MATERIALS SCIENCE
AND ENGINEERING

Proceedings of the
2nd Annual 2016 International Workshop on
Materials Science and Engineering (IWMSE 2016)

Guangzhou, China,   12 – 14 August 2016

MATERIALS SCIENCE AND ENGINEERING

Proceedings of the
2nd Annual 2016 International Workshop on
Materials Science and Engineering (IWMSE 2016)

Guangzhou, China,   12 – 14 August 2016

Editors

Roman Szewczyk

Industrial Research Institute for Automation and Measurements, Poland

Jingyu Yang

University of Maryland, College Park, USA & Shenyang Aerospace University, China

World Scientific

NEW JERSEY · LONDON · SINGAPORE · BEIJING · SHANGHAI · HONG KONG · TAIPEI · CHENNAI · TOKYO

Published by

World Scientific Publishing Co. Pte. Ltd.

5 Toh Tuck Link, Singapore 596224

USA office: 27 Warren Street, Suite 401-402, Hackensack, NJ 07601

UK office: 57 Shelton Street, Covent Garden, London WC2H 9HE

British Library Cataloguing-in-Publication Data
A catalogue record for this book is available from the British Library.

MATERIALS SCIENCE AND ENGINEERING
Proceedings of the 2nd Annual 2016 International Workshop on Materials Science and Engineering (IWMSE 2016)

ISBN 978-981-3226-50-0

Preface of IWMSE2016

Technological knowledge and industrial development are the key factors deciding on economic growth of countries and regions. From the distant past to recent times, on the base of this growth one can always find availability of necessary materials. As a result, the materials science and engineering should be located in the centre of interest of scientists, academic teachers and administration involved in economic planning.

The main aim of the second annual 2016 International Workshop on Materials Science and Engineering (IWMSE 2016) is provide a platform for scientists and engineers, to get together to share their research findings, exchange ideas and identify the future directions of R&D in materials science

The scope of the workshop covers ceramics and glasses, amorphous materials, nanomaterials and thin layers, soft magnetic materials, biomaterials, polymers, photovoltaic materials, steels, tool materials, composites, as well as functional and smart materials.

In addition, the conference also received contribution from researchers working on processes and numerical simulations; covering issues on fatigue, crack and creep resistance, corrosion and fracture mechanics, as well as non-destructive testing and reliability assessment.

In this conference, we have received over 272 high-quality papers, however, only 160 articles are included in this proceedings.

On behalf of the organization of the second annual 2016 International Workshop on Materials Science and Engineering (IWMSE 2016), we will like to express our gratitude to Collaborative Innovation Centre of Non-ferrous Metals, Henan Province, China Three Gorges University and Chongqing University for their supports. Finally, we will also like to thank all members of Advisory Committee, Academic Committee and Scientific Committee for their involvement and hard work on scientific excellence of the Workshop.

Editor

Prof. Dr Roman Szewczyk

Industrial Research Institute for Automation and Measurements, Poland

Committees of IWMSE2016

Editor

Prof., Roman Szewczyk, Industrial Research Institute for Automation and Measurements, Poland

Co-editor

Dr., Zhiyu Jiang, University of Maryland, China
Prof., Yu-Jung Huang, I-Shou University, Taiwan

The Advisory Committee

Prof., Yongchang Liu, Tianjin University, China
Prof., JianNing WEI, Jiujiang University, China
Prof., Huaili Zheng, Chongqing University, China
Prof., Feng Liu, Northwestern Poly Technical University, China
Prof., Chunxiang Li, Shanghai University, China

Academic Committee

Prof., Jingpei Xie, Henan University of Science and Technology, Luoyang, China
Keynote Speakers
Prof.,Wen-Tsai Sung, National Chin-Yi University of Technology, Taiwan
Prof., Jyh-Chiang Jiang, National Taiwan University of Science and Technology (NTUST), Taiwan
Prof., Yu-Jung Huang, I-Shou University, Taiwan
Dr., Piyas Samanta, Vidyasagar College for Women, India
Prof., Luheng Wang, Central South University, China
Dr., Merivan Sasmaz, Adiyaman University, Turkey

Honor Chair

Prof., Zhenyang Zhong, Fudan University, China
Prof. Ying Wang, Harbin Engineering University, China
Prof. WenXing Zhang, TaiYuan University of Technology, China

The Science Committee

Prof., Jiangyong Wang, Shantou University, China
Prof., Cheng bin Du, Hohai University, China
Prof., Chuan-He Tang, South China University of Technology, China
Prof., Jicheng Bai, Harbin Institute of Technology, China
Prof., Menghuai Wu, University of Leoben, Austria

Organizing Committee

Prof., Keke Zhang, Director of Henan Key Laboratory of Non-ferrous Materials Science & Processing Technology, Luoyang, China

Technical Program Committees

Prof., Wei-Bor Tsai, National Taiwan University

Prof., Wen-Tsai Sung, National Chin-Yi University of Technology

Prof., Chien-Chang CHOU, National Kaohsiung Marine University

Prof., Liu, Chuan-Hsi, National Taiwan Normal University

Prof., Jyh-Chiang Jiang, National Taiwan University of Science and Technology

Prof., Chung-Neng Huang, National University of Tainan

Prof., Yu-Jung Huang, I-Shou University

Prof., Chiung-Lin Wang, University of Illinois

Prof., Chien-Chang CHOU, National Kaohsiung Marine University

Prof., Liu, Chuan-Hsi, National Taiwan Normal University

Prof., Chi-Wai Chow, National Chiao Tung University

Associate Prof., Limin,Cao, Capital Normal University, China

Prof., Changxin Chen, Shanghai Jiao Tong University, China

Prof., M.K.BANERJEE, Southwest Petroleum University, India

Prof., Luheng Wang, Central South University, China

Dr., Merivan Sasmaz, Adiyaman University, Turkey

Prof., Ning Zhang, Beijing Union University, China

Researcher, Mohamed M. S. Wahsh, National Research Centre, Egypt

Prof., Liyuan Sheng, Peking University, China

Senior Researcher, Quanbao Ma, University of Oslo, China

Prof., Bing Han, Bei Jing Jiao Tong University, China

Prof., Felix Telegin, Ivanovo State University of Chemical Technology, Russia

Prof., Partha Pratim Das, Haldia Institute of Technology, India

Dr., Prakornchai Phonrattanasak, North Eastern University, Thailand

Dr., G.N.K.Ramesh Bapu, CSIR-Central Electrochemical Research Institute, India

Assoc. Prof., Alper Sezer, Ege University, Turkey

Assoc. Prof., Jingyu YANG, University of Maryland, China

Dr., Zhiyu Jiang, University of Chinese Academy of Sciences, China

Dr., Liang Deng, Shanghai University of Electric Power, China

Prof., Ke Wang, Huaqiao Univiersity, China

Senior Research Associate, Tianhua Xu, University College London (UCL), London

Prof., K. M. Pandey, National Institute of Technology Silchar, India

Prof., Sadaqat Jan, University of Engineering & Technology, Pakistan

Prof., Sajan Daniel George, Manipal University, India

Dr., Fabio D'Andreagiovanni, Zuse Institute Berlin (ZIB), Germany

Dr., Vasar Cristian, Politehnica University of Timisoara, Romania

Dr., Samir Hamaci, University of Angers, France

Associate Prof., Moola Mohan Reddy, Curtin University, Malaysia

Associate Prof., Sunil Karamchandani, University of Mumbai, India

Associate Prof., Bonifacio Llamazares, University of Valladolid, Spain

Dr., Roberto Suárez Antola, Energy and Mines of Uruguay, Uruguay

Dr., Andrzej Glowacz, AGH University of Science and Technology, Poland

Dr., Shahryar Jafarinejad, College of Environment, Iran

Dr., Zhi-Gang Chen, The University of Southern Queensland, Australia

Prof., Szewczyk, Industrial Research Institute for Automation and Measurements, Poland

Senior Scientist, Chris Bumby, Victoria University of Wellington, New Zealand

Senior Researcher, Buryat State University, Russia

Dr., Homero Toral-Cruz, University of Quintana Roo (UQROO), México

Prof., Antonio Politano, University of Calabria, Italy

Dr., Sayani Majumdar, Aalto University, Finland

Dr., Marc Porti, Autonomous University of Barcelona, Spain

Dr., Jianjun Pang, Zhejiang University of Water Resources and Electric Power, China

Dr., Felix Ling Ngee Leh, University Tun Hussein Onn Malaysia, Malaysia

Dr., Mariya Petrova Aleksandrova, Technical University of Sofia, Bulgaria

Dr., Togay Ozbakkaloglu, The University of Adelaide, Australia

Prof., Piotr Kulczycki, Polish Academy of Sciences, Poland

Prof., Irina Severin, Politehnica University of Bucharest, Romania

Dr., V. ANANDAKRISHNAN, National Institute of Technology, India

Dr., Chunlai Tian, State Nuclear Power Technology Corporation Research and Development Center, China

Dr., junsheng zheng, Tongji University, China

Associate Prof., Beiyue Ma, Northeastern University, China

Dr., Bo Li, Liaoning University of Technology, China

Dr., Yujie Dai, Liaoning Shihua University, China

Dr., Guo-Ying Gu, Shanghai Jiao Tong University, China

Dr., B. E. Amitha rani National Aerospace Laboratories, Bengaluru, India

Prof., Ke Dai Huazhong agricultural university, China

Dr., Qing Zhang, Henan university of science and technology, China

Prof., S K Panigrahi, Defence Institute of Advanced Technology, Maharastra, India

Dr., MOHAMMAD JAWAID, Universiti Putra Malaysia, Malaysia

Dr., Hai-Wen Li, Kyushu University, Japan

Dr., AV Santhana Babu, Indian Space Research Organisation (ISRO), India

Dr., P. Gangopadhyay, Materials Science Group, IGCAR, Kalpakkam 603 102, India

Dr., Xiaolei Su, Xi'an Polytechnic University, China

Dr., Pei-Hsing Huang, National Pingtung University of Science and Technology, Taiwan

Dr., Zawati Harun, University Tun Hussein Onn Malaysia, Malaysia

Prof., Iwahashi Tetsu, Seiwa University, Japan

Dr., Vijayalakshmi Sudhakaran, Kanpur University, India

Dr., Keming Wu, Zhengzhou University, China

Keynote Speakers of IWMSE2016

Keynote Speaker I

Electrical Characterization of Interconnection for 2.5D/3D Integration
Prof. Yu-Jung Huang

BS, MS, Ph.D, Visiting Professor, associate Professor, System Engineer, Scientist, Teaching and Research assistant

Department of Electronic Engineering, I-Shou University

Abstract

Stacking multiple dies together or connecting them via an interposer also known as "More than Moore" technologies can achieve high-performance devices and systems. Using TSV or AC-coupling circuit technology can enable designers to combine multiple chips into a high performance module. Considering the current 3D TSV technology and near future 3D applications, accurate electrical characterizations of such an advanced 3D interconnect structure is required for future system-level design and performance estimation. In this paper, electrical characteristics of stacked die structure using TSV/ACCI interconnection are presented.

Speaker's Bio

Yu-Jung Huang received the B.S. degree in material science and engineering from National Tsing Hua University, Hsinchu, Taiwan, in 1981 and the M.S. and Ph.D. degrees in electrical engineering from the University of Maryland, College Park, in 1985 and 1988, respectively. From 1988 to 1990, he was with Brimrose of America, Baltimore, MD, as a Staff Scientist for developing an infrared system. From 1990 to 1992, he was a System Engineer with Integrated Microcomputer System Inc., Dayton, OH, where he worked in the field of system design automation. In August 1992, he joined the Department of Electronic Engineering, I-Shou University, Kaohsiung, Taiwan, where he is currently a Professor. Dr. Huang also served as the Executive Secretary of Surface Mounting Technology Association, Taiwan Chapter. His current research interests are mainly in the areas of 3D IC integration, system-in-package, system-on-chip design.

Keynote Speaker II

Innovative IoT System View Based on Wireless Sensors Networks Technology
Prof. Wen-Tsai Sung

PhD, MS Editor-in-Chief, Associate-Editor

Department of Electrical Engineering, National Chin-Yi University of Technology

Abstract

IoT (Internet of Things) System is a rapidly developing area, a combination of Network, mathematics and computing technology, in order to enhance the complex sensors network and data aggregation. Traditional Wireless Sensors Networks method does not have the ability to process hung amounts sensors signals that is why the Wireless Sensors Networks design often only one cluster or one layer framework. This speech issue brings together some of the optimal fusion of innovative information technology and methods and it provides to the listeners on this issue have further improved System Integration and Applications in Wireless Sensors Networks. This will allow scientists to develop smarter process strategies for multi-sensors signals and data.

Speaker's Bio

Wen-Tsai Sung is working with the Department of Electrical Engineering, National Chin-Yi University of Technology as a professor and Vice-Dean of Academic Affairs. He received a PhD and MS degree from the Department of Electrical Engineering, National Central University, Taiwan in 2007 and 2000. He has won the 2009 JMBE Best Annual Excellent Paper Award and the dragon thesis award that sponsor is Acer Foundation. His research interests include Wireless Sensors Network, Data Fusion, System Biology, System on Chip, Computer-Aided Design for Learning, Bioinformatics, and Biomedical Engineering. He has published a number of international journal and conferences article related to these areas. Currently, he is the chief of Wireless Sensors Networks Laboratory. At present, he serves as the Editor-in-Chief in three international journals: International Journal of Communications (IJC), Communications in Information Science and Management Engineering (CISME) and Journal of Vibration Analysis, Measurement, and Control (JVAMC), he also serves as the other international journals in Associate-Editor and Guest Editor (IET Systems Biology).

Keynote Speaker III

First Principles Study of Hydrogen Production and Storage

Prof. Jyh-Chiang Jiang

B.S., PhD, postdoctoral fellow, Assistant Professor, full Professor

Chemical Engineering Department, National Taiwan University of Science and Technology (NTUST)

Abstract

In the current energy research, hydrogen, the most abundant element in the universe is considered as a "green fuel" because, it burns cleanly without emitting any environmental pollutants and is considered as a most viable energy source. H_2 can be produced from different sources, e.g., coal, natural gas, liquefied petroleum gas (LPG), propane, methane (CH_4), dry biomass, biomass-derived liquid fuels (such as methanol, C_2H_5OH, biodiesel), as well as from water. Among the liquid H_2 sources, C_2H_5OH and water are good candidates due to its low toxicity and easily available in nature etc,. Here we proposed H_2 production from steam reforming by C_2H_5OH. Steam reforming is the first step of the H_2 production process involves a light hydrocarbon reacting with steam. The second step, known as a water gas shift (WGS) reaction, the CO produced in the first reaction is reacted with steam over a catalyst to form H_2 and CO_2. We used DFT methods to investigate the mechanism of the WGS reaction on a model consisting of 3Cu atom-cluster on an $3Cu/\alpha\text{-}Al_2O_3(0001)$ surface. Here we propose three reaction mechanisms, such as redox, carboxyl and formate, have been examined and we found that the H_2 formation barrier is extremely low, 0.65 eV, on this surface. However, storage of hydrogen under appropriate conditions is the challenging problem in the modern research. In this present study we propose a new strategy in which we considered three transition metal (TM) atoms with high, medium and low hydrogen adsorption energies. These TM atoms are used to decorate the Boron doped graphene sheet and our results show that the activation energies for H atom diffusion are much smaller than the previously reported values, indicating that a fast H diffusion on this proposed surface can be achieved.

Speaker's Bio

Jyh-Chiang Jiang graduated from National Taiwan University in 1986 with a B.S. in Chemistry and received his PhD in Chemistry in 1994 from the National Taiwan University. After working as a postdoctoral fellow at IAMS, Dr. Jiang joined the faculty of National Taiwan University of Science and Technology (NTUST) in 2001 as an Assistant Professor. In 2010 he was full Professor in Chemical Engineering Department. Dr. Jiang is Executive Supervisor of the Taiwan Theoretical and Computational Science Association from 9/2014. He was the coordinator of NSC-computational chemistry group during 11/2009~1/2013. Dr. Jiang was also the Panel Member in Division of Chemistry, National Science Council during 1/2010~12/2012.

Keynote Speaker IV

Temperature Dependence Current Conduction Mechanism and Its Impact on Lifetime Prediction of Metal-Oxide-Silicon/SiliconCarbide Devices

Dr. Piyas Samanta

Assistant Professor, Ph.D., M.Sc., B.Sc., Higher Secondary, Madhyamik

Vidyasagar College for Women

Abstract

At high-field and temperature, the electronic conduction through SiO2 films on Si, SiC is due to both FN and PF emission of process-induced electron traps located below the oxide conduction band. The conduction mechanism is crucial in simulation of TDDB especially for SiO_2 film thickness down to 15 nm. The interfacial electron traps (contributing to PF current) accelerates TDDB. For 5V TTL logic, Devices having SiO2 film as thin as 12 nm can be used for 10 year projected lifetime. EEPROM devices (10-12 nm tunnel oxide) however cannot survive for 10 years at the required 1 mA/cm2 current density for writing and erasing cycles.

Speaker's Bio

Dr. Piyas Samanta graduated from Jadavpur University, Kolkata, India with first class Honors in 1989, M.Sc. (Physics) in 1991 and Ph.D in 1999. He then joined Vidyasagar College for Women as a Lecturer in Physics in 2000 and he is now an Assistant Professor in the same college. Dr. Samanta's research interest is in CMOS device physics, wide bandgap semiconductor device physics and characterization, high-k dielectric reliability. Dr. Samanta was with the Electrical and Computer Sc. Engg. Department at Hong Kong University of Sc. & Technology, Kowloon as a Research Associate from 2003 to 2005, where he worked on high-k dielectric with a few publications in peer reviewed international journals and conference proceedings. Dr. Samanta presented research papers in IWPSD in India in 1995 and 2007, CAS conference in Sinaia, Romania in 1996, IRPS at San Jose, CA, US in 2005, SSDM in Tsukuba, Japan in 2008, INFOS at Cambridge University in 2009, EDSSC in Chengdu, China in 2014. He is a reviewer of J. Appl. Phys. and Appl. Phys. Letters of Am. Inst. of Physics. He collaborated with various scientists in Taiwan, US as visiting scientists and delivered a number of invited talk in Korea and Taiwan.

Keynote Speaker V

Piezoresistive Effect of Conductive Polymer Composite
Prof. Luheng Wang
Central South University

Abstract

Conductive polymer composite is a popular functional material which can be used to develop piezoresistive sensor. The progress in the research on the piezoresistive effect of the composite is introduced, including the composition, the classification, the piezoresistivity, and the piezoresistive mechanism of the composite. Furthermore, the key technologies of the flexible piezoresistive sensor based on the composite are presented, encompassing the sensing material fabrication, the probe encapsulation, and the signal processing system design. Finally, the future important works in this field are summarized.

Speaker's Bio

Luheng Wang received the Ph.D. degree from the Department of Precision Instruments and Mechanology, Tsinghua University, Beijing, China. He is currently a professor with the School of Information Science and Engineering, Central South University, Changsha, China. His current research interest is to study the key mechanical/electrical properties and mechanisms for the novel flexible sensitive materials which possess multi-sensing-functions (e.g. pressure sensitive, temperature sensitive, magnetic sensitive, and gas sensitive, etc.), and to develop the multi-functional flexible sensor system based on the aforementioned sensitive materials.

Contents

Chapter 1
Ceramics, Glasses, Cement and Concrete

Study of the Adsorptive Behaviors of Polynaphthalene Sulphonate Superplasticizer on Alite-sulphoaluminate Cement

Xiao-Cun Liu, Tong Liu, Chang-Liang Wu, Yan-Jun Li[*]

School of Materials Science and Engineering, University of Jinan, Jinan Shandong, 250022, China
Email: mse_liyj@ujn.edu.cn

The adsorption amount of polynaphthalene sulphonate superplasticizer (NSF) on the surface of alite-sulphoaluminate cements of different mineral compositions were detected by ultraviolet-visible absorption spectrometry. The adsorptive behaviors of NSF on the surface of cement particles in the hydration system were investigated. The results showed that the adsorption amount of NSF on the cement increased with higher initial concentration; the adsorption amount and the maximum adsorption amount increased with prolonged hydration time; with a certain content of C_3S, the increased amount of calcium sulphoaluminate ($3CaO \cdot 3Al_2O_3 \cdot CaSO_4$, $C_4A_3\bar{S}$) leads to observably increased adsorption amount and the maximum adsorption amount of NSF on cement. The total adsorption amount is much higher than that of Portland cement.

Keywords: Alite-sulphoaluminate cement; mineral composition; NSF; adsorption.

1. Introduction

Alite–sulphoaluminate cement composed of main minerals of C_3S, C_2S, and $C_4A_3\bar{S}$ is a new type of energy-conserving and high-performance cement. It possesses the properties of Portland cement, and shows benefits of rapid hydration and hardening, higher early age strength and smaller volume shrinkage of hardening [1]. Polynaphthalene sulfonate superplasticizer (NSF) is a kind of non-air-entraining superplasticizer with β-naphthalene sulfonic acid formaldehyde condensation compound as the main component [2, 3]. The superplasticizers show advantages such as low cost, high water reducing rate, non-air-entraining property, good adaptability with the cement, and capability for production of high-strength, high-performance concrete. Recent studies reported naphthalene water reducing agent used in Portland cement, such as: Effect of the superplasticizers on the adding limestone, slag and other admixture cement fluidity [4]; The scattered mechanism of water-reducing agent, and revealing the regularity of dispersion effect of superplasticizer on Portland cement mineral particles [5]. There are also reported studies on the adsorptive characteristics of water-reducing agent on the hydration products AFt and AFm of early hydrating cement [6, 7]. To the best of our knowledge, there is no report about the adsorptive behaviors of superplasticizer on alite-sulphoaluminate cement. The paper used UV-visible absorption spectroscopy [8] to study the adsorption amount and behaviors of NSF on alite-sulphoaluminate cement.

2. Experimental

2.1. *Raw materials*

This experiment used commercial raw materials such as limestone, clay, phosphogypsum and fly ash and pure chemicals reagent CaF_2. The compositions of the raw materials were given in Table 1. The chemical compositions of clinkers and corresponding calculated mineral components can be seen in Table 2. The NSF (solid) was provided by in the Shandong Wenhe reagent plant (China).

---

* Corresponding author

2.2. Preparation of the samples

The mineral composition of the test clinker A, B and C has been chosen: C_3S 60%, the total contents of C_2S and $C_4A_3\bar{S}$ were ~ 30%, C_4AF and $CaSO_4$ was 5% respectively (wt. %). The raw mixtures were prepared by mixing adequate quantities of raw materials (Table 2) with the addition of 0.25% CaF_2 and grinding in a laboratory ball mill to leave a residue of ~ 4% (wt.%) on the 80 μm standard sieve. The raw mixtures ground was mixed with 8% water and pressed with 25 MPa pressure into discs using a mould with the size of 60 mm in diameter and 8 mm in thickness.

Table 1 Chemical compositions of raw materials (wt.%)

	Loss	SiO_2	Al_2O_3	Fe_2O_3	CaO	MgO	SO_3	Σ
Lime stone	41.27	4.40	0.67	0.30	51.10	1.65	—	99.39
Clay	6.96	61.44	16.58	6.85	3.26	0.74	—	95.83
Fly ash	2.37	52.11	32.69	5.55	3.12	0.56	—	96.40
Phosphogypsum	18.10	5.02	0.45	0.24	30.17	0.24	42.50	96.72

The discs were burned to clinker in an electric furnace with silicon molybdenum bars heating elements at 1320 °C for 50 min, then removed from the furnace at 1200 °C and cooled rapidly in air. The cement was made by mixing 95% clinker and 5% phosphogypsum, and grinding in a laboratory ball mill to a specific surface area of ~ 320 $m^2 \cdot kg^{-1}$.

Table 2 Practical chemical compositions and calculated mineral components of the clinkers (wt.%)

	Chemical compositions					Mineral components				
	CaO	SiO_2	Al_2O_3	Fe_2O_3	SO_3	C_3S	C_2S	$C_4A_3\bar{S}$	C_4AF	$CaSO_4$
A	63.81	23.38	4.28	1.81	3.43	56.88	25.69	6.24	5.49	4.44
B	61.82	19.94	8.18	2.17	4.69	58.98	12.76	13.56	6.61	4.94
C	60.41	18.51	10.64	2.01	5.35	57.41	9.82	18.66	6.13	4.93

2.3. Phase analysis

In order to investigate the formation of mineral phases, the content of free lime in the clinker was determined chemically using the ethanol-glycerin method. The mineral composition of the clinker was analyzed by the XRD (Cu, Kα, λ=0.154 nm), the microstructures of the clinkers were observed by scanning electron microscope (SEM).

2.4. Measurement of the adsorption amount

Superplasticizer solutions were prepared with concentrations of 600, 900, 1200, 1500 and 1800 mg/L at 25 °C. The initial concentration C_0 was determined by TU-1901 UV spectrophotometer. The sample of 2.5 g was in a beaker, and 25 mL of superplasticizers solutions with different concentrations were added. The mixture was stirred for 5 min, keeping still for 25 min, 55 min, and 85 min respectively. The clear liquid was separated, centrifuged for 30 min, and the superplasticizer solution after adsorption was obtained. The solutions were diluted to the controlled the concentration range of 20 - 40 mg/L to comply with the Beer's Law. The concentration of the solutions was measured by UV spectrophotometer, and was noted as C. The adsorption amounts were calculated with the concentration difference (C_0-C) of the solutions before and after adsorption.

3. Results and discussion

3.1. *Mineral formation of clinkers*

Ethanol-glycerin method was used to determine the content of f-CaO in clinkers A, B and C, and the result was 0.48, 0.21, and 0.41% respectively, indicating that the mineral reaction is completed at high temperature. The XRD patterns and SEM pictures of the clinkers are shown in figure 1 and figure 2. Alite–sulphoaluminate cements clinker contain main minerals: C_3S (d=0.304, 0.219, 0.176 nm), C_2S (d=0.278, 0.275, 0.261 nm), $C_4A_3\bar{S}$ (d=0.375, 0.265, 0.217 nm).

According to figure 1, the spectra of different minerals are consistent with the standard card, and the diffraction peaks are sharp and intense. According to figure 2, the crystal of different minerals is regular, clear, and uniformly distributed. The spectra and the diagram indicate the good formation of clinkers, consistent with the designed mineral components.

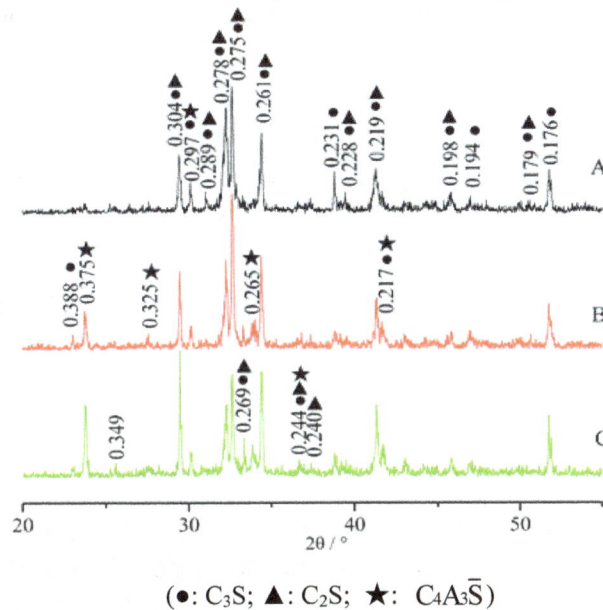

($\bullet$: C_3S; $\blacktriangle$: C_2S; $\star$: $C_4A_3\bar{S}$)

Fig. 1 XRD patterns of clinkers A, B and C

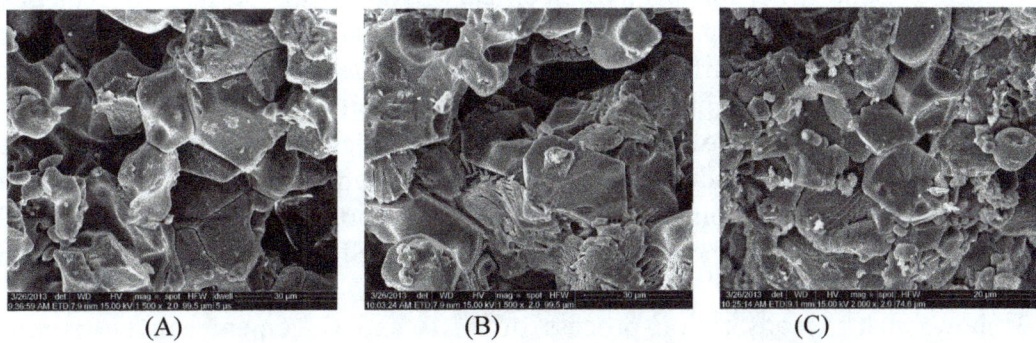

(C_3S: short pillar and hexagonal sheet; C_2S: round shape; $C_4A_3\bar{S}$: small polygon)

Fig. 2 SEM photograph of clinkers A, B and C

5

3.2. *Adsorption of NSF on alite–sulphoaluminate cement*

C_0 and C (mg/L) represent the concentrations before and after the adsorption. The adsorption amount of solute is n^s (mg/g).

$$n^s = V (C_0 - C) / m \qquad (1)$$

where V = volume of superplasticizers solutions, mL; m = mass of mineral sample, g.

Adsorption isotherm is obtained if taken n^s as the function of C (named equilibrium concentration). The adsorption equation is determined according to the shape of the adsorption isotherms to obtain the maximum adsorption amount [9]. Figure 3 present the adsorption amounts of NSF on the different cements at 30 min, 60 min, and 90 min. From figure 3, the adsorption amount of NSF on cement increases with increased initial concentration C_0 of NSF solutions. When the initial concentration C_0 is small, the superplasticizer molecules have larger dispersion, and the molecular diffusion length is longer in adsorption process. In a certain time length, the total adsorption amount is small, which is far from the maximum adsorption amount. When C_0 is larger, the dispersion of the superplasticizer molecules is less, and thus more molecules can be adsorbed on the surface of cement particles and reach the maximum adsorption amount is a shorter time length.

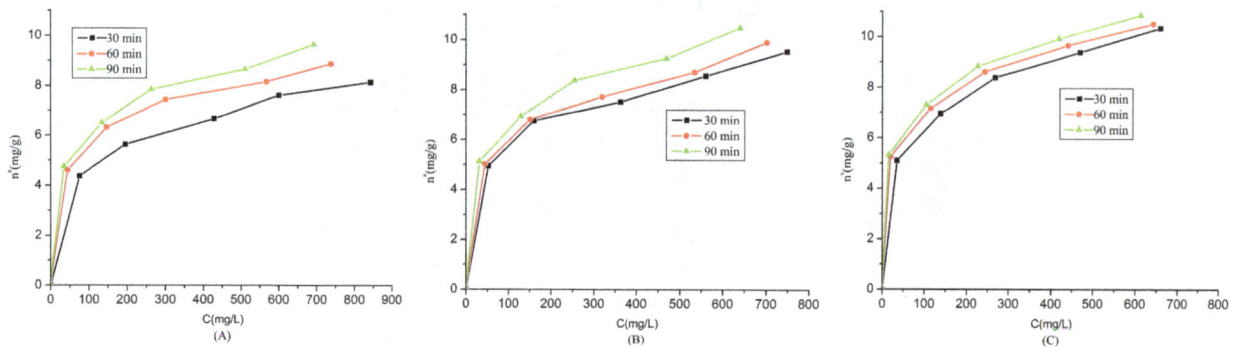

Fig. 3 Adsorption isotherm of NSF on the cement A, B and C

Figure 3 also show that the adsorption amount of NSF on cement particles is larger with longer hydration time when the initial concentrations are the same. During the hydration process, the adsorption rate of NSF molecules is more than the dissociation rate. This seems to be related to the hydration of cement. When the superplasticizer molecules are adsorbed onto cement particles, the adsorbed molecules build a net of negative electrical charge to the surface of cement particles. This phenomenon results in long range electrostatic repulsive force and short range steric repulsive force between cement particles. These dispersion effects contribute to increased fluidity of cement paste. As the hydration process continues, the specific surface area of cement particles decreases. The adsorption amount of NSF on the hydration calcium silicate and portlandite is very small. However, the hydration products AFt or AFm has much stronger adsorptive capacity [6, 7]. The dissociated NSF molecules from the surface of cement particles can be adsorbed again by the AFm or AFt molecules, and thus the adsorption amount of NSF on the cement particles increased with prolonged adsorbing time.

Figure 3 shows that the adsorption process of A, B and C cement can be expressed with Langmuir isotherm equation [10].

$$n^s = \frac{n_m^s bc}{1+bc} \qquad (2)$$

where n^s_m = maximum adsorption amount; b = adsorption constant; c = equilibrium concentration. To get the maximum adsorption amount, equation (2) can be reformed as:

$$\frac{1}{n^s} = \frac{1+bc}{n^s_m bc} \tag{3}$$

$$\frac{c}{n^s} = \frac{1}{n^s_m}\left(\frac{1}{b}+c\right) \tag{4}$$

Equation (4) is as the y = a + bx type linear equation. When changing the axis of the isotherm curves, taken C/n^s as the function of C, the linear relationship is obtained, and the slope represents the maximum adsorption amount n^s_m. Figure 4 is the C/n^s-C diagram of NSF adsorbed on A, B and C cement at 30 min, 60 min, and 90 min. According to the slopes of the straight lines in figure 4, the maximum adsorption amount n^s_m of NSF on the cements are shown in Table 3.

Table 3 showed that the maximum adsorption amount of NSF on alite–sulphoaluminate cement increased with prolonged hydration times. The results indicated that for alite–sulphoaluminate cement, as the hydration process continued, certain amount of AFm and AFt are formed with higher adsorptive capability, and facilitated the adsorption rate more than the dissociation rate of the NSF molecules. Therefore, the maximum adsorption amount increased with longer adsorption time.

3.3. *Effect of the mineral composition on the adsorption amount and maximum adsorption amount*

Figure 3 indicates that the adsorption amount of NSF on A, B and C cement increase with increased content of $C_4A_3\bar{S}$ mineral in clinkers. In order to investigate influence of the mineral composition on the adsorption amount, the content of C_3S was designed constantly, and the contents of C_2S and $C_4A_3\bar{S}$ was changed in A, B and C cement. The C_2S amount decreases correspondingly with the increased amount of $C_4A_3\bar{S}$. In a previous paper, as hydration proceeds, the adsorption amount of NSF on C_3S mineral is more than that on C_2S single mineral [5]. However, $C_4A_3\bar{S}$ mineral has much stronger adsorptive capacity than the silicate minerals C_3S and C_2S. Consequently, the superplasticizer molecules can be more absorbed on $C_4A_3\bar{S}$ mineral. The results indicated that for the different minerals composition of alite-sulphoaluminate cement, constant content of C_3S with increased $C_4A_3\bar{S}$ amount leads to the increased adsorption amount of NSF. This is related to the characteristics of clinkers mineral and the difference of hydration products.

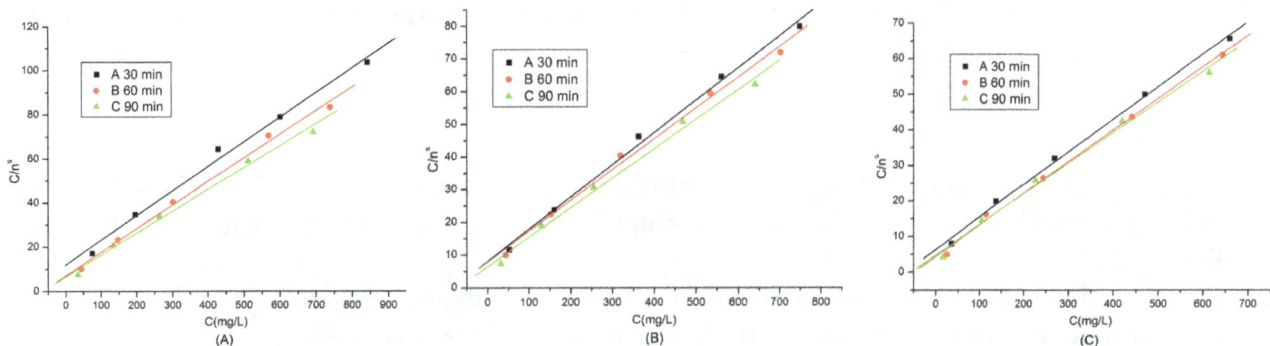

Fig. 4 $C/n^s \sim C$ curves of NSF adsorbed on the cement A, B and C at different times

7

Table 3 Maximum adsorption amount of NSF on the cement A, B and C (mg/g)

Time	A	B	C
30min	8.98	10.13	11.05
60min	9.35	10.62	11.24
90min	10.08	11.09	11.62

Table 3 also showed that the maximum adsorption amount of NSF on the cement was larger for the cement containing higher $C_4A_3\bar{S}$ mineral, while the content of C_3S was unchanged in the clinkers. Literature [9] investigated the adsorption characteristics of various NSF superplasticizers on Portland cement. Adsorptive behaviors of four types of NSF superplasticizers on Portland cement shows maximum adsorption amount of 7.57, 6.00, 6.02, and 4.87 mg/L at 30 min. In contrast, the alite-sulphoaluminate cement has stronger adsorptive capacity than Portland cement, and the maximum adsorption amount of NSF on those is 1.2 to 2.5 times as much as on Portland cement. Research [11] indicates that if the amount of $C_4A_3\bar{S}$ mineral in clinker increased, the difference between the initial fluidity of the different cement pastes is small, but the flow loss of cement paste is more. Therefore, the higher the maximum adsorption amount of NSF on alite-sulphoaluminate cement is, the worse the plasticity of cement paste. Because of the selectivity and nonuniformity of superplasticizers adsorbed on cement particles, the superplasticizer molecules will be absorbed firstly on $C_4A_3\bar{S}$ mineral and its hydration product during the hydration process. So the dissociative NSF molecules will decrease, inevitably affecting the adsorption amount of superplasticizer on the other mineral (C_3S, C_2S, and C_4AF). As a result, it will decrease the dispersion effect of superplasticizer on the cement mineral particles, and induce to worse compatibility between cement and superplasticizer, comparing with Portland cement.

4. Conclusion

With the same adsorption time, the adsorption amount of NSF on alite-sulphoaluminate cement increases with the increased initial concentration. With the same initial concentration, the adsorption amount and the maximum adsorption amount of NSF on alite-sulphoaluminate cement increases with prolonged hydration time. For the same C_3S content, as the amount of $C_4A_3\bar{S}$ mineral increased, the adsorption amount and the maximum adsorption amount of NSF on alite-sulphoaluminate cement increased. Alite-sulphoaluminate cement has much stronger adsorptive capacity than Portland cement, and the maximum adsorption amount is 1.2 to 2.5 times as much as Portland cement.

References

1. Liu Xiaocun, Li Yanjun, Zhang Ning, Influence of MgO on the formation of Ca_3SiO_5 and $3CaO \cdot 3Al_2O_3 \cdot CaSO_4$ minerals in alite–sulphoaluminate cement, Cement and Concrete Research. 32 (2002) 1125-1129.
2. Zhu Huaxiong, Synthesis and application of NSF naphthalene based high efficiency water reducer, Fuel & Chemical Processes (in Chinese). 2 (2000), 85-87, 108.
3. Ge Zhaoming, Yu Chengxing, Wei Qun, Concrete Admixture, Chemical Industry Press of China, Beijing, China, 2004.

4. O. Boukendakdji, E. H. Kena, S. Kenai, Effects of granulated blast furnace slag and superplasticizer type on the fresh properties and compressive strength of self-compacting concrete, Cement and Concrete Composites. 34 (2012) 583-590.

5. Li Yanjun, Qi Tao, Li Wenchan, Study of adsorptive behaviors of polynaphthalene sulfonate superplasticizer on C_3S and C_2S, Journal of Building Materials(in Chinese). 3 (2011), 362-365.

6. Liu Bingjing, The compatibility of cement and superplasticizer, Concrete (in Chinese). 9 (2002) 20-25.

7. J. Plank, P. Chatziagorastou, C. Hirsch, New model describing distribution of adsorbed superplasticizer on the surface of hydrating cement grain, Journal of Building Materials (in Chinese). 1 (2007) 7-13.

8. Qu Jindong, Peng Jiahui, Chen Mingfeng, Experimental methods in the research of absorption properties of concrete Superplasticizers, Concrete (in Chinese), 9 (2004) 27-28.

9. Wang Qian, Du Bing, Study on absorption performance of polynaphthalene sulfonate superplasticizer, Concrete (in Chinese). 9 (2003), 30-31.

10. Chen Jiankui, Theory and application of concrete admixture, China Planning Press, Beijing, China, 1997.

11. Qi Tao, Liu Xiaocun, Li Yanjun, Effect of mineral composition on adaptability of alite-sulphoaluminate cement with superplasticizer, Cement (in Chinese). 5 (2009) 1-4.

Influence of Calcium Sulphoaluminate Mineral Contents on the Burning and Properties of Alite-sulphoaluminate Cement Clinker

Xiao-Cun Liu, Qian-Qian Cui, Chang-Liang Wu, Yan-Jun Li[*]

School of Materials Science and Engineering, University of Jinan,
Jinan Shandong, 250022, China
[*]Email: mse_liyj@ujn.edu.cn

The contents of calcium sulphoaluminate ($C_4A_3\bar{S}$) mineral were set as a series 5-20 percent and accordingly fixed tricalcium silicate (C_3S) mineral contents at 30, 45, 60 percent respectively in alite-sulphoaluminate cement clinker. The burning and properties of clinker were investigated. Results showed that: For certain C_3S content, the clinker sintering is loose when the content of $C_4A_3\bar{S}$ mineral is 5%, and the free lime amount is low, the cement setting time is long, the early age compressive strength is low, and a significant increase in the 28 days compressive strength is observed. When the contents of $C_4A_3\bar{S}$ mineral are 10% and 15%, the free lime amount is low, the clinker sintering compacter, the main mineral formation are good, and the cements show proper setting time and high compressive strengths. When $C_4A_3\bar{S}$ mineral content is 20%, the free lime amounts are high, the setting time of the cements is shorter, the early age compressive strengths are higher, while the compressive strength growth are smaller.

Keywords: Calcium sulphoaluminate mineral; alite-sulphoaluminate cement clinker; burning properties.

1. Introduction

The calcium sulphoaluminate mineral ($3CaO \cdot 3Al_2O_3 \cdot CaSO_4, C_4A_3\bar{S}$) has the fast setting and hardening, high early age strength and hydration hardening expansion properties, etc. This kind of mineral with special properties expands the varieties of cements, such as sulphoaluminate cement and ferroaluminate cement known as the third series cement, which is mainly composed of $C_4A_3\bar{S}$ mineral [1, 2]. Some belite cements achieved a high early strength and certain practicability by introducing a proper amount of $C_4A_3\bar{S}$ mineral [3, 4]. Alite-sulphoaluminate cement can be prepared by introducing $C_4A_3\bar{S}$ into the Portland cement clinker to replace or partially for tricalcium aluminate ($3CaO \cdot Al_2O_3, C_3A$), which has not only the properties of Portland cement, but also has the excellent properties of the quick hydration and hardening, high early age strength and the volume does not shrinking or slightly expanding when hydrating [5-8]. To date, there are few reports on the influence of $C_4A_3\bar{S}$ mineral content on the burning and properties of alite-sulphoaluminate cement clinker, the related experimental researches are carried out in this paper.

2. Experimental

2.1. *Raw materials*

Commercial raw materials such as limestone, sandstone, clay, Al_2O_3 powder, phosphogypsum and the pure chemicals Fe_2O_3 and CaF_2 were used as test materials. The chemical compositions of raw materials are given in Table 1.

[*] Corresponding author

Table 1 Chemical compositions of raw materials (wt.%)

Raw material	Loss	SiO_2	Al_2O_3	Fe_2O_3	CaO	MgO	SO_3	Σ
lime stone	42.16	2.86	0.61	0.48	51.01	2.01	0.07	99.20
sandstone	2.86	85.40	8.07	2.29	0.48	0.46	0.10	99.66
Al_2O_3 powder	0.20	0.10	98.5	0.03	0.02	0.01	—	98.86
phosphogypsum	20.40	3.35	0.55	0.58	31.44	0.11	42.60	99.03

The $C_4A_3\bar{S}$ mineral contents in the cement were 5%, 10%, 15% and 20%, respectively. Changing the content of C_3S ($3CaO \cdot SiO_2$) and C_2S ($2CaO \cdot SiO_2$) accordingly, and C_3S content were designed by 30%, 45% and 60%, C_2S content was changed along with $C_4A_3\bar{S}$ and C_3S content, C_4AF ($4CaO \cdot Al_2O_3 \cdot Fe_2O_3$) and $CaSO_4$ content remained 5% respectively. The chemical and mineral compositions of clinkers are shown in Table 2.

Table 2 Chemical and mineral compositions of clinkers (wt.%)

Sample	Chemical compositions						Mineral compositions				
	SiO_2	Al_2O_3	Fe_2O_3	CaO	MgO	SO_3	C_3S	C_2S	$C_4A_3\bar{S}$	C_4AF	$CaSO_4$
A1	26.07	3.42	1.64	61.74	2.46	3.46	28.71	53.08	4.76	4.99	4.83
A2	24.41	5.84	1.64	60.42	2.38	4.10	28.74	48.29	9.59	4.98	4.83
A3	22.75	8.26	1.64	59.10	2.30	4.73	28.78	43.50	14.44	4.98	4.83
A4	21.08	10.68	1.64	57.77	2.23	5.37	28.77	38.73	19.28	5.00	4.84
B1	24.82	3.42	1.64	62.94	2.50	3.46	43.08	38.64	4.75	5.00	4.83
B2	23.16	5.84	1.64	61.62	2.42	4.09	43.13	33.85	9.58	5.00	4.83
B3	21.50	8.25	1.64	60.30	2.34	4.73	43.17	29.05	14.42	4.99	4.83
B4	19.83	10.68	1.64	58.98	2.27	5.37	43.22	24.24	19.27	4.98	4.83
C1	23.57	3.42	1.64	64.14	2.54	3.46	57.49	24.20	4.75	4.99	4.82
C2	21.91	5.83	1.64	62.82	2.46	4.09	57.54	19.40	9.58	4.98	4.83
C3	20.25	8.25	1.64	61.50	2.38	4.73	57.55	14.62	14.41	5.00	4.83
C4	18.58	10.67	1.64	60.18	2.31	5.36	57.61	9.80	19.25	5.00	4.83

2.2. Experimental process

The raw meal were prepared by mixing adequate quantities of raw materials and an amount of 0.25 % CaF_2 was added in each sample, and grinded in a laboratory ball mill to give a residue of ~10 % on the 80 μm standard sieve. The ground raw meal were mixed with water and pressed into ϕ60×8 mm discs using a mould and then dried.

The discs were burned to clinker in an electric silicon molybdenum furnace. The burned temperatures were set as 1310 °C for group A, 1320 °C for group B and C, and remained 50 minutes respectively, removed from the furnace at 1200 °C and cooled them rapidly in air.

The cements were made by mixing 95% clinker and 5% gypsum and were grinded in a laboratory ball mill to give a residue of ~10 % on the 45 μm standard sieve.

2.3. Analysis

The clinker was grinded to the particles small than 80 μm and the free lime content was determined chemically after dissolving in ethanol-glycerin.

The group C clinker was fine grinded and the SO_3 content was determined by barium sulfate quality method.

The hydration exothermic rate and heat evolution content of C1 and C3 cement hydrated for 80 h were analyzed by TAM Air isothermal calorimeter.

Group A, B and C clinker were analyzed by XRD (Cu, Kα, λ = 0.154 nm), respectively. The C3 clinker was observed by SEM, the 3 days hydration of A3, B3 and C3 samples were also analyzed by XRD, respectively. The compressive strengths were tested on 20 mm cubes prepared from cement paste with a water/cement ratio of 0.30 at 3, 7 and 28 days after 24 h curing in moist air and subsequently in water. The setting time was tested on paste with a constant water/cement ratio of 0.30 using a cylindrical mould of 40 mm diameter and 40 mm height. In order to comparing with the strength of Portland cement clinker, the strength development of Portland cement clinker was tested using the same method.

This experiment measured the weight loss of cement paste at the different temperatures by burning method, indicating the combined water content of cement paste in order to characterizing the hydration process of cement.

3. Results and Discussion

3.1. *Burning performance of clinker*

Comparing the appearance of clinkers, with an increase of $C_4A_3\bar{S}$ content, the clinkers become more compact and the hardness of the clinker gradually increases at the same C_3S content. It can be observed that the amount of liquid increases at the corresponding burning temperature with the increase of Al_2O_3 and SO_3 contents in the sample, and the clinker burning are easy. However, the density and hardness of clinkers decrease and the sintering degree reduce with the increase of C_3S content at the same $C_4A_3\bar{S}$ content. It can be known from figure 1 that free lime content of each clinker is all lower, indicating that the clinker burning are easy on the whole. Specifically, for group A, B and C samples, when the contents of $C_4A_3\bar{S}$ are 5%, 10% and 15% respectively, the free lime content in clinker is lower relatively which benefiting the clinker mineral formation. And when the $C_4A_3\bar{S}$ content is 20%, the free lime content of clinker is generally high; the clinker mineral formation becomes difficult. For group C with a high C_3S content, the free lime content of clinker is overall the highest than other groups and the clinker mineral formation is more difficult.

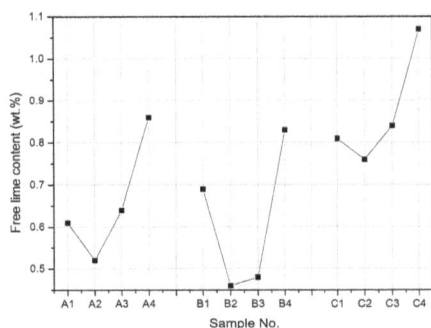

Fig. 1 Free lime content of the clinker

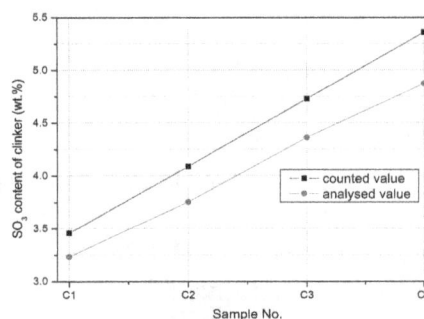

Fig. 2 SO_3 content of group C clinker

3.2. *The formation of clinker mineral*

XRD analysis of group A, B and C are shown in figure 3, it can be seen that the main minerals of each group sample are C_3S (d=3.04, 2.78, 2.61, 2.19 and 1.77Å), $C_4A_3\bar{S}$ (d = 3.75, 2.65 and 2.17Å) and C_2S (d = 2.78, 2.75, 2.61 and 2.88Å), and a small amount of C_4AF (d=7.29Å), the formation condition of clinker mineral consists with the designed mineral composition. When the designed $C_4A_3\bar{S}$ mineral content is 5% in clinker, the diffraction peaks of the mineral are not obvious.

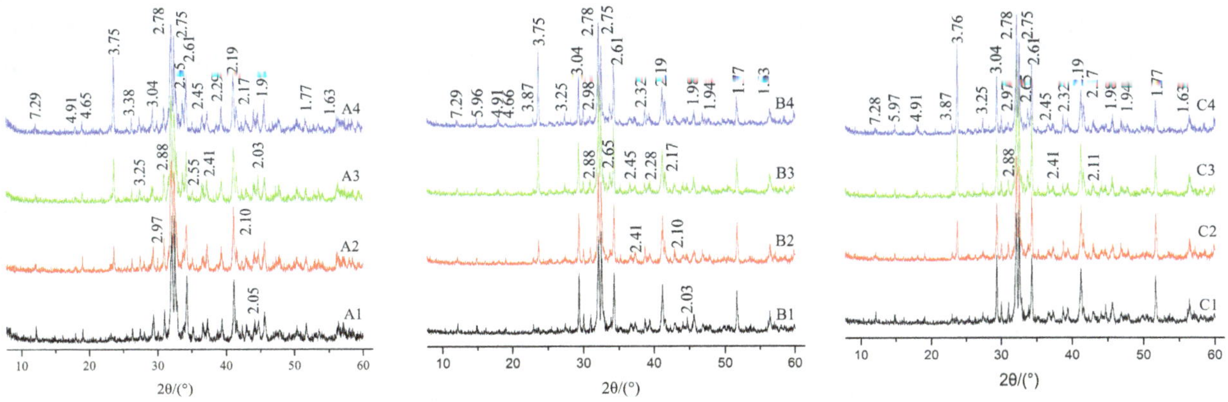

Fig. 3 XRD patterns of group A, B and C clinker

With the increase of the designed content, the intensity of the diffraction peaks of $C_4A_3\bar{S}$ mineral gradually increase and more $C_4A_3\bar{S}$ mineral is formed in the clinker. SO_3 content of group C clinker is shown in figure 2, it is known that SO_3 has a partly resolved and volatilized at high temperatures, but the loss is not much. When the designed $C_4A_3\bar{S}$ content is 5% in clinker that the diffraction peaks are not apparent, there are two reasons, one is the detection precision of the instrument not enough, another is due to SO_3 dissolving in other minerals, which reduced the amount of SO_3 that is used to form $C_4A_3\bar{S}$. The diffraction peaks of C_3S mineral in group A, B and C show that the intensity of that gradually increase with the increase of the designed content, many C_3S mineral is formed in the clinker. The overall analysis of mineral formation, C_3S and $C_4A_3\bar{S}$ mineral can well coexist in clinker in a larger content range. But the height of diffraction peaks of C_3S mineral in the group A, B and C has a certain reduction with the increase of the designed $C_4A_3\bar{S}$ mineral, the C_3S mineral content decreases, this indicates that, the high Al_2O_3 and SO_3 content hinder the C_3S mineral formation. And the effect of designed C_3S content change on the $C_4A_3\bar{S}$ mineral formation, it can not been observed from the XRD patterns.

Figure 4 shows SEM photograph of C3 clinker. It can be seen clearly, the main mineral formation is good. C_3S mineral crystal size is about 20~40 μm, $C_4A_3\bar{S}$ mineral crystal size is about 2~5 μm, C_2S mineral crystal size is about 10~30 μm.

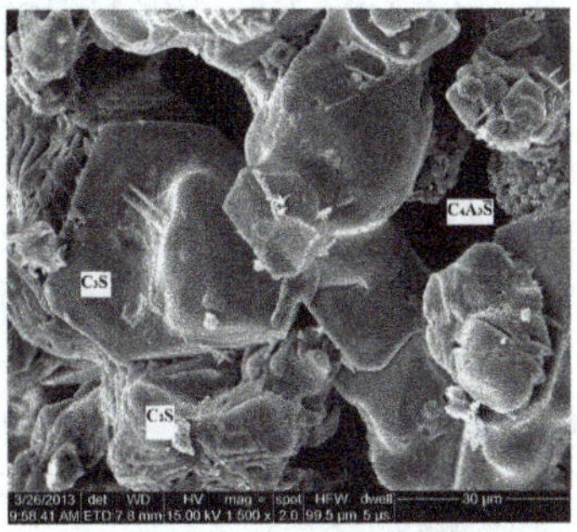

Fig. 4 SEM photograph of C3 clinker

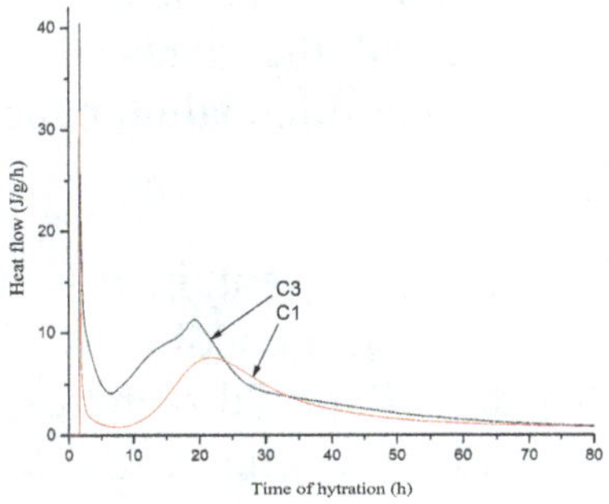

Fig. 5 Hydration exothermic rate of C1 and C3 cement

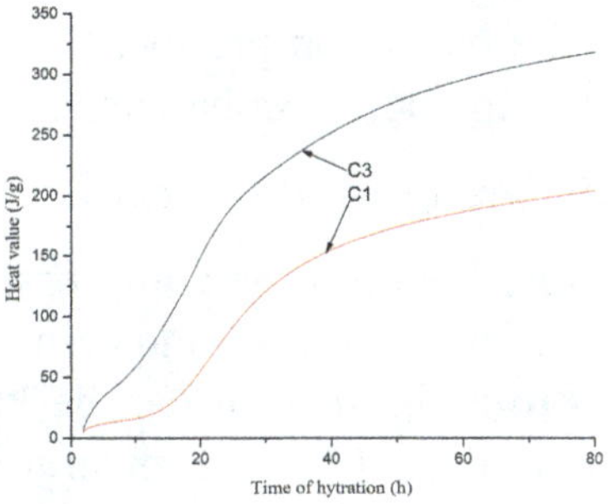

Fig. 6 Heat evolution content of C1 and C3 cement

3.3. Hydration reaction of cement

The hydration reaction of cement is an exothermic reaction and testing the cement hydration heat can be characterized the hydration process of cement. Figure 5 shows the hydration heat evolution rate of C1 and C3, having different content of $C_4A_3\bar{S}$ mineral. The content of $C_4A_3\bar{S}$ mineral is 5% in C1, and 15% in C3. The first exothermic peak of C3 cement is 40.5 J/g/h, and the first exothermic peak of C1 cement is 32 J/g/h. C1 cement exothermic peak is smaller and exothermic duration is longer than C3. After the first exothermic peak, it enters into a hydration induction period. C1 cement induction period lasts a long time than C3 cement, and it is corresponding that the C1 cement setting time is longer. At about 20 h, it comes to the second hydration exothermic peak. C3 cement exothermic peak is 12.5 J/g/h, C1 cement exothermic peak is 7 J/g/h. The hydration exothermic rate of C3 is large, the second exothermic peak is narrower, and the exothermic peak is higher. This is mainly caused by the rapid hydration and the large hydration heat evolution rate of $C_4A_3\bar{S}$ mineral.

Figure 6 shows the hydration heat evolution content of C1 and C3 cement. The heat evolution content of C1 and C3 cement are 320 J/g and 200 J/g respectively. So, the content of $C_4A_3\bar{S}$ mineral increases, leads to the cement hydration exothermic rate and the hydration heat evolution content increase.

Figure 7 shows the XRD patterns of A3, B3 and C3 cement at 3 days hydration, which can be used to analyze the hydration products of cement. The diffraction peak at d = 9.75 is ettringite (AFt), d = 8.91 is monosulphate (AFm), and d=4.92 is $Ca(OH)_2$. Therefore, the cement forms the hydration products are ettringite, AFm and $Ca(OH)_2$ at 3 days hydration, and the content of $Ca(OH)_2$ increases obviously with the increase of C_3S content.

3.4. Setting performance of cement

Figure 8 shows the setting time of group A, B and C cement. It can been seen that the initial setting time and final setting time of cement are significantly reduced with the increase of $C_4A_3\bar{S}$ mineral content for the same group of samples, and the interval between initial and final setting time also shorten significantly. This is because that $C_4A_3\bar{S}$ mineral is a rapid hardening and the early strength mineral which the hydration hardening rate is quick, the more $C_4A_3\bar{S}$ mineral, the more hydration products produce, and is the shorter setting time of the cement. This corresponds to the above analysis of the cement hydration process, the induction period duration of C1 cement is markedly prolonged compared with that of C3 cement.

However, from figure 8, the setting time of the cement is extended obviously with the increase of the designed C_3S content in clinker in general, the initial and final setting time are also prolonged obviously. This is different from Portland cement. The reason is mainly that the content of the hydration product $Ca(OH)_2$ increases with the increase of C_3S content, which results in the retarding action enhancement for aluminate and sulphoaluminum minerals with gypsum [9, 10].

3.5. Strengths of cements

Figure 9 shows the compressive strength of the cement, in order to compare, the figure also shows the strength of Portland cement under the same test conditions.

From figure 9, it can be known that the 3d and 7d compressive strength of group A samples containing 30% C_3S mineral increases with increasing of $C_4A_3\bar{S}$ mineral content, but the strength growth rate decreases for A4 sample containing 20% $C_4A_3\bar{S}$ mineral. The 28d compressive strength of A4 sample is significantly lower than that of other three samples. The 3d and 7d compressive strength of group B samples containing 45% C_3S mineral increases with the increase of $C_4A_3\bar{S}$ mineral content obviously. Although the 28d strength of B4 sample is higher than that of B1 sample, the strength growth rate decreases significantly. For group C samples containing 60% C_3S mineral, are similar to group A and B cement [11]. The hydration hardening rate of $C_4A_3\bar{S}$ mineral is fast, the hydration product ettringite are acicular crystals [9], can strengthen the structure of the cement stone, increasing the cement strength. However, when the $C_4A_3\bar{S}$ content is too many, a large amount of ettringite is formed in a short time, resulting in the partial damage of cement stone structure, leading to the decrease of the cement strength or strength growth rate.

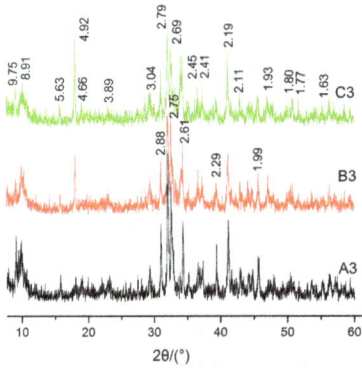

Fig. 7 XRD patterns of the cement at 3 days hydration

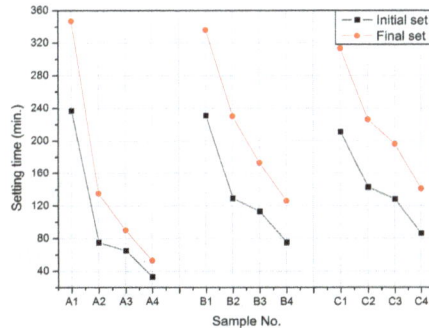

Fig. 8 Setting time of the cement

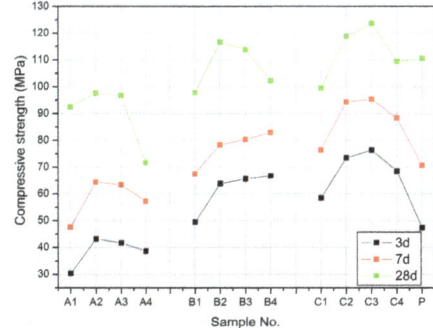

Fig. 9 Compressive strength of cement

In general, the strength of the cement is high and the strength growth rate is good when the $C_4A_3\bar{S}$ mineral content is 10% or 15%. When the $C_4A_3\bar{S}$ mineral content is 5%, the 3 d and 7 d compressive strength of the cement is lower, but the growth rate of 28 d strength is big and attains a high strength. When the $C_4A_3\bar{S}$ mineral content is 20%, the 3 d and 7 d strength are higher, but the strength growth rate of 28 d decreases. With the increase of designed C_3S mineral, the setting time of the cement prolongs, the cement strength increases and the values are higher. In contrast, the setting and hardening of alite-sulphoaluminate cement are faster, the early strength is higher, the whole strengths are also higher; but the hydrating and setting and hardening of Portland cement are slower, the early strength is lower, and the strength growth rate is larger.

4. Conclusion

The density of the clinker and the free lime content increase with the increase of designed $C_4A_3\bar{S}$ mineral. $C_4A_3\bar{S}$ and C_3S minerals can well coexist in the clinker in a large range of mineral content. The influence of C_3S mineral content change on the formation of $C_4A_3\bar{S}$ mineral is small, while the high $C_4A_3\bar{S}$ mineral content hinders the formation of C_3S mineral.

The hydration exothermic rate and heat evolution content increase with the increase of $C_4A_3\bar{S}$ mineral content, and the hydration induction period shortens. Moreover, the initial and final setting time of cement are significantly shortened with the increase of $C_4A_3\bar{S}$ mineral content, and the interval of the initial and final setting time also shorten significantly.

When the $C_4A_3\bar{S}$ mineral content is 5% in the clinker, the early strength of the cement is lower, but the strength growth rate is bigger. When the $C_4A_3\bar{S}$ mineral content is 10% or 15%, the whole strengths are higher. When the $C_4A_3\bar{S}$ mineral content is 20%, the 3 d and 7 d compressive strength are higher, but the strength growth rate of 28 d decreases.

References

1. Deng Junan, Study of sulfo-aluminate cement, Journal of the Chinese Ceramic Society (in Chinese). 1 (1982) 56-63.
2. Wang Yanmou, Su Muzhen, Third series of cement in China, China Building Materials Science and technology (in Chinese). 6 (1993) 1-5.
3. F. P. Glasser, L. Zhang, High-performance cement matrices based on calcium sulfoaluminate–belite compositions, Cement and Concrete Research. 31 (2001) 1881-1886.
4. Sui Tongbo, Liu Kezhong, Wang Jing, A study on properties of high belite cement, Journal of the Chinese Ceramic Society (in Chinese). 4 (1999) 488-492.
5. I. V. Kravchenko, T. V. Kuzhecova, V. B. Khlusov, Alite-sulphoaluminate clinker and its cement, Cement. 8 (1978) 7-8.
6. T. V. Kuzhecova, N. G. Zealishvili, D. L. Mendeleeva, Manufacture and study of modified alite-sulphoaluminate clinker, Cement. 6 (1991) 39-43.
7. S. Ma, R. Snellings, X. Li, et al, Alite-ye'elimite cement: Synthesis and mineralogical analysis, Cement and Concrete Research. 45 (2013) 15-20.
8. R. Pérez-Bravo, J. M. Compana, S. Bruque, et al, Alite sulfoaluminate clinker: Rietveld mineralogical and SEM-EDX analysis, Advances in Cement Research. 26 (2014) 10-20.
9. C. W. Hargis, A. P. Kirchheim, P. J. M. Monteiro, E. M. Gartner, Early age hydration of calcium sulfoaluminate (synthetic ye'elimite, $C_4A_3\bar{S}$) in the presence of gypsum and varying amounts of calcium hydroxide, Cement and Concrete Research. 48 (2013) 105-115.
10. G. L. Saoût, B. Lothenbach, A. Hori, et al, Hydration of Portland cement with additions of calcium sulfoaluminates, Cement and Concrete Research. 43 (2013) 81-94.
11. L. Pelletier, F. Winnefeld, B. Lothenbach, The ternary system Portland cement–calcium sulphoaluminate clinker–anhydrite: Hydration mechanism and mortar properties, Cement and Concrete Composites. 32 (2010) 497-507.

Influence of Addition 3CaO·3Al$_2$O$_3$·CaSO$_4$ Mineral on the Properties of Portland Cement

Xiao-Cun Liu, Wei-Shan Wang, Yan-Jun Li[*]

*School of Materials Science and Engineering, University of Jinan,
Jinan Shandong, 250022, China*
Email: mse_liyj@ujn.edu.cn

The influence of the C$_4$A$_3\bar{\text{S}}$ mineral addition on the properties of cement was studied, with industrial Portland cement clinker and laboratory prepared C$_4$A$_3\bar{\text{S}}$ mineral as based materials. The results indicated that the early strength of cement increased, and the late strength steadily grew with 3-5% of C$_4$A$_3\bar{\text{S}}$ mineral in Portland cement. For Portland cement with blastfurnace slag, the early and late strengths of cement increased greatly with the addition of 3-5% C$_4$A$_3\bar{\text{S}}$ mineral. For the cement in which 5% of clinker was replaced with limestone, the various age strengths of cement increased significantly with addition of 5% C$_4$A$_3\bar{\text{S}}$ mineral. For the cement in which 5% of the slag or fly ash was replaced with limestone, the strength properties of the cement improved significantly with the addition of 5% C$_4$A$_3\bar{\text{S}}$ mineral. The addition of C$_4$A$_3\bar{\text{S}}$ mineral can reduce the initial and final setting times of cement.

Keywords: Portland cement; C$_4$A$_3\bar{\text{S}}$ mineral; limestone; blastfurnace slag; cement properties.

1. Introduction

Portland cement clinker was composed by four kinds of major minerals of C$_3$S, C$_2$S, C$_3$A and C$_4$AF. The main hydration products formed in the hydration process were C-S-H, Ca(OH)$_2$ and some ettringite. This kind of cement showed low early strength and large volume of shrinkage during the hydration process [1, 2]. Addition of larger proportion of slag, fly ash and other industrial waste slag as admixtures can also lead to significant reduced strength of the cement. In recent years, new higher cementitious properties cement clinker mineral phase system, such as: alite-sulphoaluminate cement was obtained by introducing C$_4$A$_3\bar{\text{S}}$ minerals into Portland cement clinker system and replacement of C$_3$A. Alite-sulphoaluminate cement clinker was composed by four kinds of main mineral C$_3$S, C$_2$S, C$_4$A$_3\bar{\text{S}}$ and C$_4$AF. The C$_3$S and C$_4$A$_3\bar{\text{S}}$ were the mineral of high early strength, which can form hydration products of C-S-H, Ca(OH)$_2$ and larger amount of ettringite during the hydration process. This kind of cement showed higher early strength and smaller volume shrinkage or little expansion [3, 4]. When incorporated into a larger proportion of slag, fly ash, the strength of the cement was reduced to a small extension [5]. However, large amount of C$_3$S mineral formed at temperatures higher than 1400 °C, while the C$_4$A$_3\bar{\text{S}}$ mineral would decompose above 1350 °C. The coexisting temperature range of C$_3$S and C$_4$A$_3\bar{\text{S}}$ minerals was limited [6]. In addition, oxide atmosphere was required in burning process of alite-sulphoaluminate cement clinker in order to reduce the volatilization of SO$_3$, increasing the difficulty for industrial production control. Thus it is often challenging to meet the actual clinker mineral composition design requirements.

In order to introduce C$_4$A$_3\bar{\text{S}}$ mineral into Portland cement system to improve the properties of Portland cement, and to avoid difficult controlling in clinkering process of clinkers such as alite-sulphoaluminate cement clinker, this paper discusses the hydration and hardening properties of cement with addition of a certain amount of C$_4$A$_3\bar{\text{S}}$ mineral in Portland cement.

*Corresponding author

2. Experimental

2.1. Materials

This experiment used analytical pure chemical reagents of $CaCO_3$, Al_2O_3 and $CaSO_4 \cdot 2H_2O$ for synthesis of $C_4A_3\bar{S}$ mineral; Portland cement clinker was provided by a cement plant (in Shangdong China); Other raw materials were the blastfurnace slag, fly ash, limestone and phosphogypsum. The compositions of the raw materials were given in Table 1.

Table 1 Chemical composition of raw materials (wt. %)

	Loss	SiO_2	Al_2O_3	Fe_2O_3	CaO	MgO	SO_3	$\sum$
Fly ash	4.44	51.89	30.60	8.46	2.57	0.58	0.28	98.82
Phosphogypsum	18.10	5.02	0.45	0.24	30.17	0.24	42.50	96.72
Blast furnace slag	-0.85	34.74	17.26	1.58	36.67	7.53	—	96.93
Limestone	41.75	2.66	0.54	0.24	49.69	2.57	—	97.45
Portland cement clinker	0.36	21.22	5.95	3.64	62.50	2.45	0.33	96.45

2.2. Preparation of the samples

2.2.1. Synthesis of calcium sulfoaluminate (3CaO·3Al₂O₃·CaSO₄)

The composition of the controlled raw mixtures contained the following (wt.%): $CaCO_3$ (33.56), Al_2O_3 (38.19), $CaSO_4 \cdot 2H_2O$ (28.25). Considering the decomposition of $CaSO_4 \cdot 2H_2O$ during the burning process, an overdose of $CaSO_4 \cdot 2H_2O$ was used [7]. The raw mixtures were prepared by mixing adequate quantities of pure chemical reagents, grinded in a laboratory ball mill, and pressed at pressure of 25 MPa into discs using a mould of 30 mm in diameter and 5 mm in thickness. The discs were burned to clinker in an electric furnace with silicon molybdenum bars at 1300 °C for 2 h, and then removed from the furnace to cooled rapidly in air, and obtained $C_4A_3\bar{S}$ mineral.

2.2.2. Preparation of the cement

The cement was made by mixing $C_4A_3\bar{S}$ mineral, Portland cement clinker, the mineral admixtures (blastfurnace slag, fly ash and limestone) in different proportion and 5% phosphogypsum, and grinding in a laboratory ball mill to a specific surface area of $\sim$ 320 $m^2 \cdot kg^{-1}$. The proportions of cements were given in Table 2 (Groups A-F).

Table 2 Proportions of cements (wt.%)

Samples	Clinker	Phosphogypsum	Blast furnace slag	Fly ash	Limestone	$C_4A_3\bar{S}$ mineral
A0	95	5	—	—	—	—
A1	92	5	—	—	—	3
A2	90	5	—	—	—	5
A3	88	5	—	—	—	7
A4	85	5	—	—	—	10
B0	55	5	40	—	—	—
B1	52	5	40	—	—	3
B2	50	5	40	—	—	5
B3	48	5	40	—	—	7

(Continued)

Table 2 (*Continued*)

Samples	Clinker	Phosphogypsum	Blast furnace slag	Fly ash	Limestone	$C_4A_3\bar{S}$ mineral
B4	45	5	40	—	—	10
C0 (A2)	90	5	—	—	—	5
C1	80	5	10	—	—	5
C2	70	5	20	—	—	5
C3	60	5	30	—	—	5
C4 (B2)	50	5	40	—	—	5
D0	90	5	—	—	5	—
D1	87	5	—	—	5	3
D2	85	5	—	—	5	5
D3	83	5	—	—	5	7
E0 (D2)	85	5	—	—	5	5
E1	70	5	15	—	5	5
E2	60	5	25	—	5	5
E3	50	5	35	—	5	5
F0 (D2)	85	5	—	—	5	5
F1	70	5	—	15	5	5
F2	60	5	—	25	5	5
F3	50	5	—	35	5	5

2.2.3. *Measurement of the properties*

In order to investigate the formation of mineral phases, the mineral composition of $C_4A_3\bar{S}$ mineral and Portland cement clinker was analyzed by the XRD (Cu, Kα, λ=0.154 nm), the microstructures of the $C_4A_3\bar{S}$ mineral were observed by scanning electron microscope (SEM). The content of free lime in the clinker was determined chemically using the ethanol-glycerin method.

Compressive strengths were tested on 2-cm-cubes prepared from cement paste with a water/cement ratio of 0.285 at 3, 7 and 28 days after 24 h curing in moist air and subsequently in water at 20 °C. The setting time and water demand was determined at standard consistency (in accordance with Chinese standard GB1346-89).

The hydration of the cement pastes were stopped by washing with ethanol, the hydration productions were determined by XRD analysis and TG-DTA tests after 3 days hydration.

3. Results and Discussion

3.1. *Mineral formation of clinkers*

Ethanol-glycerin method is used to determine the content of f-CaO in the clinker, and the result shows only trace of f-CaO, indicating that the reaction between CaO, Al2O3, and CaSO4 is completed at high temperature. The XRD patterns and SEM pictures of mineral are shown in Figure 1 and Figure 2. According to Figure 1, the diffraction peaks of mineral are sharp and intense (d =0.492, 0.376, 0.325, 0.291, 0.265, 0.246 and 0.217 nm). From Figure 2, the crystal of minerals is regular, clear, and uniformly distributed. The spectra and the diagram indicate the good formation of the mineral.

Ethanol-glycerin method is used to determine the content of f-CaO in Portland cement clinker, and the result is 0.86%, indicating that the mineral reaction is completed at high temperature. The XRD patterns of the clinkers are shown in Figure 1. According to Figure 1, the clinker contains main minerals of: C_3S (d=0.303, 0.219, 0.176 nm), C_2S (d=0.278, 0.274, 0.260 nm), C_3A (d=0.270,

0.191, 0.220 nm) and C_4AF (d=0.277, 0.263, 0.192 nm). The diffraction peaks of different minerals are clear and sharp. This indicates the good formation of clinkers mineral and better quality clinker.

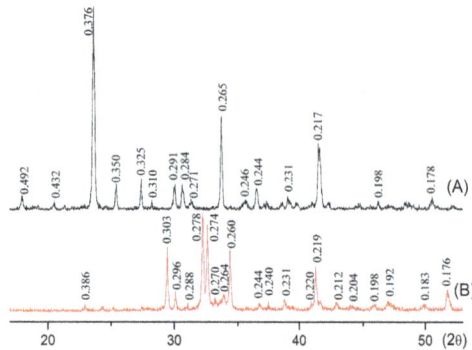

Fig. 1 XRD patterns of $C_4A_3\overline{S}$ mineral (A) and clinker (B)

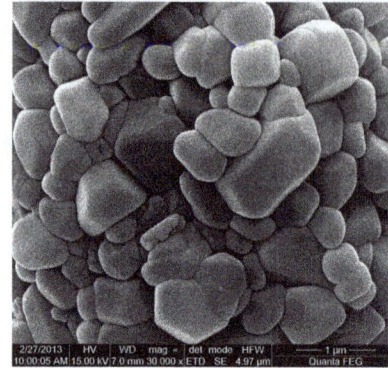

Fig. 2 SEM photograph of $C_4A_3\overline{S}$

3.2. Influence of $C_4A_3\overline{S}$ mineral on cement strength

The strength development of the cement pastes at various hydration ages are shown in Figure 3, DTA curves of the cement after 3 days hydration are shown in Figures 4. From Figure 3 (a), it is observed that 3 days strength of cement increases obviously and 7 days and 28 days strengths are steadily developed when the addition of $C_4A_3\overline{S}$ mineral is 3-5% (group A). With more than 5% of $C_4A_3\overline{S}$ mineral content, 3 days strength slightly increases, but 28 days strength decreases and is obviously lower than Portland cement (A0) without $C_4A_3\overline{S}$ mineral. This indicates that the properties of high early strength of $C_4A_3\overline{S}$ mineral is shown with a suitable addition amount to the Portland cement, and results in higher early strength of cement paste to make up the lack of lower early strength of Portland cement. In the meantime, the silicate phase in Portland clinker hydrates continually with the prolonged hydration time, and the later strength of the cement steadily increases. For the cement with addition of 40% slag (group B), 3 days strength of cement increases appreciably, and 28 days strength also increases obviously with 3-5% of $C_4A_3\overline{S}$ mineral. According to Figure 3 (b), when the addition of $C_4A_3\overline{S}$ mineral is 5%, and the addition of slag is 10-20% (group C), 3 days age strength of cement decreases slightly compared to cement C0 (without slag), but 7 days and 28 days strengths are higher than cement C0. This is attributed to the lower hydration activity of blastfurnace slag. It can decrease significantly the early age strength of the cements when a large amount is added. However, by adding a suitable amount $C_4A_3\overline{S}$ mineral, it can ensure to form more ettringite and gel at early age of hydration paste and increase early strength of the cement [5]. At the prolonged hydration times, silicate phase hydrates continually along with the hydration of the activity minerals in slag to ensure the steadily increment of the later strength of the cement.

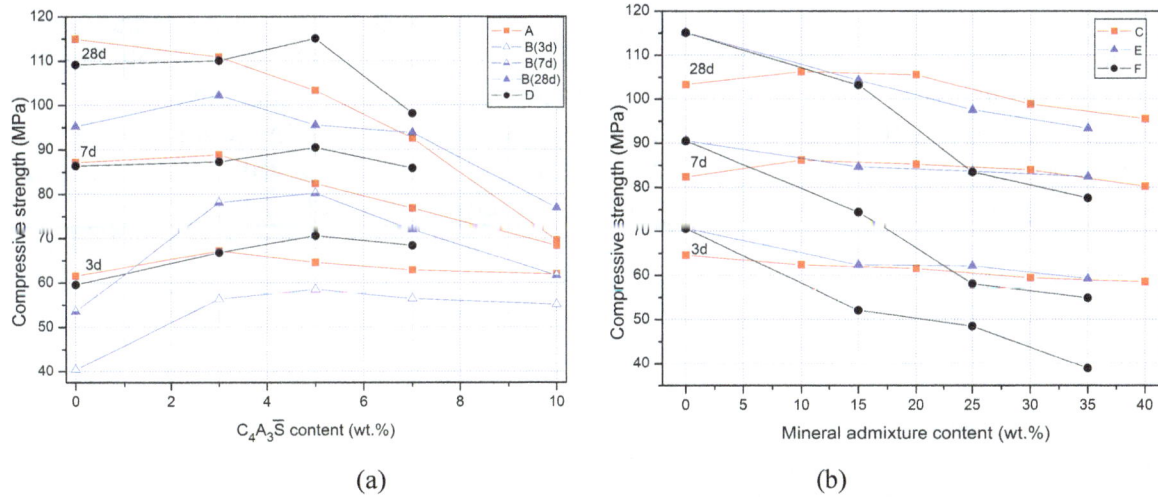

(a)　　　　　　　　　　　　　(b)

Fig. 3 Compressive strength of cement

From Figure 4, TG-DTA analysis demonstrates that the dehydration weight loss of cement A0, A1, C2 and C4 of 3 days hydration at about 116 °C are 7.85, 8.81, 7.89 and 7.21%, respectively. On the other hand, weight loss of cement A0, A1, C2 and C4 at about 460 °C of 6.10, 5.07, 4.69 and 4.93% respectively are due to the dehydration of $Ca(OH)_2$.

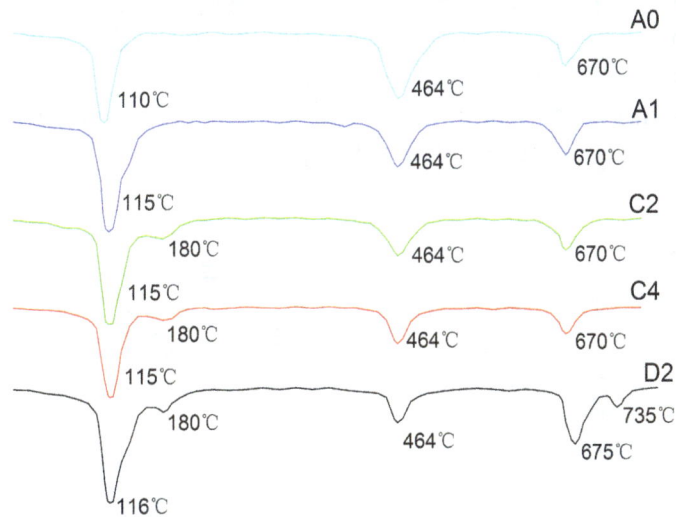

Fig. 4 DTA curves of cements after 3 days hydration

This indicates that the addition of suitable amount of $C_4A_3\bar{S}$ mineral can increase the formation of ettringite and decrease the formation of $Ca(OH)_2$. The results indicate that the presence of $C_4A_3\bar{S}$ mineral can improve the matching relationship of hydration products in the paste, increase the contents of acicular mineral ettringite and gel, and decrease the contents of sheet mineral $Ca(OH)_2$. More ettringite and gel are formed in the hydration paste even with a large addition amount of slag. Therefore, a suitable gel/crystal ratio could lead to a higher strength for hardening cement paste [8].

According to Figure 3 (a), it is also observed that various age strengths of the cement can appreciably increase with addition of suitable amount of $C_4A_3\bar{S}$ mineral in the Portland cement with 5% limestone (group D). The strength is good especially for sample with addition of 5% $C_4A_3\bar{S}$ mineral. According to Figure 3 (b), it is observed that strength properties of the cement, in which 5% of slag or fly ash is replace by limestone, are improved obviously with addition of 5% $C_4A_3\bar{S}$ mineral (group E and F). Previous research has shown that CaCO3 has multiple functions in cement: As the nucleating agent to promote the hydration of C3S; participation in the hydration reaction leading to high Ca/Si ratio of C-S-H gel and increment of hydration products in the hydration process; reduction of Ca(OH)2 grain size for more uniform distribution; participation in formation of the carbon aluminate hydrate similar to ettringite; reaction with Ca(OH)2 to form basic calcium carbonate [Ca3(CO3)2(OH)2·1.5H2O] to strengthen the structure of interfacial zone[9, 10]. From Figure 4, TG-DTA analysis demonstrates that the dehydration weight loss of sample D2 is 8.85% at about 116 °C, more than sample A1 (8.81%); 3.55% at about 460 °C, smaller than sample A1 (5.07%); and 6.87% at about 670 °C, more than sample A1 (3.52%). These results show that the coexist of the $C_4A_3\bar{S}$ mineral and limestone can promote the hydration of the minerals in clinker and mixed materials in the hydration system, increase the hydration products of acicular AFt phase and the gel *etc.*, reduce the flaky mineral Ca(OH)2 formation amount, and benefit the development of cement strength.

3.3. Influence of $C_4A_3\bar{S}$ mineral on the setting time of cement

The setting time of the cement is shown in Figure 5. From Figure 5 (a), it is observed that the initial and final setting time is shorter with the addition of the $C_4A_3\bar{S}$ mineral in the cement (groups A, B and D). Increment addition of the $C_4A_3\bar{S}$ mineral results in even shorter setting times. Comparing group A, and group D in which 5% of limestone is used instead of clinker, the initial setting time of group D is slightly longer, and the final setting times are approximately equal. Comparing group A, and group B with suitable addition amount of slag, the initial and final setting times of group B are prolonged. From Figure 5 (b), it is shown that the setting times of cement are prolonged with increment of admixtures. The setting time of cement is especially prolonged with addition of fly ash (group F).

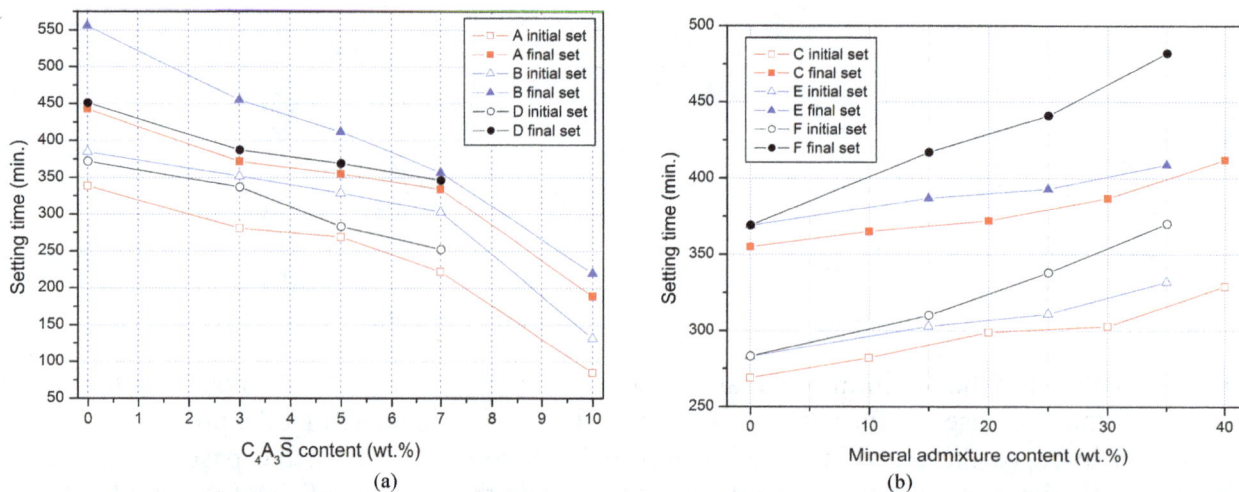

Fig. 5 Setting time of cement

The changes of setting time of the cement are related to hydration characteristics of the $C_4A_3\bar{S}$ mineral. With addition of $C_4A_3\bar{S}$ mineral in the cement, more hydration product of AFt phase is formed at early age of hydration process, and results in shorter setting times. The effects of admixtures such as blastfurnace slag, fly ash and lime stone on the setting time of the cement are consistent with some other research results of Portland cement [2].

4. Conclusion

1. The addition of $C_4A_3\bar{S}$ mineral can shorten the initial and final setting times of the cement. Increased amount of $C_4A_3\bar{S}$ mineral significantly decreased the setting times.

2. The addition of 3-5% of $C_4A_3\bar{S}$ mineral in Portland cement resulted in increased early strength and steadily growth of the late strength of the cement. Excess amount of $C_4A_3\bar{S}$ mineral (> 5%) increased the early strength of cement, but constrained the late strength development to lower levels compared to Portland cement.

3. For Portland cement with blastfurnace slag, addition of 3-5% $C_4A_3\bar{S}$ mineral increased obviously the early and late strengths. When the addition of $C_4A_3\bar{S}$ mineral was 5%, even with 10-20% of slag, the early strength of the cement was only slightly lower compared to Portland cement. However, the 7d and 28d strengths were significantly higher compared to Portland cement.

4. With the addition of 5% $C_4A_3\bar{S}$ mineral along with 5% limestone in the cement composition, the various age strengths of cement increased significantly with or without the addition of admixtures.

References

1. H. F. W. Talor, Cement Chemistry, Academic Press, London, 1990.
2. Shen Wei, Huang Wenxi and Min Panrong, Cement Technology (in Chinese), Wuhan University of Technology press, Wuhan, 1991.
3. I. V. Kravchenko, T. V. Kuzhecova, V. B. Khlusov, Alite-sulphoaluminate clinker and its cement, Cement. 8 (1978) 7-8.
4. T. V. Kuzhecova, N. G. Zealishvili, D. L. Mendeleeva, Manufacture and study of modified alite-sulphoaluminate clinker, Cement. 6 (1991) 39-43.
5. Li Yanjun, Liu Xiaocun, Zhang Ning, Study of the properties of alite-sulphoaluminate cement with blastfurnace slag, ZKG. 53 (2000) 602-605.
6. Liu Xiaocun, Li Yanjun, Shan Lianmei, Influence of ZnO and /or CaF2 on mineral formation of C3S and $C_4A_3\bar{S}$, Journal of Building Materials (in Chinese). 6 (2003) 9-12.
7. Zhang Pixing, Chen Yimin, Preparation of pure mineral $C_4A_3\bar{S}$ for structure analysis by orthogonal design, Journal of China Building Materials Academy (in Chinese). 1 (1989) 297-303
8. Li Yanjun, Liu Xiaocun, Li Shiqun, Hu Jiashan, Influence of P2O5 on the formation and hydration properties of $C_4A_3\bar{S}$, Materials Research Innovation. 4 (2001) 241-244.
9. Zhang Chunmei, V. S. Ramachandran, Influence of calcium carbonate as a fine filler on the hydration of tricalcium silicate, Journal of the Chinese Ceramic Society (in Chinese). 16 (1998) 110-117.
10. Li Yanjun, Liu Xiaocun, Influence of limestone on alite-sulphoaluminate cement properties, Bulletin of the Chinese Ceramic Society (in Chinese). 13 (1994) 64-68.

Investigation of Acoustic Emission in Grinding of 2.5D Woven Fiber-reinforced Ceramic Composites

Yu-Guo Wang[*], Chao Ding, Bin Lin

Key Laboratory of Advanced Ceramics and Machining Technology, Ministry of Education, Tianjin University,
Tianjin 300350, China
[*]Email: wyuguo@tju.edu.cn

Fiber-reinforced ceramic composites are relatively promising materials in aerospace and other high-tech areas. Machining difficulty and high cost in grinding are the main impediments to the application of those composites. In this work, the grinding mechanism of 2.5-dimensional woven quartz fiber-reinforced ceramic matrix composites was researched by experimental analysis. The acoustic emission (AE) combined with wavelet analysis was adopted to evaluate the grinding process. The effects of grinding parameters on wavelet energy were investigated. The AE energy increases with the increase of depth of cut and feed speed respectively, while it decreases with the increase of peripheral wheel speed in grinding process. The 2.5D composite is inhomogeneous and anisotropic. Different kinds of AE signals are generated when grinding parameters are changed.

Keywords: Acoustic emission; 2.5D woven ceramic composites; grinding; wavelet energy.

1. Introduction

Fiber-reinforced ceramic composites (FRCC) are widely applied in aerospace, national defense, and other high technology fields due to their superior properties [1, 2]. However, it is quite difficult to machine them because of their high hardness, brittleness and the fibers. It has been reported FRCC are strong anisotropic and inhomogeneous, which introduces many specific problems in machining process such as matrix crack, delamination, fiber pull-out, fiber breakage, fiber/matrix debonding [3,11].

Although the FRCC are usually made to the final size of the products, subsequent machining process, mainly trimming, drilling, and grinding are needed to achieve desired geometry, assembly requirements, and surface integrity. Abrasive machining of FRCC by means of grinding with diamond wheel is the most primary process to achieve high dimensional accuracy and desired surface finish [4, 5]. Wang and Lin [6] studied the grinding force and surface morphology of FRCC. It was found that grinding force were increased when the depth of cut or workpiece feed speed was increased. Surface morphology of the composites was also analyzed by scanning electron microscopy.

As a non-destructive techniques, acoustic emission (AE) has been widely used to evaluate the fracture of composite materials [7, 8, 9, 10]. In this work, the grinding mechanism of 2.5D woven quartz fiber-reinforced ceramic composites is studied by experimental analysis. The AE technology combined with wavelet transform analysis is used to evaluate the grinding process. The work will benefit to the design, manufacture and application of the composites.

2. Experimental Procedure

2.1. *Specimen*

The specimen material used in the present work was 2.5D woven fiber-reinforced ceramic matrix composites. The reinforcing material used for this investigation was quartz fibers. The matrix

---

[*]Corresponding author

material was SiO2. Quartz fibers were woven by weft yarn and warp yarn to form precast bodies by 2.5D braid technology (shallow bend-joint).

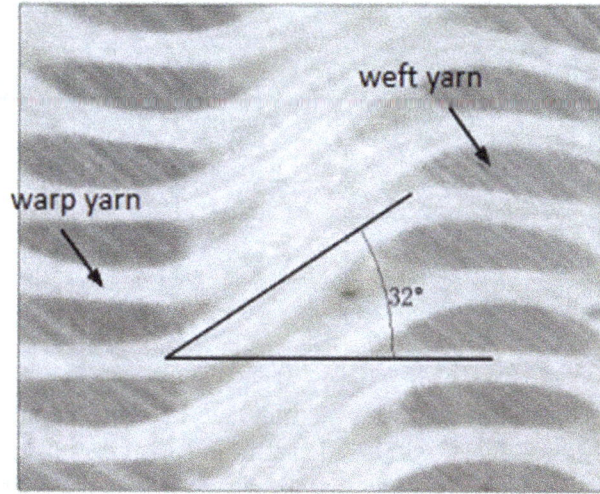

Fig. 1 Grinding surface

The fabric bodies were vacuum impregnated using colloidal silica. Then the composites were fabricated by silicasol-infiltration-sintering method. The specimen is shown in Fig. 1. Grinding orientation is perpendicular to the section of weft yarn. The grinding direction is from left to right.

2.2. Grinding experimental apparatus

The grinding was carried out on a CNC optical profile grinding machine (type MK9025). Table 1 summarizes settings of main grinding parameters. The main experimental apparatuses are shown in Fig. 2.

Table 1 Major grinding parameters

Parameters	Content
Grinding wheels	Electroplating diamond grinding wheels
Wheel diameter	100mm
Grain mesh size	100#
Coolant	Dry grinding
Grinding Condition 1	Feed speed $v_f = 4m/min$; peripheral wheel speed $v_s = 20m/s$; Depth of cut $a_e = 0.05 - 0.4mm$
Grinding Condition 2	Feed speed $v_f = 3 - 7m/min$; peripheral wheel speed $v_s = 20m/s$; Depth of cut $a_e = 0.2mm$
Grinding Condition 3	Feed speed $v_f = 4m/min$; peripheral wheel speed $v_s = 15 - 28.3m/s$; Depth of cut $a_e = 0.2mm$

25

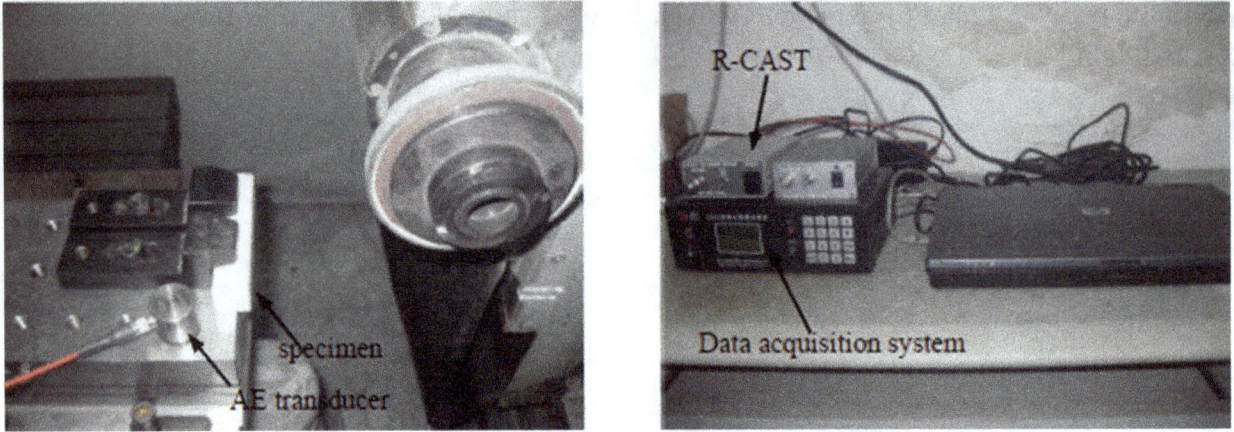

Fig. 2 Experimental apparatus

A piezoelectric AE transducer, model 1045S, manufactured by Japan Fuji Corp., was utilized. The bandwidth is in the range of 100 kHz to 1500 kHz, and the sensitivity is 51±3dB. The transducer was clamped on the platform near the specimen with screw and lock nut which prevent screw loosing. A pre-amplifier, model R-CAST, with 55dB gain and 10 kHz to 5MHz bandwidth was used for signal conditioning. The noise is 5μVrms, and the maximum output voltage is 7V.

Then the AE signals were processed by data acquisition system, model UDAQ-50612/20612, manufactured by Top Measurement & Control Technology CO.LTD. The data acquisition system was equipped with 4 channels, with a sampling rate of 20M/50MSps per channel. The raw data were acquired and performed basic data processing using the software TopView2000.

2.3. *Wavelet-based methodology for AE signals analysis*

The wavelet transform, introduced in the area of applied mathematics and signal processing, was applied to an arbitrary square-summable real function and transformed into a series of shifted and dilated sums of wavelets [7]. It has been reported that wavelet-based signal processing was a useful tool in the analysis of AE [8, 9, 10]. Wavelet transform is using wavelet functions gradually to approach object function in the particular space. Each decomposed wavelet level represents a specific sub-band which contains different information of target signal with other levels.

The discrete wavelet transform (DWT) is defined as:

$$W_f(j,k) = \int_{-\infty}^{\infty} f(t)\psi_{j,k}^* dt \qquad \psi_{j,k}(t) = 2^{-j/2}\psi(2^{-j}t - k) \tag{1}$$

Where $W_f(j,k)$ are the wavelet transform coefficients. $f(t)$ is the analyzed signal, $\psi_{j,k}(t)$ is the wavelet used for analysis, j represents the scale and determines the stretching and compressing of the wavelet, while k is referred to the time shift or translation, and * denotes the complex conjugation of a wavelet.

The inverse (or reconstruction) of wavelet transform can be expressed as:

$$f(t) = c\sum_{-\infty}^{\infty}\sum_{-\infty}^{\infty} W_f(j,k)\psi_{j,k}(t) \tag{2}$$

Here c is a factor which is irrelevant to the signals.

According to the method of wavelet transformation and reconstruction, the AE signal $f(t)$ can be decomposed into $j+1$ frequency bands on the jth level, where $c_j f(t)$, $d_j f(t) \ldots d_1 f(t)$ are the components of the jth level. In accordance with the level of the decomposed signal, the formula of decomposed wavelet energy in the jth level can be expressed as:

$$\begin{cases} E_j^c f(t) = \sum_{n=1}^{N} \left(c_j f(t) \right)^2 \\ E_i^d f(t) = \sum_{n=1}^{N} \left(d_i f(t) \right)^2, \quad i = 1, 2, \cdots j \end{cases} \tag{3}$$

where $E_j^c f(t)$ is the total energy of low-frequency signal components on level j, $E_i^d f(t)$ is the total energy of high-frequency signal components on level j.

And, the total energy of the signal is defined:

$$Ef(t) = E_j^c f(t) + \sum_{i=1}^{j} E_i^d f(t) \tag{4}$$

The AE signals collected during grinding experiments were decomposed into different wavelet levels. Each of them was related to a specific frequency range. In this work the daubechi 8 wavelet was chosen as the mother wavelet to analyze the AE signals. As the sampling frequency of AE signals was 1000 kHz in the grinding process, the frequency range which can be analyzed is 0-500kHz according to the sampling theory. The six frequency bands are separately are 0-15.625kHz, 15.625-31.25 kHz, 31.25-62.5 kHz, 62.5-125 kHz, 125-250 kHz, 250-500 kHz.

In this work, wavelet energy was used to analyze the effects of grinding parameters in grinding process.

3. Results and Discussion

In grinding process, material removal efficiency, removal mechanism and machining quality can be greatly affected by grinding parameters. Different grinding parameters may result in different fracture, which may generate different AE sources. The sources cause the different energy distribution in the different frequency bands.

Therefore, relationship between AE wavelet energy and grinding parameters can be used to analyze the machining quality in grinding process. The grinding experiments were carried out in different conditions. AE wavelet energy shown in Fig. 3, Fig. 4 and Fig. 5 was given by equation (4), summed over the six frequency bands mentioned in section 2.3.

3.1. Effects of depth of cut on AE energy

The relationship between depth of cut and AE signal energy is shown in Fig. 3. The grinding parameter is in accordance with Grinding Condition 1(shown in Table 1). It can be found that the AE signal energy increases with the increase of depth of cut, and grinding force also increases [6]. The reason is that contact arc length, total number of effective abrasive particles, and abrasive particles cutting thickness increase, which results in the increase of contact area, interaction between wheel and specimen. Therefore material removal efficiency, fibers cutting and matrix fracture increase too. AE signal sources increase, which leads to the increase of AE energy.

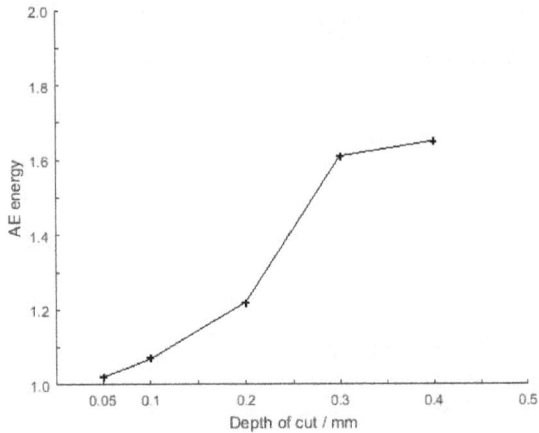

Fig. 3 Depth of cut and AE signal energy

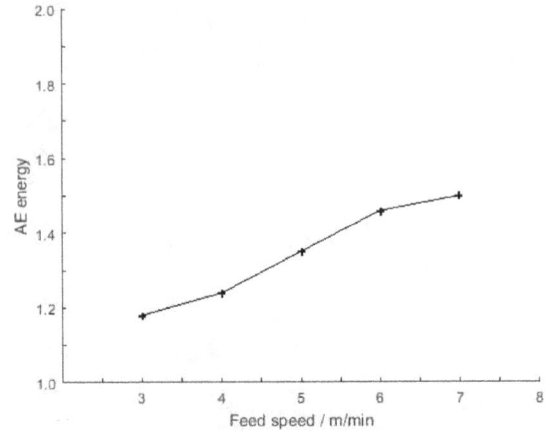

Fig. 4 AE signal energy and feed speed

3.2. Effects of feed speed on AE energy

Fig. 4 shows the effects of feed speed on AE energy under Grinding Condition 2 (shown in Table 1). It demonstrates the AE signal energy increases with the increase of feed speed, and grinding force is in the same trend [6]. Abrasive particle cutting thickness and undeformed chip increases, so the material removal rate and the total number of effective abrasive particles in unit time also increase. The increase of AE signal sources results in the increase of AE energy.

3.3. Effects of peripheral wheel speed on AE energy

Fig. 5 demonstrates the relationship between peripheral wheel speed and AE energy under Grinding Condition 3 (shown in Table 1). It can be found that the AE signal energy decreases with the increase of peripheral wheel speed. Because contact time of single abrasive particle decreases, chips deformation of fibers and matrix and material strain rate decrease with the increase of peripheral wheel speed. Therefore, the interaction between the wheel and specimen decreases, which results in the decrease of AE energy.

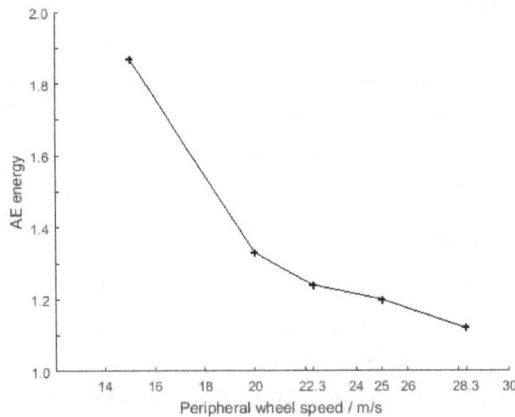

Fig. 5 AE signal energy and peripheral wheel speed

3.4. *Analysis of grinding surface morphology*

It can be seen from Fig. 6 that bundles of the fibers are fractured. There is a lot of matrix debris on the grinding surface, and there are also a large number of matrix cracks. The matrix on fibers surface fall off seriously near the fracture surface, and matrix is even separated from fibers.

Matrix cracks mainly extend along the interface of matrix and fibers, fibers and matrix appear separated simultaneously. Fiber bundles mainly bear the shearing and bending force of the abrasive particles. Fracture morphology which is flat-shape is the result of shearing force, and fracture morphology which is bevel-shape is the result of bending stress. A large number of fiber breakages in the grinding precisely cause the bundle fracture.

Fig. 6 Grinding surface morphology

4. Conclusion

The AE signals collected in grinding process of 2.5-dimensional woven quartz fiber-reinforced SiO2 composites were analyzed using wavelet analysis. The AE energy increases with the increase of depth of cut and feed speed respectively, while it decreases with the increase of peripheral wheel speed in grinding process. The 2.5D composites is inhomogeneous and anisotropic, different kinds of AE signals are generated when grinding parameters are changed.

Acknowledgment

This work was funded by the National Natural Science Foundation of China (No. 51305296).

References

1. R. Naslain, Design, preparation and properties of non-oxide CMCs for application in engines and nuclear reactors: an overview, Composites Science and Technology 64 (2004) 155–170.
2. Papakonstantinou CG, Balaguru P, Lyon RE, Comparative study of high temperature composites. Compos B. Eng 2001: 32(8), 637–49.
3. Mkaddem A, El Mansori M. Finite element analysis when machining UGF-reinforced PMCs plates chip formation crack propagation and induced damage. J Mater Des 2009. doi:10.1016/j.matdes.2008.12.009.
4. Hu, N.S., Zhang, L.C., 2003. A study on the grindability of multidirectional carbon fiber-reinforced plastics. Journal of Materials Processing Technology 140 (2003) 152–156.

5. Santiuste Carlos, Soldani Xavier, Miguélez Maria Henar, Machining FEM model of long fiber composites for aeronautical components. Composite Structures 92 (2010), 691–698

6. Wang YuGuo, Lin Bin, Research of the grinding force and surface morphology of fiber-reinforced ceramic matrix composite. Advanced Materials Research (2012), Vol. 569, 132-135.

7. Qi Gang, Barhorst Alan, Hashemi Javad & Kamala Girish, Discrete wavelet decomposition of acoustic emission signals from carbon-fiber-reinforced composites, Composites Science and Technology 57 (1997), 389-403.

8. Mareca, A., Thomasa, J.-H. Guerjouma, R. El., Damage characterization of polymer-based composite materials: Multivariable analysis and wavelet transform for clustering acoustic emission data. Mechanical Systems and Signal Processing 22 (2008), 1441–1464.

9. Gang Qi, Wavelet-based AE characterization of composite materials, NDT&E International 33 (2000), 133–144

10. E. Maillet, N. Godin, , M. R'Mili, P. Reynaud, J. Lamon, G. Fantozzi, Analysis of Acoustic Emission energy release during static fatigue tests at intermediate temperatures on Ceramic Matrix Composites: Towards rupture time prediction, Composites Science and Technology 72 (2012), 1001–1007

11. T.H. Loutas, V. Kostopoulos, C. Ramirez-Jimenez, M. Pharaoh, Damage evolution in center-holed glass/polyester composites under quasi-static loading using time/frequency analysis of acoustic emission monitored waveforms, Composites Science and Technology 66 (2006), 1366–1375

Effect of Ionic Radius on the Dielectric Properties of BaTiO$_3$-MNbO$_4$ Ceramics for Energy Storage Capacitors

Meng Wei, Ji-Hua Zhang*, Xiang-Xiang Dong, Hong-Wei Chen, Chuan-Ren Yang

State Key Laboratory of Electronic Thin Films and Integrated Devices, Collaboration Innovation Center of Electric Materials and Devices, University of Electronic Science & Technology of China, 4ᵗʰ Block 2Jianshebei Road Chenghua Area, Chengdu, Sichuan 610051, PR China
Email: jhzhang@uestc.edu.cn

The dielectric materials with high energy density and high energy efficiency are the core problem for energy storage applications. In this study, 0.9BaTiO$_3$-0.1MNbO$_4$ (M=Al, In, Y, Sm, Nd, Bi, La) ceramics are prepared by solid-state reaction method. The structure, dielectric properties and energy storage properties of the ceramics are systematically investigated. The results show that permittivity decrease firstly and then increase, lastly decrease gradually with the increased ionic radius. The differences of dielectric properties in 0.9BaTiO$_3$-0.1MNbO$_4$ systems are attributed by the effects of different ionic radius. When M=Bi, it exhibits the highest energy density of 0.797 J/cm^3 and energy efficiency of 92.5% in this study, because the second phase suppress the nonlinearity obviously. It is demonstrated that dielectric with larger P$_m$, slim hysteresis loop and lower nonlinearity, which are indispensable conditions, is a good way to improve energy density and energy efficiency in energy storage applications.

Keywords: BaTiO$_3$; dielectric properties; energy density; ionic radius.

1. Introduction

The pulse power capacitors were core component in power electronics, electric gun, directed energy weapon, active armor, and so on [1-4] Therefore, the high pulse power capacitors with high energy density dielectric materials have attracted more attentions. High energy density dielectric materials needed both high permittivity and high breakdown strength (BDS), according to the equation $W = \int E_b \, dP$ [5-7]. Nowadays, relaxor ferroelectrics and antiferroelectrics have received more attention on energy storage applications, due to larger P$_s$, small P$_r$ and moderate BDS. While most antiferroelectrics contain lead, e. g. (Pb, La)(Zr, Sn, Ti)O$_3$ [8-11] and (Pb, La)(Zr, Ti)O$_3$ [12-14], which belong to heavy metal pollution. Therefore, relaxor ferroelectrics are more suitable for energy storage applications. Recently, a new kind of materials BaTiO$_3$-BiMO$_3$ (M=Sc [15], Al [16], Y [17], Gd [18], Yb [19], etc.) ceramics were considered as potential candidates. Most of the researchers were focused on the improvement of permittivity and BDS. However, little attention was paid on the effect of different ionic radius on the dielectric properties in BaTiO$_3$-based materials. In this paper, the dielectric properties of a series materials BaTiO$_3$-MNbO$_3$ (M=Al, In, Y, Sm, Nd, Bi, La) ceramics with perovskite structures were investigated. The proportion of 0.9BaTiO$_3$-0.1MNbO$_3$ ceramics was confirmed below the solid solubility, according to the previous researches. This study could optimize an appropriate ionic radius doping in BaTiO$_3$-MNbO$_4$ system for improving energy density in energy storage applications.

2. Methods

The 0.9BaTiO$_3$-0.1MNbO$_4$ ceramics was prepared by traditional solid state sintering. BaCO$_3$(99.5%), TiO$_2$(99.99%), Nb$_2$O$_5$(99%), Bi$_2$O$_3$(99%), Al$_2$O$_3$(99.5%), In$_2$O$_3$(99.99%), Y$_2$O$_3$(99.99%), Sm$_2$O$_3$(99.5%), Nd$_2$O$_3$(99.5%), La$_2$O$_3$(99.5%) were used as raw materials. The powders were mixed according to the 0.9BaTiO$_3$-0.1MNbO$_4$ stoichiometry and ball-milled in

*Corresponding author

31

distilled water for 24h. The slurry was dried pre-sintered in oven at 1100°C for 2h. The calcined powders were ball-milled a second time for 24h. Then, dry the slurry, granulate the dried powders with 5% PVA and press the powders into pellets with diameter of 12mm and thickness of 0.2mm. The pellets were sintered in a closed crucible between 1150°C and 1250°C. The sintering temperature of different kinds of samples was confirmed by the maximum shrinkage and the minimum loss.

The structure characteristic of $0.9BaTiO_3$-$0.1MNbO_4$ ceramics were measured by using the X-ray diffractions, which could confirm the crystalline phases and the change of the lattice constant of the samples. The X-ray diffractometer (XRD, Bede QC200) was using CuKα radiation with λ=0.15406nm, which operated at 40kV and 30mA. The surface topographies of the samples were measured by scanning electron microscope (SEM, FEI Inspect-F, Holland). The temperature dependence of permittivity and loss of samples were measure from 100Hz to 1MHz and from 25°C to 300°C by using Wayne Kerr LCR Mete. The polarization-electric field (P-E) hysteresis loops were measured using a ferroelectric tester (RADIANT Precision LC, USA) at 100Hz. The changes of the capacitances with the electric field were measured at 1 kHz by using Agilent B1505A (Agilent Technologies, USA) at room temperature.

3. Results and Discussion

The XRD pattern of $0.9BaTiO_3$-$0.1MNbO_4$ (M=Al^{3+}, In^{3+}, Y^{3+}, Sm^{3+}, Nd^{3+}, Bi^{3+}, La^{3+}) ceramics are shown in Fig. 1. All the peaks fit the perovskite phase, except Bi^{3+}. $0.9BaTiO_3$-$0.1BiNbO_4$ beyond the solid solubility may because it is easy to form a second more stable phase. The lattice parameter (a), as show in Table 1, increases with increased ionic radius firstly, and then decreases, except Al^{3+}(53pm) because of smaller ionic radius for Ti^{4+}(60.5pm).

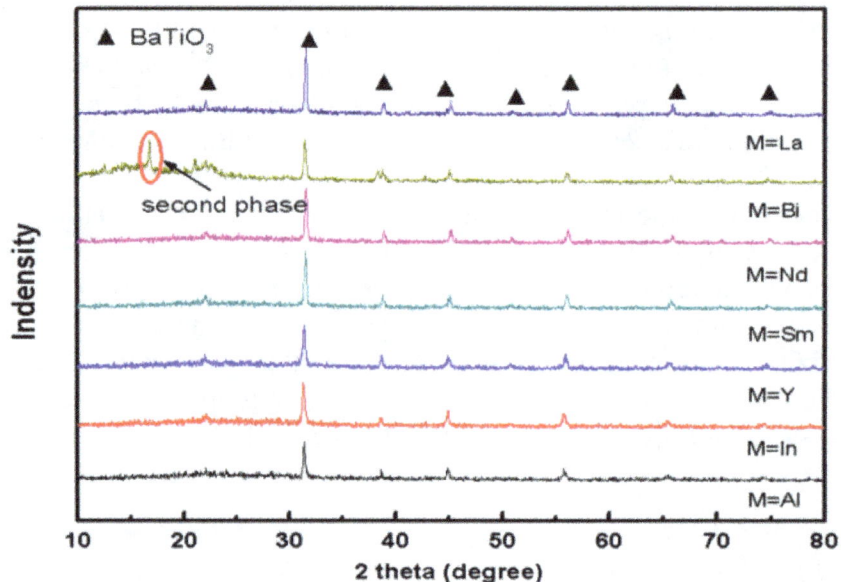

Fig. 1 X-ray diffraction profile of $0.9BaTiO_3$-$0.1MNbO_4$ ceramics

Table 1 The crystal parameter of the BT-MNb ceramics

Samples	Crystal phase	a(Å)	b(Å)	c(Å)	V(Å³)
BT-AlNb	Cubic	4.01852	4.01852	4.01852	64.89
BT-InNb	Tetragonal	3.97763	3.97763	4.03698	63.87
BT-YNb	Tetragonal	4.00264	4.00264	4.02138	63.65
BT-SmNb	Tetragonal	3.99657	3.99657	4.01576	64.14
BT-NdNb	Tetragonal	3.99344	3.99344	4.02995	64.27
BT-BiNb	Tetragonal	3.98279	3.98279	4.01962	64.14
BT-LaNb	Tetragonal	3.98141	3.98141	4.01927	63.7

The crystal phase is cubic when the substitution is smaller than Ti^{4+}, and the crystal phase change to tetragonal when the substitution is bigger than Ti^{4+}. The topography of the $0.9BaTiO_3-0.1MNbO_4$ ceramics is shown in Fig. 2.

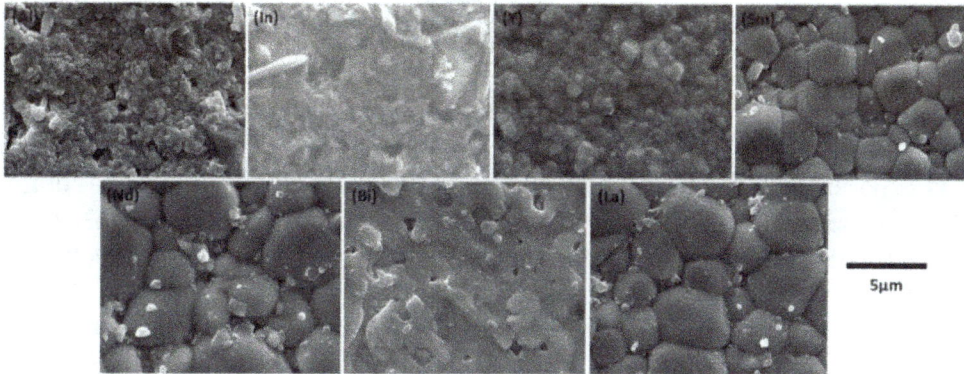

Fig. 2 Scanning electron microscopic images of thermal etched surface of $0.9BaTiO_3-0.1MNbO_4$ ceramics

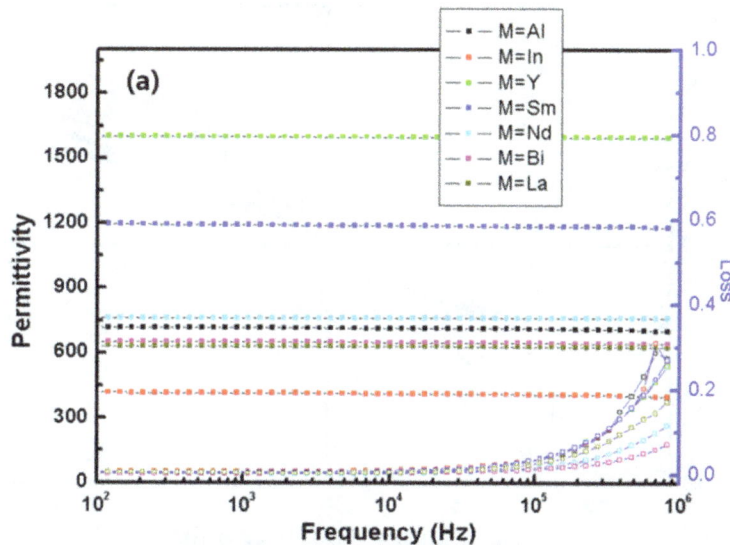

Fig. 3 (a) Dielectric constant and loss of $0.9BaTiO_3-0.1MNbO_4$ ceramics from 100 Hz to 1 MHz (at room temperature)

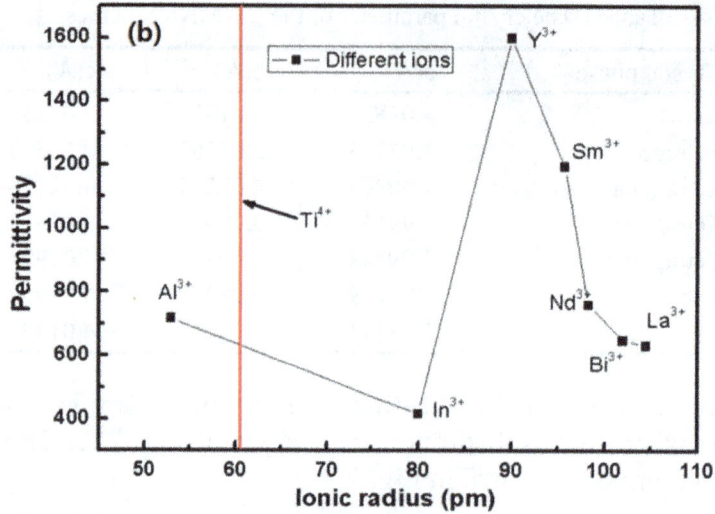

Fig. 3 (b) Relationship between dielectric constant and different ions in $0.9BaTiO_3$-$0.1MNbO_4$ systems (at room temperature with 100 Hz)

It shows that an obvious increase in grain size with increased ionic radius, which is also gradually densely sintered. Fig. 3(a) shows that the frequency dependence on the dielectric properties. It is obvious that the frequency stabilities of $0.9BaTiO_3$-$0.1MNbO_4$ ceramics are very excellent and the losses are sufficient low. The dielectric properties of the $0.9BaTiO_3$-$0.1MNbO_4$ ceramics with different ionic radius exhibit the effect of ionic radius on dielectric properties, as shown in Fig. 3(b).

When ionic radius R_i is less than 90pm, the permittivity increase with the increased ionic radius. When M=Y (ionic radius R_i= 90pm), the maximum permittivity achieve 1600. And then when R_i ＞90pm, the permittivity decrease with the increased ionic radius.

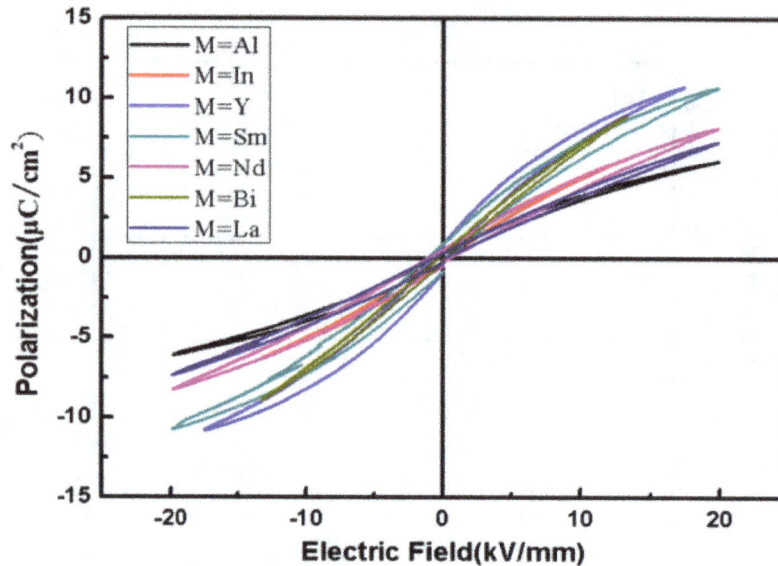

Fig. 4 Polarization-electric field relationship of $0.9BaTiO_3$-$0.1MNbO_4$ ceramics at 100 Hz

In Fig. 4, the P-E hysteresis loops of $0.9BaTiO_3$-$0.1MNbO_4$ ceramics are measured for analysis the relationship among P_{max}, P_r, energy storage density and energy efficiency, which could indicate the energy storage properties overall. The accurate values of P_{max}, P_r, energy storage density and energy efficiency are shown in Table 2.

Table 2 The properties of polarization and energy properties of $BaTiO_3$-$MNbO_4$ ceramics

	M=Al	M=In	M=Y	M=Sm	M=Nd	M=Bi	M=La
$P_m(\mu C/cm^2)$	6.1	6.281	10.764	10.724	8.221	8.915	7.343
$P_r(\mu C/cm^2)$	0.275	0.263	0.808	0.88	0.527	0.245	0.304
$W_1(J/cm^3)$	0.482	0.355	0.744	0.78	0.672	0.797	0.643
$W_2(J/cm^3)$	0.05	0.032	0.139	0.179	0.124	0.065	0.065
$\eta(\%)$	90.5	91.8	84.3	81.4	84.5	92.5	90.8
ionic radius(pm)	53	80	90	95.8	98.3	102	104.5

P_{max} increase with increased ionic radius firstly, and then decrease gradually, except Bi^{3+} which attribute to the second phase. When M=Y (90pm), P_{max} achieves the maximum value $10.764\mu C/cm^2$. Meanwhile, P_r increase with increased ionic radius firstly, and then decrease gradually. When M=Sm (95.8pm), P_r achieves the maximum value $0.88\mu C/cm^2$. What is interesting is that the energy density and energy efficiency exhibit the highest values, when M=Bi, which not fit the change of P_{max}. To explain the phenomenon, DC bias performances under C-V mode are detected. The dielectric properties exhibit continuous variations with the change in electric field, which show in Fig. 5(a, b). $0.9BaTiO_3$-$0.1AlNbO_4$ ceramics exhibit low P_m, low energy density and relative high nonlinearity, because the ionic radius of Al^{3+} is smaller than Ti^{4+}. With the increased ionic radius, the nonlinearities increase firstly, and then decrease. $0.9BaTiO_3$-$0.1BiNbO_4$ ceramics is an exception, because the second phase suppress the nonlinearity obviously, which exhibit the highest energy density and energy efficiency in this study.

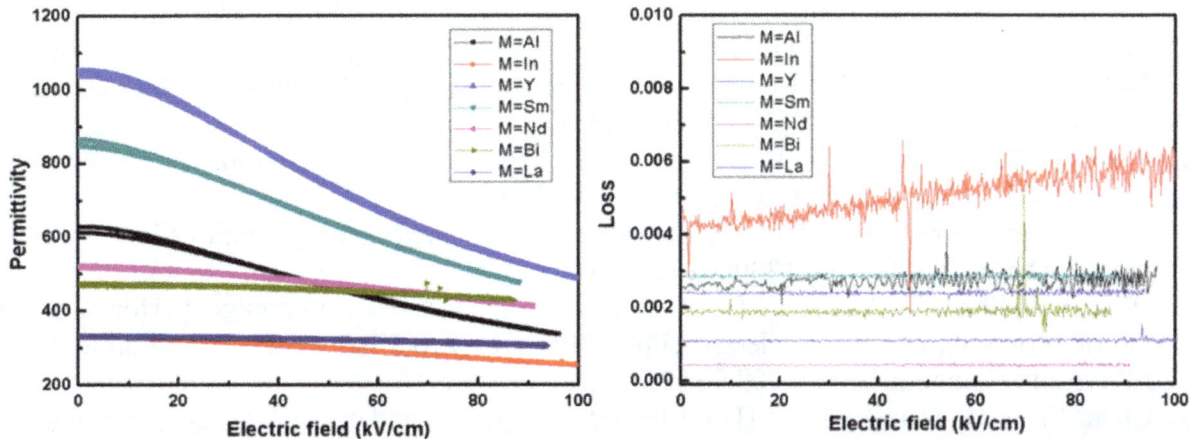

Fig. 5 Permittivity and loss versus DC electric field for $0.9BaTiO_3$-$0.1BiNbO_4$ ceramics at 1 kHz at room temperature

4. Conclusion

$0.9BaTiO_3$-$0.1MNbO_4$ (M=Al, In, Y, Sm, Nd, Bi, La) ceramics are successfully prepared by using solid-state reaction method. The structures, dielectric properties and energy storage properties are systematically investigated. The phase of $0.9BaTiO_3$-$0.1MNbO_4$ ceramics change from cubic (M=Al) to tetragonal (M= In, Y, Sm, Nd, Bi, La). With the increased ionic radius, permittivity decrease firstly, and then increase, finally decrease gradually in tetragonal phase. The differences of dielectric properties in $0.9BaTiO_3$-$0.1MNbO_4$ system are attributed to the effects of different ionic radius. With the increased ionic radius, the nonlinearities increase firstly, and then decrease. $0.9BaTiO_3$-$0.1BiNbO_4$ ceramics is an exception, because the second phase suppress the nonlinearity obviously, which exhibit the highest energy density of 0.797 J/cm^3 and energy efficiency of 92.5% in this paper. In order to gain high energy density and high energy efficiency for power pulse capacitors applications, P_m, P_r and nonlinearity should be taken into comprehensive consideration. This study could optimize an appropriate ionic radius doping in $BaTiO_3$-$MNbO_4$ system for improving energy density in energy storage applications, which possess larger P_m, slim hysteresis loop and lower nonlinearity. Therefore, the effect of ionic radius on dielectric properties should be paid more attentions.

Acknowledgment

This work was funded by the Innovation Foundation of Collaboration Innovation Center of Electronic Materials and Devices (No. ICEM2015-4002)

References

1. L. Wang, J. Liu, and J. Feng, A compact 100 kV high voltage glycol capacitor, The Review of scientific instruments, 86 [1] 014701 (2015).
2. S. Hameer and J. L. van Niekerk, A review of large-scale electrical energy storage, International Journal of Energy Research, 39 [9] 1179-95 (2015).
3. B. Huhman, J. Neri, and D. Wetz, Application of a compact electrochemical energy storage to pulsed power systems, IEEE Transactions on Dielectrics and Electrical Insulation. 22 [4] 1299-303 (2013).
4. S. A. Sherrill, P. Banerjee, G. W. Rubloff, and S. B. Lee, High to ultra-high power electrical energy storage, Physical chemistry chemical physics : PCCP, 13 [46] 20714-23 (2011).
5. T. Wang, L. Jin, C. Li, Q. Hu, X. Wei, and D. Lupascu, Relaxor Ferroelectric BaTiO3-Bi(Mg2/3Nb1/3)O3 Ceramics for Energy Storage Application, Journal of the American Ceramic Society, 98 [2] 559-66 (2015).
6. Z. Shen, X. Wang, B. Luo, and L. Li, BaTiO3–BiYbO3 perovskite materials for energy storage applications, J. Mater. Chem. A, 3 [35] 18146-53 (2015).
7. X. Hao, A review on the dielectric materials for high energy-storage application, Journal of Advanced Dielectrics, 03 [01] 1330001 (2013).
8. X. Wang, J. Shen, T. Yang, Y. Dong, and Y. Liu, High energy-storage performance and dielectric properties of antiferroelectric (Pb0.97La0.02) (Zr0.5Sn0.5−xTix)O3 ceramic, Journal of Alloys and Compounds, 655 309-13 (2016).
9. L. Chen, Y. Li, Q. Zhang, and X. Hao, Electrical properties and energy-storage performance of (Pb0.92Ba0.05La0.02)(Zr0.68Sn0.27Ti0.05)O3 antiferroelectric thick films prepared by tape-casing method, Ceramics International, 42 [11] 12537-42 (2016).
10. Q. Zhang, X. Liu, Y. Zhang, X. Song, J. Zhu, I. Baturin, and J. Chen, Effect of barium content on dielectric and energy storage properties of (Pb,La,Ba)(Zr,Sn,Ti)O3 ceramics, Ceramics International, 41 [2] 3030-35 (2015).

11. L. Zhang, S. Jiang, B. Fan, and G. Zhang, High energy storage performance in (Pb0.858Ba0.1La0.02Y0.008)(Zr0.65Sn0.3Ti0.05)O3- (Pb0.97La0.02)(Zr0.9Sn0.05Ti0.05)O3 anti-ferroelectric composite ceramics, Ceramics International, 41 [1] 1139-44 (2015).
12. Y. Zhao, X. Hao, and Q. Zhang, Energy-storage properties and electrocaloric effect of Pb(1-3x/2)LaxZr0.85Ti0.15O3 antiferroelectric thick films, ACS applied materials & interfaces, 6 [14] 11633-9 (2014).
13. Z. Hu, B. Ma, R. E. Koritala, and U. Balachandran, Temperature-dependent energy storage properties of antiferroelectric Pb0.96La0.04Zr0.98Ti0.02O3 thin films, Applied Physics Letters, 104 [26] 263902 (2014).
14. X. Hao, J. Zhai, L. B. Kong, and Z. Xu, A comprehensive review on the progress of lead zirconate-based antiferroelectric materials, Progress in Materials Science, 63 1-57 (2014).
15. H. Ogihara, C. A. Randall, and S. Trolier-McKinstry, High-Energy Density Capacitors Utilizing 0.7 BaTiO3-0.3 BiScO3 Ceramics, Journal of the American Ceramic Society, 92 [8] 1719-24 (2009)
16. M. Liu, H. Hao, Y. Zhen, T. Wang, D. Zhou, H. Liu, M. Cao, and Z. Yao, Temperature stability of dielectric properties for xBiAlO3–(1−x)BaTiO3 ceramics, Journal of the European Ceramic Society, 35 [8] 2303-11 (2015).
17. Y. Wang, Y. Pu, and P. Zhang, Investigation of dielectric relaxation in BaTiO3 ceramics modified with BiYO3 by impedance spectroscopy, Journal of Alloys and Compounds, 653 596-603 (2015).
18. G. Schileo, A. Feteira, K. Reichmann, M. Li, and D. C. Sinclair, Structure–property relationships in (1−x)BaTiO3–xBiGdO3 ceramics, Journal of the European Ceramic Society, 35 [9] 2479-88 (2015).
19. Z. Shen, X. Wang, B. Luo, and L. Li, BaTiO3–BiYbO3 perovskite materials for energy storage applications, J. Mater. Chem. A, 3 [35] 18146-53 (2015).

Discussion on Design Method of C20 Recycled Concrete Mix Proportion

Yan Wang[1], Bo Dong[1], Ai-Qin Zhang[2,*]

[1]Shandong province Hi-Speed Road and Bridge Maintenance Co., LTD. Shandong Jinan 250032

[2]Shandong Jiaotong University, Shandong Jinan 250357

*Email: zhangaq6563@163.com

In this paper, the design method of C20 recycled concrete mix proportion is studied by using recycled aggregate. The mechanical properties of C20 recycled concrete and common concrete are analyzed by contrast test. The experimental results show that the strength-distributing rule of recycled concrete is the same as that of common concrete. It follows the Bolomey Theory when the W/C is between 0.5-0.6. The volume stability of recycled concrete is shown as the expansion occurred in the early stage and dry shrinkage in the late. Its shrinkage strain is slightly higher than that of common concrete.

Keywords: Construction waste; factory recycled aggregate; surface-dry moisture content; mix proportion; mechanical property.

1. Introduction

With the continuous expansion of urban construction, infrastructure construction is increasing gradually. Construction waste has become a major and no-neglected renewable resource in urban construction. The recycling of construction waste has been taken more and more attention by people, and the research results have played an important role in the regeneration and utilization of construction waste resources in our country. But compared with foreign countries, the research and application of recycled concrete in China started late, which is about 50 years behind other countries. Due to the different sources of recycled aggregate, the broken way and treatment technology of waste concrete are different, so the research difficulty of recycled aggregate concrete is increased [1, 2], which leads to the complexity and diversification of the recycling utilization of construction waste [3, 4, 5].

Construction waste includes a wide range of muck, rubble, waste mortar, brick fragments, waste concrete block, asphalt concrete block, waste plastic, scrap metal and waste wood, etc. In this study, test material is the broken concrete of urban concrete structure demolition, and recycled aggregate is processed in the factory [6]. In view of the characteristics of high porosity, large surface cracks, uneven material and large water absorption of the recycled aggregate, an optimized design method for the mix proportion of recycled concrete is studied, and the relationship between the mixture ratio design and mechanical properties of recycled concrete is analyzed through the test of compressive strength and dry shrinkage strain indexes.

*Corresponding author

2. Recycled Aggregate of Construction Waste

Recycled coarse aggregate used in the study is mainly derived from the demolition of the modern concrete buildings.

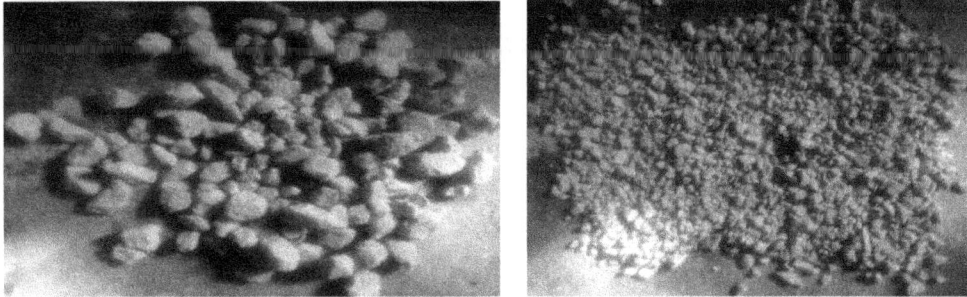

Fig. 1 Recycled coarse aggregate

After rolling and grinding process in the factory, it is separated into the specifications of 5-25mm and 5-10mm. Their external appearances are shown in Fig. 1.

Natural coarse aggregates for common concrete test and design are traditional limestone crushed stones in comparison test. The crushing value of 5-25mm aggregate is 9.6%, and its apparent density is 2680kg/m³. But the crushing value of recycled coarse aggregate is 15%, it is larger than that of the natural coarse aggregate. Its average apparent density is slightly lower, as 2580kg/m³, and the average water absorption rate is 7%, which is 94% higher than that of limestone crushed stone. After careful analysis, there are many factors, such as more pores in the concrete forming process, a lot of cracks produced in the recycled aggregate surface, the interface transition zone between aggregate and concrete, etc. These reasons lead to the reduction of the crushing capacity and the increase of water absorption in recycled aggregate.

3. The Optimized Design Method for C20 Recycled Concrete Mix Proportion

The mix proportion design adopts 42.5 ordinary portland cement, natural sand mixed with recycled sand and the recycled coarse aggregate to study on optimized mix proportion design method of C20 recycled concrete by considering the comprehensive factors of mechanical property, durability and economy. According to the characteristics of high water absorption of recycled coarse aggregate, its surface-dry condition of 24h saturation is used as the design basis. The surface-dry densities of two kinds of recycled aggregate, 5-25mm and 5-10mm, are respectively 2310kg/m³ and 2280kg/m³. Specific design method is as follows:

(1) Preliminary mix proportion design calculation:
1) Aggregate gradation design: design of the mix proportion of recycled coarse aggregate by using the principle of continuous gradation; determination of the total amount of fine aggregate by the void fraction in tamping condition;
2) Reference to the design method of mix proportion of common concrete [7], calculating the initial water cement ratio, determining the unit water consumption, choosing primary sand ratio and calculating the unit cement content;

3) Water cement ratio and unit cement content should be carried out durability checking in accordance with the temperature region of the concrete project;

4) Using absolute volume method to calculate the amount of recycled coarse and fine aggregate. Among them, the amount of recycled coarse aggregate is calculated by the surface-dry density [8];

5) Obtaining preliminary mix proportion.

(2) Adjustment of lab mix proportion:

1) According to the test results, adjusting the water cement ratio;

2) Determining the mix ratio of natural sand and recycled sand in fine aggregate by the initial mixing;

3) Mechanical index test: cube compressive strength and volume shrinkage test;

4) Frost resistance test can be carried out in cold regions;

5) Checking density to determine the lab mix proportion.

(3) The construction mix proportion can refer to the lab mix proportion.

The design method is suitable for the concrete with no water reducing agent and admixture. If they need to be added, the amount of water consumption, cement content and the dosage of admixture should be calculated according to the mix proportion design methods of common concrete and fly ash concrete.

4. Experimental Study and Analysis of C20 Recycled Concrete

4.1. C20 recycled concrete mix proportion

According to the above method, C20 recycled concrete mix proportion is designed for plastic concrete with the concrete slump 10-30mm. It does not mix water reducing agent and admixture. The proportion of recycled coarse aggregate is 5-25mm: 5-10mm=60:40, the proportion of recycled sand and natural sand is 1:2. W/C test range is 0.5-0.6, and a set of mix ratios are designed at intervals of 0.02. They are 5-25mm: 5-10mm: recycled sand: natural sand: water =480:320:235:470:175, and the amount of cement used in the four groups are 324kg, 313kg, 302kg and 292kg with W/C 0.54, 0.56, 0.58 and 0.60 respectively.

By mixing test, all of the four groups of mixture slump meet the design requirements. They have good cohesiveness and water retention except that W/C for the 0.60 groups of water retention is a bit poor and shows a slight bleeding phenomenon.

4.2. Cube compressive strength test

Four groups of length 150mm standard cube specimens are prepared with the above four groups of recycled concrete. The molding method is the vibration mode, and the vibration time is 30-60s. The curing age is 28 days with a temperature of 20±2 degrees Celsius and the relative humidity of 95% standard curing conditions [9].

The loading rate of the compressive strength test is between 0.3-0.5MPa/s. The results of the test are shown in Table 1. The relationship between cement water ratio and 28 day compressive strength is shown in Fig. 2.

Table 1 28d compressive strength test record of recycled concrete C20

W/C	$f_{cu,28}$ (MPa)									
	1	2	3	4	5	6	7	8	9	Average value
0.54	30.5	32.6	32.3	32.2	34.7	32.0	34.3	30.4	32.6	32.4
0.56	31.8	31.7	29.4	31.4	30.2	31.5	32.4	29.6	31.1	31.0
0.58	26.7	30.3	29.1	28.6	29.3	28.3	28.8	27.1	30.3	28.7
0.60	27.9	26.1	26.3	26.4	21.2	26.9	27.8	26.9	25.7	26.0

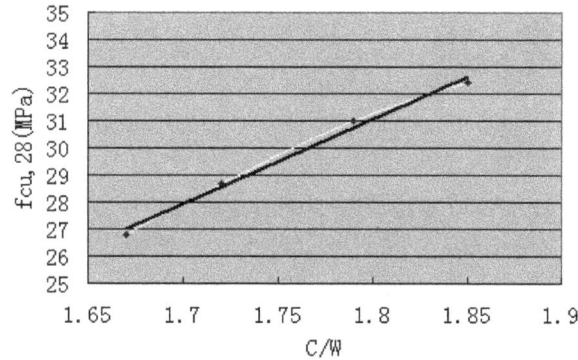

Fig. 2 Relation curve of the cement water ratio and compressive strength of recycled concrete

Known from Table 1, the mix proportions of recycled concrete reach the design strength of C20 within W/C of 0.5-0.6. It is shown in Fig. 2 that the compressive strength of recycled concrete increases with the increase of the cement water ratio when the cement strength is determined. Similar to the strength theory of common concrete, recycled concrete also follows the change of Bolomey Theory [10].

4.3. *Dry shrinkage test of recycled concrete*

In conclusion, there are two groups of recycled concrete (W/C: 0.56 and 0.58) having better comprehensive performance. Because the cement content of the recycled concrete is slightly higher in the group of W/C=0.54, it may be more obvious to the leading role of the volume stability. So the group is chosen to carry out the drying shrinkage test.

Beam type specimens with standard size 400mm*100mm*100mm are formed and maintained in accordance with the above method. Three groups of specimens are prepared, in which two groups of recycled concrete specimens and one group of common concrete specimen. At the third day after molding, the specimens are placed in the dry shrinkage test box, and dial indicators are installed and read as shown in Fig. 3.

Fig. 3 Dry shrinkage test of recycled concrete

The test cycle is 60d. Compared with the shrinkage test under the same condition of common concrete specimens of C20, test curve of dry shrinkage of recycled concrete is shown in Fig. 4.

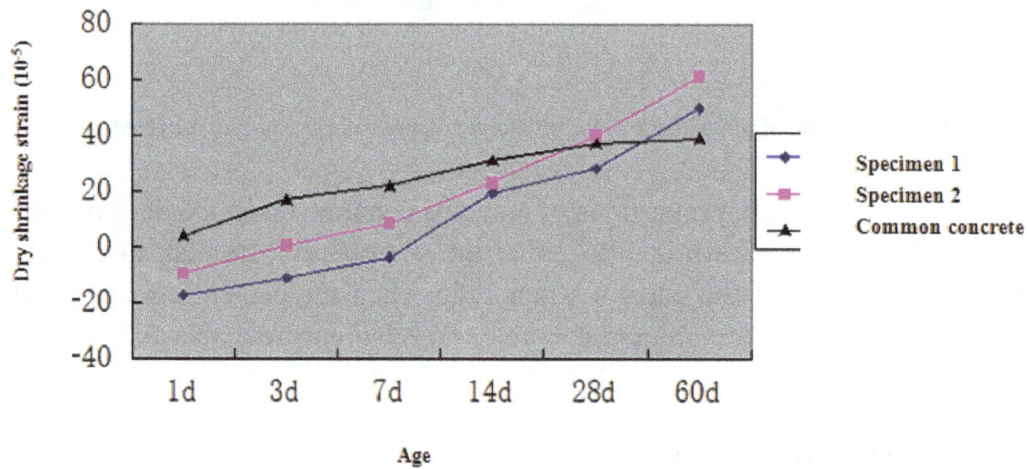
Fig. 4 Shrinkage strain curve of recycled concrete

According to Fig. 4, the early volume change of the common concrete is mainly shrinkage, but the average volume change of the recycled concrete from 1d to 7d is mainly in the expansion and gradually begins to shrink after the 7th day. The shrinkage strain of common concrete is decreased from 14d to 60d, and the volume shrinkage tends to be stable. The volume shrinkage strain of recycled concrete is close to that of common concrete, but its volume shrinkage continues increasing, and its shrinkage strain of 60 days is slightly higher than that of common concrete.

The analysis shows that the mix proportion design of recycled concrete is based on the surface-dry condition of recycled coarse aggregate, and the water absorption rate of the air dry recycled sand is relatively large. This is the main reason for the early appearance of water swelling. With the gradual hydration of cement and the continuous formation of hydration products, the volume of recycled concrete begins to shrink in the later stage. Because of losing water and the acceleration of cement hydration reaction, the dry shrinkage strain value increases. Obviously, the maintenance of recycled concrete is more important.

5. Conclusion

With the experimental study of C20 recycled concrete, the conclusions are drawn as follows:

1. Design of the mix proportion of recycled coarse aggregate by using the principle of continuous gradation, calculating the total amount of fine aggregate by the void fraction in tamping condition and determining the proportion of the two by the initial mixing.
2. Optimize design of recycled concrete using surface-dry condition and referring to lab proportion for the construction mix proportion.
3. The relationship between cement water ratio and strength of recycled concrete follows the Bolomey Theory when the W/C is 0.5-0.6.
4. Volume expansion of recycled concrete within one week after molding, and volume shrinkage after the 7th day. Its shrinkage strain is slightly higher than that of common concrete.

References

1. S. C. DENG, X. B. ZHANG, Y. S. LUO, The abandoned concrete present condition of the reborn exploitation analysis and the research outlook, Concrete, 11 (2006) 20-24.
2. W. X. LI, X. ZHANG, X. LIU, Study on the relationship of recycled aggregate replacement ratio, slurry content, apparent density and water absorption of recycled concrete, Concrete, 10 (2010) 60-63.
3. X.G. YANG, Y. C. HAN, Y. XUE. Study on the basic properties of recycled aggregates, Concrete, 2 (2012) 66-68.
4. Y. Z. XU, J.G. SHI, Analyses and evaluation of the behaviour of recycled aggregate and recycled concrete, Concrete, 7 (2006) 41-46.
5. S.C. DENG, G. X LUO, Comparision and analysis of designs of mix ratio for benchmark concrete and recycled concrete, J. of Huizhou University, 3 (2011) 9-18.
6. W. SHI, J. P. HOU, Technology and mix design on recycled concrete, Building technique development, 8 (2001)18-20.
7. Ministry of housing and urban rural development of the people's Republic of China. Regulation for design of common concrete mix (JGJ 55-2011) [S]. China Building Industry Press, 2011.
8. X. K. LI, Q. GUO, S. B. ZHAO, Experimental study on mix design of full-recycled-aggregate concrete, J. of north China institute of water conservancy and hydroelectric power, 4 (2013) 53-56.
9. W. X. LI, X. LIU, Study on influence factors of mechanical properties of recycled aggregate concrete, Architecture technology, 1 (2012) 15-17.
10. Y. J. ZHANG, S. HE, X. ZHANG, Modification of the Bolomey formula in recycled aggregate concrete, J. of building materials, 4 (2012) 538-543.

Chapter 2
Amorphous Materials, Nanomaterials and Thin Films

Thermal Analysis and Rheological Behavior of Core–shell Nanoparticles Filled with PP-EVA

Ling Liu*, Xiu-Li Wang, Shou-Lian Wei

Faculty of Chemistry and Chemical Engineering, Zhaoqing University, Zhaoqing 526061, P. R. China
Email: lingliu0813@163.com

Core-shell nanoparticles (CSN) with polystyrene (PSt)/organophilic montmorillonite (OMMT) as core and poly (butyl acrylate) (PBA) as shell were prepared through seed emulsion polymerization. The structural characteristics of CSN were examined through scanning electron microscopy and transmission electron microscopy. CSN was spherical, with an evident core-shell structure. The thermal property and rheological behavior of the polypropylene (PP)/ethylene-vinyl acetate (EVA) copolymer composite system filled with different proportions of CSN were systematically investigated by using a thermogravimetric analyzer and an RT-2000 high-pressure capillary rheometer. Results showed that temperatures corresponding to the maximum weight-loss rate and the end of weight loss for PP-EVA/CSN were higher than those of single PP-EVA and CSN when the mass fraction of CSN was 5 wt%. CSN can also improve the thermal performance of PP-EVA. The prepared composites that belonged to non-Newtonian pseudoplastic fluid and the processibility were similar to those of PP-EVA. Shear stress, apparent viscosity decreased as CSN was added; thus, CSN unlikely caused damage to the processing fluidity of PP-EVA, instead, processibility is likely improved.

Keywords: Polypropylene; butyl acrylate; styrene; organophilic montmorillonite; thermal analysis; rheological property.

1. Introduction

Since the concept of core–shell particle design is proposed, J.W.Vanderhoff and T.Matsumoto [1-3] have greatly contributed to this domain by determining the morphological characteristics and reaction mechanism of a composite latex particle with a core-shell structure. Core-shell composite nanoparticle, a composite nanostructure of high level, is an ordered structure of nanoscale formed by one nanomaterial coated in another nanomaterial through chemical bonding or other interactions [4]. With an innovative design, core-shell composite nanoparticles have many new features that cannot be found in single nanoparticles. Core-shell composite nanoparticles have considerable promising applications and receive much attention [5-9]. Polypropylene (PP) exhibits good comprehensive performance, but the application of this substance is restricted by fragility, low temperature resistance, impact resistance, flame retardancy, and compatibility with inorganic fillers [10]. Core-shell structure is possibly granted to a polymer/clay system through in-situ compound technology. The toughness and flame retardancy of PP are also likely improved by combining the evident enhancement effect of structure-controllable core--shell composite particles [11] and the flame retardancy of the polymer/clay system. Novel ideas and trials have provided for further research and development of core-shell nanocomposites.

PP-ethylene-vinyl acetate (PP-EVA) composite system (with a mass ratio of 9:1) is used to improve the interfacial compatibility of PP and core-shell composite nanoparticles. Core [polystyrene (PSt)/organophilic montmorillonite (OMMT)]-shell [poly(butyl acrylate) (PBA)] structured composite nanoparticle is synthesized through seed emulsion polymerization in an experiment. The product and PP-EVA composite system are processed using melt-blending method to achieve compatibilization and functionalization. Studies have also demonstrated the preparation and application of core-shell composite nanoparticles; however, the rheological

---

*Corresponding author

behavior of composite particle-filled polymer is rarely reported. Therefore, this study focused on the thermal stability and rheological behavior of PP-EVA/core-shell nanoparticle (CSN) composites and compared these properties with those of the PP-EVA system. This study also provided reliable data to screen and evaluate the thermal performance of materials and the reasonable selection of process condition [12].

2. Experimental

2.1. *Materials*

The following materials were used in this study: PP, T30S, Tianjin Petrochemical Company; EVA, 18% vinyl acetate content, SINOPEC Beijing Yanshan Company; OMMT, trademark i.44P, 15–20 m2/g specific surface area, USA Nanocor Company; butyl acrylate (BA), AR, Tianjin Fuchen Chemical Reagents Factory; styrene (St), AR, Tianjin Damao Chemical Reagent Factory; acrylic acid (AA), AR, Tianjin Fuchen Chemical Reagents Factory; sodium dodecyl sulfate (SDS), AR, Guangzhou Chemical Reagent Factory; polyoxyethylene octyl phenyl ether (OP-10), AR, Tianjin Fuyu Fine Chemical Co., Ltd.; and potassium persulfate (KPS), AR, Guangzhou Chemical Reagent Factory. Other reagents were commercially available products.

2.2. *Preparation of CSN*

2.2.1. *Preparation of seed emulsion*

Compound emulsifier, 1.9 g of SDS, and 0.9 g of OP-10 were added to 30 mL of distilled water and heated in a water bath at 50–60 °C. After the emulsifier dissolves completely, 40 g of St monomer as core and 1.2 g of 3% AA were added. The monomer was pre-emulsified for 30–40 min. Afterward, 20 g of BA monomer was used as shell and pre-emulsified in the same process as previously described except the amounts of reagents were altered (0.7 g of SDS, 0.3 g of OP-10, and 0.6 g of AA).

OMMT (1.0 g) and distilled water (90 mL) were added to a three-neck flask and stirred for 0.5 h. At solution temperature of 60 °C, 20% core monomer pre-emulsion and 30% aqueous solution of initiator KPS (with a concentration of 0.5%) were added successively. The remaining core monomer pre-emulsion and KPS solution were slowly added with two dropping funnels for approximately 2 h when no reflux condensation of monomer occurs. Stirring speed was maintained at 220–250 r/min. The reaction was conducted for another 1 h at a constant water bath temperature of 75 °C. PSt/OMMT seed emulsion was obtained.

2.2.2. *Preparation of CSN*

KPS solution (0.5%) and NaHCO3 (10%) were added to the three-neck flask with seed emulsion, with pH of 7–8. The shell monomer pre-emulsion was starve-fed using a constant-pressure dropping funnel at a strictly controlled dropping rate. Thus, feeding can be completed in 2 h. The reaction continued at constant temperature and stirring speed for 0.5 h and another 1 h of reaction after temperature slowly increased to 80 °C. The mixture was cooled to room temperature, and demulsified with KCl. After filtration, drying, and grinding were conducted, core-shell (PSt/OMMT)/PBA composite nanoparticles, namely, CSN particles, were obtained.

2.2.3. *Preparation of PP-EVA filled with CSN*

PP-EVA (with a mass ratio of 9:1) and CSN were premixed at mass ratios of 95:5, 90:10, and 85:15. The mixed material was placed in a Rheocord System 40 twin-screw extruder (US, HAAKE Co., Ltd.) to undergo melting extrusion. The temperature of each zone of the extruder was set as follows: first zone, 150 °C, second zone, 200 °C, third zone, 200 °C, and fourth zone, 190 °C. The rotation speed of the screw was 50 r/min. PP-EVA/CSN composite materials with different proportions of CSN were prepared, and characterized as PP-EVA-5, PP-EVA-10, and PP-EVA-15, respectively.

2.3. *Morphological characteristics of core–shell structure*

The prepared CSN emulsion was diluted for approximately 300 times. After ultrasonic dissolution was completed, one or two drops of diluent were dried and directly added to the processed silicon wafer. The surface morphological characteristics of the latex particles were observed under a field emission scanning electron microscope (Supra 55 Sapphire, Carl Zeiss, Germany). The sample was coated with gold. The diluted CSN emulsion was stained and added to the carbon film. The resulting emulsion was naturally dried and its structure was observed under a field emission transmission electron microscope (Tecnai G2 F30, FEI Company, US).

2.4. *Thermogravimetric analysis (TGA) and rheological experiment*

The thermal performance of the samples was analyzed using a DTG-60H thermogravimetric analyzer (Shimadzu, Japan) in an N2 atmosphere at a heating rate of 20 °C/min, and a heating range of 30–700°C. A Rheotens-2000 high-pressure capillary rheometer produced by German GOTTFERT Company was employed in the rheological experiment. The die diameter was 1 mm and the length-to-diameter ratio of the die was 30/1. The experiment was conducted at 200, 210, 220, and 230 °C. Data were processed in a computer.

3. Results and Discussion

3.1. *Morphological characteristics of CSN*

The scanning electron microscopy (SEM) and transmission electron microscopy (TEM) images of the prepared CSN are presented in Figure 1. CSN is a regular spherical form and particle sizes reach 70—90 nm [Figure 1 (a)]. Two particles clearly depict the core-shell structure of CSN [Figure 1 (b)]. The internal boundary of the particle is clear. The black part corresponds to the (PSt/OMMT) core, and the light part represents the PBA shell. Thus, the core–shell structure has been well formed.

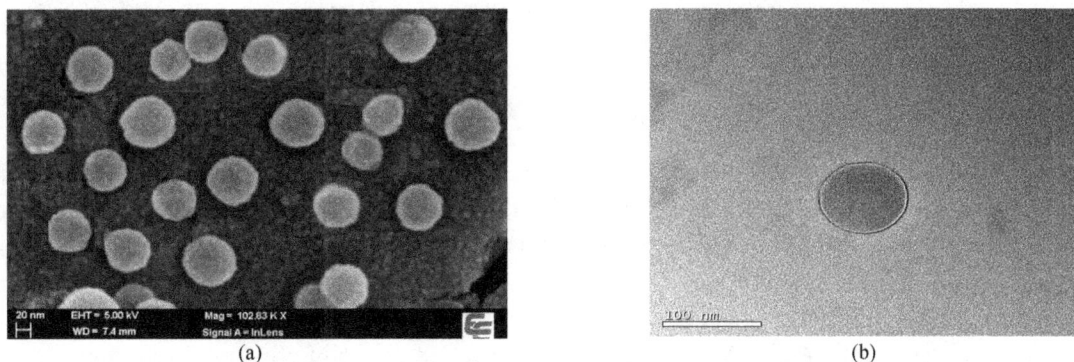

(a)   (b)

Fig. 1 SEM (a) and TEM (b) images of CSN

3.2. Thermal analysis

The thermal stability of PP-EVA, CSN and the PP-EVA-5 composite material is investigated through TGA and differential TGA (DTGA). The results are shown in Figure 2. Figures 2 (a) and (b) show that all of the three materials are characterized by one apparent weight loss step. After the initial decomposition temperature (t_d, the temperature that corresponds to 5% weight loss) [13] is reached, each material initially loses weight rapidly. t_d of PP-EVA is 419.4 °C, and temperature corresponding to the maximum rate of thermal weight loss rate (t_p) is 467.4 °C. Weight loss is terminated at 494.2 °C, and the char yield at 600 °C is 1.06%. t_d and t_p of CSN are 359.8 and 446.2 °C, respectively. Weight loss also ends at 472.1 °C, and the char yield at 600 °C is 19.15%. t_d of PP-EVA-5 composite material is 394.2 °C when 5 wt% CSN is added; t_p is higher than that of PP-EVA and CSN by 7.7 and 29.3 °C, respectively. Weight loss also ends at 502.5 °C, and the char yield at 600 °C increases by 1.72% compared with that of PP-EVA. Although the addition of CSN results in a decrease in t_d of PP-EVA, t_p, temperature at which weight loss ends, and the char yield at 600 °C increase. The acme of the thermal weight loss rate (shown in Figure b) is lower than that of PP-EVA by 20%; this result indicated that thermal decomposition rate decreases. This result also showed that the addition of CSN improves the thermal stability of PP-EVA to a certain degree. This improvement is primarily because OMMT in CSN disperses in PP-EVA system, thereby impeding and suppressing the movement of PP molecular chains [14]; thus, the thermal decomposition rate of PP-EVA composite system decreases.

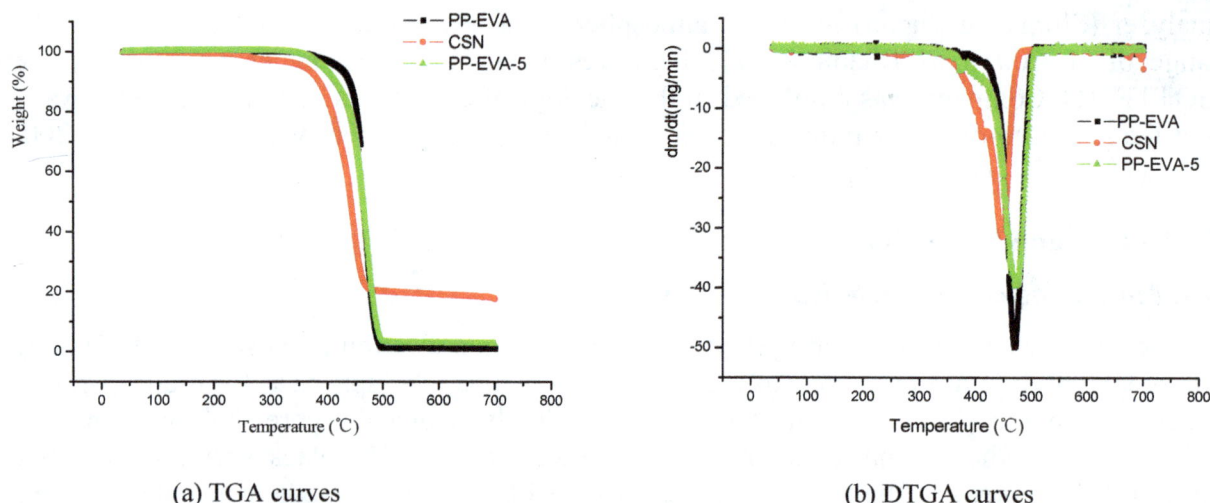

(a) TGA curves

(b) DTGA curves

Fig. 2 PP-EVA and PP-EVA-5 composites

3.3. Rheological behavior

3.3.1. Shear stress (τ) and shear rate (γ)

Figure 3 shows the rheological curves of τ and γ at 200 °C of PP-EVA and PP-EVA/CSN composites with different proportions of CSN. The curves show that the addition of CSN does not change the fluid type of PP-EVA system, which is considered as non-Newtonian pseudoplastic fluid. τ increases as γ increases. At $\lg\gamma < 3.0$ s−1 and constant γ, τ decreases as the proportion of CSN increases. At $\lg\gamma > 3.0$ s^{-1}, the curve converges and τ exhibits small amplitude of decrease; therefore, the sensitivity of PP-EVA system to the alteration of CSN content weakens.

3.3.2. Apparent viscosity (η_a) and γ

Figure 4 shows η_a-γ curves of PP-EVA and PP-EVA/CSN with different proportions of CSN at a temperature of 200 °C. η_a of the four materials decreases as γ increases. When $\gamma < 700$ s^{-1}, η_a of PP-EVA and PP-EVA/CSN drastically reduces as γ increases. As γ increases to a certain value, particularly $\gamma > 1000$ s^{-1}, the curve converges and the sensitivity of η_a to γ diminishes. Hence, change in amplitude reduces. The result shows that the PP-EVA/CSN composites are pseudoplastic melt, and the processing behavior is similar to that of PP-EVA. At low γ, molecular chains entangle, resulting in high viscosity and poor fluidity. As γ gradually increases, the molecular chains gradually disentangle along the flow direction. Thus, fluidity is improved and η_a decreases [15]. The addition of CSN greatly decreases η_a of PP-EVA system, and the amplitude is likely enhanced as the proportion of CSN increases. This result indicates that CSN can push the disentanglement and separation of PP-EVA molecular chains, which decreases the concentration of the entangled points. The interaction of particles in the matrix is shielded by the core-shell structure, which impedes particle agglomeration and reduces flow resistance. Consequently, molecular movement becomes easy and processing property of PP-EVA is improved.

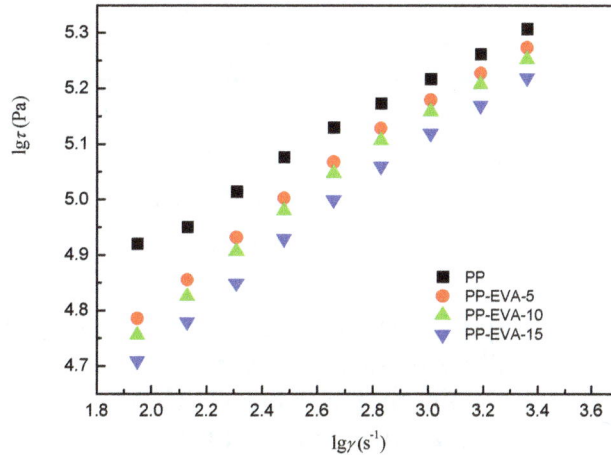

Fig. 3 Rheological curves of PP-EVA and PP-EVA/CSN composites

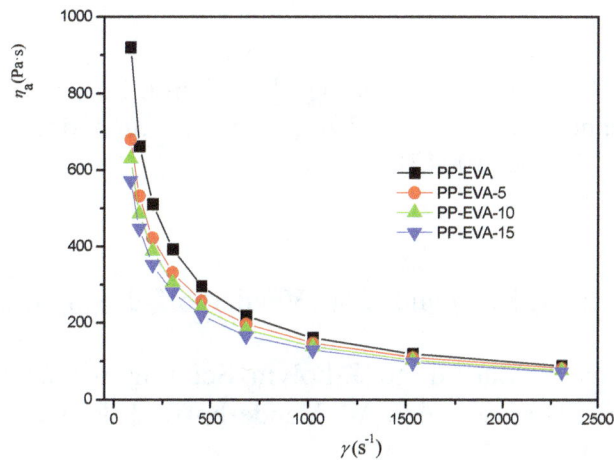

Fig. 4 Effect of γ of PP-EVA and PP-EVA/CSN on η_a

3.3.3. *Flow index (n)*

The relationship between τ and the γ of the polymer melt usually satisfies the Ostwald-de Waele power law equation $\tau = k\gamma^n$, where k is the parameter and n is the flow index or the non-Newtonian index. For Newtonian fluid, $n = 1$; for pseudoplastic fluid, $n < 1$. The difference between n and 1 indicates material viscosity dependence on γ. The $\lg\tau$- $\lg\gamma$ linear equations of PP-EVA and PP-EVA/CSN composites at different temperatures are formed. Using the slope of the line, we can obtain different n values (Table 1).

n of all materials is <1, indicating that PP-EVA and PP-EVA/CSN composites are non-Newtonian fluids. At the same temperature, n of PP-EVA is the least, suggesting that this value is most sensitive to γ. As system components change or components remain the same as the temperature changes, n of PP-EVA/CSN composites slightly varies. This result indicates that the sensitivity of this material to temperature is weak; therefore, the amount and temperature of CSN slightly affect n.

Table 1 n values of PP and PP/CSN at different temperatures

Temperature (°C)	n			
	PP-EVA	PP-EVA-5	PP-EVA-10	PP-EVA-15
200	0.283	0.349	0.356	0.367
210	0.291	0.346	0.358	0.368
220	0.314	0.358	0.369	0.370
230	0.313	0.363	0.372	0.381

4. Conclusion

In summary, (PSt/OMMT)/PBA composite nanoparticles prepared through in-situ polymerization exhibits a core-shell structure. The filling of such nanoparticles can enhance the thermal stability of PP-EVA composite system to a certain degree. Core (PSt/OMMT)-shell (PBA) nanoparticles unlikely affect the nonlinear rheological mechanism of the filled PP-EVA melt, the final composite material is pseudoplastic melt, with a similar processing behavior to that of PP-EVA composite system.

Acknowledgments

This work is funded by the Science and Technology Innovation Project of Guangdong Provincial Education Department, (Grant No. 2013kjcx0194), and the Guangdong Innovation Project for College Students (Grant No. 201410580042).

References

1. Dimonie, M. S. El-Aasser, A. Klein and J. W. Vanderhoff, J. Polym. Sci. Polym. Chem. Ed., 1984, 22, 2197.
2. M. Okubo, A. Yamada and T. Matsumoto, J. Polym. Sci. Polym. Chem. Ed., 1980, 18, 3219.
3. T.I. Min, A. Klein, M.S. EI-Aasser and J.W. Vanderhoff, J. Polym. Sci. Polym. Chem. Ed., 1983, 21, 2845.
4. J. J. Schneider, Adv. Mater., 2001, 13, 529.
5. J.H. Zeng, J. Yang and J.Y. Lee, J. Phys. Chem. B, 2006, 110, 24606.

6. J.W. Hu, J.F. Li, B. Ren, D.Y. Wu, S.G. Sun and Z.Q. Tian, J. Phys. Chem. C, 2007, 111, 1105.

7. S.L. Chai, M.M. Jin and H.M. Tan, European Polymer Journal, 2008, 44, 3306.

8. S.B. Jhaveri, D. Koylu, D. Maschke and K.R. Carter, J Polym Sci Part A-Polym Chem., 2007, 45, 1575.

9. J.J. Wang, L. Sun, K. Mpoukouvalas, K. Lienkamp, I. Lieberwirth, B. Fassbender, E. Bonaccurso, G. Brunklaus, A. Muehlebach, T. Beierlein, R. Tilch, H.J. Butt and G. Wegner. Adv. Mater., 2009, 21, 1137.

10. S.L. Zheng, Powder surface modification. Beijing: China Building Materials Industry Press, 2003, 54.

11. J. Kim, K. Lee, K. Lee, J. Bae, J. Yang and S. Hong, Polym Degrad Stab., 2003, 79, 201.

12. Q.Y. Wu, Polymer rheology. Beijing: Higher Education Press, 2002, 68.

13. D.M. Guo, X.K. Jing, J.N. Wu, Y.Q. Ou, F.Y. Zhai, L. Chen and Y.Z. Wang, Acta Polymerica Sinica, 2012, (9), 1042.

14. C.X. Zhao, G. Peng, B.L. Liu and Z.W. Jiang, Polym Res., 2011, 18, 1971.

15. J. Sun, Y.H. Song, Q. Zheng, H. Tan, J. Yu and H. Li, J Polym Sci Part B: Polym Phys, 2007, 45, 2594.

The Properties of HgCdTe/graphene Composite Thin Film with Different Layer Graphene

Mei Liu[1,2,*], Shuai Zhang[1], Dong Bi[1], Ying Shi[1], Hong Ma[1], Xue-You Xu[3]

[1]*School of Physics and Electronics, Shandong Normal University, Jinan, 250014, P.R. China*
[2]*Institute of Materials and Clean Energy, Shandong Normal University, Jinan 250014, P.R. China*
[3]*Information Research Institute, Shandong Academy of Sciences, Jinan, 250014, P.R. China*
Email: liumei@sdnu.edu.cn

The mercury cadmium telluride (MCT) thin film is grown on single- or multi-layer graphene to explore the probability of the development of a new MCT detector. The influence of graphene on the crystalline quality, surface morphology, composition and optical property of MCT thin film is studied. It is found that a single-layer graphene (SLG) almost has no effect on the quality of MCT thin film, and the MCT thin film grown on the SLG has a high crystalline quality and a smooth surface morphology. The average transmittance of the MCT thin film on the SLG/GaAs substrate is about 90% when the wavenumber is less than 1500 cm^{-1}. While the increase of the number of the graphene layer deteriorates the quality of the MCT epitaxial thin film. It proves that SLG makes it possible to fabricate a MCT devices on large-area, non-crystalline or amorphous substrates.

Keywords: Mercury cadmium telluride; thin film; single-layer graphene; multi-layer graphene.

1. Introduction

Mercury cadmium telluride (MCT, $Hg_{1-x}Cd_xTe$) is a widely used direct band gap infrared (IR) detector material. Its band gap can be adjusted by the cadmium-mercury ratio [1]. In contrast with the MCT bulk, MCT thin film with homogeneous composition, large area, high quantum efficiency and low dislocation density has been more widely used in the third generation high performance IR focal plane array detectors [2]. With the development of science and technology, there is an increasing need of smaller, lighter and folded detector IR detection devices. However, the common substrate materials for epitaxial film growth, such as Si, Al_2O_3, and GaAs, are single crystal and hard. The problem of high quality MCT thin film growth on large, low price, organic or soft substrates becomes one of the main obstacles to develop the application of MCT material in folded IR optical detectors.

Graphene is a new two-dimensional membrane which consists of single-layer carbon atoms with a hexagonal structure [3]. Single-layer graphene (SLG) has a high light transmittance (~97%) across the spectrum from UV to IR and a high carrier mobility (up to 20 m^2/Vs) at room temperature [4], and has been regarded as one of the most promising candidates for photodetectors, transparent electrode, optical displays, and so on. It has a layered structure with stable and strong covalent bonds in each layer, and weak van der Waals forces between the interlayer. So the hybrid heterostructures of graphene and epitaxial thin films can be transferred onto arbitrary substrate materials, such as metal, organic, plastic or glass [5]. In this study, we try to grow crystalline MCT thin film on graphene. To compare, MCT film is deposited on GaAs, SLG/GaAs and multi-layer graphene (MLG)/GaAs substrates, respectively. The effects of graphene film on the crystalline quality, surface morphology and optical property of MCT epitaxial films were studied. The results show a possibility of the development of the new MCT detectors.

2. Materials and Methods

SLG was grown on a thin copper foil by chemical vapor deposition (CVD) method and transferred to a GaAs substrate by PMMA. The detailed description of the growth and transfer process is

---

* Corresponding author

presented in Ref. [6, 7]. MLG was obtained by the multiple transfer method. MCT thin film was deposited by laser molecular beam epitaxy (LMBE) [8]. A 248 nm KrF laser ($\upsilon = 10$Hz, $\tau = 25$ns) was used as the excitation source. The laser energy density was 6 J/cm². Bulk $Hg_{1-x}Cd_xTe$ (x=0.14) was used as a target. The GaAs, SLG/GaAs and seven-layer graphene/GaAs substrates were located at 3 cm away from the target and held at 100 ^oC. The Ar partial pressure was fixed at 100 Pa to obtain a high quality MCT film. The deposition time is 50 min, and the thickness of the MCT layer is about 2.5 μm. Sample numbers are listed in Table 1.

Table 1 Characteristics of the MCT films grown on different substrate materials

Sample ID	Substrate	FWHM of (111) peak	Composition of Hg (1-x)	Cut off wavelength at 77K [μm]	Concentration at 77K [10^{16}cm^{-3}]
S1	GaAs	0.147	0.7496	9.57	7.036
S2	SLG/GaAs	0.143	0.7540	9.98	6.834
S3	MLG/GaAs	0.152	0.7832	10.22	6.724

The thickness and surface morphology of the MCT epitaxial film grown on different substrates were obtained by scanning electron microscopy (SEM, Zeiss Gemini Ultra-550). The elemental composition was analyzed by the energy dispersive spectroscopy (EDS) which is the accessory of SEM (Zeiss Gemini Ultra-550). Rigaku D/max-rB X-ray diffraction (XRD) spectroscopy with a Cu K_α line radiation source was used to determine the crystalline quality and preferential orientation of the MCT epitaxial layer with different substrates. Micro-Raman spectroscopy (HORIBA, LabRAM HR Evolution) with an incident laser wavelength of 532 nm was used to characterize the quality of graphene layer. The IR transmittance spectra were taken at room temperature with a Thermo Nicolet Nexus 670 Fourier Transform Infrared (FTIR) spectrometer.

3. Results and Discussion

Fig. 1(a-e) displays the surface morphologies of the MCT films on different substrates. Some obvious defects (arrowed in Fig. 1a) caused by the GaAs substrate are exhibited on the surface of S1. The crystalline particles of S2 are uniform in size and shape, while the surface of S3 is rough and made up of small particles.

Fig. 1 SEM images of the different MCT film samples. (a), (d) S1, (b), (e) S2 and (c), (f) S3. (d), (e), (f) is the local enlarged image of (a), (b), (c), respectively

The XRD patterns of the different MCT films are illustrated in Fig. 2.

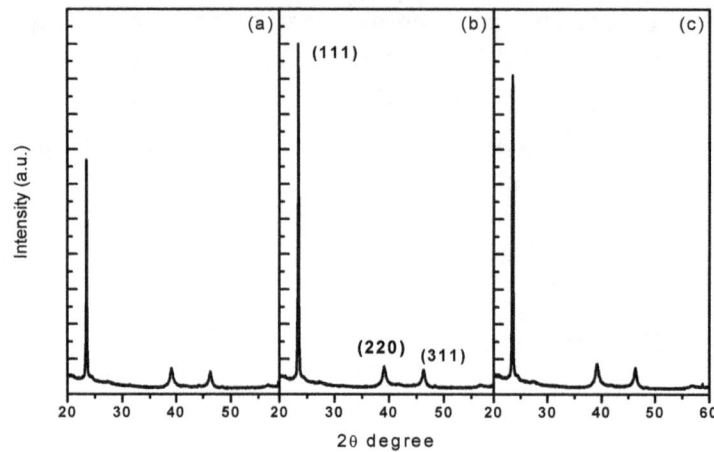

Fig. 2 Comparison of the XRD patterns of the MCT films on (a) GaAs, (b) SLG/GaAs and (c) seven-layer graphene/GaAs substrates

The results of XRD indicate that the MCT thin film grown on different substrates has a face centered cubic (fcc) structure as shown on Fig. 3(a) and the most preferential orientation is along the (111) direction which corresponds to the crystal plane with the lowest surface energy in the fcc structure.

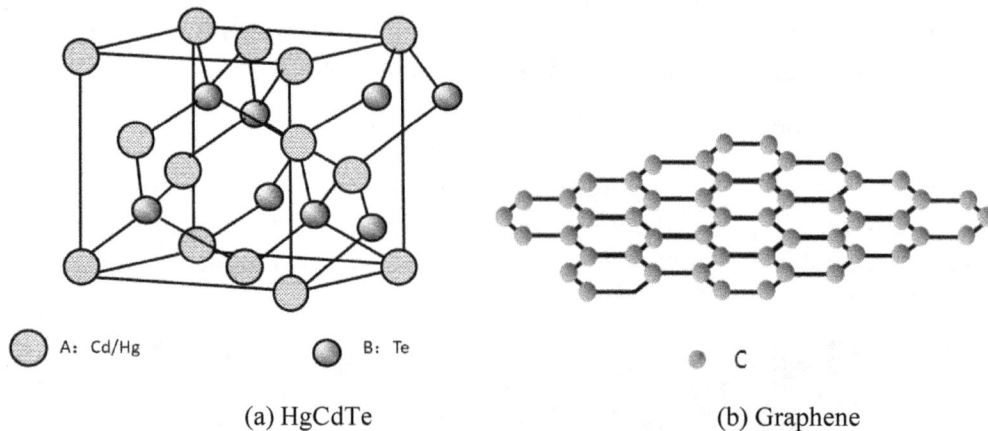

(a) HgCdTe   (b) Graphene

Fig. 3 The lattice structure of (a) HgCdTe and (b) Graphene

The XRD spectrum around (111) peak is fitted to obtain the full width at half maximum (FWHM), and the results are presented in Table 1. According to the Scherrer Formula, the average crystallite size of the MCT film deposited on the SLG/GaAs substrate is larger than that deposited on the GaAs and MLG/GaAs substrate.

SLG graphene is composed of a single layer of carbon atoms with hexagonal structure lattices (Fig. 3b). When it is transferred onto GaAs substrate, it is combined with the surface of GaAs by physical absorption. There is hardly change in the lattice constant of the single graphene membrane. The lattice mismatch between graphene and MCT is smaller than that between GaAs and MCT (a_{MCT}=6.465 Å, $a_{graphene}$=6.708 Å, and a_{GaAs}=5.653 Å). Meanwhile, the strong in-plane covalent bonds within the graphene membrane decrease the density of defects, such as holes, steps and scratches, on the surface of GaAs substrate. It increases the diffusion of the atoms on the substrate surface. Then the SLG membrane contributes to improve the crystalline quality and grain size of MCT thin film. However, the multi-transfer process to obtain the MLG sheet on the GaAs substrate inevitably introduces an increase of the impurities, wrinkles, and holes in the graphene sheet which will have an effect on the crystalline quality of MCT thin film.

The quality of the graphene sheet with different layers on GaAs substrate was characterized by Raman spectra shown in Fig. 4.

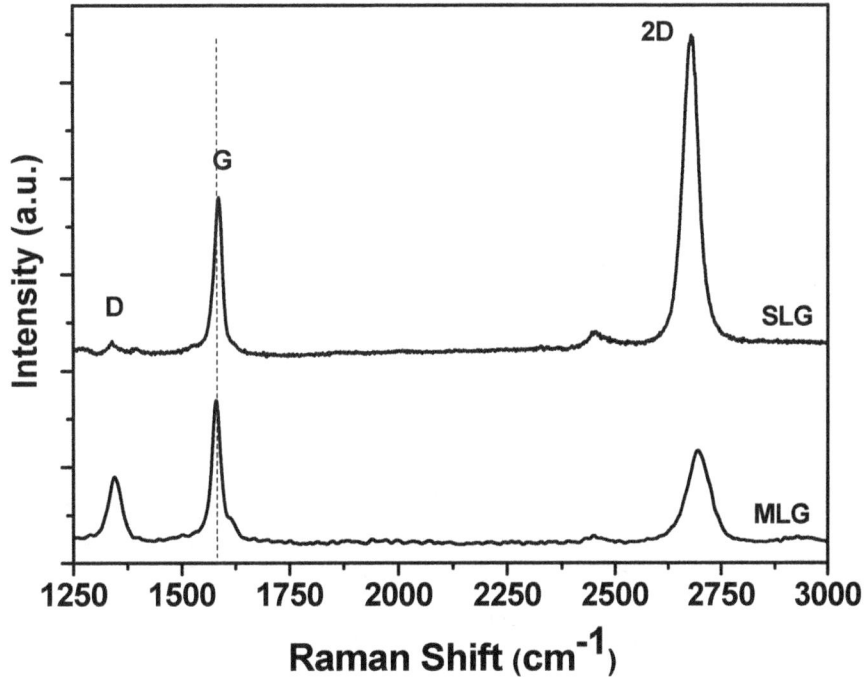

Fig. 4 Raman spectra of the graphene on the GaAs substrate with different layers

There are three main peaks: D (1345 cm^{-1}), G (1580 cm^{-1}) and 2D (2694 cm^{-1}). The G peak is an in-plane vibrational mode due to the sp^2 hybridized carbon atoms. The G peak of MLG has a red shift compare to that of SLG because of the increase of the layer number. The 2D peak is introduced by a second order overtone of the D band, and the 2D peak of MLG becomes wider and weaker than that of SLG. The D peak has a relationship with disordered carbon atoms and defects in the sample [9]. The significant D peak shown in the Raman spectrum of MLG means that there are a lot of defects in the MLG sheet which are resulted by the transfer process of the graphene. Meanwhile, during the growth of the MCT epitaxial thin film, the SLG can transfer heat to incident particles quickly owing to its high heat transfer coefficient. The particles deposited on the surface of SLG and GaAs substrate can obtain nearly the same energy from the substrate and form the crystals with similar size. While the thermal conductivity is reduced as the number of graphene layers increases. The diffusion of the particles on the MLG surface is limited because of the low temperature of the substrate. Some grains with a rough surface lead to the decrease of the crystalline quality of MCT epitaxial layer on the MLG/GaAs substrate.

Table 1 shows the mercury composition of the MCT epitaxial thin films with different substrate materials. Due to the different thermal conductivities of SLG and MLG, the interface temperature (T_i) between MCT thin film and the lower layer material (graphene or GaAs) is different ($T_{i-S1} \approx T_{i-S2} > T_{i-S3}$). The mercury atoms are more easily desorbed from the surface at a high temperature than cadmium and tellurium atoms because of its high vapor pressure. Furthermore, the defects in the MLG sheet, such as wrinkles, folds, and impurities, increase the surface diffusion barrier. It causes the number of desorbed mercury atoms to be decreased. Then the mercury composition of the MCT thin film deposited on the MLG/GaAs substrate is the largest among all the samples, and is closer to the target than that of the MCT thin film deposited on the GaAs or SLG/GaAs substrate.

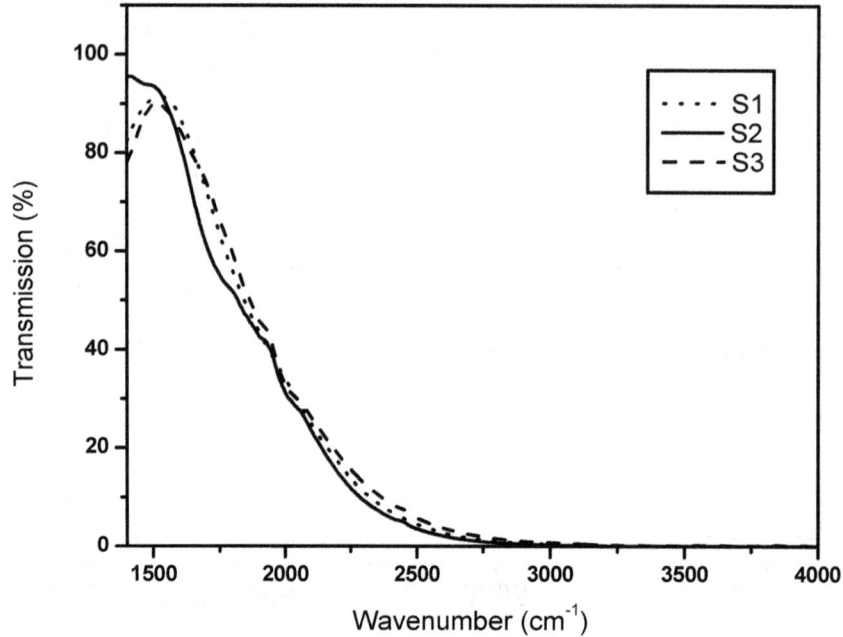

Fig. 5 IR transmission spectrum of the MCT films on different substrates

The IR transmittance spectra of the MCT films deposited on different substrates are measured with a wavenumber from 1400 to 4000 cm^{-1} (shown in Fig. 5). The incident light is completely absorbed when the wavenumber is increased to 3000 cm^{-1}. All the IR transmittance spectra of the samples have a slope in the range of 1500 to 3000 cm^{-1}. The cutoff wavelength of different samples is changed with the different cadmium concentration in samples. The cutoff wavelength and intrinsic carrier concentration of the MCT thin film grown at different substrate at 77K can be calculated according to the relationship between bandgap and cutoff wavelength of Hansen [10, 11]:

$$E_g = -0.302 + 1.93x - 0.81x^2 + 0.832x^3 + 5.35 \cdot (1 - 2x) \cdot 10^{-4} T \tag{1}$$

$$n_i = \left(5.585 - 3.82x + 1.753 \cdot 10^{-3} T - 1.364 \cdot 10^{-3} T \cdot x\right) \cdot 10^{14} E_g^{0.75} T^{1.5} \cdot e^{\frac{-E_g \cdot q}{2kT}} \tag{2}$$

where E_g is the bandgap, x is the percentage of cadmium concentration, T is temperature, n_i is the intrinsic carrier concentration, k is Boltzmann's constant, and q is the elementary electric charge. The results are presented in Table 1.

Considering that the absorption is ~2.3% for SLG, the absorption rate of the MCT thin film grown on the SLG/GaAs substrate has a little decrease when the wavenumber is less than 1500 cm^{-1} as shown in the result. While the transmittances of S1 and S3 decrease to less than 85% in this range. It has a relationship with the absorption for MLG and the quality of the MCT epitaxial thin film. The large defects on the GaAs substrate will deteriorate the mechanical properties and the adhesion of the MCT epitaxial layer. The MCT thin film deposited on SLG has a relatively stable absorption in this range because of its good surface morphology, high crystalline quality and the low light absorption of SLG (only 2.3%). While the total absorption of seven layers graphene can not be ignored, and the light reflection at the interface between the graphene layer is enhanced [12]. The properties of MLG are close to graphite. Moreover, the rough morphology of the MCT thin film grown on MLG/GaAs substrate causes the increase of the light scattering. These have an effect on the transmittance of the MCT thin film on the MLG/GaAs substrate in a long wavelength region.

4. Conclusion

The properties of MCT films deposited on GaAs, SLG/GaAs and MLG/GaAs substrate were investigated, respectively. The SLG can decrease the density of the defects on the surface of the GaAs substrate, ensure the crystalline quality, surface morphology and elemental composition, and stabilize the transmittance of MCT film. But the MLG sheet obtained by multi-transfer process leads to an increase of the wrinkles, tears and holes in the graphene layers, which will have an effect on the quality of the MCT epitaxial thin film. The results of the MCT thin film grown on SLG offer the possibility to grow the MCT thin films on large-area, non-crystalline or amorphous substrates and provide a method to fabricate some new MCT detector devices.

Acknowledgments

This work was funded by the National Natural Science Foundation of China (61307120, 11474187 and 11304186), the Project of Shandong Province Outstanding Young Scientists Research Award Fund (BS2013CL011), and the Project of Shandong Province Higher Educational Science and Technology Program (J12LA07).

References

1. P. Norton, HgCdTe infrared detectors, Opto-Electron. Rev. 10 (2002) 159-174.
2. A. Rogalski, HgCdTe infrared detector material: history, status and outlook, Rep. Prog. Phys. 68 (2005) 2267-2336.
3. K.S. Novoselovet, A.K. Geim, S.V. Morozov, D. Jiang, Y. Zhang, S.V. Dubonos, I.V. Grigorieva, A.A. Firsov, Electric field effect in atomically thin carbon films, Science 306 (2004) 666–669.
4. J.K. Wassei, R.B. Kaner, Graphene, a promising transparent conductor, Mater. Today 13 (2010) 52-59.
5. K. Chung, C.H. Lee, G. C. Yi, Transferable GaN layers grown on ZnO-coated graphene layers for optoelectronic devices, Science, 330 (2010) 655-657.
6. S.C. Xu, B. Y. Man, S.Z. Jiang, C.S. Chen, C. Yang, M. Liu, X. G. Gao, Z.C. Sun, C. Zhang, Flexible and transparent graphene-based loudspeakers, Appl. Phys. Lett. 102 (2013) 151902.
7. S. Bae, H. Kim, Y. Lee, X. Xu, J.S. Park, Y. Zheng, J. Balakrishnan, T. Lei, H.R. Kim, Y.I. Song Y.J. Kim, K.S. Kim, B. Ozyilmaz, J.H. Ahn, B.H. Hong, S. Iijima, Roll-to-roll production of 30-inch graphene films for transparent electrodes, Nat. Nanotechnol 5 (2010) 574-578.
8. M. Liu, X.Y. Xu, B.Y. Man, D.M. Kong, S.C. Xu, Effects of substrate material on carbon films grown by laser molecular beam epitaxy, Appl. Surf. Sci. 263 (2012) 362-366.
9. Frank, L. Kavan, M. Kalbac, Carbon isotope labelling in graphene research, Nanoscale 6 (2014) 6363-6370.

10. G.L. Hansen and J.L. Schmit, Calculation of intrinsic carrier concentration in Hg_{1-x} $Cd_x Te$, J. Appl. Phys. 54 (1983) 1639.

11. M. Badioli, A. Woessner, K. J. Tielrooij, S. Nanot, G. Navickaite, T. Stauber, F. J. Garcia de Abajo, F.H.L. Koppens, Phonon-mediated mid-infrared photoresponse of graphene, Nano Lett. 14 (2014) 6374–6381.

12. Y. Kumar, D.V. Shokeen, Graphene: A new beginning for semiconductor devices beyond silicon, IOSR J. Electron. Commun. Eng. 9 (2014) 51-53.

The Relationship Between Pre-annealing Time and Glass Transition Thermodynamics Parameters of Fe68Ni1Al5Ga2P9.65B4.6Si3C6.75

Teng Li, Ming-Xi Fu[*], Bin Chen, Qiang Fu, Chao Wang

College of Material Science and Engineering, Jiang Su University, Zhenjiang 212013, China

Email: fmxe@sina.com

In this paper the differential scanning calorimetry analysis (DSC) and X-ray diffraction analysis(XRD) detection methods were applied to investigate glass transition temperature T_g, onset crystallization temperature T_x, crystallization peak temperature T_p, the variation of specific heat capacity $\triangle C_{p,g}$ in the glass transition and the XRD pattern of crystallization reaction at 823K about Fe68Ni1Al5Ga2P9.65B4.6Si3C6.75 amorphous alloy, which was pre-annealed for different period (20,40,60min) at 673K $((T_g-100K)<T<T_g)$. Research results indicated that the glass transition temperature and the variation of specific heat capacity gradually increased with the extending pre-annealing time; while the onset crystallization temperature and the first crystallization peak temperature displayed a trend of gradual decrease. Besides, the XRD patterns indicated that the shapes of diffraction peaks for samples were similar, while the intensity of diffraction peaks had some slight differences. The research results were then further analyzed in terms of structural relaxation theory.

Keywords: Pre-annealing; structural relaxation; glass translation; thermodynamics parameters.

1. Introduction

Fe based amorphous alloy as a new functional material has many excellent special properties, such as good corrosion resistance, soft magnetic and so on. However, this good performance will lose in case of structural relaxation or crystallization. In thermodynamics the amorphous alloy is in the metstable state, and its free energy is higher than that of equilibrium crystal line's, so it will transform to metastable amorphous with lower energy or equilibrium state in appropriate conditions. In general, it is called structure relaxation that the amorphous transform to a more stable metastable amorphous at the low temperature; It is called the crystallization that atomic overcome barrier and rearrange into balance crystalline or metastable crystalline state at higher temperature. Chang [1-2] et al studied the amorphous materials pre- annealing and annealing crystallization and then pointed out that the structural relaxation was induced by annealing near the glass translation temperature T_g ,which formed two type atomic structure arrangement called chemical short order and topological short order. It has pointed out that [3-4] structural relaxation had effect on the nucleation and growth during crystallization; The extension of pre-annealing period made nucleation growth activation increase; Pre-annealing led to structure changing, which changed the morphology of crystallization phase. Using thermal analysis technique to measure the heat capacity change of amorphous alloy pre-annealed in different time in continuous heating process, we can investigate the structure relaxation and crystallization dynamics. In this paper,the effect of pre-annealing treatment on glass transition thermodynamics of Fe based amorphous pre-annealed below glass transition temperature was studied. The relationship between pre-annealing time and glass transition thermodynamics parameter is also indicated.

---

[*]Corresponding author

2. Raw Material and Experimental Method

Multicomponent Fe-based amorphous alloy has a larger glass-forming ability(DFA) [5].The alloy for experiment is Fe68Ni1Al5Ga2P9.65B4.6Si3C6.75 . Master alloy materials were melted with pure Fe, Ni, Al, Ga, Si and P-Fe,B-Fe alloys. The pure Fe and high purity graphite particle were melted to Fe - C alloy firstly, and then it was melted with other raw materials under an argon atmosphere in an induction furnace. The ribbon-shaped specimens 20μm and 10mm wide were prepared from these master material using a melt-spring apparatus having a copper miller with a diameter of 280mm.

The pre-annealing treatment was carried in the high temperature furnace using atmosphere heat treatment procedure at pre-annealing temperature of 673K(Tg-100 k) < T (Tg) respectively for 20, 40, and 60 min [6].

The structure of sample was analsized with D/MAXRA type diffraction instrument and the thermal parameter of it was analsized with Netzsch DSC404 type differential scanning calorimeter (DSC) instrument. In high purity argon protection, the quantity of thermal analysis was conducted for the as-quenched state and pre-annealed sample, which is heated at 10K/min rate. Glass transition temperature Tg, initial crystallization temperature Tx and crystallization peak temperature Tp were determined. At the same time the difference of specific heat capacity at Tg between supercooled liquid and glass transition are measured.

3. Results and Discussion

The XRD patterns for as-quenched Samples of Fe68Ni1Al5Ga2P9.65B4.6 Si3C6.75 amorphous alloy pre-annealed at the temperature of 673K for different period(20,40,and 60min) and crystallized at the 823K for 30min was shown in Figure 1. As was shown in the diagram, after pre-annealed for different time and crystallized, the diffraction peak shape of the as –quenched alloy was similar, but the intensity of the diffraction peak was slightly different.

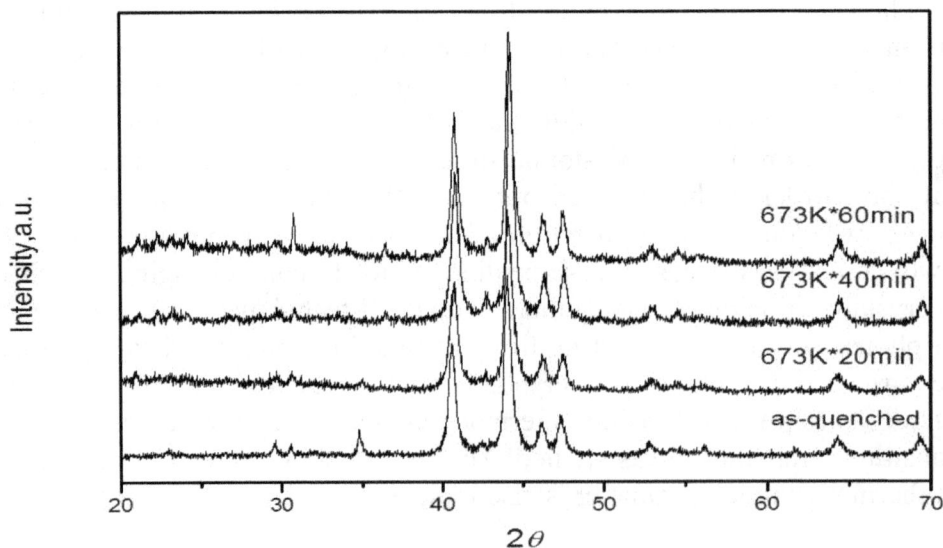

Fig. 1 The XRD patterns after crystallized of as-quenched samples and the samples pre-annealled for different time

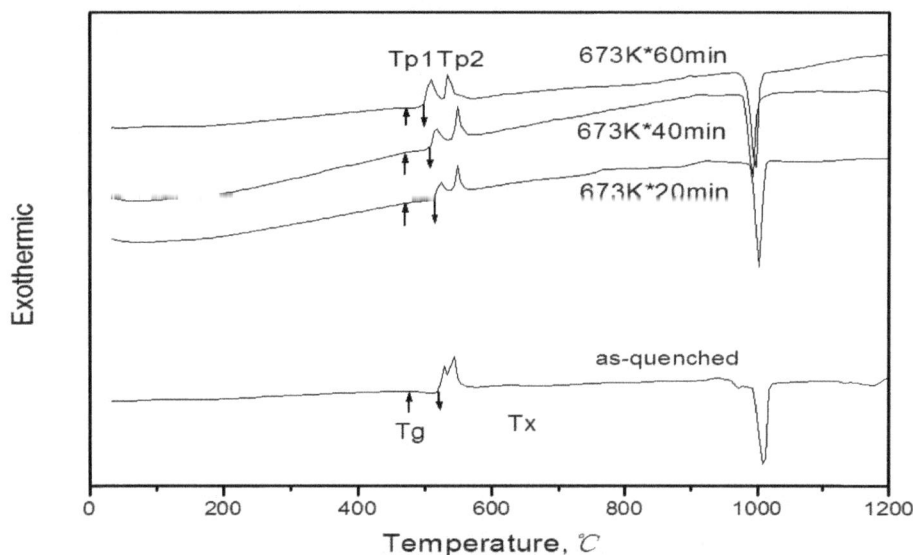

Fig. 2 The DSC curves after crystallized of as-quenched samples and the samples pre-annealled for different time

Figure 2 was the DSC curves of the amorphous alloy which was pre-annealed at the temperature of 673K at 10K/min heating rate. As what was presented in the graph, the alloys had an endothermic peak in glass transition and had a corresponding exothermic peak in supercooled liquid region in crystallization process. It was also indicated that the as-quenched state sample has amorphous structure. DSC curves of the samples after pre-annealed for different period (20,40,60min) were similar, and there existed an "enthalpy relaxation" endothermic peak near glass transition temperature. The amorphous alloy has two crystallization exothermic peak for two stage crystallization way.

Table 1 Thermodynamic parameters of samples

Number	Tg (K)	Tx (K)	Tp1 (K)	Tp1 (K)	ΔCp,g(J/g*K)
quenched state	754.4	798.9	809.1	823.2	0.8723
673K×20min	757.9	797.2	809.0	825.3	0.9006
673K×40min	763.6	791.7	803.5	827.8	0.9901
673K×60min	765.0	789.8	800.3	820.6	1.0135

Table 1 is thermodynamic parameters table of amorphous alloy pre-annealed at 673K for different time (20,40,60min).

Figure 3 indicated the relationship of alloys' Tg and the specific heat capacity change with pre-annealing time.

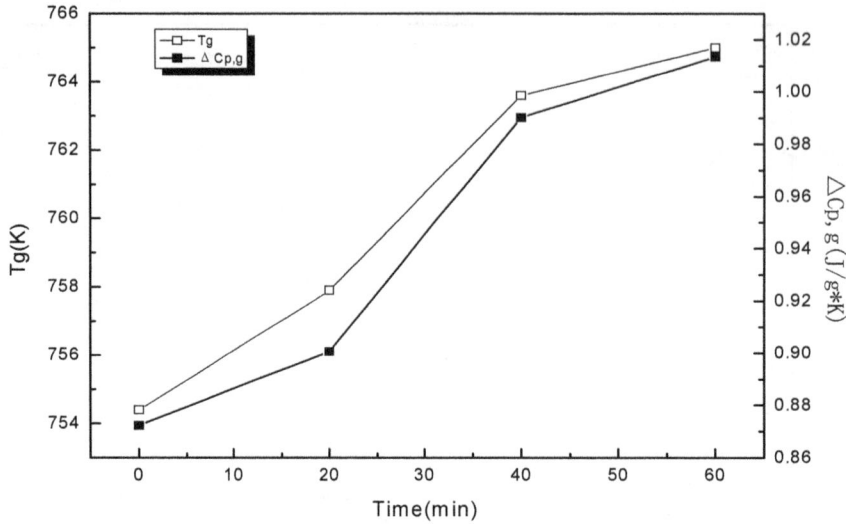

Fig. 3 The variations of Tg and ΔCp,g with pre-annealing time

From Figure 3, it can been see that with the extension of pre-annealing time the change trend of glass transition temperature Tg and specific heat capacity change ΔCp,g were general similar, which increased gradually during the initial period and increased dramatically during 20~40min, when the pre-annealing treatment was over 40min, it raised at low pace. 673K was in the high temperature structural relaxation temperature area of the amorphous alloy. In this area, the redundant free volume contained in amorphous alloy declined with the extension heat time. Because atomic rearrangement in a certain range made the part of the surplus energy gradually release, so that the amorphous alloy were in the lower energy state. Research exhibited [7] that the of structural relaxation activation energy spectrum can be divided into two regions, one is low temperature relaxation region (LTR) and the other is high temperature relaxation reign (HTR). The relaxation happened in these two regions were respectively defined as Low temperature structure relaxation and High temperature structure relaxation. Because the spectrum of low temperature structure relaxation possessed the small proportion of the whole structural relaxation spectrum, its activation energy was low, while high temperature structural relaxation was opposite. Due to structure change happened in the low temperature structure relaxation region was in local area and in short distance, which means that atomic migration and diffusion occurred independently in a small range, so structure relaxation needed lower activation energy. While in high temperature regions, atomic mainly was carrying on medium or long distance diffusion, so structural relaxation needed higher activation energy. In high temperature structural relaxation process, atomic cluster rearrangement improved the order degree of amorphous, which produced more and more orderly cluster. Chen along with other researchers [6] proposed that amorphous consisted of class liquid area and class solid area. The former one is featured big free volume and high local free energy, while the later one is opposite. According to the percolation theory and combined with the concept of free volume, Cohen[8-9] put forward the balance theory of the amorphous phase transformation and believed that each radical in amorphous alloy had certain free volume, which was called class solid cell when its volume was smaller than a certain critical value, and was called class liquid cell when its volume was bigger than the critical value. The author maintainsed that amorphous alloy consisted of these two parts. One part was class liquid cell with a large free volume and high local free energy, the other part was class solid cell that embedded among class liquid cell structure with small free volume and low local free energy. So the radical cluster appeared to be ordered. When the amorphous alloy was pre-annealed at 673K, atomic involved long range concentration rearrangement with the extending heat preservation time, which eliminated some free volume and produced more complex orderly radical cluster. As class solid cell that has little free volume and low local free energy gradually replaced class liquid cell, the number of class solid cell increased continuously and the class liquid cell number decreased, so the free energy of the whole system gradually reduced. Since the reduction of energy in amorphous alloy, the energy difference from the supercooled liquid energy was enlarged, which made the difference of specific heat capacity between glass state and supercooled liquid increased in glass transition. When the alloy in glass state transferred to the supercooled liquid state, it needed more heat quantity, whose performance was that glass transition temperature Tg gradually increased. After pre-annealed for 40min, the amorphous alloy occurred sufficient structural relaxation, therefore, the microscopic structure of the sample has been basically stabilized, capacity of exothermic absorption tended to be saturated, which result to the degree of variation for Tg and ΔCp tends to be gentle.

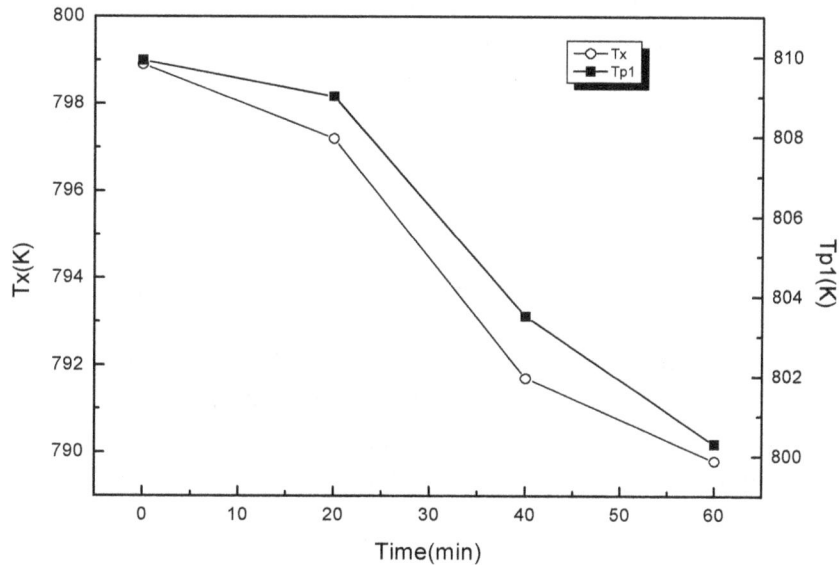

Fig. 4 Dependence of Tx and Tp1 as function of pre-annealed time

Figure 4 was the correspondence relationship of the first crystallization peak temperature Tp1, the onset crystallization temperature Tx with pre-annealing time. From the figure, the Tx and the Tp1 gradually reduced with the pre-annealing time. After the pre-annealed for different period (20, 40, 60min) at 673K, the samples heated continuously at 10K/min speed had experienced two glass transitions and crystallization process that was glass transition turns to supercooled liquid, and then the supercooled liquid turned into the crystalline state. Thermodynamically, supercooled liquid of amorphous alloy was in metastable state. However, high-temperature pre-annealing condition (673K) was conducive to the diffusion of atoms in the amorphous alloy. Moreover, atomic diffusion speed was relative high compared with low temperature pre-annealing conditions. Therefore, with the extending pre-annealing, the migration and diffusion of atoms can occure in the large area, which result in more and larger ordered atomic clusters forming and free energy and atomic structure changing. Atomic clusters were bridge that links the amorphous phase and subsequent crystal. They have an important impact on the subsequent crystallization process. For crystalline phases these ordered atomic Group induced nucleation in subsequent continuous heating process, which reduced the crystallization temperature and crystallization activation energy for initial crystallization [10]. It can be seen from fig. 1 that the crystal diffraction peak is larger by pre-annealed for 60min at 673K and the diffraction peak becomes sharper. The authors believe that ordered atomic clusters induced more core number in nucleation, which is conductive to the crystallization reaction. Seen from figure 2, the DSC curve of the amorphous alloy appears two exothermic peaks, indicating two crystallization processes. The beginning crystallization temperature refers to the temperature at the time amorphous alloy appears crystalline phase in crystallization process, while the peak temperature is the temperature at the time the collision grown crystals. Structural relaxation and crystallization of amorphous alloy is the process in which internal atomic gradually gain energy, increase mobility, reach the activated state (over the energy barrier), diffuse and appear the nucleation and growth [11]. After pre-annealing for different period (20, 40, 60min) at 673K, the amorphous happened structural relaxation. With the extending pre-annealing holding time, the mobility of atoms in the amorphous alloy increased, the structural relaxation proceed fully.

As the atoms gather and rearrange, which improve the degree of order in the amorphous, more and larger ordered atomic clusters were produced in amorphous alloy. They are as the bridge links to the amorphous phase and subsequent, and the nucleation improved in crystallization reaction. This phenomenon not only benefits to the growth of the crystal nucleus, but also promotes the cluster collision in the lower temperature, therefore temperature Tp1 reduced with the pre-annealing time.

4. Conclusion

Experiments indicate, after the amorphous alloy Fe68Ni1Al5Ga2P9.65B4.6Si3C6.75 was pre-annealed for different period (20, 40, 60 min) at the glass transition temperature Tg (673K) below Tg, microscopic atomic configuration has changed, which affect its subsequent glass transition and crystallization behavior. Its glass transition temperature Tg and the specific heat capacity $\triangle$Cp,g gradually increase with extending pre-annealing time ; while the initial crystallization temperature Tx and the first crystallization peak temperature Tp1 present a gradual decrease trend .

Acknowledgments

Project is funded by National "863(2007AA03Z548)" Jiangsu province high technology research projects (BG2007030), Provincial Department of Education Science College, and Research project (03KJD430067) Jiangsu university science and technology innovation team.

References

1. Chang Y K, Cheng Y H, Pong W F, et al. The effect of annealing time on the electronic structure of the Fe-Cu-Nb-Si-B alloys [J]. Journal of Electron Spectroscopy and Related Phenomena, 2001, 114-116:831-835.
2. Kim K B, Warren P J, Cantor B, Structural relaxation and Glass transition behavior of novel (Ti33Zr33Hf33)50(Ni50Cu50)40Al10 alloy developed by equiatomic substitution [J]. Journal of Non-crystalline Solids, 2007, 353:3338-3341.
3. Zhang G P, Liu Y, Zhang B, Effect of annealing close to Tg on notch fracture toughness of Pd-based thin-film metallic glass for MEMS applications [J]. Scripta Material, 2006, 54(5):897-901.

4. Tanga J C, Zhou L, Tan D Q, et al. Crystallization process of amorphous Fe86Zr7B6Cu1 alloy under hot isothermal pressing [J]. Journal of Alloys and Compounds, 2005, 394(1-2):215-218.

5. Fu M X, Zhao J, Zhou H J, et al. Conformation, Thermodynamic parameters and glass forming ability of bulk amorphous alloys of Fe Ni Al Ga P B Si C System [J]. Rare Metal Materials and Engineering, 2008, 37(6):970-974.

6. Chen H S, noue A, Masumoto T. Two-stage enthalpy relaxation behaviour of (Fe0.5Ni0.5)83P17 and (Fe0.5Ni0.5)83B17 amorphous alloys upon annealing [J]. J Mater Sci, 1985, 20(7): 2417-2438.

7. Wang Jinfeng, Liu Lin, Zhou Hui, et al. The bulk metallic glass Pd40Cu30Ni10 P20 structure relaxation [J]. Rare metal materials and engineering, 2005, 34(1):98.

8. Cohen M H, Grest G S. Liquid-glass transition, a free-volume approach [J]. Phys Rev B, 1979, 20(3):1077-1098.

9. Grest G S, Cohen M H. Liquid-glass transition: Dependence of the glass transition on heating and cooling rates [J]. Phys Rev B, 1980, 21(9):4113-4117.

10. Hu Y, Kou S, et al. The effect of pre-annealing temperature on Cu base bulk amorphous glass transition and crystallization [J]. Material Guide newspaper, 2007, 23(VIII):439.

11. Mei J.N, Li J.S, et al. The research of crystallize of amorphous alloy [J]. Rare metal material and engineering. 2007, 36 (7):1215.

Preparation of POSS-Polyurethane Nanocomposites by γ-Ray Irradiation

Cheng-Fei Zhou[*], Yang Liu, Wei Cao, Tong Zhai, Lian-Cai Wang

Beijing Key Laboratory of Radiation Advanced Materials, Beijing Research Center for Radiation Application, Beijing 100015, China

[*]*Email: zhou_chengfei@163.com*

In this paper, the preparation foundation of polyhedral oligomeric silsesquioxane (POSS)-polyurethane composites were reviewed. The preparation of POSS-polyurethane using polyols irradiated with γ-ray radiation was introduced. And the radiation crosslinking POSS-poly (urethane-imide) foam were discussed.

Keywords: Polyhedral oligomeric silsesquioxane; polyurethane; nanocomposite; gamma-ray irradiation; octavinylo-ctasilasesquioxane.

1. Introduction

Polyhedral oligomeric silsesquioxane (POSS) is a kind of nano-filler due to the structure characteristics of organic-inorganic polymer. Polymer modified by POSS, flame retardant properties, oxidation resistance, the glass transition temperate (T_g), thermal deformation and modulus can be improved, especially in the modulus of the significant increase. Although the study on POSS/polyurethane composites has made considerable progress [1-20], but with high energy ray radiation method to prepare, are still rarely reported.

Ionizing radiation can be used for understanding mechanism of polymerization reaction as well as confirming initiation of the polymerization process. The radiation modification has significant advantages over conventional methods, including: (i) absence of foreign matter, like initiator, catalyst, etc., (ii) radiation modification at low temperature or in solid state, (iii) the initiating radicals can be produced uniformly by γ-irradiation [21-36]. Therefore, preparation of the composites by radiation method created a new technical approach.

2. Preparation Foundation of POSS-Polyurethane Composites

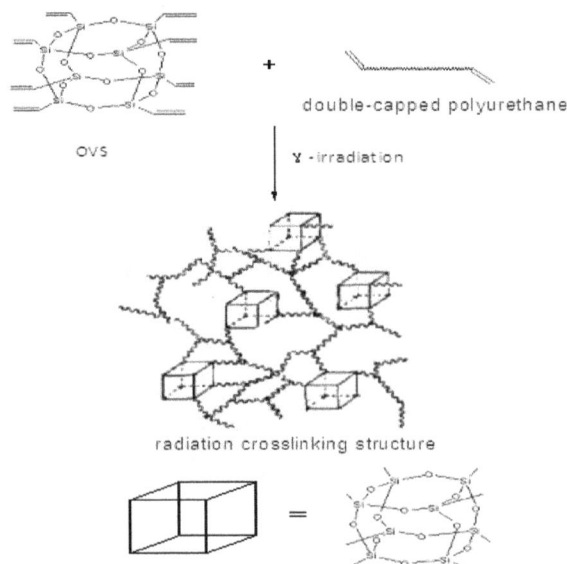

Fig. 1 Schematic diagram of OVS-polyurethane radiation crosslinked system

---

[*]Corresponding author

Radiation crosslinking technique is the most basic method for preparation of composites. In terms of polyurethane radiation crosslinking of general, has made a lot of progress [37-40]. Radiation crosslinking polymerization of polyurethane often was achieved via unsaturated bonds of molecular backbone structure, also to carry out by introduction of unsaturated terminal using end capping methods and also to enhance radiation crosslinking polymerization by adding crosslinking agent in the system.

In our study, POSS is the use of octavinyloctasilasesquioxane (OVS), mainly through the preparation of double terminated polyurethane to realize its radiation crosslinking [41], specific method was shown in figure 1.

The specific method included: (i) using 2-hydrooxyethyl methylacrylate (β-HEMA) as end capping agent, (ii) using bulk polymerization method to synthesize double terminated polyurethane prepolymer, (iii) adding crosslinking agent octavinyloctasilasesquioxane (OVS), (iv) by γ-irradiation, through free radicals react to form OVS-polyurethane crosslinked structure. The β-HEMA is a typical monomer, which are often used to radiation polymerization, here is to consider its hydroxyl groups can react with isocyanate groups, as end capping agent to use. Although the preparation of the OVS-polyurethane by radiation polymerization has been rarely reported, but there are many relevant reports.

As an example, according to the above method, the radiation crosslinking polyurethane was prepared by using polycarbonate diol and liquefied 4, 4-diphenylmethane diisocyanate (Liquefied MDI). The dose is 50kGy, and gamma radiation dose rate is 10kGy/h. The sample 1 is unirradiated polyurethane prepolymer, the sample 2 is the radiation crosslinking polyurethane elastomer without OVS, the sample 3 and sample 4 are the OVS-polyurethane radiation crosslinked samples, which adding quantity of OVS were 7% and 11%, respectively. The structure and properties of obtained samples were investigated by Fourier transform infrared spectroscopy (FTIR), X-ray diffraction (XRD), dynamic thermomechanical analysis (DMA), and thermal gravimetric analysis (TGA).

The FTIR spectra of the radiation crosslinking polyurethane are presented in Figure 2.

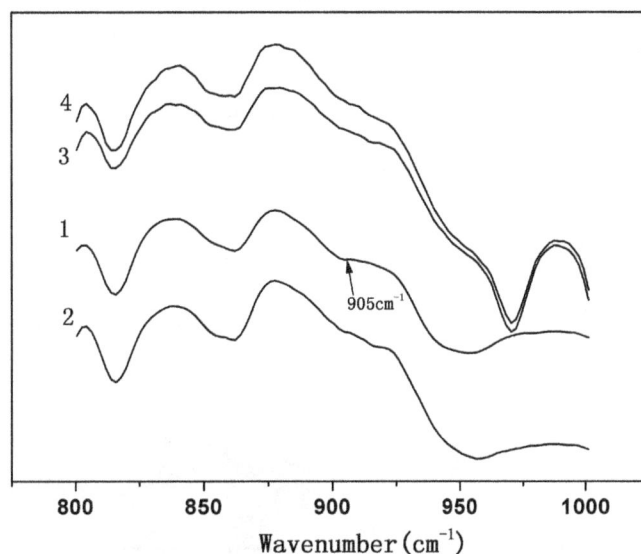

Fig. 2 FTIR spectra of unirradiated sample (1) and irradiated samples (2, 3 and 4) of polyurethane

The FTIR spectra of unirradiated sample (sample 1), appeared a weak absorption peak at 905 cm^{-1}, which can be attributed to the contribution of C=C double bond. However, in the spectra of irradiated samples (2, 3 and 4), the peak disappeared. So it is obvious that the formation of cross-linked structure is mainly due to the crosslinking reaction of the unsaturated double bonds.

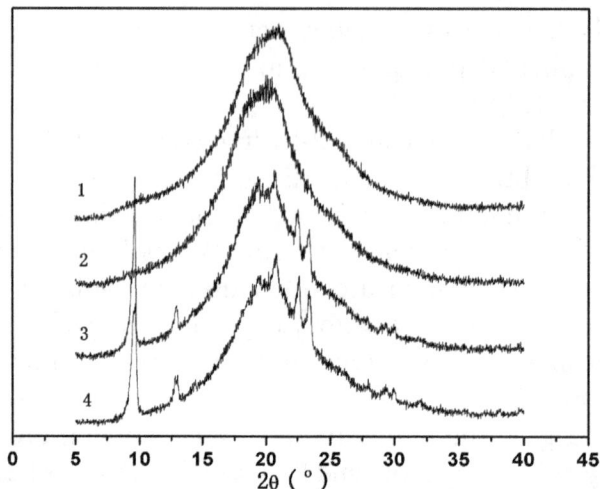

Fig. 3 XRD spectra of unirradiated sample (1) and irradiated samples (2, 3 and 4) of polyurethane

On the other hand, Figure 3 is the XRD spectra of the radiation crosslinking polyurethane. Whether linear (unirradiated sample and sample 1) or radiation crosslinking (sample 2), only one passivated diffraction peak appears and is assigned to the presence of local regular structure, meaning no obvious crystal phenomenon. However, the OVS-polyurethanes radiation crosslinking samples (3 and 4) appear sharp diffraction peaks in the diffraction angle of 7.9, 8.8, 10.9, 11.7, 18.4, 19.9, 21.9, 24.3, and 25.5, which belonged to the crystallization of OVS contained in OVS-polyurethane. In addition, Figure 4 shows how the γ-ray radiation affects the TGA curve of irradiation crosslinking polyurethane.

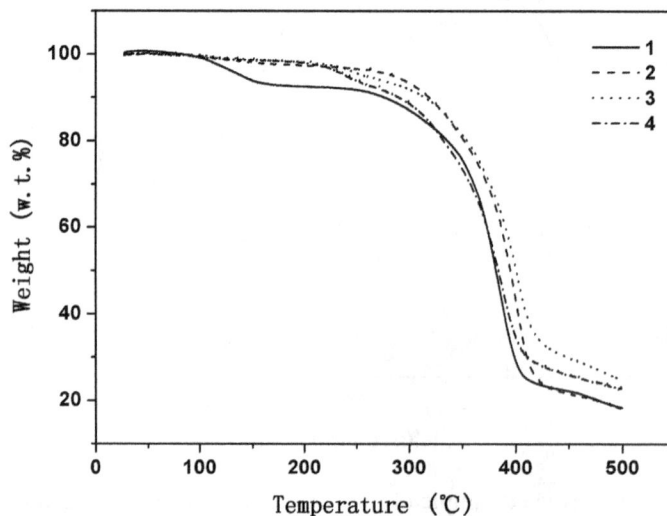

Fig. 4 TGA curves of unirradiated sample (1) and irradiated samples (2, 3 and 4) of polyurethane

Form Figure 4, the thermal stability of radiation crosslinking polyurethane is greatly improved, the temperature of 5% weight loss of sample 2, 3 and 4 are 269.6 °C 242.6 °C and 239.0 °C, respectively, but the temperature of 5% weight loss of unirradiated sample (sample 1) is only just 133.3 °C.

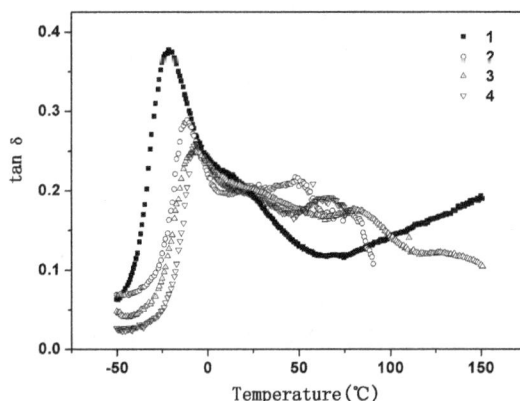

Fig. 5 Tan δ curves of unirradiated sample (1) and irradiated samples (2, 3 and 4) of polyurethane

The Tan δ curves obtained of unirradiated sample and radiation crosslinking polyurethane were shown in Figure 5. DMA charts showed that, the tan δ-T curves of the radiation crosslinking polyurethane have two tan δ peaks. From the molecular structure of polyurethane, the flexible chain segment (often called soft segment) in macromolecules presents random coil state. The soft segment and rigid chain segments (often called hard segment) of polyurethane are gathered in one block, forming the microphase separation structure. If the degree of phase separation is better, then the tan δ-T curve will appear two tan δ peaks, which belonged to soft segment and hard segment, respectively, but if between the two segments is compatibility, will become a tan δ peak. From that, the prepared radiation crosslinking polyurethane has a good degree of microphase separation. From Figure 5, it can be seen that, prior to irradiation (sample 1), only one tan δ peak from the soft segment, which is mainly the contribution of the long chain of polycarbonate diol. After irradiation (sample 2), because of the β-HEMA polymerization and formation of polymeric chain segment, it appears an obvious tan δ peak due to poly (2-hydroxyethyl methacrylate) (PHEM) hard segment, and the tan δ peak of soft segment shifted to higher temperature, which belonged to the confined effect of radiation crosslinking. After adding OVS, two tan δ peaks shifted to higher temperature, and with the increasing of adding amount of OVS, this change became increasingly evident. This should be attributed to the contribution of OVS on radiation crosslinking polymerization.

3. Preparation of POSS-Polyurethane Using Radiation Modified Polyols

Preparation of POSS-polyurethane using radiation modified polyols [42-43]. And POSS is the use of OVS. More specifically, it is that preparing the POSS-filled modified polyol by OVS and γ-ray irradiation, then POSS-filled modified polyols is used to preparation of POSS-polyurethane composites.

For the preparation of POSS-filled modified polyol, OVS and polyols were mixed, and then get by gamma-ray irradiation. The sample 1 (radiation modified polyol, OVS (25 wt%) and polyether glycol (100 wt%)), the performance test results were shown in Table 1. The results show that, the POSS-polyurethane foam has the sound absorption performance and good heat resistance.

Table 1 Determination results of performance of POSS-polyurethane foam

Properties	Determination results
Density(g/cm^3)	0.045
Ope cell rate(%)	96.7
Average acoustic coefficient (125Hz-4000Hz)	0.42
Temperature of 5% weight loss (°C)	247.4
Temperature of 15% weight loss (°C)	290.5
Temperature of 50% weight loss (°C)	336.7

POSS-filled modified polyol can also prepared by adding of OVS and other vinyl monomers (methyl methacrylate (MMA), styrene (St) and acrylonitrile (AN)), then to preparing POSS modified polyurethane foam. Sample 2 (polyol (100 wt%), OVS (15 wt%), MMA (10 wt%)), sample 3 ((polyol (100 wt%), OVS (15 wt%), St (10 wt%)), samples 4 ((polyol (100 wt%), OVS (15 wt%), AN (10 wt%)). Performance test results of foaming material were shown in Table 2. The results show that, the POSS-polyurethane foam has the sound absorption performance and good heat resistance.

Table 2 Determination results of performance of POSS-polyurethane foam

Properties	Determination results		
	Sample 2	Sample 3	Sample 4
Density(g/cm^3)	0.053	0.052	0.045
Ope cell rate(%)	96.5	96.8	97.5
Average acoustic coefficient (125Hz-4000Hz)	0.41	0.36	0.37
Temperature of 5% weight loss(°C)	272.1	283.9	283.9
Temperature of 15% weight loss(°C)	295.2	296.2	298.4
Temperature of 50% weight loss(°C)	331.1	334.5	342.4

4. Radiation Crosslinking POSS-poly (urethane-imide) Foam

Poly(urethane–imide) is a new polymeric material, it has the excellent properties of both polyurethane and polyimide, such as good mechanical properties, thermal stability, high mechanical strength, electrical insulating properties, chemical resistance, hydrolysis resistance, radiation resistance, abrasion resistance and biological compatibility, etc. [44-48]. Therefore, it can also use radiation crosslinking method for preparing POSS-poly (urethane-imide) composites. Here, specifically to discuss the preparation of POSS-poly (urethane-imide) foams.

In the preparation of POSS-poly (urethane-imide) radiation crosslinking foam, the foaming problems of the poly (urethane-imide) foam can be solved by using carbon dioxide gas (such as foaming agent H$_2$O react with isocyanate to produce CO$_2$, etc.). Here, only discuss the preparation of POSS-poly (urethane-imide) radiation crosslinking foam [49], and POSS is the use of OVS. Specifically, it was prepared by using polyester polyol, polyphenyl polyisocyanate (PAPI), and 3,3',4,4'-Benzophenonetetracarboxylic dianhydride (BTDA), and β-HEMA as end capping agent, adding crosslinking agent OVS. And by using γ-irradiation, irradiation dose is 25kGy, 50kGy, 75kGy and 100kGy, respectively, the irradiation dose rate for 10kGy/h. Also, focus on the effect of radiation dose to the structure and properties of radiation crosslinking POSS-poly (urethane-imide) foam.

First, use TGA to explore the irradiation dose on the effect of the thermal stability of radiation crosslinking POSS-poly(urethane-imide) foam, in order to facilitate comparison, select four weight loss points of 5%, 15%, 50% and 80% to determine TGA temperature, the results are shown in Table 3. The results show that, when the weight loss was 5%, 15%, 50% and 80%, respectively, the effect of irradiation dose all showed the same trend, namely, the corresponding TGA temperature increases with the increase of radiation dose, when the irradiation dose reaches 50kGy, the TGA temperature reaches the maximum, then the TGA temperature drop with the increase of irradiation dose, and the irradiation dose increased to higher, then more down low. Further analysis, the crosslinking structure between the vinyl group of POSS and terminal double bonds of poly(urethane-imide) will be formed by γ-irradiation, and the degree of crosslinking is closely related to radiation dose. If the irradiation dose is too small, the degree of crosslinking is insufficient, but if the irradiation dose is too large, the radiation degradation will increase, will cause the decrease of the degree of crosslinking, and this trend will increases with the increase of the irradiation dose.

Table 3 TGA results of radiation crosslinking poly (urethane-imide) foam

Irradiation dose (kGy)	Temperature of 5% weight loss (°C)	Temperature of 15% weight loss (°C)	Temperature of 50% weight loss (°C)	Temperature of 80% weight loss (°C)
0	178.6	212.5	486.3	642.8
25	186.2	222.3	515.1	640.8
50	193.7	232.8	596.0	717.2
75	189.1	220.6	508.6	644.6
100	186.0	218.6	504.0	646.5

Figure 6 is DMA curves of POSS-poly (urethane-imide) radiation crosslinking foam. First of all, from the relations of the irradiation dose and the elastic modulus (E'), the E' of the glassy state increased significantly with the increase of radiation dose, and when the irradiation dose was 50kGy, reached the maximum, and then, decreased with the further increase of the irradiation dose, and, the irradiation dose increased, the E' of the glass state is lower. This indicates that when the irradiation dose was 50kGy, the E' of the glassy state was the highest. This can be attributed to the radiation dose of 50kGy can be the greatest degree of crosslinking, so that the foam has a high rigidity, also displays the maximum modulus. However, in the rubbery state, the effect of irradiation dose on the E' had no evident. Secondly, from the relations of the irradiation dose and the loss modulus (E"), the peak value of E" increases obviously with the increase of dose, when irradiation dose was 75kGy, reached the maximum, and then, the irradiation dose increased again, appeared in rapid decline.

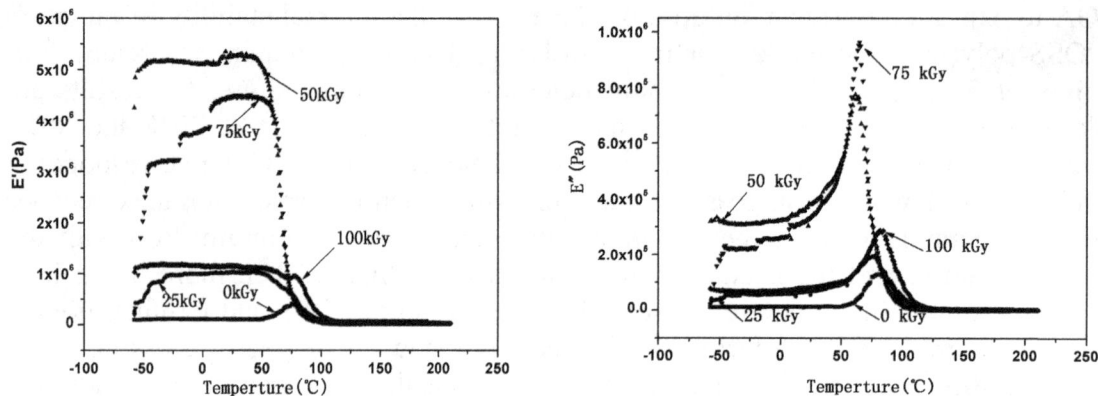

Fig. 6 DMA charts of radiation crosslinking poly (urethane-imide) foam

5. Conclusion

In recent years, POSS/polymer nanocomposite as a new material has obtained the very good development, which pay attention to the excellent performance of POSS/polyurethane nanocomposite and by people. The preparation of POSS/polyurethane composite by radiation, although in its infancy, but obviously can be prepared the POSS/polyurethane nanocomposite, and opened up a new technical approach.

Radiation method POSS/polyurethane nanocomposite preparation, the crosslinking reaction can use the end of polyurethane molecular chains and vinyl double bond in POSS to prepare, can also be prepared to achieve through the POSS-filling radiation system polyols. In addition, basically the same with other methods, POSS/polyurethane nanocomposites prepared by radiation can be used in many aspects such as foam.

References

1. Madbouly S A, Otaigbe J U, Nanda A K, et al. Pheological behavior of POSS/ polyurethane-urea nanocomposite films prepared by homogeneous solution polymerization in aqueous dispersion. Macromolecules, 2007, 40(14): 4982-4991.
2. Efrat T, Dodiuk H, Kenig S, et al. Nanotailoring of polyurethane adhesive by polyhedral oligomeric sisesquioxane (POSS). Journal of Adhesion Science and Technology, 2006, 20(12): 1413-1430.
3. Markevicius G, Chaudhuri S, Bajracharya C, et al. Polyoligomeric silsesquioxane (POSS)-hydrogenated polybutadiene polyurethane coatings for corrosion inhibition of AA2024. Progress in Organic Coatings, 2012, 75(4): 319-327.
4. James P Lewicki, Klzysztof Pielichowski, Pauline Tremblot De La Croix, et al. Thermal degradation studies of polyurethane/POSS nanohybrid elastomers. Polymer Degradation and Stability, 2010, 95(6): 1099-1105.

5. Sibdas Singha Mahapatra, Santosh Kumar Yadav, Jae Whan Cho, et al. Nanostructured hyperbranched polyurethane elastomer hybrids that incorporate polyhedral oligosilsesquioxane. Reactive & Functional Polymers, 2012, 72(4): 227-232.

6. Bliznyuk V N, Tereshchenko T A, Gumenna W A, et al. Structure of segmented poly(ether urethane)s containing amino and hydroxyl functionalized polyhedral oligomeric silsesquioxanes (POSS). Polymer, 2008, 49(9): 2298-2305.

7. Mohammad Mizanur Rahman, Eun Young Kim, Won Ki lee, et al. Effect of DMPA-clay-POSS content on thermal and mechanical properties of nanostructured ionomeric polyurethane. Journal of Nanoscience and Nanotechnology, 2010, 10(10): 6981-6985.

8. Bartlomiej Janowski, Klzysztof Pielichowski. Thermo(oxidative) stability of novel polyurethane/POSS nanohybrid elastomers. Thermochimica Acta, 2008, 478(1-2): 51-53.

9. Serge Bourbigot, Thomas Turf, Séverine Bellayer, et al. Polyhedral oligomeric sisesquioxane as flame retardant for thermoplastic polyurethane. Polymer Degradation and Stability, 2009, 94(8): 1230-1237.

10. Kim E H, Myoung S W, Jung Y G, et al. Polyhedral oligomeric silsesquioxane-reinforced polyurethane acrylate. Progress in Organic Coatings, 2009, 64(2-3): 205-209.

11. Bothe M, Mya K Y, Jie Lin E M, et al. Triple-shape properties of star-shaped POSS-polycaprolactone polyurethane networks. Soft Matter, 2012, 8(4): 965-972.

12. Samy A Madbouly, Joshua U Otaigbe, Ajaya K Nanda, et al. Rheological behavior of POSS/polyurethane-urea nanocomposite films prepared by homogeneous solution polymerization in aqueous dispersions. Macromolecules, 2007, 40(14): 4982-4991.

13. Madhavan K, Gnanasekaran D, Reddy B S R. Synthesis and characterization of poly(dimethylsiloxane-urethane) nanocomposites: Effect of (in)completely condensed silsesquioxanes on thermal, morphological, and mechanical properties. Journal of Applied Polymer Science, 2009, 114(6): 3659-3667.

14. Ralf Lach, Goerg Hannes Michler, Wolfgang Gerllmann, et al. Microstructure and indentation behavior of polyhedral oligomeric silsesquioxanes-modified thermoplastic polyurethane nanocomposites. Macromolecular Material and Engineering, 2010, 295(5): 484-491.

15. Madhavan K, Reddy B S R. Synthesis and characterization of polyurethane hybrids: Influence of the polydimethylsiloxane linear chain and silsesquioxane cubic structure on the thermal and mechanical properties of polyurethane hybrids. Journal of Applied Polymer Science, 2009, 113(6): 4052-4065.

16. Fu B X, Hsiao B S, Pagola S, et al. Structural development during deformation of polyurethane containing polyhedral oligomeric silsesquioxanes (POSS) molecules. Polymer, 2001, 42(2): 599-611.

17. Gar B Hoflund, Rene I Gonzalez, Shawn H Phillips. In situ oxygen atom erosion study of a polyhedral oligomeric silsesquioxzne-polyurethane copolymer. Journal of Adhesion Science and Technology, 2001, 15(10): 1199-1211.

18. Aravindaraj G Kannan, Namita Roy Choudhury, Naba Duba, et al. Fluoro-silsesquioxane-urethane hybrid for thin film applications. ACS Applied Materials & Interfaces, 2009, 1(2): 336-347.

19. Eric Devaux, Maryline Rochery, Serge Bourbigot. Polyurethane/clay and polyurethane/POSS nanocomposites as flame retarded coating for polyester and cotton fabrics. Fire and Materials, 2002, 26(4-5): 149-154.

20. Matthew Oaten, Namita Roy Choudhury. Silsesquioxane-urethane hybrid thin film application. Macromolecules, 2005, 38(15): 6392-6401.
21. Esmaiel Jabban, Samyra Nozari. Swelling of acrylic acid hydrogels prepared by γ-radiation crosslinking of polyacrylic acid in aqueous solution. European Polymer Journal, 2000, 36(12): 2685-2692.
22. Agnes Safrany, Barbara Beiler, Krisztina Laszio, et al., Control of pore formation in macroporous polymers synthesized by single-stepγ-radiation-initiated polymerization and crosslinking. Polymer, 2005, 46(9): 2862-2871.
23. Baljit Singh, S Kumar. Synthesis and characterization of psyllium-NVP based drug delivery system through radiation crosslinking polymerization. Nuclear Instruents and Methods in Physics Research Section B: Beam Interactions with Materials and Atoms, 2008, 266(15): 3417-3430.
24. Abd H L, Mohdy E I, Agnes Safrony. Preparation of fast response superabsorbent hydrogels by radiation polymerization and crosslinking of N-isopropylacrylamide in solution. Radiation Physics and Chemistry, 2008, 77(3): 273-279.
25. Noriaki Seko, Masao Tamada, Fumio Yoshii. Current status of adsorbent for metal ions with radiation grafting and crosslinking techniques. Physics Research Section B:Beam Interaction with Materials and Atoms, 2005, 236(1-4): 21-29.
26. Naotsugu Nagasawa, Ayaka Kaneda, Shinichi Kanazawa, et al. Application of poly(lactic acid) modified by radiation crosslinking. Nuclear Instruments and Methods in Physics Research Section B: Beam Interaction with materials and Atoms, 2005, 236(1-4): 611-616.
27. Radoslaw A Wach, Hiroshi Mitomo, Naotsugu Nagasawa, et al. Radiation crosslinking of carboxymethylcellulose of varius degree of substitution at high concentration in aqueous solutions of natural pH. Radiation Physics and Chemistry, 2003, 68(5): 771-779.
28. Nursel Pekel, Fumio Yoshii, Tamikazu Kume, et al. Radiation crosslinking of biodegradable hydroxypropylmethylcellulose. Carbohydrate Polymers, 2004, 55(2): 139-147.
29. Hennink W E, Nostrum C F. Novel crosslinking methods to design hydrogels. Advanced Drug Delivery Reviews, 2002, 54(1): 13-36.
30. Salmi A, Benfarhi S, Donnet J B, Decker C. Synthesis of carbon-polyaorylate nanocomposite materials by crosslinking polymerization. European Polymer Journal, 2006, 42(9): 1966-1974.
31. Baljit Singh, Manu Vashishtha. Development of novel hydrogels by modification of sterculia gum through radiation crosslinking polymerization for use in drug delivery. Nuclear Instruments and Methods in Physics Research Section B: Beam Interactions with Materials and Atoms, 2008, 266(9): 2009-2020.
32. Acharya A, Mohan H, Sabharwai S. Radiation induced polymerization and crosslinking behavior of N-hydroxy methyl acrylamide inaqueous solutions. Radiation Physics and Chemistry, 2002, 65(3): 225-232.
33. Jinhua Chen, Masaharu Asano, Tetsuya Yamaki, et al. Chemical and radiation crosslinked polymer electrolyte membranes prepared from radiation-grafted ETFE films for DMFC applications. Journal of Power Sources, 2006, 158(1): 69-77.
34. Andrzei G Chmielcwski, Mohammad Haji-Saeid, Shamshad Ahmed. Progress in radiation processing of polymers. Interaction with Materials and Atoms, 2005, 236(1-4): 44-54.
35. Erdener Karadag, Dursun Saragdin, Olgun Guven. Water absorbency studies of γ-radiation crosslinked poly(acrylamide-co-2,3-dihydroxybutanedioic acid) hydrogels. Nuclear Instruments and Methods in Physics Research Section B: Beam Interaction with Materials and Atoms, 2004, 225(4): 489-496.

36. Baljit Singh, Lok Pal. Radiation crosslinking polymerization of sterculia polysaccharide-PVA-PVP for making hydrogel wound dressing. Internationl Journal of Biological Macromolecules, 2011, 48(3): 501-510.
37. Roger A. Assink. Radidtion crosslinking of polyurethanes. Journal of Applied Polymer Science, 1985, 30(6): 30(6): 2701-2705.
38. Beyer G, Steckenbiegler B. Radiation crosslinked thermoplastic polyurethane. Gummi Fasern Kunststoffe (Germany), 1991, 44(11): 614-617
39. Shintani H, Akitada N. Degradation and crosslinking of polyurethane irradiated by gamma-rays. Polymer Regradation and Stability, 1991, 32(2): 191-208.
40. Azevedo E C, Chierice G O, Neto S C. Gamma radiation effects on mechanical properties and morphology of a polyurethane derivate from castor oil. Radiation Effects and Defects in Solide, 2011, 166(3): 208-214.
41. Chengfei Zhou, Wei Cao, Tong Zhai, et al. Preparation and characterization of radiation crosslinking polyurethane and its POSS-containing composites. Polyurethane Industry, 2011, 26(3): 9-12.
42. Fenin A A, Ermakov V I, Revina A A. Radiation-chemical synthesis of polymer polyols and related composites. Theoretical Foundations of Chemical Engineering, 2008, 42(5): 662-666.
43. Chengfei Zhou. The research progress of polyols modified via polymer-filled technology. Synthetic Technology & Application, 2012, 27(3): 19-23.
44. Tetsuya Kogiso, Shin-Ichi Inoue. Synthesis and properties of elastic polyurethane-imide. Journal of Applied Polymer Science, 2010, 115(1): 242-248.
45. Gnanarajan T Philip, Nasser A Sultan , Iyer N. Padmanabha,et al. Synthesis of poly(urethane-imide) using aromatic secondary amine-blocked polyurethane prepolymer. Journal of Polymer Science Part A: Polymer Chemistry, 2000, 38(22): 4032-4037.
46. Gnanarajan T Philip, Iyer N. Padmanabha. Poly(urethane-imide)s from blocked polyurethane prepolymer and pyromellitic dianhydride: effect of alkali metal alkoxides and phenoxides and substituents on the blocking agent in the polymerization reaction. Journal of Macromolecular Science, Part A: Pure and Applied Chemistry, 2001, 38(8): 807-820.
47. Hossein Behniafar. Direct synthesis of new soluble and thermally stable poly(urethane-imide)s from an imide ring-containing dicarboxylic acid using diphenylphosphoryl azide. Journal of Applied Polymer Science, 2006, 101(2): 869-877.
48. Choonkeun Lee, Iyer N. Padmanabha, Kyungwook Min, et al. Synthesis and characterization of novel poly(amide-imide)s containing 1,3-diamino mesitylene moieties. Journal of Polymer Science Part A: Polymer Chemistry, 2004, 42(1): 137-143.
49. Chengfei Zhou, Wei Cao, Tong Zhai, et al. Effect of absorbed dose on radiation crosslinking polyhedral oligomeric silsesquioxane/poly(urethane-imide) nano-composite foam. China Synthetic Rubber Industry, 2013, 36(2): 119-122.

Controlled Synthesis of Bi$_2$O$_3$ Nanomaterials with Different Morphologies by Reverse Microemulsion

Ze-Xue Li, Ya-Jun Wang[*], Hai-Yang Yu

State Key Laboratory of Explosion Science and Technology, Beijing Institute of Technology, Beijing 100081, China
Email: yajunwang@bit.edu.cn

Bi$_2$O$_3$ nanomaterials with a diversity of morphologies, such as nanoparticles, nanorods and nanoplates, were controllably synthesized successfully by a reverse microemulsion route. X-ray diffraction (XRD), field-emission scanning electron microscope (FE-SEM), and thermogravimetry differential thermal analysis (TG-DTA) were employed to characterize the obtained products. It was found that the size and morphology of bismuth oxide were affected mainly by water content (W_0), and the shape of Bi$_2$O$_3$ changed with different W_0. The common role of surfactant adsorbing on the surface of the nanoparticles led to the formation of Bi$_2$O$_3$ nanomaterials with specific morphologies. The possible formation mechanisms of these nanostructures were also discussed. With calcination temperature increasing from 275 °C to 350°C, β-Bi$_2$O$_3$ has transformed into α-Bi$_2$O$_3$ completely.

Keywords: Bi$_2$O$_3$ nanomaterials; reverse microemulsion; morphology control; mechanism.

1. Introduction

Bi$_2$O$_3$ nanomaterials, especially nanostructures with tuned morphology and dimensionality, have a range of applications due to their intrinsic size dependent properties and particular electrical, optical and fast-ion conducting characteristics [1-2]. Thus, it is of great importance to control the size and morphology of desired nanostructures.

In the past years, several ways are used to prepare Bi$_2$O$_3$ nanomaterials, such as wet chemical [3], metalorganic chemical vapor deposition (MOCVD) [4], hydrothermal [5], solution precipitation [6] and sol-gel [7] and so on. Compared with above synthesis methods, of particular note is that microemulsion route is a facile and flexible preparation method which is able control size, geometry, morphology, homogeneity and surface area of final products [8, 9]. Moreover, the synthesis process carry out at ambient temperature and pressure, and reaction solvent can be recycled. So it is a mild, low-cost and eco-friendly preparation technology.

In this paper, we disclose a method for the controlled synthesis of Bi$_2$O$_3$ nanomaterials, especially successful preparation of Bi$_2$O$_3$ nanorods. Also, we found that the size and morphology of bismuth oxide were affected by water content (W_0). The possible formation mechanisms of these nanostructures were also discussed.

2. Materials and Methods

2.1. *Materials*

TritionX-100 (chemical grade) was purchased from China Xilong Chemical Company. Bismuth nitrate pentahydrate (Bi(NO$_3$)$_3$·5H$_2$O), oxalic acid (H$_2$C$_2$O$_4$), 1-pentanol (C$_5$H$_{12}$O), n-heptane (C$_7$H$_{16}$), and nitric acid (HNO$_3$) were analytical grade and purchased from Beijing Chemical Reagent Company, China. Deionized water was used throughout the experiments.

---

[*]Corresponding author

2.2. Preparation method

In a typical preparation process, microemulsion A (RM-A): 1.77 mL of 0.5 mol/L $Bi(NO_3)_3$ solution (the concentration of solvent HNO_3 was 2.0 mol/L) was added to the mixture of 20.0 mL TritonX-100, 20.0 mL 1-pentanol and 40.0 mL of n-heptane. Microemulsion B (RM-B) had the same component as the above solution except for 1.77 mL 1.5 mol/L $H_2C_2O_4$ instead of $Bi(NO_3)_3$. Both microemulsions were left stirring vigorously for 10 min respectively to obtain an optically clear homogeneous dispersion. Then RM-B was added dropwise to RM-A with continuous stirring for 1 h. Subsequently, after standing at room temperature for 24 hour, the resulting precipitate was washed alternately with ethanol and deionized water for 3 times, then dried in a vacuum drying oven at 70 °C for 12 h. Finally, the as-obtained precursors were calcined to get the final Bi_2O_3.

2.3. Characterization

Hitachi S-4800 field-emission scanning electron microscopy (FE-SEM) was used to examine the morphology and size of the precursors and Bi_2O_3, operated on 15 kV. The X-ray diffraction (XRD) of the powder samples was examined on a Germany Bruker D8 Advance diffractometer using Ni-filtered Cu/$K\alpha$ radiation ($\lambda=1.5406$ nm) by a scanning rate of 0.02° s^{-1} in a range from 20° to 70°. Thermogravimetry differential thermal analysis (TG-DTA) was carried out by TG/DTA 6300 system on well-ground samples in atmosphere with a heating rate of 10 °C/min in the range of temperature rise (25–700 °C).

3. Results and Discussion

3.1. Composition and phase structure analysis

Fig. 1 shows a typical XRD pattern of the final sample Bi_2O_3. The peaks ($2\theta=27.47°$, 33.38°, 46.46°), marked by their indices (-121), (-202) and (041), were in line with α-Bi_2O_3 (JCPDS 65-2336). After calcined at 350 °C for 2 h, the precursors decomposed to α-Bi_2O_3. The TG-DTA curves of the precursor are shown in Fig. 2. $Bi_2(C_2O_4)_3$ decomposed to Bi_2O_3 by only one stage as shown in the DTA curve and had no weight loss after 325 °C, and finally gave the product of bismuth oxide.

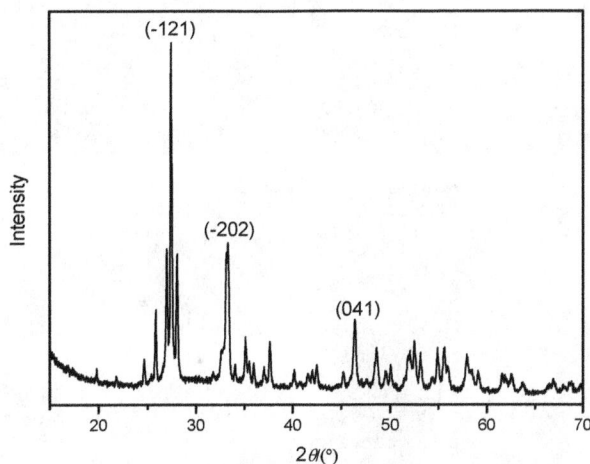

Fig. 1 XRD pattern of Bi_2O_3 calcined on 350 °C for 2 h

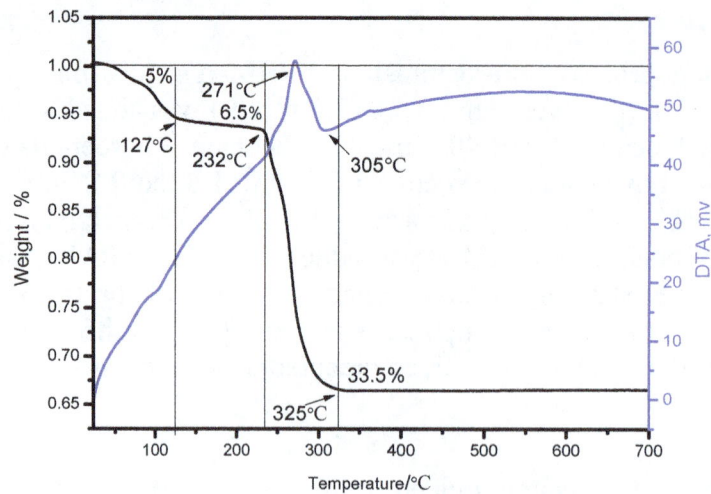

Fig. 2 TG-DTA curves of the precursors

3.2. *Influence of water content (W₀)*

(a), (b) $W_0=1$

(c), (d) $W_0=3$

(e), (f) $W_0=5$

(g), (h) $W_0=9$

(i), (j) $W_0=15$

Fig. 3 SEM images of the precursors (Left image) and Bi_2O_3 (Right image) at different W_0 values

Water content (W_0) is defined here as the molar ratio of water to surfactant. The size and morphology of droplets in microemulsion depends on water content. And when the size of the particle approaches surface of the droplet, the surfactant molecules adsorbed physically on the surface of the particle formed therein, acting as a template agent and restricting the growth of nanoparticles [9]. Water content (W_0) is a crucial factor influencing the size and morphology of Bi_2O_3 nanomaterials. Fig. 3 shows the SEM images of the products synthesized at different W_0. With the increase of water content, the morphology of the precursors and final Bi_2O_3 show extremely obvious changes, and nanobundles (Fig. 3(b)), nanorods (Fig. 3(d)), nanoplates (Fig. 3(h)) and irregular nanostucture (Fig. 3(j)) sequentially appeared.

The different microstructures were attributed to modification of the nucleation and growth kinetics, and guidance of self-assembly approach by the surfactant agent. With the change of the W_0 value, physical properties of the surfactant adsorbing on the surface of the nanoparticles would be changed. The functional surfaces and coatings result in oriented attachment and crystal growth direction. Different oriented connection route and the fusion process of reverse micelles may occur when water content (W_0) varies. With the increase of W_0, the hydrodynamic radius and the amount of the reverse micelles increase [10], which changes the nature of interfacial film [11, 12]. The interfacial film have higher fluidity due to surface tension decreasing with increasing water content, so that it is easy to fuse for the droplets in microemulsion. Then these reverse micelles collide, fuse with the nearby ones in various ways and finally grow into various nanostructures. Different water content leads to the formation of different size and morphology of the reverse micelles cavity, which determines the shape of the final product, as shown in Fig. 4.

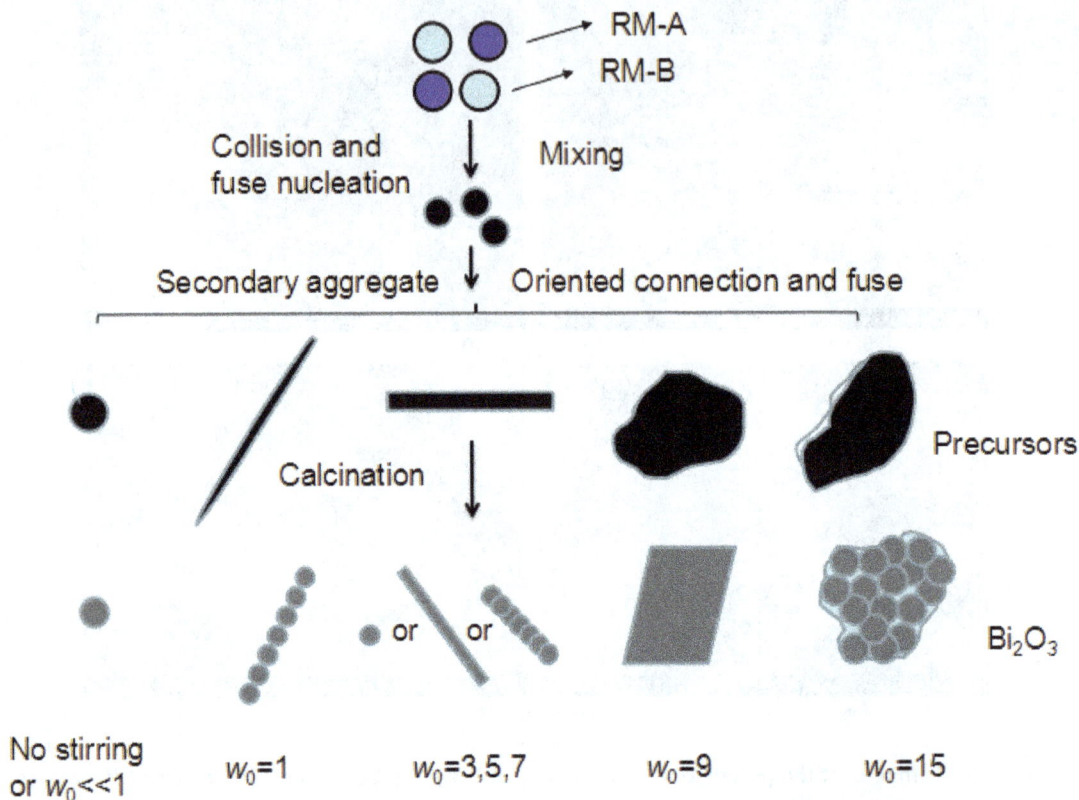

Fig. 4 Schematic diagram of formation mechanism of different structures of precursors and final Bi_2O_3 at different W_0

3.3. *Influence of calcination temperature*

And in general, bismuth oxide has four main crystallographic polymorphs denoted by α-Bi_2O_3, β-Bi_2O_3, γ-Bi_2O_3, and δ-Bi_2O_3. Crystal form is mainly affected by synthesis condition, especially the calcination temperature. As shown in Fig. 5, when kept calcination temperature at 275 °C and 300 °C, the characteristic diffraction peaks of products were in line with that of β-Bi_2O_3, and curve (b) in Fig. 5 shows that diffraction peak was more acute, which indicates that crystallization was better at 300 °C. When calcined at 325 °C, the XRD pattern shows that the as-prepared sample had both α-Bi_2O_3 and β-Bi_2O_3, in curve (c). As the calcination temperature increased to 350 °C, curve (d) shows no other peaks except those of α-Bi_2O_3, suggesting that β-Bi_2O_3 had transformed into α-Bi_2O_3 completely.

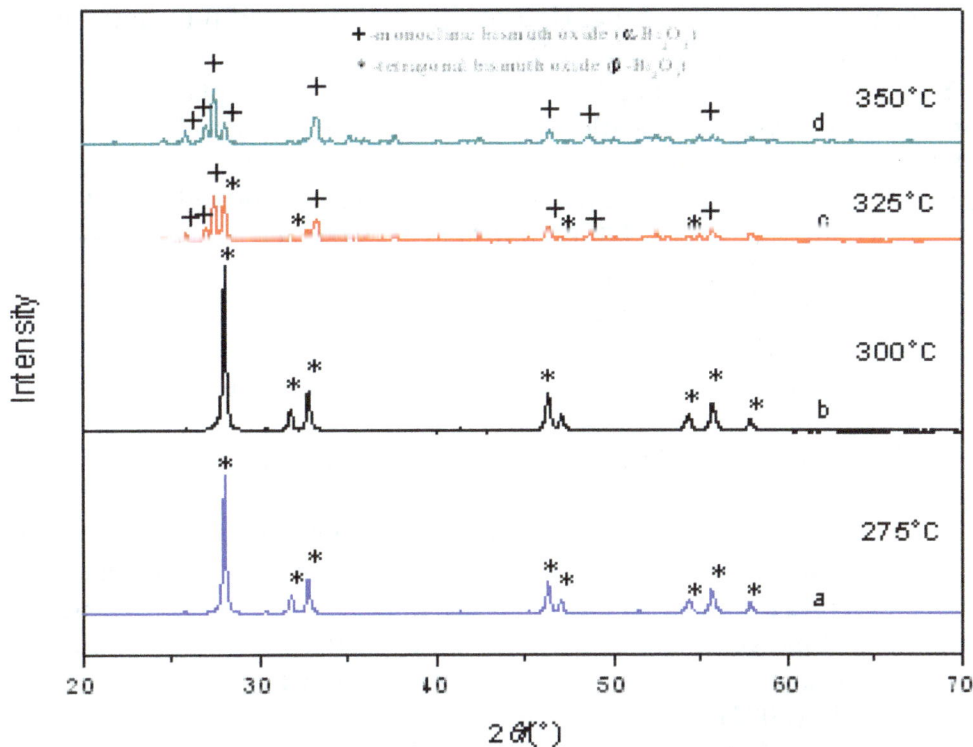

Fig. 5 XRD patterns of Bi_2O_3 at different calcination temperatures

4. Conclusion

Bi_2O_3 nanomaterials with a variety of morphologies have been obtained, such as nanoparticles, nanorods and nanoplates. Especially, there are few reports on the simple preparation of Bi_2O_3 nanorods. And it has been found that water content (W_0) plays a crucial role to the morphology and size of the final Bi_2O_3. When kept W_0 at 3 and 9, we obtained Bi_2O_3 nanorods and nanoplates respectively. Calcination temperature determines the crystal form of the Bi_2O_3. When calcination kept at 300 °C and 350 °C, β-Bi_2O_3 and α-Bi_2O_3 were obtained respectively. And self-assembly and the oriented growth mechanism were proposed for the formation of different structures.

Acknowledgment

This work is funded by the project of State Key Laboratory of Explosion Science and Technology (Beijing Institute of Technology, China) (No.YBKT16-06).

References

1. S. Tsai, K. Fung, C. Ni, Y. Chang, Stability of Bi_2O_3-based ionic conductor and its application on composite cathode, ECS Trans. 68 (2015) 867-874.
2. X. Lv, Z. Li, J. Zhang, B. Yang, A facile approach to prepare bismuth oxide nanorods for application in optoelectronic devices, Chem. Lett. 44 (2015) 97-99.
3. W. Li, Facile synthesis of monodisperse Bi_2O_3 nanoparticles, Mater. Chem. Phys. 99 (2006) 174-180.
4. H.W. Kim, J.W. Lee, S.H. Shim, Study of Bi_2O_3 nanorods grown using the MOCVD technique, Sens. Actuators, B 126 (2007) 306-310.

5. L. Liu, J. Jiang, S. Jin, Z. Xia, M. Tang, Hydrothermal synthesis of β-bismuth oxide nanowires from particles, Crystengcomm, 13 (2011) 2529-2532.

6. Y.C. Wu, Y.C. Chaing, C.Y. Huang, S.F. Wang, H.Y. Yang, Morphology-controllable Bi_2O_3 crystals through an aqueous precipitation method and their photocatalytic performance, Dyes & Pigm. 98 (2013) 25-30.

7. C. Karunakaran, P. Magesan, P. Gomathisankar, Photocatalytic activity of sol-gel derived Bi_2O_3-TiO_2 nanocomposite, Mater. Sci. Forum 712 (2012) 73-83.

8. A. Hu, Z. Yao, X. Yu, Phase behavior of a sodium dodecanol allyl sulfosuccinic diester/n-pentanol/methyl acrylate/butyl acrylate/water microemulsion system and preparation of acrylate latexes by microemulsion polymerization, J. Appl. Polym. Sci. 113 (2009) 2202-2208.

9. M.P. Pileni, Nanocrystals: fabrication, organization and collective properties, C. R. Chim. 6 (2003) 965-978.

10. M.A. Malik, M.Y. Wani, M.A. Hashim, Microemulsion method: A novel route to synthesize organic and inorganic nanomaterials, Arabian J. Chem. 5 (2012) 397-417.

11. P. Kaushik, S. Vaidya, T. Ahmad, A.K. Ganguli, Optimizing the hydrodynamic radii and polydispersity of reverse micelles in the Triton X-100/water/cyclohexane system using dynamic light scattering and other studies, Colloids Surf., A 293 (2007) 162-166.

12. M. Li, S. Mann, Emergent nanostructures: water-induced mesoscale transformation of surfactant-stabilized amorphous calcium carbonate nanoparticles in reverse microemulsions, Adv. Funct. Mater. 12 (2002) 773-779.

13. M. Li, B. Lebeau, S. Mann, Synthesis of aragonite nanofilament networks by mesoscale self-assembly and transformation in reverse microemulsions, Adv. Mater. 15 (2003) 2032-2035.

Dynamic Behavior of a Single-walled Carbon Nanotube for Nanoparticle Delivery with Surface Effect

Wei Wang[1,*], Wei-Kai Xu[2]

¹School of Civil Engineering, Shenyang Jianzhu University,
Shenyang, China
²Key Laboratory of Liaoning Province for Composite Structural Analysis of Aerocraft and Simulation
Shenyang Aerospace University, Shenyang, China,
*Email: starwei2002@163.com

As the structure sizes are in nanoscale, the surface-to-bulk energy ratio increases and the surface effects must be taken into account. Surface effect plays a key role to accurately predict the vibration behavior of nanostructures. In this paper, the dynamic behavior of a single-walled carbon nanotube (SWCNT) for nanoparticle delivery is studied. The nonlocal Bernoulli-Euler beam theory is used and the surface effect is taken into account. It is found that in the presence of the surface effects, the dimensionless displacement significantly decreases, unlike the classical solution in which the surface effect is neglected.

Keywords: Single-walled carbon nanotube; nanoparticle delivery; non-local elasticity theory; surface effect.

1. Introduction

Carbon nanotubes are expected to be one of the key structures and have many potential applications. Very recently, the hollow geometry is considered to be applied in biological devices such as nanofluid conveyance and drug delivery. These structural elements have drawn great attention in targeted drug delivery for cancer therapy.

The interaction of drug with SWCNT can be characterized as a nanobeam conveying moving nanoparticles. For example, Lee and Chang [1] presented a dynamic model of a SWCNT conveying a moving nanoparticle. In their study, the surrounding foundation stiffness and van der Waals force were all taken into account. Based on the nonlocal elasticity theory, Simsek [2] analytically studied the forced vibration of an elastically connected double-carbon nanotube system under a moving nanoparticle. Kiani [3] investigated the dynamic response of a SWCNT subjected to a moving nanoparticle by incorporating the inertial effect and friction between the nanoparticle surface and the inner surface. Kiani et al. [4, 5] respectively modeled such system based on the Rayleigh, Timoshenko, and higher-order beam theories in the context of the nonlocal continuum theory. Ghorbanpour Arani et al. [6] investigated nonlocal vibration of single-walled Boron Nitride nanotube under a moving nanoparticle. The effect of electric field, elastic medium, slenderness ratio and small scale parameter are discussed in detail. This paper opens possibilities for controlling the dynamical properties by using external electric field.

However, as the structure sizes are in nano scale, the surface-to-bulk energy ratio increases and the surface effects must be taken into account. Therefore, surface effects are important to nanostructure and many studies have been performed to investigate the surface effects on nanostructures. For example, Lee and Chang [7] used nonlocal Timoshenko beam theory to study the vibration behavior of nanotubes with surface effect. They observed that the frequency ratio increases when the surface effect is taken into account. Wang [8] formulated the dynamics of fluid-conveying nanotubes with surface effect. They found that the natural frequencies increase due to the presence of surface effect. Narendar et al. [9, 10, 11] compared the nonlocal wave properties of nanotubes and nanoplate with and without surface effect. They found that the flexural wavenumbers with surface effect become higher. Therefore, it is believed that surface effect plays a key role to accurately predict the vibration behavior of nanostructures. However, the discussion of surface effect on the dynamics of nanotube conveying nanoparticles is rather limited. In this paper, a theoretical analysis is presented and both the nonlocal size effect and surface effect are under consideration, which is different from the previous models.

2. Non-local Model for SWCNT with Surface Effect

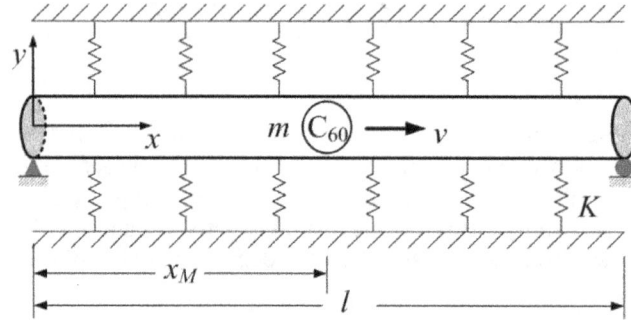

Fig. 1 A nanoparticle-conveying SWCNT embedded in an elastic medium

A schematic diagram of a nanotube conveying a moving nanoparticle is shown in Fig. 1. The nanotube has the length L, mass mc, Young's Modulus E and inertia I. tc, di and do are thickness, inner and outer diameter of the nanotube. The mass of the moving nanoparticle is m.

The one-dimensional non-local constitutive relation is [12,13,14,15]

$$M - (e_0 a)^2 \frac{\partial^2 M}{\partial x^2} = -\Gamma \frac{\partial^2 w}{\partial x^2} \qquad (1)$$

where M is the bending moment, e_0 the constant appropriate to each material and a the internal characteristic length. The value of e_0 can be identified by experiments. If $e_0 a = 0$, the relationship of Eq.(1) reduces to the classical elasticity theory. w is the transverse displacement of the nanotube. Γ is the effective flexural rigidity, which is defined as

$$\Gamma = EI + \alpha = EI + \frac{1}{8} \pi E_s (d_i^3 + d_0^3) \qquad (2)$$

where EI is the flexural rigid and α is the additional flexural rigidity induced by the surface energy in nanotubes. Es is the surface elastic modulus, do and di are the outer and inner diameters of the nanotube. It follows from Eq. 2 that surface effect is important and induces size effect in the otherwise size-independent classical elasticity.

The transverse vibration equation of the nanotube can be expressed as

$$\frac{\partial Q}{\partial x} = m_c \frac{\partial^2 w}{\partial t^2} - P \qquad (3)$$

$$Q = \frac{\partial M}{\partial x} \qquad (4)$$

where Q is the resultant shear force on the cross section, mc is the mass of the SWCNT per unite length, t is the time. P is the acting force on the SWCNT, which can be written as

$$P = mg\delta(x - vt) - Kw + H_0 \frac{\partial^2 w}{\partial x^2} \tag{5}$$

where K denotes the Winkle constant of the surrounding elastic medium. v is the moving velocity of the nanoparticle. δ is the Dirac delta function. H_0 is the surface parameter and defined by

$$H_0 = 2\tau_0(d_i + d_0) \tag{6}$$

which τ_0 is determined by the residual surface tension.

Substituting Eq. 2-Eq. 6 into Eq. 1, the governing equation of motion for a nanoparticle moving in the nanotube can be expressed as

$$(EI + \alpha)\frac{\partial^4 w}{\partial x^4} + \left[1 - (e_0 a)^2 \frac{\partial^2}{\partial x^2}\right]\left(m_c \frac{\partial^2 w}{\partial t^2} + Kw - H_0 \frac{\partial^2 w}{\partial x^2}\right)$$
$$= \left[1 - (e_0 a)^2 \frac{\partial^2}{\partial x^2}\right] mg\delta(x - vt) \tag{7}$$

In this paper, the pined-pined boundary conditions are under consideration

$$\frac{\partial^2 w(0,t)}{\partial x^2} = w(0,t) = 0, \quad \frac{\partial^2 w(L,t)}{\partial x^2} = w(L,t) = 0. \tag{8}$$

3. Analytical Solution

Standard Galerkin-type projections will be utilized in this section. Assuming the displacement expansion is

$$w(x,t) = \sum_{n=1}^{\infty} q_n(t) \sin(\frac{\lambda_n x}{L}) \tag{9}$$

where qn(t) is the generalized coordinates of the discretized system, $\sin(\frac{\lambda_n x}{L})$ is the deflection curve for the nth mode of nanotube with a simply supported boundary condition. Substituting Eq. 9 into Eq. 7 and multiplying by $\sin(\frac{\lambda_n x}{L})$ and then integrating the results from 0 to L lead to

$$\frac{d^2 q_n}{dt^2} + \omega_n^2 \left[\frac{1}{1 + \rho^2 \lambda_n^2} + \frac{k + \lambda_n^2 \hbar}{\lambda_n^4}\right] q_n = \frac{96 Q_s \omega_n^2}{(1 + \rho^2 \lambda_n^2) \lambda_n^4} \sin(\frac{\lambda_n t}{\tau}) \tag{10}$$

where

$$\rho = \frac{e_0 a}{L}, k = \frac{K + L^4}{EI + \alpha}, \omega_n^2 = \frac{\lambda_n^4 (EI + \alpha)}{m_c L^4}, Q_s = \frac{mgL^3}{48(EI + \alpha)}, \tau = \frac{L}{v}, \hbar = \frac{H_0 L^2}{EI + \alpha} \qquad (11)$$

Introduce the dimensionless velocity parameter γ

$$\gamma = \frac{v}{v_{cr}} \qquad v_{cr} = \pi \sqrt{\frac{EI + \alpha}{m_c L^2}} \qquad (12)$$

where vcr represents the critical speed of the moving nanoparticle, at which the SWCNT loses its stability by buckling, α is defined in Eq. 2. Eq. 10 is a set of ordinary differential equation and can be solved analytically, the result is

$$w(x,t) = \sum_{n=1}^{\infty} \frac{96\beta Q_s}{n^2 \pi^4 (\mu^2 n^2 - \gamma^2)} \left[\sin(n\pi \frac{t}{\tau}) - \frac{\gamma}{\mu n} \sin(\mu \frac{n^2 \pi}{\gamma} \frac{t}{\tau}) \right] \sin(n\pi \frac{x}{L}) \qquad (13)$$

where

$$\beta = \frac{1}{1 + \rho^2 \lambda_n^2} \qquad \mu = \sqrt{\beta + \frac{k + \lambda_n^2 \hbar}{\lambda_n^4}}$$

4. Numerical Results and Discussion

In this section, the dynamics of the nanotube will be numerically demonstrated. Fig. 2 illustrates the surface effect on the dimensionless displacement of a SWCNT conveying nanoparticle for $EI=7.12\times10^{-25}$ Nm^2, $E_s=5.1882$, $\tau_0=0.9108$, $\gamma=0.2$, $d_o=1.36$nm, $t_c=0.35$nm, $m_c=4.32\times10^{-25}$kg/m. $k=20$, $\rho=0.1$ and L=40nm. It is shown that, in the presence of the surface effects the dimensionless displacement significantly decreases. It indicates that the surface properties are significant for the SWCNT and cannot be neglected.

Fig. 3 shows the influence of outer diameter on the dimensionless displacement. It is found that the displacement decreases as the outer diameter do increases. The reason behind this is, as shown in Eq. 2, the effective flexural rigidity increases as the outer diameter do increases. The increasing value of the effective flexural rigidity leads to a stiffer SWCNT. Accordingly, the dimensionless vibration displacement is deceased.

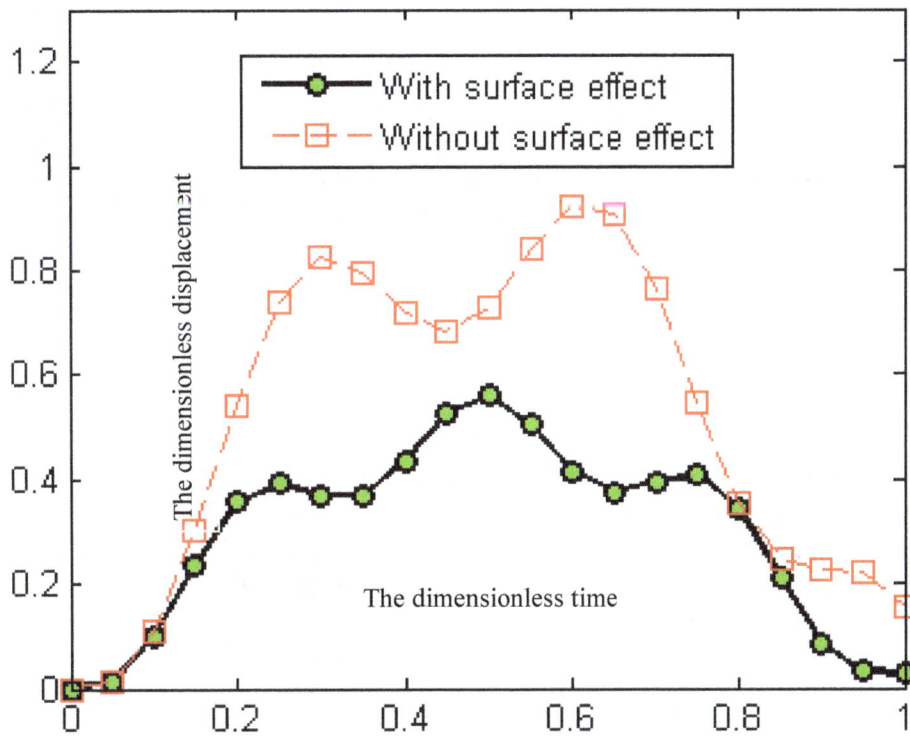

Fig. 2 The surface effects on the dimensionless displacement of a SWCNT conveying nanoparticles

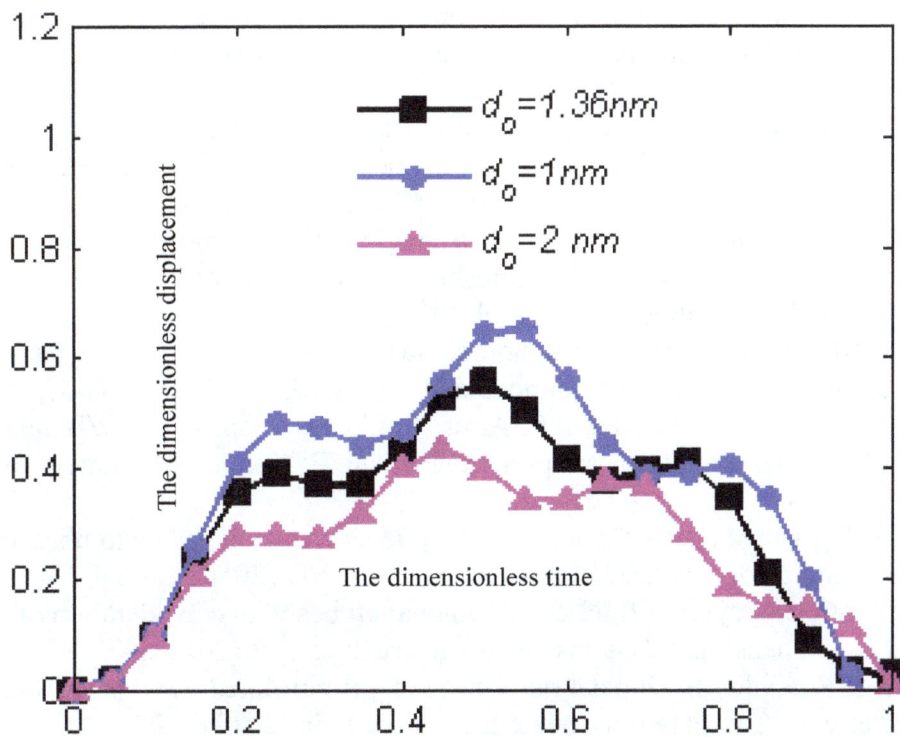

Fig. 3 The outer diameter on the dimensionless displacement of a SWCNT conveying

5. Conclusion

In the present paper, the dynamics of a SWCNT conveying a nanoparticle is investigated. The surface effect on the dynamic behavior is discussed. It is found that in the presence of the surface effects, the dimensionless displacement significantly decreases, unlike the classical solution in which the surface effect is not taken into account. Due to the surface effect, the SWCNT become stiffer. Accordingly, as the outer diameter increases, the displacement decreases. In addition, it is also found that the higher values of embedding stiffness, nonlocal parameter and the length increase lead to larger dimensionless displacements. All the conclusions obtained are expected to be useful in the precise nanoparticle delivery.

Acknowledgments

This research was funded by the following: the Youth Foundation of National Natural Science (51308357, 11302135), the Science and college and universities outstanding young scholar growth plan in Liaoning province (LJQ2015091, LJQ2014019), Natural Science Foundation of Liaoning Province of China (201602627,201602572) a and Discipline Content Education Project of Shenyang Jianzhu University (XKHY2-11).

References

1. H.L. Lee, W.J. Chang, Dynamic modelling of a single-walled carbon nanotube for nanoparticle delivery, Proceedings of the Royal Society A: Mathematical, Physical and Engineering Science, 467 (2011), 860-868.
2. M. Simsek, Nonlocal effects in the forced vibration of an elastically connected double-carbon nanotube system under a moving nanoparticle, Computational Materials Science, 50 (2011), 2112-2123.
3. K. Kiani, Longitudinal and transverse vibration of a single-walled carbon nanotube subjected to a moving nanoparticle accounting for both nonlocal and inertial effects, Physica E: Low-dimensional Systems and Nanostructures, 42 (2010), 2391-2401.
4. K. Kiani, Q. Wang, On the interaction of a single-walled carbon nanotube with a moving nanoparticle using nonlocal Rayleigh, Timoshenko, and higher-order beam theories, European Journal of Mechanics-A/Solids, (2011), 179-202.
5. K. Kiani, B. Mehri, Assessment of nanotube structures under a moving nanoparticle using nonlocal beam theories, Journal of Sound and Vibration, 329 (2010), 2241-2264.
6. A. Ghorbanpour Arani, M. Roudbari, S. Amir, Nonlocal vibration of SWBNNT embedded in bundle of CNTs under a moving nanoparticle, Physica B: Condensed Matter, 407 (17) (2012), 3646-3653.
7. H.L. Lee, W.J. Chang, Surface effects on frequency analysis of nanotubes using nonlocal Timoshenko beam theory, Journal of Applied Physics, 108 (2010), 093503.
8. L. Wang, Vibration analysis of fluid-conveying nanotubes with consideration of surface effects, Physica E: Low-dimensional Systems and Nanostructures, 43 (2010), 437-439.
9. S. Narendar, S. Ravinder, S. Gopalakrishnan, Study of non-local wave properties of nanotubes with surface effects, Computational Materials Science, 56 (2012), 179-184.

10. S. Narendar, S. Gopalakrishnan, Terahertz wave characteristics of a single-walled carbon nanotube containing a fluid flow using the nonlocal Timoshenko beam model, Physica E: Low-dimensional Systems and Nanostructures, 42 (2010), 1706-1712.
11. S. Narendar, S. Gopalakrishnan, Study of terahertz wave propagation properties in nanoplates with surface and small scale effects, International Journal of Mechanical Sciences, 64(1) (2012), 221-231.
12. A.C. Eringen. On differential equations of nonlocal elasticity and solutions of screw dislocation and surface waves, Journal of Applied Physics, 54 (1983), 4703-4710.
13. B. Arash, Q. Wang, A review on the application of nonlocal elastic models in modeling of carbon nanotubes and graphenes, Computational Materials Science, 51 (2012) 303-313.
14. A. Ghorbanpour Arani, M. Shokravi, S. Amir, M. Mozdianfard, Nonlocal electro-thermal transverse vibration of embedded fluid-conveying DWBNNTs, Journal of mechanical science and technology, 26 (2012) 1455-1462.
15. M. Mirramezani, H. Reza Mirdamadi, Effects of nonlocal elasticity and Knudsen number on fluid-structure interaction in carbon nanotube conveying fluid, Physica E: Low-dimensional Systems and Nanostructures, 44 (2012) 2005-2015.

Low Concentration Hydrogen Sensor Based on Palladium Nanoparticles Film

Peng-Cheng Huang, Gang Zhang*, You-Ping Chen

School of Mechanical Science and Engineering, Huazhong University of Science and Technology, Wuhan,China
*Email: GangZhang@hust.edu.cn

Hydrogen is promising and green energy resource, but the hydrogen is a dangerous gas. The explosion limits is 4.1% to 74.2% and the explosion power is very high, so the low concentration hydrogen detection is very important. ALD (Atomic Layer Deposition) method is used to prepare the palladium nanoparticles as the hydrogen sensing units. The palladium nanoparticle is sensitive to hydrogen. The response time and recovery time for 0.05% hydrogen are 35s and 20s respectively. The signal intensity can satisfy the detection. This work presents that the ALD preparation method can supply palladium nanoparticles for low concentration hydrogen detection.

Keywords: Hydrogen; optical fiber sensor; palladium nanoparticles.

1. Introduction

Palladium is a common material for hydrogen detection, which is highly sensitive and selective to hydrogen molecule [1]. Since the advantages of optical fiber hydrogen sensors including high sensitivity, good stability, anti-electromagnetic, and intrinsic safety, the optical fiber hydrogen sensors are studied widely [2, 3]. The Palladium film has been used as sensing unit for hydrogen detection, as its high selectivity. The palladium film could absorb the hydrogen atoms and desorb the atoms thoroughly, and above cycle is invertible. But the high concentration hydrogen will make PdH_x go through three stages from α phase, α-β phase to β phase [4]. After many absorption cycles, the sensing film would crack [5], which leads the film invalid. This paper presents a new preparation method for hydrogen sensing unit. ALD (Atomic Layer Deposition) could supply nanoparticles preparation [6, 7], which could be used for hydrogen leak detection. The optical property of the palladium nanoparticles would change after absorb hydrogen, whose variation could be used for hydrogen detection. The flammable range of hydrogen is from 4.1% to 74.2% [8], which means hydrogen is a dangerous gas. It is necessary to detect hydrogen leakage. To detect the hydrogen response in low concentration, this work present hydrogen response in 0.05%.

2. Experiment

2.1. Preparation

The substrate of the sensing unit is Si disc of diameter in 10 millimeters. Palladium(II) hexafluoroacetylacetonate (Pd) and formalin is used as precursor for nanoparticles preparation. Nitrogen is the carrier gas for the whole process. The temperature of reaction chamber is 200°C, the temperature of pipeline and precursor is 80°C and 50°C respectively. The rate of nitrogen flow is 100 sccm. The deposition time is for 2 s, the gas time is for 8 s, the whole cycle is for 10 s. The nanoparticles on substrates start to grow after 50 cycles. According to the previous experience, 100 ALD cycles are corresponding to 0.7 nm. The preparation in this work is for 400 cycles, whose thinness is about 3 nm.

---

*Corresponding author

2.2. *Optical fiber sensor system*

The dual optical path compensation is designed to be used in the sensing system. Figure 1 presents the block diagram of the optical fiber system, there are two chambers, the one is the test chamber, and the other is the reference one. The signal of the light intensity of reflection is detected by the monitor system. The 0.05% hydrogen is flowed into the test chamber, then the hydrogen is captured by Pd, the monitor system would detect the change of the sensing unit

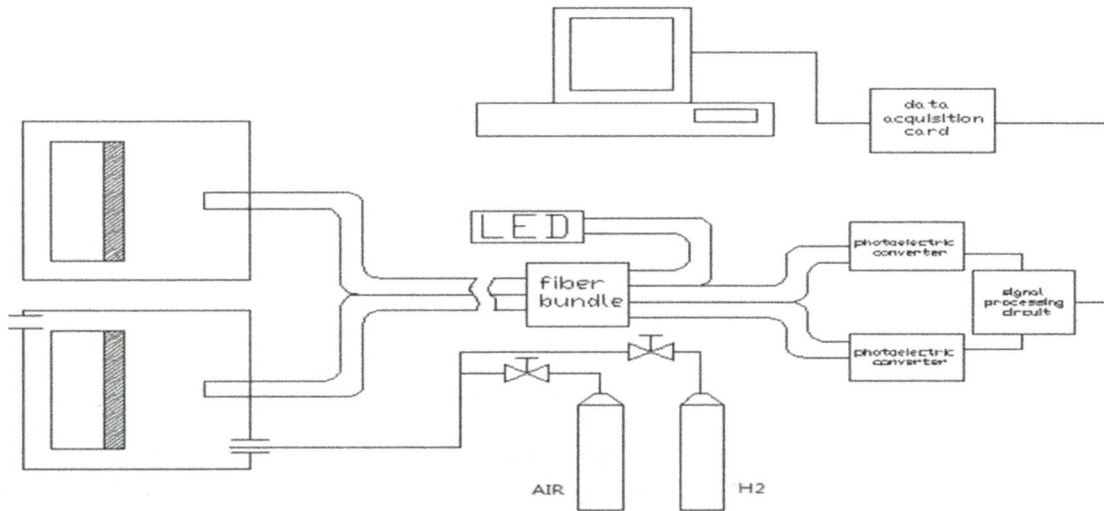

Fig. 1 Structure of detection system

3. Result and Discussion

Figure 2 shows the response and recovery of the sensing system. The test gas is 0.05% concentration hydrogen, and the balance gas is nitrogen. When the test gas is flowed into the test chamber, the optical property of the sensing unit will change. The response time of the sensor is about 35s, and the recovery time is about 20s. Since the hydrogen goes into the palladium, the optical property of the sensing film would change the light intensity of reflection increases. The reaction is showed in Equation (1). There are some signal noises during the recovery process, whose reason is that the hydrogen is not steady when hydrogen valve is closed.

$$Pd + \tfrac{x}{2}H_2 \leftrightarrow PdH_x \tag{1}$$

When the sensing film absorbs hydrogen, the palladium lattice would expand. After absorbing hydrogen the nanoparticles expand, which results in the increasing of the reflection area. The received signal would increase, so the curve in Figure 2 would rise. The output signal could be expressed as Equation (2).

$$S_{out} = -\frac{I_h}{I_r} + C \tag{2}$$

Here, I_h is signal intensity of the test signal, and I_r is signal intensity of the reference signal. C is the bias-voltage, which is a constant.

From the Figure 2, there is few zero shift in the response and recovery curve. Since the concentration is very low, the palladium phase change will not occur. This sensor is used in low concentration condition could restrain the phase change and then restrain the crack and bubble of the sensing film.

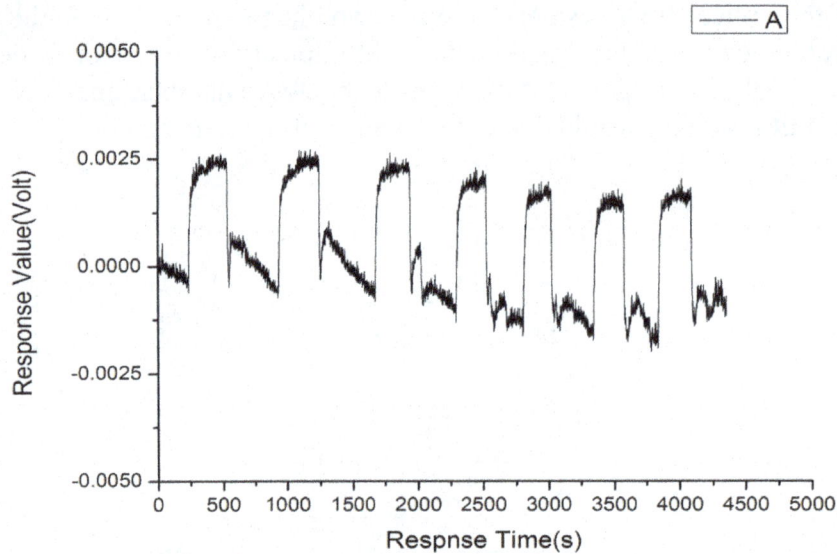

Fig. 2 Response curve of sensing unit in 0.05% hydrogen concentration

Characterization of the sample is showed in Figure 3. The AFM is used to observe the surface topography of the sample. The AFM scanned area is 5*5 micrometer and the lightspots are the palladium nanoparticles. The particles distribute on the substrate, the size of the particles are nanometers in size. The height of the nanoparticles is from 4 nm to 15 nm, the size of the particles is 4 nm to 15 nm approximately. The substrate is covered fully by palladium nanoparticles, which demonstrates that the ALD preparation method could supply the uniform distribution of the palladium particles on the SiO_2 substrate.

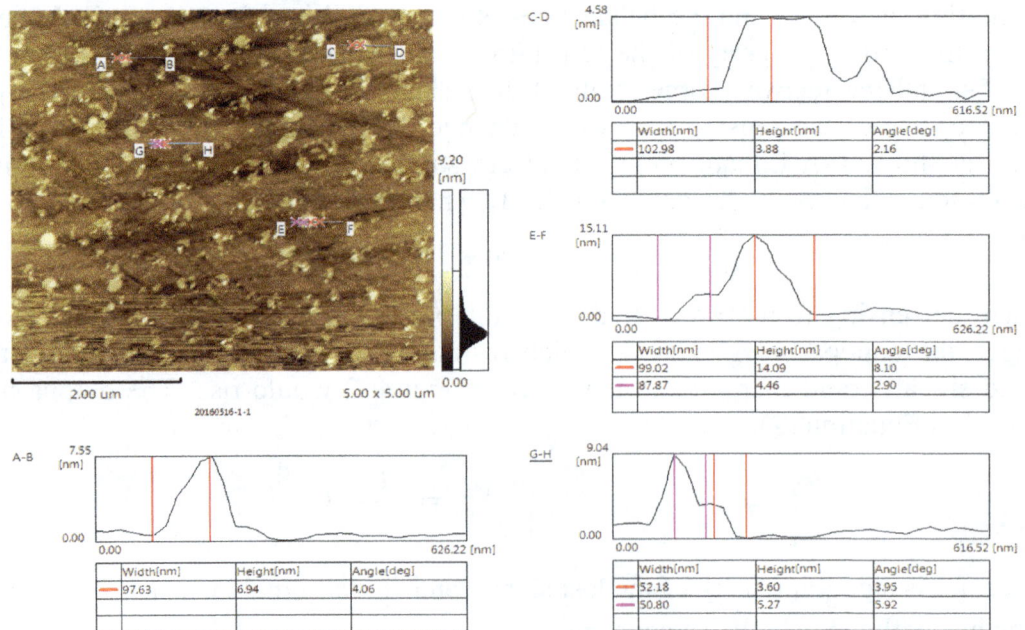

Fig. 3 AFM scan result of the sample

4. Conclusion

The ordinary optical fiber hydrogen sensor is hard to detect low concentration hydrogen. This work uses the ALD preparation method to prepare the palladium nanoparticles as the sensing film to detect the low concentration hydrogen. The sample is used to detect 0.05% hydrogen, and the response range is relatively large. The dual optical path compensation system compensates the fluctuation caused by variation of temperature and optical source. The response time and recovery time are 35s and 20s respectively. Although the response amplitude is not very large, the signal processing circuit can enhance the signal to compensate the deficiency. Since the low concentration hydrogen will not lead palladium phase change, this sensor can be used in low concentration condition. As the good performance of the sample, the sensor in this work can satisfy the detection of hydrogen leakage.

Acknowledgment

This research was funded by the Fundamental Research Funds for the Central Universities (2016YXMS269).

References

1. Hübert, T., et al., Hydrogen sensors – A review. Sensors and Actuators B: Chemical, 2011. 157(2): p. 329-352.
2. Masuzawa, S., et al., Catalyst-type-an optical fiber sensor for hydrogen leakage based on fiber Bragg gratings. Sensors & Actuators B Chemical, 2014.
3. X. Be Venot A, A.T.A.C., Hydrogen leak detection using an optical fibre sensor. Sensors and Actuators B 67 (2000). 57–67, 2000.
4. Butler, M.A. and D.S. Ginley, Hydrogen sensing with palladium-coated optical fibers. Journal of Applied Physics, 1988.
5. Chen, R., et al., A reliable and fast hydrogen gas leakage detector based on irreversible cracking of decorated palladium nanolayer upon aligned polymer fibers. International Journal of Hydrogen Energy, 2015. 40(1): p. 746-751.
6. Lu, J., et al., Porous Alumina Protective Coatings on Palladium Nanoparticles by Self-Poisoned Atomic Layer Deposition. Chemistry of Materials, 2012.
7. Goldstein, D.N. and S.M. George, Surface poisoning in the nucleation and growth of palladium atomic layer deposition with Pd(hfac)2 and formalin. Thin Solid Films, 2011.
8. Butler, M.A., Optical fiber hydrogen sensor. Applied Physics Letters, 1984. 45(10): pp. 1007-1009.

Fabrication and Adsorption Properties of Graphene Oxide Aerogel for Methylene Blue from Aqueous Solution

Li-Ping Wang[1,2,*], Ming-Yu Zhang[2,*], Min Lei[1], Shan Yang[1]

[1] *Department of Biological and Environmental Engineering, Changsha University, Changsha, Hunan, 410003, China*
[2] *Powder Metallurgy Research Institute, Central South University, Changsha, Hunan, 410083, China*
Email: misswlp@163.com, zhangmingyu@csu.edu.cn

Graphene oxide aerogel was fabricated by the vacuum freeze-drying method and characterized with scanning electron microscope (SEM), Nitrogen adsorption-desorption curve and Fourier transform infrared spectroscopy (FTIR). Then, the batch adsorption test of graphene oxide aerogel for methylene blue (MB) from aqueous solution was carried out. The experimental results showed that the developed mesoporous structure and plenty of functional groups formed in the fabricated graphene oxide aerogel. Moreover, the removal rate gradually increased from 58.4% to 99.7% with raising pH from 2.0 to 9.0. Furthermore, the adsorption isotherm was well fitted by Freundlich model and the pseudo-second-order kinetic model could well describe the adsorption process. According to the calculated thermodynamic parameters, the adsorption process was non-spontaneous and endothermic.

Keywords: Graphene oxide aerogel; fabrication; MB; adsorption.

1. Introduction

Dye wastewater is urgent to be properly treated at present, since most dyes are harmful to human being and the environment. Many previous studies about the treatment methods of dye wastewater were reported [1, 2]. Among these methods, the adsorption is the most efficient and economical way due to its simple operation, high efficiency and low cost.

Recently, graphene oxide has been paid more and more attention in adsorption field because of its high surface area and abundant oxygen-containing functional groups such as hydroxyl, epoxy groups in the basal plane, and carboxyl groups at the edges. Especially for some cationic dyes such as methylene blue (MB), graphene oxide may be a high-efficiency adsorbent in consideration of electrostatic attraction, hydrogen bond and the π–π stacking interactions of aromatic rings on the adsorbent and adsorbate. However, it is difficult to separate the graphene oxide from the solution due to its soluble property in water. In order to enhance the separability of the graphene oxide from the solution, the magnetic graphene oxide composites were prepared and their adsorption performances were investigated in some literatures [3]. However, there have been few reports referred about the adsorption of graphene oxide aerogel.

Therefore, in this paper, a graphene oxide aerogel was fabricated through vacuum freeze-drying method and its adsorption behavior for methylene blue was evaluated.

2. Materials and Methods

2.1. *Preparation of graphene oxide aerogel*

Graphite oxide was prepared by oxidation of natural flake graphite according to the improved Hummers' method [4]. Then, 2 g above-obtained graphite oxide was dissolved in the 1000 mL deionized water and sonicated for 1 h under the ambient condition. Thus, the homogeneous suspension of the graphene oxide was formed and then it was transformed into graphene oxide aerogel after vacuum freeze-drying.

---

*Corresponding author

2.2. Characterization of graphene oxide aerogel

The morphology of the graphene oxide aerogel was observed with the field emission environmental scanning electron microscope (FEI Quanta 250 FEG). The specific surface area was tested by the automatic specific surface analyzer (Quantachrome Autosorb-1), and the pore size distribution was determined by the automated pore size analyzer (Quantachrome Qudrasorb SI). The surface functional groups were analyzed by Fourier transforms infrared spectrometer (Bruker TENSOR27) MB concentrations were measured by spectrometry at the wavelength of maximum absorbance at 665 nm using UV-759 ultraviolet and visible spectrophotometer.

2.3. Batch adsorption experiments

The batch adsorption experiments of MB onto the graphene oxide aerogel were carried out at different temperatures in the aqueous medium. A certain amount of the graphene oxide aerogel was placed in a conical flask with 50 mL of MB solution, and the mixture was shaken in a temperature controlled oscillator for a time. 2 M HCl or 0.5 M NaOH was used to adjust the pH value of the solution. The amount of MB adsorption was determined according to the following equation:

$$q_e = (c_0 - c_e)V / m$$

(1)

Where C_0 and C_e (mg/L) are the concentrations of MB at initial and equilibrium, respectively; V (L) is the volume of MB solution and m (g) is the amount of used adsorbent and q_e represents the adsorption capacity (mg/g).

3. Results and Discussion

3.1. Morphology

The morphology is presented in Fig. 1. From Fig. 1a, it can be observed that the grapheme oxide aerogel has a loose and porous structure, which was formed by crosslinking and crimp of graphene oxide flake in vacuum freeze-drying. And Fig. 1b reveals that the graphene oxide aerogel has stacked layers of the sheet-like transparent structure with the smooth surface and wrinkled edges.

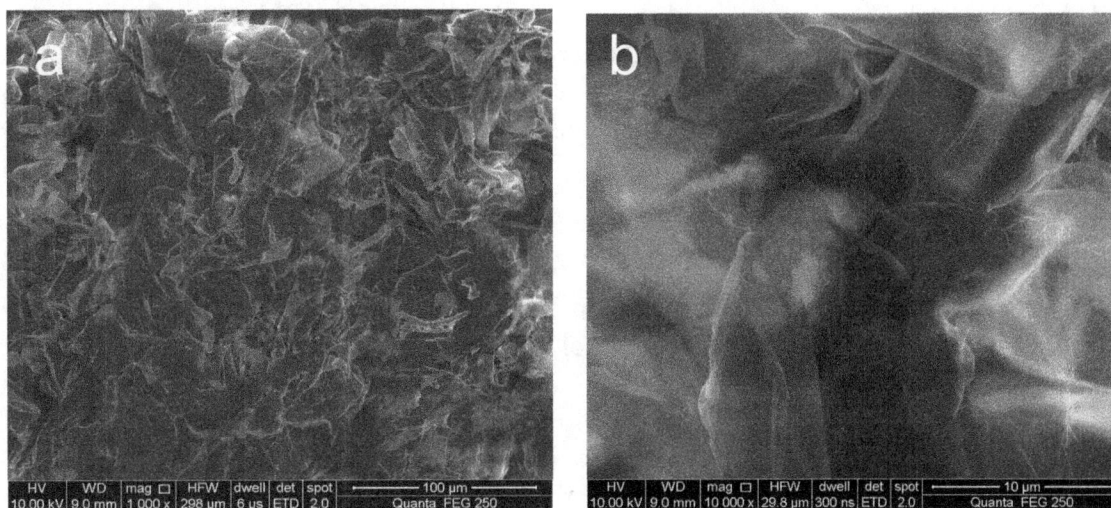

Fig. 1 SEM of graphene oxide aerogel

3.2. *Distribution of pore size and specific surface area*

N$_2$ adsorption-desorption isotherm was shown in Fig. 2a, which shows that the hysteresis loop is type H3 [5]. As we all know, the H3 is defined from the slit-shaped pores given rise to by the aggregates of plate-like particles. It is consistent with the observations in the SEM micrographs of Fig. 1. The pore size distribution graph was displayed in Fig. 2b, which reveals that the pore sizes mainly distribute in 2nm-15nm, which belong to mesoporous and the N$_2$ adsorption-desorption isotherm confirmed the result. These pores are enough large and ensure the entry of MB molecules. Moreover, by calculation, the specific surface area is 16.1 m^2/g.

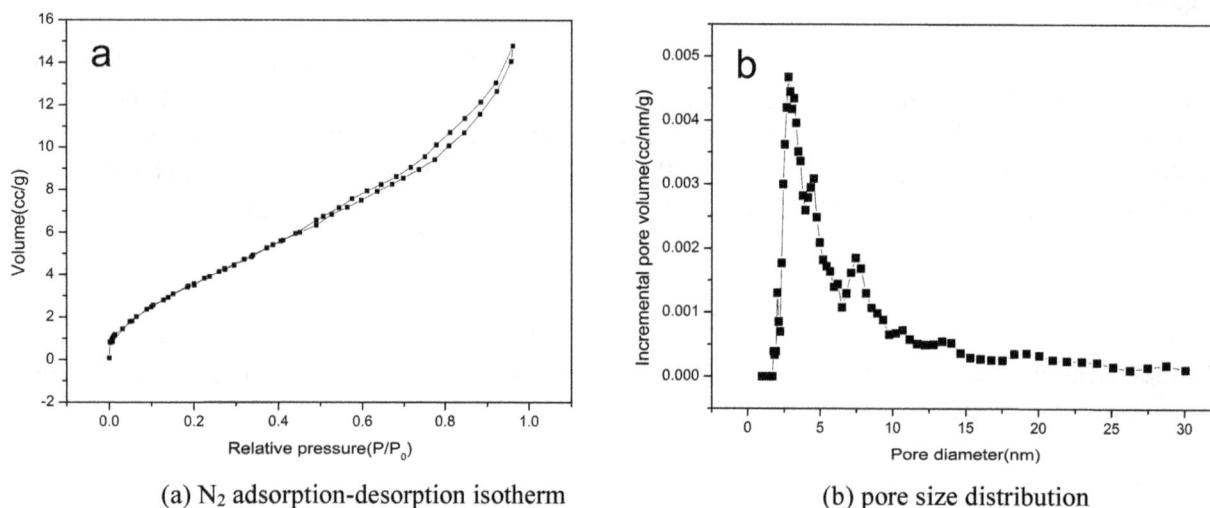

(a) N$_2$ adsorption-desorption isotherm   (b) pore size distribution

Fig. 2 N$_2$ adsorption-desorption isotherm and pore size distribution

3.3. *FTIR*

In order to clarify the surface chemical properties of graphene oxide aerogel, FTIR is made and expressed in Fig. 3. A strong characteristic band at 3383 cm^{-1} can be seen in Fig. 3, which is ascribed to O-H stretching vibrations due to surface hydroxyl groups and water. The peak at 1730 cm^{-1} is assigned to C=O stretch of carboxylic acid groups. The peak at 1620 cm^{-1} is due to the absorbance of stretching vibration of C=C bond. The peaks at 1229 and 1064 cm^{-1} may be attributed to C-OH and C-O stretching [6]. FTIR suggests that there are abundant functional groups on the surface of the graphene oxide aerogel, which can enhance the adsorption property of the graphene oxide aerogel.

3.4. *Effect of initial pH on adsorption ability*

In general, the initial pH of the solution plays an important role in adsorption process. The effect of initial pH on adsorption capacity is depicted in Fig. 4 and from Fig. 4, it can be seen that the adsorption of MB onto the graphene oxide aerogel strongly depends on the pH of the solution. With raising pH from 2.0 to 9.0, the removal rate gradually increased from 58.4% to 99.7% and the adsorption capacity increased from 145.89mg/g to 249.21mg/g. It can be attributed to that at lower pH, the predominant H$^+$ makes the graphene oxide aerogel carry positive charges and the electrostatic repulsion is strengthened between the graphene oxide aerogel and MB ions in the solution resulting in the lower removal rate. In addition, it can be concluded that the desorption process of the graphene oxide aerogel can be carried out by changing the pH of the solution.

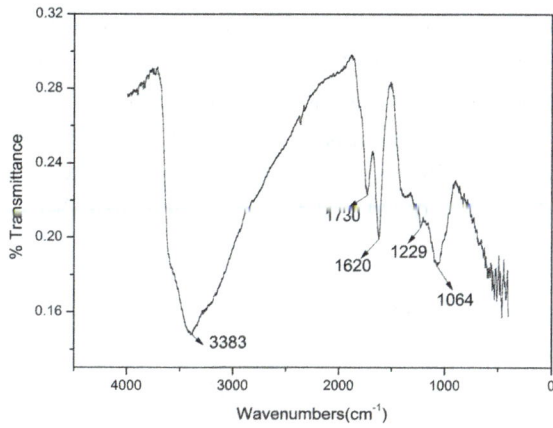

Fig. 3 FTIR of graphene oxide aerogel

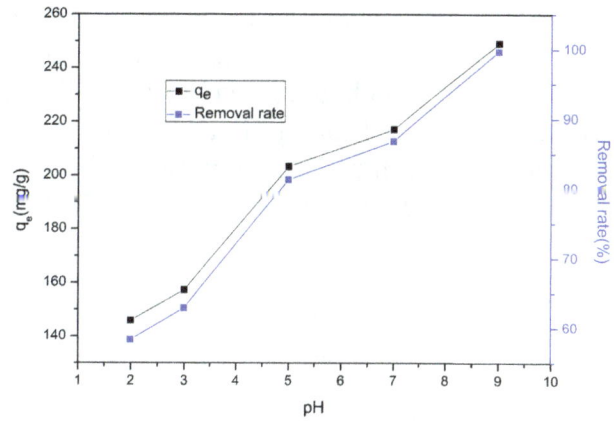

Fig. 4 Effect of pH on removal rate.(Adsorbent amount:10 mg, solution volume: 50 mL, MB concentration: 50 mg/L, shake time: 80 min)

3.5. *Adsorption isotherm*

Adsorption isotherms are important for determining the adsorption behavior of the adsorbent and assisting to reveal the adsorption mechanism much deeply. Abundant isotherm models were used to fit the experimental data, while the most conventional models are Langmuir and Freundlich isotherms models among these isotherm models. The Langmuir model assumes that the adsorption is a monolayer adsorption, while Freundlich model is an empirical model allowing for multilayer adsorption on the adsorbent.

Langmuir isotherm can be expressed by the following equation:

$$\frac{C_e}{q_e} = \frac{1}{q_0 K_L} + \frac{1}{q_0} C_e \tag{2}$$

Where q_0 and K_L are the Langmuir constants

The Freundlich isotherm can be given in the linear form by the following equation:

$$\ln q_e = \ln K_F + \frac{1}{n} \ln C_e \tag{3}$$

Where K_F and n are Freundlich constants

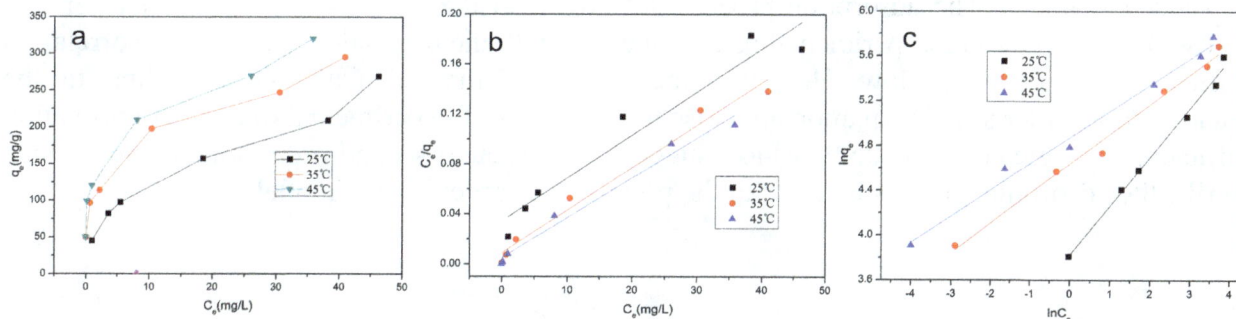

(a) Adsorption isotherm curves (b) Langmuir adsorption isotherm model   (c) Freundlich adsorption isotherm model

Fig. 5 Adsorption isotherms: (Adsorbent amount: 10 mg, solution volume: 50 mL, shake time: 80 min, pH: 7.0)

99

Fig. 5a presents the adsorption isotherms of MB onto the graphene oxide aerogel at different temperatures. Fig. 5b and 5c indicates fitted curves according to Langmuir model and Freundlich model, respectively, which are used to normalize the adsorption isotherm data. The relative parameters are listed in Table 1. It can be seen from Table 1 that the adsorption is fitted by Freundlich model better than Langmuir model, because of its higher regression coefficients, which suggests that the adsorption of the graphene oxide aerogel can be mainly considered as a multilayer adsorption process.

Table 1 Adsorption isotherm parameters of graphene oxide aerogel at different temperatures

Isotherms	Parameters	t (°C)		
		25	35	45
Langmuir	q_0 (mg/g)	289.9	293.3	312.5
	K_L (L/mg)	0.1008	0.3823	0.6349
	R^2	0.9246	0.9747	0.9771
Freundlich	K_F[mg/g (L/mg)$^{1/n}$]	45.65	103.23	130.45
	n	2.283	3.772	4.268
	R^2	0.9904	0.9878	0.9878

3.6. Adsorption kinetics

Adsorption kinetics is used to depict the adsorption rate in common. And mainly adsorption kinetics models are the pseudo-first-order kinetic model and the pseudo-second-order kinetic model [7]. The pseudo-first-order kinetic model is expressed as follows:

$$\ln(q_e - q_t) = \ln q_e - k_1 t \tag{4}$$

Where q_t (mg/g) is the adsorption capacity at time t, and k_1 (min^{-1}) is the adsorption rate constant. The pseudo-second-order kinetic model is presented by following equation:

$$\frac{t}{q_t} = \frac{1}{k_2 q_e^2} + \frac{t}{q_e} \tag{5}$$

Where k_2 [g/(mg·min)] is the rate constant of the pseudo-second-order equation.

Fig. 6a shows the effect of contact time on the adsorption of MB onto the graphene oxide aerogel and these adsorption data were fitted with the pseudo-first-order kinetic and second-order kinetic models which are shown in Fig. 6b and 6c, respectively. At the same time, the regression parameters are listed in Table 2. The adsorption process did not fit to the pseudo-first-order kinetic model because that the values of experimental q_e do not agree with the calculated q_{exp} and the correlation coefficients are relatively low. However, the values of the calculated q_e according to the pseudo-second-order kinetic equation are close to the values of experimental q_e and the correlation coefficients are greater than 0.99, which suggests the pseudo-second-order kinetic model can describe the adsorption process better than the pseudo-first-order kinetic model.

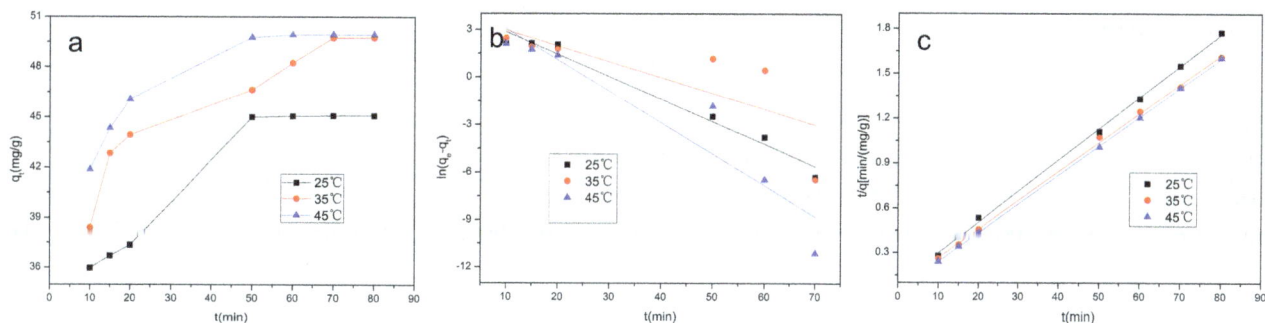

(a) Adsorption kinetics curves, (b) the pseudo-first-order kinetic model (c) the pseudo-second-order kinetic model.

Fig. 6 Adsorption kinetics: (Adsorbent amount: 10 mg, solution volume: 50 mL, MB concentration: 50 mg/L, pH: 7.0)

Table 2 Adsorption kinetic parameters of graphene oxide aerogel at different temperatures.

Kinetic models	Parameters	t(°C)		
		25	35	45
Pseudo-first-order	q_{exp}(mg/g)	72.17	51.21	153.9
	q_e (mg/g)	45.09	49.72	49.91
	k_1 (min^{-1})	0.1415	0.0996	0.1979
	R^2	0.9720	0.4829	0.8632
Pseudo-second-order	q_e(mg/g)	47.87	51.55	51.55
	k_2/[g/(mg·min)]	0.004937	0.005354	0.008448
	R^2	0.9988	0.9988	0.9998

3.7. *Adsorption thermodynamics*

The effect of temperature on MB adsorption is further explained by thermodynamic parameters. Thermodynamic parameters such as change in standard free energy (ΔG), change in standard enthalpy (ΔH), and change in standard entropy (ΔS) can be calculated by the following equations:

$$\Delta G = -RT \ln K_L \qquad (6)$$

$$\Delta G = \Delta H - T\Delta S \qquad (7)$$

Where R is 8.314 J/(mol·K), T (K) is the absolute temperature, and K_L (L/mol) is the Langmuir constant. The relation between ΔG and T is studied by linear regression analysis. The calculated thermodynamic parameters were listed in Table 3. It can be observed that the values of ΔG were positive which indicates that the adsorption process is non-spontaneous. The decrease from 56.85 to 12.01kJ/mol in the values of ΔG with the increase of temperature from 298 to 318K indicated that the adsorption process became more favorable at higher temperatures. The ΔH value of 721.7 kJ/mol suggested that the adsorption reaction was endothermic. ΔS was determined as 2.242 kJ/(mol•K), indicating the increasing randomness at the solid-solution interface during the adsorption process.

Table 3 Adsorption thermodynamic parameters for graphene oxide aerogel

T(K)	ΔG (kJ/mol)	ΔH (kJ/mol)	ΔS [kJ/(mol•K)]
298	56.85	721.7	2.242
308	24.62	721.7	2.242
318	12.01	721.7	2.242

4. Conclusion

Graphene oxide aerogel presents porous structure. Its slit-shaped pores are mainly mesoporous and specific surface area is $16.1 m^2/g$. The functional groups include O-H, C=O, C=C, C-OH and C-O. The removal rate increased from 58.4% to 99.7% with raising pH from 2.0 to 9.0. The adsorption process of MB onto the graphene oxide aerogel can be well described with Freundlich isotherm model and the pseudo-second-order kinetic model. The positive values of ΔG expressed the adsorption is non-spontaneous and the calculated ΔH value of 721.7 kJ/mol suggested that the adsorption reaction was endothermic.

Acknowledgments

This research was funded by the Nation Natural Science Foundation of China (No. 51404041), the Natural Science Foundation of Hunan Province, China (No. 2015JJ3016), Project funded by China Postdoctoral Science Foundation (2015M570690) and the Science Foundation for Postdoctoral Research of Central South University.

References

1. L P WANG, Z C HUANG, M Y ZHANG, B CHAI, Adsorption of methylene blue from aqueous solution on modified ACFs by chemical vapor deposition, Chem. Eng. J. 189-190 (2012) 168-174.
2. C Y CHEN, S H YEN, Y C CHUNG, Combination of photoreactor and packed bed bioreactor for the removal of ethyl violet from wastewater. Chemosphere. 117(2014) 494-501.
3. L L FAN, C N LUO, M SUN, X Y LI, H M QIU, Highly selective adsorption of lead ions by water-dispersible magnetic chitosan/graphene oxide composites, Colloids Surf., B. 103(2013) 523-529.
4. D C MARCANO, D V KOSYNKIN, J M BERLIN, A SINITSKII, Z SUN, A SLESAREV, L B ALEMANY, W LU, J M TOUR. Improved Synthesis of Graphene Oxide, ACS Nano. 4 (2010) 4806-4814.
5. K S W SING. Reporting physisorption data for gas/solid systems, with special reference to the determination of surface area and porosity, Pure Appl. Chem. 54 (1982) 2201-2218.
6. Z J GUO, S F WANG, G WANG, Z L NIU, J W YANG, W S WU, Effect of oxidation debris on spectroscopic and macroscopic properties of graphene oxide, Carbon. 76 (2014) 203-211.
7. Y S HO, G MCKAY. Kinetic models for the sorption of dye from aqueous solution by wood. J. Environ. Sci. Health., Part B. 76 (1998) 183-191.

A Review: The Synthesis and Application of Fe$_3$O$_4$/Graphene Materials

Lei Zhang[1], Zheng-Kang Duan[1,*], Hong-Wen Zhu, Ke Yin

[1]College of Chemical Engineering, Xiangtan University, Xiangtan 411105, Hunan, China

*Email: dzk0607@163.com

More and more scholars were committed to loading the iron oxide nanoparticles on the graphene. Shown in the studies, there were many nucleation sites on the graphene for the Fe$_3$O$_4$ particles, meanwhile, the distance between the graphene was lengthened due to the magnetic Fe$_3$O$_4$ particles, and the reunion of the graphene could be avoided. The excellent performance of the two materials was combined perfectly in this way. To facilitate the researchers work better in the future, in this paper, the synthesis of Fe$_3$O$_4$ /graphene were reviewed from its applications such as waste water treatment, magnetic targeting drugs, biosensor, and the anode materials for lithium-ion battery. Simultaneously, the prospects of the composite material were discussed.

Keywords: Wastewater treatment; magnetic targeting drugs; biosensor; anode materials; Fe$_3$O$_4$; graphene.

1. Introduction

Graphene, a plane which linked by single atomic of atoms in a crystal sp2 hybridized carbon atoms [1], [(a) is shown in Fig. 1] is the strongest material what we have found so far. Large specific surface area, jagged edges and high velocity in electron contribute it in great application potential in many fields, such as biomedical, electronics, optics, magnetism, energy storage, catalytic sensors. The excellent properties of graphene make it an ideal supporter to load all types of compounds. Until 2006, a graphene composite conductive material was prepared by Rouff et.al [2] for the first time. The graphene oxide (GO) was peeled from the graphite, then oxided [3] [4]. A layered structure was detected in the graphene oxide, there were enormous functional groups on the group layer sheet either, the chemical structure of it as seen in Fig. 1 (b). It was generally agreed that a large amount of hydroxyl and carbonyl groups were exist between the graphene oxide layers. The layer spacing of graphene was large either. These conditions provided a gainful environment for loading the inorganic nanoparticles and polymer. Since a lot of hydrophilic acidic group in the inter-layer graphene, it was easily for GO to be dispersed in water, which greatly facilitates to load mineral on it. Recent years, the development of graphene oxide composites was fantastic.

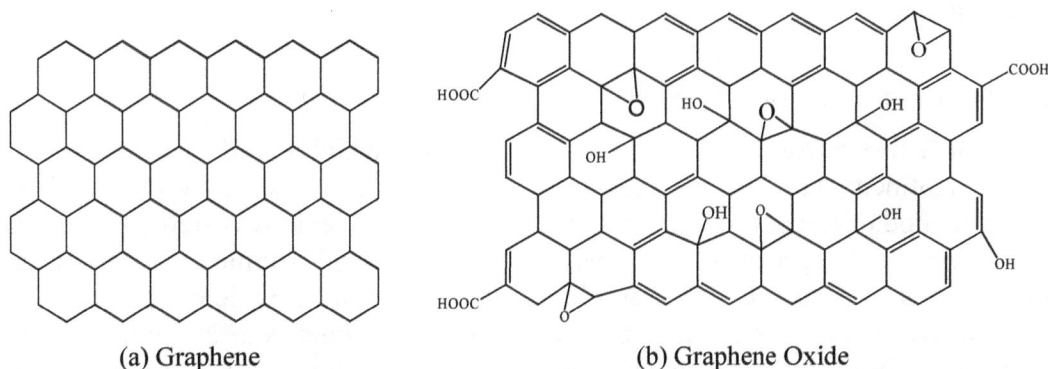

(a) Graphene   (b) Graphene Oxide

Fig. 1 The chemical structure of graphene 1 (a) and graphene oxide (b)

As the main component of magnetite, Fe$_3$O$_4$ was a traditional magnetic material [5]. The structure of the Fe$_3$O$_4$ was crystal spinel. There were many advantages and features, which had

*Corresponding author

been attracted much concerns. As gradually understanding and familiar with nanomaterials, it was precisely for people to study the size of the Fe_3O_4 material (from micron to nanometer level).The demand for new materials was more and more urgent with the various developments in various field, a lot of development was accomplished at the distribution and morphology of nanoscale iron oxide particle.

As a result of the good physical and chemical properties of Fe_3O_4 nanoparticles, shown great potential, it attracted much attention. Based on the features of super paramagnetic, the super-conductivity, inexpensive and environmentally, Fe_3O_4 was widely used in the field of electrochemistry. However, it is aggregated easily for the Fe_3O_4 particles. On the other way, the graphene would aggregate with the presence of van der Waals, which reduced the specific surface area and electrical conductivity. More and more scholars were committed to loading the iron oxide nanoparticles on the graphene. Shown in the studies, there were many nucleation sites on the graphene for the Fe_3O_4 particles, meanwhile, the distance between the graphene was lengthened due to the magnetic Fe_3O_4 particles, and the reunion of the graphene can be avoided. The excellent performance of the two materials was combined perfectly in this way. In this paper, the recent methods of synthesis and applications of Fe_3O_4/graphene were reviewed.

2. The Synthesis and Application of Fe_3O_4/graphene Materials

2.1. *The synthesis of the graphene*

Currently, most of the graphene were prepared by the methods of micro mechanical peeling, liquid lift-off, epitaxial growth, oxidation-reduction graphene oxide and chemical vapor deposition [6]. The graphene oxide was obtained by peeling the graphite and the preparation was divided into three steps: oxide, peel and disperse. The method of modified Hummers was used by most scholars to prepare the graphene because of the simple process and safe environment.

2.2. *The synthesis of Fe_3O_4/ graphene composites*

At present, there were five ways to prepare the Fe_3O_4/graphene composites: sol - gel method, micro-emulsion, precipitation, hydrothermal (solvothermal) and in situ reduction method. It will be introduced respectively at the terms of different fields for the application of Fe_3O_4/graphene composites in the later.

2.2.1. *The synthesis for wastewater treatment Fe_3O_4/graphene composites*

A certain pore structure existed in the Fe_3O_4 and graphene, which could be used as an adsorbent for wastewater treatment. The Fe_3O_4/graphene was successfully prepared by Huang Xiaomei et al [7] through the method of direct mixing, and the adsorption of arsenic was studied at the same time. In their experiments, the green vitriol was raw material, and sodium chlorate was oxidant. Experimental results showed that when graphene dispersed in iron oxide content of 55% and the pore size of the adsorbent was 55ppm, the adsorption of arsenic was 25mg/g. Compared with activated carbon or iron oxide, the advantages of high efficiency, good chemical stability and economic were showed by the composite materials. As one kind of excellent performance heterogeneous Fenton-reaction catalyst, Fe_3O_4 could be used to catalyze and decompose sulfuric acid and hydroxyl radical that produced by sulfates. It was effective for the degradation of organic pollutants. Yang Yanming et al [8] used $FeCL_3.6H_2O$ as raw material by the method of hydrothermal for loading the Fe_3O_4 on the graphene. For further study, they activated persulfate to degrade Rhodamine B wastewater with Fe_3O_4/graphene catalyst. Studies had shown that catalyze the degradation of Rhodamine B composite expressed high activity under weakly acidic conditions. The high temperature and concentration of the oxidant conditions were conducive to the degradation of Rhodamine B.

Fe_3O_4/graphene composites was prepared by the above two methods could be reunite in small scale, thus limiting its application. WeiFan et al [9] made improvements on this. A hydrothermal method was used to introduce carbon-coated Fe_3O_4 layer/graphene composites. Thereby a stable ternary composite materials was produced. The $FeCL_3.6H_2O$, sodium acetate, was used with the DEG-EG modified graphene oxide as raw materials, according to a conventional method. Then, in glucose solution the mixture must be dispersed as homogeneously as possible. After a series of treatments, the stable ternary composite materials was obtained. The resulting product was applied as an adsorbent to remove organic dye wastewater. Its adsorption properties was studied either. Compared with dual composite materials (Fe_3O_4/graphene/C), the rate of adsorption of organic dyes has been increased by 40%. At the same time, the experimental results showed that the ternary composites could exhibit not only excellent absorption properties in the water, but also in the acidic environment. They thought that the composite materials had a potential in the environmental applications.

2.2.2. *The synthesis for magnetic targeting drugs of Fe_3O_4/graphene*

Yan zhang et al [10] prepared the Fe_3O_4 nanoparticle/graphene oxide (MGO) by covalent cross-linking reaction. First, the ferromagnetic oxide nanoparticles (Fe_3O_4 - DMSA) was obtained by pyrolysis. Subsequently, it was modified by carboxyl group. Finally, it was reacted with the graphene oxide which was chemically modified with polyethyleneimine (PEI) Covalent. It was demonstrated that the amino on the GO-PEI and the carboxyl on the Fe_3O_4-DMSA were played an important role in the whole process. When the ratio between the Fe_3O_4 nanoparticles and the GO - PEI was changed in the reaction, the results showed that magnetic of the MGO composite materials was controllable. The preparation of composite material through this method had high superpara magnetism and colloid stability, which could be applied in the area of magnetic resonance imaging, magnetic targeted drug and magnetic separation.

2.2.3. *The sythesis for the biosensor Fe_3O_4/graphene*

Since the high degree of uniformity, good electrical conductivity and excellent biocompatibility of the Fe_3O_4, it was frequently used in the biosensor to be support matrix. Hui Wang et al [11] prepared an Acetyl cholinesterase biosensor, which was modified by nano Fe_3O_4/ chitosan /graphene composite membrane. It was utilized to detect the proportion of the chlorpyrifos pesticide. The graphene solution and ferroferric oxide chitosan suspension was drop on the activated electrode surface after perfectly blending, then through the reactions in a series of solutions. The electrode combined both the advantages of graphene nano materials and Fe_3O_4, such as the large specific surface area, high electron transfer properties, unique excellent biocompatibility and chitosan adhesive performance, which would provide a good combination of interface for the acetyl cholinesterase. The lowest detection limit of chlorpyrifos concentration was 0.02 mu g/L under the optimal condition of detection. This indicated that the new type of biosensor showed excellent accuracy for content of poisoning death ticks vegetables.

2.2.4. *The synthesis for the lithium-ion battery anode material of Fe₃O₄/graphene*

With the characters of low prices, abundant resources, it was a potential for Fe_3O_4 to be applied in the anode electrode materials. The theoretical specific capacity of it was 926mA / g. But as lithium battery cathode material, the changes of structure and volume of Fe_3O_4 would lead to the changes in the structure of the particles. Thus, the electrode structure was destroyed, the cycle times was lowed either. Fortunately, it could be overcome with the appearance of the graphene. More and more scholars were committed to loading the iron oxide nanoparticles.

Most of the Fe_3O_4/ graphene composites was synthesized by hydrothermal [12-15]. For example, Wei et al [16] used one-step hydrothermal synthesis of Fe_3O_4/ graphene composites. Then this compound was made into a film supercapacitor electrode without the interference of the resistance. Experimental results showed that when the proportion of graphene and Fe_3O_4 nanoparticles was 2.8: 1, the maximum ratio of electrode material capacitance 480F/g could be got. At 5A/g current density, the energy density of it was 67 W h/kg, and the power density was 5506W/kg. Lou et al [17] prepared it in glucose solution by hydrothermal method. Electrochemical tests was found that when the composite material at 1C current density after 100 cycles, the reversible capacity still maintained 808mAh/g. The Fe_3O_4/graphene nanocomposites were prepared by Lian et al [18] through the method of hydrothermal, which showed excellent performance as the same way. In 0.01-3,00V voltage range to 100mA/g current density discharge, after 40 cycles, the capacity was remained 1045mAh/g.

In order to improve the electrochemical properties of lithium battery cathode materials, a large amount of improvements to its preparation was conducted by domestic and foreign scholars.

Ming Zhang et al [19] used microwave-assisted hydrothermal method to prepare monocrystalline Fe_3O_4 nanoparticles and applied in lithium battery anode material. The anode electrode material experiments showed the wonderful properties of cycle performance and electrochemical. Capacity would be remained at 610mAh/g at 5C magnification current density. Guangmin Zhou et al [20] used the method of in-situ reduction to prepare Fe_3O_4/graphene nanocomposites. After 85 cycles, the cycle capacity was remained 950 mAh/g in the composite material. This method could effectively prevent the agglomeration of iron oxide. Reversible capacity of the electrode material and the number of cycles were improved in this way. The composite material was prepared by self-assembly using static electricity by Liu Jianhua et al [21]. The $Fe(OH)_3$ colloid was gotten in the graphene oxide (GO) dispersion. Followed by it, the $Fe(OH)_3$/GO composite structure was formed by electrostatic action. Then, through the hydrothermal reaction of the $Fe(OH)_3$/GO precursor with graphene, a three dimensional network structure was formed. Finally, the Fe_3O_4/graphene composite was calcinated, and its structural characterization and electrochemical properties were studied. At 175mA/g current density after 50 charge and discharge cycles, the composite material still had an 814mAh/g specific capacity. In 1400 mAh/g high-rate discharge capacity kept throwing at 531 mAh/g. The solvent of the reaction was important for the process. So they changed the solvent to observe it. Surprisingly found that when the solvent was a mixed solution of EG and H_2O for the hydrothermal reaction, it was possible to obtain a more uniform composite which was exhibited a higher specific capacity and better cycling stability. At 175 mAh/g current density, after 50 charge and discharge cycles, the sample capacity was remained at 947 mAh/g. In fact, the modified for the graphene played an important role for the modification of graphene to improve the properties of composite materials.

A new method to prepare iron oxide/graphene composites was invented by Linhai Zhuo et al [22]. With the decomposition of ferric nitrate in the presence of graphene oxide in the mixed solvent of CO_2-expanded ethanol, the material was prepared. Then the precursor was converted to

Fe$_3$O$_4$/graphene composites around nitrogen. The solvent was not only non-toxic but also recyclable. It could make the Fe$_3$O$_4$ uniformly and intimately supported on the graphene sheets institutions. Fortunately, after 100 charge and discharge cycles, the specific capacity of the composite material was still remained 826 mAh/g.

3. Conclusion

Depending on the different methods of preparation, the structures and properties of the resulting Fe$_3$O$_4$/graphene were different, which could be applied in different areas. In summary, it is best for the Fe$_3$O$_4$/graphene to be applied in the lithium-ion batteries. The prospect of it was particularly impressive. In future studies, it is possible to continue to improve its method of preparation, the regulation of the process parameters to optimize cycle performance and rate properties of the composites.

Acknowledgment

This research was funded by the National Natural Science Foundation of China (21576229).

References

1. C.G. Wu, D.C. D Groot, Reaction of aniline with FeOCl. Formation and ordering of conducting polyaniline in a crystalline layered host, J. Am. Chem. Soc. 117 (1995): 9229–9242.
2. F. Lernoux, B.E. Koene, Electrochemical lithium intercalation into a Polyaniline/V$_2$O$_5$ Nano composite, J. Electro. Chem. Soc. 143 (1996) 181–183.
3. T.A. Kerr, H. Wu, Concurrent polymerization and insertion of aniline in molybdenum trioxide: formation and properties of a [Poly (aniline)] 0.24MoO$_3$ Nano composite, C. Chem. Mater. 8 (1996) 2005–2015.
4. M. Lira-Cantu, P. Gómez-Romero, The organic–inorganic poly-aniline/V$_2$O$_5$ system application as high-capacity hybrid cath-odes for rechargeable lithium batteries, J. Electrochem. Soc. 146 (1999) 2029–2033.
5. N. Liao, Z. Liu, Preparation of a novel Fe$_3$O$_4$/graphene oxide hybrid for adsorptive removal of methylene blue from water, J. Macromol. Sci. A. 53 (2016) 276-281.
6. Y. Zhu, S. Murali, Graphene and grapheme oxide: synthesis, properties and applications, J. Adv. Mater. 22 (2012) 3185-3193.
7. X.M. Huang, Preparation and properties of graphene oxide and iron oxide nanoparticles composites, Gd. Chem, 02 (2012) 22-23. (In Chinese)
8. Y. Yang, Y. Leng, Fe$_3$O$_4$/Graphene activated persulfate wastewater degradation of rhodamine B, J. Environ. Manage, 04 (2014) 80-84. (In Chinese)
9. W. Fan, W. Gao, Hybridization of graphene sheets and carbon-coated Fe$_3$O$_4$ nanoparticles as asynergistic adsorbent of organic dyes, J. Mater. Chem. 22 (2012) 25108-25115.
10. Y. Zhang, B. Chen, Fe$_3$O$_4$ magnetic nanoparticles - controllable preparation, structure and properties of graphene oxide composites characterization, J. Phy. Chem. 05 (2011) 1261-1266. (In Chinese)
11. H. Wang, Y. Duan, Based on nanometer iron oxide/chitosan/graphene nano-composite modified acetylcholinesterase biosensor pesticide chlorpyrifos, Moden. Food. Sci. Tec., 02 (2016) 276-282. (In Chinese)

12. L. Lu, J. Wang, Enhanced cycling performance of nanocrystalline Fe_3O_4/C as anode material for lithium-ion batteries, J. Nanosci. Nanotechno. 12 (2012), 1246-1250.
13. T. Muraliganth, M.A. Vadivel, A. Manthiram, Facile synthesis of carbon-decorated single-crystalline Fe_3O_4 nanowires and their application as high performance anode in lithium ion batteries, Chem. Commun. 47 (2009) 7360-7362.
14. W. K. Dong, G. B. Long, A facile route towards the synthesis of Fe_3O_4/graphene oxide nanocomposites for environmental applications, Mol. Cryst. Liq. Cryst. 2014, 599 (2014) 43-50.
15. M. Ren, M. Yang, Ultra-small Fe_3O_4 nanocrystals decorated on 2D graphene nanosheets with excellent cycling stability as anode materials for lithium ion batteries, Electrochimica. Acta, 194 (2016) 219-227.
16. W. Shi，J. Zhu, Achieving high specific charge capacitances in Fe_3O_4/reduced graphene oxide nanocomposites, J. Mater. Chem. 21 (2011) 3422–3427.
17. T. Zhu, J.S. Chen, Glucose-Assisted one-pot synthesis of FeOOH nanorods and their transformation to Fe_3O_4@Carbon nanords for application in lithium ion batteries, J. Phys. Chem. C. 115 (2011) 9814-9820.
18. P. Lian, X. Zhu, Enhanced cycling performance of Fe_3O_4-graphene nanocomposite as an anode material for lithium-ion batteries, Electrochimica. Acta. 56 (2010) 834-840.
19. M. Zhang, D. Lei, Magnetite/graphene composites: microwave irradiation synthesis and enhanced cycling and rate performances for lithium ion batteries, J. Mater. Chem. 20 (2010): 5538-5543.
20. G. Zhou, D. Wang, Graphene-wrapped anode material with improved reversible capacity and cyclic stability for lithium ion batteries, Chem. Mater. 22 (2010) 5306-5313.
21. J. Liu, B. Liu, Fe_3O_4/graphene composites with a porous 3D network structure synthesized through self-assembly under electrostatic interactions anode materials of high-performance Li-Ion batteries, Acta. Phys. Chem. Sin. 30 (2014) 1650-1658.
22. L. Zhuo, Y. Wu, CO_2-expanded ethanol chemical synthesis of a Fe_3O_4@graphene composite and its good electrochemicals properties as anode material for Li-ion batteries, J. Mater. Chem. A. 1 (2013) 3954-3960.

Effect of High-temperature Treatment on the Thermal and Dielectric Properties of Nano-SiO₂/CE Resin Composites

Yan-Nan He, Zhi-Qiang Yu*

Department of materials science, Fudan University, 200433 shanghai, China
Email: yuzhiqiang@fudan.edu.cn

The nano-silica (SiO_2) was introduced into the cyanate ester (CE) resin matrix to fabricate composites by physical blending method, and then the composites were treated in different high-temperature conditions. The results of differential scanning calorimetry (DSC) indicated that the thermal properties of SiO_2 filled composites are higher than that of pure CE. But the heat resistant of them was all decreased after heat treatment. The filled composites had lower dielectric constant and loss, indicating that the introduction of SiO_2 had positive effect on the dielectric properties of CE. The high temperature treatment made the dielectric properties of CE and its composites show a trend of decrease.

Keywords: cyanate ester resin; nano-silica; thermal properties; dielectric properties; high temperature treatment.

1. Introduction

The materials applied in high temperature environments are needed to have good high-temperature properties to meet demand of elevated temperature engineering. Cyanate ester (CE) resin is a kind of high performance thermosetting resin. The excellent properties such as high thermal stability, good dielectric properties and high chemical stability of CE make it was widely used in printed circuit board, radome and so on [1, 2]. But like most thermosetting resin, the most highly crosslinking network of CE constituted by 1,3,5-triazine [3, 4] makes the CE tend to be brittle and need to be modified suitably to be able to serve mechanical or structural functions. Many attempts have been made to modify and improve the toughness and other properties of CE, and the introduction of inorganic nano-particles has been proved to be an effective method [5, 6]. However, few of them focus on the thermal stability and dielectric properties of CE resin matrix composites after high-temperature heat treatment.

In this paper, nano-SiO_2 was introduced into the CE resin matrix to prepare composites by physical blending method. Then the composites were heat-treated in different temperature for different time. The thermal stability and dielectric properties of treated composites were investigated.

2. Experimental

2.1. Materials

SiO_2 (purity of 99.5%, with the average particles size of 50nm), was purchased from the *Aladdin Chemistry Co. Ltd.* 2,2-Bis(4-cyanatophenyl)propane (analytically pure), was purchased from *Shanghai LiDa Chemical Co. Ltd.*

2.2. Composites preparation

2,2-Bis(4-cyanatophenyl)propane was molten at 150°C for 10min under the mechanical stirring. Then the nano-SiO_2 particles were added into the melt. After reaction at 150°C for 70min, 1.0wt%

---

*Corresponding author

109

of epoxy resin was added into the mixture. The product was poured into a mould after stirring 10min. The sample was vacuumized to remove the remainder bubbles. The sample was solidified at 120°C for 60min, and then raised the temperature gradually up to 200°C within 6h.

2.3. *Characterization and testing*

The DSC measurements were carried out using the DSC-Q2000 analyzer made in *American TA Company*. The dielectric properties were tested by WY2855 dielectric constant and dielectric loss analyzer made in *Shanghai WuYi Electronics Co. Ltd*. In order to test the performance of composites after high-temperature heat treatment, samples were heated at 200°C and 300°C for 15min and 60min in muffle furnace before the examination, respectively.

3. Results and Discussions

3.1. *DSC characterization of thermal properties*

The heat resistant of nano-SiO_2/CE resin composites was characterized by DSC. Fig. 1 shows the DSC curves of pure CE and its filled composites with different mass fraction particles. The T_g obtained from DSC curves is summed in Table 1.

Table 1 Tg of different mass fraction particles filled composites

Mass fraction (wt%)	0.0	1.0	3.0	5.0
T_g (°C)	296.08	306.02	302.75	294.98

As shown, the addition of nanoparticles has a influence on the heat resistant properties of CE resin matrix. Introducing different content nanoparticles to CE resin can change the glass transition temperature (T_g) of neat resin more or less. The composite with 1.0wt% addition of nanoparticles has higher T_g of 306.02°C, increased by about 10°C relative to the neat resin. The T_g decreases with the increase of the content of SiO_2 particles when the content exceed 1.0wt%. The reason can be believed that the nanosilica particles play a role of crosslinking spots in the matrix at lower loading level. The CE matrix with 1.0wt% nanoparticles has higher cross-linking degree which can obstruct effectively the resin macromolecules moving, and so it has higher T_g and presents higher heat resistant properties. The decrease of T_g with the increase of particles content is mainly resulted from the an increasing susceptibility of agglomeration of nanoparticles, whereas the agglomeration of nanoparticles as defects makes the macromolecules move easily and decreases T_g. T_g of composite with 5.0wt% nanoparticles is lower than neat CE resin. It is suggested that the dispersion of the nanoparticles in composite with 5.0wt% nanoparticles will be even poorer.

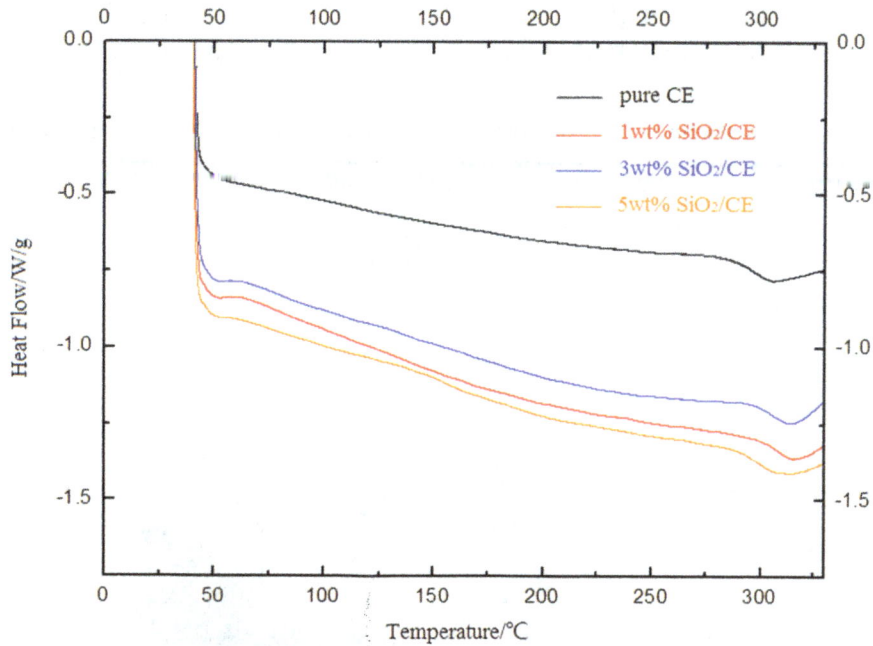

Fig. 1 DSC curves of different mass fraction particles filled composites

Fig. 2 shows the DSC curves of composites filled with 1.0wt% SiO$_2$ treated in different temperature conditions. The T$_g$ obtained from DSC curves are summed in Table 2.

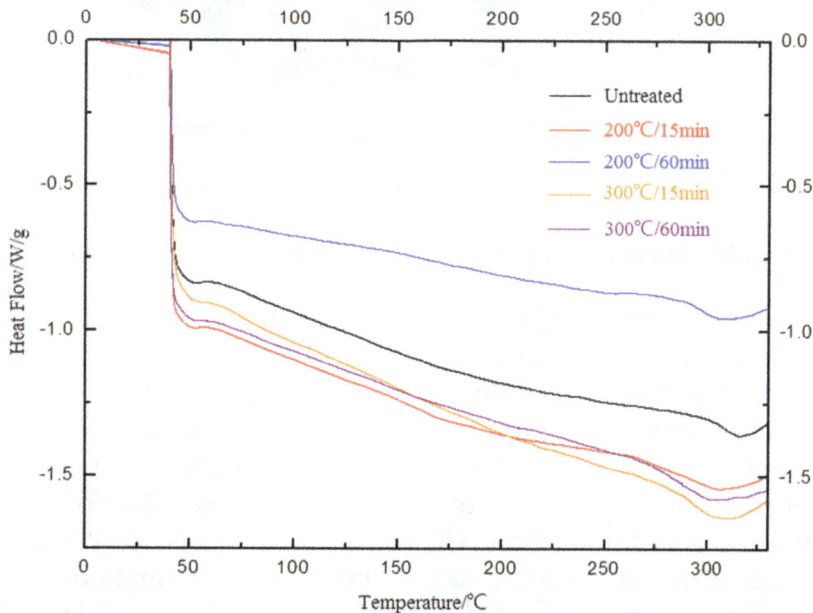

Fig. 2 DSC curves of 1.0wt% SiO$_2$/CE resin composites treated at different temperature conditions

As shown in Fig. 2, it is obvious that the baseline of DSC curves of filled composites shifts in the range from 280°C to 330°C, indicating a glass transition of sample. The T$_g$ of sample is different under different treatments of temperature and time. From Table 2, it can be seen that T$_g$

of the composites treated by high temperature is decreased clearly compared to untreated composite, and the T_g is further decreased with the increase of temperature.

Table 2 T_g of 1.0wt% SiO_2/CE resin composites treated in different temperature conditions

Treatment conditions	Untreated	200°C/15min	200°C/60min	300°C/15min	300°C/60min
T_g (°C)	306.02	293.51	296.73	291.91	287.71

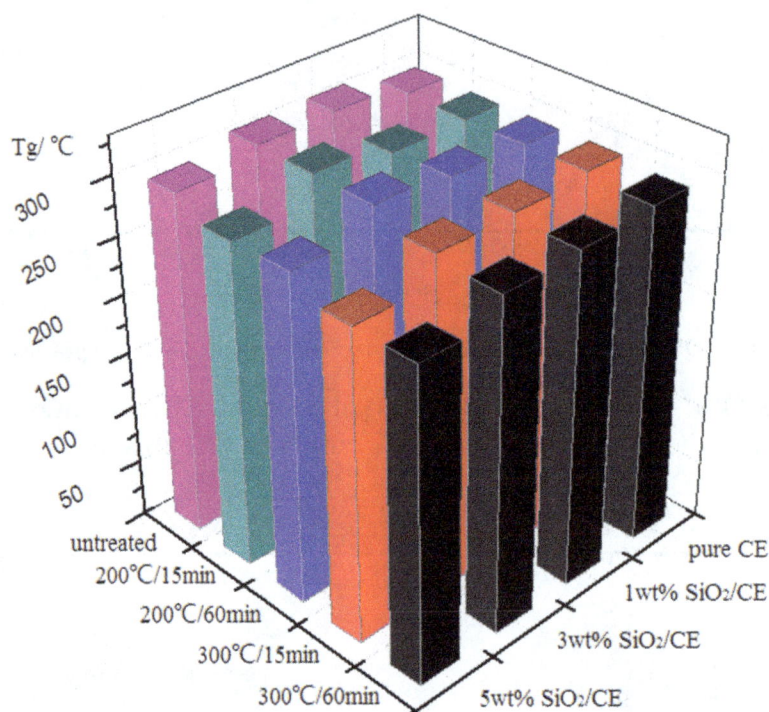

Fig. 3 The summary of Tg of samples under different treatment conditions

T_g of untreated composite is 306.02°C, whereas T_g of the composites treated at 200°C/15min and 300°C/15min is decreased by 12.51°C and 14.11°C (9.29°C and 18.31°C at 60min), respectively. It means that thermal stability of CE resin matrix composites is decreased after high-temperature heat treatment. It can be resulted from the crosslinking network system degradation and therefore the macromolecular chains are more easily to unfreeze during the heating up. In order to further verify the above conclusion on the effect of high-temperature heat treatment on the thermal stability of filled CE composites, T_g of 3.0wt% and 5.0wt% SiO_2/CE resin composites under different temperature and time treatment conditions is also investigated. Fig. 3 shows the column chart of T_g of different mass fraction particles filled composites at different treatment conditions. As shown, the variations of T_g of 3.0wt% and 5.0wt% SiO_2/CE resin composites under different temperature treatment are similar to that of 1.0wt% composite, that is to say, T_g of the both composites is all decreased with the increase of heat treatment temperature. In comparison, the composite filled with 1.0wt% SiO_2 has higher heat stability at high temperature.

3.2. Dielectric properties analysis

The dielectric constant (ε) and dielectric loss ($\tan\delta$) of different mass fraction SiO_2 particles filled composites are summed in Table 3. From Table 3, it can be seen that the dielectric constant and loss of filled composites reduce with the increase of nano-SiO_2 particles content. It can be resulted from the decrease of the polarity of composite material system as well as the density of polar groups due to the introduction of nano-SiO_2. Besides, the nano-SiO_2 has lower dielectric constant and loss. These integrated factors make the value of ε and $\tan\delta$ of composites decrease.

Table 3 The dielectric constant and loss of different mass fraction SiO_2 filled composites

Mass fraction (wt%)	1.0	3.0	5.0
ε	2.23	2.13	2.00
$\tan\delta/\times10^{-2}$	1.47	1.38	1.21

To further discuss the variation of dielectric properties of filled composites at high temperature, the ε and $\tan\delta$ of different composites treated under different conditions are investigated, and summed in Table 4 and Table 5, respectively.

It can be observed from Table 4 and Table 5 that ε and $\tan\delta$ of composites both decreased after high-temperature heat treatment compared to untreated composites. Under the condition of same temperature and time, ε and $\tan\delta$ of composites reduce with the increase of nanoparticles content. In comparison, the composite filled with 1.0wt% presents higher values of ε and $\tan\delta$. In addition, it can also be seen that ε and $\tan\delta$ of composites decrease with the increase of temperature and time at the same mass fraction. These can be explained that after high-temperature heat treatment, the structure and properties of composites would be degraded slightly [7]. On the one hand the 1, 3, 5-triazine with low polarity reduced. On the other hand, the interface interaction between SiO_2 particles and CE resin matrix were weakened. It makes the interfacial polarization reduce and therefore the values of ε and $\tan\delta$ decreased. Besides, high ratio of inorganic nano-SiO_2 has more obviously effect on the integrated performance of composite system after heat treatment. Because ε and $\tan\delta$ of SiO_2 is lower than CE resin, it makes ε and $\tan\delta$ of its filled CE composites decreased.

Table 4 The dielectric constant of composite materials under different treatment conditions

Treatment conditions	ε		
	1.0wt% SiO_2/CE	3.0wt% SiO_2/CE	5.0wt% SiO_2/CE
Untreated	2.23	2.13	2.00
200°C/15min	2.22	2.11	1.99
200°C/60min	2.21	1.99	1.97
300°C/15min	2.20	1.98	1.95
300°C/60min	2.10	1.97	1.94

Table 5 The dielectric loss of composite materials under different treatment conditions

Treatment conditions	$\tan\delta/\times10^{-2}$		
	1.0wt% SiO_2/CE	3.0wt% SiO_2/CE	5.0wt% SiO_2/CE
Untreated	1.47	1.38	1.21
200°C/15min	1.34	1.31	1.18
200°C/60min	0.97	0.79	0.69
300°C/15min	0.87	0.78	0.67
300°C/60min	0.71	0.74	0.45

4. Conclusion

The heat resistant and dielectric properties of nano-SiO_2/CE resin composites under the high temperature are researched in this paper. The introduction of nano-SiO_2 improves the heat resistant of CE resin, and the heat resistant of composites decreased with the increase of SiO_2 mass fraction due to the agglomeration of nanoparticles. After high-temperature heat treatment, the heat resistant of CE filled nanoparticles decreases owing to the 1, 3, 5-triazine in composite system degraded. Dielectric properties analysis indicate that the composites have lower dielectric constant and loss than those of CE resin, and the high temperature treatments make the dielectric constant and loss of filled composites further decreased. In comparison, the CE resin filled with 1wt% SiO_2 has better thermal and dielectric performance under high temperature. The results could be very useful for investigation and design on advanced heat-resistant and good dielectric properties polymer matrix composite materials under high temperature conditions.

Acknowledgment

This research was funded by the National Natural Science Foundation of China: (Grant no. 51273044).

References

1. M.F. Zeng, C.Y. Lu, B.Y. Wang, C.Z. Qi, Free volume hole size of Cyanate ester resin/Epoxy resin interpenetrating networks and its correlations with physical properties, Radiat. Phys. Chem. 79 (2010) 966-975.
2. L. Yuan, G.Z. Liang, A.J. Gu, The thermal and dielectric properties of high performance cyanate ester resins/microcapsules composites, Polym. Degrad. Stab. 96 (2011) 84-90.
3. Y. Lin, M. Song, C.A. Stone, S.J. Shaw, A comprehensive study on the curing kinetics and network formation of cyanate ester resin/clay nanocomposites, Thermochim. Acta. 552 (2013) 77-86.
4. Q.L. Lin, L.J. Qu, Q.F. Lu, C.Q. Fang, Preparation and properties of grapheme oxide nanosheets/cyanate ester resin composites, Polym. Test. 32 (2013) 330-337.
5. T. J. Wooster, S. Abrol, J.M. Hey, D.R. MacFarlane, Thermal, mechanical, and conductivity properties of cyanate ester composites, Composites Part A. 35 (2004) 75-82.
6. W. Qin, D.Q. Peng, X.H. Wu, J.H. Liao, Study on the resistance performance of TiO_2/cyanate ester nano-composites exposed to electron radiation, Nucl. Instrum. Methods Phys. Res., Sect. B. 325 (2014) 115-119.
7. M.L. Ramirez, R. Walters, R.E. Lyon, E.P. Savitski, Thermal decomposition of cyanate ester resins, Polym. Degrad. Stab. 78 (2002) 73-82.

Study on Stirring Auxiliary Machinery Hydroxyapatite Nanomaterials

Yong-Guang Bi[*], Xu-Si Xu, Shi-Ting Deng, Hong Yu, Xue-Wei Chen, Xue-Mei Liu, Hai-Lan Huang

School of Pharmacy, Guangdong Pharmaceutical University, Guangzhou 510006, Guangdong, China

E-mail: biyongguang2002@163.com

Papers prepared using auxiliary mechanical stirring hydroxyapatite nano-materials, and by infrared spectroscopy and X-ray diffraction of the prepared nano materials characterization, characterization results show: Infrared analysis confirmed the root samples S1 phosphate (PO_4^{-3}) and hydroxyl (OH^{-1}) presence, X-ray diffraction instrumentation products for the next hydroxyapatite crystals belonging to the hexagonal nano-HAP, sharp peak shape, crystallinity is good. Therefore, preparation of auxiliary mechanical stirring hydroxyapatite nano-materials is simple, material morphology and size controllable process parameters can be expected to provide large-scale preparation of nano-materials and reference.

Keywords: Hydroxyapatite; mechanical stirring; preparation; infrared spectroscopy; X-ray diffraction.

1. Introduction

Hydroxyapatite chemical formula ($Ca_{10}(PO_4)_6(OH)_2$, referred to as the HAP or HA, is an important biomedical materials and environmental functional materials, with a variety of important applications of thermodynamics in the generation process of the HA it will have a very big impact on their stoichiometric ratio of calcium and phosphorus if slight deviation from the standard molar ratio of 1.67, it will lead to α-type or β-type tricalcium phosphate production, many of the early research papers on HAP powder production and processing are concerned how to avoid other unnecessary calcium phosphate phase structure [1].

Hydroxyapatite with the chemical composition of human bones and teeth of the main inorganic mineral of the same, similar crystal structure, having some biological activity and good biocompatibility, in the current human hard tissue repair has been applied, and still months medical material is a hot research. Studies have shown that the crystal structure and physical and chemical properties of hydroxyapatite determine the structure of calcium ions, phosphate ions and hydroxide ions can easily be replaced by other ions, resulting in a very good ion exchange properties, can be used as ion adsorbent in drinking water as well as high levels of fluorine specific industrial waste of heavy metal ions [2]. And the porous hydroxyapatite ceramic heat resistant and moisture, humidity can be used as a semiconductor material, the higher the sensitivity.

Hydroxyapatite is an important class of bone repair materials in human applications, with excellent biocompatibility, the main component of the inorganic component of human bones

---

*Corresponding author

similar, does not contain harmful elements, and its crystal the cell calcium is necessary for the body, it is widely used in the medical field [3]. Previous basic research and clinical application of researchers demonstrated that synthetic hydroxyapatite having excellent biocompatibility after implantation of human non-toxic, non-irritating, and human adaptability better, without any adverse reaction, It can be used to repair bone tissue. Due to the low mechanical properties of dense hydroxyapatite material, so it is mainly used in non-load-bearing parts of the bone defect, and achieved better repair effect. In order to overcome the brittleness and lower strength hydroxyapatite single component, the researchers added a second in which the selective phase, including wafer, whiskers or fibers, both on the mechanical properties of materials play a reinforcing role. Hydroxyapatite has good biocompatibility and bone-inducing, so that it can be connected directly with the host bone, so the bulk material can be used as bone repair or replacement limb, as in Craniofacial Surgery to repair orthopedic reconstructive surgery as planting bone, as dental filling materials, as ophthalmic orbital implant.

Hydroxyapatite as a bulk material due to its low mechanical properties can not be used for load-bearing parts, as the coating is not limited thereto. Although current medical metallic material excellent mechanical properties, but poor biocompatibility and can release toxic metal ions often encountered in the course of failure. After the metal material and the surface of the implant is coated hydroxyapatite can enhance their biocompatibility, avoid direct contact with the metal body, but also the toxic metal ions can be masked covered up [4]. Currently using plasma spraying, electrochemical deposition, ion sputtering or the like, to enhance the biological activity of the implant in the body coated with titanium implant surface layer of hydroxyapatite coating to improve its compatibility and bone induction and reduce the failure rate of implants to improve their life.

Nano-hydroxyapatite with unique biological activity, in the medical field has been applied with the gene, the drug and the protein carrier, and in the treatment of tumors is also considerable potential. Hydroxyapatite as a drug carrier can load slowly release the drug, the drug concentration in the affected area improve yield better therapeutic effect and to reduce the concentration of drug in the blood, reducing the drug on the body parts of other health damage and poisoning. Hydroxyapatite as an inorganic component of human hard tissue, in the degradation of the human body does not produce toxic side effects [5]. With the development of nano-medicine, researchers have found that nano-hydroxyapatite inhibit cell division action, but had no effect on normal cells. Studies have shown that one characteristic of tumor cells is a strong calcium intake capability, nano calcium hydroxyapatite as a source of tumor cells has provided more than the tolerance range of the amount of calcium ions, and then produce a toxic, inhibits the growth of tumor cells . Reason may be light nano hydroxyapatite strong ability to penetrate into the interior of the tumor cells to make changes inside the cell microenvironment, influence cancer cell growth and proliferation, resulting in inhibition.

Size and particle size of the ceramic material microstructure and macroscopic properties it has a very important influence. The ceramic material is sintered after forming a ceramic powder, ceramic particles more uniform shrinkage during sintering the convergence, the smaller the particle, the smaller the defect, the strength of the obtained ceramic material are higher, further It may be some unique properties, large particles of ceramic material which is not available. In addition, the fine ceramic particles to the nano-scale, it has a large specific surface area and high chemical activity, can significantly improve the degree of density ceramic material during sintering of. Nano HA powder not only provides excellent performance in HA bulk ceramics and composite material aspects, in the treatment of cancer and it can also play its specific properties. Some scholars have been using nano-HA material for the treatment of breast cancer cells implanted in mice. The study found that when the nano-HA material particle size in the range of 20~80nm, for the inhibition of breast cancer cells in mice significantly, but the normal cells were not inhibited. In addition, physical and chemical properties, and nano-sintered HA HA are different, stronger when the particle size of about 10~100nm nano HA HA having a solubility greater than the sintering, the surface energy is higher, the ion exchange capacity than the sintered HA. Nano HA may also play a role in their cancer cell surface and therefore to which certain components and between cancer cells and play a role, inhibit cancer cell.

Adsorption is a common phenomenon in a solid to a gas or liquid adsorption of hydroxyapatite because of its special physical and chemical properties, and is currently used as an adsorbent ion exchange material, has been widely used in protein separation and purification, wastewater treatment, soil pollution governance and other fields. The crystal structure of hydroxyapatite exists in two different positions of different calcium ions, ion different radius has a strong hold, making the hydroxyapatite has a very good function of ion adsorption and ion interaction, which can be lattice developed into efficient adsorption and inorganic ion exchange material, waste water treatment and recycling is applied to the metal element. Hydroxyapatite for wastewater treatment and soil purification, the predominant mechanism include: adsorption, ion exchange, precipitation dissolution [6]. In the solution of hydroxyapatite adsorption and ion exchange occur simultaneously, namely hydroxyapatite adsorption on metal ions to the surface, and then the ion-exchange and hydroxyapatite calcium ions, are adsorbed fixed metal ions. Refers to precipitation of dissolution due to different solubility of the heavy metal ions to form insoluble heavy metal salt content or difficulty in solution in the presence of hydroxyapatite, hydroxyapatite is dissolved in this process. Liu Yu and other natural and artificial hydroxyapatite apatite wastewater treatment capacity conducted a series of experiments. The results showed that apatite adsorption and ion exchange interactions vast majority of heavy metal is better, that the main mechanism comprising: adsorption, surface complexation, dissolution-precipitation and ion exchange. Hydroxyapatite is harmful to humans, can be used in the food industry, and steady state after adsorption, no secondary pollution, the adsorbent can be used to remove fluorine in groundwater areas with high fluoride content.

This paper mechanical stirring assist preparation of hydroxyapatite nanomaterials, investigate their preparation process, and the material properties were characterized in order to provide reference for industrial scale applications.

2. Materials, Reagents and Methods

2.1. *Instrument* [7]

2.2. *Reagents* [7]

2.3. *The pre-reaction HAP sample preparation* [7]

2.4. *Mechanical agitation Nano HAP*

Equal volumes of the amount of the above-described 20mL two with a good solution, two with aqueous ammonia solution the pH of 10 to 10.5, the latter with vigorous stirring (0.030mol / L of $(NH_4)_2HPO_4$ solution) using a separator funnel slowly was added dropwise (<10mL / min) constantly stirred calcium nitrate solution, and treated with aqueous ammonia to control pH value of 10 to 10.5, the reaction temperature is room temperature 25°C. Pending completion of the addition, the stirring was continued for 1h, then the reaction product was allowed to stand for aging for about 12h, the supernatant decanted lower sediment centrifugation, thoroughly washed with deionized water, centrifugal separation three to five times, each time 15min, to remove the ammonium ion, until neutral, after which the product in an oven temperature of 80°C cross-dried 12h to give HAP powder sample denoted S1.

2.5. *Characterization of the sample* [7]

3. Results and Discussion

3.1. *FTIR results of sample S1 (see Fig. 1)*

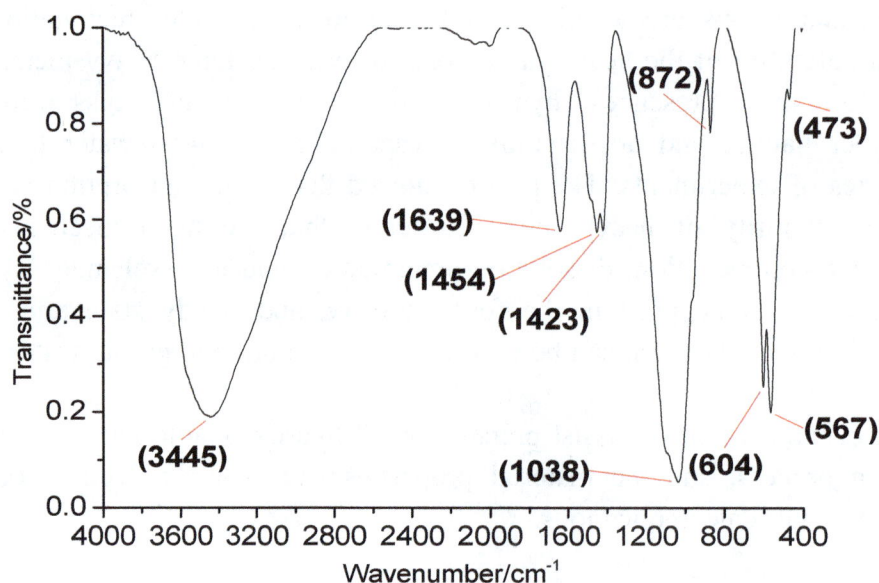

Fig. 1 FTIR spectra of the sample S1

It can be seen from Fig. 1 that 473,567,604 are the P-O bending vibration absorption peak, the absorption peak of which 567,604 at the obvious characteristics. 1038 is a P-O asymmetric stretching vibration absorption peak distinctive characteristic. It appears two peaks. it can have a preliminary determination and more content. 872,1423,1454 vibration absorption peaks, of which 1423 and 1454 appear doublet into the apatite structure is an important symbol, can explain the structure of a solid solution powder HAP part of the structure, due to the presence in the powder synthesis, aging process absorbed in the air causes less content. 1639 was adsorbed water absorption peak, 3445 is the symmetric O-H stretching vibration, infrared analysis thus confirmed the sample S1 phosphate (PO_4^{-3}) and hydroxyl (OH^{-1}) is present.

3.2. XRD results of sample S1 (see Fig. 2)

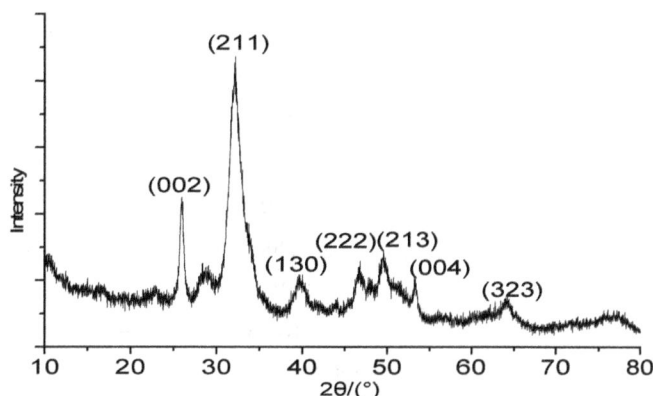

Fig. 2 XRD diffraction pattern of the sample S1

According to X-Ray Diffraction Standards JCPDS standard card PDF # 09-0432 Richard Pu hydroxyapatite, it can be seen from Fig. 2 that the product of 25.95°, 32.19°, 39.71°, 46.83°, 49.57°, 53.28°, 64.19° hydroxyapatite appears at several characteristic peaks, corresponding to HAP crystal (002), (211), (130), (222), (213), (004), (323) of crystal face diffraction characteristics, the highest peak in the figure corresponds to the (211) plane of HAP, the second peak corresponding to the (002) planes. This indicates that the product is prepared hydroxyapatite crystals belonging to the hexagonal nano-HAP, sharp peak shape, crystallinity is good. In addition to hydroxyapatite, we can also see the characteristic peak at about 77° for the trace impurity peaks, combined IR spectrum shows that the trace impurity peaks may belong carbonate; additional (002), (211), (213) high-intensity diffraction peaks of crystal face, thus indicating a higher purity of hydroxyapatite powder preparation, more content.

3.3. Gaussian fit

We have pre-prepared by a liquid ion nano-hydroxyapatite, denoted S3, ultrasonic assisted preparation of nano-hydroxyapatite, denoted by S2, will now be prepared by mechanical mixing with hydroxyapatite nanoparticles S1 and S2, S3 do FIG fitting of comparison, the XRD data Origin8.6 of three samples were calculated separately Gaussian fitting half width (FWHM), the obtained half width (FWHM) corresponding to the plane diffraction peak (211), calculate the average crystallite size according to the Scherrer formula, the results in the table below.

Table 1 The average crystallite size of HAP

Sample No.	S1	S2	S3
β(FWHM)	2.02587/180	1.65425/180	1.80477/180
2θ（211）	32.19°	32.03	31.94°
Crystal size: D/nm	12.83	15.71	14.40

(Copper target, $\lambda = 0.15418\,\mathrm{nm}$)

Table 2 Gaussian fitting Results

Model	Gauss	
Equation	$y=y0+(A/(w*sqrt(Pi/2)))*exp(-2*((x-x0)/w)^2)$	
Reduced Chi-Sqr	5729.25795	
Adj.R-Square	0.54276	
	Value	Standard Error
y0	144.8299	1.29237
x0	32.21557	0.01809
w	1.72062	0.03658
A	1273.81551	23.94168
sigma	0.86031	0.01829
FWHM	2.02587	0.04306
Height	590.69335	10.79717

4. Conclusion

Preparation of mechanical stirring hydroxyapatite material can also be seen outside the characteristic peak at about 77° for the trace impurity peaks, combined IR spectrum shows that the trace impurity peaks may belong carbonate; additional (002), (211), (213) crystal face diffraction intensity high peaks, indicating high purity hydroxyapatite powder preparation, more content. Therefore, preparation of auxiliary mechanical stirring hydroxyapatite nano-materials is simple, material morphology and size controllable process parameters can be expected to provide large-scale preparation of nano-materials and reference.

Acknowledgments

This research was funded by Guangdong Department of Water Resources Science and Technology Innovation Project, (No. 2015-20) and Guangdong Provincial Department of Education Science and Technology Innovation Project (No. 2013KJCX0109).

References

1. A. Thuault, E. Savary, J.-C. Hornez, G. Moreau, M. Improvement of hydroxyapatite mechanical properties by direct microwave sintering in single mode cavity [J]. Journal of the European Ceramic Society. 2014, 34(7): 1865-1871.

2. Li Chen, Jingxiao Hu, Jiabing Ran, Xinyu Shen, Hua Tong. Preparation and evaluation of collagen-silk fibroin/hydroxyapatite nanocomposites for bone tissue engineering [J]. International Journal of Biological Macromolecules. 2014, 65: 1-7.

3. A. Joseph Nathanael, Jun Hee Lee, D. Mangalaraj. Influence of processing method on the properties of hydroxyapatite nanoparticles in the presence of different citrate ion concentrations [J]. Advanced Powder Technology. 2014, 25(2): 551-559.

4. H. Wang, N. Eliaz, Z. Xiang, et al. Early bone apposition in vivo on plasma-sprayed and electrochemically deposited hydroxyapatite coatings on titanium alloy. Biomaterials, 2006, 27: 4192-4203.

5. Y.W Fan, K Duan, R.Z Wang. A composite coating by electrolysis-induced collagen self-assembly and calcium phosphate mineralization [J]. Biomaterials, 2005, 26: 1623-1632.

6. Y Boonsongrit, H Abe, K Sato, M Naito, M Yoshimura, H Ichikawa, Y Fukumori. Controlledrelease of bovine serum albumin from hydroxyapatite microspheres for protein delivery system. Materials Science and Engineering: 8, 2008, 148: 162-5.

7. Yong-guang Bi, Xu-si Xu. Study on Nano-hydroxyapatite assisted preparing by ionic liquids [J]. Advanced Materials Research, 2014, (1015): 501-504.

Research on Microstructure and Properties of (Y_2O_3+TiB+ TiC)/Ti-6Al-4V Composite Fabricated by Melting-casting Process

Yun-Lian Qi*, Li-Ying Zeng, Zhi-Min Hou, Wei Liu, Hua-Mei Sun, Quan Hong, Xiao-Nan Mao

Northwest Institute for Nonferrous Metal Research, Xi'an 710016, China
Email: qiyunlian@126.com

The (Y_2O_3+TiB+TiC)/Ti-6Al-4V titanium matrix composite were prepared by vacuum induction pre-melting and vacuum arc melting for 4 times melting-casting process in this paper. The composite contains element yttrium (Y) and element boron(B) and element carbon(C) in which Y content ranging from 0.1wt% to 0.15wt% and B content ranging from 0.2wt% to 0.5wt%and C content ranging from 0.5wt% to 3wt%.The microstructure and SEM images of fracture surface of composite were studied. The elastic modulus and mechanical properties of composite were tested. The results indicate that ultimate tensile strength (UTS) of composite reaches 1300MPa and increase 25% than Ti-6Al-4V titanium matrix with 0.5wt%B and 0.5wt%C; the elastic modulus of composite are 145GPa and increase 30% than Ti-6Al-4V titanium matrix with 0.5wt%B; the rare-earth oxide (Y_2O_3) improved structure of the (Y_2O_3+TiB+TiC)/Ti-6Al-4V titanium matrix composite distributed uniformly on grain boundariesβ. The microstructures have good match the ultimate tensile strength and elastic modulus with elongation containing lamellar α and grain boundaries β and the second phase TiB with bar-like and the second phase TiC with spherical-like.

Keywords: Titanium matrix composite; elastic modulus; microstructure and mechanical properties.

1. Introduction

Titanium and titanium alloys have been widely used in aerospace and shipbuilding due to its excellent corrosion resistance, favorite fatigue properties as well as fracture toughness. Specially, Ti-6Al-4V titanium alloy and its castings have found more and more applications with the development of aerospace and advanced weapons. And in recent years, some complex equipment parts made by batch casting are possible with the technology improvement of the investment casting for titanium. So, the production cycle will be shortened, the material utilization can be improved and the manufacturing cost of equipment parts can be reduced sharply by the near-net shaping technology. At the same time, advances in technologies, especially the development and progress of titanium alloy precision casting technology and the emergence of hot isostatic pressing technology (HIP), which greatly improve the quality of titanium castings and promote the application of titanium castings for engines and airplanes [1-4].

The long and thin wall titanium components are easy to be deformed during the service life because of its low rigidity, so how to improve the comprehensive performance of titanium alloys has always been concerned by many researchers. The strength and stiffness can be increased and the grain size will be refined with the adding of B, C and Y in traditional titanium alloy castings. Then, the performance of the traditional titanium alloys can be improved due to the refined grains, new, economic and practical processes can be developed according to above mentioned methods [5-7].

The microstructure and the mechanical properties of the as-cast Ti-6Al-4V bars with element B, element C and element Y are to be investigated in this paper.

---

*Corresponding author

2. Experiment Procedures and Methods

Raw materials used in this experiment were titanium sponge, aluminum beans, pure vanadium, pure B powder, pure C powder, pure Y block like and titanium foil, they were mixed in a certain proportion and were pressed into electrode block. Six kinds of Ti-6Al-4V-xB, Ti-6Al-4V-0.5B-yC and Ti-6Al-4V-0.5B-0.5C-zY titanium matrix composite, where x = 0.2 wt.%, 0.5 wt.%, y = 0.2 wt.%, 3 wt.%, z = 0.1 wt.%, 0.15 wt %, were prepared by pre-melting in induction melting furnace for one time and by melting in non consumable vacuum arc furnace for 4 times. Nominal chemical composition of titanium matrix composite is shown in Table 1.

Table 1 Nominal chemical composition of composite (*wt/%*)

No	Al	V	B	C	Y
1#	6	4	0.2	0.032	
2#	6	4	0.5	0.022	
3#	6	4	0.5	0.5	
4#	6	4	0.5	3	
5#	6	4	0.5	0.5	0.1
6#	6	4	0.5	0.5	0.15

Then the as-cast composite bars with the length of 100 mm and diameter of 8 mm were casted in a red copper mould. The bars were hot isostatic pressed (HIPed) at 920°C with the pressure of 110 MPa for 2h to reduce the cast porosity. The samples for phase composition and microstructure analysis were in length 10 mm and diameter 8 mm. The samples for mechanical property tests were in length 40 mm and diameter 8 mm.

3. Results and Discussion

3.1. *The elastic modulus of composite*

Fig. 1 shows the influence of micro-scale element B, C and Y on the elastic modulus of titanium matrix composite. From the table, it can be seen that the elastic modulus for the as-cast matrix Ti-6Al-4V alloy is 105GPa, and which is 130GPa for the composite with 0.2wt% B, the value increases 25GPa. With the adding of element boron, the elastic modulus for the composite will increase. When B content is 0.5wt%, the elastic modulus of the composite reaches a maximum value of 145GPa due to the formation of TiB reinforced phase. For the 0.5%B/TC4 composite with the element C from 0.5wt% to 3wt%, the elastic modulus of the composite is 127-130GPa, which decreases slightly. For the (0.5%B+0.5C)/Ti-6Al-4V composite with the element Y from 0.1wt% to 0.15wt%, the elastic modulus of the composite is 126-130GPa. So, Element B has a larger contribution to the increase of elastic modulus for composite, while the contribution of element Y is not obvious.

3.2. *The mechanical properties of composite*

The mechanical properties at room temperature of the as-HIPed titanium matrix composites bars are shown in Fig. 2. The tensile strength increases for the titanium matrix composite with boron content from 0.2wt% to 0.5wt%, and the elongation and reduction of area have the same increase tendency. The increment of the tensile strength is about 20%~30%, and the increase of the elongation is about 25%~30%.

The tensile strength for the composite with 0.5 wt% C greatly increases to 1305 MPa, which is only 1045MPa for 0.5B / Ti-6Al-4V composite, the increment is about 59%, but the decrease of elongation is about 50%. When the element C increases from 0.5 wt% to 3 wt%, the room temperature mechanical properties of composite casting billet is not so good and the fracture is brittle. The reason for the decrease of plasticity is mainly due to the higher volume fraction of the precipitated TiC particles with the increase of C content. Hard and brittle TiC phases gather in grain boundaries, which can counteract the effect of grain refinement, and restrict the plasticity of the alloy [7]. The tensile strength for the 0.5%B+0.5%C/Ti-6Al-4V composite with Y content ranging from 0.1wt% to 0.15wt% increases from 1225 MPa to 1232 MPa, the elongation is 6.5%~6%, and the reduction of area is 12.5%~10.5%. To sum up, the strength and plasticity of composite casting billet improves markedly because of the adding of element B. Trace element C (less than 0.5%) can increase the strength of composite casting billet, and can impair the plasticity at the same time. The elongation of composite casting billet decreases slightly adding element Y, but increase of the tensile strength is greatly.

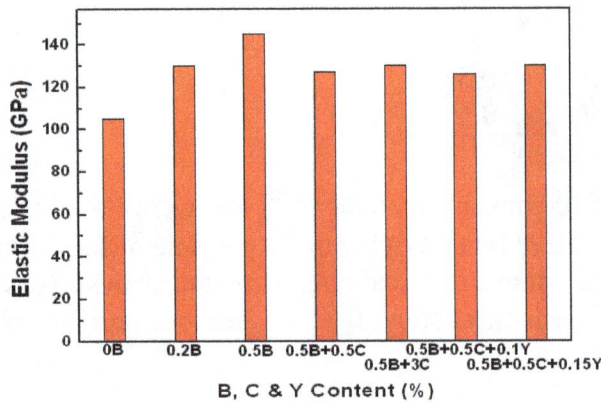

Fig. 1 Elastic modulus of titanium matrix composite

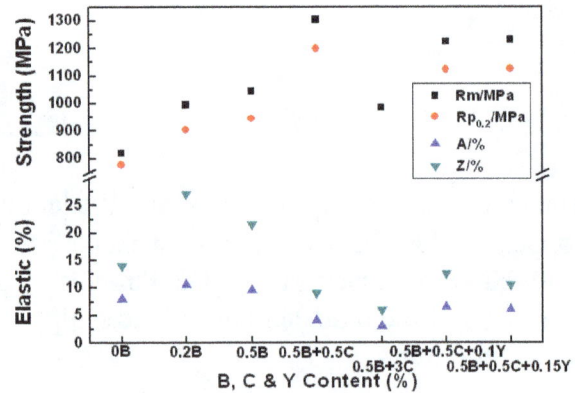

Fig. 2 Room temperature tensile properties of titanium matrix composite

3.3. *The microstructure of composite*

The microstructures of the composite with different B, C and Y content are shown in Fig. 3. From the figure, it can be seen that the microstructure are consisted of lamellar α, grain boundaries β and the second phase of TiB for composite casting billet with element B from 0.2wt% to 0.5wt%. The microstructure of composite casting billet is obviously refined because the volume fraction of the second phase of TiB increases significantly with the increment of B content. When the B element content is 0.2wt%, the morphology of α within the grains is very long and thick lathes, as shown in Figure 3a. When the B element content is 0.5wt%, the morphology of α within the grains is short and coarse lathes, as shown in Figure 3b. The microstructures of the 0.5% B/ Ti-6Al-4V composite with element C varying from 0.5wt% to 3wt% are shown in Figure 3c, d. When the C content is 0.5wt%, the morphology of α within the grains is fine needles, as shown in Figure 3c, which corresponds to the increasing strength and the reducing plasticity for the composite. When the C content is 3wt%, spherical-like second phases of TiC particles distribute evenly in the grains, as shown in Figure 3d, the plasticity for the composite is not so good due to the existence of TiC particles. The microstructures of (0.5B+0.5C)/Ti-6Al-4V composite with different Y content are shown in Figs. 3e, f. From the Fig. 1e and Fig. 3f, the grain β and lamellar α have been spheroidized adding 0.1 wt% and 0.15 wt% element Y. The spheroidized microstructures have a good match strong with plastic.

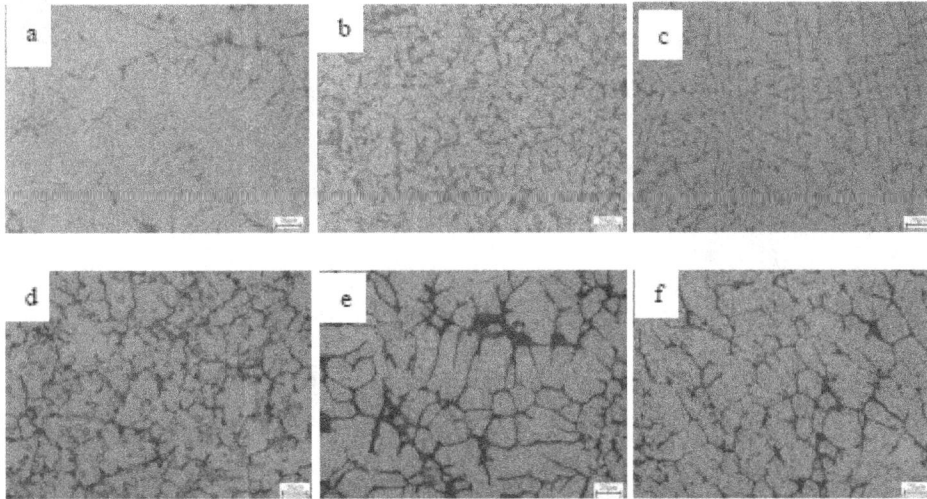

Fig. 3 The microstructure of titanium matrix composite

(a) 0.2%B, (b) 0.5%B, (c) 0.5%B+0.5%C, (d) 0.5%B+3%C,
(e) 0.5%B+0.5%C+0.1%Y, (f) 0.5%B+0.5%C+0.15%Y

The SEM microstructures of the composite with different B and C content are shown in Fig. 4a and Fig. 4b. Second phases of short bar-like TiB and spherical-like TiC can be found in Figure. When the C content increases to 3wt%, globurizing tendency of TiC particles is more obvious, and the particles also tend to float in the matrix, as shown in Figure 4b. The SEM microstructures of (0.5B+0.5C)/Ti-6Al-4V composite with different Y content are shown in Fig. 4c and Fig. 4d. The oxydum Y_2O_3 distributes uniformly in the grain boundary β.

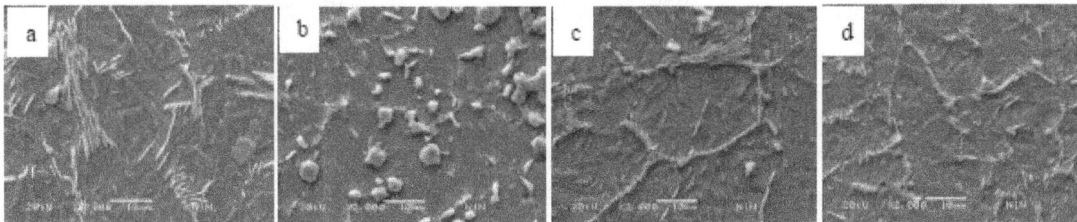

Fig. 4 SEM images of titanium matrix composite

(a) 0.5%B+0.5%C, (b) 0.5%B+3%C, (c) 0.5%B+0.5%C+0.1%Y, (d) 0.5%B+0.5%C+0.15%Y

3.4. *The tensile fracture of composite*

Fractographies of titanium matrix composite are shown in Fig. 5. From Fig. 5a, Fig. 5i, and Fig. 5k, it can be seen that the fracture surface for the composite with element B and element Y are ductile fracture, with obvious necking phenomenon. The fracture surface in Fig. 5c and Fig. 5e are more plain than that in Fig. 5a, Fig. 5i, and Fig. 5k, without obvious necking phenomenon, some small holes can also be found.

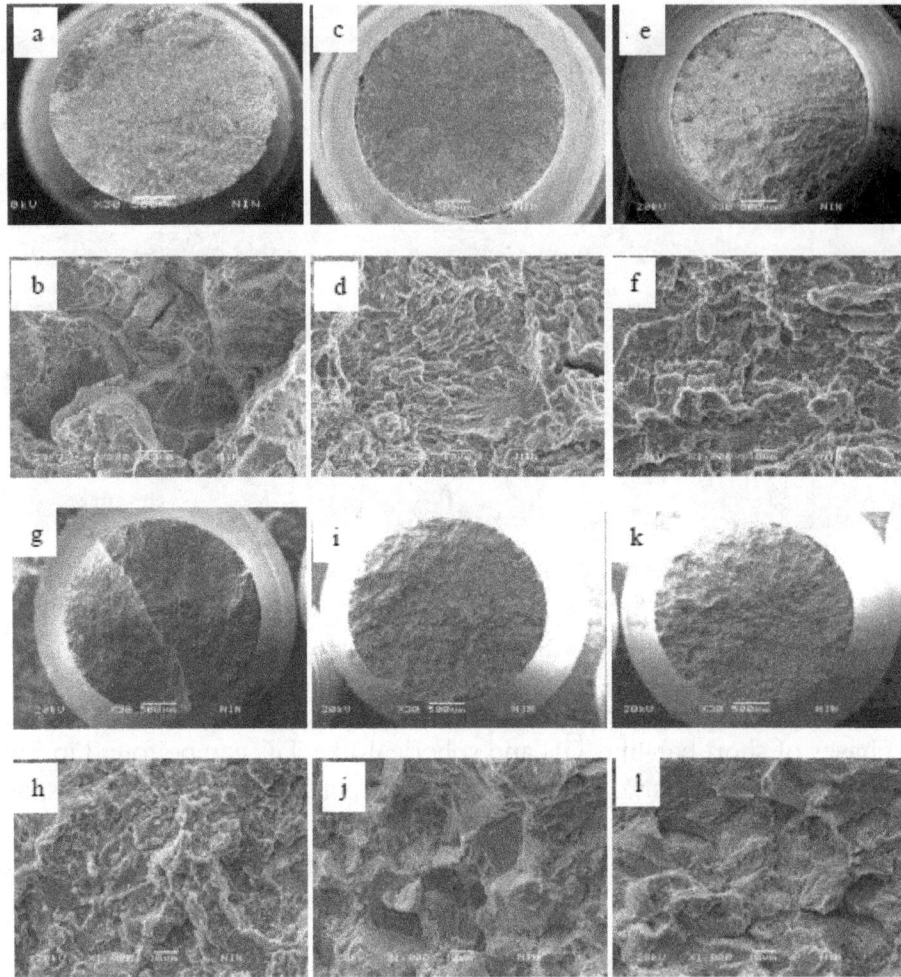

Fig. 5 SEM images of fracture surface of titanium matrix composite

a), b 0.2%B; c), d) 0.5%B; e), f) 0.5%B+0.5%C; g), h) 0.5%B+3%C;
i), j) 0.5%B+0.5%C+0.1%Y; k), l) 0.5%B+0.5%C+0.15%Y

A clear tearing ridge can be seen on the fracture surface shown in Fig. 5g. Relatively broad and deep dimples can be seen in Fig. 5b, Fig. 5j, and Fig. 5l, which corresponds to the good plasticity of composite casting billet. The fracture faces shown in Fig. 5d, Fig. 5f and Fig. 5h are relatively plain and the sizes of the shallow dimples are not uniform, which leads to the decreasing plasticity of composite casting billet. Intercrystalline fractures accompanying with secondary cracks can be found in all fracture planes, and the originated location of fracture are not obvious.

4. Conclusion

(1) Element B can greatly enhance the elastic modulus of composite casting billet, while the improving effect of element C and element Y is not apparent. When the content of boron is 0.5 wt %, the elastic modulus of the composite can reach to 145GPa.

(2) The strength and plasticity increase markedly for composite casting billet with element B, the composite has a good match between strength and plasticity. Element C can increase the strength of composite casting billet, the plasticity of composite casting billet with 3wt% element C is less than 3%, and brittle fracture can be found in the composite. The elongation of

(TiB+TiC)/Ti-6Al-4V casting billet decreases slightly adding element Y, but increase of the tensile strength is greatly.

(3) The microstructures of as-cast $(Y_2O_3+TiB+TiC)/Ti-6Al-4V$ are consisted of lamellar or spherical α, grain boundary β, rare-earth oxide Y_2O_3 and the second phases of bar-like TiB, spherical-like TiC. The rare-earth oxide Y_2O_3 distributes uniformly in the boundaryβ.

References

1. H.S. Xie, S.B. Liu, G.Q. Sun, Z.H. Wang, J. Zhao, Development and Application of Investment Casting Technology for Titanium Alloys Casting of China, Special Casting & Nonferrous Alloys. 5(2008) 462-464.
2. M. Zhang, H. Nan, D. Huang, G. P. Cao, Study of Heat Isostatic Pressing and Thermohydrogen Treatment of Titanium Alloy Castings, China Foundry Machinery & Technology. 5 (2002) 1-3.
3. P. Yan, L. Wang, J. Zhao, Z.H. Wang, C.H. Zhang, T. You, Development and Applications of the High-Strength Cast Titanium Alloy, Foundry. 5 (2007) 451-454.
4. G.Q. Su, H.S. Xie, C.H. Zhang, J. Zhao, L. Wang, Z.Q. Yu, H.Y. Liu, Microstructures and Mechanical Properties of Zti-3B Cast Titanium Alloy Material, Titanium Industry Progress. 2(2005)26-30.
5. H. Luo, Z.Q. Chen, Progress in Boron-modified Titanium Alloys, Development and Application of Materials. 4 (2010) 77-80.
6. S.Z. Zhang, J.F. Lei, S.X. Guan, Y.Y. Liu, D. Li, Influence of Heat Treatment on Mechanical Properties of High Elastic Modulus, High Strength, High Toughness and Weldable Titanium Alloys, Acta Metallurgica Sinica. 38 (2002) 74-77.
7. Z.H. Zhang, X.Z. Wang, S.L. Shang, K.W. Bai, J.Y. Shen, Influence of Processing on Elastic Modulus for a Titanium Alloy with High Strength and High Elastic Modulus, Chinese Journal of Rare Metals. 25 (2001) 19-22

Chapter 3
Magnetic Materials

Periodic Vibration Analysis of Giant Magnetostrictive Rod

Yong-An Zhu[1,2], Fan Wang[1,*]

[1]*Institute of Applied Mechanics, Jinan University, Guangzhou, P. R. China*
[2]*Infrastructure division, Jinan University, Guangzhou, P.R. China*
[*]Email: twfan@jnu.edu.cn

The axial periodic vibration of a giant magnetostrictive rod in the periodic altering magnetic field and the temperature field is analyzed in this paper. The mathematical model is established by the variational method, and the quadratic nonlinear mechanic-magnetic constitutive relationship is applied. The displacement solution and the relationship between the amplitude and the frequency are obtained by the harmonic method under the linear spring constrained boundary condition.

Keywords: Giant magnetostrictive; material, Terfenol-D; rod; nonlinear constitutive relationship; axial vibration; harmonic method.

1. Introduction

Terfenol-D is a kind of giant magnetostrictive material. Its shape can be changed with the variation of the magnetic field, which is called "magnetostrictive effect". As a new kind of smart material, it has many advantages such as large strain (about $10^{-3} \sim 10^{-2}$) and high energy conversion efficiency (about 49%~56%), which has been applied in high technology fields such as micro displacement controller, high efficiency transducer, precision sensor, sonar, damping element of aerocraft and walking robot [1-3]. There have been many research works about manufacture and mechanic- magnetic experience of Terfenol-D material. There are some research works about the vibration of giant magnetostrictive rod by using magnetoelastic theory [11]. Some researchers have studied the vibration of the rod without any constraint in the altering magnetic field by using finite element method, and some dynamic parameters such as displacement, frequency and amplitude are obtained [4-6]. Jin Ming-yu and Shang Xin-chun [7-10] have done a series of research work about the vibration of the Terfenol-D cylinder rods with the spring constraint. There is few research work about the vibration of Terfenol-D rod in the magnetic field and temperature field by now. In this paper, the axial periodic vibration of giant magnetostrictive rod in the periodic altering medium high magnetic field and temperature field is analyzed. The mathematical model including vibration controlling equations and boundary constrained conditions is established by applying the variational method, and the quadratic nonlinear mechanic-magnetic constitutive relationship is applied. The quasistatic method is used to deal with the temperature variation. After the temperature changes to a certain degree, the magnetic field is applied to the rod. The result is that there is no temperature part in the magnetization equation. The natural frequency of the axial vibration and the displacement solution are obtained by the harmonic method.

2. Mathematical Model

A magnetostrictive rod actuator as shown in Fig. 1 is considered of turn number n, current intensity $i=I\sin\omega t$, in which I and ω are the maximum value of the current intensity and frequency respectively, the rod length l, the radius of rod r, material density ρ, elastic modulus E, axial magnetostrictive coefficient d, magnetic permeability μ.

---

*Corresponding author

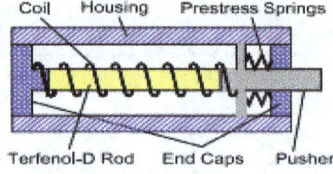

Fig. 1 Cross-section of the Terfenol-D actuator

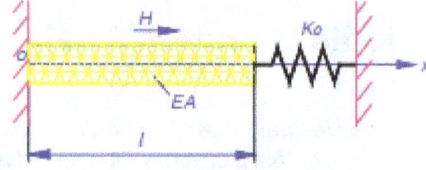

Fig. 2 Simplified mathematical model of the giant magnetostrictive rod actuator

The quadratic nonlinear constitutive relationships of the magnetostrictive material are expressed as follows:

$$\varepsilon = \frac{\sigma}{E} + mH^2 + \alpha T_d, \qquad B = \mu H + m\sigma H \tag{1}$$

Where ε and σ are the axial strain and stress respectively, B and H are the axial magnetic induction intensity and magnetic field intensity, α is the temperature coefficient, T_d is the temperature difference. The axial coordinate is x and the axial displacement is $u(x,t)$. The axial stress and the axial magnetic induction intensity are:

$$\sigma = E\varepsilon - E(mH^2 + \alpha T_d), \qquad B = \mu H + m\sigma H \tag{2a,b}$$

Where

$$\varepsilon = \frac{\partial u}{\partial x}$$

In the action of alternating current, the kinetic energy, magnetic energy and potential energy of the system, which consists of rod and spring, can be expressed as follows:

$$T = \frac{A}{2}\int_0^l \rho(\frac{\partial u}{\partial t})^2 dx, \quad \hat{H} = \int_0^l BH dx, \quad U = \frac{A}{2}\int_0^l \sigma\varepsilon dx + \frac{1}{2}k_0 u^2(l,t) \tag{3a,b,c}$$

Then, the Hamilton function quantity of the whole system is:

$$\prod[u] = \int_{t_1}^{t_2}(T + \hat{H} - U)dt \tag{4}$$

By applying Hamilton principle of least action, and applying Eqs. (1)-(3), the variational method is applied to Eq. (4), the vibration equation of the rod can be obtained,

$$\frac{\partial^2 u}{\partial t^2} - \frac{E}{\rho}\frac{\partial^2 u}{\partial x^2} = 0 \tag{5}$$

The boundary condition of the clamped end can be showed as follows:

$$u(0,t) = 0, (0 \le t \le T) \tag{6}$$

132

The boundary condition of the spring end can be showed as follows:

$$[\frac{\partial u}{\partial x} + \frac{k_0 u}{EA}]_{x=l} = mH^2 + \frac{1}{2}\alpha T_d, (0 \le t \le T) \tag{7}$$

The periodic condition can be showed as follows:

$$u(x,0) = u(x,T), \frac{\partial u(x,0)}{\partial t} = \frac{\partial u(x,t)}{\partial t} (0 \le x \le l) \tag{8}$$

Where, T is period of time.

3. Analytical Solution

According to vibration Eq. (5), boundary conditions (6) and periodic conditions (8), the following expression of the displacement solution is considered:

$$u(x,t) = \sum_{j=1}^{\infty} A_j \sin\frac{j\omega x}{p} \cos j\omega t + A_0 x \tag{9}$$

Where

$$p = \sqrt{\frac{E}{\rho}} \tag{8}$$

Obviously, Eq. (9) can satisfy vibration Eq. (5), boundary condition (6) and periodic conditions (8). By substituting Eq. (9) into Eq. (7), the results can be showed as follows:

$$\sum_{j=1}^{\infty} A_j [(\frac{j\omega}{p})\cos\frac{j\omega l}{p} + \frac{k_0}{EA}\sin\frac{j\omega l}{p}]\cos j\omega t + A_0(1 + \frac{k_0 l}{EA}) = \frac{mn^2 I^2}{l^2}\sin^2\omega t + \frac{1}{2}\alpha T_d \tag{10}$$

The following expressions are considered:

$$c_1 = \frac{mn^2 I^2}{2l^2} \qquad c_2 = \frac{1}{2}\alpha T_d \tag{11}$$

By applying the double angle formula of trigonometric function and Eq. (11), the Eq. (10) can be written as:

$$\sum_{j=1}^{\infty} A_j [(\frac{j\omega}{p})\cos\frac{j\omega l}{p} + \frac{k_0}{EA}\sin\frac{j\omega l}{p}]\cos j\omega t + A_0(1 + \frac{k_0 l}{EA}) = (c_1 + c_2) - c_1\cos 2\omega t \tag{12}$$

The following expression is considered:

$$A_0 = (c_1 + c_2)/(1 + \frac{k_0 l}{EA}) \tag{13}$$

133

According to Eqs. (12) and (13), the result can be expressed as follows:

$$\sum_{j=1}^{\infty} A_j[(\frac{j\omega}{p})\cos\frac{j\omega l}{p} + \frac{k_0}{EA}\sin\frac{j\omega l}{p}]\cos j\omega t = -c_1 \cos 2\omega t \tag{14}$$

By multiplying both ends of Eq. (14) by [cosκωt], and integrating them in [0, T], the result can be showed as follows:

$$\sum_{j=1}^{\infty} A_j[(\frac{j\omega}{p})\cos\frac{j\omega l}{p} + \frac{k_0}{EA}\sin\frac{j\omega l}{p}]\int_0^T \cos j\omega t \cos k\omega t dt = -c_1\int_0^T \cos 2\omega t \cos k\omega t dt \tag{15}$$

By considering the orthogonality of the trigonometric function:

$$\int_0^T \cos j\omega t \cos k\omega t dt = \frac{T}{2}, (k = j)$$

$$\int_0^T \cos j\omega t \cos k\omega t dt = 0, (k \neq j)$$

The results can be showed as follows:

$$A_2 = c_1 / [(\frac{2\omega}{p})\cos(\frac{2\omega l}{p}) + \frac{k_0}{EA}\sin(\frac{2\omega l}{p})] \tag{16}$$

According to Eq. (9) and Eq. (6), the displacement expression can be showed as follows:

$$u(x,t) = A_2 \sin\frac{2\omega x}{p}\cos 2\omega t + A_0 x \tag{17}$$

By substituting x=l into Eq. (17), the displacement of the rod cap can be obtained, which is showed as follow:

$$u(l,t) = A_2 \sin\frac{2\omega l}{p}\cos 2\omega t + A_0 l \tag{18}$$

If the denominator expression of Eq. (16) is assumed to be zero, the result can be showed as follow:

$$(\frac{2\omega}{p}) + \frac{k_0}{EA}tg(\frac{2\omega l}{p}) = 0 \tag{19}$$

The natural frequency of the rod can be obtained from Eq. (19).

4. Numerical Result and Discussion

A giant magnetostrictive rod with the parameter showed in Table 1 is considered as a numerical example.

Table 1 Parameter table

Name	Number	Unit
Elastic modulus E	2.65×10^{10}	Pa
Rod length l	0.25	m
Rod radius r	0.01	m
Linear stiffness of spring k_0	3.0×10^{7}	N/m
Material density ρ	9.25×10^{3}	Kg/m^3
Nonlinear magnetostrictive coefficient m	0.28×10^{-12}	m/A
Coefficient of thermal expansion α	12×10^{-6}	C^0
Current peak value I	10	A
Turn number n	120	
Current frequency f	700	Hz

The natural frequency of the rod can be obtained by the numerical calculation through Eq. (19), which is showed in Table 2:

Table 2 Frequency table

Order number	1	2	3	4	5
Frequency ω	6753.2	16568.21	26967.9	37496.11	48070.6

The maximum displacement and the minimum displacement of the rod cap can be obtained by the numerical calculation through Eq. (18), which are showed in Table 3 and Table 4 respectively.

Table 3 The maximum displacement of the rod cap

$Td(°C)$	0	0.05	0.10	0.15	0.20	0.25	0.30
$u(l)_{max}(10^{-6}m)$	1.063	1.102	1.142	1.181	1.221	1.260	1.299

Table 4 The minimum displacement of the rod cap

$Td(°C)$	0	0.05	0.10	0.15	0.20	0.25	0.30
$u(l)_{min}(10^{-7}m)$	-2.145	-1.75	-1.355	-0.961	-0.566	-0.172	-0.223

Through the expression of amplitude (16), the curves of amplitude and frequency of the rod can be obtained as follows:

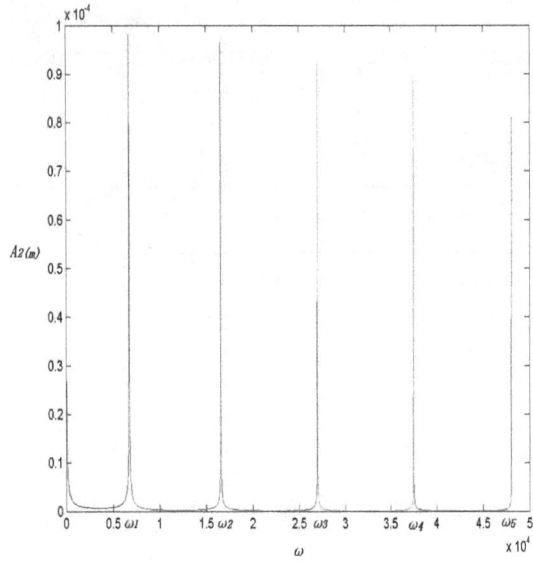

Fig. 3 Curves of amplitude versus frequency

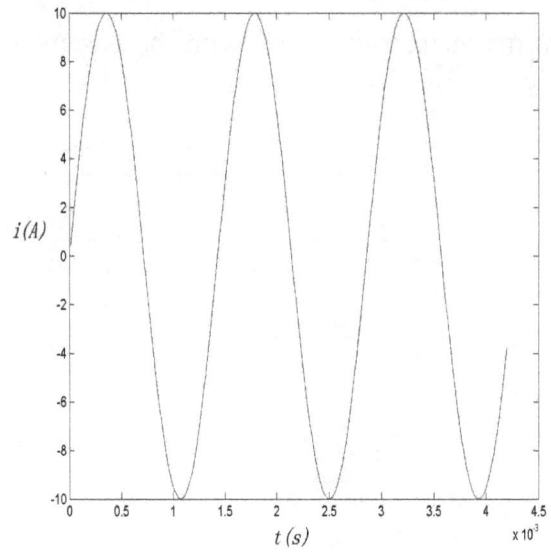

Fig. 4 Excitation current variation curve

The excitation current curve is showed in Fig. 4.

Through the expression of the displacement of the rod cap (18), the displacement responsive curves of the rod cap can be obtained:

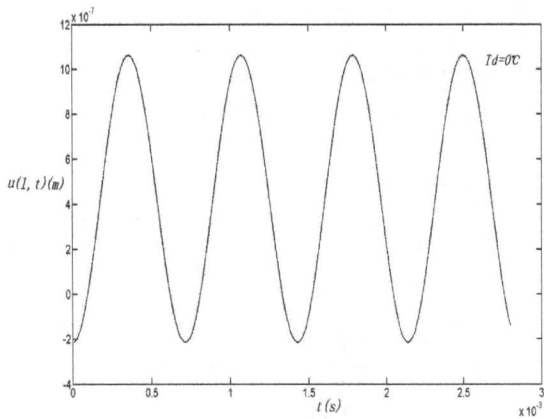

Fig. 5 The displacement responsive curves
of the rod cap(Td=0°C)

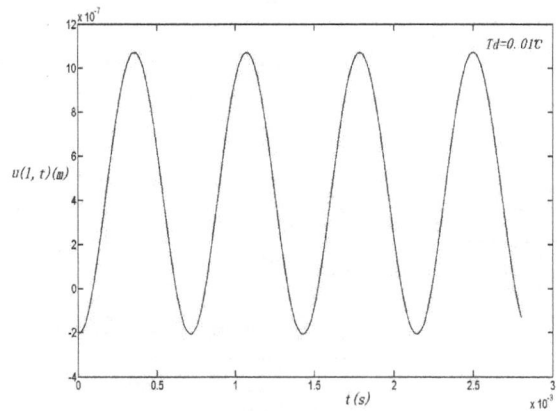

Fig. 6 The displacement responsive curves
of the rod cap(Td=0.01°C)

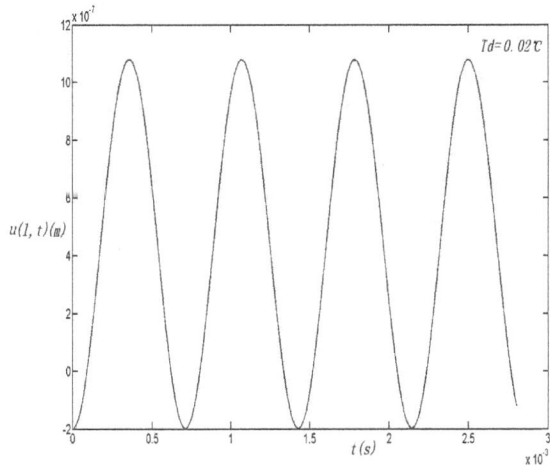

Fig. 7 The displacement responsive curves
of the rod cap(Td=0.02°C)

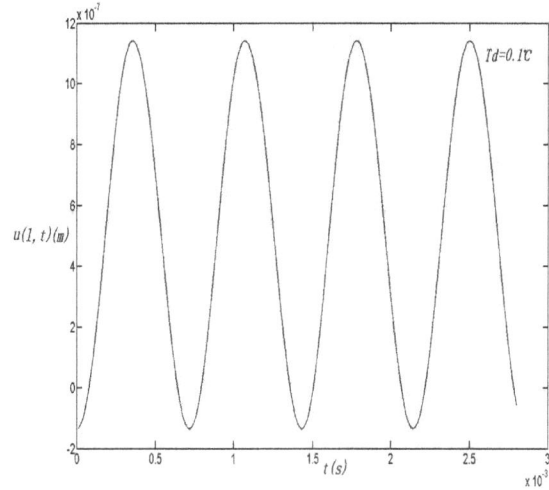

Fig. 8 The displacement responsive curves
of the rod cap(Td=0.1°C)

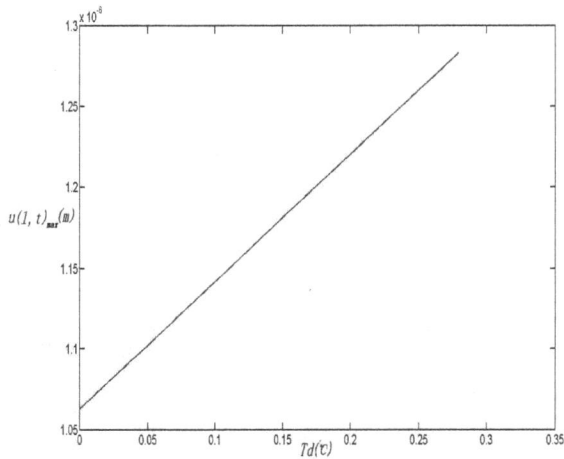

Fig. 9 Curves of the maximum displacement
of the cap versus temperature variation

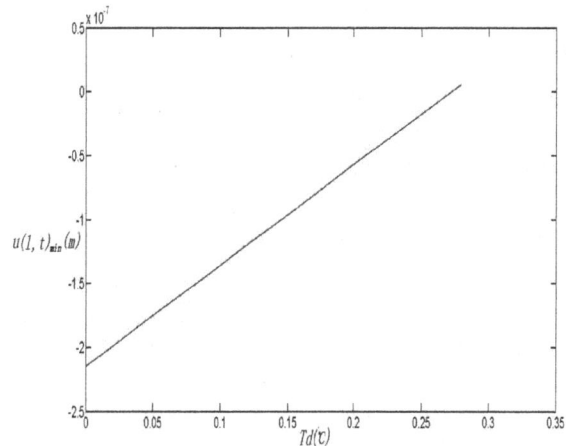

Fig. 10 Curves of the minimum displacement
of the cap versus temperature variation

1. Curves of amplitude and frequency are showed in Fig. 3. It can be found that the amplitude increases sharply when the excitation frequency approaches the natural frequency, which shows the characteristic of simple harmonic vibration.

2. The displacement responsive curves of the rod cap are showed in Fig. 5-Fig. 8. It can be found that the vibration of the rod cap shows the characteristic of periodic vibration. By comparing Fig. 5-Fig. 8 with Fig. 4, it can be found that the vibration frequency is double of the excitation current frequency, which is called *frequency-doubled effect.*

3. It can be concluded from Fig. 9 and Fig. 10 that the displacement of the rod cap increases as the temperature increases. Because of the characteristic of easily heating elongation, the output displacement of the rod cap can be increased through the increase of temperature field. The value of the maximum displacement and the minimum displacement in different temperature field are showed in Table 3 and Table 4.

5. Conclusion

The vibration of the giant magnetostrictive rod in medium high magnetic field and temperature field is harmonic vibration, and the frequency is double of the excitation current frequency. The output displacement of the rod cap can be increased as the increasement of temperature field.

References

1. Y. Yamamoto, H. Eda, T. Mori and A. Rathore, Smart Vibration Sensor Using Giant Magnetostrictive Materials, JMSE Int. J. Series C. 6(1997) 260-266.
2. M.G. Aston, R.D. Greenough, A.G.I. Jenner, W.J. Metheringham, K. Prajapati, Controlled high power actuation utilizing Terfenol-D. J. Alloys Compounds. 258 (1997) 97-100.
3. S. Ashley, Magnetostrictive actuators. Mechanical Engineering. 68 (1998) 68-70.
4. K.S. Kannan and A. Dosgupta, A nonlinear Galerkin finite-element theory for modeling magnetostrictive smart structure. Smart Mater. Struct. 6 (1997) 341-350.
5. Z. Ren, B. Ionescu and A. Razek. Calculation of Mechanical Deformation of Magnetic Materials in Electromagnetic Devices. IEE Trans. Magnetics. 31 (1995) 1893-1876.
6. M. Besbes, Z. Ren and A. Razek, Finite Element analysis of Magneto-Mechanical Coupled Phenomena in Magnetostrictive Materials. IEE Trans Magnetics. 32 (1996) 1058-1061.
7. Mingyu Jin, Xinchun Shang, Nonlinear Vibration analysis of magnetostrictive Terfenol-D rod Engineering mechanics. 6 (2002) 508-511.
8. Liping Qin, Xinchun Shang, Numerical analysis of periodic vibration of linear Magnetostrictive actuator, Engineering Journal of Wuhan University. 42 (2009) 21-26.
9. Liping Qin, Xinchun Shang, Perturbation solution for vibration of Magnetostrictive actuator. Engineering Mechanics. 26 (2009) 223-227.
10. Xinchun Shang, Liping Qin and Fan Wang. Periodic vibration analysis of magnetostrictive Actrator, International Journal of Applied Electromagnetics and Mechanics. 33 (2010) 681-688.
11. Dazhi Yang, Smart material and systems. Press of Tian Jin University. Tianjin. 2000.

Creep Analysis of Ti-600 Alloy at 600°C

Li-Ying Zeng*, Xiao-Nan Mao, Quan Hong, Yun-Lian Qi, Hang-Biao Su

Northwest Institute for Nonferrous Metal Research, Xi'an 710016, China
*Email: ZENG-ly@163.com

Creep tests were carried out on Ti-600 alloy at the temperature of 600°C, and with the stresses of 150MPa, 200MPa, 250 MPa, 300 MPa and 350 MPa, respectively. Steady state creep rate and the stress exponent n at different stresses were calculated for the alloy. Threshold stress σ_0 was introduced to get the true stress exponent p. Creep deformation mechanism was also investigated. The results indicated that the steady state creep rate will increase with the rise of stress, and the creep time will be shortened at the same time. At 600°C, the threshold stress is 76.1MPa. The value of n and p is 2.4 and 1.1 respectively for the alloy crept at lower stress region (150-200MPa); and which is 7.4 and 4.1 respectively for the alloy crept at higher stress region (200-350MPa). Constitutive equations of steady state creep rate were established for the alloy crept at 600°C. The creep deformation mechanism for the alloy is grain boundary diffusion one (Coble creep mechanism) at lower stress region, and which is dislocation climbing mechanism controlled by lattice diffusion (Weetman creep mechanism) at higher stress region.

Keywords: Ti-600 alloy; stress exponent; creep threshold stress; true stress exponent; creep mechanism.

1. Introduction

Nowadays, more and more attention has been paid on titanium and titanium alloys which can be used at 600°C or even higher temperature for longer time. Several near α titanium alloys have been developed for high-temperature applications [1, 2]. Ti-600 is a near alpha titanium alloy designed for components used in turbine engines at 600°C [1, 2], which requiring very low creep plastic strain and relatively high creep resistance [2].

In order to use these alloys in gas turbine engines for increased efficiency, certain properties like short-term strength, long-term creep strength and improved oxidation resistance have been profoundly optimized. A major factor responsible for limiting the use of these alloys at a temperature of 600°C is their poor creep resistance [3]. At present, many studies have been made on the high temperature creep property of conventional high temperature Ti alloys [4], for the creep property has significant influence on the life endurance for Ti alloys and the safety for key components [5]. Creep resistance has great relation with the stress, temperature, time and the volume fraction of β transus phases. So, it is necessary to have the related exploration on these influencing factors.

The creep process discussed in the literature for metallic alloys includes glide and climb dislocation [6, 7], grain boundary or interface sliding [8], viscous dislocation glide either dragged by jogs or by solute atoms [9] and diffusional creep [10]. These creep mechanisms in different stresses and temperature regimes are usually determined by the values of apparent activation energy and stress exponents [10].

The main deformation model is sliding and twins of dislocations at ambient temperature, while at elevated temperature, it is dislocation climb [6-9]. Dislocations can move across the plane perpendicular to the gliding plane through dislocation climb when encountering barriers, then the dislocations will not be blocked at the barriers, which can make the deformation process go on. That's a deformation softening process. It is roughly considered that the creep process is the alternating processing between the sliding and climbing of dislocations.

---

*Corresponding author

In this paper, the creep tests on Ti-600 alloy with β microstructure at 600°C and five kinds of stress levels are to be investigated, and the creep deformation mechanism is also to be discussed.

2. Experiment Materials and Procedures

A near-α high-temperature titanium alloy named after Ti-600 was used in this study, the nominal composition for the alloy is Ti-6Al-2.8Sn-4Zr -0.5Mo-0.4Si-0.1Y. A 504kg ingot was produced by electrode consumption vacuum arc furnace. The β transus temperature for the alloy is 1010°C or so. The alloy was forged at 1100°C from 440 mm starting diameter to 90 mm square cross-section. Then diameter 32 mm bars were conventionally forged. The forging was eventually rolled to diameter 12 mm bars at temperatures below 950°C. The creep samples were cut from the rolling bars and were dissolved at 1020°C, for 1 h, air cooling, then aged at 650°C for 8 h, air cooling.

Creep tests under constant tensile load in air were carried out on the specimens of 5mm gauge diameter using a RD-30 typed creep-rupture machine at 600°C with the stresses of 150MPa, 200MPa, 250MPa, 300MPa and 350MPa, respectively. Rod creep specimens were fabricated from the heat-treated coupons wire-cut electro discharge machining (EDM). The creep elongation was measured by means of one linear variable differential transformer, allowing an accuracy of 0.1%. Based on the results of the test, specimens were interrupted at the stress of 300MPa. While at other strain levels, the experiments were stopped as the tertiary stages occurred, for the secondary stages were too long and more predominant as typically observed in many other alloys with favorite creep resistance.

3. Results and Discussion

3.1. *Creep curves and steady creep states of Ti-600 alloy*

Fig. 1 shows the typical creep curves of Ti-600 alloy at 600°C with five kinds of stresses (150MPa, 200MPa, 250MPa, 300MPa, 350MPa). From the figure, it can be seen that the creep curves exhibit all the three well-defined stages, i.e. primary, secondary and tertiary creep regimes. In the primary stage, the strain rate is relatively high, but slows with increasing strain. The strain rate eventually reaches a minimum and becomes near constant. This stage is known as secondary creep. In tertiary creep, the strain rate exponentially increases with strain because of necking phenomena.

Stress is one of the important factors influencing the creep process for the alloy. Reducing the stress, the creep processing will slow up, the secondary stage in the creep curve will be extended, and the tertiary stage will even not be occurred sometimes. On the contrary, increasing the stress, the secondary stage will be shortened, the curve will transit from the primary stage to tertiary one rapidly, the intermediate zone is short; the secondary stage will not even appear completely at some condition.

As steady-state creep strain rate is one of the specific parameters to reflect the creep behavior for metal materials, the steady-state strain rates were extracted from the creep curves for the five kinds of stresses, then steady-state creep rates ($\dot{\varepsilon}_s$) can be got. By calculation, the steady creep rate for the alloy at 600°C with the stress of 150MPa, 200MPa, 250MPa, 300MPa, 350MPa is 2.78×10^{-9} s^{-1}, 5.56×10^{-9} s^{-1}, 2.78×10^{-8} s^{-1}, 9.17×10^{-8} s^{-1}, 3.72×10^{-7} s^{-1}, respectively. The data indicate that the steady-state creep rate for the alloy increases with the testing stress at 600°C, and its value is relatively low, which indicate that the alloy possess higher creep resistance.

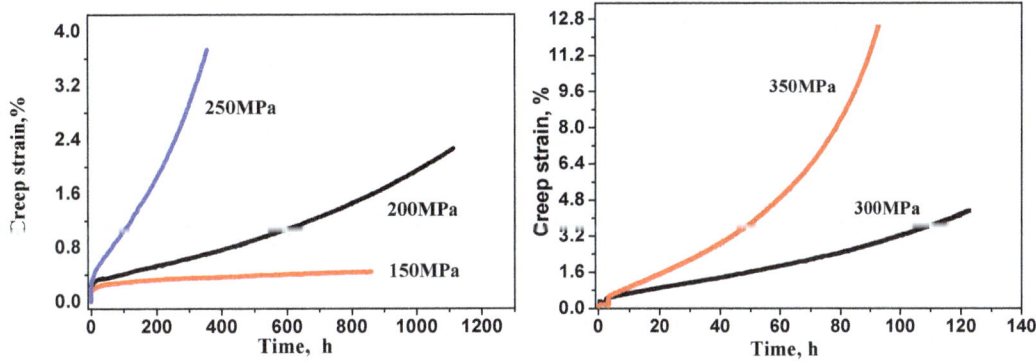

Fig. 1 Typical creep curves of Ti-600 alloy at 600°C with five kinds of stresses

3.2. Stress exponent, threshold stress and true stress for Ti-600 alloy

The combined stress and temperature dependence of steady state creep rate is frequently described by an power law of the form [10]:

$$\dot{\varepsilon}_s = A\sigma^n \exp(-\frac{Q_{App}}{RT}) \qquad (1)$$

where A is a constant, n the apparent stress exponent, R the gas constant, T the absolute temperature and Q_{App} is the apparent activation energy. When the temperature is stable, the apparent stress exponent can be described as follows through differential calculus on Equation (1):

$$n = (\frac{\partial \ln \varepsilon_s}{\partial \ln \sigma})_T \qquad (2)$$

Steady-state creep rates were plotted against applied stress on double-logarithmic scales for Ti-600 alloy, as shown in Fig. 2. Its value increase with the stress at 600°C , which has linear relation for the two parameters. Two straight lines can be found in the figure, and the slope of the line is the value of n. At 600 °C , n is calculated to be 2.4 with the stress from 150 MPa and 200 MPa, and n is 7.4 with the stress from 200 MPa and 350 MPa. High value of n has been observed for Ti-6Al-4V and IMI834 alloy [10]. The value of n equals from 5.2 to 11.3 for Ti-6Al-4V from 500°C to 600°C at the stress of 97-472MPa [10], while for IMI834 alloy from 600°C to 700°C at the stress of 150-500MPa, the value equals from 5.5 to 7.5 [11]. The present values of $n = 2.4$-7.4 at 600°C, are higher than that for α-Ti [12], which correspondence with the results for IMI834 and Ti-6Al-4V alloys.

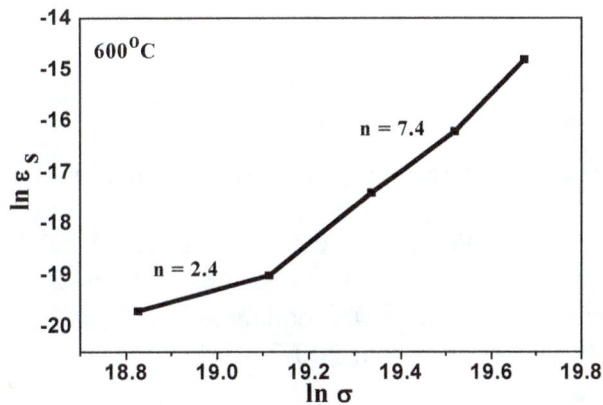

Fig. 2 Log-log plot of steady-state creep rate vs. applied stress

Reduced stress (σ-σ_0) and critical stress σ_0 were introduced to explain the relatively high stress exponent n existed in high temperature titanium alloy [10, 12]. On this basis, an attempt to rationalize these large n values, is considered that creep occurs under an effective stress (σ-σ_0) and the stress and temperature dependences of the creep rate can be written as [10, 12]:

$$\dot{\varepsilon}_s = A^* \left(\sigma - \sigma_0\right)^p \exp(-\frac{Q^*_{App}}{RT}) \tag{3}$$

where A^* is a constant, p is true stress exponent, σ_0 is critical stress, and Q_{App} are the creep activation energy for self or atom diffusion.

Assuming that the threshold stress σ_0 leads to the reason of high values of n [10], the magnitude of its value can be found for the alloy by plotting the experimental values of $\dot{\varepsilon}_s^{1/4.3}$ and σ on linear axes and extrapolating linearly to $\dot{\varepsilon}_s^{1/4.3} = 0$, as illustrated in Fig. 3. The values of σ_0 at 600°C is 76.1MPa.

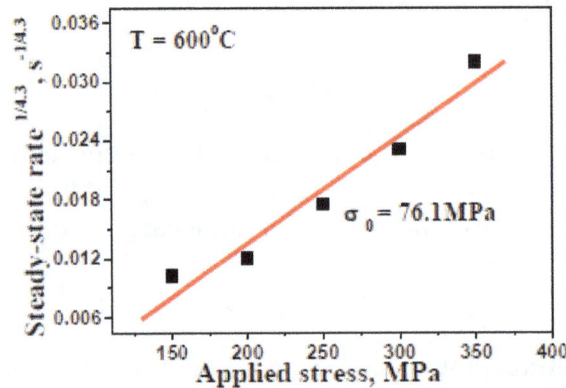

Fig. 3 The plot of $\dot{\varepsilon}_s^{1/4.3}$ against σ for Ti-600 alloy (the extrapolation of the linear regression to zero creep rate gives the threshold stress)

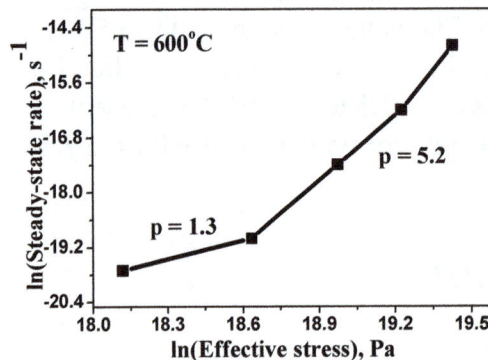

Fig. 4 The plot of steady-state creep rate against effective stress for Ti-600 alloy

The plot of steady-state creep rate against effective stress (σ-σ_0) for Ti-600 alloy at 600°C is shown in Fig. 4. By incorporating the threshold stress into analysis, the creep data are rationalized to true stress exponents of p for the alloy at 600°C equals to 1.3 at lower stress region of 150-200MPa, which equals to 5.2 at higher stress region of 200-350MPa.

3.3. Creep mechanism for Ti-600 alloy at 600°C

At present, the Dorn formula, one of the approved constitutive equations describing the relationship among the parameters of steady-state creep rate, the stress and temperature can be written as [11]:

$$\dot{\varepsilon} = A\frac{D_0 Gb}{kT}\left(\frac{b}{d}\right)^p\left(\frac{\sigma}{G}\right)^n \exp\left(\frac{-Q}{RT}\right) \tag{4}$$

where $\dot{\varepsilon}$ is steady-state creep rate, G is shear modulus, b is Burger's vector, K is Boltzmann constant, D_0 is diffusion constant, d is grain diameter, σ is applied stress, n is stress exponent, Q is diffusion activation energy, R is the gas constant, A and P is the constant related to alloy microstructures.

The grain size d of Ti-600 alloy has no marked difference in this research, which can be regarded as a constant parameter, at the same time, the shear modulus G can also be taken as a constant, so, equation (4) can be simplified as:

$$\dot{\varepsilon} = \frac{B}{T}\sigma^n \exp\left(\frac{-Q}{RT}\right) \tag{5}$$

where B is a constant, which is related to the microstructure of the alloy. Substitute the above experimental data of steady-state creep rate, stress and temperature into equation (5), parameter B, n, q can be calculated, then the steady-state creep rate constitute equation for the alloy at given temperature and stress can be got. Combined with the existing creep theory, creep mechanism at different stages for Ti-600 alloy can be analyzed. The numerical value and its changing tendency of n and p can both reflect the creep mechanism for the alloy at various temperatures and stresses [10, 13].

As mentioned above, the value of n is 2.4 for the alloy crept at 600°C with lower stress of 150-200MPa, and the value of p is approximately 1. Put the correlated experimental result into equation (5), the steady-state creep rate for the alloy at 600°C and at lower stresses can be written as:

$$\dot{\varepsilon} = \frac{3.65\times10^6}{T}\sigma \exp\left(\frac{-332700}{RT}\right) \tag{6}$$

From equation (6), it can be seen that Q is 332.7kJ/mol for the Ti-600 alloy, which can be correspondence with the activation energy of grain boundary diffusion for Ti alloys [10]. Two main diffusion creeps are Nabarro-Herring(N-H) creep and Coble creep for the alloy at high temperature low stress with the n of 1, both of which are caused by intra-crystalline diffusion or by grain boundary diffusion [11]. Match the actual creep activation energy and the diffusion activation energy for the alloy, and the reason for creating diffusion during creep process can be analyzed.

The creep activation energy for N-H creep and for Coble creep equals to the self diffusion activation energy and the grain boundary diffusion activation energy, respectively. So, it can be learned that equation (6) is the expression of mathematic model for Coble creep, and the creep for Ti-600 alloy is controlled by the process of grain boundary diffusion.

The value of n is 7.4 for Ti-600 alloy crept at 600°C with higher stress of 200-350MPa, and the value of p is 5.2. Similarly, put the correlated experimental result into equation (5), the steady-state creep rate for the alloy at 600°C and at higher stresses can be written as:

$$\dot{\varepsilon} = \frac{1.12\times10^{-14}}{T}\sigma^{5.2} \exp\left(\frac{-507500}{RT}\right) \tag{7}$$

From equation (7), it can be learned that the creep activation energy Q is 507.5kJ/mol for Ti-600 alloy, which can be correspondence with the activation energy of crystal lattice diffusion for Ti alloys [13]. The result is coincidence with Weetman creep, its creep mechanism belongs to the dislocations climb. At 600°C, the diffusion coefficient for titanium atoms increase, and the diffusion for lattices aggravate, the dislocations climb can be easily preceded. At higher stresses, dislocations climb occupy the principal position during the creep processing.

4. Conclusion

(1) The values of the threshold stresses σ_0 is 76.1MPa for Ti-600 alloy crept at 600°C with five stresses from 150MPa to 350MPa.

(2) The value of stress exponent n and the value of true stress exponent p is 2.4 and 1.1, respectively for Ti-600 alloy crept at 600°C with lower stress of 150-200MPa, the value of n and p is 7.4 and 4.1, respectively for the alloy crept at 600°C with higher stress of 200-350MPa.

(3) The constitutive equation for Ti-600 alloy crept at 600°C with lower stresses can be written as $\dot{\varepsilon} = \dfrac{3.65 \times 10^6}{T} \sigma \exp\left(\dfrac{-332700}{RT}\right)$, its creep mechanism is Coble one controlled by the process of grain boundary diffusion.

(4) The constitutive equation for Ti-600 alloy crept at 600°C with higher stresses can be written as $\dot{\varepsilon} = \dfrac{1.12 \times 10^{-14}}{T} \sigma^{5.2} \exp\left(\dfrac{-507500}{RT}\right)$, its creep mechanism is Weetman one controlled by the process of lattice diffusion, which belongs to the dislocations climb.

Acknowledgments

This work was funded by Shaanxi Province Key Science and Technology Innovation Team Program: "Titanium Alloy Research Innovation Team" (2012 KCT-23).

References

1. L.Y. Zeng, Y.Q. Zhao, Q. Hong, G.J. Yang, High cycle fatigue property of Ti-600 alloy at ambient temperature, J Alloys Comp. 509 (2011) 2081-2085.
2. L.Y. Zeng, Q. Hong, G.J. Yang, Y.Q. Zhao, Y.L. Qi, P. Guo, Tensile and creep properties of Ti-600 alloy, Trans Nonferrous Metals Soc China. 17 (2007) s522-s525.
3. R.W. Evans, B. Wilshire, Introduction to Creep, LD: Maney Publishing, London, 1993.
4. J.C. Williams, E.A.J. Starke, Progress in Structural Materials for aerospace systems, Acta Mater. 51 (2003) 5775-5781.
5. T. Matsunaga, T. Kameyama, K. Takahashi, Constitutive Relation for ambient-temperature creep in hexagonal close-packed metals, Mater Trans. 50 (2009) 2858-2864.
6. R.W. Hayes, P.L. Martin, Tension creep of wrought single phase γ TiAl, Acta Metall Mater. 43 (1995) 2761-2765.
7. H. Mishra, D.V.V. Satyanarayana, T.K. Nandy, P.K. Sagar, Effect of trace impurities on the creep behavior of a near α titanium alloy, Scripta Materialia. 59 (2008) 591-595.
8. D.A.P. Reis, C.M. Neto, C.R.M. Silva, M.J.R. Barboza, F.P. Neto, Effect of coating on the creep behavior of the Ti6Al4V alloy, Mater Sci Eng A. 486 (2008) 421-426.

9. G.B. Viswanathan, V.K. Vasudevan, M.J. Mills, Modification of the jogged-screw model for creep of γ-TiAl, Acta Mater. 47 (1999) 1399-1404.

10. M.J.R. Barboza, E.A.C. Perez, M.M. Medeiros, D.A.P. Reis, M.C.A. Nono, F.P. Neto, C.R.M. Silva, Creep behavior of Ti-6Al-4V and a comparison with titanium matrix composites, Mater Sci Eng A. 428 (2006) 319-326.

11. Owen D.M., T.G. Landon, Low stress creep behavior: an examination of Nabarro-Herring and Harper-Dorn creep, Mater Sci Eng A. 216 (1996) 20-26.

12. B. Wilshire, Observations, theories, and predictions of high-temperature creep behavior, Metall Mater Trans A. 33 (2002) 241-247.

13. A.K. Mukherjee, An examination of the constitutive equation for elevated temperature plasticity, Mater Sci Eng A. 322 (2002) 1-7.

Chapter 4
Biomaterials

Evaluation of Tanshinone IIA on Antitumor Effect and Mechanism of the Proliferation and Apotosis in Human Carcinoma Hep-2 Cells

Wen-Yi Fu[1], Jing-Hua Li[1], Run-Hong Mu[2], Ming-Cheng Li[3],*

[1]Associated Hospital, Beihua University, Jilin, Jilin, 132012, China
[2]School of Basic Medicine, Beihua University; Jilin, Jilin, 132013, China
[3]School of Laboratory Medicine, Beihua University; Jilin, Jilin, 132013, China
*Email: limingcheng1964@163.com

Tanshinone IIA (Tan IIA) is an ingredient extracted from salvia miltiorrhiza, a traditional Chinese medicine, and has been used in the therapy of cardiovascular diseases. The study was designed to investigate the antitumor effect of Tanshinone IIA (Tan IIA) at different concentrations on the proliferation and apoptosis of human laryngeal carcinoma Hep-2 cells. MTT assay revealed that Tan IIA significantly inhibited the growth of Hep-2 cells in a dose dependent manner. Flow cytometry showed Hep-2 cells became apoptotic, and the early apoptosis rate was 5.8 %, 7.9 %, 10.2 % and 20.4 %. Tan IIA may inhibit the proliferation and induce the apoptosis of Hep-2 cells in a concentration dependent manner.

Keywords: Tanshinone IIA; antitumor effect; human carcinoma Hep-2 cells.

1. Introduction

Tanshinone IIA (Tan IIA) is an ingredient extracted from salvia miltiorrhiza, a famous kind of Traditional Chinese Medicine (TCM), and has been used in the therapy of cardiovascular diseases [1]. Tan IIA may protect myocytes against oxidative stress and inflammation and has been extensively applied in the therapy of coronary heart disease and angina [2]. In Traditional Chinese Medicine, Tan IIA as a pharmacotherapeutic may also be used in the therapy of cancers [3]. Recent studies revealed that Tan IIA may reverse the malignant phenotype of cancers and compromises the migration and invasion of cancer cells. There was evidence showing that Tan IIA may inhibit the proliferation of cancer cells including breast cancer, lung cancer, osteosarcoma, liver cancer, leukemia and ovarian cancer, suggested its anti-tumor activity [4].

The World Health Organization reported that cancer was one of the major causes of death and the mortality of cancer patients in 2020 [5]. In recent year, the development of genomics and proteomics brought about the cancer therapy to a new era. In medicine, cancer is regarded as a hereditary disease. From this view, different methods are required to be introduced to the diagnosis and therapy of cancers. Generally, the anti-tumor effect of drugs is mainly ascribed to the induction of apoptosis of cancer cell. Thus, it is important to determine the influence of traditional drugs on the apoptosis in cancer cells.

Laryngeal carcinoma is a malignancy with increased expected frequency and closely related to alcohol and smoking. In developing countries, laryngeal carcinoma has been the subsequent common malignancy in men and significantly threatens the health and life of men. The mortality of laryngeal carcinoma and other life-threatening cancers in men is higher than 85 %. Thus, increasing attention has been paid to the therapy of cancers with TCM.

Epidemiological and clinical findings reveal that laryngeal carcinoma is closely related to the infection by high risk HPV (especially the HPV16 and HPV18) as the same as the cervical cancer [6]. Previous studies indicated that Tan IIA could inhibit the proliferation of cervical cancer HeLa cells infected by HPV18 and induce their apoptosis. However, the influence of Tan IIA on the HPV16 positive laryngeal carcinoma cells is still unclear and has never been reported. This study aimed to

*Corresponding author

investigate the effects of Tan IIA on the proliferation and apoptosis of HPV16 positive Hep-2 cells. Our findings may provide theoretic evidence for the therapy of laryngeal carcinoma with Tan IIA.

2. Materials and Methods

2.1. *Materials*

Tan IIA (purity with 99.2% HPLC) was acquired from Chinese Food and Drug Supervision and Management, and Dimethylsulfoxide (DMSO), L-glutamine as well as antibiotics from Sigma-Aldrich Co (St Louis, MO, USA). All reagents included in the study were of analytical grade. The Tan IIA was diluted with DMSO to 10.0 mg/L and preserved at -20°C for further use. The indicated concentrations were 0.5, 1.0, 2.0 and 5.0 mg/L respectively, and the control group was 0.9% saline solution.

2.2. *Cell culture and treatments*

Hep-2 cells lines purchased from Cell Culture Centre, Institute of Basic Medical Sciences, and Chinese Academy of Medical Sciences. Cells were cultured in DMEM containing 10 % fetal bovine serum (FBS) (Invitrogen, USA) at 37°C in a humidified environment with 5 % CO_2. The medium was refreshed once every 2-3 days and cells were passaged at 1:3 when the confluence of monolayer cells reached about 100%. Detection of cell proliferation is by MTT assay.

Hep-2 cells in logarithmic growth phase were harvested and re-suspended at a density of 1.0×107/L. Then, cells were added into 96-well plates (100μL/well). When cell adhesion was observed, the medium was removed and treated with Tan IIA at different concentrations for 24 h. Cells without Tan IIA treatment served as a blank control. There were 5 wells in each group. Cell growth was noted under an inverted microscope and photographed. Then, 20 μL of 5 mg/ml MTT (Funakoshi Co., Tokyo, Japan) was added to each well, followed by incubation in dark at 37°C for 4h. The supernatant was removed, and 150 μL of DMSO were added to each well. After incubation for 10 min under constant shaking, the absorbance was measured using an Infinite F50 microplate reader (Tacan, Mannedorf, Switzerland) at a wavelength of 570 nm.

2.3. *Detection of cell apoptosis by flow cytometry*

After Tan IIA treatment for 12 h, cells in different groups were harvested (1×10⁶/group) and then fixed with 70% cold ethanol over night at 4°C. After centrifugation at annexin A and propidium iodide (Invitrogen Life Techonology, USA), cells were incubated in dark for 30 min and then subjected to flow cytometry (Becton Dickinson, San Joe, CA, USA). Data was analyzed with WinMDI software.

2.4. *Statistical analysis*

Statistical analysis was performed with SPSS version 10.0, and data are expressed as mean ± standard deviation. When the F test showed homogeneity of variance, t test was used for comparisons between groups. A value of $P<0.05$ was considered statistical significance.

3. Results

3.1. *Effects of tan IIA at different concentrations on the morphology of Hep-2 cells*

Following treatment with Tan IIA at different concentrations, the cell morphology was observed under a microscope. There was a distinct difference in shape between Tan IIA-treated groups and the control group. In Tan IIA-treated groups with different concentrations at different time point, total volume of cells reduced (Figure 1).

Table 1 Effect of different concentrations of Tan IIA on cell viability of Hep-2 cells

Group (mg/L)	Cell proliferation (A value)	Cell growth inhibition rate(%)
0	0.89±0.12	0
0.5	0.62±0.13*	30.24±4.5*
1.0	0.46±0.07*	49.45±4.2*
2.0	0.33±0.06**	63.0±5.4**
5.0	0.26±0.03**	71.8±7.4**

Data are presented as mean±standard deviation. *$p<0.05$ compared with the control group.**$p<0.01$ compared with the control group.

Intercellular junction disappeared and detached each other. Besides, the cell growth was not in good conditions compared with the control group. The data illustrated that Tan IIA induced cell injury in a dose-and time-dependent manner (Table 1).

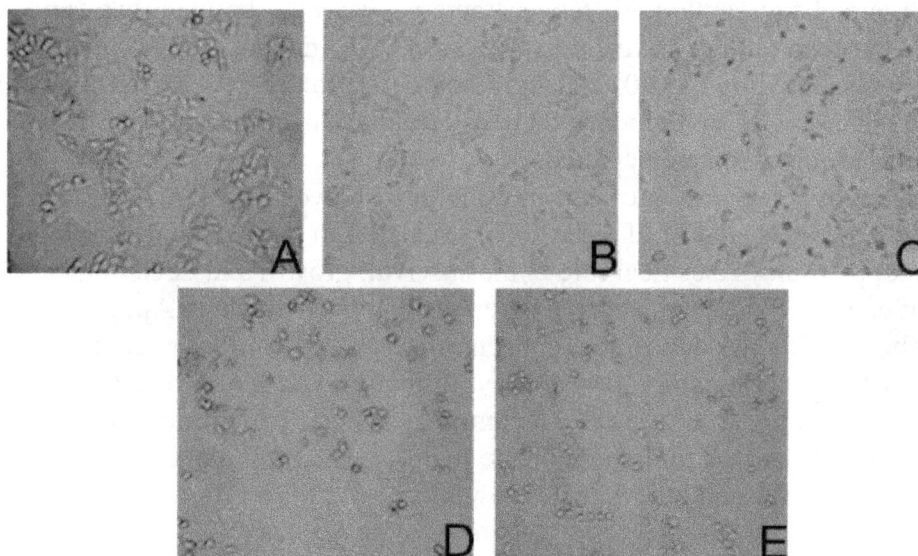

Fig. 1 Effect of different concentrations of Tan IIA on cell morphology of Hep-2 cells. A:control group; B:0.5mg/L; C:1.0 mg/L; D:2.0 mg/L; E:5.0 mg/L

3.2. *Effects of Tan IIA on the proliferation of Hep-2 cells*

After treatment with Tan IIA at different concentrations for 24h, MTT assay was performed, and absorbance was measured in different groups. Then, the number of viable cells (cell proliferation) in the control group was the highest. After Tan IIA treatment, the cell proliferation decreased significantly and the growth inhibit rate increased in a dose dependent manner. Significant difference was observed in the growth inhibition rate between Tan IIA group and the control group

(0.5-1.0mg/L Group: P<0.05; 2.0-5.0 mg/L Group: P<0.01). This suggests that Tan IIA may significantly inhibit the proliferation of cervical cancer cells in a dose dependent manner. According to the standard curve, the IC50 of Tan IIA was 1.2 mg/L in Hep-2 cells.

3.3. *Effects of Tan IIA at different concentrations on the early apoptosis of SiHa cells*

The results of flow cytometry analysis showed that the early apoptosis rate increased markedly compared with the control group after treatment with Tan IIA at different concentrations for 12h. There was a significant difference between Tan IIA-treated groups and the control group. In the control group, the early apoptosis rate of SiHa cells was 2.56±0.21 %, the early apoptosis rate in the Tan IIA-treated groups at different concentrations increased markedly (0.5 mg/L group, 5.8±0.32%; 1.0 mg/L group, 7.9±0.43%[P<0.05]; 2.0g/L group, 10.2±0.42%; 5.0 mg/L, 20.44±1.24% [P<0.01]).

4. Conclusion

Salvia miltiorrhiza is one of the most common Chinese herbs used in clinical practice and extracted from the root of *salvia*. *Salvia* has different subtypes, and their ingredients are also diverse [7]. Tanshinone (Tan) is a diethyl ether or ethanol extract of *salvia* root and the major effective ingredient of *salvia* miltiorrhiza [8]. Out of all extracts, Tan IIA has natural anti-oxidative activity. In clinical cardiovascular pharmacology, Tan IIA has anti-atherosclerotic activity and is able to reduce myocardial infarction, decrease myocardial oxygen consumption, and inhibit the thrombosis in addition to the platelet aggregation. Moreover, there is evidence showing that Tan is promising for the treatment of cancers. Studies have confirmed Tan IIA is in a position to induce the apoptosis of cervical cancer cells including HL-60 cells and K562 cells. Tan IIA possesses anti-tumor effects via directly killing cancer cells or inducing their differentiation and apoptosis. However, the exact mechanism is still required to be elucidated.

In the present study, observation of cell morphology showed Tan IIA significantly changed the morphology of Hep-2 cells, and MTT assay also revealed Tan IIA markedly inhibited the growth of Hep-2 cells. Apoptosis plays a critical role in the survival, growth and development of cells and the pathogenesis of cancers. Endogenous and exogenous pathways mediate Apoptosis.

Our results showed Tan IIA at different concentrations significantly increased the early apoptosis rate of Hep-2. Further investigation were needed to rectify Tan IIA reduced Bcl-2 expression and enhanced Bax expression, resulting in Bcl-2 family members, which induced caspase-3 activation and lead to cell apoptosis. Our findings indicate that Tan IIA may include the apoptosis of human cervical cancer Hep-2 cells and inhibit their growth.

Acknowledgments

The present project was funded by Science and Technology Development Program of Jilin Province (20090906); Science and Technology Development Program of Jilin City (2013523010); Emerging Strategic Industries and High-tech Industry Development Program of Jilin Province (2013G030).

References

1. Yang R, Liu A, Ma X, et al. Sodium tanshinone IIA sulfonate protects cardiomyocytes against oxidative stress-mediated apoptosisi through inhibiting JNK activation. J Cardiovasc Pharmacol. 51 (2008) 396-401.

2. Ren ZH, Tong YH, XU W, et al. Tanshinone IIA attenuates inflammatory responses of rats with myocardial infarction by reducing MCP-1 expression. Phytomedicine.17 (2009) 212-218.

3. Liu JJ, Zhang Y, Lin DJ, et al. Tanshinone IIA inhibits leukemia THP-1 cell growth by induction of apoptosis. Oncol Rep. 21 (2009) 1075-1081.

4. Lu Q, Zhang P, Zhang X, et al. Experimental study of the anti-cancer mechanism of Tanshinone IIA against human breast cancer. Int J Mol Med. 24 (2009) 773-780.

5. Chien SY, Kuo SJ, Chen YL and Su C. Tanshinone IIA inhibits human hepatocellular carcinoma J5 cell growth by increasing Bax and caspase 3 and decreasing CD31 expression in vivo. Mol Med Rep. 5 (2011) 282-286.

6. Su CC. Tanshinone IIA inhibits gastric carcinoma AGS cells through increasing p-p38, p-JNK and p53 but reducing p-ERK,CDC2 and Cyclin B1 expression. Anticancer Res. 34 (2014) 7097-7110.

7. Su CC. Tanshinone IIA inhibits human gastric carcinoma AGS cell growth by decresing BiP, TCTP, Mc11 and BclxL and increasing Bax and CHOP protein expression. Int J Mol Med. 34 (2014) 1661-1668.

8. Sung HJ, Choi SM, Yoon Y and An KS. Tanshinone IIA, an ingredient of Salvia miltiorrhiza BUNGE, induces apoptosis in human leukemia cell lines through the activation of caspase-3. Exp Mol Med. 1366 (1998) 151-165.

Alkaline Protease Immobilization in Alginate–Calcium Chloride Core-shell Microcapsules

Yuan Jiang[1], Ming-Di Zhang[1], Mei-Shuo Zhang[1], Song-Yi Lin[1, 2,*]

[1]Laboratory of Nutrition and Functional Food, Jilin University, Changchun, 130062, P. R. China
[2]School of Food Science and Technology, Dalian Polytechnic University, Engineering Research Center of Seafood of Ministry of Education, Dalian 116034, PR China
*Email: linsongyi730@163.com

Immobilization of enzyme into microcapsules has attracted many researchers' interest. In order to develop and optimize encapsulated rate of microcapsules mathematical model, the four variables including sodium alga acid concentration, the anhydrous calcium chloride concentration, placed time and the alkaline protease (AP) concentration were optimized by one-factor-at-a-time (OFAT). The results were as follows: sodium alga acid concentration 1.5 %, anhydrous calcium chloride concentration 3 %, placed time 1 h and alkaline protease concentration 0.75 %, placed time 62.23 min and alkaline protease concentration 0.78 %. Under this condition, the encapsulated rate of microcapsules achieves 95.63 %.

Keywords: Alkaline protease; microcapsules; encapsulated rate; response surface methodology.

1. Introduction

Proteases dominate the worldwide enzyme market, it plays an important part. Proteases perform its function by cleaving peptide bonds. The microbial proteases are mainly used in the formulations of various detergents, which contribute significantly to global enzyme sales [1]. As one of the most important enzymes, alkaline serine protease is excreted into the culture medium by strains of Bacillus licheniformis or B. pumilus [2]. These alkaline proteases have various industrial applications including detergents [3], foods, pharmaceuticals and diagnostic reagents [4]. Proteases are also used to the disposal of shellfish wastes as a waste treatment alternative [5].

The purpose of this work was to produce microcapsules of alkaline protease (AP) by Encapsulator B-395 Pro and to optimize encapsulated rate of microcapsules mathematical model by response surface methodology (RSM).

2. Materials and Methods

2.1. *Materials and instruments*

Casein, trichloroacetic acid (TCA) and all the other materials required in the experiments were purchased from Beijing Chemical Plant. Standard sample of L-tyrosine was purchased from TCI Shanghai Co., Ltd. AP was purchased from China Pangbo Biological Engineering Co., Ltd. All chemicals and reagents were analytical grade.

Encapsulator B-395 Pro was purchased from BUCHI, used for producing microcapsule. The main parts of the Encapsulator 8-395 Pro are the control unit, with the syringe pump, the electrical and pneumatic systems, and the reaction vessel.

2.2. *Microcapsules preparation*

At room temperature, sodium alga acid and AP were mixed in a 100 mL beaker with a mechanical stirrer. The magnetic stirrer (IKA, Germany) was stirred at 1500 r/min for 1 h. The 30 mL mixture was drawn into a 50 mL syringe. Washing solution was made by anhydrous calcium chloride. The

*Corresponding author

Encapsulator B-395 Pro was set as follow: stirrer 40 %, pump 5.01 mL/min, electrode 1550 V, frequency 1900 Hz and nozzle 0.20 mm. At the conclusion of the production run, the mixture was set at room temperature for a certain time. The microcapsules and liquid were separated through a sterile filter. The filtrate was collected for determination of enzyme activity. The microcapsules were separated from the outer water phase, rinsed three times with distilled water and freeze-dried (Freeze Drier, Christ ALPHA 1-2 LD plus, Germany) for 24 h (as a batch).

2.3. Determination of encapsulated rate of AP in microcapsules

The encapsulated rate of microcapsules was determined as Eq. 1.

$$Encapsulated\ rate\ (\%) = \frac{Addition\ of\ enzyme\ activity - Remain\ of\ enzyme\ activity}{Addition\ of\ enzyme\ activity} \times 100\% \quad \text{Eq. 1}$$

Protease activity was carried out by the method described by Kembhavi et al . The substrate was casein. Casein substrate was dissolved in 0.1 M sodium carbonate buffer to get one percent of solution. The reaction mixture was made up of 1 mL 100 µg/mL casein and 1mL of diluted enzyme. Then the reaction mixture was kept for 20 min at 40 °C and terminated by 2 mL of 5 % TCA. The mixture was filtrated by 0.22 µm filter membrane and measured at 680 nm. Three parallel blanks were run with each sample. In one unit of protease activity was defined as the amount of enzyme required to liberate 1 µg of tyrosine per minute under the used experimental conditions.

2.4. Effect on encapsulated rate of microcapsules through one-factor-at-a-time experiment

In our one-factor-at-a-time (OFAT) study, four independent variables were studied, including the sodium alga acid concentration (0.7 %, 1.1 %, 1.5 %, 1.9 %, 2.3 %), the anhydrous calcium chloride concentration (1.0 %, 2.0 %, 3.0 %, 4.0 %, 5.0 %), placed time (0.5, 1.0, 1.5, 2, 2.5 h) and the AP concentration (4.5 %, 3 %, 1.5 %, 0.75 %, 0.375 %). Each text changed only one variable value and other variables were fixed.

2.5. Experimental design of response surface methodology

Based on OFAT experiment, four independent variables at three levels (43) were adopted for the RSM. These variables were the sodium alga acid concentration (%), AP concentration (%), the anhydrous calcium chloride concentration (%) and placed time (h), which were labeled as X_1, X_2, X_3 and X_4 as showed in Table 1. Y, the dependent variable, was the encapsulated rate of microcapsules. The RSM was used to optimize experimental parameters including sodium alga acid concentration (1.3 %, 1.5 %, and 1.7 %), anhydrous calcium chloride concentration (2.5 %, 3 %, and 3.5 %), placed time (40 min, 60 min, and 80 min) and AP concentration (1 %, 0.75 %, and 0.6 %).

3. Results and Discussion

3.1. Effects of independent variables on encapsulated rate of microcapsules through OFAT experiment

The results were profiled in Fig. 1. The effect of sodium alga acid concentration on encapsulated rate of microcapsules was showed in Fig. 1 (A). When the sodium alga acid concentration reached a level of 1.5 %, the decrease of encapsulated rate was significantly (*P<0.01*) increased to the maximum of 95.40 % ± 0.44 %. It's effective to improve encapsulated rate of microcapsules by increasing sodium alga acid concentration suitably. The encapsulated rate significantly ($P<0.01$) with the anhydrous calcium chloride concentration increasing from 1 % to 3 % as shown in Fig. 1 (B). At the 3 % of anhydrous calcium chloride, the encapsulated rate of microcapsules reached maximum of 95.73 % ± 0.53 %. The effect of placed time on encapsulated rate of microcapsules was showed in Fig. 1 (C). At the 1 h of placed time, the encapsulated rate of microcapsules reached maximum of 95.60 % ± 0.43 %. The encapsulated rate significantly ($P<0.01$) reduced when placed time increased from 1 h to 2.5 h. Fig. 1 (D) showed the effect of different concentration of AP on encapsulated rate of microcapsules. The encapsulated rate didn't change significantly ($P>0.05$) with the AP concentration from 0.375 % to 0.75 %. However, the encapsulated rate significantly ($P<0.01$) reduced when AP concentration increased from 0.75 % to 4.50 %. Therefore, 1.5 % sodium alga acid concentration, 0.75 % AP concentration, 3 % anhydrous calcium chloride concentration and 1 h placed time was chosen as center point .

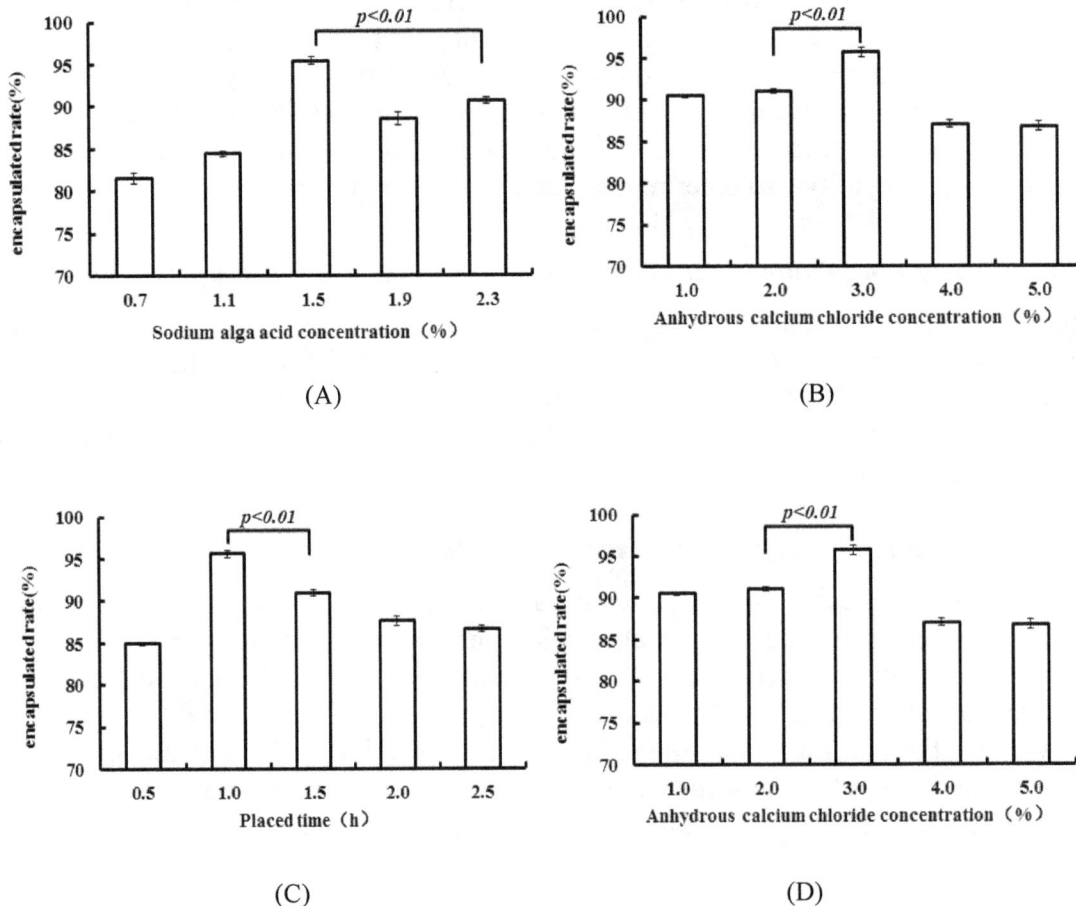

(A)   (B)

(C)   (D)

Fig. 1 Effects of variables on encapsulated rate of microcapsules through OFAT experiment

3.2. Box-Behnken design and response surface methods

The experimental dates were shown in Table 1.

Table 1 Box-Behnken design matrix and response values

Experiment number	X_1	X_2	X_3	X_4	Experimental Data (%)	Predicted Data (%)
1	-1	1	0	0	91.22±0.09	89.98
2	0	0	1	-1	86.97±0.11	85.54
3	1	-1	0	0	84.31±0.23	85.34
4	0	0	0	0	96.32±0.19	95.33
5	-1	0	-1	0	89.76±0.15	89.95
6	1	0	1	0	85.74±0.21	85.89
7	1	0	0	1	84.96±0.18	84.98
8	0	0	0	0	95.06±0.23	95.33
9	1	0	0	-1	86.31±0.27	86.54
10	0	0	1	1	87.46±0.19	86.10
11	0	0	-1	1	83.56±0.29	84.77
12	0	0	0	0	95.06±0.18	95.33
13	0	-1	0	1	84.07±0.15	84.28
14	0	0	0	0	95.06±0.15	95.33
15	-1	-1	0	0	87.45±0.27	87.87
16	1	0	-1	0	86.32±0.20	85.53
17	-1	0	1	0	85.39±0.20	86.53
18	0	1	1	0	85.23±0.19	86.78
19	0	1	-1	0	89.87±0.26	89.79
20	0	-1	0	-1	88.09±0.24	88.15
21	0	-1	-1	0	87.86±0.25	86.19
22	1	1	0	0	88.1±0.22	87.46
23	-1	0	0	1	87.13±0.19	86.77
24	-1	0	0	-1	89.95±0.28	89.80
25	0	0	0	0	95.13±0.14	95.33
26	0	1	0	-1	88.56±0.18	88.69
27	0	1	0	1	87.69±0.26	87.97
28	0	0	-1	-1	88.78±0.25	89.93
29	0	-1	1	0	86.19±0.22	86.14

The regression and coefficients of the model were shown in Table 2.

Table 2 Design matrix evaluation for the response surface quadratic model

Source	Coefficient estimate	Standard Error	Sum of Squares	Mean Square	F-Value	p-value Prob > F
Model			353.38	25.24	18.99	< 0.0001
Intercept	95.33	0.52				
X_1	-1.26	0.33	19.15	19.15	14.41	0.0020
X_2	1.06	0.33	13.44	13.44	10.11	0.0067
X_3	-0.76	0.33	7.01	7.01	5.27	0.0376
X_4	-1.15	0.33	15.85	15.85	11.92	0.0039
X_1X_2	<0.001	0.58	<0.001	<0.001	<0.001	0.9932
X_1X_3	0.95	0.58	3.59	3.59	2.70	0.1225
X_1X_4	0.37	0.58	0.54	0.54	0.41	0.5341
X_2X_3	-0.74	0.58	2.21	2.21	1.66	0.2186
X_2X_4	0.79	0.58	2.48	2.48	1.87	0.1935
X_3X_4	1.43	0.58	8.15	8.15	6.13	0.0267
X_1^2	-3.96	0.45	101.55	101.55	76.40	< 0.0001

(Continued)

Table 2 (*Continued*)

Source	Coefficient estimate	Standard Error	Sum of Squares	Mean Square	F-Value	p-value Prob > F
X_2^2	-3.71	0.45	89.12	89.12	67.05	< 0.0001
X_3^2	-4.40	0.45	125.32	125.32	94.28	< 0.0001
X_4^2	-4.35	0.45	122.49	122.49	92.14	< 0.0001
Residual			18.61	1.33		
Lack of Fit			17.37	1.74	5.61	0.0555
Pure Error			1.24	0.31		
Cor Total			371.99			

By applying multiple regression analyses on the experimental data, the final statistical model could be described by the following quadratic: $Y = + 95.32600 - 1.26333 X_1 + 1.05833 X_2 - 0.76417 X_3 - 1.14917 X_4 + 0.005 X_1X_2 + 0.94750 X_1X_3 + 0.36750 X_1X_4 - 0.74250 X_2X_3 + 0.78750 X_2X_4 + 1.42750 X_3X_4 - 3.95675 X_1^2 - 3.70675 X_2^2 - 4.39550 X_3^2 - 4.34550 X_4^2$. Where, Y is the encapsulated rate (%); X_1 is the sodium alga acid (%); X_2 is the anhydrous calcium chloride (%); X_3 is the placed time (min) and X_4 is the AP (%). The predicted values based on the second-order polynomial equation predictive data were calculated, as shown in Table 1.

The regression model was tested by ANOVA, as shown in Table 2. The lack of fit value of the model was 0.06 and the P-value of the model was significant ($P<0.01$), which indicated that the mathematical model appropriately fits. The coefficient R^2 was calculated to be 0.7258, which testified the regression model was fit . There was a high degree correlation between the experimental data and the predicted values.

The ANOVA analysis showed that the regression model well ($P<0.05$) modeled the experimental results. Therefore, the effect of the independent variables on the protease activity could be well analyzed and predicted the model of the encapsulated rate of microcapsules. As shown in the Fig. 2, 3-D response surface and 2-D contour plots were demonstrated interactions between the experimental variables and the relationship between responses and experimental variable. The contour plots presented as an ellipse, which means the interaction effect was significant. The predicted values for the optimal conditions of encapsulated rate of microcapsules: sodium alga acid concentration 1.47 %, anhydrous calcium chloride concentration 2.93 %, placed time 62.23 min and AP concentration 0.78 %. Under this condition, the encapsulated rate of microcapsules achieves 95.63 %.

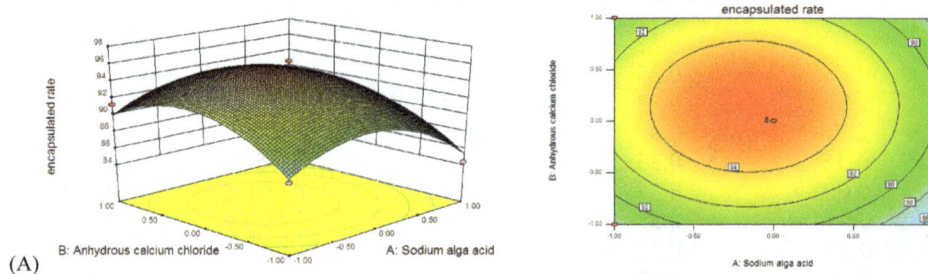

(A)

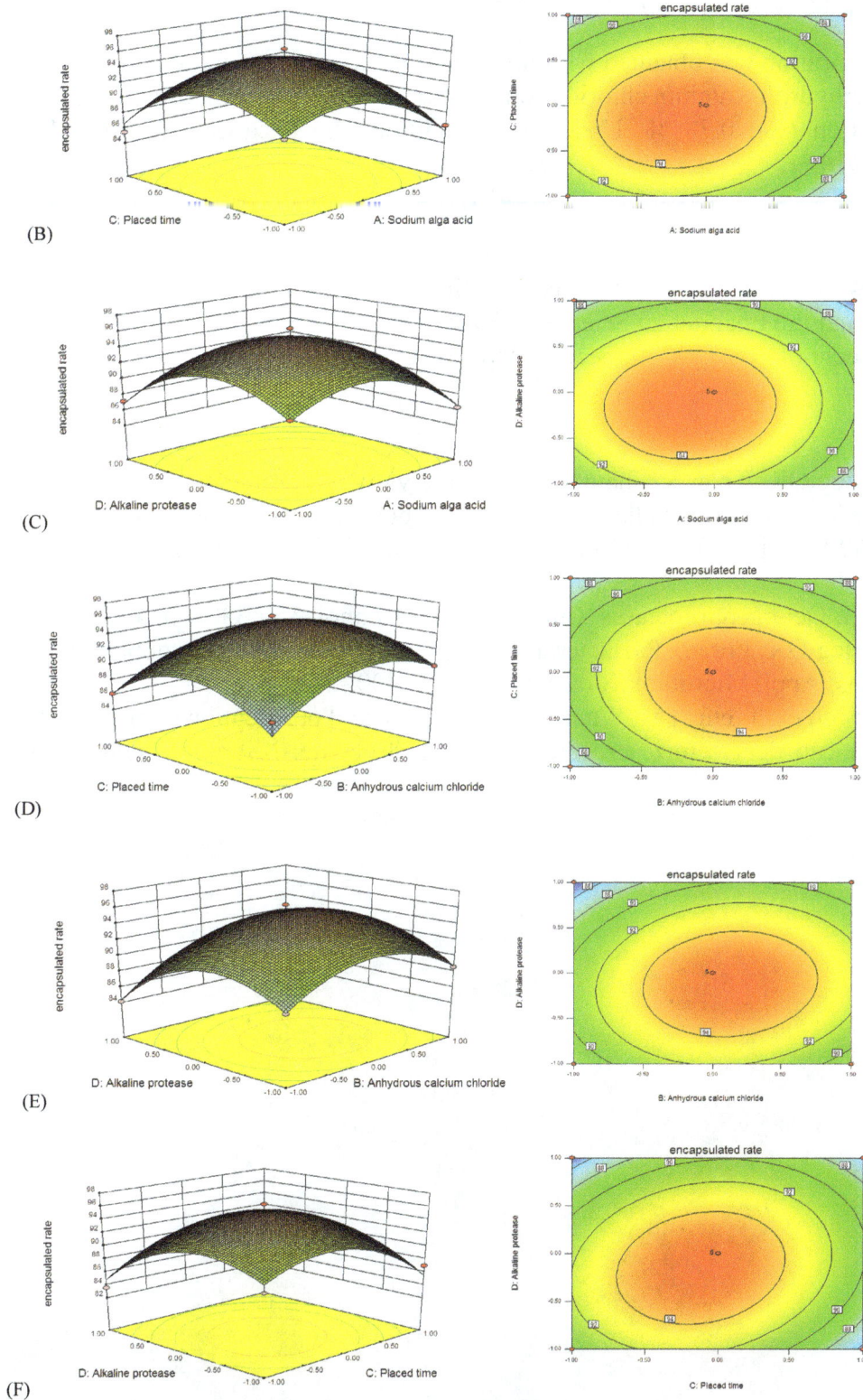

(B)

(C)

(D)

(E)

(F)

Fig. 2 Response surface plots of encapsulated rate of microcapsules

The protease activity in samples is shown as a functional interaction between either sodium alga acid concentration (A) or anhydrous calcium chloride concentration (B) or placed time (C) or alkaline protease concentration (D).

4. Conclusion

The optimal condition was obtained with the following conditions: sodium alga acid concentration (X_1) 1.47 %, anhydrous calcium chloride concentration (X_2) 2.93 %, placed time (X_3) 62.23 min and AP concentration (X_4) 0.78 %. Under this condition, the encapsulated rate of microcapsules achieves 95.6314 %.

Acknowledgments

This work was funded by the Key Projects of Jilin province Science & Technology Program (2015 0204032NY) and the Changchun Science and Technology Planning Project of Modern Agricultural Science and Technology Support Special(14NK005).

References

1. Gupta R, Beg QK, Khan S, Chauhan B. An overview on fermentation, downstream processing and properties of microbial alkaline proteases. Appl. Microbiol. Biot. 60 : 381 – 395 (2002).
2. Fogarty WM, Kelly CT. Microbial Enzymes and Biotechnology. Elsevier Applied Science, 1990 .
3. Maurer KH. Detergent proteases. Curr. Opin. Biotechnol. 15 : 330 – 334 (2004).
4. Gupta R, Beg QK, Lorenz P. Bacterial alkaline proteases: molecular approaches and industrial applications. Appl. Microbiol. Biot. 59 : 15 – 32 (2002).
5. Yang JK, Shih IL, Tzeng YM, Wang SL. Production and purification of protease from a Bacillus subtilisthat can deproteinize crustacean wastes. Enzyme. Microb. Technol. 26 : 406 – 413 (2000).

Preparation, Stability, and Cytotoxicity of Sub-micron Calcium Oxalate Monohydrate

Jian-Min Wang, Mu-Hua Wan, Jian-Ming Ouyang*

Institute of Biomineralization and Lithiasis Research, Jinan University, Guangzhou 510632, China
Email: toyjm@jnu.edu.cn

Urinary stone is one of the most common urologic diseases over the world, but its formation mechanism is not yet clearly understood. In this work, calcium oxalate monohydrate (COM) sub-micron crystal with a size of about 200 nm was prepared by complex precipitation method using NTA as complexing agent. The sample was characterized by X-ray diffraction, Fourier transform infrared spectroscopy, and Zetasizer Nano-ZS analyzer. The prepared COM crystals were nearly elliptical. The effects of initial reactant concentrations, pH in reaction system, and dispersion media etc. on morphology and size of COM crystals were investigated. Ethanol medium favored the dispersion of COM crystals. We compared the adhesion of COM to normal and injured human renal tubule epithelial cells, which indicated that injured cells can adhere to much more COM crystals. It confirmed that cell injure is the main reason of urolithiasis.

Keywords: Calcium oxalate monohydrate; sub-micron crystals; crystals adhesion; cytotoxicity.

1. Introduction

Kidney stones have been investigated for a long time. However, its formation mechanism has not yet been completely clarified [1]. Calcium oxalate monohydrate (COM) crystal is the primary inorganic constituent of kidney stones. Although calcium oxalate is usually supersaturated in renal tubule, not all the calcium oxalate crystals can form kidney stones. Usually, the calcium oxalate crystals can be carried away by the liquid in renal tubule. Because it is difficult to study the formation of kidney stone in situ, some models, such as cell models [2], are used to explore its pathological mechanism.

It is reported that high levels of oxalate and COM crystal could cause cell damage and apoptosis, inducing HK-2 cytotoxicity with concentration dependent effect [3]. Scheid et al. [4] reported that with increased oxalate concentration, free radicals in kidney cells increased, leading to further damage to LLC-PK1 cells.

In this study, sub-micron crystals COM crystals were prepared and acted with human renal tubule epithelial cells (HKC), we hope further to reveal the formation mechanism of kidney stones at molecular level and cellular level.

2. Materials and Methods

2.1. *Materials and apparatus*

All conventional reagents used were analytically pure and purchased from Guangzhou Chemical Co. (China). Human renal tubule epithelial cells (HKC) were purchased from Shanghai Cell Bank, Chinese Academy of Sciences (Shanghai, China).

Apparatus: X-L type environmental scanning electron microscope (SEM, Philips, Eindhoven, Netherlands); laser confocal microscope (LSM510 Meta Duo Scan, Zeiss, Jena, Germany); a

---

*Corresponding author

D/max-γA X-ray diffractometer (Rigaku, Japan) using Ni-filtered Cu Kα radiation (λ=1.54 Å).); FT-IR spectrometer Bruker IFS25 (Bruker Spectrospin, Wissembourg, France).

2.2. *Preparation of sub-micron calcium oxalate monohydrate (COM) crystals*

Sub-micron COM crystal was prepared by complexing precipitation method using nitrilotriacetic acid (NTA) as complex from the reaction of 0.1 mol/L $CaCl_2$ solution and 0.1 mol/L $K_2C_2O_4$ solution, at 75°C, pH 8 with fast agitation. The average dimension of COM crystals was about 200±50 nm.

COM suspension was prepared using DMEM/F12 of serum-free cell culture medium with a final concentration of 100 μg/ml.

2.3. *Adhesion of COM crystals on normal and injured renal tubule epithelial cells and zeta potential measurement*

With a cell density on average of 2×10^5 cells/mL and 1 mL/well, cells were transferred to 6-well plates and incubated with DMEM/F12 culture media containing 10% fetal calf serum for 24 h. The cells were incubated 12 h with DMEM/F12 of serum-free cell culture media to make cells for synchronizing. Then the cells were divided into two groups: ① Control group: only adding serum-free culture medium, ② H_2O_2 –induced injured cells group: serum-free culture medium containing 0.5 mmol/L H_2O_2 was added. The cells were added 1 ml COM suspension. After the adhesion of COM on cell was carried out for 6 h and 12 h, respectively, the serum-free culture medium containing of COM crystal was aspirated and the cells were rinsed twice with D-Hanks. After the supernatant solution was centrifugalized out, the left specimen sediment was redispersed in HEPES buffer (pH=7.4), and the Zeta potential of the cells was measured by Zetasizer Nano ZS.

3. Results and Discussion

3.1. *Preparation and characterization of sub-micron COM crystals*

The morphology and microstructure of the COM crystals were investigated by SEM (Fig. 1a). It could be seen that the diameters of these crystals ranged from 150 to 250 nm.

XRD pattern (Fig. 1b) revealed the phase composition of the crystals. The corresponding diffraction peaks were located at 0.594, 0.365, 0.297, and 0.235 nm, which assigned to the ($\bar{1}$01), (020), ($\bar{2}$02), and (130) planes of COM crystals [5].

Fig. 1c showed the FT-IR spectrum of the prepared sample. The main antisymmetric carbonyl stretching band ($v_{as}(COO-)$) and the secondary symmetric carbonyl stretching band ($v_s(COO-)$) of COD crystals were located at about 1618 and 1318 cm^{-1}, respectively. The five peaks at 3060~3489 cm-1 were the characteristic absorption of crystal water molecules of COM. The analytical results of Fig. 1 confirmed that the obtained sample was COM crystals.

The initial concentrations of reactants can change the size of COM. Fig. 2 showed the size distribution of COM crystals in different initial concentrations of reactants.

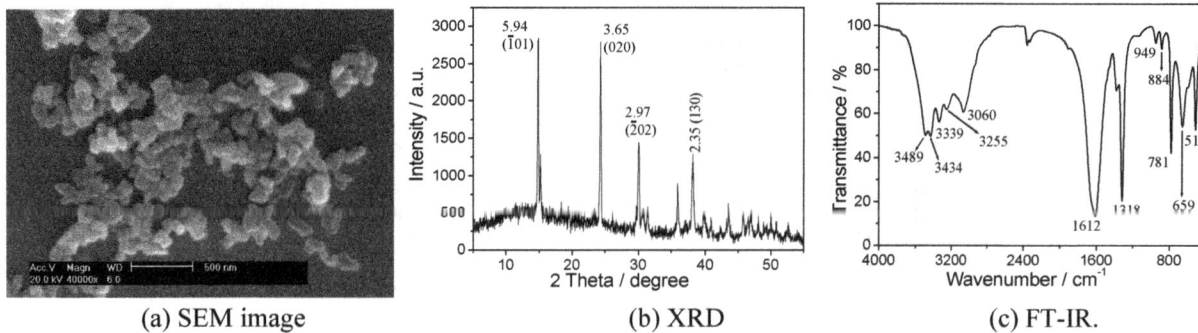

(a) SEM image   (b) XRD   (c) FT-IR.

Fig. 1 Characterization of sub-micron COM crystals

(The bar: 500 nm. Preparation conditions: c(reactant) = 0.1 mol/L; pH = 4)

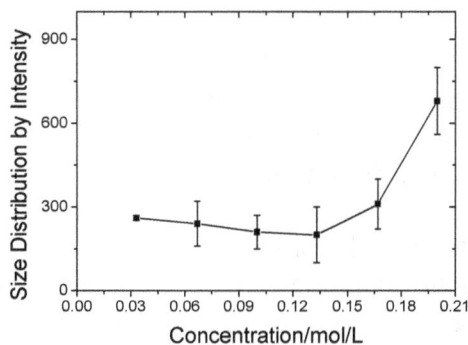

Fig. 2 The size distribution of COM crystals in different initial concentrations of reactants (pH=4, T=35°C)

3.2. *Factors affecting the stability of COM crystals*

Sub-micron COM crystals can easily aggregate into larger particles because of their high surface activity and high surface energy [6, 7]. Therefore, how to make the COM crystals dispersed homogeneously and stably was a key problem.

1) Effect of placement time

Fig. 3A showed the change of size distributions of COM crystals dispersed in ethanol. With the placement period increasing from 10 min to 30, 60, 120, and 240 min, respectively, the size distribution profiles were much similar. It indicated the sub-micron COM crystals dispersed in ethanol were stable.

2) Effect of concentration of COM suspension

The COM suspension showed a relatively high stability in the concentration ranged from 1.28 mg/mL to 5.12 mg/mL (equivalent to 10-40 mmol/L). As shown in Fig. 3B, the mean particle size was about 260 nm at a concentration of 1.28 mg/mL of COM suspension. As the concentration increased to 2.56, 3.84 and 5.12 mg/mL, the mean particle size increased to 256, 223 and 268 nm, respectively. The result further confirmed that the COM crystals had high stability.

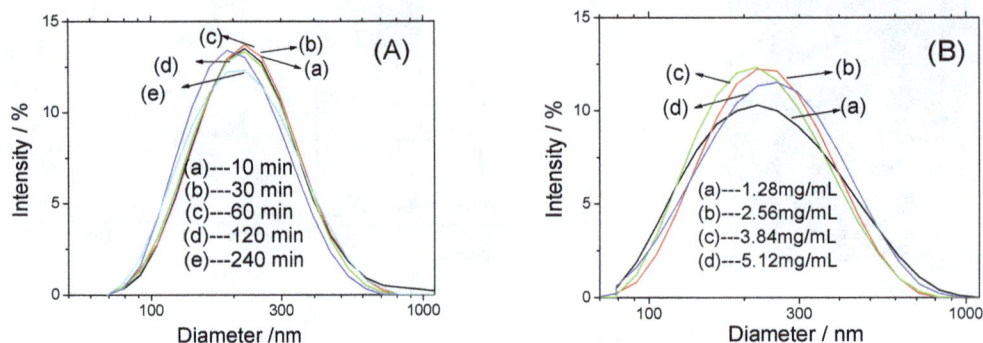

Fig. 3 Effect of standing time (A) and concentration of COM suspension (B) on size of COM crystals: Dispersion medium: absolute ethanol

3) Effect of dispersion media

The size distribution of sub-micron COM crystals in different dispersion media such as ethanol, water and ethanol-water mixture was investigated. The volume fraction of ethanol has a strong influence on the size distribution of COM crystals (Fig. 4). The observed mean size of COM remarkably increased from 210 nm to 382, 1210 and 1520 nm, respectively when the volume fraction of ethanol decreased from 1.0 to 0.8, 0.5 and 0. That is, the average size was the smallest in absolute ethanol and the largest in water. This result was attributed to the difference of many parameters of the two solvents such as the wetting ability of the media (solvation), refractive index (n20), viscosity (η20), surface tension (γ20), density, zeta potential and so on. Generally speaking, the COM particles were easily dispersed as small particles in solvent with a small viscosity due to its high shear force. The higher the refractive index of solvent has, the worse its wetting property on particles was. Then it led to a poor dispersion of particles. Although the viscosity and refractive index of water were smaller than that of ethanol, the surface tension of water was much high than that of ethanol, the dispersed particles reunited immediately. Ethanol has an appropriate viscosity and a low surface tension simultaneously, so a stable dispersion system was easily formed in ethanol. In addition, as the volume fraction of ethanol increased, the zeta potential on the surface of COM increased (Table 1). Therefore, the electrostatic repulsive among the particles increased, and thus the aggregation of particles was prevented. As a result, the particles show smaller size.

Table 1 Effect of different dispersed media on size of COM crystals

Medium	ξ (mV)	η_{20} (mPa·S)	$\gamma20$(mv·m^{-1})	n20	size of COM (nm)
Absolute ethanol	-24.67	1.17	22.27	1.3614	210±70
Ethanol solution (80%)	-19.77	-	-	-	382±110
Ethanol solution (50%)	-17.4	-	-	-	1210±340
Deionized water	-4.96	1.0019	72.58	1.3330	1520±360

[Notes] ξ: zeta potential; η_{20}: viscosity; γ_{20}: surface tension; n_{20}: refractive index.

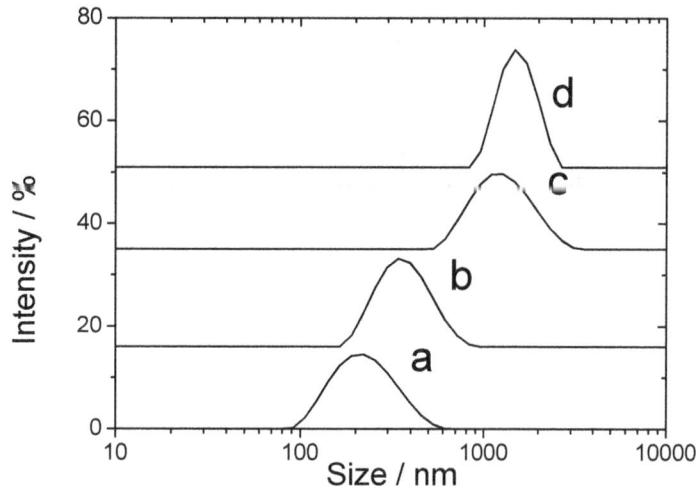

Fig. 4 Effect of dispersion media on size of COM crystals.: (a) absolute ethanol; (b) the continuous phase contains 80% (v/v) ethanol; (c) the continuous phase contains 50% (v/v) ethanol; (d) the pure water

3.3. *Adhesion of COM Sub-micron Crystals to Human Renal Tubule Epithelial Cells (HKC)*

COM was the main composition of kidney stones. The formation of COM stones included four processes: nucleation, growth, aggregation, and crystal adhesion on renal tubule epithelial cells. There were many reports about research of the previous three processes [8]. However, a few reports were seen on the adhesion of the urinary crystals to renal tubule epithelial cells [9].

The change of Zeta potential (ξ) on cell surface can evaluate the risk of urinary stone formation. Table 2 showed the ξ of the normal and injured HKC cells after adhesion with sub-micron COM crystals for different time. It could be seen that the ξ on the surface of injured cells (-46.7 mV) was smaller than that of normal cells (-24.8 mV), because the membranes of injured HKC cells were depolarized, leading the ξ of membranes decreased. The higher the injured degree of HKC cells, the smaller the Zeta potential was. As adhesion time increased from 0 h to 6 h and 12 h, the ξ of normal cells decreased significantly from -24.8 mV to -34.9 mV and -43.9 mV, respectively. However, the ξ of injured cells increased significantly from -46.7 mV to -39.8 mV and -31.9 mV, respectively. The results indicated that the adhesion or interaction between COM crystals with injured HKC was stronger than that with the normal HKC. Verkoelen et al. [10] reported that the renal tubular epithelium cells can protect themselves from crystal binding by at least two different defense mechanisms: 1) the composition of the apical membranes of polarized renal tubular cells was unfavorable for crystal attachment; 2) negatively-charged molecules in tubular fluid prevented crystal retention by covering the potential binding sites on crystal surface. However, the membranes of injured cells were depolarized, and the ξ of membranes decreased. Since a small absolute value of zeta potential will decrease the electrostatic repulsion, the adhesion of COM sub-micron crystals to injured cells was stronger.

Table 2 Zeta potential of normal and injured HKC cells after adhesion with sub-micron COM crystals for different Time (mV)

Adhesion time/h	0	6	12
Normal cell	-24.8±0.4	-34.9±0.8	-43.9±0.5
Injured cell	-46.7±0.8	-39.8±0.7	-31.9±0.9

4. Conclusion

Sub-micron COM crystals were successfully prepared. SEM showed that the COM crystals have an average size of about 200 nm. XRD and FTIR analysis indicated that the product was COM. The prepared COD sub-micron crystals had good dispersion and stability. The adhesion of sub-micron COM to injured HKC cells was much stronger than that of the normal HKC, indicating cells injure is one of the important reasons for the formation of calcium oxalate stones.

Acknowledgment

This research work was funded by the Natural Science Foundation of China (NO. 21371077).

References

1. Farmanesh S, Chung J, Sosa R D, Kwak J H, Karande P, Rimer J D, Natural promoters of calcium oxalate monohydrate crystallization. J. Am. Chem. 2014, 136(36): 12648-12657.
2. McMulkin C J, Massi M, Jones F. Tetrazoles: calcium oxalate crystal growth modifiers. CrystEngComm, 2015, 17(13): 2675-2681.
3. Chen S, Gao X, Sun Y, Xu C, Wang L, Zhou T. Analysis of HK-2 cells exposed to oxalate and calcium oxalate crystals: proteomic insights into the molecular mechanisms of renal injury and stone formation. Urol Res., 2010, 38: 7–15.
4. Thamilselvan S, Khan S R, Menon M. Oxalate and calcium oxalate mediated free radical toxicity in renal epithelial cells: effect of antioxidants. Urol Res., 2003, 31: 3-9.
5. Gan Q-Z, Sun X-Y, Ouyang J-M. Adhesion and internalization differences of COM nanocrystals on Vero cells before and after cell damage. Mater. Sci. Eng.: C, 2016, 59(1): 286-295.
6. Wei Q, Hou J, Bonnell D A. Effect of interface atomic structure on the electronic properties of nano-sized metal-oxide interfaces. Nano Lett., 2015, 15(1): 211-217.
7. Brandon M. J, Joseph A. F, Donald T. G, et al. Acute exposure to ZnO nanoparticles induces autophagic immune cell death. Nanotoxicology, 2015, 9(6): 737-48.
8. N. K. Saw, P. N. Rao, J. P. Kavanagh. A nidus, crystalluria and aggregation: Key ingredients for stone enlargement. Urol. Res., 2008, 36: 11-15.
9. Sun X-Y, Ouyang J-M, Zhu W-Y, Li Y-B, Gan Q-Z. Size-dependent toxicity and interactions of calcium oxalate dihydrate crystal on Vero renal epithelial cells. J. Mater. Chem. B, 2015, 3(9): 1864-1878.
10. C. F. Verkoelen, B. G. Van der Boom, A. B. Houtsmuller, F. H. Schroder, J. C. Romijn. Increased calcium oxalate monohydrate crystal binding to injured renal tubular epithelial cells in culture. Am. J. Physiol.-renal., 1998, 274: 958-965.

Concentration-dependent Cellular Injuries Induced by Calcium Oxalate Monohydrate and Dihydrate Crystals

Li-San Huang, Jian-Ming Ouyang[*]

Department of Chemistry, Jinan University, Guangzhou 510632, China
[*]Email: toyjm@jnu.edu.cn

This study aims to compare the cytotoxicity and adhesion of calcium oxalate monohydrate (COM) and dehydrate (COD) crystals with a size of 5 μm toward human kidney proximal tubular epithelial (HKC) cells so as to reveal the mechanism of kidney stone formation at cellular level. The measurement of cell viability and Lactate dehydrogenase (LDH) content were used to quantitatively analyze cell injury induced by COM and COD crystals; cell mortality was measured by propidium iodide (PI) staining; the adhesion of crystals on cell surface was observed by SEM. The decrease of cell viability and increase of LDH release of HKC cells caused by COM and COD were concentration-dependent in crystal concentration range of 100~1600 μg/mL. COM caused more serious injury in HKC than COD. The adhesion amount of COM was significantly greater than COD crystal. The damage of micron COM was larger than COD, and COM was more easily aggregated on HKC. The results in this paper indicated that the presence of COM crystals in urine was more likely to increase the risk of stone formation than COD crystals.

Keywords: Calcium oxalate; concentration effect; cell injury; crystal adhesion.

1. Introduction

Urine usually contains supersaturated calcium and oxalate ions that will nucleate to form calcium oxalate (CaOx) crystals [1]. CaOx is the main component of kidney stone [2], and it contains two components: calcium oxalate monohydrate (COM) and dihydrate (COD) [3]. Renal epithelial cell adhesion of COM has been widely studied, but less research has been conducted for COD crystals [4].

Cell damage was thought to be a necessary and critical in crystal adhesion process [5]. Calcium phosphate crystal was also kidney stone composition, about 20% kidney stones contained calcium phosphate crystal. It was reported that calcium phosphate showed concentration dependent relationship in cell adhesion. For most cells, the adhesion amount of COM crystal on cells was 10 times of calcium phosphate crystals when the same amount of crystals was added. COM crystal adhesion on cells also presented concentration dependent relationship for MDCK cells [6].

2. Materials and Methods

2.1. *Materials and apparatus*

(1) Materials: HKC, a line of renal proximal tubular epithelial cells of human origin, was obtained from the Shanghai Changzheng Hospital. Cell proliferation assay kit (CCK-8) and Lactate dehydrogenase (LDH) kit were purchased from Dojindo Laboratories (Kumamoto, Japan). All conventional reagents were analytically pure and purchased from Guangzhou Chemical Reagent Factory of China (Guangzhou, China).

(2) Apparatus: X-L type environmental scanning electron microscope (SEM, Phillps, Eindhoven, Netherlands), enzyme mark instrument (Safire2™, Tecan, Männedorf, Switzerland), flow cytometry (FACS Aria, BD Corporation, CA, USA).

---

[*]Corresponding author

2.2. Experimental methods

2.2.1. Preparation and characterization of calcium oxalate monohydrate (COM) and dehydrate (COD) crystals

COM and COD crystals were prepared by changing the concentration of reactants (CaCl2 and Na2Ox), reaction temperature, solvent and stirring speed. The size and crystalline phase of prepared crystals were characterized by SEM and XRD.

2.2.2. Cell test

HKC cells were cultured in DMEM culture medium containing 10% fetal bovine serum in a 5% CO2 humidified atmosphere at 37 °C. Trypsin digestion method was adopted for cell propagation. Upon reaching 80%–90% confluence, the cells were rinsed twice with PBS. A certain amount of 0.25% trypsin digestion solution was then added and maintained for 3–5 min at 37°C. Afterward, DMEM containing 10% fetal bovine serum was added to terminate the digestion. The cells were then blown well to form cell suspension.

Cell viability assay, lactate dehydrogenase (LDH) release assay, and cell death assay were carried out [2, 5].

2.2.3. SEM observation of HKC treated by COD and COM crystals

After crystals exposed to cells for 6 h, the supernatant was removed by suction, washed three times with PBS, fixed in PBS 2.5% glutaraldehyde at 4 °C for 24 h, fixed with 1% OsO4, then dehydrated in gradient ethanol (30%, 50%, 70%, 90% and 100%, respectively), eventually it was fixed with isoamyl acetate, dried under the critical point of CO2, and treated with gold sputtering, and finally observed under SEM.

3. Results

3.1. Synthesis and characterization of calcium oxalate crystals

The SEM images of COM, COD crystal and XRD spectra were shown in Fig. 1. COM crystal is a typical hexagonal (Fig. 1a), COD crystal is tetragonal bipyramid shape (Fig. 1b). Both of their average particle diameters are about 5 μm, so we define them as COM-5μm and COD-5μm.

In the XRD spectra, the diffraction peaks of COM crystals were detected at d=5.93, 3.65, 2.97, 2.49, 2.36, 2.26, and 2.08 Å (Fig. 1c), which belonged to the ($\bar{1}$01), (020), ($\bar{2}$02), (112), (130), (202), and (321) crystal faces of COM respectively. The diffraction peaks at d=6.18, 4.42, 2.78 and 2.24 Å were belonged to the (200), (211), (411) and (213) crystal faces of COD respectively (Fig. 1d). The above results indicated that the synthesized crystals are pure COM and COD crystals.

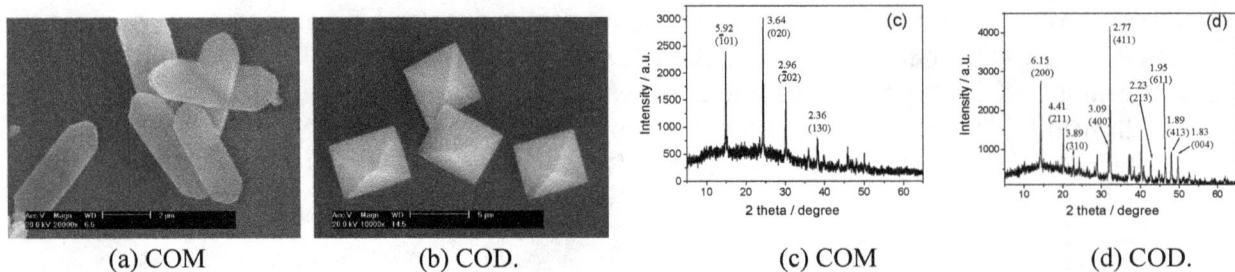

(a) COM   (b) COD.   (c) COM   (d) COD.

Fig. 1 SEM images and XRD patterns of COM and COD crystals

3.2. Cell viability and lactate dehydrogenase (LDH) release assay

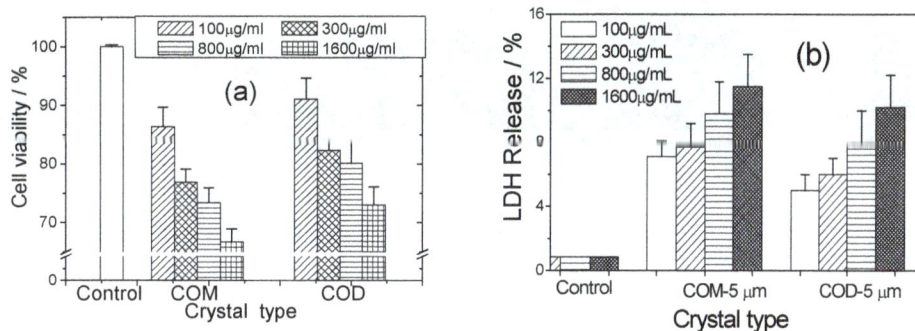

Fig. 2 Change in cell viability of HKC cells viability after exposure to different concentrations of COM and COD crystals for 6 h (a) and LDH release (b)

Fig. 2 showed that changes of cell viability and LDH release of HKC cells after interacted with different concentrations (100, 300, 800, 1600 g/mL) of COM, COD crystals. It can be seen that the changes in cell viability and LDH release induced by COM, COD crystals were concentration-dependent, that is, the greater the crystal concentration, the greater the cells were damaged. At the same crystal concentration, COM caused larger cell damage than COD.

LDH is a stable cytoplasmic enzyme which exists in all cells and releases extracellularly once the cell membrane ruptures. Therefore, the extent of the cells damage can de determined by detecting the LDH release in the supernatant of cell culture.

3.3. Propidium iodide (PI) staining after COM and COD adhered on HKC cells

PI is a commonly used fluorescent staining agent for the cell nucleus. PI cannot penetrate through normal cell membranes but can pass through terminal apoptotic cell and necrotic cell membranes and bind to DNA in the nucleus, thereby emitting red fluorescence. The red fluorescence intensity emited by PI binding with DNA was 20-30 times of the uncombined PI.

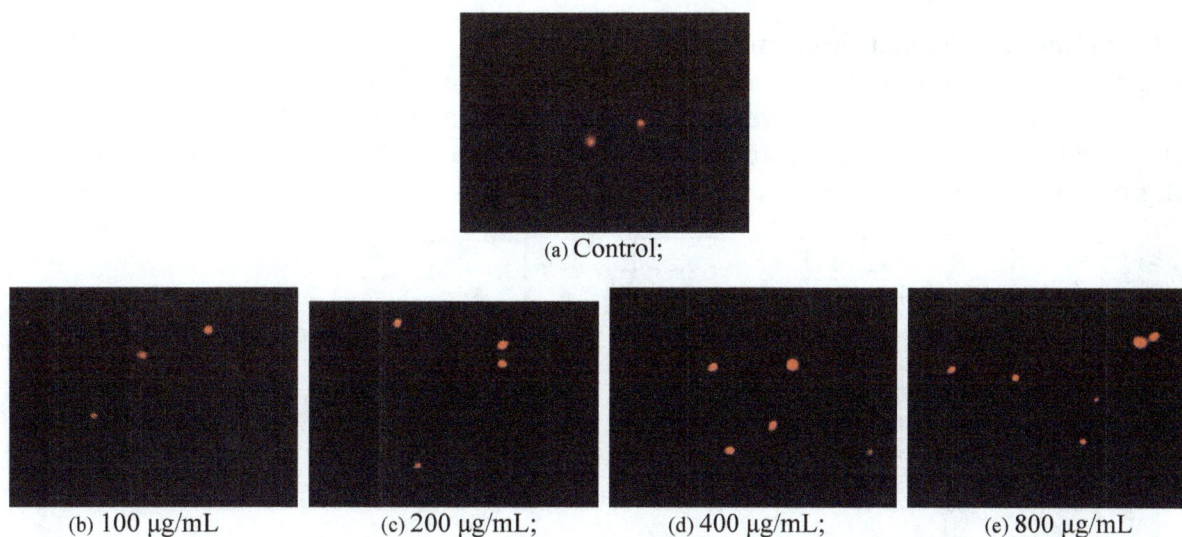

(a) Control;

(b) 100 µg/mL   (c) 200 µg/mL;   (d) 400 µg/mL;   (e) 800 µg/mL

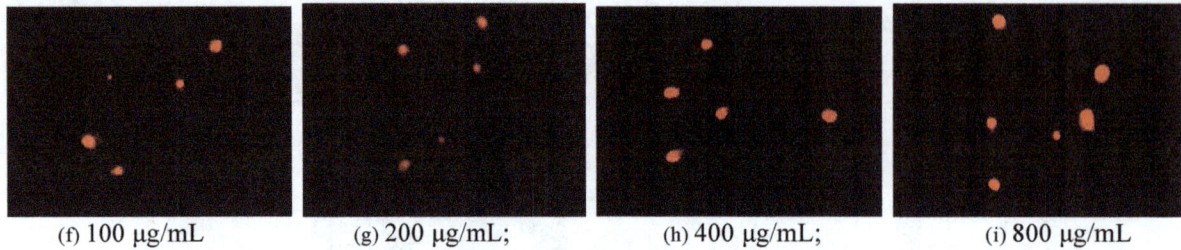

(f) 100 μg/mL　　　(g) 200 μg/mL;　　　(h) 400 μg/mL;　　　(i) 800 μg/mL

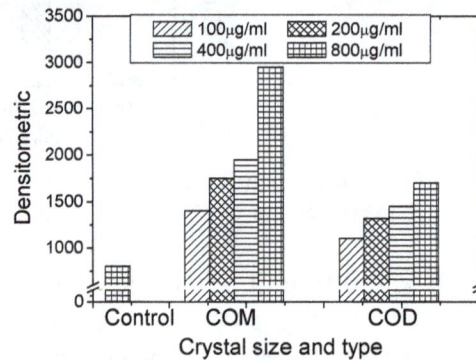

(j) Relative fluorescence intensity.

Fig. 3 Fluorescence microscope (×600) and relative fluorescence intensity of PI staining results after HKC adhered with different concentrations of COM and COD for 6 h

The results of PI staining after HKC adhered with different concentrations of COM and COD for 6 h were as shown in Fig. 3. With the crystal concentration increases, the number of PI-stained nuclei increased, and the relative fluorescence intensity gradually increased (Fig. 4B), indicating that the HKC mortality rates increased constantly. COM has larger lethal rate to HKC than COD crystal. The result is consistent with the previous detection results (Figs. 2 and 3).

3.4. *Cell death detection*

In order to analyze the cell death mechanism induced by CaOx crystals, annexin V/PI double staining was used to quantitatively analyze the cell apoptosis and necrosis by flow cytometry. Annexin V staining was used to reveal phosphatidylserine exposure (cell apoptosis) and PI was used to reveal the cell membrane integrity (necrosis). As shown in Fig. 4, cell mortality exposed in COM group (Q1+Q2+Q4) was far greater than COD group.

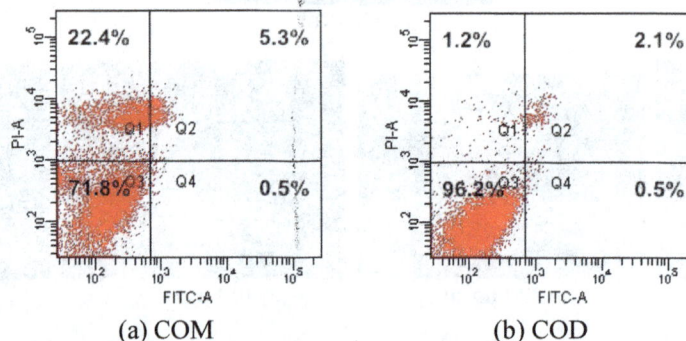

(a) COM　　　　　　　　(b) COD

Fig. 4 Flow cytometric data of cell apoptosis and death after HKC were exposed to 200 μg/mL of crystals for 6 h

The adhesion of COM crystals on HKC will induced reactive oxygen species (ROS) generation, while ROS can mediate a series of inflammation, activate many signal molecules such as p38 mitogen activated protein kinase (p38MAPK), protein N-terminal kinase (JNKp) and transcription factor NF-κB. These substances caused expression of genes and proteins and upregulation such as the monocyte chemotaxis protein-1 (MCP-1) and injured molecule KIM-1 [7]. Therefore, a series of cascade reactions resulted in cell apoptosis, cell function impaired greatly and finally cell necrosis.

3.5. *SEM detection of adhesion between HKC and COM, COD crystals*

The SEM images of HKC cells after exposure to different concentration of COM and COD are presented in Fig. 5.

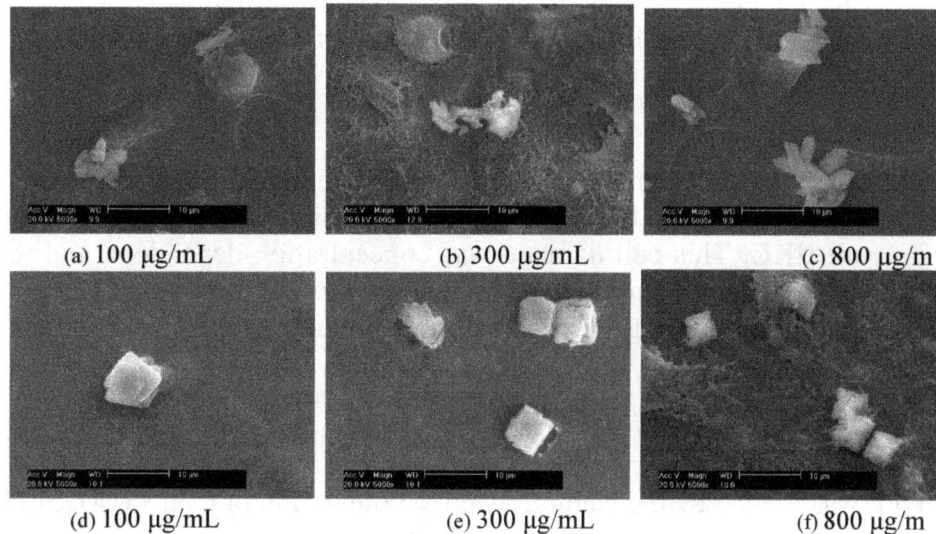

(a) 100 µg/mL   (b) 300 µg/mL   (c) 800 µg/m

(d) 100 µg/mL   (e) 300 µg/mL   (f) 800 µg/m

Fig. 5 SEM images of HKC adhered with different concentrations of COM and COD for 6 h. Bars: 10 µm

Compared to COD group, the COM crystals adhered on cell surface appeared obvious aggregation. Crystal aggregation will further damage the cell and increase the risk of stone formation. This suggests that the formation of a large number of COM crystals in the urine in a short time will cause great damage to kidney epithelial cells and thus promote the formation of stones, but the equivalent amount of COD has significantly less damage to cells than COM.

4. Discussion

From Fig. 2, we can see that COM caused greater LDH release and lower cell viability compared to COD crystal under the same concentration. Fig. 4 showed that COM induced larger HKC cells death rate. The micron-grade crystals adhered on cell surface will directly act on subcellular structure, cause cell membrane permeability change, and even rupture of the cell membrane, which induce cell necrosis. Moreover, crystals adhered on cell surface can activate cyclophilin D, cause mitochondrial collapse and induce oxidative stress [8]. NADPH oxidase is a major source of ROS generated in renal epithelial cells, oxalate will increase NADPH oxidase activity in renal epithelial cells, induce oxidative damage of cells, damage the cells integrity. The ion dynamic balance through cell membrane will be destroyed, cell membrane permeability on calcium and oxalate ions will increase, and the formed CaOx crystal can adhere to the cell membrane surface, and further damage the cell [9].

In normal cells, only neutral phospholipids are in the extracellular membrane while the negatively-charged phosphatidylserine (PS) is inside the cell membrane [10]. When renal epithelial cells are damaged, cell membrane structure will change, PS will migrate outward the membrane, CD44 and hyaluronan are also expressed on cell surface. These changes will provide effective sites for the combination with the positively-charged Ca^{2+} ions, and also promote the nucleation and growth of CaOx crystals, thus accelerate the formation of kidney stone. Wiessner et al [11] showed that PS exposed on the cell membrane surface will cause the cells to lose their polarity and destroy tight junctions between cells, which are conducive to crystal adhesion on cells.

Literature [12] showed that after oxalate and COM crystal exposed to renal epithelial cells of patients with oxalate urine, FKBP4 will increase. High concentration oxalate or calcium oxalate crystal can cause FKBP4 great expression accompanied by inhibition of TRPV5 activity and reduction of Ca2+ reabsorption, and finally lead to the formation of hypercalciuria and stones.

5. Conclusion

Micron COD and COM crystals can decrease HKC activity, increase LDH release, and increase the cell number of PI-stained nuclei, these results indicated that two kinds of crystals have toxicity and damage effect on HKC. This cell damage was concentration-dependent, and the damage of COM was significantly greater than COD. The results showed that a large number of COM crystals formed in the urine will cause great damage to kidney epithelial cells, thus promote the formation of CaOx stones.

Acknowledgment

This research work was funded by the Natural Science Foundation of China (No. 21371077).

References

1. Peng H, Ouyang J-M, Yao X-Q, Yang R-E. Interaction between submicron COD crystals and renal epithelial cells. Int. J. Nanomed., 2012(7): 4727-4737.
2. Ouyang J-M, Yao X-Q, Tan J, Wang F-X. Renal epithelial cell injury and its promoting role in formation of calcium oxalate monohydrate. J Biol Inorg Chem., 2011, 16: 405–416.
3. Martin X, Smith L H, Werness P G. Calcium oxalate dihydrate formation in urine. Kidney Int., 1984, 25: 948-952.
4. Khan A, Byer K, Khan S R. Exposure of Madin-Darby canine kidney (MDCK) cells to oxalate and calcium oxalate crystals activates nicotinamide adenine dinucleotide phosphate (NADPH)-oxidase. Urology, 2014, 83(2): 510.e1-e7.
5. Gan Q-Z, Sun X-Y, Ouyang J-M. Adhesion and internalization differences of COM nanocrystals on Vero cells before and after cell damage. Mater. Sci. Eng., C. 2016, 59: 286–295.
6. Lieske J C, Deganello S F, Toback G. Cell-Crystal Interactions and Kidney Stone Formation. Nephron, 1999, 81(1): 8-17.
7. Harrell P C, McCawley L J, Fingleton B, McIntyre J O, Matrisian L M. Proliferative effects of apical, but not basal, matrix metalloproteinase-7 activity in polarized MDCK cells. Exp. Cell Res., 2005, 303: 308-320.

8. Wiessner J H, Hasegawa A T, Hung L Y, Mandel N S. Oxalate-induced exposure of phosphatidylserine on the surface of renal epithelial cells in culture. JASN., 1999, 10: S441-5.

9. Sun X-Y, Ouyang J-M, Li Y-B, and Wen X-L. Mechanism of cytotoxicity of micron/nano calcium oxalate monohydrate and dihydrate crystals on renal epithelial cells. RSC Adv., 2015, 5: 45393–45406.

10. Thamilselvan V, Menon M, Thamilselvan S. Selective Rac1 inhibition protects renal tubular epithelial cells from oxalate-induced NADPH oxidase-mediated oxidative cell injury. Urol Res. 2012, 40: 415-423.

11. Zhang S, Su Z-X, Yao X-Q, Peng H, Deng S-P, Ouyang J-M. Mediation of calcium oxalate crystal growth on human kidney epithelial cells with different degrees of injury. Mater. Sci. Eng. C, 2012, 32: 840-847.

12. Wiessner J H, Hasegaw A T, Hung L Y, Mandel G S, and Mandel N S. Mechanisms of calcium oxalate crystal attachment to injured renal collecting duct cells. Kidney Int., 2001, 59(2): 637–644.

A Simple Kinetic Model of Alkali Metal Release during Aquatic Biomass Pyrolysis

Hua Fei[1], Jin-Ming Shi[2,*], Yang Liu[1], Jian-Hong Deng[1]

[1]Laboratory of Architectural Environment and Energy Application Engineering, Jiangxi University of Science and Technology, Ganhzhou 341000, Jiangxi Province, China;
[2]Institute of Energy, Jiangxi academy of Sciences, Nanchang 330096, China
*Email: Shijinming0012@163.com

In this work, the precipitation characteristics of alkali metal during aquatic biomass pyrolysis were studied. The results show that aquatic biomass species are important factors for K and Na releasing ratio during pyrolysis, which is related with the combined forms of K and Na in aquatic biomass and sample structure. Na precipitation rate is higher than K, which is mainly due to interfacial bond energy of Na lower than K. According to the K and Na releasing behavior of aquatic biomass during pyrolysis, the dynamical model was constructed, and the K and Na releasing characteristics predicted by this model are more satisfying in depicting the experimental data, indicating dynamical model can be applied to predict the K and Na releasing characteristics of aquatic biomass during pyrolysis.

Keywords: Aquatic biomass; pyrolysis; alkali metal; kinetic model.

1. Introduction

As a low-nitrogen, low-sulfur and carbon-neutral renewable clean energy, biomass has attracted worldwide interests, becoming research focus in the renewable energy field [1-3]. However, biomass contains significant higher amounts of alkali metal (K, Na) than other petroleum, coal et al. The low level of technology or the inappropriate process of biomass utilization not only leads to contamination, corrosion of system equipment, but also causes the metal contaminants to enter the food chain which is harmful to human health [4-6].

Pyrolysis is the first and fundamental step in combustion, gasification and other thermochemical conversion processes for biomass exploitation [7, 8], and it has been demonstrated to be an appropriate and efficient technological route for converting biomass to bio-oils, especially fast pyrolysis which prefers for biomass conversion to bio-fuels is particularly interesting [9, 10]. Thus, there are extensive investigation motivations for achieving in-depth understanding of biomass during pyrolysis for optimization and technology development.

However, numerous comprehensive studies had focused on behavior of alkali metal during biomass gasification, and only a few investigations on behavior of alkali metal during biomass pyrolysis could be found [11, 12]. Some investigations found the alkali metal were prone to be considerable parts of alkali metal would release in the form of gaseous during reaction [13-15]. Thus a well understanding of alkali metal migration during thermal conversion process is highly significant for resourceful utilization of biomass.

In this work, the precipitation characteristics of alkali metal were studied, and dynamical model was constructed according to the K and Na releasing behavior of aquatic biomass during pyrolysis. Using the model, the K and Na releasing characteristics of the reed and eichhornia crassipes pyrolysis processes were analyzed.

**Corresponding author

2. Experimental

Reed and eichhornia crassipes were used in the present study as the representatives of biomass. The samples were first crushed and sieved. Fractions in the size range of <0.8mm was used in the experiments. The small particle size used in this study ensured that temperature gradients within the samples were minimized. Elemental analysis and proximate analysis were carried out in vario EL-2 and TGA2000 (Navas Instruments, Spain), respectively. Biomass properties are shown in Table 1.

Table 1 Proximate analysis and ultimate analysis of aquatic biomass

Sample	Ultimate Analysis (wt%)				Proximate Analysis (wt%)				
	M_{ad}	V_{ad}	A_{ad}	FC_{ad}	C_{ad}	H_{ad}	N_{ad}	S_{ad}	O_{ad}
Reed	1.85	75.75	7.37	15.03	38.60	5.16	1.74	0.09	54.41
Eichhornia crassipes	1.66	75.14	18.95	4.25	45.06	5.46	0.91	0.09	48.48

Char samples were performed by using the tube furnace, which constituted a quartz tubular reactor with a length of 1600 mm and internal diameter of 45 mm and a Fourier transform infrared spectrometer. The tube furnace was first heated from room temperature to 900°C and pumped in the gas of N_2 ($\geq$99.999%) at a flow rate of 800 mL/min to provide an inert atmosphere for pyrolysis, and then moved biomass samples into heating zone rapidly. Finally, the samples were moved out of heating zone after predetermined time and cooled in N_2 atmosphere.

Alkali metal amount of samples was quantified by using inductively coupled plasma mass spectrometry (ICP-MS). Briefly, the sample was digested by 10 ml HNO_3 (68 wt%), 2 ml H_2O_2 (30 wt%) and 1 ml HF (38 wt%) mixture agents.

3. Results and Discussion

3.1. *Mathematical modeling*

Poyang Lake Basin aquatic plants are rich in K, Na and other elements, which is prone to volatilize and lead to serious problems such as deposition, agglomeration, slagging and heated side corrosion in the thermal utilization process [5, 13]. According to the precipitation characteristics of alkali metal during biomass pyrolysis, the dynamic model of alkali metal precipitation will be established. Due to a variety of forms such as alkali metal with inorganic and organic matter occurrence in aquatic biomass, it assumes that the alkali metal precipitation is one step reaction during biomass pyrolysis, and the alkali metal release in the form of NaOH and KOH in order to exclude the interference of chloride, it is expressed as

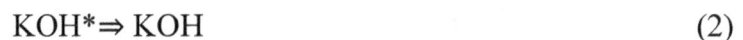

$$NaOH* \Rightarrow NaOH \tag{1}$$

$$KOH* \Rightarrow KOH \tag{2}$$

In which, KOH*, NaOH* is gas release source of alkali metal compounds respectively. Due to the content of alkali metals in biomass is a little, the assumption error of the gas release source can be negligible.

It assumes that the decomposition rate of the volatile matter is equal to the rate of precipitation during biomass pyrolysis, and reaction is carried out on the surface of the particles, the relationship between the rate of chemical reaction and the concentration can be given by

$$\frac{dX}{dt} = A_0 (1-X)^{\frac{2}{3}} \tag{3}$$

In which A_0 is reaction rare constant.

3.2. Precipitation characteristics of alkali metals

The rate of pyrolysis conversion is calculated by the following equation:

$$v_{t,p} = \frac{m_{0,p} - m_{t,p}}{m_{0,p}} \times 100\% \, c \tag{4}$$

In which $m_{t,p}$ is the sample weight at pyrolysis t time, $m_{0,p}$ is the sample weight at pyrolysis beginning. Release ratio of alkali metals is calculated by the following equation:

$$\alpha_{t,p} = (m_c - m_e)/m_c \times 100\% \tag{5}$$

Where m_e and m_c represent alkali metals amount in pyrolysed char and original sample, respectively.

The pyrolysis experiments of aquatic biomass which deriving from Poyang Lake were performed by using the tube furnace. Fig. 1 shows the precipitation characteristics of alkali metal during reed pyrolysis. As can be seen Fig. 1, the highest releasing percentage of reed during pyrolysis is more than 63% for K and Na, and Na precipitation rate were higher than K, this phenomenon is mainly due to interfacial bond energy of Na lower than K, which is consistent with Womat et al [16] studying results. Aquatic biomass species are important factors for K and Na releasing ratio during pyrolysis, which is related with the combined forms of K and Na in aquatic biomass and sample structure. Fig. 2 shows the changes characteristics of alkali metal precipitation during eichhornia crassipes pyrolysis, the highest releasing percentage of eichhornia crassipes during pyrolysis for K and Na reaches 70% and 80% respectively. Precipitation rate of Na and K for eichhornia crassipes were higher than that of reed, respectively. This phenomenon is mainly caused by the combined forms of K and Na in aquatic biomass. As can be seen Figs. 1 and 2, the release rate of Na is faster than K because the some compounds formed by Na are volatile at pyrolysis process, but C in chars is easy to react with K to form relatively less volatile compounds.

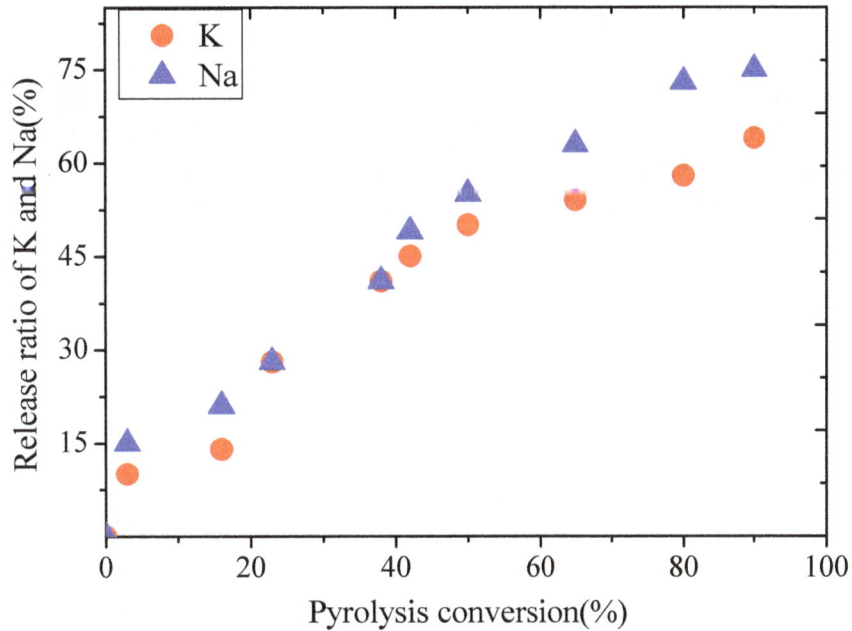

Fig. 1 Alkali metal releasing ratio of reed during pyrolysis

3.3. *Prediction of the models*

According to the K and Na releasing behavior of aquatic biomass during pyrolysis, the dynamical model was constructed, and the K and Na releasing characteristics of the reed and eichhornia crassipes during pyrolysis were analyzed by the model.

A set of experimental data are used to evaluate the model proposed in this work. A comparison can be made between the experiment data and alkali metal releasing ratio calculated by this model as shown in Fig. 3. As can be seen from Fig. 3(a), the K releasing characteristics predicted by this model have larger deviation from the experimental data at the alkali metal releasing ratio beyond 0.6.

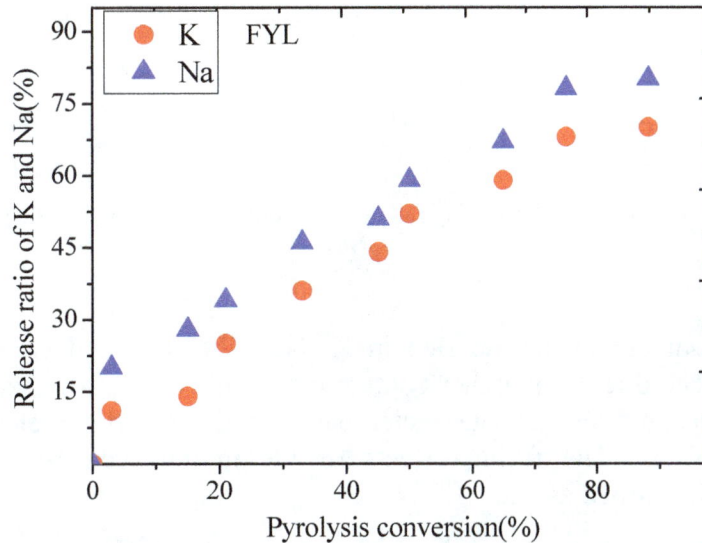

Fig. 2 Alkali metal releasing ratio of eichhornia crassipes during pyrolysis

Fig. 3(b) shows variation of the Na releasing ratio with time for reed pyrolysis under N_2 environment. At the alkali metal releasing ratio of $\alpha_{t,p}<0.6$, Na releasing ratio predicted by the model were in agreement with the experimental data, while some deviations occur at the later stage of pyrolysis. It is can be seen from Fig. 3 that the K and Na releasing characteristics are different during reed pyrolysis, and K and Na releasing ratio increase with the time. Fig. 4 shows calculation results of relationship between the times and alkali metal releasing ratio of eichhornia crassipes pyrolysis process by the model. At $\alpha_{t,p}<0.6$, the predictions of the model proposed in this work agree better with the experimental data for eichhornia crassipes pyrolysis under N_2 environment, while the simulation results of the model are poor agreement with the experiment data at the end of pyrolysis, which is probably caused by the changes of char structure during biomass pyrolysis.

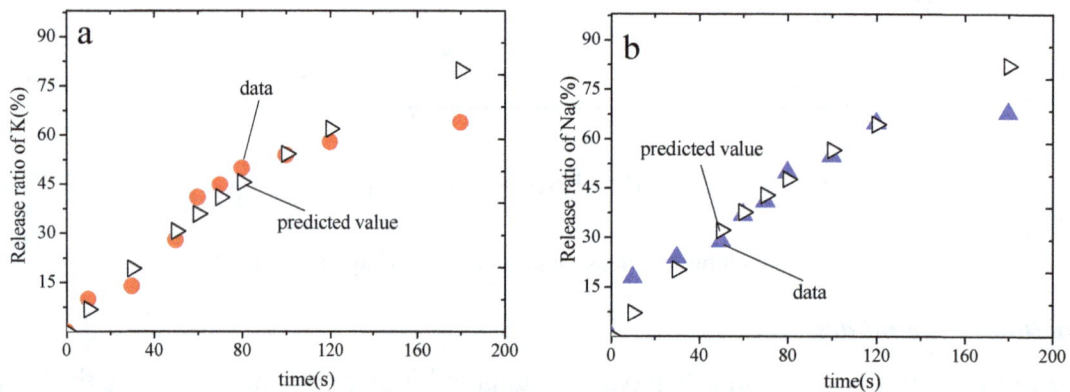

Fig. 3 Calculation results of alkali metal releasing ratio of reed pyrolysis process by the model (a: K; b: Na)

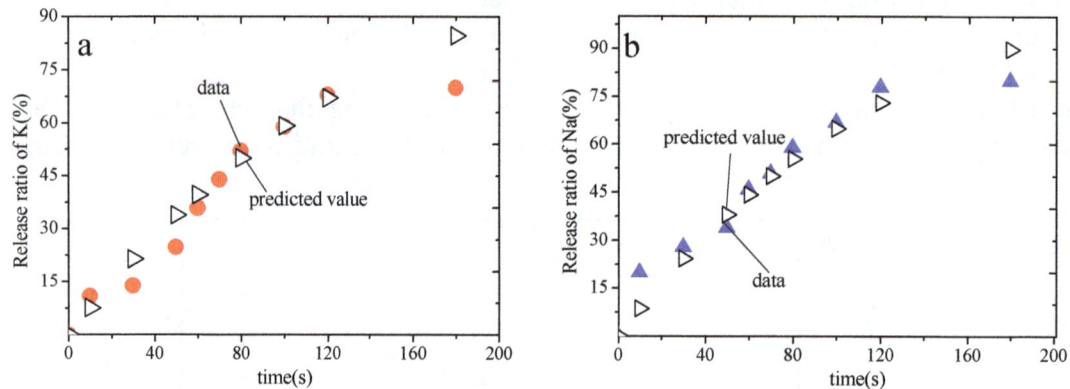

Fig. 4 Calculation results of alkali metal releasing ratio of eichhornia crassipes pyrolysis process by model (a: K; b: Na)

3.4. *Error Analysis*

Poyang Lake Basin aquatic plants are rich in K, Na. The reed and eichhornia crassipes were selected as the feedstock due to its higher content of K and Na as well as abundance in China. In this work, the dynamical model was constructed based on the K and Na releasing characteristics of biomass during pyrolysis. The fitting effects of the models can be judged by using error percentages δ_i which is defined as

$$\delta_i = \sum_{i=1}^{N} \left| X_{exp}^i - X_{pred}^i \right| \tag{6}$$

In which X_{pred}^i is predicted data point, X_{exp}^i is experimental data point.

It can be seen from Table 2 that predictions of the model are more accurate at pyrolysis time of t<120s for reed and eichhornia crassipes pyrolysis. Thereby, the model proposed in this work offers simplified mathematical equations that can be recommended as a convenient model for describing the alkali metal releasing ratio of biomass during pyrolysis.

Table 2 Error analysis of calculation results for k and Na

Conversion	Reed (%)		Eichhornia crassipes(%)	
	k	Na	k	Na
30	5.24	3.71	7.46	3.75
50	2.59	3.17	8.91	4.01
70	4.17	1.80	0.97	1
100	0.21	1.57	0.13	2.08
120	3.90	0.60	0.93	5.0
180	15.7	14.1	14.6	9.65

4. Conclusion

(1) The highest releasing percentage of reed and eichhornia crassipes during pyrolysis is more than 63% for K and Na. Na precipitation rate is higher than K, which is mainly due to interfacial bond energy of Na lower than K.

(2) Dynamical model was constructed based on the K and Na releasing behavior of aquatic biomass during pyrolysis, and the K and Na releasing characteristics predicted by this model are more satisfying in depicting the experimental data, which can be applied to predict the K and Na releasing behavior of aquatic biomass during pyrolysis.

Acknowledgments

This research was funded by the National Science Foundation (No. 51666004, 51566012), the Natural Science Foundation of Jiangxi Province: (No. 20142BAB203029, 20151BAB213025, 20151BAB206048), the Project of Jiangxi Province Education Development Fund (No.GJJ150626, GJJ150616), and the Postdoctoral Science Foundation (No.2015M571991, 2015KY37). These supports are gratefully acknowledged.

References

1. X.B. Xiao, X.l. Meng, D.D. Le, T. Takarada. Two-stage steam gasification of waste biomass in fluidized bed at low temperature: Parametric investigations and performance optimization. Bioresour. Technol. 102 (2011) 1975–1981.
2. K. Elif. Recent advances in production of hydrogen from biomass. Energy Convers. Manage. 52 (2011) 1778–1789.
3. D.M. Keown, J.I. Hayashi, C.Z. Li. Effects of volatile-char interactions on the volatilisation of alkali and alkaline earth metallic species during the pyrolysis of biomass. Fuel 87 (2008) 1187–1194.

4. I.Y. Eom, J.Y. Kim, T.S. Kim, S.M. Lee, D. Choi, I.G. Choi, J.W. Choi. Effect of essential inorganic metals on primary thermal degradation of lignocellulosic biomass. Bioresour. Technol. 104 (2012) 687–694.

5. D.M. Keown, G. Favas, J.I. Hayashiet C.Z. Li. Volatilisation of alkali and alkaline earth metallic species during the pyrolysis of biomass: Differences between sugar cane bagasse and cane trash. Bioresource Technology 96 (2005) 1570–1577.

6. H. Yu, Z. Zhang, Z. Li, D. Chen. Characteristics of tar formation during cellulose, hemicellulose and lignin gasification. Fuel 118 (2014) 250–256.

7. F.C. Borges, Z. Du, Q. Xie, J.O. Trierweiler, Y. Cheng, Y. Wan, Y. Liu, R. Zhu, X. Lin, P. Chen, R. Ruan. Fast microwave assisted pyrolysis of biomass using microwave absorbent. Bioresour. Technol. 156 (2014) 267–274.

8. H. Zhang, M. Luo, R. Xiao, S. Shao, B. Jin, G. Xiao, M. Zhao, J. Liang. Catalytic conversion of biomass pyrolysis-derived compounds with chemical liquid deposition (CLD) modified ZSM-5. Bioresour. Technol. 155 (2014) 57–62.

9. Y. Zhang, S. Kajitani, M. Ashizawa, Y. Oki. Tar destruction and coke formation during rapid pyrolysis and gasification of biomass in a drop-tube furnace. Fuel 89 (2010) 302–309.

10. A. Jensen, K. Dam-Johansen, M.A. Wójtowicz, M.A. Serio, TG-FTIR study of the influence of potassium chloride on wheat straw pyrolysis. Energy Fuels 12 (1998) 929–938.

11. O. Hirohata, T. Wakabayashi, K. Tasaka, C. Fushimi, T. Furusawa, P. Kuchonthara, A. Tsutsumi. Release behavior of tar and alkali and alkaline earth metals during biomass steam gasification. Energy Fuels 22 (2008) 4235–4239.

12. D.M. Quyn, J. Hayashi, C.Z. Li. Volatilisation of alkali and alkaline earth metallic species during the gasification of a Victorian brown coal in CO_2. Fuel Process. Technol. 86 (2005) 1241–1251.

13. T. Kowalski, C. Ludwig, A. Wokaun. Qualitative evaluation of alkali release during the pyrolysis of biomass. Energy Fuels 21 (2007) 3017–3022.

14. T. Okuno, N. Sonoyama, J. Hayashi, C.Z. Li, C. Sathe, T. Chiba. Primary release of alkali and alkaline earth metallic species during the pyrolysis of pulverized biomass. Energy Fuels 19 (2005) 2164–2171.

15. N. Shimada, H. Kawamoto, S. Saka. Different action of alkali/alkaline earth metal chlorides on cellulose pyrolysis. J. Anal. Appl. Pyrolysis 81 (2008) 80–87.

16. M.J. Womat, R.H. Hurt, N.Y.C. Yang, T.J. Headley. Structural and compositional transformations of biomass chars during combustion. Combustion and Flame 100 (1995) 131–143.

Biomass Material Processing When the Band Saw Blade Crack of Kurtosis Index Before and After the Impact

Jin-Gui Gao, Zhao-Fang Jiang*, Jian Zhang
BEIHUA UNIVERSITY
*Email: jiangzhao_fang@126.com

Taking MJ3210 cars woodworking band saw machine as the research object, using high precision under no-load and load of Beijing pop vibration analyzer and the Vibsys vibration signal acquisition, processing and analysis software of band sawing machine band saw blade transverse vibration test and the signal acquisition, and analyze the collected signal calculation, through the analysis of the kurtosis saw blade transverse vibration displacement, find the blade cracks under different conditions and the changing rule of the kurtosis. Through the orthogonal experiment analysis: the saw blade transverse vibration displacement of the most significant factor for the spindle speed followed by the tension of saw blade and feed speed. Through the analysis of the saw blade crack before and after shows that perfect band saw blade transverse vibration displacement when no-load kurtosis is below 1, and no-load crack when the K value in 3-4.8, when the load of crack band saw blade transverse vibration displacement of saw blade kurtosis between 5-6.5; K value changing with cutting trees will also change, with the increase of wood hardness, kurtosis K value will also increase, band saw blade crack length will also increase; Can use the kurtosis of interval value range to determine whether a saw blade cracks, and the extent of the crack. Thus for the first time the kurtosis theory applied to crack diagnosis of biomass material processing band saw, for processing cutting tool under the load of on-line fault diagnosis is a method and the technical basis.

Keywords: Band saw; the band saw blade transverse vibration; crack; kurtosis.

1. Introduction

Band sawing machine is widely used in wood processing enterprises of the machine tool [1]. Its quality directly affects the economic benefits of enterprises. Especially in today's situation, after our country took compulsory action, analyzing the forbidden mining engineering is especially important. The band saw blade in the fault diagnosis is still in the diagnosis of both ex ante and ex post status. The influence of improving the quality of the band saw blade faces cut greatly, due to the production practice, affected by various factors is difficult to accurately judge the band saw blade failure type and the degree of development, so it is difficult to scientifically grasp the band saw blade replacement time, it is hard to guarantee efficient application of band saw blade processing timber, is difficult to ensure the safety of band saw blade. From the author after graduation to post-doctoral study has been research to see if there is a kind of method and theory of index change can quickly diagnose the saw blade are defects [2], therefore, funded by national natural science foundation of China, after a forestry college at the university of north China woodworking machinery laboratory and birch Hua-dian Hui-bang wood industry co., Ltd., enterprise experiment, this paper intends to explore an effective method.

---

*Corresponding author

181

2. Main Experimental Instruments and Equipment

2.1. *Experimental instrument*

Instruments used in the experiment are shown in Table 1.

Table 1 Apparatus and equipments

Serial number	Device name	model	Manufacturer
1	Data acquisition instrument	WS-5942-2-50	Beijing century science and Technology Development Co., Ltd.
2	Eddy current sensor	HZ-8500	Beijing vibration instrument factory
3	Woodworking band sawing machine	MJ3210	Fred mechanical equipment manufacturing Co., Ltd.

Vibration testing system composition block diagram is shown in Figure 1.

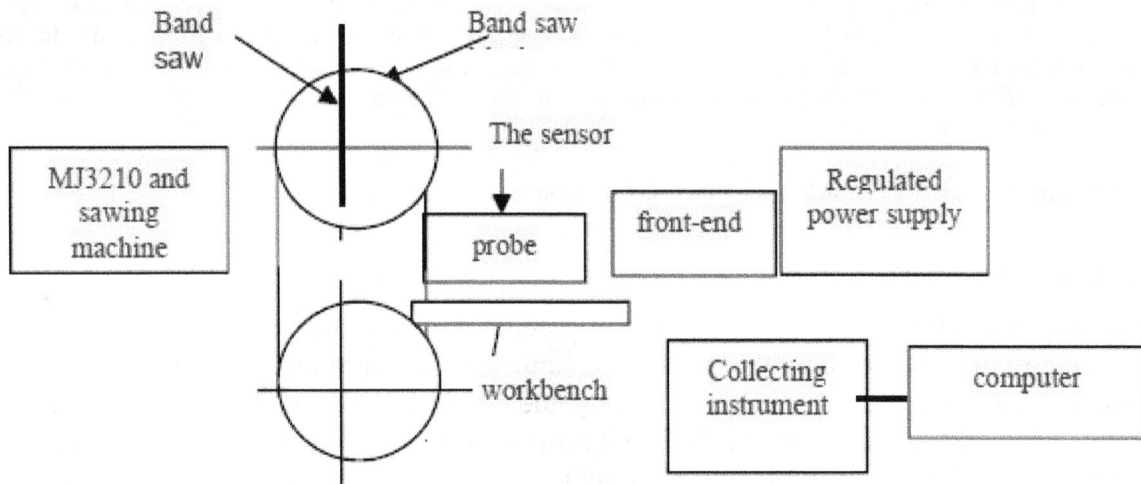

Fig. 1 Vibration testing system composition block diagram

2.2. *Equipment and parameters*

2.2.1. *Machine tool components*

The research objects of this experiment for birch Hua-dian Hui-bang wood industry co., ltd. MJ3210 type woodworking band saw. Saw wheel up and down by its structure, band saw blade, saw blade tensioning device, the power system, frame, safety devices, etc.

2.2.2. *Machine parameters*

MJ3210 cars woodworking band saw the specific parameters are as follows: Saw wheel diameter: 1000 mm blade thickness: 1.05 mm machine weight: 1000 kg saw blade width: 125 mm.

Saw wheel speed: 750 r/min; manufacturer: Hebei RuiXiang mechanical equipment manufacturing co., ltd. Machine tool appearance size: 1250 * 1000 * 2180 mm; sports car feed speed: 0-20 m/min maximum straight through the wood sawing; saw blade maximum length: 6850 mm to 6900 mm.

3. The Experimental Data and Result Analysis

3.1. Crack blade transverse vibration of orthogonal experiment

According to the analysis of the structure and movement MJ3210 woodworking band saw, spindle speed, feed speed and the blade tension as the experimental factors, the purpose is to identify the significant factors influencing the band saw blade transverse vibration displacement. 3 factors 3 levels L9 (3^4) empty under the load of the orthogonal experiment, the factors of a level are shown in Table 2.

Table 2 Orthogonal test factors and levels

Due to the grain	Level 1	Level 2	Level 3
A - saw wheel spindle speed (RPM)	700	750	800
B - feed speed (m/min)	10	15	20
C - saw blade tensioning force (N)	124	122	120

Test results as follows: the factors of primary and secondary order is given priority to, time: A - saw wheel spindle speed C - B - feed speed saw blade tension; A - saw wheel spindle speed C - B - feed speed saw blade tension as the significant factor; The optimal solution for A - saw wheel spindle speed 750 rpm, C - saw blade tensioning force of 122N, B - feed speed of 15 m/min, the thickness of the band saw blade in the following experiments are unchanged.

The experiment selects the root place to produce 1.6 mm crack band saw blade, in order to complete contrasts with the saw blade, don't change, other factors orthogonal test factors and levels are shown in Table 2.

3.2. Saw wheel spindle speed of band saw blade transverse vibration kurtosis

The band saw blade tensioning force of 122N, under the premise of the feed speed of 15 m/min, oak wood processing load. Respectively on the full blade and crack defects of the band saw blade transverse vibration data acquisition, analysis, calculation, will now and with the increase of saw wheel spindle speed saw blade transverse vibration displacement kurtosis change data into curve as shown in igure 2 [3-8].

By observing the analysis Figure 2, can be found, saw wheel spindle speed of band saw blade transverse vibration displacement kurtosis value effect is very obvious, with the increase of saw wheel speed, under no-load condition of saw blade kurtosis increase, but smaller, and the bigger kurtosis increase rate of the defective band saw blade, the range of values: no-load intact blade of kurtosis (hereinafter referred to as: K) within 0 and 1, light has a crack between 3-4 K of saw blade; Loads with the increase of saw wheel speed, band saw blade transverse vibration displacement of the kurtosis of the K value changes very fast, its value is between 5-6.

Saw wheel speed under different load effect on the band saw blade transverse vibration kurtosis

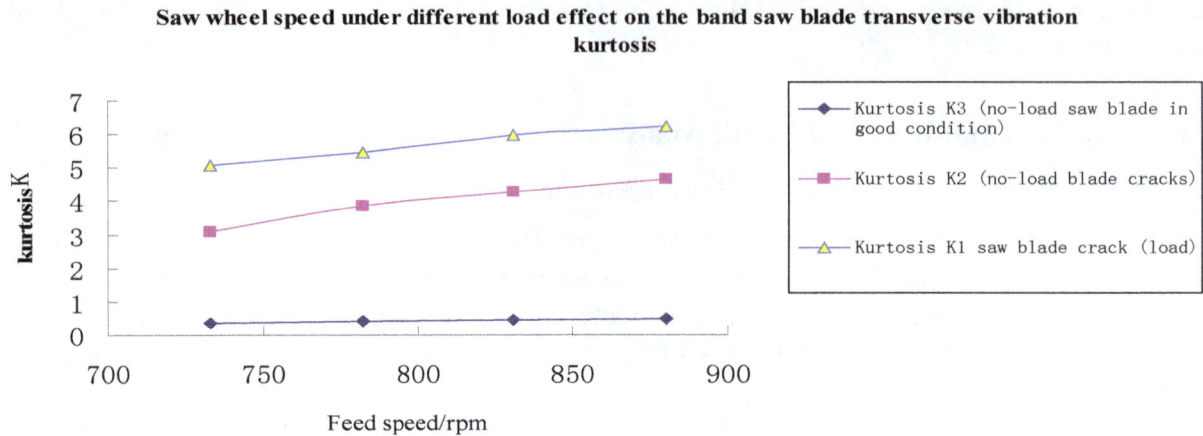

Fig. 2 Saw under different load cycle count on the effect of transverse vibration kurtosis

3.3. *Saw blade tensioning force's influence on the band saw blade transverse vibration kurtosis*

In band sawing machine saw wheel spindle speed is 750 rpm, under the premise of the feed speed of 15 m/min, the load is oak wood cutting, change the saw blade tensioning force five times. Respectively on the full blade and crack defects of band saw blade transverse vibration data acquisition, analysis, calculation, will now get with the increase of blade tension band saw blade transverse vibration of kurtosis change data into curve as shown in Figure 3.

According to the curve in Fig. 3, the band saw blade in the case of tension is not enough, will turn, transmission power is not enough, so at the time of tension is small vibration displacement kurtosis can slant. Produced by the three curves, the crack of the saw blade vibration displacement kurtosis is generally bigger than complete saw blade, K value range: no-load intact band saw blade K = 0 and 1; Light has a crack band saw blade K = 3-4.5; Load has a crack band saw blade K = 5-6.5.

The saw blade under different load tension effect on the band saw blade transverse vibration kurtosis

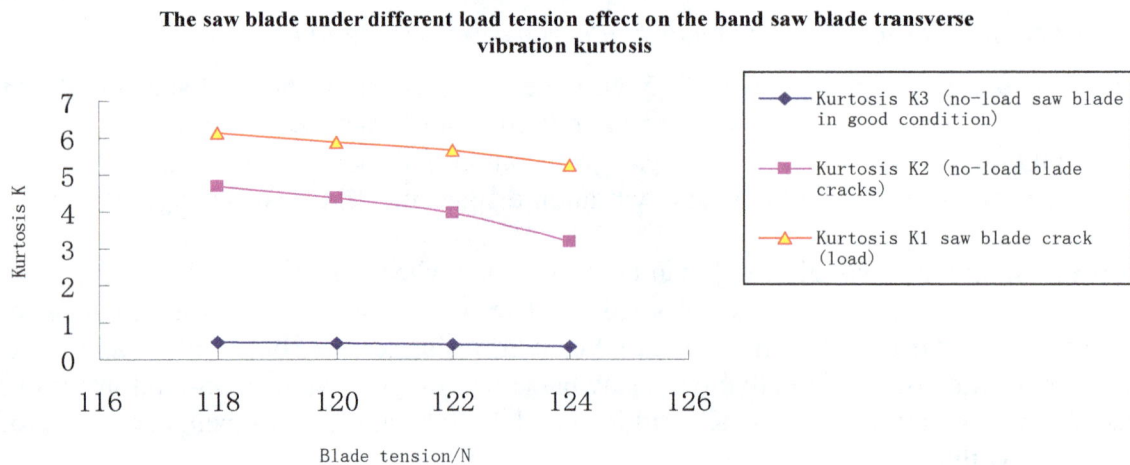

Fig. 3 Saw blade tensioning force under different load on the influence of the saw blade transverse vibration kurtosis

3.4. *Feed rate on the band saw blade transverse vibration kurtosis*

Band sawing machine spindle speed is 750 rpm, under the premise of saw blade tensioning force of 122N, five change feed speed and cutting oak wood, respectively, on the full blade and crack defects of band saw blade transverse vibration data acquisition, analysis, calculation, will now with the increase of feed speed and the change of the band saw blade transverse vibration displacement kurtosis data into a curve as shown in Figure 4.

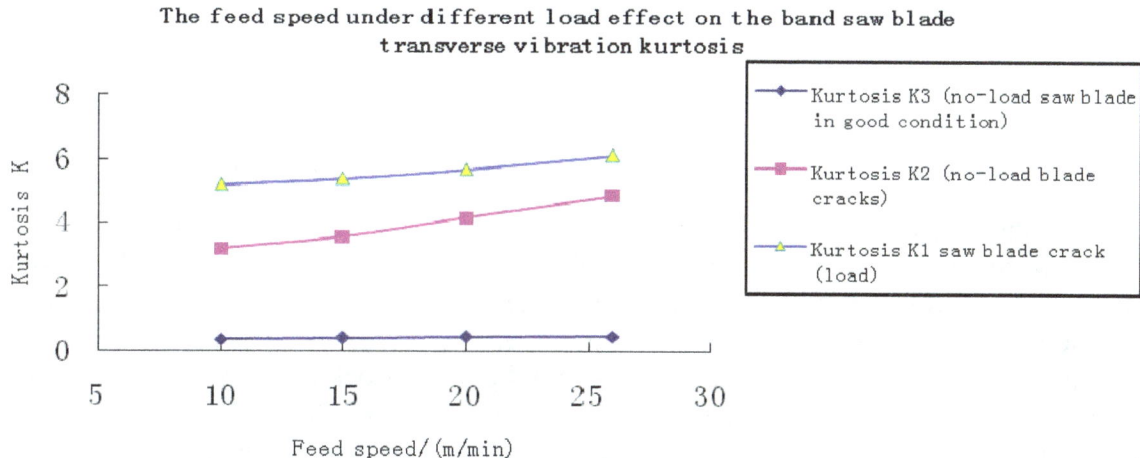

Fig. 4 The feed speed under different load effects on blade transverse vibration kurtosis

Figure 4 shows that with the increase of car speed, blade vibration displacement of kurtosis is also on the increase, but the increase rate is different, no-load intact band saw blade minimum kurtosis, no-load crack increase rate of the big band saw blade, the load has a crack band saw blade kurtosis increase rate is the largest. The scope of its K value: light in good band saw blade 0-1; Light has a crack band saw blade 3-4.8; Load has a crack band saw blade 5-6.

Sawing by the wheel speed, tension of saw blade and feed speed on the band saw blade transverse vibration kurtosis influence, in type MJ3210 sports band sawing machine normal working conditions, no-load crack and no crack band saw blade the kurtosis scope is different, respectively in no-load intact band saw blade K = 0 and 1; Light has a crack band saw blade K = 3-4.8; Under load has a crack band saw blade kurtosis than no-load crack and no crack band saw blade kurtosis is big, its scope in K = 5-6, so you can determine the band saw blade according to the scope of the kurtosis to change whether produce crack defects.

3.5. *Different tree species and crack length on the band saw blade transverse vibration kurtosis*

In determining saw wheel spindle speed 750 rpm, saw blade tensioning force of 122N, the feed speed of 15 m/min, feed speed: 15 m/min, cutting USES four kinds of wood: cypress wood, elm, oak wood, color wood; Will now under different crack length of band saw blade transverse vibration displacement of the kurtosis change data to plot in Figure 5.

185

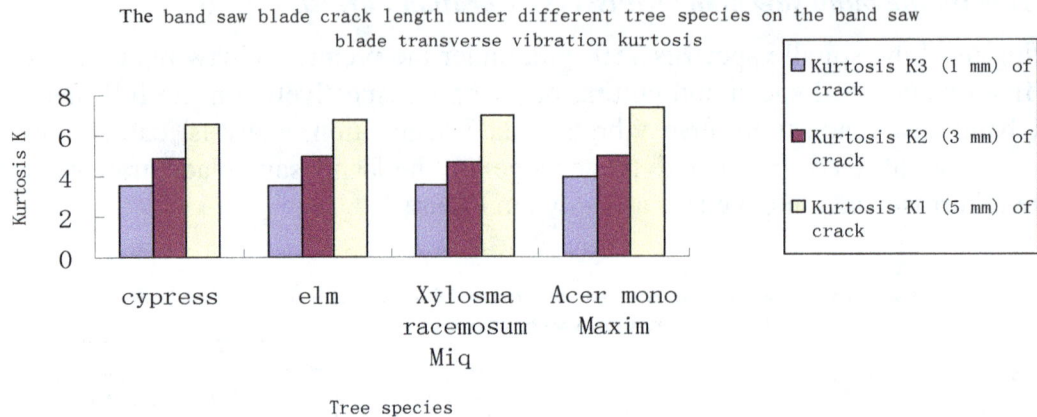

The band saw blade crack length under different tree species on the band saw blade transverse vibration kurtosis

Fig. 5 Crack length under different tree species on the influence of the saw blade transverse vibration kurtosis

Figure 5 shows that with the increase of crack length, band saw blade transverse vibration displacement of the kurtosis is increasing; With the increase of wood hardness, band saw blade with crack defects of kurtosis increase, also under the same tree species, with the increase of crack length, the kurtosis also will increase; K value range: 1 mm crack: cypress 3-3.5 elm oak 3.8 3.9 3.6 3.7 3.9-4 color wood; 3 mm crack; Cypress 4.4 4.6 4.6 4.7 oak elm 4.8 4.9 color wood, 5.0 to 5.1; 5 mm crack: 6.5 6.8 6.9 7 oak elm have 7-7.2 color wood, 7.2 to 7.4; According to the K value is also different, according to different feedstock tree species, the judge the degree of crack generation and development.

4. Conclusion

After above under no-load and load of MJ3210 cars woodworking band saw blade vibration test and analysis to the change of the kurtosis data calculation, the following conclusions:

(1) Through the orthogonal experiment analysis shows that the type of MJ3210 sports cars, woodworking band saw blade transverse vibration displacement of the most significant factor for the spindle speed, followed by the tension of saw blade and feed speed.

(2) In MJ3210 cars band sawing machine work routine work conditions, under no-load crack and no crack band saw blade the kurtosis scope is different, under no load: good band saw blade K = 0 and 1; Have a crack band saw blade K = 3-4.8; Loads a crack band saw blade kurtosis is obvious than no-load crack and no crack band saw blade kurtosis is big, the scope of its K value in 5 to 6, so you can determine whether the band saw blade according to the scope of the kurtosis value change produces crack defects.

(3) Based on kurtosis K value and type of processing raw materials species, can judge the degree of crack generation and development.

Acknowledgments

The study was funded by the national natural science fund of the People's Republic of China. National natural science fund project information: based on time series analysis of woodworking band saw blade tooth crack formation mechanism and prediction research and loss, item number (31570556).

References

1. Guo-Xi ZHU Hua-Bin WANG et al. Modern Chinese lumber production line of research. Harbin: northeast forestry university press, 1989:2-10.
2. Jin-Gui GAO. Band sawing machine mechanical tension system of the computer simulation and experimental research [D]. Harbin: the northeast forestry university, 2001
3. Hong-Bin MEI. Theory of rolling bearing vibration detection and diagnosis. Methods. System. Beijing: mechanical industry press, 1996:15 to 35.
4. Zhi-Jiang XIE. Equipment condition monitoring and fault diagnosis technology. Chongqing: chongqing university press, 1998:35-55.
5. Chun-Hao LIU. compilation. Structural vibration of rolling bearings. Bearings. (5) : 2000-42 and 43.
6. Tandon N, Nakra BC. Vibration and & monitoring techniques for the detection of defects in rolling element bearing - a review. The Shock Vibration Digest, 1992, 24 (3) : 68-92.
7. Shu-Jing ZHANG, Li JI heart. Time series analysis brief tutorial [M]. Beijing: tsinghua university press, northern jiaotong university press, 2003.9:1-100.
8. Rui-Hua XIA. The fault diagnosis of rolling bearing based on time series analysis [D]. Xinjiang, Xinjiang University, 2006.

Based on Time Series Analysis Theory of Biomass Materials Processing of Band Saw Blade Failure Prediction

Jin-Gui Gao, Hong-Gang Zhao*, Jian Zhang
BEIHUA UNIVERSITY
Email: 1125568071@qq.om

In this paper, using the theory of time series analysis, by means of econometric software Eviews8.0, based on the type gathering MJ3310A sports car band saw to produce crack defects of band saw blade transverse vibration displacement data of system analysis, time series prediction model is established, and the transverse vibration of the defective band saw blade displacement prediction and analysis, a experiment is carried out through the forecast when the band saw blade cracks on the actual value compared with the predicted value can be concluded that the accuracy of 95% and 97% respectively, error rate is smaller; After on the band saw blade crack predicted the future of the actual value compared with the predicted value, you can see that accuracy rate were 94% and 95% respectively, and less error rate. Therefore, using the prediction model can predict when the band saw blade crack and crack extension. Then using time series analysis theory for the first time for the band saw the prediction of crack generation and propagation and the safety analysis of on-line fault diagnosis for the work of band saw blade security provides a feasible theory and method.

Keywords: Band saw machine; the band saw blade transverse vibration; crack; prediction.

1. Introduction

Band sawing machine is widely used in wood processing enterprises of the machine tool [1]. Its quality directly affects the economic benefits of enterprises. Especially in today's situation, because our country compulsory after.

The core of econometric research is the design model, data collection, estimation model, inspection model, application model (structural analysis, economic forecasting, policy evaluation). Eviews is an essential tool to accomplish the above tasks. It is precisely because of the emergence of Eviews and other econometric software packages, so that the measurement of economics has made considerable progress, the development of a more practical and rigorous economic discipline. Using Eviews can quickly find a statistical relationship from the data, and use the relationship to predict the future value of the data [1].

Band saw machine has a timber produced rate high, convenient adjustment, saw straight and easy timber sawing, and in the wood processing enterprises get widely used [2]. Its quality directly affects the economic efficiency of enterprises. According to the band saw blade fault diagnosis is still in the ex ante and ex post diagnosis situation. It is very difficult to predict the band saw blade production defects and defects in the development of degree, and accurate determination of saw blade replacement time, can be by band saw blade vibration amplitude and kurtosis to predict band saw blade of the defect type and degree of development [3-9] according to my previous research results and literature. In order to predict the band saw blade of development defects, and from the analysis and forecasts information identify whether the fault is generated, from the author's Ph.D. after graduation to postdoctoral study has been in study whether there is a method and theoretical index changes, to from the predicted values can be diagnosed quickly saw the defects, therefore, with the support of the National Natural Science Foundation, through the experiment in laboratory and enterprise, this paper intends to explore a using time series analysis method for predicting the effective method of index of transverse vibration in crack band saw blade.

---

*Corresponding author

2. Major Laboratory Instruments and Equipment

2.1. *Experimental instruments*

Instruments used in the experiment are shown in Table 1.

Table 1 Apparatus and equipments

Serial number	Device name	model	Manufacturer
1	Data acquisition instrument	WS-5942-2-50	Beijing century science and Technology Development Co., Ltd.
2	Eddy current sensor	HZ-8500	Beijing vibration instrument factory
3	Woodworking band sawing machine	MJ3310A	Fred mechanical equipment manufacturing Co., Ltd.

Vibration test system block diagram shown in Figure 1.

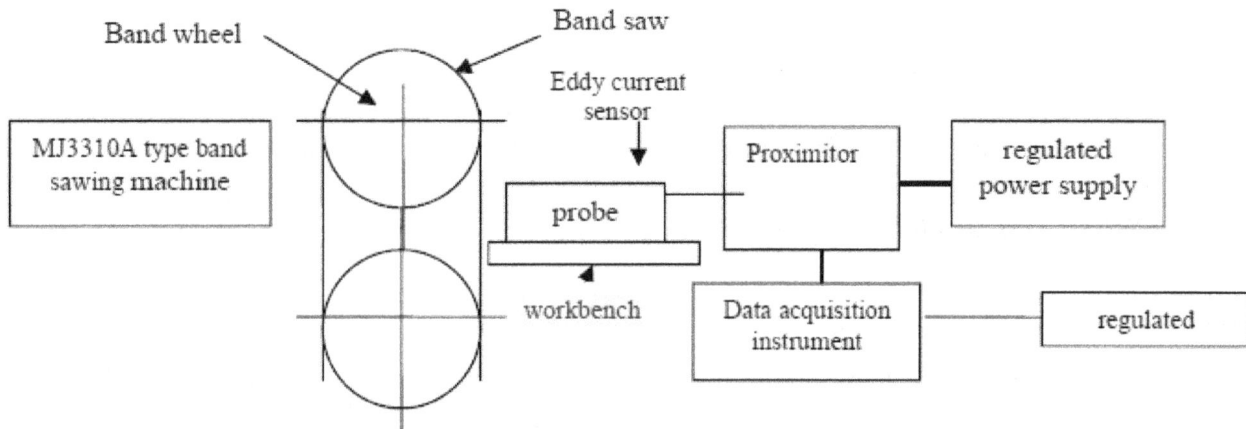

Fig. 1 Vibration testing system composition block diagram

2.2. *Device composition and its parameters*

2.2.1. *Machine tool composition*

The research object of this experiment is MJ3310A type woodwork sawing machine Hua-dian Huibang Wood Industry Co. Ltd. The structure by, saw wheel band saw blade, saw blade guide device, tensioning device, power system, frame, safety covers.

2.2.2. *Parameter of machine tool*

The following specific parameters of MJ3310A type woodwork sawing machine:

Application of saw blade thickness: 1.15mm machine weight: 1000kg band saw blade width: 125 mm; Saw wheel speed: 750r/min Machine tool appearance size: 1250*1000*2180mm.

Carriage feed speed: 0-20m/min sawing straight the largest: the maximum length of 6850mm blade: 6950mm.

2.3. *Experimental data*

3 mm band saw blade crack in the experiments, the acquisition of the band saw blade transverse vibration data as shown in Figure 2.

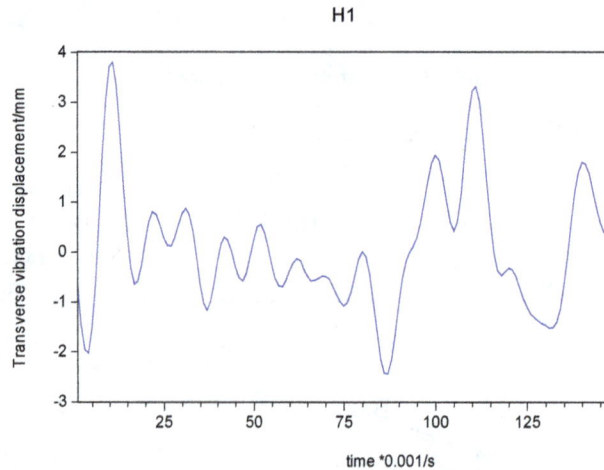

Fig. 2 Acquisition of defective horizontal band saw blade vibration displacement time history curve

3. Establishment of Model

3.1. *Stability test*

The data for the analysis and calculation of self correlation coefficient, derived from the correlation coefficient as shown in Figure 2. By Figure 2, the correlation coefficient can be seen from all around, and it is determined that the sequence is a stationary time series.

3.2. *Self correlation diagram test*

The crack band saw transverse vibration displacement time series data autocorrelation analysis are shown in Figure 3.

From Figure 2, sequence analysis results can be seen: the autocorrelation coefficient always fluctuations in around zero, to further determine the sequence for stationary time series; 2. Look at the Q statistic P values: the statistics of the original hypothesis for x 1 period, 2 period...... K since correlation coefficient was equal to 0, the alternative hypothesis for self correlation coefficient at least a is not equal to 0, so as the figure, the p value are the significant level of 5, so accept the null hypothesis that the sequence is purely random sequences, namely white noise sequence.

Date: 12/16/15 Time: 16:28
Sample: 1 149
Included observations: 149

Autocorrelation	Partial Correlation		AC	PAC	Q-Sta...	Prob
		1	0.945	0.945	135.77	0.000
		2	0.796	-0.90...	232.73	0.000
		3	0.588	0.619	286.10	0.000
		4	0.369	0.287	307.21	0.000
		5	0.180	-0.07...	312.27	0.000
		6	0.051	-0.19...	312.68	0.000
		7	-0.01...	-0.14...	312.69	0.000
		8	0.01...	0.12...	312.74	0.000
		9	0.008	0.039	312.75	0.000
		1...	0.034	0.078	312.94	0.000
		1...	0.041	0.085	313.22	0.000
		1...	0.020	0.018	313.28	0.000
		1...	-0.02...	-0.04...	313.38	0.000
		1...	-0.07...	-0.14...	314.41	0.000
		1...	-0.12...	-0.14...	317.09	0.000
		1...	-0.15...	-0.12...	321.25	0.000
		1...	-0.17...	-0.06...	326.19	0.000
		1...	-0.17...	0.002	331.37	0.000
		1...	-0.18...	0.043	336.95	0.000
		2...	-0.19...	-0.00...	343.85	0.000
		2...	-0.23...	0.023	353.55	0.000
		2...	-0.28...	-0.03...	367.62	0.000
		2...	-0.32...	-0.01...	386.60	0.000
		2...	-0.35...	0.007	408.72	0.000
		2...	-0.34...	-0.05...	429.99	0.000
		2...	-0.29...	0.009	446.16	0.000
		2...	-0.22...	-0.02...	455.28	0.000
		2...	-0.13...	0.006	458.61	0.000
		2...	-0.05...	0.028	459.10	0.000
		3...	0.007	0.028	459.11	0.000
		3...	0.033	0.013	459.33	0.000
		3...	0.030	-0.01...	459.50	0.000
		3...	0.008	-0.05...	459.51	0.000
		3...	-0.01...	-0.00...	459.56	0.000
		3...	-0.02...	0.028	459.72	0.000
		3...	-0.01...	0.019	459.79	0.000

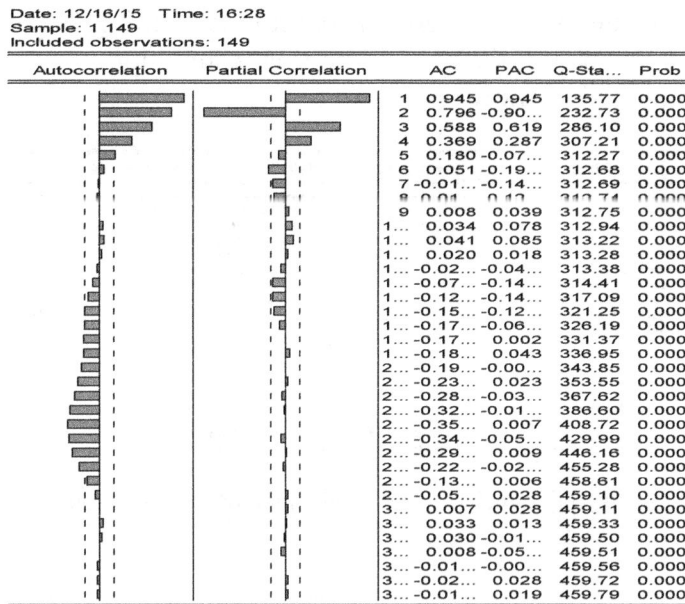

Fig. 3 Analysis of autocorrelation coefficient of defect band saw blade vibration displacement time series

3.3. *Unit root test: ADF, PP test, etc*

Unit root test is shown in Figure 4.

Augmented Dickey-Fuller Test Equation
Dependent Variable: D(H1,3)
Method: Least Squares
Date: 12/16/15 Time: 16:18
Sample (adjusted): 13 149
Included observations: 137 after adjustments

Variable	Coefficient	Std. Error	t-Statistic	Prob.
D(H1(-1),2)	-0.732164	0.101605	-7.205964	0.0000
D(H1(-1),3)	0.548842	0.096467	5.689450	0.0000
D(H1(-2),3)	1.021728	0.097494	10.47989	0.0000
D(H1(-3),3)	0.377133	0.118206	3.190479	0.0018
D(H1(-4),3)	-0.024252	0.116417	-0.208317	0.8353
D(H1(-5),3)	0.148858	0.107511	1.384575	0.1686
D(H1(-6),3)	0.282642	0.091124	3.101715	0.0024
D(H1(-7),3)	0.017401	0.085063	0.204563	0.8382
D(H1(-8),3)	0.163590	0.079331	2.062127	0.0413
D(H1(-9),3)	0.273292	0.080750	3.384417	0.0010
C	1.89E-05	0.003687	0.005117	0.9959

R-squared	0.855303	Mean dependent var		0.004888
Adjusted R-squared	0.843820	S.D. dependent var		0.108715
S.E. of regression	0.042964	Akaike info criterion		-3.380040
Sum squared resid	0.232580	Schwarz criterion		-3.145589
Log likelihood	242.5327	Hannan-Quinn criter.		-3.284765
F-statistic	74.47876	Durbin-Watson stat		2.046229
Prob(F-statistic)	0.000000			

Fig. 4 Unit root test

According to the results of 3ADF test, we can know that the unit root statistic ADF=-0.023651 is less than the ADF threshold value given by 1%-10%, so the sequence is stable and the EVIEWS value of the critical value is assumed. According to the correlation coefficient and partial correlation coefficient of tailing phenomenon, therefore the model for ARMA (P, Q) model.

3.4. *Model order*

The first step: from the sample autocorrelation coefficient and the sample partial autocorrelation coefficient, the selected process.The second step: to more accurate the model identification and more accurate judgment, value, we try several different model fitting, according to the criterion of the AIC and SiC, the final choice of the model, the final estimation results P=6, Q=3.. The fitting model parameters are estimated as shown in figure 5.

Variable	Coefficient	Std. Error	t-Statistic	Prob.
C	2.014782	0.680373	2.961293	0.0075
AR(1)	-0.929878	0.112936	-8.233690	0.0000
MA(1)	0.737137	0.075454	9.769344	0.0000
MA(6)	-1.510734	0.119855	-12.60472	0.0000

R-squared	0.737934	Mean dependent var	2.014000
Adjusted R-squared	0.700496	S.D. dependent var	21.03456
S.E. of regression	11.51158	Akaike info criterion	7.870230
Sum squared resid	2782.845	Schwarz criterion	8.065251
Log likelihood	-94.37788	Hannan-Quinn criter.	7.924321
F-statistic	19.71082	Durbin-Watson stat	1.826776
Prob(F-statistic)	0.000003		

Inverted AR Roots	-.93			
Inverted MA Roots	.98	.43+.91i	.43-.91i	-.67+.89i
	-.67-.89i	-1.24		
Estimated MA process is noninvertible				

Fig. 5 Parameters of the model parameter estimation

Prediction model

$$x_t = -0.030293366 + \frac{1 - 0.929878 * B - MA(2) * B^2 - \ldots\ldots 0.93256 MA(q) * B^3}{1 + 0.737137 * B + AR(2) * B^2 - \ldots\ldots 0.76231 AR(p) * B^9} \varepsilon_t$$

3.5. *Test for stationary and pure randomness of residuals*

The stability of residual error and the test of pure randomness are shown in Fig. 6.

From Fig. 6, we can see that ACF and PACF are not significantly different from zero, the P value of Q is far greater than 0.05, so we can think that the residual sequence is white noise sequence, the model information extraction is more adequate.

In addition, the P value of the first order parameter of the constant and the lag is very small, and the parameters are significant; the whole model is simplified, and the model is better.

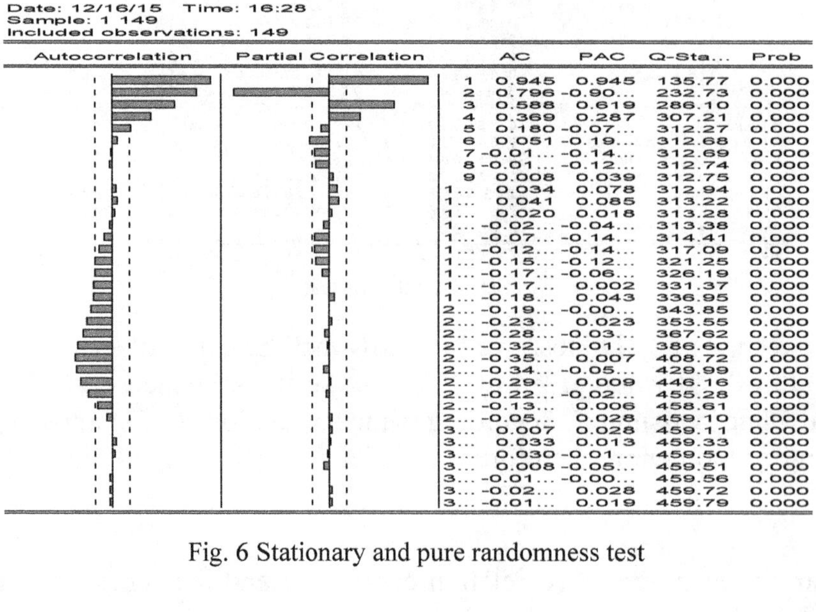

Fig. 6 Stationary and pure randomness test

3.6. *Model checking*

By using the residual (Residual) and the actual value (Actual) and the fitting value (fitted), the residual error of the model is smaller and the fitting value and the actual value of the model are in good agreement. From residual serial correlation coefficient diagram can be seen from the correlation coefficients fall within the confidence interval, and 0 no significant difference, and the sequence is random, namely the residual is a white noise sequence, the regression model was established. So the model is ideal.

3.7. *Prediction model*

To predict the transverse vibration of band saw blade by using the model, the predicted value, as Figure 7. The forecast value of the future 1.01s after the future is worth to be compared with the actual value shown in Figure 8.

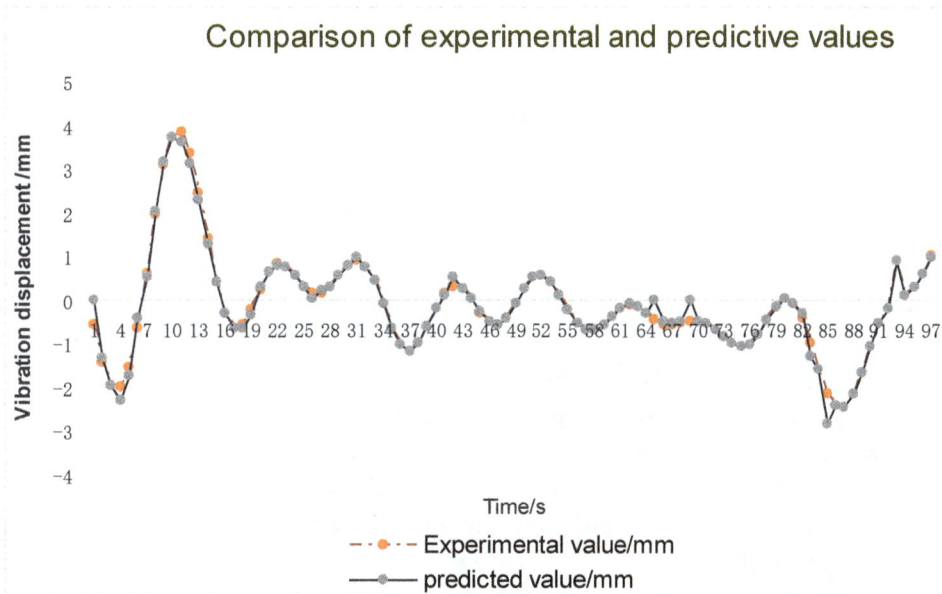

Fig. 7 Curve of forecasting

Band saw blade transverse vibration displacement

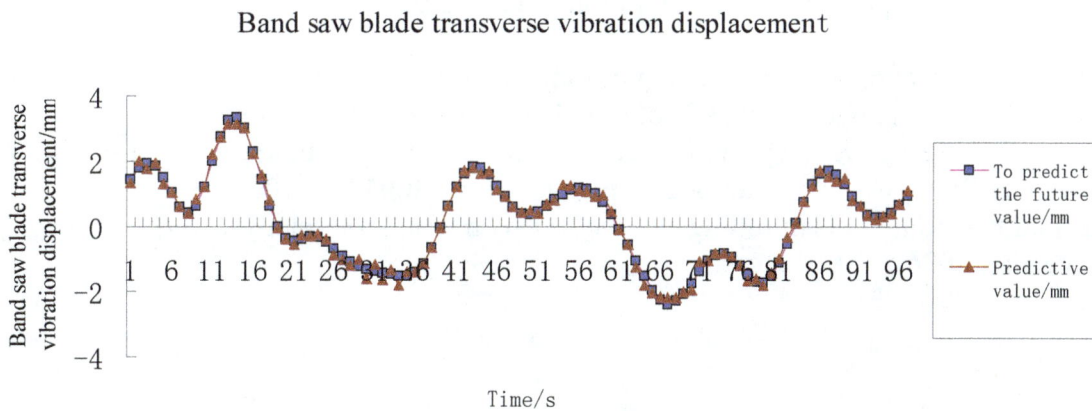

Fig. 8 Comparison of the predicted future value with the actual value

By Figure 7, can be seen through the prediction of band saw blade cracks after the actual value and the predicted value comparison, we can see that the accuracy rate of 95% and 97%, respectively, low error rate. Therefore, the use of the band saw blade can predict the crack growth after generation. From Figure 8, it is seen that, after the future to predict the band saw blade cracks after the actual value and predicted value comparison can be seen, the accurate rate was 94% and 95% respectively, low error rate.

4. Conclusion

Because the model is stationary time series model is obtained by testing and by curve fitting can see the model prediction accuracy is very high, can predict the band saw blade crack propagation. In the future, we need to optimize the relative optimal model according to the cutting tree species.

Acknowledgments

The study was funded by the national natural science fund of the People's Republic of China. National natural science fund project information: based on time series analysis of woodworking band saw blade tooth crack formation mechanism and prediction research and loss, item number (31570556).

References

1. Easy to use data analysis and Eviews application [M]. Beijing: China Statistics Press, 2003:106-132.
2. Zhu Guoxi, Wang Huabin et al. Research on Chinese modern production line [M]. Harbin: Northeast Forestry University press, September 1989: 2-10.
3. Gao Jingui, Zhang Jian, Li. Band saw blade production defects before and after the experimental analysis [J]. Fujian Forestry Science and technology, 2014 (3): 101-105.
4. Mei Hongbin. Vibration detection and diagnosis theory of rolling bearings. Methods. System. Beijing: Mechanical Industry Press, 1996:15-35
5. Xie Zhijiang. Equipment condition monitoring and fault diagnosis technology. Chongqing: Chongqing University press, 1998:35-55
6. Liu Chunhao compiler. Structural vibration of rolling bearings. Bearing.2000 (5): 42-43 Tandon N
7. Nakra BC. Vibration and acoustic monitoring techniques for the detection of defects in rolling element bearing-a review. Shock Vibration Digest.1992,24:68-92. (3)
8. Zhang Shujing, Li Xin Qi. Time series analysis concise tutorial [M]. Beijing: Tsinghua University Press, Northern Jiaotong University Press, 2003.9: 1-100
9. Xia Xia. Fault diagnosis of rolling bearing based on time series analysis [D]. Xinjiang: Xinjiang University, 2006

Fully Bio-based Isosorbide Epoxy Cured by Bio-based Curing Agents with Regulated Properties

Rui-Xue Chang, Jiang-Lei Qin*

College of Chemistry and Environmental Science, Hebei University, 180 East Wusi Road, Baoding China
Email: qinhbu@iccas.ac.cn

Isosorbide diglycidyl ether (isosorbide epoxy) was cured by maleopimaric acid and terpene maleic anhydride as co-curing agent to regulate the properties of fully bio-based epoxy resins. The non-isothermal curing kinetics for co-curing of fully bio-based epoxies with various MPA/TMA ratios was studied by differential scanning calorimetry (DSC). The dynamic mechanical analysis (DMA) was used to evaluate the mechanical properties and thermogravimetric analysis (TGA) was used to study the thermal stability of the cured resins respectively. The results showed that the fully bio-based isosorbide epoxies have comparable or even better mechanical properties to the bisphenol A (BPA) type competitor cured by petroleum based curing agents. Furthermore, the mechanical properties of these fully bio-based epoxies can be further regulated in large extent and then can be used in more areas with desired performances, showing that the fully bio-based isosorbide epoxies have great potential as substitutes for BPA type epoxies.

Keywords: Bio-based epoxy; isosorbide; curing kinetics; sustainable development; regulated properties.

1. Introduction

Because the soaring of the petroleum price and strict government regulations on biotoxicity of materials, BPA based epoxies faces great challenge because BPA has been classified as a carcinogen, mutagen, and reprotoxic (CMR) chemical in European regulations. [1] At same time, it also provides great opportunity to develop bio-based epoxy resins from sustainable resources.

Among those bio-based feedstocks, the vegetable oils and fatty oil esters are the most popular renewable feedstocks used because of its low and affordable price. [2-4] But cured epoxidized vegetable oils have long molecular chains in its structure with the epoxy rings set in the middle, which caused very low glass transition temperature (T_g) of cured resin. Rosin base epoxy exhibit good mechanical properties, but the cured rosin epoxy is extremely brittle, which limited its application. [5] Eugenol as bio-based aromatic chemical was also used to prepare bio-based epoxy and the mechanical properties of cured eugenol epoxy was comparable to BPA based epoxy. Itanic acid was also used to prepare epoxy resin cured by anhydride and crosslinker. [6] Isosorbide was used to prepare bio-based epoxies; based on its compact rigid structure, the cured epoxy exhibited a high T_g of 155 °C comparable to BPA type epoxies. [7, 8] This result shows that the sugar based isosorbide derived epoxy has great potential as substitute for BPA epoxy.

Beside epoxy itself, the weight ratio of curing agent can go up as high as 50% and can regulate the ultimate properties of the cured resin in large extent. In this research, it was revealed the T_g of MPA cured isosorbide epoxy was up to 252.9 °C but extremely fragile which; on the other hand, the TMA cured isosorbide epoxy had a T_g of 136.6 °C. The structures of the epoxy and curing agents are shown in Scheme 1. Based on different advantages of these two bio-based curing agents MPA and TMA, fully bio-based isosorbide epoxy was cured by co-curing agents of MPA and TMA with regulated mechanical properties. The curing of isosorbide epoxy by MPA/TMA was studied by FT-IR, DSC, DMA and TGA. The results showed that the T_g of these fully bio-based isosorbide epoxies can be regulated from 136.6 °C to 252.9 °C without loss of storage modulus,

*Corresponding author

which extended the application areas of the fully bio-based isosorbide epoxy as sustainable substitute for BPA epoxies.

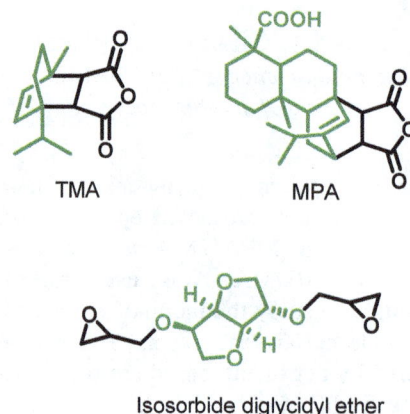

Scheme 1 Structures of isosorbide diglycidyl ether and caged curing bio-based agents

2. Experimental

Synthesis of isosorbide epoxy: The preparation of isosorbide epoxy was carried out by a two step reaction as reported by Feng et al. [9] The isosorbide was heated with allyl bromide in sodium hydroxide solution at 65 °C for 5 h to prepare isosorbide diallyl ether; then the double bond in isosorbide diallyl ether was oxidized with MCPBA in CH_2Cl_2 to give isosorbide diglycidyl ether named as isosorbide epoxy. [10]

Sample preparation: Isosorbide epoxy and curing agents with 1:1 equivalent group ratio were ground in a mortar to obtain a homogenous mixture. The MPA to TMA ratios (MPA/TMA) were varied from 0/1 to 1/0 for 6 formulations. EMID was added as catalyst with 1% to total weight of epoxy and curing agents. For the DMA specimens, the homogenous mixture was transferred into a rectangular mould with standard diameter, cured at 150 oC for 2 h and 200 oC for 4 h.

Characterizations: ^{1}H NMR spectra were recorded with a Bruker 400 MHz spectrometer at room temperature in deuterated chloroform ($CDCl_3$). Fourier transform infrared spectroscopy (FT-IR) was recorded on a Nexus 670 instrument. The samples were dissolved in CH_2Cl_2 and casted on a KBr plate for FT-IR characterization. Non-isothermal curing of the epoxies was studied on a Diamond differential scanning calorimetry under nitrogen atmosphere. The samples were heated from 35 to 280 °C at heating rates of 5, 10, 15 and 20 °C min^{-1} respectively, with all sample weights were about ~8 mg. Dynamic mechanical analysis (DMA) was performed on a TA Q800 analyzer under the single cantilever mode at a frequency of 1 Hz. The tests were scanned from room temperature to 270 °C with heating rate 3°C min^{-1}. T_g was defined as the peak temperature of tan δ curve. Thermogravimetric analysis (TGA) of cured epoxies was performed on a TA Q600 analyser under N_2 atmosphere and the heating rate of 10 °C min^{-1}.

3. Result and Discussion

3.1. *Synthesis of isosorbide epoxy*

Two-step procedure was selected in this study and pure product with an epoxy equivalent of 119 g/mol was obtained. Fig. 1 shows the ^{1}H NMR of the final product respectively, no peak between 5.00 ppm and 6.00 ppm indicated that all double bonds were oxidized into epoxy groups, and the peaks between 2.50 ppm and 3.70 ppm (peak 1, 2 and 3) showed the product was the diglycidyl ether.

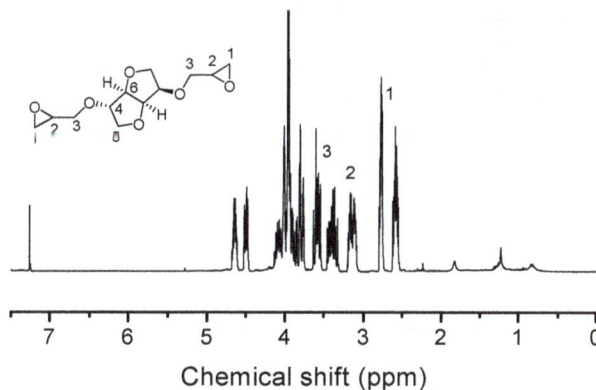

Fig. 1 ^{1}H NMR of isosorbide epoxy

3.2. Curing kinetics

MPA and TMA were prepared following reported procedure [11, 12] and used as curing agent to prepare fully bio-based epoxies. Fig. 2 shows the DSC thermograms of heating curves for isosorbide epoxy cured by various ratio of MPA/TMA as co-curing agents at 5 °C min^{-1} heating rate, and the DSC results are summarized in Table 1.

The epoxy cured by MPA showed a shoulder at low temperature of 119.8 °C because of the existence of carbonyl group. However, the epoxy cured by TMA also showed a shoulder probably because of some impurities existed in it, this is a proof showed the TMA is not pure even after column purification. With increasing ratio of TMA added into MPA as co-curing agent, the shoulder peak temperature increased gradually from 118.9 °C to 125.3 °C with the main peak almost keep intact (see Table 1). At the same time, another shoulder appeared at higher temperature range of 148.6 °C, indicated complicated post curing reaction involved in the co-curing process.

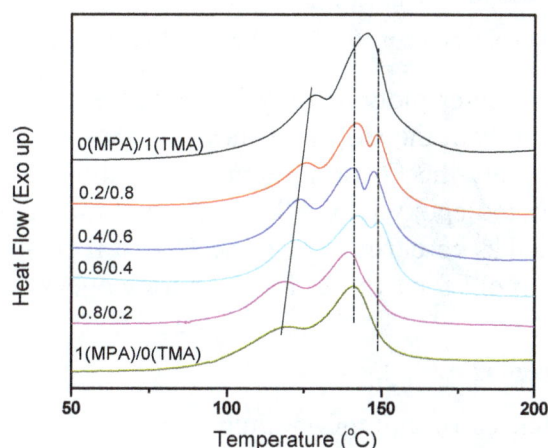

Fig. 2 Thermograms of isosorbide epoxies cured by various ratio of MPA/TMA at 5 °C min^{-1} heating rate

Table 1 DSC results of non-isothermal curing and properties of the cured epoxies

Curing agent	MPA	MPA/TMA				TMA
		0.8/0.2	0.6/0.4	0.4/0.6	0.2/0.8	
T_p(°C)	119.8(L)/141.3(H)	139.3	142.8	140.6	142.0	145.5
T_s (°C)		118.9	122.4	123.4	125.3	
E_a(kJ mol^{-1})	65.8(L)/56.2(H)	60.8	77.3	71.2	69.9	65.0
T_g	252.9	227.3	205.7	187.2	162.1	136.6
G'(GPa, 30 °C)	2.06	2.19	2.00	2.19	2.05	2.17
T_{d5}(°C)	344.6	334.7	326.1	322.7	313.3	313.8

[a] T_p was measured from the curing at a heating rate of 5 °C min^{-1}.
[b] T_{d5}: temperatures of 5% degradation.

Fig. 3 (left) shows the non-isothermal curing curves of isosorbide epoxy cured by 0.4(MPA)/0.6(TMA) at various heating rates. The Kissinger equation was widely used to calculate the curing activation energy (E_a) of curing reactions, as shown in equation 1[13].

$$\frac{d[\ln(q/T_p^2)]}{d[1/T_p]} = -\frac{E_a}{R}$$

(1)

Here T_p is the peak exothermal temperature, q the constant heating rate, E_a the activation energy of the curing reaction, and R is the gas constant with the value of 8.314.

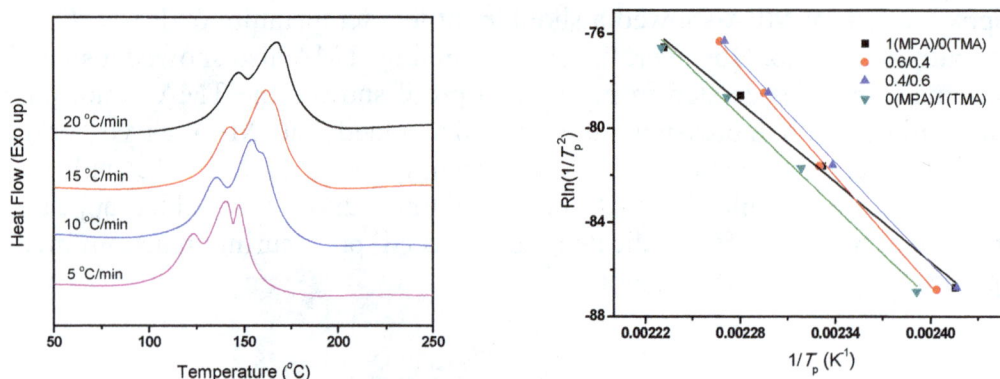

Fig. 3 DSC curves of isosorbide epoxy cured by 0.4(MPA)/0.6(TMA) at various heating rate and Kissinger plots

Fig. 3 (right) shows the Kissinger plots of fully bio-based epoxies (selected formulations) and the E_a values were determined from the slopes of $\ln(q/T_p^2)$ vs. $1/T_p$. There are two values for the MPA; for other formulations, only the E_a for main peaks are illustrated. As shown in Table 1, the E_a of the shoulder for MPA cured epoxy is 65.8 kJ mol^{-1}, and the E_a of the main peak in just 56.2 kJ mol^{-1}. With TMA was added as co-curing agent, the E_a first increased gradually to 77.3 kJ mol^{-1} and then decreased to 65.0 kJ mol^{-1} for 1 ratio TMA as curing agent.

3.3. Dynamic mechanical properties

Fig. 4 compares the dependences of storage modulus and tan δ on temperature for the epoxies cured by various ratios of MPA/TMA, and the T_g was defined as the peak temperature of tan δ curve. As shown in Fig. 4(a), the isosorbide epoxy cured by all ratio of MPA/TMA showed comparable storage modulus at glassy states (ranging from 2.00 to 2.19 GPa at 30 °C), comparable to BPA based epoxies. As for T_g of cured resins in Fig. 4(b), the TMA cured isoborbide epoxy showed a moderate T_g of 136.6 °C; with increasing ratio of MPA was added as co-curing agent, the T_g of the epoxies increased gradually to 252.9 MPA cured one. Although the storage modulus analysis indicated complicated reaction involved in the curing reactions, further analysis showed the T_g have linear relation with molar ratio of MPA/TMA. Based on its brittleness, the MPA could be a good curing agent modifier to increase the T_g of cured epoxies.

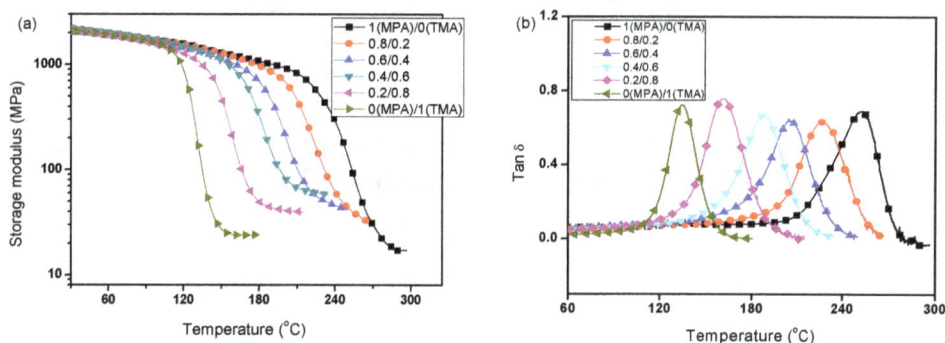

Fig. 4 Dynamic mechanical properties of MPA/TMA cured isosorbide epoxy resins

3.4. Thermal stability and of various cured epoxy resins

Fig. 5 shows the TGA curves of the isosorbide epoxy cured by various MPA/TMA ratios, the 5% weight loss (T_{d5}) as the important parameter to evaluate thermal stability are listed in Table 1. All formulations showed similar one step decomposition process with negligible char residues at 600°C. The isosorbide epoxy cured by MPA had the best thermal stability possibly because of more stable network derived from high functionality of MPA, the T_{d5} of isosorbide epoxy/MPA is as high as 344.6 °C, much higher than TMA cured one (313.8 °C). The T_{d5} of co-cured epoxies increased with increasing of MPA ratio in co-curing agent, as shown in Fig. 5 and Table 1. This result showed that the MPA as co-curing agent improved the thermal stability while increase the T_g of cured isosorbide epoxies.

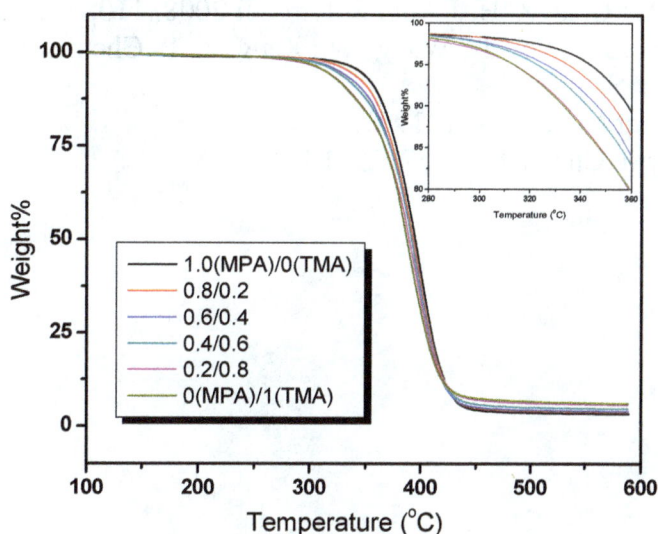

Fig. 5 TGA curves of cured isosorbide epoxies at 10 °C min^{-1} heating rate under N$_2$

4. Conclusion

Bio-based isosorbide epoxy and bio-based curing agents of TMA, MPA with anhydride group were synthesized. The non-isothermal curing kinetics of fully bio-based epoxies cured by various ratios of MPA/TMA was studied. The fully bio-based epoxy resins have regulated properties based on high rigidity of MPA and flexibility of TMA. Although MPA cured isosorbide epoxy has disadvantage of brittleness, it is a good biobased co-curing agent to increase the T_g and improve the thermal stability of fully bio-based isosorbide epoxies.

Acknowledgments

This research was funded by Scientific Research Foundation for the Returned Overseas Chinese Scholars, State Education Ministry; Science Foundation of Hebei University (y2013272) and Post-graduate's Innovation Fund Project of Hebei Province.

References

1. DIRECTIVE 2002/95/EC OF THE EUROPEAN PARLIAMENT AND OF THE COUNCIL on the restriction of the use of certain hazardous substances in electrical and electronic equipment.
2. Stemmelen, M.; Pessel, F.; Lapinte, V.; Caillol, S.; Habas, J. P.; Robin, J. J. J Polym Sci Pol Chem 2011, 49, 2434-2444.
3. Wang, Z.; Zhang, X.; Wang, R.; Kang, H.; Qiao, B.; Ma, J.; Zhang, L.; Wang, H. Macromolecules 2012, 45 (22), 9010-9019.
4. Pan, X.; Sengupta, P.; Webster, D.C. Biomacromolecules 2011, 12, 2416-2428.
5. Liu, X.; Huang, W.; Jiang, Y.; Zhu, J.; Zhang, C. *eXPRESS Polym Lett* 2012, *6(4)*, 293.
6. Ma, S.; Liu, X.; Jiang, Y.; Tang, Z.; Zhang, C.; Zhu, J. Green Chemistry 2013, 15, 245-254.
7. Chrysanthos, M.; Galy, J.; Pascault, J. P. Polymer 2011, 52, 3611-3620.
8. Lukaszczyk, J.; Janicki, B.; Kaczmarek, M. European Polymer Journal 2011, 47,1601-1606.
9. Feng, X. H.; East, A. J.; Hammond, W. B.; Zhang, Y.; Jaffe, M. Polym Adv Technol 2011, 22, 139-150.
10. Huo, L.; Gao, J.; Du, Y.; Chai, Z. H. J Appl Polym Sci 2008, 110, 3671-3677.
11. Leng, F.; Duan, W.; Xu, X.; Wei, T.; Wang, W.; Zeng, Y. P. Chemistry and Industry of Forest Products 2011, 31 (5), 65-70.
12. Bardasz, E. A. US patant No. 5066461, 1991.
13. Kissinger, H. E. Analytic Chemistry 1957, 29, 1072-1076.

Preparation of Gypsum Retarder from Hydrolyzing Waste Penicillin Mycelium by Microwave Heating Method

Meng-Meng Zhang[1,*], Feng-Qing Zhao[2]

[1]Department of Chemical Engineering, Hebei University of Science & Technology, Shijiazhuang, 050018, China
[2]Hebei Engineering Research Center of Solid Waste Utilization, Shijiazhuang, 050018, China
*Email: 820665850@qq.com

Waste penicillin mycelium (WPM) is a solid waste from the production process of penicillin in pharmaceutical industry, which is difficult to dispose because of environmental and safety problems. We treated WPM by microwave hydrolysis in alkali condition and used for gypsum retarder. Four factors were investigated: alkali concentration, solid-liquid ratio, treatment time and temperature. The optimal results for the hydrolysis process were obtained: alkali concentration 0.04 mol/L, solid-liquid ratio 1:3, treatment time 12 min, and at 85 °C. The change rule of protein content is not consistent with that of setting time from the experimental data, which suggests that the components as amino acid, citric acid and lactic acid in retarder also have the retarding effect on gypsum.

Keywords: Mycelium; gypsum retarder; microwave heating; protein.

1. Introduction

Waste penicillin mycelium is a solid waste from the production edprocess of penicillin in pharmaceutical industry. The common treating method was to make poultry feed, landfilled, or treated by incineration. However, using as poultry feed might result in antibiotic abuses and has been prohibited recently in China and other countries. Landfill and incineration would cause serious pollution to the environment. It has been found that WPM contains a large amount of proteins, fiber, enzymes and a certain amount of ergosterol, which have latent recycling value. Previous research found that the common resource utilization technology includes preparation of activated carbon directly [1], and preparation of ergosterol and chitosan by extraction process [2, 3, 4]. The former might result in wasting resource, and much NOx will be produced, causing serious pollution. The latter is more complex in process, which needs further study. Collagen protein can extend the setting time of gypsum. Based on the characteristic of rich in protein, Xiao-Qiong REN used the method of ordinary heating to extract protein from penicillin waste and prepared gypsum retarder [5]. It is pity that the hydrolysis time is long, and the retarding effect of the components was not revealed.

In this paper, we treated WPM by microwave heating method to prepare gypsum retarder, investigated the factors on the setting time of gypsum paste, and studied the retarding effect of the components in the product.

2. Experimental

2.1. Materials

The WPM was used from Huabei Pharmaceutical Co. Ltd. Coomassie brilliant blue G250 solution was prepared in laboratory [6]. Xibaipo Power Plant provided the desulfurization gypsum used in this paper. The physical and mechanical properties of the gypsum were shown in Table 1.

*Corresponding author

Table 1 Physical and mechanical properties of the desulfurization gypsum

Setting time [min]		Bending strength [MPa]	Compressive strength [MPa]	0.2 mm sieve residue	Standard consistency
Initial setting time	Final setting time				
8 min	11 min	6.8 MPa	20.5 MPa	0	0.58

2.2. *Methods*

The standard consistency, setting time were tested according to GB/T17669.3-1999 and GB/T17669.4-1999. According to the standard curve, the protein content of the retarders was tested [7]. See Figure 1.

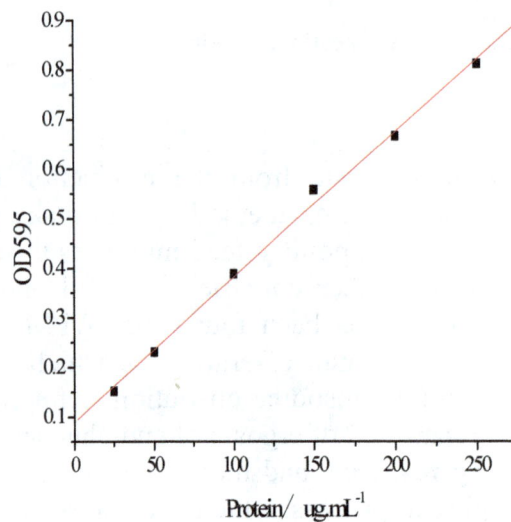

Figure 1 Standard curve of Coomassie Brilliant Bule method

Though standard curve of coomassie brilliant bule method, we can get the correlation as follows:

$$Y=0.0893+0.00293C \tag{1}$$

3. Results and Discussion

3.1. *Effect of alkali concentration*

Alkali was used to break cell wall of penicillin mycelium. Figure 2 gives the changes of protein content with different alkali concentration.

(a)   (b)

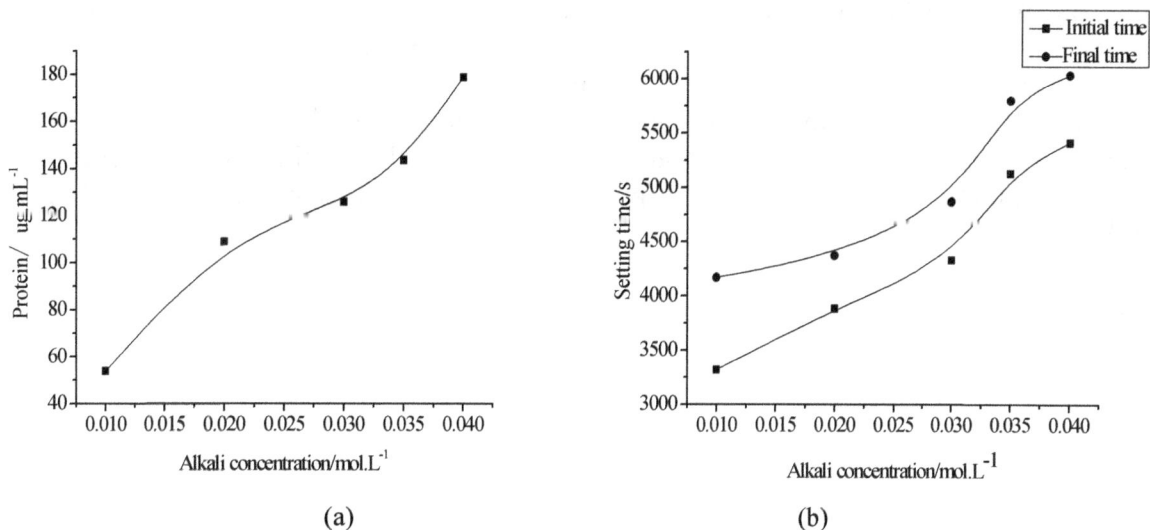

Fig. 2 Effect of alkali concentration on the protein content and the setting time

It can be known from Figure 2(a) the protein content is increased with the increase of the alkali concentration. The cell wall is unstable in the alkaline condition. The higher the alkali concentration, the more completely the cell wall breaks. Excessive alkali concentration denatures the proteins in retarder. Under the condition of high alkali concentration, the viscosity of hydrolysate increased gradually, which is hard to treat in the next step. Figure 2(b) shows that the setting time of gypsum is prolonged with the increase of the alkali concentration. It can be conclude that the main components of the retarder are protein and related materials which prolong the setting time of gypsum.

3.2. Effect of solid-liquid ratio

Solid-liquid ratio as an important factor has an important influence on the protein concentration of the retarder. See Fig. 3.

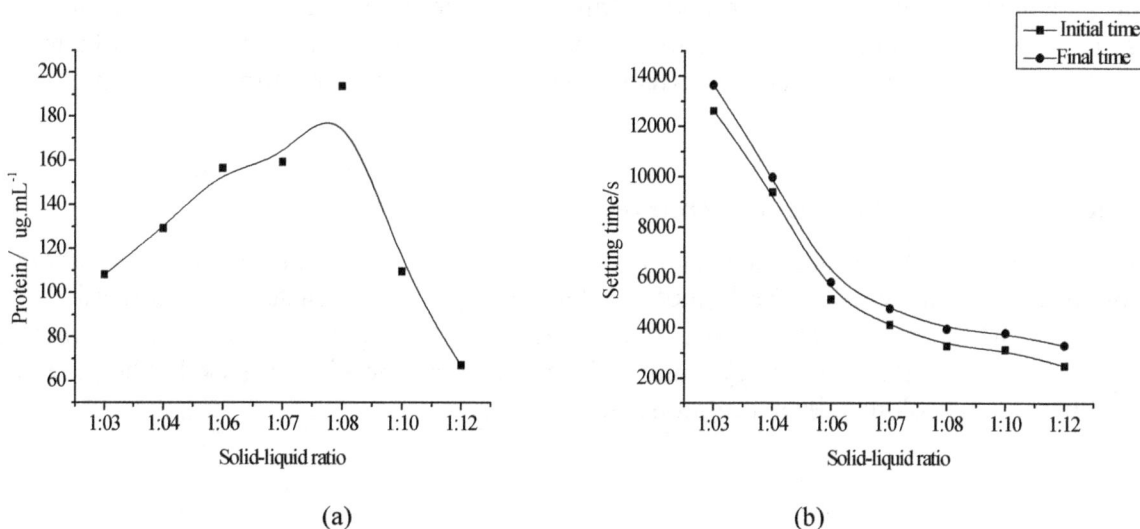

(a)   (b)

Fig. 3 Effect of solid-liquid ratio on the protein content and the setting time of gypsum

As is shown in the Figure 3(a), when the solid-liquid ratio is about 1:8, the protein content reaches peak vale. However, when solid-liquid ratio is above 1:8, the protein content declines rapidly. The reason is that the more reactive liquid, the worse the heat transfer effect. We can see from Figure 3(b), the setting time of gypsum shows a decrease trend. Compared with Figure 3(a), the rule of protein content was not consistent with that the setting time. Therefore, other components also have a retarding effect on gypsum such as amino acid, citric acid and lactic acid in retarder.

3.3. *Effect of microwave radiation time*

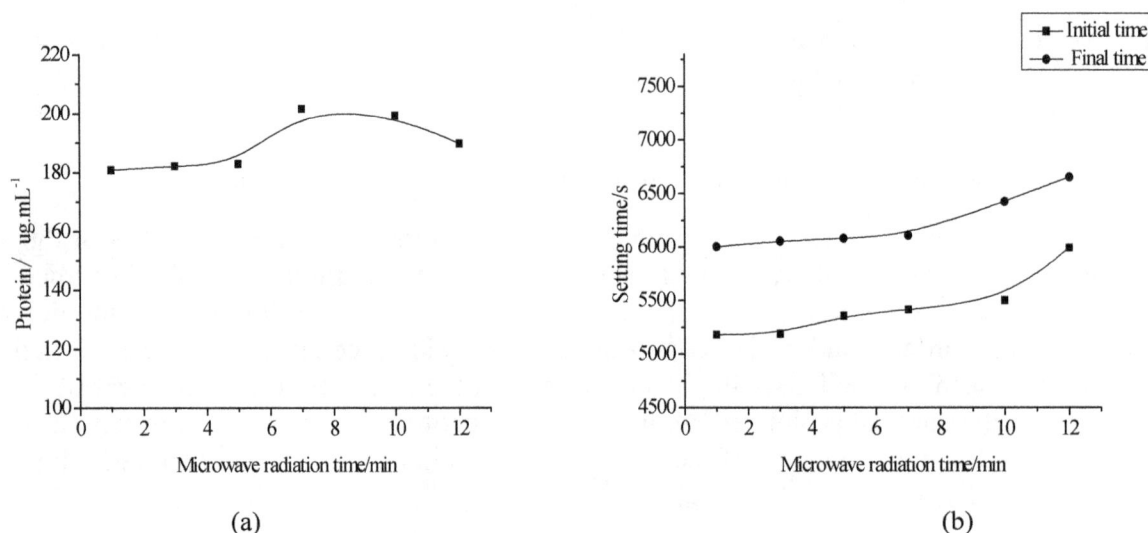

Fig. 4 Effect of microwave time on the protein content and the setting time of gypsum

Due to the microwave strengthening heating, cell wall breaking was completed in a relatively short period of time. See Figure 4(a), when microware heating time is less than 7 min, the protein content is increased slowly. However, above 7min, the protein content decreases. The setting time results are shown in Figure 4(b), which shows that the length of setting time ranged from 87 min to 98 min. Microwave heating has high-effective advantage. Therefore, time is not the main factor in this case.

3.4. *Effect of microwave heating temperature*

We can see from Figure 5(a) that the protein content reach peak vale in 80°C, however, above 80°C, the protein content decreases. The higher the temperature is, the faster polymerization rate is. Unfortunately, when the temperature was over 80°C, part of the protein will be denatured. Figure 5(b) shows that the setting time of gypsum with temperature sharply increased. The results show that it is the protein prolongs the setting time of gypsum.

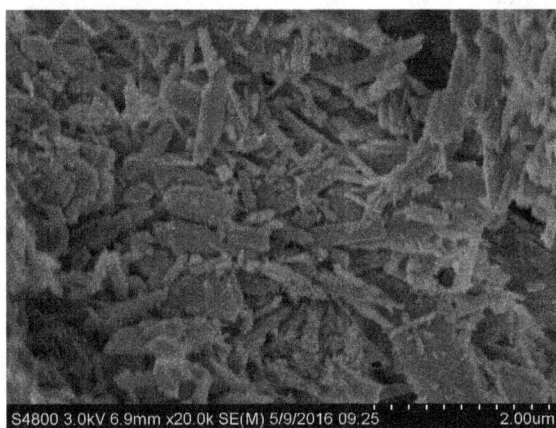

(a)

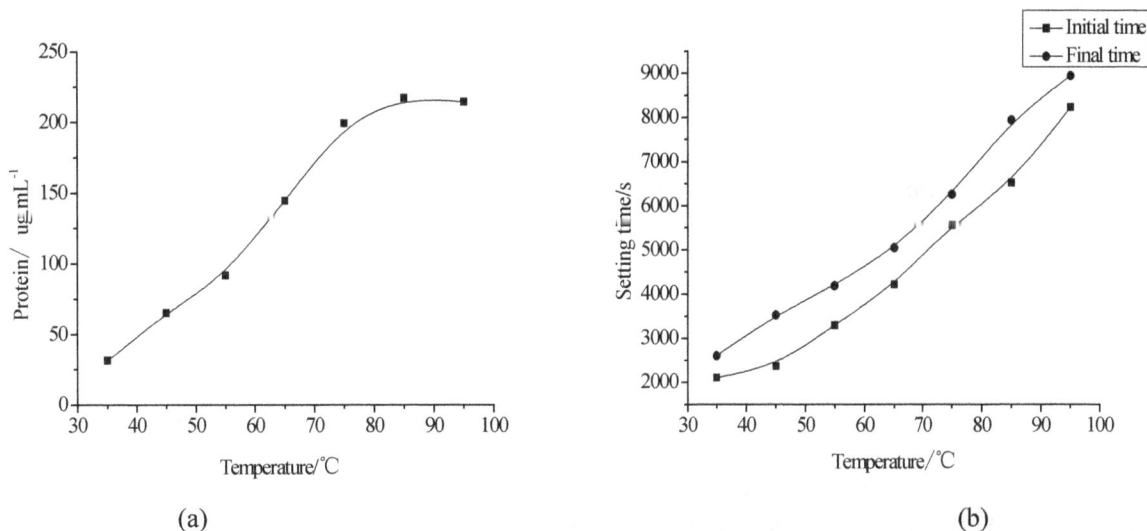

(b)

Fig. 5 Effect of microwave temperature on the protein content and on the setting time

The single factor experiment indicated that the optimum treatment conditions were alkali concentration 0.04 mol/L, solid-liquid ratio 1:3, treatment time 12 min, and the temperature 85°C.

3.5. *Microstructure of the gypsum paste*

Scanning electron micrographs of different cases, gypsum paste with different retarder no retarder, WPM retarder (at the optimum treatment conditions) are shown in Figs. 6(a)-(b).

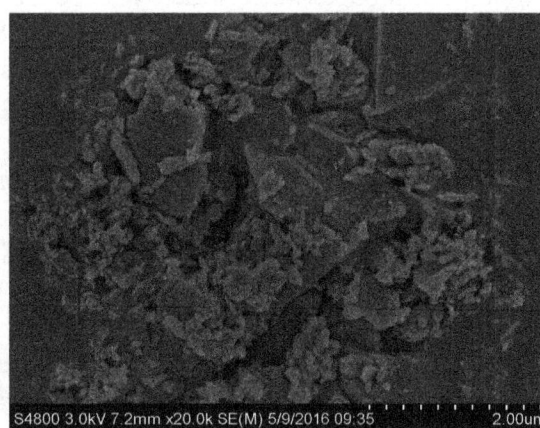

Fig. 6(a) gypsum paste without retarder

Fig. 6(b) gypsum paste with WPM retarder

Fig. 6 Scanning electron micrographs of gypsum paste with different retarder

Figure 6(a) is the micrograph of pure gypsum paste without retarder. It is shown that the crystals usually are acicular or elonmnar, which form network structure. Figure 6(b) is the micrograph of gypsum paste with WPM retarder. It is shown that the gypsum crystals become short, massive crystals generally are columnar or patches. Above all, the main component of the retarder formed colloid in water. The colloid adsorbed on the surface of the gypsum particles, preventing the nuclei formation and crystal growth, thus prolonging the setting time of the gypsum.

4. Conclusion

A kind of gypsum retarder was prepared from waste penicillin mycelium by microwave heating method. Through a series of experiments, the results were obtained: alkali concentration 0.04 mol/L, solid-liquid ratio 1:3, treatment time 12 min, and the temperature 85°C. In addition, microwave heating has the advantage of high effective and high-speed compared with ordinary hydrolysis.

By experiment, we found that the protein content was not consistent with the setting time. Other components also had a retarding effect on gypsum such as amino acid, citric acid and lactic acid in retarder.

The micro structure of the gypsum paste was studied by using SEM, and the results showed that the retarder changed the crystal growth in gypsum paste, thus the setting time of gypsum was prolonged.

Acknowledgment

We would like to show the best grateful to Qing-Qing FAN and Wen-Shu HOU, as they had given much valuable help to the experiment.

References

1. Bao-Hua ZHOU, Qin GAO, Bin GUO, et al, Optimization of activated carbon preparation from penicillin bacterial residue by response surface methodology, Journal of Hebei University of Science and Technology. 6 (2012) 554-558. (in Chinese)
2. C. F. Wang, Extraction of chitosan from waste mycelium, China Brewing. 10 (2009) 124-126. (In Chinese)
3. Ling LIU, Zheng-Qi WU, Xiong WANG, et al, Study on extraction of ergosterol from penicillin mycelium, Journal of Hubei University (Natural Science). 2 (2009) 232-235.
4. Xiao-Qiong REN, Feng-Qing ZHAO, Waste mycelium processing and resource utilization, Advanced Materials Research. 746 (2013) 58-61.
5. Xiao-Qiong REN, Feng-Qing ZHAO, Qian-Qian HU, Study on gypsum retarder from waste penicillin mycelium, Environmental Science & Technology. 37 (2014) 134-137. (in Chinese)
6. Xiao-Ping WANG, Shu-Li XING, The study on the measurement protein by Coomassie Brilliant Blue method, Tianjin Chemical Engineering. 23 (2009) 40-42. (in Chinese)
7. Wan-Yu HUANG, Wei CAO, Jing LI, et al, The study on the protein measurement by Coomassie Brilliant Blue method from fruit juice, Foods and Fermentation Industry. 35 (2009) 160-162. (in Chinese)

Oxidation of Rice Straw with Hydrogen Peroxide Solution

Gui-Zhen Gong[*], Xu Tang

College of Chemical Engineering, Xuzhou University of Technology, Xuzhou 221111, China
[*]Email: ggz72@163.com

Rice straw was oxidized with hydrogen peroxide. The reaction mixture was extracted with petroleum ether and ethyl acetate, respectively. The effects of temperature on the oxidation were investigated. Residue and extracts were analyzed with Fourier Transform infrared (FTIR) spectrometer. The results indicated that hydrogen peroxide solution can convert most of the organic matters contained in rice straw into solvent-soluble species. The research is significance in the high value added utilization of rice straw.

Keywords: Rice straw; oxidation; hydrogen peroxide.

1. Introduction

At present, fossil energy is still the main source of human production and life. But with the increasing depletion of fossil energy and serious environmental pollution, alternative energy sources have gained great concern in the past few decades [1]. Biomass, as the largest source of renewable energy available over the world, is the most promising alternative energy source. According to statistics, the yield of crop straw in China has more than 7 million tons a year, including straw 2.13 million tons, wheat straw class 1.12 million tons, corn 2.12 million tons, and other straw 2 million tons [2]. Unreasonable use, such as burning, causes serious environmental pollution. How to improve its added value has become the focus of all countries [3-8].

In the present study, the effects of different temperature on oxidation of rice straw with hydrogen peroxide (HP) were investigated. The degradation products were analyzed with Fourier Transform infrared (FTIR) spectrometer.

2. Experimental Section

Rice straw was collected from Xuzhou field, Jiangsu, China, and is denoted as RS for the convenience of description. It was prepared by washing with water and naturally drying. Then, it was pulverized into small pieces and passed through an 80-mesh sieve, followed by drying in a vacuum at 65 °C for 24 h. All the materials used in the experiments, including hydrogen peroxide (HP), anhydrous $MgSO_4$, petroleum ether (PE) and ethyl acetate (EA) are commercially purchased analytical reagents, all the liquid reagents were purified by distillation prior to use.

3. Oxidation of RS with HP

2.5 g RS sample was oxidized with 50 mL of HP solution in a 3-necked flask. The reaction was continued for 12 h at certain temperature. After completion of the reaction, the reaction mixture was filtered through a membrane filter with 0.45 um of pore size washed with distilled water until the washing become light-colored, affording filter cake (FC) and filtrate (F_1). F_1 was extracted with petroleum ether and ethyl acetate in a separatory funnel, respectively. The extractable fraction was dried over anhydrous $MgSO_4$ and then concentrated with a rotary evaporator to afford petroleum ether extract (PEE) and ethyl acetate extract (EAE), which were analyzed with FTIR. Procedure for the oxidation of RS with HP and subsequent treatments was shown in Fig. 1.

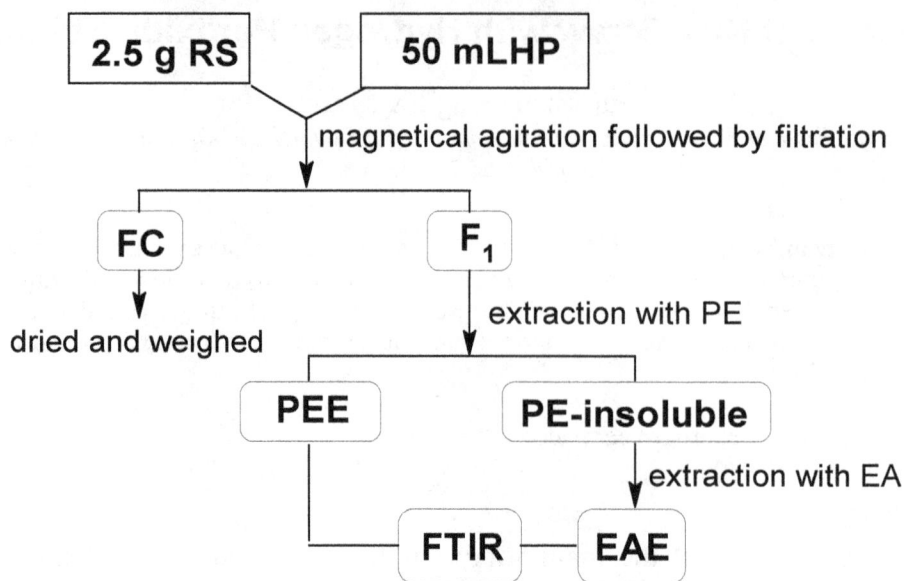

Fig. 1 Procedure for the oxidation of RS with HP and subsequent treatments

4. Results and Discussion

Fig. 2 shows the yields of residue from RS oxidized with HP under different temperature. From the Fig. 2, we can find that the yields of residue are 74.1% for 25 °C, 69.6% for 35 °C, 67.4% for 45 °C, 71.0% for 55 °C, and 71.4% for 65 °C, respectively. The results show that temperature can promote the oxidative depolymerization of rice straw in H_2O_2, and convert most of the organic matters contained in RS into small molecule substance. Degradation effect was the best at 45 °C.

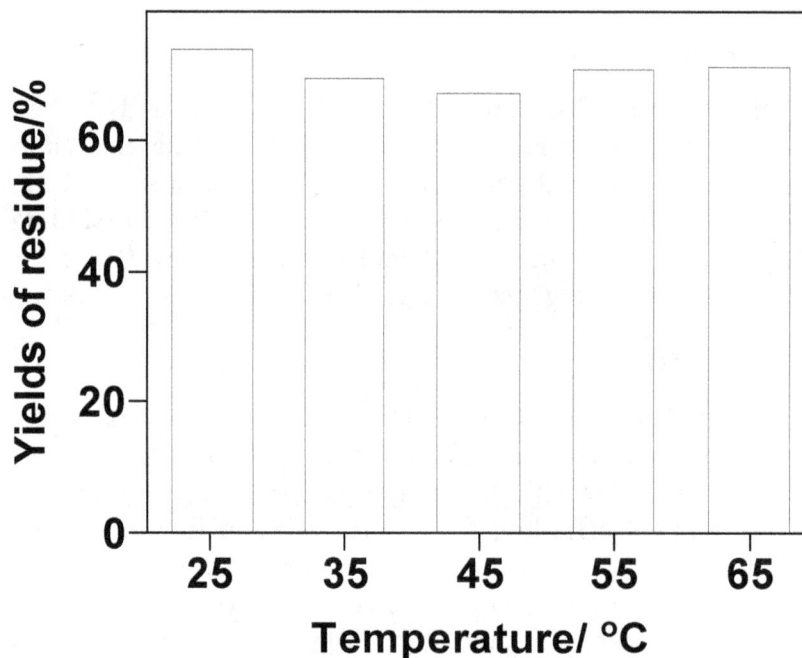

Fig. 2 Yields of residue from RS degraded with HP under different temperatures

Fig. 3 gives the assignments of FTIR absorption bands of the PEE and EAE from RS degraded with HP solution at 45 °C. The spectra of PEE and EAE appear to be slightly different, indicating a similar structure of the two extractives. The peak of –OH group stretching is at 3420 cm^{-1}. The absorption of the two kinds of extracts is strong, indicating that much OH-containing products were enriched by PE and EA. The bands at about 2928 cm^{-1} correspond to methylene and methyl stretching frequencies, which are stronger from EAE than that from PEE, indicating that more aliphatic-containing products were enriched by EA than by PE. The strong band at 1730 cm^{-1} in the spectra is attributed to carbonyl stretching., and the absorption of C-O-C vibration can be seen at around 1240 cm^{-1}, which two kinds of absorption are both stronger in EAE than PEE, revealing that EAE contain more oxygen-containing compounds than PEE.

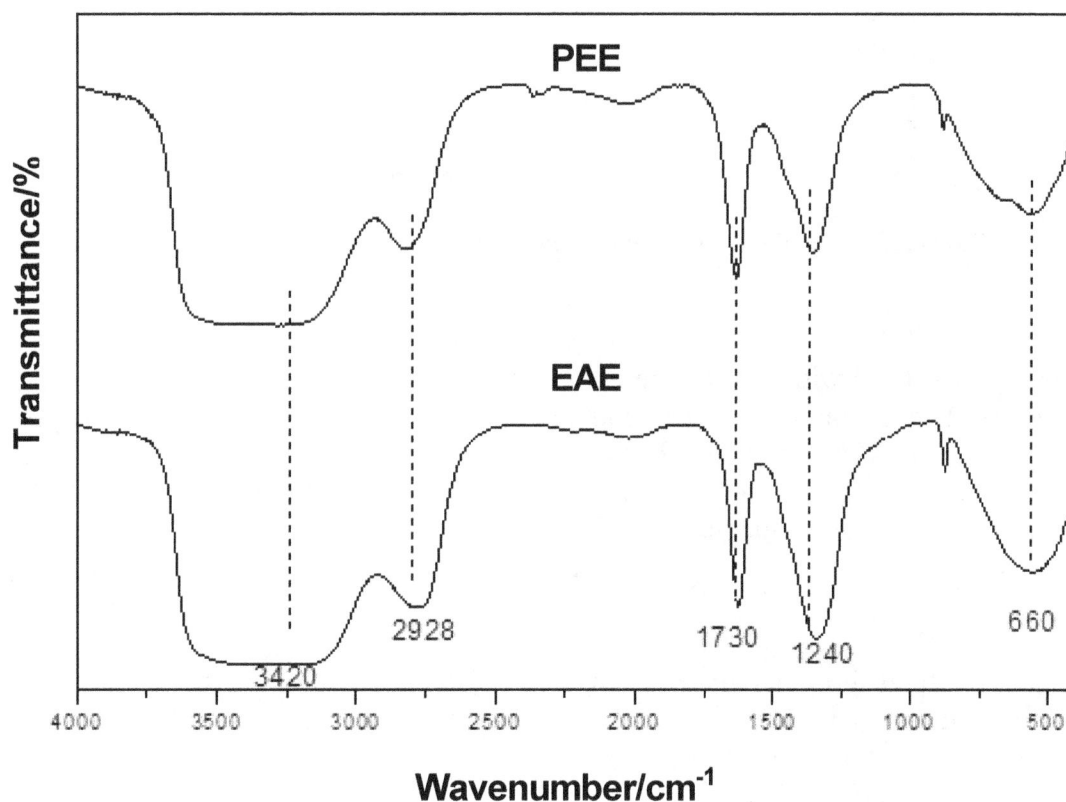

Fig. 3 FTIR spectra of PEE and EAE from RS degraded with hydrogen peroxide at 45 °C

FTIR spectra of EAE from RS degraded with HP under different temperatures are shown in fig. 4. The absorption intensity of each peak increased with the increase of temperature. The increase has become flat over 45 °C, in order to save energy. The optimum reaction temperature is 45 °C to meet the need.

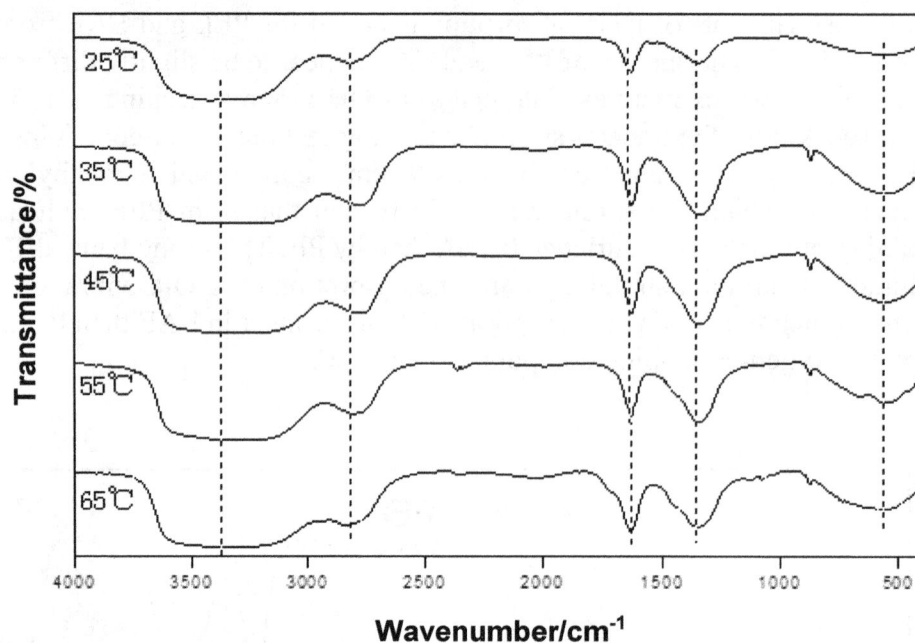

Fig. 4 FTIR spectra of EAE from RS degraded with hydrogen peroxide under different temperatures

5. Conclusion

Rice straw was oxidized with HP at different temperatures. The results indicated that HP solution can convert most of the organic matters contained in rice straw into solvent-soluble species. The yields of residue are 74.1% for 25 °C, 69.6% for 35 °C, 67.4% for 45 °C, 71.0% for 55 °C, and 71.4% for 65 °C, respectively. The products from RS oxidation with HP solution can be enriched by PE and EA. The optimum reaction temperature is 45 °C to meet the need.

Acknowledgments

This work was funded by China Building Materials Federation (2014-M3-4), and Xuzhou Information Institute (XKQ016).

References

1. M. Balat and H. Balat: Appl. Energy vol. 86 (2009), p. 2273-2282.
2. P. Di, J. Jian and L. D. Liu: Industrial Construction vol. 41 (2011), p. 57-60.
3. Z. G. Liu and F. S. Zhang: Energy Convers. Manage vol. 49 (2008), p. 3498-3504.
4. M. K. Akalin, K. Tekin and S. Karagöz: Bioresour Technol. vol. 110 (2012), p. 682-687.
5. T. Aysu and M. M. Kücük: Fuel vol. 103 (2013), p. 758-763.
6. S. Yaman: Energy Convers Manage vol. 45 (2004), p. 651-671.
7. E. Minami and S. Saka: J Wood Sci vol. 2005, 51 p. 395-400.
8. B. V. Babu and A. S. Chaurasia: Energy Convers. Manage vol. 44 (2003), p. 2135-2158.

Development and Application of a Novel Nucleic Acid Amplification Kit on Detection of MRSA

Jin-Hong Xie[1], Zhen-Bo Xu[1, 2, 3,*], Bing Li[1, 3], Lin Li[1, 3]

[1]School of Food Science and Technology, South China University of Technology, Guangzhou 510640, China

[2]Department of Microbial Pathogenesis, School of Dentistry, University of Maryland, Baltimore, Maryland, USA

[3]Guangdong Province Key Laboratory for Green Processing of Natural Products and Product Safety, Guangzhou, 510640, China

*Email: zhenbo.xu@hotmail.com

Methicillin-resistant *Staphylococcus aureus* (MRSA) now becomes a global health concern. It costs 3 d-4 d to finish MRSA detection procedure using conventional methods, also it may have false positive or false negative. Thus, developing an accurate and rapid method in MRSA detection and infection control becomes necessary. This study aimed to develop and establish a multiplex PCR assay for rapid and sensitive detection of MRSA. Four genetic loci were selected to be detected in one amplification system, *16S rRNA* of *Staphylococcus* genus, *femA* of *S. aureus*, *mecA* of methicillin-resistance and *orfX* of SCC*mec*. The PCR reaction was finished within 3 h. The specificity of multiplex PCR was brought out with the evaluation using four different kinds of reference strains including MSCNS, MRCNS, MSSA and MRSA. The diagnostic evaluation was brought out with the appliance of 33 clinical MRSA strains. The results showed that the sensitivity and diagnostic rate of multiplex PCR was 100% and without false negative or false positive. To sum up, this rapid detection method has the potential in the diagnosing and infection control of MRSA.

Keywords: MRSA; rapid detection; multiplex PCR.

1. Introduction

Staphylococcus aureus (*S. aureus*) is a kind of typical foodborne microorganism, especially methicillin-resistant *Staphylococcus aureus* (MRSA), which causes infections and brings serious public health burden. About 14 million people are infected due to foodborne microorganisms per year [1]. Infection caused by MRSA accounts for a large proportion, which has become an advanced project in food safety. [2-4]

16S rRNA gene exists in all bacteria genomes, which encodes rRNA. Confirmation of species with the detection of the variable region sequences in *16S rRNA* among different bacteria. The *femA* gene is regarded as an important gene locus existing in *S. aureus*, which possess strong phylogenetic conservation in all staphylococcal species and MRSA [5]. The *femA* gene produces a 48 kD protein acting on cell wall metabolism. The key dominant of β-lactam resistance expresses by MRSA is penicillin-binding protein 2a (PBP2a) [6], which is a modified penicillin-binding protein with low affinity for β-lactam antibiotics. The *mecA* gene exists on SCC*mec* and encodes PBP2a, because of which *mecA* gene is considered to be a molecular marker of methicillin resistance in *S. aureus*. Besides, as a highly specific gene locus, *orfX* can be integrated by SCC*mec* from 5'-terminal, which influences excision and integrating of exogenous genes.

*Corresponding author

Increasing awareness of the risk of MRSA strains and demands for capable tests of early, cost-effective, timely, and sensitive detection of staphylococci with associated antibiotic resistance determinants has made these tests an urgent demand. Diagnosed as clinical strains, MRSA have been commonly identified using standard and conventional methods as gram staining, hyaluronidase and coagulase tests, Vitek 2 automated system and the API-Staph commercial kit etc. Usually it needs 3 d- 4 d to finish the detection procedure.

Instead, PCR, especially multiplex PCR, possess significant advantages in high specificity, time-saving and high sensitivity. Though various PCR assays have been developer and applied to detect *S. aureus* and methicillin resistance, this research firstly detected four gene locus mentioned above in one PCR amplification procedure. This study aimed to developing and applying a multiplex PCR to detect MRSA rapidly.

2. Material and Methods

2.1. *Bacterial strains*

Five reference strains, including MRSA 12513 and 10864 (with *mecA*, *femA*, *16S rRNA* and *orfX* positive), methicillin-sensitive Staphylococcus aureus (MSSA) 10501 (with *femA*, *16S rRNA* and *orfX* positive, *mecA* negative), methicillin-resistance coagulase-negative staphylococci (MRCNS) 110146 (with *mecA*, *16S rRNA* and *orfX* positive, *femA* negative), methicillin-sensitive coagulase-negative staphylococci (MSCNS) 110830 (with *16S rRNA* positive, *mecA*, *femA* and *orfX* negative) were subjected to evaluation and optimization of multiplex-PCR assay. Application of the multiplex-PCR assays were further performed on a total of 33 MRSA isolates, which were isolated from The First Affiliated Hospital of Jinan University (FAHJU).

2.2. *Culturing condition and template DNA preparation*

In this section, mature methods using genomic DNA isolation kit was applied [7-9]. Genomic DNA from *S. aureus* strains used as template for PCR amplification was prepared from overnight Tryptic soy broth (TSB) cultures at 37°C with shaking. Culture was performed according to the instruction of kits (Dongsheng Biotech, Guangzhou). 0.5-2.0 ml culture was harvested in a sterile 1.5 ml micro centrifuge tube by centrifuging for 1min at 12,000 rpm. Supernatant was discarded. Resuspend pellet in 200 μl DS Buffer. Mix the mixture thoroughly with tip. Add 20 μl Proteinase K and 220 μl MS lysis buffer, mix thoroughly. Incubate at 65°C for 10 min. Add 220 μl EtOH, mix the mixture thoroughly with tip. Transfer the supernatant and floc to a new purify tube, centrifuging for 1 min at 12,000 rpm, discard filtrate. Add 500 μl PS Buffer, centrifuging for 1min at 12,000 rpm, discard filtrate. Add 500 μl PE Buffer, centrifuging for 1min at 12,000 rpm, discard filtrate. Add 500 μl PE Buffer, centrifuging for 1min at 12,000 rpm, discard filtrate. Centrifuge the tube for 3 min at 12,000 rpm, and transfer the spin column to a sterile 1.5ml centrifuge tube. Add 100 μl TE Buffer, incubating at room temperature for 2 min. At last, Centrifuge the tube for 2 min at 12,000 rpm. Remove spin column, the buffer in the micro centrifuge tube contains the highly purified DNA. Store it at -20°C.

2.3. *Primer design*

Four genetic loci were selected to differentiate MRSA、MSSA、MRCNS、MSCNS and non-Staphylococci strains. The protocol was designed to (i) detect any staphylococcal species to the exclusion of other bacterial pathogens using as an internal control, with primers corresponding to Staphylococcus-specific regions of the *16S rRNA* genes (C1: 5'-GATGAGTGCTAAGTG TTAGG-3' and C2: 5'-TCTACGATTACTAGCGATTC-3', with an expected 542 bp amplicon); (ii) distinguish between *S. aureus* and coagulase-negative staphylococci (CoNS) strains based on amplification of the *S. aureus* specific *femA* gene (F1: 5'-AAAGCTTGCTGAAGGTTATG-3' and F2: 5'-TTCTTCTTGTAGACGTTTAC-3', with an expected 823 bp amplicon) and (iii) provide an indication of the likelihood that the staphylococci present in the specimen are resistant to methicillin based on the amplification of the *mecA* gene (M1: 5'-GGCATCGTTCCAAAGA ATGT-3' and M2: 5'-CCATCTTCATGTTGGAGCTTT-3', with an expected 374 bp amplicon). (iv) detect an open reading frame existing on both sides of SCC*mec* in order to identify MRSA strain (with *mecA*+) based on the amplification of the *orfX* gene (O1: 5'-ACCACAATCMACAGTCAT-3' and O2:5'-CCCGCATCATTTGATGTG-3' with an expected 212 bp amplicon).

2.4. *Establishment of multiplex-PCR assay*

Table 1 Reaction system of multiplex PCR amplification

Component	Volume (μl)
2×UTaq PCR Master Mix	12.5
Primer M1, M2; C1, C2 (each)	1
Primer F1, F2; O1, O2(each)	1.5
DNA template	1.5
ddH₂O	1

Five reference strains were used to evaluate the multiplex PCR assay, which was carried out in a total of 25 μl reaction mixture. Components for multiplex PCR are listed in Table 1. Primers are stored at -20°C. The stock solution of primers is 100 pmol/μl. Work solution of primers is 10 pmol/μl.

Multiplex-PCR amplification was carried out using the thermal profile as follows: 94°C for 5 min, followed by 30 cycles of 94°C for 30 s, 50°C annealing for 30 s, and 72°C for 1.5 min and a final extension cycle at 72°C for 7 min. PCR amplification finished within 3 h. The amplified products (7 μL/well) were analyzed by gel electrophoresis in 1.5% agarose gels and stained with ethidium bromide for 10 min. A negative control was performed using sterile water instead of DNA template.

2.5. *Evaluation of the novel nucleic acid amplification kit*

In this section, four kinds of reference strains (Table 2) involving MSCNS, MRCNS, MSSA and MRSA, were applied to multiplex PCR amplification system to evaluate the specificity and sensitivity of multiplex PCR.

213

Table 2 Reference strains used in this study and expected loci

Reference Strains	16S rRNA	femA	mecA	orfX
MSCNS (110830)	+	-	-	-
MRCNS (110146)	+	-	+	+
MSSA (10501)	+	+	-	+
MRSA (12531 and 10864)	+	+	+	+

2.6. *Application of the novel nucleic acid amplification kit on clinical strains*

A total of 33 clinical MRSA isolates were subjected to detection by the established multiplex-PCR assay with primers pairs F1 with F2, C1 with C2, O1 with O2, and M1 with M2, and PCR amplicons were evaluated by electrophoresis as aforementioned. These experiments were replicated to ensure reproducibility.

3. Results and Discussion

3.1. *Establishment and evaluation of the multiplex-PCR*

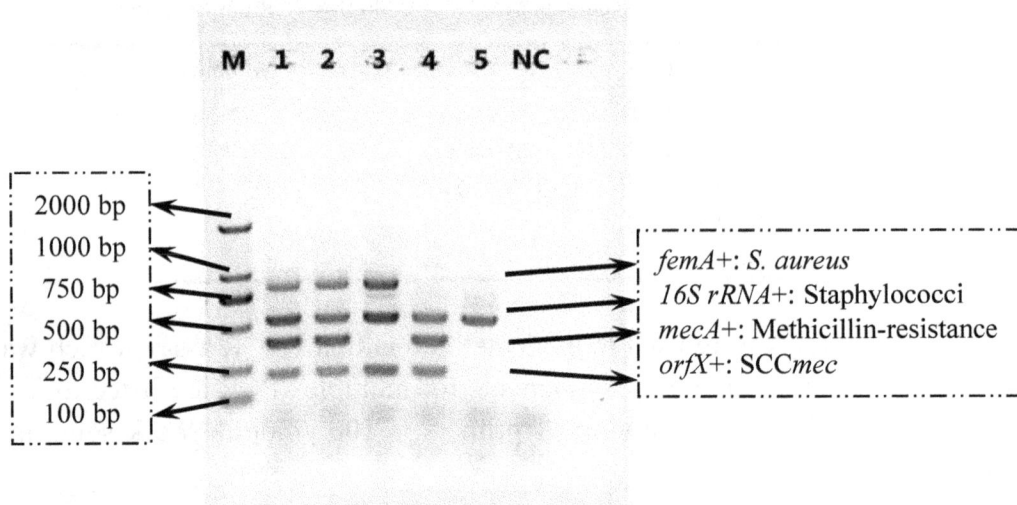

Fig. 1 Amplification results of multiplex-PCR

The specific amplification generated 4 bands on agarose gel (Fig. 1), with sizes 823 bp for *femA*, 542 bp for *16S* rRNA, 374 bp for *mecA* and 212bp for *orfX* respectively. Marker DS™ 2000 (lane M) was used for control. MRSA strain 12513 (lane 1) and strain 10864 (lane 2) showed specific amplification for *mecA*, *femA*, *16S rRNA* and *orfX*, MSSA strain 10501 (lane 3) showed positive result for *femA*, *16S rRNA* and *orfX*, MRCNS 110146 (lane 4) was detected to carry *mecA*, *16S rRNA* and *orfX*, MSCNS 110830 (lane 5) had been found to be *16S rRNA* positive, while negative control (lane NC) cannot amplified any band.

According to the amplification results mentioned above, the multiplex-PCR assay has fine resolution, not only to MRSA, but also to other Staphylococci such as MRCNS, MSCNS, and

MSSA etc. Besides, characters showed by multiplex-PCR, rapid identification and high specificity, for instance, may significantly increase efficiency both in clinical medicine and microbial safety in food industry.

3.2. Application of the novel nucleic acid amplification kit on 33 clinical strains

A total of 33 clinical MRSA strains had been subjected to the application of the multiplex-PCR detection using primers F1 with F2, C1 with C2, O1 with O2, and M1 with M2 together. Multiplex-PCR and subsequent detection by electrophoresis were performed as described previously. All 33 strains were confirmed to be MRSA, which yielded four bands as 212bp, 374 bp, 572 bp and 823 bp, corresponding to *orfX*, *mecA*, *16S rRNA* and *femA*. No false positive amplification was observed, indicating the high specificity of the established multiplex-PCR assay (Fig. 2).

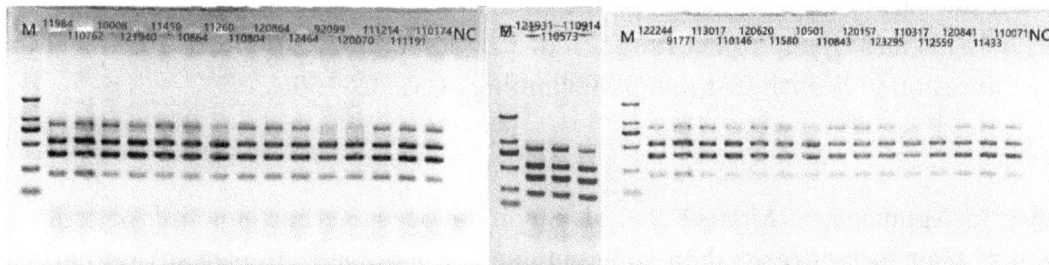

Fig. 2 The amplification results of 33 clinical strains

This study developed a combine molecular assay which was suitable for MRSA rapid detection. In addition, MSCNS, MRCNS, MSSA and MRSA were clearly discriminated by detecting 4 representative genes (*16S rRNA*, *femA*, *mecA*, *orfX*) existing in Staphylococci in one PCR amplification system. Although plenty of reports have developed multiplex PCR assay for MRSA detection, few of them have adopted internal controls to identify CoNS [10] and their methicillin-resistance. Moreover, based on the basic application, this study improved the accuracy of multiplex PCR in MRSA detection, and expanded the resolution of Staphylococci.

Contamination of food products with MRSA is an important cause of food poisoning. The presence of MRSA has been found both in China [11] and its neighboring countries [12-13]. In Europe, The Dutch Food Safety Agency sampled various kinds of meat collected from the retail trade, and MRSA was isolated from 264 (11.9%) of 2217 samples analyzed [14]. It's obvious that MRSA currently exists in various food, and possess a potential risk for health.

This presented multiplex PCR assay is practicable for culture confirmation purposes. With high specificity, time-saving and high sensitivity, PCR is quite suitable for bacteria species identification. Study by Rajan and others [15] has pointed out that MRSA rapid detection with PCR has the potential to preventing spread. Thus, with the wide-appliance, the multiplex PCR assay developed in the study has the penitential to investigate the spread of MRSA and improve both community-acquired and nosocomial infection control condition.

4. Conclusion

The novel nucleic acid amplification kit firstly used 4 genetic loci (*16S rRNA*, *femA*, *mecA* and *orfX*) to identify MRSA strains. Besides, detecting other kind of Staphylococci (MSCNS, MRCNS and MSSA) was synchronous achieved in this assay. The diagnostic accuracy was tested using 33 clinical MRSA strains and showed 100% (33/33) sensitivity. Thus, this assay possesses the potential to be an effective tool in investigation of MRSA spread and its infection control.

Acknowledgments

This work was funded by the National 973-Plan of China (2012CB720800), National Natural Science Foundation of China (31201362 & 31101278), International Science & Technology Cooperation Program (2013B051000014), National Outstanding Doctoral Dissertation Funding (201459), Guangdong Outstanding Doctoral Dissertation Funding (K3140030), Open Project Program of State Key Laboratory of Food Science and Technology, Jiangnan University (SKLF-KF-201513) and Open Project Program of Key Laboratory for Green Chemical Process of Ministry of Education in Wuhan Institute of Technology (GCP201506).

References

1. A. Rojo, A. Aguinaga, S. Monecke, et al. Staphylococcus aureus genomic pattern and atopic dermatitis: May factors other than superantigens be involved? European Journal of Clinical Microbiology & Infectious Diseases Official Publication of the European Society of Clinical Microbiology (2014) 651-8.
2. Z. Xu, L. Lin, C. Jin, et al. Development and application of loop-mediated isothermal amplification assays on rapid detection of various types of staphylococci strains. Food Research International. 47 (2012) 166-173.
3. R. You, Z. Gui, Z. Xu, M. E. Shirtliff, et al. Methicillin-Resistance Staphylococcus aureus detection by an improved rapid PCR assay. African Journal of Microbiology Research. 6 (2012) 7131-7133.
4. R. M. Klevens, M. A. Morrison, J. Nadle, et al. Invasive methicillin-resistant Staphylococcus aureus infections in the United States. Jama the Journal of the American Medical Association. 298 (2007) 1763-71.
5. R. L. Hürlimann-Dalel, C. Ryffel, F. H. Kayser, et al. Survey of the methicillin resistance-associated genes mecA, mecR1-mecI, and femA-femB in clinical isolates of methicillin-resistant Staphylococcus aureus. Antimicrobial Agents Chemotherapy. 36 (1992) 2617-21.
6. D. Lim and N. C. Strynadka. Structural basis for the beta lactam resistance of PBP2a from methicillin-resistant Staphylococcus aureus. Nature Structural Biology. 9 (2002) 870-876.
7. Y. Deng, J. Liu, B. M. Peters, et al. Antimicrobial resistance investigation on Staphylococcus strains in a local hospital in Guangzhou, China, 2001-2010. Microbial Drug Resistance. 21 (2015) 102-104.

8. R. Zhou, L. Lin, J. Su, et al. Detection of class I integron from foodborne salmonella and analysis on the drug resistance gene cassettes. Science & Technology of Food Industry. 31 (2014), 279-278.

9. B. Li, X.Liu, L. Li, et al. Molecular identification of the genotype of staphylococcus aureus biofilm. Modern Food Science & Technology. 31 (2015) 74-79

10. K.,Zhang J. A. Mcclure, S. Elsayed, et al. Novel multiplex PCR assay for characterization and concomitant subtyping of staphylococcal cassette chromosome mec types I to V in methicillin-resistant Staphylococcus aureus. Journal of Clinical Microbiology, 43 (2005) 5026-33.

11. S. Lu, Z. Tang, X. Li, et al. Prevanlence antibiotic susceptibility and enterotoxin gene patterns of Staphylococcus aureus in raw milk Nanning City. Appl Prev Med. 16 (2010) 271–274.

12. E. Shin, H. Hong, J. Park, et al. Characterization of Staphylococcus aureus faecal isolates associated with food-borne disease in Korea. Journal of Applied Microbiology. 121(2016) 277-286.

13. Y. Sato'O, K. Omoe, I. Naito, et al. Molecular Epidemiology and Identification of a Staphylococcus aureus Clone Causing Food Poisoning Outbreaks in Japan. Journal of Clinical Microbiology. 52 (2014) 2637-2640.

14. E. D. Boer, J. T. M. Zwartkruis-Nahuis, B. Wit, et al. Prevalence of methicillin-resistant Staphylococcus aureus, in meat. International Journal of Food Microbiology. 134 (2009) 52-56.

15. L. Rajan, E. Smyth, H. Humphreys. Screening for MRSA in ICU patients. How does PCR compare with culture? Journal of Infection. 55 (2007) 353-357.

Chapter 5
Polymers

Synthesis and Characterization of Polythiophenes Bearing Pyrroledione in the Conjugated Chain

Wen-Li Wang, Fu-De Liu*

School of Chemistry and Chemical Engineering, Tianjin University of Technology, Tianjin 300384, China
E-mail: liufude@tjut.edu.cn

Poly(2,5-bis[4-hexyl-thiophen-2-yl]-N-n-Octyl-3,4-thiophenedicarboximide-co-3- hexylthiophene) (PHTPD) was synthesized with $FeCl_3$ as oxidant by 3-hexylthiophene and 2,5-bis (4-thiophen-2-yl-hexyl) -N- thiophen-n-octyl-3,4-dicarboximide. The optical properties of polymer PHTPD was characterized by UV-visible absorption spectra (UV) and fluorescence spectra (FL), The electrochemical properties of polymer PHTPD was recorded by cyclic voltammetry. By calculating we obtained the optical band gap, electrochemical bandgap, HOMO/LUMO orbital energy of PHTPD. Moreover, we utilized gel permeation chromatography (GPC) to characterize its molecular weight and thermogravimetric curve (TG) to characterize its thermal properties.

Keywords: Stile reaction; polythiophene; conjugated polymer.

1. Introduction

Recently, polymer solar cells are systematically investigated due to the great potential for widespread use and the promise of high volume production at low cost [1]. With polymer poly (3-hexylthiophene) (P3HT) as donor, $PC_{61}BM$ as acceptor, the PCE of solar cells has reached 5.0% by Hiramotoin 2007. So far, the photoelectric conversion efficiency (PCE) of solar cells based double-layer film structure has already broken 8% [2]. The efficiency of organic laminated cell devices has exceeded 10% [3, 4]. Nevertheless, compared to inorganic solar cells, its process engineering is still at comparatively low levels due to its short developing time. The promotion of photoelectric conversion efficiency (PCE) is the critical means to further development and improvement of polymer solar cells [5]. The electrical and optical properties of donor materials are a critical factor to photoelectric conversion efficiency (PCE) of the organic photovoltaic cells [6]. For instance, maximum absorption wavelength increasing will improve the photon absorption efficiency [7]. The absolute value of HOMO orbital energy(E_{HOMO}) increased will enhance the open circuit voltage [8]. M.C. Scharber mapped the diagram of the photoelectric conversion efficiency of photovoltaic cells, the band gap (E_g) of the donor material and the LUMO orbit energy (E_{LUMO}) [9], it shows that the good matching between band gap (E_g) and LUMO orbit energy (E_{LUMO}) is an important factor that affect the photoelectric conversion efficiency. So the design and synthesis of new type structure of polythiophene derivatives and the determination of E_g, E_{HOMO} and E_{LUMO} is of great significance to screening power conversion efficiency material.

2. Experimental

2.1. *Materials and instruments*

Chloroform was distilled from calcium hydride under nitrogen before use. N, N-Dimethyl form amide was dried and distilled under reduced pressure. THF and toluene was distilled from Na. All other chemicals and solvents used in this work were analytical grade and purchased from commercial sources and used without further purification.

---

*Corresponding author

¹H NMR and ¹³C NMR spectra were recorded on Bruker AM - 400 spectrometer using CDCl₃ as solvent in all cases. IR analyses were performed on a Nicolet 380 spectrometer. UV-visible spectra and fluorescence spectra of the polymers were measured by HITACHI U-3310 spectrophotometer and HITACHIF-4500 fluorescence spectrophotometer, respectively. The melting points were determined on a RY-1 melting point apparatus and temperatures were uncorrected. Cyclic voltammetry (CV) measurements were conducted on an electrochemistry workstation (LK-2005A, Tianjin) with the polymer film on platinum (Pt) plate as the working electrode, Pt wire as the counter electrode, and Ag / AgCl electrode as a reference electrode with a scan rate of 50 mVs. Tetrabutylammoniumperchlorate (TBAP, 0.1 mol / L) and chloroform were used as the supporting electrolyte and solvent, respectively. The measurements were calibrated using ferrocene as the standard. TGA were conducted by Germany Netzsch TG 209F3 instrument under nitrogen atmosphere at a heating rate of 10 °C min⁻¹ from 30 °C to 1000°C. Gel permeation chromatography (GPC) measurements were performed in THF, using a Waters 2414 system (Milford, MA) equipped with a refractive index detector. Adequate molecular weight separation was achieved using three Waters Styragel columns (HT3, HT4, HT5) in series at a flow rate of 1.0 mLmin⁻¹ and a temperature of 35 °C. Calibration curves were obtained with nearly monodisperse polystyrene

2.2. Synthesis of the monomer and the polymer

The synthetic routes of the monomer and the polymer are shown in Scheme 1, and the detailed synthetic processes are as follows.

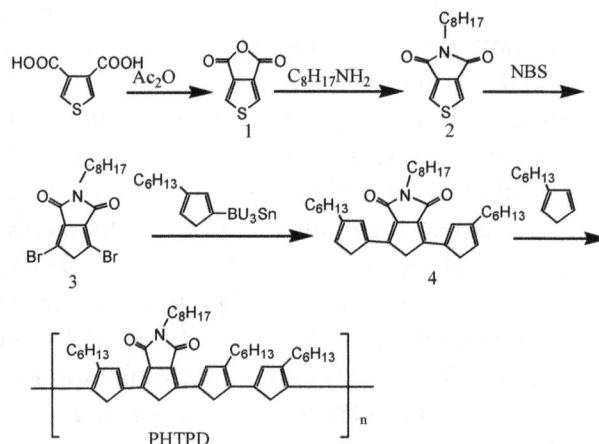

Scheme 1 Synthetic Processes of monomers and polymer

2.3. N-n-Octyl-3,4-thiophene-dicarboximide

3,4-thiophenedicarboxylicacid (0.5g 2.9mmoL) and acetic anhydride (15mL) were refluxed at 130°C overnight under nitrogen. The solution was cooled to room temperature and the solvent was evaporated with rotary evaporation. The resulting residue and n-octylamine were dissolved in toluene (15mL) and refluxed at 110°C ` `with stirring overnight. The solution was cooled to room temperature and the organic solvent was removed with rotary evaporation. Thionylchloride

(10mL) was added and refluxed at 80°C with stirring overnight. The solution was cooled to room temperature. The mixture was washed with water, extracted with dichloromethane, and finally dried with anhydrous MgSO4. The solvent was evaporated under reduced pressure and the crude product was purified by the silica gel column chromatography using petroleum ether and dichloromethane (petroleumether/ dichloromethane=1:3) as the eluents. The product was obtained as a white solid (0.66g) with a melting point of 113-120°C. Yield: 85.7%. IR (KBr, cm^{-1}): 2850-3086, 1690, 1360-1400, 1100, 750, 570. ^{1}H-NMR(CDCl3, 400MHz): 7.826(s,2H), 3.629(m,2H), 1.678(m,2H), 1.328(m,2H), 1.282(m,8H), 0.892 (t,3H).

2.4. *2,5-dibromo-N-n-Octyl-3,4-thiophenedicarboximide*

N-n-Octyl-3,4-thiop-hene-dicarboximide (0.66 g 2.4mmoL), sulfuric acid (5mL 98%) and trifluoroacetic acid (15mL) were added into a 3-neck flack (50mL). The solution was stirred at room temperature, NBS was added meanwhile. TLC was used for tracking and detection. The reaction was next quenched with water, extracted with dichloromethane, washed with water several times and then dried by anhydrous MgSO4. The solvent was evaporated under reduced pressure and the crude product was purified by the silica gel column chromatography using petroleum ether and dichloromethane (petroleumether/ dichloromethane=1:3) as the eluents. The product was obtained as a white solid (0.861g) with a melting point of 70-75°C. Yield: 80.2%. IR (KBr, cm-1): 2920, 2850, 1690, 1530, 1380, 950-1050, 750, 680.1H-NMR (CDCl3, 400MHz): 3.615 (m,2H), 1.635 (m,2H), 1.28-1.333 (m, 10H), 0.905 (t, 3H).

2.5. *4-hexyl-2-tributyltinthiophene*

Under nitrogen, 3-hexyllthiophene (5.04g, 30.0mmol) and THF (20 mL) in a 100 mL two-necked flask were cooled to -78 °C. n-Butyllithium (14 mL, 2.5 M in hexanes) was added to the solution slowly over 10 minutes, stirred for 1 h. Tributyltinchloride (11.7 g, 36mmol) was added slowly over 10 minutes and the reaction was allowed to warm to room temperature with stirring overnight. Water was added, and the aqueous phase was extracted several times with ethyl ether. The combined organic phases were dried over MgSO4, and the solvent was removed in vacuum. Vacuum distillation gave alight yellow liquid 13.5g, which was directly used for next step without further purification.

2.6. *2,5-bis[4-hexyl-thiophen-2-yl]-N-n-Octyl-3,4-thiophenedicarboximide*

2,5-Dibromo-N-n-octyl-3,4-thiophenedicarboximide(0.5g1.18mmol),4-hexyl-2-tributyltinthiophene (1.35g 2.95mmoL) and Pd(PPh3)4 (0.02 g, 0.017mmol) in toluene (20 mL) were refluxed under nitrogen. After refluxing for 48 h, the solution was cooled to room temperature .The reaction was next quenched with water, extracted with ethyl ether and finally dried with anhydrous MgSO4. The solvent was evaporated under reduced pressure and the crude product was purified by silica gel column chromatography using petroleum ether and dichloromethane (petroleum ether/ dichloromethane=2:1) as eluent. The product was obtained as a green and yellow solid(0.64g). Yield: 90.1%. IR (KBr, cm-1): 2930, 2360, 1610-1750, 1330-1470, 1070, 620, 490.1H-NMR (CDCl3,400MHz): 7.90 (d,2H), 7.047 (d,2H), 3.68 (m,2H), 2.65 (m,4H), 1.68 (m,4H), 1.55 (m,2H), 1.35 (m,6H), 1.29 (m,16H), 0.92 (m,9H).

2.7. Poly(2,5-bis[4-hexyl-thiophen-2-yl]-N-n-Octyl-3,4-thiophenedicarboximide-co-3-hexyl thiophene) (PHTPD)

FeCl₃ (0.2g 1.23mmol) and chloroform (10mL) was stirred at room temperature in a 4-neck flask under nitrogen. Under ice bath temperature and bubbling device to control the nitrogen flow, 2,5-bis[4-hexyl-thiophen-2-yl]-N-n-octyl-3,4-thiophenedicarboximide(0.2g1.23),3-hexyllthiophene(0.2g 1.19mmol) and chloroform(20mL) were added dropwise at 0-5°C. The mixture was stirred for 6 h in an ice bath. When the mixture was stirred for another 24 h at room temperature, dark grey-colored precipitate was obtained. The precipitate was extracted with dichloro, washed with water thoroughly, and then was dried over MgSO₄. Acuum distillation gave a black solid. The product was washed with methanol thoroughly. Yield: 0.36 g.

3. Results and Discussion

3.1. Optical properties

The UV-visible spectrum of polymer PHTPD in DMF solution is shown in Fig. 1. The solution absorption spectrum of polymer PHTPD shows profile with maximum peak appearing at 415 nm in Fig. 1. From the edge of the absorption spectrum (vλg=570 nm) in Fig. 1, the optical band-gap (Eg=1 240/λg) of the polymer is estimated to be 2.17eV.

The fluorescence spectrum of polymer PHTPD in DMF is shown in Fig. 2. When the excitation wavelength was fixed at 415 nm, polymer PHTPD shows the strongest fluorescence-emission at 470 nm with respect to fluorescence response range of 430-570 nm. The results show that polymer PHTPD exhibited good optical performance.

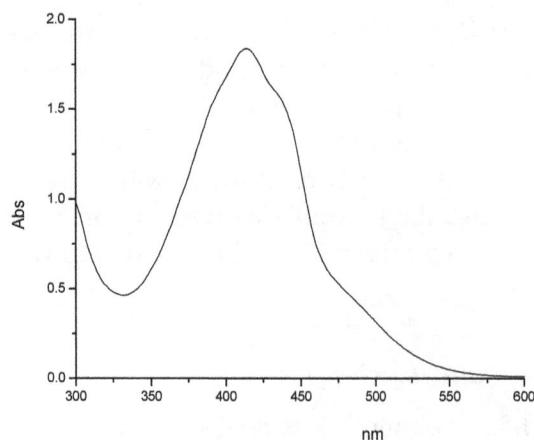

Fig. 1 UV-visible absorption spectrum of PHTPD

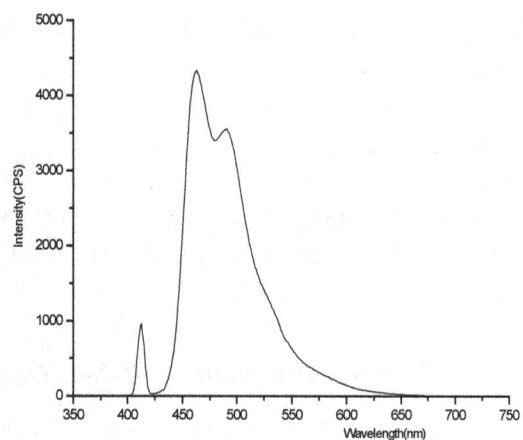

Fig. 2 Fluorescence spectrum of PHTPD

3.2. Electrochemical properties

The electrochemical property of polymer PHTPD was investigated by cyclic voltammetry (CV) to determine the energy level of its HOMO and LUMO. The HOMO and LUMO energy levels (E_{HOMO} and E_{LUMO}, respectively) of the polymer were calculated by use of the following empirical formulas, $E_{HOMO} = - (Eox + 4.4)$ eV and $E_{LUMO} = - (Ere + 4.4)$ eV, where Eox and Ere are the onset oxidation and reduction potentials, respectively.

Fig. 3 shows the onset oxidation and reduction potentials are 1.2ev and -0.56ev, respectively. From the up formulas we can estimate the E_{LUMO} was -3.84eV, meanwhile the E_{HOMO} value was found to be -5,6eV, with a electrochemical band-gap of 1.72eV.

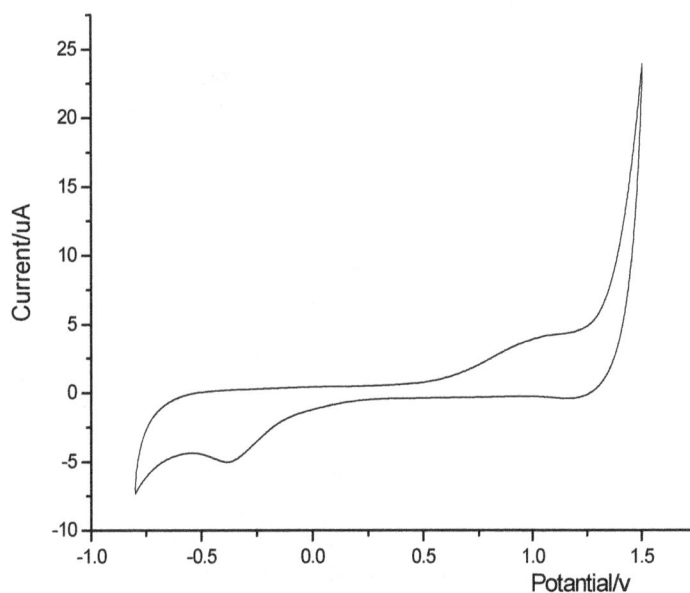

Fig. 3 Cyclic voltammetry of PHTPD

3.3. Molecular weight and thermal properties

We utilized gel permeation chromatography (GPC) to determine the molecular weight of PHTPD, and the detailed data are shown in Table 1

Table 1 shows that the molecular weight of the PHTPD reaches 16206, which is mainly due to the PHTPD with smaller steric hindrance of the straight-chain, in favor of the polymerization. By the PDI is 1.59 you can see the distribution width of PHTPD is not wide, and the relative molecular weight is relatively concentrated.

Table 1 Molecular weight of polymer PHTPD

Sample	Mw	Mn	PDI
PDTPD	25877	16206	1.59

(a) Mw—Weight average molecular weight (b) Mn—Number-average molecular weight
(c) PDI—Width of molecular weight distribution; PDI= Mw/ Mn

The thermal properties of polymer PHTPD was investigated by thermogravimetric curve (TG). The Td of polymer PHTPD, which was the thermal decomposition temperature when the percent of weightlessness was 5%, is 264.4°C. It shows that the polymer has good heat resistance, good thermal performance.

4. Conclusion

By calculating all these data, we obtain the optical band-gap (2.17ev), electrochemical band-gap (1.72ev) and HOMO/LUMO orbital energy (-5.6ev/-3.84ev) of the polymer PHTPD. These dates suggest that the band-gap and HOMO/LUMO orbital energy of the polymer is decreasing when Pyrrolediketone is introduced into the main chain of the conjugate. The decomposition temperature (264.4°C) and molecular weight (16206) show that the polymer has a good degree of polymerization and thermal stability, which is good for the conversion efficiency of PCE.

Acknowledgment

This work was funded by the National Nature Science Foundation of China (21176193) and is gratefully acknowledged.

References

1. S.H. Jin, H.J. Park, J.Y. Kim. Poly(fluorenevinylene) derivative by gilch polymerization for light emitting diode applications [J]. Macromolecules, 2002, 35: 7532–7534.
2. Z. He, C. Zhong, S. Suetal, Enhanced power-conversion efficiency in polymer solar cells using an inverted device structure [J]. Nat Photon, 2012,6 (9): 591-595.
3. M.A. Green, K. Emery, Y. Hishikawaetal. Solar cell efficiency Tables (Version 41) [J]. Progress in Photovoltaics: Research and Applications, 2013, 21 (1):1-11.
4. G. Li, R. Zhu, Y. Yang, Polymer solar cells [J]. Nat Photon 2012, 6 (3):153-161.
5. Huaxing Zhou. Rational Design of High Performance Conjugated Polymers for Organic Solar Cells [J]. Macromolecules, 2012, 01(10): 1021-1024.
6. G. Dennler, M. C. Scharber, C. J. Brabec, Polymer-fullerene bulk-heterojunction solar cell [J]. Adv. Mater, 2009, 21:1-16.
7. J.E. Carle, J.W. Andreasen, M. Jorgensen et al. Low band gap polymers based on 1,4-dialkoxybenzene, thiophene, bithiophene donors and the benzothiadiazole acceptor [J]. Solar Energy Materials & Solar Cells, 2010, 94: 774–780.

8. C.F. Lin, S.W. Liu, C.C. Lee et al. Open-circuit voltage and efficiency improvement of subphthalocyanine-based organic photovoltaic device through deposition ratecontrol [J]. Solar Energy Materials & Solar Cells, 2012, 103: 69–75.
9. M.C. Scharber, D. Muhlbacher, M. Koppe et al. Design Rules for Donors in Bulk-Heterojunction Solar Cells — Towards 10% Energy-Conversion Efficiency [J]. Adv. Mater, 2006, 18:789-794.

Characteristics of Flame Over Polyamide in High Oxygen Environments

Wen-Zhong Mi[1,2], Rui-Chao Wei[1], Shen-Shi Huang[1], Jian Wang[1,*]

[1]*State Key Laboratory of Fire Science, University of Science and Technology of China, Hefei, Anhui 230029, China*
[2]*Fire Department of Ministry of Public Security, Beijing 100054, China*
Email: wangj@ustc.edu.cn

Solid material polyamide (PA6) is used as nylon cap in tee joint mouth of oxygen bombs. Its flame characteristics are influenced by the high oxygen concentration environment in the bombs and required study. The flame characteristics of PA6 are directly related to safety of oxygen bombs. Experiments were conducted in modified oxygen index apparatus to study the effect of oxygen concentration on flame height and flame spread. Thermoplastic material polymethyl methacrylate (PMMA) and thermosetting material epoxy were selected as comparison material. Both the flame height and flame spread rate increase consistently with a power law with the increasing oxygen concentration. The results of this work provide further understanding of the characteristics of materials like PA6 in environments similar to oxygen bombs and future spacecraft.

Keywords: High oxygen concentration; flame spread rate; flame height; PA6.

1. Introduction

Oxygen is widely used in aircraft industry, medical industry and metallurgic industry. Solid material polyamide (PA6) is used as nylon cap in tee joint mouth of oxygen bombs which have the high oxygen concentration environment. Polyamide plastic has many advantages such as wide material sources, easy modification, easy processed to molding. However, the polymer is easily aged and ignited. The flame characteristics of PA6 under high oxygen concentration are different from atmosphere. Particularly important is the response of PA6 to high oxygen concentration in oxygen bombs.

Some scholars have carried on the experimental studies on combustion behaviors of materials under high oxygen concentration. Mekki [1] in Combustion/Heat Transfer Laboratory studied the effect of laminar forced flow wind-aided on flame spread over wood and PMMA in the ceiling configuration. The result shows that the flame spread rate vary as $Y_{O\infty}^{1.1}$ for wood and $Y_{O\infty}^{1.4}$ for PMMA with the increase of oxygen mass fraction. Y. H. C. CHAO and A. C. FERNANDEZ-PELLO [2] studied the fire spread speed on PMMA and found that fire spread speed vary as $V_f \sim Y_{O\infty}^1$ under low oxygen concentration environment but vary as $V_f \sim Y_{O\infty}^2$ under high oxygen concentration. Osorio [3] in Berkeley conducted the experiment to study the influences of oxygen concentration and oxidizer flow velocity on the horizontal flame spread over two fire resistant fabrics. Few studies have been conducted about polymer material PA6 which is used under high oxygen concentration.

In the process of measuring flame height, we compare PA6 with the material epoxy. When analyzing flame spread rate, we compare PA6 with the materials PMMA and epoxy. PA6 and epoxy are the thermoplastic material but epoxy is thermosetting material. We compare these three different materials trying to find the difference characters of flame spread between thermoplastic materials and thermosetting materials under high oxygen concentration. The study will provide further understanding of the characters of material like PA6 in environments similar to those of oxygen bombs and future spacecraft and provide theoretical guidance for the oxygen bombs and future spacecraft.

*Corresponding author

2. Experimental Apparatus and Materials

Modified HC-2 oxygen index apparatus is used as experimental apparatus to measure flame spread under different oxygen concentration. The experimental apparatus is presented in Fig. 1. The flame spread process and flame shape is recorded by a high speed Sony camera (HDR-XR160). The camera frame rate is set as 25 frames per second. Binary image processing technique is used to process the flame shape. The samples are 40mm in length, 4mm in diameter. For each case, the test was conducted at least 3 times to ensure repeatability.

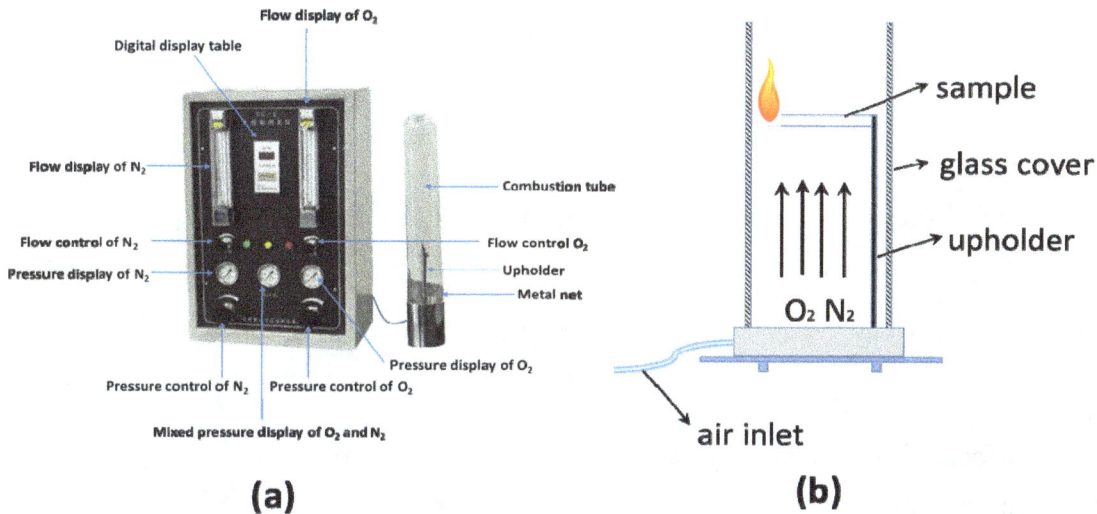

(a) HC-2 oxygen index apparatus (b) sample holder system in combustion tube
Fig. 1 Schematic of experimental apparatus

3. Result and Discussion

3.1. *Flame height*

The flame height of horizontal fire spread can be defined as the vertical distance from the upper surface of sample to the flame tip [4]. Fig. 2 is the flame shape of PA6 under 30% (a) and 35% (b) oxygen concentration at 250 frame (10s) after the fire ignition.

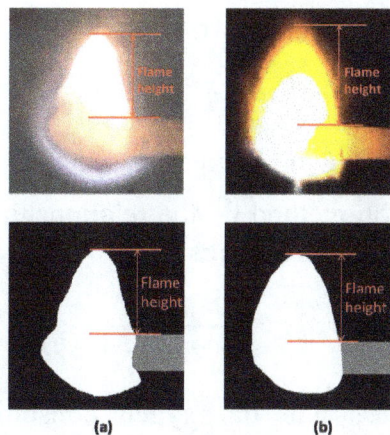

Fig. 2 Flame shape and their binary images of PA6: (a) 30% oxygen concentration (b) 35% oxygen concentration

Diffusion flame height is related to the inertial force and buoyancy. So Froude number is introduced as an analytical tool. Froude number stands for the relative importance between inertia and buoyancy. The Froude number in the field of fire is usually expressed as follows:

$$F_r = \frac{momentum}{buoyancy} = \frac{u_1}{\sqrt{gD}} \tag{1}$$

where u_1 is gas flow rate; g is acceleration of gravity; D is fuel size.

The relationship between gas evaporation velocity and mass loss rate can be expressed as

$$u_2 = \frac{\dot{m}}{\rho A} \tag{2}$$

Mass loss rate of the solid can be obtained based on the energy relationship. Mass loss rate is given as follow:

$$\dot{m} = \frac{\dot{Q}_a - \dot{Q}_l}{L_V} \tag{3}$$

where $\dot{Q}_a$ is the heat flux obtained by samples. $\dot{Q}_l$ is the heat flux used by samples to reach the gasification temperature. L_V is heat of vaporization. The heat flux that sample get can be divided into heat flux from the flame and heat flux from the external heat source.

$$\dot{m} = \frac{\dot{Q}_F + \dot{Q}_E - \dot{Q}_l}{L_V} \tag{4}$$

where $\dot{Q}_F$ is the heat flux that the samples receive from the flame; $\dot{Q}_E$ is the heat flux that the samples receive from the external radiation. The external radiation can be neglected in this study, so the value of $\dot{Q}_E$ is zero.

$\dot{Q}_F$ is a part of combustion heat, so it can be written as

$$\dot{Q}_F = \zeta \dot{q}_f \tag{5}$$

where $\dot{q}_f$ is combustion heat flux and ζ is an experimental constant.

According to oxygen consumption method [5], the relationship between flame combustion heat flux and oxygen concentration can be expressed as

$$\dot{q}_f = E\dot{V}_{298}\chi_{O_2}\left(\frac{\phi}{\phi(\alpha-1)+1}\right) \tag{6}$$

where E is heat release per mass unit of oxygen consumed ($\approx 13.1 KJ \cdot g^{-1}$); $\dot{V}_{298}$ is gas volume-flow rate under normal temperature ($25^\circ C$); χ_{O_2} is the mole percent of oxygen; α is volumetric expansion factor and ϕ is depletion factor.

Combining with (4), (5), (6), the formula of mass loss can be rewritten as

$$\dot{m} = \frac{\varsigma E \dot{V}_{298} \chi_{O_2} (\frac{\phi}{\phi(\alpha-1)+1}) - Q_l}{L_V} \tag{7}$$

Assuming that gas flow rate u_1 is equal to gas evaporation rate u_2 and combining with equations (1) (2) (7), Froude number under the influence of oxygen concentration can be written as

$$Fr = \frac{\varsigma E \dot{V}_{298} \chi_{O_2} (\frac{\phi}{\phi(\alpha-1)+1}) - Q_l}{\rho A L_V \sqrt{gD}} \tag{8}$$

$\frac{Q_l}{\rho A L_V \sqrt{gD}}$ is a small value by contrast. Thus, the Froude number is proportional to oxygen concentration.

$$Fr \propto \chi_{O_2} \tag{9}$$

In order to study the relationship between the flame height and the oxygen concentration conveniently and ignore the impact of sample size on flame height, a dimensionless flame height is defined as follows:

$$H^* = H_f / D \tag{10}$$

where H^* is dimensionless flame height; H_f is the actual flame height; D is sample diameter.

When F_r number is small enough, the dimensionless flame height meets [6]

$$H_f / D \propto F_r^n, with 1/5 \leq n \leq 1/3 \tag{11}$$

So the correlation between Froude number and oxygen concentration is obtained:

$$H_f / D \propto \chi_{O_2}, with 1/5 \leq n \leq 1/3 \tag{12}$$

Fitting the experimental data as shown in Fig. 3, we can find a similar law between dimensionless height and oxygen concentration. The dimensionless flame height is proportional to the power of oxygen concentration. But the power exponents are not in the range 1/5 to 1/3. Further testing and better parameters are needed to clarify the question. The relationship between dimensionless height and oxygen concentration is shown as follows:

$$H_{PA6} / D = 9.36 \chi_{O_2}^{1.25}, with\ 25\% \leq \chi_{O_2} \leq 50\% \tag{13}$$

$$H_{Epoxy} / D = 19.43 \chi_{O_2}^{2.31}, with\ 40\% \leq \chi_{O_2} \leq 65\% \tag{14}$$

In addition, the dimensionless flame height of epoxy is divided into two parts. When the oxygen concentration satisfies $25\% \leq \chi_{O_2} \leq 35\%$, the dimensionless flame height increases according to a linear law as a function of oxygen concentration.

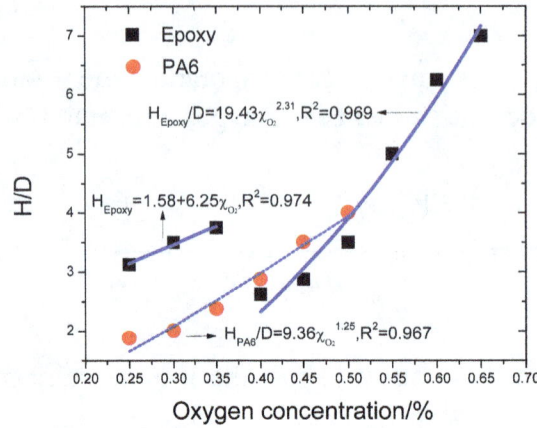

Fig. 3 Variation of dimensionless flame height with the increase of oxygen concentration, where points indicate the experimental data and curves are the best fit

This may be associated with the error of the experiment apparatus or material properties. Dripping phenomenon of thermoplastic material PA6 reduces its average flame height in a degree. Moreover, the declination of flame height with increasing oxygen concentration of the PA6 is smaller than epoxy as shown in Fig. 3.

3.2. *Flame spread rate*

Because the depth of thermal wave is less than thickness of the sample material, so the sample material we use is considered as thick material. The expression between flame spread rate and heat flux from the flame for thermally thick material developed by Quintiere [4] and Saito et al. [7] is expressed as

$$v_f \approx \frac{4(\dot{Q}_F^{"})^2 \delta_P}{\pi k \rho c (T_{ig} - T_\infty)^2} \tag{15}$$

where δ_P is the length of preheated region; $\dot{Q}^{"}$ is the heat flux of flame; k is the thermal conductivity. ρ is the density; c is the heat capacity; T_{ig} is the ignition temperature and T_∞ is the ambient temperature.

Combining (5), (6) and (15), the relationship between flame spread rate and oxygen concentration can be expressed as

$$v_f \propto \chi_{O_2}{}^m \tag{16}$$

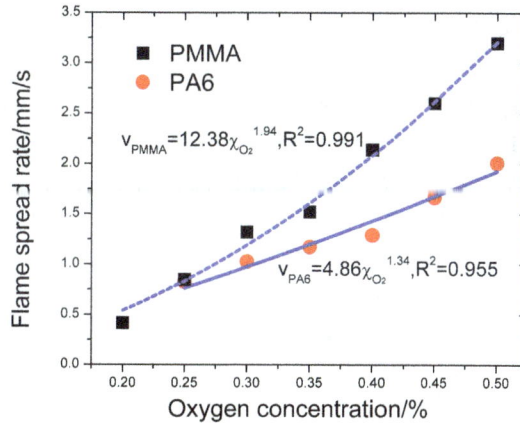

Fig. 4 Flame spread rate over PA6 and PMMA as a function of oxygen concentration. Points indicate the experimental date, and the dotted lines are the best fits

Fig. 4 describes the rate of flame spread over PA6 and PMMA as a function of oxygen concentration. The horizontal flame spread rate increases with the increase of oxygen concentration. In the experiment, thermoplastic material PMMA is selected as a comparison material to find if there exists similar law on these two thermoplastic materials. The best-fit relationship between the horizontal flame spread rate of these two thermoplastic material and oxygen concentration are as follows:

$$v_{PA6} = 4.86\chi_{O_2}^{1.34}, with\ 25\% \leq \chi_{O_2} \leq 50\%$$ (17)

$$v_{PMMA} = 12.38\chi_{O_2}^{1.94}, with\ 20\% \leq \chi_{O_2} \leq 50\%$$ (18)

where v denotes the horizontal flame spread rate. χ_{O_2} is the oxygen concentration.

4. Conclusion

The relationship between flame spread rate v_f and dimensionless flame height $H^* = H_f / D$ is established in this paper. The main result of this work is shown as follows:
1. Dimensionless flame height of PA6 increases according to a power law with the increasing oxygen concentration. Moreover, the declination of flame height with increasing oxygen concentration of the PA6 is smaller than thermosetting epoxy.
2. The horizontal flame spread rate of PA6 also increases according to a power law with the increase of oxygen concentration and the power exponent is 1.34.

Acknowledgment

This research was funded by the National Key Technology R&D Program (2013BAJ01B05). The authors deeply appreciate the support.

References

1. Mekki, K., et al. Wind-aided flame spread over charring and non-charrring solids: An experimental investigation. in Symposium (International) on Combustion. 1991.
2. CHAO, Y.H.C. and A.C. FERNANDEZ-PELLO, Concurrent Horizontal Flame Spread: The Combined Effect of Oxidizer Flow Velocity, Turbulence and Oxygen Concentration. Combustion Science & Technology, 1995. 110-111(1): p. 19-51.
3. Osorio, A.F., et al., Limiting conditions for flame spread in fire resistant fabrics. Proceedings of the Combustion Institute, 2013. 34(1): p. 2691-2697.
4. Quintiere, J.G., Fundamentals of fire phenomena. Fundamentals of Fire Phenomena, 2006.
5. DiNenno, P.J., SFPE handbook of fire protection engineering. 2008: SFPE.
6. Williams, G., Combustion theory. 1985.
7. Saito, K., J. Quintiere, and F. Williams. Upward turbulent flame spread. in Fire Safety Science-Proceedings of the First International Symposium. 1986.

Synthesis and Characterization of Waterproof Type of Waterborne Polyurethane

Xiang Li[1, 2], Ming-Yu Pan[1, 2], Chao Zhu[2], Li-Jie Ni[1, 2,*], Heng Quan[1, 2,*]

1Faculty of Chemical Engineering, Wuhan Textile University, Wuhan, Hubei, 430073, China
2Wuhan Textile University graduate (color root) workstations, Songgzi, 315600, China
Email: quanheng2002@163.com

Four types of waterborne polyurethane (WPU) have been synthesized by the reaction of toluene-2,4-diisocyanate (TDI), polytetrahydrofuran glycol (PTMG), polyester (PE3030), polyether diol (N220), Polyether triol (ZC330), hydroxyl-terminated polybutadiene (HTPB), hydroxyl silicon oil (PPC) and chain extender(EXT08). The chemical structures of WPU were characterized by FT-IR. Their properties were *measures and compared* in terms of the particle size, interfacial tension, contact angle, weight gain rate (WGR), tensile strength and elongation at break. The studies show that PPC and HTPB are introduced in. The polyether WPU with10 wt.% of EXT08 has best waterproof and better comprehensive properties.

Keywords: Waterborne polyurethane; polyether; polyester; waterproof.

1. Introduction

The eco-friendly waterborne polyurethanes (WPU) were widely used in the textile industry because it possesses properties such as storage stability, elasticity and so on [1]. However the development of the single function of WPU cannot meet the requirement of today's textile market [2]. For example, water-based system of polyurethane is of environmentally benign nature but with poor water resistance [3]. Therefore, preparing a novel WPU that has properties about waterproof, soil-release, and flame retardant etc. It is becoming a research hotspot [4].

In this paper, four types of WPU were synthesized. HTPB, PPC were introduced into WPU and the results showed that the polyether WPU modified with 10 wt.% of EXT08 had best waterproof performance and better comprehensive properties.

2. Experimental

2.1. *Materials*

Toluene-2,4-diisocyanate (TDI), polytetrahydrofuran glycol (PTMG), polyether diol (N220), Polyether triol (Z330), hydroxyl-terminated polybutadiene (HTPB), hydroxyl silicone oil (PPC) and polyester (PE3030) were all industrialized products. Dimethylolbutanoic acid (DMBA), neopentyl glycol (NPG) and acetone (ATE) were purchased from Sinopharm Chemical Reagent Co., Ltd and were all chemical pure reagent. Trimethylamine (TEA), stannous octoate and octanol are all analytical reagents. The chain extender EXT08 was prepared in lab.

2.2. *Main instruments*

Laser particle size analyzer (Nanotrac), Table model high speed centrifuge (TGL-16), Fourier transform infrared spectroscopy (Nicolet iS5), Contact angle meter (DSA20), Automatic interface tensiometer (JYW-200A), Fabric strength machine (YG(B)026H).

---

*Corresponding author

2.3. *The preparation of functional WPU*

Dehydration (polyether, 105°C-110°C, -0.1MPa, 2h); prepolymerization (IPDI, 80°C, 2h), chain extension (chain extender, 65°C, 2.5h), end-capping reagent (octanol, 1h), neutralizer (TEA,30min), emulsification by deionized water, get the WPU emulsion.

Add glycerin into the prepolymer and keep the temperature at 60°C for 1.5 h after adding octanol into HDI (Hexamethylene Diisocyanate) slowly with the temperature at 75-80°C for 2 h, get the EXT08.

2.4. *Characterization and testing*

The WPU films were characterized by FTIR on the Nicolet iS5 with the wavenumber ranged from 400 – 4000 cm^{-1}. The particle size of WPU was tested with a laser particle size analyzer (Nanotrac). The interfacial tension of WPU was obtained on an automatic interface tensiometer (JYW-200A) at room temperature. The contact angel was tested on the contact angle meter (DSA20) for 0, 1, 2, 4, 6, 8, 10 min, respectively. The tensile strength and elongation at break tests were obtained by making the films into standard dumbbell specimens which were tested according to GB/T 1040-92 standard.

To test the resistance for water, acid and alkali, WPU film was firstly cut and its quality was noted as M_1. The cut film was then immersed into the water, acid or alkali solution to be soaked for 72 h at room temperature, and the quality of the soaked film was noted as M_2. The Weight gain rate (WGR) can be calculated according to Equation (1):

$$WGR = \frac{M2\text{-}M1}{M1} \times 100\%$$

(1)

3. Results and Discussions

3.1. *Waterproof performance comparison of polyether WPU and polyester WPU*

The compositions of WPU are shown in Table 1.

Table 1 The material and structure of WPU

Number	Soft segment-1	Soft segment-2	Hard segment	Hard segment	Soft segment
A-1	PTMG		TDI	43%	57%
A-2	PE3030	PPC HTPB	DMBA NPG	38%	62%

The performance analysis of A-1 and A-2 emulsion and film are given in Table 2.

Table 2 The properties of WPU

Number	Interfacial tension (mN/m)	WGR -water%	WGR -acid %	WGR -alkali %
A-1	30.32	14.62	1.19	10.60
A-2	23.46	12.02	1.81	18.22

From Table 2, interfacial tension of A-1 was higher than that of A-2 because A-2 contains ester base, which has high cohesive energy.

Due to its hydrophilic characteristic polyether performed better than polyester. The micro-phase separation of polyether was stronger than polyester and A-1 was more hydrophilic than A-2. The ester base is destroyed easily in the alkali condition, which makes macromolecule amorphous region expanded so that water is easy to move into the inside molecule and makes hydrophilic groups form hydrogen bond with water to make polyester WPU more hydrophilic. Consideration of acid and alkali, polyether WPU had better water resistance [5]. Otherwise, if the film of polyether WPU is softer, it has litter effect on the style of the textiles.

The contact angle of WPU film of A-1 and A-2 were presented in Fig. 1. The contact angle of polyether WPU was always greater than that of polyester WPU, which suggested that polyether WPUhad better water resistance. During the WPU film formation, the micro-phase separation brought by the flexibility of polyether, help Si-O in PPC and C=C in HTPB migrate to the film surface more easily to form waterproof layer than the polyester. The uniform distribution of characteristic groups results in more hydrophilic groups are located on the surface. Therefore polyether WPU had better waterproof performance.

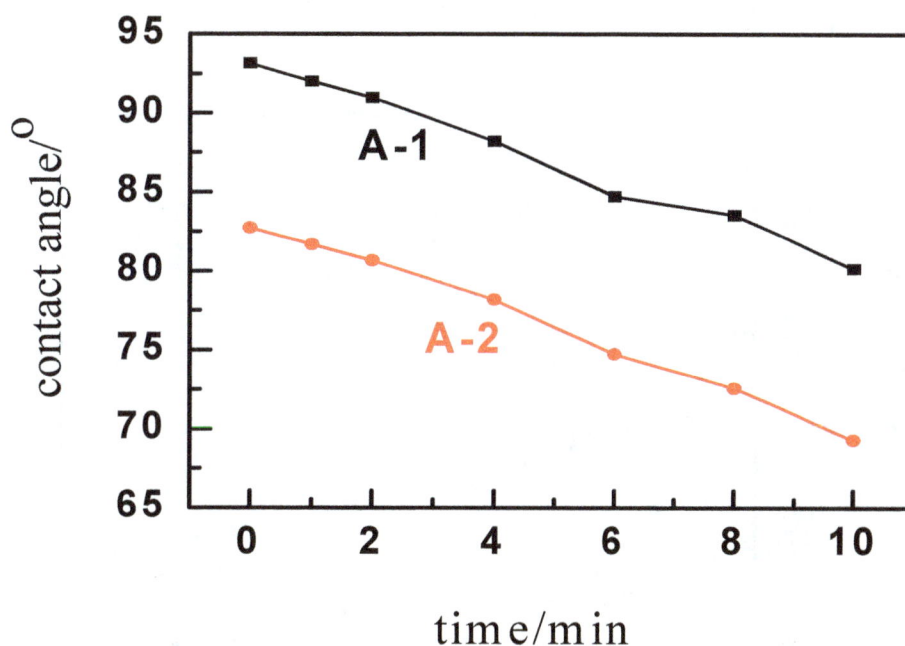

Fig. 1 The contact angle of WPU

3.2. Effects of long chain alkyl dosage on WPU film

The effective migration of waterproof segment is the determined factor to form waterproof layer during the film formation process. In order to get a better waterproof layer, a small molecule diol chain extender EXT08 containing long alkyl side chain was made and introduced into the molecule side chain. The nonionic chain extender in the A-1 recipe was replaced by EXT08, and the obtained products were named as B-1and B-2.The weight ratio of EXT08 in B-1 and B-2 were determined as 10 wt.% and 15wt.%, respectively.

The properties of B-1 and B-2 emulsions and films are given in Table 3. It shows that the interfacial tension of B-1 was greater than that of B-2. This is because the increased dosage of long alkyl chain could reduce the tension of the emulsion and increase the hydrophobic properties. On the other hand, the increasing dosage gave inferior water, acid and alkali WGR to the A-1. Although the long alkyl chain could make the hydrophobic layer more compact on the film surface, it increases the micro phase separation degree and raises bibulous WGR for a long time in the solution.

Table 3 the properties of WPU films

Number	Interfacial tension (mN/m)	WGR -water %	WGR -acid %	WGR -alkali %
B-1	21.30	12.38	1.02	10.24
B-2	20.38	15.14	1.14	10.36

The contact angles of B-1 and B-2 film are depicted in Fig. 2. Due to the introduction of long alkyl side chain, the contact angles of B-1 and B-2were more than 90° after 10 min, suggesting the waterproof property of WPU film was remarkably improved. With the EXT08 dosage increased, the benzene content increased accordingly, making the flexibility of molecular chain decreased and the planarity become better. The waterproof groups on the film surface were evenly and densely distributed, thereby improving the waterproof property. To summarize, B-1 gave the best waterproof performance.

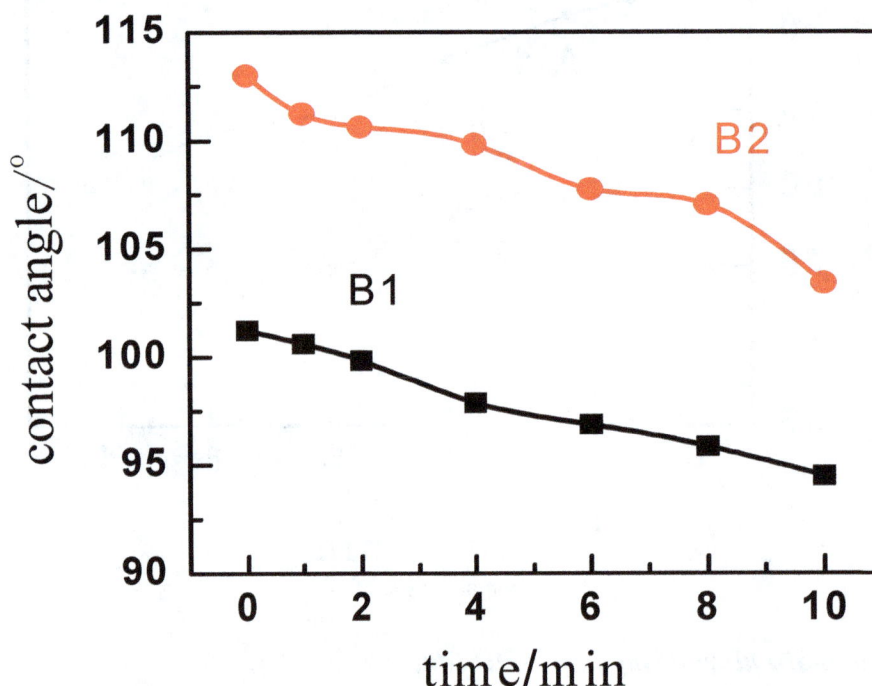

Fig. 2 The contact angle of 10 wt.% & 15 wt.% EXT08 WPU

3.3. *FT-IR analysis*

According to Figure 3, the absence of -NCO absorption peak around 2270cm^{-1} suggested no free –NCO group. The peaks, 1597cm^{-1}, 3017cm^{-1} and 1632cm^{-1}, appeared C=C bone in benzene ring, C-H and C=C bond in vinyl stretching vibration peak, respectively. These indicated the presence of C=C bond in the chain and the successful introduction of HTPB. In addition, two peaks, 1260cm^{-1} and 1000-1100cm^{-1}, appeared the symmetric deformation vibration peak of -Si-CH$_3$ and the strong stretching vibration peak of -Si-O-. These indicated that PPC was successfully introduced in the macromolecules.

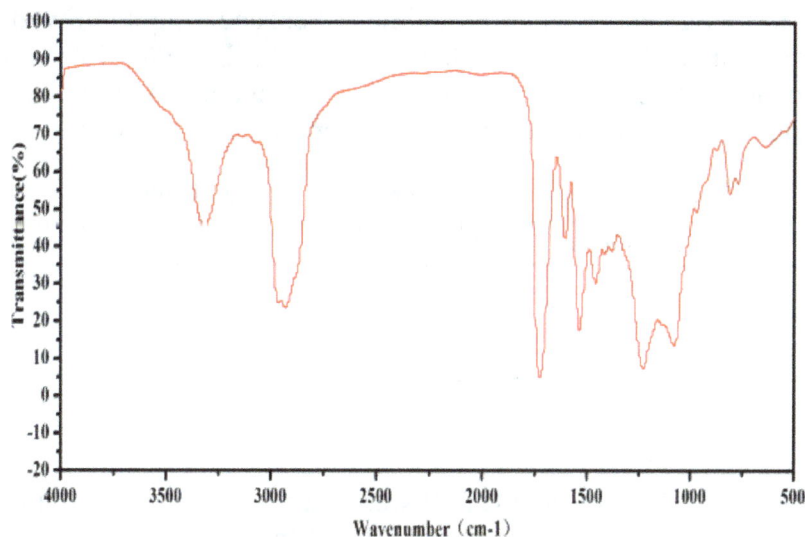

Fig. 3 FT-IR spectrogram of WPU (B-1)

3.4. Particle size analysis

From Fig. 4, the average particle size of polyether WPU was about 300nm, the distribution range was narrow, the emulsification was very good and the micro-emulsion was translucent.

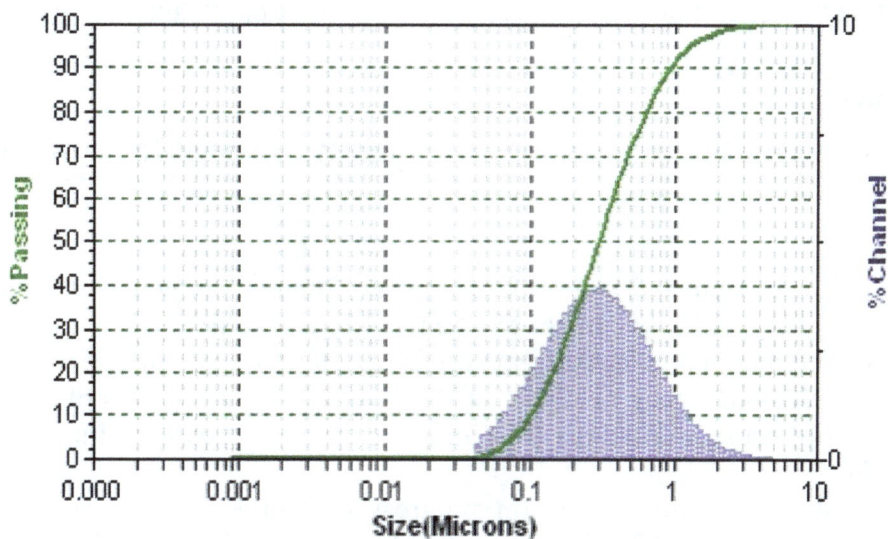

Fig. 4 The particle size of polyether WPU (B-2)

3.5. Physical properties

Functional WPU is usually used for coating and its performance requirements are very stringent. Specifically, the introduction of the waterproof property on the textiles should avoid jeopardizing the textiles themselves. Taking the tensile property as an example, if the elongation at break is less than the extensibility of fabric itself, the film will break for wearing and affect the textiles performance.

Tensile properties of the WPU films were evaluated and given in Table 4. The elongation at break was more than 4 times, and the strength increased. WPU had good flexibility and deformability, and the long alkane chain containing benzene ring improved the strength and reduced the elongation at break. B-1, which had 10 wt.% EXT08 of long chain alkyl chain extender, obtained better waterproof and its strength and elongation at break were between A-1 and B-2. So B-1 had better comprehensive performance compared to A-1 and B-2.

Table 4 The strength and elongation test of WPU films

Number	Elongation at break %	Strength MPa
A-1	480	3.6
B-1	450	3.8
B-2	440	3.9

4. Conclusion

1. FT-IR analysis indicated that PPC and HTPB were successfully introduced in WPU.
2. Waterproof of polyester and polyether WPU related to the characteristics of the groups, the distribution of hydrophobic chain in molecular and the degree of micro-phase separation. To some extent, the micro-phase separation degree determined its waterproof. Polyether WPU had higher micro-phase separation degree than polyester WPU.
3. The introduction of EXT08 modified the WPU molecular. A large number of hydrophobic chain segments provided WPU film better hydrophobicity, but made the emulsion worse and particle size bigger. Above all, when EXT08 dosage was 10wt.%, polyether WPU had better waterproof and other comprehensive properties such as strength and elongation at break.

Acknowledgment

This research was funded by the Wuhan Textile University postgraduate innovation team project.

References

1. Wei-guo Li, Ke-lin Huang, Xiao-xin Liao. Study Progress in Water-borne Polyurethane (in chinese), Technology & Development of Chemical Industry. Vol. 38 No. 11 (2009) 19-24.
2. Ge-wen Xu. Waterborne polyurethane materials (in Chinese), Beijing, 2006.
3. Jian-jun Yang, Jian-an Zhang. The Latest Progress in Design and Application of Functional Waterborne Polyurethane Coatings (in Chinese). Paint & Coatings Industry. Vol. 41 No. 3 (2011) 70-74.
4. Jun-jie Bao, Research and application progress of composite modification of waterborne polyurethane coatings (in Chinese). PU technology. Vol. 05 (2006) 76-80.
5. Hong-mei Chen. Study on improving water resistance of waterborne polyurethane (in Chinese). China adhesives. Vol. 18 No. 1 (2009) 11-15.

A MEMS-based Capacitive Flexible Film Force Sensor with Polymer Insulating Flat Dielectric

Gao-Feng Zhou*, Lu-Jun Cui, Ze-Xiang Zhao, Zheng-Feng Li, Shi-Rui Guo, Hui-Chao Shang

School of mechatronics engineering, Zhongyuan University of Technology, Zhengzhou 710049, PRC
Email: yaofabiaolunwen@163.com

Aim at detecting single-point force in biomedical field; we develop a MEMS-based flexible thin film single-point force sensor with flat interlayer structure. In terms of detecting principle of parallel plate capacitor, we deduce the linear equation between point force and output capacitance and give out relative formula on the measurement range of the developed sensor in theory. We put forward the interlayer flat structure to realize the measurement of single-point force. Through given procedures, we successfully fabricate the MEMS-based flexible thin film single-point force sensor. The results of experiments and analysis illustrate that the MEMS-based thin film single-point force sensor with flat interlayer structure is available for single-point force detection from biomedical field.

Keywords: MEMS; flexible thin film single-point force sensor; piezoelectric effect; flat interlayer structure.

1. Introduction

Tactile force sensor is used to detect the contacting force which is exerted to the measured object. Flexible thin film force sensor is flexible and thin and soft, which is being applied to biomedical field such as the measurement of body pressure distribution and biting force of teeth and so on. Other potential application of such sensor includes the sensing of organic tissue either at the end of catheter or on the fingers of an endoscope surgery. On the basis of a certain sensing mechanism, MEMS-based sensor may be divided into four types: piezoresistive, capacitive, piezoelectric and optical [1-4]. For polymer materials are mechanically robust, chemically resistant and stable, polymer thin film sensors are being widely studied for biomedical field [5-6].

Presently developed polymer sensors based on silk-screen printing technologies are being used to detect the pressure distribution between body and contacting object [7-8]. However, they are not sensitive for a single-point force because such sensor cannot be accurately placed to a single contacting point between two point-meshing objects. In the meanwhile, the tactile sensor for pressure distribution can not be utilized to detect the end of organic tissue. In the operation of MEMS device or biomedical contact there are a lot of cases of point-touching such as clamping atom or adjusting the placement of vein in body. Hence, point force sensor has to be studied and explored to adapt the case of point contacting on organic tissue such as pressure of toe and teeth biting. Claudio P. Fernandez, Per-Olof J. Glantz, Stig A. Svensson, et al explored a novel sensor for bite force determinations to detect bite force of teeth [9], and Gaofeng Zhou, Yulong Zhao, Zhuangde Jiang devised a flexible thin film single-point force sensor [10]. However, their size is too large to be successfully applied to point touching case of medical measurement due to brittleness and large volume of sensor, although there are some testifying experiments in laboratory. Micro Electro Mechanical Systems (MEMS) technologies may be utilized to make the volume of sensing device with micro scale features or with array structure. Significant developments in microscale sensors have been displayed in related literatures [11-12]. Their work shows that design and development of a contacting sensor with flexible feature is possible and available and robust.

The aim of this manuscript is to develop a MEMS-based flexible thin film single-point force sensor with flat interlayer structure. The detecting principle is narrated and the specific structure is put forward to realize the aim function. Then the range of measurement is deduced and ensured.

---

*Corresponding author

Utilizing the relevant MEMS technologies of thin film, we narrate the corresponding fabricating process and the specific steps. Finally the relevant test experiment and analysis on the developed sensor are finished to testify whether the developed sensor has the given function and whether it is applied to biomedical field.

2. Designing the Flat Interlayer Structure Based on MEMS Technology

2.1. *Detecting principle*

Parallel capacitor which consists of two parallel metal foil plates and dielectric medium is being widely applied in all kinds of detecting and converting circuit. The sensor developed is based on the capacitors of parallel plate, which diagram is seen in figure 1. The middle dielectric medium is used to separate the top and down metal foil plates. Here metal foil plates are used to transfer the sign caused by external force to the detecting circuit.

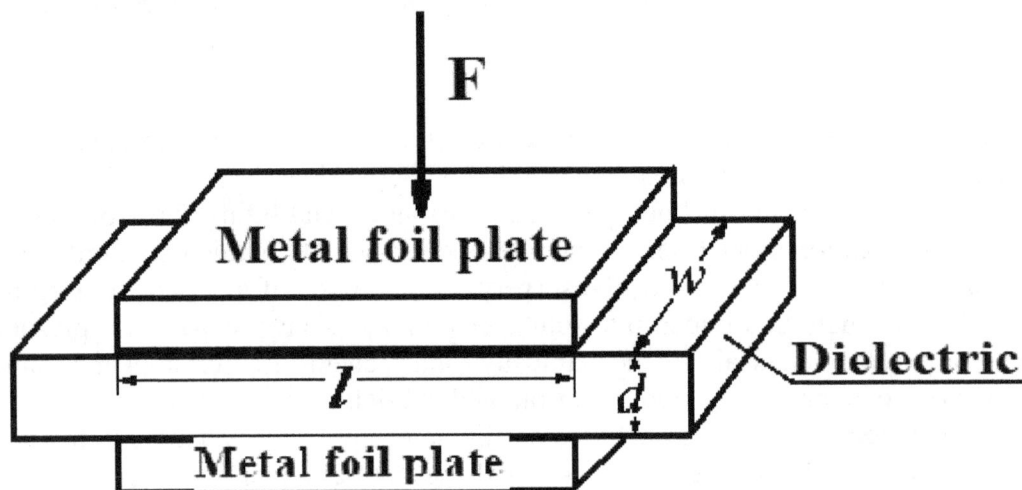

Fig. 1 Diagram of capacitive detecting principle

According to the formula of capacitor of parallel plate and the parameters in figure 1, the following equation (1) could be drawn:

$$C = \frac{\varepsilon_0 \varepsilon A}{d} = \frac{\varepsilon_0 \varepsilon}{d} \times l \times w \qquad (1)$$

Here, C —original capacitance; d —the distance between the top and down parallel plates; ε_0 —dielectric constant in vacuum; ε — the related dielectric constant of dielectric material related to vacuum; A —the area overlapped by top and down parallel plates ($A = l \times w$).

Total differential equation (2a) is gained for equation (1); then the change of capacitance may be analyzed and known:

$$\Delta C = \frac{\varepsilon_0 A}{d} \Delta \varepsilon + \frac{\varepsilon_0 \varepsilon}{d} \Delta A - \frac{\varepsilon_0 \varepsilon A}{d^2} \Delta d = \frac{\varepsilon_0 A}{d} \Delta \varepsilon + \frac{\varepsilon_0 \varepsilon}{d} \Delta A - \frac{C}{d} \Delta d \qquad (2a)$$

For the relative dielectric coefficient ε and the effective area of capacitor A are not be changed and keep constant in action, they should be zero. Then equation (2a) may be simplified to equation (2b):

$$\Delta C = -\frac{\varepsilon_0 \varepsilon A}{d^2}\Delta d = -C\frac{\Delta d}{d} \tag{2b}$$

Of course, when an external force is exerted to the metal foil plate, the distance between top and down metal foil plates will be accordingly reduced. According to the theory of material mechanics, equation (3) could be gained:

$$\sigma = \frac{F}{A} = \frac{F}{lw}; \quad \sigma = E\varepsilon' = E\frac{\Delta d}{d} \tag{3a}$$

Namely,

$$\frac{\Delta d}{d} = \frac{F}{Elw} \tag{3b}$$

Here, σ—stress under external force; E—elastic ratio of dielectric medium; Δd—the changed distance from top and down parallel metal foil plates.

After the equation (3b) is substituted into equation (2b), equation (4) could be acquired in the following:

$$\Delta C = -C\frac{F}{Elw} \tag{4}$$

In the course of measuring single-point force, the sensing point is always pressed by measured force. According to the regular direction of stress in material mechanics, a minus should be added before the value of point force. So the capacitance change ΔC will be changed into equation (5):

$$\Delta C = C\frac{F}{Elw} \tag{5}$$

Equation (5) reveals that there is a certain proportional relationship between external force and capacitance variant. External force is proportion to the changed capacitance caused by the distance change between top and down metal foil plates. That's to say, the sign of external force can be converted to the sign of capacitance change in the detecting circuit. Namely, the detecting principle in figure 1 could be used to sense the external force. Actually the capacitive sensor may be regarded as a variable capacitor in the detecting circuit.

2.2. Ensuring the range of measurement

The dielectric medium used in the developed sensor is biomedical silicon gel which has larger elasticity and resistant corrosion; hence there will be a certain distance change Δd produced by external force in figure 1. However, the distance change should be always less than the original distance between top and down parallel metal foil plates; otherwise the developed sensor will be seriously destroyed and it will lose its original function. Then the following boundary condition will be known and expressed in in equation (6a):

$$\Delta d < d \tag{6a}$$

Relative parameters are placed into inequality (6a), then the following in equation can be obtained:

$$\frac{Fd}{Elw} < d, \qquad \text{namely } F < Elw \tag{6b}$$

For any material, the stress caused by external force should not be greater than its permissible stress of dielectric medium $[\sigma]$ which could be known by looking up in the relevant material brochures. Then the following equation (7a) could be gained according to the theory of material mechanics:

$$\sigma \leq [\sigma] \tag{7a}$$

Similarly, the following expression can be also gotten:

$$\frac{F}{lw} \leq [\sigma], \qquad \text{namely } F \leq [\sigma]lw \tag{7b}$$

For strain of silicon gel is always less than one, we could know that its permissible stress $[\sigma]$ should always less than the elastic module E on the basis of the relationship between them in equation (3a). Hence, the range of measurement on the developed sensor could be ensured and acquired in equation (8):

$$0 \leq F \leq [\sigma]lw \tag{8}$$

If the measurement area is circular, the range of measurement in theory is expressed as follows:

$$0 \leq F \leq [\sigma]A = [\sigma]\pi r^2 \tag{9}$$

Here, r —the radius of measurement area; π—circumference ratio, 3.14.

2.3. Structure design

In order to sense the external force, we devised the flat interlayer structure which is seen in figure 2. The whole dimension is $50mm \times 15mm \times 500\mu m$. This component is composed of 8 components: plastic base plate, dielectric protecting plastic film, external electrode1, dielectric medium (biomedical silicon gel), and dielectric protecting film1, external electrode2, external protecting film and transparent adhering glue. Here transparent adhering glue is utilized to joint electrodes and protecting films together.

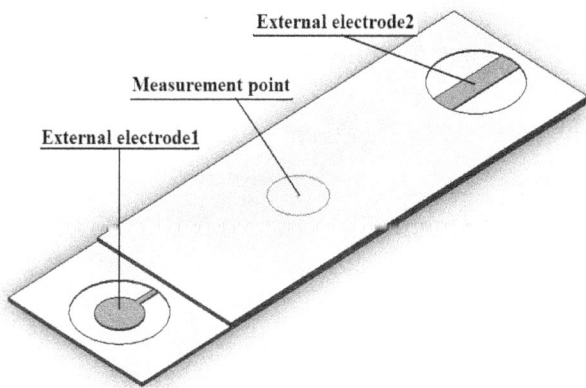

(a) 3D assembly diagram of developed sensor　　(b) Explosion diagram of given sensor

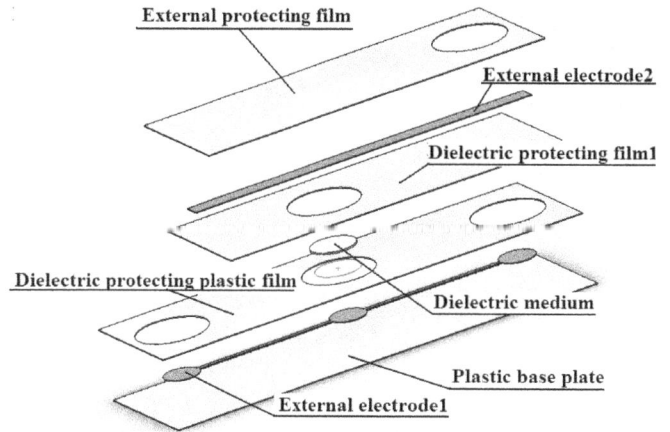

Fig. 2 Structure diagram of developed sensor

When the given sensor is connected to external circuit, external electrodes and circuit contactors keep detachable clamping connection because used protecting films can not endure high temperature welding procedure.

3. Fabricating Procedure

In the course of fabricating given sensor, thin film printing technologies and adhering procedures are utilized to fasten all metal foil electrodes and protecting films. The procedure is as follows:

Step1, the transparent protecting films are trimmed according to the dimension of design. The dimension of plastic base plate is $50mm \times 15mm \times 100\mu m$. The same things on other protecting films are done to realize their relevant function such as protecting metal foil electrodes which material are silver and aluminum apart.

Step2, circular holes are punched on the external protecting film and middle protecting films. The radius of all holes is 4.5mm. Such size is to easily realize the clamping connection in laboratory or industry.

Step3, external electrode1 (silver) is printed on the plastic base plate by the silk screen-printing technologies. The circular area on the external electrode1 is adopted to realize and display external connection and the placement of dielectric medium.

Step4, external electrode2 (aluminum) with $2mm \times 40mm \times 10\mu m$ is processed and trimmed by utilizing scissors and microscope to transfer the sign of capacitance.

Step5, all components are settled and overlapped according to the order displayed in figure 2 (b).

Step6, the adhering glue is coated onto dielectric protecting film1 and plastic base plate and dielectric protecting plastic film.

Step7, at the both sides of the middle circle from external electrode1, the dielectric glue is equably coated.

Step8, all components are pressed to fasten and make them into given sensor. Here we should guarantee that contactors from external electrode1 and external electrode2 should be enough exposed.

Step9, the developed sensor is placed and dried in a drying place for about two days.

According to above order, such sensor is fabricated as shown in figure 3. The adhering glue can be found in figure 3 for plenty of glue is coated to avoid the peel and glide between any two plastic films.

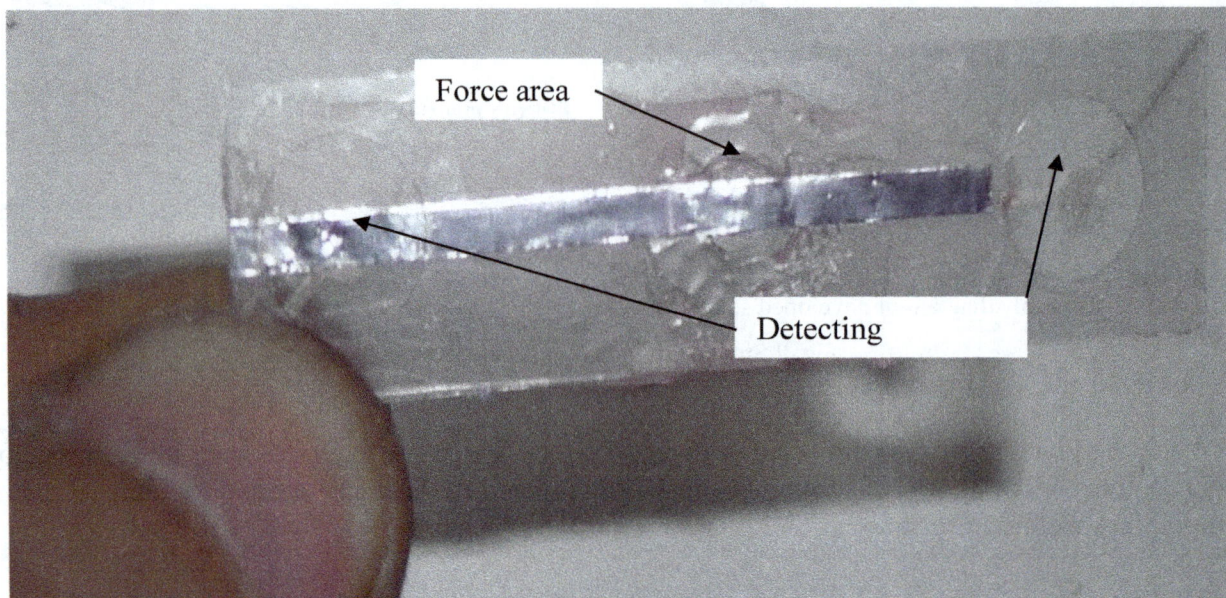

Fig. 3 MEMS-based thin film single point force sensor for biomedical field

4. Experiments

4.1. *Measurement of basic parameters*

The geometry parameters are measured as follows:

Total length: 50mm; total width: 15mm; total thickness: 0.6mm; diameter of hole in the plastic film: 9mm; the width of bottom electrode: 0.5mm; width of top electrode: 2mm; diameter of dielectric medium: 2mm; permissible stress of dielectric medium: 9MPa; range of measurement in theory: $0 < F < 113N$.

Here, the range of measurement is calculated according to the equation (9). The used tools in laboratory are listed in the following: vernier caliper, steel ruler, pair of compasses, balance, capacitance meter, wires, and contactors for wire, forceps, oil pen, slide and recording paper.

4.2. *Measurement of work curves*

In order to know the dynamic character, the developed sensor needs to be tested by actually external force. The single-point force used is respectively listed as follows: 8g, 11.8g, 14.5g, 17.5g, 44.5g, and 84.1g. The measurement curve on point force and electrical capacity is seen in figure 4. In the figure, the using unit of mass is gram, which is simplified into symbol g. The production between mass and gravity acceleration is force.

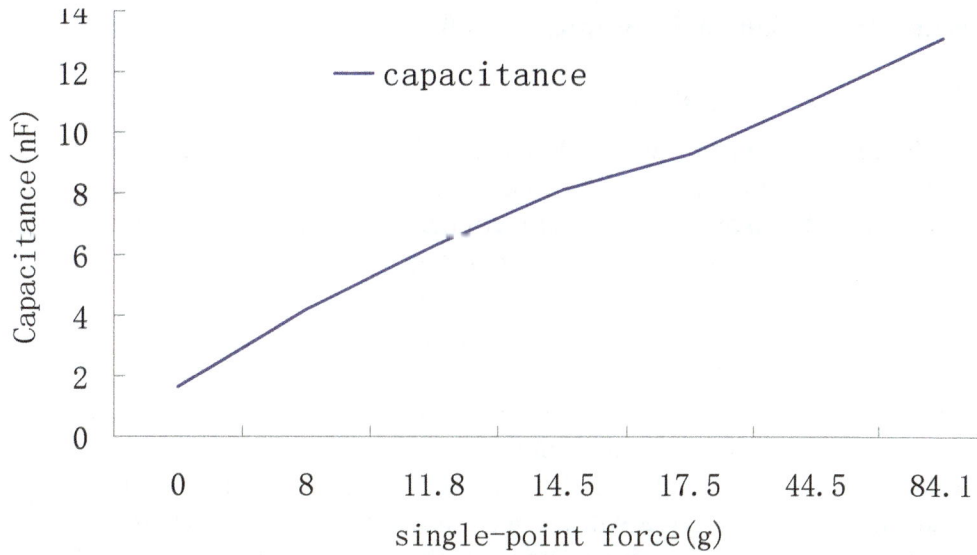

Fig. 4 Tested curve from the developed sensor

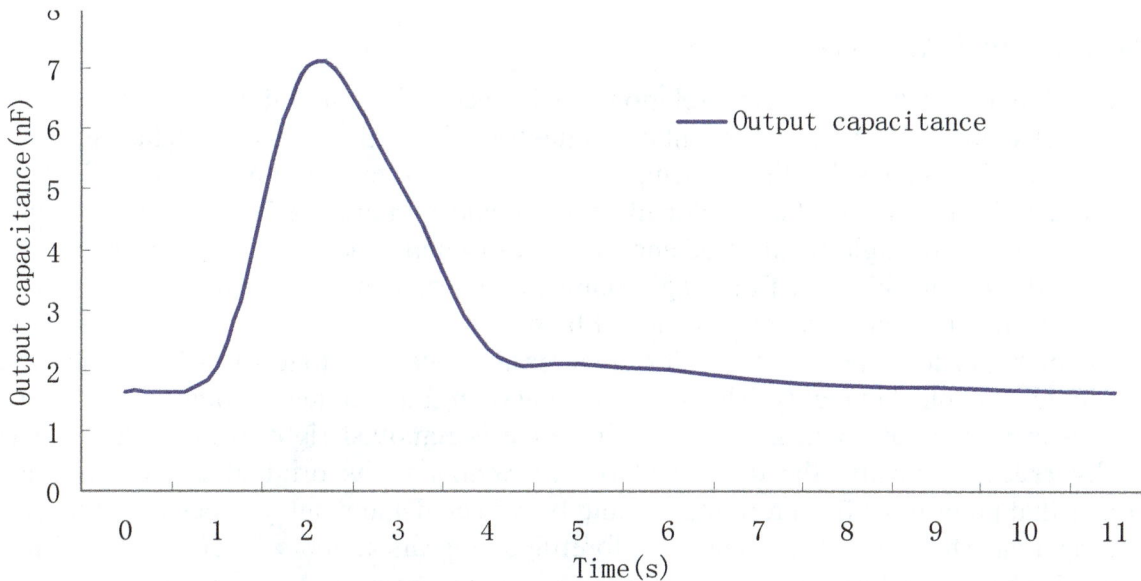

Fig. 5 Test curve on forefinger

At the same time, we also tested a random force to testify whether given device can be used to detect a single-force between forefinger and touching area. The specific method was such: step1, the developed sensor was connected to two contactors of capacitance meter; step2, the capacitance meter was turned on to observe whether the test data could be normally displayed on the screen; step3, forefinger was placed on the sensing point from the developed sensor and then right away finger was removed; step4, we observed and recorded the change of data. The test curve is seen in figure 5. This curve illustrates that given device can sense the force between forefinger and touching area.

5. Analysis and Discussion on Experiment Results

5.1. *Analysis on basic geometry parameters*

According to the information in section4.1, the developed sensor could be thin and flexible. It is suitable for the case of small single-point force in the biomedical field. The area of contactor is so large that the developed sensor could be easily connected to the detecting circuit or printing circuit board. In figure 3, we may clearly see that the developed sensor is transparent and flexible and thin. Furthermore, the detecting area and connecting area by the circular hole with 9mm is easily processed and displayed to conveniently realize point force measurement and external connection. The total thickness is 0.6mm. It's so thin that sensor hardly affects normal activities from patients or human activities. For an example, the developed sensor may be placed and settled in shoes or gloves to realize related measurement of singe-point force.

Through observing displaying photo in figure 3 and reading above geometry test parameters, we easily find that there is a larger area without any function for the MEMS-based flexible thin film single-point force sensor with flat interlayer structure. That's to say, the whole dimension of developed sensor may be reduced further to be suitable for some smaller case of point force in biomedical field.

5.2. *Analysis on the test curves*

Figure 4 shows that the displaying relationship between single-point force and capacitance is approximately linear. With the increment of single-point force, the output capacitance continues to rise. Equation (5) displays that the capacitance is proportional to single-point force which needs to be measured. That's to say, the developed sensor could correctly reflect and express the linear relationship between single-point force and output capacitance deduced by us in theory. Namely, the MEMS-based flexible thin film single-point force sensor may be available for measuring a single-point force in biomedical or mechanical fields.

Furthermore, an actual point force on forefinger is also measured to testify whether the developed sensor could be applied in reality. The displaying curve in figure 5 quickly and sharply rises. And a maximum is reached at about 2s. When point force is removed right away, output capacitance gradually reduces. Finally, the developed sensor returns to its original capacitance. Figure 5 illustrates that point force from forefinger could be reflected and displayed between forefinger and contacting area. The actual value caused by forefinger appears at about 2s. In the detecting circuit, the MEMS-based thin film single-point force sensor may be regarded as a variable capacitance. However, the exposing flaws should not be neglected in discussion and analysis.

5.3. *Discussion on the existing flaws from the developed sensor*

The developed sensor has a certain flaws through analyzing the test parameters and curves. The present sensor displayed in figure 3 is difficultly embedded into MEMS biochip because its dimension is too large to realize related slight operation in biomedical operation. Hence the dimension has to be deduced in terms of MEMS procedures. And general fabricating procedures are not used to fabricate MEMS biochip. In order to acquire MEMS biochip of single-point force sensor, part of procedures narrated in section3 needs to be adjusted besides silk-screen printing technologies.

Of course, it's very obvious that the range of measurement is too small not to detect the larger single-point force; hence the adopted material needs to be changed or adjusted to expand the range of measurement of the developed sensor. The linearity of curve in figure 4 needs be adjusted to make it further approach the expression in equation (5). The ascending time and reducing time in figure 4 are both too long for quick biomedical measurement.

By observing the measurement area of photos in figure 3, it's found that there are some bubbles produced in the course of fabricating the given sensor. How to eradicate the displaying bubbles in fabrication needs to be researched in future.

6. Conclusion

We successfully developed the MEMS-based capacitive thin film single-point force sensor and it could be used to detect the point force with interlayer flat structure in biomedical or mechanical fields. We correctly deduced the linear equation between external force and output capacitance. Furthermore, by comparison, the measurement range was calculated and confirmed in theory. The experiments results and analysis on relevant information illustrate that the MEMS-based thin film single-point force sensor could be available for single-point force from biomedical field such as measurement of toe force and finger force.

Acknowledgments

This study was funded by the Education Department of Henan Province -- (Granted No.: 14B460004, and 15A460040) and Science and Technology Bureau of Zhengzhou City (Granted No.: 131PPTGG416-5).

References

1. Takao Someya, Tsuyoshi Sekitani. Printed skin-like large-area flexible sensors and actuators [J]. Procedia Chemistry 1 (2009) 9–12.
2. G. Murali Krishna and K. Rajanna. Tactile Sensor Based on Piezoelectric Resonance [J]. IEEE SENSORS JOURNAL 4 (2004) 691-697.
3. H.B. Muhammad, C.M. Oddo, L. Beccai, M.J. Adams, et al. Development of a biomimetic MEMS based capacitive tactile sensor[J]. Procedia Chemistry 1 (2009) 124–127.
4. F. Eghtedari and C. Morgan. A novel tactile sensor for robotic applications [J]. Robotics 7 (1989) 289-295.
5. Jonathan Engel, Jack Chen and Chang Liu. Development of polyimide flexible tactile sensor skin [J]. Journal of Micromechanics and Micro engineering 13 (2003) 359–366.
6. Nathalie K. Guimard, Natalia Gomez, Christine E. Schmidt. Conducting polymers in biomedical engineering [J]. Progress in Polymer Science 8 (2007) 876-921.
7. Gaofeng Zhou, Yulong Zhao, Zhuangde Jiang. Strip double sensing layer pressure sensor for interface pressure distribution [J]. Sensor Review 29/2 (2009) 148-156.
8. Chang-Sin Park, Jongsung Park, Dong-Weon Lee. A piezoresistive tactile sensor based on carbon fibers and polymer substrates [J]. Microelectronic Engineering 86 (2009) 1250–1253.
9. Cláudio P. Fernandes, Per-Olof J. Glantz, Stig A. Svensson, Anders Bergmark. A novel sensor for bite force determinations [J]. Dental Materials 2 (2003)118-126.
10. Gaofeng Zhou, Yulong Zhao, Zhuangde Jiang. Error analysis and modification for the flexibly single-point force sensor [C]. The 4th IEEE conference on industrial electronics and applications (ICIEA2009), pp. 342-347.

11. Engel, J., et al., Polymer micromachined multimodal tactile sensors. Sensors and Actuators A 117(1) (2005) 50-61.

12. G.C. Hill, R. Melamud, F.E. Declercq, A.A. Davenport, I.H. Chan, et al. SU-8 MEMS Fabry-Perot pressure sensor [J]. Sensors and Actuators A 138 (2007) 52-62.

Synthesis and Characterization of the Hyperbranched Polyesteramide

Pei Yao[1, 2], Shu-Bai Li[1, 2,*], Yuan Liu[1, 2], Qi-Meng Zhang[1, 2]

[1]School of Chemical and Materials Engineering, Changzhou Vocational Institute of Engineering, Changzhou, China
[2]Institute of Green Technology, Changzhou Vocational Institute of Engineering, Changzhou, China
*Email: sbli@email.czie.ecdu.cn

The quasi-one-step method was used to synthesize hyperbranched polyesteramide by using diethanolamine and butanedioic anhydride as raw materials. Fourier transform infrared spectrometer (FTIR), Ubbelohde viscometer and DSC-TGA were used to identify the structure, viscosity and thermal stability of hyperbranched polyesteramide, respectively. In this investigation, the dependence of viscosity on hyperbranched polyesteramide in the presence of end-capped reagents was studied. Benzoic acid, stearic acid, crylic acid, α-methyl crylic acid and cinnamic acid as end-capped reagents can reduce the viscosity. The logarithmic value of viscosity with end-capped reagents (0.019-0.057dL/g) was much lower than the viscosity (0.146 dL/g) without end-capping reagent. In thermal analysis, T_d (thermal decomposition temperature) of hyperbranched polyesteramide was at the temperature of 316.3 °C, T_g (Glass transition temperature) was at the temperature of 270 °C. The results showed that the product exhibits low logarithmic viscosity value, good solution and relative thermal stability, which has a potential application values in rheology modifier of macromolecular materials.

Keywords: Hyperbranched polyesteramide; butanedioic anhydride; diethanolamine; quasi-one step method; synthesis.

1. Introduction

It is well known that the field of branched polymer materials is one of the hottest areas in the past ten years with the development of functional materials. It has been widespread attention from both academic and industry areas.

Hyperbranched polymer is one class of macromolecules which possesses highly branched and three-dimensional structure. Hyperbranched polymers have the excellent solubility properties and high reactive activities. However, hyperbranched polymers' properties of intensity and tenacity are not very excellent because of no twining chains in inner macromolecules [1]. More and more new functional polymers with the unique structure were prepared by modifying the end functional groups [2].

There are two methods of synthesis of hyperbranched polymers including one-step and quasi-one step methods [3]. One-step method is about preparing the hyperbranched polymer by polymerization of ABx (x>=2, A and B is the reaction functional groups.) One-step method is easy got and has a high rate of recovery [4]. But the molecular designing and synthesis of ABx are cumbersome and wasting time. Quasi-one step method is to add molecule By as the core and to prepare the hyperbranched polymer, which can be best controlled by designing the ultimate functional groups [5].

In this paper, The hyperbranched polymer was polymerization of diethanolamine and butanedioic anhydride as raw materials by quasi-one step method. The structures of hyperbranched polymers were characterized by FTIR. The thermal properties were performed by differential thermal and thermal gravimetric analyzer (DSC-TGA).

2. Materials

Chemicals and Apparatus. Diethanolanmine, dimethylacetamide (DMAc), N, N-dimethylforma -mide and butanedioic anhydride were purchased from Tedia (American International Chemical,

---

*Corresponding author

Inc. California, USA). Toluene, benzoic acid, acroleic acid, cinnamic acid, α-methylacrylic acid, octadecanoic acid and toluene-p-sulfonic acid methanoic acid, ammonia water, and ethanol were bought from Sinopharm Chemical Reagent Co. (Shanghai, China). All of these reagents were analytically pure. Differential thermal and thermal gravimetric analyzer were supplied from TA-Instruments Corporation (USA), Fourier transformation infrared spectrometer was got from Perkin-Elmer Corporation (USA). Drying cabinet，water knockout vessel, Ubbelohde viscometer and photoelectric balance were purchased from Huasheng Instrument Corporation (Guangzhou, China).

Experimental Condition of Viscosity. The viscosity was called ratio concentration logrithm viscosity [6]. It was tested by Ubbelohde viscometer. The solution of DMAc (0.005 g/mL) was prepared first. t_0 denoted the time which pure DMAc solution was flown down. t_1 denoted the time which samples need by the same way, it. So the viscosity was calculated using the formula $\mu = \ln(t_1 / t_0)/0.5(dL/g)$.

3. Experimental Methods

Synthesis of intermediate product I. Weigh 3-6 g diethanolamine quickly, add diethanolamine and succinic anhydride with the molar ratio of 1:1 into a three-neck round-bottom flask containing, DMAc(30mL) as a deliquescent reagent and the reactant mixture were stirred for 1 h with a continuous nitrogen flow at the room temperature.

Synthesis of intermediate product II. After the intermediate product I was got, a small amount of catalyst (toluene-p-sulfonic acid) and the amount of about 50 ml of azeotropic water-carrying reagent(toluene) were added into the reaction system immediately. The flask was heated by the oil bath to the temperature to 270°C, stirred slowly until back-flow was coming out. Then the water was removed by water knockout vessel, and the flask was heated all the time. When there was no water coming from the water knockout vessel, it indicated the reaction has reached the equilibrium and the intermediate product II was got finally.

Synthesis of hyperbranched polymer III. The polyester was generally prepared of the AB2 monomer with hydroxyl and carboxyl by the condensation reaction under the action of the catalyst. Adding over 10% of end-capping reagents (benzoic acid, acrylic acid, cinnamic acid, α-methylacrylic acid and stearic acid) into the reactor, when there was no water droplet generating, the reaction was finished. After that, the solvent and low molecular weight byproducts were removed by cooling and vacuum distilling and vacuum drying. Hyperbranched polymer end-capped with phenyl and vinyl was got eventually.

4. Results and Discussion

FTIR Spectra. The FTIR spectra of the product end-capped with benzoic acid and the product without end-capping were shown in Fig. 1. It can be seen from FTIR spectra that the peaks at 1721 cm^{-1} and 1731 cm^{-1} were attributed to the stretching vibration of C=O. The peak assigned to C-N appeared as a vibration absorption at 1273cm^{-1}. And the peak at 3062 cm^{-1} was attributed to the stretching vibration of C-H in the benzene ring. The results demonstrated that the end-capped reaction was occurred and the product end-capped with benzoic acid was got. The peaks at 1674-1647 cm^{-1} were attributed to the vibration of C-H in the benzene ring. Also, the wide characteristic absorption peak at 3330 cm^{-1} indicated that the product contained hydroxyl groups, which demonstrated that the product was not end-capped with some reagents, the product was a polymer with hydroxyl end groups. At the same time, due to the proportion of the whole molecule is less, the number of hydroxyl groups can be found out from the light transmission rate. Therefore, it can be judged in the final products structure, Two kinds of final products were proved to be the branched polymer with target functional groups.

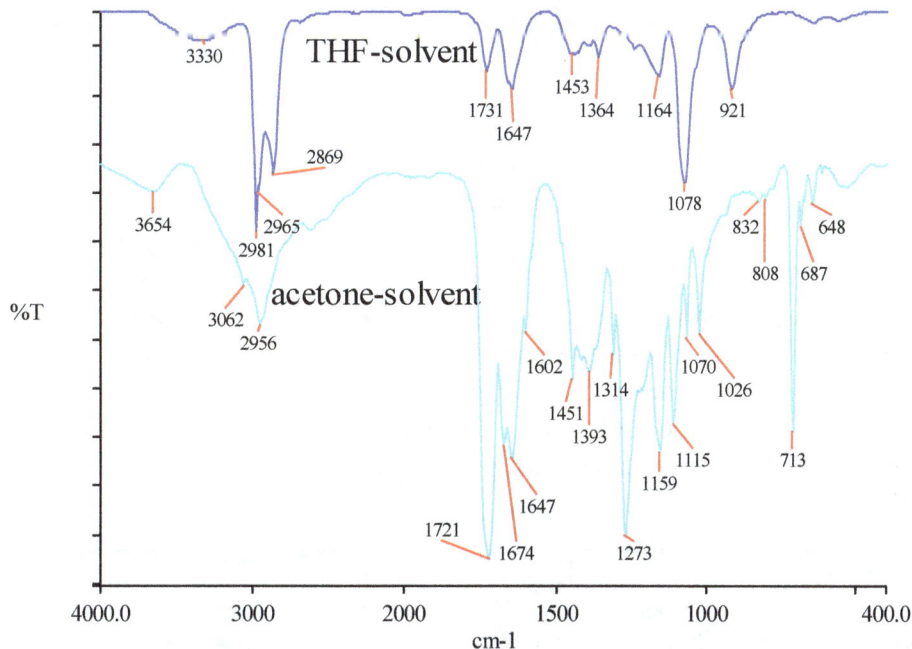

Fig. 1 FTIR spectra of the sample end-capped with benzoic acid and without end-capping

Ratio Concentration Logrithm Viscosity. The products with different kinds of end-capped reagents were dissolved in DMAc. The results were shown in Table 1. The data demonstrated that the viscosity of hyperbranched polymer with end-capped reagent was small, which illustrated that the structure of the hyper branched polymer had high symmetry [7]. However, the product without end-capping had high viscosity (0.146 dL/g). The theory of Frechet [8] showed that the higher the viscosity, the less number of degree of branching. Different end capped reagents had a certain effect on the viscosity of the samples, but the effect is not very obvious .The products end-capped with the stearic acid and benzoic acid had a large viscosity.

Table 1 Log. viscosity of the products with different end-capped reagents

End-capped reagents	Logarithmic viscosity (dL/g)
uncapped	0.146
benzoic acid	0.057
stearic acid	0.063
crylic acid	0.028
α-methyl crylic acid	0.031
cinnamic acid	0.019

Thermal Properties. Fig. 2 shows the thermal properties results of sample end-capped with benzoic acid. There were three stages of decomposition process marked in the product end-capped with benzoic acid. The weight loss of first stage before 158 °C was due to water evaporation in the sample. Volatilization of the low molecular weight (such as solvent molecules and small molecular by-products) accounted for 7.0% of the total weight of the gross weight. The temperature between

158-316.3°C was the second stage. Due to polycondensation reaction under high temperature, the weigh was lost. The main cause of the third stage after 316.3°C was backbone chain scission process of the polymer. Therefore the T_d was 316.3°C. The results implied that the polymer has a fine thermal performance, is expected to be used in the application of the rheology modified.

In DSC curve, T_g of the sample end-capped with benzoic acid was about 270°C. The obvious endothermic peak appeared at about 330°C was the transformation for viscous flow of the absorption, so we can find that the hyperbranched polymer T_m (melting point) was 330°C. The solidifying temperature of the sample end-capped with benzoic acid was determined for about 353°C by the exothermic peak temperature.

In Fig. 3, it shows the thermal properties results of sample without end-capping. The decomposition process of the product without end-capping was similar to the decomposition process of the product end-capped with benzoic acid, T_d was 334.6°C, T_g was 304°C, and the solidifying temperature was about 330°C, and the DSC curve had no obvious endothermic peak [9].

T_d of the product end-capped with benzoic acid is smaller than the product without end-capping, mainly because ester fracturing is easier than hydroxyl fracturing. T_g is one characteristic of chain motion. The properties of terminal groups had some influence on the value of T_g. The bigger polarity of the terminal group, the higher T_g is. Because the polarity of hydroxyl group is stronger than benzene group, T_g of the product without end-capping is a little higher than the product end-capped with benzoic acid.

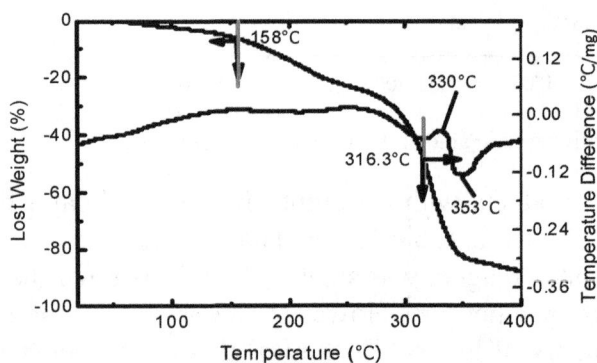

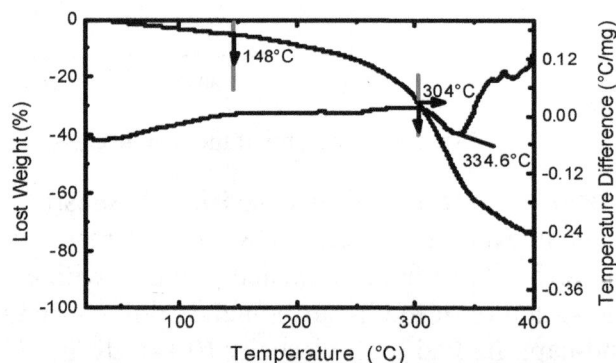

Fig. 2 DSC-TGA of the sample end-capped with benzoic acid.   Fig. 3 DSC-TGA of the sample without end-capping

Solubility. The solubility of the samples(end-capped with benzoic acid and without end-capping) were tested, and the results were shown in Table 2. The experimental results showed that hyperbranched polymer end-capped with benzoic acid was easily soluble in acetone, pyridine and tetrahydrofuran (THF). The hyperbranched polymer without end-capping was soluble in pyridine and tetrahydrofuran. As can be seen from Table 2, the solubility of the products was mostly followed by the rule of Similar compatibility [10].

Table 2 The solubility of products

Solvent	End-Capped products	Unend-capped products
Ketone	+	+-
Ethanol	+-	+
Carbon tetrachloride	+-	+-
THF	+	+
Methyl acetate catalyzed	+-	+-
Hot benzene	+-	+-
Pyridine	+	+

+ dissolve   - indissolve   +- tiny dissolve

5. Conclusion

The hyperbranched polyesteramide end-capped with benzoic acid was got by quasi-one step method, the process was simple and well controlled by polycondensation. The hyperbranched polyesteramide can be used in a wide range of application prospects because of the excellent solubility, low viscosity and well thermal stability.

Acknowledgments

This work was funded by universities in Jiangsu province of outstanding young teachers and principals of overseas training program, Jiangsu Qinglan Project, fund of JiangSu high school brand construction project chemical engineering (PPZY2015B178), the Natural Science Foundation of the Jiangsu Higher Education Institutions (No. 16KJB530005).

References

1. I. Gadwal, S. Binder, M. C Stuparu, et al. Dual-reactive hyperbranched polymer synthesis through proton transfer polymerization of thiol and epoxide groups, J. Macr. 47 (2014) 5070-5080.
2. X.H Wang, Y.Q Fu, L.F Ren, et al. Synthesis and characterization of the PTMG aliphatic hyperbranched polyurethane, J. Funct. Mater. 44 (2013) 289-293 (In Chinese).
3. C. Gao, D.Y Yan, Synthesis of Hyperbranched Polymers Functional Materials by End Capping, J. Chinese Science Bull. 45 (2000) 1145-1150 (In Chinese).
4. H.Y Wei, W.F Shi, Structural Characteristics, Syntheses and Applications of Hyperbranched Polymer, Chem. J. Chin. Univers. 22 (2001): 338-342 (In Chinese).
5. H. Zhao, Y. J. Luo, H. X. Song, The Advance in Reseaches on Hyperbranched Polymers(II) The Senthetic Methods of Hyperbranched Polymers, J. Therm. Res. 5 (2004) 34-38.
6. H.G. Kou, W. F. Shi, Study of Photopolymerization Characteristics of Hyperbranched Methacrylated Poly (amine-ester)s, J. Acta. Polymer. Sin., 5 (2000) 554-558.
7. L. M. Tang, H. You, Y. Li Synthesis and Curing of Acrylate Term Inated Hyperbranched Polymer, J. Tsinghua Univers., 12 (2003) 1613-1615 (In Chinese).
8. J. M. J. Frechet, C. J.Hawker, Gitsov et al. Hyperbranched polymer, J. Macromol. Sci. Pure. Appl. Chem., A33 (1996) 1399-1425.
9. B. I. Voit, A. Lederer, Hyperbranched and highly branched polyer architectures-synthetic strategies and major characterization aspescts, J. Chemic. Rev., 109 (2009) 5924-5973.
10. S. Gamier, A. Laschewsky, New amphiphilic diblock copolymers: surfactant properties and solubilization in their micelles, J. Langm., 22 (2006) 4044-4053.

Evaluation Character of Polymer Matrix Composite Materials by Ultrasonic Method

Xing-Guo Wang[*], Wen-Lin Wu, Zheng-Lin Chen, Yong-Gang Xie, Shi-Chong Jian

*School of Mechanical and Electronic Engineering, Jingdezhen Ceramic Institute,
Jingdezhen 333403, China*
[*]*Email: xgwang@yeah.net*

A new evaluation method for polymer composite materials dynamic viscoelasticity by ultrasonic wave method is provided based on the ultrasonic wave propagation theories in complex number range and the more theories. Considering the ultrasonic wave propagation characteristics in polymer composite materials with viscoelasticity, the mathematics model of ultrasonic wave method is built according to the attenuation coefficient and the loss tangent $\tan \delta$, and the various rubber samples are discussed by both experiment analysis and numerical simulation, as a result experiment and simulation data are fitted extremely. A nondestructive evaluation system is presented for the ultrasonic wave method for the polymer composite materials dynamic viscoelasticity. This is a new method for researching on the polymer composite characteristic and on line nondestructive evaluation, which can be also used for the evaluation of polymer material with viscoelasticity.

Keywords: Rubber material; ultrasonic method; attenuation coefficient; loss tangent $\tan \delta$; dynamic viscoelasticity evaluation.

1. Introduction

As the important ingredient of automobile tire, rubber is the only material that cars contact with the ground. The friction between the tire and the ground is the only external force for braking, acceleration and cornering. Tribological property used to evaluate the property of rubber productions is an important parameter. In order to improve the tribological property and wear resistance of the rubber tires, people tend to manufacture rubber by filling modification, so that not only can the physical and mechanical property of rubber be changed, but the tribological property can also be changed [1, 2]. Therefore, it is one of the most critical links for driving safety to study friction characteristics of automobile tire. Rubber with low modulus of elasticity and high viscoelasticity is a polymer material, and it is very sensitive to temperature, pressure, scroll speed, and other factors. So the friction mechanism of the rubber is a very complex problem and is very difficult to research [3-5]; for researching tribological property of the rubber, Varieties of the laboratory equipment and test methods have been used, as a result, the theory of the friction mechanism of the rubber have made a great progress [6-8]. According to these theories, we can gain the link between the friction characteristics the rubber and viscoelasticity [9].

2. Ultrasonic Echo Method Underside

Ultrasonic bottom echo method is shown in Figure 1. U_{A0} is the reflected wave from the bottom surface of delay material, U_A is the reflected wave from the interface between testing material and the delay material, and U_B is the reflected wave from the bottom surface of the test material. If there is no the couplant between the sensor and the delay material, as well as, between the delay material and test material, then it is difficult to amend reflection and transmission error of interface between the probe and the delay material and between the test material and the delay material, therefore we used water as couplant in this experiment. Figure 2 is a block diagram of an ultrasonic analysis system back reflection method. Signal is emitted by the pulser, the reflected signal is received by the

[*]Corresponding author

receiver after the A / D conversion and then we analyze these process data. If the bottom echo amplitude of delay material referred to U_{A0}, echo signals amplitude of the interface between delay material and the test material referred to U_A, the echo signal amplitude between the test material and the bottom referred to as U_B, the echo signal amplitude U_{A0}, U_A and U_B are applied with method of fast Fourier transform, the amplitude of frequency spectrum are referred to as A_0, A and B, respectively, then, the attenuation coefficient of the test material can be obtained according to the following equation [6].

$$\alpha(f) = \frac{1}{2h} \ln\left(\frac{A_0(f)^2 - A(f)^2}{A_0(f)B(f)}\right) \tag{1}$$

In the above formula, f is the frequency, h is the thickness of the testing material. $\alpha(f)$ is the attenuation coefficient of the testing material and that can reflect the relationship between attenuation coefficient and changeable frequency when ultrasonic propagate in the medium. The phase velocity V_p can be obtained according to the plural real part and the imaginary part of the frequency spectrum of various each echo.

$$V_p(f) = \frac{2h\omega}{\tan^{-1}\frac{Im[B(f)]}{Re[B(f)]} - \tan^{-1}\frac{Im[A(f)]}{Re[A(f)]} + 2N\pi + \omega T} \tag{2}$$

In the above formula, f is the frequency in the ultrasonic propagation medium, ω is the angular frequency ($\omega = 2\pi f$), T is the time difference of extracting the waveform of U_A and U_B when it apply FFT to the ultrasonic echo signals called U_A from surface and echo signal called U_B from bottom. According to the ultrasonic wave propagation theory in plural ranges, when the ultrasonic wave propagate in the medium, the elastic modulus is expressed in the form of plural due to the presence of dielectric loss:

$$L = L' + iL'' \tag{3}$$

$$\tan\delta = \frac{L'}{L''} \tag{4}$$

In which, L' is storage modulus of the medium
L'' is loss modulus of the medium
$\tan\delta$ is loss tangent of the medium

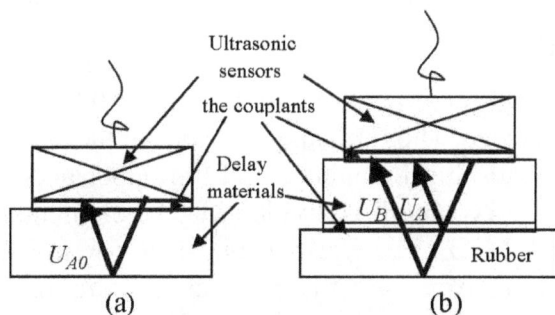

Fig. 1 Ultrasonic echo signal schematic diagram   Fig. 2 Ultrasonic wave experimental system

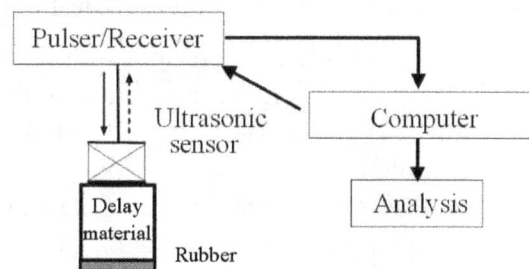

It is assumed $\alpha Vp/\omega \ll 1$, the medium's the storage modulus, loss modulus and the loss tangent can be obtained by the following equation:

$$L' = \rho V_p^2 \tag{5}$$

$$L'' = \frac{2\alpha \rho V_p^3}{\omega} = \frac{2\alpha V_p}{\omega}L' \tag{6}$$

$$\tan \delta = \frac{L''}{L'} = \frac{2\alpha V_p}{\omega} \tag{7}$$

The characteristic of the rubber polymer composites viscoelastic can be evaluated by using the above formula.

3. Laboratory Equipment and Materials

Ultrasonic testing equipment is materials testing equipment USH-B produced by Toshiba. The main technical performance of the device is a rectangular pulse waveform, the operating frequency range is 70 kHz ~ 15MHz pulse transmitter and receiver, the sampling frequency is 100MHz, a resolution is 8bit, 10ns minimum sampling interval of A/D converter. In order to improve the resolution of S/N ratio and the vertical resolution, this experiment extracts mean value for 16 waveforms.

The ultrasonic sensor with longitudinal wave is produced by Parametrics with a diameter of 12.7mm, the center frequency of 1MHz.

Table 1 Experimental material physics parameters

Test material	Hard-ness [HA]	Glass transition temperature [°c]	Velocity [m/s]	density [kg/mm³]
EPX-46	41	-66	1463	0.98
OR4Si	45	-50	1546	1.077
ORHS4Si	48	-24	1723	1.106
B4Si	45	-105	1520	1.059

The testing materials are four different rubbers, which are most commonly used for tire. The material physical parameters are shown in Table 1. The filler of various rubbers are silicon. The glass transition temperature can be obtained by the experiment of dynamic mechanical analysis. The sizes of the testing materials are $80 \times 80 \times 10$ mm³ rectangular parallelepiped, the delay materials are PMMA acrylic resin whose size of $40 \times 40 \times 40$mm³ cube.

4. Numerical Simulation of Ultrasonic Propagation

Analytical model of ultrasonic propagation is shown in figure 3, the size of the sensor, the resin and the testing sample in the figure is identical to the actual size. The right and left sides of the analytical model for PML (Perfect Matching Layer) is fully absorbing boundary, and the upper and lower bounds is completely elastic support boundary. Each wavelength is divided into 80 elements. The incident wave is sine waves whose frequency is 1 MHz. The various physical parameters of analytical analysis are shown in Table 2. According to Hooke's law, the movement equations of the ultrasonic propagation can be expressed availably by equations (8) under the binary plane strain state in isotropic medium. The first item of the equations in the left corresponds to longitudinal wave, the second corresponds to shear wave. u and v in the equations (8) is displacement of x and y, respectively, ρ is density.

$$\left.\begin{aligned}\frac{\partial \sigma_{xx}}{\partial x}+\frac{\partial \tau_{xy}}{\partial y}&=\rho\frac{\partial^2 u}{\partial t^2}\\\frac{\partial \sigma_{yy}}{\partial y}+\frac{\partial \tau_{xy}}{\partial x}&=\rho\frac{\partial^2 v}{\partial t^2}\end{aligned}\right\}\tag{8}$$

The computer is used to calculate ultrasonic propagation echo response in the medium, R1 is ultrasonic echo response on the surface of the material in experiment, and R2 is ultrasonic echo response on the bottom of the material. Finally, According to the method mentioned in section 1, the data of the impulse response is dealt with, and then the analysis results are compared with the experimental results

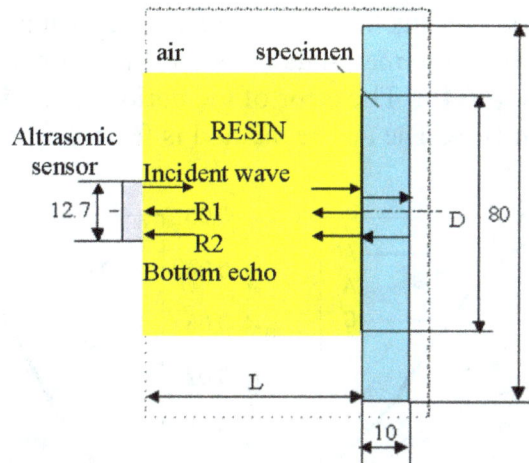

Fig. 3 Ultrasonic propagation analytic model

Table 2 Physic parameters by analytic model

Material name	resin	EPX-46	OR4Si	ORHS4Si	B4Si	Air
Density [kg/m³]	1170	980	1077	1106	1059	1.3
Longitudinal wave velocity [m /s]	2703	1463	1546	1723	1520	340
Shear wave velocity [m /s]	1358	780	780	780	780	0
Attenuation coefficient [dB/cm]	0	3.28	5.98	17.68	0.89	0
Incident wave	P0 [1+sin (ωt-π/2)]					

5. Result and Analysis

Taking ultrasonic experimental testing and numerical simulation results in EPX - 46 materials as an example, as shown in figures 4~6. By the method of the ultrasonic bottom echo (figure 3), echo signal U_{A0}, U_A and U_B have been measured, and then these echo signal U_{A0}, U_A and U_B have been analyzed by the method of the frequency spectrum analysis. it means that amplitude of echo signal

U$_{A0}$, U$_A$ and U$_B$ have been analyzed by the FFT transform process, its amplitude spectrum spectrogram A$_0$, A and B is shown in figure 4. The relationship between the attenuation coefficient and frequency can be obtained according to the power spectrum of A$_0$, A and B and equation (1), and then according to formula (2), (5) and (6), the relationship between storage modulus and frequency, as well as between loss modulus and frequency can be obtained the spread of the ultrasonic wave in the test materials rubber, the result is shown in figure 5. Finally, the relationship between loss tangent and frequency can be obtained via equation (7). The relationship between attenuation coefficient and loss angle tangent are shown in figure 6. According to power spectrum analysis results in figure 4, the center frequency of the experiment results and analytic results be shifted from 1 MHz to 0.88 MHz, the reason is internal sensor coupling. Here, taking out loss tangent whose centre frequency is 0.88MHz after the migration and attenuation coefficient α. The loss tangent of four kinds of rubber materials in the center frequency of 0.88 MHz is presented in Table 3, attenuation coefficient α values of four kinds of the rubber materials (EPX - 46, ORH4Si, OR4Si, B4Si) in the center frequency of 0.88 MHz is shown in Table 4. The parameters of the experimental and analytical results from Table 3 shows that loss tangent of four kinds of materials have the same order, the order is as follows: ORH4Si > OR4Si > EPX - 46 > B4Si. The parameters of the experimental and analytical results from Table 4 also shows that the attenuation coefficient α values of the four measured composite materials have the same order and the order is as follows: ORH4Si > OR4Si > EEPX - 46 > B4Si. The error of the various materials experimental results and the analytical results are within 10%. The above method is feasible through result analysis.

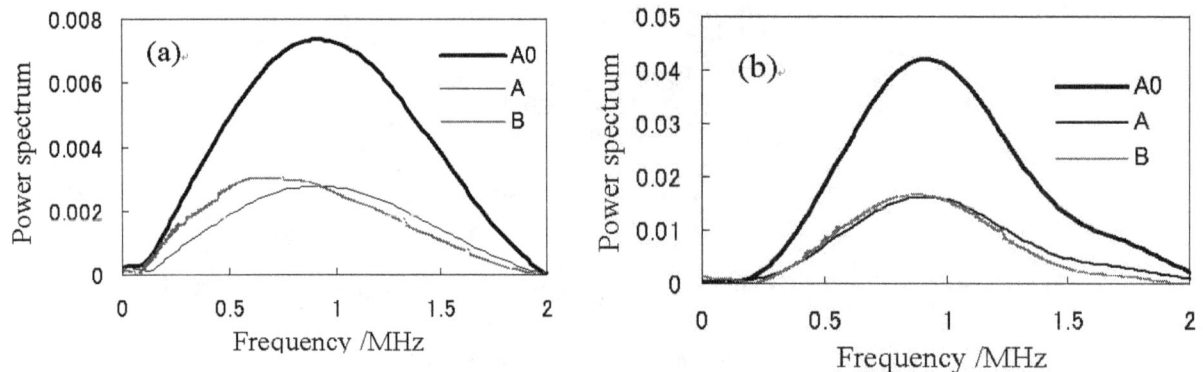

Fig. 4 Comparison of EPX-46 power spectrum (a) Experimental result (b) analytic result

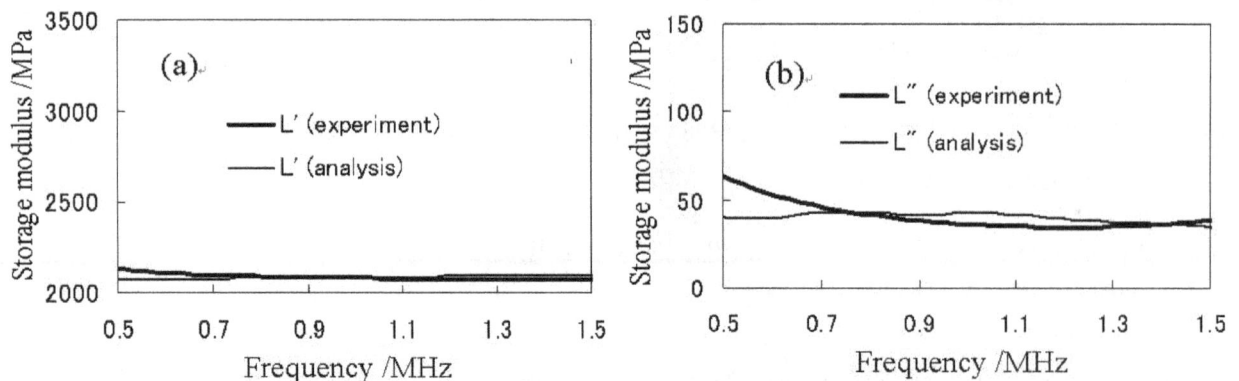

Fig. 5 Experimental result and analysis result for EPX-46 (a) Relationship between frequency and storage modulus, (b) Relationship between frequency and loss modulus

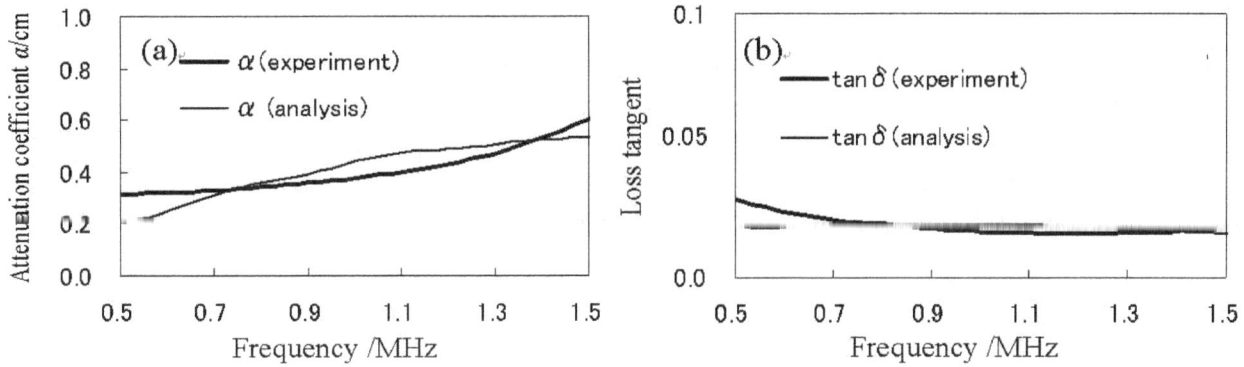

Fig. 6 Experimental and analytic result for EPX-46 (a) Relationship between frequency and attenuation coefficient, (b) Relationship between frequency and loss tangent

Table 3 Experimental and analytic result of loss tangent tanδ

Experimental materials	Ultrasonic testing	Numerical analysis	Error
EPX-46	0.018	0.0198	9.2%
OR4Si	0.034	0.0357	4.6%
ORHS4Si	0.105	0.108	3.0%
B4Si	0.005	0.0056	9.7%

Table 4 Experimental and analytic result of attenuation coefficient α

Experimental materials	Ultrasonic testing	Numerical analysis	Error
EPX-46	0.377	0.402	6.31%
OR4Si	0.688	0.744	7.53%
ORHS4Si	2.036	2.228	8.62%
B4Si	0.102	0.114	9.68%

6. Conclusion

According to the relationship between the attenuation coefficient and the group velocity, both the attenuation coefficient α and loss tangent tanδ of ultrasonic propagation in rubber polymer composite materials are solved both by experimental analysis and numerical analysis, which is based on the ultrasonic bottom echo method. The two results have been compared, it showed that the experiment results and analytic results fit very well, the error of them is within 10%. The dependence of two results of attenuation coefficient and loss tangent frequency is also the same. Thus the evaluation method, which is suitable for the dynamic and viscoelastic polymer composite materials, has been established.

Acknowledgment

This project is funded by National Natural Science Foundation of China (Grant No. 51305184 and No. 51565020).

References

1. Kai YU, Li LIU, Qiang YU, et al. Study on properties of crumb tire rubber and waste PE modified bitumen, J. Chinese Journal of Environmental Engineering, 2010, 4(3):689-692. In Chinese.
2. Nejad F M, Aghajani P, Modarres A, et al. Investigating the properties of crumb rubber modified bitumen using classic and SHRP testing methods, J. Construction & Building Materials, 2012, 26(1):481–489.
3. Sharp R S, Gruber P, Fina E. Circuit racing, track texture, temperature and rubber friction, J. Vehicle System Dynamics, 2016:1-16.
4. Persson B N J. Role of Frictional Heating in Rubber Friction, J. Tribology Letters, 2014, 56(1):77-92.
5. Selig M, Lorenz B, Henrichm D, et al. Rubber Friction and Tire Dynamics: A Comparison of Theory with Experimental Data, J. Tire Science & Technology, 2014, 42(4):216-262.
6. Buck O. Characterization of a propagating crack by ultrasonic techniques, J. Molecular Microbiology, 2011, 81(2):327–339.
7. Xing-Guo WANG, Jun-Jie CHANG, Ying-Chun SHAN, et al. Measurement of attenuation of ultrasonic propagating through the thin layer media with time delay spectrum, Chinese Journal of Mechanical Engineering (CJME), 2010, 23, (1), 129-134. In Chinese.
8. Xing-Guo WANG, Jun-Jie CHANG,Ying-Chun SHAN, et al. Testing of Characteristic of Rubber Thin Layer by Ultrasonic Echo Signal, Chinese Journal of Mechanical Engineering (CJME), 2008, 44, (10), 114-117. In Chinese.
9. Moore D F. The Friction of Pneumatic Tires [M]. New York: Elsevier Scientific, 1975:4.

Chapter 6
Photovoltaic Materials

Photoelectric Response of TiO₂ Thin Films Sensitized by CuInS₂ Nanocrystals

Jian-Bo Yin[1, 2,*], Qi-Zheng Dong[1, 2]

State Key Laboratory of Advanced Processing and Recycling of Non–ferrous Metals,
Lanzhou University of Technology, Lanzhou, 730050, China
School of Material Science and Engineering, Lanzhou University of Technology, Lanzhou, 730050, China
Email: jianbery@163.com

In this article, we presented the preparation of copper indium disulfide ($CuInS_2$; CIS) nanocrystals (NCs) by solution process as the sensitizers for microporous titania (TiO_2) thin film photoelectrodes. The CIS NCs ranging from 7 to 15 nm in diameter were synthesized by one-step thermolysis method. The results of UV-visible absorption spectra show that the as-prepared CIS NCs have strong absorption in visible light range (400 nm-1000 nm), and the band gap energy is 1.43 eV. Under illumination of simulated AM 1.5 G at one sun intensity, the thin films exhibit sensitized behavior and enhanced photoelectrochemical properties. The photocurrent of the CIS sensitized TiO_2 is three times higher than that of pure TiO_2.

Keywords: Titania; copper indium disulfide; sensitizers.

1. Introduction

Solar cells have drawn much attention due to the rise in energy consumption and environmental concern. Semiconductor-sensitized solar cells based on the nanocrystals of narrow band gap semiconductors as sensitizers have attracted more attention for their potential low cost alternatives to existing silicon cells. TiO_2 with a band gap of 3.2 eV, is one of the most significant semiconductor materials extensively applied in the photovoltaic field because of its high chemical stability and a long lifetime of photon generated carriers. However, the energy conversion efficiency of TiO_2 is low mainly because TiO_2 absorbs solar light only in the UV region, which comprises a small portion of the solar spectrum [1-4]. To extend the spectral response range, various methods have been employed and dye-sensitized and NCs-sensitized TiO_2 photoelectrodes are two strategies generally adopted by researchers [5, 6]; while the NCs-sensitized TiO_2 photoelectrodes has attracted more attention in virtue of the advantages of low cost and durable character [7-9]. Consequently, short band-gap and high optical absorption coefficient semiconductors are studied by lots of researchers. CIS is a low cost direct band gap semiconductor material with a band gap energy range of 1.20-1.50 eV and an optical absorption coefficient of 1.0×10^5 cm⁻¹. The n-type or p-type conductivity of CIS can be easily adjusted by controlling the molar ratio of the compositional elements [10, 11]. So far, several approaches have been employed to synthesize CIS NCs, such as solvothermal method, precursor decomposition method, hot injection techniques [12-14]. But the synthesis of CIS NCs with above methods normally need complex processes and toxic compounds, and the aggregation of the synthesized CIS NCs with high surface energies embarrasses the application in sensitizing photoelectrodes.

In here, the synthesis of hexagonal wurtzite polytypism CIS NCs with one-step thermolysismethod was presented, and we had decided on anatase structural TiO_2 thin films instead of rutile due to its optical activity of anatase much higher than that of rutile. Under illumination of simulated AM 1.5 G at one sun intensity, the sensitized thin films exhibit sensitized behaviors and enhanced photoelectrochemical properties.

---

*Corresponding author

2. Experimental Section

2.1. Chemicals

Copper iodiode (CuI) and polyethylene glycol were purchased from Sinoparm chemical reagent corporation; indium (III) chloridetetrahydrate (InCl$_3$·4H$_2$O), titra-n-butyl titanate (Ti(OC$_4$H$_9$)$_4$) and thiourea were pursed from Shanghai chemical reagent corporation; thioglycollic acid, tetramethylammonium hydroxide, Sodium sulfide hydrate, Sodium sulfite, methanol anhydrous, diethanolamine, ethanol absolute, ethyl acetate, hexane and acetonitrile are all purchased from Tianjin chemical regent corporation; Cetylamine (Chengdu) was purchased and used as-received.

2.2. Synthesis of P-Type CIS NCs

Typical synthesis of CIS NCs was performed as follows: CuI (0.228 g, 1.2 mmol) and InCl$_3$·4H$_2$O (0.294g, 1mmol) and thiourea (0.152g, 2.0mmol) were added to 20 mL cetylamine in a 50 mL three-necked bottle at 120 °C under the protection of N$_2$ After reaction for 10 minutes for nucleation. The system was subsequently heated to 240 °C and reacted for 60 minutes under N$_2$ flow. After being cooling naturally, the NCs were separated by adding 100 mL absolute ethanol. The NCs are washed by excess absolute ethanol for two more times and dispersed in a methanol solution of thioglycollic acid (120 mM) and tetramethylammonium hydroxide (150 mM) to exchange cetylamine molecules adsorption on the surface of NCs for 60 minutes to activate the optical activity. The CIS NCs precipitated with an ethyl acetate-hexane (1/4, v/v) solution and were redispersed in methanol.

2.3. Preparation of CIS NCs sensitized TiO2 thin films

TiO$_2$ thin films were prepared by dip-coating method and the process is as follow: 5 mL diethanolamine and 10 mL Ti(OC$_4$H$_9$)$_4$ were added into 30 mL absolute ethanol in a 100 mL beaker with vigorous stir. Then, 5 mL (0.2g/mL) polyethylene glycol solution were dipped into the Ti(OC$_4$H$_9$)$_4$ / ethanol solution, the purpose of which is to shape porous surface on titania thin films to increase the specific surface area of titania thin films. The mixed solution was continued stirring vigorously for 30 minutes. The dip coating of TiO$_2$ thin films was carried out on a dip coater with a dip velocity of 1mm/s, and a hold time of 300 s. And dip for 10 times to coat films on the FTO (20×10 mm). Subsequently, the TiO$_2$ substrates were calcined at 500 °C for 120 minutes with a heating rate of 2 °C/min, and cooled down naturally to the room temperature. For attaching the linker molecules to the TiO$_2$ surface, the TiO$_2$ thin films were immersed in an acetonnitrile solution of thioglycollic acid (1.2 M) and sulfuric acid (0.15 M) for 12 h; after that, the TiO$_2$ thin films were immersed in the CIS NCs-methanol solution for 36 h to entrap the NCs sensitizers according to the literature [10].

2.4. Characterization

High resolution transmission electron microscope (HRTEM, Hitach H-7500, Japan) was used to explore the microstructure of the CIS NCs. Scanning electron microscope (FESEM, JSM-6701F, Japan) was used to detected the microstructure of the thin films. The crystal structure of the samples was characterized by power X-ray diffraction (XRD, Rigaku D/MAX-2500, Japan) using Cu Kα radiation (λ=0.15418 nm) at the scanning speed of 1.2°/min. The component analysis of the NCs was measured by electron probe microanalyser (EPMA-1600, Rigaku). UV-vis absorption spectra were recorded using a LabTech (China) spectrophotometer within the wavelength range of

300-1000 nm. Energy dispersive X-ray analysis (EDAX, GENESIS, America) was employed to characterize the state of TiO_2 thin films sensitized by CIS NCs. Photoelectrochemical reactions of the sensitized TiO_2 electrodes were carried out in an aqueous solution of Na_2S (0.1 M) and Na_2SO_3 (0.02M) served as the electrolyte, and photoelectrochemical properties were recorded under illumination with a solar simulator (AULTT, China) at 100 mW/cm^2 (AM 1.5) using an electrochemical analyzer (CHI 660, China).

3. Results and Discussion

Fig. 1 shows HRTEM images of the as-prepared CIS NCs.

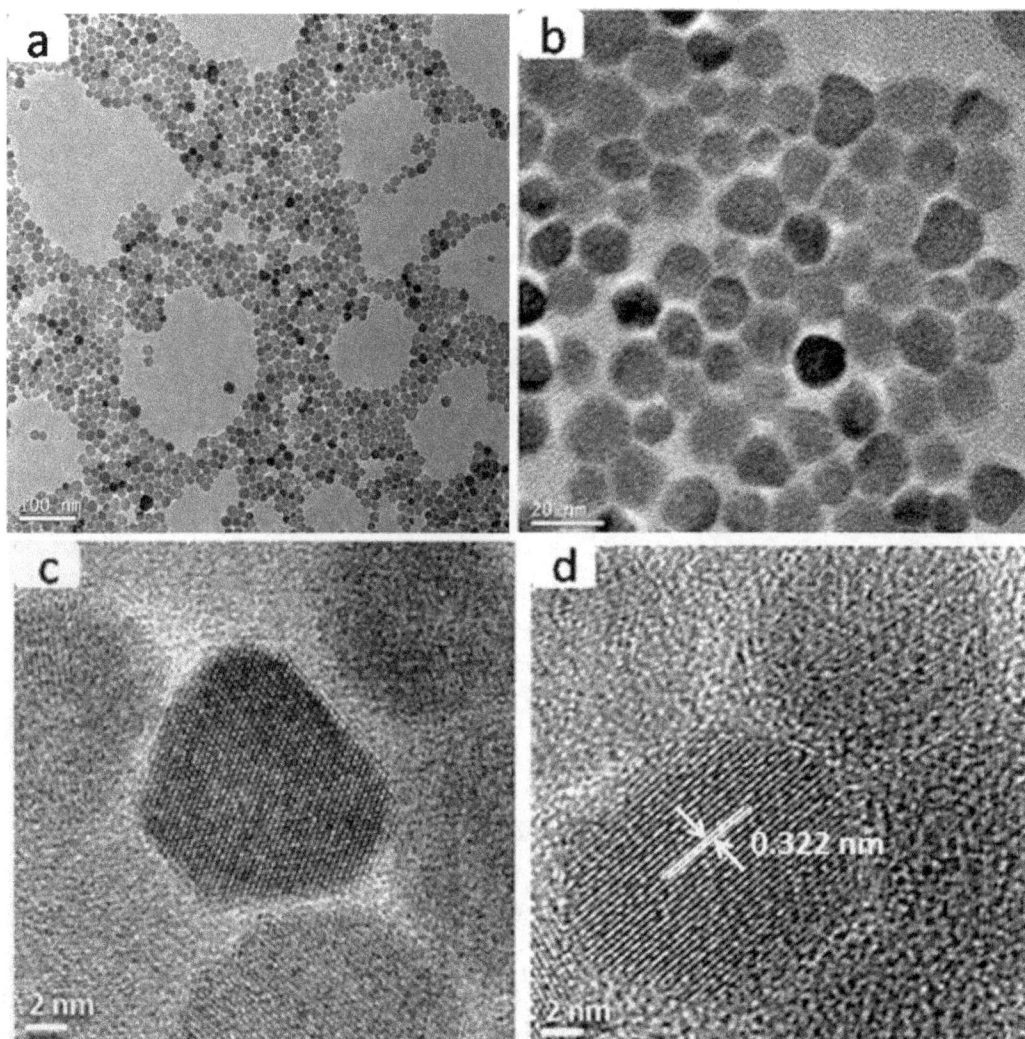

Fig. 1 (a) HRTEM Image of the as-prepared CIS NCs dispersed in cyclohexane,
(b, c and d) Magnified HRTEM Images of (a).

It can be seen that the CIS NCs are monodispersed in cyclohexane (Fig. 1a) and they are irregularly spherical with a diameter range of 7 to 15 nm (Fig. 1b). The magnified HRTEM images indicate that the as-prepared CIS NCs are highly crystallized (Fig. 1c and d). The lattice spacing in Fig. 1c corresponds to a crystal with a hexagonal structure of (001) planes, and that in Fig. 1d

267

corresponds to (002) planes [15]. Fig. 2 shows the XRD patterns of the as-prepared CIS NCs. It indicated the characteristic of a hexagonal wurtzite polytypism structure according to the literatures [16, 17]. For it is a new structure, there is no standard JCPDS card database can be employed. The average grain size of the CIS NCs was calculated about 12.1 nm according to Scherrer Equation, which matches the HRTEM results well.

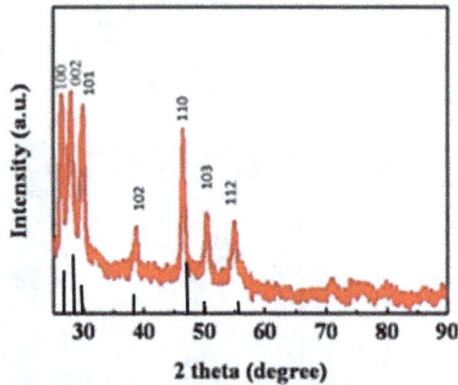

Fig. 2 XRD pattern of CIS NCs

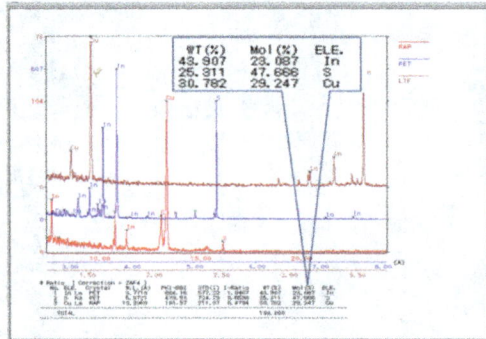

Fig. 3 Component analysis results of EPMA

EPMA analysis shows that the as-prepared CIS NCs is composed of Cu, In and S with an average composition ratio of approximate to 1.2 : 1 : 2 (Fig. 3), and this doesn't matches the normal ratio of CIS. The formation of Cu-riched CIS, which is assigned to p-type semiconductors [9], was due to the redundant composition leading to intrinsic defect structures in crystal. The UV-vis absorption spectrum of CIS NCs is shown in Fig. 4. It shows a wide absorption in the whole visible range which results in a black color of the material. From the plot of $(\alpha h\upsilon)^2$ versus $h\upsilon$ (Fig. 4b), band gap energy of 1.43 eV for CIS NCs, which were evaluated from the absorption spectrum by extrapolation of the plot of $(\alpha h\upsilon)^2$ versus $h\upsilon$ to the x axis, is within the range of CIS band gap energy (1.2 to 1.5 eV), reported in literature [18, 19, 20].

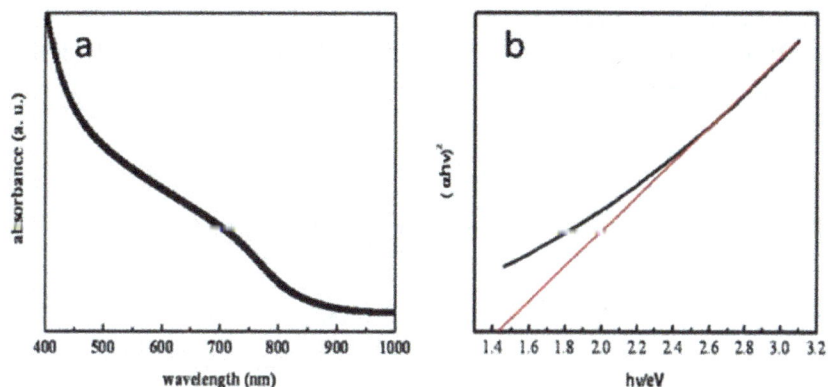

Fig. 4 (a) Optical absorption spectra of synthesized CIS NCs in cyclohexane.
(b) Plots of (αhv) 2 against photon photo energy (hv) for CIS NCs. The photon energy (hv) region for the plot was selected based on the region of significant absorbance (400-800nm) in the absorption spectra. The value estimated is 1.43 eV.

Fig. 5 displays the SEM image of microporous TiO_2 thin films. It can be seen that the thin film is 1.0 μm in thickness and the micropores size is smaller than 400 nm. The typical XRD pattern indicates that the TiO_2 thin film is anatase structure (Fig. 6), which takes on high optical activity in UV region.

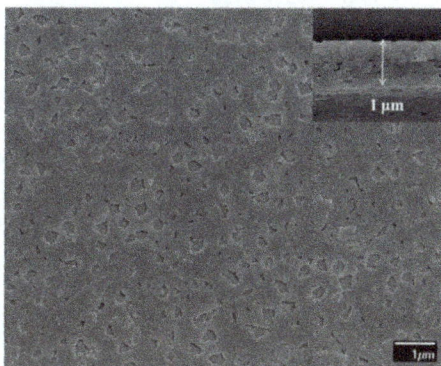

Fig. 5 FESEM image of microporous TiO_2 thin films, inset is the cross-section of TiO2 thin film

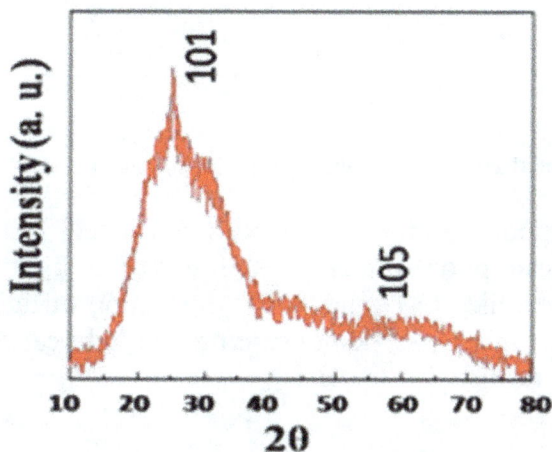

Fig. 6 XRD pattern of TiO_2 thin films

Fig. 7 shows the sensitized process of the CIS NCs sensitized TiO_2 photoelectrodes. The hydroxyl groups (-OH) on the surface of TiO_2 react with acidic groups (-COOH) of thioglycollic acid. Then, the sulfhydryl groups (HS-) react with cuprous ion (Cu^+) and indium ion (In^{3+}) of CIS.

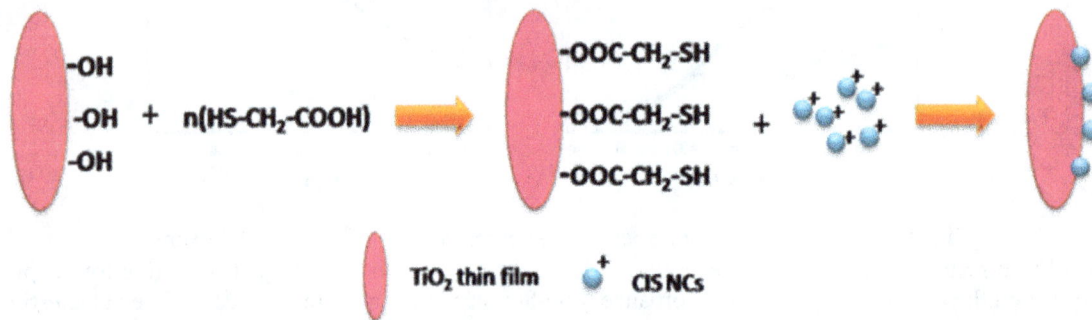

Fig. 7 The images of sensitized process

The binding reaction attracts enough quantitative CIS NCs on the surface of TiO_2 films can be proved by the EDAX results (Fig. 8), the atomic ratio of Cu and In is no change, but the percentage of S is high which attributes to the (HS⁻) groups. The redundant O can attribute to the (FTO) glass substrate (SnO_2 and SiO_2).

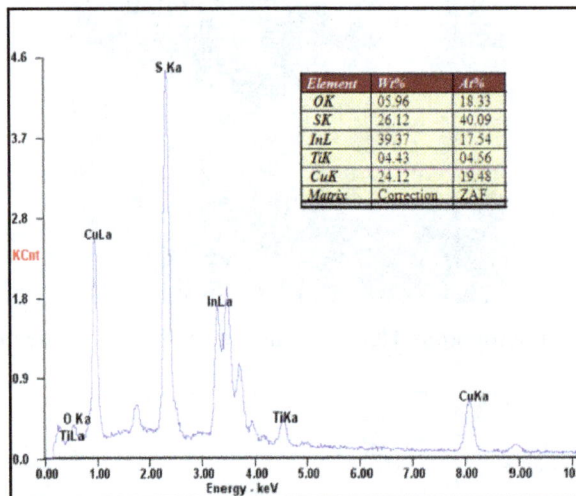

Element	Wt%	At%
OK	05.96	18.33
SK	26.12	40.09
InL	39.37	17.54
TiK	04.43	04.56
CuK	24.12	19.48
Matrix	Correction	ZAF

Fig. 8 EDAX image of sensitized TiO_2 thin film after sensitized in CIS NCs-methanol solution for 36h

Fig. 9 presents the absorption spectra of CIS NCs sensitized TiO_2 (S-TiO_2) thin film. After sensitized by CIS NCs, the absorption spectra red shift to visible light; and this result corresponds to Fig. 10, which shows the cyclic "ON" and "OFF" formation of light-induced photocurrents in the system for sensitized TiO_2 thin films, demonstrating that light causes photoelectron generation and reaction of photoelectrochemistry.

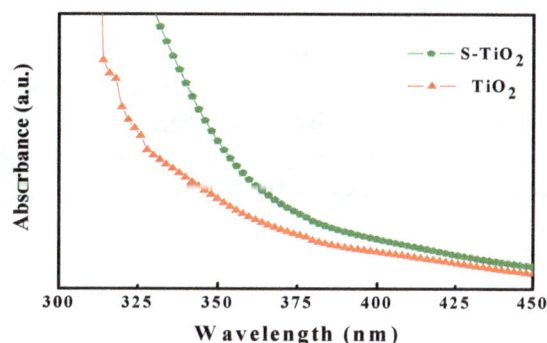

Fig. 9 UV-vis absorption spectra of CIS NCs sensitized TiO_2 thin film for 36 h.

In Fig. 10, the photocurrent of S-TiO_2 is higher than that of TiO_2, The relatively low photocurrent indicates that TiO_2 absorbs solar light only in the ultraviolet region, and the relatively high photocurrent indicates that S-TiO_2 absorbs solar light not only in UV region, but also in visible region. It is interesting to note that the sensitized TiO_2 photoelectrodes shows a high photocurrent generation of 0.40 mA cm^{-2}, which is three times higher than that of pure TiO_2 films (0.12 mA cm^{-2}).

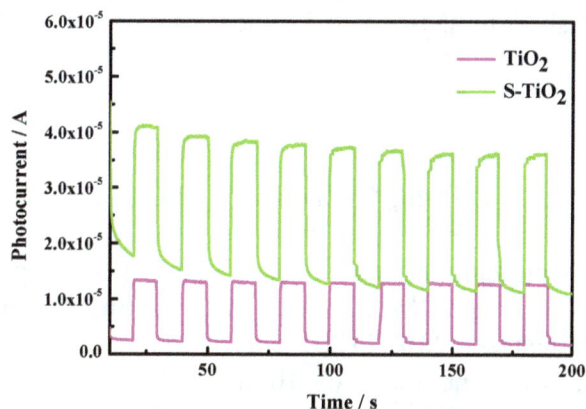

Fig. 10 The current responses of sensitized TiO_2 thin films photoanodes under illumination with AM 1.5 G at full sun intensity (100 mVcm^{-2}) at daytime without any shied in the room in 0.1 M Na_2S and 0.02 M Na_2SO_3 electrolyte solution with an illumination area of 1 cm^2, and sensitized TiO_2 thin films as the photoelectrode, Pt as the cathode and (Ag/AgCl) as reference electrode.

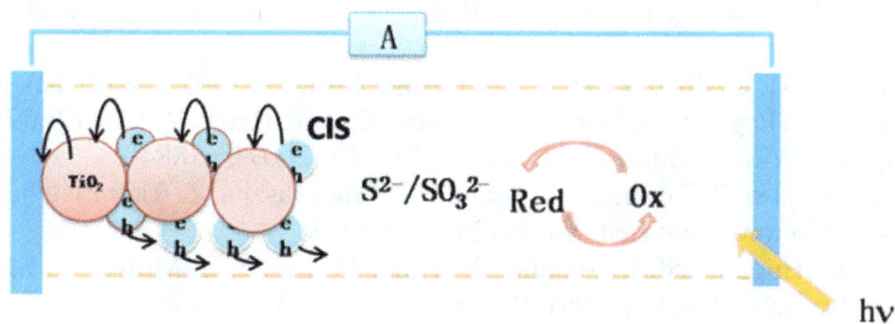

Fig. 11 The principle image of photocurrent generation of CIS NCs sensitized TiO_2

The principle of photocurrent generation is shown in Fig. 11. Under illumination, TiO_2 absorbs UV light which is only 3% of solar spectra, and charges separate from the surface. Charges inject into anode, and holes flow into the (S^{2-}/SO_3^{2-}) electrolyte solution, which is followed by collection of charges at the electrode surface, and the redox function of the solution ensures the regeneration of CIS NCs. After sensitized by CIS NCs, low bandgap material CIS NCs absorb visible light (42% of solar spectra), and much more photogenerate-electrons inject into TiO_2, and holes into electrode solution to generate photocurrent. This is why the enhanced photoelectrochemical properties can be detected.

4. Conclusion

CIS NCs were successfully synthesized by one step thermolysis method. The nanoparticles are 7 to 15 nm in diameter and hexagonal structural with a band gap of 1.43 eV. The CIS nanocystals and TiO_2 thin films were bounded by the thioglycollic acid molecules. The characteristic of current responses of the TiO_2 thin films sensitized by CIS NCs shows high sensitized behavior and enhanced photoelectrochemical properties, and the photocurrent of the CIS sensitized TiO_2 is three times higher than that of pure TiO_2.

Acknowledgments

This work is funded by the Gansu Provincial Youth Science and Technology Fund Projects (1506RJYA093) and National Natural Science Foundations of China (Grant No. 51402142 and No. 21301084).

References

1. J.B. Joo, Q. Zhang, I. Lee, M. Dahl, F. Zaera, Y. Yin,. Adv. Funct. Mater. Vol. 22 (2012), p. 166.
2. E. Li, N. Wang, H. He. Nanoscale. Research. Lett. Vol. 11 (2016), p. 1.
3. A.B. Wijeratne, D.N. Wijesundera, M. Paulose, I.B. Ahiabu, W.K. Chu, O.K. Varghese, K.D. Greis, ACS appl. mater. interface. Vol. 7 (2015), P. 11155.
4. C. Anderson, A.J. Bard, Phys. Chem. B. Vol. 101 (1997), p. 2611.
5. M. M. Momeni, Y. Ghayeb. J. Solid Stat. Electrochem. 2015, P. 1-7.
6. E.L. Unger, S.J. Fretz, B. Lim, G.Y. Margulis, M.D. McGehee, T. D. P. Stack,. Phys. Chem. Chem. Phys. Vol. 17 (2015), p. 6565.
7. J. Zhao, J. Wu, F. Yu, X. Zhang, Z. Lan, J. Lin. Electrochim. Acta Vol. 96 (2013), p.110.
8. X. Zeng, J. Bao, M. Han, W. Tu, Z. Dai. Biosens. Bioelectron. Vol. 54 (2014), p. 331.
9. N. S. Makarov, H. McDaniel, N. Fuke, I. Robel, V. I. Klimov, J. physical chem. lett. Vol. 5 (2013), p. 111.
10. Y. Shi, F. Xue, C. Li, Q. Zhao, Z. Qua, X. Li. Appl. Surf. Sci. Vol. 258 (2012), p. 7465.
11. F.M. Courtel, R.W. Paynter, B. Marsan, M. Morin. Chem. Mater. Vol. 21 (2009) p. 3752.
12. W. Du, X. Qian, J. Yin, Q. Gong. Chem. Eur. J. Vol. 13 (2007), p. 8840.
13. J.J. Nairn, P.J. Shapiro, B. Twamley, T. Pounds, R. Wandruszka, T. Rick Fletcher, M. Williams, C. Wang, M.G. Norton. Nano Lett. Vol. 6 (2006) p. 1218.
14. C. Czekelius, M. Hilgendorff, L. Spanhel, I. Bedja, M. Lerch, G. Muller, U. Bloeck, D.S. Su, M. Giersig. Adv. Mater. Vol.11 (1999), P. 643.
15. B. Koo, R. N. Patel, A.K. Brian. Chem. Mater. Vol. 21 (2009), p. 1962.

16. M. Kruszynska, H. Borchert, J. Parisi, J. K. Olesiak. J. Am. Chem. Soc. Vol.132 (2010). p. 15976.
17. D. Pan, L. An, Z. Sun, W. Hou, Y. Yang, Z. Yang, Y. Lu. J. Am. Chem. Soc. Vol. 130 (2008), p. 5620.
18. M.G. Panthani, V. Akhavan, B. Goodfellow, J.P. Schmidtke, L. Dunn, A. Dodabalapur, P.F. Barbara, B.A. Korgel. J. Am. Chem. Soc. Vol. 130 (2008) p. 16770.
19. Y. Chen, X. He, X. Zhao, M. Song, X. Gu. Mater. Sci. Engin. B Vol. 139 (2007) p. 88.
20. P.V. Kamat. J. Phys. Chem. C Vol.112 (2008), p. 18737.

Effect of Doped-CeO$_2$ on Structure and Ultra-violet Absorption Property of ZnO Thin Film

Yun-Yun Xu*, Tao Zhang, Zhen-Rong Lin

Airforce Logistics College, Xuzhou 221000, China

*Email: xuyunyun713@163.com

High quality ZnO thin films doped with CeO$_2$ was prepared by RF magnetron sputtering technique. The influence of CeO$_2$ on the structure and optics absorption property was studied by XRD apparatus and UV-Visible spectrophotometer. The results show that doped CeO$_2$ has affected developing ways of crystal grains and UV absorption property of ZnO. The films' UVA absorption is enhanced. The slope of the absorption margin increased and the absorption edge obviously moved to short wave direction. In addition, the breadth of the absorb peak increased and the absorption intensity improved.

Keywords: RF magnetron sputtering; ZnO thin film; doped; ultra-violet absorption.

1. Introduction

ZnO thin film is an ultraviolet semiconductor optoelectronic device material with great potential application value. The ZnO thin films doped with certain elements has piezoelectricity, varistors, gas sensor, photoelectric and transparent conductive properties [1-4]. Recent studies show that ZnO thin films doped with Sb$_2$O$_3$ and Bi$_2$O$_3$ has low voltage varistors, which is excellent in the development of low voltage varistor's potential. Ultraviolet detector can be made by use of light conductive properties and wide band gap of ZnO [5-9]. Study on ZnO grain boundary character has more, but the research on the metal oxide doping on the phase structure of ZnO main crystal effect are few. Electronic information and atomic state crystals can be learned by ultraviolet absorption spectrum and the change of crystal structure will cause the UV absorption peak of the transfer and change. In this paper, the effect of doped CeO$_2$ on UV-Vis optical absorption property of ZnO thin film has been discussed.

2. Experimental

In the experiment, ZnO thin films were prepared on 15×15×1 mm glass sheets by JGP 500 R.F. magneto sputtering apparatus. The glass substrates were dried and then quickly placed into the sputtering chamber after thoroughly ultrasonic cleaned in ethanol and deion water. The targets with different percent of CeO$_2$, 1%, 3%, 5%, 10%, 15% (wt%), were made of the mixing ZnO and CeO$_2$ powders with highly purity of 99.95% and were placed 45 mm on the top of the substrate. The target and substrate were pre-sputtered for 5 minutes to get rid of the impurities. In the process, the basic vacuum of the chamber, the pressure of sputtering gas mixed of 20sccm Argon and 20sccm Oxygen, the sputtering powder, and the substrate temperature is 2.0×10-4Pa,1.0Pa, 80W, and 300°C, respectively. The five kinds of Ce-ZnO thin films are marked as S1-S5. ZnO thin

*Corresponding author

film's structural and their optical properties are studied by D8 Advanced X Ray Diffraction, (Cu kα, λ=0.15418nm) and 6010 Ultraviolet-visible spectrophotometer with the wavelength ranging from 200 to 900 nm.

3. Results

Fig. 1 is the XRD spectra of pure and CeO_2 doped ZnO thin films. There is only one strong ZnO (002) diffraction peak, whose intensity is lower than that of pure ZnO thin film.

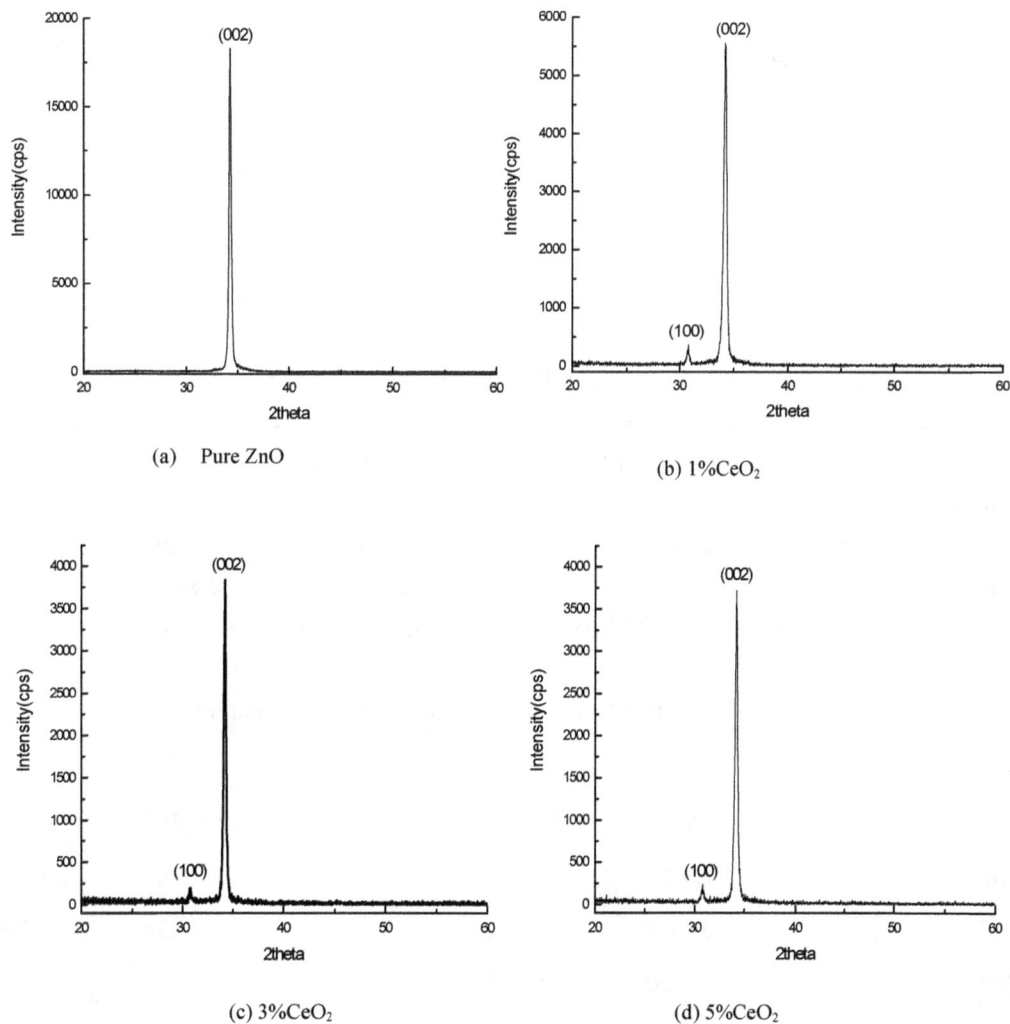

(a) Pure ZnO

(b) 1%CeO_2

(c) 3%CeO_2

(d) 5%CeO_2

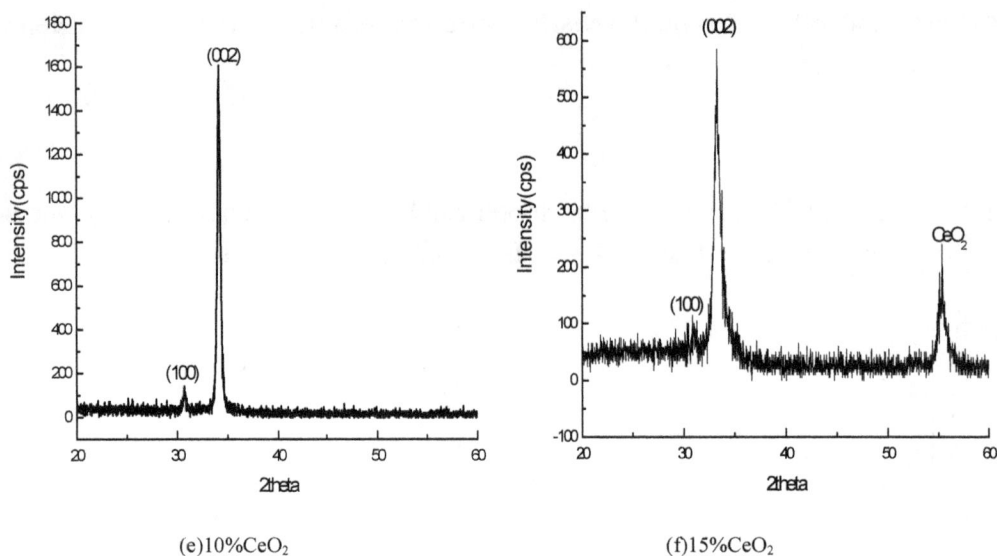

(e)10%CeO$_2$ (f)15%CeO$_2$

Fig. 1 XRD spectra of pure and CeO$_2$ doped ZnO thin films

Moreover, when the CeO$_2$ content is more than 1%, diffraction spectrum also appeared on the ZnO (100) crystal face diffraction peaks, the intensity and with the increase of CeO$_2$ content, the corresponding increase, but the content of CeO$_2$ reached 15%, the diffraction spectrum appeared on the diffraction peaks of CeO$_2$.

According to the calculated strain and grain size of thin film D, C-axis along C-axis stress. Results are shown in Table 1. All samples of D values are larger than the standard value d$_0$ of bulk materials, indicating the direction of the C axis in the films always has stress.

Table 1 XRD data analysis of doped ZnO thin film samples

Sample	Pure ZnO	S$_1$	S$_2$	S$_3$	S$_4$	S$_5$
d/Å	2.61418	2.61978	2.62172	2.62190	2.62397	2.69083
σ/MPa	2.123	3.099	3.438	3.470	3.830	15.4855
2θ/°	34.2532	34.199	34.1729	34.1706	34.1428	33.2694
B/°	0.248	0.250	0.257	0.291	0.334	0.710
D/nm	35.037	25.553	33.803	29.853	26.008	12.206

Table 1 shows that stress always exist in ZnO thin films. After CeO$_2$ doping, due to the role of Ce^{4+}, the lattice distortion is further increased, and the stress resulting in the films also increased manifold. In addition, with the increase of Ce content in the films, the grain size of the films gradually decreases. This shows that in the films of Ce inhibit the growth of ZnO grains to the speed and degree of orientation. ZnO doped with CeO$_2$, because Zn, Ce, O three elements could not form chemical compounds, while the Ce^{4+} radius of 0.80 Å, so a portion of the Ce^{4+} will enter the crystal lattice of ZnO instead of Zn^{2+}. When adding a small amount of CeO$_2$, Ce^{4+} mainly entered the ZnO lattice position, caused a small expansion of the ZnO lattice. ZnO (002) increase in the interplanar lattice spacing, increased from 2.61418Å to 2.62397Å, due to the increase of lattice distortion caused the stress in the film is also a corresponding increase in. With increasing of CeO₂ doping amount, cause further relaxation of the ZnO lattice, (002) the interplanar lattice spacing is increased, reaching 2.69083 Å. When the CeO₂ content is over 1%, many defects in the films make the mixing direction of ZnO crystal grains, such as (100).

Because the ZnO grain has different growth direction, inhibiting ZnO grain (002) orientation length rate, resulted in the gradual decrease of grain size. When the CeO₂ content further increased up to 15%, CeO₂ phase will be in ZnO grain boundary, the generation of CeO₂ also significantly inhibited the ZnO grain oriented growth, the intensity of the diffraction peaks decreased, peak widths increase. Due to the differences of Sb^{3+} and Ce^{4+} on the size and the binding energy with Zn^{2+}, their degree of influence on ZnO is different. Radius of Ce^{4+} ions is greater than Sb^{3+}, and when the two defects exist as ions, due to the size of Zn^{2+} and Sb^{3+} more closely, so easier to exchange Zn^{2+}. While the Ce^{4+} size is larger, so the probability of replacement of Zn^{2+} is relatively small, resulting in the lattice distortion caused by the same Sb^{3+} content greater than by Ce^{4+}.

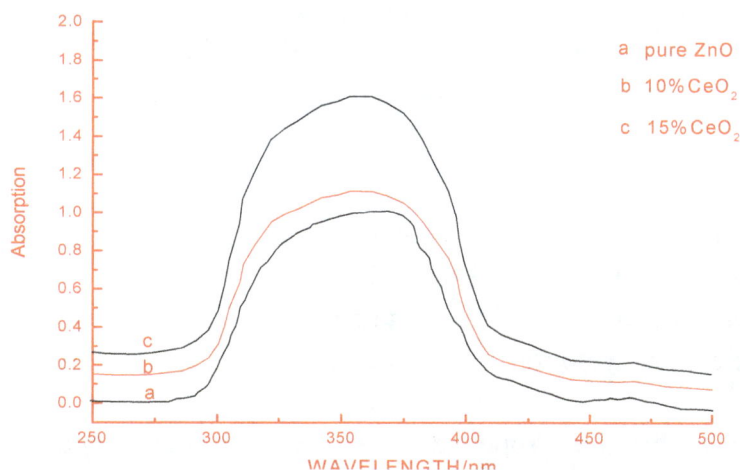

Fig. 2 The ultraviolet absorption spectra of pure and CeO₂-doped ZnO thin films

The spectra of pure and CeO₂ doped ZnO thin films are shown in Fig. 2. From the plot we can see that spectrum shape without big changes and the absorption peak is consistent. The main differences are as follows: first, after 15% CeO₂ doped ultraviolet absorption capacity of ZnO thin film is significantly enhanced, and absorption intensity increases obviously. Second, the absorption edge is slightly shifted to short wavelength and the slope of absorption edge had small increase. Third, there is a slight increase in the width of the absorption peak. 10% CeO₂ doped samples has the similar change, but the absorption intensity is not big improvement compared with the pure ZnO. We can see that CeO₂ doped ZnO film can enhance the original band ultraviolet absorption of ZnO thin film.

4. Discussion

Because of the doped Ce^{4+} increases the exciton concentration in the films transitions, the intensity of absorption peak increase. Therefore, CeO_2 is an effective material to enhance ZnO film ultraviolet absorption ability. We can know from the previous analysis, because the ion size of Ce and Sb are different, the lattice distortion is vary, resulting in the absorption edge of CeO_2 doped samples was lower than the Sb_2O_3 doped. In addition, UV absorption abilities of CeO_2 and Sb_2O_3 also have the difference. The test shows that the Sb_2O_3 ultraviolet band has a good optical absorption and optical absorption of CeO_2 mainly concentrated in two bands. This leads to significant differences between the two kinds of samples UV absorption strength.

5. Conclusion

1) By mixing CeO_2 powder in the target, to prepare CeO_2 doped ZnO thin films by magnetron sputtering technique, Ce exists in the form of substitutional atoms, ZnO still C axis preferred orientation growth.
2) The electron binding energy of ZnO increases after CeO_2 doped.
3) ZnO thin film's UV absorption property is improved after introducing CeO_2. The absorption intensity increases in some sort. Equal amounts of Sb_2O_3 doped ZnO thin film is better than CeO_2 doped ZnO film on ultraviolet absorption improvement ability.

References

1. E.M. Bachari, G. Baud, S.B. Amor, M. Jacquet, Structural and optical properties of sputtered ZnO films [J], Thin Solid Films, 2003, 348: 165-172
2. T. Inukai, M. Matsuoka and K. Ono, Characteristics of zinc oxide thin films prepared by r.f. magnetron-mode electron cyclotron resonance sputtering[J], Thin Solid Films, 2005, 257: 22-27
3. R. Ondo-Ndong, G. Ferblantier, Properties of RF magnetron sputtered zinc- oxide thin films [J], Journal of Crystal Growth, 2003, 256: 130-135
4. J. Molarious, J. Kaitila, Piezoelectric ZnO films by r.f. sputtering [J], Journal of Materials Science: Materials in Electronics, 2003, 14: 431-435
5. Horio N, Hiramatsu M, Yoko T. [J]. J Am Ceram Soc, 2003, 81(6): 1622-1632
6. Yaniu HUANG, Meidong LIU, Yike ZENG, et al. [J]. Materials Science and Engineering B, 2001, 86: 232-236
7. Jin-zhong WANG, Xiao-tian YANG and Bai-jun ZHAO, Semiconductor Optoelectronics, 2008, 23:426(in Chinese)
8. H. Suzuki, and R.C. Bradt, J Am Ceram Soc, 2009, 78:1354.
9. Yan-qiu HUANG, Mei-dong LIU, Zhen LI and Yi-Ke ZENG, Journal of Functional Materials, 2010, 33: 653 (in Chinese).

Effect of Interfacial Properties on the Band Gaps of One-dimensional Phononic Crystal

Li Wang[1,a], Xi-Qiang Liu[1,b,*], Gui Zhang[1,c], Bi-Li Wang[1,d]

[1]*College of Sciences, PLA University of Science and Technology, Nanjing, China, 211101*
*Email: liuxiqiang2010@163.com

The effect of interfacial properties on the band gaps is studied when shear horizontal (SH) wave obliquely incident on a one-dimensional phononic crystal, based on the spring interface model. When the interface phase is thin and light (the mass density is small), the inertia effect can be neglected. In this situation, the interface phase is usually substituted by distributed springs. The boundary conditions for spring interface model are obtained. The dispersion relation is derived using the transfer matrix method and Bloch theorem. The numerical simulation is performed for the combination of Al and Epoxy. The band structures corresponding to real wave vectors are computed in the case of different interfacial adhesion degrees. As a result, it shows that the width of the band gap increases when the interfacial adhesion degree decreases. The localized modes appear when the interface is almost unbonded. In addition, the wave propagates through the lattice with a small attenuation near the localized mode frequency.

Keywords: Phononic crystal; spring interface model; band gaps; localized modes.

1. Introduction

Phononic crystals are composite materials with periodic structures [1]. Similar to the phontonic crystal, phononic crystals are characterized by bandgap effect, negative refraction effect, wave guide effect and defect effect etc. These characteristics have a wide range of applications in the field of vibration reduction and noise reduction, wave guiding and wave filtering, new acoustic devices designing etc.

Band gap is one of the primary properties which is concerned by many reserchers. However, the studies on the solid / solid phononic crystals mostly adopted the continuous boundary conditions, i.e. the assumption of continuous displacement vector and stress vector at the boundary. In fact, the interface layer/interphase exists between different materials due to different processing technology and environment [2], even many kinds of flaw and damage appear. They have great influences on the propagation of elastic wave. Some vibration isolation and noise reduction facilities may be devised by particular design of the interface conditions. So, the effect of interfacial properties on the band gap should be identified firstly. Wei [3] investigated the influence of gradient profile on the band gap of two-dimensional phononic crystal via the use of multiple scattering method. Wang [4] calculated the bandgap of in-plane waves in nanoscale phononic crystals taking account of surface/interface effects by the use of DtN mapping method. The transfer matrix method [5] is usually used to calculate the energy band gap structure of one dimensional phononic crystal, in this case, the wave motion equation is very simple, and the analytical dispersion relation can be obtained.

This paper discussed the effect of interfacial properties on the band gap of one-dimensional two-component phononic crystal using transfer matrix method base on spring interface model. The paper is arranged as follows. Section two lists relevant elastodynamics equations for completeness. Section three is devoted to the theoretical derivation of the dispersive relation based on spring interface model. In section four, the bandgaps are computed for various interface adhesive degrees in consideration of Al and Epoxy combination.

2. SH Wave Propagating in 1-D Phononic Crystal

We consider one-dimensional two-component phononic crystal. The geometry is depicted in figure 1, A and B are homogeneous materials, the mass density is denoted by ρ_j, the material constant is denoted by μ_j, the velocity of SH wave is $c_j = \sqrt{\mu_j/\rho_j}$, the thickness of material A in a single is a_j, $a = a_1 + a_2$ is the lattice constant ($j = 1,2$ denotes two different materials A and B, respectively).

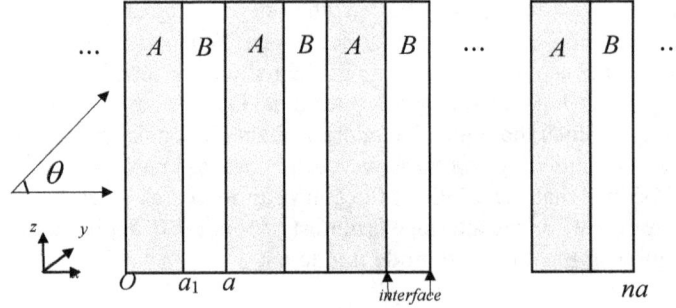

Fig. 1 The schematic of 1-D phononic crystal

Assume the time harmonic SH wave propagates obliquely in plane xoz, θ is the angle between the wave vector and the positive direction of x axis, SH wave satisfies the wave motion equation

$$u_{tt} = c^2\left(u_{xx} + u_{zz}\right), \tag{1}$$

Where $u = u(x,z,t)$ is the displacement. k_3 is the wave number along z direction and it keeps invariant according to the Snell law in acoustics [6], so the displacement can be assumed as

$$u_{nj}(x,z,t) = \left(P_{nj}^+ e^{\alpha_j(x-na)} + P_{nj}^- e^{-\alpha_j(x-na)}\right)e^{i(k_3 z - \omega t)}, \tag{2}$$

Where, $\alpha_j = \sqrt{k_3^2 - \left(\omega/c_j\right)^2}$ denotes the wave number along x direction, P_{nj}^+ and P_{nj}^- denote that the wave propagate along the positive direction and the negative direction of x axis, respectively. $i\left(=\sqrt{-1}\right)$ represents the imaginary unit. According to the relation between displacements and stress components, the stress expressions are

$$\sigma_{nj} = \rho_j c_j^2 \frac{\partial u_{nj}}{\partial x} = \alpha_j \rho_j c_j^2 \left(P_{nj}^+ e^{\alpha_j(x-na)} - P_{nj}^- e^{-\alpha_j(x-na)}\right)e^{i(k_3 z - \omega t)}, \tag{3}$$

For convenience, the displacements and the stress can be expressed as

$$\mathbf{W}_{nj}(x) = \begin{bmatrix} u_{nj}(x) \\ \sigma_{nj}(x) \end{bmatrix}. \tag{4}$$

3. The Dispersive Equation Based on Spring Interface Model

The function of the interphase is to link different material components together. If the interphase is thin enough and the mass density is small, then, the inertial effect of the interphase can be ignored. The interphase can be modeled as spring. The boundary conditions of spring model can be expressed as [7]

$$[\mathbf{t}] = \mathbf{0}$$

$$[\mathbf{u}] = \mathbf{F} \cdot \mathbf{t} = \begin{bmatrix} F_n & & \\ & F_{s1} & \\ & & F_{s2} \end{bmatrix} \begin{bmatrix} t_n \\ t_{s1} \\ t_{s2} \end{bmatrix} \tag{5}$$

Where, $[\cdot]$ denotes the jump of physical quantity, $[\mathbf{t}] = \mathbf{t}^+ - \mathbf{t}^-$ denotes the jump of traction, $[\mathbf{u}] = \mathbf{u}^+ - \mathbf{u}^-$ denotes the jump of displacement. $\mathbf{t}^+, \mathbf{t}^-$, $\mathbf{u}^+$ and $\mathbf{u}^-$ denote the traction and displacement at both sides of the interphase. $\mathbf{F}$ is called flexible matrix of the spring. When the thickness of interphase is enough thin, the asymptotic analysis leads to $F_n = h/(\lambda_s + 2\mu_s)$, $F_{s1} = F_{s2} = h/\mu_s$, where h is the thickness of interphase, λ_s, μ_s are the material constants of interphase. The main diagonals of matrix $\mathbf{F}$ are called flexible coefficients that describe the imperfect degree of the interface when changing between 0 and $+\infty$. The very small flexibility coefficient indicates the interface is close to perfect, and the very big flexibility coefficient indicates the interface is close to unbonded.

Taking into account SH wave incidence, the boundary conditions of spring interface can be expressed as

$$\mathbf{W}_{n1}(na + a_1) = \mathbf{FW}_{n2}(na + a_1), \tag{6}$$

$$\mathbf{W}_{n1}(na) = \mathbf{FW}_{n-1,2}(na), \tag{7}$$

where, $\mathbf{F} = \begin{bmatrix} 1 & F_s \\ 0 & 1 \end{bmatrix}$, $F_s = h/\mu_s$.

The boundary conditions of spring interface (6), (7) lead to

$$\mathbf{F}_1 \mathbf{K}_1 \mathbf{\Psi}_{n2} = \mathbf{H}_1 \mathbf{\Psi}_{n1}, \tag{8}$$

$$\mathbf{F}_1 \mathbf{K}_2 \mathbf{\Psi}_{n-1,2} = \mathbf{H}_2 \mathbf{\Psi}_{n1}, \tag{9}$$

where,
$$\mathbf{F}_1 = \begin{bmatrix} 1 & F_s \\ 0 & 1 \end{bmatrix}, \qquad \mathbf{K}_1 = \begin{bmatrix} e^{\alpha_2 a_1} & e^{-\alpha_2 a_1} \\ F_2 e^{\alpha_2 a_1} & -F_2 e^{-\alpha_2 a_1} \end{bmatrix}, \qquad \mathbf{H}_1 = \begin{bmatrix} e^{\alpha_1 a_1} & e^{-\alpha_1 a_1} \\ F_1 e^{\alpha_1 a_1} & -F_1 e^{-\alpha_1 a_1} \end{bmatrix},$$

$$\mathbf{K}_2 = \begin{bmatrix} e^{\alpha_2 a} & e^{-\alpha_2 a} \\ F_2 e^{\alpha_2 a} & -F_2 e^{-\alpha_2 a} \end{bmatrix}, \ \mathbf{H}_2 = \begin{bmatrix} 1 & 1 \\ F_1 & -F_1 \end{bmatrix}, \ \boldsymbol{\Psi}_{nj} = \begin{bmatrix} P_{nj}^+ \\ P_{nj}^- \end{bmatrix}, \ F_j = \alpha_j \rho_j c_j^2, \ j = 1, 2.$$

The relation between the nth cell and $(n-1)$th cell can be obtained from equations (8) and (9)

$$\boldsymbol{\Psi}_{n2} = \mathbf{T} \boldsymbol{\Psi}_{n-1,2}, \tag{10}$$

where, $\mathbf{T} = \mathbf{K}_1^{-1} \mathbf{F}_1^{-1} \mathbf{H}_1 \mathbf{H}_2^{-1} \mathbf{F}_1 \mathbf{K}_2$ is the transfer matrix, $|\mathbf{T}| = 1$ can be easily proved.

Considering the periodicity along x direction, the following relation can be obtained according to **Bloch theorem** [8]

$$\boldsymbol{\Psi}_{n2} = e^{ika} \boldsymbol{\Psi}_{n-1,2}, \tag{11}$$

and the matrix eigenvalue problem can be obtained from equations (10) and (11)

$$\left| \mathbf{T} - e^{ika} \mathbf{E} \right| = 0, \tag{12}$$

where, $\mathbf{E}$ is the two order unit matrix. After tedious derivation, the relation between wave vector k and frequency ω can be obtained through solving the eigenvalue problem (12)

$$\cos(ka) = \cosh(a_1 \alpha_1) \cosh(a_2 \alpha_2) + \frac{1}{2}\left(F + \frac{1}{F}\right) \sinh(a_1 \alpha_1) \sinh(a_2 \alpha_2)$$

$$- \frac{1}{2} F_1 F_2 F_s^2 \sinh(a_1 \alpha_1) \sinh(a_2 \alpha_2) \tag{13}$$

where, $F = F_1 / F_2$.

4. Numerical Results and Discussion

Consider a phononic crystal composed of Al and Epoxy, the lattice constant is $a = 0.15m$, $a_1 : a_2 = 1:1$, $\theta = 0.5$. The material constants are given in Table 1.

Table 1 Material parameters of Al and Epoxy

Material	Mass density $\rho\ (kg \cdot m^{-3})$	shear modulus μ (Gpa)
Al(A)	2799	26.8
Epoxy(B)	1180	1.59

Without loss of generality, the shear modulus of the adhesive material is selected as $\mu_s = \dfrac{\mu_1 + \mu_2}{10}$, the thickness of the adhesive layer is selected as $h = a_1/10$, and the flexibility coefficient is $F_s = F_0 \cdot \dfrac{h}{u_s}$, thus we can use a dimensionless parameter F_0 to describe the bonding degrees, the range of F_0 is $[0, +\infty)$, and $F_0 = 0$ denotes the perfect bonding interface, $F_0 = \infty$

denotes the fully debonding interface. The band structures corresponding to real wave vectors for different interface adhesive degrees are shown in figure 2, and (a)(b)(c)(d) correspond to $F_0 = 0,5,10,100$ respectively.

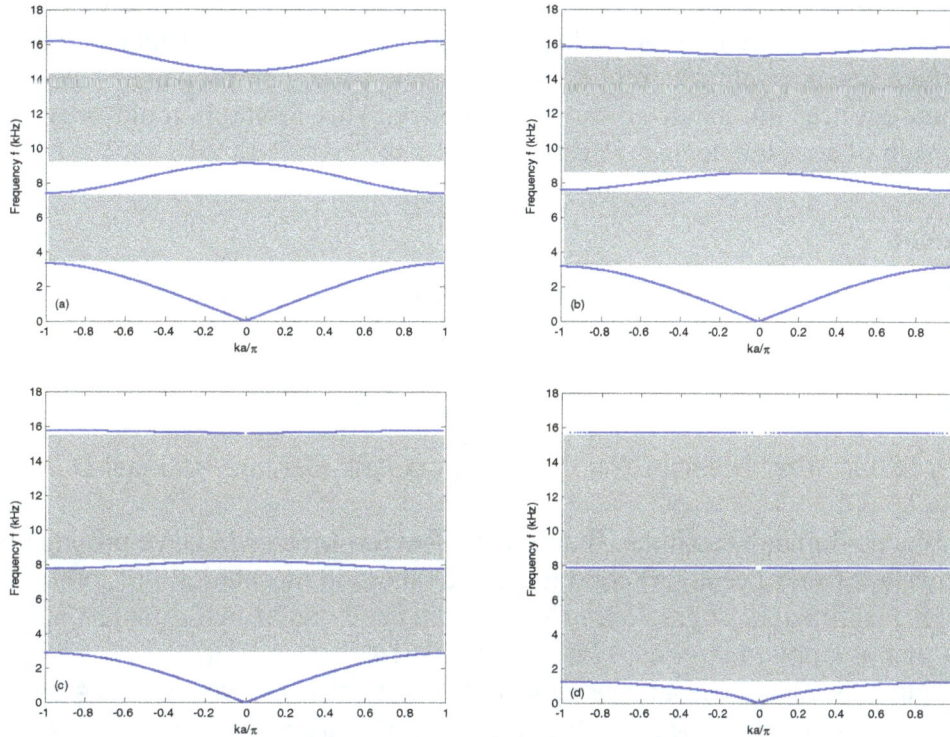

Fig. 2 Band structures corresponding to real wave vectors for different interface adhesive degrees

The curve in the figure is called the dispersion relation curve, on which each point corresponds to a kind of propagation mode of the elastic wave. The shadow part indicates the band gap range. The band gap range and width for different interface adhesive degrees are shown clearly in table 2. It can be seen that two band gaps appear in the range of 0-15.8kHz. The band gap width of the phononic crystal become wide with the increase of F_0, i.e. the decrease in the adhesive degree of the interface.

The interface can be considered as a defect when the interface becomes almost fully debonding, and the localized modes can be observed obviously. The localized mode between the first and second band gaps exists near 7.869kHz. This phenomenon can be explained by the localized resonant mechanism when the interfaces between two solids become almost fully separated.

Table 2 Bandgap range and bandgap width for different interface adhesive degrees

F_0	Bandgap range(kHz)	Bandgap width(kHz)
0	(3.330,7.390)	4.060
	(9.114,14.473)	5.360
5	(3.164,7.583)	4.419
	(8.574,15.366)	6.792
10	(2.866,7.741)	4.875
	(8.198,15.619)	7.421
100	(1.230,7.865)	6.635
	(7.872,15.733)	7.861

5. Conclusion

The analytical dispersion relation of one-dimensional two-component phononic crystal with spring interface was derived by the use of transfer matrix method and Bloch theorem. The interfacial properties have obvious influences on the band gap characteristics. The band gap width becomes wide with the decrease of the interface adhesive degrees. The localized modes can be observed when the interface becomes almost fully debonding. The wave can propagate through the lattice with less attenuation near the localized mode frequency. This research could provide theoretical basis for the design of acoustic devices in future.

Acknowledgment

The work is funded by the Advanced Research Foundation of PLA University of Science and Technology.

References

1. Xisen Wen, Jihong Wen, Dianlong Yu. Phononic crystals. Beijing: National Defence Industry Press, 2009, 53-57.
2. Yuesheng Wang, Guilan Yu, Zimao Zhang, et al. Review on elastic wave propagation under complex interface (interface layer) conditions. Advances in mechanics, 30(2000) 378-390.
3. Cai B, Wei P J. Influences of gradient profile on the band gap of two-dimensional phononic crystal. Journal of Applied Physics, 110 (2011) 103514.
4. Zhen N, Wang Y S, Zhang C. Bandgap calculation of in-plane waves in nanoscale phononic crystals taking account of surface/interface effects. Physica E: Low-dimensional Systems and Nanostructures, 54 (2013) 125-132.
5. Zhenguo Zhang,Ying Liu. Study on sound insulation of 1D two-component phononic crystal based on the theory of sound transmission. Materials Review B. 27 (2013):155-158.
6. Gonghuan Du, Xiufen Gong. Fundamentals of Acoustics. nanjing: Nanjing University Press. 2001, 135-140.
7. Mal A K, Xu P C. Elastic waves in layered media with interface features. Elastic Wave Propagation, 1989, 67-73.
8. Kun Huang, Ruqi Han. Solid state physics. Beijing: Tsinghua University Press. 2003, 154-157.

Chapter 7
Functional and Smart Materials

Enhanced Mechanical Properties and Cell Affinity of Chitosan Gels by Blending with Silk Fibroin

Hui Wu[1], Jing-Wan Luo[1], Zu-Wei Luo[1], Yi Jiang[1], Guo Chen[1], Yi Zhang[2], Ming-Zhong Li[1,*]

[1]*National Engineering Laboratory for Modern Silk, College of Textile and Clothing Engineering, Soochow University, No. 199 Ren'ai Road, Industrial Park, Suzhou 215123, China;*
[2]*Department of Burns and Plastic Surgery, Affiliated Hospital of Nantong University, Nantong 226001, China;*
[*]*Email: mzli@suda.edu.cn*

Blending with another biocompatible natural polymer is a potential way to enhance the mechanical properties and cell affinity of chitosan gels. In this study, chitosan (CH)/silk fibroin (SF) hybrid gels were fabricated by initiating gelation using β-glycerol phosphate at 37°C. Statistical analysis results from SEM images showed that the internal pore diameter of CH/SF gels with blending ratio of 75/25 was about 93.9 μm, and a number of micro fibers with the size of several micrometers were observed within the gels. Mechanical measurements showed that the compressive strength of CH/SF gels with blending ratio of 75/25 was about 4.39 kPa which were significantly bigger than that of CH gels, and it showed viscoelasticity characteristics of the elastic materials by rheological tests. Bone marrow mesenchymal stem cells were incubated on CH/SF blend gels for 3 days. Results from SEM and LSCM observation showed that blending CH with SF was beneficial for the cell attachment, spreading and proliferation. These results indicated that the CH/SF gel is expected to be useful in tissue engineering.

Keywords: Chitosan; silk fibroin; blend gels; BMSCs.

1. Introduction

Chitosan (CH), a natural cationic polymer, no specific adhesion ability to cell [1], due to its excellent biocompatibility, injectability, thermosensitivity and degradability [2], has been currently receiving a great deal of interest and has been proposed for biomedical applications, especially used as tissue engineering scaffolds for cartilage, skin, vessel and bone regeneration [3]. A cellular scaffold of tissue engineering, in addition to being biocompatible, should be a biomaterial device with physical and mechanical properties such as strength, elasticity and stiffness that match those of the target tissues and surrounding tissues [4, 5]. The compressive strength of CH gels can be enhanced by impregnation with crystal grains [6], staple fibers [7] and nanotubes [8]. Blending with another biocompatible natural polymer is a potential way to improve mechanical properties and biocompatibility of CH gels [2, 9].

Silk fibroin (SF), a natural fibrous protein, has been used as a potential biomaterial for biomedical and biotechnological fields because of its super biocompatibility, low inflammatory reactions and tunable biodegradation rate [7, 10], especially the excellent cytocompatibility and tensile strength [11]. SF materials can support the attachment, proliferation and differentiation of primary cells and cell lines. It has been effectively used as tissue engineering scaffolds for cartilage, bone, skin, nerve and blood vessel regeneration [12].

In the present study, CH/SF hybrid gels were prepared by blending chitosan with silk fibroin under the initiation of β-glycerol phosphate (β-GP). The compressive strength and viscoelasticity of CH/SF gels were measured and compared with CH gels. Bone marrow mesenchymal stem cells (BMSCs) were seed on CH gels and CH/SF gels for *in vitro* culture, the cell affinity of CH gels and CH/SF gels were observed by SEM and LSCM.

2. Materials and Methods

2.1. *Preparation of silk fibroin solution*

Raw *bombyx mori* silk fibers (Haian, China) were boiled for 30 min in an aqueous solution of 0.05 (wt)% Na_2CO_3 at 98~100°C for three times and then rinsed in distilled water to remove sericin. The extracted fibers were subsequently air dried and dissolved in 9 M LiBr solution at 60 ± 2°C for 1 h with stirring. The cooled solution was dialyzed against distilled water in cellulose tubes (MWCO 9–12 kDa) for 4 days, and then diluted to 2 wt% with deionized water. The resulting SF solution was stored at 4°C after filtration.

2.2. *Preparation of CH/SF gels*

Chitosan (degree of deacetylation 95.8%, sigma) was dissolved in 0.1 N acetic acid to prepare 2.0 wt% chitosan stock solution. Beta-glycerophosphate (β-GP) powder (sigma) was dissolved in distilled water. The prepared chitosan solutions and silk fibroin solutions were mixed in the ice baths at various chitosan/silk fibroin mass ratios (w/w) of 100/0, 87.5/12.5, 75/25, 62.5/37.5, 50/50 and an appropriate amount of pre-cooled 40 wt% β-GP solution was added dropwise to the above mixture, respectively, to obtain a mixed solution containing 8 wt% β-GP. The resulting solutions were injected to the 10 ml centrifugal tubes, respectively. After incubation at 37°C, CH/SF gels with various blend ratios were obtained.

2.3. *Scanning electron microscopy (SEM)*

For SEM observation, selected gel samples were frozen at -80°C and freeze-dried in a lyophilizer (ALPHA1-4LSC, CHRIST, Germany) for 48 h. The dried gels were cut with a sharp blade to expose internal microstructure and sputter coated with platinum-gold. The morphologies of the gels were observed with SEM (HITACHI S-4800, Japan). Each pore area (S_1, S_2 ...S_i....S_n) in the visual field was calculated according to the limits of each pore in bitmaps and the number of image points in the whole image. The pore diameter d_i (μm) of each pore and average pore diameter $\overline{d}$ (μm) are given by Eqs. (1) and (2), respectively [13]. For each CH/SF scaffolds minimum 100 pores were examined.

$$d_i = \sqrt{\frac{4s_i}{\pi}} \qquad (1) \qquad\qquad \overline{d} = \frac{\sum_{i=1}^{n} d_i}{n} \qquad (2)$$

2.4. *Compression strength measurement*

The above-prepared CH/SF solutions with the ratios of 100/0, 87.5/12.5, 75/25, 62.5/37.5, 50/50 containing 8wt% β-GP were injected into 24-well culture plates, respectively, and incubated at 37°C for 15 min to transform into gels. Cylinder-shaped gel samples with a diameter of 11 mm and a height of about 12 mm were prepared using a hollow sample punch. Compressive stress of gel samples was determined in unconfined compression mode at room temperature using an Instron-3365 mechanical testing system (INSTRON, America). Samples were compressed at a constant deformation rate of 0.5 mm/min until reaching 50% strain. Compressive strength of gel samples was determined with a constant strain of 50%. All gel samples were measured in triplicate.

2.5. Rheological tests

After been incubated for 24 h at 37°C, the rheological properties of the gel samples with various ratios were measured on an AR 2000 rheometer (TA Instrument, New Castle, DE). The evolutions of log (G'), log (G") and tanδ with the frequency varying from 0.01 to 500 rad/s were determined while the strain was fixed at 1%. It was defined as frequency f=0 rad/s when the temperature reached the desired temperature (37°C).

2.6. Isolation and cultivation of BMSCs

The 2-week-old SD rats were provided by the animal center of Soochow University. The animal experiments were approved by the Jiangsu Province in experimental animals management rules ([2008] No. 26). BMSCs were isolated as described [14]. The bone marrow was flushed out from the femur and resuspended in DMEM/F-12 (HyClone, Logan, UT) culture medium supplemented with 10% fetal bovine serum (FBS; HyClone, Logan, UT) and a 1% Penicillin-Streptomycin solution (invitrogen). The cells were cultured in a humidified incubator at 37°C with a 5% CO_2 atmosphere. The culture medium was replaced every 3 days and the cells were passaged upon approaching confluence. Only passages 4 of the BMSCs were used for experiments.

2.7. Morphology of BMSCs cultured on CH/SF gels

CH/SF gels with ratios of 100/0, 75/25, 50/50 were immersed in phosphate buffer saline (PBS, 0.05 M, pH 7.4) for a week, freeze-dried for 48 h and sterilized. BMSCs were seeded into the gels at a density of $1×10^5$ cell/ml in DMEM with 10% FBS. The cell-seeded gels were incubated at 37°C in a humidified atmosphere for 4 h to adhere to the gels before the addition of complete medium into each well. Cells were incubated for 3 days at 37°C and 5% (v/v) CO_2 in a humidified incubator. After rinsing three times with PBS, cells were fixed with 4% (v/v) paraformldehyde in PBS for 30 min, rinsed three times with PBS and permeabilized with 0.2% Triton X-100 in PBS for 5 min and then immersed with 2% (v/v) BSA in PBS (corning) for 30 min at room temperature. The cells were stained with 5μg/ml FITC-Phalloidin (Sigma-Aldrich) for 2 h at room temperature. The samples were rinsed three times with PBS for observation by LSCM (FV1000, OLYMPUS, Japan).

BMSCs were seeded into CH gel and CH/SF gels. After cultured for 3 d, all samples were rinsed three times with PBS, fixed with 2.5% (v/v) glutaraldehyde for 4 h at 4°C and rinsed three times with PBS. The fixed samples were dehydrated with series of ascending graded ethanol (50%, 70%, 90% and 99.7%) for 5 min and then dried with hexamethyldisiloxane (HMDS, Sigma-Aldrich) for 3 min. The dried samples were sputtered coated with gold for 90 s before SEM observation.

3. Results

3.1. Morphology of gels

The porous structure of CH gels and CH/SF gels with blending ratios of 75/25, 50/50 was monitored by SEM as shown in Fig. 1. CH gel (Fig. 1A) showed an irregular porous structure. Compared with CH gels, due to the blending with SF, the porous structure of CH/SF gels obviously changed (Figs. 1B, 1C). The pore wall of CH/SF gels displayed lamellar structure which formed by a number of fibers with the diameter of several micrometers, which played an important role in maintaining the integrity of the gels and inter-connection of pores. Initial pore sizes of the gels were statistically calculated based on SEM images. The internal average diameter of the CH gel was about 90.7 ± 1.6 μm, and those of CH/SF gels with ratios of 75/25, 50/50 were 93.9 ± 2.4 μm and 114.8 ± 4.8 μm, respectively. The pore size in CH/SF gels were slightly enlarged compared with CH gel.

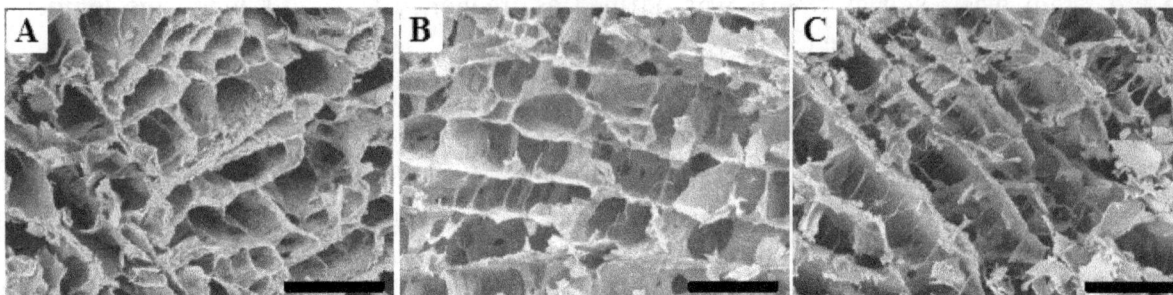

Fig. 1 SEM images of CH/SF gels. CH/SF ratios: (A) 100/0, (B) 75/25, (C) 50/50. Scale bars: 200 μm.

3.2. Compression properties

Results of compression tests for CH gel and CH/SF gels with various ratios are shown in Fig. 2. Stress-strain profiles (Fig. 2A) showed that no significant differences of stress from 0 to 10% strain were exhibited between all the CH/SF gels. Increasing the strain from 10% to 50% led to an increase in stress of CH/SF gels and the differences between the gels were expanded. The increasing rate in stress of CH/SF gel with blending ratio of 75/25 was the quickest, while that of the CH/SF gel with blending ratio of 50/50 is the lowest. The increasing rates in stress of CH/SF gels with blending ratios of 100/0, 87.5/12.5, 62.5/37.5 were similar, less than that of CH/SF gel with blending ratio of 75/25 and higher than that of CH/SF gel with blending ratio of 50/50. It is suggested that SF could enhance the stress of CH gel when the proportion of SF was at about 25%. Compressive strength of CH/SF gels at 50% strain was shown in Fig. 2B. It showed that compressive strength of CH/SF gel with ratio of 75/25, about 4.39 ± 0.29 kPa, was significantly bigger than that of CH gels and CH/SF gels with blending ratios of 87.5/12.5, 62.5/37.5 and 50/50 ($p<0.05$).

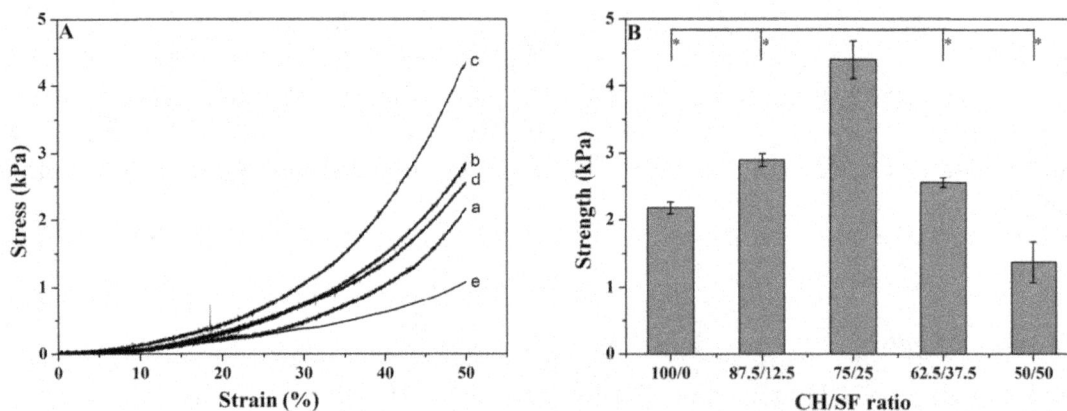

Fig. 2 Compression stress-strain curves of CH/SF gels (A) and strength at 50 % strain (B). CH/SF ratios: (a) 100/0, (b) 87.5/12.5, (c) 75/25, (d) 62.5/37.5, (e) 50/50. *p< 0.05.

3.3. Rheological characteristics

The relationships that log G' and log G'' of CH gel and CH/SF gels with ratios of 75/25, 50/50 versus sweep frequency are shown in Fig. 3A.

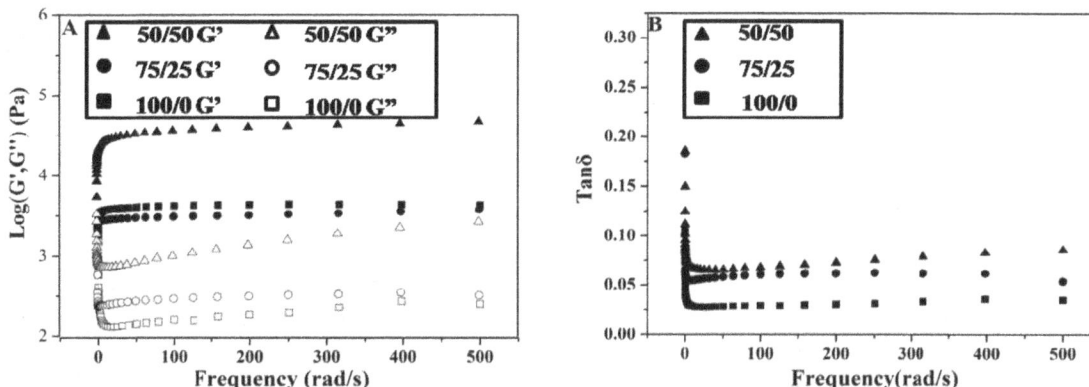

Fig. 3 Rheological measurements of CH/SF gels (after formed 24 h) (A) frequency dependence of log G' and log G'';
(B) frequency dependence of phase angle Tanδ.

The storage modulus (G') of CH gel and CH/SF blend gels were bigger than their loss modulus (G''), and no great fluctuations of both G' and G'' were observed. It suggested that all the three CH/SF gels were elastic materials with stable structure. Fig. 3B showed the relationship that phase angle (Tanδ) of CH gel and CH/SF gels vary with sweep frequency. Tanδ is a physical parameter to evaluate viscoelasticity of materials. When Tanδ is less than 1%, the gel is elastic (G'>G''), otherwise the gel is viscous. Fig. 3B showed Tanδ of CH gel and CH/SF gels were less than 0.3, suggested that CH gel and CH/SF gels with blending ratios of 75/25, 50/50 are elastic materials.

3.4. Growth of BMSCs on gels

Fig. 4 showed LSCM images of cultured BMSCs on CH gels and CH/SF blend gels at day 3. Less cells, scattered in the CH gels, were observed in CH gel (Fig. 4A). Many cells were observed in CH/SF (75/25) gels (Fig. 4B) and the number of cells obviously increased compared with CH gels (Fig. 4A). Fig. 4C showed that hundreds of cells distributed in CH/SF (50/50) gels and the number of cells was increased significantly compared to CH gels and CH/SF (75/25) gels. The results suggested blending with SF could significantly improve the microenvironment of CH gel for cell growth and proliferation.

(A) CH/SF ratios: 100/0,   (B) CH/SF ratios: 75/25,   (C) CH/SF ratios: 50/50.

Fig. 4 LSCM images of BMSCs cultured on CH/SF gels for 3 days. (Scale bars: 100 μm.)

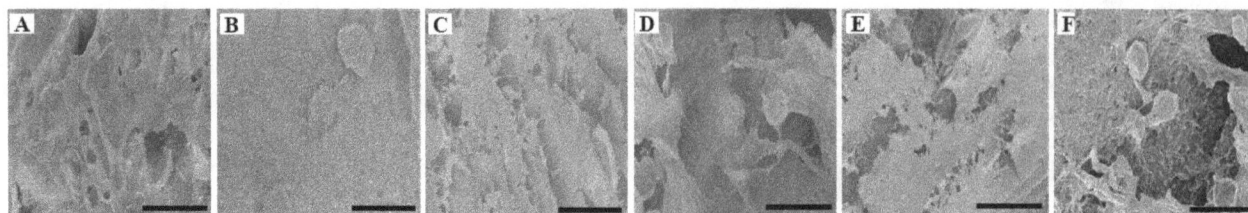

(A) CH gel   (B) CH gel   (C) 75/25 CH/SF gel (D) 75/25 CH/SF gel   (E) 50/50 CH/SF gel (F) 50/50 CH/SF gel.
Scale bars: (A, C, E) 200 μm, (B, D, F) 20 μm.

Fig. 5 SEM images of BMSCs cultured on CH/SF gels for 3 days.

Fig. 5 showed the morphology of cultured BMSCs on CH gel and CH/SF gels with blending ratios of 75/25, 50/50 at day 3. Cells cultured on the CH gel showed almost rounded shape, anchored to the pore walls inside gel through little number of filopodia and not fully extended (Figs. 5A, 5B). Cells on the CH/SF gel (Figs. 5C, 5D) with blending ratio of 75/25 showed more filopodia and lamellipodia than on pure CH gel, anchored to the pore walls closely and fully spreaded on the pore walls. The cells on CH/SF gel with blending ratio of 50/50 (Figs. 5E, 5F) showed fully spreaded morphology, extended lamellipodia and filopodia and a large quantity of extracellular matrix networks which secreted by cells were observed.

4. Discussion

In the present study, the sol-gel transformation of CH/SF blends was initiated by β-GP. The sol/gel transition mechanism of CH/SF system can be briefly described as follow: (1) elevated temperature (4-37°C) induced proton transfer between the cationic amine groups in CH/SF blends and the anionic phosphate groups in β-GP, result in reduction of electrostatic repulsion between CH/SF blends; (2) the increase of CH/SF interchain hydrogen bonding as a consequence of reduction of electrostatic repulsion; (3) the hydrophobic interactions between CH/SF molecules were enhanced by the structuring action of glycerol on water [15]. The above effective hydrogen bonding and hydrophobic interactions made the CH/SF chains entangled one another. As a result, the gel was formed.

CH is a cationic polymer, while SF, isoelectric point is about 4.5, is a protein with negative charge on the surface [16]. After the blending of CH with SF, the two polymers would be bound through intermolecular electrostatic interactions. Fig. 2 showed that the compressive strength of CH/SF gel with ratio of 75/25 was significantly bigger than that of CH gel, it suggested that the binding of CH with SF through intermolecular electrostatic interactions could reduce the electrostatic repulsion between the CH chains, increase the binding force between CH chains through interchain hydrogen bonds and hydrophobic interactions, strengthen the ability of the CH/SF aggregates to resist the external force and effectively enhance the compressive strength of CH gel. However, when the proportion of SF in the CH/SF blend system was higher than 50% (the ratio of CH/SF was up to 50/50), the electrostatic repulsion between CH/SF aggregates increased as a consequence of the introduction of excess negative charged SF, and the compressive strength of CH/SF (50/50) gels was lower compared with CH/SF (75/25) gels.

Different from the variation regularity of compressive strength with the changes of CH/SF ratios, Fig. 4 and Fig. 5 showed that with the increase of SF proportion in the CH/SF blends, the formed gels would be more beneficial for the cell attachment, spreading and proliferation. BMSCs were seeded into the CH/SF gel with blending ratio of 50/50, after cultured for 3 d, the cells could fully spread and secrete a large quantity of extracellular matrix networks (Figs. 5E, 5F), it suggested that blending CH with SF could significantly improve the cell affinity of CH gel.

5. Conclusion

CH/SF blend gels at various ratios initiated by β-GP were fabricated at 37°C. The compressive strength of CH/SF gels with blending ratio of 75/25 were significantly bigger compared with CH gel and the gels showed viscoelasticity characteristics of the elastic materials. Large of micro fibers were observed in the internal pore walls inside CH/SF gels. SEM and LSCM observation for BMSCs cultured on CH/SF gels suggested that blending CH with SF was beneficial for the cell attachment, spreading and proliferation. These CH/SF gels have potential to be tissue engineering scaffolds for cartilage and adipose regeneration.

Acknowledgments

This work was funded by the National Nature Science Foundation of China (313709 68): Natural Science Foundation of Jiangsu Province (BK20131177), College of Natural Science Research Project of Jiangsu Province (12KJA430003) and Jiangsu Province Science and Technology Support Program (BE2013734).

References

1. C.M. Lehr, J.A. Bouwstra, E.H. Schacht, H. E. Junginger, *In vitro* evaluation of mucoadhesive properties of chitosan and some other natural polymers. Int. J. Pharm. 78 (1992) 43-48.
2. H. Park, B. Choi, J. Hu, M. Lee, Injectable chitosan hyaluronic acid hydrogels for cartilage tissue engineering. Acta Biomater. 9 (2013) 4779-4786.
3. R.A.A. Muzzarelli, Chitins and chitosans for the repair of wounded skin, nerve, cartilage and bone. Carbohyd. Polym. 76 (2009) 167-182.
4. T.J. Keane, S.F. Badylak, Biomaterials for tissue engineering applications. Semin. Pediatr. Surg. 23 (2014) 112-118.
5. F.M. Chen, X. Liu, Advancing biomaterials of human origin for tissue engineering. Prog. Polym. Sci. 53 (2016) 83-168.
6. B. Ma, A. Qin, X. Li, X. Zhao, C He, Bioinspired design and chitin whisker reinforced chitosan membrane. Mater. Lett. 120 (2014) 82-85.
7. Mirahmadi F, Tafazzoli-Shadpour M, Shokrgozar M A, Bonakdar S, Enhanced mechanical properties of thermosensitive chitosan hydrogel by silk fibers for cartilage tissue engineering. Mate. Sci. Eng. C. 33 (2013) 4786-4794.
8. S. Chatterjee, M.W. Lee, S.H. Woo, Enhanced mechanical strength of chitosan hydrogel beads by impregnation with carbon nanotubes. Carbon 47 (2009) 2933-2936.
9. L. Wang, J.P. Stegemann, Thermogelling chitosan and collagen composite hydrogels initiated with β-glycerophosphate for bone tissue engineering. Biomaterials 31 (2010) 3976-3985.
10. J. Melke, S. Midha, S. Ghosh, K. Ito, S. Hofmann, Silk fibroin as biomaterial for bone tissue engineering. Acta Biomater. 31 (2015) 1-16.
11. Z. Shao, F. Vollrath, Surprising strength of silkworm silk. Appl. Polymer. Sci. 82 (2001) 1928-1935.
12. F. Mottaghitalab, H. Hosseinkhani, M.A. Shokrgozar, C. Mao, M.Y. Yang, M. Farokhi, Silk as a potential candidate for bone tissue engineering. J. Control. Release 215 (2015) 112-128.

13. Y.H. Cheng, S.H. Yang, W.Y. Su, Y.C. Chen, K.C. Yang, W.T.K. Cheng, S. Wu, F.H. Lin, Thermosensitive chitosan–gelatin–glycerol phosphate hydrogels as a cell carrier for nucleus pulposus regeneration: an in vitro study. Tissue Eng. Part A. 16 (2009) 695-703.

14. R. You, X. Li, Y. Liu, G. Liu, S. Lu, M. Li, Response of filopodia and lamellipodia to surface topography on micropatterned silk fibroin films. J. Biomed. Mater. Res. A. 102 (2014) 4206-4212.

15. J. Cho, M.C. Heuzey, A. Bégin, P.J. Carreau, Physical gelation of chitosan in the presence of β-glycerophosphate: the effect of temperature. Biomacromolecules 6 (2005) 3267-3275..

16. Q. Cheng, T.Z. Peng, X.B. Hu, F.Y. Catherine, Ion recognition and analytical application of a fibrion modified electrode. Chinese Chem. Lett. 15 (2004) 1473-1476.

Air Glow Discharge Plasma Reduction to Synthesize Pd/C Catalyst for Suzuki Coupling Reaction

Yang Liu, Xue-Feng Bai*

Institute of Petrochemistry, Heilongjiang Academy of Sciences, Harbin 150040, China
*Email: tommybai@126.com

Pd nanoparticles supported on activated carbon were synthesized by air glow discharge plasma reduction method (Pd/C-AirP) without any chemical reduction reagents and protective agents. The Pd/C-AirP catalysts were characterized by N_2 adsorption-desorption, X-ray diffraction (XRD) and transmission electron microscope (TEM) analysis. The results suggested that Pd nanoparticles with an average particle size of 23.3 nm are formed in Pd/C-AirP by plasma reduction for 15 min (Pd/C-AirP-15). The formation of PdO is found for Pd/C-AirP prepared by plasma reduction for 30 min. A 97% yield for the Suzuki coupling reaction of 4-bromotoluene with phenylboronic acid using EtOH/H_2O (1:1) as solvent and K_2CO_3 as base in the presence of 0.5 mmol‰ Pd/C-AirP-15 was obtained. The Pd/C-AirP-15 could be recovered and yield was dropped to 85% for the third cycle.

Keywords: Air glow discharge plasma reduction; Pd/C catalyst; Suzuki coupling reaction.

1. Introduction

Transition metal palladium catalyzed Suzuki coupling reactions of aryl halides with organic boric acid have been regarded as one of the most important and powerful tools for the selective formation of carbon-carbon bonds to synthesize biphenyl compound such as agrochemicals, natural product, pharmaceuticals and advanced functional materials [1-4].

Recently, supported Pd nanocatalysts with nano-size have drawn a great deal of attention due to their high activity and reusability. Numerous researches have been made to develop effective method for the synthesis of highly active supported Pd nanocatalysts. The chemical reduction is one of the most commonly reported preparation methods, however, the chemical reduction reagents and protective agents are necessary during the chemical reduction process [5-6]. Some researchers propose [7-8] that glow discharge plasma with energetic electrons and activated radicals can realize the preparation of metal nanoparticles without any reduction and protective agents, and the prepared nanoparticles exhibit high activity, providing a valid route for preparing supported metal nanocatalysts.

In this paper, the supported Pd/C catalysts were prepared by air glow discharge plasma reduction method. The pore structure, crystal structure and particle size of the Pd/C-AirP were studied. The catalytic properties of Pd/C-AirP for the Suzuki coupling reaction of 4-bromotoluene and phenylboronic acid were examined.

2. Experimental

2.1. *Characterization*

N_2 adsorption-desorption isotherms were measured at 77K on a Quantachrome Autosorb Gas Sorption analyzer. The Brunauer-Emmett-Teller (BET) method was applied to calculate the surface areas of activated carbon and Pd/C-AirP catalyst. X-ray diffraction (XRD) patterns of samples were recorded on a Bruker D8 advance diffractometer using a Cu Kα (λ=0.15418 nm) radiation with a fixed power source (40 kV, 40 mA) and 2θ range from 20° - 80°. Transmission electron microscopy (TEM) images were performed on a FEI Tecnai G2 S-TWIN Transmission electron microscope operating at 200 kV.

*Corresponding author

2.2. Preparation of Pd/C-AirP catalysts

A commercially available coconut shell activated carbon was crushed and sieved. The activated carbon (200-300 mesh) was chosen and pretreated with concentrated HNO_3 for 12 h, and then washed with distilled water to adjust the pH of the filtrate to 6-7 and dried at 383 K. The obtained samples were used as supports. The aqueous solution of Na_2PdCl_4 was prepared by dissolving $PdCl_2$ and $NaCl$ into deionized water by ultrasonic treatment. The aqueous solution of Na_2PdCl_4 was added into activated carbon and impregnated for 12 h. The obtained Na_2PdCl_4/C was denoted as activated carbon-Im and placed in the apparatus of plasma reactor. Then glow discharge plasma was generated using air as plasma forming gas by applying a high discharge voltage under the system pressure of 80 Pa. The 5 wt.% Pd/C-AirP catalysts were prepared by plasma reduction for 15 min and 30 min. The obtained catalysts were denoted as Pd/C-AirP-15 and Pd/C-AirP-30, respectively.

2.3. General procedure for the Suzuki coupling reactions

The 4-bromotoluene (1.0 mmol), 1.5 mmol phenylboronic acid and K_2CO_3 (2 mmol) were added into a mixture of 12 mL $EtOH/H_2O$ (1:1) and stirred at 60°C. To this was added 0.5 mmol‰ of Pd/C-AirP-15 catalyst, the reaction was carried out for 30 min. The resulting mixture (5 mL) was added to 0.2 mol/L sodium hydroxide solution (5 mL) and extracted with ethyl acetate (10 mL). The combined organic layers were collected, and dried in air to obtain the product. Evaluation results of products derived from the analysis of the HPLC.

3. Result and Discussion

N_2 adsorption-desorption isotherms of the Pd/C-AirP catalysts are shown in Fig. 1 and the pore structure parameters are listed in Table 1. As shown in Table 1, there is no obvious difference for the pore diameters of activated carbon and the Pd/C catalyst prepared by plasma, revealing that the pore structure is not damaged by air glow discharge plasma. The pore volume of activated carbon and activated carbon-Im are 0.876 cm^3/g and 0.694 cm^3/g, respectively. The decrease suggests that Pd^{2+} absorbed on the channels of support. However, the pore volume of Pd/C-AirP-15 is increased to 0.780 cm^3/g. A possible reason for this is that the particles with large particle size are formed and moved from the channel to surface of support.

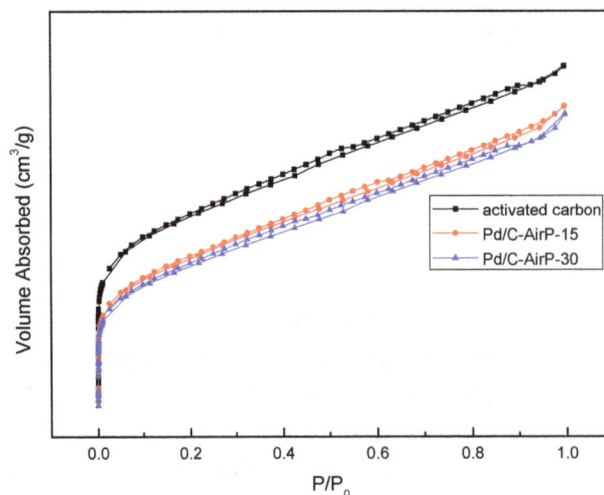

Fig. 1 N_2 adsorption-desorption isotherms of activated carbon and Pd/C-AirP

Table 1 Pore structure parameters of activated carbon and Pd/C-AirP

Sample	$S_{BET}[m^2/g]$	$V_{HK}[cm^3/g]$	$D_{HK}[nm]$
Activated carbon	1197	0.876	0.508
Activated carbon-Im [a]	909	0.694	0.508
Pd/C-AirP-15	957	0.780	0.502
Pd/C-AirP-30	918	0.763	0.502

S_{BET}: calculated by BET method; V_{HK}, D_{HK}: pore volume and pore size calculated by HK method

[a] Activated carbon-Im: activated carbon after impregnation

The XRD patterns of activated carbon, Pd/C-AirP-15 and Pd/C-AirP-30 are presented in Fig. 2. All patterns show two broad diffraction peaks at 24.6° and 43.8°, indicating the carbon structure of activated carbon. The peaks with 2θ of 40.1°, 46.6°, 68.1° are attributed to the (111), (200) and (220) planes of Pd nanoparticles, revealing the formation of Pd (0). Otherwise, a weak and broad peak at 33.5° indicates the presence of PdO for Pd/C-AirP-30.

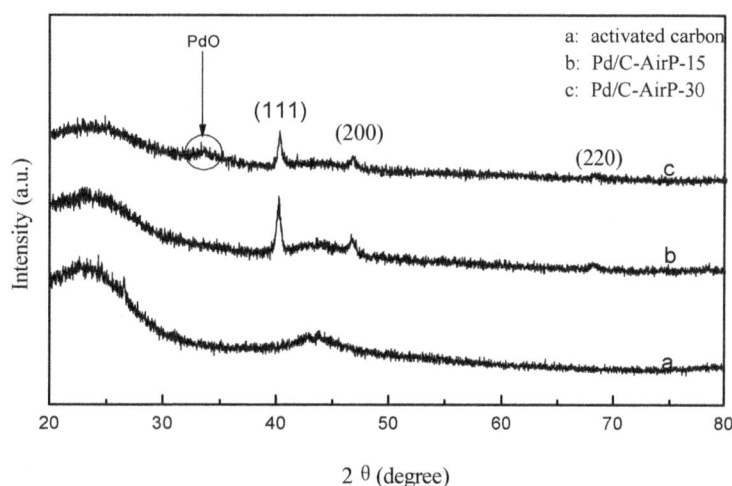

Fig. 2 XRD patters of Pd/C-AirP catalysts

Fig. 3 shows the TEM images of Pd/C-AirP-15 and Pd/C-AirP-30. From the images, the Pd nanoparticles are clearly visible in all Pd/C-AirP catalysts. Compared Pd/C-AirP-15 with Pd/C-AirP-30, the Pd/C-AirP-15 has a smaller particle size range from 4.8 nm-47.9 nm (an average diameter of about 23.3 nm) than Pd/C-AirP-30 with the largest size of 68.6 nm. This result suggests that it is not conducive for the formation of Pd nanoparticles with small particle size by prolonging reduction time.

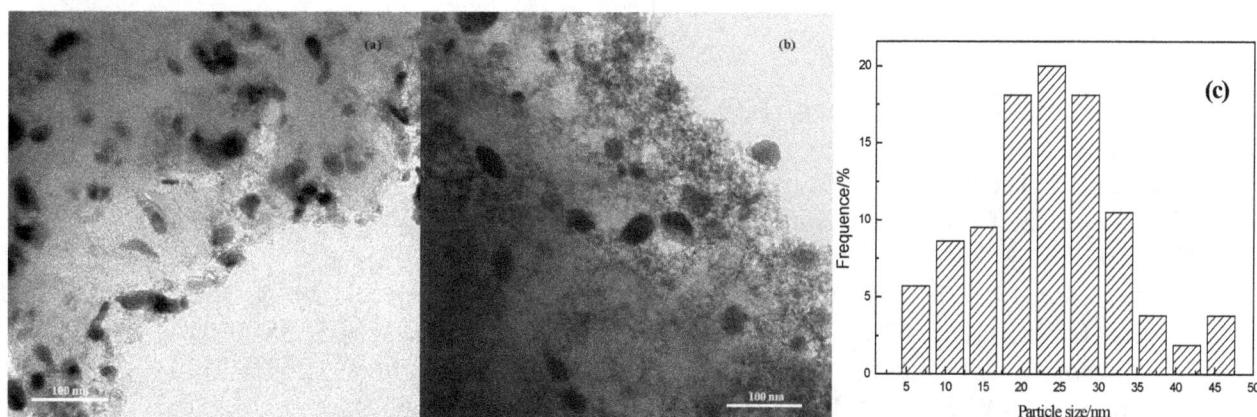

Fig. 3 TEM images of Pd/C-AirP-15 (a), Pd/C-AirP-30 (b) and particle size distribution of Pd/C-AirP-15 (c)

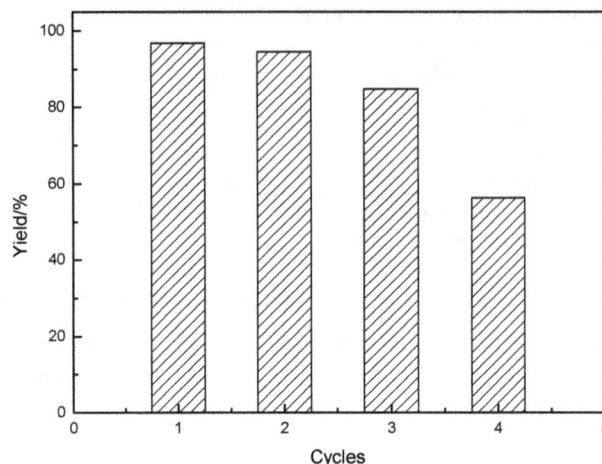

Fig. 4 The recycling of Pd/C-AirP-15 for the Suzuki coupling reaction

The reusability of Pd/C-AirP-15 was tested for Suzuki coupling reaction of 4-bromotoluene and phenylboronic acid using EtOH/H$_2$O (1:1) as solvent and K$_2$CO$_3$ as base in the presence of 0.5 mmol‰ catalyst. As shown in Fig. 4, a 97% yield is obtained for the first run, and the yield is reduced to 85% for the third run. The activity of Pd/C-AirP-15 drops in the fourth cycle yielding 56%.

4. Conclusion

In conclusion, Pd/C-AirP catalyst was prepared by air glow discharge plasma reduction method for different discharge time. Pd nanoparticles with an average particle size of 23.3 nm are obtained for Pd/C-AirP-15. The formation of PdO is found for Pd/C-AirP-30. A 97% yield was reached for the Suzuki coupling reactions of 4-bromotoluene with phenylboronic acid catalyzed by 0.5 mmol‰ Pd/C-AirP-15 catalyst, and reused for 4 times with an obvious loss of activity.

Acknowledgment

This work was funded by the National Natural Science Foundation of China (Grant No.21276067), NSFC-RFBR (Grant No.214111301884), Program of International S&T cooperation (Grant No.2013DFR40570) and Science Foundation of Heilongjiang Academy of Sciences.

References

1. M.C. Kozlowski, B.J. Morgan, E.C. Linton, Chem. Soc. Rev. 38 (2009) 3193-3207. Reference to a book:
2. A. Evitt, L.M. Tedaldi, G.K. Wagner, Chem. Commun. 48 (2012) 11856-11858.
3. B. Yalcouye, S. Choppin, A. Panossian, F.R. Leroux, et al. Eur. J. Org. Chem. 28 (2014) 6285-6294.
4. S.M. McAfee, J.S.J. McCahill, C.M. Macaulay,et al. RSC Adv. 5 (2015) 26097-26106.
5. F. Wang, C.H. Li, H.J. Chen, et al. J. Am. Chem. Soc. 135 (2013) 5588-5601.
6. P.M. Uberman, L.A. Pérez, S.E. Martin, et al. RSC Adv. 4 (2014) 12330-12341.
7. Y.T. Chen, H.P. Wang, C.J. Liu, et al. J. Catal. 289 (2012) 105-117.
8. M.B. Zhang, X.L. Zhu, X. Liang, et al. Catal. Commun. 25 (2012) 92-95.

Development of Al-Si-Cu-Zn-Mn Filler Metal for Brazing 3003 Aluminum Alloy

Zhong-Li Dong[1], Xue Luo[2], Xiao-Qiang Li[2,*], Jing-Mao Li[2], Ming Nie[1], Qing Xiao[2]

[1]Electric Power Research Institute of Guangdong Power Grid Corporation, Guangzhou 510640, China

[2]National Engineering Research Center of Near-net-shape Forming for Metallic Materials, South China University of Technology, Guangzhou 510640, China

*Email: Lixq@scut.edu.cn

The study is concerned with developing a low-melting-point filler metal for brazing 3003 aluminum alloys. For this purpose, a series of designed Al-Si-Cu-Zn-Mn filler metals were prepared and evaluated in the light of melting characteristics and spreading area. Al-25Cu-7.5Si-5Zn-1Mn alloy was picked out as contrasting filler. Based on the orthogonal test, the composition of Al-20Cu-7.5Si-10Zn-1Mn filler metal was then optimized. Its solidus and liquidus temperatures of 454.0 °C and 521.9 °C were lower than those of the contrasting filler, and the spread ability was slightly better. The mechanical properties, micro structure and fracture morphology of the joints brazed for 20 min by the two types of filler metal were investigated subsequently. The results showed that the shear strength of the joint brazed with the optimized filler was higher in the brazing temperature range of 540-580 °C, compared to Al-25Cu-7.5Si-5Zn-1Mn filler. The maximum strength was obtained at the brazing temperature of 570 °C, being 70.8 MPa. The fracture surface located in the central brazed layer and was characterized by the mix of local ductility and main cleavage, which is favorable to improve the joint strength.

Keywords: Brazing; Al-Si-Cu-Zn-Mn filler; 3003 aluminum alloy.

1. Introduction

3003 aluminum alloy has been widely used in the manufacturing of heat exchangers due to its favorable thermal conductivity, excellent formability as well as good corrosion resistance. In industrial application, the bonding problem with aluminum alloys has been a serious consideration. Among a variety of bonding techniques, brazing has been recognized as a reliable method for joining aluminum components [1-3]. The method is also suitable to precise bonding. Compared with flame brazing, salt bath brazing and vacuum brazing, air furnace brazing is operation-convenient and cost-effective, for which only a universal heating furnace is often needed. However, traditional aluminum based filler metal based on eutectic Al-12Si alloy has a working temperature above 590°C, so it is difficult to ensure the good quality of brazed joints and the precise dimension of brazed parts, resulting in a restricted application. Particularly for the air exposure of furnace brazing, high brazing temperature easily induces severe oxidation of joints and weakens bonding. A series of low-melting-point filler metals for brazing aluminum alloys have been developed by adding copper, germanium, zinc, or tin into Al-Si alloys. For instance, Suzuki et al. introduced a eutectic Al-4.2Si-40Zn filler metal with a melting point of 535 °C, which has good hot processibility [4]. Nevertheless, the high content of zinc easily causes the dissolution of aluminum substrate. Chang developed a Al-12Si-20Cu alloy with liquidus temperature of 535 °C [5]. Sharma found that the liquidus temperature of Al-12Si-20Cu decreased to 531.6 °C by further adding 0.05 wt% ZrO_2 nanoparticles [6]. Niu et al. explored novel Al-9.5Si-10Ge-15Zn filler, of which the melting temperature range is 505.2-545.1 °C and the solidus-liquidus interval is only 39.9 °C [7]. However, the addition of expensive Ge limits its application. Therefore, developing a low-melting-point filler metal has a very important significance to achieve a low-cost and high-quality brazed joint of 3003 aluminum alloy. In this study, a series of Al-Si-Cu-Zn-Mn alloys are evaluated in the light of brazing characteristics, and a proper low-melting-temperature filler metal is developed to braze 3003 aluminum alloy. Additionally, the joint reliability is carried out

---

*Corresponding author

299

by measuring the joint shear strength as well as characterizing the microstructure and fractography.

2. Experimental

A series of Al-Si-Cu-Zn-Mn filler metals with various compositions designed by orthogonal design method were prepared by melting in graphite crucibles. The Al-Si alloys were molten first at 750 °C for 5 min. Then the temperature rose up to 1000 °C and subsequently Cu, Zn, Mn and rare earth (RE) were added into the molten aluminum alloy followed by refining for 3 min. They were finally cast in a stainless steel mold. The cast filler metals were machined into columnar shape with weight of 0.2 g for spreading test and rolled into foils with thickness of 2.0 mm for brazing. 3003 aluminum alloy as the base metal were processed into plates with dimensions of 40 mm×40 mm×3 mm for spreading test and 30 mm×15 mm×3 mm for brazing, as shown in Fig. 1. Prior to brazing, both the surfaces of filler metals and bonding surfaces were ground with SiC paper down to grade 2000 and coated with QJ201flux. The brazing process was conducted in an air furnace at 540-580 °C for 20 min. Spreading test was carried out at 580 °C for 10 min. The solidus and liquidus temperatures of the filler metals were determined by differential scanning calorimetry with heating rate of 10°C/min under argon atmosphere. The metallographic analysis of the filler metals and brazed joints was carried out on an optical microscope after metallographic preparation and etching in 0.5 vol.% HF for 30 s. The phases were identified with X-ray diffraction diffractometer and energy dispersive spectrometer. The shear strength of lap joints was measured at room temperature with a constant crosshead displacement speed of 0.5 mm/min. In order to ascertain reproducibility, at least three specimens were tested for each brazing condition. Fractography of the joints was characterized by scanning electron microscopy.

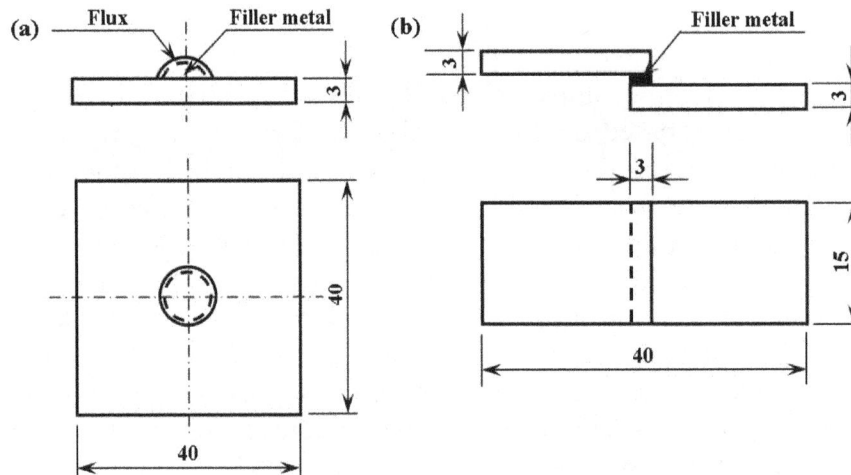

Fig. 1 Schematic illustration of specimen for spreading test (a) and brazed joints (b) Unit: mm

3. Results and Discussion

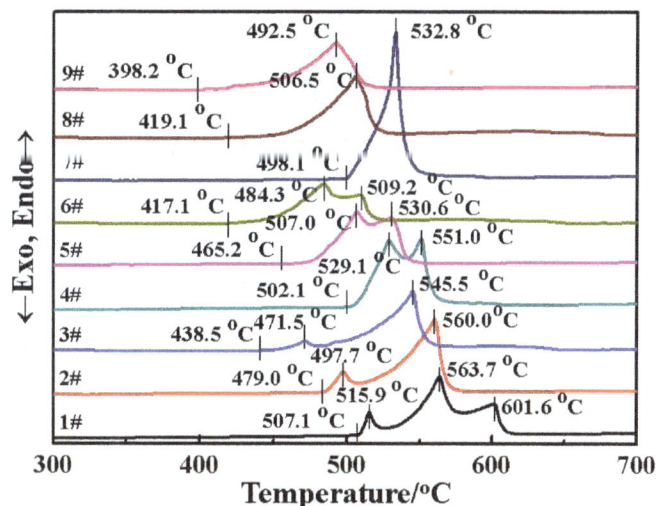

Fig. 2 DSC curves of filler metals

Fig. 2 shows the DSC curves of nine types of designed filler metals, whose compositions are listed in Table 1. Two or three endothermic peaks are seen in the DSC curves of fillers 1-6 in the range of 300-700 °C, while fillers 7-9 have only one endothermic peak. It can be inferred that there are different phases in fillers 1-6. For all the designed fillers, the main elements are Al, Si, Cu and Zn, and the contents of Mn and RE are much lower. Taking Al-5Cu-5Si-5Zn-1Mn (filler 1) for example, the three endothermic peaks in the DSC curves are considered to likely correspond to the Al-Si-Cu ternary eutectic point, Al-Cu binary eutectic point and Al-Si binary eutectic point at 527 °C, 542 °C and 577 °C, respectively. The addition of Zn and Mn cause the endothermic peaks to shift. An ideal filler metal generally possesses narrow melting range and only one endothermic peak in its DSC curve till entirely melting, which is beneficial to spreading and gap-filling properties.

From Fig. 2 and Table 1, we can see that Al-25Cu-7.5Si-5Zn-1Mn alloy (filler 7) has just one endothermic peak and the most narrow melting range. Additionally, its melting temperature is markedly depressed by adjusting the contents of Cu, Si and Zn alloying elements, with the solidus and liquidus temperatures of 498.1 °C and 532.8 °C. Therefore, filler 7 is considered to be an appropriate filler. Although some designed alloys have lower melting temperatures, their melting ranges is much wider, which is unfavorable to brazing. For example, the melting ranges of Al-15Cu-7.5Si-25Zn-1Mn alloy (filler 6) and Al-25Cu-5Si-25Zn-1Mn alloy (filler 9) widens to 92.1 °C and 94.3 °C. By means of orthogonal design and range analysis, it is concluded that the change of Cu and Zn contents have bigger influence than Si on the solidus and liquidus temperatures as well as melting range of filler metal. The solidus and liquidus temperatures of filler metal visibly decrease with the increase of Cu content from 5 wt% to 25 wt%. The effect of Zn is similar to Cu on the solidus and liquidus temperatures. With the Cu content increasing from 5 wt% to 15 wt%, the melting range of filler metal decreases, while a further increase instead causes the melting range to widen. Different from Cu, increasing Zn content in the investigated range induces the melting range to widen monotonously. As for Si, the solidus temperature first rises and then descends with its content increasing from 5 wt% to 10 wt%. On the contrary, the liquidus temperature first descends and then slightly rises, which is similar to the change of melting temperature. From Table 1, filler 7 exhibits good spreadability on the surface of 3003

301

aluminum plate and the spreading area at 580 °C gets 547.2 mm^2. The value is just lower than 554.8 mm^2 of filler 6 and 574.3 mm^2 of filler 9. The range analysis shows that the increase of Cu and Zn contents is apt to enhance the spreadability.

Table 1 Solidus temperature and liquidus temperature of filler metals and their spreading areas on 3003 aluminum alloy

Filler number	Chemical composition[wt%]						Melting temperature[°C]			Spreading area[mm^2]
	Cu	Zn	Si	Mn	RE	Al	T_s	T_l	ΔT	
1	5	5	5	1	0.1	bal	507.1	601.6	94.5	36.2
2	5	15	7.5	1	0.1	bal	479.0	560.0	81.0	419.2
3	5	25	10	1	0.1	bal	438.5	545.5	107.0	470.5
4	15	5	10	1	0.1	bal	502.1	551.0	48.9	424.2
5	15	15	5	1	0.1	bal	465.2	530.6	65.4	482.7
6	15	25	7.5	1	0.1	bal	417.1	509.2	92.1	554.8
7	25	5	7.5	1	0.1	bal	498.1	532.8	34.7	547.2
8	25	15	10	1	0.1	bal	419.1	506.5	87.4	322.6
9	25	25	5	1	0.1	bal	398.2	492.5	94.3	574.3

*T_s and T_l indicate the solidus and liquidus temperatures of filler metal, respectively. Melting range $\Delta T = T_l - T_s$.

Based on the orthogonal test, the composition of filler metal is optimized to Al-20Cu-7.5Si-10Zn-1Mn (marked filler 10). The corresponding melting temperature and spreading area at 580°C is listed in Table 2. Filler 10 has solidus temperature, liquidus temperature, melting range and spreading area of 454.0 °C, 521.9 °C, 67.9 °C and 548.0 mm^2, respectively. Compared with filler 7, the values of filler 10 are more suitable as filler except melting range. The microstructure of each filler metal studied in this paper consists of grey θ (Al$_2$Cu) phase, dark silicon particles and white Al base solid solution containing Cu, Si, Zn and Mn alloying elements. It is identified by OM, XRD and EDS methods.

Table 2 Melting temperatures of Al-20Cu-7.5Si-10Zn-1Mn filler metal and its spreading areas on 3003 aluminum alloy.

Filler number	Chemical composition[wt%]						Melting temperature[°C]			Spreading area[mm^2]
	Cu	Zn	Si	Mn	RE	Al	T_s	T_l	ΔT	
10	20	10	7.5	1	0.1	bal	454.0	521.9	67.9	548.0

Fig. 3 displays the RT shear strength of joints brazed at different temperatures for 20 min with fillers 7 and 10. The strength of joint brazed with filler 10 is higher in the brazing temperature range of 540-580 °C. For filler 10, the maximum strength appears at 570 °C and is 70.8 MPa, about 53% of 3003 aluminum alloy. The maximum is 51.9 MPa with filler 7, which appears at the brazing temperature of 560 °C. The joints brazed with the two types of filler metals have similar microstructure.

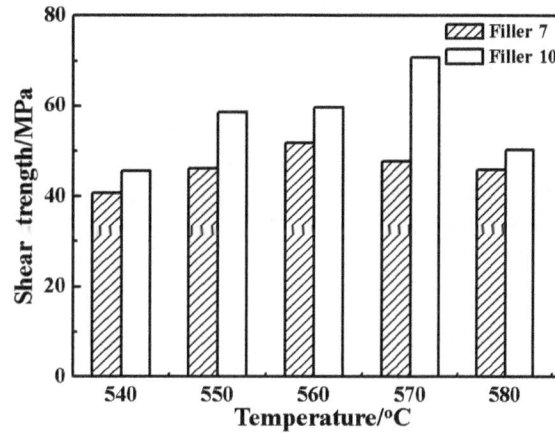

Fig. 3 RT shear strength of joints brazed at different temperatures with fillers 7 and 10, respectively

Typical brazed joint consists of central brazed layer (zone I) and two diffusion zones adjacent to 3003 base material (zone II), as shown in Fig. 4. The central brazed layer is composed of α(Al) solid solution, θ(Al$_2$Cu) intermetallic compound, massive/acicular silicon phase and fine AlFeMnSi phase. The diffusion zone mainly consists of α(Al) solid solution. From Fig. 4b, the microstructure similar to that of central brazed layer can be seen in the local diffusion zone for filler 10, which may improve the joint strength in a certain extent. However, a big volume fraction of eutectic microstructure in diffusion zone weakens the joint strength, like the joint brazed at 580 °C for 20 min with filler 10. Low brazing temperature such as 540 °C will cause an entire gap-filling and insufficient element-diffusion between filler metal and base material, resulting in a relatively low joint strength.

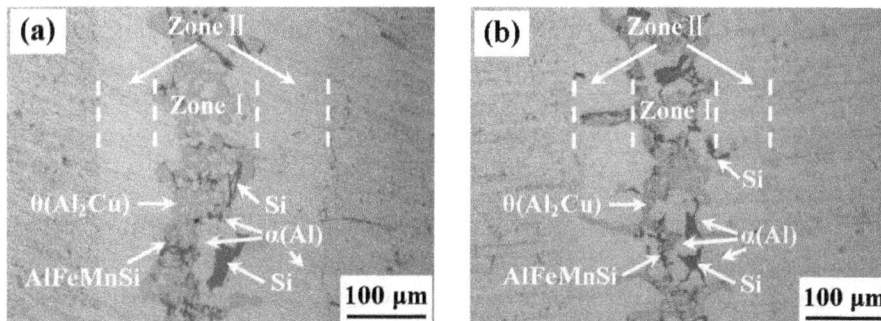

Fig. 4 Microstructure of the joints brazed at 570 °C for 20 min with filler 7 (a) and filler 10

The shear fracture surfaces of the joints brazed for 20 min with filler 10 migrates with the brazing temperature. It lies in the interface between 3003 alloy and brazed seam when the brazing temperature is 540 °C. For 550 °C and 560 °C, it locates between zones I and II.

With the advent of maximum joint strength responding to 570 °C brazing temperature, the shear fracture surface lies in zone I as shown in Fig. 5. Nevertheless, higher brazing temperature of 580 °C makes the fracture surface return to the interface of zones I and II. Appropriate brazing temperature not only enhances the gap-filling of filler but also facilitates the element diffusion between filler metal and base material and improves the microstructure of brazed joint. The shear fracture surface of the joint brazed at 570 °C exhibits the mix of local ductility and main cleavage.

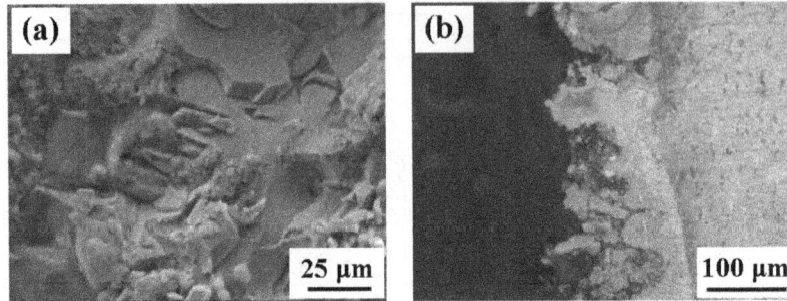

Fig. 5 SEM graphs of the shear fracture surface (a) and the cross-section (b) of the joint brazed at 570 °C for 20 min with filler 10.

4. Conclusion

A series of designed Al-Si-Cu-Zn-Mn filler metals were evaluated in the light of solidus temperature, liquidus temperature, melting range and spreading area. Amongst these filler metals, Al-25Cu-7.5Si-5Zn-1Mn alloy was outstanding, owing to the low melting temperature and high spreading area, especially the narrow melting range of 34.7 °C. Based on the orthogonal test, the composition of Al-20Cu-7.5Si-10Zn-1Mn filler metal was optimized to braze 3003 aluminum alloy. Its solidus and liquidus temperatures as well as spreading area were respectively 454.0 °C, 521.9 °C and 548.0 mm^2, which were better than those of Al-25Cu-7.5Si-5Zn-1Mn filler. The shear strength of joint brazed for 20 min with the optimized filler was higher in the brazing temperature range of 540-580 °C. The maximum strength appears at 570 °C, being 70.8 MPa. The fracture surface located in the central brazed layer and was the mix of local ductility and main cleavage.

Acknowledgments

This work was funded by the Technology Program of Southern Power Grid Corporation (No. GDKJ00000081), and the Research Project of Special Furnishment and Part (No. XZJQ-B1120560).

References

1. L.C. Tsao, W.P. Weng, M.D Cheng, Brazeability of a 3003 aluminum alloy with Al-Si-Cu based filler metals, J. Mater. Eng. Perform. 11 (2002) 360-364.
2. W. Dai, S.B. Xue, F. Ji, J. Lou, B. Sun, S.Q. Wang, Brazing 6061 aluminum alloy with Al-Si-Zn filler metals containing Sr, Int. J. Min. Metall. Materal. 20 (2013) 365-370.
3. W. Dai, S.B. Xue, J.Y. Lou, S.Q. Wang, Development of Al-Si-Zn-Sr filler metals for brazing 6061 aluminum alloy, Mater. Design. 42 (2012) 395-402.
4. K. Suzuki, M. Kagayama, Y. Takeuchi, Eutectic phase equilibrium of Al-Si-Zn system and its applicability for lower temperature, J. Jan. I. Light Met. 43 (1993) 533-538.
5. S.Y. Chang, L.C. Tsao, T.Y. Li, T.H. Chuang, Joining 6061 aluminum alloy with Al-Si-Cu filler metals, J. Alloy Compd. 408 (2009) 174-180.
6. A. Sharma, M.H. Roh, D.H. Jung, J.P. Jung, Effect of ZrO$_2$ nanoparticles on the microstructure of Al-Si-Cu filler for low-temperature Al brazing applications, Metall. Mater. Trans. A. 47 (2016) 510-521.
7. Z.W. Niu, J.H. Huang, H. Yang, S.H. Chen, X.K. Zhao, Preparation and Properties of a Novel Al-Si-Ge-Zn Filler Metal for Brazing Aluminum, J. Mater. Eng. Perform. 24 (2015) 2327-2334.

Comparative Research of Different Admixtures in Cement-based Materials

Mei-Juan Rao[1], Shu-Hua Liu[2,*], Jian-Peng Wei[2]

[1]Changjiang River Scientific Research Institute, Wuhan, 430010, China
[2]State Key Laboratory of Water Resources and Hydropower Engineering Science, Wuhan University, China
[*]Email: shliu-job@163.com

With the rapid development of material science nowadays, adding mineral admixture to concrete can not only improve the utilization of industrial by-products, but can be the most efficient method to improve the mechanical property and the durability of cement-based materials. The purpose of this paper is to study the mechanical property of cement-based materials by adding various active mineral admixtures. The adding of glass powder was detrimental to the development of specimen's early strength, but the late pozzolanic reaction increased the strength significantly. Besides, the adding of limestone powder (uni mixed) obviously improved the early strength of the specimens. The incorporation of glass powder had little effect on the hydration product category of the gelling system but had a greater impact on the generated amount. The steel slag powder and limestone played a role of filling at the early gelling system. The nucleus effect of the limestone powder also promoted early cement hydration. In addition, the late hydration of limestone powder produced single carbon hydrated calcium aluminate.

Keywords: Concrete; mineral admixtures; mechanical property; microstructure.

1. Introduction

Concrete in civil engineering in modern times is the most versatile and the largest amount of construction materials [1-3]. Research into the role played by the deserted mineral admixtures in cement-based materials will ensure the re-use of industrial waste and also increase their scope and level. Subsequently, more environmentally friendly and efficient composite cement-based materials could be developed [4-6]. In recent years, focus on the different mineral admixtures in cement group about the filling effect and the active role of the law and its impact. Such experiments have proposed a variety of estimation methods, such as intensity effect factor method, compressive strength ratio method [7], pozzolanic activity diagram method, lime absorption value method and the specific strength. The scholars' experiments have also laid a good foundation for research on the use of mineral admixtures in cement-based materials [8]. National and international studies have shown that adding finely ground glass powder (GP), limestone powder (LP), and steel slag powder (SP) into concrete will improve the concrete's mechanical properties, performance, and durability in varying degrees by improving the hydration environment, the micro-aggregate effect, and the pozzolanic effect. On the contrary, limestone powder is an inert material; its main component, $CaCO_3$, manifests by improving early compressive by filling pores, and by accelerating the hydration process of the complex binder in the early curing ages. Steel slag powder, composed mainly of C_2A and C_3A, are similar to clinkers; they have potential hydraulic properties, and reaction activity that is significantly lower than that of clinkers.

This paper is based on a variety of mineral admixtures mixed with cement-based material and the impact on the macro and micro performance. The paper also discusses the complex method of new admixture which provides a theoretical basis for the study of new admixture in concrete.

---

[*]Corresponding author

2. Experimental

2.1. *Raw materials*

Table 1 Physical properties of cement (wt, %)

Materials	Density (g/cm³)	Standard water consumption (%)	Fineness		Stability	Setting time(h:min)	
			80μm Sieving Residue (%)	Specific Surface (m²/kg)		Primary setting	Final setting
Cement	3.1	25.6	1.4	437	qualified	3: 25	5: 51

Cement 42.5 from Huaxin Cement Co., Ltd (Hubei Province, China) was used. The physical and chemical properties of cement and admixtures are shown in the Table 1, 2, 3 and 4.

Table 2 Physical properties of limestone powder

45μm Sieving Residue (%)	Specific Surface (m²/kg)	Water demand ratio (%)	Density (g/cm³)	Activity Index	
				7d	28d
16.0	457.9	98	2.69	64.1	69.0

Table 3 Physical properties of glass powder and steel powder

Materials	Density (g/cm³)	Fineness	
		80μm Sieving Residue (%)	Specific Surface (m²/kg)
Glass powder	2.35	75	230
Steel powder	3.29	90	265

Table 4 Chemical properties of cement and admixtures (wt, %)

Name	SiO_2	Al_2O_3	CaO	Fe_2O_3	MgO	BaO	F	SO_3
Cement	21.25	2.91	63.09	3.24	0.68	-	-	3.36
Glass powder	55.75	10.64	6.60	0.28	1.01	0.43	0.94	0.27
Limestone powder	0.73	0.20	30.08	0.06	19.38	-	-	0.01
Steel slag powder	14.11	3.51	42.39	17.45	6.60	0.10	-	0.50
Name	K_2O	Na_2O	P_2O_5	MnO	Cl	CeO_2	ZnO	Loss
Cement	1.12	0.31	0.17	0.04	/	/	/	3.52
Glass powder	0.54	9.92	0.03	0.29	0.31	1.10	0.30	11.9
Limestone powder	0.05	-	0.03	-	0.01	-	-	49.47
Steel slag powder	0.13	0.22	1.63	2.96	0.06	-	0.03	8.72

Fig. 1 SEM images of cement and admixtures (a-cement; b-glass powder; c-limestone powder; d-steel powder)

2.2. *Test method*

Molded paste specimens 40mm × 40mm × 40mm, test pieces with a hammer to crack after standard conservation to 28d and 90d. The clean bean size from the central portion of the broken specimen sample block, were then soaked in absolute ethanol to terminate the hydration process. The test pieces were dried at 60 °C for 2-3h before XRD and TG-DT testing, and were dried in an oven at 80 °C for 2-3 hours before SEM testing, followed by placing the dried test piece in a vacuum coating machine for dehumidifiers and plating before testing again.

We designed a series of net pulp tests which was about Single or compound mineral admixture. The tests were based on Pu Si cement and mineral admixture as cementitious materials. The net plasma test studied the effect of mineral admixture on mechanical properties by controlling the replacement rate of the mineral admixture instead of the cement (15%, 30%, 45%) , Three kinds of mineral admixtures (glass powder, limestone powder, slag powder mixed with the equivalent) mixed with these variables.

Table 5 Mix proportion of pastes

Samples	W/C	Content (%)	Cement (g)	Glass powder (g)	Limestone powder(g)	Steel slag powder(g)	Water (g)
NPC	0.4	0	600	0			240
G1	0.4	15	510	90	—	—	240
G2	0.4	30	420	180	—	—	240
G3	0.4	45	330	270	—	—	240
L1	0.4	15	360	—	90	—	240
L2	0.4	30	320	—	180	—	240
L3	0.4	45	280	—	270	—	240
S1	0.4	15	400	—	—	90	240
S2	0.4	30	360	—	—	180	240
S3	0.4	45	320	—	—	270	240

Notification: G-glass powder, L-limestone powder, S-steel slag powder

3. Results and Discussion

3.1. *Contribution rates of the hydration activity effect of mineral admixture*

To evaluate the activity of the glass powder, an evaluation methodology of the activity index of the mineral materials is adopted. A relative index is obtained which is called the hydration activity contribution rate (Pa)., Pa can be used to represent the degree of contribution of the hydration activity effect of the auxiliary gel materials to concrete.

$$P_a = \frac{R_{sp}}{R_{sa}} = \frac{R_{sa} - R_{sc}}{R_{sa}} = 1 - \frac{R_{sc}}{R_{sa}} = 1 - \frac{R_c * q_0}{R_a * 100}$$

Table 6 Contribution rates of the hydration activity effect of mineral admixture

Samples	3d	7d	28d	90d
G1	-0.25	-0.38	-0.26	-0.10
G2	-0.31	-0.30	-0.36	0.04
G3	-0.12	-0.09	-0.15	0.06
L1	0.26	0.25	-0.06	0.01
L2	0.13	0.14	0.04	0.11
L3	0.08	-0.13	-0.20	0.12
S1	0.14	0.25	0.04	0.14
S2	0.10	0.14	-0.08	0.16
S3	-0.10	-0.01	-0.19	0.07

According to the above methodology, the contribution rates of the hydration activity effect of mineral admixture in the test samples of all groups are calculated. Table 6 shows the calculated the results. From the table it is obvious that glass powder, limestone powder, steel slag content, age and the combined methods all have great importance on the contribution rate of hydration activity effect on strength.

The contribution rates of the hydration activity effect of glass powder increased with the age which indicates that pozzolanic reaction of glass powder is stronger later than early age. The contribution rates of the hydration activity effect of limestone powder are bigger at early age so that the early strength development would be worse when the content of limestone powder is too large.

3.2. XRD analysis

The content of 30% samples was taken as the characteristic specimen samples in each group (Fig. 2 and Fig. 3). Compared with the NPC group and G group, the pozzolanic reaction of glass powder is very small, mainly at late age. This is because the pozzolanic reaction of the glass powder needs to occur under certain excitation conditions such as alkali-activated, thermal and physical stimulation, etc.

It should be paid attention that $C_3A \cdot CaCO_3 \cdot 11H_2O$ could be seen in samples adding limestone powder. $C_3A \cdot CaCO_3 \cdot 11H_2O$ is nearly not found at 28d but obvious at 90d which determined by the content of C3A in cement. When adding with slag powder, the content of CH would increase because of the hydration activity of active constituent in slag powder which would form CH.

At the early age of hydration (before 28d), the incorporation of different mineral admixtures has no big influence on the changes in the kinds of the systematic hydration products, but the number of hydration products changed obviously. With the increase in age, groups mixed with glass powder have impressively dropped in the content of $Ca(OH)_2$ at the age of 90d. This is because at the later age of hydration, alkali concentration of the pore solution gradually grows up activating the pozzolanic activity and consuming partial $Ca(OH)_2$. In these test samples, the unhydrated clinkers such as dicalcium silicate and tricalcium silicate have lower diffraction peak compared to the other groups showing that the hydration of cement is relatively sufficient.

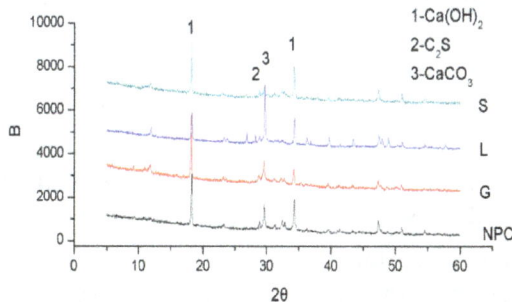

Fig. 2 XRD results of samples at 28d

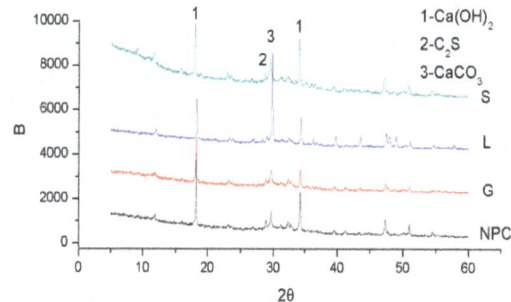

Fig. 3 XRD results of samples at 90d

3.3. SEM analysis

Take 30% of the content for samples from each group as characteristics specimens of the group. Similar to pure cement specimens, a large number of fibrous CSH is observed from the figure, sometimes showing forms of lath rod, tube, or roll foil sheet, and so on. Contrary to the pure cement specimens at the age of 28 days, the structure of slurry mixed with glass powder becomes significantly dense and porosity is further reduced making it difficult to observe and identify the structure and morphology of the crystal under the electron microscope. The reason is ground fine glass powder plays a role of filling at the early stage. The dual cementing system of cement and glass powder with more reasonable graded distribution improves the bulk density of the composite binder. Therefore, the pores are reduced compared with pure cement test block. The slurry at the age of 90 days has basically formed a compact whole, which plays an important role in improving the slurry structure of the late stage.

Fig. 4-Fig. 6 SEM images of samples adding glass powder, slag powder and limestone powder

As shown in Fig. 5, there are a large number of needle rods, mesh CSH gel, and hexagonal flake calcium hydroxide crystal in slurry mixed with limestone powder. The hydration products of cement attach to the surface of calcite act as a nucleating effect. This effect promotes early hydration of cement, increases the system's early strength, but slows the late strength development of cement and glass powder dual cementing system which is consistent with the strength test. It is worth noting that some tiny cracks related to the morphological characteristics of calcite are found. The cracks are easily produced on the surface due to smooth fracture of limestone powder particles. This is also one of the reasons for which the late strength of the test block mixed with limestone powder is weakened. The CSH gel is mostly type II mesh and III CSH gel at the age of 90 days. The phenomenon that the surface of the limestone has been eroded proves the involvement of the limestone powder in post-hydration reaction. The structure of the reaction product is a bit dense.

The microstructure of the slurry mixed with steel slag powder (uni mixed) is relatively complete (see Fig. 6), and the hydration degree is not high enough. The enrichment of hydration products that can be seen on the surface of the slag for grinding the fine steel slag plays a role of filling at the early stage. The generated CSH gel basically fills pores of the hydrate at the age of 90 days. The surface of the steel slag powder has been seriously corroded and is covered with CSH gel, indicating the occurrence of hydration reaction at the late stage of the steel slag. The calcium hydroxide crystal is gradually surrounded by other hydration products to form a dense structure with the surroundings.

4. Conclusion

(1) The adding of glass powder was detrimental to the development of specimen's early strength, but the late pozzolanic reaction increased the strength significantly. Also, the adding of limestone powder (uni mixed) obviously improved the early strength of the specimens. For instance, the strength at the age of 7 days was significantly greater than that of pure cement test block. The specimens with 15% of added steel slag powder had a good contribution to the early and late strength of the specimens. As a result the strength at the age of 7 days or 90 days was higher than that of pure cement block.

(2) The incorporation of glass powder had little effect on the hydration product category of the gelling system but had a greater impact on the generated amount. The steel slag powder and limestone played a role of filling at the early gelling system. The nucleus effect of the limestone powder also promoted early cement hydration. The later hydration of active

substances in the steel slag powder such as C_2S and C_3S also generated calcium hydroxide. In addition, the late hydration of limestone powder produced single carbon hydrated calcium aluminate.

Acknowledgments

This material is based upon work funded by the central non-profit scientific research fund for institutes (No. CKSF2016003/CL) ; China Postdoctoral Science Foundation (No. 2015M582213).

References

1. Johnston C D. Waste Glass as Coarse Aggregate for Concrete. Journal of Testing & Evaluatio, 1974, 2(5): 344-350.
2. Jin W, Meyer C, Baxter S. Glascrete – concrete with glass aggregates [J]. ACI Materials, 2000, 97: 208-213.
3. Entec Consulting Ltd. Report on Ontario blue box material recovery facilities [R]. 2007.
4. Meyer C, Baxter S. Use of Recycled Glass for Concrete Masonry and Blocks [R], Columbia University Final Report to New York State Energy Research and Development Authority, 1997,15.
5. Meyer C and Baxter S. Use of Recycled Glass and Fly Ash for Precast Concrete [R], Columbia University Final Report to New York State Energy Research and Development Authority, 1998,18.
6. Bazant Z P, Zi G, Meyer C. Fracture mechanics of ASR in concretes with waste glass particles of different sizes [J]. ASCE J. Eng. Mech., 2000, 126: 226-232.
7. Motz H, Geiseler J. Products of Steel Slags an Opportunity to Save Natural Resources [J]. Waste Management, 2001, (2): 2-8.
8. J.N. Murphy, T.R. Meadowcroft, et al. Enhancement of the cementitious properties of steelmaking slag [J]. Canadian Metalurgical Quarterly, 1997, 36(5): 331-335.

Functional Treatment Effect of Formaldehyde-free Fixing Agent CN Applied on Silk Fabric

Lei Xu[1,*], Rong Zhang[2], Jun Zhang[1], Ran Tao[1]

[1]No. 287, Xuefu Road, Suzhou   Institute of Trade & Commerce, Shi-hu International Education Park, Suzhou city, Jiangsu Province, China;

[2]No.69, Wenqu Road, Suzhou Institute of Inspection on Fiber, Suzhou, Jiangsu Province, China

*Email: 1981_xl@163.com

As a new type of dyeing-fixing agent, CN can be applied to silk fabric with satisfied anti-crease effect. Based on the multi-analysis of performance to fixing-agent CN and effect of anti-crease in silk fabric, the optimum parameters of CN application in silk fabric can be reached. The appropriate processing is as following: the optimum dosage of CN is 30g/L; the fabric is treated at 60°C for 30min. After drying, it is baked at 150°C for 3min. The dry crease recovery angle can be reached at grade 4 of treated silk crepe de Chine, and the wet crease recovery angle can be reached at grade 3. With the use of formaldehyde-free fixing agent CN, the anti-crease of silk fabric can be improved effectively with little influence on the shade.

Keywords: Anti-crease; fixing agent; silk; acid dyes.

1. Introduction

Silk fabric is a kind of comfortable textile with good wearing performance. The clothing made of silk are popular in consumers which possess good performance such as moisture absorption, permeability and soft handle. But poor anti-crease function of silk is limited the application of silk, especially after wearing for a long time. In traditional anti-crease functional treatment, the release of formaldehyde will be produce environment problems. And the handle of fabric also will be affected by the fixing agent. Nowadays, the polycarboxylic acid anti-crease fixing agent can be improved the effect of crease-resistance in silk, but the high cost restricted its application [1]. So some new types of anti-crease agent are developed in the finished treatment.

Fixing agent CN, applied after acid dyeing of silk fabric, can be carried out crosslink reaction by itself. Amount of amino, hydroxyl and carboxyl groups in the molecular chain of agent CN are reacted by valence bond cross linking with the groups such as hydroxyl and carboxyl in prime molecular chain of silk fiber. So the reticular structure is formed in the amorphous region of silk fiber and the color fastness of acid dyes also can be improved. By application testing, with the usage of new type of fixing-agent CN, the color fastness can be improved effectively and anti-crease properties are promoted also.

2. Experiment

2.1. Materials and instruments

Material: 03 silk crepe de Chine

Drugs: agent (industrial grade), acid red FG (industrial grade), sodium sulfate (Hua Xuechun), acetic acid (Hua Xuechun) and time phosphoric acid sodium (chemical pure.

Instruments: YG541B fabric fabric crease recovery tester (Hongda Textile Instrument Co., Ltd.), WSB-2 digital Bai Duyi (Xin Rui Instrument Instrument Co., Ltd.), YG026 fabric strength machine (Hongda Textile Instrument Co., Ltd.), HH-8 thermostatic water bath (Honghua Instrument Co., Ltd.), Y571B colour fastness to rubbing tester (Hongda Textile Instrument Co., Ltd.), Datacolor 600 test color matching instrument (Datacolor Trading (Shanghai) Co., Ltd.),

---

*Corresponding author

JA1003 electronic precision balance (Hongda Textile Instrument Co., Ltd.), M6 sample type dryer (Fuda Dyeing & Printing Machinery Co., Ltd.).

2.2. Experiment methods

1) Anti-crease finishing process [2, 3]

   The finishing solution is prepared with bath ratio 1:20. The silk sample is dipped into the solution for a period of time. Then the sample is dried at 100°C and baked for complete dye fixation.

2) Dyeing process

   The dyeing prescription is as follows:

   Dye concentration: 4% (o.w.f)

   Acetic acid: 5ml/L

   Sodium sulfate: 5g/L

   Bath ratio: 1:40

   Dyeing process of silk crepe is as follows:

   After preparing the dyeing solution, the temperature is increased to the 50°C. Dip the silk sample and increase the temperature to 90°C as the speed of 2°C /min. Then keeping on dyeing for 30 min, the sample is taken out for washing and drying.

3) Test standard

   1. Test for color fastness to rubbing

      According to the national standard GB/T3920-2008

   2. Test for color fastness to washing

      According to national standard GB/T3921-2008

   3. Test for color fastness to perspiration

      According to national standard GB/T3922-1995

   4. Rating color standard card

      Rating color-changed standard card: GB250-1995

      Rating color-stained standard card: GB250-1995

   5. Whiteness

      After folded in 4 layers, the sample is tested in Model WSB-2 whiteness measurement and obtained the average value.

   6. Wrinkle recovery angle

      According to the national standard GB/T3819-1997, by the vertical testing, the angle obtained in warp +weft direction.

   7. Breaking strength

      According to the national standard GB/T3923.1-1997, the remaining of breaking strength is calculated as follows:

      Remaining of breaking strength (%) = Breaking strength after treatment/ Breaking strength before treatment * 100%

2.3. Results and discussions

Compared the performance of treated-silk sample with the usage of CN by testing value of anti-wrinkle effect, the optimum process parameter can be reached. The process respects and wrinkle recovery angle test method respects to experimental methods.

3. Applied Parameters of Agent CN

3.1. *Dosage of agent CN*

With different dosage of agent CN, the samples are treated at 60°C for 30 min. After drying, baked at 140 °C for 4 min, the anti-wrinkle angle (wrap + wept), whiteness and breaking-strength remaining of silk crepe de Chine are be measured. By compared the different results, the optimum dosage of agent CN can be obtained.

From Table 1, it is obviously that the wrinkle-recovery angle of untreated sample is compared low. Increasing the dosage of agent CN, the wrinkle-recovery is increased correspondingly at dosage of 30g/L. When the dosage is higher than 30g/L, the wrinkle-recovery angle is decreased conversely. And the remaining breaking strength is gradually decreased with the agent CN using.

Table 1 Test results with different dosage of agent CN

CN dosage (g / L)	Crease recovery angle (°)		Breaking strength retention (%)		White degree
	Dry (warp + weft)	Wet (warp + weft)	Warp	Weft	
0	215	186	100	100	87.6
10	222	192	93.8	93.0	86.2
20	232	203	92.7	90.2	86.3
25	241	212	90.1	88.2	86.5
30	252	231	88.4	87.5	85.7
35	252	232	87.5	86.7	85.5
40	253	230	87.2	85.9	85.0

The reason of the phenomena is mainly because the self-polymerization of agent CN in the surface of silk fabric. The reaction between hydroxyl groups, carboxyl groups in the silk fiber and carboxyl groups and amino groups in the agent CN, and the carboxylic acid amines are produced. The crosslinking reaction will be further carried out in carboxylic acid amines and fibroin, and a layer of network structure is formed in the surface is silk fiber. The acid dyes are included in the layer, which attributed to the color fastness improvement and anti-wrinkle effect increasing. With the dosage of agent CN increasing, the anti-wrinkle effect will be increasing correspondingly until at the optimum dosage of 30g/L. The retention rate of the breaking strength decreased with the increasing used amount of agent CN, which mainly due to the reducing of molecular chain slip because of crosslinking of agent CN and silk fibroin. It is little influence on the whiteness. Based on the above-mentioned factors, the optimum dosage of agent CN is 30g/L.

3.2. *Anti-wrinkle finishing time*

In the dosage of agent CN 35g/L, treat it at 60 °C for different finishing time. After drying, baked at 140 °C for 4 min, the anti-wrinkle angle (wrap + wept), whiteness and breaking-strength remaining of silk crepe de Chine are measured. By compared the different results, the optimum finishing fixation time of agent CN can be obtained.

Table 2 Test results in different finishing time of agent CN

Table 2 Test results in different finishing time of agent CN

Finishing time (min)	Crease recovery angle (°)		Breaking strength retention (%)		White degree
	Dry (warp + weft)	Wet (warp + weft)	Dry	Wet	
10	231	207	92.5	91.4	88.1
15	240	209	91.5	89.8	87.4
20	245	212	89.0	88.2	86.5
25	250	223	88.3	87.2	86.8
30	255	233	88.1	87.0	85.9
35	254	232	87.3	86.4	84.9
40	256	231	87.2	86.1	84.5

From Table 2, it can be confirmed that with increasing the finishing time, the wrinkle-recovery angle of silk is increased correspondingly and the maximum angle is obtained at the finishing time of 30 min. Keep on prolonging the finishing time, there are little change in the wrinkle-recovery angle. As the finishing time goes by, the crosslinking of agent CN and silk fiber is deeper and deeper, and the reaction between the hydroxyl groups, carboxyl groups in the silk fiber and carboxyl groups and amino groups in the agent CN will be fixed until to a equilibrium value. After this value, finishing time prolonging cannot improve the anti-wrinkle effect. The retention rate of the breaking strength also decreased with the increasing the finishing time. It is because that the molecular chains slip for crosslinking of agent CN and silk fibroin. The whiteness decreases because of more and more deposits. Based on the above mentioned factors, the optimum finishing time is 30 min.

3.3. Baking temperature

In the dosage of agent CN 35g/L, treat it at 60°C for 30 min finishing time. After drying, baked at different baking temperature for 4 min, the anti-wrinkle angle (wrap + wept), whiteness and breaking-strength remaining of silk crepe de Chine are measured. By compared the different results, the optimum baking temperature of agent CN can be obtained.

Table 3 Test results in different baking temperature of agent CN

Curing temperature (°C)	Crease recovery angle (°)		Breaking strength retention (%)		white degree
	Dry (warp + weft)	Wet (warp + weft)	Dry	Wet	
120	233	210	88.4	88.5	87.8
130	245	216	88.2	88.1	86.5
140	253	229	87.9	87.6	85.9
150	255	231	87.2	86.8	85.3
160	255	231	86.5	85.7	84.8
170	252	227	85.2	84.0	82.2

From Table 3, when the baking temperature is high than 120 °C, it is found that the wrinkle-recovery angle of silk is climbed up and then decline. The reason is for the reaction between hydroxyl groups, carboxyl groups in the silk fiber and carboxyl groups and amino groups in the agent CN goes fatherly with baking temperature increasing. At baking temperature of 150 °C, the best anti-wrinkle effect can be reached. When the baking temperature exceeds 170 °C, the molecular crosslinking can be broken by the high temperature, the effect of anti-wrinkle will be declined as well. The whiteness grade is down obviously at baking temperature at 160 °C because of flavescent silk itself. The handle also is influenced in high temperature. Comparing all the above mentioned factors, the optimum baking temperature is at 150 °C.

3.4. *Baking time*

In the dosage of agent CN 35g/L, treat it at 60°C for 30 min finishing time. After drying, baked at 150°C for different baking time, the anti-wrinkle angle (wrap + wept), whiteness and breaking-strength remaining of silk crepe de Chine are measured. By compared the different results, the optimum baking temperature of agent CN can be obtained.

From Table 4, it can be seen that the wrinkle-recovery angle of silk is climbed up and then decline. According to the increased anti-wrinkle angle, it is mainly because that the degree of cross-linking with silk fiber and agent CN are deepened. The maximum anti-wrinkle angle is baking time for 4 min. With baking time goes by, the high temperature will cause the cross-linking broken between the agent CN and silk fiber, also the degree of handle is down. Considering the similar anti-wrinkle effect of baking time for 3 min and 4 min, and no obvious difference in whiteness, the optimum baking time is for 3 min.

Table 4 Test results for different baking time of agent CN

Curing time (min)	Crease recovery angle (°)		Breaking strength retention (%)		White degree
	Dry (warp + weft)	Wet (warp + weft)	Warp	Weft	
1	237	213	93.4	92.8	87.7
2	249	225	91.7	89.5	86.6
3	255	231	89.4	88.3	86.0
4	256	232	86.5	84.8	85.1
5	252	226	83.6	82.3	83.8

4. Conclusion

(1) Applied in the fixation treatment of silk fabric, the agent CN can be increase the anti-wrinkle performance effectively, and little influence on the shade.
(2) To obtain the best dye fixation and appropriate anti-wrinkle results, the optimum process is as follows: the 30g/L dosage of agent CN, at 60°C for 30 min finishing time. After drying, bake it at 150°C for 3 min. After treatment by the agent CN, the grade of anti-wrinkle angle is up to level 4, and the grade of wet crease recovery angel can be reached to level 3.

Acknowledgments

This material is based on work funded by the Natural Science Foundation of Jiangsu Province, China (grant no. BK2013024), Science & technology project 2016 of Jiangsu province quality and Technical Supervision Bureau, China, National Special Foundation of Cocoon Silk Development Project 2016, Jiangsu Province, China.

References

1. Xu Lei, Zhang Rong. Research development of functional non-formaldehyde wrinkle resistantfinishing of textiles [J]. Silk, 2015, 05:26-35.
2. U. Sewekow. How to meet the requirements for eco-textiles [J]. Textile Chemists and Colorist, 1996 (28): 21-27.
3. Xu Lei, Chen Lin. Research and Application of Anti crease finishing agent sericin for silk fabric [J]. Silk, 2014, 12:20-23.

Influence on the Blockage of Diamond Grinding Wheel by Carbon Chain Characteristic of Additives

Zhi-Yuan Wu[1,*], Shu-Hui Wang[2], Kai-Wen Ji[3], Xiu-Jian Tang[1], Xin-Li Tian[1]

[1]*National Key Laboratory for Equipment Remanufacturing, Academy of Armored Force Engineering, Beijing,China*
[2]*Department of Scientific Research, Academy of Armored Force Engineering, Beijing,100072 ,China*
[3] *northwest institute of technology, Xi'an, 710065, China*
*Email: Wu_zhiyuan20021@163.com

The blockage of grinding wheel is the key obstacle against the industrial application of the grinding fluid dedicated for ceramics. With a strong focus on the study of carbon chain, this paper explores the rule of blockage of grinding fluid prepared by organics with different number of carbons. Experimental result shows that: for low-carbon organics, the blockage area on grinding wheel increases with the increase of number of carbons, while for long-carbon the blockage area on grinding wheel decreases with the increase of number of carbons. The optimal anti-blocking effect can be realized only when the number of carbons of organics reaches a certain number, mainly on account of that long-carbon organics can form large oil film thickness, which can prevent grinding collection in grinding area and blockage of grinding wheel.

Keywords: Silicon nitride ceramic; diamond grinding wheel; blockage; carbon chain.

1. Introduction

With the development of science and technologies, people's research on engineering ceramics becomes increasingly deep. They are widely used in various fields of production and life[1], with more and more obvious excellent performance such as high rigidity, light weight, and good wear and heat resistance. At present, the processing methods of ceramics mainly include ELID mirror grinding, ductile regime grinding, super-high speed grinding, precise abrasive belt grinding, precise grinding wheel grinding, etc.[2]. However, there are some problems such as high processing cost, high processing conditions, and processing difficulty during processing, which hinders the development of engineering ceramics seriously[3]. Dedicated grinding fluid is a good method to assist the processing of engineering ceramics, but blockage is still a key problem urgent to be solved in this field[4]. This paper mainly researches the relation between number of carbons of organics and blockage of grinding wheel, and analyzes the production rule and mechanism of blockage of grinding wheel, laying a solid foundation for further practice and theoretic research on the formula of grinding fluid.

2. Experimental Conditions

Experimental material: silicon nitride

Organics: normal alkanes with the number of carbons of 9-16, normal organic acid with the number of carbons of 3-9, normal organic alcohol with the number of carbons of 3-9

Experimental equipment: XD-250AH machine tool

Grinding parameters: the moving speed of machine tool is the maximum gear; the feed rate of machine tool is 0.03 mm; the concentration of emulsion is 10%; the flow velocity of emulsion is 500 ml/h, and the grinding wheel speed is 3,000 r/min

3. Influence on Blockage by the Change of Carbon Chain Length

3.1. *Influence on blockage by the increase of carbon chains of alkanes*

Alkanes are hydrocarbons of which the carbon atoms are mutually connected through single bond, and other valence bonds are saturated by hydrogen atoms[5]. In industrial application, most of engine oil, emulsion, etc. are prepared mainly with alkanes. Therefore, the influence on the blockage of diamond grinding wheel by alkane organics has extremely important application value. In order to eliminate the influence on test by volatilization and heterogeneous characteristic, normal long-carbon alkanes are selected for the experiment, including decane, undecane, dodecane, tridecane, tetradecane, pentadecane, and cetane.

The experimental result is shown as Fig. 1:

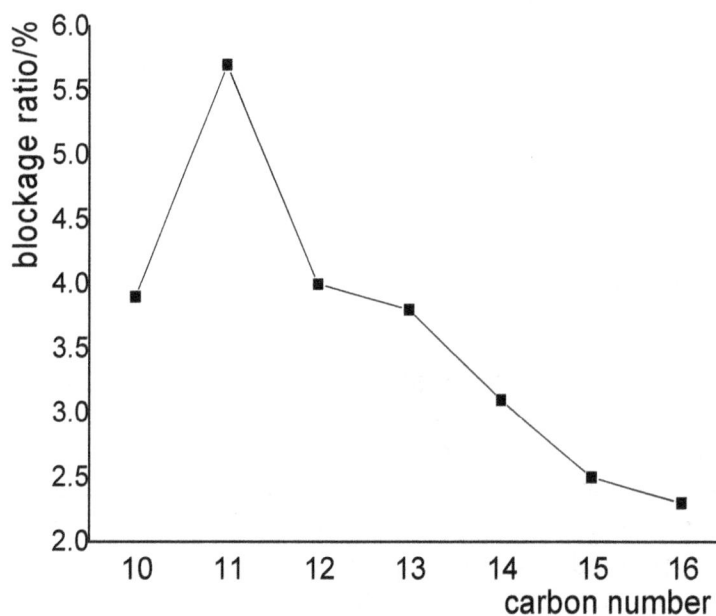

Fig. 1 Relation between number of carbons and blockage ratio of alkanes

It can be seen from Fig. 1 that, when number of carbons is increased from 10 to 11, the grinding wheel is more probably blocked. When number of carbons is over 11 with the increase of carbons, the blockage ratio of grinding wheel reduces rapidly, with disciplinary monotonic decreasing tendency. Thus it can be seen that, when the number of carbons of normal alkanes is above a certain value, increasing the number of carbon can restrain blockage of grinding wheel better.

3.2. *Influence on blockage by the increase of carbon chains of organic acid*

Organic acid is organic compounds of which the molecules contain carboxyl [5]. The organic acid selected for experiment is in normal type, with number of carbons of 3-9. Organic acid with above 9 carbon atoms is not applied, as it is mostly solid and insoluble in water. The experimental result is shown as Fig. 2:

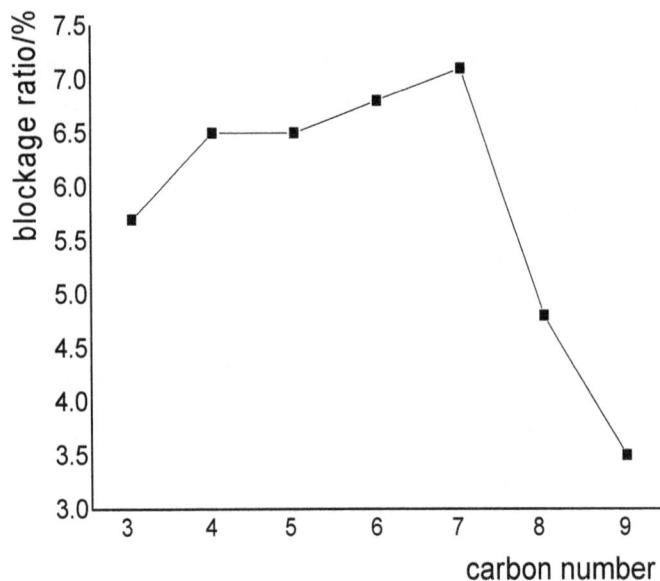

Fig. 2 Relation between number of carbons and blockage ratio of organic acid

It can be seen from Fig. 2 that, the blockage area of low-carbon organics increases with the increase of number of carbons. When number of carbons is above 7 (heptylic acid), the blockage area decreases with the increase of number of carbons, through which the blockage is restrained greatly. What is noteworthy is that due to the participation of carboxyl, the turning point between increase and decrease of the blockage of grinding wheel reduced from the 11 carbon atoms of alkanes to 7 carbon atoms. Although alkanes with equivalent number of carbons cannot be involved in the experiment, this phenomenon reflects the participation of carboxyl is beneficial to restraining the blockage of grinding wheel.

3.3. *Influence on blockage by the increase of carbon chains of organic alcohol*

Organic alcohol is organic compounds of which the molecules contain hydroxyl [5]. The organic alcohol selected for experiment is in normal type, with number of carbons of 3-9. The experimental result is shown as Fig. 3.

It can be seen from Fig. 3 that, when number of carbons is below 6, the blockage area of grinding wheel increases with the increase of number of carbons. When the number of carbons is above 6, the blockage area reduces rapidly with the increase of number of carbons. This rule basically conforms to the function rules of alkanes and organic acid. However, the participation of hydroxyl in alcohol makes the turning point acquired when number of carbons is above 6 in alcohol experiment, and the blockage area less than alkanes and acid.

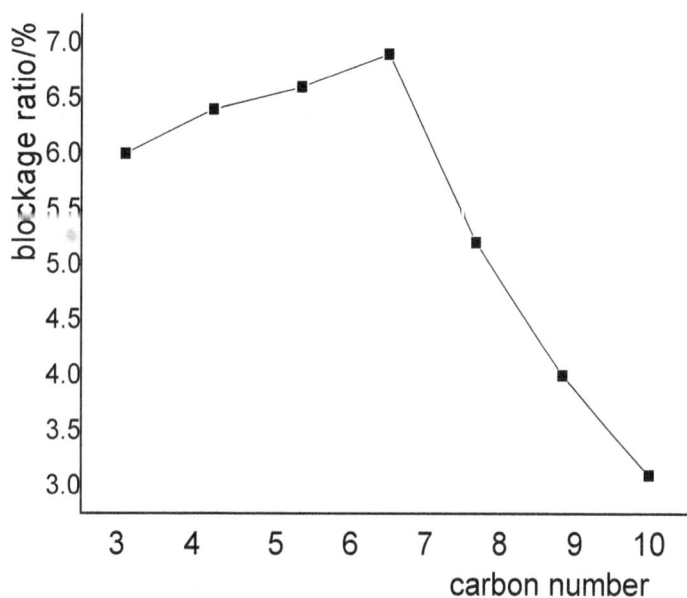

Fig. 3 Relation between number of carbons and blockage ratio of alcohol

Above all, during the application of different organics, the blockage area of grinding wheel will increase at first and then reduce, with the increase of number of carbons. However, number of carbons of different organics occurring turning point differs greatly.

4. Analysis of Anti-blocking Mechanism

4.1. *Discussion of correlation of number of carbons to the optimal anti-blocking organic*

Characteristic points of above experiments are summarized as shown in Table 1:

Table 1 Characteristic points of different organic materials

Organic	Turning point	The minimum number of carbons of blockage below 4.5%
Alkanes	11	12
Acid	7	8
Alcohol	6	9

According to the investigation on the turning points occurring reduction of blockage area of different reagents, the turning points of alkanes, alcohol, and acid occur at the number of carbons of 11, 6 and 7 respectively. The participation of carboxyl or hydroxyl in acid or alcohol makes the realization of acid and alcohol. The stronger the polarity of acid is, the less the demanded number of carbons is, which shows that additives with stronger polarity have larger anti-blocking capacity. This rule can also be found through comparison, with the limit of blockage ratio below 4.5%. Blockage of grinding wheel can be restrained under 4.5% when the number of carbons of organic acid is 8.

4.2. *Influence on grinding blockage by oil film thickness*

All organic acid, organic alcohol and alkanes applied to the experiment are normal organics. When they are adsorbed on the surface of ceramics, the adsorption mode is shown in Fig. 4:

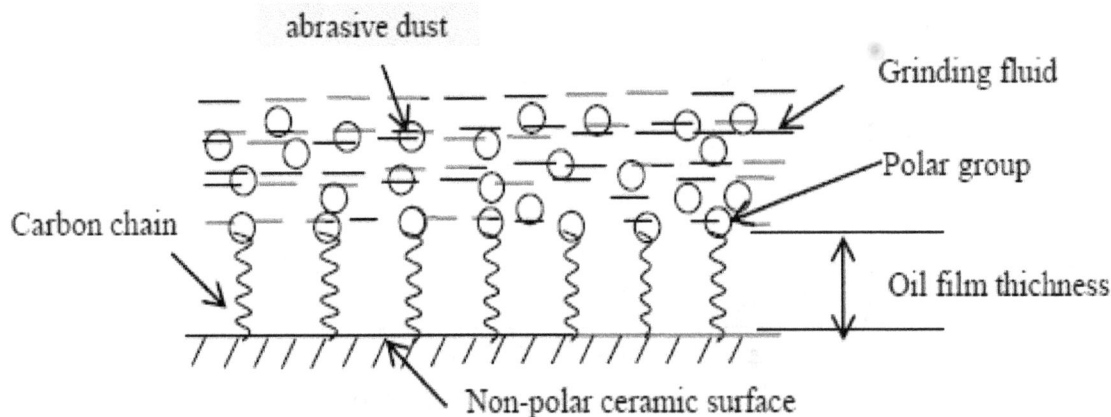

Fig. 4 Adsorption model of additives on non-polar ceramic surface

It can be easily seen from Fig. 4 that, the properties of polar water and non-polar ceramics are opposite, with different attractions and repulsion interactions on organic groups with different polarities and non-polar organic groups. Therefore, normal organics in the experiment are vertically adsorbed on ceramic surface vertically. Under that adsorption mode, the oil film thickness depends on the length of carbon chain of organics. Organics with larger number of carbons can form thicker oil film, which possesses stronger buffer function against wear debris, thus it is more difficult for wear debris to generate extrusion and cause blockage in the grinding zone between ceramic and grinding wheel, which reduces the occurrence of blockage of grinding wheel.

5. Conclusion

1. The influence on blockage by the increase of carbon chain length of organics includes two stages: when the length of carbon chain is short, the blockage of grinding wheel increases with the increase of the length of carbon chain. When the length of carbon chain is above a certain value, the blockage of grinding wheel reduces with the increase of the length of carbon chain.
2. The common characteristic of different organics acquiring the optimal anti-blocking effect is long carbon chain.
3. The anti-blocking effect of organics relates to the thickness of oil film formed on ceramic surface by them. The larger the thickness of the formed oil film, the better the anti-blocking effect after addition.

Acknowledgment

This work was funded by National Natural Science Foundation of China (Project numbers: 51275527 and 51475474).

References

1. Melanie Lee, Bo Wang, K. Li, et al. New designs of ceramic hollow fibres toward broadened applications. Journal of Membrane Science, 2016, 503: 48-58

2. H. Tang, Z.H. Deng, Y.S. Guo, J. Qian, D. Reynaerts. Depth-of-cut errors in ELID surface grinding of zirconia-based ceramics. International Journal of Machine Tools and Manufacture, Volume 88, January 2015, Pages 34-41.

3. Renjie Ji, Yonghong Liu, Yanzhen Zhang, et al. Machining performance of silicon carbide ceramic in end electric discharge milling. International Journal of Refractory Metals and Hard Materials, 2011, 29(1): 117-122

4. Tian Xinli, Ji Kaiwen,Wu Zhiyuan, et al. Analysis of factors affecting clogging in difamond wheel griinding of Si3N4 ceramic. [J]. Journal of Academy of Armored Force Engineering. 2015, 06(2): 39~42

5. Ji Maozhi, Peng Song, Ge Zhenghua. Organic chemistry; [M], Science Press, 2013: 01-10.

The Blockage of Diamond Grinding Wheel with Normal Acid and Alcohol Additive

Zhi-Yuan Wu[1,*], Shu-Hui Wang[2], Kai-Wen Ji[3], Jun-Wei Yang[1]

[1]National Key Laboratory for Equipment Remanufacturing, Academy of Armored Force Engineering, Beijing, China
[2]Department of Scientific Research, Academy of Armored Force Engineering, Beijing, 100072, China
[3]Northwest institute of technology, Xi'an, 710065, China
*Email: Wu_zhiyuan20021@163.com

The grinding fluid dedicated for ceramics is one of the effective methods to realize efficient grinding of ceramics. The blockage of grinding fluid dedicated for ceramics is researched in this paper. On basis of the fundamental principles of organic chemistry, the influence from changes of organic acid, organic alcohol carbon chain and functional group on the blockage of grinding wheel are investigated respectively in this paper. According to the experiment result: the blockage area of grinding wheel increases along with the increase of low-carbon chain acid and alcohol, but for the long-carbon chain, the blockage area on grinding wheel decreases with the increase of the number of carbons; the influences of adding functional group are highly related to the hydrophile-lipophile (HLB value), which is finally formed by organic matters and has regular changes.

Keywords: Ceramics; blockage; carbon chain; functional group.

1. Introduction

Featured by excellent physicochemical properties, the engineering ceramics are widely used in all areas [1]. As the major means to process ceramics, grinding is of high difficulty and cost and thus restraining the large-scale application of ceramic materials [2]. The grinding fluid is an effective way to improve the economical efficiency of ceramic processing, but it faces many bottlenecks due to short development history [3]. As one of the key problems, the grinding wheel may be blocked during the grinding process of diamond grinding wheel [4]. Functioning as the special atomic or atomic group determining the chemical property of molecules [5], the functional group has decisive influences on the application characteristics and results of organic matters. In this paper, experiment is made to check the influences from characteristics of organic matter functional group on blockage of grinding wheel to explore the relevant laws and lay the theoretical basis for preparing new grinding fluid.

2. Experimental Conditions

Experimental material: silicon nitride

Organics: normal alkanes with the number of carbons of 9-16, normal organic acid with the number of carbons of 3-9, normal organic alcohol with the number of carbons of 3-9

Experimental equipment: XD-250AH machine tool

Grinding parameters: the moving speed of machine tool is the maximum gear; the feed rate of machine tool is 0.03 mm; the concentration of emulsion is 10%; the flow velocity of emulsion is 500 ml/h, and grinding wheel speed is 3,000 r/min

Emulsion preparation method: OP10 and Span80 are used as emulsifier in the experiment to implement emulsification to the paraffin so as to prepare basic emulsion. To reduce the influences from emulsifier on experiment, the total quality of emulsifier should be lower than 5% of paraffin. Experiment may be started by directly adding the proportioned additive into the basic emulsifier for even mixing, since all organic matter additives are soluble in water.

3. Influences on Blockage by Functional Group Difference

The physicochemical properties of organic compounds are determined by its structure. The difference of structure may lead to changes of carbon chain and can be influenced by the functional group, which is manifested by its type, quantity and position, on basis of which deep analysis will be implemented as follows.

3.1. *Influence on blockage by type of functional group*

Functional group functions serve as special atomic or atomic group determining the chemical property of molecules [5]. In this part, the influences from adding of different functional groups on blockage are researched. Comparison and analysis are implemented by combining the experiment data above and the analysis result is as shown in Fig. 1.

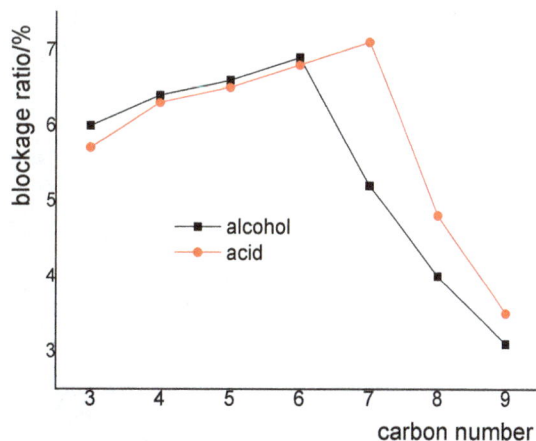

Fig. 1 Comparison of blockage ratio for organic acid and alcohol along with number changes

According to Fig. 1, when number of carbons is 3-6, the blockage ratio of organic acid is slightly lower than that of organic alcohol; when number of carbons is 6-9, the blockage ratio of organic acid is slightly higher than that of organic alcohol.

3.2. *Influences on blockage by number of functional groups*

The experiments above show that the blockage of grinding wheel can be alleviated by effectively using the polar functional group. The methylacetic, malonic acid and tricarballylic acid, with number of carbons of 3 and number of functional groups of 1, 2 and 3 respectively, are used in further experiment in order to check the influences from number of functional groups on blockage of grinding wheel. The experiment result is as shown in Fig. 2.

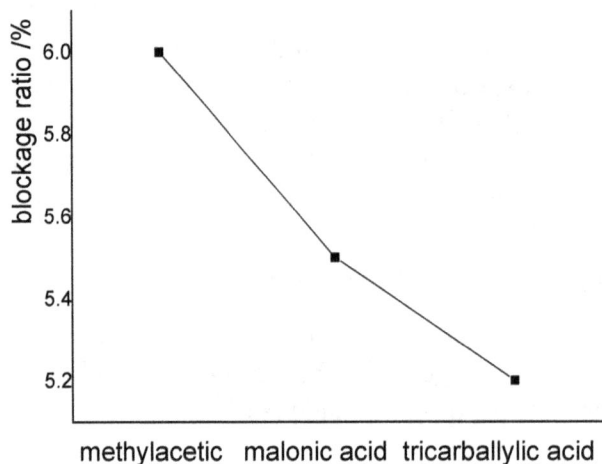

Fig. 2 Relation among methylacetic, malonic acid and tricarballylic acid

According to Fig. 2, the experiment result for methylacetic, malonic acid and tricarballylic acid has monotonic decreasing, which means, the organic acids added have better improvement effects along with the increasing of number of functional groups.

3.3. *Influence on blockage by position of functional group*

In this part, research is carried out regarding the position of polar functional group. Propanol, 1, 2-propanediol, 1, 3-propanediol, and glycerol are selected for the experiment.
The experimental result is shown as Fig. 3:

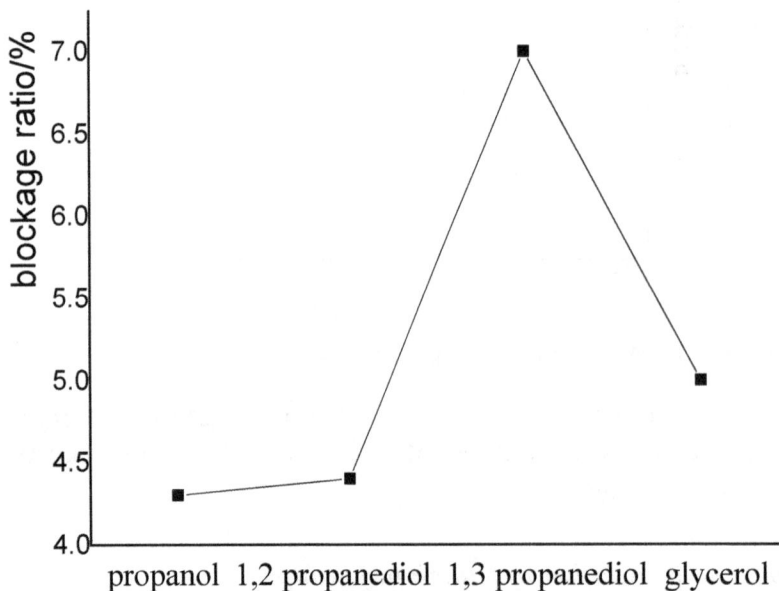

Fig. 3 Blockage ratio among propanol, 1, 2-propanediol , 1, 3-propanediol, and glycerol blockage ratio

According to Fig. 3, except for 1, 3-propanediolwhich is a peak, the improvement effect of the other three organic alcohols gets worse with the increase of functional groups.

4. Influence on Blockage by Organic Interaction on Blockage

The affinity between applied organics and ceramic surface, and the condition of formed oil film has extremely important influence on the blockage of grinding wheel [6]. In this section, the conception of HLB value is introduced to research relevant phenomenon. As the equilibrium value of hydrophilic and lipophilic properties of substances, HLB value is the balance of the two opposite groups which are hydrophilic and lipophilic respectively in molecules on size and power. The scope of HLB value is from 1 to 40. The lower the HLB value, the better its lipophilic property. Conversely, the higher the HLB value, the better its hydrophilic property. Therefore, HLB value can reflect the increase and decrease of the polarity of organics themselves and their affinity on solid surface. The HLB value of organics used for experiment can be calculated through structural factorization method, of which the specific arithmetic is shown as Formula 1:

HLB=7+$\sum$ (cardinal number of hydrophilic groups)+(cardinal number of lipophilic groups)   (1)

The HLB value of above organics can be calculated through Formula 1. The relation between HLB value and blockage ration is shown as Fig. 4.

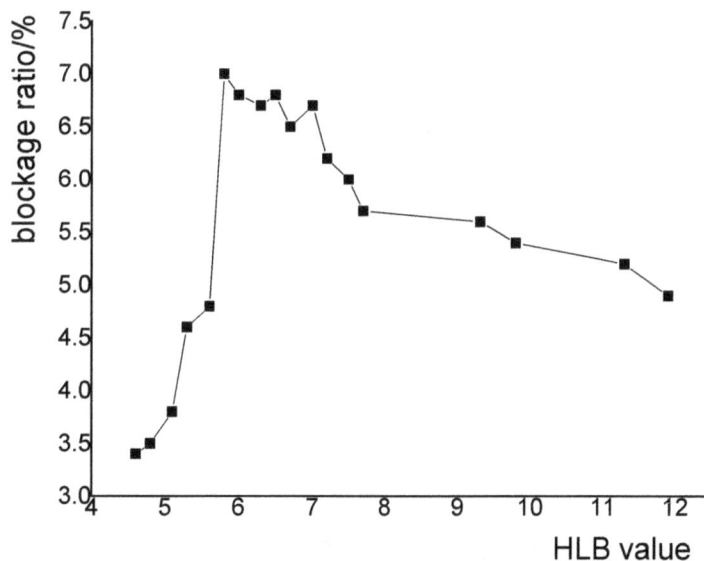

Fig. 4 Influence on blockage ratio by the change of HLB value

According to Fig. 4, as to non-polar silicon nitride ceramics, there is obvious correlation between the anti-blocking effect of organics and their HLB value. When the value of substance is lower than 5.8, the blockage ratio increases gradually with the increase of blockage ration. When the HLB value is above 5.8 or even larger than 7, the blockage ratio reduces obviously with the increase of blockage ration.

The best anti-blocking additives in Fig. 4 are organic matters with low HLB, i.e. decyl alcohol and decanoic acid. As the common characteristics, they have the longest carbon chain in organic matters. By referring to their HLB value in Fig. 4, it can be deduced that the molecular structure characteristics for the optimal anti-blocking organic matters of non-polar silicon nitride ceramics are: long carbon chain and low HLB value (low organic matter molecular polarity). This result is consistent with the similarity attraction principle, indicating that, the absorption effects of organic matters are of great importance for the realization of anti-blocking performance of additive.

By reviewing the two inflection point materials, hexyl alcohol and heptanoic acid, in the two curves in Fig. 1 and referring to the table, it can be found that the HLB value is 2.1 for –COOH of organic acid and 1.9 for –OH of organic alcohol. The HLB value of carboxyle is higher, but the number of carbons of corresponding heptanoic acid is one more than alcohol. Based on calculation of Formula 1, the HLB value of two inflection points is 6.0 and 5.8 respectively and they are quite close, which means, the application effects of organic matters for certain ceramic materials are not dependant on the type of organic matter functional group, but on the comprehensive characteristics of molecules composed by carbon chain and functional groups.

5. Conclusion

1. The influences on blockage by different lengths of organic matter carbon chain can be divided into two stage: when carbon chain length is short, the blockage of grinding wheel will be aggravated along with the increase of carbon chain length; when carbon chain length is higher than certain value, the blockage of grinding wheel will be alleviated along with the increase of carbon chain.
2. The influence on blockage by functional group depends on the final organic matters formed. When organic matters have low hydrophile (HLB value is lower than 5.8), the blockage of grinding wheel will be aggravated along with the increase of hydrophile of functional group. If this value is exceeded, the blockage of grinding wheel will be alleviated along with the increase of hydrophile of functional group.

Acknowledgement

This work was funded by National Natural Science Foundation of China (Project number: 51275527 and 51475474).

References

1. Jie Li, Liang Fang, Hao Luo, et al. Structure and Microwave dielectric properties of a novel temperature stable low-firing $Ba_2LaV_3O_{11}$ ceramic. Journal of the European Ceramic Society, 2016, 36(8): 2143-2148
2. Tang H, Deng Z H, Guo Y S, Qian J, D Reynaerts. Depth-of-cut errors in ELID surface grinding of zirconia-based ceramics International[J]. Journal of Machine Tools and Manufacture, 2015, 88(2): 34~41.
3. C. Piazzoni, M. Blomqvist, A. Podestà, G. Bardizza, M. Bonati, P. Piseri, P. Milani, C. Davies, P. Hatto, et al. Nanocomposite TiN films with embedded MoS_2 inorganic fullerenes produced by combining supersonic cluster beam deposition with cathodic arc reactive evaporation [J]. Applied Physics A. 2008, 26(2): 31~38.
4. Tian Xinli, Ji Kaiwen, Wu Zhiyuan, et al. Analysis of factors affecting clogging in diamond wheel grinding of Si_3N_4 ceramic. [J]. Journal of Academy of Armored Force Engineering.2015,06(2): 39~42

5. Ji Maozhi, Peng Song, Ge Zhenghua. Organic chemistry; [M], Science Press, 2013: 01-10.
6. Nettikadan S R, Johnson J C, Mosher C, et al. Virus particle detection by solid phase immunocapture and atomic force microscopy[J]. Biochemical and Biophysical Research Communications, 2013, 42(2): 33~45.

Comparative Study on the As-cast Microstructures and Mechanical Properties of AZ91D and AM60B Alloys by Addition of Nd

Zheng-Hua Huang[1,*], Dong-Fu Song[1], Wen-Jun Qi[1], Yang-De Li[2], Wei-Rong Li[2]

[1]Guangdong Institute of Materials and Processing, Guangdong Academy of Sciences, Guangzhou 510650, China
[2] DongGuan EONTEC Co. Ltd., DongGuan 523662, China
*Email: zhhuang@live.cn

As-cast microstructures and phase compositions of AZ91D-xNd and AM60B-xNd (x=0~2.5) alloys were investigated. Meanwhile, the tensile mechanical property and impact toughness were tested. With increasing the Nd content, β-$Mg_{17}Al_{12}$ phase originally precipitating by discontinuous network or small dispersed block begins to break and decrease gradually. At the same time, fine rod-shaped or small block new phase Al_3Nd with high thermal stability increases gradually. The tensile mechanical property at ambient temperature is enhanced gradually for AZ91D-xNd alloys; however, it is firstly reduced obviously and then enhanced gradually for AM60B-xNd alloys. The maximum value of tensile strength σ_b and elongation δ can reach 225 MPa, 11.0% and 230 MPa, 16.5% for AZ91D-2.17Nd and AM60B-2.32Nd alloys, respectively. Addition of Nd can enhance the tensile mechanical property at elevated temperature for AZ91D alloy, but it cannot ameliorate that for AM60B alloy. Meanwhile, it is better for the former series alloys. Impact toughness is enhanced gradually for AZ91D-xNd alloys, while for AM60B-xNd alloys, it is reduced significantly at the start and then remains almost unchanged, which is still better.

Keywords: AZ91D alloy; AM60B alloy; Nd modification; microstructure; mechanical property.

1. Introduction

Magnesium alloy, one of the most important structural materials having high specific strength, is used widely in the automotive, communicated, electronic and aerial industries [1]. AZ91D and AM60B alloys are two of the most widely used commercial die-casting heat-resistant magnesium alloys, however the mechanical property is limited at ambient temperature. Meanwhile, the strength and creep resistance are relatively poor at elevated temperature because the main strengthening phase β-$Mg_{17}Al_{12}$ exhibits the low thermal stability. Then they are difficult to be fabricated as the components used under the elevated temperature for long time. Researchers had ameliorated the microstructures and enhance the mechanical properties of Mg-Al series alloys by precipitating the phases with high thermal stability such as Al_2Ca, Al_2Sr, Mg_2Si and Al-RE through adding alkaline earth metal [2-4], Si [5, 6] and rare earth RE [7-13]. Rare earth with the unique atomic electron and chemical property can purify the alloy melt, ameliorate the microstructure and enhance the property effectively. Compared with other rare earth metals, the effects of Nd on the microstructures and mechanical properties of Mg-Al series alloys have been rarely studied yet. At the same time, an alloy as a structural component should possess not only the sufficient static strength but also the superior dynamic property, since the failure of component is usually due to the dynamic loading against the foreign object during the service. Impact toughness, an important dynamic mechanical property for engineering structural materials, can affect the performance and structural reliability significantly. However, the investigation on the impact toughness of magnesium alloy is relatively few yet [14-16]. Therefore, the present paper is focused on studying the evolution in the as-cast microstructure, tensile mechanical property and impact toughness of AZ91D and AM60B alloys by adding 0~2.5%Nd respectively, and present the role of second phase on strengthening mechanism.

---

*Corresponding author

2. Experimental

Alloy ingots were prepared by melting AZ91D, AM60B alloy ingots and Mg-30wt.% Nd master alloy in an electric resistance furnace under the mixed atmosphere of CO_2 and SF_6. When the melt temperature of alloy ingot reached 1003 K, the master alloy was added into the melt. Then the melt was stirred for twice within an hour to ensure the compositional homogeneity. After adding the refine agent, the melt was held at 1023 K for 20 min. When the temperature cooled to 988 K, the melt was poured into the wedge permanent mold with a preheated temperature of 523 K, and then as-cast samples were obtained.

The Nd content in the two series alloys was measured by inductively coupled plasma analyzer (ICP, JY Ultima2). All specimens were etched with 4 vol.% HNO3 in ethanol. Microstructural observation was carried out on optical microscope (OM, Leica DM IRM) and scanning electron microscope (SEM, JEOL JXA-8100), respectively. Phase analysis was carried out on X-ray diffractometer (XRD, D/MAX-RC) with Cu Kα radiation. Tensile test at the temperature of 298 K and 423 K was performed on material test machine (DNS200) at a rate of 2 mm/min. Standard un-notched impact specimen with dimension of 10×10×55 mm3 was tested on the pendulum impact testing machine at ambient temperature.

3. Results and Discussion

Figures 1 to 4 show the optical and SEM graphs of as-cast alloys, respectively. The results show that the evolution in the as-cast microstructures of AZ91D and AM60B alloys with increasing the Nd content is different.

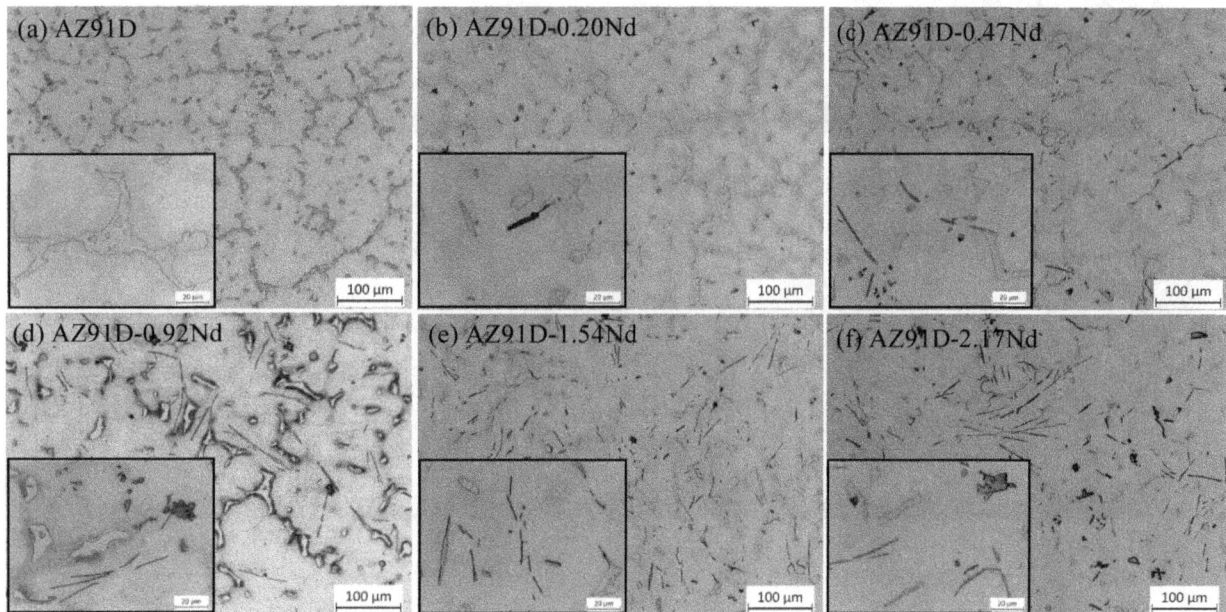

Fig. 1 OM graphs of as-cast AZ91D-xNd alloys. Illustrations are the graphs with high magnification

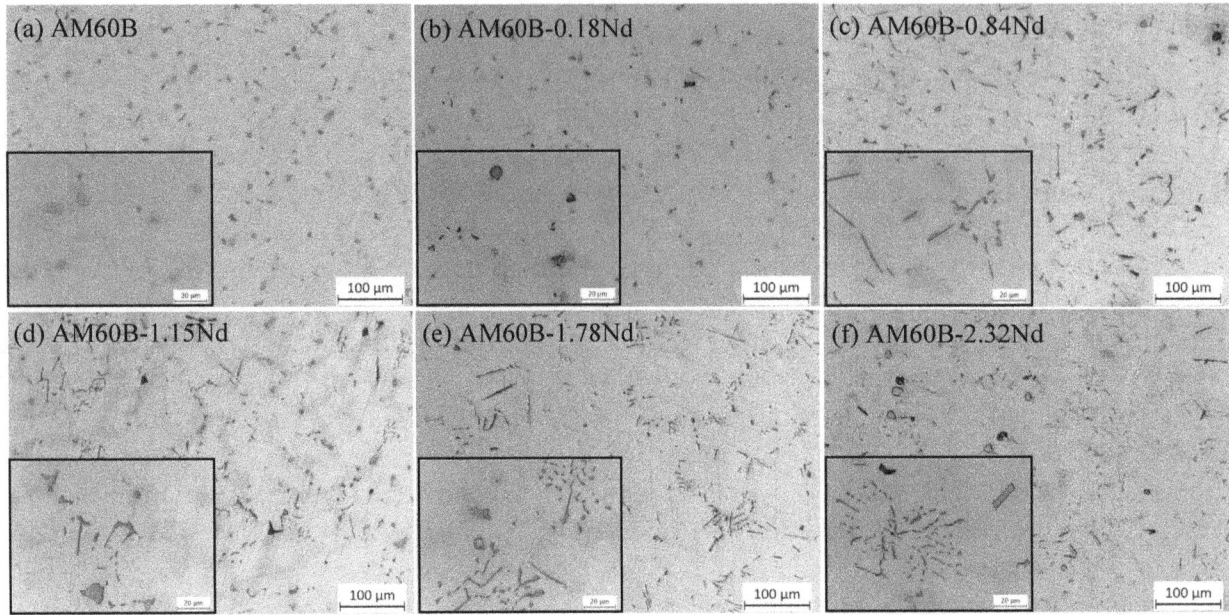

Fig. 2 OM graphs of as-cast AM60B-xNd alloys. Illustrations are the graphs with high magnification

Fig. 3 SEM graphs of as-cast AZ91D-xNd alloys

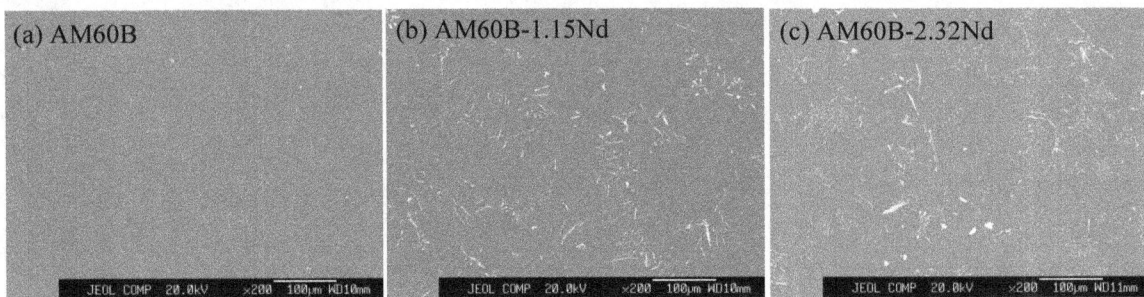

Fig. 4 SEM graphs of as-cast AM60B-xNd alloys

As-cast AZ91D alloy exhibits the coarse dendrite and is mainly composed of α-Mg matrix and many β-Mg$_{17}$Al$_{12}$ phase precipitating along grain boundary discontinuously (see Figs. 1a and 3a), while the β-Mg$_{17}$Al$_{12}$ phase among the as-cast AM60B alloy exhibits few dispersed small block (see Figs. 2a and 4a). With increasing the Nd content, the β-Mg$_{17}$Al$_{12}$ phase begins to break and decrease gradually. Meanwhile, the precipitated fine rod-shaped or small block compounds increase gradually, and the fine rod-shaped compound occupies the majority.

Table 1 EDS results of as-cast AZ91D-xNd and AM60B-xNd alloys (atomic percent, at.%)

Alloy	Position in Figs. 5 and 6	Mg	Al	Zn	Mn	Nd
AZ91D	1	62.86	35.30	1.84	-	-
	2	95.22	4.78	-	-	-
AZ91D-0.92Nd	1	57.63	33.66	1.57	-	7.14
	2	12.49	60.91	-	-	26.60
	3	66.94	31.41	1.65	-	-
	4	97.88	2.12	-	-	-
AZ91D-2.17Nd	1	4.20	64.31	-	-	31.49
	2	9.48	59.12	-	-	31.40
	3	70.67	27.77	1.56	-	-
	4	95.08	4.90	-	-	0.02
AM60B	1	90.43	8.53	-	1.04	-
	2	72.27	27.73	-	-	-
	3	94.04	5.96	-	-	-
AM60B-1.15Nd	1	72.95	21.72	-	-	5.33
	2	70.07	29.93	-	-	-
	3	96.93	3.07	-	-	-
AM60B-2.32Nd	1	48.87	36.26	-	-	14.87
	2	8.19	62.20	-	-	29.61
	3	72.81	27.19	-	-	-
	4	96.91	3.09	-	-	-

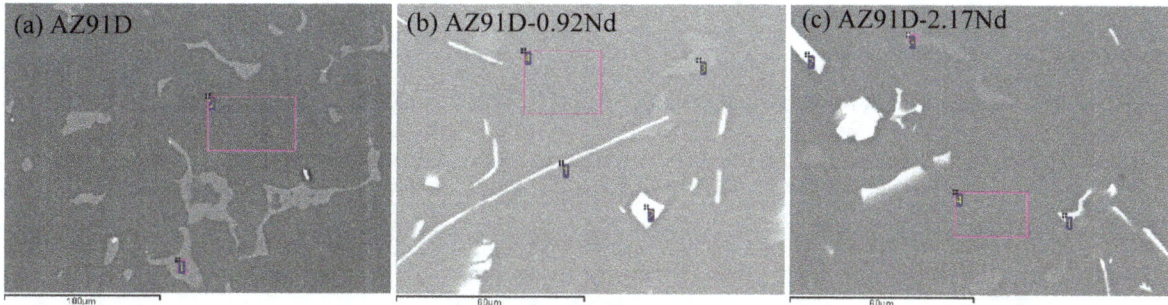

Fig. 5 EDS spectra of as-cast AZ91D-xNd alloys

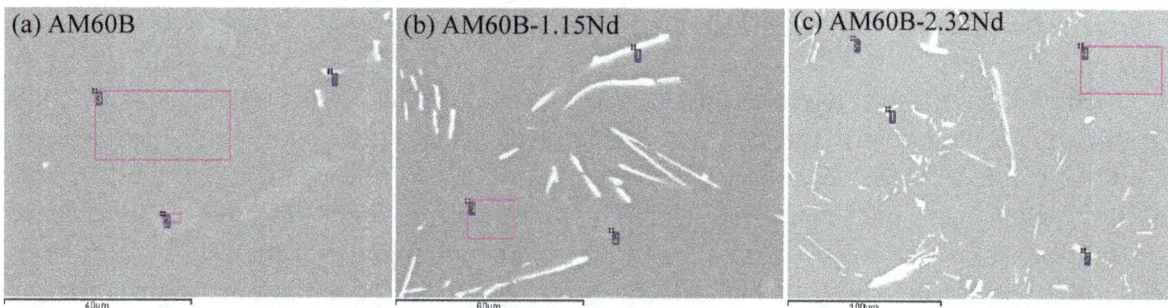

Fig. 6 EDS spectra of as-cast AM60B-xNd alloys

Figures 5 and 6 show the EDS spectra of as-cast alloys, and then the results are listed in Table 1. As-cast microstructures of the two base alloys main composite of α-Mg and Mg-Al phases i.e. β-Mg$_{17}$Al$_{12}$.

Meanwhile, few Mn-containing phase can be observed (see spectrum 1 in Fig. 6a). However, other fine rod-shaped or small block compounds can be observed for all as-cast Nd-containing alloys.

In order to determine the phase composition after adding the Nd content, the comparison between the as-cast XRD spectra of the two base alloys and the two alloys containing the maximum Nd content was made (see Fig. 7). XRD spectra of the two base alloys consist of the peaks of α-Mg and β-Mg$_{17}$Al$_{12}$ phases, while the peak of Al$_3$Nd phase can be observed among the XRD spectra of AZ91D-2.17Nd and AM60B-2.32Nd alloys. Combined with the EDS results, the fine rod-shaped or small block compounds are considered as Al$_3$Nd phase, which exhibits higher thermal stability than β-Mg$_{17}$Al$_{12}$ phase.

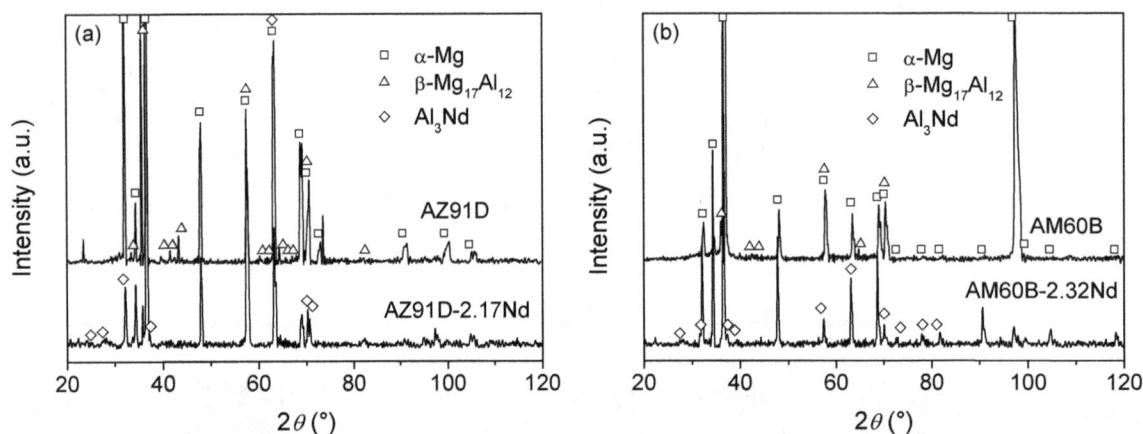

Fig. 7 XRD spectra of as-cast AZ91D-xNd and AM60B-xNd alloys

Al-RE, Mg-RE or Mg-Al-RE compounds may be formed when rare earth metal RE is added into Mg-Al series alloys. Degree of difficulty in forming compounds between different elements can be judged by electronegativity difference $\Delta\chi$. The greater the value of $\Delta\chi$ is, the larger the binding force is and then the easier the formation of compounds is [17]. Electronegativity χ is 1.31, 1.61 and 1.14 for Mg, Al and Nd respectively [18], thus Nd-Al exhibits the larger value of $\Delta\chi$ (0.47) than Nd-Mg (0.17), which indicates the greater binding force between Nd and Al. From the thermodynamic point of view, small block Al$_3$Nd not Mg-Nd or Mg-Al-Nd phase precipitates on the priority when Nd is added into AZ91D and AM60B alloys respectively, which has been confirmed by XRD and EDS results.

Rare earth Nd, a surface active element, can be adsorbed on the tip during the growth process of β-Mg$_{17}$Al$_{12}$ phase, and then its growth can be inhibited. Thus, the β-Mg$_{17}$Al$_{12}$ phase becomes to break and be refined. At the same time, the formation of Al$_3$Nd phase can consume the partial Al atoms in the alloy melt, and then the β-Mg$_{17}$Al$_{12}$ phase can be reduced undoubtedly. During the process of alloy solidification, Al$_3$Nd phase can form eutectic together with α-Mg and β-Mg$_{17}$Al$_{12}$ phase through Nd participating in the eutectic reaction. Al$_3$Nd phase is pushed into the growth interface, and then the growth of the dendrite is hindered. Therefore, as-cast microstructures of the alloys can be refined slightly.

Table 2 shows the tensile mechanical property and impact toughness of as-cast alloys. It is seen that the evolution with increasing the Nd content is also different. With increasing the Nd content, the tensile mechanical property and impact toughness at ambient temperature increase gradually for AZ91D-xNd alloys. Tensile strength σ_b, elongation δ and impact toughness α_{nK} gradually increase from 182 MPa, 7.0% and 9 J/cm^2 for AZ91D alloy to 225 MPa, 11% and 16 J/cm^2 for AZ91D-2.17Nd alloy respectively, with the improvement amplitude of 24%, 36% and 78% respectively. When the test temperature increases to 423 K, σ_b first obviously increases from 176 MPa for AZ91D alloy to 189 MPa for AZ91D-0.20Nd alloy, and then gradually increases to 179 MPa for AZ91D-2.17Nd alloy, which is still slightly higher than that for AZ91D alloy.

Table 2 Tensile mechanical property and impact toughness of as-cast AZ91D-xNd and AM60B-xNd alloys

Alloy	298 K		423 K		α_{nK} (J/cm^2)
	σ_b (MPa)	δ (%)	σ_b (MPa)	δ (%)	
AZ91D	182	7.0	176	15.5	9
AZ91D-0.20Nd	190	8.5	189	15.0	9
AZ91D-0.47Nd	199	8.5	186	18.0	11
AZ91D-0.92Nd	210	10.5	185	14.0	13
AZ91D-1.54Nd	215	10.5	182	12.5	14
AZ91D-2.17Nd	225	11.0	179	16.5	16
AM60B	210	14.0	167	20.0	24
AM60B-0.18Nd	185	11.0	160	18.0	17
AM60B-0.84Nd	182	10.5	144	11.5	17
AM60B-1.15Nd	205	12.0	164	18.5	19
AM60B-1.78Nd	215	13.5	156	21.5	19
AM60B-2.32Nd	230	16.5	162	28.5	18

AM60B alloy exhibits the more excellent tensile mechanical property and impact toughness at ambient temperature than AZ91D alloy, where σ_b, δ and α_{nK} can reach 210 MPa, 14.0% and 24 J/cm^2 respectively. However, its tensile mechanical property at elevated temperature is lower. When few Nd content (0.18~0.84%) is added, the tensile mechanical property and impact toughness decrease significantly at ambient temperature. With the further increase of Nd content (1.15~2.32%), the tensile mechanical property gradually increases at ambient temperature, where σ_b and δ can reach 230 MPa and 16.5% for AM60B-2.32Nd alloy respectively, with the improvement amplitude of 10% and 18% respectively. Meanwhile, the impact toughness remains almost unchanged at the range from 18 J/cm^2 to 19 J/cm^2, which is still better than that for AZ91D-xNd alloys. When the test temperature increases to 423 K, the change trend in the tensile mechanical property of the alloys is on the whole consistent with that at the ambient temperature, however that does not excel than that for AM60B alloy.

In summary, addition of Nd can enhance the tensile mechanical property and impact toughness at ambient temperature significantly and the tensile mechanical property at elevated temperature slightly for as-cast AZ91D alloy. Addition of high Nd content can improve the tensile mechanical property at ambient temperature for as-cast AM60B alloy, but does not eliminate the impact toughness and elevated tensile mechanical property.

With increasing the Nd content, the discontinuous network β-Mg$_{17}$Al$_{12}$ phase begins to break and decrease gradually for AZ91D-xNd alloys. Meanwhile, the fine rod-shaped or small block compound Al$_3$Nd increases gradually and as-cast microstructure is refined slightly. Thus, the tensile mechanical property is enhanced significantly at ambient temperature owing to the dispersion strengthening and grain-refinement strengthening. When few Nd content is added (0.20%), lots of discontinuous network β-Mg$_{17}$Al$_{12}$ phase as well as few Al$_3$Nd new phase lead to a significant enhancement in the elevated tensile mechanical property. Although the gradually increasing Al$_3$Nd phase can play a role in pinning the grain boundary with the further increase of Nd content, the rapid decrease in β-Mg$_{17}$Al$_{12}$ phase can play a negative role. Therefore, the elevated tensile mechanical

335

property decreases gradually and still is higher than that of AZ91D base alloy owing to the relatively many second phase.

AM60B alloy exhibits the higher tensile mechanical property at ambient temperature owing to the dispersion strengthening resulting from the small block β-Mg$_{17}$Al$_{12}$ phase; however it does the lower elevated tensile mechanical property due to few β-Mg$_{17}$Al$_{12}$ phase. When few Nd content is added (0.18~0.84%), the significantly decreasing β-Mg$_{17}$Al$_{12}$ phase and few Al$_3$Nd phase result into decreasing the tensile mechanical property obviously. When the Nd content increases to 1.15~2.32%, the Al$_3$Nd phase increases significantly, and then the tensile mechanical property is enhanced gradually which does not still exceed that of AM60B alloy because of the lack of β-Mg$_{17}$Al$_{12}$ phase. It is seen that the morphology, size, distribution and volume fraction of second phase as well as the thermal stability can affect the tensile mechanical property of magnesium alloy at elevated temperature significantly. The second phase should exhibit the many quantity, high thermal stability and precipitation by continuous network.

Compared with the elastic deformation during the impact process, plastic deformation reflected by the crack initiation and propagation is predominant for magnesium alloy. Among the internal affecting factors, grain refinement can make the impurity concentration on grain boundary decrease, and then the brittle fracture weakens. Impediment of grain boundary is large when plastic deformation crosses from one grain to another grain. Meanwhile the change in the sliding direction can consume more energy when crack passes through grain boundary. Therefore, alloys with fine grain should exhibit high impact toughness. Compared with the matrix, second phase generally precipitating along grain boundary may become the place of crack initiation more easily owing to the un-synchronous plastic deformation. The harmful effect can be weakened effectively by few fine dispersed second phase [14]. With increasing the Nd content, the discontinuous network β-Mg$_{17}$Al$_{12}$ phase begins to break and decrease gradually and the fine rod-shaped or small block compound Al$_3$Nd increases gradually for AZ91D-xNd alloys. But, the volume fraction of the second phase decreases gradually on the whole. Thus, the impact toughness is enhanced gradually. AM60B alloy exhibits the better impact toughness owing to the few small block β-Mg$_{17}$Al$_{12}$ phase. The second phase on the whole increases gradually and the as-cast microstructure is refined slightly with increasing the Nd content. Therefore, the impact toughness first decreases significantly and then remains almost unchanged.

4. Conclusion

With increasing the Nd content, β-Mg$_{17}$Al$_{12}$ phase originally precipitating by discontinuous network or dispersed small block begins to break and decrease gradually. Meanwhile, fine rod-shaped or small block new phase Al$_3$Nd increases gradually. With increasing the Nd content, the tensile mechanical property is enhanced gradually at ambient temperature for AZ91D-xNd alloys, while it is first reduced obviously and then enhanced gradually for AM60B-xNd alloys. The maximum value of σ_b and δ can reach 225 MPa, 11.0% and 230 MPa, 16.5% for AZ91D-2.17Nd and AM60B-2.32Nd alloys, respectively. Addition of Nd can enhance the tensile mechanical property at elevated temperature for AZ91D alloy; however it cannot ameliorate that for AM60B alloy. Meanwhile, that for the former series alloys is better than that for the latter series alloys. Impact toughness is enhanced gradually for AZ91D-xNd alloys, while it first is reduced obviously and then remains almost unchanged for AM60B-xNd alloys, which is still excellent than that for the former series alloys. The morphology, size, distribution and volume fraction of second phase can affect the tensile mechanical property and impact toughness of magnesium alloy significantly.

Acknowledgments

This work was funded by the Guangdong Key Laboratory of Metal Toughening Technology and Application (2014B030301012), Guangzhou Key Laboratory of Advanced Metal Structural Materials (201509010003), High Technology Industrialization Project of Guangdong Province (2013B010102021 and 2013B010102024), Project on the Integration of Industry, Education and Research of Guangdong Province (2014B090903016), Guangdong and Hong Kong Generic Technology Bidding Project (2013B010138001), International Science and Technology Cooperation Project of Guangdong Province (2014A050503002), Hong Kong, Macao and Taiwan Science and Technology Cooperation Project (2014DFH50050) and Technology Innovation Project of Science and Technology Small and Medium Enterprise of Guangdong Province (2016A010120024).

References

1. I.J. Polmear, Magnesium alloys and applications, Mater. Sci. Technol. 10(1994) 1-14.
2. Q.D. Wang, W.Z. Chen, X.Q. Zeng, Y.Z. Lu, W.J. Ding, Y.P. Zhu, X.P. Xu, M. Mabuchi, Effects of Ca addition on the microstructure and mechanical properties of AZ91magnesium alloy, J. Mater. Sci. 36 (2001) 3035-3040.
3. X.G. Min, W.W. Du, F. Xue, Y.S. Sun, Analysis of EET on Ca increasing the melting point of $Mg_{17}Al_{12}$ phase, Chin. Sci. Bull. 47(2002) 1082-1086.
4. Y. Lou, X. Bai, L.X. Li, Effect of Sr addition on microstructure of as-cast Mg-Al-Ca alloy, Trans. Nonferrous Met. Soc. China, 21(2011) 1247-1252.
5. G.Y. Yuan, Z.L. Liu, Q.D. Wang, Y.P. Zhu, W.J. Ding, Microstructure refinement of Mg-Al-Zn-Si alloys, Mater. Lett. 56(2002) 53-58.
6. Y.Z. Lu, Q.D. Wang, X.Q. Zeng, Y.P. Zhu, W.J. Ding, Behavior of Mg-6Al-xSi alloys during solution heat treatment at 420°C, Mater. Sci. Eng. A 301(2001) 255-258.
7. K.M. Asl, A. Tari, F. Khomamizadeh, The effect of different content of Al, RE and Si element on the microstructure, mechanical and creep properties of Mg-Al alloys, Mater. Sci. Eng. A 523(2009) 1-6.
8. H. Yokobayashi, K, Kishida, H, Inui, M. Yamasaki, Y. Kawamura, Enrichment of Gd and Al atoms in the quadruple close packed planes and their in-plane long-range ordering in the long period stacking-ordered phase in the Mg-Al-Gd system, Acta Mater. 59(2011) 7287-7299.
9. H.T. Son, J.S. Lee, D.G. Kim, K. Yoshimi, K. Maruyama, Effects of samarium (Sm) additions on the microstructure and mechanical properties of as-cast and hot-extruded Mg-5 wt%Al-3 wt%Ca-based alloys, J. Alloys Compd. 473(2009) 446-452.
10. K.J. Li, Q.A. Li, X.T. Jing, J. Chen, X.Y. Zhang, Effects of Sb, Sm and Sn additions on the microstructure and mechanical properties of Mg-6Al-1.2Y-0.9Nd alloy, Rare Met. 28(2009) 516-522.
11. S.C. Zhang, B.K. Wei, Q.Z. Cai, L.S. Wang, Effect of mischmetal and yttrium on microstructures and mechanical properties of Mg-Al alloy, Trans. Nonferrous Met. Soc. China, 13(2003) 83-87.
12. J.H. Zhang, P. Yu, K. Liu, D.Q. Fang, D.X. Tang, J. Meng, Effect of substituting cerium-rich mischmetal with lanthanum on microstructure and mechanical properties of die-cast Mg-Al-RE alloys, Mater. Des. 30(2009) 2372-2378.

13. J.H. Zhang, Z. Leng, S.J. Liu, M.L. Zhang, J. Meng, Structure stability and mechanical properties of Mg-Al-based alloy modified with Y-rich and Ce-rich misch metals, J. Alloys Compd. 509(2011) L187-L193.

14. Z.H. Huang, W.J. Qi, J. Xu, Effect of microstructure on the impact toughness of magnesium alloys, Trans. Nonferrous Met. Soc. China, 22(2012) 2334-2342.

15. J.S. Liao, M. Hotta, K. Kaneko, K. Kondoh, Enhanced impact toughness of magnesium alloy by grain refinement, Scripta Mater. 61(2009) 208-211.

16. M.Vedani, Microstructural and impact toughness properties of a magnesium AM60B die cast alloy, Key Eng. Mater. 188(2000) 129-138.

17. Z.H. Huang, X.F. Guo, Z.M. Zhang, Effects of Ce on damping capacity of AZ91D magnesium alloy, Trans. Nonferrous Met. Soc. China, 14(2004) 311-315.

18. J.A. Dean, Lange's Handbook of Chemistry, 15th ed., McGraw-Hill, New York, 1999.

Activated Carbon/Polyvinyl Formal Sponge as a Carrier for Laccase Immobilization for Dyes Decolourization

Wei Ma, Bo-Kai Cao, Sai-Nan Xu, Da-Ming Chen, Yong Chen[*]

State Key Lab of Marine Resource Utilization in South China Sea; Hainan Provincial Key Laboratory of Research on Utilization of Si-Zr-Ti Resources, College of Materials Science and Chemical Engineering, Hainan University, China

[*]Email: ychen2002@163.com

The activated carbon/polyvinyl formal (AC/PVF) sponge was prepared for the immobilization of laccase, exhibiting high adsorption, good bio-affinity and recyclability. Compared to the free laccase, immobilized laccase exhibited a higher affinity to the substrate, a significantly improved stability and a better cycleability with 51% and 35% of its initial activity after 7 and 9 cycles, respectively. Thus, it is a promising method for the laccase immobilization with a high catalytic activity.

Keywords: Laccase; immobilization; AC/PVF sponges; dye waste water.

1. Introduction

Water pollution has been one of the greatest threats to human survival. Industrial waste water, especially dye waste water is the main source of water pollution. Moreover, most of the dyes are difficult to decompose and their degradation products are toxic to our environment [1]. So, many efforts were made to solve these problems and develop to various new physical, chemical, and biological technologies [2-7]; among these methods, bio-treatments have attracted significant attention due to the environmental friendless and high efficiency [8]. Laccase, a copper-containing polyphenol oxidase, can catalyze the degradation of several organic pollutants and was widely used in the treatment of industrial wastewater and the decolorization of dyes, due to the high efficiency, specificity and mild properties of enzyme-catalyzed reactions [9, 10]. Nevertheless, its applications are still limited because of the instability and poor cycle performance. The immobilization of enzymes on supports is an effective method to overcome these limitations [11, 12]. The suitable carrier materials are key factor for the enzyme immobilization. In addition to the adsorption of toxic and harmful substances because of the large surface areas and high pore volume, activated carbon is also used for loading enzymes [13, 14]. However, the use of activated carbon or laccase alone has the problems of poor recycling and possible secondary pollution. Polyvinyl formal (PVF) containing formyl and hydroxyl functional groups exhibits an excellent affinity to various enzymes [15]. Herein, the activated carbon/polyvinyl formal (AC/PVF) sponges were prepared for the laccase immobilization by combining the advantages of AC and laccase, which significantly improve the stability of enzymes in the acidic region. Congo red, which is widely presented in industry waste water, especially in the textile industry and difficult to biodegrade due to its complex constituents and chemically inertness, was selected for evaluating the effectiveness of wastewater treatment.

2. Experimental

First, 10g of PVA was dissolved in 140 mL of deionized (DI) water with stirring for 1h at 100 °C. Then starch solutions (10g of starch in 20 mL DI water), 10g of AC, 24mL of dilute H2SO4 (9%) and 8mL of a HCHO solution (40%) were successively added with strong stirring at 70 °C. After acetalization and hydrolysis reaction for 12h at 55 °C followed by thoroughly washing with DI water and dried at 60 °C, AC-PVF sponges were obtained. Fourier-transform infrared spectra (FTIR) of the pellets were then recorded on a Bruker Tensor 27 spectrometer.

AC-PVF sponges (0.1 g) were immersed in 30 mL of a laccase solution (1.5 g/L in a $C_2H_4O_2$-Na_2HPO_4 buffer solution) with stirring at 300 rpm for 12 h at room temperature followed by washing with DI water. To determine the activity of laccase, the 1.5 g/L of laccase in 1mL 1.5 g/L $C_2H_4O_2$-Na_2HPO_4 buffer solution (pH 5.5), and 0.5 mM of ABTS in 29 mL 1.5 g/L $C_2H_4O_2$-Na_2HPO_4 buffer solution (pH 5.5) under stirring for 5 min at room temperature. After incubation at 30°C for 25min, the absorbance was recorded at 420 nm on a UV-vis spectrophotometer using 1 mL deactivated laccase (1.5 g/L) under the same experimental conditions as those of the blank control group. For immobilized laccase, the experimental method was the same as that of free laccase. The activities of laccase were defined by Eq. 1 and Eq. 2.

$$Average\ activity = 1000X/t \tag{1}$$
$$Specific\ activity = X1/X2^*100\% \tag{2}$$

Here, X is the absorbance at 420 nm, t is the reaction time, X1 and X2 represent the average activities of immobilized and free laccase, respectively.

The effects of pH on the performance of immobilized laccase were investigated in solutions with pH ranging from 1.1-1.5 (KCl-HCl buffer solution) and 2.2-6.0 ($C_2H_4O_2$-Na_2HPO_4 buffer solution) at room temperature. The effects of temperature on enzymatic activity were also measured at the temperature range from 10 to 50°C at pH 4 ($C_2H_4O_2$-Na_2HPO_4 buffer solution). The decolourization was performed in a reaction mixture containing 100 mg/L of congo red in 29 mL of $C_2H_4O_2$-Na_2HPO_4 buffer solution (pH 5) and 5 mL of $C_2H_4O_2$-Na_2HPO_4 buffer solution (pH 5). In comparison, the decolourization by free laccase was also provided. The congo red concentration in solution was measured by a UV-vis spectrophotometer at 482 nm using 1 mL laccase (1.5 g/L) under the same conditions.

3. Results and Discussion

The AC-PVF sponges exhibited a 3D hierarchical porous structure with uniformly distributed pores as sown in Fig. 1a. Their specific surface area and pore volume reached to 1243.6 m^2/g and

0.69 cm³/g, respectively, according to the nitrogen adsorption/desorption measurement. The detail was described in literature [16]. The macropores not only serve as tunnels for the diffusion of dye molecules but also provide adsorption sites and storage space for dye molecules. The meso- and micro-pores play a role in adsorbing pollutant molecules.

Fig. 1b shows the FTIR spectra of immobilized laccase and AC-PVF. The functional groups of -CH₂-(2922.3 cm⁻¹ and 2861.6 cm⁻¹), O-H (3343.0 cm⁻¹), C=O (1647.1 cm⁻¹), and C-O-C (1020.3 cm⁻¹) simultaneously appeared in Fig. 1b. The presence peak of C=O (1647.1 cm⁻¹) is indicative of residual aldehyde groups after acetalization and hydroxylamine reaction. By comparison, the existence of the N-H stretching vibration (1569.9 cm⁻¹) indicated the doping of laccase in AC-PVF composites by the covalent attachment of the aldehyde and hydroxyl groups on the surface of AC-PVF and the primary amino groups of laccase. Hence, AC-PVF could be an excellent matrix for the laccase immobilization, caused by its hierarchical porous structure as well as the availability of adequate enzyme binding sites on its surface.

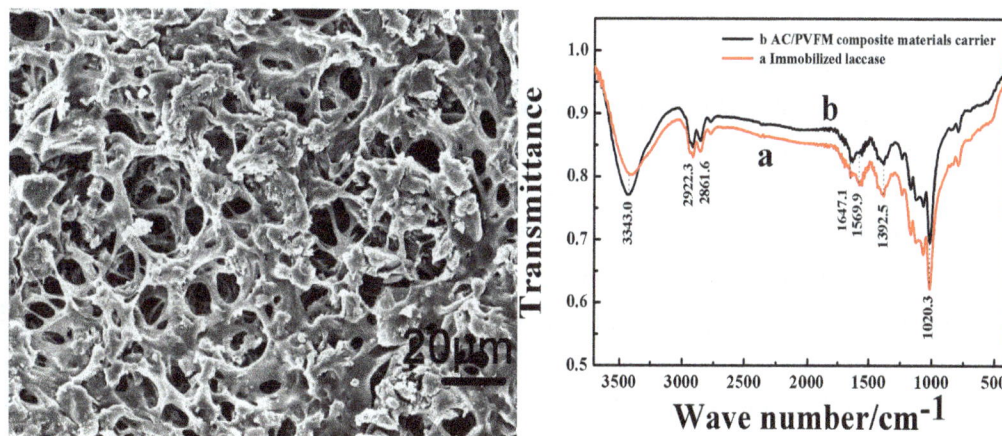

Fig. 1 (a) SEM images and (b) FTIR spectra of AC-PVF composites

Fig. 2 shows the Lineweaver-Burk plots of free and immobilized laccase. The Km value for immobilized laccase was found to be 0.34 mmol/L, far lower than that of free laccase (8.7 mmol/L). The decreased value for immobilized laccase indicates a significantly higher affinity for the substrate than that of free laccase, which is attributed to 3D diffusion channels and synergistic effect of AC and laccase.

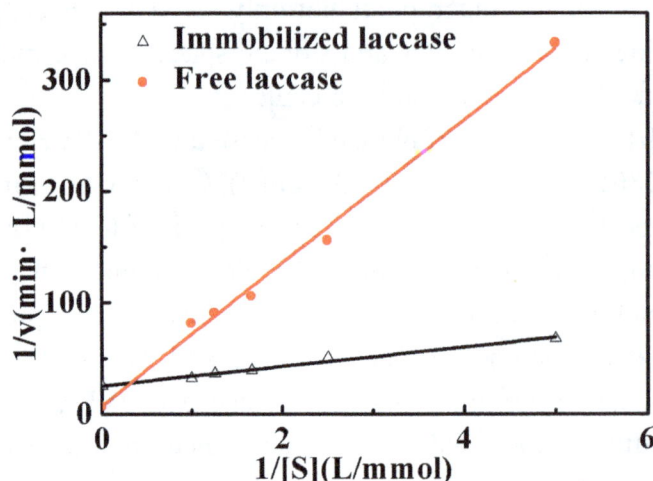

Fig. 2 Lineweaver-Burk plots of free and immobilized laccase

The effects of pH on free and immobilized laccase are shown in Fig. 3a. It was found that free laccase exhibited the maximum catalytic activity at a pH of 4. In contrast, the maximum catalytic activity of immobilized laccase was at a pH of 1.1. This shift is attributed to the fact that pH influences the molecularity of substrates and changes the dissociated state of the active center of enzymes, thereby significantly increasing the reactivity. Meanwhile, the covalent bonding between the laccase molecules and AC-PVF sponges protected the space conformation of laccase molecules and improved the stability of the enzyme against structural denaturation [3, 7, 17, 18]. Also, as shown in Fig. 3a, the activity of immobilized laccase was significantly better than that of free laccase from pH 1.1 to 6, which is ascribed to not only the adsorption and channels provided by the hierarchical porous structure, but also synergetic catalysis between laccase and AC, leading to the decrease in activation energy for the degradation [19].

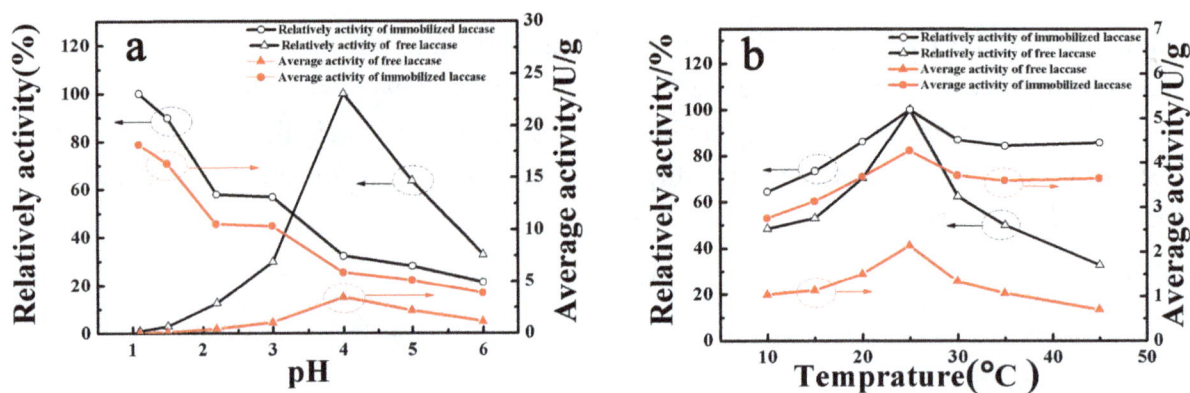

Fig. 3 (a) Effect of pH and (b) temperature on the activity of laccase

Fig. 3b shows the effect of temperature on the activity of free and immobilized laccase. For both free and immobilized laccase, when the temperature reached an optimum value, the activity of laccase increased to the maximum level. The highest activity for free laccase or immobilized laccase was achieved around 25°C. However, the thermal stability of immobilized laccase was superior to that of free laccase. This enhancement can be understood by considering the stronger bonding between laccase and the carrier, caused by the presence of AC-PVF composites, which contained containing a large amount of free hydroxyl and aldehyde groups. This bond prevented the conformational denaturation of the enzyme at high temperatures [20]. Meanwhile, laccase also maintained a significantly higher activity after immobilization.

The main advantage of the immobilized enzyme was its recyclability. As shown in Fig. 4a, the immobilized laccase retained around 51% and 35% of its initial activity after 7 and 9 cycles, respectively. Duo to the presence of micro- and meso-pores, the AC-PVF composites can absorb pollutant molecules quickly. Coupled with the catalytic role of laccase, they can simultaneously perform the adsorption and decomposition of pollutant molecules. With the increase of cycling times, pollutant molecules accumulated in the micro- and meso-pores result in the gradual decrease in the laccase activity, caused by enzyme inactivation attributed to continuous use and enzyme leaching [7], inconsistent with the previous studies [21, 22].

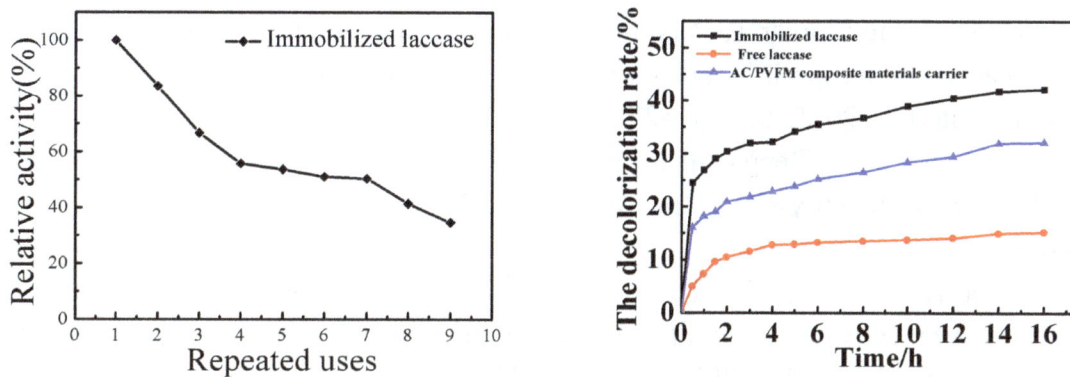

Fig. 4 (a) Recycleability and (b) decolorization rate of immobilized laccase

The decolorization rates were 15.2% for free laccase, 32.1% for laccase immobilization on AC-PVF composites, and it reached up to 42.3% after 16h (after 16h, the decolorization rates were 42.3% for laccase immobilization on AC-PVF, 32.1% for AC-PVF composites composite carrier, and 15.2% for free laccase, respectively) . The catalytic efficiency of immobilized laccase for congo red was better than that of free laccase. During degradation, immobilized laccase still exhibited a relatively high activity, indicating that the 3D hierarchical porous structure of AC-PVF composites exhibit good adsorb ability for congo red.

4. Conclusion

In this study, 3-D hierarchically porous AC-PVF composites were prepared and immobilized laccase for dye wastewater treatment. The activity reached 53.70 U/g, 3.84 times higher than that of free laccase under the same conditions. Compared to free laccase, immobilized laccase significantly improved stability and recyclability, particularly, a significantly higher activity in a strong acidic environment even at pH 1.1. The immobilized laccase retained around 51% and 35% of its initial activity after 7 and 9 cycles, respectively. The congo red removal rate by immobilized laccase reached up to 42.3%, higher than that of free laccase (15.2%).

Acknowledgments

This study was funded by the NSFC (Grant No. 51362009, and Grant No. 51162006), Key Science & Technology Project (ZDXM2015118), International Science & Technology Cooperation Program of Hainan (KJHZ2015-02) and Science and Technology Development Special Fund Project (ZY2016HN07).

References

1. Benigni R, Giuliani A, Franke R, Gruska A. Quantitative structure-activity relationships of mutagenic and carcinogenic aromatic amines. Chemical Reviews. 2000; 100(10): 3697-714.
2. Arulkumar M, Sathishkumar P, Palvannan T. Optimization of Orange G dye adsorption by activated carbon of thespesia populnea pods using response surface methodology. Journal of Hazardous Materials. 2011; 186(1): 827-34.
3. Spadaro JT, Gold MH, Renganathan V. Degradation of azo dyes by the lignin-degrading fungus Phanerochaete chrysosporium. Applied and Environmental Microbiology. 1992; 58(8): 2397-401.
4. Sathishkumar P, Arulkumar M, Palvannan T. Utilization of agro-industrial waste jatropha curcas pods as an activated carbon for the adsorption of reactive dye remazol brilliant blue R. Journal of Cleaner Production. 2012; 22(1): 67-75.
5. Kabra AN, Khandare RV, Govindwar SP. Development of a bioreactor for remediation of textile effluent and dye mixture: A plant-bacterial synergistic strategy. Water research. 2013; 47(3): 1035-48.
6. Dos Santos AB, Cervantes FJ, van Lier JB. Review paper on current technologies for decolourisation of textile wastewaters: perspectives for anaerobic biotechnology. Bioresour Technology. 2007; 98(12): 2369-85.
7. Sathishkumar P, Kamala-Kannan S, Cho M, Kim JS, Hadibarata T, Salim MR, et al. Laccase immobilization on cellulose nanofiber: The catalytic efficiency and recyclic application for simulated dye effluent treatment. Journal of Molecular Catalysis B: Enzymatic. 2014; 100:111-20.

8. Li L, Dai W, Yu P, Zhao J, Qu Y. Decolorisation of synthetic dyes by crude laccase fromRigidoporus lignosusW1. Journal of Chemical Technology & Biotechnology. 2009; 84(3):399-404.

9. Liu Y, Zeng Z, Zeng G, Tang L, Pang Y, Li Z, et al. Immobilization of laccase on magnetic bimodal mesoporous carbon and the application in the removal of phenolic compounds. Bioresour Technology. 2012; 115:21-6.

10. Chivukula M, Renganathan V. Phenolic azo dye oxidation by laccase from pyricularia oryzae. Applied and Environmental Microbiology. 1995; 61(12): 4374-7.

11. Sathishkumar P, Chae JC, Unnithan AR, Palvannan T, Kim HY, Lee KJ, et al. Laccase-poly(lactic-co-glycolic acid) (PLGA) nanofiber: Highly stable, reusable, and efficacious for the transformation of diclofenac. Enzyme and Microbial Technology. 2012; 51(2): 113-8.

12. Cristóvão RO, Tavares APM, Brígida AI, Loureiro JM, Boaventura RAR, Macedo EA, et al. Immobilization of commercial laccase onto green coconut fiber by adsorption and its application for reactive textile dyes degradation. Journal of Molecular Catalysis B: Enzymatic. 2011; 72(1-2): 6-12.

13. Karnib M, Kabbani A, Holail H, Olama Z. Heavy metals removal using activated carbon, silica and silica activated carbon composite. Energy Procedia. 2014; 50:113-20.

14. Durán-Jiménez G, Hernández-Montoya V, Montes-Morán MA, Bonilla-Petriciolet A, Rangel-Vázquez NA. Adsorption of dyes with different molecular properties on activated carbons prepared from lignocellulosic wastes by taguchi method. Microporous and Mesoporous Materials. 2014; 199: 99-107.

15. Sharma SK, Suman, Pundir CS, Sehgal N, Kumar A. Galactose sensor based on galactose oxidase immobilized in polyvinyl formal. Sensors and Actuators B: Chemical. 2006; 119(1):15-9.

16. Ma W, Xu S, Chen K, Guo Y, Chen Y. Preparation and oil absorption performance of sponge-like activated carbon/organic composites. New Carbon Materials, 2015, 3(5): 425-431.

17. Xu R, Zhou Q, Li F, Zhang B. Laccase immobilization on chitosan/poly (vinyl alcohol) composite nanofibrous membranes for 2,4-dichlorophenol removal. Chemical Engineering Journal. 2013; 222: 321-9.

18. Secula MS, Suditu GD, Poulios I, Cojocaru C, Cretescu I. Response surface optimization of the photocatalytic decolorization of a simulated dyestuff effluent. Chemical Engineering Journal. 2008; 141(1-3):18-26.

19. Calvino-Casilda V, López-Peinado AJ, Durán-Valle CJ, Martín-Aranda RM. Last decade of research on activated carbons as catalytic support in chemical processes. Catalysis Reviews. 2010; 52(3): 325-80.

20. Osma JF, Toca-Herrera JL, Rodríguez-Couto S. Biodegradation of a simulated textile effluent by immobilised-coated laccase in laboratory-scale reactors. Applied Catalysis A: General. 2010; 373(1-2): 147-53.

21. Areskogh D, Henriksson G. Immobilisation of laccase for polymerisation of commercial lignosulphonates. Process Biochemistry. 2011; 46(5):1071-5.

22. Da Silva AM, Tavares APM, Rocha CMR, Cristóvão RO, Teixeira JA, Macedo EA. Immobilization of commercial laccase on spent grain. Process Biochem. 2012; 47(7): 1095-101.

Thermal Property of Waterborne Ultrathin Vac-veova Latex Intumescent Fire Retardant Coatings for Structural Steel

Yun-Chun Xia

School of Civil Engineering, Anhui Jianzhu University
No.856, Jinzhai South Road, Hefei City, Anhui Province, 230022, P.R. China
wxiayc@126.com

The Vac-veova latex coatings was a good type of film-forming material in intumescent fire retardant coatings, it had good properties in high thermal stability, so it was very suitable to serve as the film-forming material of IFR coatings for structural steel to improve its fire resistance. According to the research results, the different ratio between each component had obvious difference for fire prevention effect, in the APP/PER/MEL/EG for IFR coatings, the best ratio of APP/PER/MEL/EG was 5.25:2:2:1.25, and the best content ratio between IFR and Vac-veova latex emulsion was 35/25. According to the chosen component ratio in IFR, its intumescent ratio could reach 22.68%, the fire-resistant time reached 149s. After burnt out, the residual weight reached 47.12% at 75kW/m².

Keywords: Thermal property; waterborne; ultrathin; Vac-veova latex coatings.

1. Introduction

In recent years, structural steel is widely used for many buildings, but it is poor for fire resistance. To protect it from damage in fire, fire retardant coatings is one of the most effective protection methods. Solvent-type ultrathin fire retardant coatings for structural steel contain a large amount of VOC and other toxicities, and they are easily polluted in environment [1, 2]. Waterborne ultra-thin fire retardant coatings, which based on waterborne polymer as film-forming and water as dispersion medium, are friendly to environment. It has the advantages in little VOC, low energy consumption and less harm to human body [3]. Compared with the other types of fire retardant coatings, it can meet with the need of efficient fireproof and building decorative appearance [7, 8]. Therefore, the ultrathin fire retardant coatings of structural steel have the properties of green environmental protection.

2. Experimental Reagents and Emulsion Preparation

2.1. *Experimental reagents*

Vinyl acetate-vinyl ester of versatic co-polymer emulsion (called Vac-veoVa emulsion, solid content ≥50%, made by myself). APP (its polymerization degree n≥1000), PER (weight content 98%), MEL (weight content 99%), EG (400meshes), TiO_2, chlorinatedparaffin and other auxiliaries.

2.2. *Basic recipes*

The basic recipe mainly included vinyl acetate-vinyl ester of versatic acid (Vac-Veova) emulsion, it was 25~30wt% in coatings. For the intumescent fire retardant, IFR included APP, MEL, PER and EG, it was 30~35wt%. In IFR, the content ratio of APP/MEL/PER/EG was 5:2:2:1.25. Wetting dispersant was DP-18, it was 0.2~0.5wt%, water was 18~22wt%, methyl silicone oil was taken as the de-foaming agent, it was 0.4~0.5wt%, film-forming agent propylene glycol phenyl was 1.2~1.5wt%. The plastcizier chlorinatedparaffin (CP-52) was 3wt%. The compound pigments and fillers included 0.5~1.0wt% Nano-ZrO_2, 8~10wt%TiO_2 and 2.5~3.0wt% ZnO, the de-foaming agent was the compound of Tributyl phosphate and NOPC8034L, and their ratio was 1:4 in weight, it was 0.2~0.3wt% in IFR [4]. Leveling agent BMC-1001S was 0.3~0.4wt%, and the wetting dispersant Bermawet-1000 was 0.2~0.3wt%. The coupling agent TMC-114 was 0.1wt%, and the reinforcer phenolic fiber was 0.05~0.1wt%, antifreeze ethanedio was 0.1~0.2wt%, stabilizer PVA was 0.2~0.3wt%, coalescent DBG was 4~5wt%, and the other auxiliaries was 0.5~1wt%.

3. Preparations of Coatings and Experimental Steel Samples

Pre-Treatment for Raw Materials:

For powder materials, they should be grinded, and then they were sieved by 200meshes standard sieve shaker for ensuring the uniformity in coatings.

Preparation of Vac-Veova Latex Film:

200g Vac-veova emulsion was put into a PTFE vessel, dried and filmed for 3 days at ambient temperature, then it was dried in an oven at 60°C for 2 days, Vac-Veova latex film was obtained [5].

Modification for Water Dispersion of Nano-ZrO2:

The modified surface of nano-ZrO_2 was dispersed in water, 0.4% isopropanol was added into it, dispatching the pH-value to 9~10 with strong aqua ammonia, and it was dispersed by a 200W ultrasonic cleaner for 1 hour, the water dispersion of nano-ZrO_2 was obtained [6].

Batching and Pre-Mixing:

The modified nano-ZrO_2 was put into Vac-Veova latex emulsion, which was dissolved in Vac-Veova latex film by methanol. At first, stirring and mixing it at 60 rpm for 5 minutes, and then keep 30 rpm stirring for 10 minutes, after that, they were put into the grinding vessel. According to the formula, APP, PER, MEL, TiO_2, chlorinatedparaffin and some other auxiliaries were added in order, then stirring and dispersing at 60 rpm for 20 minutes, the pre-compound was obtained. During the pre-mixing, water was put by twice, the first was 2/3, after all of the powders were premixed, the rest water could be added, and the compound consistency should be adjusted.

Grinding and Dispersing:

After wetting dispersant was put into the premixed mixture and grinding the premixed mixture for 60 to 90 minutes at high-speed. Then EG was put into the premixed mixture at the late stage, stirring the mixture at low-speed (20 rpm) for 5~10 minutes so that the mixture mixing in well.

Diluting and De-foaming:

Some water and de-foaming agent were put into the pre-mixture, stirring at 30rpm until no bubbles discharged. Adding thickener to adjust its viscosity to 4000~5000cP.s, the waterborne ultrathin Vac-Veova latexes fire retardant coatings (WUVLFRC) for structural steel product was obtained after a series filtration.

4. Fluid Deformation Capacity

Dynamic rheological double logarithmic curves of a polymer Vac-Veova latex film at 200°C was showed in Fig. 1. The melt complex of inorganic filler/Vac-Veova steady shear viscosity was tested at 20°C, the range of shearing rate was from 10 s^{-1} to 5000 s^{-1}, and the capillary was 1mm in diameter and length of 30mm.

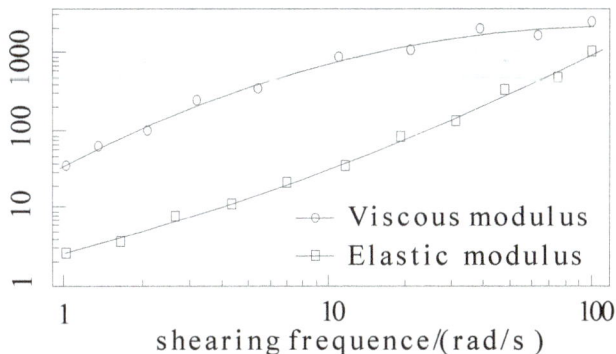

Fig. 1 Elastic and viscous modulus

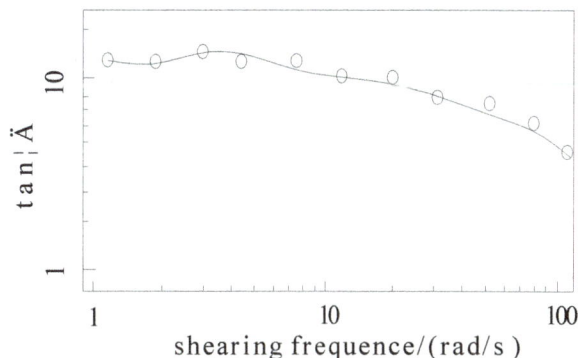

Fig. 2 Loss factor (tanδ) at 20•

The elasticity and viscous modulus increased when the shearing frequency increased. The melt polymer was main the elastic flow when the elasticity modulus was larger than its viscous modulus. However, the melt polymer was main the viscous flow when the viscous modulus was larger than its elasticity modulus.

5. Combustion Property

The photos of WUVLFRC before and after combustion were shown in Fig. 3, and its SEM was shown in Fig. 4.

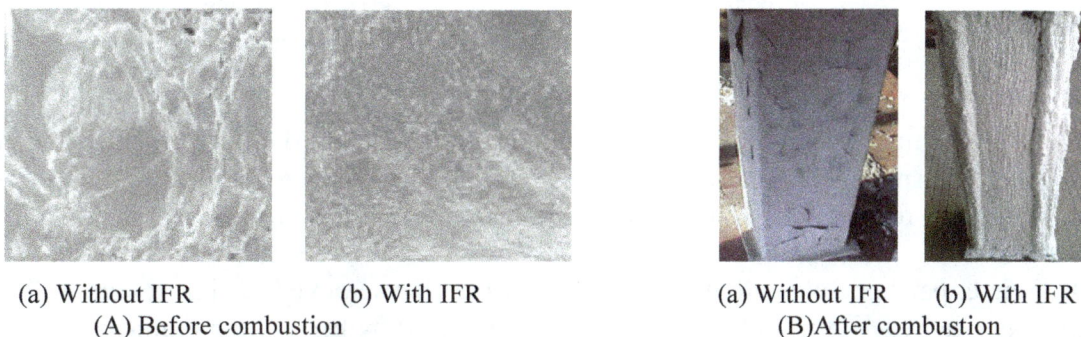

(a) Without IFR   (b) With IFR
(A) Before combustion

(a) Without IFR   (b) With IFR
(B) After combustion

Fig. 3 The photos of before and after combustion

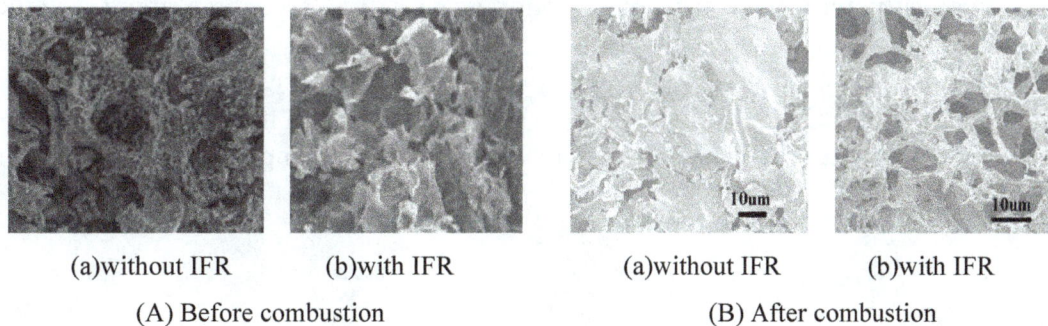

(a) without IFR   (b) with IFR

(a) without IFR   (b) with IFR

(A) Before combustion   (B) After combustion

Fig. 4 SEM of before and after combustion (×1000)

When with no IFR, the coatings cracked at high temperature, and the small particles fell off, and they had the structural defects on perforations. For the coatings with IFR, it mainly composed of bulky and particulate residues, its expansion layer surface formed a lot of closed cell, expanded fireproof, it had no perforations and cracks, and the expansion layer was dense, it was conducive barrier to the external heat and oxygen, and it could prevent the internal organics from being oxidized, it played a good role in protection for structural steel in fire.

6. Fire Resistance of WUVLFRC

In order to verify the performances of the coatings, both the made specimens of square-type and round-type structural steel were tested in fire.

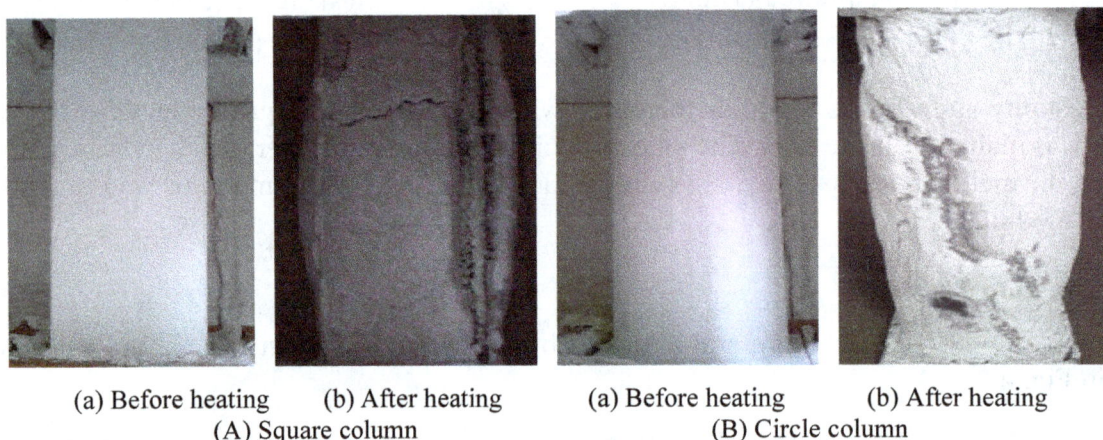

(a) Before heating   (b) After heating   (a) Before heating   (b) After heating
(A) Square column   (B) Circle column

Fig. 5 Photos of test results

In the test, the temperature of heating-up fit for ISO-834 Standard, and controlled temperature specification of test specimen were: For, when its temperature reached 650°C at single-point or the average temperature reached 600°C, the test stopped heating. The test results were shown in Fig. 5.

In test, their expansion rates were over 40 times, but they cracked during foaming. Their fire resistance time reached 67.1 minutes. In addition, when acted by fire at high temperature, the crack was different between a square and a round specimen structural steel with the expansion coatings, the crack on square structural steel was smaller, but it was relative larger on a round structural steel. However, there was little difference between their expansion ratios.

For the coatings, when it was analyzed by DSC/TG analyzer in air atmosphere and its temperature was from 600°C to 800°C, the heating rate was 10°C/min, the TG and DTG were shown in Fig. 6. When heating rate was 75kW/m^2 on CONE, its backside temperature was shown in Fig. 7.

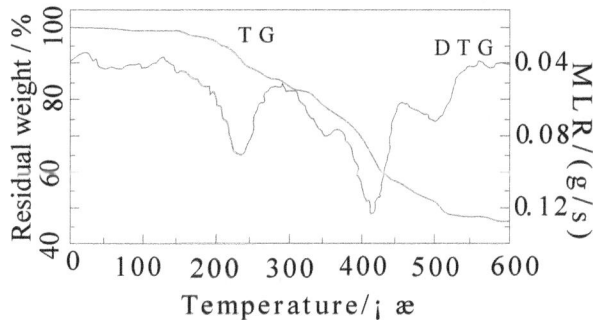

Fig. 6 TG and TDG

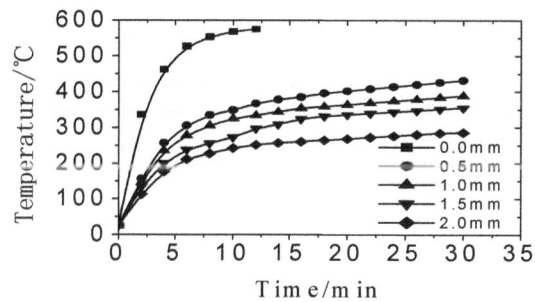

Fig. 7 Back temperature at different thickness

According to Fig. 6, the thermal weight-loss mainly included the next four stages:

1) Stable heated stage. Thermo-gravimetric curve was from room temperature to 210°C, weight loss was about 5%. It mainly volatilized the additives and other small molecule volatile compounds.

2) Foaming expansion and carbonization stage. It was divided two stages of weight-loss, the first was from 210°C to 280°C, and the second was from 280°C to 460°C, its total weight-loss was approximately 40%. After 280°C, coatings started to expand. The heated chlorinated paraffin, APP and TiO_2 started to prolyse, and it released HCl, NH_3 and steam, they made the molten coatings start to swell.

3) Loss of carbon. Temperature was from 460°C to 750°C, weight loss was about 8%. The prolysed products of APP further reacted with TiO_2 and released steam, burning charcoal was oxidized to CO_2 and escaped from the carbonization layer.

4) Inorganic layer. Temperature was from 750°C to 900°C, intumescent layer remained about 47wt% inorganic skeleton, it was with loose structure and poor adhesion, but it plays a very important role in fireproof and heat insulation. In Fig. 7, when the brushed thickness of coatings was only 2.0mm, the backside temperature of the coatings was 276°C.

The prolysis characteristic temperature of Vac-Veova latex film included the temperature ($T_{-5\%}$) weight-loss at 5%, the temperature of weight-loss at 50% ($T_{-50\%}$) and weight-loss at 90% ($T_{-90\%}$). According to Table 1, $T_{-5\%}$ was at 292°C, $T_{-90\%}$ was at 508°C, the difference between weightloss start temperature and thermal weightloss end temperature was at 216°C. For Vac-Veova latex fire retardant coatings, the temperature $T_{-5\%}$ was at 237°C, and $T_{-90\%}$ was at 486°C, their difference was 249°C. That showed that Vac-Veova latex coatings had a smaller average prolysis rate and higher thermal stability. The prolysis process of Vac-Veova latex film was divided into three stages. Unsaturated olefin carbon chain and the later carbonization products made the Vac-Veova latex film appear a turning on TG at 415°C, within the range of 400~550°C, there was a higher thermal stability. $T_{-50\%}$ of Vac-Veova latex film was only 347°C, but $T_{-50\%}$ of Vac-Veova latex coatings reached 486°C.

Table 1 TG analyses of polymer film and fire retardant coatings

material	$T_{-5\%}$/°C	$T_{-50\%}$/°C	$T_{-90\%}$/°C	Residual weight/% (600°C)
polymer film	292	347	508	0.15
fire retardant coatings	237	486	/	47.12

According to Table 2, the best content ratio between intumescent fire retardant (IFR) and Vac-veova latex emulsion in WUVLFRC was 35/25.

Table 2 Steel backside temperature and intumescent ration of coatings

IFR/ vac veova	backside temperature/°C	intumescent ratio	Fire- resistant time	feature of intumescent layer
50/10	442	22.15	102	surface was cracking off
45/15	278	21.23	116	particles was obvious
40/20	272	22.74	142	particles was a little obvious on surface
35/25	256	22.68	149	surface was dense
30/30	281	18.56	124	surface was dense
25/35	325	16.12	76	surface was dense
20/40	427	11.03	32	surface was dense

Table 3 Fire retardant contents in orthogonal experiment

No.	APP/wt%	PER/wt%	MEL/wt%	EG/wt%
A1	16	8	8	3
A2	18	8	4	3
A3	18	6.5	6.5	4
A4	20	6	6	3
A5	20	8	4	3
A6	20	6	4	5
A7	22	5	5	3
A8	22	5	3	5
A9	25	4	4	3

Table 4 The orthogonal results of intumescent fire retardant system

No.	backside temperature/°C	intumescent ration	consistency	structure strength /MPa	Burning time /s
A1	12	7	9	7	21
A2	11	8	8	8	16
A3	18	9	10	10	6
A4	9	7	8	8	15
A5	7	5	6	7	21
A6	11	7	8	9	25
A7	8	9	9	8	29
A8	5	4	5	4	31
A9	3	3	3	3	32
average1	26.32	27.36	21.56	15.78	
average2	24.35	29.54	23.45	26.12	
average3	12.63	28.75	20.76.67	23.31	
range	13.68	18.23	2.36	11.15	

When the testing indexes were the backside temperature of steel plate, the expansion ratio of coatings layer, the internal dense of intumescent layer and the structure strength of intumescent layer to determine the comprehensive performance of fire retardant coatings, which could reach for fire tests after 30 minutes. Compared all the parameters, the best internal dense of intumescent layer and the highest structure strength of intumescent layer were 10.

The influence order of experimental factors on the properties of fire retardant coatings was APP>EG>PER>MEL. The best intumescent fire retardant system could be obtained from the average values of the experimental factors: factor 1 was at level 2, factor2 was at level 1, factor 3 was at level 2, and factor 4 was at level 2. So the best proportion for APP: PER: MEL: EG = 5.25:2:2:1.25, it was the same as that of the chosen experimental formula.

7. FTIR Spectra of Burnt Residuals

The FTIR spectrum of burnt residue of WUVLFRC was shown in Fig. 8.

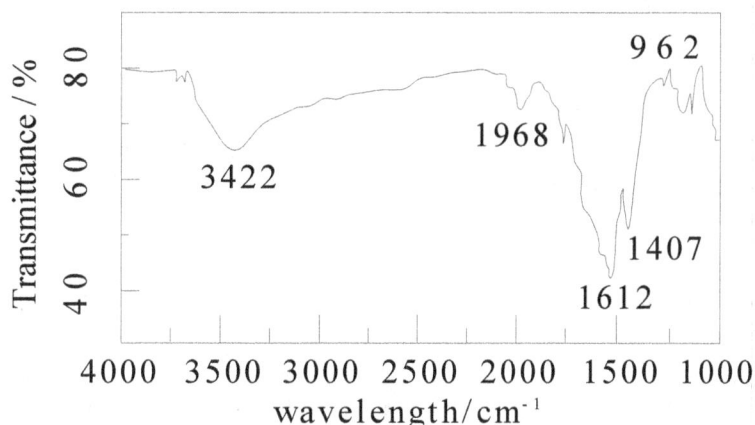

Fig. 8 FTIR spectra of residue of WUVLFRC

The Characteristic peak in Fig. 8 was mainly belong to vac-veova polymers, such as APP, PER, MEL, $1968cm^{-1}$ C=O and $1612cm^{-1}$ C-O-C characteristic peak abated, vac-veova polymer began to pyrolyse. In position of $1407cm^{-1}$, P-O-C structure compounds appeared. P-O-C keys were produced by dehydration reaction of APP and PER, de-amination reaction of APP and MEL and the removed side chain reaction of APP with vac-veova polymers. At the spectra of $3422cm^{-1}$, C-H characteristic peak appeared, the fireproof coatings had begun to produce hydrocarbons. At $962cm^{-1}$, Ti-O characteristic peak was significant, that was the intumescent layer mainly composed of TiO_2 and other inorganic substances. $3420cm^{-1}$ was its water spectra in the residue when it absorbed from air.

8. Conclusion

Waterborne ultrathin Vac-Veova latex IFR coatings for structural steel were a new type of environmental, friendly and high efficient heat insulation coatings. The prepared Vac-Veova latex IFR coatings had the high intumescent ratio (22.68). In fire, its intumescent layer was complete, no punch and cracking off. The coatings with Vac-Veova latex emulsion as film-forming material was better than other types of film-forming materials in IFR coatings in fire prevention and physical properties. In the above APP/PER/MEL/EG for IFR system, the different ratio between each component had obvious difference for fire prevention. For the compound of APP/PER/MEL/EG intumescent system, the best ratio of APP/PER/MEL/EG was 5.25:2:2:1.25.

Acknowledgment

This work was funded by the National Natural Science Foundation of China (Grant No. 51478002).

References

1. Duquesne S, Intumescent paints: fire protective coatings for metallic substrates, Surface & Coatings Technology. 180-181 (2004) 302-307.
2. Duquesne S., Magnet S. and Jama C, Thermoplastic resins for thin film intumescent coatings towards a better understanding of their effect on intumescent efficiency, Polymer Degradation and Stability. 88 (2005) 63-69.
3. Wang Z. Y., Han E. H. and Ke W., Effect of acrylic polymer and nanocomposite with nano-SiO_2 on thermal degradation and fire resistance of APP-DPER-MEL coating, Polymer Degradation and Stability. 91 (2006) 1937-1947.
4. Delobel R., Thermal behaviors of ammonium polyphosphate-pentaerythritol and ammonium pyrophosphate-pentaerythritol intumescent additives in polypropylene formulations, Journal of Fire Sciences. 8 (1990) 85-108.
5. Bourbigot S., Le Bras M., Delobel R, Synergistic effect of zeolite in an intumescence process: study of the carbonaceous structures using solid-state NMR, Journal of the Chemical Society, Faraday Transactions. 92 (1996) 3435-3444.
6. Bourgeat-Lami, E., Espiard, Ph., Guyot, A., Poly (ethyl acrylate) latexes encapsulating nanoparticles of silica: 1. Functionalization and dispersion of silica, Polymer. 36 (1995) 4385-4389.
7. Wang G. A., Cheng W. M., Tu Y. L., Characterizations of a new flame-retardant polymer, Polymer Degradation and Stability. 91 (2006) 3344-3353.
8. Chen X. L., Hu Y., Jiao C. M., Preparation and thermal properties of a novel flame-retardant coating, Polymer Degradation and Stability. 92 (2007) 1141-1150.

Effects of Superfine Grinding Method on Some Physicochemical Properties of Mung Bean Starch

Gui-Xiang Zhang[1], Zheng-Hong Hao[2,*], Bing-Wen Zhang[1]

[1] Department of food science and nutrition, Business school, University of Jinan, Jinan, China
[2] Department of food science and engineering, Shandong Agriculture Engineering College, Jinan, China
*Email: zhenghao227@163.com

Four types of Mung bean starch micronized powders were prepared by pulverizing its crude powder through superfine grinding for different periods of time (10, 20, 30 and 40 min). The physicochemical properties of the micronized powders and the crude powder were then compared and investigated. Particle sizes, granule morphology, bulk density, water solubility, water holding capacity and starch retro gradation degree were studied in this article. With increased superfine grinding time, the particle size increased which was different from other reports. The results of granule morphology proved this point. Bulk density decreased compared with controls. Water solubility and water holding capacity increased after superfine grinding. As for retro gradation degree, it gradually increased with the extension of time and the maximum value was attained at 4~10h for micronized starch which was significantly different from the controls. Superfine grinding of Mung bean starch had some significant characteristics and might be of potential use in the food industry.

Keywords: Superfine grinding; crude mung bean starch; mung bean superfine powders; physical property.

1. Introduction

As one of the most commonly consumed food legumes, mung beans (*vigna radiate L.*) have been used as noodles, starch gels, flour pastes and ground-up mung beans in China. Starch is the main component (about 56.9%) of mung bean which plays an important role in its quality.

Although native mung bean starch is an excellent raw material in food industry, it possesses some undesirable properties limiting its use [1]. Its poor solubility in cold water and strong texture stability makes it difficult to use widely in food and bio-medical applications. Chemical modifications are one of the most common techniques performed by oxidation, esterification, etherification, acid hydrolysis and cross-linking [1, 2] to improve its properties. Physical methods such as heat-treatment, microwave radiation, heat-moisture treatment are also applied in other studies [3].

Superfine grinding technology is a new type of food processing which has shown outstanding properties such as high solubility, dispersion, adsorption, chemical activity and fluidity [2, 4]. It had been considered that superfine grinding had the functions to decrease particle size and improve reactive surface to the greatest extent possible, and it had been considered to be less energy consuming than traditional mechanical grinding with respect to the increase of surface area [5, 6].

Nowadays, superfine grinding technology has become the focus of research conducted in many countries [7] reported in fiber [8] and pharmaceutics [9] as well as foodstuffs [10]. It has shown a high potential for many other commercial applications. However, little information is available regarding of superfine grinding treatment on mung bean starch.

Therefore, the present work was carried out to investigate the superfine grinding technology on crude mung bean starch (CMBS) and to obtain mung bean superfine powders (MBSP). The relationship of superfine time with powder size was discussed in this article. Moreover, some basic physical properties of MBSP including water solubility, water holding capacity, and starch retrogradation degree were also studied and compared to CMBS in this paper.

*Corresponding author

2. Materials and Methods

2.1. *Sample preparation*

CMBS was kindly supplied by Yantai Shuangta Food Co. Ltd., Yantai, Shandong Province, and China. CMBS was ground to superfine powder at room temperature for 10, 20, 30 and 40 min to obtain four different superfine powders, which were named as MBSP-A, MBSP-B, MBSP-C and MBSP-D in this work. The color of all the samples was white.

2.2. *Chemicals*

All chemicals used in this research were of analytic reagent.

2.3. *Particle size analysis*

The particle size distribution of crude powder and four superfine powders were determined by a Laser particle size analyzer (Beckmann, LS13320, USA) in 95% ethyl alcohol suspension.

2.4. *Granule morphology*

The surface morphology of CMBS and MBSP was observed using a Scanning Electron Micrographs (SEM; Hitachi, Japan). The samples were sprinkled on double-sided tape, fixed to an aluminum stub, and then coated with gold. The images were taken at an accelerating voltage of 10 Kv.

2.5. *Bulk density*

The bulk density (g/ml) was determined as Zhao et al [11]. Different types of powders were filled in a 10 ml volumetric flask (W_1) up to the mark and were weighed (W_2) separately. The bulk density of powders (d_0) was calculated as follows:

$$d_0 = (W_2 - W_1)/10 \qquad (1)$$

Where W_2 was the total weight of the mung bean powder and flask, and W_1 was the weight of the flask only. The experiments were repeated five times and the measurement of each sample was repeated three times.

2.6. *Water holding capacity (WHC)*

Water holding capacity was determined in triplicate by adopting the method of Anderson [12]. Firstly, the weights of cleaned centrifuge tubes (M) were measured, and 0.5g (M_1) of each sample was poured into it. Water (M_2) was added to disperse the powder with a powder/water ratio of 0.05/1 (w/w) at ambient temperature. The dispersion was incubated in a water bath at 60°C for 10, 20, 30, 40 and 50 min separately and then they were placed in cold water for 30 min, followed by centrifugation for 15 min at 4000 r/min. The supernatant liquid was removed and the centrifuge tubes with the powders (M_3) were weighed again. The formula to calculate water holding capacity (WHC) is as follows:

$$WHC \ (g/g) = (M_3 - M - M_1)/M_1 \qquad (2)$$

2.7. *Water solubility (WS)*

Water solubility was determined in triplicate by adopting the method of Zhao et al [11] with a slight modification. Starch sample (1.00g, W_1) was added to a centrifuge tube, then 20 ml of distilled water was added and heated at 50°C for 30 min with continuous stirring, cooled to room temperature in ice water and centrifuged at 3000 rpm for 20min. The supernatant was dried in an evaporating dish in water bath, then oven-dried at 105°C until constant weight (W_2) was obtained. This procedure was repeated at 60, 70, 80 and 90°C using the above-described protocol. All tests were repeated three times. Water solubility (WS) was determined by the following formula:

$$WS\% = (W_2/W_1) \times 100 \qquad (3)$$

2.8. Starch retrogradation degree

1g (M) of samples weighed accurately was added into 100 ml of distilled water, heated to thorough gelatinization using a home-made induction cooker and then cooled to room temperature. The volume of mixture was tested and adjusted to 100ml by adding distilled water, shook well and placed at room temperature. The bed volume of the sample (V_1) was recorded after every 1 hour. The degree of starch retrogradation (RD) was calculated as follows:

$$RD\% = V_1/100 \qquad (4)$$

3. Results and Discussion

3.1. Results of particle size analysis

The particle size distribution of CMBS and MBSP obtained by laser particle size analyzer was shown in Fig. 1.

It showed that superfine grinding couldn't reduce the size of mung bean starch. In contrast, the mean size of MBSP-A, B, C and D was increased with the extension of time. At the same time, the SD value was enhanced significantly along with longer superfine grinding time. This maybe because of the influence of superfine movements and the concentration of inner energy, some molecular particles was reformatted and assembled together, and could also be due to the expression of surface forces, namely van der Waals, magnetic, and electrostatic forces, leading to the agglomerate phenomenon of the powder. It is reported that by using superfine grinding technology, a narrower and more uniform particle size distribution could be obtained in green tea powders, silver carp bone powder and ginger powder et al. [2, 4]. It is just different in mung bean starch in this study.

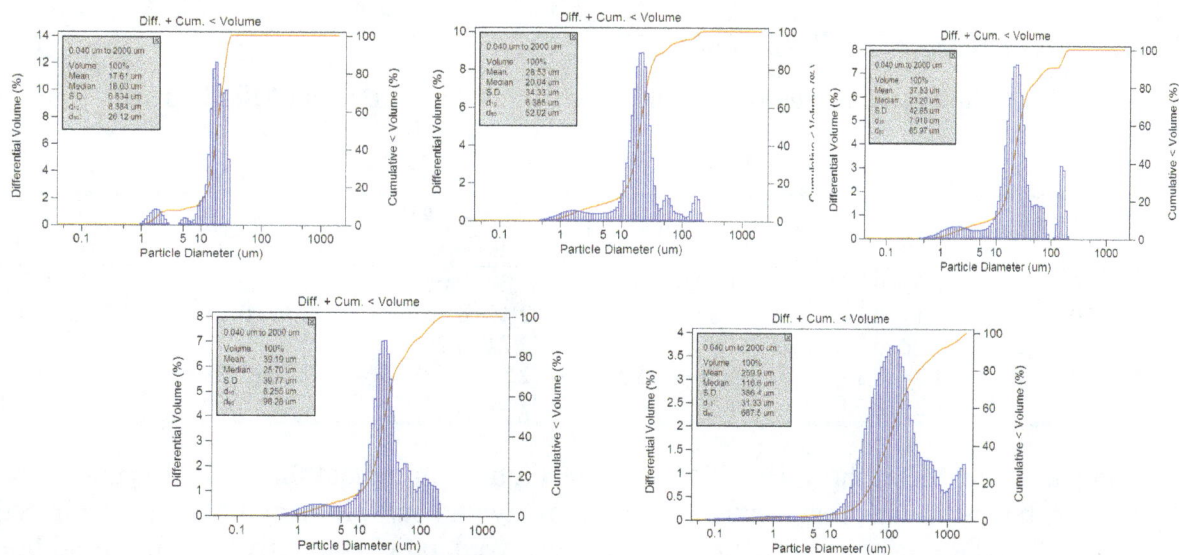

Fig. 1 The particle size distribution of CMBS and MBSP

3.2. Microphotographs of CMBS and MBSP

The microphotographs showed the morphology of fragmented mung bean starch granules (Fig. 2). As shown, with the longer superfine grinding time on the mung bean starch, the lager particle was obtained; the effect of superfine grinding time on mung bean starch particle size was negative. It was possibly attributed to the particle aggregation as revealed from the SEM images. Mechanical damage was a transformation from an ordered to a disordered structure via the breakage of intermolecular bonds. At the start, the particle of CMBS was dispersed and smooth. When broken into MBSP, the combination of flattening and aggregation resulted in various shapes of mung bean particles, as could be seen in Fig. 2. After milling, mung bean starch became rough and the particle size was increased considerably, which was accordant with laser particle size analyst.

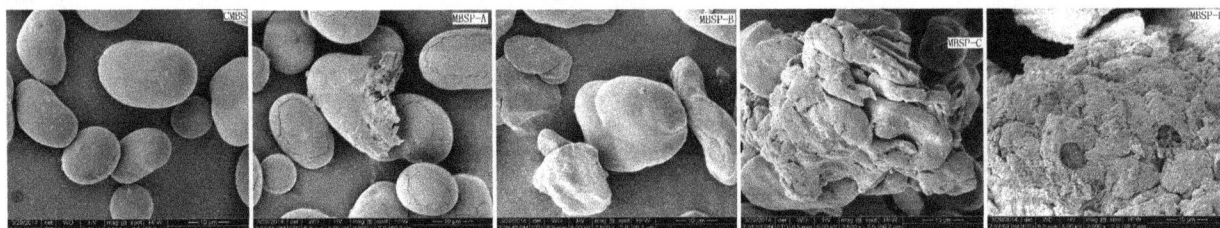

Fig. 2 SEM images of different sized mung bean starch

3.3. Results of bulk density

The bulk density of the mung bean starch fractions ranged from 5.44×10^{-2} to 6.07×10^{-2} g/ml seen in Table 1. The bulk density of MBSP with a larger particle size (Figs. 1 and 2) was smaller than CMBS. There was no very significant difference ($p < 0.05$) in bulk density among MBSP with the different size compared to CMBS.

Table 1 Bulk Density of CMBS and MBSP

Starch	CMBS	MBSP-A	MBSP-B	MBSP-C	MBSP-D
Bulk density ($\times 10^{-2}$ g/mL)	6.07±0.03	6.06±0.15	5.88±0.12	5.44±0.03	5.87±0.05

3.4. Results of WHC of CMBS and MBSP

The effect of temperature on the water holding capacity of CMBS and MBSP is shown in Table 2.

Table 2 WHC of CMBS and MBSP

Time(min)	WHC(g/g)				
	CMBS	MBSP-A	MBSP-B	MBSP-C	MBSP-D
10	1.03±0.08	1.59±0.06	1.87±0.05	2.26±0.07	2.39±0.06
20	1.17±0.14	1.84±0.11	2.17±0.15	2.61±0.13	2.51±0.18
30	1.28±0.21	1.92±0.24	2.21±0.28	2.64±0.22	2.65±0.26
40	1.34±0.15	1.86±0.28	2.61±0.31	2.80±0.25	2.97±0.21
50	1.33±0.32	1.93±0.36	2.40±0.23	3.02±0.19	3.24±0.31

Compared to CMBS, the WHC of MBSP was higher throughout the studied temperature range. This might be due to the great changes of morphology and particle size of MBSP which lead to the increase of surface energy, specific surface area, void ratio and active point .In addition, the disruption of starch crystal lattice structure and the dissociation of double helical structure promote the bound of water molecules and free hydroxyl of starch molecules greatly and therefore ,the granule could swell to the maximum capacity.

3.5. Results of water solubility (WS)

Water solubility of CMBS and MBSP are summarized in Table 3.

Table 3 WS of CMBS and MBSP

Temperature(°C)	WS(%)				
	CMBS	MBSP-A	MBSP-B	MBSP-C	MBSP-D
50	18.12±0.98	19.11±0.86	21.27±1.06	28.16±0.99	36.23±1.01
60	19.14±0.67	19.22±0.73	21.22±1.01	28.43±0.87	42.12±0.69
70	19.34±0.56	19.57±0.31	22.36±0.85	30.44±0.67	41.32±0.59
80	20.21±0.46	20.57±0.23	23.38±0.49	30.12±0.34	42.44±0.67
90	20.67±0.35	20.76±0.56	23.67±0.45	30.34±0.66	41.89±0.86

From the results, we can see that water solubility of the MBSP increased compared to CMBS. Increment in water solubility was due to the fact that MBSP had the minimal relative molecular weight and bigger superficial area, which increased in the course of the water sucking as a result of easier water absorption .On the other hand, the presence of holes and channels inside MBSP starch enhances the binding of starch molecular and water molecular, thus increasing hydrophilicity of the starch.

3.6. Results of retrogradation degree (RD)

The retrogradation degree (RD) of CMBS and MBSP were studied and the results are listed in Fig. 3.

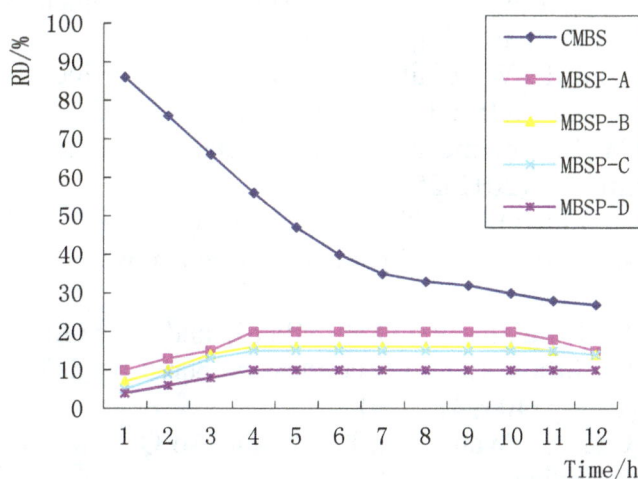

Fig. 3 Results of Retro gradation Degree(RD) of CMBS and MBSP

The results clearly show that there is a sharp decline of CMBS at the first 7 hours with the RD value from 86% to 35%, yet with a close statement from 8 to 12 hours and the RD value was on the verge of 30%. Compared to CMBS, the RD value of MBSP was gradually increased with longer time, and the maximum value was attained at 4~10h. Retrogradation degree increasing is related to the reorganization of starch chains and the re-agglomerating of functional groups such as hydroxyl of glucose units during aging which is characterized by increase of crystallinity which is contrary to gelatinization.

4. Conclusion

Superfine grinding of mung bean starch had some significant characteristics: the water solubility and water holding capacity of mung bean starch was increased after superfine grinding, and had a good retrogradation degree. But it had no or little influence on bulk density, and with the extension of superfine grinding time, the particle size was enlarged which was different from other reports. The superfine ground mung bean starch showed some special physical properties that might be of potential use in the food industry.

References

1. Maisa B., Xu X.Y., Si Y.P., A. Hydamaka, et al., Effect of oxidation and esterification on functional properties of mung bean (Vigna radiata (L.) Wilczek) starch, Eur. Food Res. Technol. 236(2013) 119-128.
2. Wu G. Ch., Zhang M., Wang, Y. Q., K. J. Mothibe, et al., Production of silver carp bone powder using superfine grinding technology: suitable production parameters and its properties, J. Food Eng.109(2012) 730-735.
3. Uthumporn U., Shariffa Y. N., Karim. A. A., Hydrolysis of native and heat-treated starches at sub-gelatinization temperature using granular starch hydrolyzing enzyme, Appl. Biochem. Biotechnol. 166(2012) 1167–1182.
4. Zhao X. Y., Yang Z. B., Gai G. Sh., Effect of superfine grinding on properties of ginger powder, J. Food Eng., 91(2009) 217–222.
5. Jin Sh. Y., Chen H. Zh., Superfine grinding of steam-exploded rice straw and its enzymatic hydrolysis, Biochem. Eng. J., 30(2006) 225-230.
6. Zhang, Z.Z, Superfine-grinding technology: Principle, Equipment and Application, 1st ed., Chemical Industry Press/ Putnam, Beijing, 1998.
7. Zhao X. Y., Ao Q., Yang L. W., et al., Application of superfine pulverization technology in biomaterial industry, J. Taiwan Inst. Chem. Eng., 40(2009) 337-343.
8. Liu X., Gu Sh. J., Xu W. L., Thermal and structural characterization of superfine down powder, J Therm. Anal. Calorim., 111(2012)259-266.
9. Zhao X. Y., Du F.L., Zhu Q. J., Qiu D.L., Yin W.J., and Ao Q., "Effect of superfine pulverization on properties of A stragalus membranaceus powder," Powder Technol., vol. 203, pp. 620–625, 2010.
10. Zhang Z. P., Song H. G., Peng Zh., Luo Q.N., Ming J., and Zhao G.H., "Characterization of stipe and cap powders of mushroom (lentinus edodes) prepared by different grinding methods," J. Food Eng. ,vol.109, 109, pp. 406–413, 2012.
11. Zhao X.Y., Chen J., Chen F., Wang X., Zhu Q., and Ao Q., "Surface characterization of corn stalk superfine powder studied
12. Anderson R.A., Water absorption and solubility and amylograph characteristics on roll-cooked small grain products, Cereal Chem., 59(1982) 265–269.

Micro Structure and Mechanical Properties of Chitosan-Corn Starch-Methyl Cellulose 3-layer Film

Li-Yan Wang, Jian Wang, and Xue-Jun Liu*

College of Food Science and Engineering, Jilin Agricultural University, Changchun, 130118 China
Email: liuxuejun63@163.com

This experiment was mainly prepared for investigating mechanical properties and characterization of the chitosan-corn starch-methyl cellulose 3-layer film, in comparison to the single chitosan, starch and methyl cellulose film respectively. Through single factor experiment, we obtained the film deposition conditions. The structural characterization and mechanical properties were observed and measured by the scanning electron microscope (SEM) and universal tensile machine. Experiments indicate the prepared conditions of 3-layer film are 50g of concentration 2% (w/w) chitosan solution, 10g corn starch solution (5% w/v) and methylcellulose at a concentration of 2% (w/v) for 10g. The results showed that all plasticized films surfaces exhibited homogeneous structures, and the three-layer structure of the chitosan-corn starch-methyl cellulose 3-layers film could be observed clearly. However, the mechanical properties of the three-layer film are not significantly increased compared with other controls groups. When will the layer stack destroy the original network structure still needs further validation.

Keywords: Micro structure; mechanical; chitosan-corn starch-methylcellulose films.

1. Introduction

The environmental concern over non-renewable and non-biodegradable petrochemical-based plastic packaging is growing. It has been a hot study spot that biodegradable materials from renewable sources would take place of traditional packaging materials [1]. The key points of a sustainable food system are to maintain food quality, to improve safety, to limit wastes and reduce storage losses [2]. Using two or more polymers has gradually become an important approach to develop new materials that exhibit properties for a better performance. Thus, biodegradable materials have been widely studied to replace (partially substitute) petroleum-based plastics, such as chitosan, starch, methylcellulose, gelatin, cellulose etc.

Chitosan is a natural macromolecule polysaccharide from fish or shrimp waste obtained by the deacetylation of chitin, so chitosan is a derivative of chitin. Chitosan is abundant in the nature [3]. Chitosan, β-(1-4)- linked D-glucosamine and N-acetyl-D-glucosamine [4-5] have some advantages of being non-toxic, antibacterial, biodegradable, renewable etc. Chitosan-based films have been used to pack fruits [6-7], vegetables [8] and meats [9-11].

Starch is a natural polymer. It is a mixture of linear amylose molecules and branched amylopectin molecules. Starch is the main raw material for preparing an edible film because of its non-toxic residue, degradation, wide source, high yield, low cost and transparency [12]. Because of these advantages, the starch has been widely used. The effect of oil lamination between plasticized starch layers on film properties is examined. Monolayer starch films were used as control groups. The results indicated that the addition of rapeseed oil significantly reduces water vapor and oxygen permeability [13]. However, due to poor mechanical properties, the influence of surface esterification with alkenyl succinic anhydrides on mechanical properties of corn starch films was investigated using dodecenyl succinic anhydride (DDSA) and octenyl succinic anhydride (OSA) [14]. The films modified with DDSA were more rigid and stronger, while the films modified with OSA were more flexible and ductile.

Methylcellulose (MC) is one of the derivatives of cellulose. It can be produced from cotton cellulose, wood etc. with great film-forming property. MC-based films have been provided with good mechanical and gas barrier properties, which can contribute to increasing the shelf-life of products [15]. Ploy Klangmuang and Rungsinee Sothornvit have a study about hydroxypropyl methylcellulose-based composite films with beeswax and nanoclay, which evaluates the films on barriers, sorption isotherm and mechanical properties [16].

The objective of this work is to research mechanical properties and micro structure of chitosan-corn starch-methylcellulose 3-layers film in comparison to others single films.

2. Materials and Methods

2.1. *Materials*

Chitosan, methyl cellulose (Sinopharm Group Co., Ltd., Shanghai), starch (Gouffault, Beijing), glacial acetic acid (AR, Beijing Chemical Plant). Electric heated water bath (HW.SY21-K, Beijing Changfeng Instrument Company), electric oven blast (101A-2E, Shanghai Experimental Instrument Factory Co., Ltd.), hygrometer (CEM DT-172, Shenzhen Huasheng Chang Machinery Industrial Co., Ltd.), electronic analytical balance (BSA124S-CW, Sartorius scientific instruments (Beijing) Co., Ltd.), magnetic stirring, ultrasonic cleaner.

2.2. *Films preparation*

2.2.1. *Chitosan film preparation*

A chitosan solution (2% w/w) was prepared by dissolving chitosan in acetic acid solution (1% v/v) at room temperature with magnetic stirring at 5krpm for 20min, placed in water bath heated to 50°C for 30 min, and blend once every 5 min. Then, power it with ultrasonic oscillation for 20 min, to remove air bubbles in the liquid chitosan film. Allow it to stand for 12 h. The 50g chitosan solution is then poured directly on a 93 mm glass dish, and dried at 70°C. Balance it at 75% humidity (RH) for 48h and strip the film (C50,) at last.

2.2.2. *Corn starch film preparation*

Firstly, corn starch solution (5% w/v) was prepared by dissolving chitosan in distilled water. The solution was stirred at 5krpm for 20 min. Secondly, keep stirring and heating at 95°C for 60 min, then undergo ultrasonic treatment for 20 min to remove any remaining insoluble particle and air bubbles. Thirdly, dump 10g corn starch in a petri dish and dry at 25°C. Lastly, get the corn starch film (S10).

2.2.3. *Methylcellulose film preparation*

The films were prepared by the casting method. Methylcellulose film solution (2%, w/v) was prepared by dissolving it in distilled water at 50°C for 30min with stirring. The film forming was blended using ultrasonic at room temperature for 20min to degas air. The films were prepared by dumping an amount of 10g solution on petri dish and drying at 70°C in a blast electric oven. Methyl cellulose films (M10) of uniform thickness is obtained at 75% RH.

2.2.4. *Chitosan-corn starch-methyl cellulose 3-layers film preparation*

Preparation of three-layer film, it is mainly that the underlying film need to be dried well, the next layer can be extended stream. The first layer of 3-layers film is chitosan solution, which can be prepared according to the above described method. The second layer of film that prepared by 10g starch solution were then cast. Wait for it to dry, pour 10g methyl cellulose solution on it to get the third layer. The corresponding drying temperature was of 70°C, 25°C and 70°C. The duration of time was 8h, 12h and 3h respectively. Thus, the thin film chitosan-corn starch-methyl cellulose 3-layers film (CSM70) was prepared.

2.3. *Scanning electron microscopy (SEM)*

The film samples were glued to the sticky adhesive conductive aluminum stage, after use gold spray SSX-500 (Shimadzu, Japan) SEM surface and section morphology were observed. Films, which had been previously kept in lyophilizer (-80°C) for 48 h, were fixed onto stubs and coated with a gold layer for 45s in 20 mA.

2.4. *Mechanical properties*

The films should be equilibrated at RH of 33%, 55%, 75%, 95% for 48 h before the test. Tensile tests were carried out using a universal testing machine (Model QJ 210, Shanghai, Qingji, China) at a crosshead speed of 5 mm/min. A load cell of 100N was used. Each test film is measured at least five parallel samples and the average values were taken.

2.5. *Statistical analyses*

Date were analyzed through an analysis software of variant (ANOVA) by means of the *SPSS*. Significance was defined at P < 0.05. All data are presented ± as mean standard deviation.

3. Results and Discussion

3.1. *SEM analysis*

Figure 1 show that the four film surfaces were intact with no rupture phenomenon.

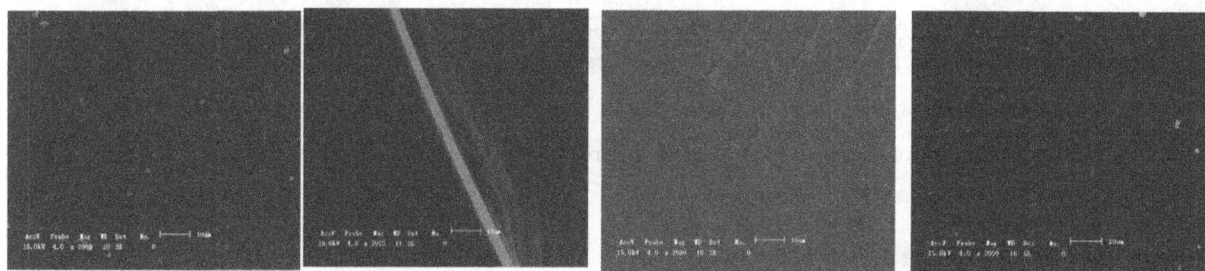

(a) Chitosan film   (b) Corn tarch film (S10).   (c) Methyl cellulose film (M10)   (d) Three-layer film (CSM70)

Fig. 1 Surface scanning electron micrographs of different films

The figure of a small bright spot, which may be present on the surface, is due to the impurity or dust in the air. In Figure 1(c), the slightly wrinkled surface, which might be in the drying process, shows shaking vessels and results in an uneven film surface. As the vessel bottom of the film is not flat, the film thickness is uneven. This phenomenon is more significant after lyophilization.

3.1.1. *SEM images of the cross sections of the films*

From the figure 2, a single-layer film of fine uniform appearance and structural stability is formed. Single film of chitosan and starch film form uniform and compact structure, while methyl cellulose film is fluffier. It may exhibit strong absorption characteristics, which can be the follow-up test of high solubility in methyl cellulose film. Figure (d) shows that the bionic layered film has a three-layer structure significantly, in line with the test expectations.

However, the film was too thin. It could not be fixed on the table and get a complete cross-section, so that the state of binding 3-layer film between each two cannot be clearly observed. The follow-up research issues are to be resolved.

| (a) Single chitosan film | (b) the single starch film | (c) single methyl cellulose film | (d) Three-layers film |

Fig. 2 Cross-sectional scanning electron micrograph of different films

Figure 3 is a scanning electron micrograph of a layered observation biomimetic layer of plastic wrap. Figure 3 shows that each layer separately presented three films but does not depict the status of the binding between the layers.

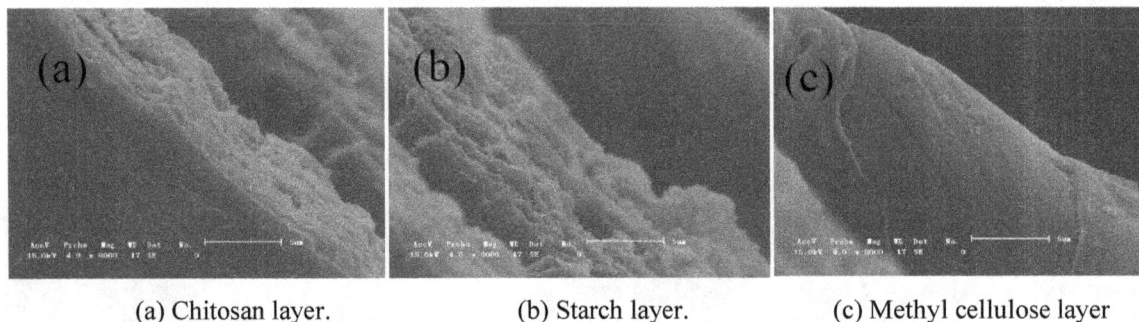

| (a) Chitosan layer. | (b) Starch layer. | (c) Methyl cellulose layer |

Fig. 3 Three-tier film scanning electron micrographs (hierarchical observation)

3.2. *Mechanical properties*

For films used in food packaging, the mechanical property is one of the decisive factors. Table 1–Table 4 shows the mechanical performance data of three-layer film (CSM70) with the control groups (C50, S10, M10) at 95%, 75%, 55%, 33% RH.

According to the tables, the tensile strength of the C50, S10 and M10 films were higher than CSM70 film. After superimposed, the film has not improved its mechanics, or even has a significant drop phenomenon. This may result in each layer of the thin film for dissolution. The destruction of its network structure results in a loose structure so that the mechanical performance goes bad. Studies have shown that the mechanical properties of starch and chitosan film mixing could be improved, which was due to hydrogen bonding interactions between starch and chitosan molecules to increase compatibility. However, the micro-phase separation is still unavoidable, affecting the mixed film to further improve the mechanical properties.

Table 1 Mechanical properties of four films at 95% relative humidity (RH)

Films	Tensile Strength (MPa)	Elongation at Break(%)	Young's Modulus (MPa)
M10	38.43±3.11ab	20.96±2.45c	102.07±17.21b
S10	39.80±2.89a	9.39±1.04d	246.76±29.93a
C50	31.21±2.98bc	89.26±7.84a	32.62±3.60c
CSM70	11.93±2.33d	30.27±9.29b	20.39±1.45c

Values are given as mean ± standard deviation. Different letters in the same column indicate significantly different ($p < 0.05$) in *SPSS*.

From Table 1 and Table 2, the tensile strength of the film CSM70 decreased significantly compared with the control groups, which may be due to each film forming material in the environment of different water absorption capacity, resulting in deformation between the layers of the film with different sizes, thus leading to the rival force between the layers and destroying the original compact structure.

Table 2 Mechanical properties of four films at 75% RH

Films	Tensile Strength (MPa)	Elongation at Break (%)	Young's Modulus (MPa)
M10	28.70±1.66a	16.49±2.81c	92.96±2.27a
S10	27.64±1.84a	5.43±2.91d	52.38±3.36c
C50	16.62±1.37c	52.61±1.92a	66.90±3.71b
CSM70	10.37±1.22d	28.46±2.68b	14.48±0.90d

Values are given as mean ± standard deviation. Different letters in the same column indicate significantly different ($p < 0.05$) in *SPSS*.

As shown in Table 3 and Table 4, the tensile strength of the film CSM70 has no significant difference compared with the control groups. This may be due to the reduced humidity in the environment compared with Tables 1 and 2. The decrease of water-absorbing material to reduce the gap between the layers shrinks the gap extending between the two layers, thus seriously damage did not occur on structure of the CSM70 film.

Table 3 Mechanical properties of four films at 55% RH

Films	Tensile Strength (MPa)	Elongation at Break(%)	Young's Modulus (MPa)
M10	38.19±2.93bc	18.93±2.07b	103.79±7.01c
S10	44.01±2.83ab	11.03±0.32c	1111.34±104.72a
C50	36.45±6.32c	38.09±7.05a	37.82±3.06cd
CSM70	36.24±4.43c	17.06±2.85bc	47.40±3.15cd

Values are given as mean ± standard deviation. Different letters in the same column indicate significantly different ($p < 0.05$) in *SPSS*.

Table 4 Mechanical properties of four films at 33% RH

Films	Tensile Strength(MPa)	Elongation at Break(%)	Young's Modulus (MPa)
M10	25.72±9.78a	14.55±5.59ab	274.56±37.57b
S10	46.46±6.45b	5.10±1.19ab	536.63±51.04a
C50	35.07±3.97ab	16.95±6.37ab	53.12±6.52cd
CSM70	29.57±1.59b	36.17±5.37b	0.93±0.23d

Values are given as mean ± standard deviation. Different letters in the same column indicate significantly different ($p < 0.05$) in *SPSS*.

From the tables, elongation at break of the S10 film is worst compared to others, mainly because S10 films without the aid of an emulsifier were more brittle, and will not likely to have fiber-like filaments that increase the elongation at break. This is a deficiency of the starch in the edible film packaging. As shown in Tables 1, 2 and 3, the CSM70 film has value between C50, S10 and M10 in terms of elongation at break. It described all materials of CSM70 film have demonstrated their performance, leading to a combined effect of the mechanical properties of the CSM70 film.

Young's modulus is a physical description of the solid material to resist deformation. Young's modulus of the CSM70 films has no significant change compared with the control group. It mainly affected tensile strength and elongation at break.

4. Conclusion

This study suggested that for the preparation of three-layer film, the optimum concentrations of chitosan, methyl cellulose, corn starch were 2%, 5% and 2%, the drying temperature were 70 °C, 25 °C and 70 °C, drying time were 8h, 12h and 3h. The entire film surface is flat and stable structure. CSM70 film can be clearly observed with three-layer structure, but the specific binding state between the layers needs further experimental observations. The CSM70 film has poor tensile strength compared with the control group because each layers of the CSM70 film layer is on the dissolution. The destruction of its network structure results in a loose structure, leading to the poor mechanical performance. Another speculation is humidity effect on the mechanical properties of the CSM70 film, which still needs validation. The mechanical property of the film as a packaging material is important so that it needs to be improved for the further.

Acknowledgment

This research was funded by the Project of Scientific Research Start-up Fund of Jilin Agricultural University of China (2015019).

References

1. J. Bonilla *et al.*, Effects of chitosan on the physicochemical and antimicrobial properties of PLA films, J. Journal of Food Engineering. 119 (2013) 236-243.
2. Nasreddine Benbettaïeb et al., Release of coumarin incorporated into chitosan-gelatin irradiated films, J. Food Hydrocolloids. 56 (2016) 266-276.

3. Li Yan Wang et al., Preparation and characterization of active films based on chitosan incorporated tea polyphenols, J. Food Hydrocolloids. 32 (2013) 35-41.

4. Jawhar Hafsa et al., Physical, antioxidant and antimicrobial properties of chitosan films containing Eucalyptus globulus essential oil, J. LWT - Food Science and Technology. 68 (2016) 356-364.

5. E. Genskowsky et al., Assessment of antibacterial and antioxidant properties of chitosan edible films incorporated with maqui berry (Aristotelia chilensis), J. LWT - Food Science and Technology. 64 (2015) 1057-1062.

6. Merve Duran et al., Potential of antimicrobial active packaging'containing natamycin, nisin, pomegranate and grape seed extract in chitosan coating' to extend shelf life of fresh strawberry, J. Food and Bioproducts Processing.98 (2016) 354-363.

7. Chun Mei Tian., Study on property of cassava starch/chitosan edible blend films and application in fresh-cut pineapple, South china University of Tropical Agriculture. Danzhou, china, 2007.

8. Gordana D. Jovanović et al., Antimicrobial activity of chitosan coatings and films against Listeria monocytogenes on black radish, J. microbiologia. 89 (2016) 1-9.

9. Danial Dehnad et al., Thermal and antimicrobial properties of chitosan nanocellulose films for extending shelf life of ground meat, J. Carbohydrate Polymers. 109 (2014) 148-154.

10. Giselle Pereira Cardoso et al., Selection of a chitosan gelatin-based edible coating for color preservation of beef in retail display, J. Meat Science. 114 (2016) 85-94.

11. Storage K.V. Reesha et al., Development and characterization of an LDPE / chitosan composite antimicrobial film for chilled fish storage, J. International Journal of Biological Macromolecules. 79 (2015) 934-942.

12. K. Sudharsan et al., Production and characterization of cellulose reinforced starch (CRT)films, J. International Journal of Biological Macromolecules 83 (2016) 385-395.

13. Ewelina Basiak et al., Effect of oil lamination between plasticized starch layers on film properties, J. Food Chemistry 195 (2016) 56-63.

14. Lili Rena et al., Influence of surface esterification with alkenyl succinic anhydrides on mechanical properties of corn starch films, J. Carbohydrate Polymers 82 (2010) 1010-1013.

15. Tian Zhong et al., Effects of ultrasound treatment on lipid self-association and properties of methylcellulose/stearic acid blending films, J. Carbohydrate Polymers 131 (2015) 415-423.

16. Ploy Klangmuang, Rungsinee Sothornvit, Combination of beeswax and nanoclay on barriers, sorption isotherm and mechanical properties of hydroxypropyl methylcellulose-based composite films, J. LWT - Food Science and Technology 65 (2016) 222-227.

Study on Hot In-plant Recycling Mixture Modified by Buton Rock Asphalt

Qing-Qing Lu

Tongji University, Key Laboratory of Road and Traffic Engineering of the Ministry of Education, Shanghai, P.R.China
Shanghai Road and Bridge (Group) Co.Ltd., No.36 Guoke Road, Yangpu District, Shanghai, P.R.China
Email: loo_chey@hotmail.com

Buton rock asphalt (BRA) is usually used as modifier applied to bitumen and bituminous mixture. In this paper, the performance of hot in-plant recycling mixture modified by buton rock asphalt are evaluated using rutting test, immersion Marshall test and freeze-thaw split test. The modification mechanisms of buton rock asphalt are studied by element analysis, scanning electron microscope analysis and infrared spectrum analysis. The test results showed that buton rock asphalt could improve the rutting resistance and the ability to anti-water damage of hot in-plant recycling mixture. The combined actions of rock asphalt and RAP asphalt lead to the improvement of high-temperature stability. The high-temperature softening and miscellaneous clusters adsorption of buton rock asphalt reinforce the bonding of asphalt and aggregate, and the rock asphalt powder absorbs asphalt to increase binder viscosity and reduce flow ability, thus improving the moisture susceptibility of asphalt mixture.

Keywords: Buton rock asphalt; hot recycling mixture; performance; mechanism.

1. Introduction

BRA (BRA) is a kind of natural mineral resources, as the asphalt and asphalt mixture additive, in recent years, more and more application in asphalt pavement carried out. It is a chemical modification agent of one of the alternative materials. Numerous studies [1, 2, 3, 4] revealed that BRA on asphalt modification mechanism, and show that the rock asphalt can significantly improve the ability of anti-rutting, anti-water damage of asphalt mixture and resistance to fatigue. And the plant hot-mix recycled mixture which uses recycled asphalt pavement material (RAP) is in response to the current green environmental protection politics and energy saving construction with low carbon conception. The study of [5, 6] showed that the plant hot-mix recycling mixture had better high temperature stability, but the ability to resist water damage was weakened. If the use of rock asphalt on the plant hot-mix recycling mixture, as the advantages of complementary, it should improve the water stability of recycling mixture as well as the high temperature performance.

When the plant is mixed with heat recycled mixture, the RAP material which is the proportion of all mixture 10%~50% should be heated separately, and the temperature is not more than 120 °C [7]. In order to ensure material temperature to meet the construction requirements, the total mixture heating temperature for new aggregates (proportion of 50%~90%) must be improved, the general is 190~200°C or more. BRA belongs to hard asphalt, and its observed softening point is more than 180°C. Consequently, in the case of low mix temperature, BRA is more used as organic filler and its modification effect can't be fully reflected in the mixture performance [8]. Only in high temperature can BRA show its advantages effectively. Therefore, BRA and hot recycled mixture have the same demand for high temperature. It is feasible and effective to combine the two. This paper selects 100 mesh (particle size<147 μm) BRA of rock asphalt modified hot in plant recycling mixture experimental study.

2. Mix Design and Performance Tests

BRA modified hot-mix plant recycling asphalt mixture preparation. First step is to blend and stir the mixture of BRA and aggregate evenly at a high temperature and make rock asphalt soft enough to attach the aggregate. Then after adding rap mixing, hot mix asphalt should be put in. And the last step is to add powder mixing. In this paper, the infrared temperature gun was used during the mixing process to promise correct mix temperature. The test results are shown in Table 1. In this experiment, the amount of rock asphalt is 0.94% weight of the aggregate, and the 70[#] asphalt was chosen as base bitumen.

Table 1 Test results of mixing temperature

Main processes	Temperature at each step /°C		
	AC-20	AC-20+30%RAP	AC-20+40%RAP
Preheating temperature of RAP	-	121	120
Preheating temperature of aggregate	180	196	194
Preheating temperature of Zhonghai-70	146	146	144
Preheating temperature of mix tank	150	157	157
Temperature after mixing of Buton Asphalt and aggregate	-	194	191
Temperature after RAP mix	-	160	168
Temperature after ZhonghaiI-70 mix	160	157	161
Temperature after ore powder mix	155	156	157

Marshall Compaction method [9] was used for asphalt mixture ratio design, and rutting test was conducted to determine the performance of high temperature stability of mixture, and both of the immersion Marshall Test and freeze thaw split test were implemented to figure out the water stability of mixture. Test results are shown in Table 2. The asphalt aggregate ratio (mass fraction) is the weight of three types of asphalt rap in old asphalt, including new adding asphalt, bitumen coming from BRA and RAP material contained asphalt. DS means mixture dynamic stability (dynamic stability), MS shows mixture immersion residual stability (residual Marshall Stability and TSR reveals mixture freeze-thaw splitting strength ratio thaw splitting strength ratio.

Table 2 Test results of mixtures

Mixtures	Asphalt-aggregate rate /%	DS/ (times·mm-1)		MS/%	TSR/%
		60°C	70°C		
GAC-20	4.3	1427	618	87.5	83.5
GAC-20+30%RAP	4.2	2281	744	84.3	86.4
GAC-20+30%RAP+9%BRA	4.3	4468	1488	88.4	92.5
GAC-20+30%RAP+12% BRA	4.3	4821	1819	91.6	94.6
GAC-20+30%RAP+15% BRA	4.4	5526	2188	97.1	94.6
GAC-20+30%RAP+18% BRA	4.7	5887	3257	96.5	98.1
GAC-20+40%RAP	4.2	4256	2316	88.2	82.2
GAC-20+40%RAP+9% BRA	4.2	4582	2881	90.4	88.4
GAC-20+40%RAP+12% BRA	4.4	5291	3419	93.5	94.6
GAC-20+40%RAP+15% BRA	4.5	6141	3479	95.4	97.6
GAC-20+40%RAP+18% BRA	4.8	6561	4012	97.0	98.3

Compared with the conventional asphalt mixture, BRA hot in-plant recycling mixture shows better performance. With the addition of rock asphalt, hot recycling mixture's bitumen content increased, and the high-temperature property and water stability were improved significantly by the combination of BRA and RAP. Performance of the mixture has been far more than F40-2004 JTG "highway asphalt pavement construction technical specifications" [10] in the modified asphalt mixture. Especially when the rock asphalt content reached 15%~18%, the basic and the current widely used SBS modified asphalt mixture performance was flat. But the addition of rock asphalt will increase the optimum asphalt content, which leads to a higher cost. Considering the price of materials and the performance of the mixture, the amount of rock asphalt should not exceed 15%.

3. BRA Modification Mechanism

3.1. *Element analysis*

BRA powder has a large specific surface area, whose density is small, little more than water, strong adsorption, and chemical stability characteristics. As the difference in elemental composition contrast BRA with ordinary asphalt, CHN automatic elemental analyzer elements of the BRA, base asphalt Zhonghai-70, Taiwan - 70, Gaofu - 70, and SK-70 is analyzed respectively, and the results are shown in Table 3. The molar ratio of hydrogen to carbon (NH/NC) is a parameter that reflects the saturation degree of organic compounds. When NH/NC is large, meaning that chemical structure contains more saturated components, it can make the asphalt colloid flocculation and sedimentation, thus unable to form a stable structure; when NH/NC is small, the structure contains more aromatic and alkyl chain, asphalt facilitates the formation of a gel type stable structure [11-12].

Table 3 Results of elemental analysis

Element name	Element content /%				
	BRA	ZhonghaiI-70	Taiwan-70	Gaofu-70	SK-70
C	64.34	83.52	86.76	85.36	84.55
H	6.82	10.68	7.03	9.36	9.86
N	0.78	0.45	0.26	0.74	0.31
S	8.06	5.10	5.24	3.88	4.53
NH/NC	1.24	1.53	0.96	1.27	1.38

Compared with the 4 kinds of base asphalt, BRA carbon and hydrogen content is low, but the higher content of atomic nitrogen, sulfur and other clutter. NH/NC in a lower value Heteroatom which contains strong polar functional groups could produce strong adsorption on the surface of the rock, to improve the anti-stripping and adhesion ability and anti-water damage ability of the mixture. NH/NC results also show that the rock asphalt could provide strong anti-aging ability because of its stable structure.

3.2. *Scanning electron microscope analysis*

For further study on rock asphalt and set material attachment ability to the surface of the state, the Japanese Olympus company EVO18 scanning electron microscope(SEM) was used to observe the structure of 100 mesh BRA particles and the interfacial conditions of rock asphalt adhesion, as shown in Figure 1 and Figure 2. The test parameters were set as EHT=10.00kv, WD=7.5mm, Mag=1000X for Figure 1 and EHT=10.00kV, WD=7.5mm, Mag=80X for Figure 2.

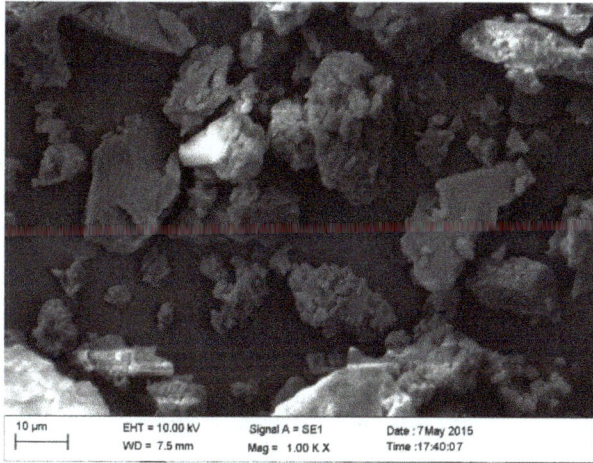

Fig. 1 SEM picture of BRA

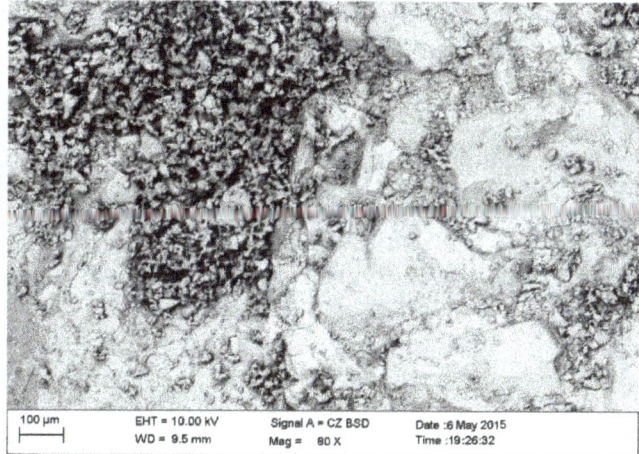

Fig. 2 SEM picture of BRA adhering to aggregate

Figure 1 shows the particle sizes of BRA powder are not homogeneous, distributing in the range of 10μm~150μm. Large particles have smooth surface and cleavage clear, and fine particle clusters attached on the larger particles due to their small size. Rock asphalt has large surface area and strong asphalt absorption. Figure 2 shows the interfacial conditions of BRA powder blending with high-temperature aggregates. The BRA powder becomes softening, adhering and covering the surface of the aggregate. As a medium of aggregate and asphalt, the rock asphalt also absorbs asphalt at the same time, effectively improving the bonding of asphalt and aggregate. At the same time, parts of the rock asphalt powder disperse to asphalt to improve the asphalt viscosity, reduce liquidity, increase asphalt film thickness, and thus the water stability of the mixture is improved.

3.3. Infrared spectrum analysis

For research BRA chemical bonds and functional group composition, the German Bruker Corporation Vector33 infrared spectrometer on BRA for analysis and detection, as shown in Figure 3.

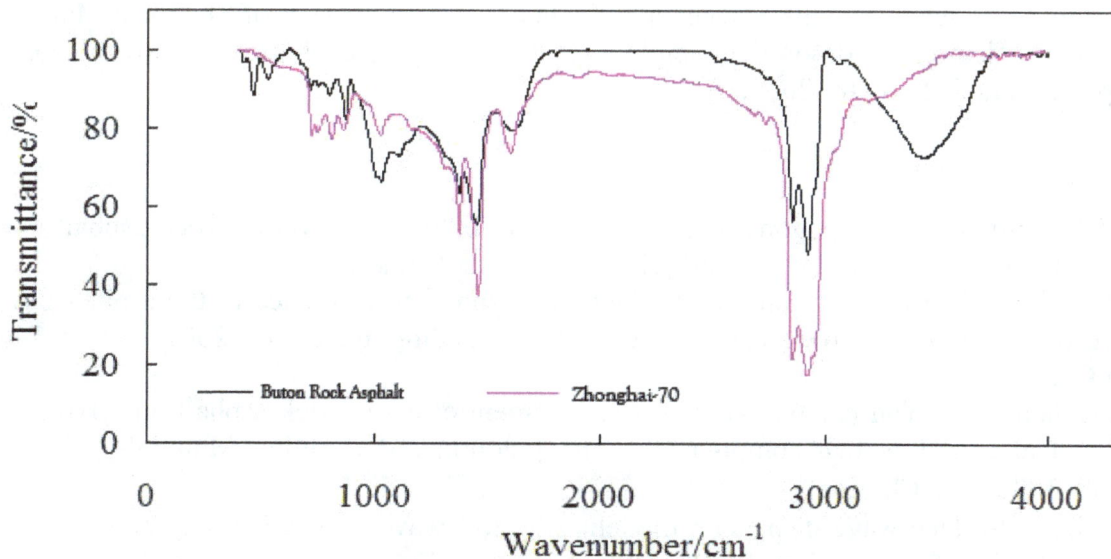

Fig. 3 Infrared spectrum of BRA and Zhonghai-70

As shown in Figure 3 BRA and asphalt together with the absorption peak for 700~900cm^{-1} at C-H bending vibration peaks, 1029cm^{-1} at S = O stretching vibration peak, 1374 cm^{-1} and 1453 cm^{-1} C-H deformation vibration peak, 2858 cm^{-1} and 2928cm^{-1} at C-H stretching vibration peak. BRA belong to the same asphalt base material, which means with ordinary asphalt compatibility is good, won't appear mutually exclusive segregation phenomenon, which is rock asphalt can used in a direct way. BRA characteristic absorption peak for 450~600cm^{-1} o-si-o bending vibration peaks and 3435cm^{-1} at - OH stretching vibration peak, which O-Si-O absorption peak of the ash of BRA bring. Containing silicon oxygen bond material temperature resistance and hardness are higher, which BRA can be one of the reasons of the hard asphalt. Meanwhile, the asphalt of RAP also belongs to hard asphalt. And under the coaction of rock asphalt and RAP asphalt, the high-temperature performance of the mixture is improved after plant mixing.

4. Conclusion

1. BRA modified of hot mix plant mixed recycled material will not affect the production energy consumption and greatly improve the performance of mixes, with characteristics of complementary advantages, the technology on the process and properties of the feasible.
2. BRA of hot mix plant recycling asphalt mixture is studied and on conventional asphalt mixture effect is consistent, which can also improve the mixture high temperature stability and water stability. Due to the superposition of the BRA and RAP, the high temperature stability of the mixture is especially obvious. As showed in infrared spectrum of BRA and Zhonghai-70, the joint function of the rock asphalt and the old asphalt has led to the high temperature stable performance.
3. BRA performance meets to the high temperature of mixes; and its high temperature softening point and miscellaneous clusters adsorption role together to improve the bonding properties of asphalt and aggregate; rock asphalt has larger area and absorb a large number of asphalt can increase the viscosity limit the flow and improve the performance of mixture.

Acknowledgment

This research was funded by the Science and Technology Commission of Shanghai Municipality. "Research and demonstration on the complete set technology of ecological road with sponge city idea" (project number: 16DZ1202000)

References

1. ZHOU Fu-qiang, Zhou bigong and Li baoguo et al. "the study on the rock-asphalt modified asphalt application". Highway. 2006(12):140-142. (in Chinese)
2. FAN Liang, SHEN Quan-jun and ZHANG Yan-yan. The influence to the asphalt pavement modified with nature rock asphalt. Journal of building materials. 2007, 10(6): 740-744. (in Chinese)
3. LI Li-han, SUN Yan-na, WANG Fei. Enhancement effect of rock asphalt on mixture of soft asphalt and hard asphalt compound binder[J]. Journal of Building Materials, 2013, 16(6): 1087~1091．(in Chinese)
4. Safani, J. Surface wave dispersion modeling by full-wavefield reflectivity and inversion for shallow subsurface imaging. Ph.D. thesis, School of Engineering, Kyoto University, Japan. 2007

5. FANG Yang, LIU Yu, ZHANG Guo-min. Research on high temperature stability of central plant hot recycling mixtures with high content of RAP [J]. Highway, 2013, 4: 99~102. (in Chinese)

6. Walubita, L.F. Comparison of Fatigue Analysis Approaches for Predicting Fatigue Lives of Hot-Mix Asphalt Concrete (HMAC) Mixtures. PhD Dissertation, Texas A&M University, College Station, Texas, 2006.

7. JTG F41-2008 Technical specifications for highway asphalt pavement recycling [S]. Beijing: China Communications Press，2008．(in Chinese)

8. Siswosoebrotho, B. I., Kusnianti, N., Tumewu, W. Laboratory evaluation of lawelebuton natural asphalt in asphalt concrete mixture. In Proceedings of the Eastern Asia Society for Transportation Studies. Vol. 5, No. 5, pp. 857-867.2005

9. JTG E20-2011 Standard test methods of bitumen and bituminous mixtures for highway engineering[S]. Beijing：China Communications Press, 2011.(in Chinese)

10. JTG F40-2004 Technical specifications for construction of highway asphalt pavements [S]. Beijing：China Communications Press，2004．(in Chinese)

11. TIAN Song-bai．Hot spot review of oil processing technology [M]. Beijing: Chemical Industry Press，2012. (in Chinese)

12. ZHA Xu-dong, BAI Lu, WANG Wei. Research on BRA modified asphalt performance. Transportation Science and Engineering, 2009, 25(1):10-13. (in Chinese)

Experimental on Mechanical Properties of Mixed Sand Concrete

Chun Li

The GuangXi Communications Research Institute, Nanning, Guangxi China
Email: 27977930@qq.com

Natural sand is becoming increasingly scarce, manufactured sand is always with unsatisfactory grading. In this paper, the natural sand concrete and the manufactured sand concrete were prepared to the mixed sand concrete. The mechanical properties of natural sand concrete, manufactured sand concrete and mixed sand concrete are studied in this paper. Test results showed that: mixed sand concrete composed of manufactured sand concrete and natural sand concrete with appropriate proportion has good working performance and mechanical properties, which may meet the technical requirements of concrete.

Keywords: Manufactured sand; concrete; mechanical properties; mixed sand.

1. Introduction

With the continuous increase of the scale of infrastructure construction, the demand for concrete is increasing day by day. Sand is one of the main raw materials of concrete. Natural sand is becoming increasingly scarce, with soaring prices. The exploitation of natural sand is a serious damage to the ecological environment of the river bed. Manufactured sand is rich in resources, which can be the nearest mining, with low transportation cost which cause its market price only 1/3 of natural sand. But the manufactured sand has poor gradation, high stone powder content and the poor wearing resistance, which limit its application [1, 2]. In order to better utilize the manufactured sand to reduce the engineering cost and to protect the environment, in this paper, the natural sand concrete and the manufactured sand concrete were mixed into mixed sand concrete with a certain proportion. It is also studied the working performance and mechanical properties of these three kinds of concrete in this paper to provide reference for the application of manufactured sand in engineering.

2. Raw Materials and Test Methods

2.1. *Raw materials*

1) Cement: P.O42.5 ordinary portland cement, Physical properties of the cement are shown in Table 1.

Table 1 Cement physical property index

Fineness	Initial final setting time (min)		Bending strength (MPa)		Compressive strength (MPa)	
80μm Square hole sieve (%)	The initial setting	The final setting	3D	28D	3D	28D
2.8	237	306	4.9	8.9	25.8	47.1

2) Coarse aggregate/manufactured sand and natural sand performance index are shown in Table 2.

Table 2 Coarse aggregate/manufactured sand and natural sand performance index

Coarse aggregate (limestone)		Natural sand		Manufactured sand		
Apparent density (g/cm³)	Maximum particle size(mm)	Apparent density (g/cm³)	Fineness modulus	Apparent density (g/cm³)	Fineness modulus	The content of stone powder (%)
2.715	31.5	2.664	2.71	2.705	2.79	10.7

3) Addition agent: Type CNF-1B high efficiency water reducing agent (volume 1.5%) Water reducing rate to take water reduction rate of 18%~20% and the solid content is 40%.

2.2. *Experiment design and test method*

At first, manufactured sand concrete and natural sand concrete were configured according to their respective proportioning ratio. The concrete design label was 5.0, the design of slump was 20~50mm, and the design of bulk density was 2400kg/m³and 2450kg/m³. The gravel content ratio of 20% fine gravel, and coarse gravel of 80%. Then mixed these two kinds of concrete in different proportions, ratio (manufactured sand concrete: river sand concrete) was: 1:9/2:8/3:7/4:6/5: 5/6:4/7:3/8:2/9:1. The final molding size of beam specimen was 150mm * 150mm * 550mm, then test the flexural strength of concrete for 7 days and 28 days with standard curing.

Table 3 Manufactured sand concrete mix ratio

Sand rate SP(%)	Theoretical mix ratio	Per m³ concrete material dosage /kg				
	c:s:g:w	Cement	Natural sand	Coarse aggregate	Water	Water reducer
37	1:2.0:3.41:0.43	351	702	1196	151	5.76

Table 4 Natural sand concrete mix ratio

Sand rate SP (%)	Theoretical mix ratio	Per m³ concrete material dosage /kg				
	c:s:g:w	Cement	Natural sand	Coarse aggregate	Water	Water reducer
36	1:1.90:3.37:0.4	360	683	1213	144	5.76

2 Experimental Results and Analysis

2.1. *Mechanical properties of natural sand concrete and manufactured sand concrete*

Test results of working performance and flexural strength of natural sand concrete and manufactured sand concrete are shown in Table 5 and Table 6.

Table 5 Test results of working performance and flexural strength of manufactured sand concrete

Slump (mm)	Concrete state	7D Bending strength (Mpa)	28D flexuaral strength (Mpa)
45	Better adhesion	5.4	6.0

Table 6 Test results of working performance and flexural strength of natural sand concrete

Table 6 Test results of working performance and flexural strength of natural sand concrete

Slump (mm)	Concrete state	7D Bending strength (Mpa)	28D flexuaral strength (Mpa)
55	Good adhesion	5.7	6.4

Table 5 and Table 6 show that, the working performance and flexural strength of natural sand concrete are better than that of manufactured sand concrete. The 7D flexural strength of natural sand concrete and manufactured sand concrete is 5.7 and 5.4 respectively, and 28D bending tensile strength is 6.4 and 6 respectively. The 7D strength of the manufactured sand concrete is lower than that of the natural sand concrete 0.3Mpa, 28D strength lower 0.6MPa.The main reason is the higher content of stone powder and its bad gradation [3, 4].

2.2. *Mechanical properties of mixed sand concrete*

Experimental results of mechanical properties of mixed sand concrete are shown in Table 7.

Table 7 Test results of working performance and flexural strength of mixed sand concrete

Number	Proportion	Slump (mm)	Concrete state	7D Bending strength (MPa)	28D Bending strength (MPa)
1	1:9	45	Slight bleeding	5.8	6.4
2	2:8	45	Better adhesion	5.9	6.5
3	3:7	50	Good adhesion	6.0	6.6
4	4:6	55	Good adhesion	6.2	6.8
5	5:5	50	Better adhesion	5.9	6.5
6	6:4	45	Slight bleeding	5.7	6.3
7	7:3	40	Slight bleeding	5.5	6.2
8	8:2	35	bleeding	5.4	6.0
9	9:1	40	Slight bleeding	5.6	6.1

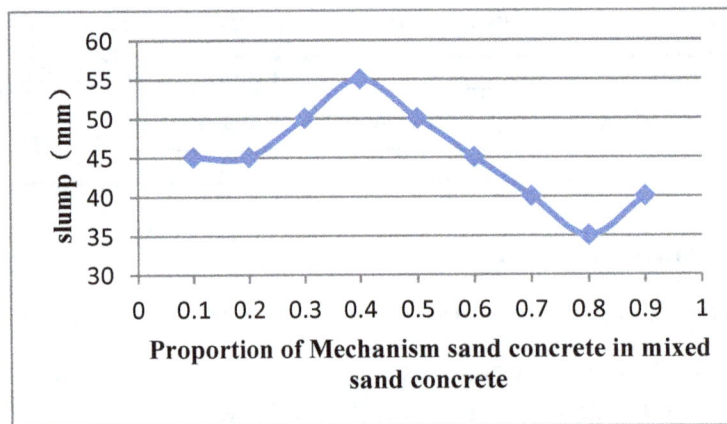

Fig. 1 Effect proportion of mixed sand concrete sand concrete on slump of concrete

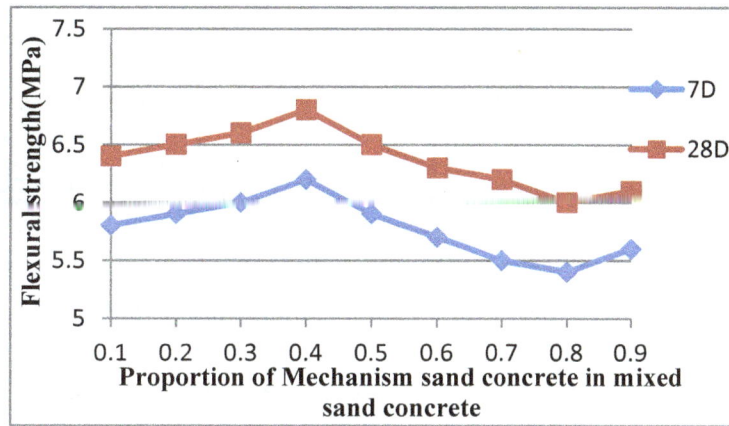

Fig. 2 The Flexural Strength of Concrete Effect of manufactured sand concrete in mixed sand concrete

From analyzing the experimental results of Table 7 and Figure 2, slump and flexural tensile strength of mixed sand concrete changed with the increase of the proportion of manufactured sand concrete, the trend first increased and then decreased. When the sand concrete proportion is small (less than 40%), with the increasing content of sand manufactured concrete, the working performance of the mixed sand concrete has been improved gradually. When the manufactured sand concrete: natural sand concrete proportion is 4: 6,the working performance and the working strength are both the best, The 7D flexural strength is higher than that of pure natural sand concrete and pure manufactured sand concrete with 0.5MPa and 0.8MPa higher respectively, the 28D flexural strength is higher than that of pure natural sand concrete and pure manufactured sand concrete with 0.5MPa and 0.8MPahigher respectively, On the one hand is because of the high content of stone powder in manufactured sand ,which can lubricate and improve the flow performance of mixed sand, at the same time to fill the powder mixed sand concrete void and increase the density of concrete structure. On the other hand, manufactured sand and natural sand mixed in appropriate proportion can be embedded and squeezed between each other, which make up the defects of the manufactured sand concrete poor gradation and improve the working performance and mechanical properties of concrete. When the manufactured sand concrete ratio is more than 40%, with the increasing ratio of manufactured sand concrete, the working performance and flexural tensile performance of mixed sand concrete shows a trend of first decrease. When manufactured sand concrete ratio is too high, the content of stone powder in mixed sand will increase, the excessive high content of stone powder added the water absorption of mixed sand concrete, making the concrete become sticky. The high content of stone powder caused excessive concrete bad gradation, Coarse aggregate ratio decreasing, the skeleton function of coarse aggregate reducing, at the same time increasing the unit powder aggregate surface area, more hydration products needed to provide cohesion, the package ability of cement slurry reducing, which has resulted in the decrease of the working performance and the bending strength [5, 6].

3. Conclusion

1. Because of the high content of stone powder and the bad gradation in manufactured sand, the working performance and flexural strength of the manufactured sand concrete is inferior to that of natural sand concrete;
2. When the proportion of manufactured sand concrete with mixed sand concrete is low, with the increase of the proportion, the working performance and flexural strength of mixed sand concrete presents an upward trend. When the proportion is more than 40%, the working performance and the flexural strength of the mixed sand concrete are decreased with the increase of the ratio of the manufactured sand concrete;
3. When the proportion manufactured sand concrete: natural sand concrete is 4: 6, the working performance and mechanical properties of mixed sand concrete can reach the best.

Acknowledgment

This work was partially funded by the Guangxi traffic science and technology project.

References

1. P.P. Li, H. Jiang, J.B. Xiong, et al, Influence of composite sand on workability and durability of C60 grade marine concrete, J. Concrete. 7 (2013) 92-94. In Chinese
2. B. Li, G. Ke, M. Zhou, Influence of manufactured sand characteristics on strength and abrasion resistance of pavement cement concrete, J. Construction & Building Materials. 25 (2011) 3849-3853.
3. E. Özgür, k Marar, Effects of limestone crusher dust and steel fibers on concrete, J. Construction & Building Materials. 23 (2009) 981-988.
4. C.W. Gao, J.H. Zhang, Application of mixed sand to concrete with low and medium intensity, J. Concrete. 288 (2008) 87-91. In Chinese
5. G.D. Tan, D.C. Liu, Application of mixed sand in C50 as-cast finish concrete, J. Concrete. 9 (2007) 60-62. In Chinese
6. Y.H. Fang, Y. Gu, Q. Kang, et al, Utilization of copper tailing for autoclaved sand–lime brick, J. Construction & Building Materials. 25 (2011) 867-872.

Study on Preform Design of Multi-stage Forging for Connecting Rod to Upgrade Material Retention Ratio

Huu-That Nguyen, Quang-Cherng Hsu*, Chun-Hung Liu

Department of Mechanical Engineering, National Kaohsiung University of Applied Sciences, 415 Chien Kung Road, Sanmin District, Kaohsiung 80778, Taiwan, R.O.C.
Email: hsuqc@kuas.edu.tw

In this research, three different preform types (i.e., type-1, type-2 and type-3) were designed and simulated by finite element method (FEM) based on DEFORM-3D software. By way of comparison between preform types, the required forging load and die stress was significantly improved in both the second and final stage. Among the three types of forging preform investigated, the type-2 was found to be the optimum. Compared to type-1, the material retention ratio for type-2 significantly increased to 83.7%. The percentage of flash volume for type-2 was only 16.3%, which yields a significant economic benefit. Eventually, a validation test was carried out to verify dimensional accuracy of the connecting rods in the manufacturing process. The result indicated that the maximum errors of dimension between simulated and actual forging parts for type-1 and type-2 were 0.09 mm and 0.18 mm, respectively. These errors were found to be insignificant. Therefore, the proposed preform design (type-2) could be considered to apply for manufacture of connecting rod by the closed die hot forging method.

Keywords: Reverse engineering; multi-stages forging; connecting rod; material retention ratio.

1. Introduction

The connecting rod is one of the main components in every internal combustion engine of car, motorcycle, trucks etc [1]. It is a coupling link that connects the piston to the crankshaft to convert linear, reciprocating motion into the rotary motion in an internal combustion engine [2]. During the operation process of the engine, the connecting rod is subjected to complicated loads, which are tensile, compression and bending load [3].

In order to ensure the rigidity, strength, ductility as well as resistance to fatigue, the connecting rod is often manufactured from different materials. In general, the connecting rod is made of steel in internal combustion engines. However, some previous researchers have afforded to use alternative lighter materials such as titanium, aluminum, composites based on aluminum for the higher performance engines, but not yet commonly utilized due to cost of expensive [2].

It is well known that the connecting rod is a mass product, so that the selection of optimal method to manufacture these parts is of vital technological and economic importance. Recently, with computer aided finite element simulation, several researchers have discussed issues related to the forging simulation. For instance, Grass et al. [4] researched the hot forging of a connecting rod by 3-D simulation. Fuertes et al. [5] reported the optimization of preform design in forging of aluminum alloy connecting rods by FEM based on DEFORM-3D.

To enhance the material retention ratio, many previous studies have provided the forging of connecting rods without flash. Andrzej et al [6] presented the theoretical and experimental studies on the flashness die forging of the aluminum alloy connecting rod when three-slides forging press is utilized. Vazquez and Altan [7] concentrated on an investigation of material flow and the preform design for flashness cold forging of aluminum connecting rod. Their result noted that the amount of flash was approximately 5%. However, the stringent requirements of this forging were that the volume of initial billet must be equal to the volume of the die cavity and the dimensional tolerance must be very close. In addition, compared to the flash die forging, the flashness forging brings a higher tooling cost due to complex die cavities in use.

*Corresponding author

On the other hand, Khaleed et al. [8] carried out the optimization of preform geometry when flash cold-forging of connecting rod without under-filling. They revealed that the material waste to flash was about 21.16%. However, the amount of the flash volume was still relatively large. This can lead to an increase in the production cost of connecting rod.

In order to fill these gaps, the research work focused on the multi-stage preform design for closed die hot forging of AISI 4140 alloy steel connecting rod upgrading material retention ratio.

2. Research Methodology

In this research, three different forging preform types were designed and numerically simulated by FEM based on DEFORM-3D software. Each of them consists of three main stages (i.e., first stage, second stage, and final stage) as shown in Figure 1. First, the design of both initial billet and die geometries for the type-1 were conducted through the reverse engineering process. Next, a validation test was carried out to verify the accuracy of the design process. Finally, in order to enhance the material retention ratio, a new design for type-2 and type-3 was employed.

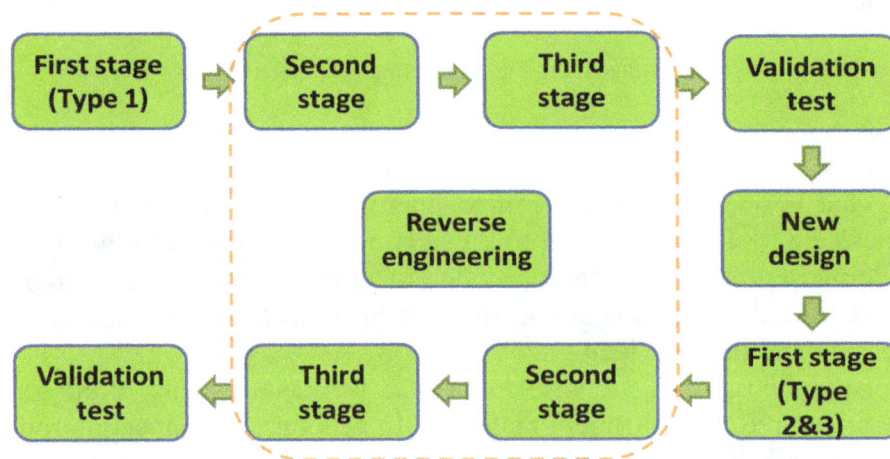

Fig. 1 Research procedure

2.1. *Reverse engineering*

Reverse engineering is a process that can achieve a geometric CAD model from measurements acquired by non-contact or contact scanning technique of an existing physical model [9]. It consists of many steps such as data acquisition, pre-processing, triangulation, feature extraction, segmentation and surface fitting, and application of CAD/CAM/CAE tools [10]. In order to do these tasks, the forged parts were measured by using a 3D scanner ATOS-I which was shown in Figure 2a. After scanning, the 3D CAD models were obtained as shown in Figures 2b-c.

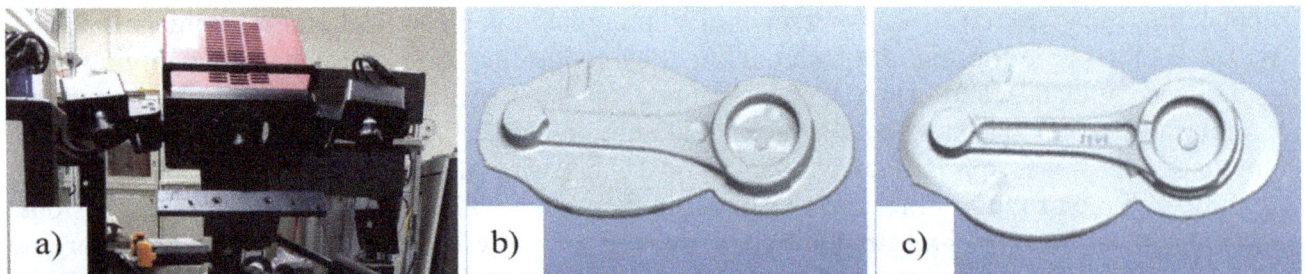

Fig. 2 Illustration of reverse engineering: a) 3D scanner ATOS-I, b) and c) 3D CAD model
in the second stage and in the final stage

2.2. *Design of forging dies and initial billets*

Based on the proposed CAD models, the forging dies were designed by using Pro/ENGINEER software. For forging preform, in the present study, three billet types were designed for closed hot forging of a connecting rod. In order to assure the die fill, the specifications of the billets were achieved through the assumption that the volume of the billet was more than the volume of die cavity or the final forged product [8]. From this assumption, the geometry and the dimension of the initial billets were designed as given in Figure 3.

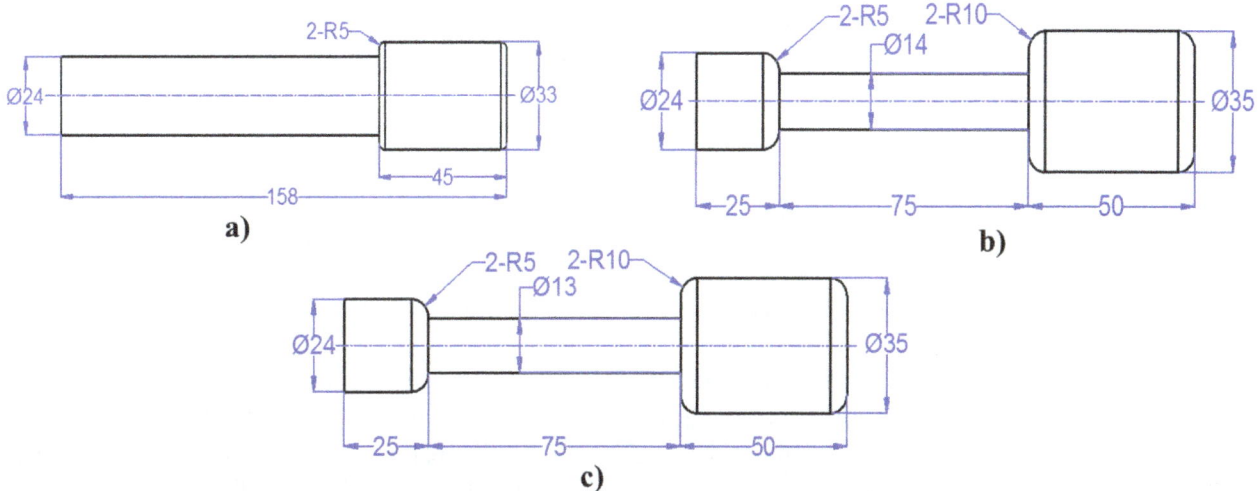

Fig. 3 Geometry and dimension of preforms: a) Type-1, b) Type-2 and c) Type-3

According to Figure 3, the volumes of the billet types (type-1, type-2, and type-3) were 90309.0 mm^3, 66476.9 mm^3, and 65014.5 mm^3, respectively. Based on the die cavity volume, the volume of connecting rod was determined to be 55645.5mm^3.

2.3. *Forging simulation based FEM*

In this study, the simulation of the forging process was carried out using FEM based on DEFORM-3D software. The boundaries and the material properties of the forging dies and preforms were completely listed in Table 1.

Table 1 The boundaries and the material properties of the forging dies and preforms

Parameters	Unit	Die	Preform
Material type		AISI H13	AISI 4140
Number of elements		-	90000
Temperature	[°C]	200	1000
Friction coefficient		0.3	0.3
Cycle time	[Cycles/sec]	1.5	-

3. Results and Discussion

3.1. *Analysis of forging load*

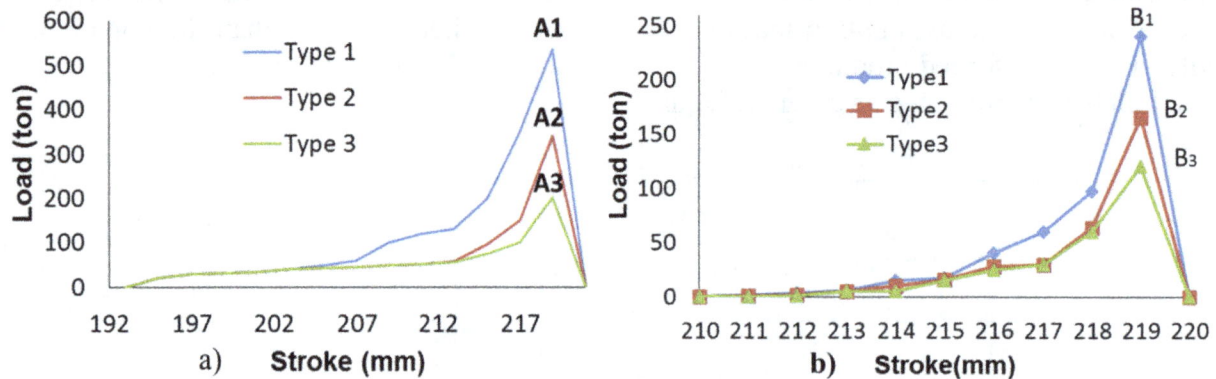

Fig. 4 Predicted forging loads for types of preform: a) At first and second stage, b) At final stage

Figure 4 shows the relationship of the predicted forging load with the stroke length for three different preforms. Figure 4a shows that the forging loads are similar in the first stage, but there is a remarkable difference of forging force between each kind of preforms in the second stage. In this stage, the maximum forging force is discovered to be for type-1 (536 tons) followed by type-2 (339 tons) and, finally, by type-3 (201 tons). This can be explained that the optimization of preform geometry lead to a decrease in flash volume. Thus, the forging forces are correspondingly reduced [5]. Similarly, as can be observed in Figure 4b, it indicates that the required loads to form the connecting rod in final stage for type-1, type-2 and type-3 are (240.8; 165.9; and 120.7) tons, respectively.

3.2. *Analysis of forging die stress*

In this research, the stress values on dies are obtained using FEM simulation at seven different positions on each forging die as shown in Figures 5a-b. The stress graphs of dies for three forging performs in the final stage are also illustrated in Figures 6a-b. From these figures, they show that most of the investigated positions, the stress of dies for type-3 is found to be least. This can be understood that optimization of preform sizes brings a decrease in required forging forces. Therefore, the amount of stress on dies are correspondingly reduced.

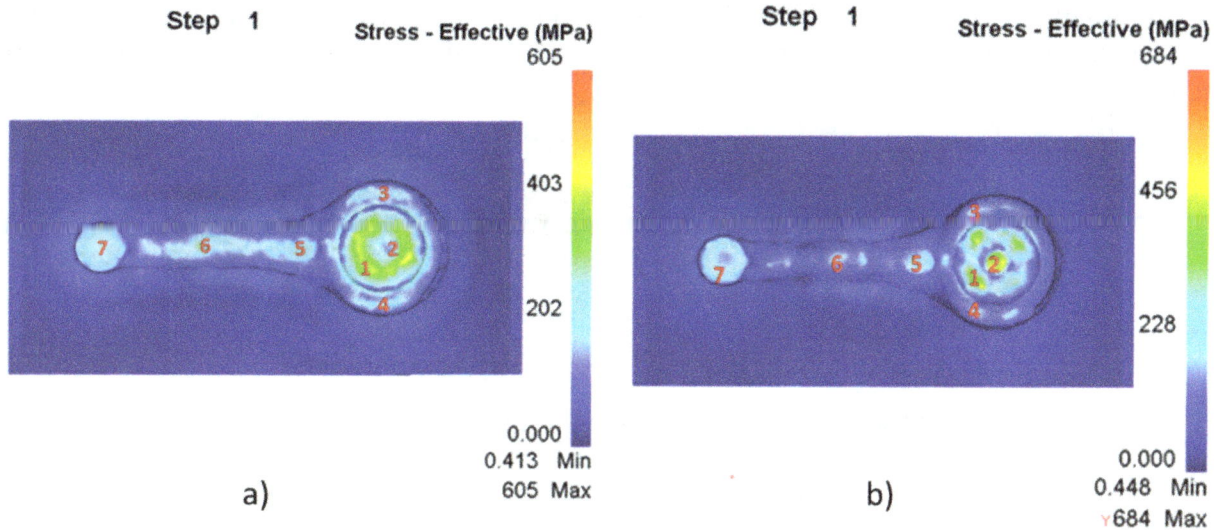

Fig. 5 Investigated positions of stress on dies in final stage for type-2: a) Upper die;b) Lower die

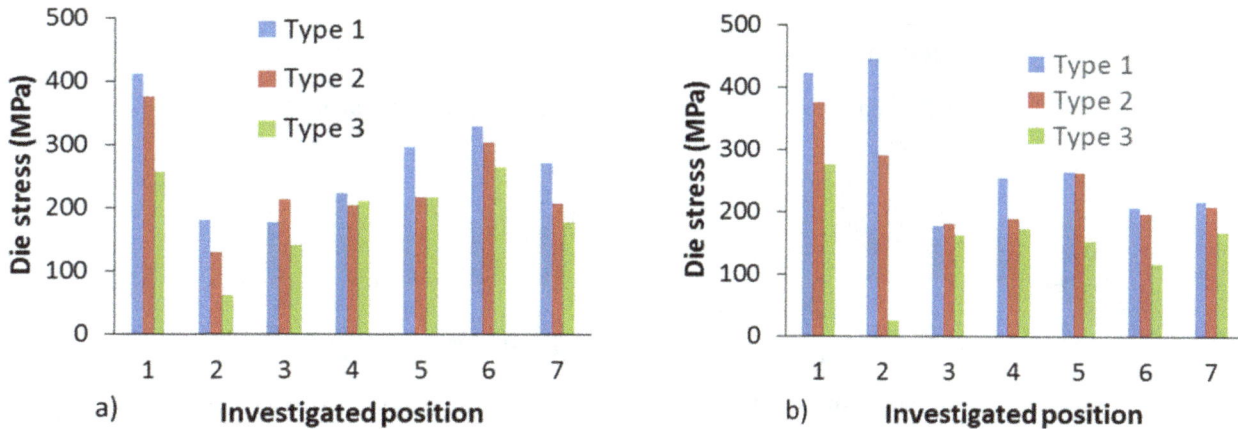

Fig. 6 Plots of die stress in the final stage: a) Upper die stress; b) Lower die stress

3.3. *Material retention ratio*

After simulation, the geometries of the connecting rod in the second and final stage are illustrated in Figures 7-9. Study of Figure 7, it shows that the flow of metal inside die cavities is very good, but the material retention ratio only obtains about 61.6%. In this case, the volume of flash is relatively high. This leads to an increase in the production cost of connecting rod. Figure 8 indicates that the die fill not only completes but also saves material in the forging process in both two stages. Compared to type-1, the material retention ratio of type-2 significantly increased to 83.7%. As can be seen in Figure 9, it shows that the material retention ratio reaches about 85.5%. However, the filling of die cavities is incomplete after simulation in the second stage and the final stage, as shown in the red circle in Figure 9. Therefore, in this research, the type-2 is found to be the optimum.

In order to clarify this research work, a relative comparison can be conducted with other author. Khaleed et al. [8] reported that the material waste to flash was about 21.16% when the flash cold-forging of connecting rod without under-filling. This means that an amount of the material for the formation of the flash in closed die cold forging is relatively large. The current study reveals that the flash volume has only been about 16.3%. This can be explained because the specifications of proposed preform are selected appropriately. Thus, this study yields a significant economic benefit.

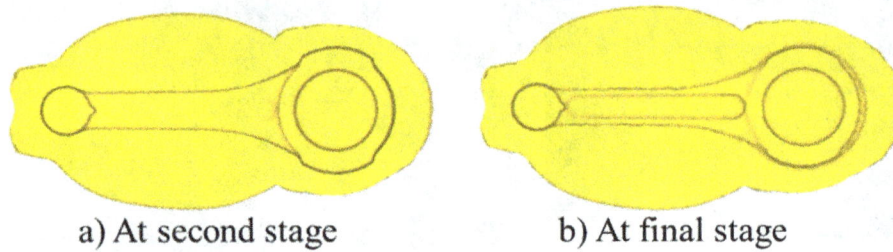

a) At second stage   b) At final stage

Fig. 7 Forged shapes of type-1 with material retention ratio of 61.6%

a) At second stage   b) At final stage

Fig. 8 Forged shapes of type-2 with material retention ratio of 83.7%

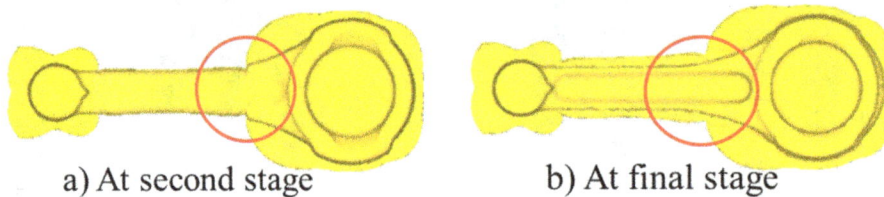

a) At second stage   b) At final stage

Fig. 9 Forged shapes of type-3 with material retention ratio of 85.5%

3.4. *Validation experiment*

In this section, the validation test is conducted for both type-1 and type-2. The check of the forged part size is accomplished by the reverse engineering process as mentioned above. The valid results reveal that the maximum dimensional errors between simulated and actual parts for type-1 and type-2 were 0.09 mm and 0.18 mm, respectively, as shown in the red circle in Figure 10, and Figure 11. It can be concluded that these values are found to be very small. Thus, the simulations of the multi-stage forging for manufacture of connecting rod by the closed die hot forging method are reasonably accepted.

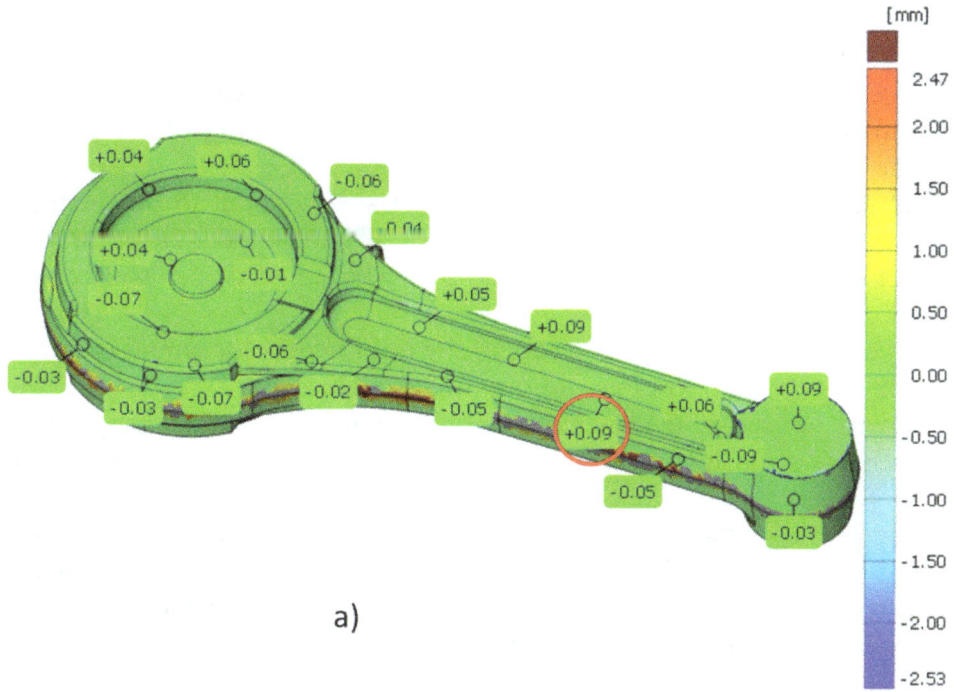

Fig. 10 Validation test of forged parts: a) Type-1

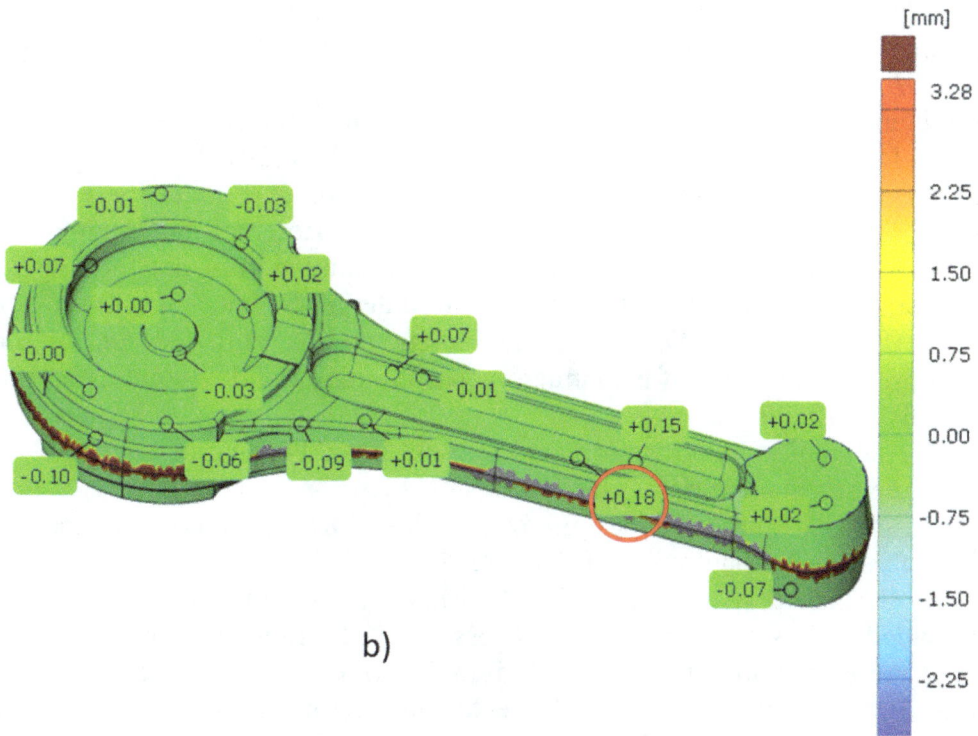

Fig. 11 Validation test of forged parts: b) Type-2

4. Conclusion

This paper has presented the study of the preform design of multi-stage forging for the connecting rod to upgrade material retention ratio when closed die hot forging of AISI 4140alloy steel. It could be concluded as follows:

At the second stage, the maximum forging load was discovered to be for type-1 (536 tons) followed by type-2 (339 tons) and, finally, by type-3 (201 tons). Similarly, the required loads to forge the connecting rod inthe final stage for type-1, type-2 and type-3 were (240.8; 165.9; and 120.7) tons, respectively.

Most of the investigated positions, the stress of the forging dies for type-3 is found to be least.

Among the three types of forging preform investigated, the type-2 was found to be the optimum. Compared to type-1, the material retention ratio for type-2 significantly increased to 83.7%. The percentage of flash volume for type-2 was 16.3%, which yields a significant economic benefit.

By reverse engineering process, the geometry and dimension of the connecting rod for both type-1 and type-2 were verified. The result shows that the maximum dimensional errors between simulated and actual forged parts for type-1 and type-2 in the final stage were 0.09 mm and 0.18 mm, respectively. These errors were found to be insignificant.

References

1. M. Plancak, D. Vilotic, O. Luzanin, I. Kacmarcik, A. Ivanisevic, D. Movrin, et al., A review of the possibilities to fabricate connecting rods, Annals of the Faculty of Engineering Hunedoara-International Journal of Engineering. 11 (2013).
2. L. K. Vegi and V. G. Vegi, Design and analysis of connecting rod using forged steel, International Journal of Scientific & Engineering Research. 4 (2013) 2081-2090.
3. P. D. Toliya, R. C. Trivedi, and N. J. Chotai, Design and finite element analysis of aluminium-6351 connecting rod, International Journal of Engineering Research and Technology. (2013).
4. H. Grass, C. Krempaszky, and E. Werner, 3-D FEM-simulation of hot forming processes for the production of a connecting rod,Computational Materials Science. 36 (2006) 480-489.
5. J. Fuertes, J. León, C. Luis, D. Salcedo, I. Puertas, and R. Luri, Design, optimization, and mechanical property analysis of a submicrometric aluminium alloy connecting rod,Journal of Nanomaterials. 2015 (2015).
6. G. Andrzej, P. Zbigniew, S. Grzegorz, and T. Arkadiusz, Forging of connecting rod without flash,Steel Research International. 81(2010) 358-361.
7. V. Vazquez and T. Altan, Die design for flashless forging of complex parts, Journal of Materials Processing Technology. 98 (2000) 81-89.
8. H. Khaleed, Z. Samad, A. Othman, M. A. Mujeebu, A. Badarudin, A. Abdullah, et al., Computer-aided FE simulation for flashless cold forging of connecting rod without underfilling, Arabian Journal for Science and Engineering. 36 (2011) 855-865.
9. A. Kumar, P. Jain, and P. Pathak, Reverse engineering in product manufacturing: An overview, DAAAM International Scientific Book. (2013).
10. B. Bidanda and Y. A. Hosni, Reverse engineering and its relevance to industrial engineering: A critical review, Computers & industrial engineering. 26 (1994) 343-348.

Chapter 8
Tool Materials and Special Alloys

Preparation of W-20Cu-0.5Co Alloy by Milling and Spark Plasma Sintering and Its Arc Ablation Performance

Zhong-Li Dong[1], Mo Guan[2], Xiao-Qiang Li[2,*], Xue Luo[2], Ming Nie[1]

[1]Electric Power Research Institute of Guangdong Power Grid Corporation, Guangzhou 510640, China

[2]National Engineering Research Center of Near-net-shape forming for Metallic Materials, South China University of Technology, Guangzhou 510640, China

*Email: Lixq@scut.edu.cn

W-20Cu-0.5Co alloy were prepared by milling and spark plasma sintering. Property tests such as hardness measurement bend test and arc ablation as well as microstructural characterization such as scanning electron microscope and transmission electron microscope all have shown that the appropriable milling and SPS process improve the properties of the material. A near-fully dense alloy with 200 nm W grains and homogenous microstructure was obtained by sintering at 1060 °C for 5 min under 50 MPa of 20 h milled powders. It possessed hardness of 48 HRC, bend strength of 573.7 MPa, electric conductivity of 37.5 ICAS% and good arc erosion resistance. Further elongated milling time could enhance the hardness and bend strength, but the conductivity lowered badly.

Keywords: W-Cu alloy; spark plasma sintering; mechanical property; arc ablation.

1. Introduction

W-Cu alloys have been widely used in the manufacture of electric contact parts, welding electrodes and thermal management devices owing to high thermal and electric conductivities, good high-temperature behavior, outstanding arc erosion resistance and low thermal expansion coefficient [1-2]. Because of the negligible mutual solubility in the W-Cu system and the extremely big melting temperature difference of the components, W-Cu powder compacts show very poor sinterability, even if by liquid phase sintering above the melting point of Cu phase [3]. W-Cu material is conventionally produced by infiltration of liquid copper into a porous tungsten piece sintered previously. However, this method has prominent shortcoming, which pronely produces defects including pores, coarse grains, and phase agglomeration in the microstructure. As a result, the electric and mechanical properties will weaken sensitively [4, 5]. It has been known that an activated sintering process, which is to add a small amount of metal such as Co, Ni and Zn, can increase the sinterability of W-Cu powders [5, 6]. Nevertheless, the addition of these elements may deteriorate the electrical and thermal properties of W-Cu alloys, so the amount of these elements must be strictly controlled. Nanocrystalline powders can also enhance the sinterability and produce a homogenous and utrafine/fine grained microstructure after sintering. High energy milling is considered an attractive and simple process for the synthesis of nanocrystalline powder mixtures. Especially for immiscible W-Cu alloys, milling facilitates the embedment of fine W grains in copper and the evolution of forced solid solution. It has been reported that W-Cu nanocomposite powder with good sinterability can be fabricated by high energy milling [7]. Compared with conventional sintering techniques, spark plasma sintering (SPS) is more suitable to consolidate nanocrystalline powders, due to a series of advantages of rapid heating rate, spark plasma effect, lower sintering temperature and so on. Resultantly, a highly dense and fine grained material can be easily obtained by SPS of milled powders [8]. The aim of the present investigation is to explore the manufacture of high-property ultrafine-grained W-Cu alloy by the addition of 0.5 wt% Co to W-20wt% Cu and the process routine of milling plus SPS. The influences of milling time, sintering temperature and sintering pressure on the microstructure and properties of sintered alloys are discussed in detail.

*Corresponding author

2. Experimental

Elemental W (purity≥99.9%, particle size of 4.5 μm), Cu (purity≥99.9%, particle size of 12.5 μm) and Co (purity≥99.9%, particle size of 2.0 μm) were used as raw powders. The powders were weighed accurately in a desired stoichiometric composition of W-20Cu-0.5Co (wt%), then milled in a planetary ball mill (QM-3SP2, Nanjing NanDa Instrument Plant, China) with a rotation speed of 269 r/min under argon gas protection. Tungsten carbide milling balls and stainless steel vials were used. The diameter of the balls was 10 mm and the ball to powder weight ratio was 10: 1. The milling was reversed every 30 min, with an acceleration and deceleration time of 10 s. 30.0 g powders every time were moved into a graphite die with an internal diameter of 20.4 mm and a wall thickness of 15 mm. All the sintering experiments were conducted on a Spark Plasma Sintering System (Dr. Sinter Model SPS-825, Sumitomo Coal Mining Co. Ltd., Japan) with a pulse sequence consisting of twelve pulses (with a pulse duration of 3.3 ms) followed by two periods of zero current. In sintering, the residual cell pressure was ≤ 6 Pa and a constant pressure was applied. An optical pyrometer, focused on the bottom of a small blind hole (with a diameter of 2 mm and a distance of 7.5 mm away from the inner wall of the female die) which was drilled in the female die wall at the same height as the center of the powder compact, was used to measure and adjust the temperature. The heating from room temperature to 600 °C was controlled by a preset heating program and completed within 4 min. From 600 °C to 900 °C, it took three minutes. Above 900 °C, the powder compacts were heated at a rate of 50 °C/ min to the objective sintering temperature. The sintering time was 5 min.

The sintered density was calculated by Archimedes' method. Phase identification was conducted by an X-ray diffractometer (XRD, D/Max-IIIA, Rigaku Co., Japan) using Cu Kα radiation, and the morphology of microstructure was observed by high-resolution scanning electron microscopy (SEM, Nova Nano 430, FEI, USA) and high-resolution transmission electron microscopy (HRTEM, JEM-2010, JEOL, Japan). Hardness was tested on Rockwell hardness tester (TH320, Beijing Time High Technology Co., China). The bend strength was measured at room temperature using a universal test machine (CMT5105, SANS Co. Ltd., China) with a constant crosshead displacement speed of 0.5 mm/min. Electric conductivity was measured by direct current double-arm bridge (QJ19, Shanghai Electric Apparatus Co., China). The sintered samples were machined to rods of Ø5 mm diameter for arc ablation tests. Arc ablation was done on a self-made apparatus based on TIG welding device, with voltage of 9 KV and current of 6 A in vacuum of about 0.01MPa, and the arcing times is 200.

3. Results and Discussion

Figure 1 shows the SEM photographs of W-20Cu-0.5Co composite powders milled for different times. With the milling time increasing, the distribution of tungsten and copper phases as well as the powder size becomes more and more homogeneous. It is attributed to the welding and fracturing effects from milling. Meanwhile, the average size of composite powders gradually increases with milling time, mainly because of the excellent deformability of Cu and the low mutual solid solubility of W-Cu system. The longer the milling time is, the more serious the deformation of Cu W and Co phases is. Large plastic deformation induces severe lattice distortion and a large number of dislocations in the component phases. Workhardening effect will intensify and hasten the fracture and spheroidization of powders. After 20 h milling, a dynamic balance of welding and fracturing comes up. The powder size thus tends to be constant. The dislocation cells and sub-grains are also gradually developed because of the accumulated deformation in milling. Thus the W and Cu grains refine finally and even a forced solution occurs. It makes the diffraction peaks of W and Cu gradually lower and wider with the milling time increased, as shown in Fig. 2. Because the content of Co is very little, the corresponding diffraction peaks is neglectable.

Fig. 1 SEM photographs of powders milled for 0 h (a), 10 h (b), 20 h (c) and 40 h (d)

Figure 3 shows the effect of sintering temperature on relative density for the unmilled and 40h milled powders (dwell time of 5 min and sintering pressure of 50 MPa). In the range of 900 °C to 980 °C, the relative sintered density rises quickly. In contrast to the 40 h milled powders, the unmilled powders are more easily densified, owing to the lack of workhardening from milling. Further increasing the sintering temperature, the densification rate slows down, especially for the unmilled powders. After sintering at 1020 °C and 1060 °C, the relative density of 40 h milling presents higher than that of unmilling. It is associated with the activated sintering of finer grains, higher dislocation density and more homogenous element distribution. The relative densities respectively go up to 98.3 % and 93.6% after 1060 °C sintering. When the sintering temperature exceeds 1060 °C, copper is prone to squeeze out. So the sintering temperature of 1060 °C is suitable.

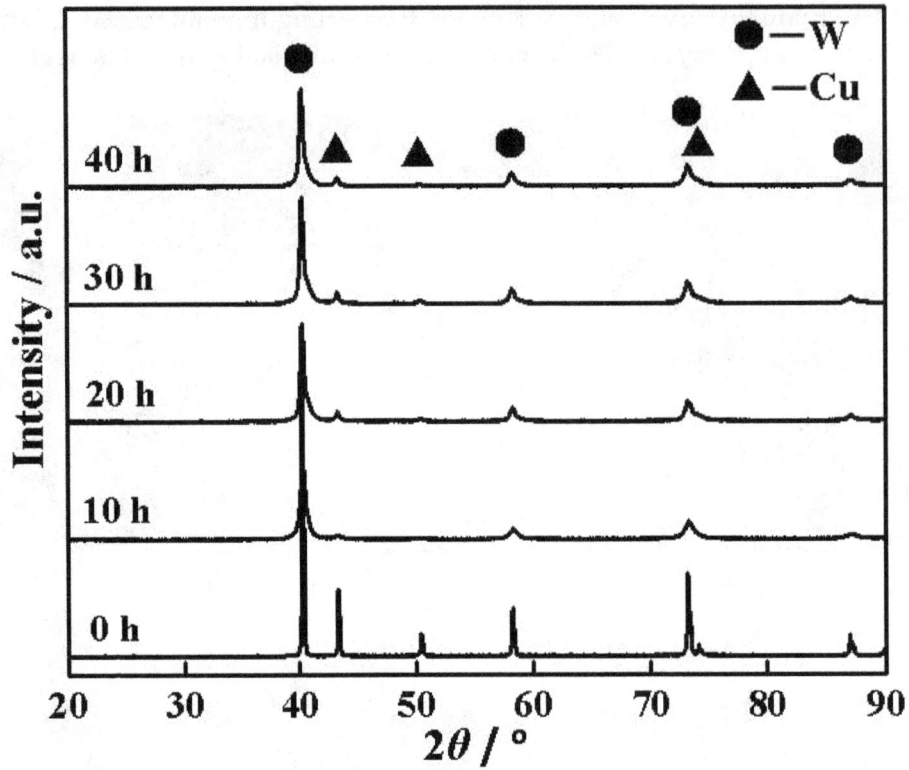

Fig. 2 XRD patterns of the composite powders milled for various times

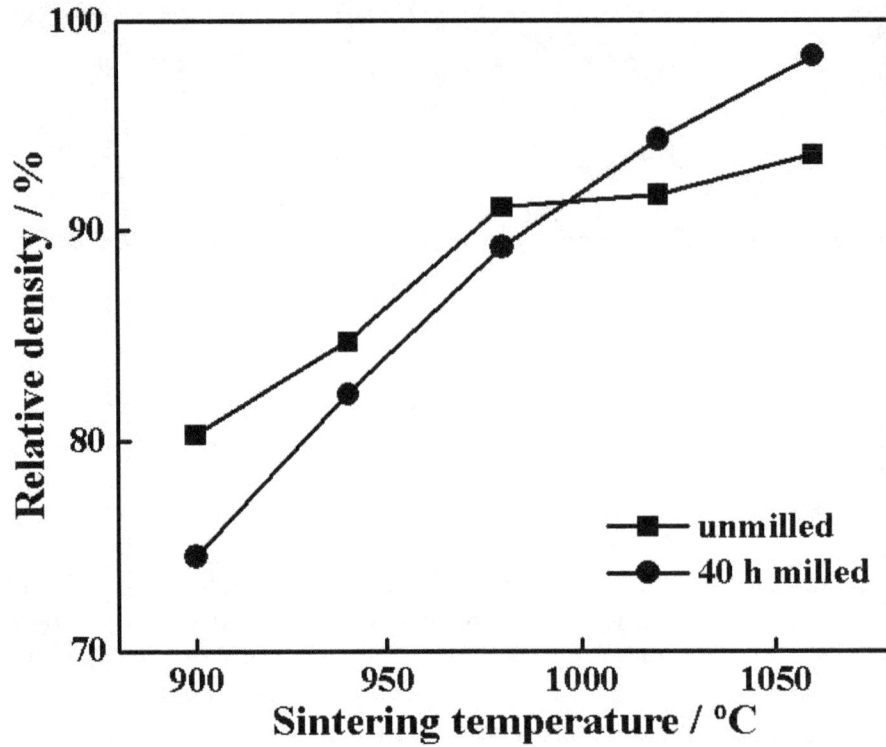

Fig. 3 Effect of sintering temperature on relative density (sintered under 50 MPa)

Figure 4 manifests the effect of milling time on relative density and electrical conductivity of sample sintered at 1060 °C for 5 min and under 50 MPa. Obviously, milling facilitates the sintering densification. After milling for 10 h, the relative sintered density reaches 98.7%. 30 h milling yields the maximum relative sintered density of 99.5%. However, further milling does not benefit to improve the relative sintered density and even produces more micropores in the sintered sample, because the corresponding powder size is too little. For W-20Cu-0.5Co sintered materials, their properties such as hardness, bend strength and electrical conductivity are determined not only by grain size and phase distribution but also by porosity, micro-defect and phase connectivity. The sample sintered from mixed powders isn't dense enough and there are many very tiny pores in it. Meanwhile, the microstructure is coarse. So its hardness and bend strength are low, being 10 HRC and 288.7 MPa, respectively (show in Fig. 4 and Fig. 5). However, its electrical conductivity is good (46.6 ICAS%), mainly owing to the coarse grained microstructure. With the milling time increasing, the porosity and pore size of sintered sample decrease, and the microstructure becomes finer. Resultantly, the hardness and bend strength keep rising with milling time. For 40 h milling, they go up to 55 HRC and 823.1 MPa. On the contrary, the electrical conductivity exhibits a downward tendency. The value of 40 h milling is just 20.8 ICAS%. Lower porosity is helpful to higher electrical conductivity. As is described above, further milling causes a finer microstructure of sintered sample. Consequently, grain boundary and phase interface in unit volume increase after sintering. In addition, a higher solubility perhaps remains because of the solution forced by long time milling and the short time sintering. The electron scattering effect of grain boundary, phase interface and micro-defects will weaken electric conductivity. The conductivity change in this study is the concurrent result of the two counteractive aspects of milling. The comprehensive performance of sample sintered from 20 h milled powders is good, with the hardness, bend strength and conductivity of 48 HRC, 573.7 MPa and 37.5 ICAS%, respectively.

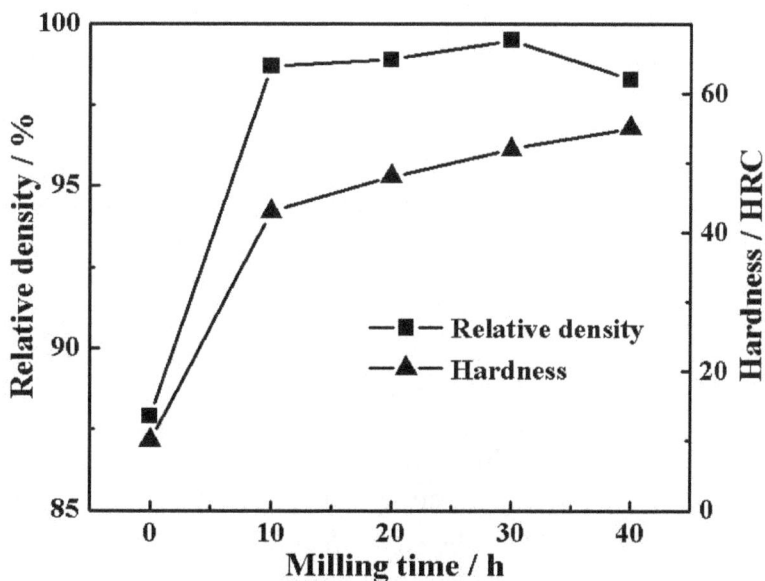

Fig. 4 Effect of milling time on relative density and electrical conductivity (sintered at 1060 °C and under 50 MPa)

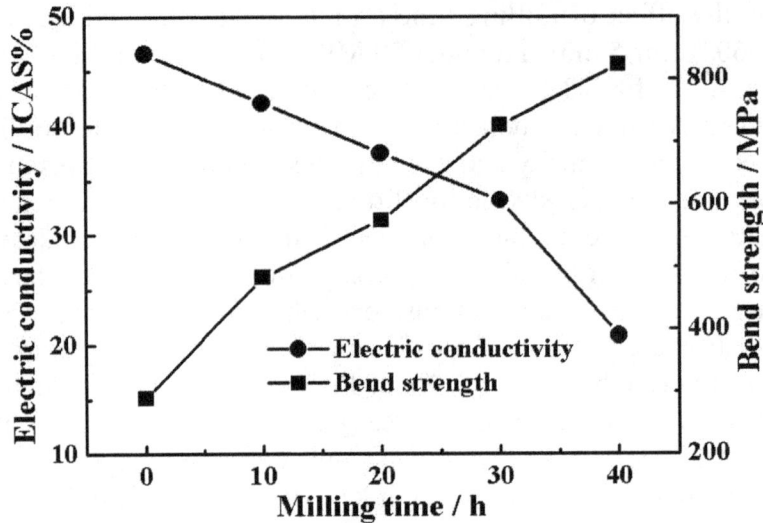

Fig. 5 Effect of milling time on bend strength and electrical conductivity (sintered at 1060 °C and under 50 MPa)

Although the adjustment range of sintering pressure in SPS is narrow owing to the strength limit of graphite die, its influence is also noteworthy. The increase of sintering pressure is favorable to the densification of powders by enhancing the deformation of the constitute phases and improving the liquidity and penetrability of liquid phase such as copper. For example, when 20 h milled powders are sintered at 1060 °C for 5 min, the relative sintered density and hardness increase with sintering pressure in the range of 20 MPa to 50 MPa, as shown in Fig. 6. After the pressure exceeds 40 MPa, the ascent rates of relative density and hardness sharply slow down. Based on Fig. 6, we can expect that both the relative density and the hardness will have a slight increase even if the sintering pressure rises above 50 MPa. It is a pity that Cu between W phases is more likely to squeeze out in practice, owing to high pressure. Although the decrease of sintering temperature can prevent copper from overflowing in SPS, low temperature is a great disadvantage to sintering, as shown in Fig. 3. Considering these, 50 MPa is an optimal sintering pressure.

Figure 7 displays the microstructure of sample sintered at 1060 °C for 5 min under 50 MPa from 20 h milled powders. The fracture surface contains lots of dimples and tear edges, which means the fracture is ductile. The copper phase is distributed well among W grains. It is attributed to the good separation and Cu encapsulation of W grains by milling. The tungsten grains are very fine and about 200 nm in diameter. Though a little amount of tiny pores are observed in the sintered sample, the size scale of pores is below 0.2 μm. It proves that a dense and superfine grained W-20Cu-0.5Co alloy can be obtained by appropriate milling and SPS process. Resultantly, the sample sintered from 20 h milled powders shows good comprehensive properties such as hardness of 48 HRC, bend strength of 573.7 MPa and electric conductivity of 37.5 ICAS%. W-Cu alloys are usually used as electrode, which suffers from the high temperature, chemical erosion and mechanical denudation of arc. Cu will evaporate and splash under arc, resulting in formation of micropores and surface crack. Then pores and crack hasten arc ablation.

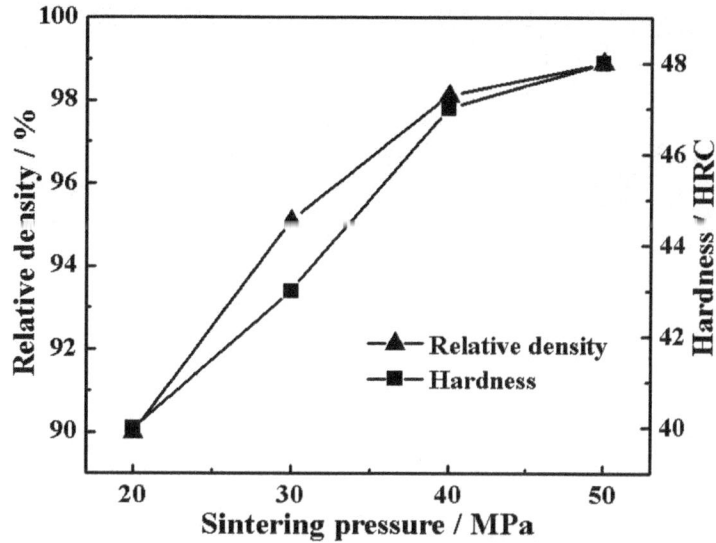

Fig. 6 Effect of sintering pressure on relative density and hardness (20 h milled powders, sintered at 1060 °C)

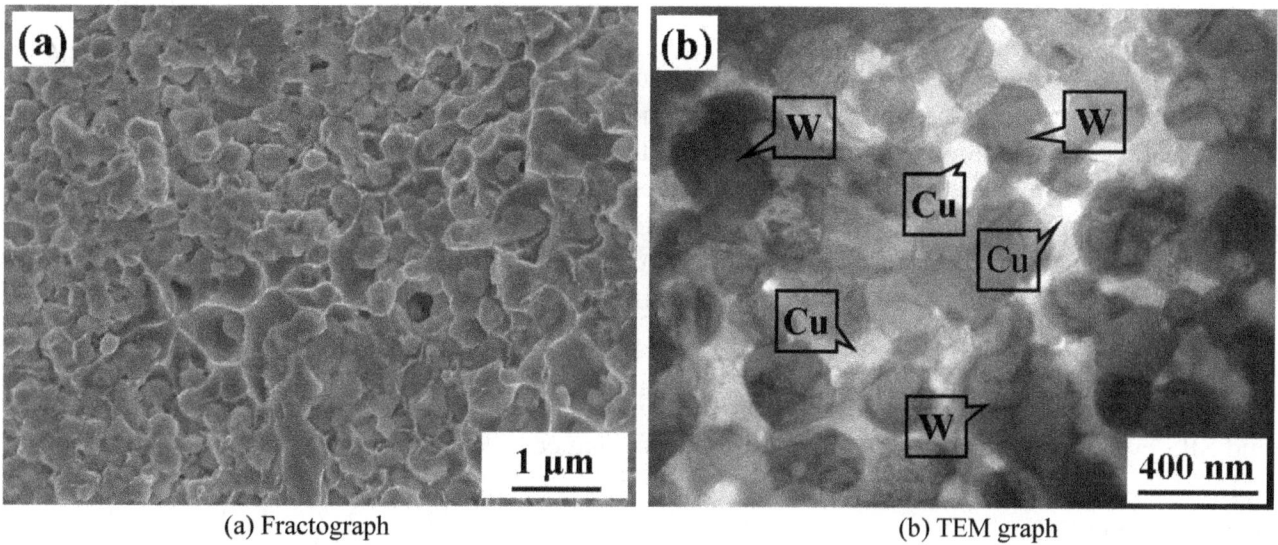

(a) Fractograph   (b) TEM graph

Fig. 7 The sample sintered at 1060 °C under 50 MPa from 20 h milled powders

Fig. 8 shows the arc ablation morphology of anode samples sintered at 1060 °C under 50 MPa. The ablation crater of 20 h milling is shallow and small compared with unmilling. Obviously, milling improves the arc erosion resistance. In the sintered sample from 20 milled powders, Cu phase evenly distributes in the W skeleton and is fine in size scale. So, the liquid Cu formed in arcing is more likely absorbed in the gaps of superfine W grains and the arc ablation is weakened.

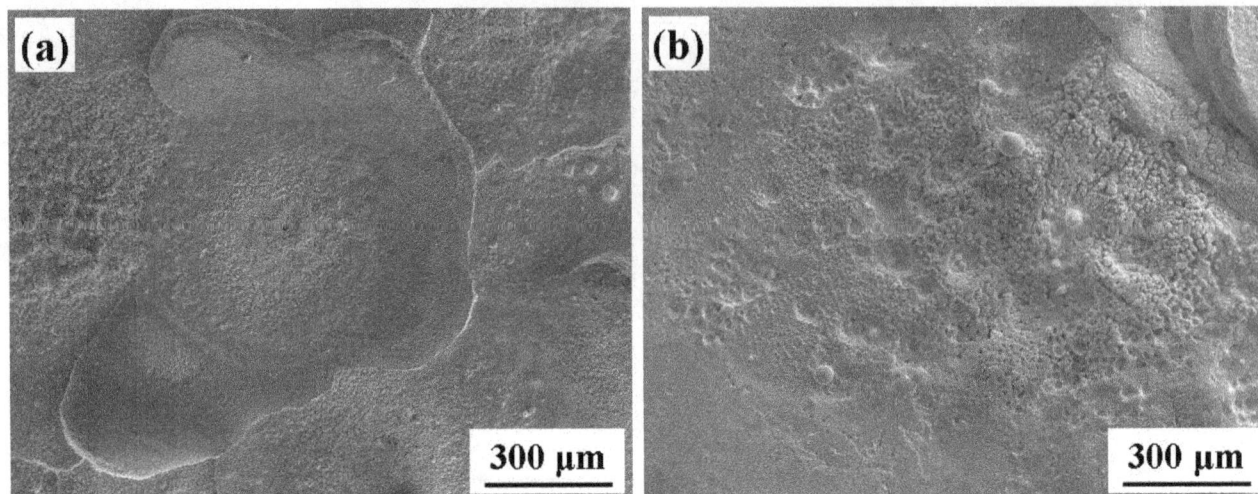

(a) From powders without milling   (b) milled for 20 h (b).

Fig. 8 Arc ablation morphology of anode samples sintered at 1060 °C under 50 MPa

4. Conclusion

W-20Cu-0.5Co alloy with high properties and superfine grains could be prepared by milling and SPS. Sintering temperature, sintering pressure and milling time determined the microstructure and properties of sintered alloy. Increasing sintering temperature and pressure was beneficial to obtain the high dense alloy. The relative density, hardness and bend strength increased with milling time. But long time milling such as 40 h caused a slight decrease in relative density. Differently, the electrical conductivity deteriorated by milling, owing to the microstructure fining. A near-fully dense W-20Cu-0.5Co alloy with 200 nm W grains and homogenous microstructure was obtained from 20 h milled powders by sintering at 1060 °C for 5 min under 50 MPa. Its hardness, bend strength and electric conductivity was 48 HRC, 573.7 MPa and 37.5 ICAS%, respectively. It is also affirmed that the dense, homogenous and superfine grained microstructure caused by milling and SPS evidently improved the arc ablation performance.

Acknowledgments

This work was funded by the Technology Program of Southern Power Grid Corporation (No. GDKJ00000081) and the Advanced Research Fund of DOD (No. 9140A18070114JW16001).

References

1. X.X. Wei, J.C. Tang, N. Ye, H.O. Zhou, A novel preparation method for W-Cu composite powders, J. Alloy Compd. 661 (2016) 471-475.
2. L.L. Zheng, J.X. Liu, S.K. Li, G.H. Wang, W.Q. Guo, Investigation on preparation and mechanical properties of W-Cu-Zn alloy with low W-W contiguity and high ductility, Mater. Design. 86 (2015) 297-304.
3. J.L. Johnson, Activated liquid phase sintering of W-Cu and Mo-Cu, International Journal of Refractory Metals & Hard Materials, 53 (2015) 80-86.
4. Y.P. Li, S. Yu, Thermal-mechanical process in producing high dispersed tungsten-copper composite powder, Int. J. Refract. Met. H. 26 (2008) 540-548.

5. P.A. Chen, G.Q. Luo, Q. Shen, M.J. Li, L.M. Zhang, Thermal and electrical properties of W-Cu composite produced by activated sintering, Mater. Design. 46 (2013) 101-105.

6. S.W. Kim, S.I. Lee, Y.D. Kim, I.H. Moon, High temperature compressive deformation and fracture characteristics of the activated sintered W-Ni compacts, Int. J. Refract. Met. H. 21 (2003) 183-192.

7 D.R. Li, Z.Y. Liu, Y. Yu, E.D. Wang, The influence of mechanical milling on the properties of W-40 wt.%Cu composite produced by hot extrusion, J. Alloy Compd. 462 (2008) 94-98.

8. Y.Q. Ye, X.Q. Li, K. Hu, S.G. Qu, Y.Y. Li. Effects of alloy composition on microstructure and mechanical properties of iron-based materials fabricated by ball milling and spark plasma sintering, Metall. Mater. Trans. A. 46 (2015) 476-487.

The Effect of Cr and Mo Addition on the Mechanical Properties and Microstructures of the Hot-dip Galvannealed DP Steels

Ying-Hua Jiang*, Shuang Kuang, Hua-Sai Liu, Chun-Qian Xie, Hua-Xiang Teng

Shougang Research Institute of Technology, Beijing 100043, China
Email: yinghuajiang@163.com

In comparison to the Cr-added hot-dip galvannealed DP steel with fully-recrystallized and equiaxed grains, the Mo-added hot-dip galvannealed DP steel shows small quasi-polygonal grains. The Mo-added hot-dip galvannealed DP steel shows better mechanical properties than Cr-added hot-dip galvannealed DP steel. The increased yield and tensile strengths in the Mo-added steel hot-dip galvannealed DP are mainly caused by grain refinement and martensite transformation. It is recognized that Cr and Mo suppress cementite transformation in the steels, but Cr is not as strong as that of Mo. Compared with Cr-steel, discontinuous yielding in the stress-strain curve did not appear in the Mo-steel because the martensite volume fraction was greater than the 5%.

Keywords: Hot dip galvannealed DP steel; chromium; molybdenum; mechanical properties; microstructure.

1. Introduction

DP steels, with their hard phase islands (martensite) embedded in a soft phase (ferrite), have unique properties such as high strength, low yield-to-tensile strength ratio, high initial work hardening rate, bake hardenability [1-2]. Therefore, DP steels were widely applied to great numbers of automotive parts. Recently, the automotive industry has been required to ensure the corrosion resistance of automotive component [3]. To meet this demand, hot-dip galvannealed DP steel was developed. However, compared with a continuous annealing line (CAL), in a continuous galvanizing line (CGL), the cooling rate was limited during galvanizing and galvannealing treatment, it is difficult to obtain DP microstructure [4]. Hence, the alloying elements Cr and Mo were added to improve the hardenability of a sheet to obtain DP microstructure [5].

However, the effect of Cr and Mo addition on the mechanical properties and microstructures of the hot-dip galvannealed DP steels have not in detail studied yet. The present study, therefore, investigated the microstructure and tensile properties of the hot-dip galvannealed DP steel to gain a better understanding of the Cr and Mo role in the hot-dip galvannealed DP steel.

2. Experimental Procedure

In the present study, the hot-dip galvannealed DP steels were produced using a laboratory annealing simulator. The low Si –low C materials are expected to improve coatability and spot weldability [6]. The low-C grades needed to add appropriate the alloying elements (Cr, Mo etc.) to improve the hardenability of a sheet to obtain DP microstructure. The chemical compositions of the steels used for this study were provided in Table 1. Experimental ingots made about 50kg in the laboratory by vacuum induction furnace. They were homogenized at 1200°C for 1hours and then hot rolled to 3mm hot band. Finishing temperatures were kept above 870°C and coiling temperatures were kept at 670°C, then air cooled to ambient temperature. Then, steel sheets are cold rolled to 1.2mm thick. The heat treatment cycle fully simulated the industrial CGL. The cold rolled steel sheets were annealed at a range between 780°C and 820°C for 70sec. After they were cooled at 460°C and dipped in a molten zinc pot at 455°C, they were then galvannealed at 520°C. In order to distinguish each phase (martensite, ferrite etc.) the microstructure, the average grain size and the martensite

---

*Corresponding author

volume fraction of steels were investigated after etching using Nital and Le Pera solution by an optical microscope.

Table 1 The chemical compositions of the steels used for this study (wt%)

	C	Si	Mn	P	S	Al	Cr	Mo	N
Cr-steel	0.07	0.15	1.3	≤0.02	≤0.007	0.035	0.5	-	≤0.005
Mo-steel	0.07	0.15	1.2	≤0.02	≤0.007	0.035	-	0.25	≤0.005

The mechanical properties of the steels were examined using tensile test machine. The precipitates of steels were observed by transmission electron microscopy (TEM).

3. Experimental Results

The optical micrographs of the Cr-steel and Mo-steel were shown in Fig. 1. The microstructure of trial steels consists of mainly ferrite and martensite with some cementite. In comparison to the Cr-steel with fully-recrystallized and equiaxed grains, the Mo-steel showed small quasi-polygonal grains. The average grain sizes on the ASTM scale for the Cr-steel and Mo-steel were 11.2μm and 12.2μm, respectively. Compared with Cr-steel, Mo-steel has the lower cementite volume fraction and higher martensite volume fraction. The martensite volume percentages of the Cr-steel and Mo-steel were 4.1% and 5.4%, respectively. This is because Mo addition effectively suppresses cementite formation so as to ensure more C contributed to the improvement of martensite formation [5].

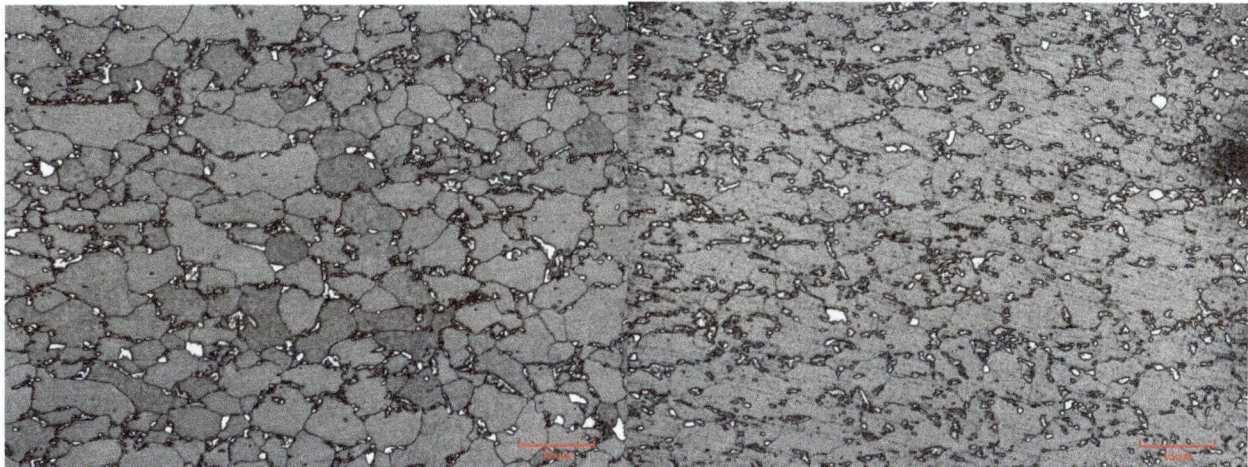

Fig. 1 Optical micrographs of the Cr-steel and Mo-steel

The TEM micrograph and EDX analysis result of Cr-steel was shown in Fig. 2. EDX analysis confirmed that the precipitates are mostly large Fe-C-Cr-based compound carbides.

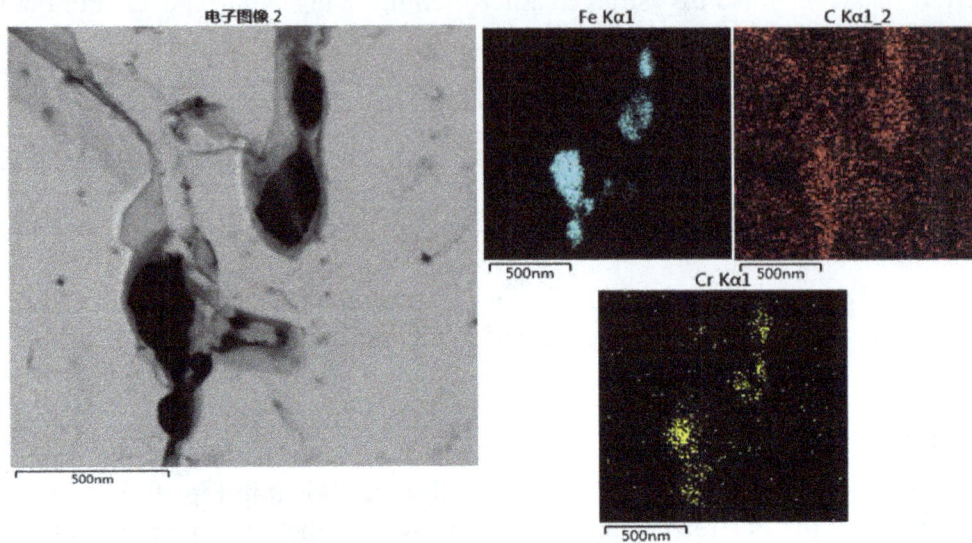

Fig. 2 TEM micrograph and EXD analysis result of the Cr-steel

The TEM micrograph and EDS analysis result of Mo-steel were shown in Fig. 3. EDS analysis confirmed that the precipitates are mostly fine Mo2C carbides. Therefore, the fine Mo2C precipitates resulted in the grain-refinement in Mo-steel.

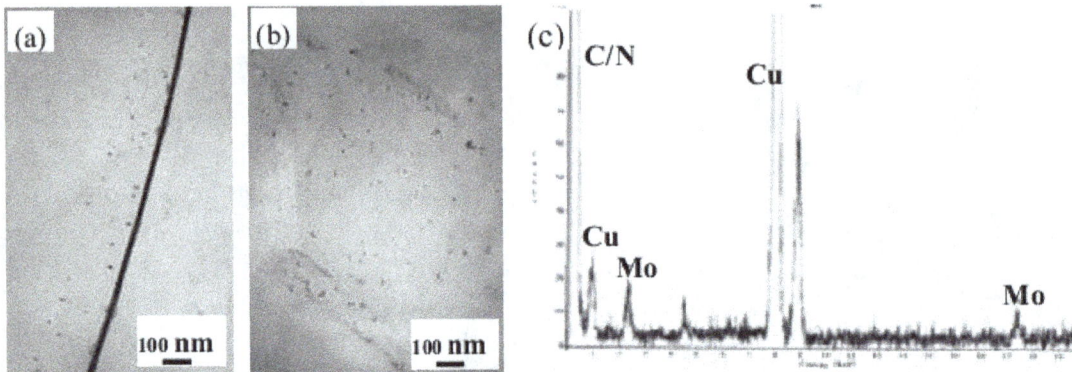

Fig. 3 TEM micrograph and EDS analysis result of the Mo-steel

Figure 4 shows the mechanical properties of Cr-steel and Mo-steel for all annealing temperatures. Compared with Cr-steel, Mo-steel shows higher yield strength and tensile strength. And the elongation for Mo-steel is the same as that of Cr-steel, reach to 36.5% at the same condition. Increasing the yield strength is obtained with decreasing the annealing temperature. In contrast, increasing the tensile strength is obtained with increasing the annealing temperature. The tensile strength and yield strength were attributed to grain-refinement hardening combined with martensite transformation hardening.

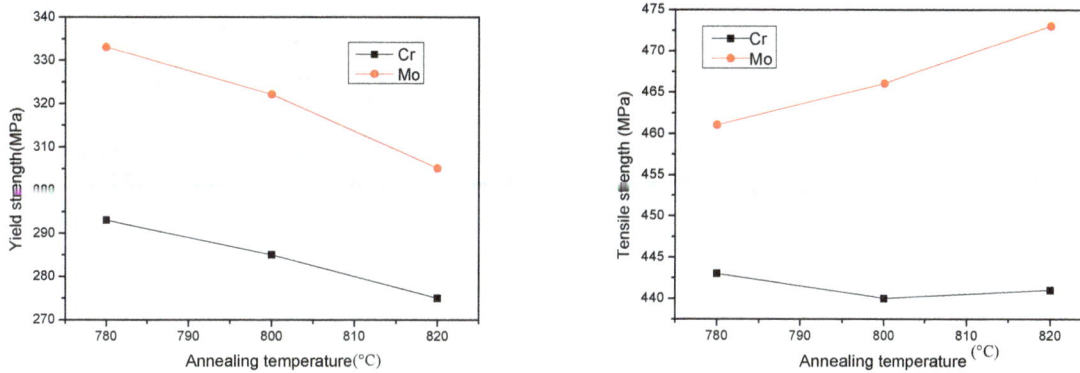

Fig. 4 Mechanical properties of Cr-steel and Mo-steel for all annealing temperatures

Figure 5 shows the strain-stress curves of Cr-steel and Mo-steel. Compared with Cr-steel, discontinuous yielding in the stress-strain curve did not appear in the Mo-steel because the martensite volume fraction was greater than the 5%. It has been reported that dual-phase steels should have at least ~5% of martensite in the microstructure for continuous yielding behavior without temper rolling [7].

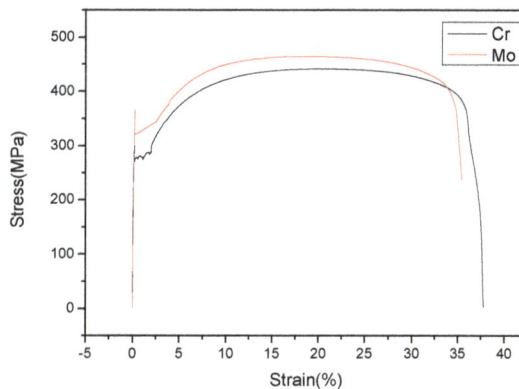

Fig. 5 The strain- stress curve of Cr-steel and Mo-steel

4. Conclusion

The effect of Cr and Mo addition on the mechanical properties and microstructures of the hot-dip galvannealed DP steels were investigated. In comparison to the Cr-added hot-dip galvannealed DP steel with fully-recrystallized and equiaxed grains, the Mo-added hot-dip galvannealed DP steel showed small quasi-polygonal grains. This is due to the grain refinement of fine Mo2C precipitates. Increasing the tensile strength and decreasing the yield strength were obtained with increasing the annealing temperature. The Mo-added hot-dip galvannealed DP steel shows better mechanical properties than Cr-added hot-dip galvannealed DP steel. The increased yield and tensile strengths in the Mo-added steel hot-dip galvannealed DP were mainly caused by grain refinement and martensite transformation. It is recognized that Cr and Mo suppress cementite transformation in the steels, but Cr is not as strong as that of Mo. Compared with Cr-steel, Mo-steel has no yield point elongation. This is contributed to a higher martensite volume fraction of Mo-steel than 5%.

References

1. Jiang Haitao, Tang Di, Mi Zhenli, Journal of Iron and Steel Research. 19 (2007) 1-6.
2. Hamed Asgari Moslehabadi, Galvannealing of Dual Phase Steels, McMaster University, Canada, 2012.
3. Zhang Liyang, Li Jun, Zuo Liang, Iron Steel Vanadium Titanium. 25 (2004), 15-19.
4. Byoung-Jin Kim, Young-Hee Kim, The Asia-Pacific Galvanizing Conference 2009.
5. Seong-Woo Kim, Jong-Sang Kim, The Asia-Pacific Galvanizing Conference 2009.
6. Kazunori Osawa, Yoshitsugu Suzuki, Shungo Tauaka, Kawasaki steel report. 34 (2002) 59-65.
7. Woochang Jeong, Met. Mater. Int. 20 (2014) 49-53.

Technics Study on Polycrystalline Diamond Cutting Tools Dry Turning Ti-6AL-4V Alloy Base on Orthogonal Experimental Design

Yun-Hai Jia[1,2,*], Yue Sun[1]

[1]Beijing Institute of Electro-machining, Beijing, China 100191
[2]Beijing Key Laboratory of Electrical Discharge Machining Technology, Beijing, China 100191
Email: jyh308401@sina.com.cn

Titanium alloys find wide application not only in the aerospace industry, but also for bio-medical applications. The machinability of titanium alloys is impaired by their high temperature chemical reactivity, low thermal conductivity and low modulus of elasticity. PCD represents a substitute tool material for turning titanium alloys due to its high hardness, wear resistance and thermal stability. For determination of suitable cutting parameters in dry turning Ti-6AL-4V alloy by PCD cutting tools, the samples, 300mm in length and 100mm in diameter, were dry machined in a lathe. The suitable turning parameters, such as cutting speed, feed rate and cut depth were determined according to workpieces surface roughness and tools flank wear base on orthogonal experimental design. The experiment showed that the cutting speed in the range of 120–160 m/min, the feed rate is 0.15 mm/rev and the depth of cut is 0.15mm, ideal workpiece surface roughness and little cutting tools flank wear can be obtained.

Keywords: Polycrystalline diamond cutting tools; Ti-6AL-4V alloy; dry turning; technics study; orthogonal experimental design.

1. Introduction

Titanium alloys find wide application in many industries. Titanium after being alloyed with aluminum, vanadium and other elements is highly suitable to be used in aircraft, naval ships, armor plating, missiles and aircrafts, due to their unrivalled and unique combination of high strength-to-weight ratio and high resistance to corrosion. Most importantly, it resists the crack growth and creep elongation even at high temperatures. Ti-6Al-4V is the mostly used alloy in aircraft applications, almost 50% of all alloys [1, 2]. The machinability of titanium alloys is impaired by their high temperature chemical reactivity, low thermal conductivity and low modulus of elasticity [2]. Notching at the tool nose, flank and crater wear, chipping and catastrophic cutting tool failure are some of the common failure modes occurring when machining titanium. Ti-6Al-4V is generally difficult to machine at cutting speeds of over 30 m/min with high-speed steel tools and over 60 m/min with cemented tungsten carbide tools [3]. The development of new tool materials in the past ten years, such as Polycrystalline Diamond (PCD), makes precision hard turning possible, which provides surface roughness, dimensional and shape tolerances similar to those achieved in grinding. The thermal conductivity for PCD ($\lambda \approx 400$ W/m K) is roughly four times greater than that of tungsten carbide ($\lambda \approx 100$ W/m K), and PCD is significantly harder (approximately 6,000 HV) than carbides (approximately 2,500 HV). The softening temperature (hot hardness) of PCD is higher than that of other commercially available cutting materials, indicating better performance at elevated temperature [4]. Some Research showed that PCD represents a substitute tool material for turning titanium alloys and that it produced a better workpiece surface integrity in finish turning operations.

There have been many valuable research results from Ti-6AL-4V alloy machining. Nurul Amin et al. studied the effectiveness of PCD and compared it to uncoated tungsten carbide–cobalt inserts machining Ti6Al4V. The authors concluded that PCD inserts can be used effectively up to cutting speeds of 160 m/min, as the wear rate is relatively low and the amount of metal removal per unit of

*Corresponding author

tool life is acceptable [5]. In Gert Adriaan Oosthuizen paper, a fine-grain polycrystalline diamond (PCD) end mill tool was tested, and its wear behavior was studied. The performance of the PCD tool has been investigated in terms of tool life, cutting forces, and surface roughness. The PCD tool yielded longer tool life than a coated carbide tool at cutting speeds above 100 m/min. A slower wear progression was found with an increase in cutting speeds. Whereas the norm is an exponential increase in tool wear at elevated speeds [6]. In literature [7], the author focuses on the influence of cutting tool edge preparation, cutting speed and feed rate on the tool performance and workpiece's surface integrity in dry turning of Ti-6Al-4V alloy using PCBN inserts. The results show, by increasing the cutting speed and feed rate resulted in tool life reduction. Cutting with honed edge insert at cutting speed of 180 m/min has shown very little wear. Rosemar B et al. investigate the behavior of PCD tools when machining Ti-6AL-4V alloy at high speed conditions using high pressure coolant supplies. Increase in coolant pressure tends to improve tool life and reduce the adhesion tendency. The authors concluded that adhesion and attrition are the dominant wear mechanisms when machining at the cutting conditions investigated [8]. Goutam Devaraya Revankar et al. deals with the investigation on machining of Ti–6Al–4V using polycrystalline diamond tool under different coolant strategies, namely dry, flooded and MQL. Taguchi technique has been employed and the optimization results indicated that MQL lubricating mode with cutting speed of 150 m/min, feed rate of 0.15 mm/rev, nose radius of 0.6 mm and 0.25 mm depth of cut is necessary to minimize surface roughness and dry mode with cutting speed of 150 m/min, feed rate of 0.15 mm/rev, nose radius of 0.6 mm and 0.75 mm depth of cut is necessary to maximize surface hardness [9].

The aim of this investigation was to study the suitable turning parameters, such as cutting speed (v), feed rate (f) and depth of cut (a_p), which were determined according to workpieces surface roughness (W_{Ra}), average cutting tool flank wear (VB) when using PCD cutting tools dry turning Ti-6AL-4V alloy.

2. Experimental Equipment and Method

Workpiece materials selected for investigation were the Ti-6AL-4V alloy rod with the composition given in Table 1. The size of the workpiece used for experimentation was round rods with dimension 100mm diameter and 300mm long.

Table 1 Composition of workpiece ($\Phi100\times300$ mm) material (%)

Workpiece material	C	AL	V	Si	Fe	N	H	O	Ti
Ti-6AL-4V	0.1	6.2	4	0.15	≤0.25	0.05	0.0125	≤0.2	Others

The physical and mechanical properties of the Ti-6AL-4V alloy were listed in Table 2. They were machined by PCD cutting tools (Compax 1300 produced by DI Corporation) in dry turning in a lathe (CAK6150).

Table 2 Physical and mechanical properties of the Ti-6AL-4V alloy

Density (g/cm^3)	Melting point (°C)	Specific heat capacity (J/Kg °C)	Thermal conductivity (W/mK)	Electric resistivity (ohm cm)	Fracture Toughness (MPam$^{1/2}$)	Elastic modulus (GPa)	Hardness (HB)
4.43	1649	526.3	6.7	0.000178	75	113.8	334

The cutting tool geometry parameters were shown in Table 3. The TR240 surface roughness tester, 19JPC-V universal tool maker's microscope and S-4800 scanning electron microscope (SEM) are selected as experimental analyzer.

Table 3 Cutting tool geometry parameters

Rake angle $\gamma_0(°)$	Flank angle $\alpha_0(°)$	Blade angle $\lambda s(°)$	Edge angle $kr(°)$	Nose radial r_ε (mm)	Vice chamfering width b_{r1}(mm)	Vice chamfering angle $\gamma_{01}(°)$
3	8	0	50	0.5	0	0

Surface roughness is used as the critical quality indicator for the machined surface. Formation of a rough surface is a complicated mechanism involving many parameters. The quality of the workpiece (either roughness or dimension) is greatly influenced by the cutting conditions, tool geometry, tool material, machining process, workpiece material and tool wear during cutting.

Use orthogonal experiment design to analyze relationships between the turning parameters and workpiece surface quality. The combinations of the factors at their different levels and the corresponding measured values of the surface roughness generated after the machining operations are given in Table 4.

Table 4 Orthogonal experimental design and experiment results (continue turning 10 minute)

S/N	Cutting speed v (m/min)	Feed rate f (mm/rev)	Depth of cut a_p (mm)	Workpiece surface roughness W_{Ra} (um)	Flank face wear VB (mm)
1	120	0.10	0.10	0.31	0.010
2	120	0.15	0.15	0.33	0.011
3	120	0.25	0.20	0.78	0.100
4	160	0.10	0.15	0.27	0.012
5	160	0.15	0.20	0.35	0.025
6	160	0.25	0.10	0.38	0.029
7	200	0.10	0.20	0.41	0.031
8	200	0.15	0.10	0.34	0.025
9	200	0.25	0.15	0.55	0.080

3. Study on Relationship between Turning Parameters and Workpiece Surface Roughness

The values of workpiece surface roughness were compared in Fig. 1(a) on the condition of the feed rate of 0.15 mm/rev, cutting speeds from 40 to 200 m/min, three different depths of cut 0.10, 0.15 and 0.20mm. From this figure, it can be known that with increasing of cutting speed, workpiece surface roughness values decreased until a minimum value was reached, after that surface roughness values increased. The lowest average value of workpiece surface roughness got obtained at 160 m/min cutting speed. The values of workpiece surface roughness were compared in Fig. 1(b) in the condition of depth of cut 0.15, feed rate from 0.05 to 0.40 mm/rev, three different cutting speeds 120, 160 and 200 m/min.

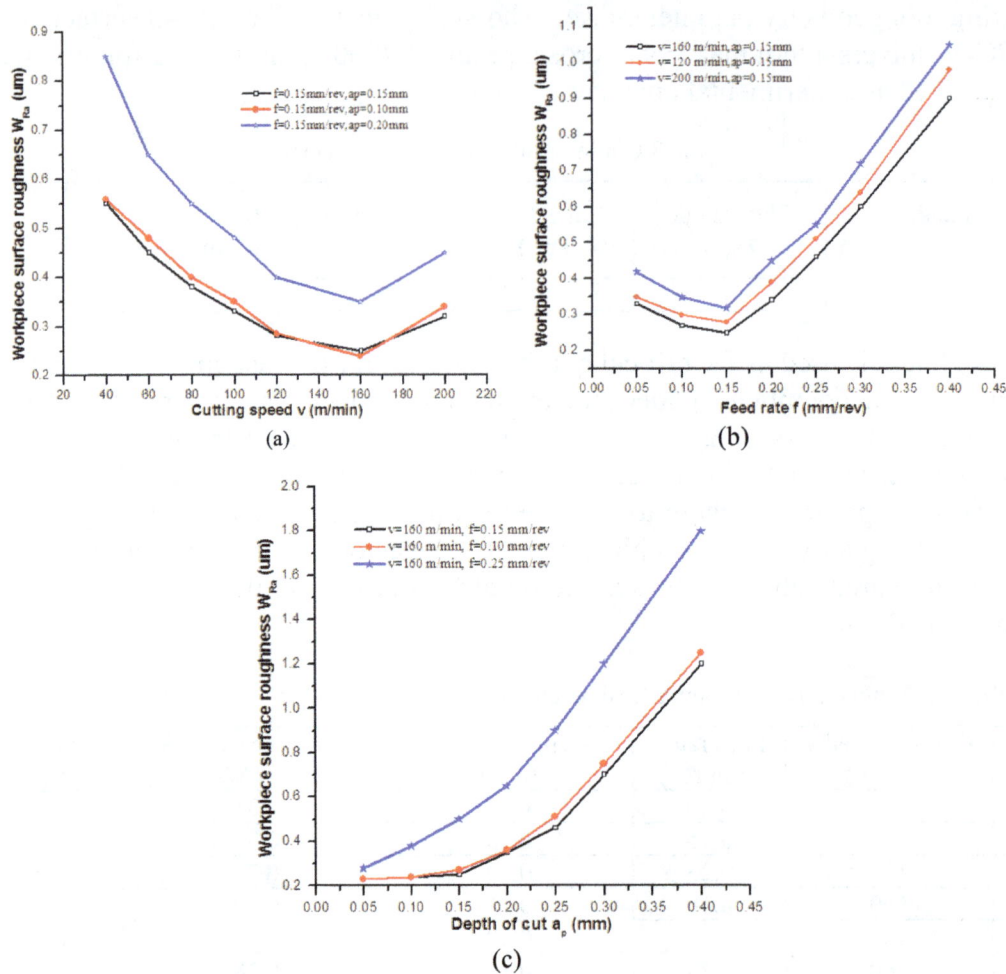

Fig. 1 Relation curve between turning parameters and workpiece surface roughness

It can be seen in this figure that workpiece surface roughness with the increase of cutting depth, slightly reduced then increased, because of the low elastic modulus of titanium alloy, so when the cutting depth is less than 0.05mm, processing produced the spring-back deformation, so that the processing quality rather than cut deep 0.15mm. With the further increase of the cutting depth, the dynamic cutting force also increases, which leads to the deformation of the titanium alloy and the vibration of the machining system, and the surface roughness increases. Fig. 1(c) shows the relationship between depths of cut with workpiece surface roughness. From this figure, it can be known that with increasing feed rate, workpiece surface roughness values increase.

4. Research on Cutting Tool's Flank Wear

Fig. 2(a) shows the worn surface of PCD cutting tools rake face after using PCD continue dry turning Ti-6AL-4V alloy rod ten minutes. Fig. 2(b) shows the worn surface of PCD cutting tools flank face after using PCD continue dry turning Ti-6AL-4V alloy rod ten minutes.

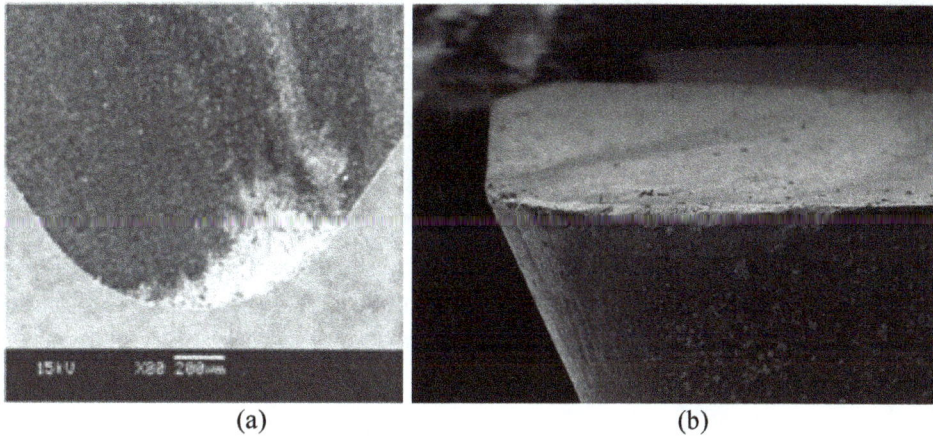

Fig. 2 SEM of PCD cutting tools face

The effects of cutting speed and feed rate values on the cutting tools flank wear were given in Figs. 3(a) and (b). From Fig. 3(a), it can be seen that flank face wear slow vary with the cutting speed increase when cutting speed is below 160 m/min, after that flank wear huge increase. From Fig. 3(b), it can be seen that flank face wear slow increase in the feed rate increase when the feed rate is below 0.15 mm/rev, after that flank wear huge increase.

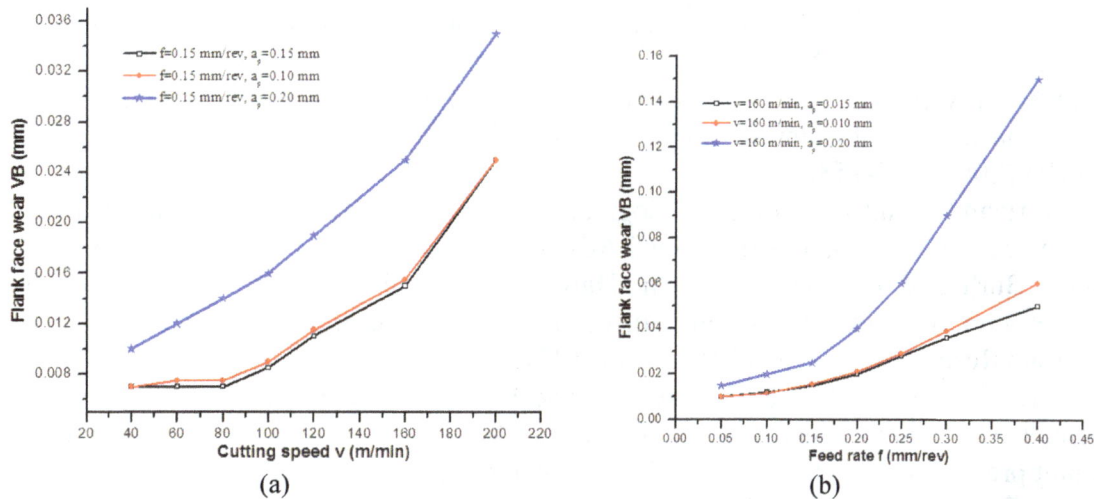

Fig. 3 Relation curve between turning parameters and cutting tools flank wear

5. Conclusion

Through the test and analysis of the above, the conclusion can be obtained as follows:

1. The surface roughness was more sensitive to variation in cutting speed than the depth of cut when the feed rate is below 0.20 mm/rev.
2. The effect of feed rate on surface roughness was consistently higher than that of cutting speed and depth of cut during PCD cutting tools dry turning Ti-6Al-4V alloy workpiece.
3. The effect of feed rate and depth of cut on flank wear was higher than that of cutting speed during PCD cutting tools dry turning Ti-6Al-4V alloy workpiece.

4. Good surface quality can be achieved at low feed rate and depth of cut with high cutting speed in using PCD cutting tools dry turning Ti-6Al-4V alloy. It can be obtained ideal workpiece surface roughness and little cutting tools flank face wear, the cutting speed in the range of 120–160 m/min, the feed rate is 0.15 mm/rev and the depth of cut is 0.15mm.

Acknowledgments

This work is funded by Beijing Natural Science Foundation the Grant No. 3162013 and Beijing Academy of Science and Technology Innovation Team Project No.IG201504N. The authors would also like to thank the anonymous reviewers whose comments greatly helped in making this paper better organized and more presentable.

References

1. V. A. Joshi. Titanium Alloys: An Atlas of Structures and Fracture Features, CRC Press, London, 2006.
2. Ezugwu E, Bonney J and Yamane Y. An overview of the machinability of aero engine alloys, Journal of Materials Processing Technology, 134 (2003) 233-253.
3. Rahman M, Wang Z-G, Wong Y-S. A review on high-speed machining of titanium alloys, JSME Int J, 49 (2006) 11–20.
4. Ezugwu E, Bonney J, Da Silva Rosemar B, and Cakir O. Surface integrity of finished turned Ti-6Al-4V alloy with PCD tools using conventional and high pressure coolant supplies, International Journal of Machine Tools and Manufacture, 47 (2007) 884-891.
5. Nurul Amin A, Ismail A and Nor Khairusshima M. Effectiveness of uncoated WC-Co and PCD inserts end milling of titanium alloy-Ti6Al-4V, Journal of Materials Processing Technology, 192-193 (2007) 147-158.
6. Gert Adriaan Oosthuizen, Guven Akdogan and Nico Treurnicht.The performance of PCD tools in high-speed milling of Ti6Al4V. Int J Adv Manuf Technol, 52 (2011) 929–935.
7. Yanuar Burhanuddin, Che Hassan Che Haron and Jaharah A.Ghani. The Effect of Tool Edge Geometry on Tool Performance and Surface Integrity in Turning Ti-6AL-4V Alloys. Advanced Materials Research, 264-265 (2011) 1211-1221.
8. Rosemar B. da Silva, Alisson R.Machado and Emmanuel O. Ezugwu etc. Tool life and wear mechanisms in high speed machining of Ti-6AL-4V alloy with PCD tools under various coolant pressures. Journal of Materials processing Technology, 213 (2013) 1459-1464.
9. Goutam Devaraya Revankar, Raviraj Shetty, Shrikantha Srinivas Rao and Vinayak Neelakanth Gaitonde. Analysis of Surface Roughness and Hardness in Titanium Alloy Machining with Polycrystalline Diamond Tool under Different Lubricating Modes. Materials Research, 17 (2014) 1010-1022.

Effect of Heat Treatment on Microstructure and Properties of Laser Welded Joint of 304 Stainless Steel Plate

Guo-Lin Guo[1, 2,*], Peng Liu[1], Li Yang[1], Jun Dai[1], Yao-Cheng Zhang[1]

1School of Mechanical Engineering, Changshu Institute of Technology, Changshu 215500, China
2Jiangsu Key Lab of Recycling & Reuse Technology for Mechanical and Electronic Products, Changshu 215500, China
*Email: gguolin@163.com

In this paper, 0.6mm thick stainless steel plates were welded by YAG pulse laser machine. According to the performance parameters of 304 stainless steel and the preliminary experiments, welding parameters were determined. Laser scanning speed is 36mm/min. After welding, different heat treatment processes were carried out on 304 stainless steel samples. Experimental results show that by water cooling at 1000°C for 60min for welding samples, the microstructure of the weld center is fine and uniform equiaxed grains. The weld edge is columnar crystal that perpendicular to the fusion line and extends to the weld center. The hardness of the weld center is 216.39HV. The tensile strength of welded joint is 375MPa. It is much higher than that of the other heat treatment processes. By the heat treatment process, weld quality of the samples is nice, the interface of the joint achieve metallurgical bonding and has higher mechanical properties.

Keywords: 304 stainless steel; laser welding; heat treatment; microstructure; mechanical properties.

1. Introduction

304 austenitic stainless steel has good mechanical properties at room temperature and excellent corrosion resistance. So it is widely used in aircraft, turbine blade, nuclear power station, gas turbine, petrochemical, pipeline and so on [1]. For different needs, it is commonly used in heat treatment system: solid solution treatment, stress relief treatment and stabilization treatment. In this paper, in order to improve the mechanical properties of welded joint of austenitic stainless steel, heat treatment is carried out for 304 stainless steel laser welding joint, the effect of different heat treatment processes on microstructure, hardness and tensile strength of welded joint is analyzed, to provide reference for the development of heat treatment process after austenitic stainless steel is welded by laser machine.

2. Experimental Materials and Methods

2.1. *Experimental materials*

The experimental material is 304 stainless steel. It is machined to the dimension of 60mm×40mm×0.6mm. Its chemical composition (mass fraction, %) is shown in Table 1. Experimental materials are grinding, polishing and cleaning with acetone after grouped, then stainless steel plates are welded.

Table 1 Chemical composition of 304 stainless steel (wt-%)

C	Si	Mn	Cr	Ni	S	P
≤0.07	≤1.0	≤2.0	17.0~19.0	8.0~11.0	0.03	0.035

2.2. *Experimental methods*

304 stainless steel plates are welded with multi-function laser welding machine of HAN SLASER WF300. The welding parameters are shown in Table 2.

Table 2 Laser welding parameters

Table 2 Laser welding parameters

Parameters	Laser pulse power (kW)	Pulse width (ms)	Scanning speed (mm/min)	Pulse frequency (Hz)	Defocusing distance (mm)
Numerical value	6	5	36	13.6	0

The welding process is shown in Fig. 1.

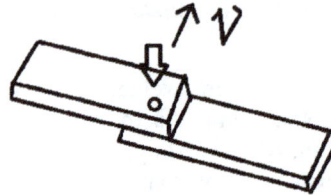

Fig. 1 Laser welding process

Heat treatment is carried out for welding samples, which are cut off by wire cutting for microstructure observation, hardness test and tensile test of welded joint. Heat treatment processes of the welding sample are shown in Table 3.

Table 3 Heat treatment processes for welding samples

Process number	Solid solution treatment			Annealing		Stress relief treatment	
	Heating temperature (°C)	Holding time (t)	Cooling Methods	Heating temperature (°C)	Holding time (t)	Heating temperature (°C)	Holding time (t)
1	1100	0.5	Water cooling	850	2	300	2.5
2	1100	0.5	Air cooling	850	2	300	2.5
3	1000	1	Water cooling	850	2	300	2.5
4	1000	1	Air cooling	850	2	300	2.5
5	900	1.5	Water cooling	850	2	300	2.5
6	900	1.5	Air cooling	850	2	300	2.5

3. Experimental Results and Analysis

3.1. *Microstructure analysis of the joint of different heat treatment processes*

3.1.1. *Morphology of the joint*

The joint morphology of welded samples after heat treatment is shown in Fig. 2. As can be seen from the figure, differences of the joint morphology are small, presenting the shape of U. There is a clear division at the junction of the weld and the base materials. There is an obvious stratification at the bottom of the weld, which is caused by the laser pulse. There is no obvious stratification in the upper part of the weld because of the uniformity of the microstructure after heat treatment.

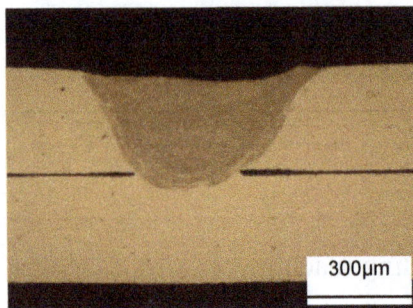

(a) Joint morphology of process 1   (b) Joint morphology of process 2   (c) Joint morphology of process 3

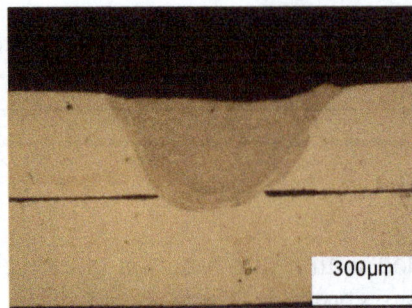

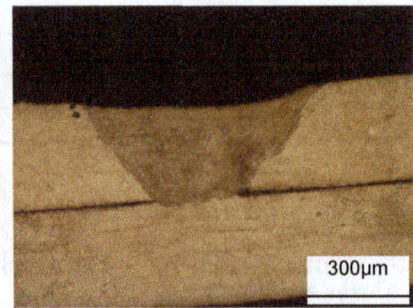

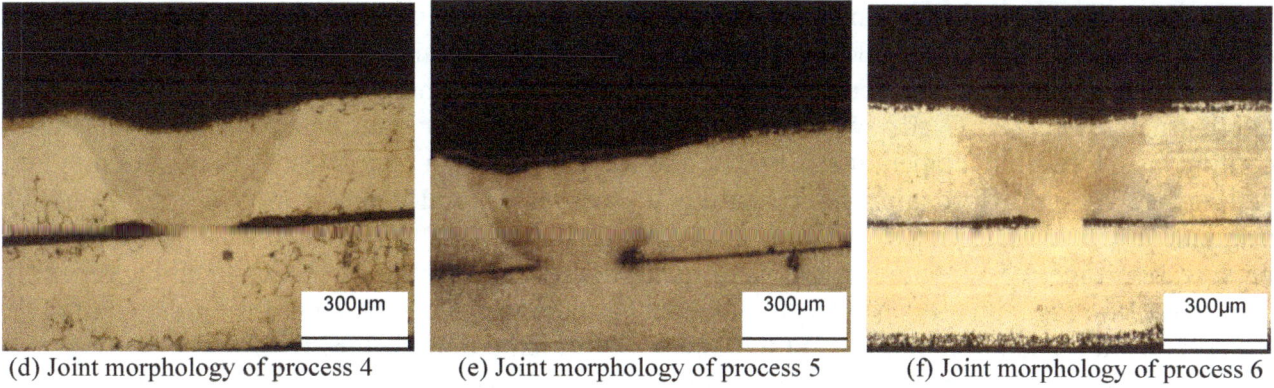

(d) Joint morphology of process 4   (e) Joint morphology of process 5   (f) Joint morphology of process 6

Fig. 2 Joint morphology of different heat treatment processes

3.1.2. *Microstructure of the fusion zone*

Microstructure of the joint fusion zone after different heat treatment processes is shown Fig. 3. Grains size in the base metal is significantly larger than that in the weld zone. The fusion zone appears columnar crystals growing perpendicular to the boundary, because surface free energy of the fusion zone is lower and heat dissipate rapidly vertical to the boundary. So firstly nucleating in the boundary of fusion zone and growing to the weld center along the opposite direction with heat dissipation, finally forming columnar crystal. Because during cooling of solid solution treatment, the edge of the weld zone exists a great undercooling and the heterogeneous nucleation, a large number of nuclei form in the center of the weld zone, growing in every direction. Growing grain crystal soon meet, it is difficult to continue to grow, thus the grains show fine equiaxed in shape in the weld center.

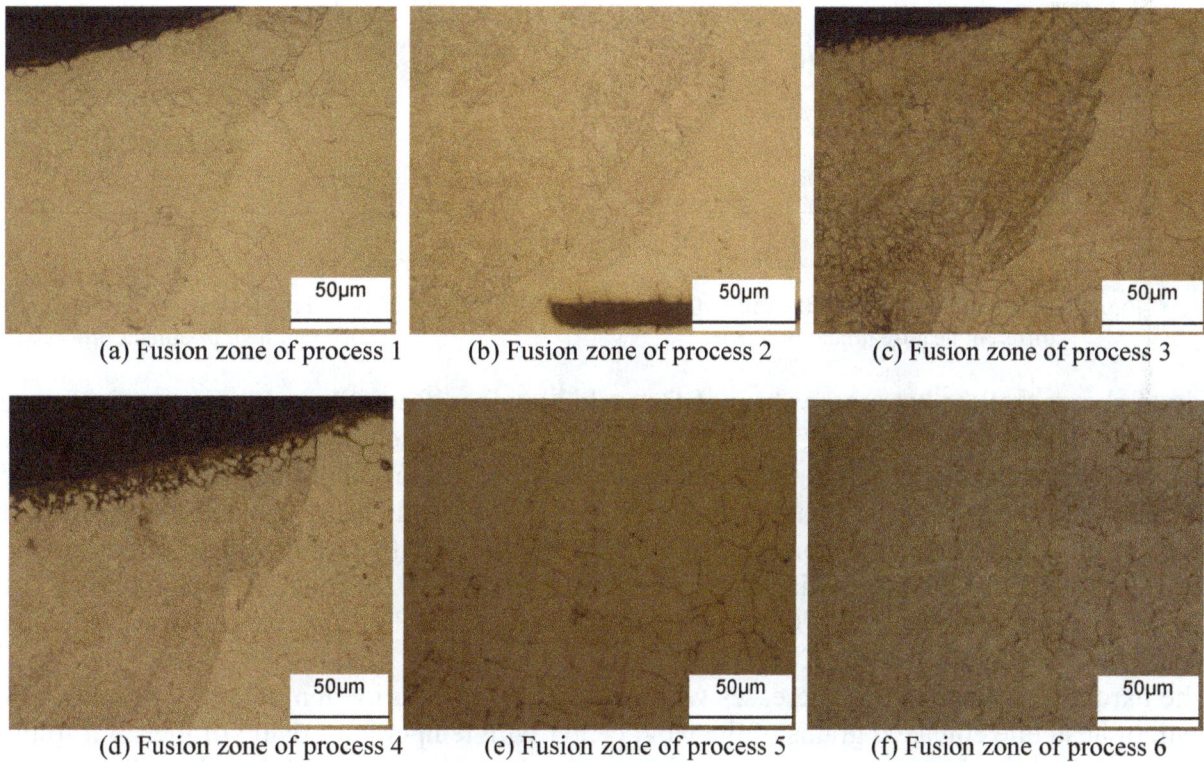

(a) Fusion zone of process 1   (b) Fusion zone of process 2   (c) Fusion zone of process 3

(d) Fusion zone of process 4   (e) Fusion zone of process 5   (f) Fusion zone of process 6

Fig. 3 Microstructure of fusion zone of different heat treatment processes

411

The more fine austenitic organization can be seen from Fig. 3 (c), After solid solution treatment at 1000 °C and water cooling, Rapid cooling of austenite, restraining the precipitation of ferrite and reducing proeutectoid ferrite content, cause fine grains. Fig. 3 (a) is fusion zone microstructure at 1100°C and water cooling. Its grains is bigger than that of 1000°C and water cooling, as the heating temperature is higher and the grain growth rate is faster, the final grain size is larger [2-5]. Rapid quenching after heating above 950°C will remove all carbides from the original structure. Heated to 1100°C, dissolving part or all of the ferrite, it depends on heat preservation time, composition of the weld metal and content of the ferrite. Quenching generally uses water cooling, and air cooling can make the carbide precipitation again during the cooling [4]. So organization has better plastic and toughness at the same temperature with water cooling than that of air cooling, As can be seen from the Figs. 3 (e) and 3 (f), the microstructure is fine austenite and ferrite. 912°C is the temperature of α-Fe to γ-Fe transition, the temperature of 900°C is in the vicinity of A_3 line at solid solution treatment of 900°C, the transformation of austenite is not sufficient, there may also be ferrite. The lower the heating temperature, the slower the growth rate of austenite and the finer the grain size.

3.2. *Microhardness analysis of the welded joint*

Microhardness test was performed by HVT-1000 micro-hardness tester every 0.1mm along transverse section of the joint, the loading force was 100g, and the duration time was 15s. The hardness distribution of the joints in different heat treatment processes is shown Fig. 4.

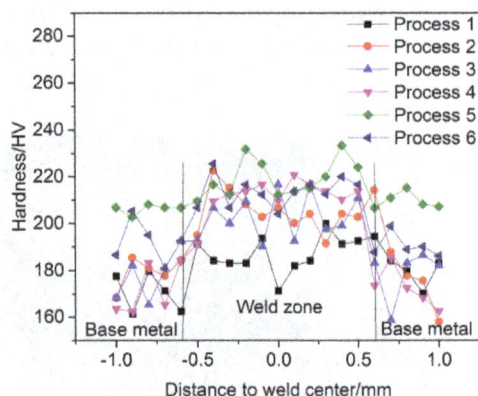

Fig. 4 Hardness distribution of the joint of different heat treatment processes

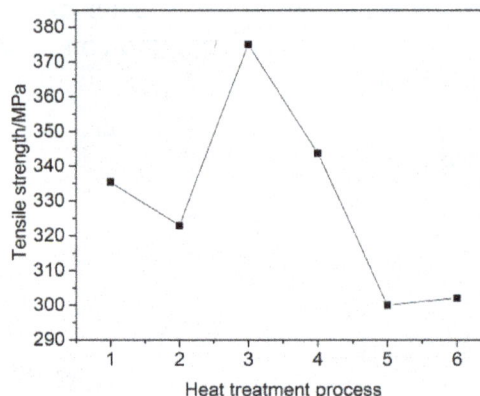

Fig. 5 Tensile strength of the joint of different heat treatment processes

Fig. 4 shows that the hardness values of the weld zone in different heat treatment processes are higher than that of the base metal, because the weld zone is fine equiaxed grain. But the hardness of the weld center is the lowest in the whole weld zone because of incomplete annealing at 850°C after solid solution treatment. Heat treatment at 650~900°C leads to $M_{23}C_6$ and σ-phase to form rapidly in the weld center, the former can cause sensitizing of the welding parts, the latter can reduce the toughness of the welding parts to embrittlement [6, 7]. The hardness of the joint is the highest by water cooling at 900°C for 90min, reaching 219.3HV, and hardness of the joint is lowest by water cooling at 1100°C for 30min, reaching 186.9HV.

The hardness of fusion zone decreases with the increase of solution temperature. Because of the growth of austenite equiaxed grains of the weld center with temperature rising of the solid solution treatment, which leads to hardness reduce. From Fig. 4, the hardness values of weld zone by air cooling at 1100°C for 30min and at 1000°C for 60min are higher than that of weld zone by water cooling at same temperature. The cooling rate of water cooling is faster than that of air cooling and has more internal stress than air cooling [8-10].

The hardness of weld zone by water cooling at 900°C for 30min is higher than that of air cooling, but the hardness fluctuation is greater than air cooling at 900°C because of fast cooling rate of water cooling and more ferrite composition segregate during cooling.

3.3. *Analysis of tensile strength of the welded joint*

The tensile strength of the welded joint in different heat treatment process is shown in Fig. 5. Heating temperature, holding time and cooling mode of the solid solution treatment are the important factors to determine the strength of the weld. The temperature of 900°C is in the vicinity of the A_3 line, there may also be ferrite, resulting in the strength decreasing of the weld. At 1100°C, fully austenitic transformation and grains coarsening lead to reducing of the strength of the weld. At 1000°C, heating temperature is higher than the A_3 line, in the lower temperature range of the austenite zone, the microstructure is fine austenite. So, the tensile strength at 1000°C and water cooling is highest, reaching 375MPa, much higher than other heat treatment processes.

4. Conclusion

The austenitic organization are fine and uniform by the solid solution treatment of 1000°C for 60min and water cooling, because of rapid cooling of austenite, restraining the precipitation of ferrite and reducing proeutectoid ferrite content.

The hardness of weld zone is the highest by water cooling at 900°C for 90min, reaching 219.3HV, and it is lowest by water cooling at 1100°C for 30min, reaching 186.9HV.

The tensile strength of the joint reaches 375MPa by water cooing at 1000°C for 60min, much higher than that of other heat treatment processes.

Acknowledgments

This work was funded by Science & Technology Program of Suzhou City: (No. SGZ2013125), the Open Fund Program of Jiangsu Province Key Laboratory of Recycling Technology of Mechanical & Electrical Products (No. KY301201402-2), the National Natural Science Foundation of China (No. 51401037), the Science and Technology Program of Jiangsu Province of China (No. BK20141228).

References

1. Zhenyu Gu. Effects of different heat treatment methods on grain boundary characteristics and intergranular corrosion resistance of 304 stainless steel [D]. Nanjing: Nanjing University of Science and Technology, 2013. (in Chinese)
2. S. Yang, Z. J. Wang, H. Kokawa. Grain boundary engineering of 304 austenitic stainless steel by laser surface melting and annealing [J]. J Mater Sci, 2007 (42): 847~853.
3. Terada M., Saiki M., Costa I., et al. Microstructure and intergranular corrosion of the austenitic stainless steel 1.4970 [J]. Journal of Nuclear Materials, 2006, 358:40~46.
4. Qingrong Zhang, Bing Yang, et al. Effects of heat treatment on microstructure and properties of 0Cr18Ni9J]. Hot Working Technology, 2013, 42(8):184-186. (in Chinese)
5. Guolin Guo, Li Yang. The investigation on wears resistance of Fe-based alloy cating by argon arc cladding [J]. Applied Mechanics and Materials, 2012, 217-219: 1247~1250.

6. Harish Kumar, P. Ganesh, Rakesh Kaul, et al. Laser welding of 3mm thick laser-cut AISI 304 stainless steel Sheet [J]. Journal of Material Engineering and Performance, 2006, 15:23-31.

7. Qinyi Shi, Yuren Yan, et al. Study on heat treatment process of austenitic stainless steel [J]. Science Technology and Engineering, 2011, 11(24): 5910-5913. (in Chinese)

8. Miao Jin, Xingang Liu, Lu Gao, et al. 316LN Dynamic Recrystallization and Microstructure Evolution [J]. Advanced Science Letters, 2012, 12:398~401.

9. Zhongqi Cui, Yaochun Qin. Metallography and Heat Treatment [M]. Beijing: Mechanical Industry Press, 2007. (in Chinese)

10. [10] Jin Miao, Lu Bo, Liu Xingang, et al. Static recrystallization behavior of 316LN austenitic stainless steel [J]. Journal of Iron and Steel Research, International, 2013, 20(11): 67~72.

Effect of Line Energy Density on Microstructure and Tensile Properties of Cobalt-based Alloy by Laser Direct Metal Deposition

Jin-Bao Li, Shuo Shang, Xiao-Rui Zhang, Chang-Sheng Liu[*]

Key Laboratory for Anisotropy and Texture of Materials, Ministry of Education, Northeastern University, Shenyang, Liaoning, 110816, China
[*]Email: csliu@mail.neu.edu.cn

Direct Laser Metal Deposition (DMD) processes can be utilized to generate functional parts providing an opportunity to generate complex shaped or functionally graded. In this paper, the relationship between energy density and microstructure of cobalt base alloy during DMD process has been discussed mainly. The microstructure, properties and preparation mechanisms of the specimens were studied using metallographic microscopy, scanning electron microscopy and X-ray diffraction. The result shows that relative density of the deposited builds is more than 99.40%. With the decrease of energy density, the relative density is reduced, the grain size and secondary dendritic space are also decrease. The average tensile strength and yield strength is 860.27 MPa and 572.4 MPa, respectively. The average elongation is 19.97%. The deposited specimens show obviously anisotropy at tensile property due to columnar grain microstructure throughout interlayer.

Keywords: Direct deposition; cobalt-based alloy; microstructure; tensile property.

1. Introduction

Nowadays, with the improvement of people's living standard as well as the promotion of life expectancy, the demand for biomedical materials is increasing rapidly. The medical cobalt base alloy, has been widely used in the field about medical implant material, for its excellent mechanical properties, such as excellent wear resistance, corrosion resistance, good biocompatibility[1]. Cobalt base alloy has been regarded as the most typical biomedical metallic materials with wide application scope currently [2,3]. In order to meet individuals' requirements, most biomedical materials need to realize personalized design and production. However, due to the cobalt base alloy is difficult to mechical processing, traditional methods such as forging or casting bring about troubles like high cost and long productive cycle when fabricating production with complex needs of customization.

Direct metal deposition(DMD) is an addictive manufacturing(AM) technology, which has ability to fabricate a wide range of metal components with a complex geometry, strating from metal powder [4]. During DMD process, it combines computer-aided design and a melt pool generted by laser radiation. Simultaneously, raw material is injected into the melt pool protected by nitrogen. By moving the laser beam or worktable, a good metallurgical bonding preparation will be acquired. Prosthetic applications are particularly well suited for processing by means of DMD due to their complex geometry, low volume and strong individualization [5].

As the advantages noted above, a number of previous efforts have been undertaken to develop Cobalt based alloy AM processes. A Cobalt based alloy specimens fabricated by DMLS have excellent hardness values due to ε-lamellae grown on the $\{111\}\gamma$ planes that restricts the dislocations slip in the γ (fcc) phase[6]. It is also reported that LENS is uesd to design and fabricate CoCrMo alloy based novel porous structuer for loading bearing implants. The novel sturcture can potentially eliminate the long standing issues such as stress-shielding, poor interfacial bond between the host tissue and the implant, and wear induced bone loss, and its modulus between 33 to 43 GPa[7] The microstructures, which are a dominant factor for influencing mechanical properties, were also reported on SLM Co–Cr builds [8].

---

[*]Corresponding author

In this paper, metallic components of a Cobalt based alloy produced by the DMD technique were investigated in order to correlate energy density with the corresponding microstructure and machanical property. To this aim, X-ray diffraction analysis, tensile test, electron microscopy observations and Energy dispersive spectrometer were performed on the spencimens.

2. Experimental Procedure and Materials

2.1. *Experimental materials*

Plates of dimension 100mm×50mm×12mm were machined from stainless steel 316L and used as substrate material. The substrate was polished, cleaned with acetone and drying before the deposition runs so as to improve substrate surface laser absorptivity and remove contaminants, respectively. The gas atomized powder was produced by Shenyang Research Institute of Nonferrous Metals used as raw materials for the direct metal deposition process.

Table 1 The chemical composites of the Co21 alloy in wt.%

Element	Co	Cr	Ni	Fe	C	Mo	W	Si
Content(wt.%)	Balance	26.25	3.5	1.21	0.22	6.45	0.08	0.4

The particle size of the powder is in the range of 50-150μm, and the chemical compositions are shown in Table 1.

2.2. *Experimental procedure*

The experiments was carried out in a fabrication system machine, fitted with single-mode continuous ytterbium doped fiber optic laser operating at 1064 nm wavelength produced by GSI corporation (maximum power of 1000W). A coaxial annular powder delivery and 4-axis mechanical work station operation by CNC system is used to deposit the thin wall. Argon gas is used as a shielding and powder carrier gas. Its values is 6L/min and 8L/min, respectively. The specimens were section, mounted, ground and etched with solution ($CuSO_4$:HCl=4.05g:20ml) for SEM (JSM-6510A) and OM (OLYMPUS-GX71) investigation. Phases presented in DMD specimens were identified using XRD (PW3040/60) with Cu Kα generated of 40kV and 40mA and a scanning speed of 4 deg/min. The process parameters showed in Table 2.

Table 2 The process parameters of direct metal deposition process

Level	1	2	3
Power(W)	500	600	700
Scanning speed (mm/min)	540	600	660
Feeding rate (g/min)	30.1	36.1	41.5

2.3. *Tensile property test*

The schematic drawing of tensile test specimen as per GB-T 228-2002 standard is shown in Fig. 1a. Specimens are test on Universal Material Testing Machine(AG-X100kN) produced by DAOJIN with a loading speed 0.5mm/min. All tests were conducted at room temperature. As shown in Fig. 1b and Fig. 1c, two types of deposition pattern are used to investigate the anisotropy of tensile property.

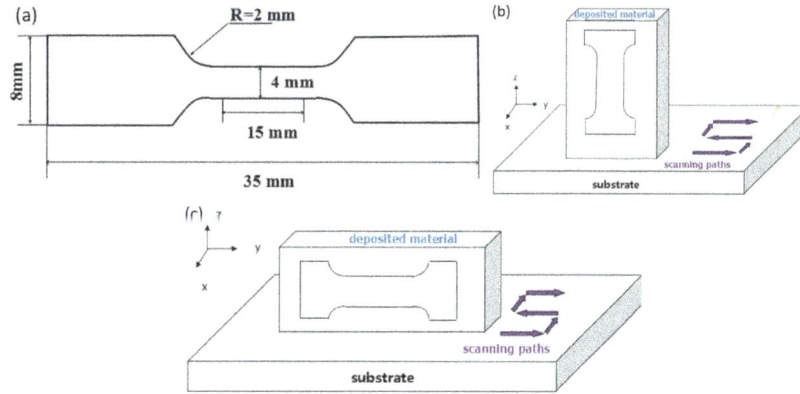

(a) The design drawing (b) sepecimen of vertical deposited direction (c) sepecimen of paralled deposited direction

Fig. 1 The schematic drawing of tensile test specimen

3. Results and Discussion

3.1. *Microstructure*

Fig. 2 shows the microstructure of deposition specimens at transverse cross-section (XOY) and longitudinal cross-section (YOZ) observation. The fine particles are observed in the OM images on the transverse cross-section (XOY) like Fig. 2(a). A SEM observation revealed that the fine particles presents regular equiaxed grain like Fig. 2(b). In the longitudinal cross-section (YOZ) paralleled to the deposition direction(Fig. 2(c)), coarse columnar grain which throughout the inter layer and fine grain zone were observed. It is also clearly seen that the fusion line between two layers, which were formed by laser scan. Fig. 2(d) shows that with the deposition process, some columnar grain growth direction angle between former layer and latter layer is 90 degrees, due to re-melting phenomenon. In addition, the columnar grain direct is not always parallel building direction because of the solidification of melt pool lags behind the laser beam, so that thermal gradient have a lean. This indicates the equiaxed grain are grown along the building direction.

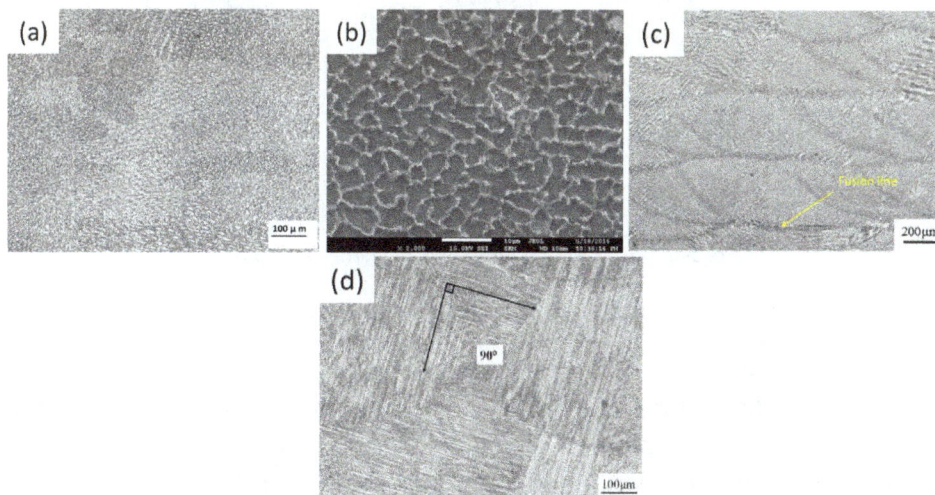

(a) Transverse cross-section 100X   (b) Transverse cross-section 2000X (c) Longitudinal cross-section 50X   (d) Longitudinal cross-section 100X

Fig. 2 The microstructure morphology of specimens

Fig. 3 shows the OM images of specimens in different line energy density, which calculated by the following equation(1).

$$E_V = P/V \tag{1}$$

Where P indicates the laser power (W), V indicates the scanning speed (mm/min), line energy density is the unit of J/mm.

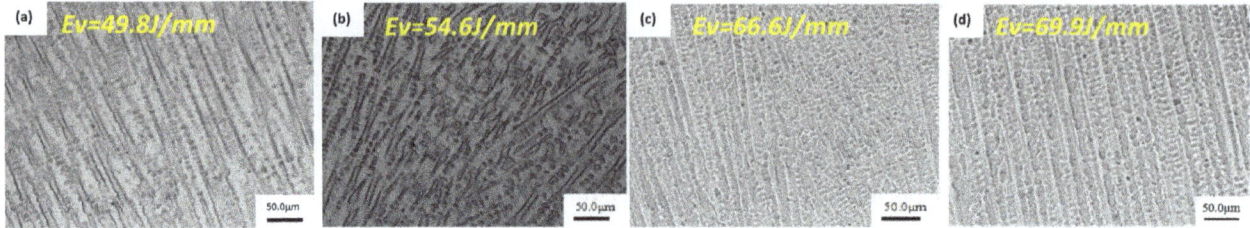

Fig. 3 OM images of grain size and secondary arm space on different energy density

It is easy to find that the average grain size and secondary dendrite arm space are obviously reduced with the increased of line energy density. Because of laser direct deposition is a rapid solidification process, the higher line energy density lead to great cooling rate. According to solidification theory, dendrite arm space is determined by the cooling conditions of solid-liquid surface. The stronger the heat diffusion ability, the smaller the influent area of the latent heat for every dendrite arm, so that the dendrite spacing get smaller, and the great cooling rate will suppress dendrite space grow up.

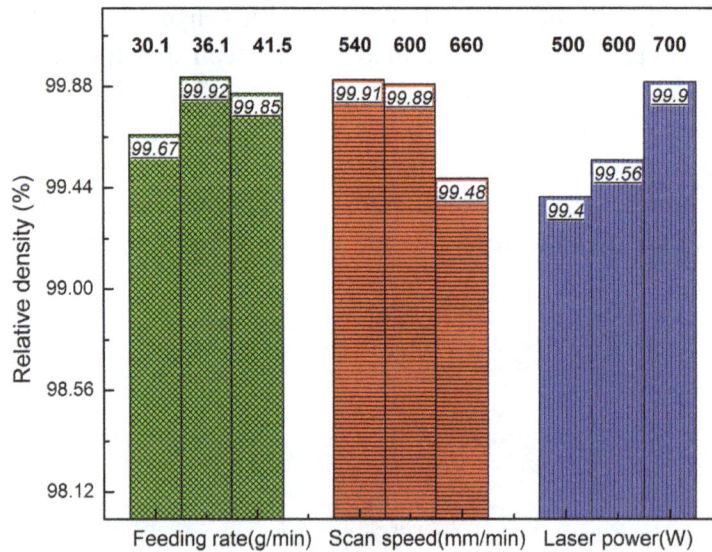

Fig. 4 The relative density of deopsitied specimens at various parameters

As shown in Fig. 4, relative density of the specimens is more than 99.40%. The specimens fabricated by laser direct deposition have higher relative density comparing with selective laser melting and powder metallurgy preparation[9]. The bar graph (Fig. 4) revealed that the increase of laser power and the decreases of scanning speed, which lead to a drop in the line energy density. Therefore, powder can`t fully melt and the porosity of parts preparation increase. Furthermore, when Feeding rate is 36.1g/min, the specimen`s relative density is higher than that of others, because of the energy density and the raw material powder were not match in other parameters. The relative density has line relationship with sufficient input energy. When the feeding rate is

30.1g/min and add up to 36.1g/min, its density is 99.67% and soar to 99.92%. If the raw powder so excessive that laser energy density cannot melted all of them, it lead to the porosity formed and the decreased of relative density.

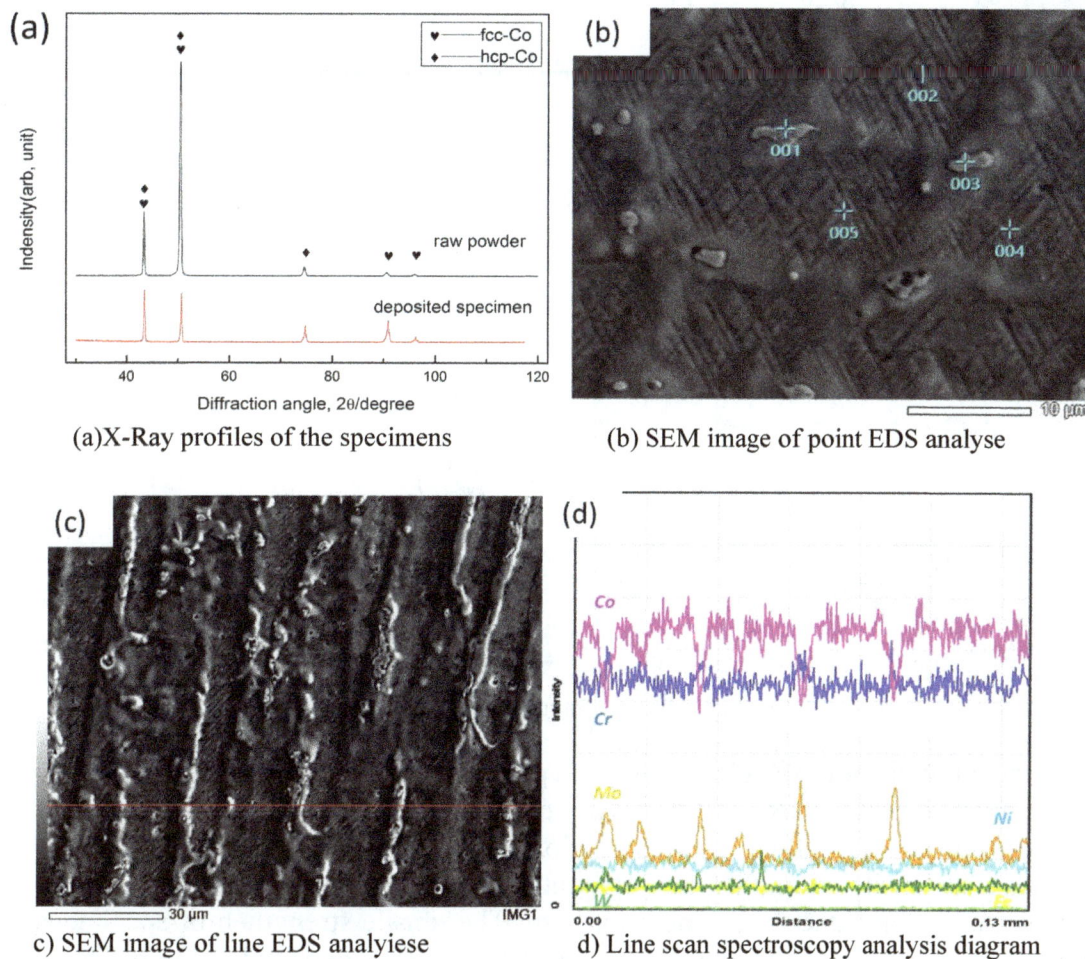

(a) X-Ray profiles of the specimens

(b) SEM image of point EDS analyse

c) SEM image of line EDS analyiese

d) Line scan spectroscopy analysis diagram

Fig. 5 The X-ray profiles of the raw powder and DMD builds.

Fig. 5(a) shows the X-ray profiles of the raw powder and DMD builds. The builds major phase is high temperature stable γ phase, which has a face-centered cubic structure, due to the γ phase have no time to transformed at rapid cooling rate and existence of Ni element which are the role of stable cubic structure. In addition, this phenomenon reveals that Cr and Ni formed supersaturation solid solution. Fig. 5(b) shows SEM observation of DMD builds and point analyses were performed at the interdendritic (1,3) and dendritic (2,4,5).

Combining with Table 3, this indicates the Cr and Mo contents at the interdendritic higher than those at the dendritic. On the contrary, it is notable that the Co contents decline obviously at the boundary of dentritic, which has a white contrast in Fig. 5(c). Therefore, it is clealy that dendritic is easy to corrode due to less Cr contents than interdendritic which shows sunken observation and Cr, Mo element enriched at dendritic boundary, as Fig. 5(d) shows.

Table 3 The compsition list of piont EDS analyises

Element	Si(wt.%)	Cr(wt.%)	Fe(wt.%)	Co(wt.%)	Ni(wt.%)	Mo(wt.%)	W(wt.%)
1	0.22	47.49	0.59	29.37	0.4	21.75	0.18
3	0.41	34.53	0.59	38.68	0.57	24.88	0.41
2	0.15	30.26	1.4	62.42	1.35	4.02	0.4
4	0.11	30.67	1.32	63.01	1.13	3.56	0.2
5	0.12	31.18	1.37	61.61	1.25	4.08	0.2

3.2. *Mechanical properties*

Fig. 6a shows stress-strain curve of deposited specimens at various energy density and those mechanical porperties summarised in Table 4.

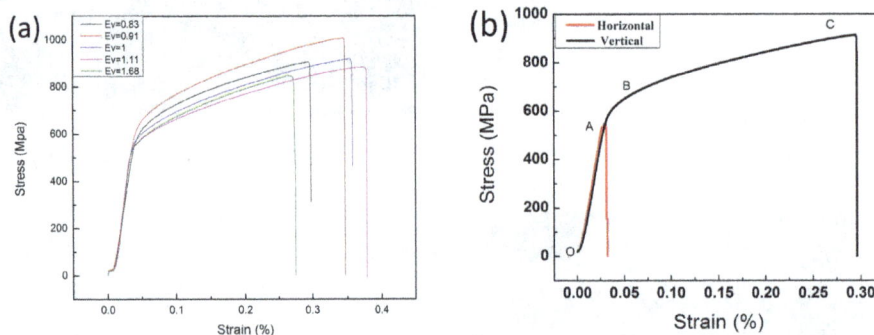

(a)   various energy density   (b)   different tensile direction

(b)   Fig. 6 Stress-strain curves of specimens on

Its typical curve did not exist obviously yield phenomenon and is characterized by high working hardening similar to continuous excessive type curve, which happend at steel material, some times. The average tensile strengthen and elongation is 860.27MPa and 19.97%, respectively. It demonstrates that tensile strengthen and elongation of specimens fabricated by direct deposition were higher than as-cast due to its fine grain size. The tensile strengthen of specimens increased with line energy density at high value, excepting for the energy density is 1J/mm. This reveals that high energy density make the grian size small and promote directional columnar grain formed. Otherwise, when feeding rate is too large to melt by laser, the porosity will be generated as defects which reduce mechanical properties. Fig. 6b shows stress-strain curve of deposited specimens on vertical and parallel to the building direction. The BC line segment clearly reflects the working hardening process. It is noteworthy that the tensile strengthen of horizontal specimen much lower than the longitudinal specimen and shows brittle fracture, because of columnar grain throughout interlayer is not paralleled tensile direction and not contribute to elongation property.

Table 4 Mechanical porperties of the deposition specimens and as-cast alloy

Specimen	Energy density (J/mm)	Yield Strength (Mpa)	Tensile strength (Mpa)	elongation (%)
1	0.83	591	905	20.4
2	0.91	605	1003	28.71
3	1	533	845	18.71
4	1.11	562	882	31.36
5	1.68	571	917	26.82
As-cast		450	655	8.0

Fig. 7 shows surface morphology of specimens after tensile test. The fracture surfaces of vertical specimen shows the dark grey contrast and serrated observation at low magnification. The dimple-like parttens were observated and have a large quantity in Fig. 7(b). This indicates that the verctical specimens are ductile fracture with great elongation, because of the tensile direction of verctical specimen was near <0 0 1>, the slip system of {1 1 1}<1 1 0> works during plastic deformation [10] and also contributed to fine grain structure. The horizontal specimen shows the cleavage surface observation and tearing ridge taken as typical brittle fracture features in Fig. 7c.

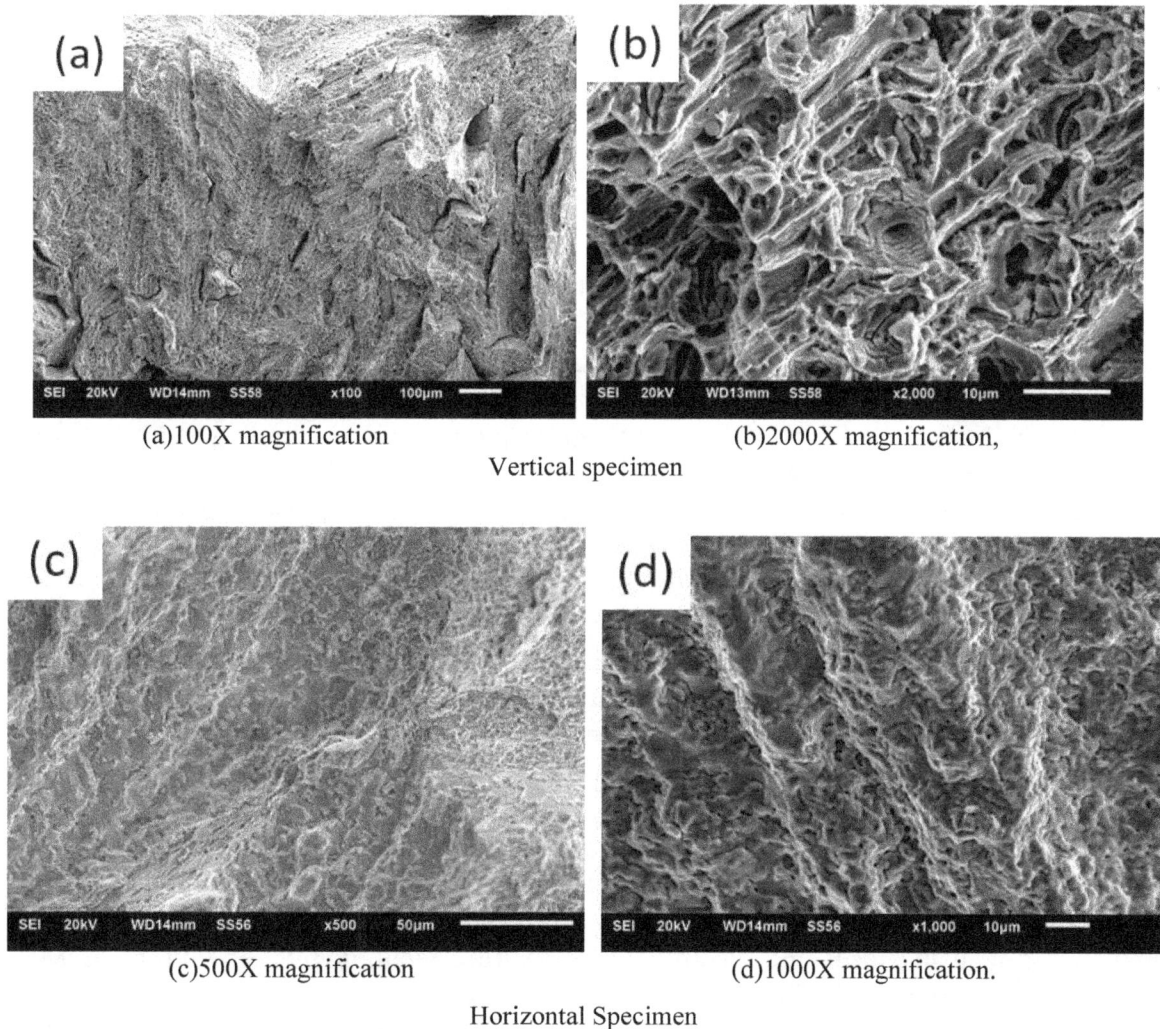

(a)100X magnification   (b)2000X magnification,

Vertical specimen

(c)500X magnification   (d)1000X magnification.

Horizontal Specimen

Fig. 7 The surface morphology of tensile fracture.

4. Conclusion

1. Relative density of the specimens fabricated by direct metal deposition is more than 99.40% and its relative density drops with the decreased of line energy density.

2. The microstructure of DMD builds consist of fine equiaxed grain and colunmar grain. Higher energy density lead to cooling rate increased so that the grain size and secondary dendritic space become smaller.

3. The DMD builds shows excellent mechanical properties as the value of average tensile strength arrive at 860MPa and reflects obviously the working hardening process. The deposited specimens show anisotropy at tensile property when the energy density in a large value, the specimens ,which paralleled deposited direction will form columnar grain microstructure throughout interlayer and make its tensile strength higher the casting part. Otherwise, the specimen shows brittle fracture.

Acknowledgments

This research was funded by the NSFC – Liaoning joint fund key projects (U1508213), GUANGDONG frontier and key technological innovation special funds (2015B010122001) and the ministry of education of institutions of higher learning basic scientific research business expenses project (N130810001).

References

1. G.D. Janaki Ram, C.K. Esplin, B.E. Stucker, Microstructure and wear properties of LENS deposited medical grade CoCrMo, J. Mater. Sci. Mater. Med. 19 (2008) 2105–2111.
2. Chunlei Qiu, G.A. Ravi, Moataz M. Attallah, Microstructural control during direct laser deposition of a β-titanium Alloy, J. Materials and Design. 81 (2015) 21-30
3. M.B. Nasab, M.R. Hassan, B.B. Sahari, Metallic biomaterials of knee and hip — a review, Trends Biomater, J. Artif. Organs. 24 (2010) 69–82.
4. Scott M. Thompson, Linkan Bian, Nima Shamsaei, et al., An overview of Direct Laser Deposition for additive manufacturing; Part I: Transport phenomena, modeling and diagnostics, J. Additive Manufacturing. 8 (2015) 36-62.
5. I. Gibson, L.K. Cheung, S.P. Chow, W.L. Cheung, S.L. Beh, M. Savalani, S.H. Lee, The use of rapid prototyping to assist medical applications, J. Rapid Prototype. 12 (2006) 53–58.
6. G. Barucca, E. Santecchia, G. Majni, et al., Structural characterization of biomedical Co–Cr–Mo components produced by direct metal laser sintering, J. Materials Science and Engineering C. 48 (2015) 263-269.
7. Félix A. España, Vamsi Krishna Balla, Susmita Bose, Amit Bandyopadhyay, Design and fabrication of CoCrMo alloy based novel structures for load bearing implants using laser engineered net shaping, J. Materials Science and Engineering: C. 30 (2010) 50–57.
8. Murr, L.E., Gaytan, S.M., Ramirez, D.A., Martinez, et al, Metal fabrication by additive manufacturing using laser and electron beam melting technologies, J. Journal of Materials Science & Technology. 28 (2012) 1–14.
9. Guifang Sun, Rui Zhou, Jyotirmoy Mazumder, Evaluation of defect density, microstructure, residual stress, elastic modulus, hardness and strength of laser-deposited AISI 4340 steel, J. Acta Materialia. 84 (2015) 172-189.
10. Atsushi Takaichi, Suyalatu, Takayuki Nakamoto, et.al., Microstructures and mechanical properties of Co–29Cr–6Mo alloy fabricated by selective laser melting process for dental applications, J. journal of the mechanical behavior of biomedical materials. 21 (2013) 67-76.

The Modal Analysis of Large Diameter Percussive Reverse Circulation Drill Bit

Xiao-Qin Shen, Huan-Huan Li*, Bo Wang, Qi-Liang Cheng

School of Mechanical and Electronic Engineering, Shandong Jianzhu University, Jinan 250101, China

Email: lhh890426@126.com

Large diameter percussive reverse circulation drill bit is the foremost part of the percussive rig. It is mainly used in the construction of large diameter bored pile foundation. It also has some problems, such as short service life, easy damage, difficult to test etc. The theory of the finite element modality analysis is applied to research the dynamic characteristic of the drill. It aims to get references for dynamic analysis and optimization design of the drill. A modal analysis of drill bit is made in the cases of free modal and node constrain based on introducing some of the drill bit's failure mode in this paper. The modal vibration mode and natural frequency are obtained in these two cases through simulation. The node of drill bit's maximum deformation is also found out and analyzed. The simulation results are compared with the actual damage drill. It indicates that they are both consistency. The results provide basis for further optimization design and precision of large diameter percussive reverse circulation drill bit.

Keywords: Large diameter percussive reverse circulation drill bit; modal analysis; mode shape.

1. Introduction

Large diameter drilling is a drilling technique which uses mechanical and broken rock cutting tools to break the rock. Impact of reverse circulation drilling is one of the large diameter drillings, which is a combination of the traditional percussion drilling technology and the continuous loop wire rope discharge technology [1-3]. Large diameter impact reverse circulation drilling rig is widely used in geological exploration, which can be used in complex strata such as floating pebble and hard rock [4]. As a matching drill bit of a large diameter impact drilling rig, there are still many problems about large diameter percussive reverse circulation drill bit. Because of the large diameter of the drill bit, the structure is easy to break, sticking and so on. The larger the diameter of the drill bit is, the easier it is to break, and the shorter life it is [5]. It is difficult to get the data through the general test because of large size and weight. The large diameter drill bit weighs 8 tons whose diameter reaches to 2.5m. It is difficult and expensive to carry out a prolonged fatigue test for different ground conditions. Therefore, it is difficult to carry on the research through the traditional experiment. The structural design of the drill bit needs to be further optimized. Although there are a lot of problems about large diameter percussive reverse circulation drill bit, its related research is very little. At present, the method of finite element analysis is used to analyze the fatigue failure of other small drill bits, which has certain reference significance for the analysis of the drill bit.

The damage form of the drill is weld cracking, rib fracture. The damage of the drill bit directly affects the normal operation of the construction. Study of bit failure problem, finding out the bit of pixels, the modal analysis of the drill bit are very significant for further optimization, improved application and development of the drill bit. So the modal analysis of the drill bit is carried out on the cases of free modal and node constrain in view of the short life ,the big volume of the drill bit and the difficult test and soon on. The analysis results are compared with the failure of drill bit in actual working conditions, which have laid a good foundation for the further optimization design of the bit.

*Corresponding author

2. Failure Mode of Drill Bit

The drill bit is matched with a large diameter impact reverse circulation drill, which is mainly used in the construction of large diameter bored pile foundation. The structure of construction stratigraphic is diverse and complex. Common ground soil includes sand, gravel layer and soft rock layer [4]. But the complex stratum such as drift gravel, hard rock formations are very common. The drilling works in low efficiency when working in complex strata such as floating pebble. It is easy to cause the sticking, ramp-hole, bit falling off, holes collapse and other accidents. Those seriously affect the normal operation of the construction so that the drill bit is very serious destroyed. The damage of drill bit can generally be divided into two categories: one is the damage of the bit body, the broken neck, crack and whole out; the second is the damage of hard alloy tooth, that is, drop, broken teeth and wear [6, 7].

3. Modal Analysis of Drill Bit

3.1. *Theory of modal dynamics analysis*

The finite element analysis method is a common method of computational tools and state analysis for a linear constant system, which has N degrees of freedom. Under the action of external force F (t), the basic equation for vibration is following:

$$[M]\{\ddot{X}(t)\} + [C]\{\ddot{X}(t)\} + [K]\{X(t)\} = \{F(t)\} \tag{1}$$

Interpretation of the equation :
$[M]$ is the mass matrix of the elastic system; $[C]$ is the damping matrix;

$[K]$ is the stiffness matrix; $\{\ddot{X}(t)\}, \{\dot{X}(t)\}$ and $\{X(t)\}$ are vector matrix of acceleration vector,

the velocity vector and displacement vector matrix; $\{F(t)\}$ is the vector matrix Dynamic incentive load vector.

$[C]$ and $\{F(t)\}$ are zero in eq.1 [8], so it can be simplified to:

$$[M]\{\ddot{X}(t)\} + [K]\{X(t)\} = 0 \tag{2}$$

Since the free vibration of any elastomer can be decomposed into a series of superposition which is composed of harmonic vibration. According to the superposition theorem, the solution of eq.2 is the following formula:

$$\{X(t)\} = \{X_0(t)\} \sin \omega t \tag{3}$$

Further, substitute eq.2 into eq.3, you will acquire the result:

$$([K] - \omega^2[M])\{X_0\} = 0 \tag{4}$$

The amplitude of each node is not all zero in structural free vibration. So the equation(4) must be satisfied the following requirement if it make the eq.4 true:

$$[K] - \omega^2[M] = 0 \qquad (5)$$

Eq.5 is N-order algebraic equation about ω^2, which can acquire N intrinsic frequency through solving this equation. For each intrinsic frequency, you would acquire the amplitudes of each node, which is the damping of the structure through solving eq.4.

Any structure has its inherent frequency, which has been associated with certain forms of vibration when the inherent frequency of some structure is activated. It is called vibration mode that an inherent frequency corresponds to a kind of vibration mode. Large diameter percussive reverse circulation drill bit will set a certain impact frequency in actual work. We should be aware of the inherent frequency of the drill bit in order to prevent form happening resonance and causing the damage of the drill bit. In this paper, a simplified three-dimensional model of drill bit is made of modal analysis by calculating the first-order natural frequency of the vibration mode.

3.2. *Establishment of model*

Firstly, the 3D software of SolidWorks is used to model the drill bit, and then the model is imported into workbench for meshing generation.

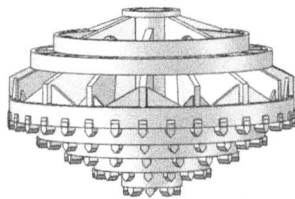

| Fig. 1 The 3D model of drill bit | Fig. 2 The finite element model of drill bit |

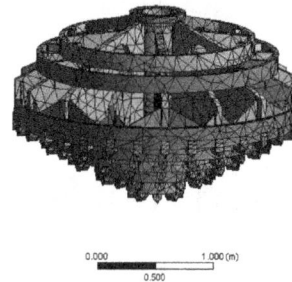

In order to simplify the calculation, the lifting structure of the drill bit is removed in the calculation process, but the main body part of the drill bit is retained. The drill bit is a welded structure, so the strength of the weld seam is similar to that of the base metal. The drill is considered as a whole for the simplified calculation without considering the weld. The actual drilling bit of the bulk material is Q235.The density of the material is 7800kg/m³; the elastic modulus is $2.1 \times 10^{11} N/m^2$; Poisson's ratio is 0.28. In this method, the drill bit is divided by the global mesh, and the tetrahedral element is adopted. The mesh is divided into 96809 nodes and 50475 elements. The simplified 3D model and finite element model of the drill bit are shown in Fig. 1 and Fig. 2.

Table 1 Frequency Response Chart of Drill Bit

Order /(i)	1	2	3	4	5	6	7	8
Frequency /(Hz)	0	1.41e-4	4.41e-4	1.26	1.28	1.31	213.27	235.5

3.3. *Modal analysis*

After bit dynamics model is built, the modal analysis of the bit is carried out under the free states and the constraint condition of the entities drill bit. The results obtained are compared with the actual failure of the drill bit.

3.3.1. *Modal analysis of drill bit under the free state*

The modal analysis of the drill bit is carried out under the condition of complete freedom without any restriction. Because the bit is under the complete free state, the first six order vibration of drill bit is rigid body motion, which is the rigid body mode of the bit. The natural frequency of drill is close to 0, and the frequency value grow significantly from the seventh order, and the related scholars take the seventh order as the first order mode of the drill bit [9, 10]. This research method is also adopted in this paper.

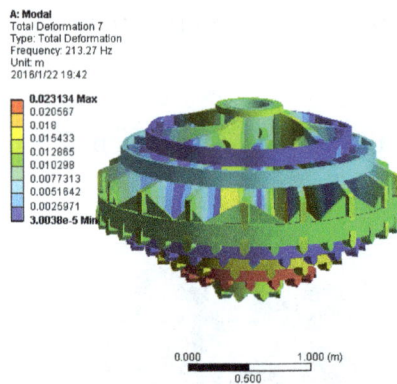

Fig. 3 The first mode of drill bit

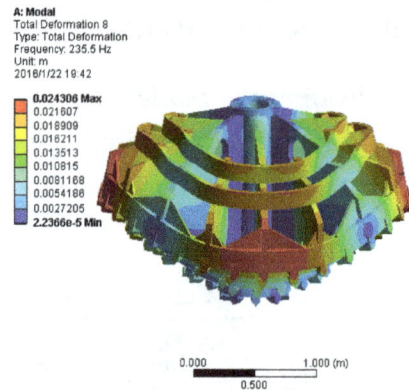

Fig. 4 The second mode of drill bit

The frequencies of the drill bit under the complete free state are shown in Table 1. The first two order modes are shown in Fig. 3 and Fig. 4.

In Fig. 5, the deformation of the first order mode of the drill bit can be seen that the deformation of the inner ring 1 and 2 of the drill bit is larger than those of the other parts of the drill bit. It explains that the impact force of the inner ring 1 and 2 of the drill bit is larger than those of the other inner ring. According to the actual damage of the drill bit (Fig. 6), most of the damage points of the drill bit is in the inner ring 1 and 2 of the drill bit .The simulation results are consistent with the actual situation. The results verify the feasibility of the simulation results, besides the modal analysis of the drill bit lays the foundation for the further optimization design of the drill bit.

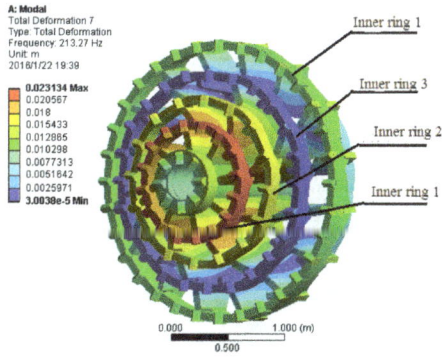

Fig. 5 The section of first order mode shape

Fig. 6 Actual damage of drill bit

3.3.2. *Modal analysis of drill bit under constraint condition*

In the actual working condition, the drill bit is suspended by the steel wire rope, and the punching operation is completed by the gravity action of the bit itself. In the case of no impact working, the drill bit can't move up and down, in addition, it can't move around. At this point, the equivalent of the drill bit is fully constrained. The fixed constraint is applied to the top of the bit, and then the solution is solved. After solving, the solution of the post-processing results can be seen. The frequency of each order is shown in Table 2.The first six order modes are shown in Fig. 7.

Table 2 The First 6 Orders of the Drill Bit under Fixed Constraints

Order/(i)	1	2	3	4	5	6
Frequency f/(Hz)	25.84	46.62	46.725	207.93	208.25	219.47

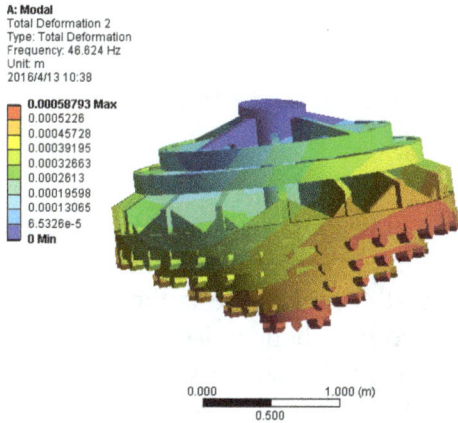

(a) The First Order Mode

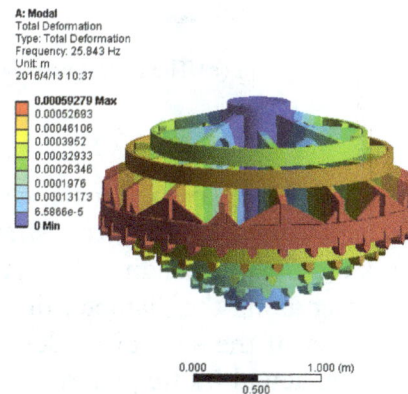

(b) The Second Order Mode

Fig. 7 The first six order modes

427

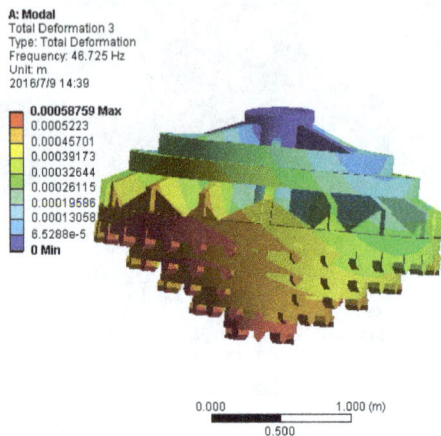

(c) The Third Order Mode

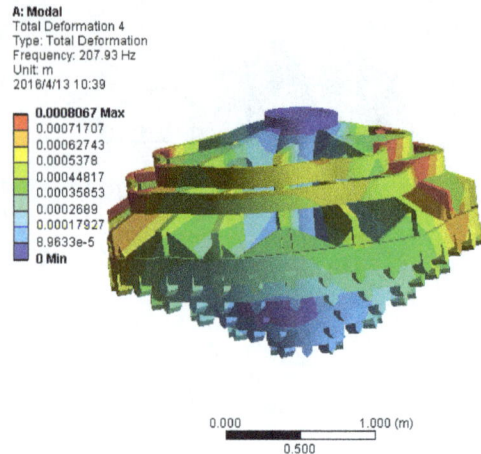

(d) The Forth Order Mode

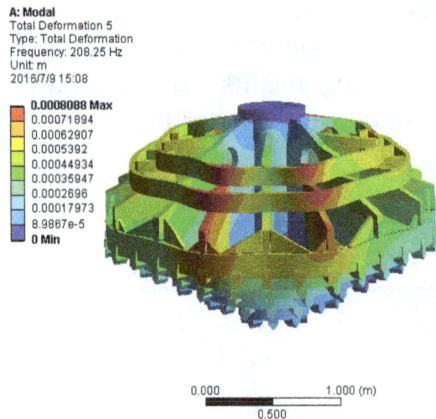

(e) The Fifth Order Mode

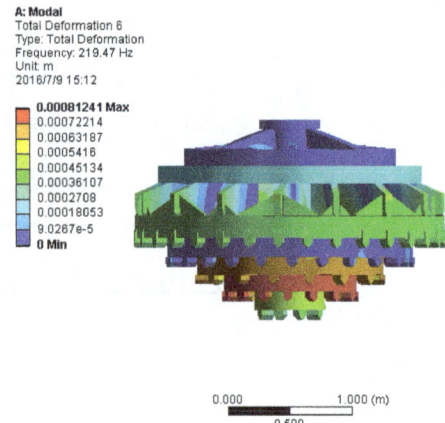

(f) The Sixth Order Mode

Fig. 7 (*Continued*)

From the above six pictures, it can be seen that connecting plates and ribbed plates distort along the diameter direction under constraint condition. Those are the first order mode. The second order mode is the overall bending deflection of the drill bit. The forth order mode is the inner ring and the top ring along the diameter direction inward bending. Core tube bends along the axial of drill bit. The rest of the several modes are no different than the original one. Low order frequency is close to the actual bit frequency. The impact frequency is 0-30 bit / min .The inherent frequency of the bit is much smaller than that obtained by the simulation 25Hz, in the actual impact process. So the resonance does not occur.

4. Conclusion

In this paper, the modal analysis of large diameter impact reverse circulation bit is carried out under the condition of free mode and full constraint. The first few modes of the drill bit are obtained in the two cases, respectively. Damaged points which reflect a bit deformation modes of the drill bit are compared to damage of drill bit under actual working condition. Results show that:

1. Under the complete free state, the first order mode of the drill bit is obtained. The inner ring 1 and 2 suffer more force. The same as the actual damage is.
2. Under the constraint condition, the natural frequency of the drill bit is 25 Hz, which is much larger than the impact frequency of the drill bit. Resonance wouldn't occur. It indicates that the drill bit works safely in impact frequency.
3. The first mode represented as connecting plates and ribbed plates distorting along the diameter direction. After constraints are imposed, the rigidity of drill bit increases. The drill bit distorts after adding excitation. So in order to increase the impact strength of the drill, it needs add a connection plate and stiffened plate fin.

The results of finite element simulation are consistent with the results of real data. It solves the technical problems which are not easy to test because of the large size of the large diameter impact reverse circulation drill bit. It lays the foundation for the further optimization and improvement of the large diameter impact reverse circulation drill bit. It is beneficial to the further development and application of large diameter impact reverse circulation drill bit.

Acknowledgments

This project was funded by A Project of Shandong Province Higher Educational Science and Technology Program（J14LB05）, PhD Start-up Fund of Shandong Jianzhu University (XNBS1014).

References

1. Zhen-Ya LI Status Quo and Trend of Percussive Reverse Circulation Pile Hole Drill Rigs in China [J]. Exploration Engineering, 2001, (1): 57-59.
2. Li-Jun SU. Brief Introduction of Large Caliber Boring Technique and Bore Bit Type [J]. Zhejiang Hydrotechnics, 2001, (5): 53-54.
3. Xu-Ming LIU. An Initial Discussion on the Development of Engineering Boring Machine for Large Diameter Pile Foundation [J]. Minerals Resources and Geology, 200, 15(S1): 563-566.
4. Ding-Cheng HU. CFZ-1500 Impact Reverse Circulation Drilling Rig in Complex Strata [J]. Geology and Exploration, 36(2): 35-36.
5. Jing-Dong SI, Wei HE, Zu-Hao JIANG. Application of Percussion Reverse Circulation Drilling Method for Large Diameter Pile Hole Construction[J]. Western Exploration Engineering, 2005, (11): 172-173.
6. Xiang-Dong LIU. Research on Main Geometric Parameters of PDC Bit Influencing On the Mechanical Behavior [D]. Xi'an: Xi'an University of Science and Technology, 2009
7. Ya-Lin CAO. An Approach to the Cause of Failure of Large Diameter Long-hole Drilling Tools [J]. Mining Research and Development, 2000: 33-34.
8. T. Wang, O. Celik. A Frequency and Spatial Domain Decomposition Method for Operational Strain Modal Analysis and its Application [J]. Engineering Structures, 2016, (114): 104-112.

9. Li-Gang CAI, Shi-Ming MA, Yong-Sheng ZHAO, Zhi-Feng LIU. Finite Element Modeling and Modal Analysis of Heavy-duty Mechanical Spindle under Multiple Constraints [J]. Journal of Mechanical Engineering, 2012, 48 (3): 165-173.

10. Tian-Biao YU, Xue-Zhi WANG, Peng GUAN, Wan-Shan WANG. Modal Analysis of Spindle System on Ultra-high Speed Grinder[J]. Journal of mechanical engineering, 2012, 48 (17): 183-187.

Effect of AlMnMgFe Deoxidization on Microstructure and Mechanical Property of Low Carbon and Low Alloy Steel

Dong-Ping Zhan[1,*], Guo-Xing Qiu[1], Yang-Peng Zhang[1], Zhou-Hua Jiang[1], Hui-Shu Zhang[2]

School of Metallurgy, Northeastern University, Shenyang 110004, China
School of Metallurgy Engineering, Liaoning Institute of Science and Technology, Benxi, 117004, China
Email: zhandp1906@163.com

In order to study the impact of Mg deoxidation on the inclusions and properties of Low Carbon and Low Alloy Steel, five heats of 5.72%Mg-50.5%Al-7.92Mn-Fe Alloy (AlMnMgFe) deoxidation trails were done during a 150 tons BOF taping at a steelmaking plant. Optical microscope (OM) was used to test the diameter and number of the inclusions. SEM and EDS were used to examine the types of the inclusions. Then the microstructures and properties of the steels were tested at the room temperature. The results show that, by using AlMnMgFe as the deoxidizer, the inclusions, microstructures and properties of the low carbon and low alloy steel Q195 are improved. There are some spherical $MgO \cdot Al_2O_3$ inclusions in the AlMnMgFe deoxidation heats. The diameters of 97.01 percent inclusions are smaller than 3 μm in the AlMnMgFe deoxidization steel, and it is higher than that of Al-killed steel. The mean yield strength, mean tensile strengths and mean elongation of the AlMnMgFe deoxidization steels are higher than Al-killed steels.

Keywords: Steel; Mg deoxidization; inclusion; microstructure; property.

1. Introduction

Knowing that non-metallic inclusions affect the steelmaking processing and properties of steel, efforts for their control, removal or modification have increased in the past few years. Different inclusion control or utilization technologies for clean steelmaking are being continuously developed to fulfill the ever increasing demands on material properties [1, 2]. The addition of calcium to the aluminum killed steel promotes partial reduction of the Al_2O_3 inclusions, giving rise to the formation of liquid calcium aluminates with low melting point and spherical morphology [3-6]. The previous work studied the effect of AlMnCa additions on the deoxidization and modification of Al_2O_3 inclusions in a low carbon and low silicon steel and AlMnFe is used to be the contrastive deoxidizer [7]. Some other alloys such as barium [8, 9] or magnesium [10-14] were tried to improve the inclusions in steel. 430 stainless steel treated by Mg-Al alloy showed better properties than the contrast steels [15]. In present work, the inclusions and properties of a low carbon and low alloy steel Q195 deoxidized by 5.72%Mg-50.5%Al-7.92Mn-Fe alloy were investigated.

2. Plant Trail Method

2.1. *Industrial trial process description*

The industrial trials have been done at a steelmaking plant in China. The molten steel, having a temperature of ~1650°C, was tapped from 150 tons Basic Oxygen Furnace (BOF). During the normal taping, ferrosilicon, ferromanganese, ferro-aluminium and some calcium aluminates synthetic slag were added into the ladle. For the five tested heats, the ferro-aluminium were replaced by 5.72%Mg-50.5%Al-7.92Mn-Fe Alloy (AlMnMgFe) for deoxidation. Subsequently, the ladle was stirred by bottom argon blowing to promote the reaction in the molten steel, the growth and separation of inclusions from the steel into the slag.

---

*Corresponding author

Table 1 Compositions of the tested steels [wt%]

AlMnMgFe deoxidization steel					Al-killed steel					
Heat No.	C	Si	Mn	P	S	Heat No.	C	Si	Mn	P	S
1	0.02	0.01	0.18	0.019	0.013	1	0.04	0.04	0.26	0.023	0.013
2	0.03	0.01	0.19	0.019	0.015	2	0.06	0.01	0.16	0.010	0.015
3	0.08	0.03	0.23	0.012	0.017	3	0.05	0.03	0.20	0.019	0.012
4	0.04	0.02	0.22	0.014	0.013	4	0.06	0.03	0.19	0.017	0.013
5	0.04	0.02	0.23	0.023	0.020	5	0.05	0.03	0.21	0.020	0.011

Finally, the molten steel was cast to slab and then hot rolled. The compositions of the AlMnMgFe deoxidation steels Q195 are listed in Table 1.

2.2. *Sample tested method*

Steel samples were taken from ladle and hot rolled plate. The size, quantity and distribution of inclusions were investigated by using the Axio Imager M2m Optical microscope (OM) under the magnification of 500 times. The morphology and compositions of inclusions were analyzed by the SSX-550 scanning electron microscope (SEM) and X-ray energy dispersive spectrometer (EDS). The specimens were etched with the reagent which contains 1 grams of picric acid, 5 ml of hydrochloric acid and 100 ml anhydrous alcohol, then observed the microstructure by scanning electron microscope. According to the requirements of GB/T228-2002 and GB/4338-84, specimens were processed into size of 4 mm ×10 mm × 86 mm. The tensile tests were carried out by the AG-X100kN electronic almighty material experiment machine at room temperature. The tensile speed was 1.5mm/min.

3. Results and Discussion

3.1. *Effect of AlMnMgFe deoxidation on inclusion*

During deoxidation, the aluminum and/or magnesium will react with oxygen in the steel as Eq. 1 or Eq. 2, and then Al_2O_3 or $MgO \cdot Al_2O_3$ are formed. The typical inclusions in the steel samples taken from ladle are shown in Fig. 1 and Fig. 2. It is shown that, there are some Al_2O_3 or Al_2O_3 cluster in the normal Al-killed steel. For the AlMnMgFe deoxidation heats, there are some spherical $MgO \cdot Al_2O_3$ inclusions.

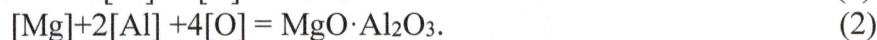

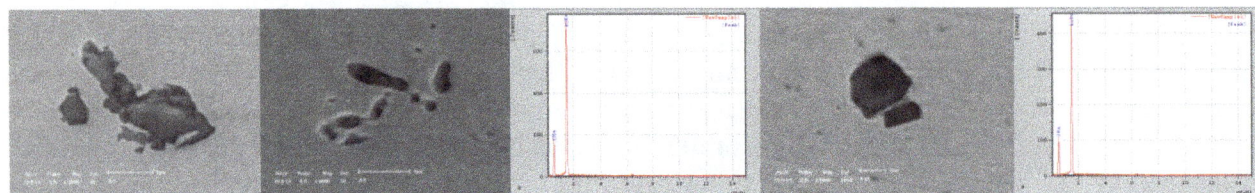

$$2[Al] +3[O] = Al_2O_3. \tag{1}$$
$$[Mg]+2[Al] +4[O] = MgO \cdot Al_2O_3. \tag{2}$$

(a) Al_2O_3 or Al_2O_3 cluster inclusions in ladle   (b) Al_2O_3 in hot rolled plate

Fig. 1 Typical inclusions in the Al-killed steel samples (without Mg)

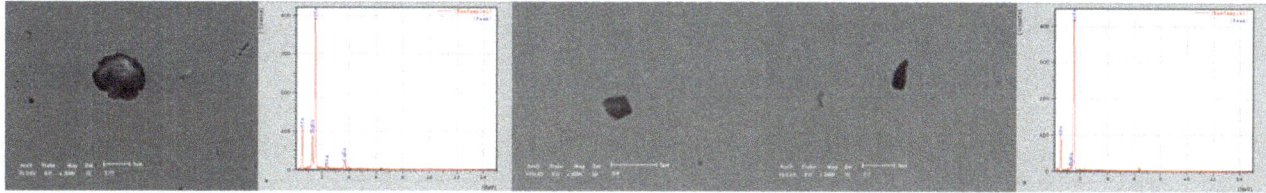

(a) MgO·Al2O3 inclusions in ladle   (b) MgO·Al2O3 in hot rolled plate

Fig. 2 Typical inclusions in the AlMnMgFe deoxidation steel samples (with Mg)

The results tested by metallographic microscope show that the mean inclusion diameter are 0.97 μm and 1.34 μm for the AlMnMgFe deoxidization steel(with Mg) and the Al-killed steel (without Mg), and the total inclusion numbers are 210 and 343 per square millimeter, respectively. Fig. 3 shows the distribution of inclusions in the steel with and without Mg deoxidization. It is shown that, the diameters of most of inclusions are less than 5μm. The inclusions with diameter smaller than 3 μm account for 97.01 percent in the AlMnMgFe deoxidization steel, while it is 94.46 percent in the Al-killed steel. The addtion of Mg in steel can decrease the size of inclusion obviously [11]. For the two steels without refined by Ladle Furnace (LF), there are some inclusions with diameter larger than 5 μm.

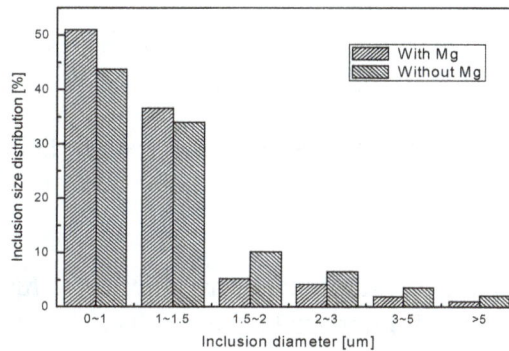

Fig. 3 Inclusions in the steel samples taken from ladle

3.2. *Effect of AlMnMgFe deoxidation on microstructures of tested steels*

Microstructures of the hot rolled plates are shown in Fig. 4. The addition of Mg in steel can reduce the grain size obviously. The grain size of the AlMnMgFe deoxidization steel is smaller than that of the Al-killed one. The maximum grain size is less than 30 μm for the former steel, while it is more than 100 μm for the later steel.

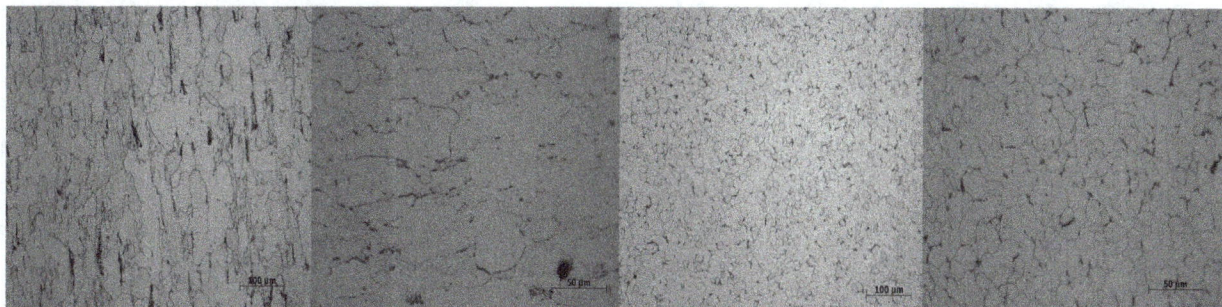

(a) Without Mg   (b) without Mg   (c) with Mg   (d) with Mg

Fig. 4 Microstructure of the tested steels

3.3. *Effect of AlMnMgFe deoxidation on properties of tested steels*

The mechanical properties of the tested steels are shown in Table 2.

Table 2 Mechanical properties of the tested steels at room temperature

	AlMnMgFe deoxidization steel			Al-killed steel		
	Yield strength [MPa]	Tensile strength [MPa]	Elongation [%]	Yield strength [MPa]	Tensile strength [MPa]	Elongation [%]
1	284	342	48.5	285	368	40
2	283	344	46	288	339	42.5
3	312	375	45.5	280	330	41
4	278	370	43	287	348	40.5
5	306	356	46	285	342	42
Mean	292.6	357.4	45.8	285	345.4	41.2

It can be seen from Table 2 that because the grains of the AlMnMgFe deoxidization steels are finer than those of the Al-killed steels, and the particle distribution is more uniform and dispersive, the mean yield strength, mean tensile strengths and mean elongation of the five AlMnMgFe deoxidization steels increases 2.67 percent, 3.47 percent and 11.17 percent, respectively as compared to the Al-killed steels. The result is obedient to Hall-Petch relation in Eq. 3.

$$\sigma_s = \sigma_0 + Kd^{-1/2}. \tag{3}$$

Where, σ_s is yield strength. σ_0 is tensile strength. K is constant. D is the grain size.

4. Conclusion

By using AlMnMgFe as the deoxidizer, the inclusions, microstructures and properties of the low carbon and low alloy steel Q195 are all improved. The results are as follows:

1. Inclusions in Al-killed steel are Al_2O_3 or Al_2O_3 cluster, while they are $MgO \cdot Al_2O_3$ inclusions in AlMnMgFe deoxidized alloy. The diameter of 97.01 percent inclusions is smaller than 3 μm in the later steel.
2. The grain size is less than 30 μm for AlMnMgFe deoxidized steels, while it is more than 100 μm for Al-killed steel.
3. The mean yield strength, mean tensile strength and mean elongation of the AlMnMgFe deoxidized steels increase 2.67 percent, 3.47 percent and 11.17 percent, respectively, as compared to that of its Al-killed counterparts.

Acknowledgments

This research was funded by National Science Foundation of China (51574063) and Fundamental Research Funds for the Central Universities (N150204012, N152306001) and Program for Liaoning Excellent Talents in University [LJQ2015056].

References

1. P Kaushik, M Lowry, H Yin and H Pielet. Inclusion characterization for clean steelmaking and quality control. Ironmaking and Steelmaking, 39, (2012), 284-300.

2. Lachmund H, Xie Y. High purity steels: a challenge to improved steelmaking processes. Ironmaking and Steelmaking, 30, (2003), 125-129.

3. Dawson S. Tundish nozzle blockage during the continuous casting of aluminum-killed steel. Iron & Steelmaker, 17, (1990), 33-39.

4. Katsuhiro S. Reaction mechanism between alumina immersion nozzle and low carbon steel. ISIJ International, 34, (1994), 802-809.

5. Luls Truebra Jr. Nozzle clogging during the continuous casting of aluminum-killed steel. Rolla: University of Missouri-Rolla. (2003).

6. Holappa L, Hämäläinen M, Liukkonen M, et al. Thermodynamic examination of inclusion modification and precipitation from calcium treatment to solidified steel. Ironmaking and Steelmaking, 30, 2003, 111-115.

7. ZHAN Dong-ping, ZHANG Hui-shu, JIANG Zhou-hua. Effects of AlMnCa and AlMnFe Alloys on Deoxidization of Low Carbon and Low Silicon Aluminum Killed Steels. Journal of Iron and Steel Research International, 15, (2008), 15-18.

8. Han Jian-huai, Lu Ping; Wang Zhong-ying. Deoxidization process of molten steel with barium alloy. Journal of Iron and Steel Research, 16, (2004), 18-22. (In Chinese)

9. V P Kirilenko, A Y Zaslavskii, V A Golubtsov, et al. Improving wheel steel by means of barium-based modifiers. Steel in Translation, 39, (2009), 1078-1083.

10. Yang J, Kiwabara M, Sakai T. Simultaneous desulfurization and deoxidation of molten steel with in situ produced magnesium vapor. ISIJ Int, 47, (2007), 418~426.

11. Ohta H, Suito H. Characteristics of particle size distribution of deoxidation products with Mg, Zr, Al, Ca, Si/Mn and Mg/Al in Fe-10% Ni alloy. ISIJ Int, 46, (2006), 14~21.

12. Pervushin G V, Suito H. Effect of primary deoxidation products of Al_2O_3, ZrO_2, Ce_2O_3 and MgO on TiN precipitation in Fe-10% Ni alloy. ISIJ Int, 20, (2001), 748~756.

13. Ohta H, Suito H. Precipitation and dispersion control of MnS by deoxidation products of ZrO_2, Al_2O_3, MgO and MnO-SiO_2 particles in Fe-10% Ni alloy. ISIJ Int, 46, (2006), 480~489.

14. Kimura S, Nakajima K, Mizoguchi S. Bahaviour of alumina-magnesia complex inclusions and magnesia inclusions on the surface of molten low-carbon steels. Process Metallurgy and Materials Processing Science, 32, (2001), 79~85.

15. JIANG Zhou-hua, ZHUANG Ying, LI Yang, LI Shuang-jiang. Effect of Modification Treatment on Inclusions in 430 Stainless Steel by Mg-Al Alloys. Journal of Iron and Steel Research International, 20, (2013), 6-10.

Influence of Annealing Treatment on the Microstructure and Properties of the Laser Welded Joint of Ti-Ni Alloy

Yu-Hua Chen[*],Wen-Ming Cao, Shu-Han Li ,Yang-Yang Yu

*National Defense Key Disciplines Laboratory of Light Alloy Processing Science and Technology,
Nanchang Hangkong University, Nanchang 330063, China*
Email: ch.yu.hu@163.com

Ti-Ni alloy of 0.2mm thick was welded by Nd: YAG laser. In order to obtain the welded joint of TiNi alloy with good shape memory effect (SME), two kinds of welding methods were studied: vacuum annealing treatment of cold-rolled base metal before welding and vacuum annealing treatment of welded joint after welding. The results show that there is no obvious loss of SME in the joint welded with cold-rolled base metal and annealed after welding, and the phase transformation process is basically similar with the annealed base metal. Shape recovery ratio of the joint welded with cold-rolled base metal and annealed after welding is 98. 5%. But welded joint made by that kind of method suffers serious loss of tensile strength. The tensile strength of the joint welded with cold-rolled base metal and annealed after welding is 67.5% of annealed base metal. The grains in weld seam center and weld edge of the joint welded with annealed base metal before welding are coarser than the joint welded with cold-rolled base metal and annealed after welding.

Keywords: Shape memory alloy; laser welding; annealing treatment; microstructure and properties of welded joint.

1. Introduction

Ti-Ni shape memory alloy (SMA) is a kind of new functional material which has special shape memory effect (SME), super elasticity (SE) and good bio-compatibility [1], it has broad application prospects in precision instrumentation, aerospace, medical equipment and other areas [2-3].

However, successful applications of any novel material not only hinge on its inherent characteristics, but also depend on solving problems of welding technologies. Delobelle et al. [4] studied on electrical resistance welding to join NiTi tubes. He had established a technique for the design and creation of self-joined NiTi material architectures from stacked tubes. Barcellona et al. [5] investigated the feasibility of friction stir welding process to join NiTi shape memory alloys, microscopy observations of Friction Stir Processed (FSP) material have been used to highlight processed zone microstructures and the austenitic and martensitic transformation temperatures of the processed material were investigated using a stress applied method. A post processing thermal treatment has been applied in order to investigate the possibility to recover the partial losing of the shape memory capability of material subsequent the FSP. Chan et al. [6] studied the effect of fibre laser welding parameters and their interactions upon the weld bead aspect ratio of nickel–titanium thin foil and studied the SME and corrosion of welded joint. Khan et al. [7] and Falvo et al. [8] investigated the effects of the laser Nd: YAG welding on the functional properties of the NiTi SMA.

But, the SME of TiNi alloy could be got by thermal treatment or mechanical treatment such as annealing treatment, aging after solid solution treatment and heat and stress cycle.

In order to obtain the welded joint of TiNi alloy with good shape memory effect (SME), two kinds of methods were studied in this paper: vacuum annealing treatment of cold-rolled base metal before welding and vacuum annealing treatment of welded joint after welding.

*Corresponding author

2. Experimental Method

Ti-Ni SMA sheets of 0.3mm thickness (Ti49.4Ni50.6 in atomic percent) were used as the base metal in this investigation. The sheets were produced by cold rolling with a thickness reduction of about 22%. In order to remove the surface layer of oxide produced during the cold rolling process, the TiNi sheets were degreased with propanone and soaked 15 minute in the mixture solution of hydrofluoric acid and nitric acid, and then washed with clear water and weathered. After these treatments, the thickness of the sheets was 0.2 mm. The dimensions of the welding specimens were 60 mm×40 mm×0.2 mm.

Two kinds of methods were used, the one method is welded with base metal annealed at 500°C in vacuum for 1 hour before welding(annealing treatment before welding), the other is welded with cold-rolled base metal and the welded joint was annealed at 500°C in vacuum for 1 hour(annealing treatment after welding). The welding processes of these two methods were all carried out using a Nd: YAG laser source (SL-80, average power is 80 W). In order to protect the welding zone, a special clamping fixture was used to protect both frontal surface and back surface of the welded joint. For all experiments, the spot diameter, welding rate, argon flow rate, laser power, pulse frequency, pulse width is fixed at 0.3 mm, 0.3mm/s, 5L/min, 16.8W, 4Hz and 2.3ms respectively.

Tensile samples of every group of welding parameters were made by spark cutting and tensile properties of welded joint were tested on tensile testing machine (INSTRON 5540). Scanning fracture morphology of tensile sample was observed by electronic microscope (Quanta 200). Metallographic specimen was made along cross direction of the welded joint and etched by mixture acid solution of HF and HNO3. The microstructure and element distribution were studied by metallographic microscope (4XB-TV).

The SME of welded joint and annealed base metal was tested by DSC measurements which were performed at temperatures ranging from -60°C to 200°C under a cooling and heating rate of 10°C/min.

3. Results and Discussion

3.1. *Phase transformation behavior*

The essence of SME of TiNi SMA is the thermo- elastic martensitic transformation, so the phase transformation temperature and order of annealed TiNi shape memory alloy base metal and the welded joints got by the two methods were tested by DSC, and the results are shown in Fig. 1.

It can be seen from Fig. 1a, there is a phase transformation heat release peak during the heating process of base metal, which indicates that transformation from martensite phase (B2) to austenite phase (B19') occurs. There is no R phase transformation during the heating process of base metal. During the cooling process, a phase transformation heat release peak also occurs nearby 43.81°C before the transformation from austenite phase to martensite phase, which indicates that transformation from B19'phase to R phase occurs. At 60.75°C, transformation from R phase to B2 phase occurs.

The DSC curve of the joint made by the first method (welded with annealed base metal before welding) is shown in Fig. 1b. There are two phase transformation heat release peaks during the heating process, which each correspond to the phase transformation of B2→R and R→B19. But the two heat release peaks do not separate completely and almost overlap each other. One step phase transformation of B19 ' →B2 occurs during the cooling process.

(a) Annealed base metal

(b) The joint welded with annealed base metal before welding

(c) The joint welded with cold-rolled base metal and annealed after welding

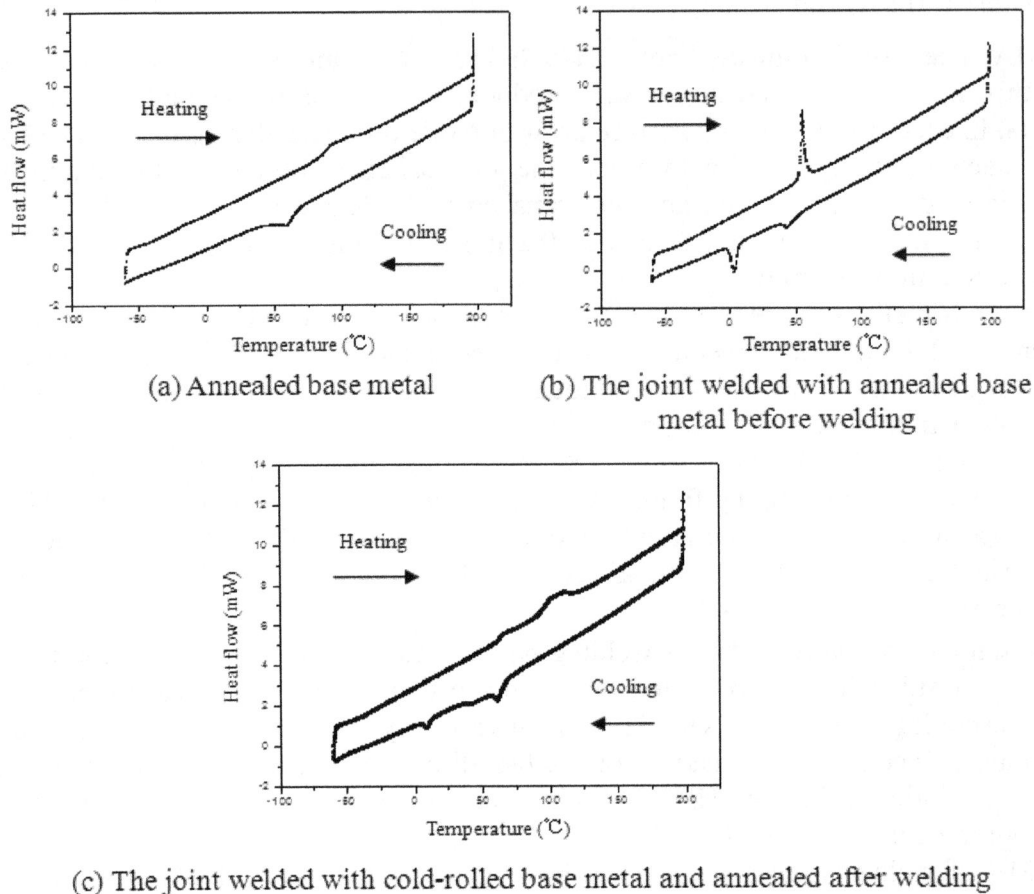

Fig. 1 DSC curves

The DSC curve of the joint made by the second method (welded with cold-rolled base metal and annealed after welding) is little different with the annealed base metal as shown in Fig. 1c. There are also two phase transformation heat release peaks during the heating process, which indicates two steps phase transformation of B2→R and R→B19 '. During the cooling process, there are also two steps phase transformation of B19 ' →B2 and R→B2 occur. The points of R phase transformation and martensite phase transformation drift compared with the annealed base metal because of the new phases such as Ti_3Ni_4 precipitating in the welded joint.

As a whole, the phase transformation process of the welded joint made by the second method is basically similar with the annealed base metal, which indicates that there is no obvious loss of SME.

3.2. Shape recovery ratio

Bending test (Test principle is showing in Fig. 2) was used to test the shape recovery ratio of the base metal and welded joints. Test result is showing in Fig. 3, and the deformation angle after recovery was measured.

In formula (1) and Fig. 2, θ, θ_1 and θ_2 is the initial bending angle ($\theta=90°$), deformation angle after elastic recovery and deformation angle after the test sample was put in water whose temperature is higher than the austenite phase transition temperature of TiNi alloy. The calculating results are showing in Table 1. The shape recovery ratio of the joint welded with cold-rolled base metal and annealed after welding is 98. 6%, this is close with the annealed base metal. The shape recovery ratio of the joint welded with annealed base metal before welding is much lower than the annealed base metal.

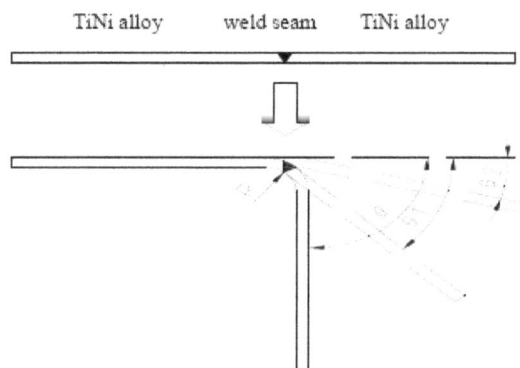

Fig. 2 Sketch map of bending test

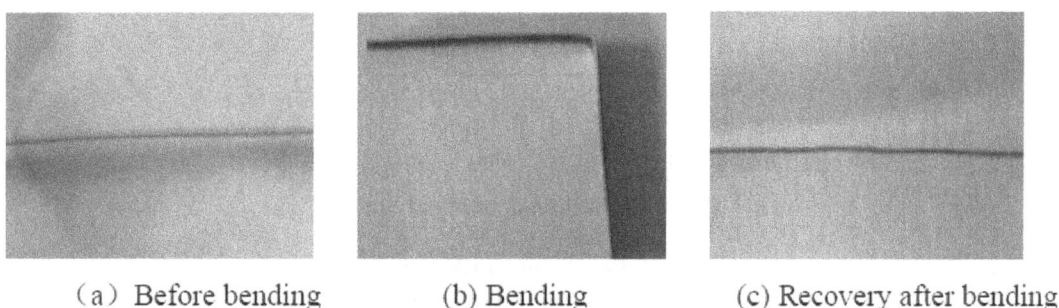

(a) Before bending　　　　(b) Bending　　　　(c) Recovery after bending

Fig. 3 Bending test of annealed base metal

The shape recovery ratio is calculated according to formula (1).

$$\eta = \frac{\theta_1 - \theta_2}{\theta_1} \times 100\% \qquad (1)$$

Table 1 Results of shape recovery ratio test

Test sample	$\theta_1/°$	$\theta_2/°$	$\eta/\%$
a	75	0	100
b	65	23	67
c	70	1	98.5

Notes:　a—— Annealed base metal,
　　　　b—— The joint welded with annealed base metal before welding,
　　　　c—— The joint welded with cold-rolled base metal and annealed after welding

3.3. Tensile properties

The stress–strain curves are shown in Fig. 4 and the tensile strength is listed in Table 2. From Fig. 4 and Table 2, we can see that the tensile strength of annealed base metal is 1183MPa and the fracture strain is 7.7%. The tensile strength and the fracture strain of the joint welded with annealed base metal before welding are 600MPa and 5.2% which are lower than annealed base metal.

The tensile strength and the fracture strain of the joint welded with cold-rolled base metal and annealed after welding are 752 MPa and 6.4%, which are higher than the joint welded with annealed base metal before welding but lower than annealed base metal. The tensile strength of the joint welded with cold-rolled base metal and annealed after welding is 67.5% that of annealed base metal. Welded joint made by this method suffers serious losses of tensile strength.

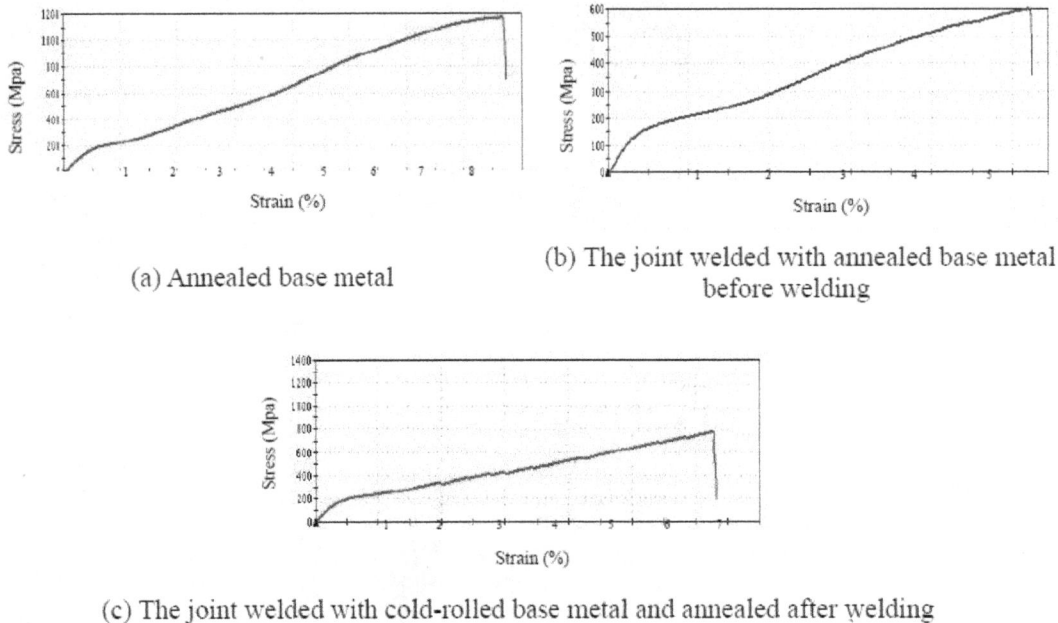

(a) Annealed base metal

(b) The joint welded with annealed base metal before welding

(c) The joint welded with cold-rolled base metal and annealed after welding

Fig. 4 Stress-strain curves

Table 2 Results of tensile strength test

Test sample	Tensile Strength δ_b (MPa)	Fracture strain ε (%)
a	1183	7.7
b	600	5.2
c	752	6.4

Notes: a——Annealed base metal,
      b—— The joint welded with annealed base metal before welding,
      c——The joint welded with cold-rolled base metal and annealed after welding

3.4. *Microstructure*

Microstructure in the weld seam center and microstructure in the weld edge of the joint welded with annealed base metal before welding and the joint welded with cold-rolled base metal and annealed after welding are shown in Fig. 5 and Fig. 6. From Fig. 5 and Fig. 6 we can see, microstructures in welded seam center are fine equiaxed crystals, but the microstructure of the joint welded with annealed base metal before welding (Fig. 5a) is coarser than the joint welded with cold-rolled base metal and annealed after welding (Fig. 5b).

There is almost no obvious coarse grain heat affected zone at the edge between welded seam and base metal. Because the energy density of laser welding is large and the affection time on the welded joint is very short, only very small zone on the welded joint is heated and the cooling velocity of welded joint is quick, so the coarse grain heat-affected zone is not obvious.

The directivity of columnar crystals in the weld edge of the joint welded with annealed base metal before welding (Fig. 6a) is stronger than that of the joint welded with cold-rolled base metal and annealed after welding (Fig. 6b).

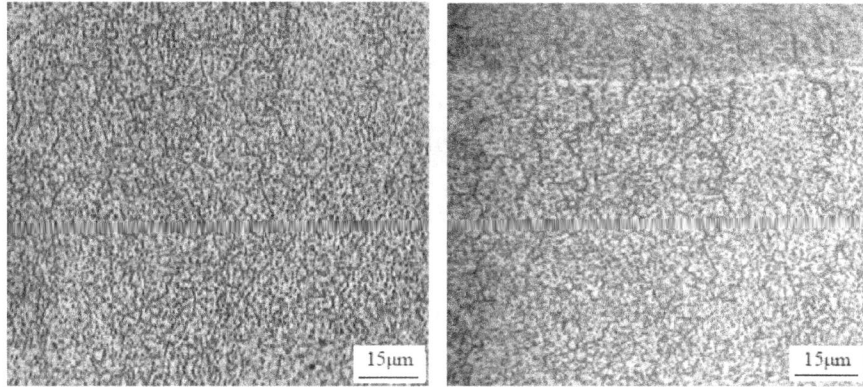

(a) The joint welded with annealed base metal before welding

(b) The joint welded with cold-rolled base metal and annealed after welding

Fig. 5 Microstructure in the weld seam center

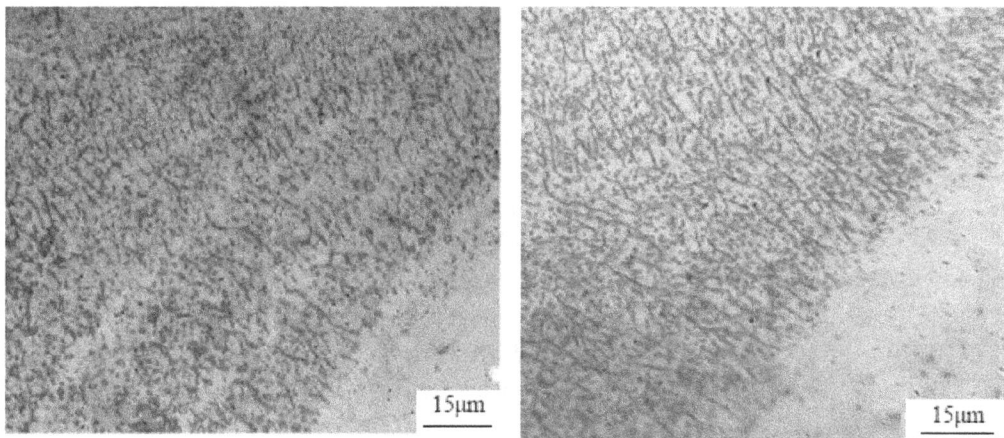

(a) The joint welded with annealed base metal before welding

(b) The joint welded with cold-rolled base metal and annealed after welding

Fig. 6 Microstructure in the weld edge

4. Conclusion

1. There is no obvious loss in the joint welded with cold-rolled base metal and annealed after welding, and the phase transformation process is basically similar with the annealed base metal.

2. Shape recovery ratio of the joint welded with cold-rolled base metal and annealed after welding is 98.5%. But welded joint made by that kind of method suffers serious losses of tensile strength. The tensile strength of the joint welded with cold-rolled base metal and annealed after welding is 67.5% of annealed base metal.

3. The grains in weld seam center and weld edge of the joint welded with annealed base metal before welding are coarser than the joint welded with cold-rolled base metal and annealed after welding.

Acknowledgments

This research was funded by National Natural Science Foundation of China: (No. 51565040), Aviation Science Funds of China (2014ZE56016), Science and Technology Planning Project of Jiangxi Province (20114BAB206006) and Frontier Science Research Project of Educational Commission of Jiangxi Province (KJLD14055).

References

1. Otsuka K, Ren X, Recent developments in the research of shape memory alloys, Intermetallics, 1999, 7(5): 511-528.
2. Gong W H, Chen Y H, Ke L M, Microstructure and properties of laser micro welded joint of TiNi shape memory alloy, Transactions of Nonferrous Metals Society of China, 2011, 21(9): 2044-2048.
3. Chen Y H, Ke L M, Huang Y D, et al, Laser butt welding of TiNi shape memory alloy sheet, Transactions of the China Welding Institution, 2010, 31(8): 37-40. (in Chinese)
4. Delobelle V, Delobelle P, Liu Y, et al, Resistance welding of NiTi shape memory alloy tube, Journal of Materials Processing Technology, 2013, 213: 1139-1145
5. Barcellona A, Fratini L, Palmeri D, et al, Friction stir processing of Niti shape memory alloy: microstructural characterization, International Journal of Material Forming, 2010, 3(1): 1047-1050.
6. Chan C W, Man H C, Laser welding of thin foil nickel-titanium shape memory alloy, Optics and Lasers in Engineering, 2011, 49(1): 121-126.
7. Khan M I, Zhou Y, Effects of local phase conversion on the tensile loading of pulsed Nd: YAG laser processed Nitinol, Materials Science and Engineering: A, 2010, 527(23): 6235-6238.
8. Falvo A, Furgiuele F M, Maletta C, Laser welding of a NiTi alloy: Mechanical and shape memory behaviour, Materials Science and Engineering: A, 2005, 412(1): 235-240.

Microstructure and Property of A7N01P-T5 Aluminum Alloy Joint by MIG Welding

Hui-Jin Zheng[1], Hao-Bo Liu[2], Shang-Lei Yang[2,*]

[1]CSSC, Shipbuilding Technology Research Institute, Shanghai 200032, China
[2]School of Materials Engineering, Shanghai University of Engineering Science, Shanghai 201620, China
*Email: yslei@126.com

The A7N01P-T5 aluminum alloy was welded by MIG welding in test, and the microstructure and mechanical property of joint were investigated. The results of experimentation show that the feature of the A7N01P-T5 aluminum alloy MIG welding joint is perfect with a gentle transition. The microstructure of fusion metal in A7N01P-T5 joint is as-cast grain. The microstructure of fusion zone consists of the fine columnar grains. It is rolling structure in the parent metal. In HAZ, the rolling structure is disappearing. The η' ($MgZn_2$) strengthening phases were precipitated in the grain boundary of the A7N01P-T5 aluminum alloy. There is no precipitation phase in welding metal of A7N01P-T5 joint. The fine η' transitional phases of the primeval parent metal are turned to the coarse η ($MgZn_2$) stable phases of the HAZ. The tensile strength and the percentage elongation after fracture of A7N01P-T5 welding joint is below the parent metal. The zone of fracture is located in the welding seam. The fracture microstructure of the A7N01P-T5 parent metal is the mixed-type fracture. The fracture microstructure of the A7N01P-T5 welding metal is the typical ductile fracture. The microhardness of the welding seam is lowest in the A7N01P-T5 joint.

Keywords: Aluminum alloy; welding joint; microstructure; mechanical property.

1. Introduction

Wrought aluminum alloy has wide application in shipbuilding and marine engineering, aerospace and automobile industries due to their superior properties compared to casting aluminum alloy. Base on the principle of design and produce of high-speed vehicle, the large-scale extruded shape of advanced aluminum alloy with higher strength and better ductile is developed [1, 2]. The long life and safe running of high-speed train also depends on the quality of welding joint. The high temperature of arc can affect the microstructure and mechanical property of A7N01P-T5 aluminum alloy. The welding technique of aluminum alloy is significant to ensure the service life of vehicle [3-5]. The high-speed train always adopts welding construction of aluminum alloy with large length and width, because of high strength, well extrusion and corrosion resistance. The A7N01P-T5 aluminum alloy is suitable for producing large-scale extruded-shape material using in carriages structure. This kind of alloy has been used to produce high-speed trains in Japan, Germany, China, and so on. The A7N01P-T5 alloy is Al-4.5Zn-1.5Mg-0.5Mn aluminum alloy with extruding and quenching online. The aging treatment system, the strengthening mechanism and the effect of welding process on mechanical properties of A7N01P-T5 alloy must go into further. So the advanced welding technologies are significant to getting ahead of fabrication technique for high-speed train, and realization of light-weight, high-speed and well-safe for the rail transit.

2. Experimental

2.1. *Material*

The experimental aluminum materials are the A7N01P-T5 alloy, which condition of heat treatment are T5. The filler metal is ER5356 welding wire of Al-Mg alloy. The chemical composition of A7N01P-T5 aluminum alloy and ER5356 filler metal is shown in Table 1.

*Corresponding author

Table 1 The chemical composition of the A7N01P-T5 aluminum alloy and the ER5356 filler metal [%]

	Al	Mg	Si	Zn	Fe	Cu	Mn	Cr	Ti	V	Zr
A7N01P-T5	base	1.0-2.0	≤0.3	4.0-5.0	≤0.35	≤0.20	0.2-0.7	≤0.30	≤0.20	≤0.10	≤0.25
ER5356	base	4.5-5.5	≤0.25	≤0.10	≤0.40	≤0.10	0.05-0.2	0.05-0.20	0.06-0.20	--	--

2.2. *Welding process*

The welding coupons of 300mm×100mm×4mm with V-shape groove (70°) butt joint were polished prior to welding in order to remove oxide films and oil stain, and were subsequently cleaned by acetone. The welding parameters of MIG are shown in Table 2.

Table 2 The welding parameter in the experiments

Diameter of filler wire [mm]	Welding current [A]	Arc voltage [V]	Welding speed [mm/s]	Ar gas flow [L/min]
1.2	120-150	16-18	6-8	15-20

2.3. *Test methodology*

The mechanical property of joint was tested according to standard of GB/T228-2002, using an AG-10KNA tensile test machine with strain rate of 0.001mm/s. The hardness of welding joint was measured using a HV-1000 tester, with a load of 0.98N being applied for 10s. The microstructure was observed by using a XJL-03 microscope, and the sample was etched by using hydrofluoric acid of 0.5%. Morphology of welding joint was studied by using a JEM-100CX TEM and a JEM-6700F SEM.

3. Results and discussion

3.1. *Feature of welding seam*

The appearance of MIG welding joint is shown in Fig. 1. The feature of joint and fusion of welding seam are perfect with a gentle transition. The welding defect do not found such as cracking, pore, and so on.

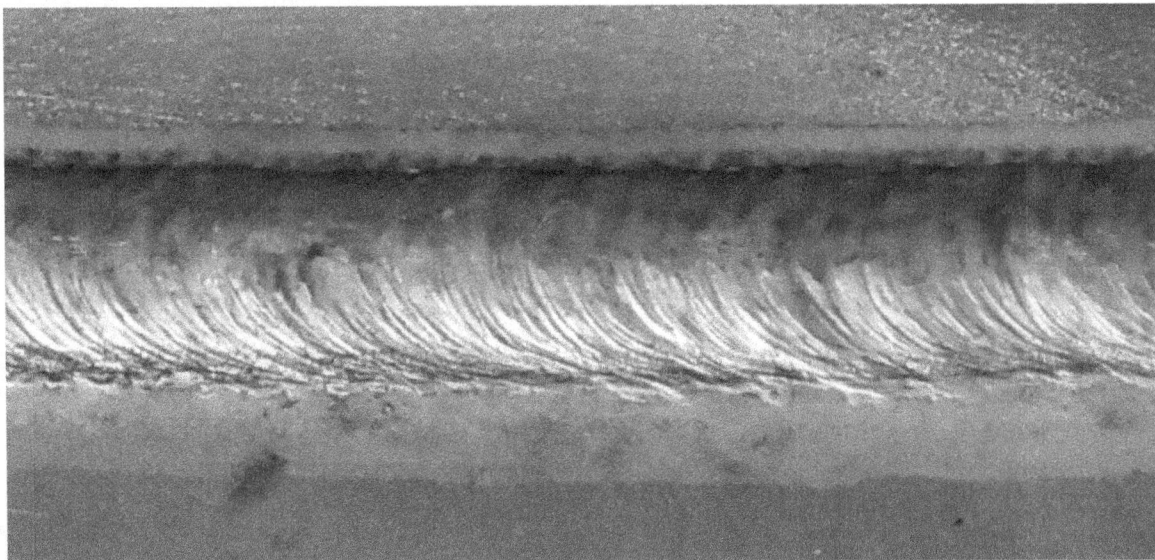

Fig. 1 Appearance of the welding joint

3.2. *Microstructure*

The microstructure in the center of welding seam is shown in Fig. 2a. The welding seam is mostly consists of fused Al-Mg filler wire of ER5356. The microstructure of fusion metal is as-cast grain. The microstructure of fusion zone (fusion line) in A7N01P-T5 joint is shown in Fig. 2b. It indicates that there is a narrow region near the edge of the welding seam, which consists of the fine columnar grains.

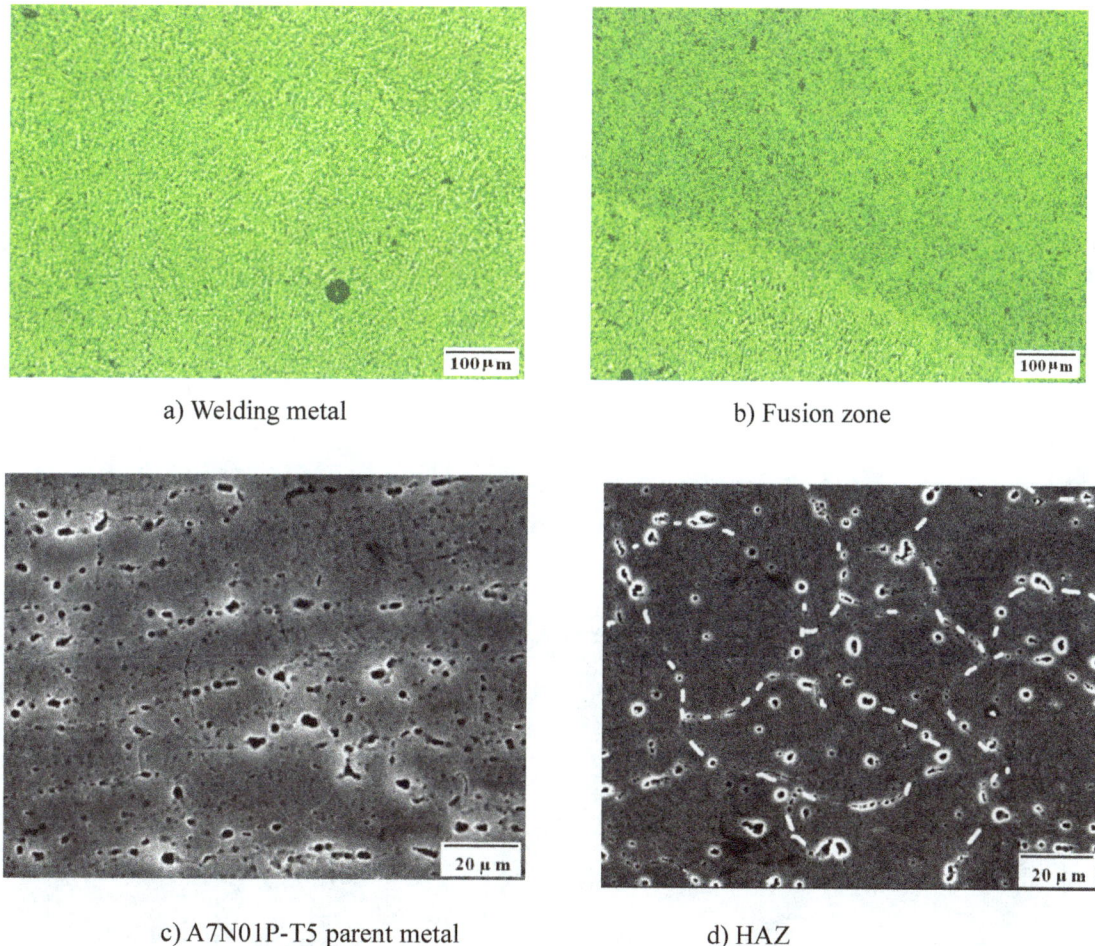

a) Welding metal

b) Fusion zone

c) A7N01P-T5 parent metal

d) HAZ

Fig. 2 The microstructure of A7N01P-T5 aluminum alloy welding joint

The microstructure of A7N01P-T5 parent metal is shown in Fig. 2c. The microstructure of the base metal is rolling structure. The strengthening of A7N01P-T5 aluminum alloy depends largely on η' ($MgZn_2$) precipitation phases. The η' phases can be observed in parent metal, which distribution dispersedly in α-Al matrix structure. However, in the welding metal, no precipitation phase was found. Because of the fusion metal is Al-Mg alloy, the super-saturation of α-Al solid solution is not very high, the precipitation tendency of second phase is low, resulting of non-obvious precipitation phase in the welding metal.

The microstructure of the heat affected zone (HAZ) in A7N01P-T5 joint is shown in Fig. 2d. The rolling structure in HAZ is disappearing, the grain become coarsen, and the strengthen phase decreasing, comparison with parent metal.

3.3. *Mechanical property*

The mechanical property of A7N01P-T5 aluminum alloy MIG welding joint was tested as shown in the Table 3. The results indicate that tensile strength and percentage elongation after fracture of A7N01P-T5 welding joint is below the parent metal. The zone of fracture is located in the welding seam.

Table 3 Mechanical properties of the welding joints of the A7N01P-T5 aluminum alloy

Material	Tensile strength σ_b /MPa	Non-proportional extension $\sigma_{0.2}$ /MPa	Percentage elongation δ /%
A7N01P-T5	426	339	12.6
ER5356	≥265	--	--
A7N01P-T5 Joint	267	--	6.2

According to the macro-fracture morphology, the A7N01P-T5 parent metal fractures in the middle of the specimen, the angle between the fracture plant and the tensile direction is 45°, and there is no obvious necking before failure. The fracture of the A7N01P-T5 welding sample occurs in the welding seam. The facture face is rough and dark, and there is obvious necking before failure.

3.4. *Fracture*

The SEM of the fracture in the A7N01P-T5 aluminum alloy was shown in the Fig. 3a. The fracture microstructure of the parent metal is composed of dimples and cleavage planes, which is characteristic of a mixed-type fracture.

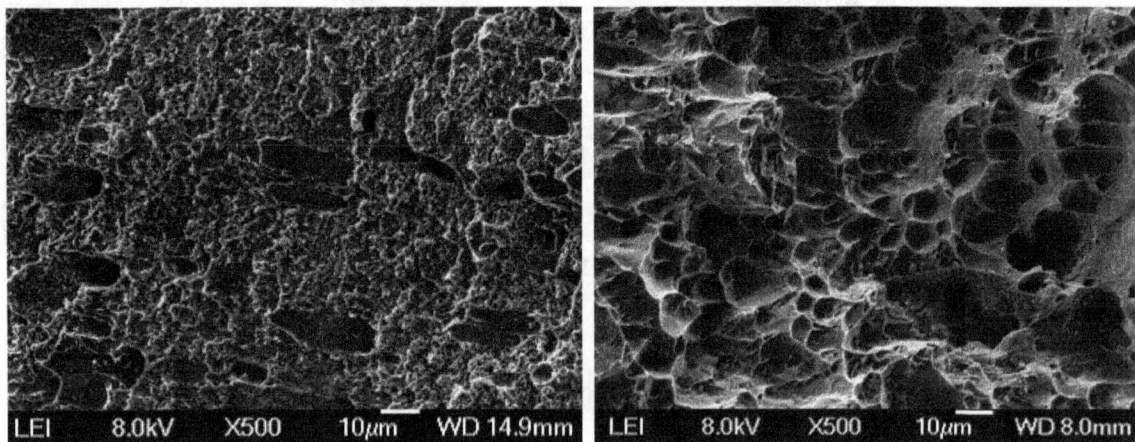

a) A7N01P-T5 parent metal   b) Welding metal

Fig. 3 SEM of the fracture in the A7N01P-T5 aluminum alloy-welding joint

The SEM of the fracture in the A7N01P-T5 joint was shown in the Fig. 3b. The fracture microstructure of the welding metal consists of dimples, which is a typical micro-void coalescence fracture, and it present ductile fracture.

3.5. *Microhardness*

The microhardness of the A7N01P-T5 welding joint is shown in Fig. 4.

446

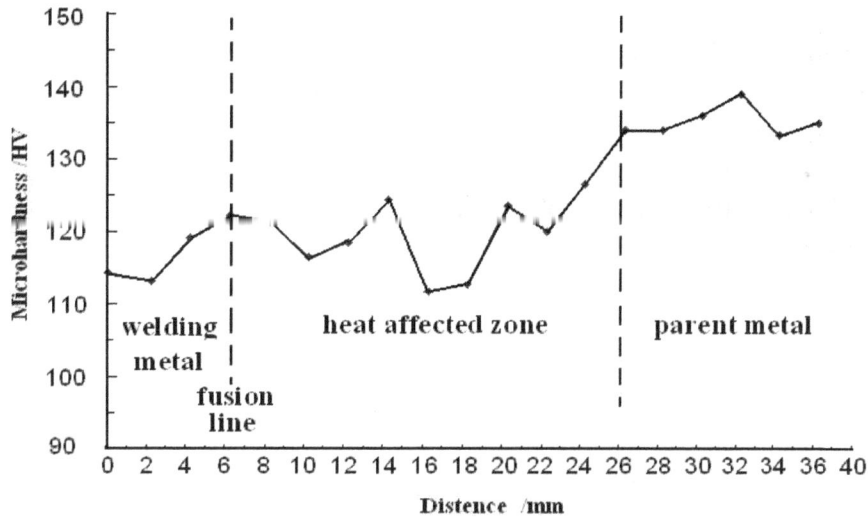

Fig. 4 Microhardness of the A7N01P-T5 welding joint

The microhardness of the welding seam is lowest in the A7N01P-T5 joint, which is about HV115. The microhardness increases with increasing of the distance to the welding seam. The microhardness of the A7N01P-T5 parent metal far away the welding seam is highest, which is about HV135. The microhardness of HAZ is higher very much than the welding seam, and a little lower than the parent metal.

4. Conclusion

1. The feature of the A7N01P-T5 aluminum alloy MIG welding joint is perfect with a gentle transition. There is no welding defect such as cracking, pore and so on.
2. The microstructure of fusion metal in A7N01P-T5 joint is as-cast grain. The microstructure of fusion zone consists of the fine columnar grains. It is rolling structure in the parent metal. In HAZ, the rolling structure is disappearing, the grain become coarsen, and the strengthen phase decreasing, comparison with parent metal.
3. The tensile strength and the percentage elongation after fracture of A7N01P-T5 welding joint is below the parent metal. The zone of fracture is located in the welding seam. The fracture microstructure of the A7N01P-T5 parent metal is composed of dimples and cleavage planes, which is the mixed-type fracture. The fracture microstructure of the A7N01P-T5 welding metal consists of dimples, which is the typical ductile fracture.
4. The microhardness of the welding seam is lowest in the A7N01P-T5 joint. The microhardness of the A7N01P-T5 parent metal far away the welding seam is highest. The microhardness of HAZ is higher very much than the welding seam.

Acknowledgments

This work was funded by Natural Science Foundation of Shanghai (Grant No. 14ZR1418800) and the Shanghai automobile industry science and Technology Development Fund (Grant No. 1404). These supports are gratefully acknowledged.

References

1. G.J. Xu, L.M.Zhong and H. Wang: Chinese Journal of Lasers Vol. 41 (2014), p. 1, (In Chinese).
2. D.J. Yan, X.S. Liu and H.Y. Fang: The Chinese Journal of Nonferrous Metals Vol. 22 (2012), p. 3313, (In Chinese).
3. M. Nicolas and A. Deschamps: Metallurgical and Materials Transactions A: Physical Metallurgy and Materials Science Vol. 20 (2015), p. 1437.
4. W.H. Tuo, S.L, Yang and W.T, Yang: Materials Review Vol. 24 (2015), p. 105, (In Chinese).
5. J.Yan, M. Gao and X.Y. Zeng: The Chinese Journal of Nonferrous Metals Vol. 12 (2009), p. 2012, (In Chinese).

The Study of Preparation and Properties of TC21 Titanium Alloy Powder for Laser 3D Printing

Dong Wang[1], Sui-Yuan Chen[1, 2,*], Kuai-Kuai Guo[2], Wen-Qian Zhang[2], Jing Liang[1], Changsheng Liu[1, 2]

[1]Key Laboratory for Anisotropy and Texture of Materials, Ministry of Education, School of Materials and Engineering, Northeastern University, Shenyang 110819, Liaoning, China

[2]Northeastern University (An Shan) Laser Application Research Institute, Anshan, Liaoning, China

*Email: chensy@smm.neu.edu.cn

TC21 titanium alloy powder for Laser 3D Printing was prepared by electrode induction melting gas atomization (EIGA). The morphology, particle size, microstructure and physical properties of TC21 powder were studied by Optical Microscopy (OM), Scanning electronic Microscopy (SEM), X-Ray Diffraction (XRD). The results showed that TC21 powder is spherical, surface structure are composed of dendritic crystals and a small amount of cellular crystal. The main phases are α', β and metal compound such as $AlTi_3$, Ti_2AlNb, Zr_3Al. The content of oxygen in TC21 alloy powder is 0.1 wt.%, powder bulk density is 2.74 g/cm^3, particle size is from 10 μm to180 μm. The fluidity of the 53-180 μm powder is 24.36 S/50g, hollow ball rate is less than 2%. The TC21 alloy powder prepared by this method can meet the request of the laser 3D printing.

Keywords: Laser 3D printing; TC21 titanium alloy powder; EIGA.

1. Introduction

Along with the progress of science and technology, and the performance requirements of the aircraft, advanced titanium alloy materials have been evolved towards the direction of damage tolerance titanium alloy with high fracture toughness and low crack growth rate [1]. TC21 alloy is widely used in aerospace structure because of its good matching of strength, ductility, fracture toughness and crack growth rate [2]. Using laser 3D printing to rapid prototyping TC21 titanium alloy makes the microstructure of the formed parts small, and its properties are similar to the forgings, the characteristics of TC21 powder for laser 3D printing are short period, high flexibility, low oxygen content, and low hollow ball rate.

The characteristic of titanium powder for laser 3D printing is the key factor to determine the titanium and titanium alloy powder products. Laser 3D printing metal powder mostly relies on imports. While the high performance aerospace titanium alloy powder for laser 3D printing is the key component to achieve laser 3D printing material foundation. The properties of the powder such as the structure of the powder, the distribution of the particle size, the fluidity, the oxygen content and the defect rate, have important influence on the performance of the laser forming parts. Therefore, the research on the preparation technology and characteristics of titanium alloy powder using for laser 3D printing has important scientific and application value [3].

So far, the research of laser 3D printing titanium alloy powder mainly focused on the TC4 alloy [4-6]. Therefore, in order to satisfy more applied area, other titanium alloy powder is urgently in need. Based on the characteristics of complex composition and outstanding performance of TC21 alloy, in this paper, we use the laboratory independent research and development of electrode induction melting atomization equipment to study on the preparation of TC21 titanium alloy powder for laser 3D printing. The preparation process, the microstructure of the alloy powder and the powder characteristics were systematically studied to provide basic research work for 3D laser printing of TC21 titanium alloy.

---

*Corresponding author

2. Experimental Materials and Methods

The raw material is $\Phi50\times1000$ mm TC21 Titanium alloy bar, as an induction electrode for electrode induction melting gas atomization, one end of the titanium rod is processed into a conical tip of 40-50 degree, the other end was processed a clamping slot whose shape is a semicircle.

In the whole process of preparation, the first step is pumping vacuum, when the vacuum degree reaches 3.0×10^{-3} Pa, the whole equipment is filled with argon gas, the electrode titanium rod is placed in an inert gas environment and controlled the distance of conical tip of titanium rod and chamber nozzle between 5cm and 7cm. As rotating and cutting magnetic line, the electrode rod is being heated, and the tip of the electrode rod continuously melt to be liquid drip, then they go through the gas nozzle center to be sharply blasted by low temperature, high speed, high pressure spraying argon and rapidly solidify into tiny spherical powder. The power of induction melting atomization process is about 52-64 kW; the electrode rotation speed is about 6r/min, the rate of decline is 650 µm/s, the melting chamber and the aerosolization chamber pressure difference 0.02 MPa. At last, the TC21 alloy powder was collected into the vacuum bag, then sieved and placed into the vacuum container.

The particle size distribution was analyzed by laser scattering particle size distribution analyzer, the rate of powder hollow ball was observed by OLYMPUS-GX71 type inverted optical microscope (OM). Shimadzu-SSX-550 scanning electron microscope (SEM) was used to observe the surface morphology and the degree of roundness of the powder. The phase analysis was carried out using the Japanese X type SmartLab-9000 ray diffraction (XRD), the oxygen content of TC21 titanium alloy powder was determined by TCH-600 nitrogen oxygen analyzer, and the bulk density ratio and fluidity of the titanium alloy were measured by the HYL-102 Holzer flow meter.

3. Experimental Results and Analysis

3.1. Composition analysis of TC21 titanium alloy powder

Table 1 is TC21 titanium alloy powders composition analysis prepared by the electrode induction melting atomization method. Controlling the content of N, H and O in titanium alloy is an important way to control the properties of titanium alloy. As Table 1 showed that the oxygen content of the powder is 0.1 wt.%, increased by 0.02 wt.% in the process of preparing powder, it proves that the oxygen content increases little in the equipment and process for preparing TC21 titanium alloy powder.

Table 1 Chemical composition analysis of TC21 titanium alloy powder

Element	Al	Sn	Zr	Mo	Cr	Nb	Fe	C	N	H	O	Ti
Content (wt.%)	6.62	2.35	2.23	2.93	1.59	1.85	0.025	0.015	0.003	0.004	0.1	Bal.

3.2. Powder particle size

Table 2 shows the relationship between power and fine powder (1-180 µm) rate. With the power increased and the process parameters remain unchanged, the rate of fine powder increases and then decreases. This is because the higher the power, the greater the degree of superheat of the droplets powder. Fine powder increases with the degree of superheat increased, which is due to the volume expansion of the liquid when the temperature rises and the molecular spacing increases, weakening the body phase intermolecular forces on the surface layer of molecules. At the same time, with the increasing superheat, molten liquid metal viscosity and surface tension of the droplet will gradually decrease, which is conducive to the formation of fine droplets impact argon. When the power reaches at 61 kW, with power continues increased, the rate of fine powders is decline. This is because when the temperature increases, the droplet solidification time becomes longer. In the process of whereabouts, small droplets into large droplets solidified to become part of the meal, another part of the solidification front hit the wall forming flakes.

Table 2 Fine powder ratio under different power

Power (kW)	52	55	58	61	64
Fine powder quality (g)	647.9	666.9	684.0	647.9	704.1
Fine powder ratio (wt.%)	31.49	35.55	37.19	41.39	34.67

Fig. 1 is the TC21 alloy powder particle size. Fig. 1 (a) is a cumulative mass distribution of TC21 titanium alloy powder whose size is less than 180 μm. It can be seen that 90% of the cumulative mass of powder particle size is about 174 μm. Fig. 1 (b) is independently TC21 titanium powder particle size distribution, it can be seen the powder particle size mainly between 10-180 μm. The range of powder particle size is wide, mainly because the high speed blowing argon blowing molten droplets broken, the gas flow field is not stable, which lead to broken droplet size uneven. The range of particle size distribution is wide, but it still fits the normal distribution.

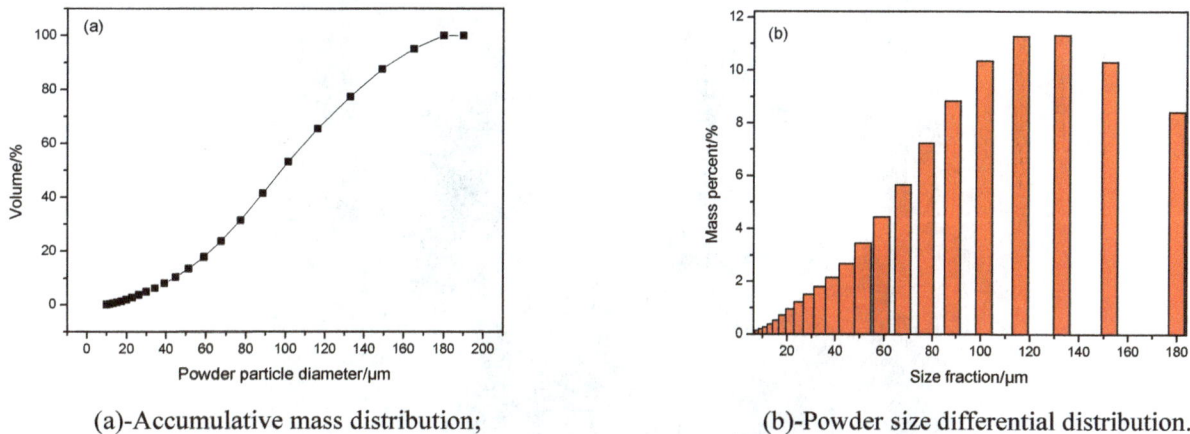

(a)-Accumulative mass distribution;   (b)-Powder size differential distribution.

Fig. 1 TC21 alloy powder particle size

3.3. Microstructure of powder

Fig. 2 is a SEM photograph of powder particle surfaces. Fig. 3 is a transverse cross section of the powder particles of SEM photograph. As can be seen from these two photos, both surface and inside of the powder is original β grain equiaxed structure, the secondary dendrite along the grain boundary to the grain interior development. The secondary dendrite in the powder is more developed than on the surface. This is because the cooling rate of the powder surface is faster than that of the inside. The emergence of dendrite shows that although the cooling rate has been very fast, there will still be dendritic segregation.

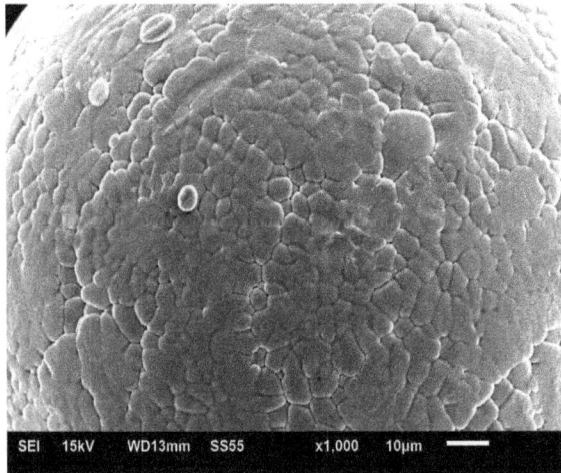

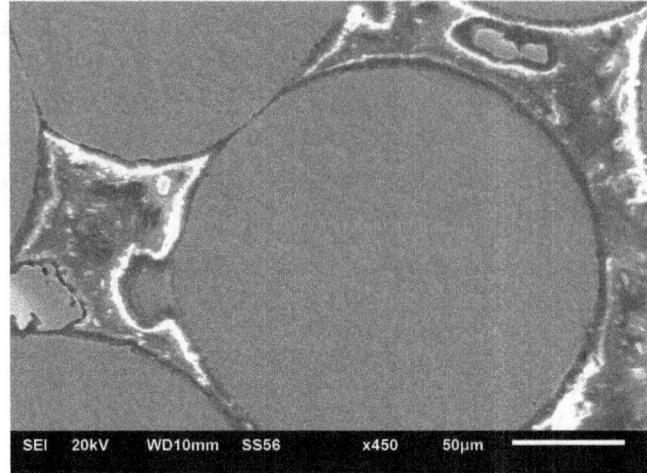

Fig. 2 SEM of powder surface   Fig. 3 SEM of powder cross-sectional surface

Fig. 4 is a photomicrograph of powder under optical microscope, matrix phase is β, there is a lot of martensite needles α' in the powder.

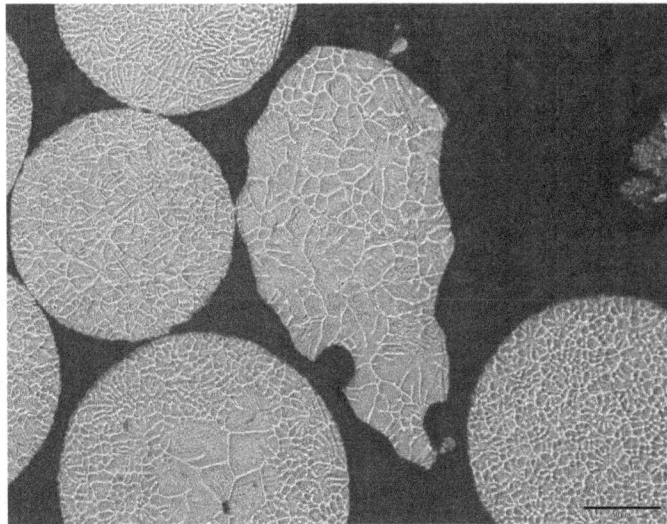

Fig. 4 The OM photo of the powder

3.4. *Analysis of defects in powder hollow ball*

Fig. 5 is the OM photos of powder hollow ball. As it showed that the alloy powder hollow ball can be seen in the hollow rate of less than 2% at the powder of 61 kW. Hollow ball formed when high-speed mechanical blowing argon into the molten metal droplets aggregate expanded in the process of milling. The main effects to reduce the hollow ball parameters in the electrode induction melting aerosolized milling process are gas pressure and power. However, when the pressure is small, it will easily cause insufficient powder crushing and over-sized particle diameter of powder. The powder size for 3D laser printing is mainly between 20-180 μm, so it is undesirable to reduce the pressure. When the gas pressure constant, the higher the induction coil power, the larger the droplet is, the area and the probability of the effect of argon and droplet are increased, it is easy to form a hollow ball. However, the induction power is too small, the stick cannot be melted or melting rate too slow, the production efficiency is low, thus increasing the cost of production.

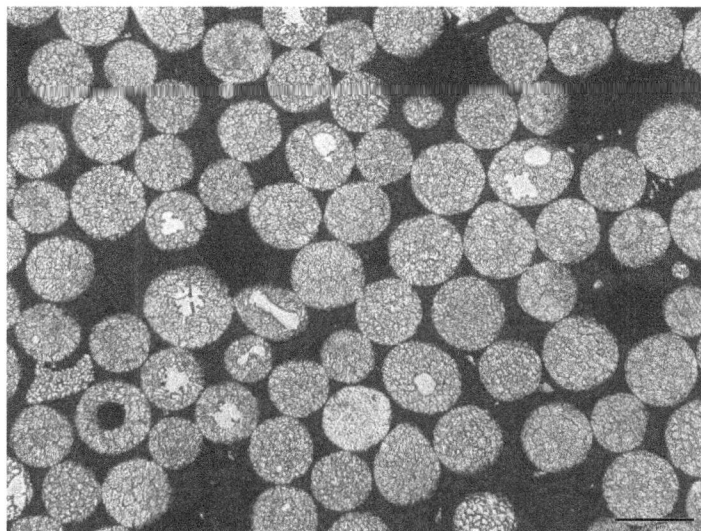

Fig. 5 The OM photo of hollow powder

3.5. *Powder phase analysis*

The XRD pattern of TC21 titanium alloy powder is shown in Fig. 6. The powder phase is mainly composed of α' and β. Martensite needles α' is a kind of metastable phase, it is due to the rapid solidification of molten droplets and β is the matrix phase of the alloy powder. There are still some metal compounds such as $AlTi_3$, Zr_3Al, and Ti_2AlNb. These metal compounds enhance the strength and toughness of TC21 titanium alloy.

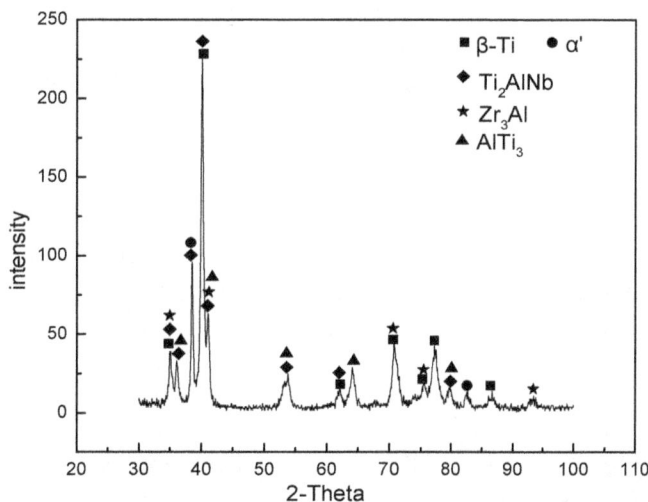

Fig. 6 XRD pattern of TC21 titanium

3.6. *Fluidity and bulk density of powder*

In laser 3D printing, particle size of 53-180 μm powder is uesd for laser direct metal deposition (LDMD) and 10-53 μm powder for Selective laser melting deposition (SLM). The fluidity of TC21 alloy powder directly impact on structure and performance of laser 3D printing parts, only good fluidity can meet the requirements of laser 3D printing metal parts without defects. The fluidity bulk density of TC21 alloy powder is shown in Table 3, with decreasing size of the powder, the better the fluidity of the powder.

Table 3 The fluidity and bulk density of TC21 titanium alloy powder.

Particle (μm)	Fluidity(s/50g)	Bulk density(g/25cm³)
150-180	24.37	68.58
106-150	21.93	67.70
53-106	21.39	66.44
53-180	21.19	69.20

When the size of powder particle less than 53 μm, regardless of percussion stabbed with a spoon or funnel, the powder will not flow down, no explanation has fluidity. Mixing powder with 53-180 μm has highest bulk density of 69.20 g/25cm3, this is because the powder with small particle size can go into the large gap among powder with large size, and bulk density is even greater.

4. Conclusion

TC21 titanium alloy powder for laser 3D printing was prepared by Electrode induction melting aerosolized under the condition of process parameter. The morphology of the prepared powder was spherical, and hollow ball rate is less than 2%. The surface structure of powder is composed of dendritic crystals and a small amount of cellular crystal. The main phases of TC21 alloy powder are composed of α', β and metal compound such as $AlTi_3$, Ti_2AlNb, Zr_3Al. The content of oxygen of TC21 alloy powder is 0.1 wt.%, powder bulk density is 2.74 g/cm^3, particle size is from 10 μm to 180 μm. Mixing powder with 53-180 μm has highest bulk density of 69.20 $g/25cm^3$, the fluidity of the 53-180 μm powder is 24.36 S/50g. The TC21 alloy powder prepared by this method can meet the request of the laser 3D printing parts.

Acknowledgments

This work was funded by Science and Technology Plan Project of Liaoning Province: (2014221006), the Fundamental Research Funds for the Central Universities (N130810002), the National Key Research Project (2016YFB1100201).

References

1. Z. H. Huang, H. L. Qu, C. Deng, J. C. Yang, Development and Application of Aerial Titanium and Its Alloys, Materials Review. 25 (2011) 102-107.
2. Y. Q. Zhao, H. L. Qu, L. Feng, H. Y. Yang, H. Li, Y. N. Zhang, H. C. Guo, D. K. Huang, Research on high strength, high toughness and high damage-tolerant titanium alloy-TC21, Titanium Industry Process. 21 (2004) 22-24.

3. H. W. Xie, L. M. Zou, X. Liu, C. Le, Y. X. Cai, The situation of preparation technology of spherical titanium powders. Materials Research and Application. 8 (2014) 78-82.
4. J. H. Zheng, Preparation and properties of TC4 titanium powder under high pressure, Yanshan University, 2011.
5. B. M. Wei, L. M. Tai, C. Z. Chi, Study on Technology of Boronizing on TC4 Titanium Alloy by Solid Powder Method. Hot Working Technology, 44 (2015) 201-204,
6. Y. Huang, H. P Tang, J. Chen, F. Y. Zhang, Y. F. Cui, Y. J. Wu, Research of rapid laser deposition by different appearance titanium and titanium alloy powder. Applied Laser, 44 (2005), 81-83+120.

A Composite of Carbon Fiber Reinforced Nylon PA6 for the Military

Can-Duo Shen[1,*], Qing Xue[2], Sheng Zhu[2]

[1]*Institute of quartermaster equipment, Beijing, China, 100010*
[2]*Academy of armored forces engineering, Beijing, China, 100072*
*Email: shencanduo@126.com

Utilizing the special properties of carbon fiber (CF), nylon PA6 resin was filled with that by employing the technique of mechanical alloying, and the effects of carbon fiber on the mechanical property of the nylon PA6 were studied. From the results, filling nylon PA6 resin with CF could effectively enhance the tensile strength, the elastic modulus, the flexural strength and the flexural modulus for the material. In addition, the material shrinkage was also significantly reduced. The enhancing effects varied as CF content changed. Through comparative experiments, the optimal proportion of the addition was determined. Replacing the conventional iron back pad of military outdoor pot by this composite would not only largely reduce the load and improve the motility for the soldiers, but also ensure a better battlefield adaptability and comfort.

Keywords: Carbon fiber; nylon PA6; reinforced strength; military using.

1. Introduction

Carbon fiber (CF) features many advantageous properties, such as high specific strength, high abrasion tolerance, high fatigue tolerance, low thermal expansion coefficient, excellent self-lubricating property and etc. As a result, CF has become one of the most important reinforcing materials in the recent years. One most common application of such reinforcement is the CF reinforced (CFR-) resin composites [1]. Being a well-performing engineering plastic, nylon still has some shortages, e.g., high water absorption, low dimensional stability. Besides, its strength and hardness are much lower than metal. In order to overcome these shortages, nylon material could be reinforced by carbon fiber or other fibers to improve the property [2, 3]. In the recent years, carbon fiber reinforced materials has developed very rapidly, and the resultant composites integrated the excellence from both, such as increased hardness and rigidity comparing to the unreinforced nylon, decreased high temperature creep, significantly increased thermal stability, high abrasion tolerance, excellent fire resistance and etc. With better battlefield adaptability and comfort, the composite is proved to be a great substitute which could replace Q235 back pad of conventional PLA outdoor pot.

2. Experiment Method

2.1. Preparation of CFR-PA6 sample

CFR-PA6 was prepared by uniformly mixing 59-94% PA6 with 5-40% surface-treated carbon fiber and about 1% other auxiliaries in an extruder. The temperatures on the front, middle and back parts of the feed cylinder (barrel) were 250°C, 260°C and 270°C. The temperature of the die was 260°C. The prills were extruded at a screw rotation speed of 80 r/min and a die pressure of 0.5-2.0 Mpa. Afterwards, the sample was molded by an injection molding machine with a 1st-zone temperature of 220°C, a 2nd-zone temperature of 230°C, a 3rd -zone temperature of 260°C, a 4th-zone temperature of 270°C and a 5th-zone temperature of 270°C. The temperature of the nozzle was 290°C, and the injection pressure was 0.5-3.0 Mpa [4, 5].

**Corresponding author

2.2. *Properties tests*

The tensile strength and elastic modulus were determined in accordance with GB/T 1040.1-2006Plastics-Determination of tensile properties-Part 1: General principles. The flexural strength and the flexural modulus were determined in accordance with GB/T 9341-2008 Plastics - Determination of flexural properties. The impact strength was determined in accordance with GB/T 1451-2005 Fiber-reinforced plastics composites-Determination of Charpy impact properties. The Rockwell hardness was determined in accordance with GB/T3398.2-2008 Plastics - Determination of hardness - Part 2: Rockwell hardness.

3. Experimental Data

3.1. *On the mechanical properties of CFR-PA6*

PA6 was reinforced by CF with different proportions. The mechanical properties were shown in Table 1. From Table 1, the tensile strength, elastic modulus, flexural strength, flexural modulus and hardness of CFR-PA6 all showed a certain degree of improvement as compared to pure PA6. In particular, the improving magnitudes of the tensile strength, elastic modulus, flexural strength and flexural modulus were relatively larger.

Table 1 Comparison of the mechanical properties of CFR-PA6 and pure PA6

Items	Pure PA6	CFR-PA6 (CF mass ratio)			
		5%	10%	20%	30%
Tensile Strength [Mpa]	72	95	125	155	176
Elastic Modulus [Mpa]	3100	6035	8025	10075	12355
Flexural Strength [Mpa]	110	145	170	185	205
Flexural Modulus [Mpa]	2500	4565	8700	10145	13455
Rockwell Hardness	95	98	105	115	118
Impact Hardness [kJ/m^2]	did not break	did not break	45	28	25

As CF content increased from 0% to 40%, the tensile strengths of CFR-PA6 were shown in Fig. 1, reaching a maximum when the amount of CF reached 30%.

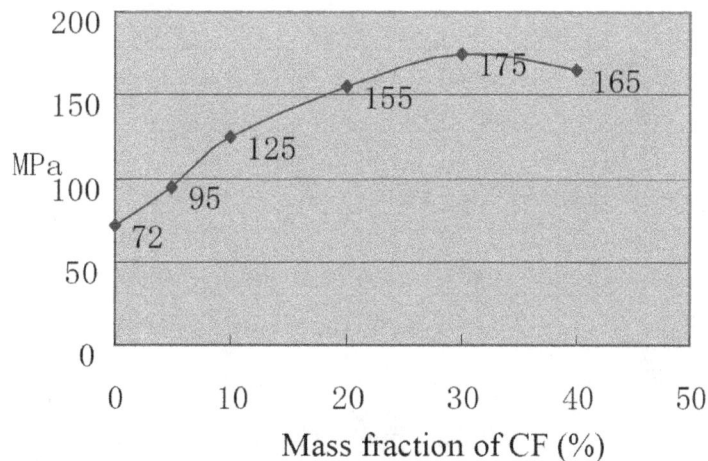

Fig. 1 The correlation between tensile strength and CF content

As CF content increased linearly from 0% to 40%, the flexural strength of CFR-PA6 also depicted an increasing trend, and the increment tended to be slow when the amount of CF reached 30%. As shown in Fig. 2.

As CF content increased linearly from 0% to 40%, the hardness of CFR-PA6 increased within a certain range, as shown in Fig. 3.

It was because CF was a highly rigid material. When CF was compounded into PA6 matrix, its high strength helped withstand the stress, and transmit the stress by making use of the elasticity of the matrix resin and the caking property of the fiber [6]. The features of this composite were as follow: Under the stress, the strain of the fiber turned to equalize the strain of the matrix resin. However, matrix resin had a much lower elastic modulus than CF and it promoted plastic yielding [7]. Therefore, when the strains of CF and the matrix resin were the same, the stress in CF was much higher than that in matrix resin, making those fibers with the cleft break first. Sequentially, as such breakage was blocked by the plastic flow in the matrix resin. These broken fibers still withstand the same load as the unbroken fibers at the end near the break site [8].

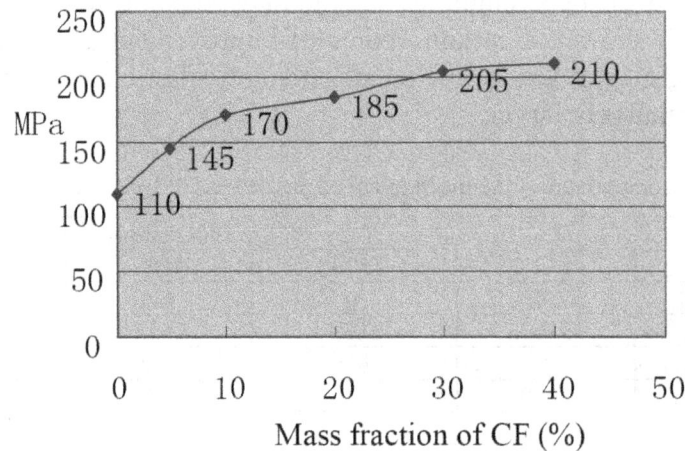

Fig. 2 The correlation between flexural strength and CF content

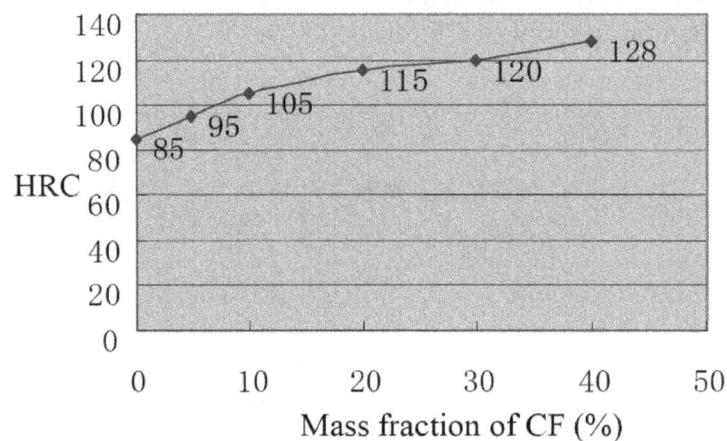

Fig. 3 The correlation between hardness and CF content

Another reason for the composite reinforcement was that the matrix cracking was inhibited by matrix resin. Through the shearing effect, the soft matrix resin stopped the cracks from propagating vertically. The propagation deflected, causing a large part of breakage was used to offset the adhesive force from the matrix resin to the fiber, and thus resulting in an inhibition on the craze throughout the whole CFR-PA6 volume. Hence, the composite acquired a much higher resistance to the formation, growth and breakage of the craze, as well as the propagation of the cracks [9]. As a result, the mechanical properties of CFR-PA6 improved to a large extent.

3.2. *Analysis of electron microscope*

When CF and PA6 are mixed and prilled in the double-screw extruder, a higher draw ratio of CF should be ensured to maximize effect of CF enhancement as long as CF is well dispersed within PA6. It required a rational combination of the feed, plastify shear and mill components of the screws in the double-screw extruder [10, 11]. From the experimental results, under a particular screw-component combination, CF was evenly distributed within PA6, and the maximum length of CF could reach 0.5mm. The distribution of CF length within PA6 complied with the distribution pattern of the common reinforced fibers. As shown in Fig. 4.

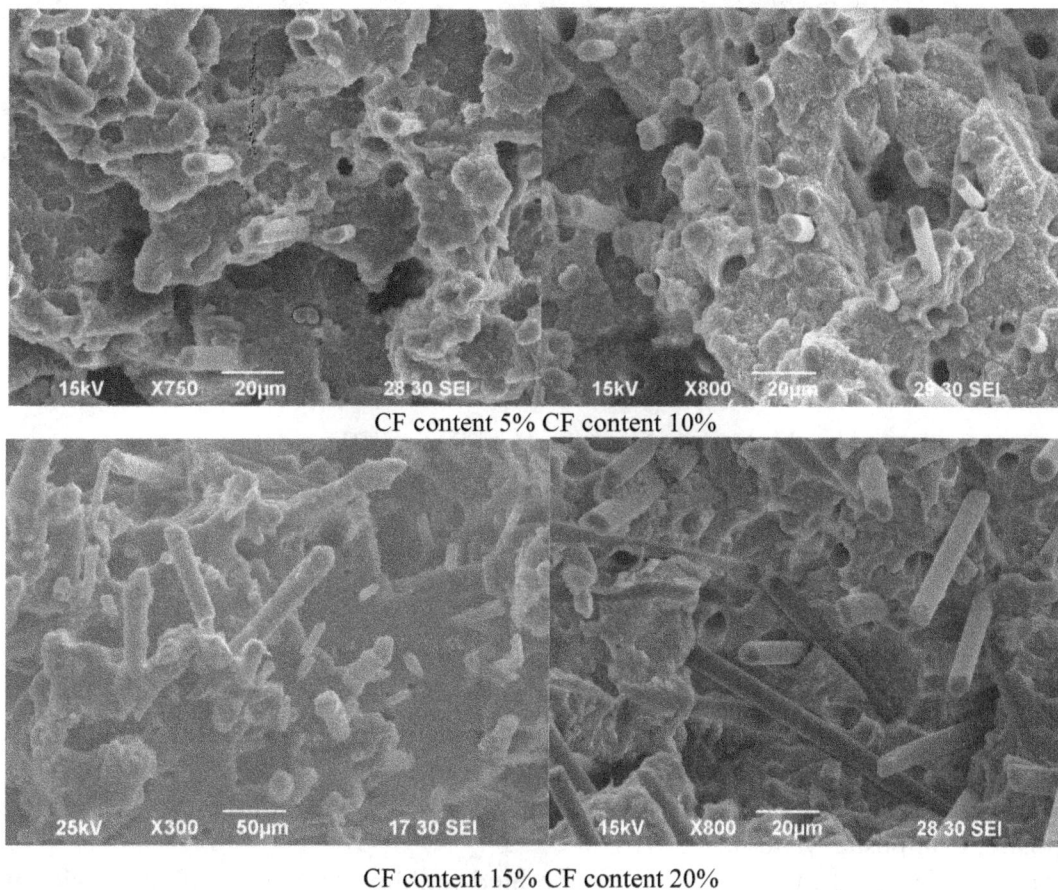

CF content 5% CF content 10%

CF content 15% CF content 20%

Fig. 4 Micrograph of CFR-PA6 with different CF content

4. Conclusion

1. As CF content increased, the tensile strength, elastic modulus, flexural strength, flexural modulus and hardness of CFR-PA6 also increased within a certain range.
2. As CF content increased, the impact strength of CFR-PA6 material decreased. When CF content was 30%, the material had the most outstanding mechanical properties.

3. In conclusion, CFR-PA6 material is a substitute which could replace steel Q235 in the manufacture of the back pad of military outdoor pot. As shown in Fig. 5, the various mechanical specifications met the requirements of military usage and the weight was reduces by 58%, leading to a large increase in soldier's motility.

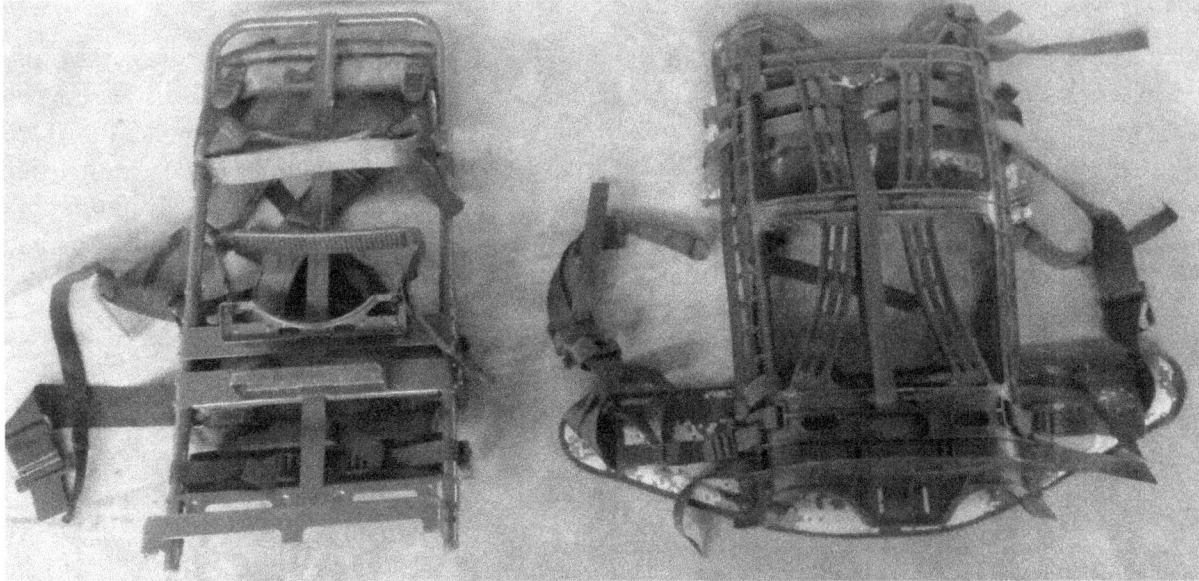

Fig. 5 Comparisons between Q235 and CFR-PA6 in making the back pad

Acknowledgment

This research was funded by the National Postdoctoral Science Foundation.

References

1. Z.H. Li, Carbon fiber composites and their applications, Application of engineering plastics. (1998).
2. L.G. Xia, A.J. Li, Study on carbon fiber surface treatment and its effects on carbon fiber resin interface, Material Review. (2006).
3. Y.Y. Zheng, Study on the interfacial-layer design and the structural properties of fiber reinforced nylon 6 composite, Journal of Fuzhou University. (2005).
4. X.J. Bi, Z.H. Li, Z.H. Tang, W. Li, Design of molding technique for stress-based resin composite, Composite Material. (2010).
5. Y. Xiao, Y.Y. Shi, L.L. Chang, Study on the bind-molding technique of composite materials, Electromachining and Mold. (2007).
6. Z.Z. Kang, X. Wang, Discuss on technique of Soft mold/Vacuum Assisted Resin Infusion flanged semisphere, Fibrous composite material. (2011).
7. C.Z. Wang, Study on the preparation of C-type carbon fiber and the mechanism of the cathodic oxidation surface treatment, Beijing University of Chemical Technology, 2010.
8. M. Zhang, Study on key influential factors on the interfacial binding strength of carbon fiber reinforced resin composite materials, Journal of Shandong University. (2010).

9. R.J. Zheng, Study on the coating agent for carbon fiber, Journal of Changchun University of Technology, 2002.
10. M.C. Liu, Mold development for fiber reinforced composite materials, Application of engineering plastic. (2005).
11. G. Du, Study on Design and Integral Manufacture of Composite Thrust Cylinder, National University of Defense Technology, 2007.

Synthesis and Photochromism Studies of 1-[2-Methyl-5-Phenyl-3-Thienyl]-2-[2-Methyl-5-(3-Aldehyde-4-Hydroxy)-3-Thienyl] Perfluorocyclopentene

Rui-Min Lu, Lu Huang, Shi-Qiang Cui[*], Shou-Zhi Pu

Jiangxi Key Laboratory of Organic Chemistry, Jiangxi Science and Technology Normal University, Nanchang 330013, P.R. China
[*]*Email: cuisq2006@163.com*

A new unsymmetrical photochromic diarylethene compound, namely, 1-[2-methyl-5-phenyl-3-thienyl]-2-[2-methyl-5-(3-aldehyde-4-hydroxy)-3-thienyl] perfluorocyclopentene was synthesized and its properties were investigated in detail, including photochromic reactivity kinetics and concentration effect. The results showed that the compound exhibited remarkable photochromism, changing between colorless and blue in solution, respectively. What is more, the kinetic experiments illustrated that the cyclization/cycloreversion process of this compound was determined to be the zeroth/first reaction.

Keywords: Diarylethene; photochromism; fluorescence.

1. Introduction

In the past several decades, there has been considerable interest in the photochromic compounds due to their wide applications in various optoelectronic devices, the term 'diarylethene derivatives' describes the reversible transformation of the chemical species by absorption of photoirradiation between two isomers that have distinguishable absorption spectra. These derivatives are promising materials for the manufacture of advanced materials due to their excellent thermal stability, remarkable fatigue resistance, rapid response and high photocyclization quantum yields. [1-5]

Furthermore, their fluorescence can be reversibly modulated by alternating the irradiation with UV and visible light. So far, many DAE fluorescence chemosensors with various functional groups have been designed and synthesized. [6-11]

As we all know, diarylethenes can reversibly transform from the open-ring isomers to the close-ring isomers upon irradiation at an appropriate wavelength of light, and their photochemical properties mainly depend on several factors, such as the categories of heterocyclic moieties, substituents and extension of π-conjugation. [12]

It is a hot topic in current chemical research to develop multiresponsive switching diarylethene molecules that integrate several switchable functions into a single molecule in order to miniaturize the components of machinery and electronics down to the molecular level. [13] In order to achieve the aim, an available approach is to introduce some functional substituents into the skeleton of diarylethene systems. For example, Matsuda et al. reported that the photochromic reactivity of a dithienylethene with a (4-pyridyl) ethynyl group could be effectively modulated by trifluoroacetic acid and diethylamine. [14]

In recent years, the development of effective fluorescent sensors with instantaneous response and high sensitivity has gained a lot of interest because of their applications in medical, environmental, and biology. [15-17] Due to the fact that metal ions are crucial for all types of organisms, development of fluorescent chemosensors for specific metal ions has become an active research area. [15]

A challenge in the development of photochromic molecular switching technology is the search for more complicated systems that integrate several switchable functions into a single molecule, which could enable complex logic-gate operations and multimode data storage with increased information density. However, multi-responsive fluorescent diarylethenes sensitive to different external stimuli [18-20] have been rarely hitherto reported.

As described above, herein, a novel unsymmetrical diarylethene 1-[2-methyl-5-(2-methoxyl-phenyl)-3-thienyl]-2-[2-methyl-(5-ethynyl) trimethylsilane-3-thienyl] has been synthesized and we not only examined its photochromic properties but also studied fluorescence properties. The results indicated that the diarylethene has good photochromic properties and fluorescence switching properties. The photochromic reaction of 1a is shown in Scheme 1.

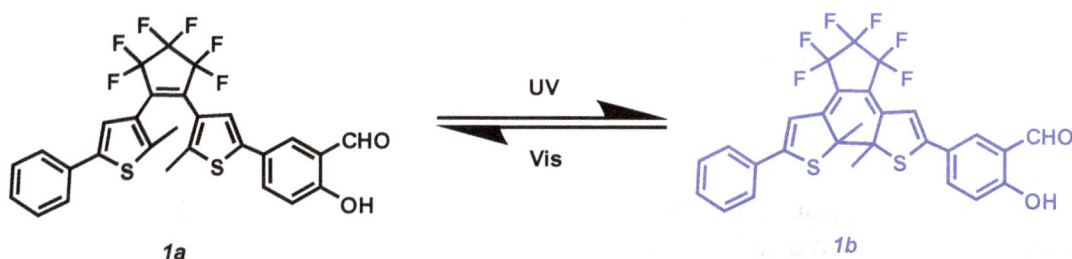

Scheme 1 Photochromism of 1a

2. Experiments

Synthesis of *1a*

Diarylethene *1a* was synthesized according to the route described in Scheme 2. The precursor diarylethene *1a* was synthesized by the reported method. Firstly, aldehyde of compound *1* was protected. Then we get *2*. Under an argon gas atmosphere, compound *3* was dissolved in THF and *n*-butyl lithium acetonitrile solution was added at -78 °C. Stirring was continued for 30 min at this low temperature, octafluorocyclopentene was added and the mixture was stirred for 1 h at this temperature. After extracting with diethyl ether and evaporation in vacuum, the residue was purified by column chromatography on silica gel (petroleum ether) to obtain *4*. Compound *5* was prepared by reacting compound *2* and *4* in dry THF at -78°C, then through purified by column chromatography on silica gel (petroleum ether), and *5* was obtained, after deprotection we get *6*. Compound *6* dissolved in anhydrous CH_2Cl_2 was added to BBr_3 at 195 K under an argon atmosphere. After stirring for 0.5 h at this temperature, the mixture was warmed to room temperature and stirred for 2 h. The crude product was purified by column chromatography on silica gel using petroleum ether/ethyl acetate (15:1) as the eluent to give compound *1a*.

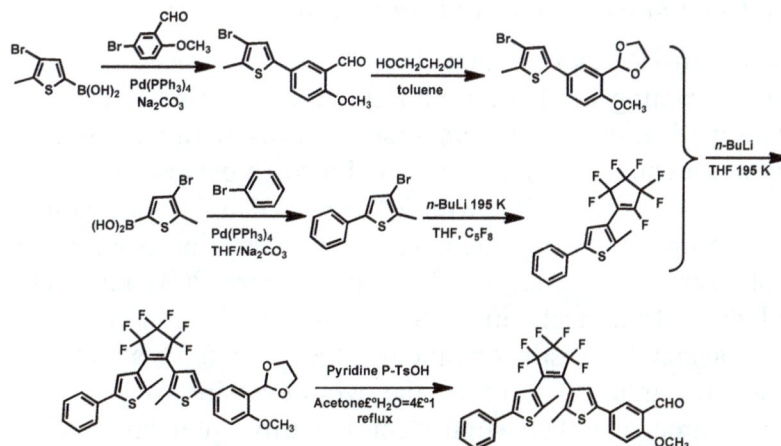

Scheme 2 Photochromism of 1a

463

3. Results and Discussion

3.1. *Photochromism of the diarylethene 1a*

Absorption spectra were monitored with an Agilent 8453 UV/Vis spectrophotometer. The asymmetric diarylethene 1a exhibits good photochromic properties and can be toggled between its colorless ring-open and colored ring-closed forms by alternate irradiation with appropriate wavelengths of light. The absorption spectral changes in acetonitrile are shown in Figure 1. The open ring isomer has an absorption maximum at 288 nm, which was arisen from $\pi \rightarrow \pi^*$ transition. Upon irradiation with 297 nm UV light, the colorless acetonitrile solution of compound 1o turned blue, in which absorption maximum was observed at 587 nm. The blue color is due to the formation of the closed ring isomer. When the blue solution was irradiated with visible light ($\lambda > 450$ nm), the color could return back to colorless and its spectrum became the same as that of original one, indicating compound 1a returned to the initial open ring isomer. The coloration/discoloration cycle could be repeated more than 50 times and the photostationary spectrum was almost the same as that of the colored isomer. This indicates a high conversion from the colorless to the colored isomers by irradiation with 297 nm light.

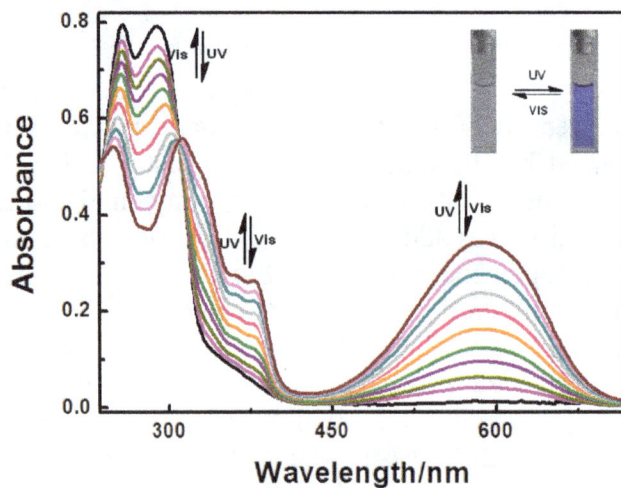

Fig. 1 Absorption spectral and color change of 1a in acetonitrile (2.0×10^{-5} mol L^{-1})

3.2. *Photochromic reaction kinetics in acetonitrile solution*

The photochromic cyclization/cycloreversion kinetics of 1a in acetonitrile were determined by UV-Vis spectra upon alternating irradiation with UV and appropriate wavelength visible light at room temperature. The cyclization and cycloreversion curves of 1a were shown in Figure 2(A) and Figure 2(B) respectively. It can be seen that the relationships between the absorbance and exposal time have good linearity upon irradiation with 297 nm UV light suggesting that the cyclization processes of 1a belong to the zeroth order reaction when open-ring isomer changed to closed-ring isomer. The slope of every line in Figure 2(A) and Figure 2(B) represents the reaction rate constant (k) of diarylethene 1a in acetonitrile. So all k of cyclization/cycloreversion process (k_{o-c}, 10^{-3}) of diarylethene 1a can be easily obtained, which are 6.54 s^{-1} and 1.08 s^{-1} in solution, respectively. As shown in Figure 2(B), during the cycloreversion of 1b, the relationship between $-\log(Abs)$ and exposal time also behave perfect linearity, indicating that the cycloreversion process belong to the first order reaction.

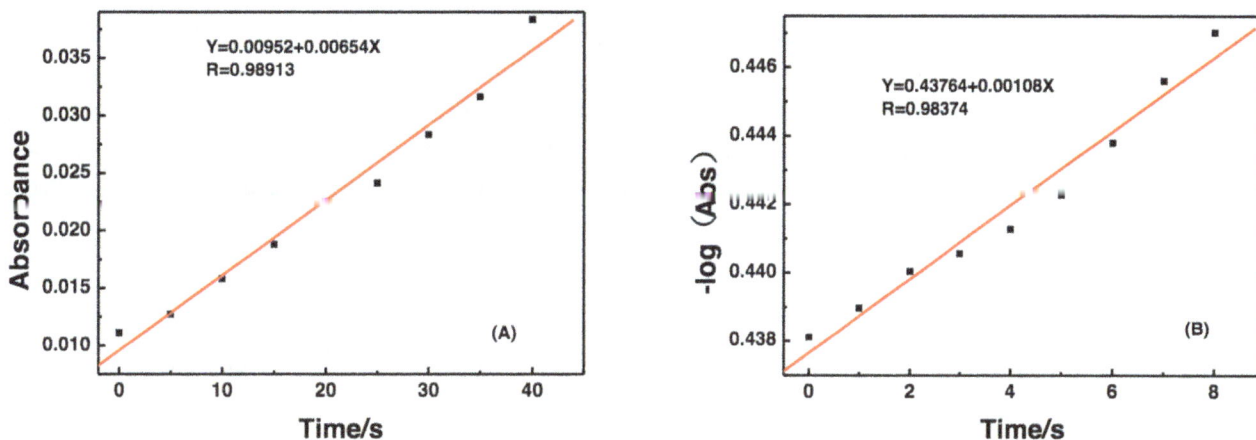

Fig. 2 The cyclization kinetics (A) and cycloreversion kinetics (B) of 1 in acetonitrile (2.0×10^{-5} mol L^{-1})

3.3. *Fluorescence of diarylethene 1a*

The fluorescence properties in acetonitrile solution of the compound 1a were measured using a Hitachi F-4600 spectrophotometer, and the breadths of excitation and emission slit were selected 10 nm and 20 nm, respectively. Figure 3 shows the fluorescence spectral changes of 1a in acetonitrile solution upon irradiation with 297 nm light at room temperature. As shown in Fig. 3, it could be clearly seen that the acetonitrile solution of 1a exhibited relatively strong fluorescence at 561 nm when excited at 360 nm. The fluorescence intensity decreased along with the photochromism from open-ring isomers to closed-ring isomers upon irradiation with 297 nm UV light, due to producing the non-fluorescence closed-ring isomer. The back irradiation by appropriate wavelength visible light regenerated its open-ring isomer and recovered the original emission intensity. The reversible changes of the emission intensity of diarylethene 1a are useful for application as the fluorescence switches.

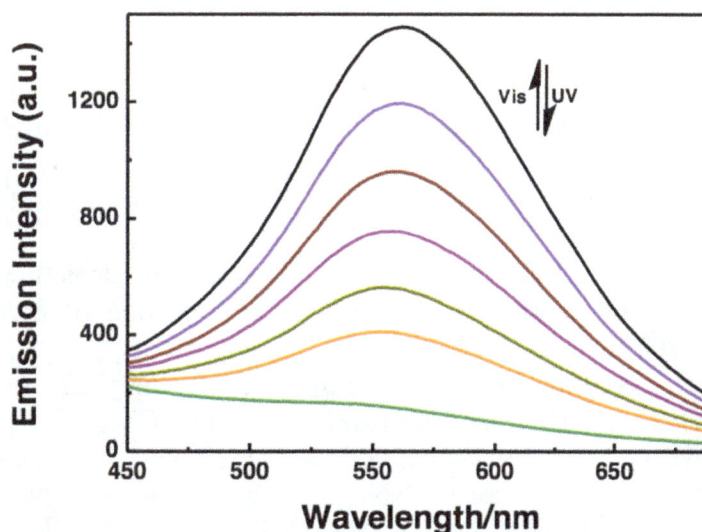

Fig. 3 Fluorescence spectra of diarylethene 1a in acetonitrile (2.0×10-5 mol L-1) upon irradiation with 297 nm light at room temperature

465

4. Conclusion

A novel unsymmetrical diarylethene was synthesized to investigate its photochromism, kinetic and fluorescence. Diarylethene 1a exhibited predominant photochromism and a relatively strong fluorescence switches along with the photochromism from open-ring isomers to closed-ring isomers in hexane. The present results are useful for the design of efficient photoactive and excellent characteristic diarylethene compounds. Furthermore, the compound also functioned as a fluorescence switch.

Acknowledgment

The authors are grateful to the Science Funds of Natural Science Foundation of Jiangxi Province (20151BAB203019) for funding this work.

References

1. H. Tian and S. Wang, Photochromic bisthienylethene as multi-function switches, Chem. Commun. 8, 781 (2007).
2. H. Tian and S.J. Yang, Recent progresses on diarylethene based photochromic switches, Chem. Soc. Rev. 33, 85 (2004).
3. M. Irie, Diarylethenes for memories and switches, Chem. Rev. 100, 1685 (2000).
4. T. Shiozawa, M.K. Hossain, T. Ubukata, Y. Yokoyama, Ultimate diastereoselectivity in the ring closure of photochromic diarylethene possessing facial chirality, Chem. Commun. 46, 4785 (2010).
5. W.H. Zhu, X.L. Meng, Y.H. Yang, Q. Zhang, Y.S. Xie, H. Tian, Bisthienylethenes containing a benzothiadiazole unit as a bridge: Photochromic performance dependence on substitution position, Chem. Eur J. 16, 899 (2010).
6. L. Giordano, T.M. Jovin, M. Irie and E.A. Jares-Erijman, Diheteroarylethenes as thermally stable photoswitchable acceptors in photochromic fluorescence resonance energy transfer (pcFRET), J. Am. Chem. Soc. 124, 7481 (2002).
7. H.C. Ding, G. Liu, S.Z. Pu and C.H. Zheng, Multi-addressable fluorescent switch based on a photochromicdiarylethene with triazole-bridged methylquinoline group, Dyes. Pigm. 103, 82 (2014).
8. S.Z. Pu, H.C. Ding, G. Liu, C.H. Zheng and H.Y. Xu, Multiaddressing Fluorescence Switch Based on a New Photochromic Diarylethene with a Triazole-Linked Rhodamine B Unit, J. Phys. Chem. C.118, 7010 (2014).
9. T. Fukaminato, T. Doi, N. Tamaoki, K. Okuno, Y. Ishibashi, H. Miyasaka and M. Irie, J. Am. Chem. Soc.133, 4984 (2011).
10. Q. Zou, X. Li, J. Zhang, J. Zhou, B. Sun and H. Tian, Highly selective fluorescence turn-on chemosensor based on naphthalimide derivatives for detection of copper (II) ions, Chem. Commun. 48, 2095 (2012).
11. J. He, T. Wang and H. Zeng, Ion-induced cycle opening of a diarylethene and its application on visual detection of Cu2+ and Hg2+ and keypad loc, J. Mater. Chem. C.2, 7531-7540 (2014).
12. T. Darwish, R. Evans, M. James, N. Malic, G. Triani, T. Hanley, J. Am. Chem. Soc. 132, 10748 (2010); (b) M. Taguchi, T. Nakagawa, T. Nakashima, T. Kawai, J. Mater. Chem. 21, 17425 (2011); (c) Y.C. Jeong, J.P. Han, Y. Kim, E. Kim, S.I. Yang, K.H. Ahn, Tetrahedron. 63, 3173 (2007).

13. A.J. Myles, T.J. Wigglesworth, N.R. Branda, An ab initio simulation of a dithienylethene/phenoxynaphthacenequinone photochromic hybrid, Adv. Mater. 15, 745 (2003).
14. K.J. Yumoto, M. Irie, K.J. Matsuda, Control of the photoreactivity of diarylethene derivatives by quaternarization of the pyridylethynyl group, Org. Lett. 10, 2051 (2008).
15. V. Amendola, L. Fabbrizzi, F. Forti, M. Licchelli, C. Mangano, P. Pallavicini, A. Poggi, D. Sacchi, A. Taglieti,Light-emitting molecular devices based on transi-tion metals, Coord. Chem. Rev. 230, 273 (2000).
16. D.A. Leigh, M.A.F. Morales, E.M. Perez, J.K.Y. Wong, C.G. Saiz, A.M.Z. Slawin, et al, Patterning through controlled submolecular motion: rotaxane-based switchesand logic gates that function in solution and polymer films, Angew. Chem. Int. Ed. 44, 3062 (2005).
17. C. Caltagirone, P.A. Gale, Anion receptor chemistry: highlights from 2007, Chem. Soc. Rev. 38, 520 (2009).
18. N. Soh, K. Yoshida, H. Nakajima, K. Nakano, T. Imato, T. Fukaminato, et al, A fluorescent photochromic compound for labeling biomolecules, Chem Commun. 520, 6 (2007).
19. T. Fukaminato, T. Doi, N. Tamaoki, K. Okuno, Y. Ishibashi, H. Miyasaka, et al, Single-Molecule fluorescence photoswitching of a diaryletheneperylenebisimide dyad: non-destructive fluorescence readout, J Am Chem Soc. 133, 4984 (2011).
20. W.J. Tan, X. Li, J.J. Zhang, H. Tian. A photochromic diarylethene dyad based on perylene diimide, Dyes Pigm. 89, 260 (2011).

Chapter 9
Composites

Influence of Aggregate Gradation Segregation on the Water Stability of Asphalt Mixes

Chao Geng, Yao-Dong Su[*]

Key Laboratory for Special Area Highway Engineering of Ministry of Education,
Chang'an University, Xi'an, Shannxi, P.R.China 710064
[*]Email: 327499098@qq.com

In order to analyze the influence of gradation segregation on the water stability of Asphalt Mixes, this paper designs five group of aggregate gradation by varying coarse aggregates or fine aggregates, and analyzes the relations between aggregate gradation and asphalt mixtures water stability. The results show that aggregate gradation segregation decreases the water stability of asphalt mixtures, and coarse aggregates segregation has more adverse effects than fine aggregates segregation.

Keywords: Gradation segregation; water stability; asphalt mixture.

1. Introduction

The amount of aggregate in hot mix asphalt (HMA) consists over 90% of the total volume. Hence, such properties of aggregate as gradation and size definitely affect the quality of asphalt mixtures for pavement. Aggregate gradation, which is one of the most important factors to resist pavement distress, is the distribution of particle sizes which is normally expressed in percentage of the total weight [1].

The optimization of aggregate gradation is advantageous for economic and technical reasons. The most well-liked and well-known methods of aggregate gradation include: i) using two different segments of aggregate (i.e. Fine aggregates (FA) and Coarse aggregates (CA)), ii) using total aggregate gradation that is combined aggregate gradation [2]. However, the segregation of asphalt mixture was not an avoided problem on the process of pavement, and the asphalt mixtures segregation can be very harmful to the performance of the asphalt mixture.

This study focuses on the influence of gradation segregation on the water stability of asphalt mixture. The Marshall test and Split test were performed to evaluate the water stability of asphalt mixtures of different gradation designed.

2. Test Materials

2.1. *Raw materials*

The municipal special 90# asphalt was adopted for test asphalt and the main technical indicators of it were determined as shown in Table 1[3].

Table 1 Technological properties of asphalt

Test terms	Requirements	Test results
Penetration degree（25°C, 100g, 5s）/0.1mm	80～100	86.1
Ductility（5cm/min，10°C）/cm	≮30	＞ 150
Softening point（R&B）/°C	≮44	49

The aggregate consisted of coarse limestone, fine limestone, and limestone powder. The main technical indicators of them are listed in Table 2.

[*]Corresponding author

Table 2 Physical properties of aggregate

Test terms	Requirements	Test results
Crushing value/%	≯28	14.3
Los Angeles abrasion loss/%	≯30	14.3
Adhesion to asphalt	≮4 level	4 level
Percentage of Flat-elongated/%	≯15	10

2.2. *Mixture composition design*

Referring to the domestic and foreign gradation design methods, Gradation 0# was chosen the design gradation, and 1~4# were the other four types of aggregate gradations designed by varying coarse aggregates and fine aggregates respectively, as shown in Table 3.

Table 3 Aggregate gradation

Gradation number	Mass fraction of aggregates passing sieve pore(mm)/%										
	19	16	13.2	9.5	4.75	2.36	1.18	0.6	0.3	0.15	0.075
0#	100	95.0	84.0	70.0	48.0	34.0	24.5	17.5	12.5	9.5	6.0
1#	100	95	92	70	34	27	20	17.5	12.5	9.5	6.0
2#	100	95	88	70	41	30	22	17.5	12.5	9.5	6.0
3#	100	95	84	70	48	34	18.5	9	7	5.5	5.0
4#	100	95	84	70	48	34	18.5	13.5	9.5	7	5.5

Referring to the asphalt content design method, OAC=4.5% was determined by Marshall Test. Based on the optimum asphalt-aggregate ratio for gradation 0#andtheory of effective thickness of asphalt, the optimum asphalt content of gradations1~4# was determined, The optimum asphalt-aggregate ratios of five types gradations are shown in Table 4.

Table 4 Asphalt content

Gradation number	0#	1#	2#	3#	4#
Optimum Asphalt-aggregate ratio	4.5	4.35	4.41	3.31	3.81

3. Test Methods

Test specimens are made by large scale Marshall compaction test at 140~150°C，and the size is ø152.4mm×95.3mm. Split test is conducted at 25°C, and the loading rate is 50mm/min. The splitting tensile strengths of specimens after freeze-thaw cycling and not freeze-thaw cycling are measured. The intensity ratio of freeze-thaw split is an index to evaluate the water stability of different gradation [4, 5].

4. Test Results and Discussion

In order to study the influence of aggregate gradation segregation the water stability of asphalt mixtures, the cleavage strength not freeze-thaw cycling and after freeze-thaw cycling of gradations 0~4# were determined as shown in Figure 1.

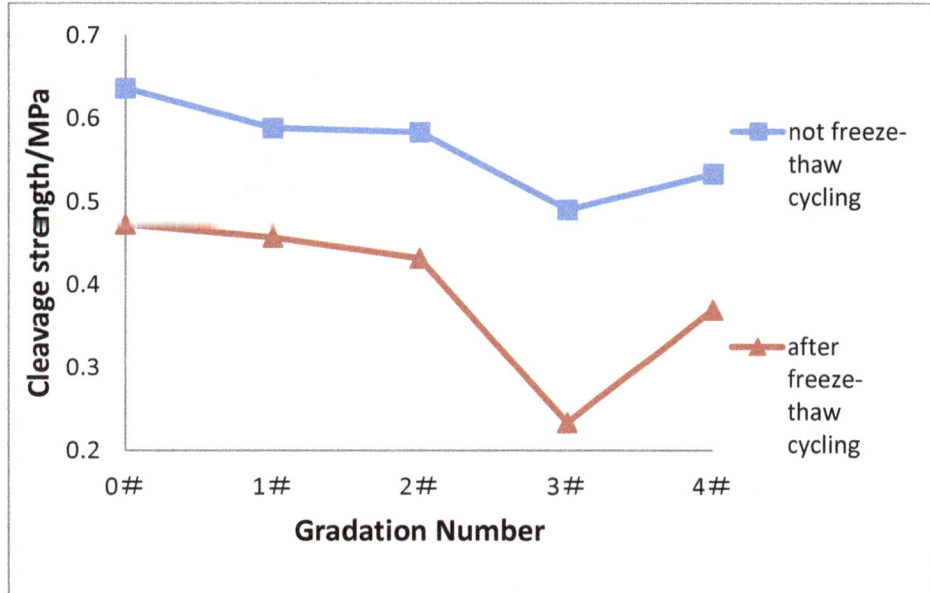

Fig. 1 Relation of gradation and cleavage strength

Tensile strength ratio of gradations 0~4# were determined as shown in Figure 2.

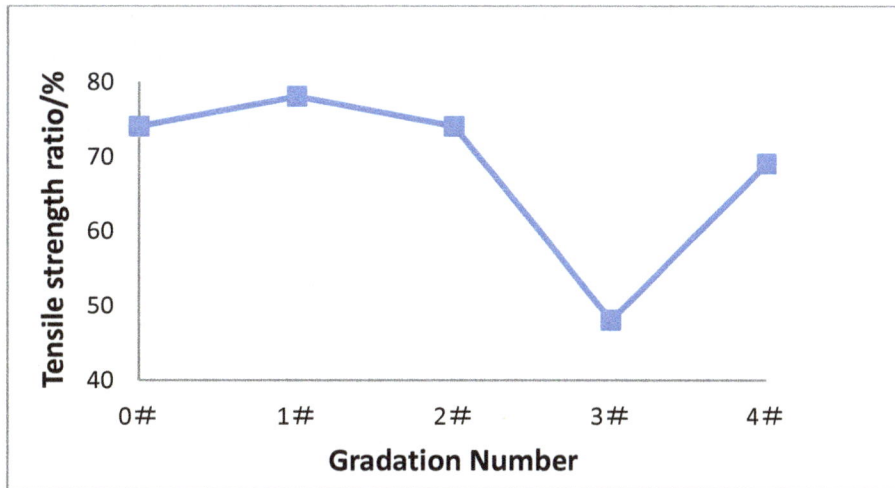

Fig. 2 Relation of gradation and cleavage strength

From the Figure 1, it can be seen that under the same condition of modeling and testing, on contrast of gradation 0#, the cleavage strength of gradation 1~4# decreases, no matter after freeze-thaw cycling and not freeze-thaw cycling.

Figure 2 shows the TSR values of asphalt mixture, compared to the design gradation 0#, gradation 1# and 4# have little change, but for gradation 3# and gradation 4#,the TSR values has a rapid reduction , one of the greatest reduction of 35%.

Compared to designed gradation 0#, the coarse aggregate of gradation 1# and 2# increases, the fine aggregate below 2.36mm almost not changes, the water stability of mixture is not changed; for gradation 3# and gradation 4#, the fine aggregate below 2.36mm decreases but the coarse aggregate almost not changes, the water stability also has a great change, and with the reduction of the fine aggregate content, the value of TSR decreased.

5. Conclusion

Aggregate gradation is one of the most important factors to the pavement performance of asphalt mixture. In this study on the influence of aggregate gradation on the water stability of pavement, the large-scale Marshall Test, freeze-thaw split test were conducted. As shown in the results, it is found that the main factors influence the water stability of the mixture is fine aggregate content. After asphalt mixture segregation happened, once the fine aggregate content changed, the influence to water stability is very significant. Even though there are larger changes in the content of coarse aggregate, as long as the fine aggregate material does not change significantly, the ability of anti-water damage showing certain stability.

References

1. Der-Hsien Shen, Ming-Feng Kuo, and Jia-Chong Du, Properties of gap-aggregate gradation asphalt mixture and permanent deformation, Construction and Building Materials 19 (2005) 147–153.
2. A.E. Hunter, G.D. Airey, and A. C. Collop, Aggregate Orientation and Segregation in Laboratory Compacted Asphalt Samples, in TRB, National Research Council, Washington D.C.,2004.
3. Feng Ma, Chao Zhang. "Road Performance of Asphalt binder Modified with Natural Rock Asphalt". Advances in Chemical, Material and Metallurgical Engineering, vl, 2729-2733.
4. Qi-lin Fu, Shuan-fa Chen, "Influence of Aggregate Gradation on Pavement Performance of Open-graded Large Stone Asphalt Mixes". ICCTP 2010: Integrated Transportation Systems— Green•Intelligent•Reliable, vl, 3133-3140.
5. M. Stroup-Gardiner, E. R. Brown, et al. Segregation in Hot-Mix Asphalt Pavements. Report 441[C]. Washington, D.C.: National Academy Press, 2000.

Influence of Gradations Design on Segregation of Asphalt Mixtures

Chao Geng, Xiao-Yu Chang*

Key Laboratory for Special Area Highway Engineering of Ministry of Education,
Chang'an University, Xi'an, Shannxi, P.R.China 710064
Email: 709236976@qq.com

In order to compare the different degrees of the aggregate segregations for asphalt mixtures with different gradation types during construction, in this paper, two kinds of asphalt mixtures with different gradation types are selected, and the granulometric composition conditions at different positions on cross section after spreading were analyzed. After the mixtures were rolled and molded, their construct depths and density were measured at different positions on cross section. Analysis result shows that, on the same condition, compared with S type features 0 gradation, near the maximum density curve of 1 graded aggregate segregation degree is smaller.

Keywords: Gradation types; asphalt mixture; segregation degree.

1. Introduction

Aggregate segregation of asphalt mixture is common in pavement construction process [1, 2]. Due to the influence of segregation of asphalt, service life of asphalt mixture pavement is greatly reduced, and on serious segregation, asphalt pavement service life will be reduced by more than 50%. It can be said that most of the early damage of the pavement is originated from unevenness of asphalt mixture, which has become one of the important factors to determine the quality of asphalt pavement [3, 4]. If the mixture can't maintain a certain stability of the construction, the indoor performance parameters obtained in the actual pavement can't be achieved, and the design will lose the meaning [5, 6].

This paper relies on the construction of Henan Shang Zhou expressway. Field studies the influence of gradation types on segregation of asphalt mixtures.

2. Experiment

2.1. *Selection of grading*

AC-25 type asphalt mixture design gradation is selected as the gradation0. Adjust the maximum density curve according to the 0.45 power as the gradation 1. Fig. 1 is the grading curve of two kinds of gradation.

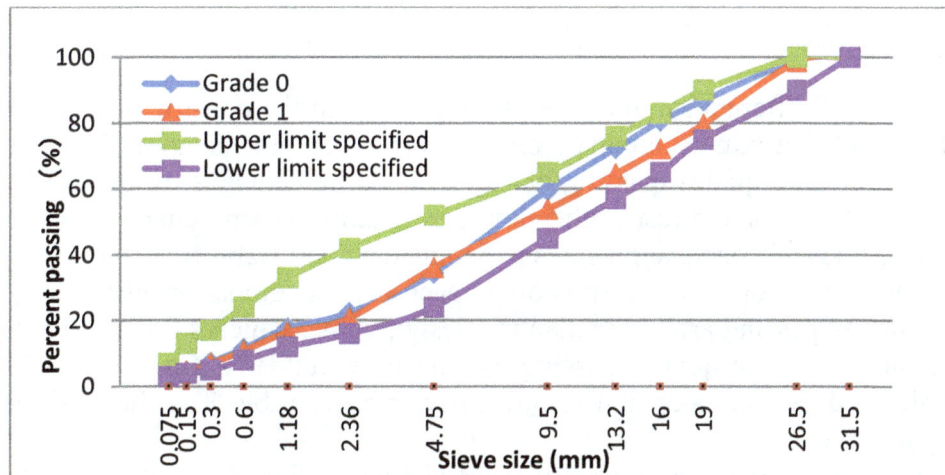

Fig. 1 Grading curve of two kinds of gradation

Table 1 is the percent passing of gradations.

Table 1 Percent passing of gradations

Sieve size (mm)	0.075	0.15	0.3	0.6	1.18	2.36	4.75
Grade 0	3.76	4.71	7.33	11.5	17.73	22.42	34.49
Grade 1	3.73	4.63	7	10.75	16.42	20.74	36.08
Upper limit specified	7	13	17	24	33	42	52
Lower limit specified	3	4	5	8	12	16	24
Sieve size (mm)	9.5	13.2	16	19	26.5	31.5	
Grade 0	59.54	72.22	80.66	86.83	99.15	100	
Grade 1	53.78	64.53	72.11	79.62	98.68	100	
Upper limit specified	65	76	83	90	100	100	
Lower limit specified	45	57	65	75	90	100	

2.2. Sampling test of mixture

The production of the mixture in the center of the paving machine (position 1), spiral distributor 1/4 (position2) and the edge (position 3) were sampled, and the degree of variation of the gradation was analyzed by the extraction test. Table 2 is the position of grade 0 gradation analysis table.

Table 2 The positions of grade 0 gradation analysis table

	Sieve size (mm)	31.5	26.5	19	16	13.2	9.5	4.75
	Grade 0	100	99.15	86.83	80.66	72.22	59.54	34.49
1	percent passing （%）	100	100	87.9	80.5	71.7	60.1	33.9
	variation	0	-0.8	-1.1	0.1	0.5	-0.6	0.6
2	percent passing （%）	100	100	86	78.9	71.6	58	30.8
	variation	0	-0.8	0.9	1.7	0.6	1.5	3.7
3	percent passing （%）	100	95.2	79.6	69.5	59.4	45.9	25.4
	variation	0	3.9	7.3	11.1	12.8	13.7	9
	Sieve size (mm)	2.36	1.18	0.6	0.3	0.15	0.075	
	Grade 0	22.42	17.73	11.5	7.33	4.71	3.76	
1	percent passing （%）	21	16.5	10.8	7.6	4.5	3	
	variation	1.5	1.2	0.7	- 0.2	0.2	0.7	
2	percent passing （%）	19.6	15.5	10.2	7.2	4.6	3.3	
	variation	2.8	2.2	1.3	0.1	0.1	0.5	
3	percent passing （%）	17.3	14	9.3	6.5	3.9	2.7	
	variation	5.1	3.7	2.2	0.9	0.8	1.1	

From the analysis of different sampling position grading, in the construction of production, the fluctuation of 0 graded mixtures is small on the whole. To the position 1, for example, the thickness of the aggregate quality passing changes in the percentages except 2.36mm, 1.18mm mesh changes is 1.5%, 1.2%, the rest of the sieve size quality passing changes in the percentages were less than 1%, fine and coarse aggregates maintained a good stability.

For position 3 on the edge of the spreading machine, the gradation change is larger, from 26.5mm sieve, percent passing are lower than the design value, which 4.75mm, 9.5mm, 13.2 mm, 16 mm continuous five mesh quality passing by rate in about 10%. Compared with the design grade, the whole gradation composition is obviously thicker. Table 3 is the position of grade 1 gradation analysis table.

Gradation 1 curve has a maximum density curve characteristics. Through sampling of different position of gradation analysis, the different positions of mixture gradation changed, but the change is mainly concentrated in 2.36mm particle diameter or more of the coarse aggregate part, for 2.36mm sieve particle size and each of the following file mesh gradation change smaller, variation of average is less than 0.5%.

Compared to gradation 0, it is not too big in the position 3 and position 1 in the process of paving mixture, which shows that the change of the mixture gradation is mainly the performance of the mixture itself and has no obvious relation with the interference of the external factors.

Table 3 The positions of grade 1 gradation analysis table

	Sieve size (mm)	31.5	26.5	19	16	13.2	9.5	4.75
	Grade 1	100	98.68	79.62	72.11	64.53	53.78	36.08
1	percent passing（%）	100	100	89.6	80.9	71.2	55.9	31.9
	variation	0	-1.3	-10	-8.8	-6.6	-2.1	4.2
2	percent passing（%）	100	100	86.8	72.5	65.6	53.4	31.2
	variation	0	-1.3	-7.2	-0.4	-1.1	0.4	4.9
3	percent passing（%）	100	92.9	79.9	74.2	69.1	53.4	31.8
	variation	0	5.8	-0.3	-2.1	-4.6	0.4	4.3
	Sieve size (mm)	2.36	1.18	0.6	0.3	0.15	0.075	
	Grade 1	20.74	16.42	10.75	7	4.63	3.73	
1	percent passing（%）	21.5	16.7	10.6	7.3	4.7	3.3	
	variation	-0.7	-0.3	0.2	-0.3	-0.1	0.4	
2	percent passing（%）	20.9	16.6	11	7.6	5.3	3.9	
	variation	-0.2	-0.2	-0.2	-0.6	-0.7	-0.1	
3	percent passing（%）	20.8	16.1	10.4	7.3	4.9	3.5	
	variation	-0.1	0.3	0.3	-0.3	-0.2	0.2	

2.3. Field test

After the completion of roller compaction, the structure depth and density test of three positions were carried out at the point of each test section. Table 4 is the field test results.

Table 4 Field test results

Test item	Average depth of construction (cm)			Average density(g/cm^3)		
Gradation / Position	1	2	3	1	2	3
Grade 0	20.9	20.48	17.25	2.345	2.31	2.182
Grade 1	19.75	20.49	17.41	2.456	2.457	2.422

For the gradation 0, each test section, position 3 sand diameter are less than other points, and the average diameter is 17.25 cm. Compared with position 1, it changes 3.65cm, the edge location of the structure is significantly greater than the middle position. Along with the increase of the depth of structure, the density of the mixture has a decreasing tendency. The density of position 3 decreased by 0.163 g/cm^3 compared with position 1.

After mixture segregation occurred, the gradation and pavement characteristics occur great changes. The field test structure depth and density gradation are consistent with numerical analysis results, position 1, 2 sampling location gradation analysis results change small compared with design gradation, gradation of position 3is obvious partial coarse compared with design gradation, and it consistent with the actual structure depth and density tests. This shows that mixture by spiral distributor after delivery occurred obvious segregation and actual segregation position mainly in the stalls shop machine spiral distributor edge.

For the gradation 1, edge position of the depth of the surface structure is larger, the density is small. Compared position 3 with 1, average value of shop sand diameter reduced 2.34cm and the average density reduced 0.034 g/cm^3. The difference decreased compared with gradation 0.

From the point of view of a single section, in the investigation of section, shop sand diameter of position 3 has no significant reduction in contrast to position 1. Diameters of some sections are even more than position 1. According to the gradation analysis, gradation change in the spread of the process is not too large. The segregation degree is decreased compared to gradation 0. The occurrence of segregation mainly affected by own combination of the particle gradation in the mixture design.

3. Conclusion

For the main conclusion of gradation 0 evaluation: overall stability is better, and segregation occurs mainly at the edge of paver paving and the degree of segregation is more serious. The main features of gradation 1: the innate stability of gradation is poor, but in construction the impact is less.

On the same condition, compared with S type features 0 gradation, near the maximum density curve of 1 graded aggregate segregation degree is smaller.

References

1. LI Yan-chun, MENG Yan, ZHOU Li-wei, et al. Grey relation degree analysis of influence factors on asphalt mixtures voids [J]. China Journal of Highway and Transport, 2007, 20 (1): 30-34.
2. ZHAO Zhan-li, ZHANG Zheng-qi, HU Chang-shun. Influence of gradation on ant-i skidding performance of asphalt pavement [J]. Journal of Chang'an University: Natural Science Edition, 2005, 25(1): 6-9.
3. Transportation Research Board National Research Council. Segregation in Hot-Mix Asphalt Pavements [M]. Washington DC: National Academy Press, 2000.
4. Transportation Research Board National Research Coucil. Guide for mechanistic-empirical design of new and rehabilitated pavement structures[R]. Washington DC: Transportation Research Board National Research Coucil, 2004.
5. PENG Yong, SUN Lijun. Relation of homogeneity and performance variation of asphalt mixture [J]. China Journal of Highway and Transport, 2006, 19(6): 30-34.
6. FENG Zhong-xu, YAO Yun-shi, FENG Jian-sheng. Rolling segregation of hot asphalt mixture [J]. Journal of Chang'an University: Natural Science Edition, 2006, 26(3): 96-99.

Microstructure and Mechanical Properties of SiCp/Cu Composite Prepared by Powder Injection Molding

Yi-Qiang He [1, 2,*], Jun-Jie Li[2], Hai-Sheng Zhou[2], Li-Chao Feng [1, 2]

[1]Jiangsu Marine Resources Develepment Research Insititue, Lianyungang, Jiangsu 222005, China
[2]College of Mechanical Engineering, Huaihai Institute of Technology, Lianyungang, Jiangsu 222005, China
*Email: ant210@126.com

SiCp/Cu composite reinforced with 5vol%, 10vol% and 15vol% SiC particles were prepared by powder injection molding (PIM) technology respectively. Surface microstructure, microhardness and tensile strength of the composite were investigated. Scanning electron microscope (SEM) and Energy Dispersive Spectrometer (EDS) were used to studied tensile fracture surface. The results show that the composite prepared on the condition of hydrogen atmosphere sintering does not contain oxygen element. SiC particles distribute uniformly in Cu matrix. With the increase of SiC content, the microhardness of the composite increases while its tensile strength increases first and then decreases. Crack sources of the composite during tension process mainly includes: cracking of the Cu matrix in vicinity of the SiC particles and debonding of an interface of the SiC particles and the Cu matrix. For 15vol% SiC copper-based composite, it is suitable for wear areas for low stress because of wear resistance and some toughness.

Keywords: Cu matrix; composite; particle reinforcement; powder injection molding.

1. Introduction

Powder injection molding, as an advanced powder metallurgy technology, is mainly used for the production of components of small-scale and complex-shaped, alloyed and relatively high requirements of material performance, large demand and difficultly process using conventional methods at a low manufacturing costs [1]. It applies to all of the material can be made powder, including steel, stainless steel, carbide alloys and high-density tungsten alloys and other traditional materials used in powder metallurgy process, intermetallic compounds, cobalt-based alloys, titanium alloys, metal matrix composites, and magnetic materials can also be used [2, 3, 4, 5]. In recent years, there are a growing number of powder injection molded copper-based materials [6, 7, 8, 9], including the powder injection molding of Cu, W-Cu, Mo-Cu, SiCp/Cu, etc. The focus of this study is to obtain optimization of process parameters by researching the influences of process parameters for the various stages to the quality of SiCp/Cu composite prepared by powder injection molding, while the study of hardness and wear resistance of components prepared by powder injection molding.

In this paper, microstructure, microhardness, tensile strength and wear behaviors of SiCp/Cu composites prepared by powder injection molding were investigated.

2. Materials and Methods

Spherical copper of the average particle size of 20μm and ordinary α green SiC powder of the average particle size of 10μm were used. The SiCp/Cu composites were prepared by powder injection molding includes powder preparations, feedstock preparation, injection molding, degreasing and sintering, as shown.

SiC powder was first pretreated and then premixed Cu powder and SiCp powder by ball mill. 5vol%, 10vol% and 15vol% SiCp in mixed powder was studied. Binder (55%PW+40%LDPE+5%SA) and premixed powder were kneaded in kneading machine (NH-5 type kneader). Powder loading was 50%. Feeds were heated and then injected in forming mold cavity by injection

---

*Corresponding author

molding machine (SA600 type) after sufficient kneading and granulation. Process parameters of injection molding are given in Table 1. Preform injection was designed in tensile specimen shape in order to facilitate performance testing of injection molding.

Table 1 Process parameters of injection molding

SiC$_P$ content (vol %)	Injection temperature/(°C)	Injection pressure/(M Pa)	Injection rate/(g·s^{-1})
5	180	120	70
10	180	140	75
15	180	160	80

A two-step method for degreasing, solvent degreasing before hot degreasing was used for preform injection degreasing. The solvent was $C_2HCl_3(65\%)+C_2H_6O(20\%)+C_3H_6O(15\%)$. Degreasing temperature was 35°C. Time was 15h. Hot degreasing was in single-chamber vacuum sintering furnace (WZDS-20 type) with maximum temperature of 490°C. Preform injection was sintered in sintering furnace (WZDS-20 type) after degreasing with the protection of hydrogen. The heating rate of sintering was 5°C/min. The maximum temperature of sintering was 1050°C, 2h insulation.MH-5 type micro hardness tester was used to test the micro hardness of the sintered parts surface. CMT5205 type computer controlled electronic universal (tensile) testing machine was used to test the tensile strength of the sintered parts. JSM-6480 type electron microscope was used to observe the microstructure and tensile fracture morphology of the sintered parts. EDS was used to determinate the elements of fracture and wear surface.

3. Results and Analysis

3.1. *Surface morphology of sintered parts*

The microstructure of sintered SiC$_P$/Cu composite (SiC$_P$ volume content of 15%) parts surface is shown in Fig. 1.

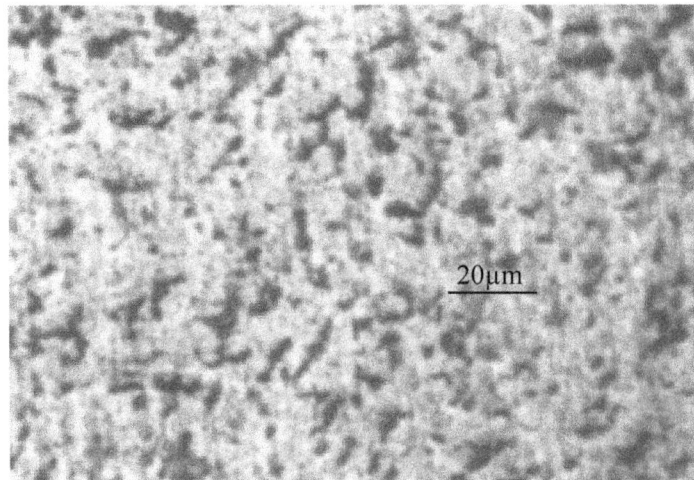

Fig. 1 Microstructure of PIM SiC$_P$/Cu composite

This shows SiC particles are more evenly distributed in the Cu matrix without significant agglomeration. It indicates that the effect of milling premix and kneading powder with binder of Cu powder and SiC powder in the process of powder injection molding is good.

3.2. *Microhardness and tensile strength of sintered parts*

The microhardness and tensile strength of PIM SiC$_P$/Cu composite with different content of SiC$_p$ are given in Table 2. It's clear to see that the microhardness of sintered parts increased with the increase of the content of SiC$_p$. Specifically, SiC$_p$ volume fraction increases from 5% to 15%, its hardness is increased from 150HV to 214HV. When SiC$_p$ volume content is from 5% to 15%, its tensile strength is between 254MPa and 291MPa. And with the increase of the content of SiC$_p$, the tensile strength of sintered parts presented trends of decreases after increases. The tensile strength of 10vol% SiC$_p$ is the maximum.

Table 2 Variation of microhardness and tensile strength with volume fraction of SiC$_p$

SiC$_p$ content (vol %)	5	10	15
Microhardness (HV)	150	195	214
Tensile strength/(M Pa)	254	291	278

For SiC$_p$/Cu composite, the strength of the composite material is improved due to the second phase enhanced dislocation, the hinder of dislocation motion, grain refinement, internal stress etc. interface caused by the addition of SiC$_p$ particles.10% SiC$_p$ content of composites compared with 5% SiC$_p$ content of the composite material, the strengthening effect of the above mentioned causes is more obvious because more SiC$_p$ particles are distributed in the copper matrix, thus improving the tensile strength. However, SiC$_p$ and copper matrix are weak interfacial bonding in SiC$_p$/Cu composite, thus cracks preferentially form at the interface, expansion, eventually lead to fracture when the material is subjected to tensile deformation. This is a cause that the tensile strength of 15% SiC$_p$ composite is lower than that of 10%SiC$_p$ composite. Because during the stretching process, copper matrix produces stress concentration near SiC$_p$ particles, easily forms crack source. When the content of SiC$_p$ is higher, relatively small distance between particles makes crack propagation path shorter. Crack is easily to further expand by SiC particles, so that close pores connect to macroscopic crack and then lead to brittle fracture. Discontinuous SiC particles produced restrictive to plastic flow of matrix, so it formed torn ribs around SiC particles.

Interfacial bonding status and strength of matrix and reinforcement is one of the main factors affecting the composite reinforcement effect. This is why many researchers preliminarily coated Cu or Ni on SiC particles when preparing SiC$_p$/Cu composites. It improves the interfacial bonding of the composite material and the material properties. It can also take this way, coated SiC powder, when in the further research about PIM prepared SiC$_p$/Cu composites.

3.3. *Morphology of tensile fracture*

The morphology tensile fracture of PIM sintered SiC$_p$/Cu composite of 10vol% SiC$_p$ is shown in Fig. 2.

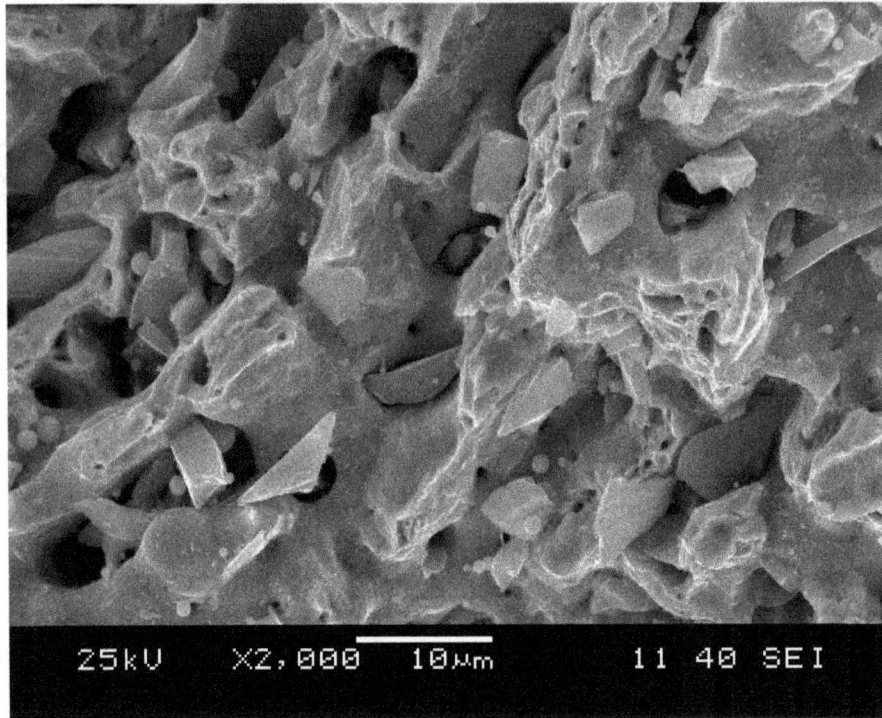

Fig. 2 Morphology of tensile fracture of PIM SiC$_p$/Cu composite

Obviously, Cu, the continuous phase distribution, covers SiC particles and there are SiC$_p$ in its dimple. There are three kinds of fracture mode in the stretching process of SiC$_p$/Cu composite. They are tough break of copper matrix, debonding of SiC particles with copper matrix and fracture reinforcement of SiC particles. Can also be seen from Fig. 2, SiC particles and Cu matrix have a good combination. Stress concentrates on SiC particles and nearby Cu matrix because of large modulus difference between SiC particles and Cu matrix in the stretching process. When the stress exceeds the ultimate strength of Cu matrix around SiC particles or the interfacial strength of SiC$_p$/Cu, it appears the tear strip because of crack of matrix near SiC particles and SiC particles are pulled out because of debonding of SiC particles with copper matrix. The crack source are mainly crack of matrix near SiC particles and debonding of SiC particles with copper matrix because the interfacial strength of the composite is not enough to make SiC particles to be pulled off.

3.4. *EDS of tensile fracture*

The fracture of tensile specimen was analyzed to determine the elements of the composite. The test results are shown in Fig. 3. It is found that the sample contains only Cu, Si and C elements and no O element. This explains that hydrogen fully restored the oxide of the copper surface in the sintered process. So it is feasible to select the sintering in a reducing atmosphere of hydrogen.

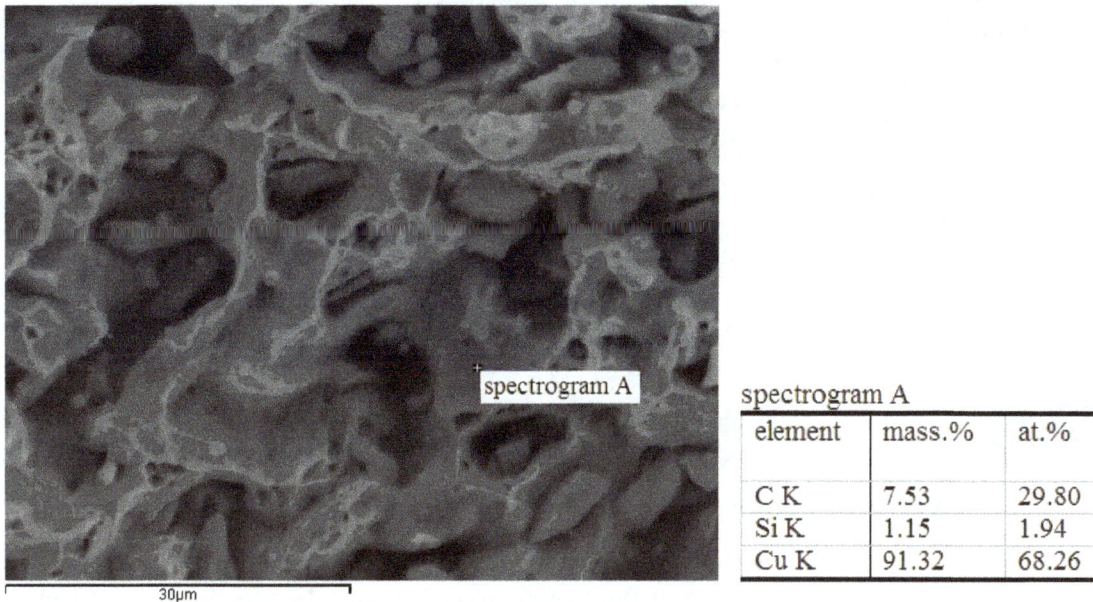

spectrogram A

element	mass.%	at.%
C K	7.53	29.80
Si K	1.15	1.94
Cu K	91.32	68.26

Fig. 3 EDS analysis of tensile fracture of samples

3.5. *Friction and wear properties*

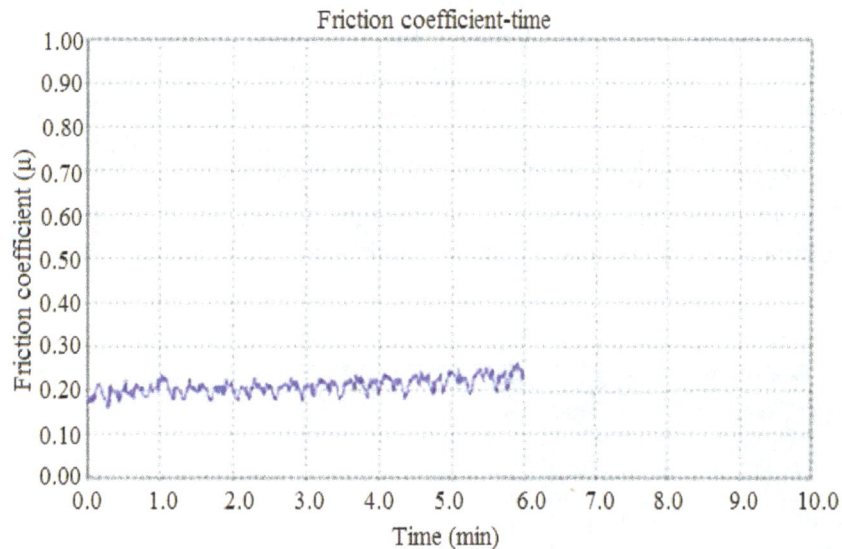

Fig. 4 The relationship between the friction coefficient and time

Couplings for wear in this experiment were 40Cr steel. Under conditions of normal load of 10N, a line speed of 13188mm/min and duration of 6min. The tested friction coefficient of 15vol% SiC sample is as shown in Fig. 4. Because of the same experimental conditions and the friction coefficient of 40Cr steel was 0.55, it indicates that the friction coefficient of 15vol% SiC composite of copper-based is lower than that of 40Cr steel and the copper-based composite have a relatively excellent friction properties.

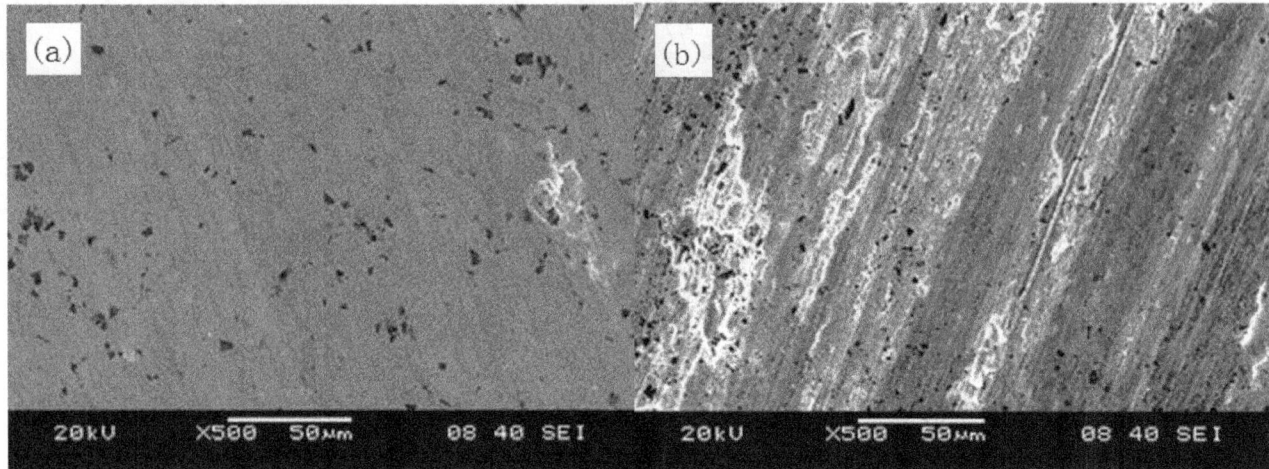

Fig. 5 Morphology of worn surface of (a) 10vol% and (b) 15vol% SiC$_p$/Cu composite

SiC particles are embedded in the copper matrix and have protective effect on the matrix because of the resistance of the plowing of dual surface was strong. In Fig. 5 (a), in the process of sliding, the adhesive and plastic deformation degree of worn surface of 10vol% SiC$_p$/Cu composite is light. In Fig. 5 (b), a lot of furrow-like grooves are formed on the worn surface of 15vol% SiC$_p$/Cu composite. There are abrasive wear characteristics. SiC particles are embedded in the copper matrix with the plowing effect of the dual surfaces. It generates plowing off and forms debris. Debris will adhere to the surface of the steel pipe. The friction between composite and steel will change into the worn between composites. The resistance of the plowing of dual surface and wear of composite is better with the increase of SiC particles. The protective effect of the matrix is reinforced. A 20μm particle size SiC particles which have better wear resistance was used in the testing process. When the SiC particles are too large, it may cause the three-body wear after falling from the matrix into contact surface of friction. Thus, the larger SiC particles reinforce composites but reduce wear. For 15vol% SiC composite, its strength decrease, but its wear resistance is better. For 15vol% SiC copper-based composite, it is suitable for wear areas for low stress because of wear resistance.

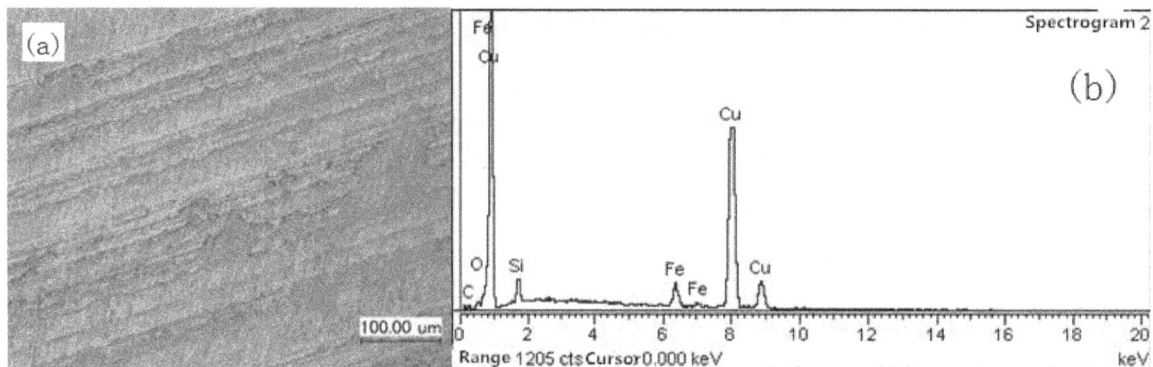

Fig. 6 (a) Morphology and (b) EDS of worn surface of SiC$_p$/Cu composite

The surface of dual steel loop is plowed by SiC particles on the surface of composite in the normal force. As shown in Fig. 6 (a), the elements of the wear surface of the composite in addition itself contains Cu, Si, C, but also contains elements such as Fe. As shown in Fig. 6 (b), it produces iron-rich particles. These particles mix with composite debris on the contact surfaces of friction. Pressed under repeated external force, it produces a dense layer of mechanically mixing on the surface of composite. It improves the wear resistance of the surface of composite.

4. Conclusion

1. SiC_p volume content of 5% to 15%, the microhardness of PIM SiC_p/Cu composite increases with increasing SiC_p content, and the tensile strength of sintered parts present trends of decreases after increases.
2. The crack source of tensile fracture of PIM SiC_p/Cu composite is mainly crack of matrix near SiC particles and debonding of SiC particles with copper matrix.
3. Sintering in atmosphere of hydrogen can eliminate the problems caused by the oxidation of copper. For 15vol% SiC copper-based composite, it is suitable for wear areas for low stress because of wear resistance and some toughness.

Acknowledgments

The research described in this publication was funded by the Natural Science Foundation of China (No.51004050), the Natural Science Foundation of Jiangsu Province (No.BK20141250), Natural Science Foundation of Jiangsu Colleges and Universities: (No. 14KJB430005), Natural Science Foundation of Lianyungang: (No. CG1418 and No. CXY1404), Postgraduate research and Innovation Project of Jiangsu Province (KYLX15_1485) and The Major R&D Project of Jiangsu Province (BE2015100).

References

1. Fayyaz A, Muhamad N, Sulong AB, Rajabi J, Wong YN, Fabrication of cemented tungsten carbide components by micro-powder injection moulding, J. Master Process Technol. 214 (2014) 1436-1444.
2. Javad Rajabi, Hafizawati Zakaria, Norhamidi Muhamad, Abu Bakar Sulong, Abdolali Fayyaz, Farication of miniature parts using nano-sized powders and an environmentally friendly binder through micro powder injection molding, J. Microsystem Technologies. 21 (2015) 1131-1136.
3. Chang Kyu Kim, Chang-Yong Son, Dae Jin Ha, Tae Sik Yoon, Sunghak Lee, Nack J. Kim, Microstructure and mechanical properties of powder-injection-molded products of Cu-based amorphous powders and Fe-based metamorphic powders, J. Materials Science and Engineering A. 476 (2008) 69-77.
4. Shu, G.J, Hwang, K.S, Pan, Y.T, Improvements in sintered density and dimensional stability of powder injection-molded 316L compacts by adjusting the alloying compositions, J. Acta Materialia. 54 (2006) 1335-1342.
5. Abdullahi, Choudhury, Azuddin, Process Development and Product Quality of Micro-Metal Powder Injection Molding, J. Materials & Manufacturing Processes, 30 (2015) 1377-1390.
6. Qu Xuanhui, Advance in Research of Power Injection Molding, J. MATERIALS CHINA. 29 (2010) 42-47.
7. Liu Manmen, Liu Jie, Xie Ming, Research Progress of Metal Injection Molding in Copper-based Materials, J. Materials Review. 25 (2011) 85-88.

8. Moballegh L, Morshedian J, Esfandeh M, Copper injection molding using a thermoplastic binder based on paraffin wax, J. Materials Letters. 59 (2005) 2832-2837.
9. Nan Hai, Qu Xuanhui, Fang Yucheng, POWER INJECTION MOLDING OF Mo/Cu ALLOY ELECTRONIC PACKAGING, J. Powder Metallurgy Industry. 14 (2004) 1-5.

The Design and Experimental Study of Surface Wetting System Based on Metal-Plastic Molding

Lu Zhang, Ning-Ning Gong*, Lin-Chao Dong, Xi-Ping Li

College of Engineering, Zhejiang Normal University, Jinhua, 321004, China

*Email: gnn@zjnu.cn

Featuring advantages of high strength, small quality and easy molding of complex construction, etc, high strength metal-polymer composites are increasingly popular in fields including aerospace, automobile manufacturing, communication and so on. This paper firstly discusses the direct injection molding principle of metal-polymer composites, and then analyzes the effect of metal surface micromorphology and the surface tension on the interface bonding strength of metal-polymer composites. In line with the injection molding principle of metal-polymer composites and on the basis of the resting drop method, the composition of system used to measure the contact angle of the polymer melt on different metal surfaces is studied and produced. The aluminum alloy samples are fabricated by using the physical sandblasting. The affecting law of the melting temperature, the surface temperature of aluminum alloy and the surface micromorphology on the contact angle variation of the PBT polymer melt on the metal surface are studied. The result, obtained from the experiment, has definite guiding significance in improving the mechanical property of interface bonding of metal-polymer composites and optimizing the injection molding technology of metal-polymer composites.

Keywords: Metal-polymer composites; wetting behaviors; contact angles; surface tension.

1. Introduction

The direct injection integrated molding products of metal-polymer organically blend thin-wall features of metal forming and complex structural of plastic parts molding. The light metal thin-wall parts are firstly pressed and surface treated. Making use of the special pattern, the molten polymer, under high pressure, is injected into the mold cavity to combine with the metal. Finally, after cooling stage, the product is obtained in the shape consistent with pattern cavity and the direct injection molding of metal-polymer is completed [1]. Compared with traditional methods of bonding metal plastic products, it features short production process and strong bonding property, and so on.

The metal surface micromorphology and the surface tension are particularly important to the direct injection molding of metal-polymer composites [2]. They can greatly affect the bonding strength of the direct injection molding of metal-polymer composites. By using the sandblasting treatment, various surface unprocessed are formed on the metal, leading to different contact angle of the polymer melt on the metal surface. If the contact angle is greater than 90°, the melt will be blocked when it flows on metal surface. Otherwise, if the contact angle is less than 90°, it is favorable for the melt flowing on metal surface. Therefore, it is very necessary to measure the contact angle of the melt on different roughness of the metal.

2. Design of Surface Wetting System

The main structure of the device (as the Figure 1) includes micro injection molten cavity, microcontroller, high-speed camera and working platform, and so on.

The polymer melt, under the high temperature molten state is extremely unstable and likely to be oxidized by air affecting the stability of the experiment, so, nitrogen is injected into the transparent glass shield. For the plastic molten dripping fluently in the experiment process, the pressure control valve is mounted in the micro injection molten cavity to control the pressure within

---

*Corresponding author

it. Also, a built-in propeller is installed in the micro injection molten cavity for driving the dropping of plastic molten drops. Circulating devices are equipped in the equipment so as to realize the ventilation in the inner cavity of the transparent glass shield and keep the uniformity of temperature in the inner cavity.

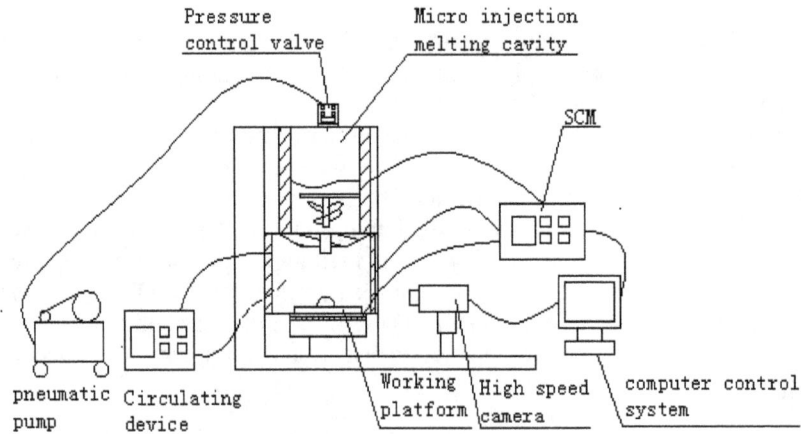

Fig. 1 Schematic diagram of device structure

The system (as the Figure 2) is composed by the main engine, thermocouple, cooling pipe, temperature sensor and A/D converter. The SCM can detect and control the working platform, the inner part of micro injection molten cavity and inner air temperature spots within the transparent glass shield through numerous temperature sensors in real time. The number of temperature sensors is determined by the accuracy of the system and the size of the device.

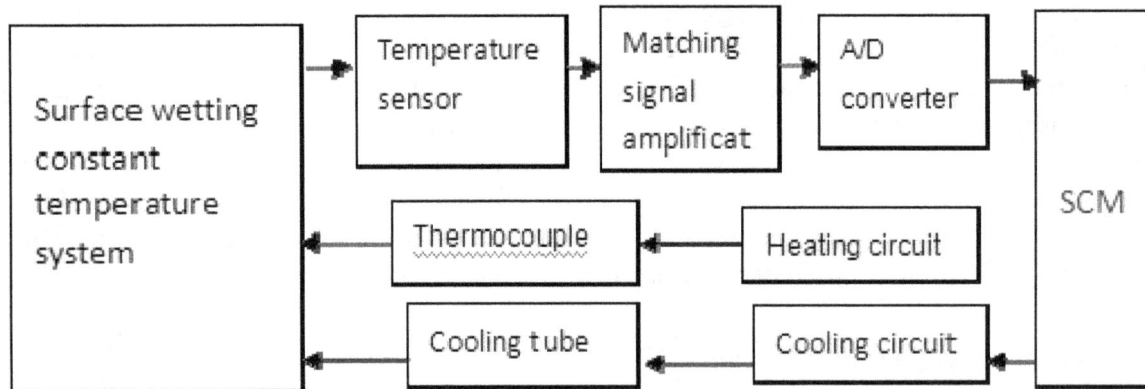

Fig. 2 Design diagram of surface wetting constant temperature system

The focus of design is the comparing control program inside the SCM [3], and the flow chart is as the Figure 3.

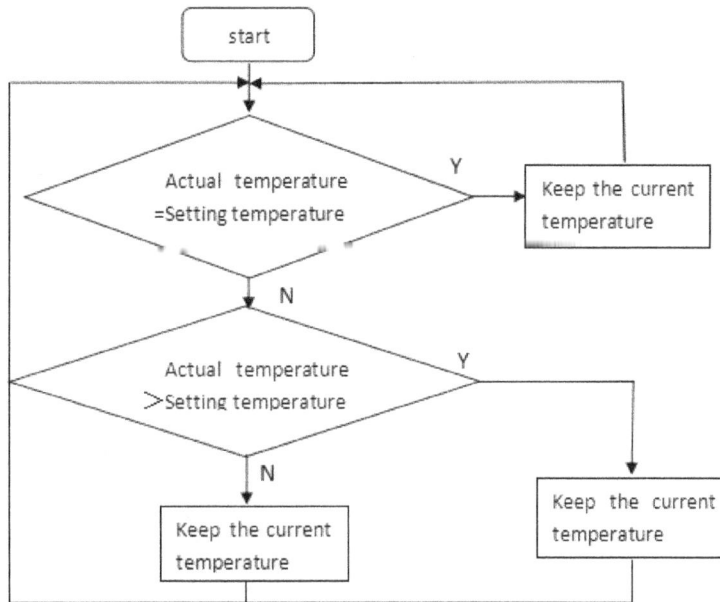

Fig. 3 Flow chart of the program

The emphasis of the system is to monitor and control to the temperature inside the micro injection molten cavity and on the working platform. When the actual inner temperature is lower than the set temperature of the system, the heating circuit is started and the thermocouple begins to work. When the temperature reaches the set temperature range, heating is stopped; when the actual inner temperature is higher than the set temperature of the system, the cooling circuit is started and the cooing pipe begins to work, and when the temperature reaches the set temperature range, heating is stopped. The temperature control system based on the SCM reaches higher control accuracy to the variation of temperature.

3. The Experimental Study Based on Metal-Plastic Molding

The resting drop method is adopted this time to calculate the contact angle.Fabricates aluminum alloy samples making use of physical sandblasting , and studies the affecting law of the melt temperature, the surface temperature of aluminum alloy and the surface micromorphology on the contact angle variation of the PBT polymer melt on the metal surface.

3.1. *Experimental materials*

The engineering plastic, polybutylene terephthalate (PBT), is used in the experiment, featuring high heat resistance, toughness and low cost, etc. [4]. Due to the small shrinkage of PBT, and the high accuracy and less distortion of the molded product, the molded structure of the interface between metal and PBT shall not be damaged by the shrinkage of cooling. As the great flowing property and easy filling of small holes, PBT is easy to mold in industrial application. PBT is a kind of semicrystalline enginnering plastic featuring more obvious melting point with the temperature range of $225 \sim 235°C$, and the decomposition temperature of PBT is around $280°C$. So, the maximum temperature for molding shall be lower than $270°C$. The actual molding temperature of PBT commonly is between $240 \sim 260°C$ [5]. Therefore, four temperatures, 230, 240, 250, 260, are selected for the experiment by PBT.

The substrate material of the experiment adopts Al6061 with the thickness of 2 mm. 6061 aluminum alloy is the aluminum alloy product with high quality featuring great corrosion resistance and excellent oxidation effect. Upon the requirement of the experiment, the strength requirement for needed aluminum sheet samples is not high, but the requirement for keeping of geometrical shape after machining is higher and the requirement for the flawless of the contact surface with plastic is increased, so Al6061 is a relatively suitable sample material [9, 11].

Physical sandblasting method is adopted in the experiment to produce aluminum alloy sample, as the Figure 4, and 4 kinds of brown fused alumina, respectively 45#, 80#, 120# and 180#, are taken as the sandblasting material. The particle size is respectively about 2mm, 1mm, 0.5mm, and 0.3mm, and the pressure is 0.1MP.

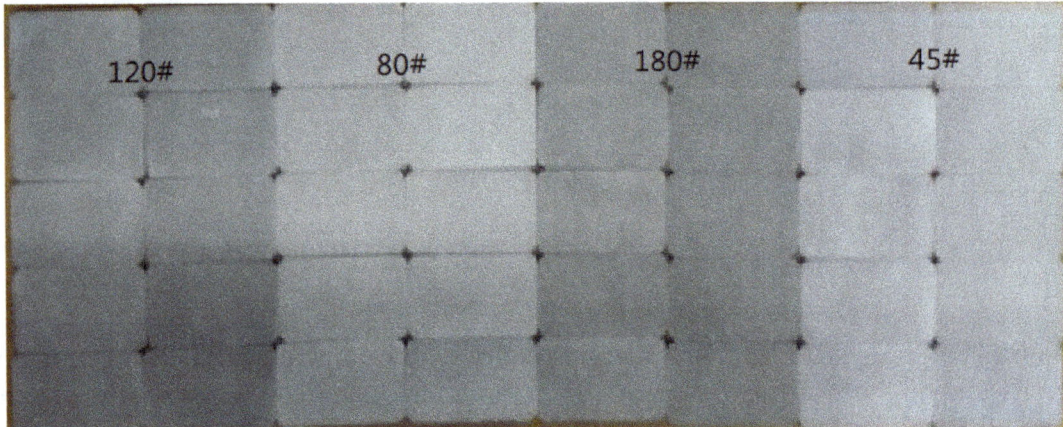

Fig. 4 The sample after sandblasting

3.2. *The process and data of measuring contact angle experiment*

This experimental study falls into three trials. The first trial is the aluminum sheet with different melt temperatures and same surface micromorphology and the variation of contact angles under the temperature f the aluminum sheet. Figure 5 shows the aluminum sheet adopting 120#0.1 and the contact angle presents under different melt temperatures when the temperature of aluminum sheet is 60°.

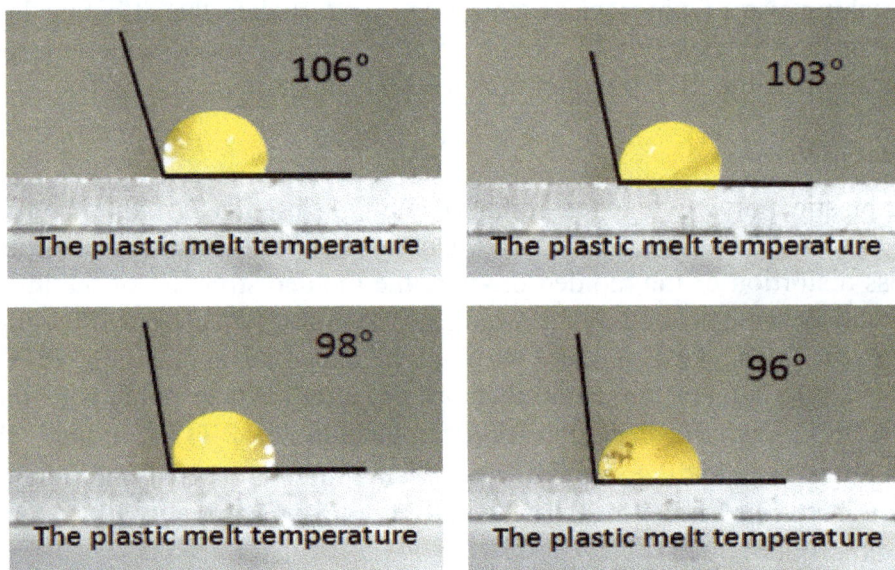

Fig. 5 Contact angles presented at different melt temperatures

Figure 6 displays the relation between the static contact angle of the injection material PBT melt and the met temperature in nitrogen atmosphere [6-8].

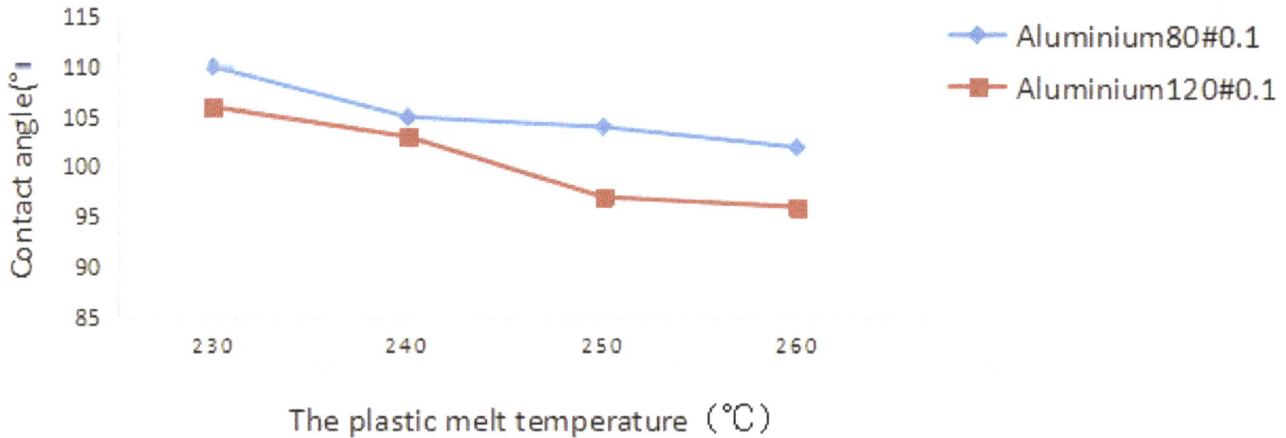

Fig. 6 The variation diagram of contact angles presented at different melt temperatures

The second trial is the aluminum sheet with different temperature of aluminum sheets and same surface micromotphology and the variation of contact angles at same melt temperature. Four temperatures of the aluminum sheet, 25°C, 60°C, 80°C, 100°C and 120°C are selected in the experiment. Figure 7 shows the relation between the static contact angle of injection material PBT melt and the substrate temperature in nitrogen atmosphere.

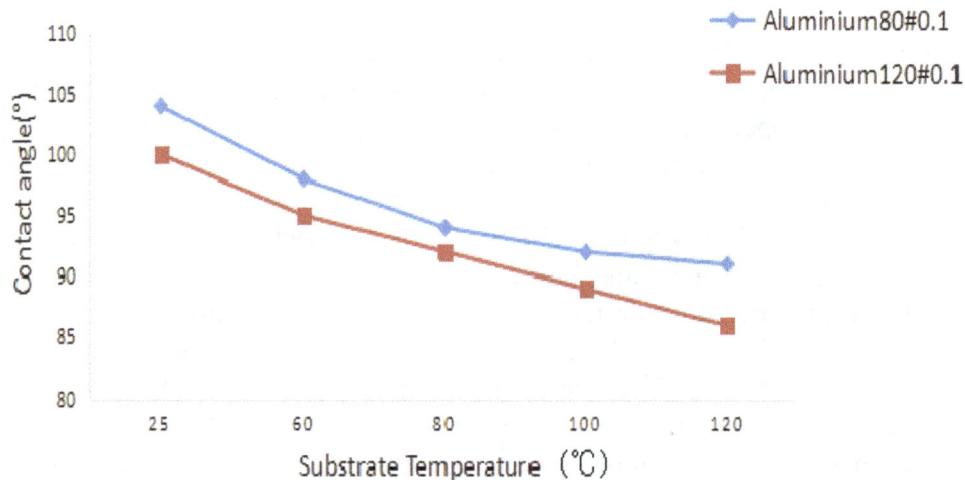

Fig. 7 The variation diagram of the contact angle at different substrate temperature

The third trial is the aluminum sheet with different surface micromorphology and the variation of the contact angle at the same temperature of aluminum sheet and same melt temperature. Four kinds of aluminum sheets with different surface micromorphology are 45#0.1, 80#0.1, 120#0.1, and 180#0.1. Figure 8 shows the relation between the static contact angle of injection material PBT melt and the aluminum sheet with different surface micro -structure in nitrogen atmosphere.

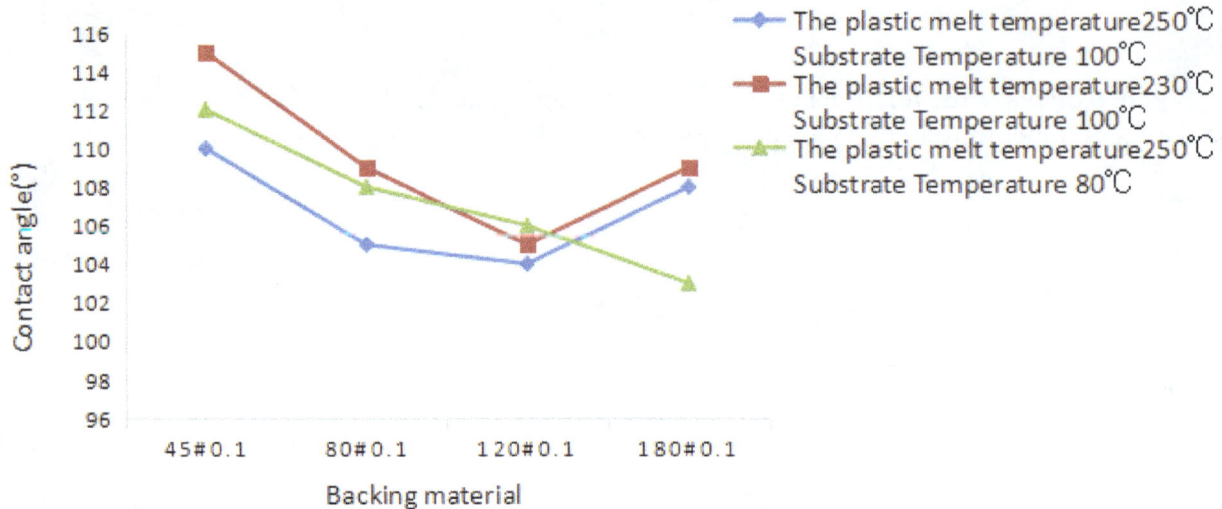

Fig. 8 The variation diagram of the contact angle presented by different substrate materials

3.3. *The analysis of experimental results*

The experiment studies the interfacial wetting property of PBT plastic. The contact angle of PBT plastic is tested in four different aluminum alloy samples and at different substrate temperature by the resting drop method. It researches the effect law of melt temperature, the surface temperature of aluminum alloy substrate and the surface micromorphology on the variation of the contact angle of PBT polymer melt on metal surfaces. The following conclusions are obtained through the calculation and analysis of the experiment and the figure intuitively:

1) For the same polymer PBT plastic, under the situation of same aluminum alloy substrates and same substrate surface temperature, the three-phase static contact angle is decreased gradually and the wetting property gets better as the increasing of melt temperature. It indicates that in a certain degree, the temperature rising of the melt can reduce the resistance of melt flow from the surface tension [10].

2) When the same melting polymer PBT plastic is injected in the aluminum sheet with same surface treatment, the substrate temperature is different, which leads to different contact angles. Therefore, the substrate temperature is one of factors affecting the surface tension of mold injection molding. From the above analysis, it can be seen that the substrate temperature is increased in a certain degree, the contact angle of melt is smaller, and the fluid motion is drove by the surface effect.

3) When the same PBT plastic, at the same temperature, is injected in four aluminum alloy sheets with different surface treatment, 45#0.1, 80#0.1, 120#0.1, and 180#0.1, the size of the contact angle of same polymers are different. From the variation chart of the contact angle, it can be seen that with the aluminum sheet of 120#0.1, the contact angle is generally smaller than with other aluminum sheets. The fluid flow resistance is smaller, the fluidity is better correspondingly, and it is more favorable to the injection molding processing of metal-polymer. Therefore, the surface micromorphology of substrate materials is one of factors affecting the surface tension of mold injection molding. The surface tension is different when the plastic injection molding is done on the surface with different treatments. For the injection molding of metal-polymer, the metal with appropriate surface

micromorphology shall be selected, and that is to say the metal surface shall be treated appropriately [12-15].

4. Conclusion

Featuring advantages of high strength, small quality and easy molding of complex construction, etc, high strength metal-polymer composites is increasingly active in fields including aerospace, automobile manufacturing, and communication, and so on. The wetting between polymer melts and metal surfaces is the common phenomenon in the injection processing, determining the bonding strength of metal-polymer composites to a great extent. The experiment studies the interface wetting property of PBT plastic. Through measuring and analyzing of the surface contact of metal-polymer composites, it studies the wetting property and various experimental parameters of metal-plastic. The interface contact angle of metal-polymer and the fluid flow resistance are smaller, the fluidity is better correspondingly, and it is more favorable to the injection molding process of metal-polymer. The result of the experiment has a certain guiding significance to improve the mechanical property of interfacial bonding of metal-polymer, such as the bonding property of plastic-aluminum injection plastic on the metal surface, and optimize the injection molding technology of metal-polymer composites. It also provides a reference for the follow-up optimization of the direct injection molding process of metal-polymer, and has practical value in the engineering field.

Acknowledgments

The research work was funded by Natural Science Foundation of Zhejiang Province (Grant No. LQ14E050003), China Postdoctoral Science Foundation (Grant No. 2014T70579), National Natural Science Foundation of China (Grant No. 51305405), and National Training Programs of Innovation and Entrepreneurship for Undergraduates (Grant No. 2015R404033).

References

1. J. Giboz, T. Copponex and P. Mele: J. Micromech. Microeng. Vol. 17 (2007), PP. 96-109.
2. D. Yang: *Study on Injection Molding of Polymer Melt Surface Effect and Panel Micro Devices* (Ph.D., Dalian University of Technology, China 2011), pp. 35-37. (In Chinese)
3. W.Q. Li, L.Z. Yu and L.X. Qiang : J. Chin J Electron Vol. 35 (2013), PP. 45-48.
4. Y.B. Bei, Y. Jun, L.Q. Ping, et al.: J. Plastic Industry Vol. 38 (2010), PP. 14-16.
5. Z. Zhao, S.Y. Hui and D. Gance: J. J East China Univ Techno Vol. 34 (2008), PP. 360-363.
6. J.Z. Gao: *Plastic Molding Process and Mold Design* (China Machine Press, China 2013), pp. 9 8-107. (In Chinese)
7. Z. Yuan, W. Huang: J. Colloid&Interf Sci Vol. 267 (2003), pp. 155-159.
8. K. Ahlh, T. Wadewitz and J. Winkelmann: J. J Chem Eng Data Vol. 48 (2003), pp. 1500-1507.
9. K.J. Kyo: J. Compos Sci Technol Vol. 25 (2000), pp. 22-24.
10. T. Nose: J. Macromolecules Vol. 28 (1995), pp. 3702-3706.
11. T. Zhang: J. Aeronaut Manuf Technol Vol. 15 (2013), pp. 32-35.
12. J.H. Phelps, A.I. Abdel-rahman and V. Kunc: J. Composites Part A Vol. 51 (2013), pp. 11-12.
13. C.Y. Xia, L. Hua: J. Guangdong Che Ind Vol. 41 (2014), pp. 53-54.
14. J. Chen, D. Zhao and X. Jin: J. Compos Sci Technol Vol. 97 (2014), pp. 41-45.
15. H.L. Luo, G.Y. Xiong and C.Y. Ma: J. Mater Design Vol. 64 (2014), pp. 294-300.

Effect of Sintering Temperature on Properties of Al$_2$O$_3$/Al Cermets

Rui-Hua Wang[1], Xiu-Qin Wang[2], Jie-Guang Song[1,*], Ming-Han Xu[1], Yao-Qi Li[1], Jian-Qing Liu[1], Ting-Ting Xia[1], Chong Wu[1], Han-Xing Yan[1]

[1]*Engineering & Technology Research Center for Materials Surface Remanufacturing of Jiangxi Province, School of Mechanical and Materials Engineering, Jiujiang University, Jiujiang 332005, China*
[2]*Library, Jiujiang University, Jiujiang 332005, China*
*Email: songjieguang@163.com

Alumina ceramics with good mechanical and corrosion resistance are the ones of the most widely used engineering ceramics. The aluminum has high strength, high conductivity and high plasticity etc. so that aluminum ceramics are used in more and more industries. In this paper, the mass fraction of 25% Al$_2$O$_3$ powder and the mass fraction of 75% Al powder were mixed in the blender. Mixer speed is100r/min with mixing time of 3.5 h. Forming, sintering and a series of processes for preparing the alumina/aluminum metallic ceramic materials, through performance testing and analysis, found that the density of the sample firstly increased and then decreased with the increase of sintering temperature. A melting point is close to the sintering temperature and the density of the cermet can be made relatively high. When the sintering temperature is about 600°C and 700°C, the macro performance of sample is better. The cermet is sintered at 700°C and its microstructure is relatively better.

Keywords: Cermets; alumina; sintering temperature; powder metallurgy method.

1. Introduction

Alumina is the main means of α- alumina ceramic, is currently the largest production volume, most widely used application surface ceramic materials, functional materials can be used as the substrate and the encapsulation of integrated circuits, as a structural material in ceramic cutting tools, dies, other high-temperature bearings used, wear-resistant, corrosion-resistant structural materials [1-2], but α- alumina ceramic brittle low flexural strength, toughness is poor, heavily influenced by the application of alumina moieties to improve the brittleness and increase strength, usually alumina composite ceramic support or a ceramic composite. One way is to add available titanium, iron, aluminum and other metallic toughness particles (metal) [3]. In order to meet the increasingly high performance materials, alumina particles, whiskers, fibers, and other just added to obtain a high specific strength, excellent heat resistance, an aluminum composite material of the conductive material [4].

Alumina-aluminum cermet usually powder metallurgy sintering, the sintering temperature is generally higher than the melting point of the metal but below the melting point of the ceramic [5-6]. Alumina ceramic powder can be joined together after the molten aluminum metal, so that both alumina ceramic and metal-ceramic aluminum advantages and features of both ways. In this article, the densification mechanism of Al$_2$O$_3$/Al metallic ceramics prepared via powder metallurgy method were investigated, which lay a base for preparing high-performance metallic ceramic products [7-8].

Cermet hard phase is composed of a ceramic or metal alloy binder phase consisting of structural materials. Cermet while maintaining the high-strength ceramics, high hardness, wear resistance, high temperature, oxidation and chemical stability characteristics, but also has good plasticity and toughness of the metal [9-10]. Cermet is a kind of very important tool and structural materials, which are extremely versatile, involving almost all areas of the various departments of the national economy and modern technology, the development of industry and improved productivity plays an important role. In promoting, for metal ceramic art materials has become a very important research branch [11-12].

2. Materials and Experimental

The raw materials used in the experiment were from Tianjin Kermel Chemical Reagent Co., Ltd. purity Al powder and Shanghai 54 Chemical Reagent Factory production analytically pure Al_2O_3 powder, alumina ceramic ball milling media. The mass fraction of 25% Al_2O_3 powder and the mass fraction of 75% Al powder were mixed in the blender, mixer speed 100r/min, mixing time of 3.5h, mixing with an agate ball mill. The mixed powder is loaded Φ30mm good mold, applying pressure to the hydraulic machine to obtain a sample. Pressure forming is 40MPa, dwell time of 5min. Placed in a vacuum sintering furnace sintering, the sintering temperature was 600 °C, 650 °C, 700 °C, 800 °C, 900 °C, holding time are 1h. After the sample after sintering, finished with analyzing balance measure the density, take a small piece, sonication and dried after treatment were sprayed gold, followed by SEM observation of the interface structure.

3. Results and Discussion

From Table 1, the density of the sample first increased and then decreased with the increase of sintering temperature. Since the melting point of aluminum is 660 °C, a melting point close to the sintering temperature and sintering, the density of the cermet can be made relatively high.

Table 1 Effect of sintering temperature on density of Al_2O_3/Al cermets

Sample number	Sintering temperature (°C)	Holding time (h)	Density (g/cm^3)
1#	600	1	1.665
2#	650	1	2.084
3#	700	1	1.935
4#	800	1	1.568
5#	900	1	Unable to test

Seen from Fig. 1, the surface of sample (a) does not overflow formed aluminum ball, the sample looks relatively the rules and no cracking. The surface of sample (b) formed a small amount of aluminum ball due to spilling aluminum. The sample looks relatively the rules and no cracking. The surface of sample (c) shows a slight aluminum overflow formed aluminum ball, it looks relatively the rules, and there is no cracking. The surface of sample (d) shows aluminum large group overflow to form the large aluminum ball, aluminum ball began to reunite the outer surface began to crack. The aluminum at the sample (e) reunites more seriously with serious overflow, which makes the body destruction. From the above analysis, the rule can be obtained the degree of spilled aluminum is increased with the increase of sintering temperature.

Because the sintering temperature is increased, the degree of melting of aluminum is increased, the surface tension is reduced, the mobility of molten aluminum is increased, and the green body is sintered in the vacuum environment, aluminum flowing out the green body through the pores between the Al_2O_3 particles to produce alumina balls on the body surface. From the above analysis, it shows that, when the sintering temperature of 600°C and 700°C, the macro performance of sample is better.

The scanning microscope can be seen from Fig. 2, the samples (c) is most dense tissue, the samples (b) is followed, the tissue density of sample (c) and (d) are minimum and most organizations loose, aluminum melted and serious spill in the sintered sample at 900°C,the sample cannot be detected.

Fig. 1 Effect of sintering temperature on the outside shape of Al2O3/Al cermets

The metal Al particles were larger and continuous distribution from SEM image, the Al_2O_3 particles are not contiguous and relatively small, which is shown in Fig. 3, Al is the matrix phase, the Al_2O_3 particles is reinforced phase, the density of the sample changes with increasing the sintering temperature, because Al melting point is 660°C, Al_2O_3 melting point is 2050°C. Mixing Al powder and Al_2O_3 powder are sintered at 700°C, Al powder completely melted, and there is no Al_2O_3 powder, which binds to binding mode between powders, densification is not combined with osteoporosis. Molten Al coated the Al_2O_3 powder surrounded to form the dense cermet. Al content in the case of sufficiently high can wrap Al_2O_3 to form a more continuous distribution block, which results in a higher density of the organizational structure.

By spectrometer detector scanning electron microscope samples can be seen in Fig. 3, when the sintering temperature is 700°C, the spherical particle shape for the rule of alumina particles, the other is aluminum metal. Alumina ceramic particles in the metal Al particles can serve to enhance the effect, especially in a series of performance under mechanical action can be improved. Through the above analysis and the experimental results, the cermet is sintered at 700°C, its performance is relatively better.

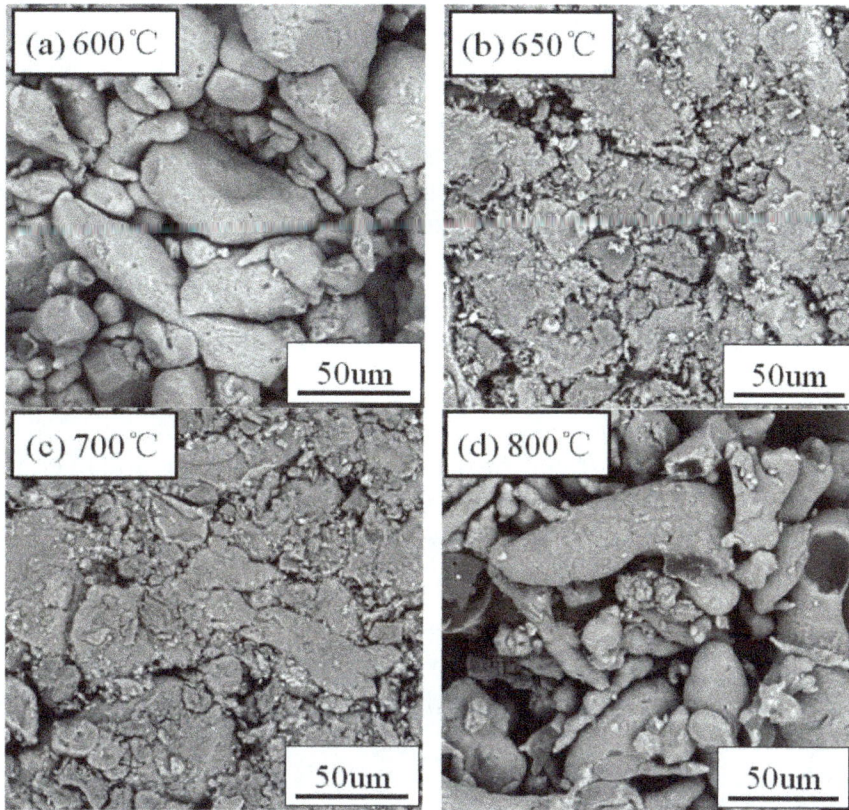

Fig. 2 Effect of sintering temperature on the microstructure of Al_2O_3/Al cermets

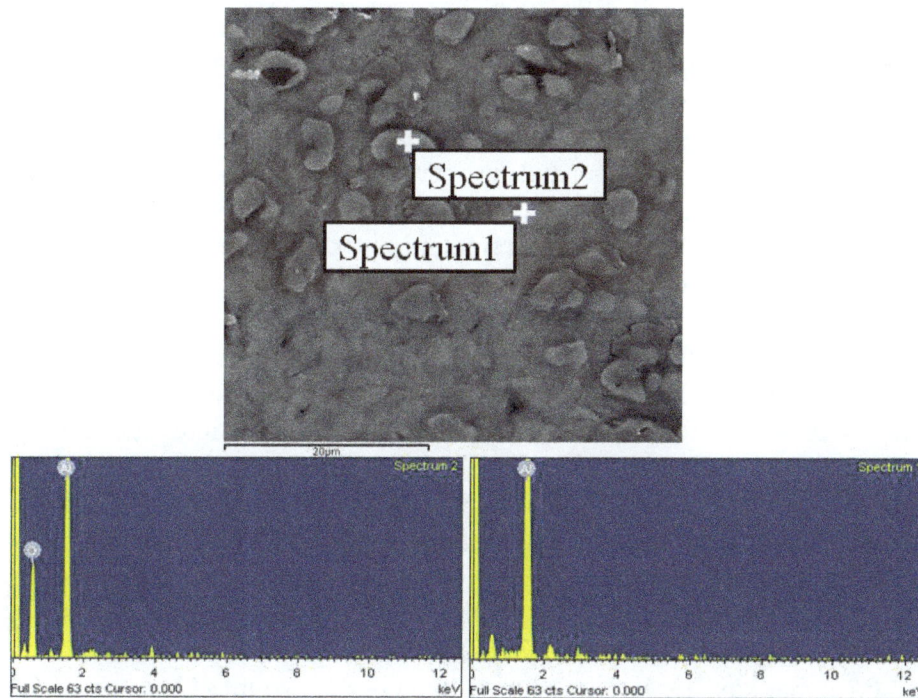

Fig. 3 EDS of sintered Al_2O_3/Al cermets at 700°C for 1h

4. Conclusion

In this paper, the mass fraction of 25% Al₂O₃ powder and the mass fraction of 75% Al powder was mixed in the blender, mixer speed 100r/min, mixing time of 3.5h, forming, sintering and a series of processes for preparing the alumina/aluminum metallic ceramic materials, through performance testing and analysis can be found the density of the sample first increased and then decreased with the increase of sintering temperature, a melting point close to the sintering temperature and sintering, the density of the cermet can be made relatively high. When the sintering temperature is of 600°C and 700°C, the macro performance of sample is better. The cermet is sintered at 700°C, its microstructure is relatively better.

Acknowledgments

The authors are thankful for the fund provided by the Science and Technology Fund of Educational Department of Jiangxi Province, China (KJLD12096), the Science and Technology Fund of Jiujiang University, China (2015LGYB12) and the Teaching Reform Fund of Jiujiang University, China (XJJGYB-16-10).

References

1. G. Bilir, J. Liguori, Laser diode induced white light emission of γ-Al₂O₃ nano-powders, J. Lumin. 153 (2014) 350-355.
2. J.M. Fang, X.Y. Huang, X. Ouyang, X. Wang, Study of the preparation of γ-Al₂O₃ nano-structured hierarchical hollow microspheres with a simple hydrothermal synthesis using methylene blue as structure directing agent and their adsorption enhancement for the dye, Chem. Eng. J. 270 (2015) 309-319.
3. C. Ottone, V.F. Rivera, M. Fontana, K. Bejtka, B. Onida, V. Cauda, Ultralong and mesoporous ZnO and γ-Al₂O₃ oriented nanowires obtained by template-assisted hydrothermal approach, J. Mater. Sci. Tech. 30 (2014) 1167-1173.
4. J.H Xu, K. Bandyopadhyay, D. Jung, Experimental investigation on the correlation between nano-fluid characteristics and thermal properties of Al₂O₃ nano-particles dispersed in ethylene glycol-water mixture, Int. J. Heat Mass Trans. 94 (2016) 262-268.
5. S. Singh, R. Singh, Effect of process parameters on micro hardness of Al-Al₂O₃ composite prepared using an alternative reinforced pattern in fused deposition modelling assisted investment casting, Rob. Comp. Integ. Manuf. 37 (2016) 162-169.
6. M. Ashida, Z. Horita, Effects of ball milling and high-pressure torsion for improving mechanical properties of Al-Al₂O₃ nanocomposites, J. Mater. Sci. 47 (2012) 7821-7827.
7. H. R. Derakhshandeh, Effect of ECAP and extrusion on particle distribution in Al-nano-Al₂O₃ composite, Bull. Mater. Sci. 38 (2015) 1205-1212.
8. M.Z. Mehrizi, R. Beygi, G. Eisaabadi, Synthesis of Al/TiC-Al₂O₃ nanocomposite by mechanical alloying and subsequent heat treatment, Ceram. Int. 42 (2016) 8895-8899.
9. J.Y. Xu, B.L. Zou, S.Y. Tao, M.X. Zhang, X.Q. Cao, Fabrication and properties of Al₂O₃-TiB₂-TiC/Al metal matrix composite coatings by atmospheric plasma spraying of SHS powders, J. Alloy. Compd. 672 (2016) 251-259.
10. A. Poulia, P.M. Sakkas, D.G. Kanellopoulou, G. Sourkouni, C. Legros, Chr. Argirusis, Preparation of metal–ceramic composites by sonochemical synthesis of metallic nano-particles and in-situ decoration on ceramic powders, Ultrason. Sonochem. 31 (2016) 417-422.

11. G. Miranda, M. Buciumeanu, S. Madeira, O. Carvalho, D. Soares, F.S. Silva, Hybrid composites Metallic and ceramic reinforcements influence on mechanical and wear behavior, Comp. Part B Eng. 74 (2015) 153-165.
12. G.Q. Xie, D.V.L. Luzgin, F. Wakai, H. Kimura, A. Inoue, Microstructure and properties of ceramic particulate reinforced metallic glassy matrix composites fabricated by spark plasma sintering, Mater. Sci. Eng. B 148 (2008) 77-81.

Design and Study of the Frangible Composite Covers

Yong Feng[1,*], Guo-Dong Yan[1], Zhen-Qin Xu[1], Zhong-Li Zhang[2]

[1] *School of Mechanical Engineering, Nanjing Institute of Technology, Nanjing 211167, China*
[2] *School of Mechanical Engineering, Henan University of Science and Technology，China*
*Email: fengyong007@sina.com

The aim of this study is to design and analyse the impact properties of frangible covers. Two type frangible covers made of rigid polyurethane (RP) as base material and epoxy resin (ER), glass reinforced fibre plastic (GRFP) as base material are developed. The methods of finite element analysis (FEA) and impact experiment are used to verification of material and structure. FEA and experiment results show clearly that the feasibility of two type frangible covers. Meanwhile, from the purpose of application, The II type frangible cover is proven to be better than I type because of its regular segments which have less influence on other rockets launch.

Keywords: Frangible cover; self-locking structure; finite element analysis; rigid polyurethane; epoxy resin.

1. Introduction

The seal cover used at both ends of rocket launcher was one of the key technologies to make sure rocket weapon in the quick reaction capability. According to the requirements of operational condition and tactical index, seal cover should be broken at the moments of rocket projectile flying away rocket launcher to ensure rocket projectile pass normal without fuze damage. In the past, mechanical open mode and explosive open mode were the main modes applied in seal cover [1]. However, both open modes had their respective defects. Mechanical open cover needs to be opened by hydraulic system was easy to cause mechanical failure. Explosive open cover was easy to make electronic components and explosive devices failure due to its relative complexity in reload and maintenance. In recent years, light frangible seal cover has been widely used in rocket weapon, such as Russian SAM-6 vertical launching system, American M270 rockets and HIMARs rockets. This type seal cover can be broken in the case of smaller pressure because of its reserved weak areas in structure, which could cause regular shape and consistent size fragments. Moreover, the frangible cover assembled easily with rocket launcher at flange by bolts can quickly improve the combat troops response ability.

As to frangible seal cover, the chosen of materials is the key factor to the frangible cover development. Generally, only non-metallic materials can be chosen, such as plastic, glass, ceramic etc. A plastic as base material foamed frangible cover is applied in China HQ-6 surface to air missile. Some auxiliary materials are added into the base material. Meanwhile, the out surface of cover is coated with silicon rubber material to improve performance of corrosion resistance, moisture-proof and insulation. American experts [2] proposed a new frangible cover structure by using the method of bonding double layer or multilayer GRFP membrane with ER, which can control the broken tracks for making the notch on a layer to coincide with the torn direction. In addition, another frangible cover made of a low density ceramics like modified inorganic material was presented with the characteristics of good tightness, long life, easy molding and low cost [3].

Two new frangible seal covers are presented in this paper by using the topology self- locking structure. The impact properties of both type covers made of RP and ER are contrastive analyzed by using the non-linear dynamic finite element method. Finally, impact experiments are carried out to discuss the frangible cover fracture mechanism. A good agreement is obtained after comparison

*Corresponding author

between the simulation results and experiment results, which show the validation of material and structure.

2. Structure Design

2.1. *Type frangible cover*

According to the operation conditions and performance of a certain type rocket weapon, the structure of I type frangible cover [4] is developed as shown in Fig. 1.

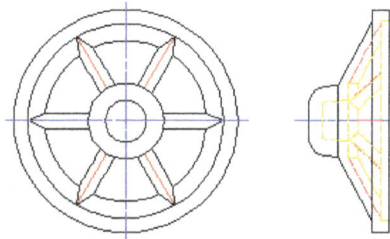

Fig. 1 I type frangible cover

The upper cover is a cylindrical bulge with inner hole, which is beneficial to reduce the impact effect on rocket fuze. The central cover is a conical structure with six 90° taper grooves on inner wall, which is beneficial to reduce the gas flow impact and also easy to cause cover broken. The lower cover is a cylindrical platform used for sealing with launcher end face. The RP is chosen to be base material. The copper foil is pasted on inner wall for electromagnetic shielding. Meanwhile, the out surface of cover is coated with silicon rubber material to improve performance of corrosion resistance, moisture-proof and insulation.

2.2. *Type frangible cover*

Professor Y. Estrin and Doctor A. V. Dsykin [5] proposed a new method of using the regular arrangement of same size and shape substructure to obtain a topology self-locking composite structure. In this structure, each independent substructure is fixed by the surrounding substructures, without the help of bonding materials. Experimental results showed that the self-locking structure has higher strength, good toughness and impact resistance, especially can prevent cracks stretch effectively. As shown in Fig. 2, a topology self-locking structure is used in II type frangible cover as the support layer which can be broken into regular fragments while striking [6]. The GRFP is used in supporting layer which consists of eight substructures, including a substructure on the top of arc, six watermelon section-like substructures and a substructure used to fix with launcher. The continuous seal layer is made of ER, which is bonded with each supporting layer substructures.

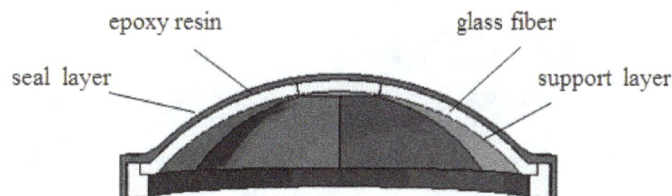

Fig. 2 II type frangible cover

3. Numerical Simulation

3.1. *I type frangible cover simulation*

- **Simulation model**

The ABAQUS/EXPLCIT is utilized for impact simulations [7]. It is developed based on the dynamic explicit Lagrangian formulation. The mesh model is established as shown in Fig. 3, in which the hexahedral element C3D8R is used. Moreover, the cover is meshed with four elements in thickness for taking account of broken detail. The initial velocity of rocket projectile is 12 m/s. The terminal time of simulation is set 7 ms.

The material property of the frangible cover (RP) is assumed to be a closed cell foam material model. The relationship between pressure P and stress σ, stress σ and volumetric strain VS of RP are shown in Fig. 4, other material properties for all components are summarized in Table 1. In order to predict the segments separation and the form of the fractured surface during impacting process, fracture initiation and propagation should be considered. The maximum elongation strain fracture criterion is used to estimate if and where a segment separation occurs. The formula is expressed as follows:

$$\sigma_1 - \mu\left(\sigma_2 + \sigma_3\right) = \sigma_b \tag{1}$$

While σ1, σ2 and σ3 are the principal stress of element, σb is the tensile strength of material, μ is the Poisson ratio of material. Therefore, a fracture initiates at any point in the material where the principal stress expression reaches the tensile strength.

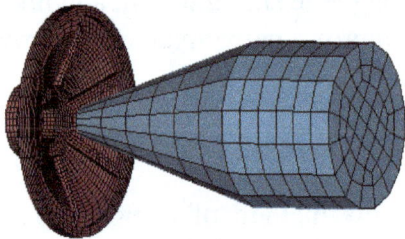

Fig. 3 I type frangible cover mesh model

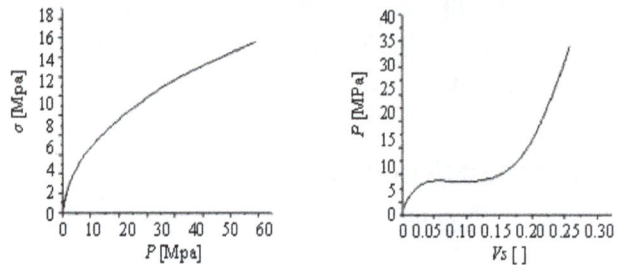

Fig. 4 Relationship curve of RP a) P vs. σ, b) P vs. VSv

Table 1 Basic meterial properties of aluminum and polyurethane

	Elasticity modulus GPa	Shear modulus GPa	Poisson ratio	Density Kg/m³	Yield strength MPa
aluminum	70	/	0.34	2700	253
polyurethane	0.2	0.05	0.4	350	/

- **Simulation results**

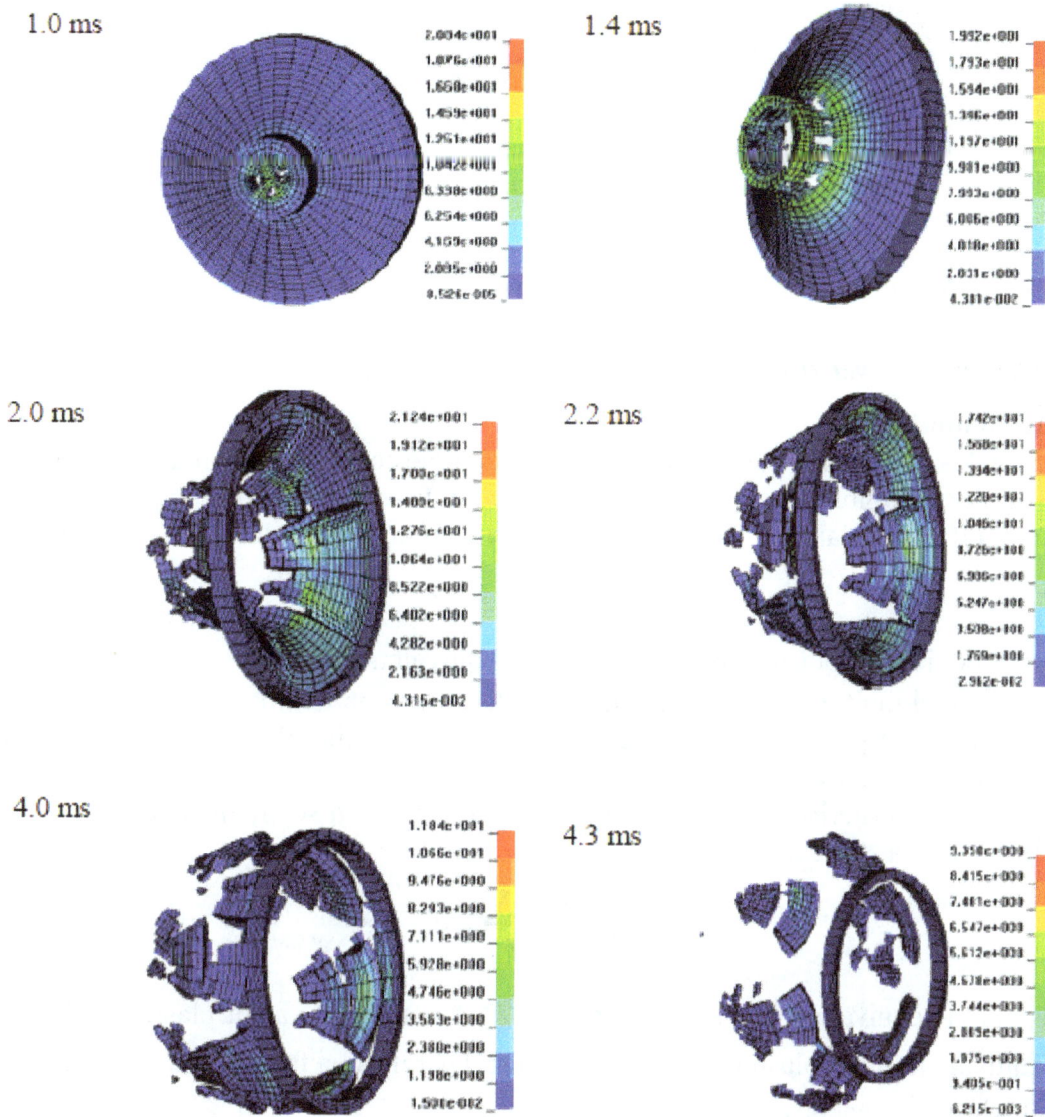

Fig. 5 I type frangible cover broken process [Mpa]

Fig. 5 shows the variation of stress at different time in the broken process, the maximum stress is 21.24 MPa occurred at 2.0 ms. The detail of impact process is investigated. When 1.0 ms, fracture is occurred at the top cover. When 1.4 ms, fracture is occurred at the root of cylindrical bulge, then the whole cylindrical bulge is broken. When 2.0 ms, cracks are occurred at conical structure along the taper grooves, then the whole conical structure is broken into six segments. When 2.2 ms, cracks are extended to the root of conical structure along the taper grooves, then each segment is broken due to the squeeze of rocket projectile. When it is 4.5 ms, six segments are separated from the cylindrical platform. The entire process takes less than 4.5 ms. Fig. 6 shows the variation of projectile kinetic energy E at different time in whole process. Four significant strikes are occurred. After these strikes, the kinetic energy of projectile is reduced from 7056 J to 7031.42 J. The total energy loss is about 0.35%.

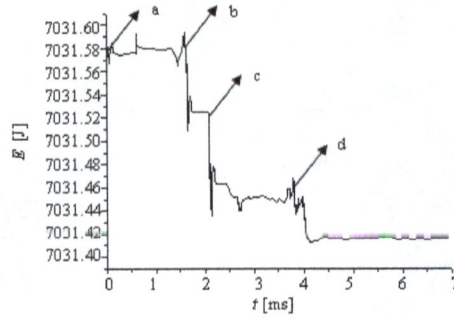

Fig. 6 Projectile kinetic energy at different time (a,b,c,d are the significant strike points)

3.2. *II type frangible cover simulation*

- **Simulation model**

The mesh model is established as same as the I type frangible cover, as shown in Fig. 7. The material properties of the frangible cover (ER) and projectile fuze (aluminum) are assumed to be a Johnson-Cook (JC) material model [8, 9]. The equivalent strength formula is expressed as follows:

$$\sigma = (A + B\varepsilon_p^n)(1 + C \ln \dot{\varepsilon}) \tag{2}$$

While ε_p is equivalent plastic strain, $\dot{\varepsilon} = \dot{\varepsilon}_p / \dot{\varepsilon}_{p0}$ is equivalent strain rate, $\dot{\varepsilon}_{p0}$ is reference strain rate, A is static yield limit, B is strain hardening modulus, n is hardening exponent, C is coefficient of strain rate. Material properties are summarized in Table 2. the glass reinforced fiber material properties of supporting layer are summarized in Table 3.

The shear fracture criterion is used to estimate if and where a segment separation occurs. The formula is expressed as follows:

$$\omega = \frac{\overline{\varepsilon}_0^{pl} + \sum \Delta \overline{\varepsilon}^{pl}}{\overline{\varepsilon}_f^{pl}} \tag{3}$$

While $\overline{\varepsilon}_0^{pl}$ is the equivalent initial plastic strain, $\Delta \overline{\varepsilon}^{pl}$ is the equivalent plastic strain increment, $\overline{\varepsilon}_f^{pl}$ is the equivalent initial plastic strain while failure. Therefore, a fracture initiates at any point in the material where ω reaches 1.

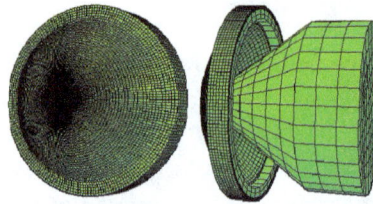

Fig. 7 II type frangible cover mesh model

Table 2 Constitutive model parameters of aluminum and ER

	A(MPa)	B(MPa)	N	C	$\dot{\varepsilon}_{p0}$
aluminum	265	426	0.34	0.015	1
ER	72	156	0.82	0.0007	1

Table 3 Basic material properties of GRFP

Elasticity modulus Gpa	Poisson ratio	Longitudinal tensile strength MPa	Longitudinal compression strength MPa	Transverse tensile strength MPa	Transverse compression strength MPa
Ea=11 Eb=10 Ec=3.2	ubc=0.27 uca=0.1 ucb=0.1	200	250	350	150

- **Simulation results**

Fig. 8 shows the variation of stress at different time in the broken process, the maximum stress is 85.18 MPa occurred at 1.0ms.

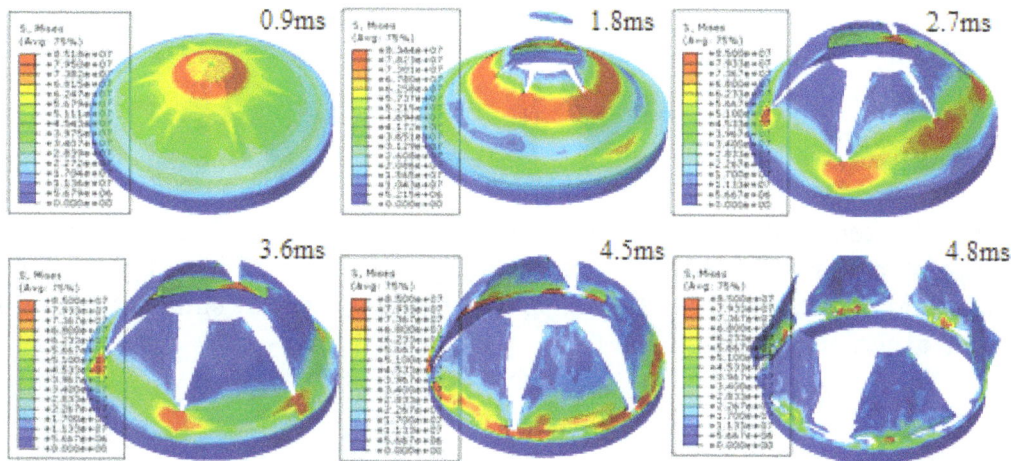

Fig. 8 *II type frangible cover broken process* [Pa]

The detail of impact process is investigated. When 0.9ms, projectile impacts the top arc substructure, then the maximum impact stress is occurred at the top of sealing layer accompanied with the enlargement of top arc substructure. When 1.8ms, fracture is occurred at the top of sealing layer, then a hat shape like segment is formed. When 2.7ms, the radial cracks are occurred at the junctions of substructures, then the cracks are expanded outwards. When 3.6ms, petal like segments are formed, then the sealing layer is broken into six regular segments as same as the substructures shape of supporting layer. When 4.8ms, the six segments are separated from cylindrical platform. The entire process takes about 5ms

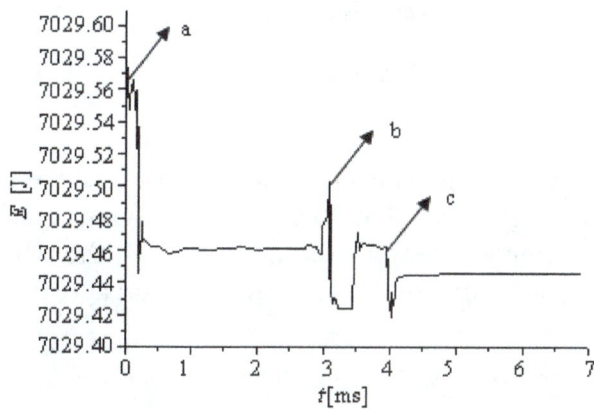

Fig. 9 Projectile kinetic energy at different time

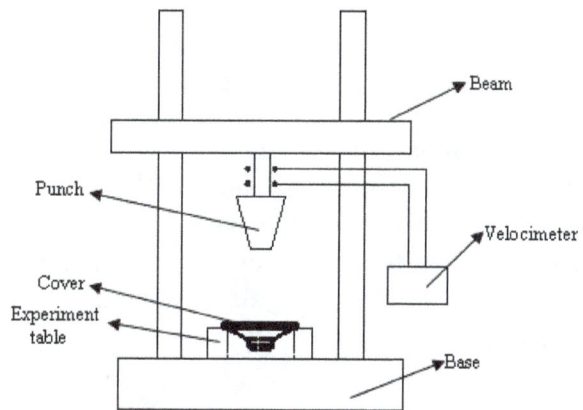

Fig. 10 Impact test bench sketch

Fig. 9 shows the variation of projectile kinetic energy E at different time in the broken process. Three significant strikes are occurred in whole broken process. After these strikes, the kinetic energy of projectile is reduced from 7056 J to 7029.42 J, the total energy loss is about 0.38%.

4. Experimental Validation

4.1. *Experiment setup*

The modified drop-weight test device is used in experiment, as shown in Fig. 10. The impact energy can be adjusted by regulation of drop hammer weight and impact speed. The punch is made according to the rocket fuze shape. Meanwhile, a simple experiment table is made to fix frangible cover. The impact speed is obtained by adjusting the height of drop hammer fixed on a beam, which can be measured using velocimeter.

4.2. *Experimental results*

- **I type frangible cover**

Fig. 11 shows the broken results. When the impact energy is equal to 13.8J as Fig. 11 (a), the top of cylindrical bulge is broken away. When the impact energy is equal to 27.5J as Fig. 11 (b), the whole cover is broken into several irregular segments. The broken status of simulation results is good agreement to the experiment results.

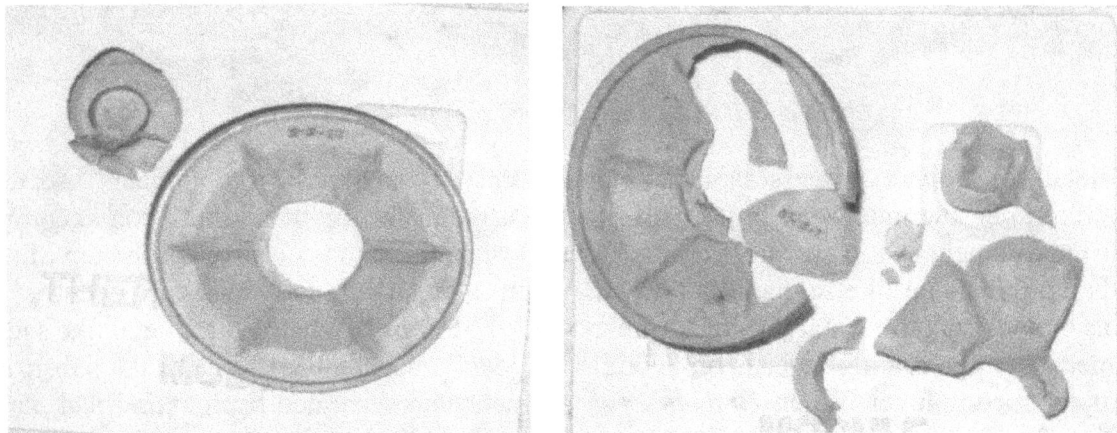

(a) Energy is equal to 13.8J   (b) Energy is equal to 27.5J

Fig. 11 Fracture photographs of I type frangible cover

- **II type frangible cover**

Fig. 12 shows the broken results of II type cover. When the impact energy is equal to 6.28J, the top structure of cover is broken away. When the impact energy is equal to 10.05J, some obvious cracks are occurred at the substructure junctions. When the impact energy is equal to 17.87J, the whole cover is broken into six regular segments. The broken status of simulation results is good agreement to the test results.

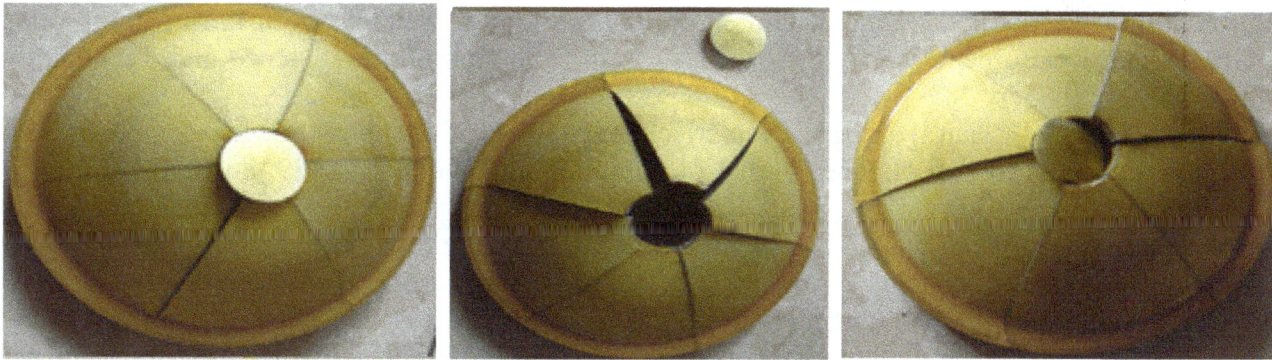

| (a) impact energy is 6.28J | (b) impact energy is 10.05J. | (c) impact energy is 17.87J |

Fig. 12 Fracture photographs of II type frangible cover

5. Conclusion

Two frangible covers are presented in this paper. The impact properties of two type frangible covers are contrastive analyzed by using the non-linear dynamic finite element method. Impact tests are carried out to discuss the fracture mechanism of frangible cover. A good agreement is obtained after comparison between the simulation results and test results, which show the validation of material and structure. With some modification, the theories and methods used in this paper can be presented to general public. From the analysis in this paper, the below conclusions are obtained:

1. The maximum stress of two type frangible covers are respectively 21.24 MPa and 85.18 Mpa less than the yield stress of fuze material, which Suggests that the frangible structures validation.
2. From simulation and experiment results, a good agreement is obtained, which Suggests that the simulation validation.
3. From the purpose of application effects, The II type frangible cover is better than I type because of its regular segments which have less influence on other rockets launch.
4. The reduced projectile kinetic energy is 26.58J while the II type frangible cover be struck, which is bigger than 24.58J of the I type frangible cover. But, it has no effect on the attacking accuracy.

Acknowledgments

This work was funded by the Natural Science Foundation of Jiangsu Province of China (NO: BK20131341 and BK20150728) and the National Natural Science Foundation of China (No. 51 405234).

References

1. Boeglin P.H (1982). Plate-glass fitted with an explosion-cutting device. US Patent, No.4333381.
2. Bell Re (1992). Missile weapon system. US Patent, No. 523990.
3. Doane WJ (1985). Frangible flies through diaphragm for missile launch canister. US Patent No. 4498368.
4. Zhang Z.L., Yu C.G., Ma D.W. (2008). Numerical simulation and experimental analysis of a crisp airproof lid under impact. The Explosion and shock waves Journal, Vol. 28, no. 1, p. 62-66. (in Chinese)

5. Dyskin A V, Estrin Y, Kanel-Belov A J (2003). A new principle n design of composites materials: Reinforcement by interlocked element. The Composites Science and Technology Journal, Vol. 63, p. 483-491.
6. Xu Z.Q, Le G.G, Han B. (2010). Numerical analysis of a directional fracture composite seal cover subjected to impact. The Journal of Vibration and shock, Vol. 29, no. 4, p. 77-80. (in Chinese)
7. Wu J.H, Wang W.T., Kam T.Y. (1999). External failure pressure of a trangible laminated composite canister cover. The Proc inst Mech Eng Journal, Part G, p. 187-195.
8. QIAN Yuan, ZHOU Guangming, HE Weidong, YU Dianjun (2012). Design and Analysis of the Weak Structure for Frangible Composite Canister Covers. The journal of Acta Aeronaticaet Astronautica Sinica., Vol. 33, No. 3, p. 487-493. (in Chinese)
9. Sun Zhibin, Wang Xinfeng, Zhou Guangming (2012). Specified direction separation of circle composite frangible cover. The journal of Nanjing University of Aeronautics & Astronautics, Vol. 44, No. 6, p. 803-808. (in Chinese)

Design Optimization for Layered Banded Materials with Multi-phase Microstructures

Wei Wang[1,*] , Wei-Kai Xu[2]

[1]School of Civil Engineering, Shenyang Jianzhu University, Shenyang, China
*[2]Key Laboratory of Liaoning Province for Composite Structural Analysis of Aerocraft and Simulation,
Shenyang Aerospace University, Shenyang, China*
[*]Email: starwei2002@163.com

Design and optimization of bandgap materials is essential in many processes. The purpose of this paper is to propose a method for achieving the optimized layered banded materials with multi-phase microstructures. In general, the characteristic of the periodic materials can be designed by controlling the materials layout within the microstructures. By using an appropriate design variables and the GA method, the dynamic responses of layered materials are studied, and the design of multi-phase layered elastic material is presented in the paper. Finally, the topologies of periodic multiphase microstructures are obtained. Two cases studies were presented in the paper, and the results show that 3-phase material can obtain quite better designs on the basis of fewer layers, whether a maximized stopbands or a strong attenuation. There will be almost a whole bandgap cover the frequency domain. The authors think the paper is a novel one and the presented designs will be useful for generating periodic materials that could be used for shock/sound isolation.

Keywords: Topology design; multi-phase materials; layered banded material; genetic algorithm component.

1. Introduction

It is well known that there exist an important dispersion-related characteristic of periodic materials, named stopband and passband. This frequency-banded response has become of considerable interest in periodic material, especially since these materials have many practical applications across multi-disciplinary [1, 2]. Then the design and optimization for the bandgap feature and properties at desired bands became the focal points in many works [3].

By controlling the materials layout and the ratio of their properties within a unit cell, the desired composite with special bandgap characteristics could be designed [4-7]. In this paper, the topology optimization of layered 3-phase materials is investigated. The technique of genetic algorithms is chosen because of the advantage that it can operate discrete variables well and converge to a near-global optimality. Two cases of prescribed objectives have been studied, and numerical examples show that the systematic approach can provide satisfactory design results with respect to multi-phase microstructures.

2. Theory

2.1. *Dispersive wave motion for layered materials*

Consider a multi-layered medium as shown in Figure 1, the governing equation for longitudinal wave propagation in the x direction in an arbitrary layer j is:

$$E_j \frac{\partial^2 u(x,t)}{\partial x^2} = \rho_j \frac{\partial^2 u(x,t)}{\partial t^2} \tag{1}$$

---

[*]Corresponding author

where ρ_j, E_j are density and Young's modulus, and the velocity is $c_j = \sqrt{E_j/\rho_j}$. Let x^{jL} and x^{jR} denote the position along the x-axis of the left and right boundaries of layer j, d_j is the thickness of the layer, then the displacement u and stress σ at x^{jL} can be related to those at x^{jR}:

$$\begin{bmatrix} u(x^{jR}) \\ \sigma(x^{jR}) \end{bmatrix} = \mathbf{T}_j \begin{bmatrix} u(x^{jL}) \\ \sigma(x^{jL}) \end{bmatrix} \tag{2}$$

where

$$\mathbf{T}_j = \begin{bmatrix} \cos(k_j d_j) & (1/Z_j)\sin(k_j d_j) \\ -Z_j \sin(k_j d_j) & \cos(k_j d_j) \end{bmatrix} \tag{3}$$

is the transfer matrix in the jth layer, and k_j, $Z_j = \rho_j c_j k_j$ are the wavenumber and the acoustic impedance, respectively. Thus the cumulative transfer matrix of an n-layered composite shown in Fig. 1 is $\mathbf{T} = \mathbf{T}_n \mathbf{T}_{n-1} \cdots \mathbf{T}_1$. Therefore, the dispersive wave motion can be resulted in the eigenvalue problem:

$$\left[\mathbf{T} - \mathbf{I} e^{ikd} \right] \begin{bmatrix} u(x^L) \\ \sigma(x^L) \end{bmatrix} = 0 \tag{4}$$

While this problem is solved for the dispersion curves, the bandgap can be determined fully.

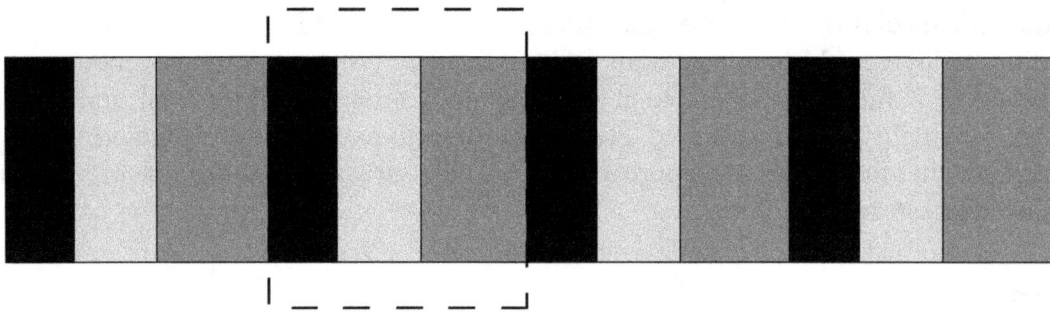

Fig. 1 The sketch of the layered material

2.1. *The design variables of multi-phase materials*

Making reference to Hussein [5] and Xu [7], the configuration of unit cell can be depicted by the number of layers and the thickness of each layer. However, due to the difficulty of performing a optimization problem that a variable depends on another one, an alternative binary variable $\mathbf{b} = (b_1, b_2 \cdots b_l)$ (with a constant dimension l) was introduced by assuming that a unit cell is divided into l imaginary "divisions" or "slots", each of which can be filled with different material, i.e. "0" for phase1 and "1" for phase 2. Furthermore, we will extend the vector $\mathbf{b}$ to play the same role in the multi-materials problems. Based on the concept of Solid Isotropic Material Penalization (SIMP) method for multi-phase materials [8], we introduce a pseudo-density variable:

$$(x_t, y_t) = \begin{cases} (0,0) & phase1 \\ (0,1) \, or \, (1,0) & phase2 \\ (1,1) & phase3 \end{cases} \tag{5}$$

Then the unit cell which is divided into l divisions can be expressed as:

$$\mathbf{a} = ((x_1, y_1), (x_2, y_2) \cdots (x_t, y_t) \cdots (x_l, y_l)) \tag{6}$$

This is a vector of binary variable with dimension 2l. Define:

$$b_t = sum(x_t, y_t) = \begin{cases} 0 & phase1 \\ 1 & phase2 \\ 2 & phase3 \end{cases} \tag{7}$$

The unit cell can be expressed by the improved variable $\mathbf{b}$. Certainly, a 4-phase optimization problem can be performed if we distribute different materials to $(1,0)$ and $(0,1)$.

Similar to the study by Hussein [5], define a logical operator XOR:

$$XOR(b_t, b_{t+1}) = \begin{cases} 0 & if \, b_t = b_{t+1} \\ 1 & if \, b_t \neq b_{t+1} \end{cases} \tag{8}$$

and a vector $\mathbf{s} = (s_1, s_2 \cdots s_{l-1})$, where $s_t = XOR(b_t, b_{t+1})$. Thus, the total number of layers can be determined by:

$$n = \sum_{t=1}^{l-1} s_t + 1 \tag{9}$$

Introduce $\mathbf{q} = (q_0, q_1 \cdots q_n)$, where $q_0 = 0, q_n = l$ and $q_j (j = 1, 2 \cdots n)$ to be the sequential number of the end slice of jth material layer. The thickness and the phase of jth layer can be expressed as:

$$d_j = (q_j - q_{j-1})\frac{1}{l}, \quad j = 1, 2 \cdots n \tag{10}$$

$$m_j = b_{q_j}, \quad j = 1, 2 \cdots n \tag{11}$$

From Eq.6-Eq.11, we can see that while the binary vector $\mathbf{a}$ is determined, the number of the layers, the thickness and the material phase of every layer are determined, i.e. the configuration of the unit cell can be depicted completely by the binary vector $\mathbf{a}$. Take the binary vector $\mathbf{a}$ as decision variable, the optimization problem can be formulated as a zero-one integer programming problem, which is suitable for the GA method.

3. Case Study

In the section, two case studies are considered by the following assumptions: The three component materials are chosen as PMMA (for 2), rubber (for 1) and aluminium (for 0), respectively. The moduluses and the densities are shown in Table 1. The total thickness of the unit cell is 0.2m. The dimension of the variable **b** is chosen to $l = 40$. Without loss of generality, it is also assumed that the first layer in unit cell is always PMMA.

Table 1 Parameters of Component Materials

	PMMA	**Rubber**	**Aluminium**
Modulus (GPa)	5.28	0.02758	70.9
Density(kg/m³)	1200	1200	2830

3.1. Case I: Maximization of the percentage of stopbands

The aim in this case is to find the construction of the unit cell consisting of the three materials which has Maximum bandwidth of negative wavenumbers. The fitness function can be used in the following form Hussein[5]:

$$F = \frac{\int_{f_{min}}^{f_{max}} H(\xi_{real}(f)) df}{f_{max} - f_{min}} \times 100\% \qquad (12)$$

where $\xi = \xi_{real} + i\xi_{imag} = k*d$ denote the wavenumber and $H(\xi_{real})$ is a hard limit function defined as:

$$H(\xi_{real}(f)) = \begin{cases} 1 & if \ \xi_{real} > 0 \\ 0 & if \ \xi_{real} < 0 \end{cases} \qquad (13)$$

f_{min}, f_{max} denote the lower and upper limits of the range respectively. In the case we select as 0 and 50 kHz. Thus, the optimization problem can be formulated mathematically as:

$$\begin{aligned} &find \quad \mathbf{b} = (b_1, b_2 \cdots b_l) \\ &min \quad F(\mathbf{b}) \\ &subject\ to \quad b_t = 0\ or\ 1 \end{aligned} \qquad (14)$$

For a single-objective optimization, a standard GA method had been employed. The results and the topological configuration of the cell are shown in Table 2 and Fig. 2.

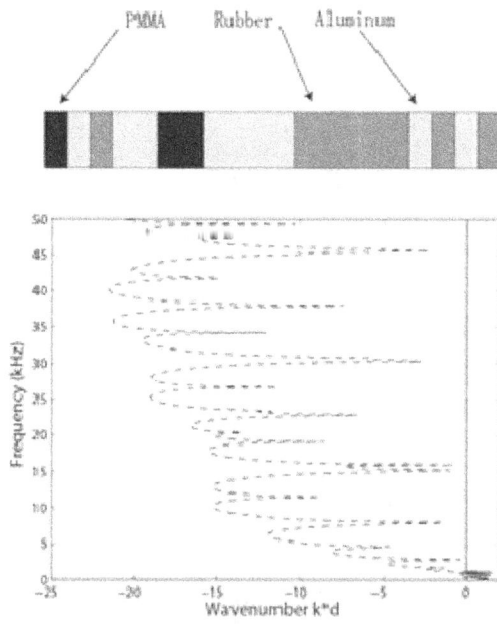

Fig. 2 Case I: the optimal topological configuration and the dispersion curves.

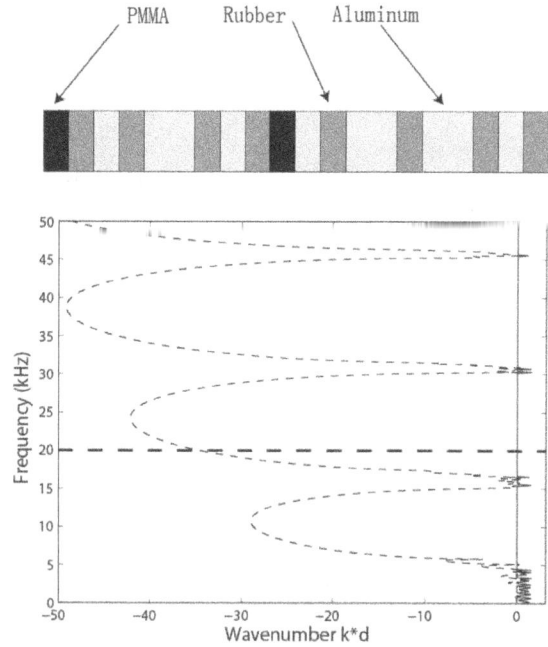

Fig. 3 Case II: the optimal topological configuration and the dispersion curves.

The solid lines denote the real part of the wavenumber, which mean the passband, and the broken lines denote the imaginary part which shown a stopband domain. From these results we can find that while we extend a 2-phase material to a 3-phase, the percentage of stopbands can be increased to 98.4%, almost a whole bandgap cover the frequency domain.

Table 2 Layer Thicknesses and Phase Properties for Case I

	layer 1	layer 2	layer 3	layer 4	layer 5	layer 6
Thickness	0.005	0.005	0.005	0.01	0.01	0.02
Phase	2	0	1	0	2	0
	layer 7	layer 8	layer 9	layer 10	layer 11	
Thickness	0.025	0.005	0.005	0.005	0.005	
Phase	1	0	1	0	1	

3.2. Case II: Maximized attenuation at a specified frequency

In general the strength of spatial attenuation of an incident wave at a stopband frequency is exponentially related to the value of corresponding imaginary wavenumber. Thus, the aim in this case can be chosen as the follow:

$$F = \xi_{imag}\left(f\right) \tag{15}$$

Consider the special frequency $f = 20kHz$ and substitute the Eq.15 for the objective function in Eq.14, the results and the topological configuration of the cell are shown in Table 3 and Fig. 3. The goal in this case is -35.59, a quite strong attenuation. Similar to the results in the work of

513

Hussein (2016), the maximized attenuation at the special frequency is not centered on around $f = 20kHz$.

Table 3 Layer thicknesses and phase properties for case II

	layer 1	layer 2	layer 3	layer 4	layer 5	layer 6
Thickness	0.005	0.005	0.005	0.005	0.01	0.005
Phase	2	1	0	1	0	1
	layer 7	layer 8	layer 9	layer 10	layer 11	layer 12
Thickness	0.005	0.005	0.005	0.005	0.005	0.01
Phase	0	1	2	0	1	0
	layer 13	layer 14	layer 15	layer 16	layer 17	
Thickness	0.005	0.01	0.005	0.005	0.005	
Phase	1	0	1	0	1	

4. Conclusion

In conclusion, the topology optimization technology is introduced to design potential layered elastic bandgap materials with multi-phase microstructures for desired properties. Two cases studies were considered

1) Case I: maximizing the percentage of stopbands within a specified frequency range and

2) Case II: maximizing the attenuation at a specified frequency.

The results shown that while the number of component materials is three, we can get a design exhibiting stopbands on almost the whole frequency range in case I. For case II there would be a quite strong attenuation at the desired frequency. The presented design will be useful for generating periodic materials that could be used for shock/sound isolation.

Acknowledgments

This research was funded by the following: the Youth Foundation of National Natural Science (51308357, 11302135), the Science and college and universities outstanding young scholar growth plan in Liaoning province (LJQ2015091, LJQ2014019), Natural Science Foundation of Liaoning Province of China (201602627, 201602572) and subject cultivation plan of Shenyang Jianzhu University (XKHY2-11).

References

1. Jensen, J.S., Phononic band gaps and vibrations in one- and two-dimensional mass–spring structures, J. Sound Vib., 266 (2003), 1053–1078.
2. Kushwaha, M.S., Halevi, P., Dobrzynski, L. and Djafari-Rouhani, B., Acoustic band structure of periodic elastic composites, Phys. Rev. Lett., 71 (1993), 2022–2025.
3. Kushwaha, M.S., Classical band structure of periodic elastic composites, Int. J. Mod. Phys. B, 10 (1996), 977–1094.
4. Hussein, M.I. Hulbert, G.M. and Scott, R.A., Dispersive elastodynamics of 1D banded materials and structures: analysis, J. Sound Vib., 289 (2005), 779-806.

5. Hussein, M.I., Hamza, K., Hulbert, G.M., Scott, R.A. and Saitou, K., Multiobjective evolutionary optimization of periodic layered materials for desired wave dispersion characteristics, Struct. Multidisc. Optim., 31 (2006), 60–75.

6. Hussein, M.I. Hulbert, G.M. and Scott, R.A., Dispersive elastodynamics of 1D banded materials and structures: design, J. Sound Vib., 307 (2007), 865–893.

7. Xu, W.K., Wang. W., Yang, T.Z., Multi-Objective Optimization of Layered Elastic Metamaterials with Multiphase Microstructures，J. Vib. Acoust., 135，(2013), p. 041010.

8. Bendsoe, M.P. Sigmund, O., Material interpolation schemes in topology optimization, Arch. Appl. Mech., 69 (1999), 635-654.

Microstructure and Mechanical Properties of TiC-Mo$_2$C-Fe Cermets by Vacuum Sintering Process

Zhen-Gang Li[1], Zong-De Liu[2,*], Bo Peng[2]

[1]*Science and Technology College, North China Electric Power University, Baoding 071051, China*
[2]*School of Energy Power and Mechanical Engineering, North China Electric Power University, Beijing 102206, China*
*Email: lzd@ncepu.edu.cn

TiC-6%wt.Mo$_2$C-Fe cermets fabricated by vacuum sintering technology were investigated. The microstructure of the cermets was analyzed by scanning electron microscopy (SEM), with attached energy dispersive spectroscopy (EDS) microprobe and by X-Ray diffraction (XRD). The bending strength and wear properties were taken by three point bending strength test and wet sand rubber wheel abrasion test respectively. The results showed that the cermets exhibit a typical core-rim structure, the core phase was undissolved TiC particles, while the rim was (Ti,Mo)C phases formed through Molybdenum atoms substituting Ti atoms randomly. The main wear mechanism of the cermets is shedding of hard phase, rather than a furrow wear or adhesive wear. The morphologies for the fracture surface of the cermets showed that the fracture mechanism is typical brittle fracture.

Keywords: TiC-Fe cermets; microstructure; mechanical properties; Mo$_2$C.

1. Introduction

TiC-Fe cermets are widely used in many applications requiring high resistivity to wear and thermal fatigue erosion e.g. plastic industries, hot work industries, aircrafts, tools, jigs and fixtures due to their good wear resistance, high-temperature hardness, low friction coefficient, chemical stability and superior thermal deformation resistance. [1-5] As a result, TiC-based cermets have attracted researchers' wide attention and a growing interest has been given to the use of TiC-based cermets. TiC-based cermets have several advantages as compared to conventional cermets (WC–Co), possessing poor oxidation resistance and plastic deformation at high temperature, which prohibit their application at elevated temperatures. [6-9]

Numerous elements (Fe, Cr, Co, Ni, Mo, and (Ni-Mo)) can be used as a binder for TiC-based cermets, [10-15] but TiC-Fe cermets are most commonly used because of their low costs and good mechanical properties. However, TiC-Fe cermets inherent brittleness at low temperature, thus many researchers have attempted to improve the mechanical properties of TiC-based cermets.[16,17] In general, the mechanical properties of cermets depend on chemical composition, grain size, microstructure, and sintering temperature. Recently researches indicated that Mo can improve the mechanical properties of TiC-based cermets [18, 19] and the wettability between reinforced phase TiC and metal binder phases. [20, 21] Molybdenum is generally added as Mo$_2$C in the raw powders. The toughness of the cermets will decrease with the increase of the molybdenum content, and according to Li's research, [18] Mo can decrease the hardness when its content exceeds 10wt-%. Previous studies [22, 23] indicated that the bending strength, impact toughness and wear resistance of the samples with 6wt-% Mo are better than the samples with 4wt.% and 8wt-% Mo content. In view of this, the aim of this study is to fabricate TiC-6wt-%.Mo$_2$C-Fe cermets and to investigate its microstructure and mechanical behaviors.

---

*Corresponding author

2. Experimental Procedure

2.1. *Preparation of sample*

Vacuum sintering was used for the preparation of samples. The composition of the raw materials is listed in Table 1.

Table 1 Composition of starting powders

Powders	TiC	Mo_2C	Mn	Ni	Fe
(wt-%)	40	7	9	5	Balance

The experimental process is shown in Fig. 1. The milled powders were pressed in a die under a pressure of 100 MPa and cold compacted by isostatic pressure (350 MPa). The green compacts were sintered at 1440°C for 30 minutes in the vacuum of 10^{-3}-10^{-4} Pa.

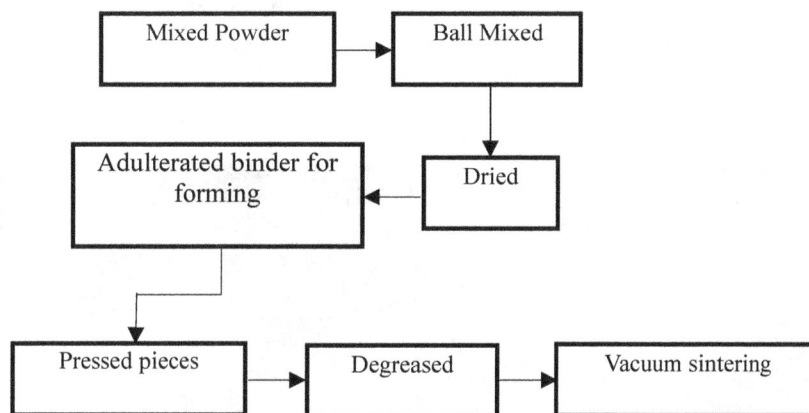

Fig. 1 Schematic diagram of the sintering process

2.2. *Microstructure observation*

The characterization of microstructure of the cermets was performed with JEOL-7001F scanning electron microscope (SEM) in back scattered electron (BSE) imaging mode, equipped with an attached energy dispersive spectroscopy (EDS) microprobe. A type of Rigaku D/Max-rB X-Ray diffractometer with Cu-Kα radiation operating at 45kV and 40mA was used to analyze the phase of the cermets.

2.3. *Microhardness and bending strength tests*

Microhardness measurements were taken by a FM-300 microhardness Vickers tester with a load of 500g applied for 15s. An average hardness value (based on ten indentations) was determined. Three point bending strength tests were performed with an omnipotence material testing machine model WDW-10E, under displacement control. The three test specimens were machined with size of 30×5×5 mm and the span is 24mm.

2.4. *Impact toughness and wear tests*

Impact toughness tests were carried out with an instrument impact machine. Samples were also prepared for wet testing using a MLS-225 wet sand rubber wheel abrasion tester. The tests were conducted at room temperature under a load of 70N and 240 rev/min rotation speed of rubber wheel. Three tests were done for reproducibility of the results. The worn surfaces were observed by SEM in order to analyze the wear mechanism.

3. Results and Discussion

3.1. *Microstructure analysis*

Back scattered electron images (BSE) of the cermets (none-etched) are presented in Fig. 2. The microstructure of the TiC-Fe cermets included binder and grains. Ceramic particles are distributed evenly on the metal binder with approximately maximum size of 8*um* and minimum size of 1*um*. As described in several previous studies, [24-26] the cermet materials exhibit a typical core-rim structure (core is dark and rim is grey dark) embedded into a metal binder (white zone) network and no porosity is observed. From the figures, there are three determinant phases identified in the micrographs based on the contrast level of each phase.

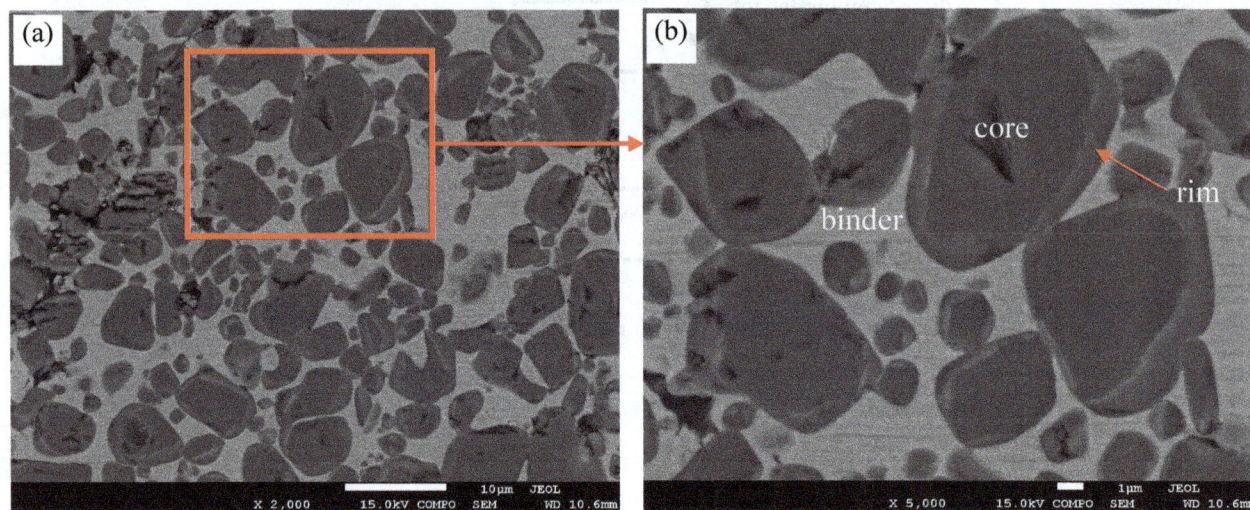

(a) Polished sample showing at ×2000 times;   (b) Magnification showing at ×5000 times

Fig. 2 Back Scattered Electron images of the TiC-Fe cermets

The corresponding results of EDS spectrometer exhibit that the black core is composed of only Ti and C two elements with Ti/C weight ratio 4.09/1 and molar ratio 1/1, so we can believe the black core is TiC phase. The grey dark rim contains Ti and C, as well as a small number of Mo. This accounts for some Mo atoms replace Ti atoms statistically and form $(Ti_{1-x}Mo_x)C$. The white binder is mainly Fe with some Ni and a little amount of Mn in solution with it. From the above analysis, the three phases are TiC, $(Ti_{1-x}Mo_x)C$ solid solution and Fe-based metal binder respectively, as is identified by the XRD results shown in Fig. 3.

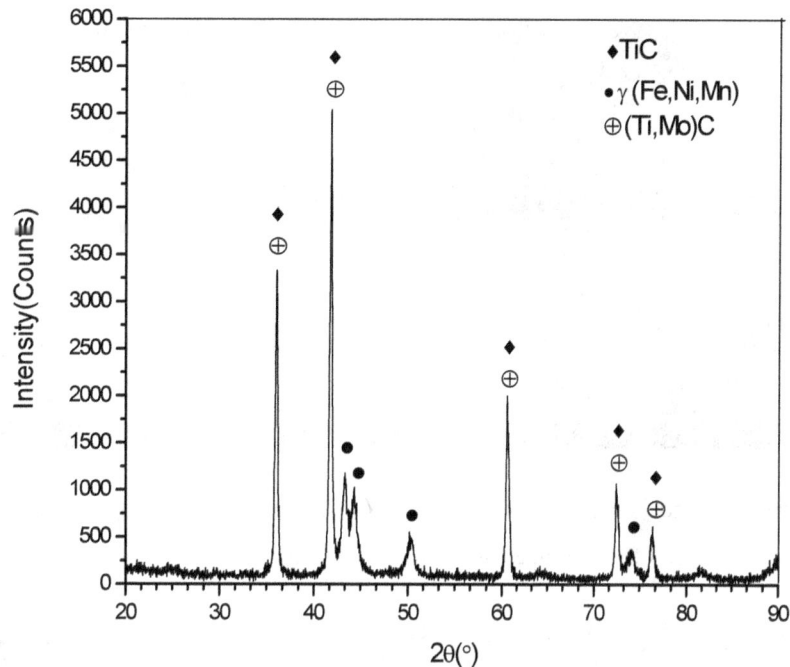

Fig. 3 X-Ray diffraction pattern of the cermets

The initial TiC particles, which were not completely dissolved during sintering appear as black cores. The core is none-dissolved TiC particle and the rim (i.e. ring coated phase) is $(Ti_{1-x}Mo_x)$ C, which is formed during sintering process via dissolution–reprecipitation. It is precisely because of the formation of the $(Ti_{1-x}Mo_x)$ C rim between the TiC particles and metallic binder, the wettability of the metal binder on the carbide phase is improved, which resulting in a decrease in detrimental microstructure defects and an increase in the interphase bond.

3.2. Mechanical properties

The average microhardness of the TiC-Fe cermets is measured to be 1040HV. The bending strength is up to 1310 MPa and the average impact toughness is $2.84J/cm^2$. The addition of Mo results in a decrease in carbide phase contiguity and an increase in the interface bonded strength, while the addition of Mn and Ni results in an increase in the strength of binder phase by solution strengthening. The thickness of the rim will increase with the addition of Mo. The increase in rim thickness can be a result of decreasing mechanical properties of the cermets because of the existence of tensile stress at the core–rim interface. Therefore, rim thickness has been moderated with the use of Mo. Therefore, the addition of Mo quantity cannot be too much, which is below 10%wt. mostly [4, 27].

Fig. 4 is SEM images of the worn surface under different magnification. Obvious traces of wear were not observed on the worn surface but different size of shed pits can be clearly seen left off due to the hard phase loss. Examination of the wear zones revealed that the wear mechanism of the metal ceramic sintering specimen was mainly shedding of hard phase, rather than a furrow wear or adhesive wear.

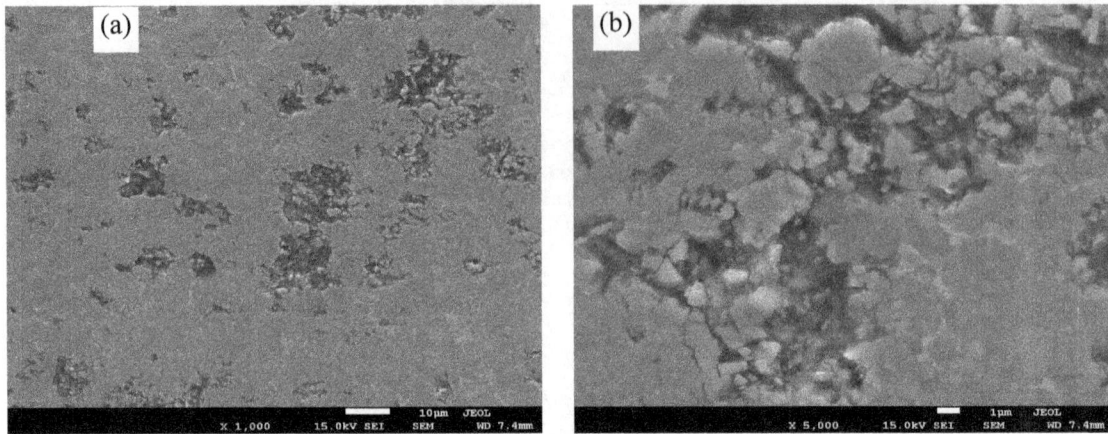

(a) Wear craters formed at the surface of the sample;   (b) Brittle fracture of the bigger phases.

Fig. 4 SEM images of the worn surface of the TiC-Fe cermets

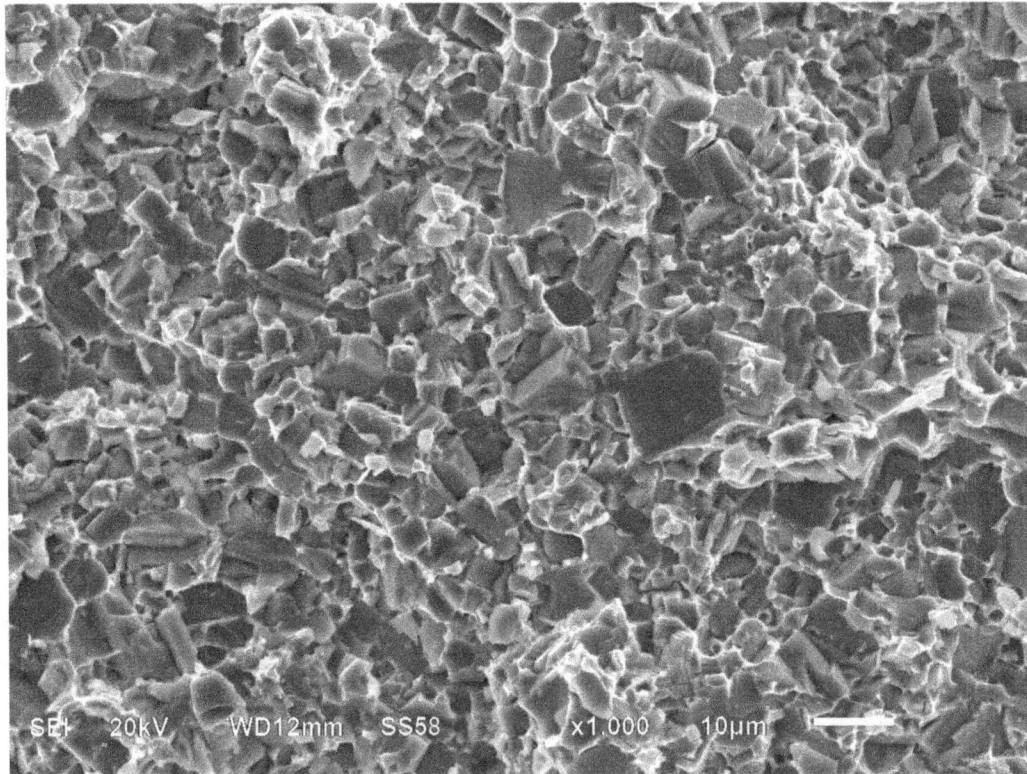

Fig. 5 SEM images of the fracture morphology of the TiC-Fe cermets

Some cracks or loose clusters within cermets can be used as a stress source, which exacerbating the ceramic phase loss. So, increasing the bonding strength of ceramic phase and matrix is the basic way to improve the wear performance, at the same time the high density of ceramic phases and low porosity are beneficial to wear resistance of the cermets.

Fig. 5 is SEM images of the appearance of fracture after impact toughness under different magnification. From the picture, we can see the sample is given priority to with the intergranular fracture. In addition, we can observe the torn edges are not obvious, and as well as plastic deformation of the binding phase. All this revealed that the fracture of the TiC/Fe cermets is the typical brittle fracture.

4. Conclusion

1) TiC-based cermets were prepared by vacuum sintering technology using Fe as the binding materials with additions of 6%wt. Mo_2C. The cermet materials exhibit a typical core-rim structure. The core is none-dissolved TiC particle and the rim is $(Ti_{1-x}Mo_x)C$, which is formed during sintering process via dissolution–reprecipitation with the same crystal structure as TiC.

2) Additions of molybdenum can increase the bonding strength between TiC particles and Fe binders. The microhardness and bending strength of the TiC-Fe cermets are up to 1040 HV and 891 MPa respectively.

3) The main wear mechanism of the cermets was shedding of hard phase, rather than a furrow wear or adhesive wear. Through observation fracture morphology, the fracture of the TiC/Fe cermets belongs to the typical brittle fracture.

Acknowledgments

This paper was funded by the National Natural Science Foundation of China (Grant No. 11372110). The authors are grateful to Mr. Zhi-Jian Bao for his contribution into samples preparing and to Dr. Hui-Na Ma for the support with SEM images.

References

1. Y. Zheng, W. Xiong, W. Liu, W. Lei and Q. Yuan, Effect of nano addition on the microstructures and mechanical properties of Ti(C, N)-based cermets, Ceram. Int. 2005, 31, 165–70.

2. A. Pyzalla, C. Genzel and W. Reimers, Thermal residual microstresses in steel-NbC particulate composites studied by X-ray and neutron diffraction, Mater. Sci. Eng., 1996, 21, 130–138.

3. M. Jones, A. J. Horlock, P.H. Shipway, D.G. McCartney, J.V. Wood, A comparison of the abrasive wear behavior of HVOF sprayed titanium carbide- and titanium boride-based cermet coatings, Wear. 2001, 251, 1009–1016.

4. A. Rajabi, M. J. Ghazal, J. Syarif and A. R. Daud, Development and application of tool wear: A review of the characterization of TiC-based cermets with different binders, Chem. Eng. J. 2014, 255, 445–452.

5. Q. Yang, W. Xiong, S. Li, H. Dai and J. Li, Characterization of oxide scales to evaluate high temperature oxidation behavior of Ti (C, N)-based cermets in static air, J. Alloy Compd. 2010, 506, 461–467.

6. Y. Li, N. Liu, X. Zhang and C. Rong, Effect of WC content on the microstructure and mechanical properties of (Ti, W)(C, N)–Co cermets, Int. J. Ref. Met. Hard Mater. 2008, 26, 33–40.

7. Y. Wu, J. Xiong, Z. Guo, M. Yang, J. Chen, S. Xiong, et al, Microstructure and fracture toughness of Ti ($C_{0.7}N_{0.3}$)–WC–Ni cermets, Int. J. Refract. Met. Hard Mater. 2011, 29, 85–89.

8. Y. Zheng, S. Wang, M. You, H. Tan and W. Xiong, Fabrication of nanocomposite Ti (C, N)-based cermet by spark plasma sintering, Mater. Chem. Phys. 2005, 92, 64–70.

9. X. Shi, M. Wang, Z. Xu, W. Zhai and Q. Zhang, Tribological behavior of Ti_3SiC_2/(WC–10Co) composites prepared by spark plasma sintering, Mater. Des. 2013, 45, 365–376.

10. N. Liu, C. Han, H. Yang, Y. Xu, M. Shi, S. Chao and F. Xie, The milling performances of TiC-based cermet tools with TiN nanopowders addition against normalized medium carbon steel AISI 1045, Wear, 2005, 258, 1688–1695.

11. G. Upadhyaya, Materials science of cemented carbides–an overview, Mater. Des. 2001, 22, 483–489.

12. W. Zhang, X. Zhang, J. Wang and C. Hong, Effect of Fe on the phases and microstructure of TiC–Fe cermets by combustion synthesis/quasi-isostatic pressing, Mater. Sci. Eng. 2004, 381, 92–97.

13. F. Arenas, C. Rondón and R. Sepúlveda, Friction and tribological behavior of (Ti,V)C–Co cermets, J. Mater. Process. Technol. 2003, 143, 822–826.

14. S. Cardinal, A. Malchere, V. Garnier and G. Fantozzi, Microstructure and mechanical properties of TiC–TiN based cermets for tools application, Int. J. Refract. Met. Hard Mater. 2009, 27, 521–527.

15. J. Pirso, M. Viljus, K. Juhani and M. Kuningas, Three-body abrasive wear of TiC–NiMo cermets, Tribol. Int. 2010, 340–346.

16. Bhaskar UK, Pradhan S, One-step mechanosynthesis of nano structured Ti (C_xN_{1-x}) cermets at room temperature and their microstructure characterization, Mater. Chem. Phys. 2012, 134, 1088–1096.

17. M. Liang, J. Xiong, Z, Guo, W. Wan and G. Dong, The influence of TiN content on erosion–corrosion behavior of Ti (C, N)-based cermets, Int. J. Refract. Met. Hard Mater. 2013, 41, 210–215.

18. Y. Li, N. Liu, X. Zhang and C. Rong, Effect of Mo addition on the microstructure and mechanical properties of ultra-fine grade TiC–TiN–WC–Mo$_2$C–Co cermets, Int. J. Refract. Met. Hard Mater. 2008, 26, 190–196.

19. X. Guo, Y. Niu, L. Huang, H. Ji and X. Zheng, Microstructure and tribological property of TiC–Mo composite coating prepared by vacuum plasma spraying, J. Therm. Spray Technol. 2012, 21, 1083–1090.

20. H. Dai, J. Li, F. Zhai, X. Cheng and Y. Wang, Effect of molybdenum on the microstructure and mechanical properties of TiC-Fe cermets, J. Adv. Mater. Res. 2012, 557-559, 205-208.

21. D. Mari, S. Bolognini, G. Feusier, T. Cutard, C. Verdon, T. Viatte and W, Benoit. TiMoCN based cermets Part I. Morphology and phase composition, Int. J. Refract. Met. Hard Mater. 2003, 21, 37–46.

22. J. Russias, S. Cardinal, Y. Aguni, G. Fantozzi, K. Bienvenu and J. Fontaine, Influence of titanium nitride addition on the microstructure and mechanical properties of TiC-based cermets, Int. J. Refract. Met. Hard Mater. 2005, 23, 358–362.

23. I. Hussainova, Effect of microstructure on the erosive wear of titanium carbide-based cermets, Wear. 2003, 255, 121–128.

24. M. Zhang, Q. Yang, W. Xiong, et al, Effect of Mo and C additions on magnetic properties of TiC-TiN-Ni cermets, J. Alloy Compd. 2015, 650, 700–704.

25. M. Liang, W. Wan, Z. Guo, J. Xiong, G. Dong, X. Zheng, et al, Erosion–corrosion behavior of Ti (C, N)-based cermets with different TiN contents, Int. J. Refract. Metal. Hard. Mater. 2014, 43, 322–328.

26. S. Park, S. Kang, Toughened ultra-fine (Ti, W)(CN)–Ni cermets, Scripta. Mater. 2005, 52, 129–133.

27. A. Rajabi, M. J. Ghazali and A. R. Daud, Chemical composition, microstructure and sintering temperature modifications on mechanical properties of TiC-based cermet – A review, Mater. Des. 2015, 67, 95–106.

Effect of Microwave Curing Process on Molding Properties of Carbon Fiber Composite

Nai-Shu Zhu*, Hua-Dong Gong, Xin-Li Kong, Jian-Liang Zhang, Jing-Yang Zhou

National Key Laboratory for Disaster Prevention & Mitigation of Explosion & Impact, PLA University of Science and Technology, China

Email: zns2000@163.com

A study on microwave curing process of carbon fiber composite prepared using epoxy adhesive with different parameters has been undertaken. The molding performance of carbon fiber composite materials could be improved by adjusting the processing parameters according to dynamic mechanical thermal analysis. When microwave power was 300W and curing time was 15 minutes, the highest stiffness of composite could be obtained, furthermore, its glassy state storage modulus E′ was up to 5000 MPa. However, when the composite was overheated, such as 400W, its stiffness was lost. And the glass transition temperature of composite processed at 200W was 10 ℃ lower than that of composite at 300W. It was indicated that appropriate power of microwave process was not only conducive to wetting carbon fiber with adhesive resin but also to preventing the material from being overheated. It has been shown that strong interface bonding between carbon fiber and resin was one of the important factors that could produce good processibility for composites.

Keywords: Carbon fiber; microwave; dynamic mechanical thermal properties; composite.

1. Introduction

Carbon fiber composites are widely used in industry, construction and other fields due to their high specific strength, high specific modulus and excellent design performance. In general, carbon fiber composites are formed by normal temperature or hot pressing process, during which there is a long curing period and requires high manufacturing cost. And microwave curing, as a new kind of composite curing technologies that enjoys the advantages of fast heating, even heating, selective heating and pollution-free heating [1] was expected to solve the problems in traditional thermal curing methods.

At present, domestic researches in this field have just started [2, 3], among which static tensile, flexural properties for characterization were used. In this paper, dynamic thermal mechanical analysis (DMA) and morphology analysis was used to study the effect of microwave curing process on the molding properties of the materials.

2. Test

2.1. *Materials*

Polyacrylonitrile carbon fiber plain weave papers were used (3k), whose radial and weft direction tensile rigidity are 80 N·mm⁻¹. Homemade rubber-toughened epoxy resin adhesive; modified amine curing agent; filler; and auxiliaries.

2.2. *Sample preparation method*

Surface pretreatments of carbon fiber cloth were presented in reference [4]. The epoxy resin was coated on carbon fiber cloth by hand scraping method, and three layers of epoxy resin were coated on the carbon fiber cloth, and were molded into thin-sheet samples under 200W/20 minutes, 300W/15 minutes, and 400W/10 minutes respectively.

---

*Corresponding author

2.3. *Testing and characterization*

The dynamic mechanical properties of thin-sheet composite samples were tested and analyzed by the American dynamic mechanical spectrum TA2980 which used single cantilever mode, 1Hz test frequency and 5°C/min of rising temperature rate. The morphology of the sample's interface constituted by carbon fiber and resin was observed by using S-4800 electronic scanning microscope.

3. Results and Discussion

3.1. *Effect of microwave curing process on the storage modulus of composite materials*

Stiffness and damping of carbon fiber composite materials can be measured by DMA at a certain temperature, frequency, stress or strain level and the interface binding capacity of composite materials could be reflected effectively [5]. From Fig. 1, it could be observed that when curing process for microwave power was 300W and curing time was 15 minutes, maximum stiffness of material could be achieved and its glassy state storage modulus E′could reach 5000 MPa. When the microwave curing power was 200W and the curing time was extended to 20 minutes, the stiffness of carbon fiber composite material would be greatly reduced, and the glassy state storage modulus was only 2580MPa. When the microwave curing power was 400 W and the curing time was shortened to 10 minutes, the carbon fiber composite material has lost its stiffness due to being overheated. The resin in the composite material was a kind of dielectric, which was produced by the polar functional group in the system under the action of alternating electric field. The dielectric loss was related to the temperature. For a fixed frequency, if the temperature was too low, the viscosity of the medium tended to become too large, the polarization process tended to become too slow, and the dielectric loss tended to become small. During the initial reaction stage, crosslinking reaction occurred within the system to produce rigid molecular chain network structure which embedded adjacent active functional groups to prevent them from spreading, thus hindering the further reaction between carbon fiber surface and resin, polarizing functional groups within resin itself and leading to the decline of material stiffness.

In addition, although microwave curing might occur within a few seconds or tens of seconds in theory, a certain amount of heat was needed to initiate reaction at the beginning of the curing since the curing of the epoxy resin was an exothermic reaction. Once the reaction was triggered, it would rely on the power released by the reaction to proceed. Meanwhile if microwave heating was further implemented, an overheated system temperature would appear, causing the worsening of the whole process and overheating of composite materials.

3.2. *Effect of microwave curing process on the glass transition temperature T_g of composite materials*

From Fig. 2, it could be observed that overheated carbon fiber composite materials didn't possess heat-resistance capacity under 400W/10 minutes. When the microwave power of the carbon fiber composites was low, its T_g was greatly reduced under 200W/20 minutes. Decreased T_g was caused by low microwave power, negative chain reaction molecular activity and few cross-linking points at the beginning of the reaction.

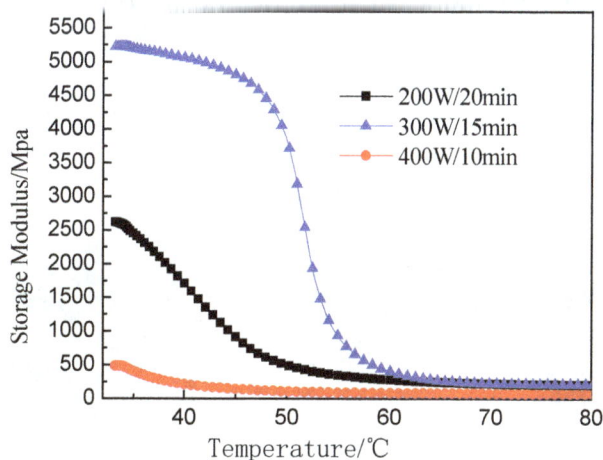

Fig. 1 Variation of storage modulus of composite materials with temperature in different microwave curing power

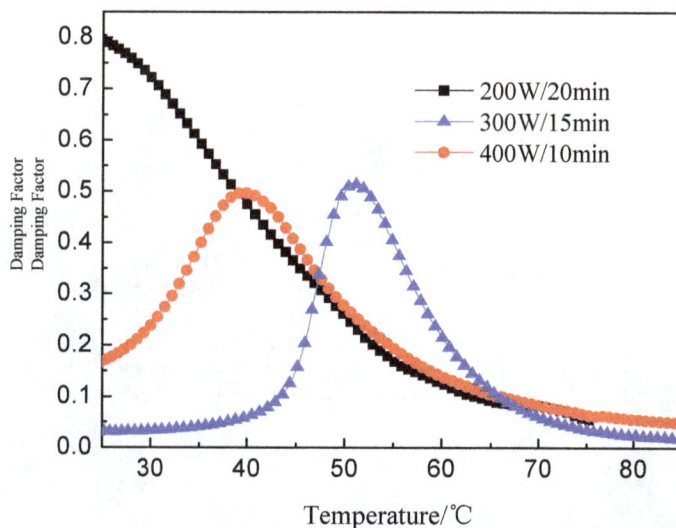

Fig. 2 Variation of damping factor of composite materials with temperature in different curing power

Cross-linking of resin has occurred in the adjacent region of the molecular chain of different degree of aggregation and caused the appearance of local rich epoxy clusters, due to the presence of free resin with epoxy terminated chain and these easy moving chain units resulted in incomplete cross-linking structures, leading to decreased curing degree and T_g reduced to 40°C.

3.3. *Effect of microwave curing process on the interfacial morphology of composites*

As shown in Fig. 3-1, holes would appear on the interfacial area between fiber and resin of overheated carbon fiber composite (400 W/10 min), which was because under high power microwave curing of composite materials the temperature of the system rose too quickly in a relatively short time before the gas generated in the reaction process could be discharged to the outside of the system. This has become a structural defect. In contrast, if microwave curing power was rather low (200W/20 minutes), as shown in Fig. 3-2, the infiltration between resin and fiber was not ideal, and cracks were expected to appear between the interfacial area of fiber and resin. As shown in Fig. 3-3, when the microwave power was 300W/15 minutes, it was conducive to the resin and carbon fiber infiltration, and wouldn't cause the overheating of material, molecular chain damage and bond failure, therefore the interface was firmly bonded.

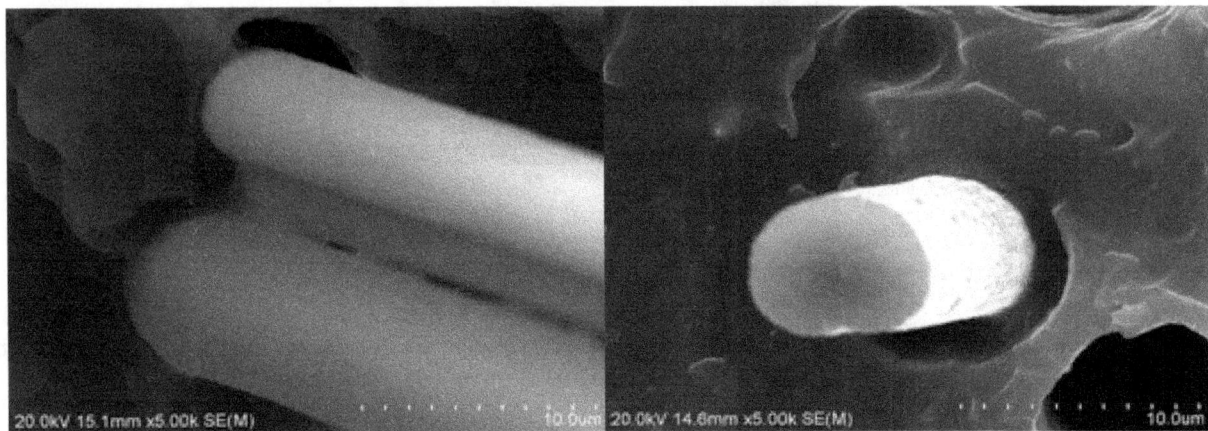

Fig. 3-1   Interfacial morphology of composite materials (400 W/10 min)

Fig. 3-2   Interfacial morphology of composite materials (200W/20min)

Fig. 3-3   Interfacial morphology of composite materials (300W/15minutes)

4. Conclusion

(1) The dynamic mechanical performance of carbon fiber composite material was good if microwave curing process was adopted under 300W/15minutes, and its glassy state storage modulus E' could reach 5000 MPa. And if power was up to 400W, the material could lose its stiffness. When the power was reduced to 200W, the glass temperature was lower than that of the 300W treatment by 10°C.

(2) Selected the appropriate microwave power, which was conducive to the adhesion and infiltration of resin and carbon fiber, and would not cause the overheating of material. Scanning electron microscopy showed that good interfacial bonding between carbon fiber and resin was an important cause of gaining carbon fiber composites with good performances.

Acknowledgment

This work was funded by the National nature science foundation of China (51505497).

References

1. Shining Ma, Battlefield emergency maintenance technology of equipment, National Defend Industry Press, Beijing, 2009.
2. Renli Ma, Xinlong Chang, Yingqiang Liao, Xiaojun Zhang, Mechanical Properties of NOL Rings of Carbon Fiber /Epoxy Composites Cured by Microwaves. Polym. Mater. Sci. Eng. 32, 3 (2016) 96-101. In Chinese.
3. Asif Shah, Yonghui Wang, Hao Huang,ed,Microwave absorption and flexural properties of Fe nanoparticle/carbon fiber/epoxy resin composite plates. Compos. Struct. 131 (2015) 1132–1141.
4. Naishu Zhu, Shining Ma, Xiaofeng Sun, Xi Chen, Effect of Plasma Treatment and Curing Condition on Dynamic Mechanical Properties of Carbon Fiber Composite. China Surf. Eng. 23, 5 (2010) 59-63. In Chinese.
5. Meili Guo, Dynamic mechanical thermal analysis of Polymer and Composites, Chemical Industry Press, Beijing, 2002.

Study on Mechanical Properties of Micro Injection Molding UV Curing Material

Wei-Jia Kang[1], Jian-Yun He[1], Zhen-Wen Liu, Xidan Luo, Peng-Feng Hua, Xue-Tao He[1,*]

School of Beijing University of Chemical Technology, Beijing, China100029
Yinglan Laboratory of Advanced Polymer Processing
*Email: 995216959@qq.com

A new method for the fabrication of polymer products with complex micro features is proposed in this laboratory, which is a new method of micro injection molding UV curing. Light curing composite prepolymer, monomer type and content has great influence on the mechanical properties of the cured product, such as tensile modulus, tensile strength and elongation at break, and also has influence on the physical properties of product, such as hardness .This study found that tensile strength, hardness of the light curing formula containing three functional monomers is higher than the light curing formula containing mono functional monomer, and the tensile strength, hardness increases with the increase of trifunctional monomer content. Elongation at break of light curing formula containing three functional monomers is lower than the light curing formula containing mono functional monomer, and the elongation at break decreases with the increase of trifunctional monomer content. The experimental results provide some reference value for the preparation of light cured products with different performance requirements.

Keywords: UV light curing; micro injection; molding; microfluidic chip; mechanical properties; hardness.

1. Introduction

Fine structure parts (micro parts) are highly demanded in national high-tech fields, playing an important role in the field of optical communication, image transmission, Bio Medical, information storage and aerospace [1], becoming one of the hotspots in the research of advanced manufacturing technology in the world. There are many kinds of processing methods of polymer micro products. The polymer micro injection molding has become a hot spot of research because of the efficiency of the micro injection molding and the diversity of the products [2]. Figure 1 is a typical polymer micro injection part. Using light (UV or visible) as energy, the process that the fast transition of liquid substance with chemically reactive activity into solid state is called light polymerization (photo curing) process [3-4]. In recent years, with the development of laser 3D printing and light curing molding technology, photochemical manufacturing technology is more and more applied to the field of material forming [5-6].

Because liquid photocurable resin is small molecular structure, only under the condition of light, can it be polymerized quickly, so it has a good fluidity and filling properties, and has broad prospects for development in the field of precision molding of fine products [7]. In view of the problems existing in the injection molding of polymer micro parts, a new method of micro injection molding is put forward in this paper. The light curing material is prepared with monomer, pre polymer and light initiator. There are many kinds of light curing oligomers, and their properties are different. Light cured micro injection part's mechanical properties are determined by oligomer structure. Monomer is one of the essential components in the light curing system. It not only can reduce the system viscosity, but also is involved in the curing process, affecting various properties of the cured film after curing [8]. According to the number of reactive groups in each molecule, the reactive diluent can be divided into single functional group, double functional group, and multifunctional reactive diluent. The functional degree affects the crosslinking structure of the cross-linked films, which affects the mechanical and physical properties of the cross-linked films [9]. The number of polar groups affects the cohesive energy of

---

*Corresponding author

the cured films, which affects the mechanical properties of the cross-linked films [10]. The spline die is made based on standard ISO527-1-1BA, the different proportion of the light curing material are stirred evenly and after being static, injected into the self-made spline die. Then close the die, light illumination using UVLED lamp light, remove the spline after complete curing. The results of this study will provide some reference value for improving the mechanical and physical properties of the products with fine structure, which are made by photochemical manufacturing.

2. Experimental

2.1. *Main experimental materials*

- **Prepolymer:** Polyurethane acrylate (PUA1), Heshan City, the li zhi men trade Co., ltd.;
- **Monomer:** three hydroxyl methyl propane three acrylic acid (TMPTA), Changshu Heng Rong Trading Co., ltd.;
- **Monomer:** ethyl acrylate (EOEOEA), Changshu Heng Rong Trading Co., ltd.;
- **Photo initiator:** 2- hydroxy -2- methyl -1- phenyl -1- acetone (1173), Changshu Heng Rong Trading Co., ltd..

2.2. *Experimental instruments*

- **Adjustable UV cold light source:** ULAMP- II, the intensity of light is 0~1000mw/cm2, Shenzhen City, You Lan Pu technology Co., ltd.;
- **Analytical balance:** measurement accuracy of 0.1mg, FR124CN, USA Ohaus Inc.;
- **Type A shore hardness tester:** Shanghai Lu Chuan Measuring Tool Co., ltd.;
- **Type D shore hardness tester:** Shanghai Lu Chuan Measuring Tool Co., ltd.;
- **Spline mold:** self- made, as shown in Fig. 2;
- Universal testing and stretching machine.

2.3. *Experimental process*

The prepolymer (PUA1) and monomer (EOEOEA), (TMPTA) respectively prepared according to Table 1 and Table 2.

Table 1 Formula 1-containing different percentages of TMPTA monomer concentration

Monomer TMPTA	Pre-polymer PUA1	Light curing initiator 1173
20%	80%	3%
30%	70%	3%
40%	60%	3%
50%	50%	3%

Table 2 Formula 2- different percentages of EOEOEA monomer concentration

Monomer EOEOEA	Pre-polymerPUA1	Light curing initiator 1173
20%	80%	3%
30%	70%	3%
40%	60%	3%
50%	50%	3%

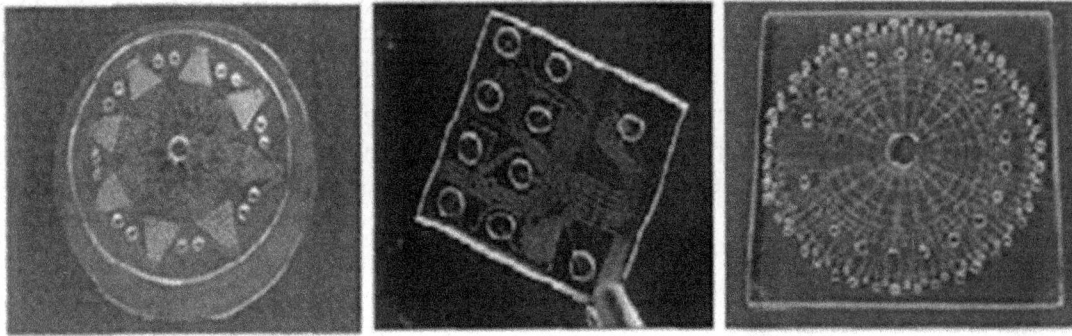

Fig. 1 Microfluidic

The chemical molecular structures of various experimental drugs used in the experiment are shown in Figs. 3, 4 and 5. The sample size using ISO527-1-1BA standard and its schematic diagram are shown in Fig. 6.

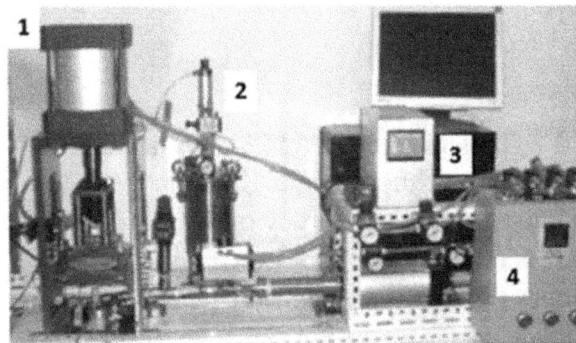

Fig. 2 Self-made mold (1-die; 2-, 3- feeding system, 4- control section)

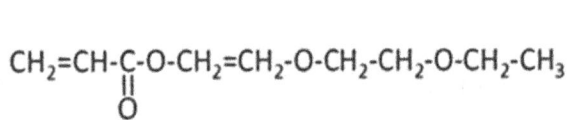

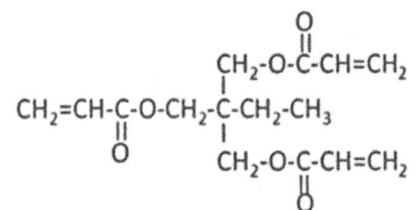

Fig. 3 Chemical molecular structure of monomer EOEOEA   Fig. 4 Chemical molecular structure of monomer TMPTA

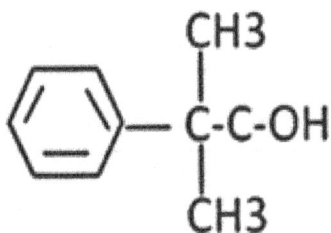

Fig. 5 Chemical molecular structure of light curing initiator 1173

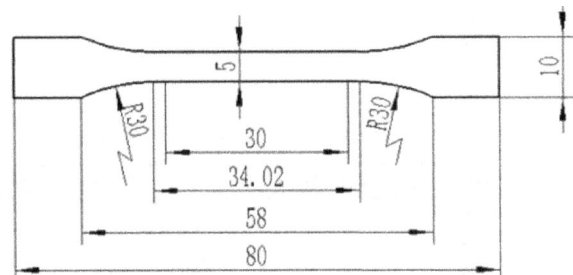

Fig. 6 Schematic diagram of spline (ISO527-1-1BAstandard)

2.4. Test conditions

Test speed 10 mm/min, based on the standard: V=50 mm/min, G20 mm, temperature of 24 degrees, humidity 30%.

2.5. Test results

The tensile strength, elastic modulus, elongation at break and hardness of the test in Table 1 and 2 are shown in Figures 7, 8, 9, 10 respectively.

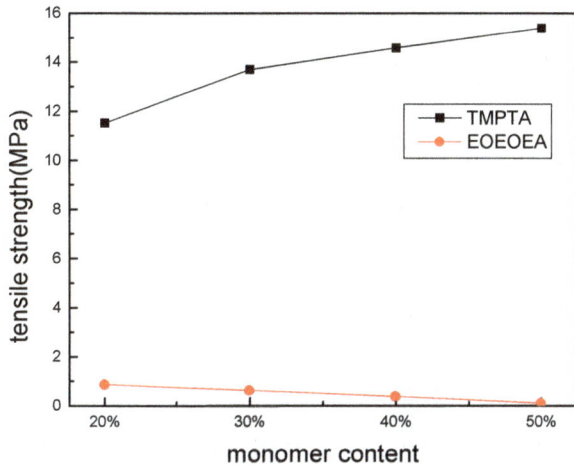

Fig. 7 Relationship between tensile strength and monomer content

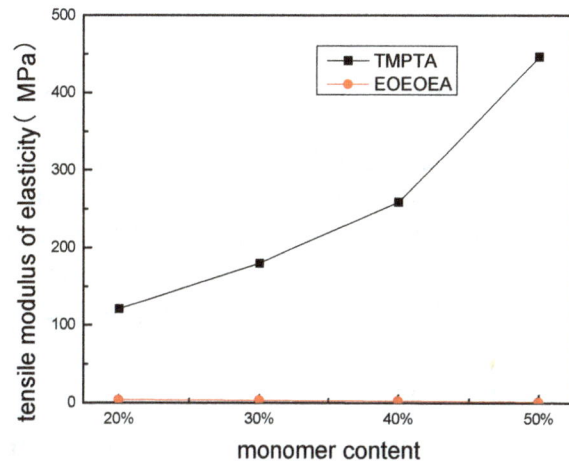

Fig. 8 Relationships between tensile elastic modulus and monomer content

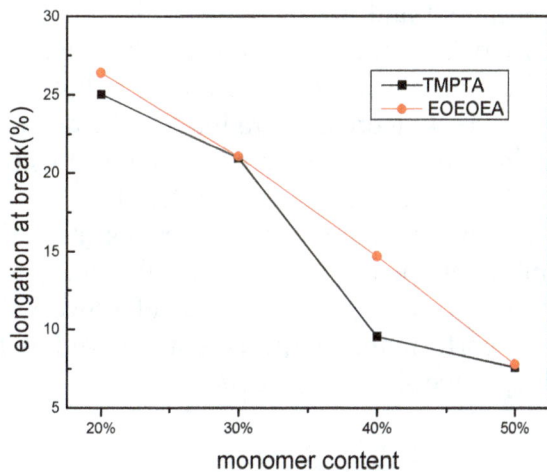

Fig. 9 Relationship between elongation at break and monomer content

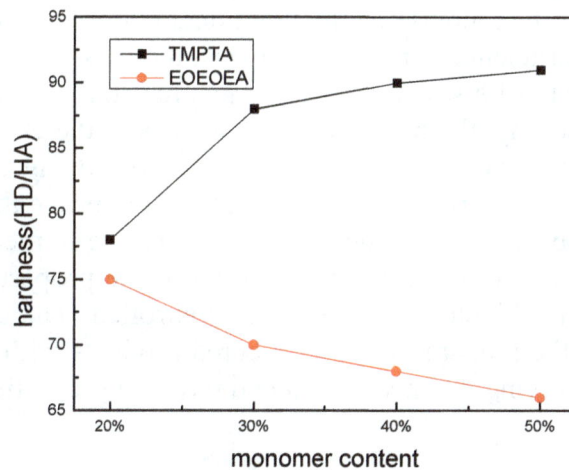

Fig. 10 Relationship between hardness and monomer content

The tensile strength, elastic modulus and hardness of the material are increased with the increase of TMPTA content and the elongation at break decreases with the increase of TMPTA content in 7, 8, 9, and 10. Because TMPTA is three functional groups reactive diluent by the molecular structure of TMPTA, and when the active diluent exists with functional degree of more than or equal to 2, cross-linked polymer network is obtained after curing, a high degree of

531

crosslinking network structure is obtained after PUA1 and TMPTA are cured. So with the increase of TMPTA, the cross-linking degree of the crosslinking curing membrane reticular structure increased and the polar groups were more, the mechanical properties of the cross-linked films were enhanced. However, with the increase of cross link density, the brittleness of cured films increased correspondingly, so the elongation at break decreased gradually.

The tensile strength, elastic modulus, hardness and elongation at break of the material decreased with the increase of EOEOEA content. Because EOEOEA is single functional active diluent from the molecular structure of the EOEOEA, it can only get linear polymer after mono functional active diluent curing and crosslinking does not occur. So with the increase of EOEOEA, the degree of crosslinking of the network structure of the cured film after curing with PUA1 and EOEOEA was decreased, and the polar groups were less, the molecular mass of EOEOEA is small. And with the increase of EOEOEA, the relative molecular mass of the cured film decreases, the interaction between the molecules decrease.

The tensile strength, elastic modulus and hardness of the cured films of formula 1 were higher than those of the formula 2 after curing, but the elongation at break of the cured film of formula 1 after curing was lower than that of the cured film of formula 2. Because the relative molecular mass of TMPTA is larger than that of EOEOEA, and TMPTA's polarity is larger, the crosslinking density of the formula 1 after curing was greater than that of the formula 2.

3. Conclusion

With the ever increasing demand for polymer micro products, the efficiency of the micro injection molding has become the focus of research and development, in particularly the UV light polymerization process. In this study, we put forward a proposal to fabricate polymer products using micro injection molding UV curing. From the results we had obtained, it was shown that with the increase of the functional degree of the active diluent and the increase of the amount of multifunctional reactive diluents, the crosslinking density of the cured materials can be improved. The tensile strength, elastic modulus and hardness of the light curing material increase, but at the same time the brittleness also increases, the elongation at break decreases gradually. The decrease of the functional degree of the active diluent and the increase of the amount of single function diluent can reduce the brittleness and improve the flexibility of the light curing material.

So as to obtain the light curing coating to meet performance requirement, the appropriate active diluents were selected according to the property requirements of the coating, and the amount of active diluent was reasonably controlled. The experimental results provide some reference value for the preparation of light cured products with different performance requirements as well as for improving the mechanical and physical properties of the products with fine structure.

Acknowledgment

The work was funded by the National Nature Fund: 51573017.

References

1. Jiang Bingyan, Xie Lei. The development status and prospect of the injection molding machine [J]. Plastic industry, No. 2, pp. 31-34, 2008.

2. Wang Xiaohua, Yang Weimin, Xie Pengcheng. Eighteenth International Plastics and rubber exhibition special report (K2010) - injection molding new technology. China plastic, Vol. 25, No. 1, pp. 110-115, 2011.

3. Nie J, E Andrzejewska, J F Rabek, et al. Effect of peroxides and hydroperoxides on the camphorquinone-initiated photopolymerization [J]. Polymer Chemistry Physics, 1999, 200: 1692-1701.

4. Muh K, J Marquardt, J E Klee, et al. Bismethacrylate-based hybrid Michael-Addition reactions [J]. Macromolecules, 2001, 34: 5778-5785.

5. Wang Wenjun, Zou Yingquan. Progress in the application of light curing coatings [J]. Information recording materials, 2004.05 (1).

6. Liu Yaxiong, Wang Qian, Jiankang He, et al. Study on the precision of the rapid casting process of customized titanium alloy prosthesis [J]. Journal of mechanical engineering, 2014, 50 (6): 75-80.

7. HSU C. Method for manufacturing optical film involves rotated cylindrical roller to press UV curable material when UV curable material is cured: USA US2013032959-A1 [P], 2012-10-12.

8. Xie Pengcheng, Changle, Jiao Zhiwei, et al. A light curing injection molding: China, 201320223941.2[P], 2013-07-17.

9. Wei Jie, Kim Yangzhi. UV curing coatings [M]. Beijing: Chemical Industry Press. 2013. 71-77.

10. Lu Yang Bin, Bao Qing Wang, Zhang Yan, Liu Ren, Zhang Sheng Wen, Liu Xiao Ya UV curing polycarbonate type polyurethane acrylate synthesis and performance [J]. Chemical new materials, 2012, 40 (1): 72-76. (11) 37-39.

Chapter 10
Fatigue, Crack and Creep Resistance, Corrosion and Fracture Mechanics

Corrosion Resistance of 7075 Aluminum Alloy Surface by Dry Turning and MQL Machining

Hong-Jie Pei[*], Lin-Feng Chen, Shao-Feng Chen, Gui-Cheng Wang

Institute of Precision Engineering, Jiangsu University, Zhenjiang, P.R.C
[*]Email: hjpei@ujs.edu.cn

7075 high strength aluminum alloy is easily to be corroded in atmosphere. During the mechanical manufacturing process, 7075 aluminum alloy component surface was obtained through dry and MQL machining method. In this paper, with the same cutting parameter (400m/min cutting speed, 0.02 mm/rev feed rate and 0.06 mm depth of cut), both dry and MQL machining experiments were carried out to turn Φ30mm 7075 aluminum alloy rod. After that, salt spray tests were conducted for three cycles, each period was 72 hours. Then, the corrosion state of surface morphology was viewed, and meanwhile the corrosion area, corrosion pits quantities, the degree of corrosion damage, the average depth and average diameter of corrosion pits were counted. The study results show that the corrosion of aluminum surface with MQL machining is more serious than with dry machining.

Keywords: 7075 aluminum alloy; MQL; dry cutting; corrosion.

1. Introduction

7075 aluminum alloy is an Al–Zn–Mg–Cu–Zr alloy developed to obtain a good combination of high strength, high resistance to stress corrosion cracking and good fracture toughness, particularly in thick sections [1]. In order to reveal the corrosion mechanism of this material, many scholars have finished a lot of research works and achieved some good performances. Zhou et al. analyzed the corrosion behavior of aluminum alloy 7075-T6 dependent of the thin electrolyte layers in 1 M sodium sulfate solution using cathodic polarization, electrochemical impedance spectroscopy(EIS), scanning electron microscopy (SEM) and X-ray photoelectron spectroscopy (XPS) [2]. Reda et al. investigated the corrosion behavior of three-point-loaded 7075 Aluminum alloy specimens after 7 days of immersion in 3.5% NaCl solution, using modified electrochemical cell [3]. Sabelkin et al. observed the transition of corrosion pit to crack in 7075-T6 under ambient laboratory and saltwater environments [4]. Meng et al. investigated the corrosion fatigue crack growth behaviors of 7075 alloy in seawater [5]. Yue et al. applied Nd-YAG laser surface treatment under two different gas environments, air and nitrogen, on 7075-T651 aluminum alloy with the aim of improving the stress corrosion cracking resistance of the alloy [6].

The aluminum alloy surface used in previous study was not obtained through the mechanical machining. So the corrosion state of machined component would be different from the actual conclusion. At present, the main machining methods of aluminum alloy are dry cutting and Minimum Quantity Lubricant (MQL) machining [7]. Compared with dry cutting, MQL technology could get the better surface quality and meanwhile make the tool service life longer, so it has more and broader application prospects. Therefore, the research on the corrosion of aluminum alloy surface under both cutting conditions.

2. Experimental Set-up and Scheme

2.1. *Experimental set-up*

The cutting experiments were performed with SB-CNC ultra-precision machining center. The lubricating device was MQL spray system produced by Accu-Lube Company. The salt spray corrosion tests were performed on the F-60C salt spray corrosion test chamber. The measurement

---

[*]Corresponding author

instrument was OlympusDSX500 optical digital microscope produced. The cutting experiments adopted SandvikCCGX 09 T308-AL H10 cutting blade produced. The relief angle of cutting tool was 7°.Workpiecerodof 7075 high strength aluminum alloy was with 30mm in diameter.

2.2. Experimental scheme

The single factor design method was used in the turning experiment. Experimental parameters were cutting speed v, feed rate f and depth of cut. These three parameters remained constant, and the values were 400 m/min, 0.02 mm/rev and 0.06 mm respectively. In addition, two different machining methods (Dry and MQL) were applied in the turning experiments. For the MQL turning, the cutting fluid flow rate and air flow rate were both set to fixed values based on the literature [8], the position of nozzle was selected along the major flank surface and the stand-off distance was 20mm.

The pH value of salt water was measured by pH paper with four colors. The sodium hydroxide solution and hydrochloric acid solution was used to adjust the spray pH value. According to the standard of GB/T10125-1997, the temperature of salt spray corrosion test chamber was set to 35±2°C, the settling velocity of salt spray was set to 2mL/80cm²/h and the concentration of sodium chloride was set to 50±5g/L. The solution should be heated properly to remove the carbon dioxide which had a certain impact on the pH value of the solution. During the experimental process, the test chamber should not be opened except for some special cases.

The parameters of salt spray test chamber were set according to the standard of GB/T10125-1997.The salt-spray corrosion experiments were carried out in the chamber environment of wet and dry cycles. The total corrosion time was divided into three periods, and each period was about 72 hours (spraying for 48 hours, drying for 24 hours).

2.3. Evaluation method of surface corrosion damage

The appearance and distribution of the corrosion area were the important effect factors of corrosion damage. DuQuesnay considered the maximum depth of the corrosion pit as an evaluation criterion of corrosion damage [9]. However, the maximum depth of the corrosion pit was not easy to be obtained due to the measurement limitation, so the average depth of corrosion damage D and the degree of corrosion damage DOP were used as the parameters to evaluate the corrosion damage in this study. The average depth of corrosion damage was the average value of the measured depth of corrosion pits by the 3D shape measurement method. The formula is as follows:

$$D = \frac{1}{n}\sum_{i=1}^{n}Di \tag{1}$$

Where D is the average depth of corrosion pits, Di is the depth of each pit, and n is the number of pit. The calculated average depth D was simply used to evaluate the degree of damage along the depth direction.

According to the definition in the literature [10], the degree of corrosion damage DOP is the ratio between the total area of all corrosion pits and the original surface area. The formula is as follows:

$$DOP = \frac{1}{S}\sum_{i=1}^{n}Si \tag{2}$$

Where *DOP* is the degree of corrosion damage, *S* is the original surface area, S_i is the surface area of each pit, and *n* is the number of pits. According to the definition in the literature [11], a round or oval shaped pit represented the real area of the pit S_i. When the pit was close to a circle, S_i can be expressed as formula (3); when was close to an ellipse, S_i can be expressed as formula (4).

$$S_i = \pi d_i^2 / 4 \tag{3}$$

$$S_i = \pi d_{i1} d_{i2} / 4 \tag{4}$$

Where d_i is the maximum diameter of pit, d_{i1} is the long axis length of the pit and d_{i2} is the short axis length of the pit. This paper applied the threshold segmentation method to measure the *DOP* value.

2.4. *Corrosion surface pretreatment*

In order to measure the area and depth of corrosion pits after the end of each cycle expediently, the corrosion products on the workpiece surface should be treated. Accordance with the national standard GB/T16545-1996, chemical and electrolytic cleaning method was applied to remove the corrosion products. The workpiece was placed in the concentrated nitric acid (HNO_3, $\rho = 1.42g/Mol$, mass fraction was 71%) and soaked for five minutes; after that, a large amount of water was used to remove the nitric acid, and then the acetone was applied to clean the surface.

The nitric acid could not only remove the corrosion products but also protect the surface from being damaged due to the passivation effect. So, this method could effectively remove the newly formed corrosion products, which was appropriate for the subsequent analysis.

3. Experimental Results

3.1. *Corrosion surface morphology analysis*

Fig. 1 shows the corrosion morphology of workpiece surface in different cycles under the dry cutting and MQL cutting conditions. After 72 hours, the substrate surface lost the original metallic luster, and a small amount of round and black spots appeared on the entire surface area. This corrosion phenomenon was defined as pitting corrosion. Because aluminium alloy is easily passivated metal, the pitting corrosion occurs in the corrosive medium containing chloride or chloride ions [12]. With the increase of corrosion time, the depth of pits became deeper and deeper, the quantity of pits also increased, and the area of corrosion pits was also constantly expanding. After 144 hours, it was found that the depth of pits became deeper, the width of pits became larger, the number of pits increased significantly and distribution of pits was relatively uniform. After 216 hours, spalling phenomenon was observed, especially on the MQL cutting condition.

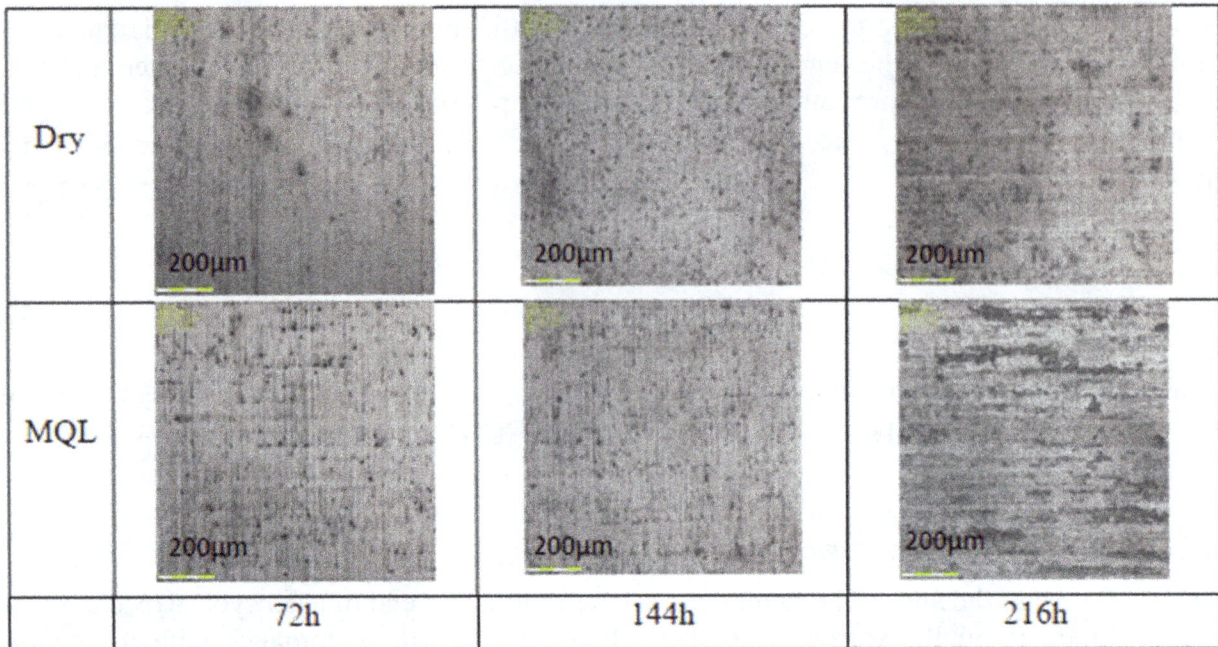

Fig. 1 The corrosion morphology photos

The corrosion morphology photos are shown as Fig. 1 and the statistics results of corrosion morphology parameters were counted with the OLYMPUS image analysis software. As the corrosion time increased, the surface area of corrosion pits, the quantity of pits, the average depth of pits were all increased. Under the dry cutting condition, when comparing with the previous one, the corrosion damage increases by 43.36% and 42.8% respectively, and on the MQL cutting condition, the corrosion damage increased by 70.64% and 20% respectively. 72 hours later, the corrosion area under the MQL cutting condition is 3.5% smaller than that under the dry cutting condition, and the number of corrosion pits was also 24.6% less than the number under the dry cutting condition. After 144 hours or 216 hours, both corrosion area and degree of corrosion damage under MQL condition were larger than that under the dry cutting condition; but the number of corrosion pits under MQL condition was less than that of dry cutting. The number of corrosion pits with larger surface area and diameter were more and spalling phenomenon was found under the MQL condition.

Fig. 2 and Fig. 3 respectively shows the ratio of all kinds of corrosion pits with different area to the total corrosion pit area and the ratio of the number of these pits to the total number under the MQL cutting and dry cutting condition after 144 hours. From the Fig. 2 and Fig. 3, the quantity of corrosion pits was decreasing with the increase of the pit area and the variation trend was as the reciprocal curve. However, the area distribution of corrosion pit was as basin type curve, which indicated that the pitting corrosion mainly contained a lot of small corrosion pits and few of large corrosion pits. Under the dry cutting condition, less than $20\mu m^2$ pit area accounted for 28.67% and its number accounted for 62.96%. Underthe MQL cutting condition, less than $20\mu m^2$ pit area accounted for 20.50% and its number accounted for 57.25%. Besides, for the range of pit area more than $20\mu m^2$, the percentage of them was higher, especially for the pits area more than $100\mu m^2$, the percentage of pit number and pit area were both increasing remarkably.

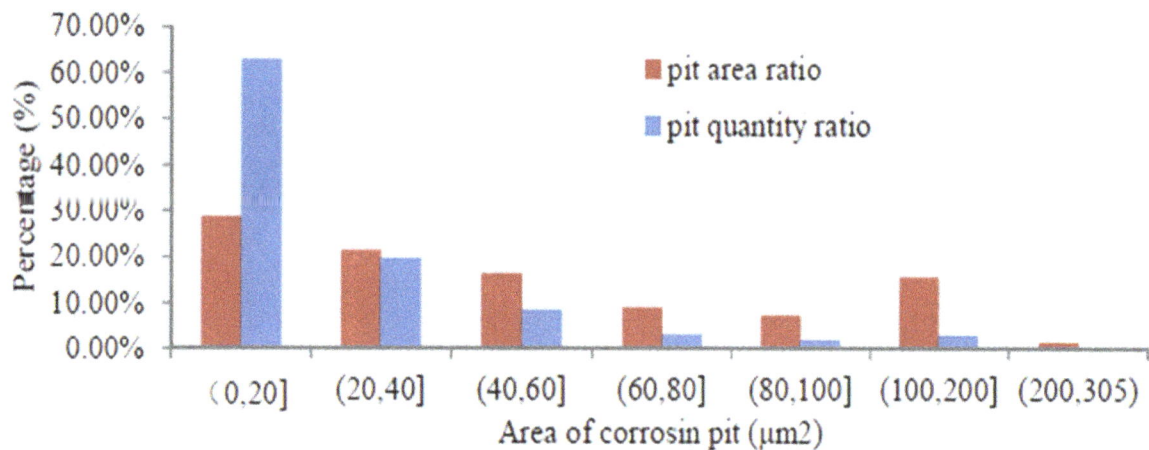

Fig. 2 (Dry，v=400m/min, f=0.02mm/rev, a_p=0.06mm，Corrosion time 144 hours)

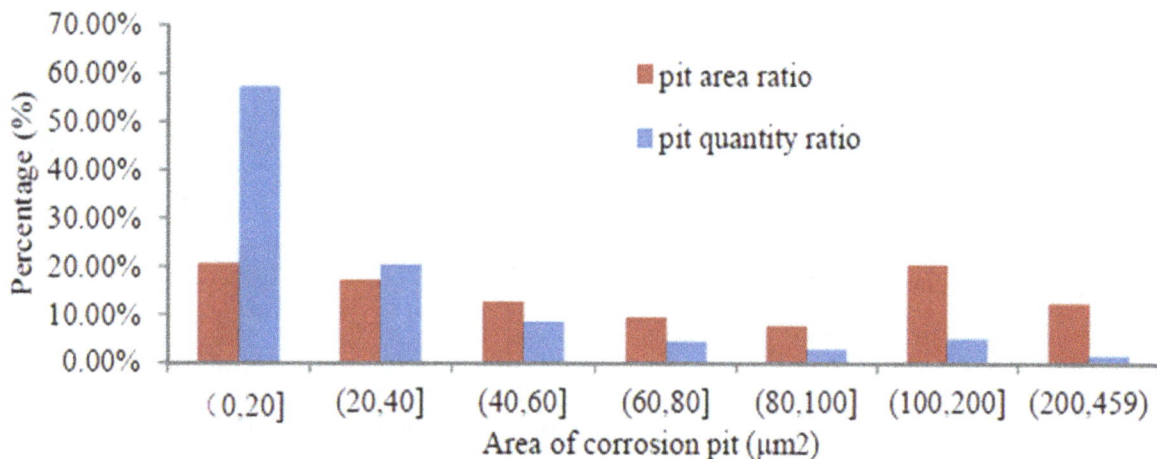

Fig. 3 (MQL，v=400m/min, f=0.02mm/rev, a_p=0.06mm，Corrosion time 144 hours)

3.2. *The degree of corrosion damage*

Fig. 4 shows that the degree of corrosion damage as a whole increased with increase of corrosion time for both drying and MQL machining. Besides, the corrosion area on the workpiece surface was smaller within the first two corrosion periods, but as the corrosion time increases, the corrosion degree was more serious due to the effect of passive film. When the corrosion time increased, the passive film was damaged and the corrosion damage was aggravated. Finally, the degree of corrosion damage of workpiece machined under the dry cutting condition was lower than that under the MQL condition in the same period. Especially in the second cycle, the above-mentioned phenomenon was more obvious, which revealed that the cutting fluid, to a certain extent, promoted the corrosion.

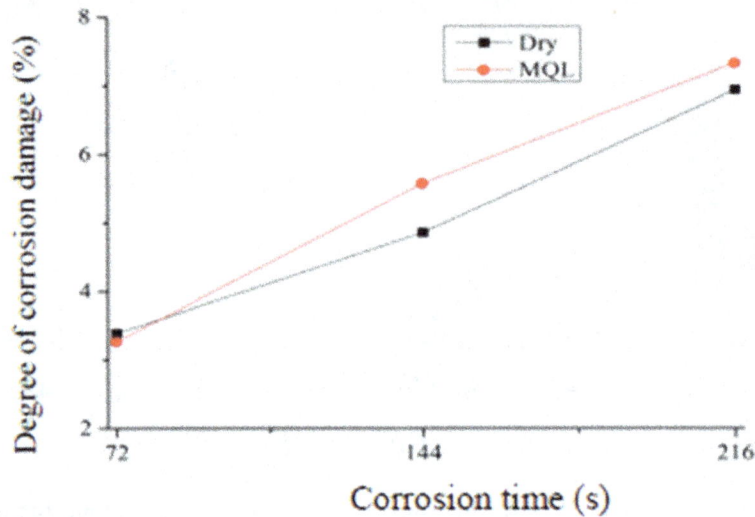

Fig. 4 Variation of the degree of corrosion damage with corrosion time

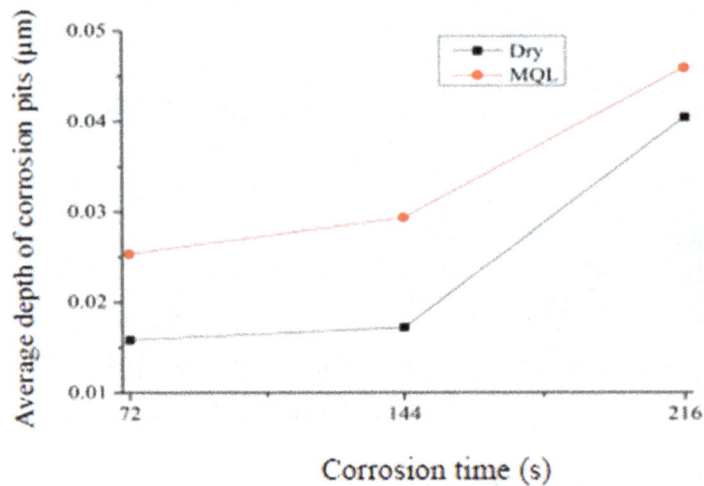

Fig. 5 Variation of the average depth of corrosion pits with corrosion time

3.3. *The average depth of corrosion damage*

Fig. 5 shows that the variation of average depth of corrosion pits with corrosion time for both drying and MQL machining. It was found that the average depth of corrosion damage gradually increased with increase of corrosion time. In addition, the average depth value under the dry cutting was less than that under the MQL machining condition for the same corrosion cycle, which revealed that the surface corrosion was even worse under the MQL condition.

3.4. *The average diameter of corrosion pit*

As shown in Fig. 6, the average diameter of corrosion pits became larger with the increasing corrosion time, which showed that the corrosion phenomenon was severer at the longer corrosion period. Meanwhile, it was also found that the average diameter of corrosion pits under the MQL machining condition was larger than that on the dry cutting condition.

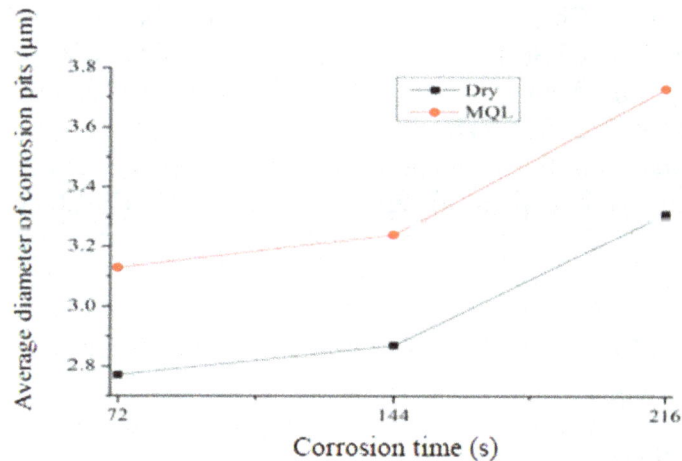

Fig. 6 Variation of the average diameter of corrosion pits with corrosion time

4. Conclusion

1. Pitting corrosion was the main corrosive form in the early and middle term of the salt spray test. The lamellar spalling phenomenon was found in later stage for the 7075 aluminum alloy surface machined on the dry and MQL cutting condition.
2. Compared with the corrosion damage in dry cutting, the corrosion damage in MQL cutting was more serious.
3. The quantity, superficial area, average depth and average diameter of corrosion pit all increased with increase of corrosion time.

Acknowledgments

This study was funded by the National Science, and Technology Major Project of China: (No.2013ZX04009031), Science and Technology Innovation Fund of Jiangsu Province (No.BC2014202), and Graduate Student Research and Innovation Plan of Jiangsu Province (No.CXZZ12-0658)

References

1. Yu-Hua ZHANG, Shu-Cai YANG, Hong-Zhi JI. Microstructure evolution in cooling process of Al−Zn−Mg−Cu alloy and kinetics description [J]. Transactions of Nonferrous Metals Society of China, 2012, 22(9): 2087−2091.
2. H.R. Zhou, X.G. Li, J. Ma, C.F. Dong, Y.Z. Huang. Dependence of the corrosion behavior of aluminum alloy 7075 on the thin electrolyte layers. Materials Science and Engineering B, 162 (2009) 1–8.
3. Y. Reda, R. Abdel-Karim, I. Elmahallawi. Improvements in mechanical and stress corrosion cracking properties in Al-alloy 7075 via retrogression and reaging. Materials Science and Engineering A 485 (2008) 468–475.
4. V. Sabelkin, V.Y. Perel, H.E. Misak et al. Investigation into crack initiation from corrosion pit in 7075-T6 on ambient laboratory and saltwater environments. Engineering Fracture Mechanics 134 (2015) 111–123.
5. Xiangqi MENG, Zhuoying LIN, Feifei WANG. Investigation on corrosion fatigue crack growth rate in 7075 aluminum alloy. Materials and Design 51 (2013) 683–687.

6. T.M. Yue, L.J. Yan, C.P. Chan. Stress corrosion cracking behavior of Nd-YAG laser-treated aluminum alloy 7075. Applied Surface Science 252 (2006) 5026–5034.

7. Weinert K, Inasaki I, Sutherland J W, et al. Dry Machining and Minimum Quantity Lubrication [J]. Annals CIRP, 2004, 53(2):511-537.

8. Wenjie ZHENG. Study on the characteristics of flow field during MQL turning and its application fundament [D]. Zhenjiang: Jiangsu University, 2011 (In Chinese).

9. Duquesnay D L, Onhill P R, Britt H J. Fatigue crack growth from corrosion damage in 7075-T6511 aluminium alloy on aircraft loading [J]. International Journal of Fatigue, 2003, 25(3): 371-377.

10. Paik J K, Jae M L, Ko M J. ultimate shear strength of plate elements with pit corrosion wastage [J]. Thin-Walled Structures, 2004, 42: 1161-1176.

11. Ruren ZHANG. 3D surface topography and its corrosion-resistant research of milling aeronautic aluminum alloy [D]. Jinan: Shandong University, 2012 (In Chinese).

12. Turnbull A, McCartney L.N,Zhou S. Modeling of the evolution of stress corrosion cracks from corrosion pits, Scripta Materialia, 2006, 54(4): 575-578.

Experimental Study on Fatigue Strength of Super-high Strength Sucker Rod

Wen-Bin Cai

Petroleum Engineering Academy, XI'an shiyou University, XI'an, Shanxi, 710065, China

Email: peter_acai@sina.com,

The super-high strength sucker rod is widely used in Chinese oil companies. The fatigue limit is one of most important features of super-high sucker rod. The fatigue strength experiment and reliability analysis were carried out according to petroleum industry standard. The alternating stress loading cycles before sucker rod fatigue failure were obtained under three different load levels, and then the experiment data were counted by probability statistic. The P-S-N formula and fatigue limit were calculated based on the experiment data, which applies theoretical basic to sucker rod design for oil lifting.

Keywords: Super-high strength sucker rod; fatigue strength; reliability; probability statistics; fatigue limit.

1. Introduction

The super-high strength sucker rod is one kind of sucker rod with higher tensile strength and fatigue strength, which is suitable to excessive load, no corrosion or micro corrosion oil wells. There are many super-high sucker rod manufactures in China. Super-high sucker rod is used widely in most of Chinese oil companies [1, 2] but the deep research on super-high sucker rod fatigue features is not enough. Empirical value of fatigue limit which has been used since decades ago is still used for the sucker rod design [3, 4] so it is necessary to do research on super-high sucker rod fatigue characters.

2. Experimental Principle and Experimental Procedure

During the experiment the PLG-300C electromagnetic resonance high frequency fatigue testing machine was used to supply load to the sucker rod, the loading frequency is 150Hz, and there are three different stresses are used (500MPa, 540MPa and 600MPa), sucker rod suffer alternating stress during the experiment, the alternating stress is sinusoidal waveform as is showed in Fig. 1, σ_a is stress amplitude. The fatigue limit was record for each sucker rod and each stress level.

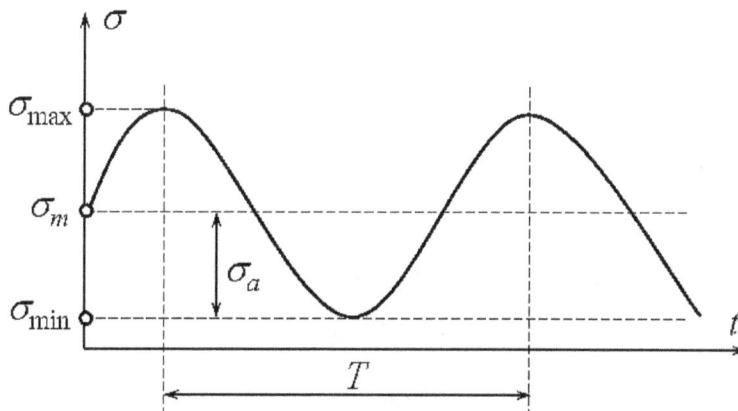

Fig. 1 Alternationg stress supplied by testing machine

The super-high sucker rod samples are from 5 manufactures, according to SY/T 5029-2013, there are two different sucker rod diameters which are 19mm and 22mm, the length of sucker rod is

500±50mm, the shape of sucker rod sample is showned in Fig. 2. the sucker rod samples is shorter than real sucker rod, but the mechanical properties are the same as real sucker rod.

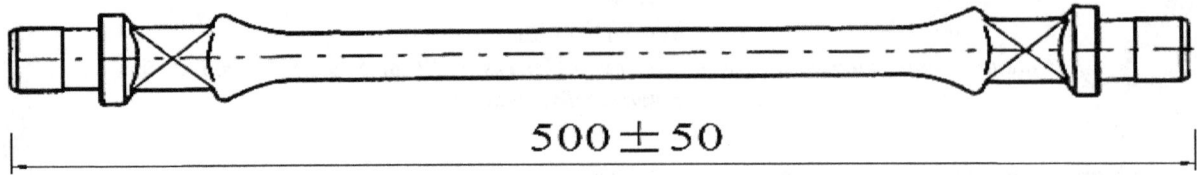

500±50

Fig. 2 The shape of super-high strength sucker rod

3. Experimental Results

The experimental results are showed in Table 1, there are three groups of sucker rod samples, each group has 5 sucker rod samples and a different maximum stress, and for example, group 1 have 5 sucker rod samples number from 1 to 5, its maximum stress is 500MPa. Table 1 shows the alternating stress loading cycles before sucker rod fatigue failure and the failure position.

Table 1 Fatigue failure experimental results of super-high strength sucker rod

No.	diameter (mm)	Max. stress (MPa)	Stress ratio	Stress loading cycles	Failure position
1	22.23	500	0.1	465674	stress groove
2	22.32	500	0.1	492523	wrench square
3	22.38	500	0.1	482672	wrench square
4	22.37	500	0.1	556596	wrench square
5	22.32	500	0.1	516241	stress groove
6	19.19	540	0.1	439481	thread
7	19.18	540	0.1	406807	thread
8	19.11	540	0.1	427751	thread
9	19.20	540	0.1	415322	thread
10	19.09	540	0.1	362733	thread
11	18.95	600	0.1	250950	thread
12	19.18	600	0.1	334635	wrench square
13	18.85	600	0.1	287449	thread
14	19.02	600	0.1	241548	wrench square
15	19.08	600	0.1	301519	thread

4. Experimental Data Normal Distribution Test

The Kolmogorov-Smirnov test method was used to test experimental data whether follow the normal distribution or not. Suppose data follow normal distribution, we could get $F_0(x_i)$ value for each data. Compare $F_0(x_i)$ to empirical distribution function $F_n(x_i)$, the test statistic D_n could be calculated by Eq. 1. Compare D_n to critical value $D_{n,a}$ (the value of $D_{n,a}$ could get from Table 2).

Table 2 Kolmogorov-Smirnov test value of $D_{n,a}$

α \ n	0.40	0.20	0.10	0.05	0.04	0.01
5	0.369	0.447	0.509	0.562	0.580	0.667
10	0.268	0.322	0.368	0.409	0.422	0.487
20	0.192	0.232	0.264	0.294	0.304	0.352
>20	$0.87/\sqrt{n}$	$1.07/\sqrt{n}$	$1.22/\sqrt{n}$	$1.36/\sqrt{n}$	$1.37/\sqrt{n}$	$1.63/\sqrt{n}$

If $D_n \leq D_{n,a}$, then the experimental data follow normal distribution, otherwise the experimental data don't follow normal distribution.

$$D_n = \sup_{-\infty < x < \infty} |F_n(x) - F_0(x)| = \max\{d_i\} \quad (i=1,2,3,4..., n) \tag{1}$$

$$d_i = \max\left\{F_0(x) - \frac{l-1}{n}, \frac{l}{n} - F_0(x_i)\right\} \quad (i=1,2,3,4..., n) \tag{2}$$

Table 3 shows the value of $F_n(x_i), F_0(x_i), d_i$ and fatigue limit at the Max. stress is 500MPa, D_n could be find by Eq. 1, its value is 0.3939, from Table 2, $D_{n,a}$ value is 0.562 when a=0.05, n=5, because $D_n \leq D_{n,a}$, the experimental data follow normal distribution. We found that each group of test data with different maximum stress follows normal distribution.

Table 3 Calculation value when the maximum stress is 500MPa

No.	stress loading cycles	$F_0(x_i)$	$F_n(x_i)$	d_i
1	465674	0.0934	0	0.1066
2	492523	0.2148	0.2	0.1856
3	582672	0.7939	0.4	0.3939
4	556596	0.6551	0.6	0.1449
5	586241	0.8133	0.8	0.1867

5. P-S-N Formula Set Up

P-S-N curve is the relationship between stress level S and fatigue limit N of the certain failure probability P. The straight line style of sucker rod P-S-N formula is showed as Eq. (3).

$$\lg N_P = A_p + B_p \lg S \tag{3}$$

A_p, B_p could be found by Eq. (4):

$$\begin{cases} A_p = \dfrac{1}{n}\sum_{i=1}^{n} \lg N_{pi} - \dfrac{B_p}{n}\sum_{i=1}^{n} \lg S_i \\[3mm] B_p = \dfrac{\displaystyle\sum_{i=1}^{n} \lg S_i \lg N_{pi} - \dfrac{1}{n}(\sum_{i=1}^{n} \lg S_i)(\sum_{i=1}^{n} \lg N_{pi})}{\displaystyle(\sum_{i=1}^{n} \lg S_i)^2 - \dfrac{1}{n}(\sum_{i=1}^{n} \lg S_i)^2} \end{cases} \tag{4}$$

The cumulative probability density distribution function $F(x, \mu, \sigma)$ of a certain reliability :

$$F(x, \mu, \sigma) = \frac{1}{x\sigma\sqrt{2\pi}} e^{-(\ln x - \mu)^2 / 2\sigma^2} \tag{5}$$

If $Z = \dfrac{\ln x - \mu_y}{\sigma_y}$, then we get Eq. (6):

$$F(Z) = \frac{1}{\sqrt{2\pi}} \int_{-\infty}^{\ln x - \mu_y} e^{\frac{1}{2}z^2} \, dz = \phi(\frac{\ln x - \mu_y}{\sigma_y})$$

(6)

Eq. (7) shows the relationship between failure probability P and $F(Z)$:

$$P = 1 - F(Z)$$

(7)

From Eq. (6) and Eq. (7) we get:

$$1 - \phi(\frac{\ln x - \mu_y}{\sigma_y}) = \phi(\frac{\mu_y - \ln x}{\sigma_y}) = P$$

(8)

The average of fatigue life μ_y, variance σ_y^2 and standard deviation σ_y could be found by standard normal distribution table for a certain value of failure probability P. Table 4 shows the canculated value of average value, variance and standard deviation of logarithmic fatigue life for each stress level. Table 5 shows the value of logarithmic fatigue life of different reliability.

Table 4 Average value, variance and standard deviation of logarithmic fatigue life

Max. Stress (MPa)	μ_y	σ_y^2	σ_y
500	5.6259	0.00097	0.0311
540	5.4988	0.00124	0.0352
600	5.2906	0.00202	0.04492

The value of logarithmic fatigue life for different reliability is showed in Table 5.

Table 5 The value of logarithmic fatigue life (lgN) of different reliability

lgN / P	500MPa	540MPa	600MPa
99%	5.5547	5.4183	5.1877
90%	5.5908	5.4592	5.2399
70%	5.6095	5.4801	5.2668
50%	5.62595	5.4988	5.2906

Table 6 shows the P-S-N formula for different reliability:

Table 6 P-S-N formula of different reliability

P	S-N formula
99%	lgN=18.1334-4.6579lgS
90%	lgN=17.613-4.452lgS
70%	lgN=17.3473-4.3468lgS
50%	lgN=17.1108-4.2533gS

6. Fatigue Limit Calculation

The fatigue limit is the maximum stress when stress loading cycles is more than 10^7 before sucker rod failure in the fatifue strength experiment. That is when the stress level decrease to a certain value σ_r, the fatigue limit N reaches infinite value, and the S-N curve tend to be a horizon, we defined σ_r as fatigue limit.

According to P-S-N formula, from Eq. (3) we get formula (9):

$$N = 10^{(A_p + B_p \lg S)} \tag{9}$$

If $N=10^7$, fatigue limit σ_r is:

$$\sigma_r = S = 10^{\frac{7-A_p}{B_p}} \tag{10}$$

Converting allowable stress $[\sigma_{-1}]$ could be canculated by Eq. (11):

$$[\sigma_{-1}] = (\sigma_r - \sigma_m)/(1 - \frac{\sigma_m}{\sigma_b}) \tag{11}$$

Where σ_r is the fatigue limit of stress ratio r, MPa; σ_m is average value of alterating stress, MPa; σ_b is the tensile strength of sucker rod, MPa.

Table 7 shows the canculated value of fatigue limit of different reliability P when the stress ratio is 0.1.

Table 7 Fatigue limit of different reliability

P	99%	90%	70%	50%
σ_r (MPa)	245.5	242.0	240.1	238.3
$[\sigma_{-1}]$ (MPa)	119.5	117.5	116.4	115.4

7. Conclusion

Based on 15 super-high sucker rods fatigue strength experiment, alternating stress loading cycles before sucker rod fatigue failure under 3 different load levels were obtained, the failure position shows the thread is more easily failure than any other parts of sucker rod, data distribution test shows that all the experimental data followed normal distribution, the value of logarithmic fatigue life of different reliability is calculated, P-S-N formula fatigue limit is also calculated, when the reliability P are 99%, 90%,70% and 50% , the value of fatigue limit σ_r are 245.5MPa, 242.0MPa, 240.1MPa and 238.3MPa respectively.

Acknowledgement

This research was funded by the scientific research plan projects of Shanxi Education Department (No. 2013JK0859)

References

1. Hein N W, Hermanson D E. A New Look at Sucker Rod Fatigue Life. Spe Annual Technical Conference & Exhibition, 1993: 26-27
2. Bianca de C.Pinheiro, Ilson P. Pasqualino. Fatigue analysis of damaged steel pipelines undercyclic internal pressure. International Journal of Fatigue, 2009, 31(5): 962-973
3. Qing-xue Huang, Jian-mci Wang, Li feng MA, et al. Fatigue Damage Mechanism of Oil Film Bearing Sleeve. Journal of Iron and Steel Research, International, 2007, 14(1): 60-63
4. Lu Z, Xiang Y, Liu Y. Crack growth-based fatigue-life prediction using an equivalent initial flaw model. Part II: Multiaxial loading. International Journal of Fatigue, 2010, 32(2): 376–381

Effect of Na₂S₂O₃ and Na₂S on the Corrosion Behavior of Q235 Steel in Sodium Aluminate Solution

Chao-Yi Chen[1, 2], Xia-Qiong Yang[1, 2], Jun-Qi Li[1, 2,*], Man Zhang[1, 2]

[1]*School of Material and Metallurgy, Guizhou University, Guiyang 550025, China;*

[2]*Guizhou Province Key Laboratory of Metallurgical Engineering and Process Energy Saving, Guiyang 550025, China*

Email: jqli@gzu.edu.cn

Corrosion behavior of Q235 steel in sodium hydroxide solutions with $S_2O_3^{2-}$ and different concentration of S^{2-} were investigated by immersed corrosion and electrochemical corrosion tests. Combined with polarization curve method, impedance spectrum method, SEM and EDS discussed the corrosion mechanism of Q235 steel. Results show that compared with only contains sodium thiosulfate in the sodium aluminate solution, corrosion behavior of Q235 steel soaking in the sodium thiosulfate and sodium sulfide coexisting in the sodium aluminate solution has been inhibited, S^{2-} and $S_2O_3^{2-}$ work together to play a passive role, mainly because of $S_2O_3^{2-}$ react with S^{2-} to generated SO_3^{2-}. Therefore, S^{2-} is thiosulfate corrosion inhibitors. With the increase of concentration of Na_2S, the corrosion rate showed a trend of first decrease and then increase, which is due to the sample surface forming loose iron sulfide corrosion products, but as a result of loose corrosion products were oxidized into dense oxide corrosion products, corrosion rate decrease.

Keywords: Sodium aluminate solution; Q235 steel; sodium sulfide; sodium thiosulfate; polarization curve; impedance spectrum.

1. Introduction

Shortage of high-grade bauxite resource puts a strain on aluminum production. Abundant high sulfur bauxite are not available for better application [1-3]. Sulfur of ore enter solution in various forms (S^{2-}, $S_2O_3^{2-}$, SO_3^{2-}, SO_4^{2-}) and cause serious corrosion to equipment in Bayer process. Many researches about corrosion of sulfur were conducted, but most of them focused on corrosion of S^{2-} [4-7]. Some researchers [8] investigated corrosion behavior of 16Mn steel in the sodium aluminate solution with S^{2-}, the low concentration of S^{2-} will inhibit corrosion while high concentration of S^{2-} accelerate corrosion [9-15]. Peterman and others [16, 17] used the weightlessness method, electrochemical method, and methods of analytical chemistry, found thiosulfate and sulfide can accelerate corrosion when they exist alone, when both exist in the solution, the corrosion rate of the sample decreases instead, therefore $S_2O_3^{2-}$ and S^{2-} has the coupling effect in promoting both passivation. In view of this, in order to better reveal the coexistence of $S_2O_3^{2-}$ and S^{2-} in sodium aluminate solution of low carbon steel corrosion mechanism, this paper explored corrosion behavior and mechanism of Q235 steel in sodium aluminate solution with $S_2O_3^{2-}$ and S^{2-} simultaneously.

2. Experiment

2.1. *Experimental material and solution*

Experimental material is Q235 steel, its chemical composition is shown in Table 1. Using analytical pure NaOH and Al(OH)₃ to prepare aluminate solution, using analytical pure Na₂S·9H₂O and Na₂S₂O₃·5H₂O adjust the mass concentration of $S_2O_3^{2-}$ and S^{2-}.

Table 1 Chemical composition of Q235 steel (wt%)

Element	C	Si	Mn	P	S	O	Fe
Mass fraction /%	0.2	0.6	0.3	0.0	0.2	1.3	97.4

---

*Corresponding author

2.2. *Autoclave static corrosion*

Autoclave static corrosion experiments were conducted at 110°C. Samples are Q235 steel with a size of 10mm×10mm×3mm. Grind the samples with SiC sandpaper of 180#, 240#, 360#, 600#, 800# and 1000# by degrees to make their surface smooth. Rinsed samples with alcohol and dried them with cold wind before use. Each sample was weighed by electronic balance and measured by Vernier caliper before experiments. After experiments, samples were washed with deionized water, and corrosion products were removed before weighing again.

2.3. *Electrochemical corrosion*

Electrochemical corrosion experiments were conducted on VSP electrochemical workshop at 65°C with a conventional three-electrode system. Counter electrode was platinum electrode, reference electrode was saturated calomel electrode (SCE) and sample with 100mm² work area was work electrode. Work electrode was sealed with epoxy resin except work surface. Poteniodynamic polarization tests were conducted from -2V vs. SCE to -1V vs. SCE with sweep rate of 2mV/s. Impedance spectroscopy tests of sine wave excitation signal amplitude of 5mV, frequency range is 100 kHz~10 mHz. All the electrode potentials are relative to potential of SCE.

3. Results and Analysis

3.1. *Micro morphology and energy spectrum analysis*

Fig. 1 shows SEM of Q235 steel after corrosion 9days in sodium aluminate solution with 0, 3, 4, 6g/L S^{2-} and 3g/L $S_2O_3^{2-}$ respectively. From 50 times micro morphology, substrate surface of Q235 steel is covered with complete protective film in sodium aluminate solution containing 0 g/L S^{2-}. When concentration of S^{2-} reached to 3 g/L, localized pitting occurs on the surface of substrate obviously. When S^{2-} is 4g/L, the pitting phenomenon shows a trend of expansion. When S^{2-} is 6g/L, the substrate surface damage range is relatively small, but surface pitting has deepened. From 3000 times micro morphology, corrosion product film and substrate adhesion is very poor with 0g/L S^{2-}. When S^{2-} increased to 3g/L, corrosion products are very loose and uniform. But when S^{2-} concentration increased to 4g/L, loose corrosion products dropped and formed local corrosion. When S^{2-} is 6g/L, substrate surface is relatively smooth with few groove corrosion pits.

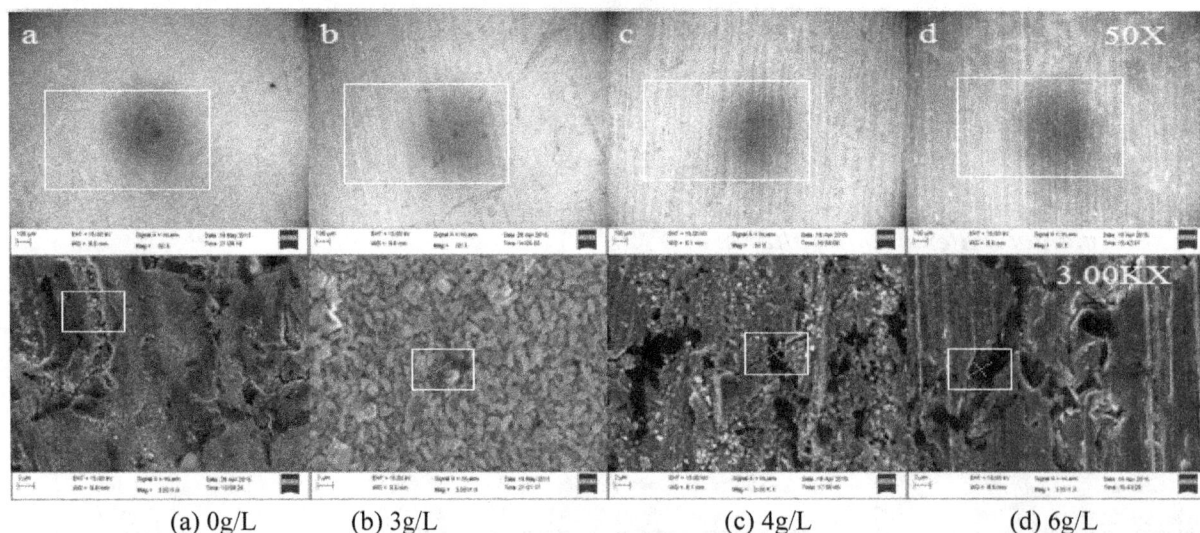

(a) 0g/L   (b) 3g/L   (c) 4g/L   (d) 6g/L

Fig. 1 SEM of Q235 steel in different concentration of S^{2-}

In order to further study concentration of S^{2-} on the influence of corrosion products, EDS analysis results are shown in Table 2. Corrosion products of Q235 steel after corrosion 9 days in sodium aluminate solution with sulfur is mainly composed of oxide of iron. With increasing of S^{2-} concentration in the solution, the O content of corrosion products showed a trend of decline, S content showed a trend of decline after rising first, which is due to the high temperature and alkali condition, oxide of iron can react with sulfur and make sulfide generate on the solubility substrate surface, undermines the integrity of the substrate surface of iron oxide.

Table 2 EDS analysis results of Q235 steel in different concentration of S^{2-} (wt%)

S^{2-}-$S_2O_3^{2-}$	C	O	Al	Si	S	Mn	Fe
0-3	0.14	28.24	1.75	0.11	0.13	1.17	68.47
3-3	2.72	13.17	1.09	0.16	2.16	0.66	80.04
4-3	5.46	11.06	0.51	0.51	5.23	0.27	76.96
6-3	3.78	10.52	0.55	0.20	4.10	0.95	79.91

3.2. Corrosion weightlessness analysis

Fig. 2 shows corrosion weightlessness of Q235 steel in NaAlO$_2$ solution with 3g/L $S_2O_3^{2-}$ and 0, 3, 4, 6g/L S^{2-}. In solution with 3g/L $S_2O_3^{2-}$ and 0g/L S^{2-}, corrosion weight loss is biggest, but corrosion weightlessness shows obvious downward trend with increasing S^{2-}. When S^{2-} increases from 3g/L to 4g/L, corrosion weightlessness increases obviously but still below the weight loss with 0g/L S^{2-}. When concentration of sodium sulfide increases to 6g/L, corrosion rate reduces again. It is consistent with the results of the micro morphology. When sodium aluminate solution only contains sodium thiosulfate, the effect of corrosion on Q235 steel is bigger, but when sulfide and thiosulfate coexist in solution corrosion weightlessness is reduced. It shows that the sulfide and thiosulfate work together at the same time can play a certain passivation effect.

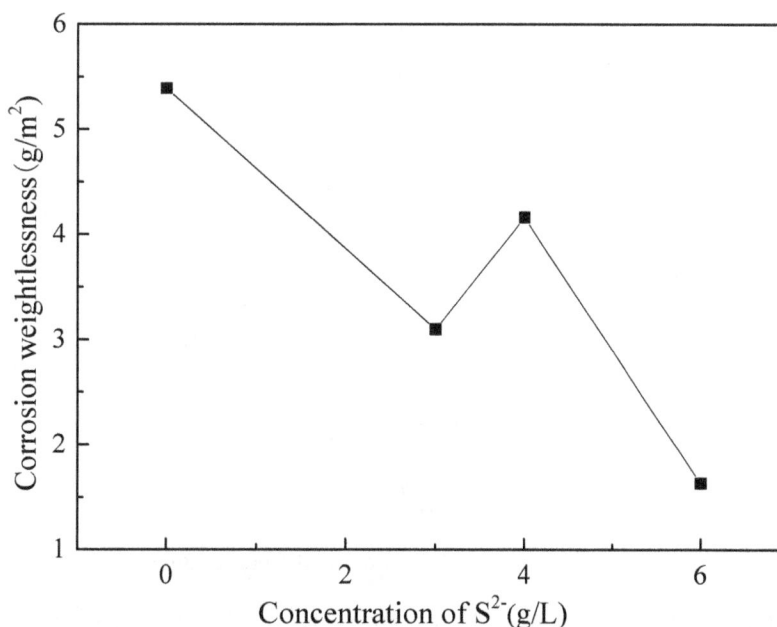

Fig. 2 Corrosion weightlessness curve of Q235 steel in different concentration of S^{2-}

3.3. *Polarization curve analysis*

Polarization curves of Q235 steel corrosion 9days in $NaAlO_2$ solution with 3g/L $S_2O_3^{2-}$ and 0, 3, 4, 6g/L S^{2-} are shown in Fig. 3. When concentration of S^{2-} increased from 0g/L to 3g/L, corrosion current density significantly decrease. Corrosion current density showed a trend of first increases and then decrease when S^{2-} increased from 3g/L to 6g/L. A layer of deep black loose black film generate on electrode surface after immersion corrosion in solution with 3g/L $S_2O_3^{2-}$. With increasing concentration of S^{2-} to 3g/L, the electrode surface corrosion products are firming tan, indicate on the corrosion inhibition, when the concentration of S^{2-} was 4g/L and 6g/L, the electrode surface corrosion products from the pale yellow to dark color, indicate that corrosion degree began to increase. Cathodic polarization curves in different concentration of Na_2S are very similar, manifest that the cathodic reaction are similar and the anodic reaction are significant activation and passivation phenomenon, later into the passivation zone. When the potential is higher than 0.5V, current density increases sharply, oxygen evolution reaction take place as follows: $O_2+2H_2O+4e=4OH^-$. Below the corrosion potential -1.2V, the hydrogen evolution reaction take place as follows: $2H_2O+2e=2OH^-+H_2\uparrow$; -1.2V to 0.75V: anode in passivation transition zone, the anode passivation transition region has three limit peak, different concentration of thiosulfate anodic peak potential limit, illustrate the anode electrode reaction is the same;-0.75V: substrate surface stability of the resulting passivation membrane; From -0.75V to 0.75V basic into passivation state: stable passivation membrane generated on the surface of the substrate and restrain the anodic dissolution process, therefore, anode ionization rate decreases, and the corrosion rate is decreased.

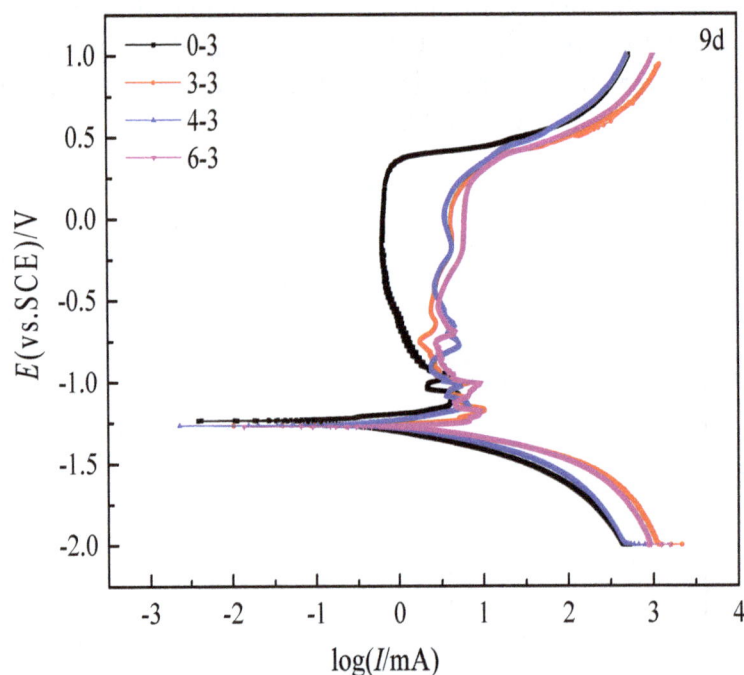

Fig. 3 Polarization curves of Q235 steel in the different concentration of S^{2-}

Corrosion current density i_{corr}, anodic Tafel slope β_a, cathodic Tafel slope β_c are shown in Table 3. By visible $\beta_c > \beta_a$, illustrate the anodic reaction control corrosion process. With the increase of concentration of Na_2S in sodium aluminate solution, the corrosion current showed a trend of decrease, accelerate the formation of Q235 steel surface passivation membrane, slowed the anodic dissolution reaction.

Table 3 Fitting data of polarization curves in the different concentration of S^{2-}

Concentration of S^{2-}(g/L)	E_{corr}/mV	I_{corr}/μA	β_a/mV	β_c/mV
0	-1262.26	3612.4557	121.4	142.3
3	-1233.02	2882.092	89.6	127.4
4	-1170.18	2901.902	112.8	132.8
6	-1258.05	827.477	110.1	118.3

3.4. *Impedance spectrum analysis*

Fig. 4 shows EIS of Q235 steel corrosion 9days in $NaAlO_2$ solution with 3g/L $S_2O_3^{2-}$ and variable S^{2-}. Ac impedance spectra of sample under the Nyquist diagrams include two semicircular arc, which corresponds to the corrosion product layer of high frequency impedance signal, the low frequency is the corrosion reaction substrate Q235. With the increase of the concentration of S^{2-}, the capacitive reactance arc radius first increases and then decreases, and the corrosion rate increase after first decrease gradually. Fig. 5 shows the equivalent circuit diagram of EIS, the equivalent circuit diagram in accordance with the experimental data. R_s as medium resistance in the figure, the Q235 steel corrosion EIS dispersion effect is very strong, the constant phase element Q instead of capacitance element, Q_r, Q_{ct} are constant phase components, which represent corrosion products sediment resistance and electric double layer capacitors respectively. R_r and R_{ct} are the resistance of the corrosion product film and the charge transfer resistance. The reaction mechanism is as follows:

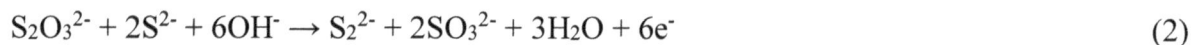

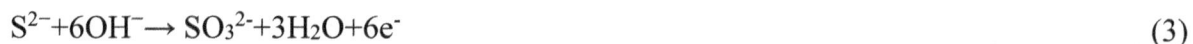

$$S_2O_3^{2-} + 6OH^- \rightarrow 2SO_3^{2-} + 3H_2O + 4e^- \tag{1}$$

$$S_2O_3^{2-} + 2S^{2-} + 6OH^- \rightarrow S_2^{2-} + 2SO_3^{2-} + 3H_2O + 6e^- \tag{2}$$

$$S^{2-} + 6OH^- \rightarrow SO_3^{2-} + 3H_2O + 6e^- \tag{3}$$

Thiosulfate is activator of corrosion. When thiosulfate is reduced, the passivation ability is bigger, such as Eq.1. S^{2-} is thiosulfate corrosion inhibitors, such as Eq.2. S_2^{2-} will be generated by partial oxidation S^{2-} and make the S^{2-} content dropped significantly, electrochemical system plays a supply electronic role, for the anodic dissolution reaction to play the role of a block of iron, the corrosion rate showed a trend of decline, as is shown in Eq.3.

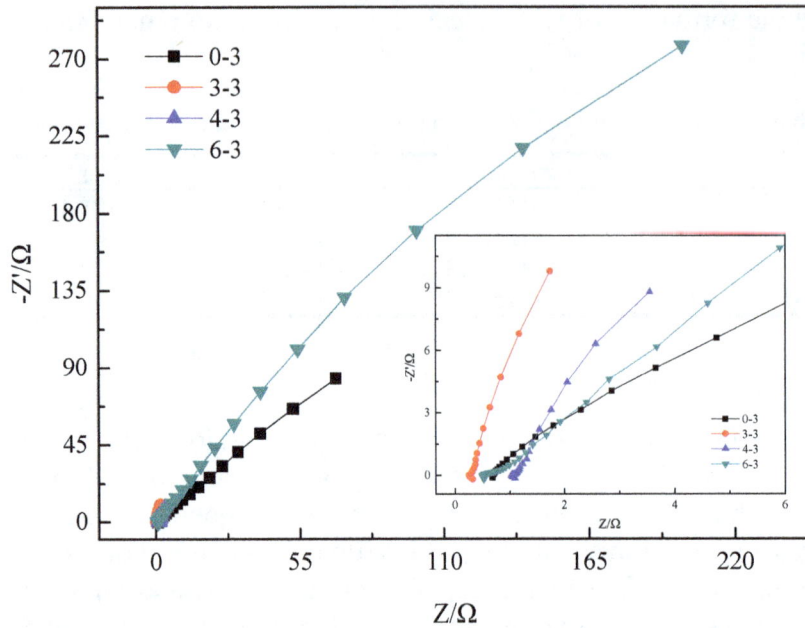

Fig. 4 Nyquist curves of Q235 steel in different concentration of S^{2-}

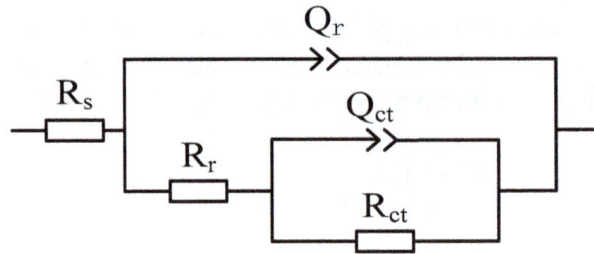

Fig. 5 Equivalent circuit diagram of Nyquist curves

4. Conclusion

The maximum corrosion weightlessness of Q235 steel is 5.39g/m² after corrosion 9 days in sodium aluminate solution with 3g/L $S_2O_3^{2-}$ and 0g/L S^{2-}. With the addition of sulfide, corrosion weightlessness reduced instead. Compared to the corrosion behavior of Q235 steel in sodium aluminate solution with only 3g/L thiosulfate, corrosion gets slighter with increasing S^{2-} in sodium aluminate solution with 3g/L thiosulfate, suggesting that S^{2-} and $S_2O_3^{2-}$ work together to play a role in passivation. When the concentration of S^{2-} from 3g/L up to 6g/L, corrosion weightlessness of Q235 steel is maximum when $S_2O_3^{2-}$ is 3g/L and S^{2-} is 4g/L in sodium aluminate solution. With the increase of concentration of Na2S, the corrosion rate of sample showed a trend of first increase and then decrease, at this time sulfide plays a corrosion activator.

Acknowledgments

This project was funded by the National Natural Science Foundation of China (No. 51574095, 51474079), KY (2015)334, The Industrial Projects Guiyang Municipal Science and Technology Bureau ([2012103]69, [2012205]64).

References

1. Y.Z. Jia, J.W. Wang, J.Q. Li and C.S. Lv. The Chinese Journal of Process Engineering, 2013, Vol. 13, pp. 801-806. "In Chinese"
2. S.X. Li and R. Akid, Engineering Failure, 2013, Vol. 34, pp. 324-334. "In Chinese"
3. X.Q. Yang, C.Y. Chen, J.Q. Li and B.L. Quan. Surface Technology, 2015, Vol. 44, pp. 80 182. "In Chinese"
4. H. Luo, C.F. Dong, K. Xiao and X.G. Li. Applied Surface Science, 2011, Vol. 258, pp. 631-639.
5. X.B. Li, C.Y. Li and T.G. Qi. The Chinese Journal of Nonferrous Metals, 2013, Vol. 23, pp. 829–835. "In Chinese"
6. Q.L. Xie, W.M. Chen and Q.P. Yang. Corrosion, 2014, Vol. 70, pp. 842-849.
7. L. Freire, M.J. Carmezim, M.G.S. Ferreira and M.F. Montemor. Electrochimica Acta, 2011, Vol. 56, pp. 5280-5289.
8. Q.L. Xie and W.M. Chen. The Chinese Journal of Nonferrous Metals, 2013, Vol. 23, pp. 3462–3469. "In Chinese"
9. I. Betova, M. Bojinov, O. Hyökyvirta and T. Saario. Journal of Electroanalytical Chemistry, 2011, Vol. 654, pp. 52-59.
10. K. R. Chasse and P. M. Singh. Metallurgical and Materials Transactions A, 2013, Vol. 44, pp. 5039–5053.
11. S.S. Xin, M.C. Li, Corrosion Science, 2014, Vol. 81, pp. 96-101.
12. J. Zhang, Corrosion Science, 2009, Vol. 51, pp. 1207-1227.
13. D.W. Shoesmith, P. Taylor, M.G. Bailey and B. Ikeda. Electrochimica Acta, 1978, Vol. 23, pp. 903-916.
14. F. Wenger, S. Cheriet and B. Talhi. Corrosion Science, 1997, Vol. 39, pp. 123.
15. A. Wensley. Corrosion, 2000, Vol. 589, pp. 26 – 31.
16. Q.L. Xie, W.M. Chen. Corrosion Science, 2014, Vol. 86, pp. 252 - 260.
17. Y. Li, J. Wu, D. Zhang, Y. Wang and B. Hou. Journal of Solid State Electrochemistry, 2010, Vol. 14, pp. 1667-1673.

Residual Stress and Radial Thermal Deformation during Hard Turning Bearing Ring

Lan-Ying Xu[1], Qiang Wu[1,*], Meng-Yang Qin[1], Bang-Yan Ye[2]

[1]Guangdong Polytechnic Normal University, Guangzhou 510635, Guangdong, China
[2]South China University of Technology, Guangzhou 510640, Guangdong, China
*Email: 510635wuqiang@163.com

The thermal deformation depends on the shape and material of being machined parts, it is also affected by the surface residual stress of being machined parts. Based on the principle of elastic mechanics and thermodynamics, this paper establishes thermal deformation mathematical model of bearing ring based on the residual stress and quantitatively calculates the radial deformation produced by the residual stress, the thermal deformation error is less than 3.8%, the calculation results and the experimental data are in good agreement with the mathematical model.

Keywords: Generator; bearing; residual stress; thermal deformation.

1. Introduction

Bearing is the key component of machine, while it is also the weak part. Its failure is one of the main faults of machine. Study on surface quality of bearing has become a research focus for experts and scholars in the field of cutting processing. Surface residual stress state has a great influence on the performance, such as reliability and stability of machine parts. Theoretical research and practical application results indicate that it can improve the fatigue strength and corrosion resistance of parts by controlling or adjusting the finished surface stress state [1, 2]. Umbrello D established a hybrid finite element method-artificial neural network approach for predicting residual stresses [3, 4]. Atsushi et al. adopted light press of sheet metal edge for reducing residual stress generated by laser cutting considering mechanical properties and intensity of residual stress [5], Arshpreet et al. investigated on surface residual stress distribution in deformation machining process for aluminum alloy Arshpreet [6], Saurabh et al. analyzed modeling of residual stresses in orthogonal machining of AISI4340 Steel [7]. These above scholars all analyzed the residual stress and its effect on surface quality. Based on the research results of the above scholars this paper studies the effect of surface residual stress on thermal deformation to establish mathematic model of thermal deformation, and deduces a more accurate calculation method of the thermal deformation.

2. Residual Stress and Radial Thermal Deformation Model

In order to calculate thermal deformation considering residual stress, first of all it should be to determine the residual stress distribution. At present, calculating residual stress accurately is more difficult in theory, because the factors influencing the distribution of residual stress is too much, In this paper, first of all, Supposing both ends of the bearing is free, temperature field is on the Z axis of symmetry and distributes along the radial direction, the value of temperature is only related to the radial size of ring and has nothing to do with the position of the Z axis. Establishing cylindrical coordinates of bearing (shown in figure 1), for convenience of calculation, assume that axial displacement of any point (signed A) within bearing is zero, we analysis the relationship between radial displacement and temperature of point A, when the temperature changes, the radial displacement of point A is determined, then thermal deformation information of bearing ring can be obtained.

---

*Corresponding author

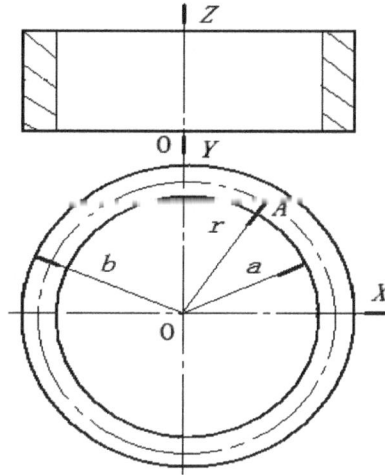

Fig. 1 Cylindrical coordinates of bearing ring

Owing to temperature is on the Z axis of symmetry, the shear stress and shear strain are all zero. According to the related literature [8], we can get solution of the plane strain for the cylinder, it is shown as follows:

$$
\begin{cases}
\sigma_r = \dfrac{-E}{1-\mu} \cdot \dfrac{\lambda}{r^2} \int_a^r t(r)r\,dr + \dfrac{E}{1+\mu}\left[\dfrac{C}{1-2\mu} - \dfrac{D}{r^2}\right] \\[2mm]
\sigma_\theta = \dfrac{E}{1-\mu} \cdot \dfrac{\lambda}{r^2} \int_a^r t(r)r\,dr - \dfrac{\lambda E t(r)}{1-\mu} + \dfrac{E}{1+\mu}\left[\dfrac{C}{1-2\mu} + \dfrac{D}{r^2}\right] \\[2mm]
\sigma_z = \dfrac{-\lambda E t(r)}{1-\mu} + \dfrac{2\mu E C}{(1+\mu)(1-2\mu)}
\end{cases}
\tag{1}
$$

$$
u_0 = \frac{1+\mu}{1-\mu} \cdot \frac{\lambda}{r} \int_a^r t(r)r\,dr + Cr + \frac{D}{r}
\tag{2}
$$

Where: E — The elastic modulus of bearing materials.

λ — The linear expansion coefficient of bearing materials.

μ — Poisson's ratio of bearing materials.

t — The changing rule of the temperature. $t = t(r)$

σ_r — radial stress of the bearing ring.

σ_θ — tangential stress of the bearing ring.

σ_z — axial stress of the bearing ring.

u_0 — The radial displacement of the bearing ring.

C and D are the integral constant; they are determined by boundary conditions.

For cylindrical parts, surface residual stress theoretically should be divided in three directions, which can be expressed as the tangential stress σ_θ, axial stress σ_z and radial stress σ_r [9]. In general, when the cylinder wall is thin, the radial residual stress σ_r is very small, and in this case the tangential residual stress σ_θ is the main part. As a result, we could only consider the effect of tangential residual stress σ_θ in order to simplify the calculation when adopting the boundary condition to determine the integral constant C and D in this paper.

Assuming that tangential stresses of bearing ring and bearing outer ring are $\sigma_{\theta a}$ and $\sigma_{\theta b}$ respectively, take them in the equation (1) above, and get the integral constant values of C and D, then substitute them in the equation (2) above to get the value of μ_0.

In order to meet the conditions of the free end, based on the above solution, we apply a pair of uniform axial force σ_z on both ends of cylinder surface to make whole axial force of the end be zero, at this time due to the influence of axial stress σ_z, radial displacement need to add a term $\dfrac{\mu\sigma_a r}{E}$ in right end of equation (3), that is

$$\mu = \mu_0 - \frac{\mu\sigma_a r}{E} \tag{3}$$

The rule of material's elastic modulus changing with temperature can be expressed as follows:

$$E = E_0\left[1 + \lambda_E t(r)\right] = \frac{E_0}{1 - \lambda_E t(r)} \tag{4}$$

where: E_0 — material's elastic modulus when temperature is t_0, take $t_0 = 0$

λ_E — The temperature coefficient of material's elastic modulus, it is far less than 1, and it often is negative value.

Assuming that temperature of bearing ring is $t(a)$, while temperature of bearing outer ring is $t(b)$, diameter of inner ring is a, while diameter of outside ring is b, then the temperature distribution inside the bearing is expressed: $t(r) = t(a) + \dfrac{t(b) - t(a)}{\ln(b/a)} \ln(r/a)$, take it into above equation (4) to acquire E.

The residual stress of surface layer is the local stress, it only effects local deformation of the surface layer, therefore, when calculating the deformation of the inner surface, it does not need to consider the influence of the outer surface residual stress $\sigma_{\theta b}$, similarly, when calculating the deformation of the outer surface, without considering the influence of the inner surface residual stress $\sigma_{\theta a}$. Take equation (3) and (4) into equation (2) to acquire calculation mathematical model of thermal deformation of axial parts taking into account residual stress, they are as follows:

$$\mu_a = \frac{1 - \lambda_E t(a)}{E_0} \bullet \left\{1 + \frac{\mu(b^2 + a^2)}{b^2 - a^2}\right\} \bullet a\sigma_{\theta a} + \lambda a t(a) + \frac{(1+\mu)b^2\lambda}{(1-\mu)(b^2 - a^2)}\left[a(1 - 2\mu) - 1\right]\bullet\left[t(b) - t(a)\right] \tag{5}$$

$$\mu_b = \frac{1 - \lambda_E t(b)}{E_0} \bullet \left[1 - \frac{\mu(b^2 + a^2)}{b^2 - a^2}\right] \bullet b\sigma_{\theta b} + \lambda b t(a) + \frac{(1+\mu)b\lambda}{(1-\mu)(b^2 - a^2)}\left[b^2(1 - 2\mu) - a\right]\bullet\left[t(b) - t(a)\right] \tag{6}$$

From equation (5) and equation (6) can be seen that the bearing radial deformation due to temperature change consists of two parts, the first part is produced by the surface residual stress, namely it is the first term on the right side of equation above, and it is relative with the size and direction of residual stress and bearing diameter; the second part is generated by the thermal expansion of the material, it includes the second item and the third item on the right side of equation above, from the analysis above, because of the influence of the surface residual stress, we can understand that the actual thermal deformation of bearing ring is not equal to deformation only generated by the thermal expansion.

3. Experimental Verification

3.1. *Residual stress measurement*

The parameters of bearing ring are as follow: r= 85 mm, R = 117.4 mm, B = 41 mm, the material of bearing ring is Gr15 steel. Its elastic modulus is 210 Gpa, Poisson's ratio is 0.3, the density is 7800 kg/m^3. The chemical component and proportion of Gr15 steel are shown in Table 1.

Table 1 Chemical component and proportion of Gr15 steel/%

elements	C	Mn	P	S	Cr	Ni	Si
Quality	0.95~1.05	0.2~0.4	<0.027	<0.02	1.3~1.65	<0.3	0.15~0.35

This experiment adopts the numerical control lathe and PCBN tool, after tool is adjusted to the appropriate location, the machine feed automatically. Cutting parameters are shown in Table 2.

Table 2 Cutting parameters

Category	Lubrication type	cutting tool	Cutting speed $v_c(m/s)$	Feed speed $f(mm/r)$	Cutting depth $a_p(mm)$
Fine turning	Dry cutting	PCBN	0.8	0.08	0.1

We measure residual stress in bearing ring surface through X-ray diffractometer type of ADVANCE (shown in Figure 2). From each 90 degree of the bearing surface, we select four different points to measure the residual stress and take their average value as the calculated value to be used later. Among these measurement results (shown in Table 3), positive value represents residual tensile stress, negative value represents residual compressive stress. It can be seen from the measured data that residual stress values are basically negative which express compressive stresses, and its value is about 200MPa.

Table 3 Average residual stress of machined surface

Sample	Residual stress /MPa			
	Point 1	Point 2	Point 3	Point 4
1	-198	-207	-189	-210
2	-190	-200	-201	-195
3	-200	-203	-187	-197
4	-196	-203.3	-192.3	-200.6

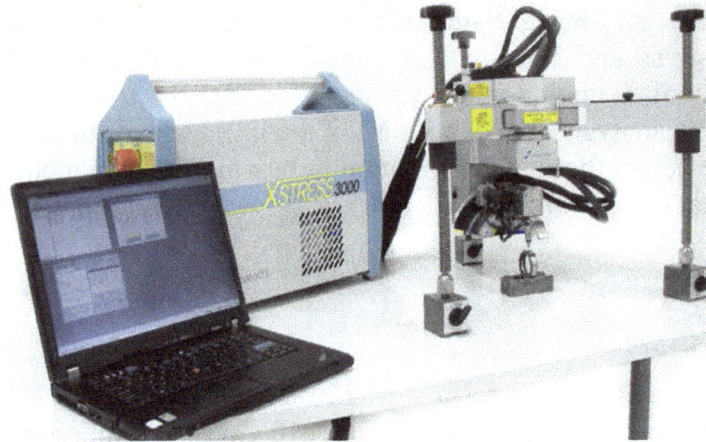

Fig. 2 Residual stress test

Fig. 3 Measurements of radial thermal deformation

3.2. *Measuring radial thermal deformation caused by residual stress*

By adopting infrared thermometer the temperature data acquisition can be real-time during and after the completion of the turning bearing sleeve, the measurement of thermal deformation is also carried out at the same time as the measurement temperature is measured. The thermal deformation produced by residual stress during processing bearing ring can be measured by Dial indicator, In the measuring process, the measured bearing ring rotates slowly, the dial indicator measuring head moves linearly along the axis direction, and passes through the whole outer ring and inner ring surface, write down maximum and minimum reading of the dial indicator pointer, take the difference between the two readings of measured elements as radial total run-out tolerances. The measured values and the calculated values of thermal deformation calculated by the above formula (6) are shown in Table 4.

Table 4 The thermal deformation and temperature of bearing ring

No.	Inner ring temperature/°C	Outer ring temperature /°C	thermal deformation of outer ring /um	
			Calculated value	Measured value
1	139.1	163.1	91.32	88.38
2	127.1	149.0	78.24	75.35

(*Continued*)

562

Table 4 (*Continued*)

No.	Inner ring temperature/°C	Outer ring temperature /°C	thermal deformation of outer ring /um	
			Calculated value	Measured value
3	118.2	136.1	61.19	64.79
4	103.3	124.6	56.48	55.90
5	98.3	108.0	40.56	38.56
6	87.2	95.4	30.68	28.80
7	75.7	88.7	21.99	22.49
8	63.9	70.1	13.83	14.32
9	58.3	63.1	11.30	10.87
10	51.7	54.5	6.98	4.89
11	32.9	36.7	3.51	2.35
12	26.1	27.4	1.21	1.03

By comparing thermal deformation measurement value of the bearing ring and the numerical calculation value derived from the above formula (6), it can be seen that both results approximately equal, which shows that the mathematical model of radial deformation derived from residual stress is correct.

4. Conclusion

Through the analysis above, precision measurement result of thermal deformation for the mechanical parts not only depends on parts material thermal expansion but also depends on its internal residual stress. This paper deduces relationship among radial thermal deformation and material thermal expansion and surface residual stress of bearing ring parts, and obtains the approximate equation calculating the radial thermal deformation. The maximum error of this formula is 3.8%. The calculation formula is of great significance for precise measurement of the radial thermal deformation. Further study should use a mechanical method or other thermal methods on adjusting machined surface residual stress to control the radial deformation quantitatively and accurately.

Acknowledgments

This project is funded by National Natural Science Foundation of China: (No. 51375101), Natural Science Foundation of Guangdong Province of China (No. 2014A030313638 and 2015A030313673) and Guangdong Science and Technology Program (No. 2014A010104014)

References

1. Hua J, Rajiv S, Cheng X M: Effect of feed rate, work-piece hardness and cutting edge on subsurface residual stress in the hard turning of bearing steel using chamfer plus hone cutting edge geometry. Materials science engineering A. Vol. 394 (2005), p. 238.
2. Madariaga A, Esnaola J.A., Fernandez E., Arrazola P.J., Garay A., Morel F: Analysis of residual stress and work-hardened profiles on Inconel 718 when face turning with large-nose radius tools, International Journal of Advanced Manufacturing Technology, Vol. 71 (2014), p. 1587
3. Umbrello D, Hua J, Shivpuri R.: Hardness based flow stress for numerical modeling of hard machining AISI 52100 bearing steel. Materials Science and Engineering A. Vol. 374 (2004), p. 90.

4. Umbrello D, Ambrogio G, Flice L, Shivpuri R.: A hybrid finite element method-artificial neural network approach for predicting residual stresses and the optimal cutting conditions during hard turning of AISI 52100 bearing steal. Materials and Design, Vol. 29 (2008), p. 873.
5. Atsushi Maeda, Yingjun Jin, Takashi Kuboki: Light press of sheet metal edge for reducing residual stress generated by laser cutting considering mechanical properties and intensity of residual stress, Journal of Materials Processing Technology. Vol. 225 (2015), p. 178.
6. Arshpreet Singh, Anupam Agrawal: Investigation of surface residual stress distribution in deformation machining process for aluminum alloy Arshpreet, Journal of Materials Processing Technology Vol. 225 (2015), p. 195.
7. Saurabh Agrawal, Suhas S. Joshi: Analytical modeling of residual stresses in orthogonal machining of AISI4340 Steel, Journal of Manufacturing Processes. Vol. 15 (2013), p. 167.
8. Huang Yan: *Engineering mechanics of elasticity*. Beijing: (Tsinghua university press, Beijing 1982).
9. Zhu Xinghua: *Grinding principle*. Beijing (Mechanical Industry Press, Beijing 1988).

Power Consumption of Groove-textured Tools in Dry Milling of Titanium Alloys

Ze Wu[*], You-Qiang Xing, Peng Huang

Department of Mechanical Engineering, Southeast University, Nanjing 211189, China
[*]*Email: wuze@seu.edu.cn*

The groove-textured tools were fabricated by making textures on the rake faces and filling them with molybdenum disulfide. Dry milling of Ti-6Al-4V alloys was carried out with the groove-textured tools and conventional tools for comparison. Results show that the groove-textured tools can reduce the power consumption by 5% or so. The radial width of cut, the cutting speed as well as the axial depth of cut all have statistical and physical effect on the power consumption per unit volume in dry milling of Ti-6Al-4V alloys, while the feed per tooth seems to have no significant effect on that.

Keywords: Power consumption; Ti-6Al-4V alloy; dry milling; Taguchi method.

1. Introduction

Surface texturing is an effective way to improve the performance of cutting tools. Song et al. [1] fabricated micro-holes on the rake and flank faces of cemented carbide turning inserts using a micro-electronic discharge machining, and analyzed the effect of micro-holes on the mechanical properties of cutting tools by finite element analysis. Sugihara et al. [2] developed milling cutters with periodical stripe-grooved surfaces on their rake face formed using femtosecond laser technology, which displayed high crater wear resistance in cutting of medium carbon steel. Meanwhile, their study also indicated that the wear resistances of the textured rake face had s strong correlation with the width of convex and concave area of the textures. Enomoto et al. [3] reported that a kind of nano/micro-textured surface significantly improved anti-adhesiveness at the tool-chip interface in face milling of aluminum alloys, which was attributed to the functions of surface textures, namely retaining cutting fluid and reducing actual contact area between the tool and chips. Kümmel et al. [4] indicated that the built-up edge (BUE) of the tools could be stabilized in dry straight turning of carbon steel by applying a dimple texture on the rake face, which was accompanied with a better wear behavior compared to the untextured cutting tool. Xing et al. [5] fabricated a kind of nano/micro-scale textured Al2O3/TiC ceramic cutting tool, which could reduce the vibration for a stable cutting and produce more uniform surface quality in dry turning of AISI 1045 hardened steel. The surface texturing can also exhibit superiority by combining with coating technology. Viana et al. [6] fabricated micro textures on the surface of cemented carbide inserts (ISO K grade) and then coated them with TiAlN and AlCrN. As a result, by using the textured tools, the delamination of coating was suppressed in face milling of compacted graphite cast iron, which resulted in a prolonged tool life.

However, literature review shows the lack of study on milling of Ti-6Al-4V alloys with surface textured tools, especially the study of power consumption in milling operation with the improved tools. In the present study, surface textures were fabricated on the rake faces of cemented carbide inserts by laser beam machining to form so-called groove-textured tools. Dry milling of Ti-6Al-4V alloys was carried out with the groove-textured tools and conventional tools for comparison. This study aims at investigating the feasibility of surface texturing in improving cutting performance

---

[*]Corresponding author

when dry milling of Ti-6Al-4V alloys, especially the effect of so-called groove-textured tools on power consumption, which has gradually become the focus of attention in machining operation.

2. Experimental Procedures

Cemented carbide (WC/Co) was selected as cutting tool material for this study. A fiber laser was used to fabricate the textured grooves on the rake face of the milling inserts. The focal length and scanning field of the scanner lens are 65 mm and 45×45 mm^2, respectively. Machining was accomplished in air with the average operating voltage of 15 V, the working current of 20A and the processing speed of 5 mm/s. The width of the textured groove is about 50 μm and the depth is about 100 μm. Molybdenum disulfide was filled into the textured grooves.

Dry milling of titanium alloys (Ti-6Al-4V) was conducted on a vertical computer numerically controlled machining center DAEWOO ACE-V500, which had a 15-kW drive motor and a maximum spindle rotation speed of 10,000 rpm. The axial rake angle , radial rake angle, and major cutting edge angle of the used tool holder were 20°, -5°, and 45°, respectively. In all the tests, only one of the teeth was used in order to avoid the effects induced by small differences between the teeth and keep the cutting condition constant. Up milling was selected in the present study. Firstly, single factor tests were conducted to investigate the cutting performance of so-called groove-textured tools. In the single factor tests, the cutting speed v ranging from 200 to 360 m/min in step of 40 m/min was adopted by fixing radial width of cut a_e, axial depth of cut a_p, and feed per tooth f_z at 50 mm, 1 mm, and 0.1 mm/tooth, respectively. Furthermore, Taguchi method was applied to investigate the influence of cutting parameters on power consumption. A commercial power sensor was used to measure the power consumption in the milling process.

3. Results and Discussion

Fig. 1 shows the comparisons of power consumption of the used milling machine between the operations with different cutting tools (radial width of cut a_e=50 mm, axial depth of cut a_p =1 mm, and feed per tooth f_z=0.1 mm/z). It can be seen from Fig. 1 that the power consumptions of the used milling machine increase gradually when the cutting speeds increase, either for the groove-textured tool or for the conventional one. For the milling operation with the conventional tool, the power consumptions increase from 4.41 kW to 6.17 kW as the cutting speeds increase from 200 m/min to 360 m/min. Moreover, the power consumption for operation with the groove-textured tool is lower than that with the conventional one under the same cutting condition. For example, the power consumption for operation with the conventional tool is 5.16 kW obtained at the speed of 280 m/min. However, the value is 4.89 kW for operation with the groove-textured tool. Calculation indicates that the using of the groove-textured tools can reduce the power consumptions by 5% or so. The reduction of power consumptions may be due to the reduced cutting forces which are obtained by using the groove-textured tools.

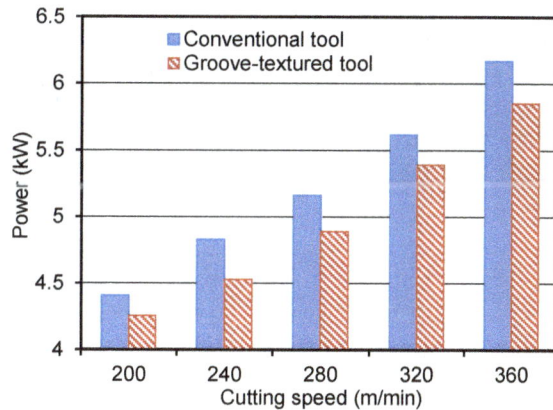

Fig. 1 Comparisons of power consumption of the machine between the operations with different cutting tools

In cutting process, the power consumption per unit volume can be expressed as power consumption divided by material removal rate. Fig. 2 illustrates the power consumptions per unit volume as a function of cutting speeds with the so-called groove-textured tool and the conventional one for comparison. It is obvious that the power consumptions per unit volume present a decreasing tendency with the increase of cutting speeds, which is opposite to variation of the power consumption. Moreover, the power consumption per unit volume for operation with the groove-textured tool is lower than that with the conventional one at the same cutting speed.

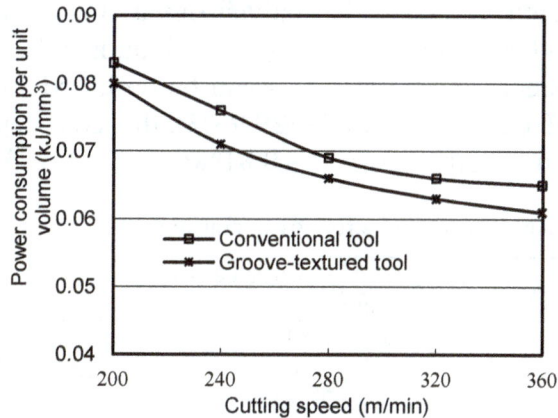

Fig. 2 Power consumptions per unit volume for operations with different cutting tools

The aforesaid experiment results indicate that the using of the groove-textured tools can reduce the power consumptions in milling process. In order to further investigate the influence of cutting parameters on power consumptions per unit volume, orthogonal experiment was conducted in milling of Ti-6Al-4V alloys by the so-called groove-textured tools. The experiment was conducted as per the standard orthogonal array. In the present investigation, an L16 orthogonal array which has 16 rows and 5 columns was chosen. The experiment consists of 16 tests and the columns were assigned with parameters. The first column was assigned to cutting speed, second column was assigned to feed per tooth, third column was assigned to axial depth of cut, fourth column was assigned to radial width of cut, and the last column was assigned to test error. The experiment was conducted as per the orthogonal array with level of parameters. The test results were subject to the analysis of variance.

Table 1 Orthogonal array of cutting parameters and the corresponding results of power consumption per unit volume

Test	Cutting speed, v (m/min)	feed per tooth, f_z (mm/z)	axial depth of cut, a_p (mm)	radial width of cut, a_e (mm)	Error term	Power consumption per unit volume, W (kJ/mm³)
1	200	0.05	0.5	15	1	0.082
2	200	0.1	1	30	2	0.074
3	200	0.2	1.5	50	3	0.068
4	200	0.3	2	80	4	0.063
5	240	0.05	1	50	4	0.076
6	240	0.1	0.5	80	3	0.071
7	240	0.2	2	15	2	0.083
8	240	0.3	1.5	30	1	0.079
9	280	0.05	1.5	80	2	0.071
10	280	0.1	2	50	1	0.075
11	280	0.2	0.5	30	4	0.081
12	280	0.3	1	15	3	0.092
13	320	0.05	2	30	3	0.065
14	320	0.1	1.5	15	4	0.078
15	320	0.2	1	80	1	0.063
16	320	0.3	0.5	50	2	0.069

The orthogonal array of parameters and the corresponding experiment results are shown in Table 1. Moreover, the results of analysis of variance for power consumption per unit volume are presented in Table 2. The seventh column of the analysis of variance for the power consumption per unit volume (see Table 2) indicates the percentage contribution (P) of each factor on the total variation indicating their influence on the result. It can be observed from Table 2 that the radial width of cut (P = 62.35%), the cutting speed (P = 29.96%), the axial depth of cut (P = 4.86%), the feed per tooth (P = 2.02%) and the error term (P = 0.81%).

Table 2 Analysis of variance for power consumption per unit volume

Source of variances	Sum of squares ($\times 10^{-6}$)	Degree of freedom	Variance ($\times 10^{-6}$)	Test F	F	Percentage of contribution (%)
Cutting speed	296	3	98.7	37	5.39 [a]	29.96
Feed per tooth	20	3	6.7	2.5	9.28 [b]	2.02
Axial depth of cut	48	3	16	6	29.5 [c]	4.86
Radial width of cut	616	3	205.3	77		62.35
Error	8	3	2.7			0.81
Total	988	15				100

[a] 90% confidence level. [b] 95% confidence level. [c] 99% confidence level.

According to Table 2, the cutting speed and the radial width of cut all produce significant level within the reliability interval of 99%, in other words, the cutting speed and the radial width of cut have high remarkable effect on the power consumption per unit volume in milling of Ti-6Al-4V alloys. The axial depth of cut has certain effect on the power consumption per unit volume at the reliability interval of 90%. However, the feed per tooth has no effect on the power consumption per unit volume at the reliability interval of 90%. Conclusively, the radial width of cut, the cutting speed as well as the axial depth of cut all have statistical and physical effect on the power consumption per unit volume in dry milling of Ti-6Al-4V alloys, while the feed per tooth seems to have no significant effect on that.

In cutting process, the friction forces between the chip and rake face vary linearly with the tool-chip contact area and the average shear strength. Firstly, surface textures on the rake face of the groove-textured tools can decrease the tool-chip contact areas compared with the conventional tools. Secondly, molybdenum disulfide may be released and smeared on the friction surface of the groove-textured tools under high cutting temperature for its high thermal expansion coefficient. The shear strength of molybdenum disulfide is very lower than that of the cemented carbide, as a result, the friction forces can be reduced with the application of the groove-textured tools. These may be the main reasons why the power consumption is saved in milling process with the groove-textured tools.

4. Conclusion

The using of the groove-textured tools can reduce the power consumptions by 5% or so. The cutting speed and the radial width of cut all have certain effect on the power consumption per unit volume within the reliability interval of 99%. The axial depth of cut has certain effect on the power consumption per unit volume at the reliability interval of 90%. However, the feed per tooth has no effect on the power consumption per unit volume at the reliability interval of 90%. The mechanism for improved performance of the groove-textured tools can be mainly interpreted as their self-lubricating function.

Acknowledgments

This work was funded by the National Natural Science Foundation of China: (Grant no. 51405080), the Natural Science Foundation of Jiangsu Province and Chinese Postdoctoral Science Foundation (Grant no. 2014M561547).

References

1. W. L. Song, J. X. Deng, H. Zhang, P. Yan, J. Zhao, X. Ai, Performance of a cemented carbide self-lubricating tool embedded with MoS_2 solid lubricants in dry machining, J. Manuf. Process 13 (2011) 8–15.
2. T. Sugihara, T. Enomoto, Crater and flank wear resistance of cutting tools having micro textured surfaces, Precis. Eng. 37 (2013) 888–896.
3. T. Enomoto, T. Sugihara, Improvement of anti-adhesive properties of cutting tool by nano/micro textures and its mechanism, Procedia Eng. 19 (2011) 100–105.
4. J. Kümmel, D. Braun, J. Gibmeier, J. Schneider, C. Greiner, V. Schulze, A. Wanner, Study on micro texturing of uncoated cemented carbide cutting tools for wear improvement and built-up edge stabilization, J. Mater. Process Technol. 215 (2015) 62–70.
5. Y. Q. Xing, J. X. Deng, J. Zhao, G. D. Zhang, K. D. Zhang, Cutting performance and wear mechanism of nanoscale and microscale textured Al_2O_3/TiC ceramic tools in dry cutting of hardened steel, Int. J. Refract. Met. Hard Mater. 43 (2014) 46–58.
6. R. Viana, M. S. F. Lima, W. F. Sales, W. M. Silva, A. R. Machado, Laser texturing of substrate of coated tools - Performance during machining and in adhesion tests, Surf. Coat. Technol. 276 (2015) 485–501.

Effects of Solution Treatment on the Microstructure and Corrosion Resistance of Mg-Zn-Zr-Ce Biomedical Magnesium Alloy

Shao-Fan Lei[1], Jiu-Ba Wen[1,2,*], Ya Liu[1], Jun-Guang He[1,2]

[1]Henan University of Science and Technology, School of Materials Science and Engineering NO.263 Kaiyuan Road, Luoyang, 471003, CN

[2]Collaborative Innovation Center of Nonferrous Metals of Henan Province, NO.263 Kaiyuan Road, Luoyang, 471003, CN

*Email: wenjiuba12@163.com

Effects of solution treatment on the Mg-2Zn-0.4Zr-0.6Ce (wt.%) biomedical magnesium alloy have been studied through SEM, immersion test, electrochemical measure. The result shows, compared with as-cast alloy, the grain becomes coarse and the most of second phases dissolve for the quenched alloys. Meanwhile, solution treatment gives an enhancement in the corrosion resistance and electrochemical properties, which is attributed to the reduction of second phases and homogenization of alloying element. The corrosion rate of quenched alloys firstly decreases with the increasing temperature of solution treatment, when the temperature is up to 460 °C, the corrosion rate increases slightly. And the optimal corrosion rate is acquired at 450 °C and is about 0.7293 mm·a^{-1}. Electrochemical measure shows: with the increasing temperature of solution treatment, the diameter of capacitive loop firstly increases and then decreases, I$_{corr}$ first decreases and then increases.

Keywords: Biomedical magnesium alloy; heat treatment; corrosion resistance.

1. Introduction

Magnesium alloys have become the most promising candidate for biomedical application in 21th century mainly due to its adequate mechanical properties, good biocompatibility and biodegradability [1]. However, the poor corrosion resistance of magnesium alloy is the great obstacle to reach its clinical application. Rapid corrosion of magnesium alloy would cause the alkalization in the vicinity, and also result in subcutaneous hydrogen accumulation and premature loss of mechanical integrity [2]. Therefore, it is of great interest to enhance the corrosion resistance of magnesium alloys for their biomedical application. The corrosion mechanism of magnesium alloy has been studied extensively. Usually, the potential of second phase is nobler than that of magnesium matrix. During degradation, the second phases act as the cathode site, the magnesium matrix acts as anodic site. Then magnesium matrix is attacked by the micro-galvanic cell which is built up between the second phases and the matrix, but the second phases are protected. Tao Li et al [3] reports that the corrosion resistance of single-phase Mg-1.5Zn-0.6Zr is better than that of AZ91D because of the absence of the micro-galvanic cell. Y Lu et al[4] investigates the influence of aging time on the corrosion resistance of Mg-3Zn and the result indicates that aging treatment would deteriorate the corrosion resistance due to more micro-galvanic cells. Take the previous studies into consideration, the alloy with single-phase or less second phases may exhibit lower corrosion rate. Therefore, solid solution treatment is the conventional process to enhance the corrosion resistance of magnesium alloys.

In initial period, effects of Ce content on microstructure and performance of the Mg-2Zn-0.4Zr-xCe alloys were investigated. According to the results, the optimal performance was obtained with the addition of 0.6% Ce. Here, Mg-Zn-Zr is one biomedical magnesium alloy which possesses good biocompatibility and low cytotoxicity, and Ce was chosen as an additional rare earth element to improve the mechanical properties and corrosion resistance by considering availability and effects on the health. In this work, one novel kind of Mg-Zn-Zr-Ce was fabricated. This study is

---

*Corresponding author

aimed to enhance the corrosion resistance of Mg-2Zn-0.4Zr-0.6Ce (nominal composition) through heat treatment.

2. Experimental Procedures

2.1. *Materials and microstructure*

The Mg-2Zn-0.4Zr-0.6Ce alloy ingots were cast with Mg (purity ≥ 99.93%), Zn (purity ≥ 99.93%), Mg-25%Zr, Mg-25%Ce as raw materials. The alloys were melted at 720 °C for 5 min and protected by the mixed gas atmosphere of SF_6 and CO_2. The solution treatment was carried out at different temperature ranging from 430 to 470 °Cfor 10 h protected by N_2 atmosphere. Then the ingots were quenched in the water of 65 °C. The microstructure of the alloys were examined by scanning electron microscopy(SEM) in conjunction with energy dispersive spectrum(EDS). All the SEM specimens were polished to mirror surface.

2.2. *Immersion test and electrochemical measures*

Specimens were polished with emery papers up to 2000 grit. In the case of the weight loss rate tests, the specimens with the size of $\Phi18$ mm×5 mm were exposed to the SBF (Simulated Body Fluid) for 5 days. The ratio of degradation solution volume to the surface area of specimen is 30 ml/cm². The SBF is composed of 8.0 g/L NaCl, 0.14 g/L $CaCl_2$, 0.4 g/L KCl, 0.35 g/L $NaHCO_3$，0.1 g/L $MgCl_2 \cdot 6H_2O$, 1.0 g/L Glucose，0.06 g/L $MgSO_4 \cdot 7H_2O$, 0.06 g/L Na_2HPO_4, and 0.06 g/L KH_2PO_4. During the test, the temperature of the solution was kept at 37 °C using water bath, and the immersion solution was renewed every 24 h. After immersion test, the corrosion products were removed using the chromic acid (200 g/L Cr_2O_3+10 g/L $AgNO_3$), Then the specimens cleaned ultrasonically in alcohol, dried in open air. The dried specimens were weighed and the corrosion rates were calculated by 3 specimens at each state.

Electrochemical measures were carried out on NOVA Autolab (AUT84580). The specimens were molded into epoxy resin with only one side, with area of 1 cm² exposed for the test. The work surface was ground with emery papers up to 2000 grit. Three-electrode system was adopted, namely, a saturated calomel electrode (SCE) as a reference electrode, a graphite electrode as a counter electrode and the specimen as a working electrode. The specimens were exposed in SBF for 1 h in order to make sure the stabilization of the open circuit potential. OPC was recorded with an immersion time of 5 mins. EIS measure was tested with the frequency ranging from 0.1 Hz to 100 kHz and the voltage amplitude was 5 mV. The potentiodynamic polarization was performed at a scanning rate of 1 mV/s from -0.25 V in the cathodic direction to +0.4 V in the anodic direction based on OPC.

3. Result and Discussion

Fig. 1 shows the SEM image of the as-cast and quenched alloys. It could be clearly seen that the microstructure is composed of α-Mg and second phase, the grains become coarser, and the second phases dissolve gradually for the quenched alloys. When the temperature of solution treatment is below 450 °C, there are still a few of semi-continuous second phases along the grain boundary. But, with the increasing of temperature, the second phases appearing semi-continuous network shape begin to dissolve. According to the result of EDS in Fig. 2, the undissolved second phase is composed of Mg, Zn, Zr and Ce element. Therefore, the undissolved second phase is probably

Mg-Zn-Ce compound. It is reported that the Mg, Ce-containing compound possessed the high-temperature stability, and a small amount of Zn can dissolve into such compound, such as (Mg, Zn)$_{11}$Ce and (Mg, Zn)$_{12}$Ce[5]. Thus there still exist undissolved second phases at triple junction after solution treatment.

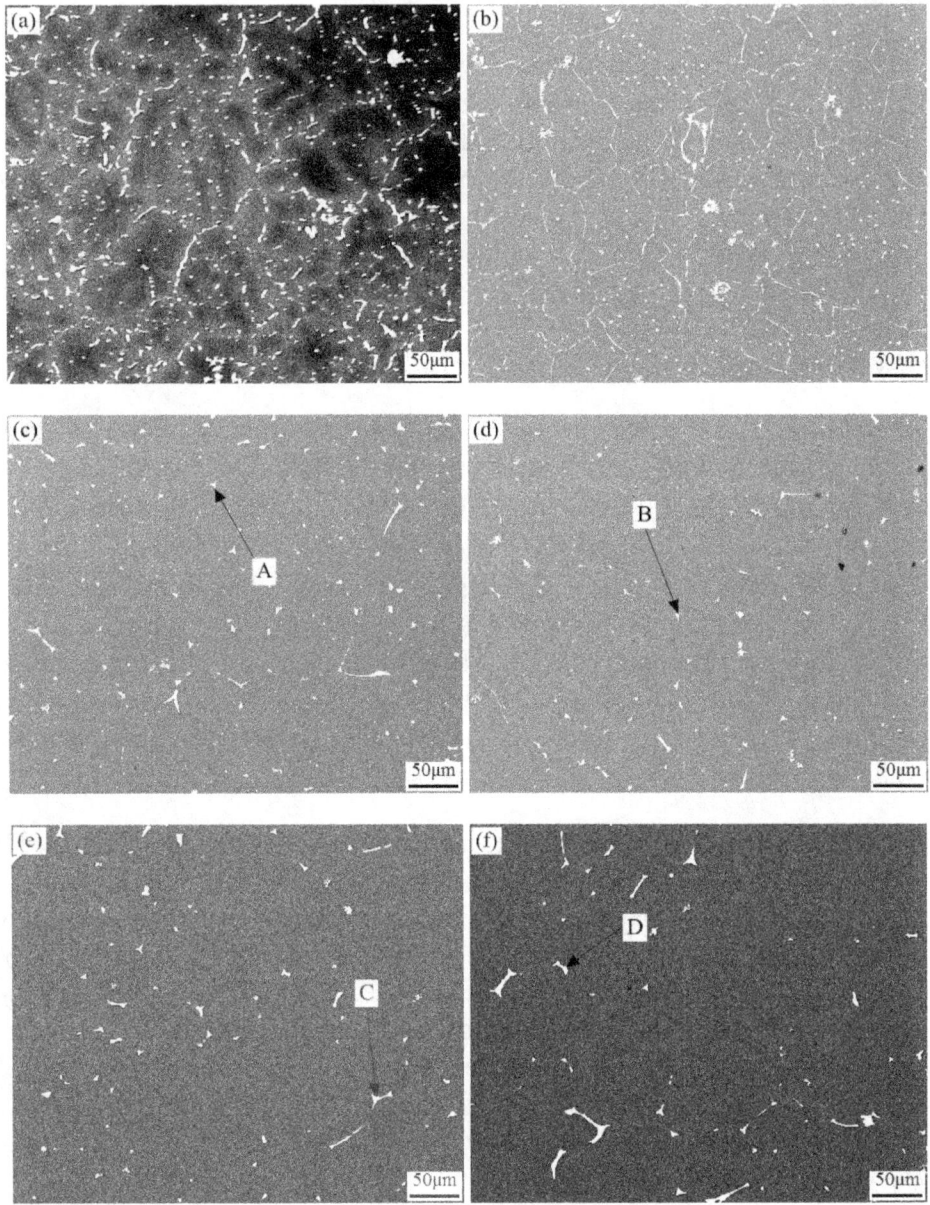

Fig. 1 The SEM image of the as-cast and quenched Mg-2Zn-0.4Zr-0.6Ce alloy

(a) as-cast alloy, (b) 430 °C, (c) 440 °C, (d) 450 °C, (e) 460 °C, (f) 470 °C

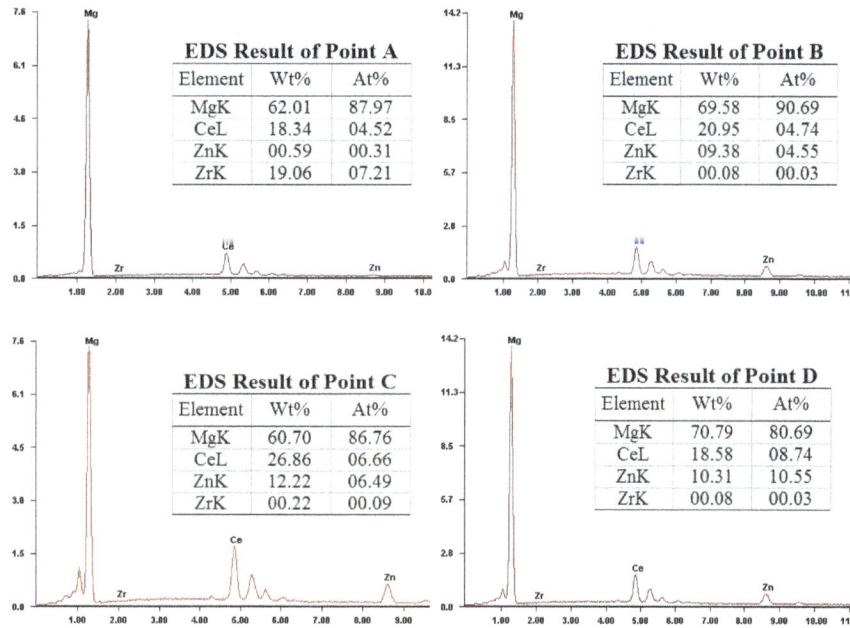

EDS Result of Point A		
Element	Wt%	At%
MgK	62.01	87.97
CeL	18.34	04.52
ZnK	00.59	00.31
ZrK	19.06	07.21

EDS Result of Point B		
Element	Wt%	At%
MgK	69.58	90.69
CeL	20.95	04.74
ZnK	09.38	04.55
ZrK	00.08	00.03

EDS Result of Point C		
Element	Wt%	At%
MgK	60.70	86.76
CeL	26.86	06.66
ZnK	12.22	06.49
ZrK	00.22	00.09

EDS Result of Point D		
Element	Wt%	At%
MgK	70.79	80.69
CeL	18.58	08.74
ZnK	10.31	10.55
ZrK	00.08	00.03

Fig. 2 The EDS result of point A, B, C, D in Fig. 1

Fig. 3 shows the corrosion rate of the Mg-2Zn-0.4Zr-0.6Ce alloys quenched at various temperatures. It demonstrates that the corrosion rate of the quenched alloys is lower than that of the as-cast alloy (about 1.142 mm·a^{-1}). When the temperature of solution treatment increases, the corrosion rate firstly decreases and then increases. And the optimal corrosion resistance of Mg-2Zn-0.4Zr-0.6Ce alloy is obtained at 450 °C, and is about 0.7293 mm·a^{-1}. Compared with as-cast alloy, the increase of the corrosion rate reaches 35.6%. That is to say, the solution treatment enhances the corrosion resistance and Mg-2Zn-0.4Zr-0.6Ce alloy quenched at 450 °C for 10 h possesses better corrosion resistance. According to the SEM microstructure of those alloys in Fig. 1, the solution treatment results in the reduction of second phase and diffusion of alloying element. And Mg-2Zn-0.4Zr-0.6Ce alloy quenched at 450 °C displays more homogeneous distribution of second phase and less second phases, so the alloy undergoes the slighter corrosion. It is clear that the enhancement in corrosion resistance is ascribed to the reduction of second phase and homogenization of alloying element [3].

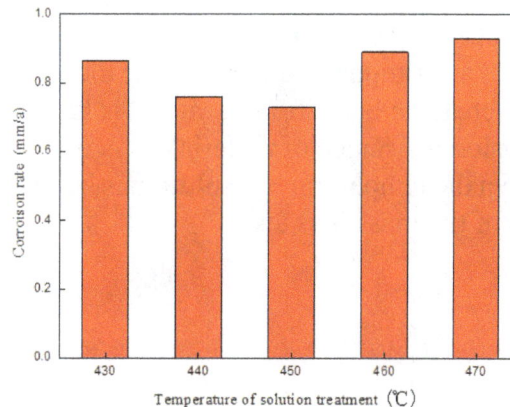

Fig. 3 The corrosion rate of the quenched Mg-2Zn-0.4Zr-0.6Ce alloy

Fig. 4 illustrates the OPC of the as- cast alloy and the alloy quenched at 450 °C versus immersion time in the SBF.

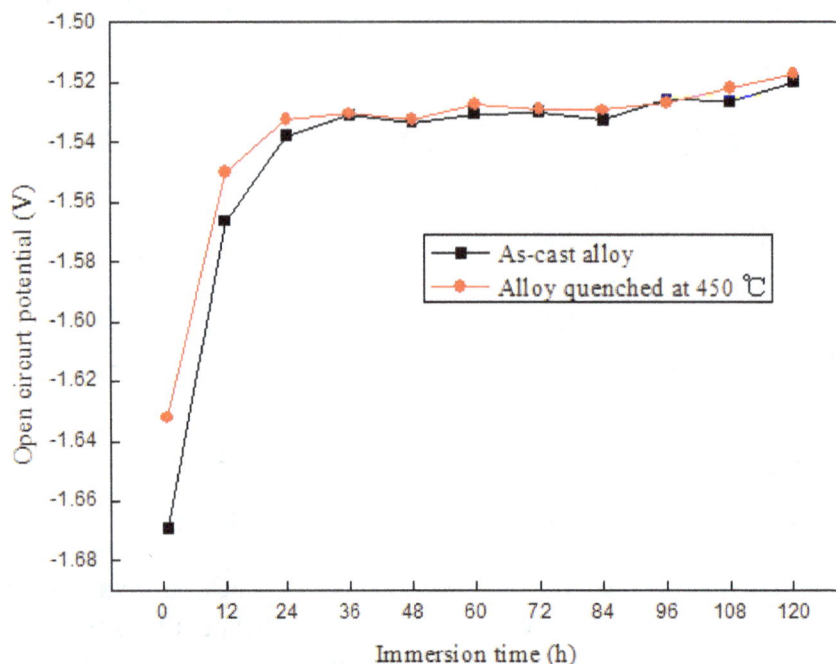

Fig. 4 Open circuit potential versus immersion time in the SBF

As it is shown, at the initial stage, the OPC of the alloys under two conditions shifts toward positive value rapidly, and then trends to be the constant OPC value for -1.529 V after immersion for about 24 h. Also, it is seen that the OPC value of the quenched alloy is more positive at the initial stage. Considering the OPC value represents the corrosive tendency of the material, the result indicates that the as-cast alloy more easily undergoes corrosion than those quenched at 450 °C. However, after the alloys immersed in the SBF for a period of time, the corrosive tendency of the alloys under two conditions not only decreases, but also represents no obvious distinction. The constant value of OPC can be related to the formation of the corrosion layer. After immersion for 24 h, the protective layer has been acquired, which prevents the matrix away from further corrosion.

The Nyquist plot of the quenched alloys acquired after immersion for 1 h is displayed in Fig. 5. As it is showed, the Nyquist plots are similar in shape, but the diameters of loop are different, which indicates the same corrosion mechanism but different rate. It is known that the larger capacitive arc diameter means the better corrosion resistance of material [6]. As is showed in the Fig. 5, the diameter of capacitive loop increases when the temperature of solution is below 450 °C, then decreases, which is in conformity with the result of the mass loss test.

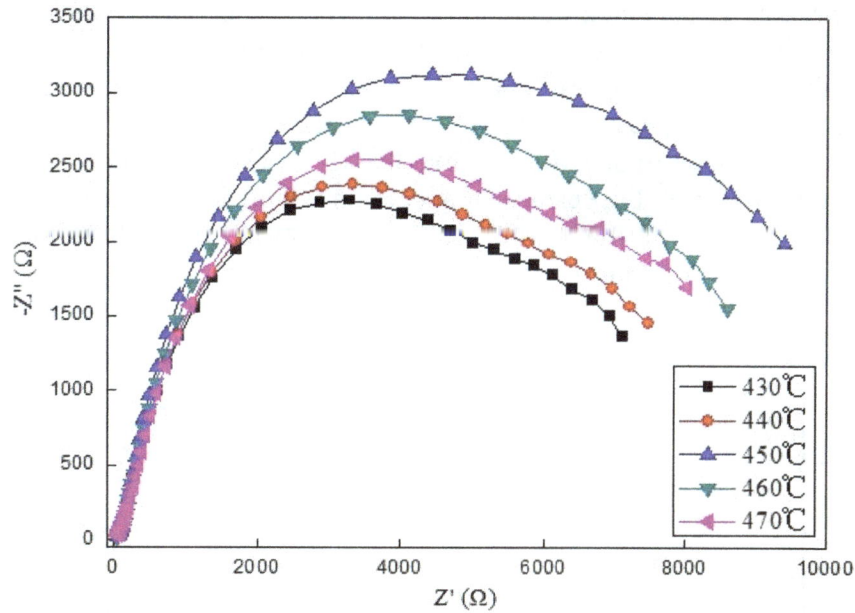

Fig. 5 EIS of Mg-2Zn-0.4Zr-0.6Ce alloy after solution treatment at different temperature

The polarization curve of the quenched alloys is showed in Fig. 6(a).

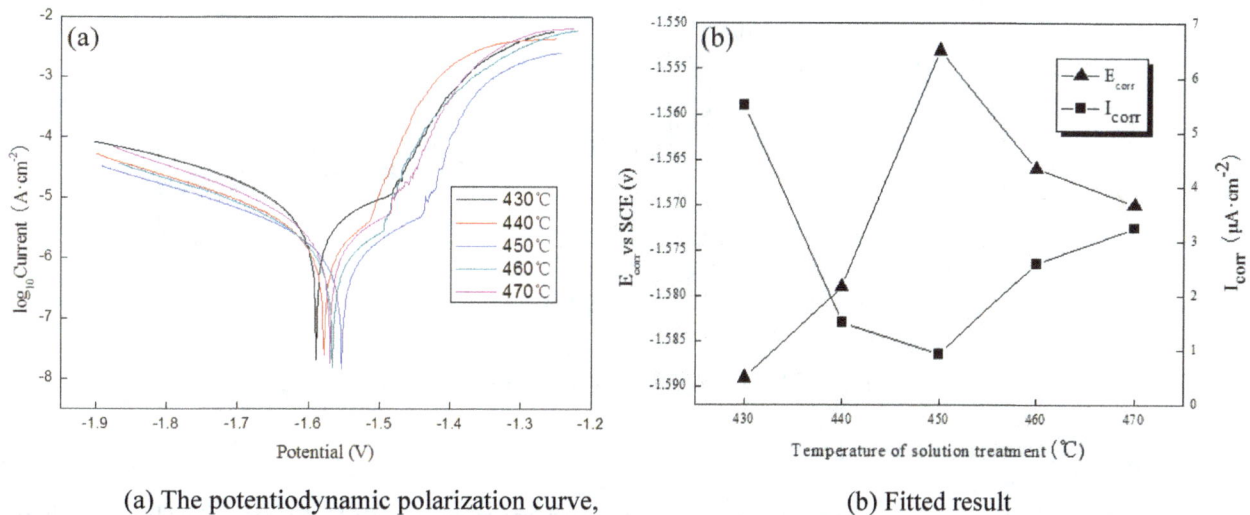

(a) The potentiodynamic polarization curve,

(b) Fitted result

Fig. 6 The potentiodynamic polarization curve and fitted result of Mg-2Zn-0.4Zr-0.6Ce alloy after solution treatment at different temperature

For the magnesium alloys, the cathodic polarization represents the hydrogen evolution and the anodic polarization represents the dissolution of magnesium. It is seen that the curves of the tested alloys are similar in shape, and there exists a step in anodic polarization area, which demonstrates the passivation of alloy occurs in the SBF [7]. The fitted result of polarization curves is showed in Fig. 6(b), and it is seen that the E_{corr} first shifts toward the positive value and then shifts to the negative potential, I_{corr} first decreases and then increases with the temperature of solution treatment increased. The lower the I_{corr} is, the better the corrosion resistance is [8]. As the result shows, the best corrosion resistance is obtained at 450 °C, which is in agreement with the result of the mass loss test.

The corrosive product is collected after immersion for 120 h. The XRD pattern of corrosion product is displayed in Fig. 7. It is clear that there exist the obvious diffraction peaks of Mg and Mg(OH)$_2$. Here, the Mg element originates from specimen. What's more, the XRD pattern demonstrates slight diffraction peaks of (Ca, Mg)$_3$(PO$_4$)$_2$ and Ca$_{10}$(PO$_4$)$_6$(OH)$_2$ (HA). Therefore, the corrosion layer on the surface of specimen is mainly Mg (OH)$_2$, (Ca,Mg)$_3$(PO$_4$)$_2$ and (HA). The corrosion product depositing on the surface of specimen would exhibit partially-protective ability due to decreased water access to magnesium substrate [9]. So the OCP shifts towards the positive value and then stabilizes at a constant.

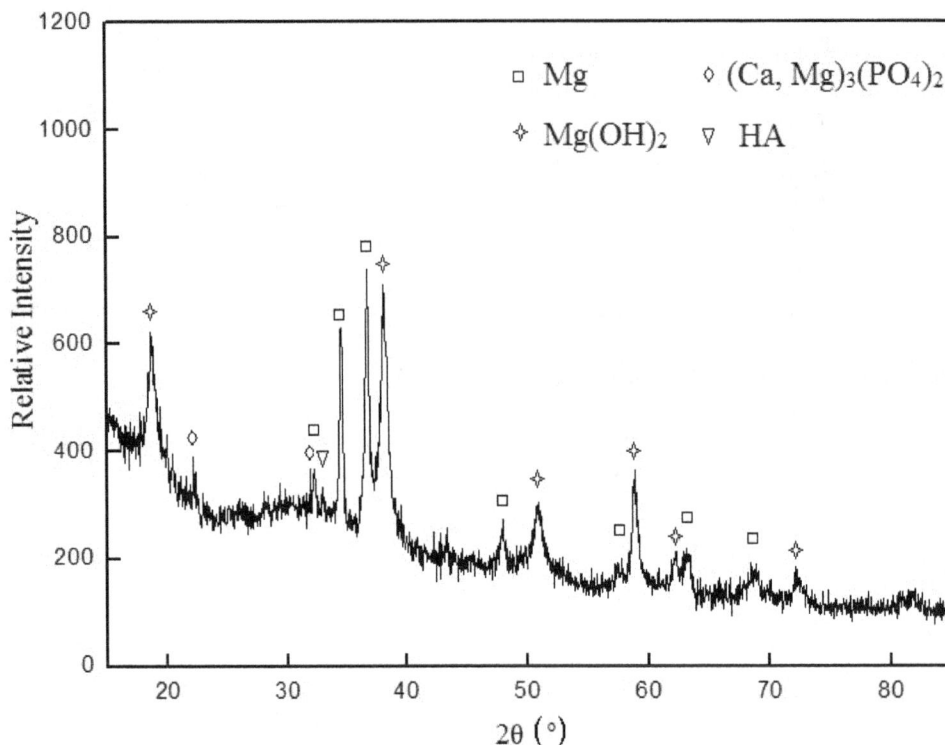

Fig. 7 The XRD pattern of corrosion product

4. Conclusion

For the quenched alloys, the grains become coarse and most of second phases dissolve. With the increasing temperature of solution treatment, the corrosion rate of Mg-2Zn-0.4Zr-0.6Ce quenched alloy firstly decreased, when the temperature is up to 460 °C, the corrosion rate rises slightly. The optimal corrosion resistance is acquired at 450 °C and the corrosion rate is about 0.7293 mm·a^{-1}.

Acknowledgments

The research is funded by Collaborative Innovation Center of Nonferrous Metals, in Henan Province, University Science Innovation Team of Henan Province: (2012IRTSTHN008) — New Nonferrous Metal Material and Major Program of Education Department of Henan (15A430024)

References

1. J.M. Seitz, R. Eifler, J. Stahl. Characterization of Mg-2Nd alloy for potential applications in bioresorbable implantable devices, J. Acta Biomaterials. 10 (2012) 3852-3864.
2. M.P. Staiger, A.M. Pietak, J. Huadmai, et al. Magnesium and its alloys as orthopedic biomaterials: a review, J. Biomaterials. 9 (2006) 1728-1734.
3. Tao Li, Yong He, Hailong Zhang. Microstructure, mechanical property and in vitro biocorrosion behavior of single-phase biodegradable Mg-1.5Zn-0.6Zr alloy, J. Journal of Magnesium and Alloys. 2 (2014) 181-189.
4. Y. Lu, Bradshaw A R, Y.L.Chiu. The role of precipitates in the bio-corrosion performance of Mg–3Zn in simulated body fluid, J. Journal of Alloys and Compounds. 25 (2014) 345-352.
5. HUANG Ming-li, LI Hong-xiao, DING Hua, BAO Li. Intermetallics and phase relations of Mg-Zn-Ce alloys at 400 C[J]. Transactions of Nonferrous Metals Society of China. 3 (2012) 539-545.
6. Yingwei Song, En-Hou Han, Dayong Shan, Chang Dong Yim. The role of second phases in the corrosion behavior of Mg–5Zn alloy, J. Corrosion Science. 7 (2012) 238-245.
7. M. Ascencio, M. Pekguleryuz. An investigation of the corrosion mechanisms of WE43 Mg alloy in a modified simulated body fluid solution: The influence of immersion times, J. Corrosion Science. 9 (2014) 489-503.
8. WANG Yong Ping, HE Yao Hua, ZHU Zhao Jin, et al. In vitro degradation and biocompatibility of Mg-Nd-Zn-Zr alloy, J. Materials Science. 17 (2012) 2163-2170.
9. Peng-Wei Chu, Emmanuelle A, Marquis. Linking the microstructure of a heat-treated WE43 Mg alloy with its corrosion behavior, J. Corrosion Science. 10 (2015) 94-104.

Investigation on the Strength and Fatigue of AC-14 High Holding Power Anchor

Wei-Guang Zhang[1], Jin-Tai Li[1,*], Guo-Yan Peng[2], Yuan Li[3], Wen-Xian Tang[1]

1. School of Mechanical Engineering, Jiangsu University of Science and Technology, China

2. Chongqing Productivity Council, China

3. Chongqing Qianwei Science & Technology group Co., Ltd, China

*Email: 1371567303@qq.com

AC-14 high holding power anchor is one of the best recognized by the world. The strength and fatigue characteristics of AC-14 high holding power anchor directly related to the safety of the ship. Taking 3097mm fluke width of AC-14 high holding power anchor as an example, according to the tensile test of the anchor in the anchor technology condition of 548-1996 GB/T, we established the stress distribution model of the anchor and combined with the finite element analysis to carry on the static analysis and fatigue analysis. The results show that the maximum stress of anchor is 207.5MPa lying on the fluke, this value is less than the yield limit of material 230MPa ; The maximum displacement of the anchor is 15mm appearing in the fluke tip, this value is very small; The fatigue life of the anchor is 3.5×10^{13} times, it's very long. So the structural strength, rigidity and life of the anchor meet the requirements of the ship.

Keywords: AC-14 high holding power anchor; finite element analysis; fatigue analysis.

1. Introduction

The high holding power anchor is widely used in large and medium-sized ships. Ships in the wind, waiting for the tide, narrow road U-turn and other emergency situations should thrown the high holding power anchor down into the water to resist wind and wave forces [1-2]. In the use of the anchor, the damage can directly affect the safety of the ship and the crew [3]. Therefore, it is very important to study the safety of the high holding power anchor.

At present, the research on the high holding power anchor is mainly focused on the research of the grasping force and the performance of the anchor at the bottom of the sea, and the research methods are mostly real ship test [4-5]. In the aspect of finite element analysis, the strength analysis of the OFFSHORE ST anchor is carried out by Lei Lin et al [6]. The anchor finite element simulation of clay pull is carried out by Yang Guanghui et al [7]. The finite element simulation of lifting ship anchor is carried out by Tang Wenxian et al [8]. But single time finite element analysis can't determine whether the anchor is in use or not after many years of working, also can't predict what problems will arise. So, combine the strength analysis and fatigue analysis can be able to explain exactly where the problem lies, which would make our design more reasonable, also further for us save the cost.

*Corresponding author

Therefore, this paper takes AC-14 high holding power anchor as an example, through the finite element analysis method, analysis AC-14 high holding power anchor's Strength and fatigue.

2. Theoretical Model of the AC-14 High Holding Power Anchor

2.1. Geometric model of the AC-14 high holding power anchor

This paper selects 3097mm fluke width of AC-14 high holding power anchor as the research object, the application of SolidWorks software to create the geometric model as shown in Fig. 1.

The anchor geometry roots in the national standard GB/T 3972-2005 "AC-14" high holding power anchor. In addition, due to the complex overall structure of the model, the chamfer of modeling is ignored.

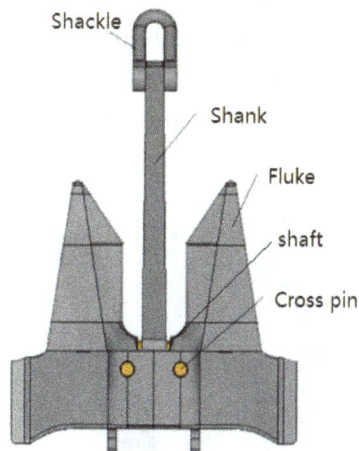

Fig. 1 AC-14 high holding power anchor

2.2. Mechanical model of the AC-14 high holding power anchor

The external load of the high holding power anchor can be reference to "anchor technology condition" in the national standard GB/T 548-1996 and "the tensile test of anchor in material and welding specification". As shown in Fig. 2, the tensile load is 987kN. One tension point is at the end of the anchor shackle, the other is at the place where L/3 is far from flukes tip.

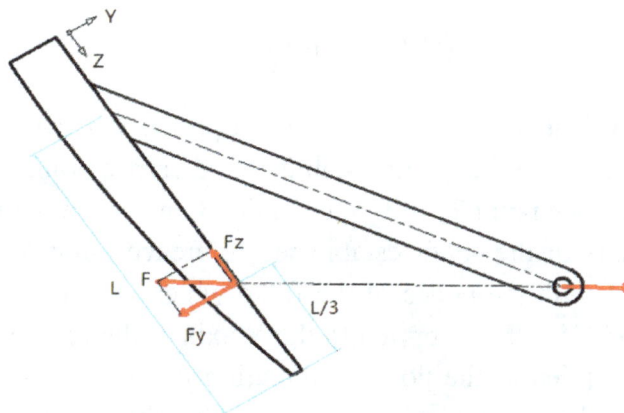

Fig. 2 Schematic diagram of Hall anchor tension test

In analysis, the whole coordinate system is based on the intersection point between the anchor line and the axis of the shaft. Fluke tip points the direction of Y axis. In this global coordinate system, the angle between the fluke and the shank is 35°, the angle between the tension force and the shank is 14.24°, the angle between the tension loading direction and the X-Y plane is 41.76°, so the force of the Y, Z axis is:

$$F_y = 987\sin(49.24) = 657.3537KN \ (1) \qquad F_z = 987\cos(49.24) = 736.244KN \ (2)$$

2.3. *Statics analytical model of the AC-14 high holding power anchor*

This paper uses software to mesh the model. The anchor element type is C3D4, the number of the element is 307758, the number of the nodes is 72308, the shank element type is C3D4, the number of the element is 36180, and the number of the nodes is 42159. In addition, the shaft can be replaced by the MPC hinge unit in consideration of its simple cylinder shape. The material of fluke and shank both is cast steel, whose elastic modulus is $2.1 \times 10^5 MPa$, and the Poisson's ratio is 0.3.

Fig. 3 AC-14 high holding power anchor mesh model

The anchor's static analysis is in Abaqus. The contact property between the fluke and the shank is defined as small sliding contact. this paper established reference point 1 at where is L/3 far from flukes tip, established reference point 2、 reference point 3 and reference point 4 at the position of three equal parts in the axis of the shaft, established reference point 5 at the upper shank hole. Coupling reference point 1 with the nodes of the fluke which is L/3 far from flukes tip and near the shank and a force of 657353.7N is applied to the Y axis of the reference point 1, and the other force of the 736244N is applied to the positive direction of the Z axis of the reference point 1. Couple reference point 3 and reference point 4 respectively with reference point 2. The shank and fluke's nodes which are in contact with the shaft are respectively connected with the corresponding reference point 2、 the reference point 3 and the reference point 4. The nodes in the upper hole of the shank are coupled with the reference point 5, and their x, y and z degree of freedom are restricted. The whole model only has a static load analysis step.

2.4. *Fatigue analysis model*

AC-14 high holding power anchor's frequency of use is very high. Its strength analysis is difficult to determine whether the anchor can meet the strength requirements accurately. So, it is necessary to use the fatigue analysis software Femfat to carry on fatigue analysis of AC-14 high holding power anchor.

The material of fluke and shank both is cast steel, the strength limit of it is 430MPa, and the yield limit is 230MPa. The flexural strength and pulse strength of the material are calculated by the software, which are 200.5MPa and 378.4MPa, respectively. This paper applied 90000 times symmetrical cyclic stress to the AC-14 force anchor.

3. Results Analysis and Discussion

3.1. *The analysis and verification of geometric model of the AC-14 high holding power anchor*

In the geometric model of the AC-14 high holding power anchor, this paper apply 657353.7N force to the negative direction of the Y axis, and apply 736244N force to the direction of Z axis in the reference point 1. Results shows that, in the reference point 5, the maximum support force of the Y axis is 657354N, the maximum support force of the negative Z direction axis is 736244N. This result is in agreement with the mechanical model of the AC-14 high holding power anchor.

3.2. *Static analysis for the AC-14 high holding power anchor*

Because the material cast steel, this paper uses the first strength theory to check, that is theory of maximum tensile stress. The maximum stress value of the fluke is 207.5MPa appearing at the contact position between the fluke and the shaft, which is mainly due to contact property between the fluke and the shank is line contact. 207.5MPa is less than the strength limit of the material 430MPa, but also less than the yield limit of the material 230MPa, so the strength of the fluke meets the requirements. But it's very close to the yield limit of the material, in order to ensure that the anchor is not easy to be damaged, the position between the fluke and the shaft should have been done surface heat treatment.

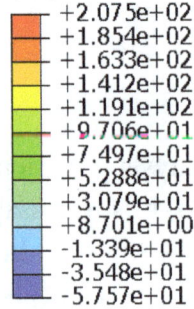 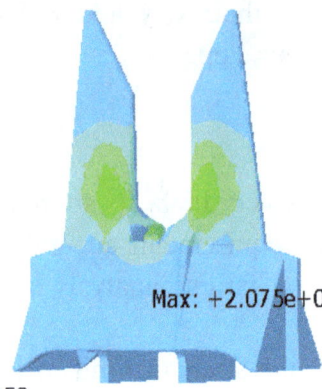 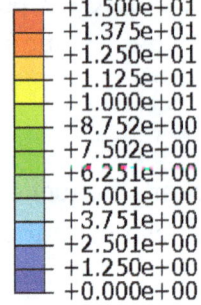 

(a) Fluke stress contour   (b) Fluke displacement contour

Fig. 4 Stress and strain contour of anchor fluke

Fig. 4- (b) shows that the maximum displacement of fluke is 15mm, appearing in the tip of it, This value is small to the overall size of the AC-14 high holding power anchor, so the fluke's deformation is little, its stiffness meets the requirements.

As shown in Fig. 5, its stress and strain contour of the shank. Fig. 5- (a) shows that the maximum stress of shank is 98.19MPa appearing at the contact position between the fluke and the shaft. The maximum stress value of the shank is 98.19MPa is less than the strength limit of the material 430MPa, but also less than the yield limit of the material 230MPa, so the strength of the shank meets the requirements.

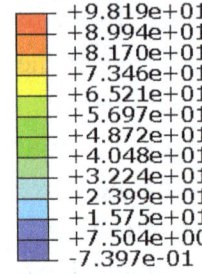 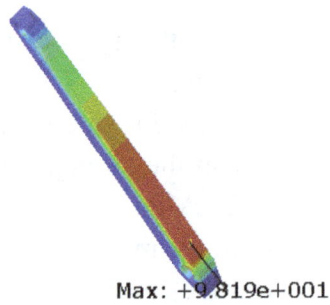 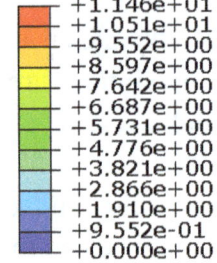 

(a) Shank stress contour   (b) Shank displacement contour

Fig. 5 Stress and strain contour of anchor Shank

Fig. 5- (b) shows that the maximum displacement of shank is 11.46mm, appearing in the bottom of it, this value is small to the overall size of the AC-14 anchor, so its stiffness meets the requirements.

3.3. *Fatigue analysis for the AC-14 high holding power anchor*

As shown in Fig. 6, the minimum fatigue reciprocal of damage value of fluke is 3.938×10^8, appearing in the bottom of the fluke, this is because the stress there is more concentrated than some other parts of the fluke, it indicates that this position is the dangerous position. According to the formula 3: fatigue life= Stress cycle times/ damage value. The fatigue life of the anchor is calculated, which is 3.5×10^{13} times. The fluke's fatigue life is very long. It's not easy for the fatigue damage to occur, which meets the requirements of frequent use of ships. But the dangerous position should been strengthened.

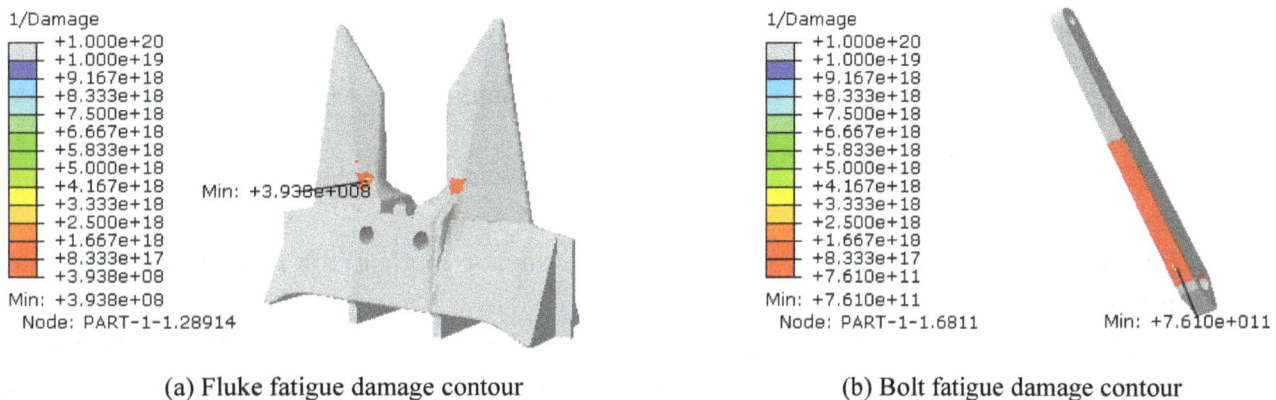

(a) Fluke fatigue damage contour

(b) Bolt fatigue damage contour

Fig. 6 The reciprocal contour of fatigue damage value of fluke

As shown in Fig. 6-(b), the minimum fatigue reciprocal of damage value of shank is 7.61×10^{11}, appearing in the contact position of the fluke and the shank. According to the formula 3 as the same: fatigue life= Stress cycle times/ damage value. The fatigue life of the shank is calculated, which is 6.8×10^{16} times. The shank's fatigue life is very long. It's not easy for the fatigue damage to occur, which meets the requirements of frequent use of ships.

4. Conclusion

This paper take 3097mm fluke width of AC-14 high holding power anchor as Example, through the establishment of three-dimensional model, mechanical model, the static analysis, fatigue analysis, we get the specific conclusions as follows:

The finite element analysis results shows that, in the reference point 5, the maximum support force of the Y axis is 657354N, the maximum support force of the negative Z direction axis is 736244N. This result is in agreement with the mechanical model of the AC-14 high holding power anchor. The results verify the mechanical model.

Through the static analysis, we know that the maximum stress of the anchor is 207.5Mpa, which is less than the material yield limit 230MPa, so the anchor's strength meets the requirements and is not easy to damage, but it's very close to the yield limit of the material, in order to ensure that the anchor is not easy to be damaged, the position between the fluke and the shaft should have been done surface heat treatment. The maximum displacement of the anchor shows in fluke tip, and the value is 15mm. This value is very small to the anchor's size, so the rigidity of the anchor is in line with the requirement and is not easy to deformation.

Through fatigue analysis, we know that the fatigue life of the anchor is 3.5×10^{13} times, so the fatigue life of the anchor is very long and meet the requirements of frequent use of ships. But the most damaged position is the bottom of the fluke, and the damage is more concentrated than some other parts of the fluke, where is the dangerous position and should been strengthened.

References

1. Xia Zhongguo. Ship structure and equipment [M]. DaLian: Dalian Maritime University Press, 1998: 58-61.
2. Jiang Zhiqiang, Yu Yang. History and Current Situation of the Development of Modern Ship Anchor [J]. China Water Transport: 2013, 13(2): 120-123.
3. Xu Peijin, Yu Rongyuan. New exploration of the anchor method for large ship[J]. Marine Technology: 1994, 02:15-16
4. Lee JH, Seo BC, Shin HK. Experimental Study of Embedding Motion and Holding Power of Drag Embedment Type Anchor (DEA) on Sang Seafloor [J], Journal of the Society of Naval Architects of Korea, 2011, 48(2): 183-187
5. Shin HK, Seo BC, Lee JH. Experimental Study of Embedding Motion and Holding Power of Drag Embedment Type Anchor on hard and soft Seafloor [J], International Journal of Naval Architect and Ocean Engineering, 2011, 3: 193-200.
6. Lei lin, Wang Zhixiang, Zhang Ming, Zhang Guopeng. Analysis on ST OFFSHORE Anchor 3D modeling & FEA [J]. Journal of Chongqing jiaotong university (natural science). 2010, 29(2): 303-306
7. Yang Guanghui, Xu Jian, Duan Menglan. Modeling and Analysis for Anchor Process of Hall Anchor under Clay Condition [J]. Journal of graphics. 2015, 36(2): 193-197.
8. Tang Wenxian, Wu Wenle, Zhang Jian, Wang Xiaorong, Sun Ze, Li Jintai. Resistance Analysis of Lifting Anchor Groundbreaking Process for Mooring System [J]. Ship & ocean engineering. 2015, 44 (6): 31-36.

Studies on Stress Concentration and Fatigue Damage for Ferromagnetic Material Based on Permeability Testing Technology

Mei-Fang Yang*, Shang-Kun Ren, Zhen-Yan Zhao

Key Laboratory of Nondestructive Testing of Ministry of Education, Nanchang Hangkong University, Nanchang 330063, Jiangxi, China
Email: 815138862@qq.com

Permeability testing technology is an evaluation method which based on the electromagnetic induction principle which states that induced voltage is proportional to the change in rate of flux in the closed magnetic circuit of probe to test the permeability change of specimen. It can detect various changes related to the permeability in a certain area of the component with high precision such as stress concentration, fatigue damage, aging and decay, etc. Taking Q235 and 45 # steel as examples, the relationships between the detection signal and the tensile stress, residual stress and fatigue damage were studied from experiments. The results show that permeability testing technology can effectively measure the stress state of rod material specimen and the maximum stress of rod material specimen which had been suffered before. We can calculate the maximum stress by measuring the residual stress after stress was applied to the rod material specimen. The detection sensitivity of the fatigue damage was lower than the stress concentration. The detection sensitivity of fatigue damage of the low carbon steel Q235 was greater than the medium carbon steel 45 # steel. The research indicated that the permeability testing technology has a broad application prospects.

Keywords: Non-destructive testing; permeability detection; residual stress; fatigue damage.

1. Introduction

In modern industry, a large number of steel components are used in the aerospace, railway, power, pressure vessels and other industries. Steel components will undergo stress concentration and fatigue damage due to stress, fatigue loading. The internal workings of the media or the outside working environment may lead to stress corrosion, function of aging and fatigue fracture during use which may cause major accidents, ultimately leading to disasters affecting countries and its people [1-2]. Therefore, in the non-destructive testing of the component, detecting the stress concentration and fatigue damage of critical areas quickly, accurately and conveniently has great significance in the prevention of the fracture failure of components and the occurrence of a major disaster [3-4]. Magnetic permeability testing technology is a kind of evaluation method to detect stress concentration and fatigue damage [5-6]. This method has very high detection sensitivity. It not only can detect the ferromagnetic material degree of stress concentration and the fatigue damage condition of the specimen but can also detect martensite-austenite transformation, mechanics, toughness-brittleness transition, dislocation defect density, cementite ferrite and pearlite transformation, and grain interior and grain boundary transition, etc. in the detection of ferromagnetic materials [7-8].

2. Detection Principle

Detect probe structure is shown in Figure 1. The probe structure comprises a detection coil and an excitation coil, which are winded on a tubular plastic frame, and is suitable for detecting rod ferromagnetic material specimen. When testing, the test specimen was put into the tube of the sensor, and the excitation coil was connected with an AC voltage, and a closed magnetic circuit was formed by the external environment through the magnetic field of ferromagnetic specimen. Component of the stress concentration caused by permeability changes causes a closed magnetic circuit. Magnetic resistance and flux change. By detecting the signal induced, permeability changes in the specimen is measured. Thus specimen stress distribution and the degree of fatigue damage was measured.

A constant voltage source was selected as the source of excitation. According to the principle of electromagnetic induction, the output signal of the detection coil changes with the change of magnetic flux density on tube center region. By the induction coil voltage value can reflect changes in the permeability of ferromagnetic materials.

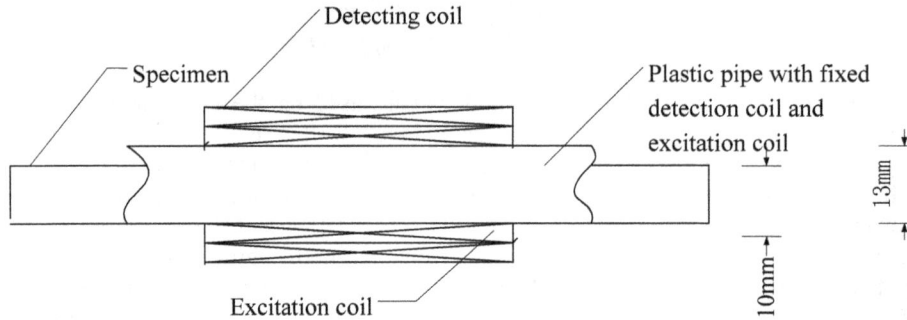

Fig. 1 Probe sensor structure diagram

The method realizes the detection of stress concentration and fatigue damage of ferromagnetic specimens. The number of turns of the detection coil was **N1**, and the number of turns of excitation coil was **N2.** The permeability of the tube part of the sensor center was composed of a closed magnetic circuit and an external environment. The average equivalent sectional area of the test piece was **S**, and the magnetic flux density was **B**. In the equivalent closed loop, the induction signal (ε) of detection coil can be expressed as:

$$\varepsilon = N_1 S \frac{dB}{dt} = N_1 S \frac{dB}{dH} \frac{dH}{dt} = N_1 S \mu \frac{dH}{dt} \tag{1}$$

permeability of test pieces is **μ**, and $\mu = \frac{dB}{dH}$. $u = U \sin \omega t$ is the excitation voltage power supply voltage, $H = cN_2 \frac{U \sin \omega t}{Z'}$ is excitation magnetic field, input impedance is **Z'**, **c** is a constant that is related to the number of turns and the structure of the excitation coil, therefore:

$$\varepsilon = N_1 S \mu \frac{dH}{dt} = N_1 N_2 S \mu c U \omega \cos \omega t / Z' \tag{2}$$

It can be seen that the detection signal (ε)is direct proportional to magnetic permeability(μ),It can be a direct response to the test specimen permeability. By detecting the sensor signal, determining changes in the permeability of the specimen, the ferromagnetic member of stress concentration and fatigue damage status is determined.

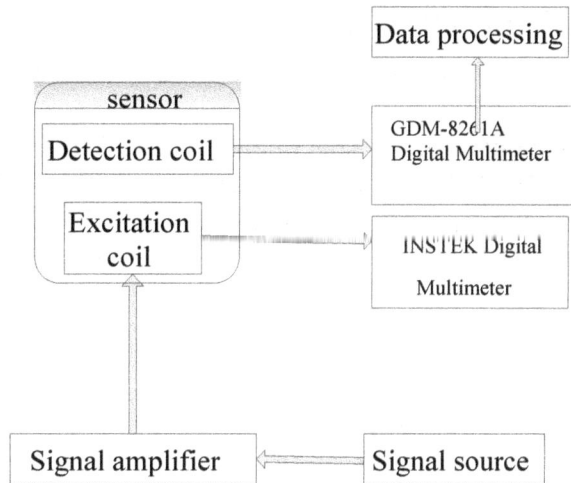

Fig. 2 Test device structure diagram

The excitation signal frequency of the sensor was 250Hz. the excitation voltage was 5V. The excitation coil and the detection coil both had 400 turns and the excitation signal waveform was a sine wave. All the test specimens were annealed at the test. In the test, specimen was fixed on a stretcher and the sensor was fixed on the specimen. The excitation coil was connected to a sinusoidal AC voltage of 5V and the detection coil was connected to GDM-8261A Digital Multimeter. Test device structure diagram is shown in Figure 2.

3. Test Equipment and Test Methods

Tensile test using tensile test equipment DW-100 electronic tensile testing machine at normal temperature state for carbon structural steel Q235 and medium carbon steel 45# was conducted at constant velocity tensile loading. The maximum tensile force could not exceed the maximum force 100kN. The loading rate was 2mm / min.

The fatigue test electronic equipment INSTRON 8801 fatigue testing machine for Q235 and 45 #steel was carried out at normal temperature stress ratio 0.1, sine wave, loading frequency 15Hz, static load 15.4kN, and dynamic load 12.6kN. The maximum tensile strength test fatigue testing machine could not exceed the maximum force 100kN and maximum load frequency couldn't exceed 50Hz.

To ensure the accuracy of test results, each of the three test materials were tested and each material dimensions were consistent. Each specimen was conveniently labeled with the label number to distinguish them clearly.

4. Test Results and Analysis

Static load stretching and unload stretching tests on Q235 and 45 # steel were carried out. Static load stretching determines the detection signal change with the tensile loads, which is getting the trend of permeability detection signal when the tensile loads are increased at a constant rate. However, unload stretching determines the relation between the detection signal and residual stress of ferromagnetic specimen by testing the permeability detection signal after unloading when a certain tensile stress was applied to the ferromagnetic specimen. The relation between the detection signal and residual stress of ferromagnetic specimen was studied.

4.1. *Yield strength and tensile strength of specimens*

Yield strength and tensile strength of the specimen can be obtained by measuring the stress-strain curve of the specimen.

Figure 3 is the stress-strain curve of the specimens. Figure 3 (a) shows that the yield strength of Q235 specimen is 248MPa (20.9kN) and tensile strength is 388MPa (32.7kN). Figure 3 (b) shows that the yield strength of 45 # specimen is 382MPa (30kN) and tensile strength is 603MPa (47.4kN). The yield strength of Q235 steel is 248MPa, which is close to nominal yield strength, 235MPa. Yield strength of 45 # steel is 382MPa, which is close to nominal yield strength of 355MPa.

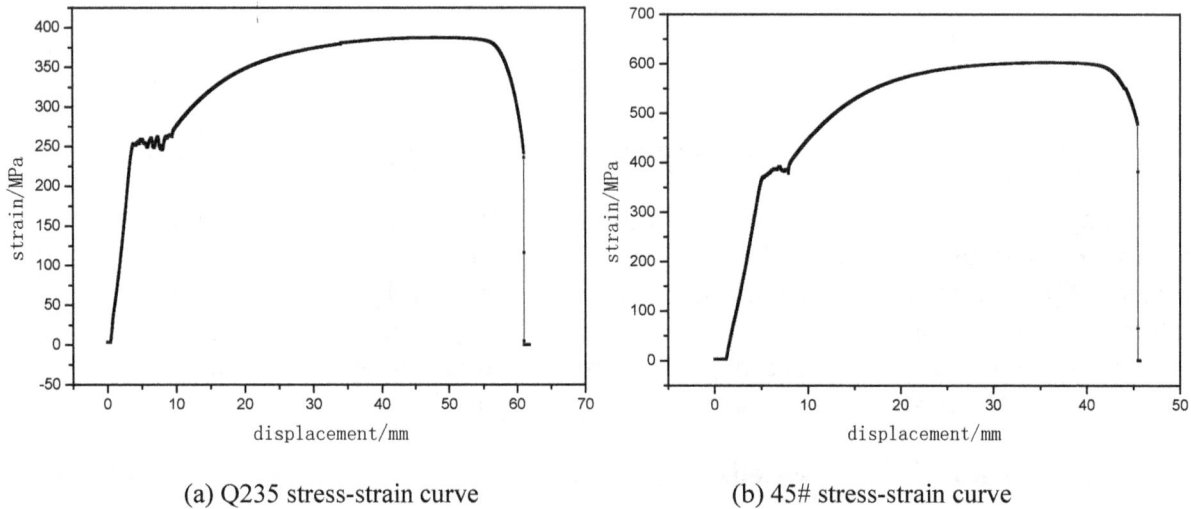

(a) Q235 stress-strain curve	(b) 45# stress-strain curve

Fig. 3 Stress -strain curves of specimens

4.2. *The relationship between the detection signal and the tensile stress*

The sensor was fixed on the specimen surface and the specimen was fixed on stretcher. With stress increasing constantly, the relationship between detection signal and the tensile stress is shown in Figure 4.

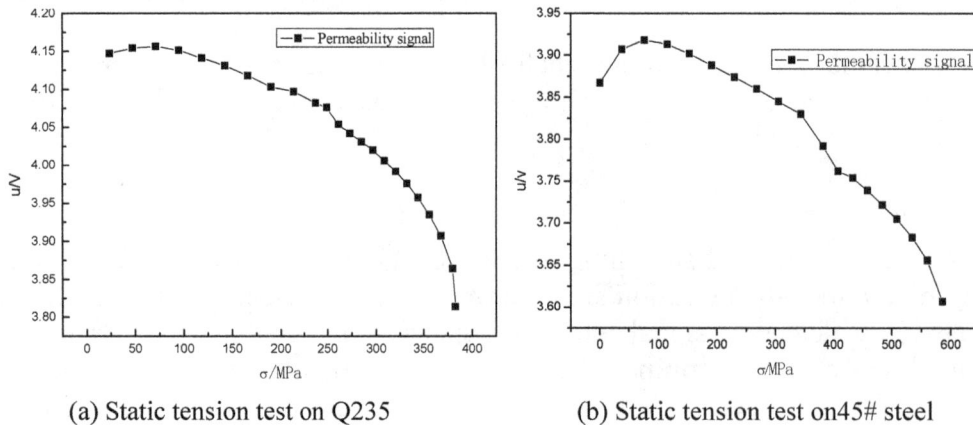

(a) Static tension test on Q235	(b) Static tension test on45# steel

Fig. 4 The relationship between the detection signal and the tensile stress

Figure 4 shows the variation of the detection signal with the change of the specimen tensile stress. Fig. 4(a) and Fig. 4(b) respectively show the relationship between the detection signal of Q235 and 45# steel specimens with changes in the tensile stress. Fig. 4 (a) shows that the detection signal is slightly increased in the early stage of the tension and is reduced slowly when the stress was lower than the yield strength (248MPa). When tensile stress is within the range from 248MPa to 388MPa, the detection signal is reduced rapidly. Fig. 4 (b) shows that for the stress within the range of 75MPa, the signal is slightly increased. Within the range from 75MPa to 382MPa, the detection signal is decreased slowly with the increase in stress. In the range from 350MPa to 600MPa, the detection signal is reduced quickly with the increase in stress.

Fig. 4 shows the detection signal is related to stress state of the specimen. The change in detection signal reflects the change in permeability. Fig. 4 shows that the variety of the magnetic permeability can be judged through the stress state of the specimen. The absolute amount of change in the signal of Q235 is 0.33V and the relative change in signal could reach 8%. For 45 # steel, the absolute signal variation is 0.3V with the relative change in signal could reach 7.6%.

4.3. *The relationship between the detection signal and the tension residual stress*

In a manner similar to the static tensile test, the test component was fixed to the machine. With stress increasing constantly, at each step the detection signal values of load tension and unload tension was recorded.

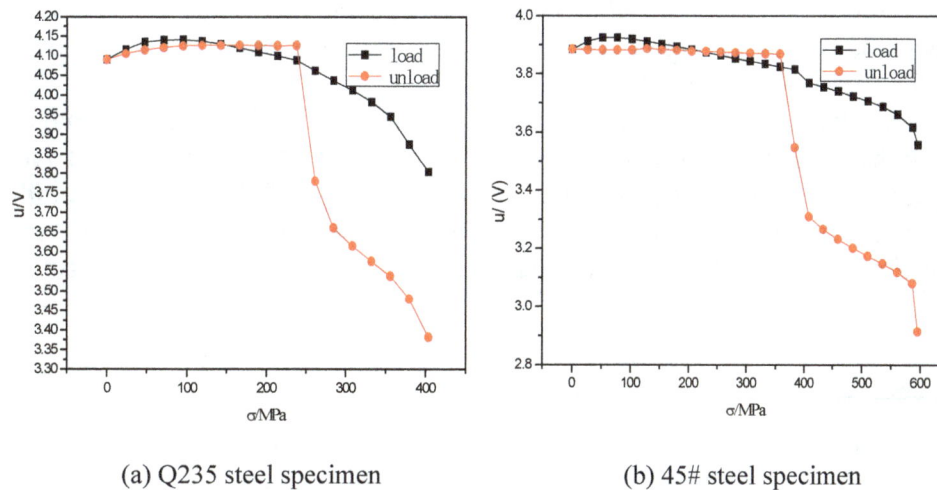

(a) Q235 steel specimen   (b) 45# steel specimen

Fig. 5 detection signal changing with tension and residual stress

Fig. 5 shows the relationship between the detection signal and the change of the residual stress of the tension and the loads. The Fig.5 (a) and (b) separately indicate the changing relation of test signals of Q235 and 45 # steel according to the stress and residual stress change.

Fig. 5 shows that the detection signal of the residual tensile stress is more sensitive and more suitable for the detection of the specimen before the stress situation. Throughout the residual stress in the force during the test, the absolute amount of change in the signal of Q235 was 0.8V and the relative signal change amounted up to 20%. The absolute signal variation of 45# steel was 1.0V and the relative signal change reached 26%.

Figure 5 (a) and Figure 5 (b) show that the stress state of 45# steel and Q235# are very different from the stress and the residual stress state. The detection signal varies greatly with the increase in the residual stress and the change in the detected signal, which is obviously larger than the stress. Description permeability caused by the residual stress is greater than the tension.

Experimental results show that detection signal independent of the residual stress remains substantially constant in elastic deformation such that tensile stress is less than the yield strength to Q235 steel and 45# steel material. When the tensile stress is greater than the yield strength, the detection signal decreases obviously with the increase in the residual stress.

Fig. 5 (a) shows that there is a dramatic change in the detection signal of the residual stress when the yield point is near 250MPa due to structural disorder increased by the elastic deformation of the material into the plastic deformation. When the tensile stress was greater than 300MPa, the performance of the detection signal saw a slow decline until the fracture caused by residual stress. This stage is the necking stage and the sample begins to undergo uneven plastic deformation, atomic slip blocks the accumulation of defects increases the stress even more. 45# steel and Q235 both have similar, regular pattern.

4.4. *The effect of fatigue damage on the detection signals*

It is similar to static load and the data was measured accordingly by subjecting the specimen to a hundred thousand cycles until the specimen was pulled off.

Fig. 6 shows the relation between the detection signal and the number of fatigue cycles, Fig. 6 (a) and Fig. 6 (b) separately indicate Q235 and 45#. Fig. 6 (a) indicate the regular of a low cycle fatigue, the maximum tensile strength of 28kN was greater than the yield strength of 20.9kN; Fig. 6 (b) reaction was the regular of high cycle fatigue of 45#, the maximum tensile strength of 28kN was lower than the yield strength 30kN.

As it can be seen from Fig. 6, the detection signal sensitivity of fatigue damage detection is lower than the stress concentration detection. Throughout the test for detection of fatigue damage process, the signal of relative variation of Q235 was 0.13V. The relative change could reach 3.3%. The signal of relative variation of 45# was 0.035V. The relative change could reach 1%. Tests showed that the detection sensitivity of detection techniques was related to the type of steel and carbon content.

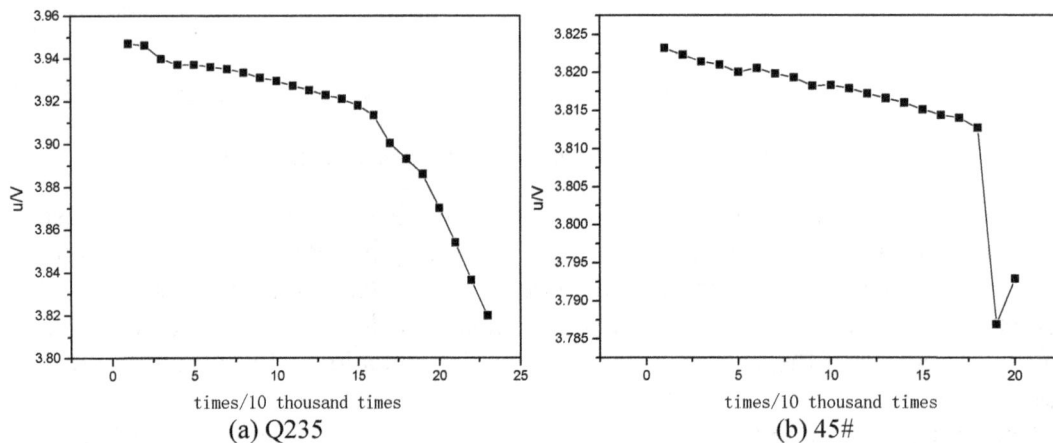

(a) Q235   (b) 45#

Fig. 6 The relationship between the detection signal and the number of cycle fatigue

Fig. 6 (a) shows that for the fatigue before the 160 thousand cycles, the detection signal slowly decreases, the signal change of 33mV, the detection signal variation of 25% was caused by 75% of cycles of fatigue. When the number of times of fatigue ranged from 160 thousand times to 220 thousand times, detection signal was sharply decreased, the signal variation reached 97mV, the detection signal variation of 25% is caused by 75% of cycles of fatigue. It can be observed that the detection technique is feasible for fatigue damage detection and residual life assessment of Q235.

Fig. 6 (b) shows that for the fatigue before the 180 thousand cycles, the detection signal decreases slowly, the signal change of 10mV, the detection signal variation of 29% is caused by 95% of cycles of fatigue. When the number of times of fatigue range from 180 thousand times to190 thousand times, the detection signal decrease sharply, the signal variation can reach 25mV.The detection signal variation of 75% is caused by 71% of cycles of fatigue. When the signal due to the significant changes in the specimen is close to rupture, this detection technology is not suitable for the fatigue damage detection and residual life assessment of 45# steel specimen.

5. Conclusion

Permeability detection technique is a foreseeing detection method of ferromagnetic member with high accuracy, which can predict stress concentration condition, the degree of fatigue damage and aging degeneration features in some areas based on the changes of magnetic induction intensity in the probe in which closed magnetic circuit detects specimen permeability changes, then explores the resulting microscopic structural changes in the components. Through the experimental study of Q235 and 45# steel specimens, the following conclusions can be made:

1. Based on the permeability measurement technique, the stress state of the rod material specimen can be effectively measured.
2. Based on the remaining residual stress, we can determine the maximum stress the component had gone through and the stress condition before. When the stress exceeds the elastic limit, the experimental results showed that it has higher detection sensitivity for the maximum stress detection component had received than the directly measured specimen stress.
3. Fatigue damage detection sensitivity is under the stress concentration detection sensitivity for the detection technology. Fatigue damage detection sensitivity is related with the kind of steel and the carbon content of the steel. The detection sensitivity of Q235 steel fatigue damage is higher than the detection sensitivity of 45# steel samples.

References

1. Yang E, Li L M, Chen X, Magnetic field aberration induced by cycle stress [J]. Journal of magnetism and magnetic materials, 2007, 312 (1): 72-77.
2. S K Ren, et al. Studies on stress-magnetism coupling effect for 35 steel components [J]. Insight: Non-Destructive Testing and Condition Monitoring, 2010, 52(6): 305-309.
3. S K Ren, et al. Influences of environmental magnetic field on stress magnetism effect for 20 steel ferromagnetic specimens [J]. Insight: Non-Destructive Testing and Condition Monitoring, 2009, 51(12): 672-675.
4. Dong Lihong, Xu Binshi, Dong Shiyun, et al. Variation of stress-induced magnetism signals during tensile testing of ferromagnetic steels [J]. NDT&E International, 2008, 41:184-149.
5. I Tomas, O Stupakov, J Kadlecova, O. Perevertov Magnetic adaptive testing-low magnetization, high sensitivity assessment of material modifications. Journal of Magnetism and Magnetic Materials, 2006, 304(2): 168-171.

6. G Vertesy, T Uchimoto, I Tomas, T Takagi. Nondestructive characterization of ductile cast iron by magnetic adaptive testing. Journal of Magnetism and Magnetic Materials, 2010, 322(20): 3117-3121.
7. I Tomas. Magnetic Adaptive Testing of Non-magnetic Properties of Ferromagnetic Materials. Czechoslovak Journal of Physics, 2004, 54, (4): 23-26.
8. G Vertesy, I Tomas, I Meszaros. Non-destructive indication of plastic deformation of cold-rolled stainless steel by magnetic adaptive testing. Journal of Magnetism and Magnetic Materials, 2007, 310(1): 76-82.

Influence of Clay Content on Incipient Motion and Erosion Features of Artificial Filling Clay

Qiang Zhang

College of Hydrodynamic and Ecology Engineering, Nanchang Institute of Technology, China
Email: zhangqiang8812@163.com

In order to study the influence of clay content on incipient motion and erosion features of artificial filling clay, a combination of methods of data analysis and experiment was used. The factors of incipient shear stress and erosion rates were analyzed according to the soil experiment results, which considered the influence of clay content. The two dimensionless formulas of the shear stress and erosion rate were put forward, which used multiple linear regression method to determine the coefficient and index. According to the results, when the clay content is different, the incipient shearing stress will increase as the clay content of soil samples increases, and the erosion rate will decrease as the clay content of soil samples increases.

Keywords: Clay content; artificial filling soil; incipient shear stress; erosion rate.

1. Introduction

In recent years, there were many achievements on the topic of incipient motion and erosion of sediment up to now, some scholars begun to concentrate on the incipient motion and erosion of cohesive soil, but their researches focused on cohesive undisturbed soil and cohesive sediment. Incipient motion of artificial filling clay is significantly different from the cohesionless granular sediment, as for cohesionless granular sediment, the starting unit of its incipient motion is generally in the form of single-particle starting, nevertheless, for dam construction soil, its incipient motion proceeds in the form of clump or conglobation by virtue of the intergranular cohesive force. Until now a unified scientific explanation for incipient motion and erosion mechanism hasn't been put forward. Therefore, the research into incipient motion and erosion of artificial filling clay is of great theoretical significance.

In order to study the influence of clay content on incipient motion and erosion features of artificial filling clay, a combination of methods of data analysis and experiment were used. The factors of incipient shear stress and erosion rates were analyzed according to the soil experiment results, which considered the influence of clay content. The two dimensionless formulas of the shear stress and erosion rate were put forward, which used multiple linear regression method to determine the coefficient and index.

2. Experiment on Incipient Motion and Erosion Features

2.1. *Experimental equipment*

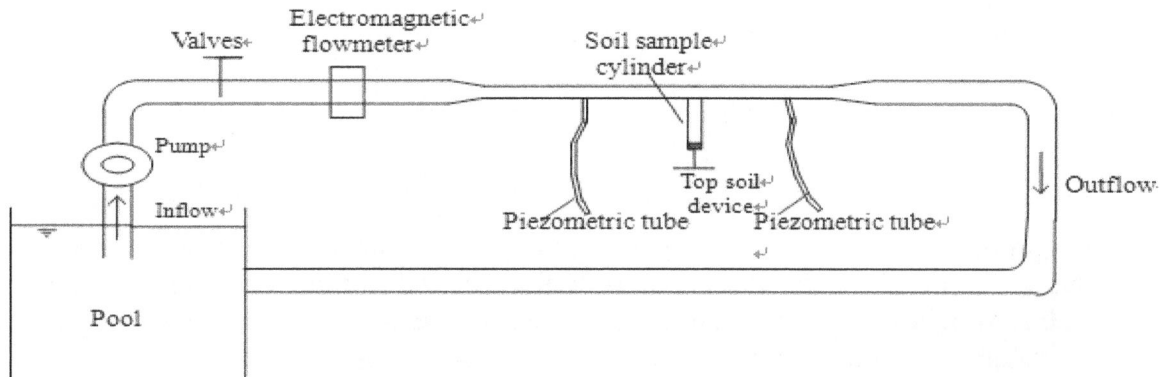

Fig. 1 The schematic diagram of the experimental equipment

In the past experimental investigations on the incipient and erosion rate of the clay, sample used is usually naturally deposited sediment [1]. Because of the difference between the filled clay and the naturally deposited sediment, it is necessary to study the incipient and erosion rate of the filled clay. The schematic diagram of the experimental equipment is shown in Figure 1 [2].

In Figure 1, a 200cm closed pipe with rectangular section (12cm wide and 3cm high) is used. During the experiment, the soil sample was placed in the cylinder with an internal diameter of 7cm, and piezometric tubes are installed to measure the pressure difference, thus the shear stress can be obtained. Distance between the two piezometric tubes is 120cm. The device below the cylinder that pushes up the soil is a rubber piston, and a screw is adopted to control the rubber piston to adjust the level of the sample in the cylinder. Flow discharge can be read on the electromagnetic flowmeter, and the maximum average velocity of the pipe is 3m/s. The experiment was carried out in the laboratory.

2.2. Experiment of incipient shear stress

2.2.1. Soil samples

This experiment mainly adopted two kinds of the soil. One kind of soil contained more viscous soil, because the kaolin clay content was very high. This kind of cohesive soil was collected from the construction site, which was known as the loess. Another kind of soil contained less viscous soil, and the color was black. This kind of cohesive soil was collected from the mountain Luojia, which was known as the black soil. Before the experiment, the soil sample was dried and crushed into powder, and the grading of the two kinds of soil sample was measured by sieving method and sedimentation method. The sieving method was adopted when the particle size is above 0.062 mm, and the sedimentation method was adopted when the particle size is below 0.062 mm. In order to increase the experiment number and make full use of the soil sample, the two kinds of soil samples were mixed in accordance with the mass ratio of 1:1, 2:1 and 1:2 three, which makes a total of five groups of the soil sample of different viscosity. The characteristics values of the soil samples are shown in Table 1 and the grading curve of the soil samples are shown in Figure 2. In Table 1, the clay particle means that the particle size is below 0.005 mm. In Figure 2, there are five kinds of soil samples.

Table 1 The characteristics values of the soil samples

Soil sample number	Loess (%)	Black soil (%)	Clay content (%)	Median size(mm)	Heterogeneous coefficient
A	100.00	0	19.01	0.061	3.407
B	66.67	33.33	14.95	0.081	2.947
C	50.00	50.00	12.93	0.089	2.733
D	33.33	66.67	10.90	0.098	2.549
E	0	100	6.85	0.114	2.068

2.2.2. Experiment procedure

The soil was properly dried, crushed, and placed in the clean container, and added proper amount of water and stirred evenly. Then it was put aside for a period of time and evenly layered rolled with the 32.0kg weight rolling roller on the soil. The dry density measured with the method of drying after rolling, and the sample was obtained by the sample box. The sample box which contains test soil samples was topped in the soil sample cylinder by using of top soil device.

The soil samples should be firstly deposited in the soil sample cylinder. After incipient motion pump, it needs to regulate valves to control the discharge. The soil sample will be pushed until the surface is exposed to the bottom of rectangular flume when the whole rectangular flume water is full of water, and the flow was stable. Slowly adjust the valve, so that the flow velocity in the pipe increased slowly to observe the start of test soil samples. When the test soil samples reach the start standard, note the pipe discharge, and the incipient motion of the soil shear stress can be obtained by calculating.

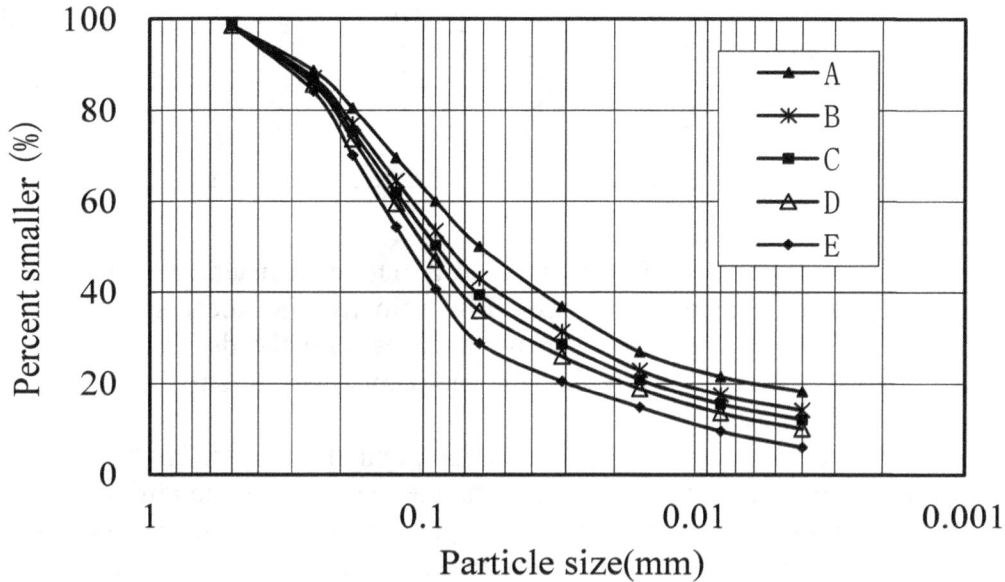

Fig. 2 The particle size distribution curve of soil samples

2.2.3. Analysis of experiment result

The 25 groups of soil samples were experimented in accordance with the operating steps, and the corresponding starting flow in the experiment was recorded to calculate the incipient shear stress according to the derived shear stress formula.

Table 2 The results of incipient shear stress

Soil samples	Clay content (%)	Dry density (g/cm³)	Water content (%)	Discharge (m³/h)	Flow velocity (m/s)	Incipient shear stress (measure)	Incipient shear stress (calculation)
	19.008	1.285273	25.27783	4.48	0.346	0.588579	0.6025
	19.008	1.333381	25.0945	4.71	0.363	0.642473	0.633604
A	19.008	1.445778	24.58278	4.93	0.380	0.695906	0.695942
	19.008	1.493839	24.13344	5.04	0.389	0.723305	0.719343
	19.008	1.51023	23.79773	5.08	0.392	0.733381	0.72697
	14.954	1.227328	25.7842	4.33	0.334	0.554526	0.554549
	14.954	1.356985	24.6532	4.68	0.361	0.635329	0.641574
B	14.954	1.410894	23.87448	4.86	0.375	0.678706	0.671391
	14.954	1.464426	22.99766	4.94	0.381	0.698378	0.698481
	14.954	1.497053	21.83533	4.99	0.385	0.710795	0.713968
	12.927	1.258611	24.13206	4.45	0.343	0.581699	0.574772
	12.927	1.342687	23.40883	4.66	0.360	0.630585	0.62948
C	12.927	1.383852	23.14716	4.72	0.364	0.644862	0.652973
	12.927	1.43089	22.9021	4.86	0.375	0.678706	0.677815
	12.927	1.48709	20.56913	4.93	0.380	0.695906	0.705193

(*Continued*)

Table 2 (*Continued*)

Soil samples	Clay content (%)	Dry density (g/cm³)	Water content (%)	Discharge (m³/h)	Flow velocity (m/s)	Incipient shear stress (measure)	Incipient shear stress (calculation)
D	10.899	1.28717	24.74981	4.51	0.348	0.595494	0.590507
	10.899	1.345533	23.59277	4.68	0.361	0.635329	0.626871
	10.899	1.427568	23.5053	4.84	0.373	0.673826	0.671522
	10.899	1.445322	22.5684	4.86	0.375	0.678706	0.680405
	10.899	1.457631	21.9634	4.91	0.379	0.690973	0.686424
E	6.845	1.278696	24.94596	4.41	0.340	0.57258	0.57404
	6.845	1.323	24.64233	4.53	0.350	0.600123	0.602057
	6.845	1.341789	24.5091	4.6	0.355	0.616445	0.613154
	6.845	1.365143	23.21409	4.64	0.358	0.625856	0.626384
	6.845	1.42523	22.07142	4.77	0.368	0.656864	0.657977

The specific experimental data is shown in Table 2. The data listed in Table 2 contains clay content, dry density, water content, discharge, flow velocity and shear stress. When the clay content is constant, the incipient shearing stress increases with the dry density of soil samples decreases. When the clay content is different, the incipient shearing stress increases as the clay content of soil samples increases.

So far, the equilibrium conditions of sediment incipient motion are mainly three types: the equilibrium conditions of the horizontal direction, the equilibrium conditions of the vertical direction, and the equilibrium conditions of the torque. The equilibrium of force is suitable for slipping, and the equilibrium of torque is suitable for rolling. The incipient motion of soil micro-aggregate is similar with the incipient motion of sediment, so the previous research of incipient motion equilibrium conditions of granular sediment can be referenced [3]. In order to maintain a harmony of dimension, the following calculation expression is proposed with a comprehensive consideration of the influencing factors [4, 5].

$$\tau_c = k\gamma RS^m \left(\frac{\rho_d - \rho}{\rho} \right)^n \tag{1}$$

The concrete steps are as follows: firstly log the equation on both sides, and the problem is transformed into data fitting of the n data points. Then use SPSS statistical analysis software as to derive regression coefficients. Finally, the artificial filling clay shear stress formula is that:

$$\tau_c = 0.00821\gamma RS^{0.040} \left(\frac{\rho_d - \rho}{\rho} \right)^{0.323} \tag{2}$$

Where, ρ_d is the soil dry density; ρ is the fresh water density; S is the clay content; τ_c is the incipient motion shear stress of soil; γ is the water unit weight; R is the hydraulic radius.

2.3. Experiment of erosion rate

2.3.1. Soil samples

Erosion rate test was still conducted in a closed rectangular conduit, where the original incipient motion trials were conducted. The test soil sample was the same to the incipient motion shear stress trials.

2.3.2. *Experiment procedure*

Put the well-prepared soil sample into the cylinder and top it by using of top soil device. Start pump to regulate the flow discharge, observe the incipient motion situation on the surface of the soil and fix the flow discharge until the soil began to start. When the soil sample gradually collapse with the water erosion, timely adjust the soil sample in order to keep the surface of it with the same height of the flume bottom. Use the scale on the cylinder to read the height of erosion and record the erosion time with a stopwatch. Read the electromagnetic flow meter to get the discharge and divide the cross-sectional area of the rectangle pressurized water tank to get the velocity, and then the flow shear stress can be calculated according to the formula. Finally, the erosion rates are obtained with different groups, different water content of soil samples and different discharge conditions.

2.3.3. *Analysis of experiment result*

The soil samples were experimented in accordance with the operating steps. The erosion height, erosion time and discharge were recorded to calculate the incipient shear stress according to the derived shear stress formula. The specific experimental data is shown in Table 3, and the data listed in Table 3 contain clay content, dry density, water content, discharge, flow velocity, Water shear stress, incipient shear stress and erosion rate. When the clay content does not change, the erosion rate decreases as the dry density of soil samples increases. When the clay content is different, the erosion rate decreases as the clay content of soil samples increases.

Table 3 The results of erosion rate

Soil samples	Clay content (%)	Dry density (g/cm³)	Water content (%)	Discharge (m³/h)	Flow velocity (m/s)	Water shear stress (N/m²)	Incipient shear stress(N/m²)	Erosion rate (mm/s)
A	19.008	1.285273	25.27783	7.9	0.610	1.588238	0.588579	0.017
	19.008	1.285273	25.27783	11.5	0.887	3.064005	0.588579	0.044
	19.008	1.285273	25.27783	15.3	1.181	5.049848	0.588579	0.083
	19.008	1.445778	24.58278	8.2	0.633	1.695284	0.695906	0.014
	19.008	1.445778	24.58278	12.33	0.951	3.461415	0.695906	0.039
	19.008	1.445778	24.58278	16.8	1.296	5.947843	0.695906	0.078
	19.008	1.51023	23.79773	8	0.617	1.623588	0.733381	0.011
	19.008	1.51023	23.79773	11.6	0.895	3.110783	0.733381	0.033
	19.008	1.51023	23.79773	15.1	1.165	4.934895	0.733381	0.061
B	14.954	1.227328	25.7842	8.84	0.682	1.933569	0.554526	0.028
	14.954	1.227328	25.7842	12.58	0.971	3.585167	0.554526	0.061
	14.954	1.227328	25.7842	16.3	1.258	5.641526	0.554526	0.111
	14.954	1.410894	23.87448	8.98	0.693	1.987476	0.678706	0.019
	14.954	1.410894	23.87448	12.15	0.938	3.373469	0.678706	0.042
	14.954	1.410894	23.87448	16.38	1.264	5.69007	0.678706	0.081
	14.954	1.497053	21.83533	8.98	0.693	1.987476	0.710795	0.017
	14.954	1.497053	21.83533	12.15	0.938	3.373469	0.710795	0.039
	14.954	1.497053	21.83533	16.23	1.252	5.599196	0.710795	0.072
C	12.927	1.258611	24.13206	7.43	0.573	1.426588	0.581699	0.017
	12.927	1.258611	24.13206	11.75	0.907	3.181519	0.581699	0.050
	12.927	1.258611	24.13206	15.08	1.164	4.923462	0.581699	0.089

(*Continued*)

Table 3 (*Continued*)

Soil samples	Clay content (%)	Dry density (g/cm³)	Water content (%)	Discharge (m³/h)	Flow velocity (m/s)	Water shear stress (N/m²)	Incipient shear stress(N/m²)	Erosion rate (mm/s)
	12.927	1.383852	23.14716	7.93	0.612	1.598808	0.644862	0.017
	12.927	1.383852	23.14716	11.65	0.899	3.134286	0.644862	0.039
	12.927	1.383852	23.14716	15.54	1.199	5.189285	0.644862	0.078
	12.927	1.48709	20.56913	7.57	0.584	1.473961	0.695906	0.011
	12.927	1.48709	20.56913	11.35	0.876	2.994408	0.695906	0.033
	12.927	1.48709	20.56913	16.25	1.254	5.611276	0.695906	0.075
	10.899	1.28717	24.74981	8.16	0.630	1.680839	0.595494	0.022
	10.899	1.28717	24.74981	12.34	0.952	3.466329	0.595494	0.056
	10.899	1.28717	24.74981	15.78	1.218	5.330347	0.595494	0.094
	10.899	1.427568	23.5053	8.75	0.675	1.899251	0.673826	0.019
D	10.899	1.427568	23.5053	12.42	0.958	3.505751	0.673826	0.044
	10.899	1.427568	23.5053	16.2	1.250	5.581097	0.673826	0.081
	10.899	1.457631	21.9634	7.75	0.598	1.535841	0.690973	0.011
	10.899	1.457631	21.9634	12.08	0.932	3.33953	0.690973	0.039
	10.899	1.457631	21.9634	16.33	1.260	5.659709	0.690973	0.078
	6.845	1.278696	24.94596	7.4	0.571	1.416523	0.57258	0.017
	6.845	1.278696	24.94596	10.83	0.836	2.758469	0.57258	0.044
	6.845	1.278696	24.94596	14.5	1.119	4.59687	0.57258	0.083
	6.845	1.341789	24.5091	8.37	0.646	1.757267	0.616445	0.022
E	6.845	1.341789	24.5091	11.52	0.889	3.073336	0.616445	0.044
	6.845	1.341789	24.5091	15.63	1.206	5.241993	0.616445	0.089
	6.845	1.42523	22.07142	8.3	0.640	1.731629	0.656864	0.017
	6.845	1.42523	22.07142	11.7	0.903	3.157865	0.656864	0.042
	6.845	1.42523	22.07142	15.2	1.173	4.99223	0.656864	0.072

Considering the shear stress is the main factor of soil erosion, many researchers determine the relationship between the erosion rate and the flow shear stress and the critical shear stress through on-site or laboratory experiments [6-8]. In order to maintain a harmony of dimension, the following calculation expression is proposed with a comprehensive consideration of the influencing factors.

$$E = k\sqrt{gR}S^{l}\left(\frac{\rho_d - \rho}{\rho}\right)^{m}\left(\frac{\tau - \tau_c}{\tau_c}\right)^{n} \qquad (3)$$

Coefficients are got by using of SPSS statistical analysis software, so that the erosion rate formula of artificial filling of cohesive soil is:

$$E = 0.0000213\sqrt{gR}S^{-0.034}\left(\frac{\rho_d - \rho}{\rho}\right)^{-0.165}\left(\frac{\tau}{\tau_c} - 1\right)^{1.041} \qquad (4)$$

Where, E is the soil erosion rate; S is the clay content; ρ_d is the soil dry density; ρ is the fresh water density; τ_c is the soil shear stress, τ is the flow shear stress; R is the hydraulic radius.

Conclusion

When the clay content is the same, the incipient shearing stress will increase as the dry density of soil samples decreases, and the erosion rate will decrease as the dry density of soil samples increases.

When the clay content is different, the incipient shearing stress will increase as the clay content of soil samples increases, and the erosion rate will decrease as the clay content of soil samples increases.

Acknowledgments

This research is funded by the Funding Projects of Jiangxi Provincial, Department of Education: (GJJ14761), the Funding Projects of Jiangxi Provincial Department of Science and Technology: (20161BAB216108) and the Non-profit Industry Financial Program of MWR: (201401039).

References

1. Cao, S.Y. and Du, G. H. (1986). Experimental study of erosion and deposition of cohesive soil. Journal of Sediment Research, Issue 4, pp 73-82. (in Chinese)
2. Wang, J., Tan, G. M. and Shu, C. W. (2014). Experimental study on inception of consolidated cohesive sediment. Journal of Sediment Research, Issue 6, pp 25-29. (in Chinese)
3. Qian, N. (1983). Mechanics of sediment transport. Beijing: Science Press. (in Chinese)
4. Shu, C. W., Wang, J. and Tan, G. M. (2007). Influence of dry bulk density on incipient motion and erosion of cohesive deposits. Engineering Journal of Wuhan University, Volume 40, Issue 1, pp 25-28. (in Chinese)
5. Chang, L. Y. and Chen, Q. (2012). Progress in contact scouring research. Advances in Science and Technology of Water Resources, Volume 32, Issue 2, pp 79-82. (in Chinese)
6. Zhang, Q., Wang, Y. and Chen, C. B. (2012). Experiment on incipient motion and scour features of artificial filling clay. Advances in Science and Technology of Water Resources, Volume 32, Issue 6, pp 75-78. (in Chinese)
7. Lv, P., Tan, G. M. and Wang, J. (2008). Study on the incipient motion velocity of cohesive sediment after deposition and consolidation. China Rural Water and Hydropower, Issue 2, pp 56-58. (in Chinese)
8. Liu, J. (2011). Mechanism of seepage contact scours between cohesionless soil layers. Advances in Science and Technology of Water Resources, Volume 31, Issue 3, pp 27-30. (in Chinese)

Chapter 11
Nondestructive Testing and Reliability Assessment

A Study of AA2219 Plate Friction Stir Welding Features with Different Initial Tempers

Jing-Wen Feng[1, 2, 3], Li-Hua Zhan[1, 2, 3,*], Yong-Lun Song[4]

[1]Condition Key Laboratory of High Performance Complex Manufacturing, Central South University, Changsha 410083, Hunan, China;

[2]School of Mechanical and Electrical Engineering, Central South University, Changsha 410083, Hunan, China;

[3]2011 Collaborative Innovation Center, Central South University, Changsha 410083, Hunan, China;

[4]College of Mechanical Engineering and Applied Electronics Technology, Beijing University of Technology, Beijing 100124;

*Email: yjs-cast@csu.edu.cn

This paper conducted a friction stir welding (FSW) experiment on 2219 aluminum alloy plates with two different initial tempers of T87 and solution treatment (W), and artificial aging treatment(S) was carried out for the welded 2219 aluminum alloy at W to obtain the welds at the two conditions of AA219T87+FSW and AA2219W+FSW+S. By the means like metalloscopy scanning, tensile test and micro hardness test, metallographic structure, mechanical performance and hardness distribution rule of the two welds were compared and analyzed. Results showed that there were different defect forms of FSW welds at the two conditions; although the two different defects emerged at the advancing side, the weld defects of 2219 aluminum alloy at T87 were relatively scattered with a concentration in the welding nugget zone close to the advancing side and they had different sizes of defects. On the weld condition of AA2219+W+FSW+S, defects were concentrated at the bottom of weld close to advancing side. Compared with base metal, the mechanical performance of welding materials on the two conditions dropped. The welds of base metal at T87 dropped to 67.6% for welding tensile strength, 40.4% for yield strength and 37.4% for ductility, while the welds of base metal at the condition W+FSW+S dropped to 70.6% for welding tensile strength, 68.8% for yield strength and 25.9% for ductility. Average welding nugget hardness at T87 was lower than W+FSW+S process. The hardness of the upper part of the nugget's weld section was obviously higher than that of the lower part and more evenly distributed. This research is about to provide important support for the welding processes of manufacturing large structures.

Keywords: Friction stir welding; different initial tempers; morphologies; mechanical performance.

1. Introduction

Thanks to the sound weldability, anti-stress corrosion and mechanical performance as well as fracture toughness of 2219 aluminum alloy within the temperature scope of -250~+250°C, it has gained wide popularity within the aerospace industry [1]. For 2219 aluminum alloy welding, several welding methods, such as MIG, TIG, EBW and VPPA, used to be adopted in foreign countries.

FSW is short for friction stir welding, which is a new type of solid connection technology invented by TWI in 1990s and has gained a quick global popularity [2, 3]. FSW can solve the defects of pores and cracks in traditional welding methods, and significantly uplift joint hardness, thus widely applied to produce high-quality aluminum alloy welded joints [4, 5]. At present, most researches focus on the influence of welding parameters upon FSW features. H1 Jin, Y.S. Sato and S. Benavides et al. studied the microstructural evolution of weld [6-13]; M. Song, T. Nishihara and P. Colegrove et al. studied the distribution of remaining stress on weld [14,15]; T.U. Seidel, Y. Li, A.P. Reynolds, K. Colligan, B. London et al. explored the plastic flow of welding zone materials [16-20]; H.J. Liu, M. W. Mahoney, M.G. Dawes, T. Hashimoto, P.S. Pao et al. studied the mechanical

*Corresponding author

performance of joint [21-26]. However, few researches involve the influence of materials at original condition upon FSW, especially the influences of the same aluminum alloy at different heat treatment conditions [27]. With constant upgrading of aviation and aerospace equipment, the molding of large structural members like rocket storage tank wall plate and fairing cannot be completed at one stroke, for it demands multiple procedures, among which butt joint welding or even heat retreatment after heat treatment of aluminum plate is dispensable. Therefore, studying various welding features of 2219 aluminum alloy at different heat treatment conditions after FSW can help us understand the FSW mechanism therein so as to make clear which condition can deliver the optimal performance after welding. In this research, AA219T87+FSW and AA2219+W+FSW+S are chosen for experiment so as to compare the microstructure, defect, tensile property and hardness distribution of joints.

2. Experimental Procedure

The experiment adopted 6mm-thickness T87 aluminum alloy rolled plates and solution-treatment aluminum alloy rolled plate as experiment material, whose chemical components are as shown in Table 1.

Table 1 Chemical components of 2219 aluminum alloy

Main elements	Cu	Mg	Mn	Si	Fe	Ni	Zr	Ti	Al
Mass fraction/%	5.24	0.028	0.27	0.042	0.13	0.03	0.14	0.065	Bal.

And the mechanical properties of the materials at different conditions are as shown in Table 2.

Table 2 Mechanical performance of different condition 2219 aluminum alloy

Material condition	Tensile strength/Mpa	Yield strength/Mpa	Ductility/%
O condition	160.5	327.1	36.3
T6 condition	433.2	304.3	20.7
T87 condition	466.5	386.2	18.2

The weldment size was 300x150x6mm. The base metal surface and joint face in touch with shaft shoulder were polished with sandpaper, the oxidation film was removed and the welds were cleaned with acetone. Welding was conducted by CPM120 FSW welding machine produced by SIEMENS as shown in Fig. 1. The stirring head was in the shape of threaded trigone, and the length for stirring rod was 5.5mm, the diameter at the bottom was 7.4mm, the diameter at the top was 5.5mm, and the diameter for shaft shoulder was 18mm. The welding speed was 180mm/min, with a rotation speed of 600r/min and a shaft shoulder under draught of 5.9mm/min. The welding method was single way butt welding. Before welding, the two aluminum alloy plates to be welded were fixated on a clamp. The stirring head rotated in a high speed and gradually drilled into the joints. When the stirring head shaft shoulder pressed tightly the aluminum plate, we completed the welding along the joints.

Fig. 1 Friction stir welding picture

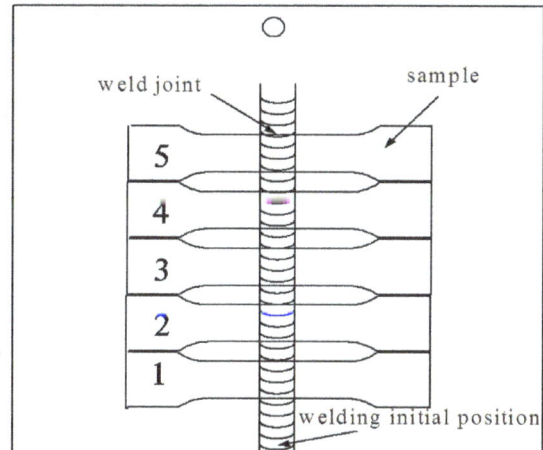

Fig. 2 Sampling positions

After welding, samples were in a perpendicular direction to the weld as shown in Fig. 2. Phased array ultrasound nondestructive inspection was conducted on the samples to detect and mark the size and position of defects. Typical welded joints were chosen to make metallographic samples to be polished. Then Keller's agent (hydrofluoric acid 1mL, hydrochloric acid 1.5mL, nitric acid 2.5mL and water 95mL) was adopted for submergence. Olympus DSX500 optical microscope was used to observe joint microstructures. Under 9.8N pressure and parameter of 15s, Duramin-10 microhardness tester was used to measure the microhardness distribution of welded joints from advancing side to retreating side, including base metal zone, heat engine influence zone and welding nugget zone, with an interval of 1mm. Tensile test was conducted on SUST CMT-5000 electronic tensile tester to measure the mechanical performance. Before this, samples were polished, cleaned and free from oxidation film and scratches.

3. Results and Discussions

3.1. *Morphologies and defects of welds*

Typical metallographic structures of AA2219T87+FSW and AA2219 solution treatment+FSW+ aging weld are shown in Fig. 3 and Fig. 4.

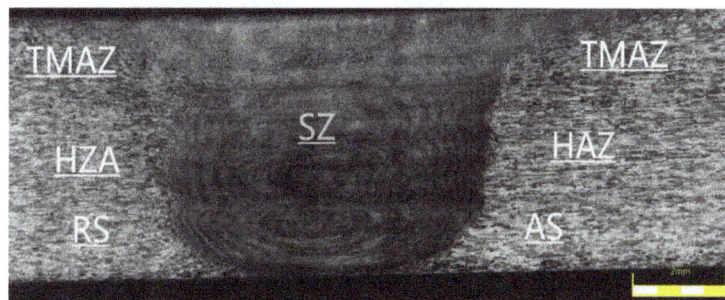

(a) Non-defect microstructure metallographic panorama

605

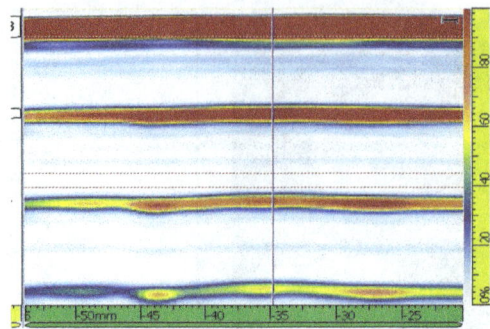

(b) non-defect weld ultrasound phased array nondestructive testing scanogram

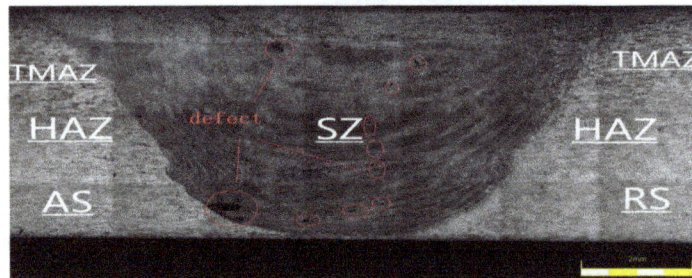

(c) defect microstructure metallographic panorama

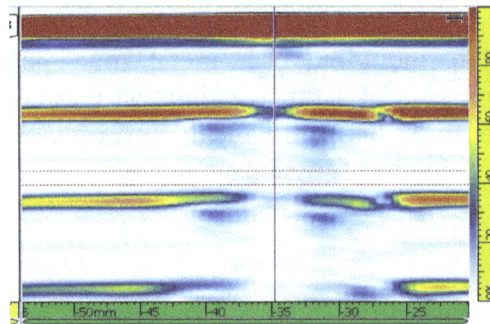

(d) defect weld ultrasound phased array nondestructive testing scanogram

Fig. 3 AA2219T87+FSW weld section

Through comparing Fig. 3(c) and Fig. 4(c), it can be discovered that the original condition of materials has a significant influence upon the position and size of structure defect. Although defects of both emerge at the advancing side, they are relatively scattered at T87 with different sizes and all closing to welding nugget. In welding structure with an original condition of solution treatment, defects concentrate at the bottom of weld with closeness to advancing side.

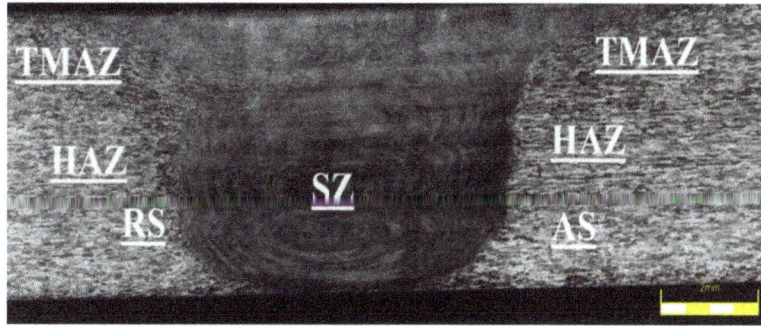

(a) Non-defect microstructure metallographic panorama

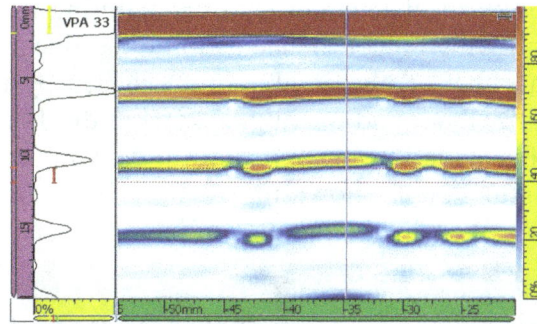

(b) Non-defect weld ultrasound phased array nondestructive testing scanogram

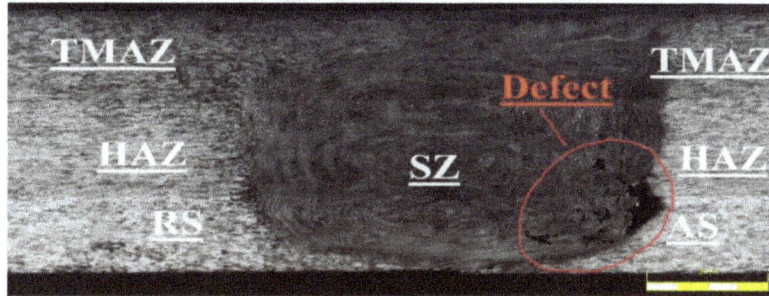

(c) Defect microstructure metallographic panorama

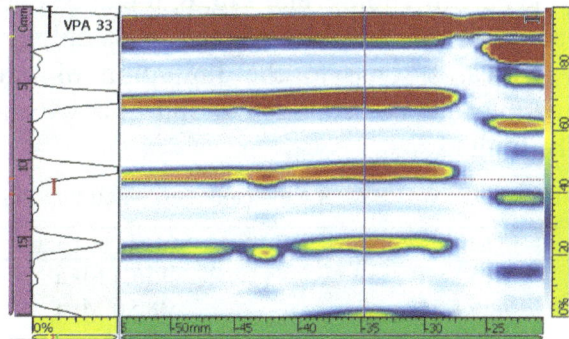

(d) Defect weld ultrasound phased array nondestructive testing scanogram

Fig. 4 AA2219 solution treatment+FSW+aging heat treatment section

3.2. Microstructures of different zones in FSW welded joints

For the two welds, the common feature lies in that material plastic flow behavior around stirring head has a significant influence upon the mechanical property of joint structure in FSW. The onion rings in the joint section of Fig. 3 and Fig. 4 fully display the plastic metal flow. Fig. 5 presents four subzones of welded joints, namely base metal zone, welding nugget zone, heat machine influence zone and heat influence zone. Fig. 5(a) is metallography of base metal zone far away from weld, with an orderly arrangement of particles along the rolling direction. Fig. 5(b) is welding nugget zone (SZ). Due to the effect of mechanical stirring of stirring rod as well as partial high temperature caused by intense friction, dynamic recrystallization occurs to structure, with hardening constituent particles scatted in it. Compared with other zones, the difference of this one lies in that it is most influenced by mechanical effect of rod, during which recrystallized grains are broken due to stirring before growing bigger, hence forming the fine and equiaxed grain structures. Under the adhesive attraction effect of plastic aluminum caused by intense stirring, there occur partial breakup and adhesive growth around the nugget zone in the TMAZ. However the structures of other parts undergo significant deflection and deformity, which also leads to reverting and recrystallization under heat circulation as shown in Fig. 5(c). HZA is free from the stirring effect, but it also undergoes coarsening of original structures of base metal as shown in Fig. 5(d).

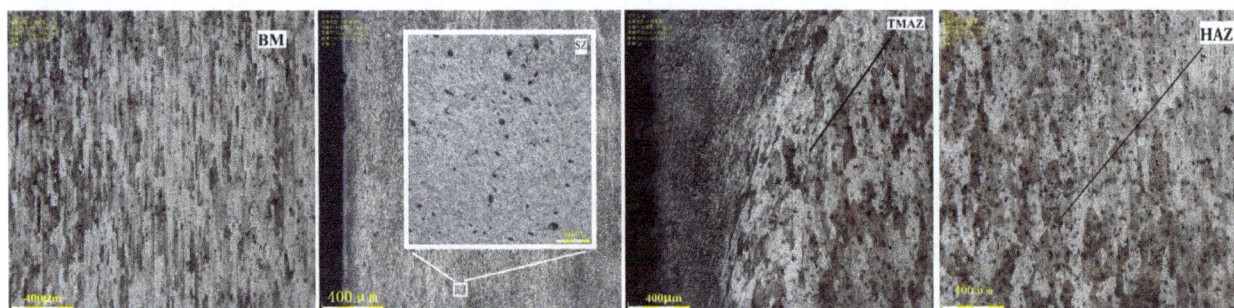

(a) base metal zone   (b) welding nugget zone   (c) HMAZ;   (d) HAZ

Fig. 5 Microstructures of different zones in FSW welded joints

3.3. Mechanical property of weld

Base metal mechanical properties of AA219T87+FSW and W+FSW+S welds are as shown in Table 3. Through comparing the data in Table 3 and Fig. 6, it can be known that original materials have different influences upon property. With the change of sampling positions, the mechanical property of weld gradually increases. Due to the limitation of equipment (the clamp size 300x300mm), large-scale experiment is not conducted, and thus we can only obtain the optimal mechanical property within the limited size.

Table 3 AA219T87+FSW and AA2219+W+FSW+S welds and corresponding base metal mechanical property

Material condition	Tensile strength	Yield strength	Ductility
2219 T87 +FSW	315.8MPa	156.4 MPa	6.7%
2219 T87	466.8 MPa	386.6 MPa	17.9%
T87 +FSW /T87	67.6%	40.4%	37.4%
2219W+FSW+AGING	305.9 MPa	209.7 MPa	5.3%
2219 T6	433.2 MPa	304.5 MPa	20.7%
W+FSW+AGING/T6	70.6%	68.8%	25.9%

From Fig. 6, it can be seen that for all processes and methods, the mechanical property from starting to end gradually increases. People have grasped the rule that welding speed causes heat input changes. In FSW, heat input is a process in which temperature change, structural transition, stress strain and metal flow interact. The level of heat input can not only influence temperature of weld but also plastic deformity of materials. The stirring rod starts welding at room temperature, and the welding materials will undergo temperature increase as time goes on. As welding continues, welding materials and welding head both reach a high temperature. Therefore, the total heat input increases, weld is fully stirred, material flow becomes better, and materials tend to fill lower gaps, hence delivering better weld structure moldability and avoiding defects. Therefore, with the moving of stirring head, the molding of welds and joint structure and performance will be influenced, i.e., weld performance is closely related with the distance to starting point of the weld.

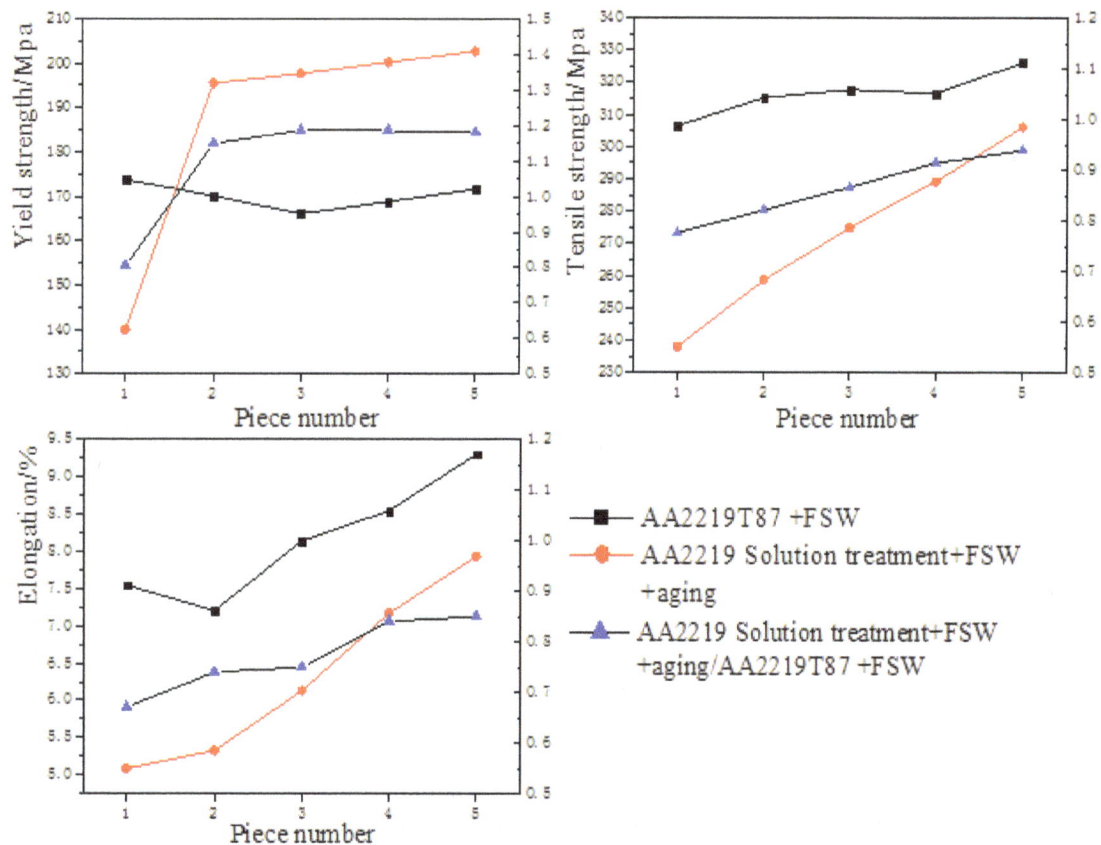

Fig. 6 Mechanical property comparison between 2219T87+FSW and 2219W+FSW+aging

Meanwhile, from Fig. 6, it can be seen that in terms of tensile strength and ductility, the weld at T87 outperforms that at the condition of W; but in yield strength, the weld at W condition is 120% of that at T87 condition, with this level maintaining along the direction of joint (the first error point ignored). For ductility, the weld at W condition is 65%-85% of that at T87 condition, which also follows the rule of increasing along the welding direction.

3.4. *Hardness distribution inside welds*

Figs. 7(a) and 7(b) present Vickers hardness distribution diagrams of weld section along three lines in Fig. 7(c) with two point intervals of 1mm. According to Table 2, the mechanical performance of 2219T87 is higher than T6, but in Fig. 7, the average hardness of the former is lower than the latter, indicating that FSW molding has severely weakened the strength of 2219 aluminum alloy with an original condition of T87. For Figs. 7(a) and 7(b), there is the rule of W-shape distribution of hardness, indicating that there is an obvious weakening zone in joints. The mechanical property and crack positions of joints are related with joint microstructure, but the lowest value of microhardness in solution treatment+FSW+aging appears in HAZ of advancing side and in HAZ of retreating side at T87. In welding nugget zone, the hardness of (a) and (b) peaks at the top and decreases to the direction of the bottom.

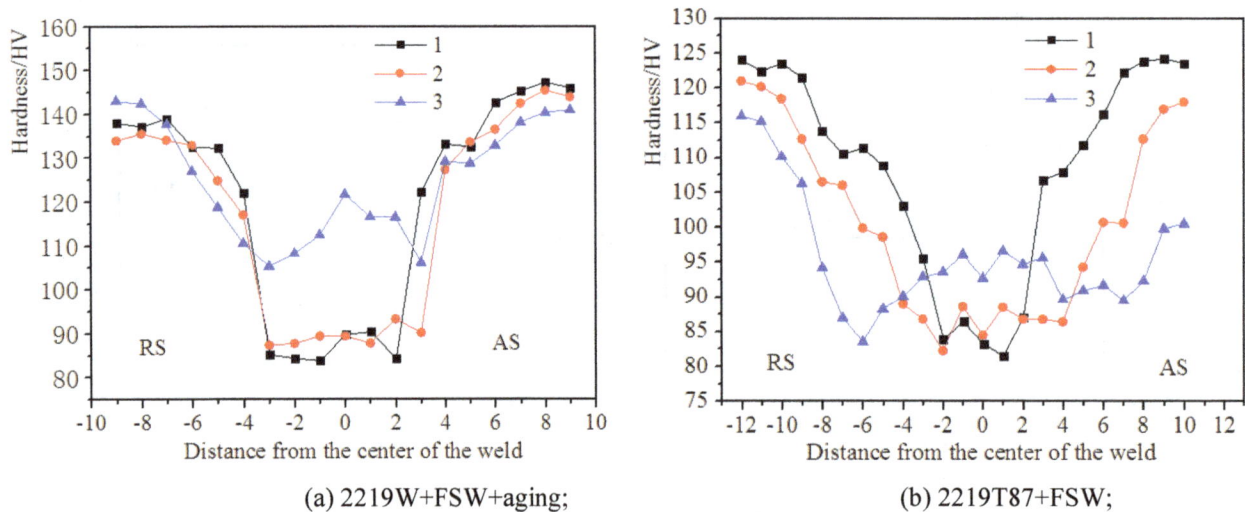

(a) 2219W+FSW+aging;

(b) 2219T87+FSW;

(c) 1, 2 and 3 stand for the line of the hardness test of weld section

Fig. 7 Weld section microhardness distribution

In the direction of thickness, there exists big difference in hardness between the upper and lower parts of joints in solution treatment, which is opposite with the above hardness rule. These two phenomena are result of different heat input mechanisms. Heat input at the stirring head can be divided into two parts, namely heat produced by shaft underdraught and heat produced by welding rod friction. In welding, the level of heat input determines the molding quality of weld. The two parts of heat input close to surface all participate in heat production, causing enough heat input and thus producing better molding. In the welding nugget zone, only detection rod participates in heat production, causing insufficient heat input and thus leading to poor molding and hardness.

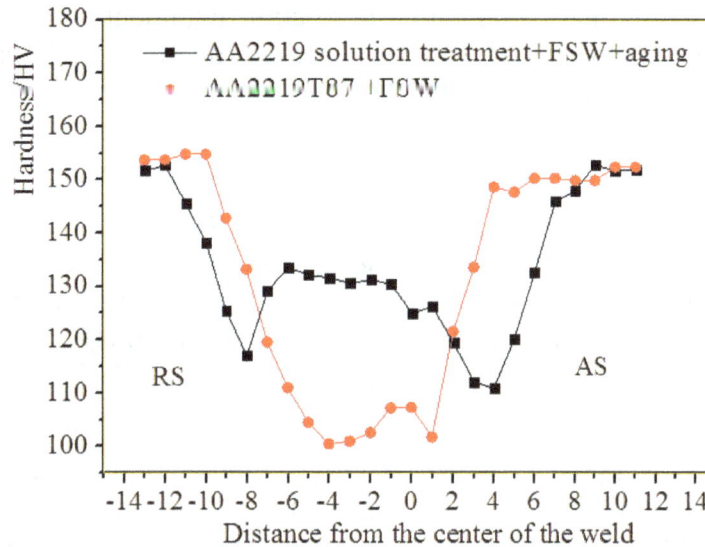

Fig. 8 weld surface microhardness comparison between 2219T87+FSW and 2219 solution treatment+FSW+aging

From Fig. 8, it can be seen that the weld surface hardness distribution of materials at two original conditions shows W-shape distribution. In vertical welding direction, there is a small wave length in joint microstructure distribution for solution treatment, small difference between joint softening zone and base metal zone. The hardness is even at the RS, about 130HV-135HV. The hardness at the AS has a higher fluctuation than that at the RS, ranging 110HV-130HV. The weld surface at 2219T87 condition witnesses big microstructure hardness fluctuations at the vertical welding direction, with the minimal value appearing at the RS, which is consistent with the result in Fig. 5.

4. Conclusion

1. Initial temper of 2219 aluminum alloy has an obvious influence upon microstructure of welding. At the condition of T87, defects tend to distribute at the whole advancing side, with big ones emerging at the bottom and there are obvious interfaces between nugget zone and base metal zone. At the condition of solution treatment, defects concentrate on the bottom of advancing side, with vague interface between nugget zone and base metal zone.
2. Initial temper of 2219 aluminum alloy has an obvious influence upon mechanical property of weld. The tensile strength at solution-treatment condition is 80%-93% of that at T87 condition, and it increases along the direction of welding. The ductility at solution treatment condition is 65%-85% of that at T87 condition, which is consistent with the rule of increasing along the direction of welding. However, the yield strength at solution treatment condition is 120% of that at T87 condition and maintains this level along the direction of welding.
3. Initial temper of 2219 aluminum alloy has an obvious influence upon microhardness of weld. AA2219+W+FSW+S and AA219T87+FSW display W-shape hardness distribution rule both inside and outside, with obvious softening zone at joints. Mechanical property and crack

positions of welds are related with microstructure of joints, but the minimal microhardness of former appears at the HAZ at advancing side while the latter is at retreating side. In welding nugget zone, the hardness of both peaks at the top while decreases to the bottom. The biggest difference between upper and lower parts at solution treatment condition is 30HV, but on horizontal level, the fluctuation is smaller, 10HV. Therefore, T87 enjoys relatively low hardness in welding joint zone with slight fluctuations.

Acknowledgments

This research was funded by the National Basic Research Program of China (2014CB046602), Grants from the Project of Innovation-driven Plan in Central South University (2015CX002), the National Natural Science Foundation of China (51235010) and the Specialized Research Fund for the Doctoral Program of Higher Education of China (201201621110003).

References

1. Narayana, G. Venkata, et al. "Fracture behaviour of aluminium alloy 2219–T87 welded plates." Science & Technology of Welding & Joining 9.2 (2013): 121-130.
2. Thomas, W. M, and E. D. Nicholas. "Friction stir welding for the transportation industries." Materials & Design 18.4 (1997): 269-273.
3. Thomas, Wayne Morris, et al. "improvements relating to friction welding." CA, EP0615480. 1995.
4. C.J. Dawes, W.M. Thomas, Weld. J. 75 (4) (1996) 41.
5. K.E. Knipstrom, B. Pekkari, Weld. J. 76 (9) (1997) 55.
6. H. Jin, S. Saimoto, M. Ball, P.L. Threadgill, Mater. Sci. Technol. 17(12) (2001) 1605.
7. Y.S. Sato, M. Urata, H. Kokawa, K. Ikeda, M. Enomoto, Scripta Mater. 45 (1) (2001) 109.
8. S. Benavides, Y. Li, L.E. Murr, D. Brown, J.C. Mcclure, Scripta Mater. 41 (8) (1999) 809.
9. G. Liu, L.E. Murr, C.S. Niou, J.C. Mcclure, F.R. Vega, Scripta Mater. 37 (3) (1997) 355.
10. Y.S. Sato, H. Kokawa, M. Enomoto, S. Jogan, Metall. Mater. Trans. A 30 (9) (1999) 2429.
11. M. Song, R. Koracevic, Proceedings of the Fourth International Symposium on Friction Stir Welding, Utah, USA, TWI Ltd., May, 2003.
12. T. Nishihara, Y. Nagasaka, Proceedings of the Fourth International Symposium on Friction Stir Welding, Utah, USA, TWI Ltd., May, 2003.
13. P. Colegrove, M. Painter, D. Graham, T. Miller, Proceedings of the Second International Symposium on Friction Stir Welding, Gothenburg,Sweden, TWI Ltd., June, 2000.
14. M. James, M. Mahoney, D. Waldron, Proceedings of the First International Symposium on Friction Stir Welding, CA, USA, TWI Ltd., June, 1999.
15. C.D. Donne, E. Lima, J. Wegener, A. Pyzalla, T. Buslaps, Proceedings of the Third International Symposium on Friction Stir Welding, Kobe, Japan, TWI Ltd., September, 2001.
16. T.U. Seidel, A.P. Reynolds, Metall. Mater. Trans. A 32 (11) (2001) 2879.
17. Y. Li, L.E. Murr, J.C. Mcclure, Scripta Mater. 40 (9) (1999) 1041.
18. A.P. Reynolds, T.U. Seidel, M. Simonsen, Proceedings of the First International Symposium on Friction Stir Welding, CA, USA, TWI Ltd., June, 1999.
19. K. Colligan, Weld. J. 78 (7) (1999) 229.
20. B. London, M. Mahoney, W. Bingel, M. Calabrese, D. Waldron, Proceedings of the Third International Symposium on Friction Stir Welding, Kobe, Japan, TWI Ltd., September, 2001.

21. H.J. Liu, H. Fujii, M. Maeda, K. Nogi, Proceedings of the Third International Symposium on Friction Stir Welding, Kobe, Japan, TWI Ltd., 27–28 September, 2001.

22. H.J. Liu, H. Fujii, M. Maeda, K. Nogi, J. Mater. Sci. Technol. 20 (1) (2004) 103.

23. M.W. Mahoney, C.G. Rhodes, J.G. Fiulintoff, R.A. Spruling, W.H. Bingel, Metall. Mater. Trans. A 29 (7) (1998) 1955.

24. M.G. Dawes, S.A. Karger, T.L. Dickerson, J. Przyoatek, Proceedings of the Second International Symposium on Friction Stir Welding, Gothenburg, Sweden, TWI Ltd., June, 2000.

25. T. Hashimoto, S. Jyogan, K. Nakada, Y.G. Kim, M. Ushio, Proceedings of the First International Symposium on Friction Stir Welding, CA, USA, TWI Ltd., June, 1999.

26. P.S. Pao, E. Lee, C.R. Feng, H.N. Jones, D.W. Moon, Proceedings of the Fourth International Symposium on Friction Stir Welding, Utah, USA, TWI Ltd., May, 2003.

27. C. Juricic, C.D. Donne, U. Drebler, Proceedings of the Third International Symposium on Friction Stir Welding, Kobe, Japan, TWI Ltd., September, 2001.

Statistical Analysis on Welding Parameters of 316L Stainless Steel Diffusion Welding Joint Based on the Weibull Distribution

Zi-Liang An, Li Zhang, Mei-Fang Hou [*]

Shanghai Institute of Technology, Shanghai 201418, China

[*]Email: cmfhou@sit.edu.cn

Based on an orthogonal experimental design method, optimal welding process parameters are studied on 316L stainless steel diffusion welding joint. Three factors, i.e. welding temperature, pressure and holding time, are distinguished. The optimal parameter set is obtained to have welding temperature of 1100°C, welding pressure of 10MPa and holding time of 3 hrs. At the same time, a statistical estimation is performed for the tensile strength data of the welding joint round bar samples following three Weibull distributions. Estimated statistical parameters of the distribution are with a position parameter of 85.5242, scale parameter of 64.8969 and shape parameter of 1.0351, respectively. Average value of the strength data is 150 MPa.

Keywords: 316L stainless steel; orthogonal experimental; Weibull distribution.

1. Introduction

The fusion welding is one welding method in which two pieces being welded are compressed tightly together under the vacuum or protection atmosphere when the temperature is below the melting point of the base material. When pressure is applied, the two welding surfaces produce micro plastic deformation so that solid metallurgical connection is formed after certain time of holding. The study on normal temperature performance of fusion welding joint has been paid greatly attention to between 316L stainless steel and heterogeneous material. However, little report is given on diffusion welding connection of homogeneous material of 316L stainless steel, particularly on the design criteria for high temperature performance and high temperature strength of the joints [1-3]. Therefore, orthogonal test design approach is adopted in this paper to study the diffusion welding connection process of homogeneous material of 316L stainless steel, in which the influence of welding temperature, welding pressure and holding time are investigated on high temperature and tensile performance of 316L stainless steel joint. Purpose is to get the optimal welding process parameters. Meanwhile, Weibull distribution is used to predict separately its tensile strength at high temperature.

2. Design of Orthogonal Test

2.1. *Design technology*

Orthogonal test design is one important method to study multiple factors and multiple levels in which some representative points were selected for test from a comprehensive test based on a fraction principle of factor design to make statistical analysis of the results [4]. Basic element is to design an orthogonal table so as to ensure that the tests with minimum number can get all the information affecting the performance parameters in the total factor test. To reduce the test number and cost, small part of factor combination is applied from all possible process parameters of 316L stainless steel diffusion welding to ensure that all the information affecting performance parameters is enclosed in all the tests with minimum number of tests [5].

---

*[*]Corresponding author

2.2. Test sample

For benefiting to produce work hardening, the surface in the piece to be welded is controlled to be roughness of 0.8um and plainness of 0.02. For avoiding the error due to difference of mechanical processing and welding batch number, two $\phi70\times50$mm 316L stainless steel round rods are used to make butt diffusion welding ends and these rods are pickled and degreased before welding [6].

3-factor and 3-level orthogonal test design method is adopted. The three factors are welding temperature, welding pressure and holding time, as shown in Table 1.

Table 1 316L stainless steel test sample level of orthogonal test

Factor	Level		
	1	2	3
Welding temperature（℃）	980	1050	1110
Welding pressure （MPa）	4	7	10
Holding time （H）	1	3	5

By previous experiments, the range of each parameter is controlled as: welding temperature of 980 -1100°C, welding pressure of 4 - 10MPa and holding time of 1 - 5 hours. During the welding, the vacuum in the furnace chamber is maintained at 1.33*10-3Pa. The welding test is conducted on the FJK-2 vacuum diffusion welding machine according to the parameter combination sequence in Table 2.

Table 2 Orthogonal experimental parameters and tensile strength at normal temperature

Sample ordinal	Welding temperature (°C)	Welding pressure (MPa)	Holding time (hrs)	Tensile strength Yi (MPa)	Y_i^2
	1	2	3		
1	980	4	1	102.4	10490
2	980	7	3	122	14880
3	980	10	5	159.7	25500
4	1050	4	3	123.9	15350
5	1050	7	5	160.8	25860
6	1050	10	1	149.1	22230
7	1100	4	5	140	19600
8	1100	7	1	87.8	7710
9	1100	10	3	300	90000
I j	384.1	366.3	339.3	$\sum_{i=1}^{9} Y_i$	$\sum_{i=1}^{9} Y_i^2$
IIj	433.8	370.6	605.9		
IIIj	587.8	668.8	460.5		
R j	203.7	302.5	266.6	1345.7	229624

Tensile tests at 550°C are performed on the Shimadzu material tester with a temperature control in high temperature furnace according to national standards GB/T 228-2002 (entitled as Metallic Materials-Tensile Testing at Ambient Temperature) and GB/T 4338-2002 (entitled as Metallic Materials-Tensile Testing at Elevated Temperature). The gauge length of tensile sample is $\phi5\times30$mm and the weld is at the middle position of the sample [7-10]. The test result is given in a mean value of three times of tests.

2.3. *Test results and analysis*

The orthogonal test arrangement and tensile test results are shown in Table 2 in which I, II and III represent respectively level I, level II and level III of each factor, j represents No.j row; I j, II j and III j represent respectively the sum of indexes of the corresponding No.j row; range Rj represents the difference between the maximum value and the minimum value in the mean value of indexes of the factor, which can be applied to judge the influence level of a factor on the joint performance. The higher the range is, the bigger the influence will be. Table 2 shows R2>R3>R1, which indicates that factor 2 (welding pressure) has the largest influence, factor 3 (holding time) is weaker and factor 1 (welding temperature) is weakest.

3. Failure Distribution

3.1. *Distribution types*

Based on the tensile strength data of test sample of 316L stainless steel diffusion welding, K-S test method is applied to perform goodness of fit test for exponential distribution, normal distribution and Weibull distribution [11] and the test results are shown in Table 3.

Table 3 Fitted results of the distribution of high temperature tensile strength

Distribution type	Test statistic	Assumed critical value	Conclusions
Exponential distribution	0.4160	0.4300	Refuse
Normal distribution	0.3362	0.4300	Obey
Two parameter Weibull distribution	0.3005	0.4300	Obey
Three parameter Weibull distribution	0.2346	0.4300	Obey

According to the data in Table 3, the tensile strength data of 316L stainless steel diffusion welding test sample at high temperature follows the normal distribution and Weibull distribution, refuses to follow exponential distribution. Three-parameter Weibull distribution is wide applied in statistical analysis of life data [12]. Therefore, it is specially studied in the present paper.

3.2. *Weibull distribution*

The cumulative failure probability function and probability density function of the three parameter Weibull distribution can be given as, respectively:

$$F_{(t)} = 1 - \exp[-(\frac{t-\gamma}{\eta})^{\beta}], t \geq \gamma > 0 \tag{1}$$

$$f_{(t)} = \frac{\beta}{\eta}(\frac{t-\gamma}{\eta})^{\beta-1} \exp[-(\frac{t-\gamma}{\eta})^{\beta}], t \geq \gamma > 0 \tag{2}$$

Where γ is the position parameter, and when $\gamma=0$, the functions are for two-parameter Weibull distribution; η is the scale parameter and β is the shape parameter.

3.3. *Parameter estimations*

Currently there are many parameter estimation methods for the Weibull distribution, such as maximum likelihood estimation, least square method, graphic estimation, linear regression estimation, etc. The graphic estimation method requires large number of samples and low fitting; the least square method and maximum likelihood estimation is suitable for complete samples, no-replacement fixed number truncation and fixed time situations. Generally the parameter estimation with maximum likelihood method is more accurate than with least square method while the latter is easier to be calculated than the former. The data used in this paper is from a situation of fixed number truncation, so that the maximum likelihood estimation is more suitable. The present parameter estimation is solved with the help of MATLAB software. The results are shown in Table 4.

Table 4 Weibull distribution fitting of tensile strength at high temperature

Weibull distribution	Test statistic	Assumed critical value	Position parameter	Scale parameter	Shape parameter	Correlation coefficient
Two parameters	0.3006	0.4300	0	168.2803	2.6273	0.9056
Three parameters	0.2013	0.4300	85.5242	64.8969	1.0351	0.9546

It seems that the fit by the three-parameter Weibull distribution fitting is better than by the two-parameter Weibull distribution. The cumulative distribution probabilities by the three-parameter Weibull distribution is drawn with the help of MATLAB software and given in Fig. 1.

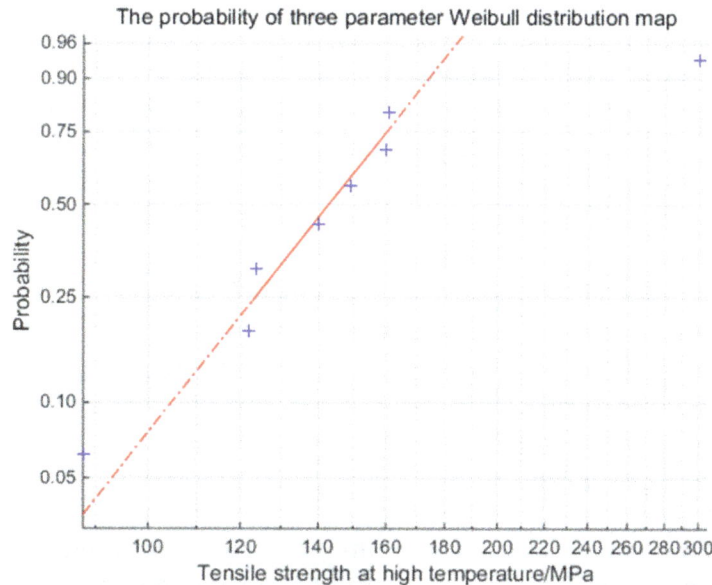

Fig. 1 Cumulative distribution probability of the three parameter Weibull distribution for the tensile strength data of 316L stainless steel diffusion welding joint under high temperature

617

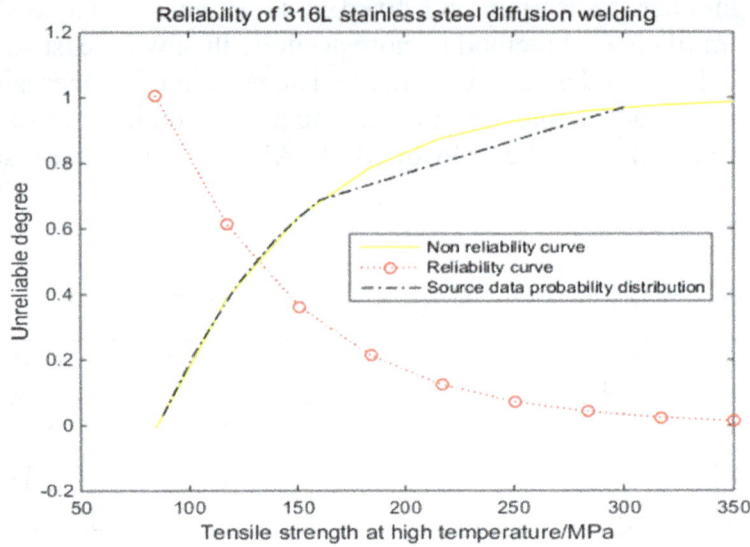

Fig. 2 Probability curves for the tensile strength data of 316L stainless steel diffusion welding joint at high temperature

3.4. *Reliability analysis*

By the analysis above, the cumulative distribution probability for the tensile strength data of 316L stainless steel diffusion welding joint is given in the following, by the three-parameter Weibull distribution, as:

$$F_{(t)} = 1 - \exp\left[-\left(\frac{t - 85.5242}{64.89691}\right)^{1.0351}\right] \tag{3}$$

Relative probability curves are shown in Fig. 2. Average tensile strength can be predicted by a formulation as:

$$MTBF = \int_0^{+\infty} R_{(t)}dt = \gamma + \eta * \Gamma\left(\frac{1}{\beta} + 1\right) \tag{4}$$

The predicted average tensile strength value of the present 316L stainless steel diffusion welding joint is 150 MPa.

4. Conclusion

Following basic conclusions can be reached after analysis and study above:

1. The orthogonal test design approach is successfully applied to the combination design of tensile tests of 316L stainless steel diffusion welding joint and the minimum number of samples is optimized to reduce the test cost.

2. The analyzed results show that the influence levels of factors on the tensile strength of 316L diffusion welding point at high temperature ordered in sequence are: the welding pressure, holding time and welding temperature, respectively, from larger to smaller. The optimal parameter set is welding temperature of 1100°C, welding pressure of 10MPa and holding time of 3 hrs.

3. The present statistical parameters for the tensile strength data of $\phi70\times50$mm 316L stainless steel welding joint round bar samples following three Weibull distribution are with a position parameter of 85.5242, scale parameter of 64.8969 and shape parameter of 1.0351, respectively. Average value of the strength data is 150 MPa.

References

1. Nishi H, Araki T, Eto M. Diffusion bonding of alumina dispersion-strengthened copper to 316L stainless steel with interlayer metals [J]. Fusion Engineering and Design, 1998, 39-40: 505-511.
2. Nishi H, Kikuchi K. Influence of brazing conditions on the strength of brazed joints of alumina dispersion-Strengthened copper to 316 stainless steel [J]. Journal of Nuclear Materials, 1998, 258-263: 281-288.
3. Kliauga A M, Travessa D, Ferrante M. Al2O3/Ti interlayer/AISI304 diffusion bonded joint: Microstructural Characterization of the two interfaces [J]. Materials Characterization, 2001, 46(1): 65-74.
4. Yang De. Experiment Design and Analysis [M]. Beijing: China Agriculture Press, 2002. 171-171.
5. Huang ChunYue, Zhou DeJian, Wu ZhaoHua. Study on the Relationships between Solder Joint Process Parameters and Reliability of Plastic Ball Grid Array Component Based on the Orthogonal Experiment Design [J]. Electronic Journal, 2005, 33(5): 788-792.
6. An ZiLiang, Xuan FuZhen, Tu ShanDong. Study on Mechanical Properties and Diffusion Bonding Parameters of 316L Stainless Steel [J]. Journal of Shanghai Institute of Technology, 2012,12(4): 257-260.
7. Somekawa H, Higashi K. The optimal surface roughness condition on diffusion bonding [J]. Materials Transactions, 2003, 44(8): 1640-1643.
8. Xuan FuZhen, Zhang Bo, Li ShuXin. Numerical simulation of the resistance method for the evaluation of micro hole defects in diffusion welded joints [J]. Welding Journal, 2007, 28(4): 9-12.
9. Han WenBo, Zhang KaiFeng, Wang GuoFeng. Study on superplastic forming and diffusion bonding process of Ti-6Al-4V alloy with multi plate structure [J]. Journal of Aeronautical Materials, 2005, 25(6): 29-32.
10. Zhu Hanliang, Zhao Bing, Li Zhiqiang, et al. Superplasticity and superplastic diffusion bonding of a fine-grained TiAl alloy [J]. Materials Transactions, 2005, 46(10): 2150-2155.
11. He ZhengFeng. MATLAB Probability and mathematical statistics analysis[M]. Beijing: Machinery Industry Press, 2012.
12. Ling Dan. Research on Wwibull Distribution and Its Applications in Mechanical Reliability Engineering [D]:[Doctoral Dissertation of University of Electronic Science and technology of science and technology]. Chengdu: University of Electronic Science and Technology of China, 2010.

Thermal Error Predictive Model of Motorized Spindle Based on Self-recurrent Wavelet Neural Network

Jian Yin[1, 2,*], Ming Li[1]

1 Shanghai Key Laboratory for Mechanical Automation and Robot, Shanghai University, 200240, P. R. China;
2 Mechanical Department, Tongling University, Anhui 244000, P. R. China
*Email: yinjianshanghai@163.com

A great challenge in improving the machining accuracy of high speed machine center is to establish accurate thermal error models for motorized spindle as its thermal errors are the main sources of inaccuracy. With the rising of the rotation speed, the spindle's temperature and thermal error increase gradually. In this paper, a new approach to derive an effective mathematic thermal model for motorized spindle is presented. The thermal errors accumulated are the combination of thermal distortions from different components with different thermal characteristics. The thermal deformation is a nonlinear procedure due to the variational working condition. By taking into consideration the thermal-elastic characteristics, a dynamic self-recurrent wavelet neural network is applied to capture the dynamics in order to assure thermal error predictive model accuracy. The structure of this model determines its dynamic characteristic with memory feedback loop. To evaluate the performance of proposed model, a verification experiment is carried out. The predictive results show the proposed model can improve the accuracy of motorized spindle effectively.

Keywords: Thermal error; motorized spindle; thermal-elastic; self-recurrent wavelet neural network.

1. Introduction

Today, high precision manufactured products can increase their competitive and added value. Improvement the accuracy of machine tool is the key aim of manufacturing industrial. Although improvement individual part precision of the machine tool can get a better performance, the cost is very high. Therefore, the error compensation approach has received widely attention as a way to improve machine tool accuracy cost-effectively [1]. In error compensation process, an artificial value that was directly measured or predicted from error model is feedback to the control system to offset errors through some suitable algorithm. It is an effective and reliable method to improve the manufacturing accuracy.

Modeling, error components measurement and identification and compensation implementation algorithm are three main sections involved in error compensation approach. Understand the reasons that error source affect the accuracy of machine tool is a useful procedure in compensation. It is very difficult to use a mathematical model to describe machine tool errors exactly and to process them technically. In manufacturing process, many random disturbances factors originate from various sources related to cutting process, thermal deformation affected the accuracy of machine tool. In general, machine tool errors consist of geometric errors, thermal errors, force induced errors, dynamic errors, etc. The errors caused by thermal deformation have the same order of magnitude or higher than the errors due to inaccuracy caused by geometric errors. As estimated by A.Mottu, "50% to 60% of the errors in precision machine result from thermal errors" [2]. Therefore, compensation for thermal errors can efficiently improve machine tool accuracy. The error predictive model is the key problem in error compensation system. The effectiveness of an error compensation system highly depends on the accuracy and robust of the thermal error predictive models.

Recently, high speed machining has meet the rapid development in aerospace and mould manufacturing industrial, which can improve material removing rates and enhance the finished products surface quality. To achieve higher speed, motorized spindle with integrated driving motor

---

is introduced. Spindle is one of the most critical components of machine tool which provide accurate axis rotation to drive the tool make an exact relative motion to workpiece. Obviously the higher rotating speed and the very high heat loss of integrated motors would make the thermal errors of spindle more complicated and serious. Although careful arranging thermal symmetry structure and appropriate heat sources layout, adding some cooling system, etc, thermal error can be reduced considerably. But it is inevitable that some residual thermal errors still exist which will reduce the accuracy of machine tool. Therefore, it is a vital task to compensate the thermal deformation errors of a motorized spindle [3, 4].

It's almost impossible to measure the thermal error directly between the tool tip and workpiece during machining process because of the discharged chips and the splashed cooling fluid. In order to get the values of thermal error, many methods such as finite element methods, the neural network method, and multiple regression approach have been used to predict them. The regression model using a least squares estimation method is employed to describe the thermal deformation for simple structure. The finite difference and finite element models are not suitable to predictive thermal errors for the values of the internal heat sources were difficult to measure and the heat boundary conditions were also too complicated to determine. In recent years, different types neural network have been employed in the thermal error modeling of motorized spindle [5, 6, 7]. For the neural network method and multiple regression approach, only the current temperature information is used as the inputs for error model. These static models are not inadequate since the previous time variables aren't considered in the model. Thermal effects have the characteristic of memory of the previous temperature. The thermal errors in a spindle are not only determined by the current temperature information but also influenced by the previous thermal status [8, 9, 10]. The hysteresis of thermo-elastic and the nonlinearity of the temperature fields in spindle are not considered in these models. As the spindle thermal error has a lagging characteristic compared with the temperature measured in certain key locations, the time independent static models are not suitable for predicting the thermal error of motorized spindle. The thermal error model of motorized spindle must have the time-dependent dynamic properties. Not only the current temperature information but also the previous temperature information is taken as the inputs for error model [11, 12, 13].

This paper begins with a review of thermal errors of motorized spindle. In section 2, on the basis of theoretic analyze of motorized spindle impacted by heat sources, the traits of lag between temperature response and thermal deformation were obtained. The design of the SRWNN to solve the thermal error prediction problem for the motorized spindle was proposed in section 3. Here the self-recurrent wavelet neural network (SRWNN) was used as the model to predict thermal error for dynamic system. In section 4, a verification experiment was implemented on a motorized spindle and the results showed that the SRWNN has high prediction accuracy. Some conclusions were drawn in section 5.

2. Overview the Heat Resources and the Thermo-elastic Characteristic of the Spindle

There are four main components of a motorized spindle which are driving motor, spindle-bearings system, tool holder mechanism, cooling and lubrication system as showed in Fig. 1. The motor is integrated into the spindle between the front and the rear bearings and drive the spindle directly. So the spindle can get higher rotation speed than the traditional spindle which is droved by belts or gears. This is why motorized spindle are very common in high speed machining. However, the high speed rotation and the integrated motor also introduce large amount of heat which required precisely regulated cooling, lubrication. The internal heat sources of motorized spindle are the driving units and power transmission system, including motors, bearings, cooling and lubrication oil and the machining process. The driving motor and the front and rear bearings are three main internal heat sources in the motorized spindle. Some researches indicated that 60% of the power input to a machine tool is dissipated in the driving units and power transmission system. The heat generated in motorized spindle is detrimental to geometric accuracy of spindle. Transfer of heat away from motor components is critical.

In order to ensure the motorized spindle accuracy running at high speed, it is necessary to cool down the motor. A circuit cooling water jacket is designed in the spindle housing which not only accommodates all the parts but also contributes to the cooling and lubrication of the spindle. As more heat was generated from bearings, the spindle bearings must be lubricated and cooled down correctly. So the bearings are individually cooled by oil/air lubrication. As a result, the thermal behaviors become very difficult to predict.

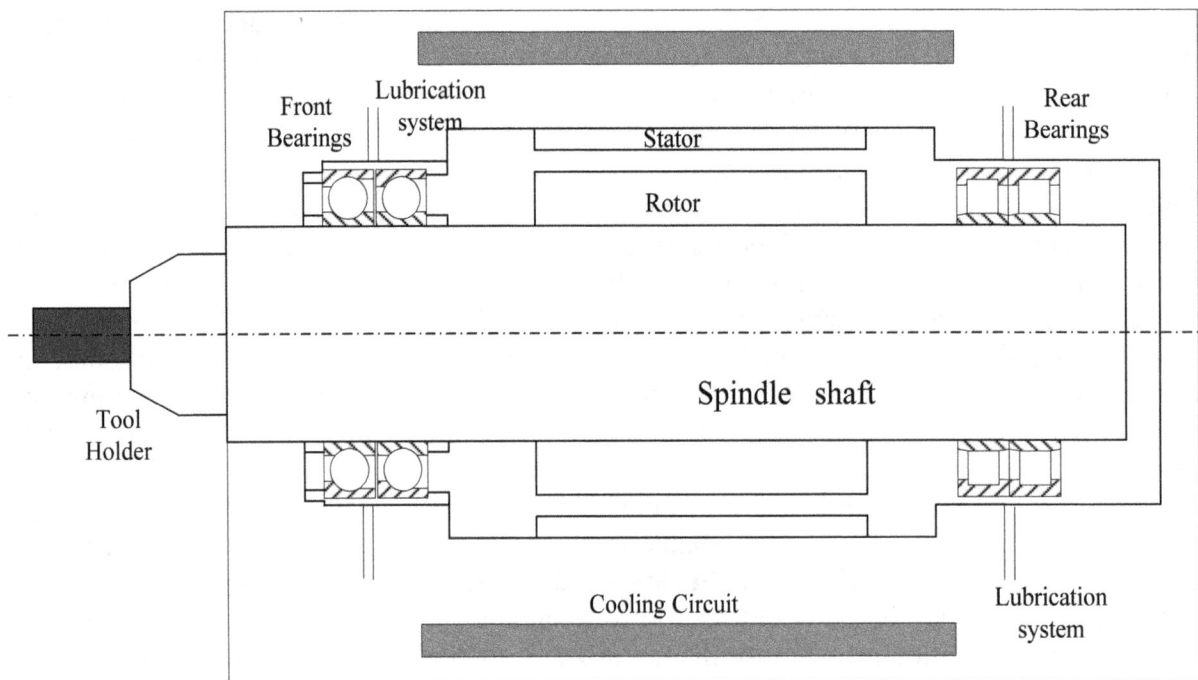

Fig. 1 Overview the structure of a motorized spindle

Thermal effects have the characteristic of memory of the previous temperature. The errors in a spindle are not only determined by the current temperature but also influenced by the preceding thermal status[9]. Those static models are not inadequate because the time-variable temperature isn't considered in the model. The hysteresis of thermo-elastic and the nonlinearity of the temperature fields in spindle are not considered.

The thermal error in the spindle is not only determined by the current temperature but also influenced by the preceding thermal status. A dynamic and nonlinear function should be used to predictive the thermal deformation error of the spindle.

3. Dynamic Modeling for Thermal Error of the Motorized Spindle

As there are different heat sources in different positions in the spindle, it will cause non-uniform temperature field. On the other hand, the different shape and dimension spindle components have different thermal characteristics. The induced thermal errors are nonlinear and time varying. The definite mathematical model to estimate the thermal error is difficult to establish. In recent years, different type artificial neural networks have been used in thermal error modeling to analysis the relationship between thermal state and deformation. However, these modeling approaches are essentially static neural network modeling methods for only considering the current temperature. The predictions usually cannot satisfy the accuracy and robust requirement of the thermal error compensation system.

The model in this paper is been used to estimate the thermal deformation of a motorized spindle operated in different selected working conditions with a few experimental data. It means, given a series of observed values of a system, the model can be trained to learn the essential property of the system and hence calculate an expected value for a given input. Also, some papers successfully applied neural network to the model predictor. But it has some drawbacks such as accuracy, slow convergence speed, which come from its inherent characteristics. Therefore, a new model, self-recurrent wavelet neural network (SRWNN) which combines the properties of dynamics of recurrent neural network and the fast convergence of wavelet neural network (WNN), is proposed to solve the prediction problem for complex dynamic systems.

The proposed SRWNN model has a mother wavelet layer composed of self-feedback neurons. As the self-feedback neuron can store previous information of the network, it can capture the response properties of the dynamic system. So the modified feature made the SRWNN well suit to apply for the complex dynamic system. For the SRWNN has less wavelet nodes than the WNN, its structure is simple than that of WNN.

The structure of SRWNN is described first[11]. Multiple temperature inputs and a single axis-deformation output are considered here. A schematic diagram of the proposed SRWNN structure is shown in Fig. 2, which has N_i inputs, one output and $N_i \times N_w$ wavelons. The model is composed of four layers which are input layer, wavelons layer, product layer and output layer. The input layer accepted the input variables first and then transferred them to the second layer. The second layer is wavelons layer. This layer of neurons is consisted of wavelons, whose input parameters include the wavelet dilation and translation coefficients. Each node of the dynamic wavelon consisted of a mother wavelet and a self-feedback loop. This layer consisted of j wavelons.

In this paper, a Morlet function, $\varphi(x) = \cos(5x)\exp(-\frac{x^2}{2})$ as a mother wavelet function is selected. A wavelet φ_{ij} of each node is derived from mother wavelet function as equation 1:

$$\varphi_{ij}(z_{jk}) = \varphi(\frac{u_{jk} - m_{jk}}{d_{jk}}), \ z_{jk} = \frac{u_{jk} - m_{jk}}{d_{jk}} \tag{1}$$

Where m_{jk} and d_{jk} are the translation factor and the dilation factor of the wavelet function respectively. The subscript j and k indicates the k input term of the j wavelet.

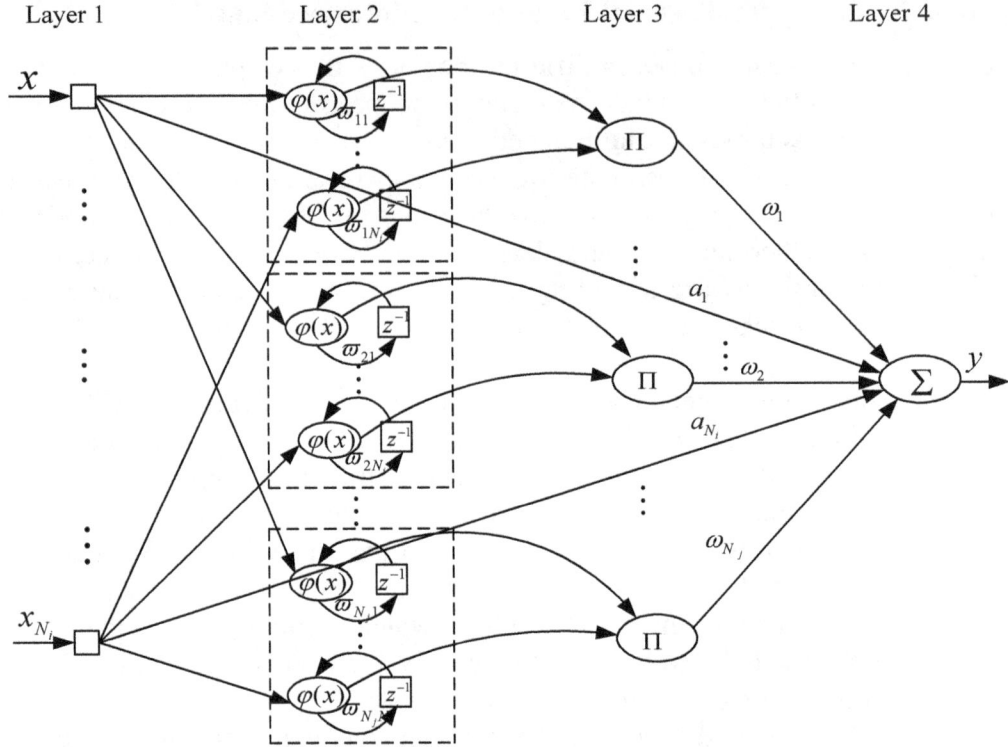

Fig. 2 Structure of the SRWNN

The inputs of the layer for moment n can be indicated as equation 2:

$$u_{jk}(n) = x_k(n) + \varphi_{jk}(n-1) \cdot \theta_{jk} \tag{2}$$

Where θ_{jk} is the weight of the self-feedback loop. The input of this layer contained the memory parameter $\varphi_{jk}(n-1)$, which can store the previous information of the model. That means the dynamic of the variables can maintain for the next sample moment. Here θ_{jk} is a factor represent the rate of information storage. These are apparent dissimilar characteristic between the WNN and the SRWNN. The layer 3 is a product layer. In this layer all the results of the nodes are the production of the wavelons, and they are showed as equation 3:

$$\varphi_j(X) = \prod_{k=1}^{N_j} \varphi(z_{jk}) \quad = \prod_{k=1}^{N_j} \left[-(z_{jk}) \exp\left(-\frac{1}{2}(z_{jk})^2 \right) \right] \tag{3}$$

The layer 4 is output layer. The result of the layer is a linear summation of outputs calculated from the layer 3. Moreover, the output result directly accepted the input variables from the first layer simultaneously. So the output of SRWNN is composed of results from each self-recurrent wavelet and the direct input variables are shown as equation 4:

$$y(n) = \sum_{j=1}^{N_w} w_j \Phi_j(X) + \sum_{k=1}^{N_i} a_k x_k \tag{4}$$

624

Where w_j is connection weight value between product nodes and output node and a_k is the connection weight between the input nodes and the output nodes. W in equation 5 is the weight vector of SRWNN.

$$W = \begin{bmatrix} a_k & m_{jk} & d_{jk} & \theta_{jk} & w_j \end{bmatrix}^T \qquad (5)$$

Where the initial values of adjusting parameters a_k, m_{jk}, d_{jk} and w_j are given in the range of [-1 1], but $d_{jk} > 0$. When the initial value of $\theta_{jk} = 0$ meant there is no feedback value in the initial status.

4. Experiments

For most thermal error compensation system, some parameters such as spindle speed or temperature at certain key positions in machine tool influence errors are the first considered problem in error models. Due to the complexity structures and changeable thermal conditions of motorized spindle, temperature sensors need to be mounted on the significantly influential positions. Then the better accuracy can be obtained from the model.

4.1. *Thermal error measurement*

The spindle used in experimentation is a type G30 spindle from CYTEC industries. The spindle is rated at 34 KW maximum power and 25,000 rpm maximum speed, corresponding to 1.5 million DN. In this paper, four thermal sensors were required and their locations were relatively easier to be determined. Four temperature sensors were attached to the motorized spindle front bearing, rear bearing, motor and the spindle body and four temperature variations were the most correlation analysis of temperature and thermal deformation for the dynamic system. In this experiment, four temperature sensors were mounted in the spindle in order to analyze temperature variations at heat source locations. The experiment equipments were composed of a CNC machining center and PT-100 thermal resistances, sensing units and a double ball-bar. PT-100 thermal resistance sensor was used to measure the temperature in the motorized spindle, because of high accuracy and finely calibration property. A double ball-bar was used to measure the thermal deformation of spindle. The circuit tests were exploited for identification of the thermal error of spindle using a space geometry analytical algorithm. The thermal errors of axial spindle are shown in Fig. 3. The spindle was operated at 5000rpm first for an hour. When the thermal condition was stable, then the speed changed to 10000rpm for thirty minutes and at last the speed reached to the 15000rpm.

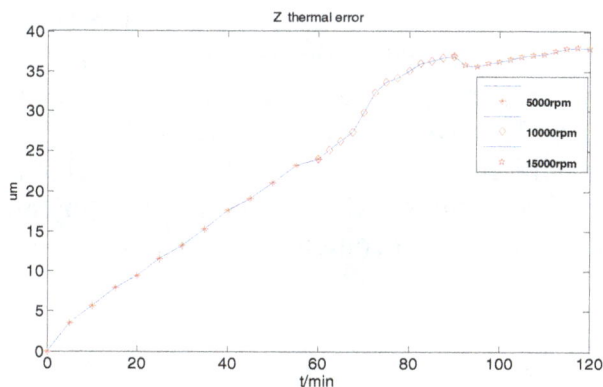

Fig. 3 Measurements of Z-axial thermal error

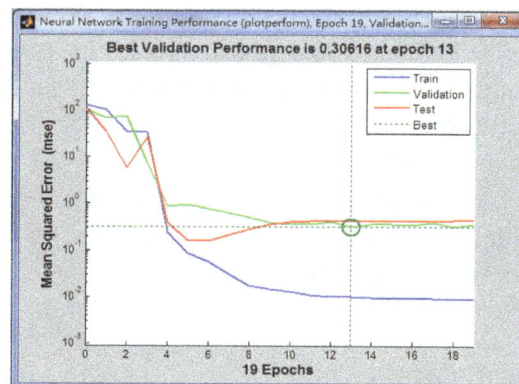

Fig. 4 Training in Matlab environment

4.2. *Thermal error forecast with SWNN and compared with different predictive models*

The SRWNN model was trained in the MATLAB environment as showed in Fig. 4. The gradient descent algorithm was selected to function-approximation, where network was trained against the data collected from the measurements of the spindle. Compared with the test results, it can be found that the SRWNN is an accuracy model to describe the behavior of thermal error of motorized spindle as shown in Fig. 5.

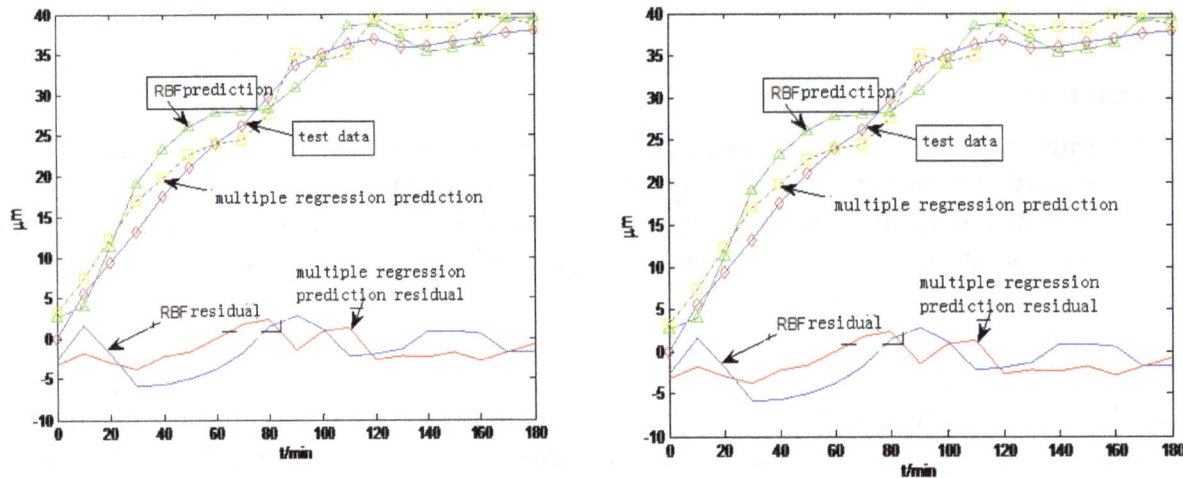

Fig. 5 SRWNN prediction compared with test data and RBF, MR model

In this paper, three methods are used to predictive the thermal error including multiple regression analysis, RBF artificial neural network and SRWNN model. As shown in Fig. 5, it can be observed that the SRWNN model predictive the thermal error much better than the static model RBF network model and MR model.

5. Conclusion

Improving the accuracy of machine tool is an essential content in manufacturing industrial. Spindle thermal error compensation is an effective way to improving the machine's accuracy. Through the theoretical study of the dynamic characteristics of motorized spindle thermo-elastic system, a new self-recurrent wavelet neural network was developed based on the investigation of the nonlinear and non-stationary effects of thermo-elastic system. This modeling methodology has been proven to be very effective and has great advantage over the conventional static model in terms of model accuracy. Model performance evaluation on the experiments of spindle thermal deformation shows that the SRWNN model is superior to the RBF neural network and multiple regressions in giving consistent predictions under a variety working conditions. The feasibility of using an SRWNN to predict the thermal error of a motorized spindle has been demonstrated. In all, the modeling method presented in this paper is applicable and effective for practical operation. It can notably improve the precision of the thermal model.

Ackowledgment

This work was funded by the Research Foundation of Anhui Province. (No. KJ2016A885)

References

1. Bryan J. International status of thermal error research [J]. CIRP ANN-Manuf Tech: 1990, 39(2): 645-656

2. Ko J K, Gim T W, Ha J Y. Particular behavior of spindle thermal deformation by thermal bending [J]. International Journal of Machine Tools and Manufacture, 2003, 43: 17-23

3. J. Jedrzejewski, Z. Kowal. High-speed precise machine tools spindle units improving [J]. Journal of Materials Processing Technology. 2005, vol 162-163: 615-621

4. Yang S, Yuan J, Ni J. The Improvement of thermal error modeling and compensation on machine tools by neural network [J]. International Journal of Machine Tools and Manufacture, 1996, 36(4): 527-537

5. Yang H, Ni J. Adaptive model estimation of machine tool thermal errors based on recursive dynamic modeling strategy [J]. International Journal of Machine Tools and Manufacture, 2005, 45: 1-11

6. Yang H, Ni J. Dynamic neural network modeling for nonlinear, nonstationary machine tool thermally induced error [J]. International Journal of Machine Tools and Manufacture, 2005, 45: 455-465

7. John M. Fines, Arvin Agah. Machine Tool Positioning Error Compensation Using Artificial Neural Networks [J]. Engineering Applications of artificial Intelligence. 2008, 21: 1013-1026

8. Kang Y, Chang C W, Huang Y R, et al. Modification of a neural network utilizing hybrid filters for the compensation of thermal deformation in machine tools[J]. International Journal of Machine Tool and Manufacture, 2007, 47(2): 376-387

9. Chang C F, Chen J J, Chen T R. A theory thermal growth control techniques of high speed spindles [J]. In: Proc of Int Conf on Advances in Electronics and Microelectronics. Washington, 2008, 96-101

10. Zhiyong Yang, Minglu Sun, Modified Elman network for thermal deformation compensation modeling in machine tools[J]. Int J Adv Manuf Technol 2011, 54: 669-676

11. Sung Jin Yoo, Bae Park, Yoon Ho Choi. Stable Predictive Control of Chaotic Systems Using Self-Recurrent Wavelet Neural Network[J]. International Journal of Control, Automation, and Systems. 2005, Vol. 3: 43-55

12. Guo Fan, Jianguo Yang. Orthogonal polynomials based thermally induced spindle and geometric error modeling and compensation [J]. Int J Adv Manuf Technol 2013, 65: 1791-1800

13. Gauracy Ameta, Shawn Moylan. Investigating the role of geometric dimensioning and tolerancing in additive manufacturing[J]. Journal of Mechanical Design 2015. vol. 137: 111706

Developmental Characteristics and Controlling Factors of Fractures of Volcanic in Yingcheng Formation of Xujiaweizi Fault Depression

Bing-Yang Lv[1, 2,*], Lei Gong[1, 3], Bo Liu[1]

1 Science and Technology Innovation Team on Fault Deformation, Sealing and Fluid Migration, Northeast Petroleum University, Daqing 163318, Heilongjiang

2 PetroChina Daqing Oilfield Company, Oil Recovery plant No.10 Daqing Oilfield Corp. Ltd, Daqing, 166405, Heilongjiang

3 Postdoctoral Programme in Daqing Petroleum Administration Bureau, Daqing 163453, Heilongjiang

*Email: 863721223@qq.com

Most scholars have reached a consensus on this point that there is a very close relationship between the development characteristics and controlling factors of fractures and gas accumulation. Thus, the study of volcanic fracture characteristics and controlling factors is significant. We selected Xujiaweizi fault depression Yingcheng as the study area. There are many pores and primary fractures in the volcanic reservoirs of Yingcheng Formation in the Xujiaweizi fault depression, but the connectivity is very poor. The development degree of tectonic fractures determines the reservoir quality and probability of hydrocarbon accumulation. In order to illuminate the relationship between fractures and natural gas accumulation, using data of cores, image logs and experimental analysis, researches are conducted firstly on the fracture genetic types, characteristics, controlling factors. Among them, secondary tectonic fractures are dominant. The distribution of tectonic fractures is controlled by lithology, lithofacies and fault in the plane, while cyclicity exists in the longitudinal direction which is controlled by unconformity.

Keywords: Fractures; development characteristics; controlling factors; volcanic; Xujiaweizi fault depression; Yingcheng formation.

1. Introduction

The rift Lower Cretaceous Yingcheng Formation of Xujiaweizi fault depression in the northern part of SongLiao Basin mainly develops volcanic strata. It has received high-yield industrial gas flow in exploration more than 50 wells and shows good exploration prospects [1, 5]. In recent years, exploration results show: Yingcheng Formation rock stratum is fracture-porosity reservoir. Fractures have important implications on reservoir stimulation, gas migration and accumulation. Distribution of fractures controls the gas reservoirs enrichment and single well productivity. Therefore, the study of volcanic reservoir fracture has important significance in guiding volcanic reservoir exploration and development. In this paper, we analyze the development characteristics and controlling factors of fracture formation in the study area using data of cores, image logs and experimental analysis.

2. Fracture Developmental Characteristics

Different scholars have different classification schemes on fractures [3, 4]. Most scholars' classification schemes reflect the different aspects on different research purposes. Usually in the study of the formation and describe of fractures divided into three categories according to geological: primary fracture, secondary fracture and artificially induced fractures. The other can also be classified according to the degree of opening of cracks, mechanical properties and fracture surface morphology. For quantitative description of fracture, parameters are mainly from the fracture size (density), width mechanical properties, filler and tendency.

According to the 2449m core of 29 wells and imaging logging data of 89 wells in the study area, Yingcheng Formation reservoir fractures can be divided into primary and secondary fracture. Primary fracture includes condensate shrinkage cracks, fried cracks and so on. They are irregular,

---

*Corresponding author

often lenticular or network distribution. At the same time they are usually small-scale cracks and are generally filled with calcite or chlorite. Most of the fractures are invalid. Secondary fracture mainly includes tectonic and weather-leaching fractures. Among them, the tectonic fractures are the main types of fractures. They are widely distributed, with good directionality and regularity in a variety of volcanic rock. They appear packaged in group with a large scale. They usually cut through the entire core, filling weakly. Only a small part of tectonic fractures are mineral filler and filling half the majority, more than 80 percent of tectonic fractures are effective fractures (Fig. 1a).

Analysis of the occurrence of cracks from core orientation and paleomagnetic imaging logging data, Yingcheng Formation of Xujiaweizi fault depression reservoir fractures mainly NEE-SWW, NNE-SSW, NW-SE and near EW direction four groups. Fracture dip focus on middle and high angle fractures oriented (Fig. 1b). Calculated according to the measurement and imaging core logging, single well fracture linear density mainly in the 1.0 to 6.0 fractures per meter, part of the well-stage fracture well developed 33.9 fractures per meter; Hydrodynamic crack width is mainly distributed in 50 ~ 400μm, average 267.5μm. After dissolution, hydrodynamic width up to 8.17mm; Fracture faces mainly in the rate of 0.1% to 1.0%, average of 0.54%, about 10 percent of total porosity reservoir (average 5.8%), it is accounted that fractures are an important part of Yingcheng volcanic reservoirs space. Fracture permeability mainly in 10 to 100mD. It is 1 to 3 orders of magnitude of the matrix permeability reservoir. Fractures are reflected in the main channel of volcanic reservoirs of natural gas migration.

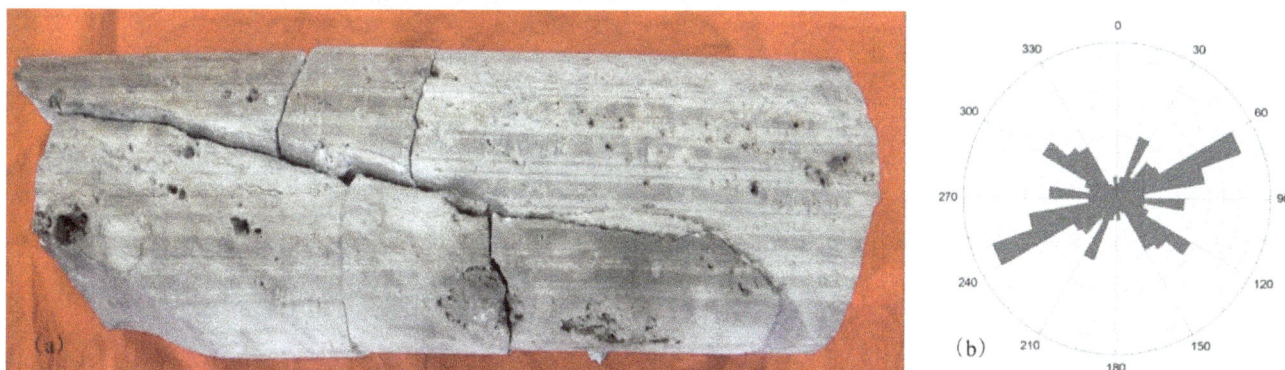

Fig. 1 Tectonic fracture and rose diagram of fracture strikes

3. Controlling Factors of Fractures

Through the study of formation mechanism of volcanic fracture, we know that the type of volcanic rock, formative period, during the eruption and the form in which the construction site are static controlling factors of fractures and also are the basis for the formation of fracture. Its response to dynamic control factor determines the distribution of fractures in different periods. There are three main aspects:

1) Different lithology fractures developing differently. Since the difference of composition, structure and construction, lithology becomes different, thus physical and chemical composition become different as well, making it a response to the effect of the late diagenesis. Hence, structure has become more different. Finally it forced the fractures have different levels of development (Fig. 2). From the general situation is concerned, in accordance with the same conditions as stress, if the rock is brittler, it is more prone to rupture [2, 4, 6]. From rhyolite to andesite to basalt, the degree of development of fractures gradually increased; from tuff to

volcanic breccia to subvolcanics, the degree of development of fractures gradually increased. Daqing Xujiaweizi, Chen Jianwen statistics of fractured volcanic rocks in the case reached the same conclusion.

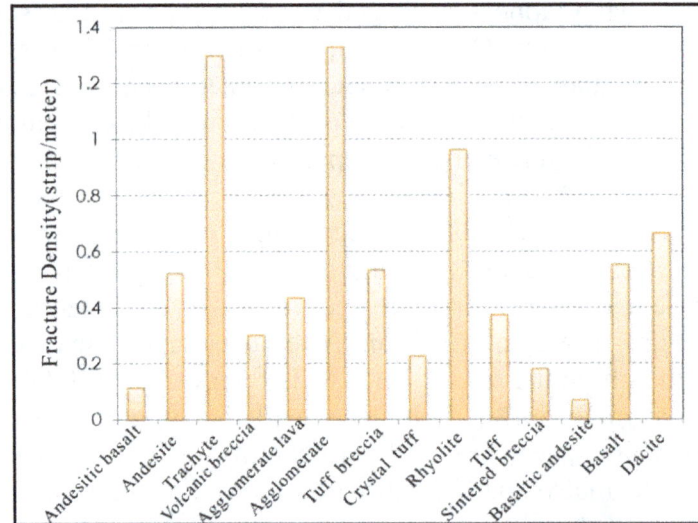

Fig. 2 Different lithology fractures' density distribution

2) The effect of volcanic facies control. Through scanned image of field outcrops and drilling cores we know that volcanic fractures are commonly development, but in different phases band the degree of fracture development is not the same in Songliao Basin (Fig. 3).

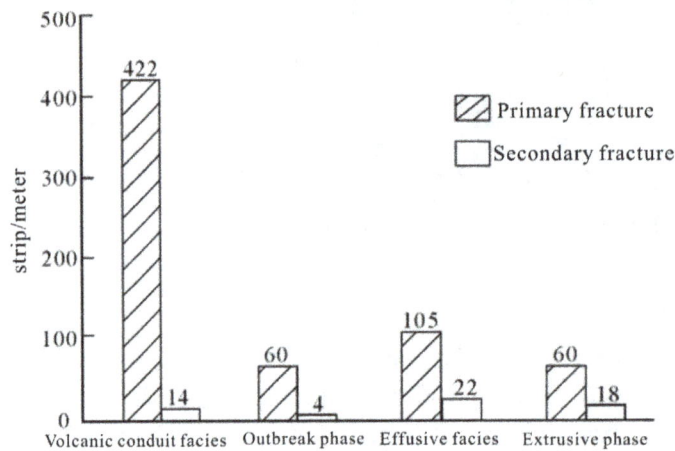

Fig. 3 Volcanic rocks associated with fracture lines

3) The degree of development of fractures in structural positions control. It is seen that construction site of fracture development degree of control plays a certain role from the mechanism of fracture formation, especially with the faults and folds associated fracture distribution in space with certain regularity. Summarized as follows:

a) Fractures develop well in turning end portion of the shaft portion of minor anticline
Since various types of volcanic facies zone tensile cracks the top of the anticline relatively development, so often in the small amplitude anticline structure of the terminal

portion of the shaft portion turn, become a good fracture zone. The anticline tectonic fractures in the wings due to stress relationship are more rudimentary. If oil and gas migration and accumulation of fracture formation is consistent with the peak period, oil preferentially occupy fracture, the calcite crystal growth will be inhibited and stop growing, filling the role of weak with a high degree of connectivity fracture, at the same time become effective oil and gas fractures; Thus the lower part of the structure is similar to crack, and tectonic fractures may be formed in the bottom surface of the rock.

b) Fracture intersection, broken ends, fracture turning end and curved convex portions often easily development of tectonic fractures

Because tectonic movements cause stress perturbations around faults in tectonic fractures, local tectonic stress concentration occurs around faults. So, fracture intersection, broken ends, fracture turning end and curved convex portions often easily develop tectonic fractures (Fig. 4).

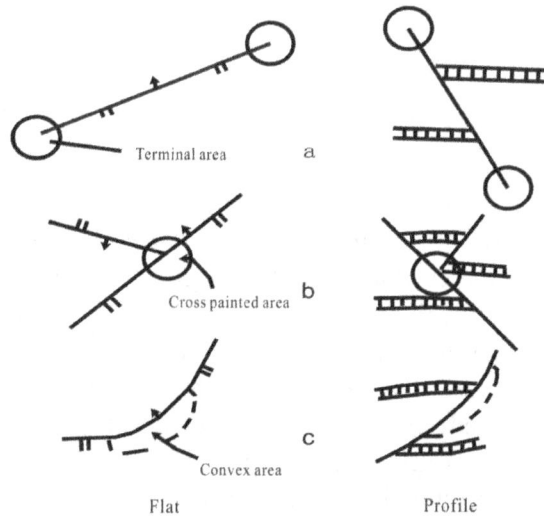

Fig. 4 Schematic of cracks prone areas near fault

c) Fracture develops differently near different parts of the fault

Stress perturbations generate on both sides of the fault zone within a certain range. Tension fractures parallel to fault as well as shear fractures oblique fault account for most of the band; other, in the cross-fault, multi-branch and knee stress concentration sites, these sites also fractures most developed areas. In the local area, in a certain area near fault often have many fractures and fractures density decreases away from the fault (Fig. 5). Nearby fault easily develop of tectonic fractures due to stress concentration. According to the literature, there is a positive correlation between fracture zone width and telescopic amount faults. Regional and deep fault caused fracture width is often from 5 to 10km, while some faults can be up to 40km.

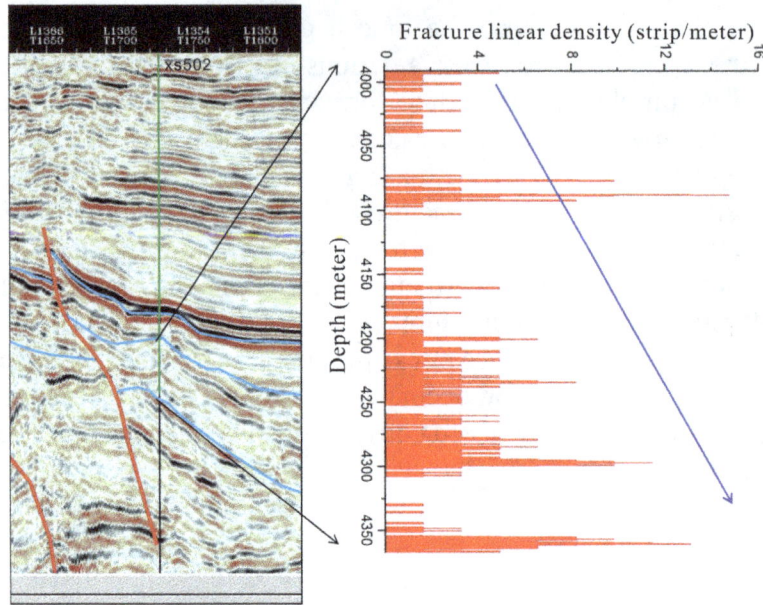

Fig. 5 Xu Shen 502 crack linear density variation with depth figure

Based on the core observation and imaging logging data analysis, distribution of Yingcheng volcanic reservoir fractures is mainly controlled by lithology, lithofacies, tectonic position and unconformity etc. There is obvious regularity on the longitudinal and flat surface. According to the statistics, cracks in rhyolite and andesite develop most (an average of 15.6 fractures per meter, up to 33.9 fractures per meter), followed by basalt and hydrothermal breccia, etc. There is less fracture in tuff. In terms of facies, overflow with cracks most developed, followed by is the outbreak of the phase, volcanic sedimentary facies fracture development is weak. On the plane, fracture development is obviously controlled by faults. In the vicinity of Xu Zhong and Xu Dong strike-slip faults, cracks develop obviously. With increasing distance from the fracture, the number of cracks begins to reduce. In the longitudinal direction, distribution of reservoir fracture has cyclicity characteristic. We use unconformity as a starting point. With increasing depth, there are regular changes on fracture type and degree of development. Less than 10m from the unconformity, corroded fissure and tectonic fractures are the main types. The degree of fracture development is high with a good connectivity; the following is tectonic fractures segment. Fracture development is declined, but its size is greater than the upper portion.

4. Conclusion

(1) There are two types of fractures in Xujiaweizi depression Yingcheng Formation reservoir. They are primary and secondary fractures. Secondary tectonic fractures occupy an important position. Tectonic fractures is an important reservoir space and the main channel of natural gas lateral migration

(2) Tectonic fractures are controlled by lithology, facies and fracture in the plane. In the longitudinal direction tectonic fractures is mainly controlled by the unconformity, at the same time it has the characteristics of the distribution cycle.

Acknowledgments

This study is funded by the China Postdoctoral Science Foundation: (Grant No.2015M581424), National Natural Science Foundation of China: (Grant No.41572126) and Natural Science Foundation of Heilongjiang Province: (Grant No.41417837-8-13145). At the same time, full of thanks to the 2nd Annual International Workshop on Materials Science and Engineering and other staff members.

References

1. CAI Dongmei, SUN Lindong, QI Jingshun, et al. Reservoir characteristics and evolution of volcanic rocks in Xujiaweizi fault depression [J]. Acta Petrolei Sinica, 2010, 31(03): 400-407.
2. Carcione J M. Constitutive model and wave equations for linear, viscoelastic, anisotropic media [J]. Geophysics, 1995, 28(02): 537-548.
3. Chen Huanqing, Hu Yongle, Jin Jiuqiang, et al. Multiple sets of information synthesized to describe fractures of volcanic reservoir: Taking volcanic reservoir of the Member 1 of Yingcheng Formation in Xudong Area of Xushen Gas Field as an example [J]. Earth Science Frontiers, 2011, 18(2): 294-303.
4. Ren Desheng. Research on fracture formation mechanism and forecasting in volcanic rock of Song Liao basin [D]. Chang Chun: Ji Lin University, 2004.
5. WANG Jinghong, ZOU Caineng, JIN Jiuqiang, et al. Characteristics and controlling factors of fractures in igneous rock reservoirs[J]. Petroleum Exploration and Development, 2011, 38(06): 708-715.
6. Zhang ErHua, Jiang ChuanJin, Zhang YuanGao, et al. Study on the formation and evolution of deep structure of Xujiaweizi fault depression [J]. Acta Petrologica Sinica, 2010, 35(01): 149-157.

Research on the Relationship Between the Wheel Tread Wear and Stability of 209 Bogies

Wen-Xue Li

CRRC Changchun Railway Vehicles CO., LTD. No.435 Qingyin Road, Changchun, Jilin, China
Email: liwenxue@cccar.com.cn

To confirm the relationship between the wheel tread wear of 209 bogie and the stability of the bogie, choose the passenger cars operated on Qingzang Line to measure and calculate the wheel tread wear and equivalent conicity; measure the acceleration signals on the three axles x, y, z of floor above the bogie by the portable test device, analyze the bogie's instability condition by time frequency and frequency spectrum. According to the instability distribution condition of the bogie and the equivalent conicity of every wheel, analyze the relationship between the wheel tread wear and the stability of the bogie, grasp primarily the wheel's limitation condition based on the conicity management. The result shows: the wheel tread wear of 209 bogie with disc brake and surface cleaning device is smaller than that with tread brake; 209 series bogie's instability condition and the equivalent conicity distribution reflect the corresponding relationship between the large equivalent conicity and the bogie's instability condition; it is suggested to choose UIC519 with 0.5 equivalent conicity as the limitation volume of 209 series bogie's wheel to the conicity management.

Keywords: Equivalent conicity; stability; 209 bogie.

1. Introduction

The wheel wear is a common problem existed in the railway field. The change of the wheel surface appearance affects the wheel-rail contact geometrical relationship directly and then affects the train's dynamical property. At present, the process mode of the wheel wear is to repair in China, but the frequent repairing of the wheel must cause high repairing costs, however, not repairing in time will cause the decrease of vehicle's operation property and may cause security problems if serious. In order to make a proper wheel repairing strategy, researching the relationship between the wheel wear and the train's operation dynamical property is the premier problem [1].

Scholars around the world research more about the wheel wear and high-speed trains' operation dynamical property. Kalkerj [2] proposed that the wheel rail contact fatigue is mainly caused by the small contact spots and large wheel rail lateral force. The small wheel rail contact area is mainly because after the surface is worn, the wheel rail conformal degree decreases and lead to non-ideal wheel-rail contact geometrical relationship; the large wheel rail lateral force is mainly because after the surface periphery is worn, the train steering capability decreases. Li Yan [3] aimed at one of the cars in the train to measure the wheel surface appearance in the fixed time by WP-D surface appearance inspection device, summarized the surface appearance wear rules and features and stimulated the influence of the wheel surface wear to the train's dynamical property in the 5 groups of wheel wear working conditions in the multi-rigid-body dynamical software. Wang Yijia [4] carried on the line track test based on the influence of the worn surface appearance to the wheel-rail contact geometrical relationship and the train's operation property, according to the angular difference with the change of surface wear, she built a dynamical stimulation model which considered the wheel-rail nonlinear and suspend force nonlinear features at the same time, in general, she chose three typical types of bogies and researched comparatively the influence of the equivalent conicity of primary wheel rail and worn stable wheel rail to the critical speed and snakelike operation stability. Chen Rong [5] compared the wheel rail contact conditions of different surface appearance and stimulated the influence of the contact geometrical relationships of different surface appearances to the derail security and operation stability.

Based on the relationship between the wheel surface wear of 209 bogies and its stability, there are few research achievements now. This article is mainly aimed at the 209 bogie trains on Qingzang Line, analyzing the relationship between the wheel surface wear of 209 bogies and its stability by measuring the wheel surface wear conditions and its bogie instability condition.

2. Vehicle Conditions

The tested vehicle is25G-type passenger cars operated on Qingzang Line, 18 cars in total, the operating mileage is 20 thousand km, including 6 cars of 209T bogies and 2 cars of 209P bogies. The main research object of this article is 209 bogie, so it is only aimed at 209P bogie and 209T bogie of this train for analysis.

209T bogie adopts the tread brake device; 209P bogie replaces the disc brake with the tread brake based on 209T bogie and installs surface cleaning devices.

3. Wheel Tread Wear and Equivalent Conicity

3.1. *Wheel tread wear*

Fig. 1 showed that the surface wear consumption against 209 bogies among the tested trains, from this diagram, the average wheel surface wear consumption against 209T bogies is 1.37, that against 209P bogies is 0.865; so the average wheel surface wear consumption against 209T bogies is 58% more than that against 209P bogies.

It can be seen that compared with 209T bogies with tread brake, 209P with disc brake and surface cleaning device can reduce the wheel surface wear consumption effectively.

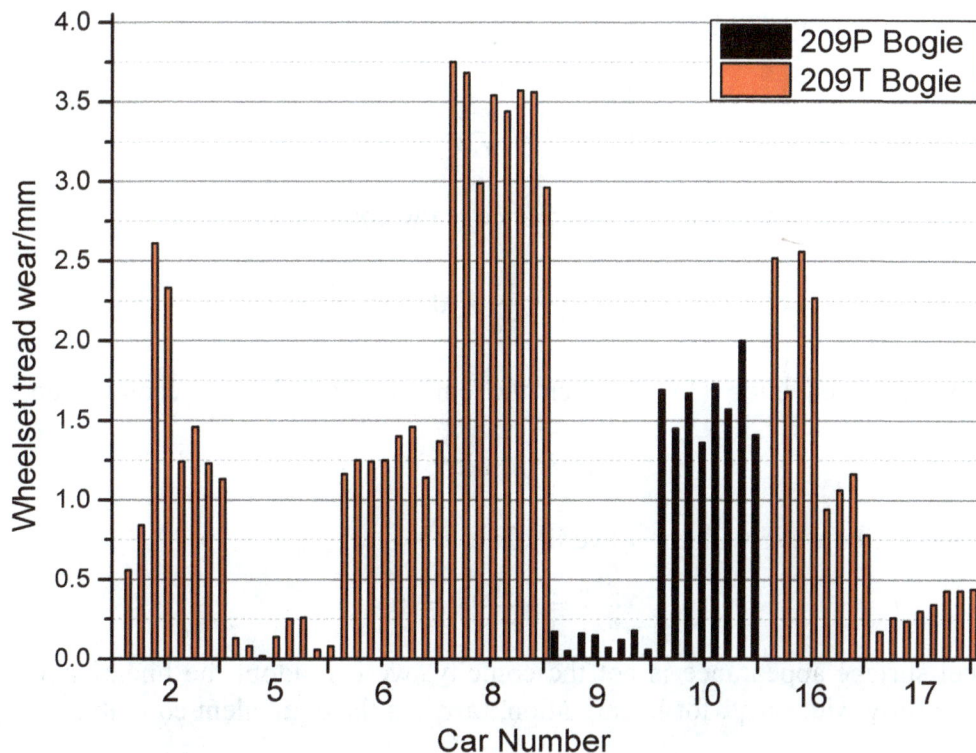

Fig. 1 Wheel tread wear of 209 bogie

3.2. Equivalent conicity

Due to the factors of wheel rail wear and wheel rail deviation, many parameters of wheel rail contact geometrical calculation have large discreteness, so it is needed a simple and proper parameter to evaluate the wheel rail contact geometrical relationship. The equivalent conicity as the wheel rail contact linear index is widely used to represent the wheel rail contact geometrical features [6]. The substance of the equivalent conicity is that, a non-conicity surface wheel set is equal to a conicity surface wheel set under each lateral amplitude. Its equivalent basis is the wave lengths of both motion trails are the same [7].

There are many methods to calculate the equivalent conicity, in this article, it uses the equivalent conicity calculation method of UIC519 [8] standard.

The free wheel operating on the track can be described as the differential equation [9]:

$$\ddot{y} + \frac{v^2}{er_0}\Delta r = 0 \tag{1}$$

In the equation, y is wheel set lateral displacement; $\ddot{y}$ is wheel set lateral accelerated speed; e is contact span; r_0 is nominal rolling radius; Δr is wheel diameter difference; v is wheel set forward speed.

Assuming v is a constant. It doesn't affect the equivalent conicity calculation, that is:

$$v = \frac{dx}{dt} \tag{2}$$

In the equation, x is the longitudinal displacement of the wheel set on the track. So, put Eq. (2) and Eq. (3) into Eq. (1), we can get

$$\begin{cases} \frac{dy}{dt} = v\frac{dy}{dx} \\ \frac{d^2y}{dt^2} = v^2\frac{d^2y}{dx^2} \end{cases} \tag{3}$$

Assuming the angle of wheel surface appearance is the conicity y,

$$\frac{d^2y}{dx^2} + \frac{\Delta r}{er_0} = 0 \tag{4}$$

So the differential equation (1) becomes the constant coefficient second order differential equation:

$$\Delta r = 2y\tan\gamma \tag{5}$$

The result is the sine wave with the wave lineλ:

$$\frac{d^2y}{dx^2} = \frac{2\tan\gamma}{er_0}y = 0 \tag{6}$$

If the wheel surface appearance is not the conicity, we can adopt the linearization method, in Eq. (7), replace $\tan\gamma$ with $\tan\gamma_e$ for linearization, $\tan\gamma_e$ is the equivalent conicity.

$$\lambda = 2\pi\sqrt{\frac{er_0}{2\tan\gamma}} \tag{7}$$

The initial condition of Eq. (4) is:

$$\begin{cases} y = y_0 \\ \dfrac{dy}{dx} = 0 \\ x = 0 \end{cases} \quad (8)$$

By the given initial value y_0 integral, the cyclical movement can be exported the wheel set with the features of amplitude $2y$ and wave length λ. Calculate the equivalent conicity by Klingel formula:

$$\tan\gamma_e = \left(\frac{\pi}{\lambda}\right)^2 2er_0 \quad (9)$$

Calculate its equivalent conicity by the wheel set surface tested data of the tested trains, see Fig. 2 attached.

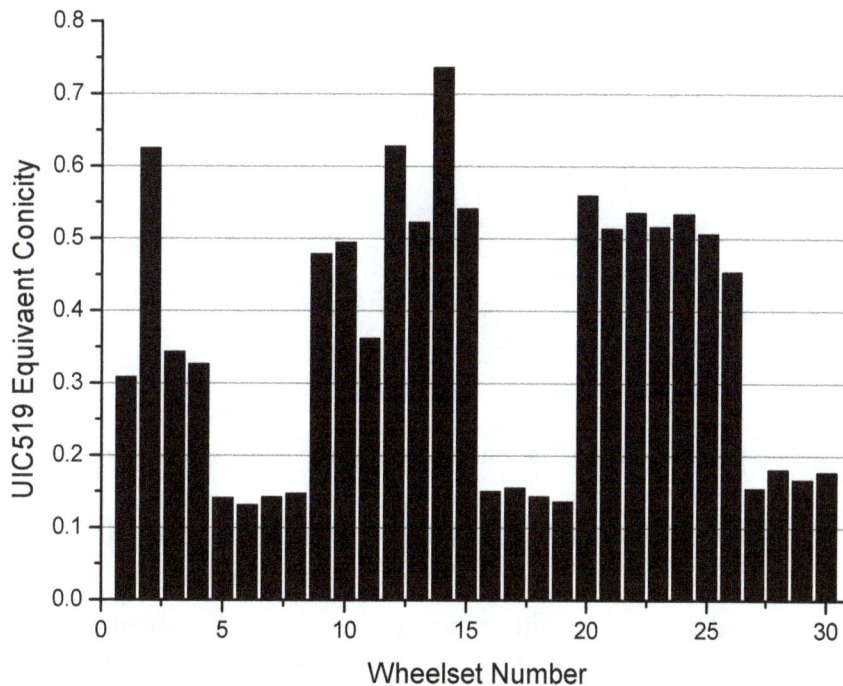

Fig. 2 The equivalent conicity of 209 bogie

4. The Method of Determining Bogie Instability

Measure the acceleration signals on the three axis x, y, z of floor above bogie by the portable test device, by analyzing the time frequency and frequency spectrum of the acceleration signals judge whether the corresponding bogies are unstable or not. The judgment method is: the 3-10Hz frequency features of floor lengthways (x axis) and horizontal(y axis) are very obvious, that is to be judged that the bogie is on the unstable condition.

From Fig. 3, the 3-10Hz frequency features of floor lengthways (x axis) and horizontal(y axis) are very obvious, so the bogie corresponding to this signal is unstable.

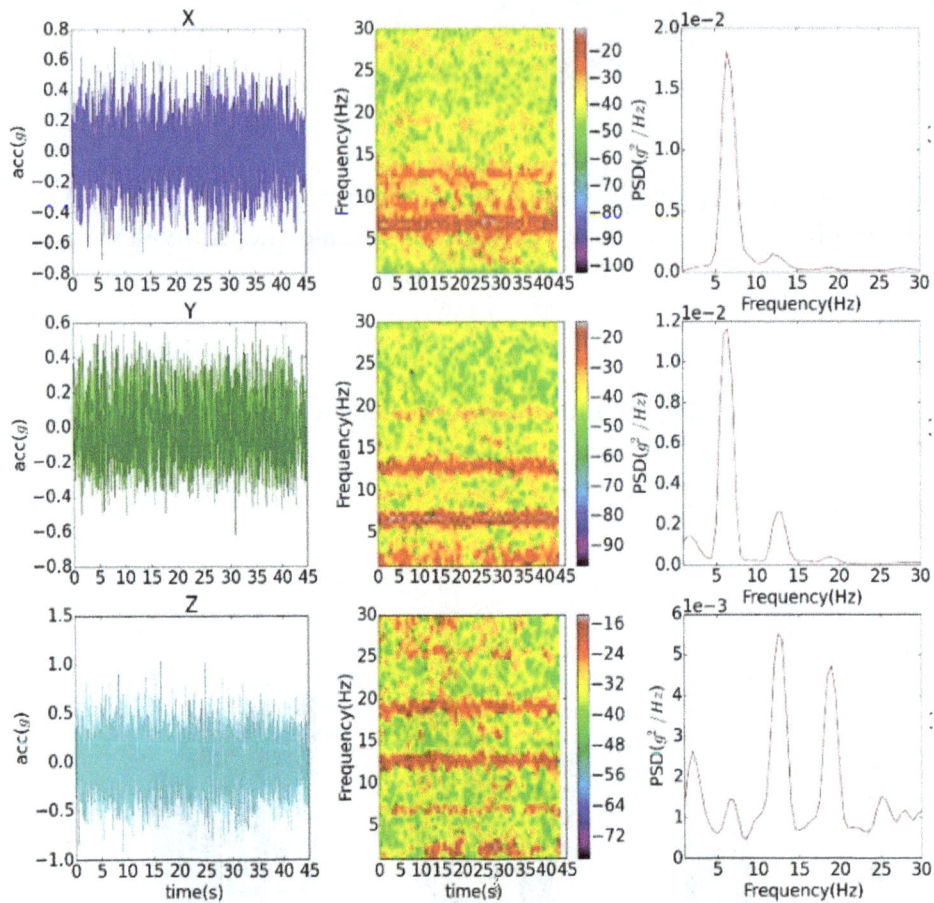

Fig. 3 The time frequency and frequency spectrum of acceleration signals on the three axis x, y, z of floor

5.   The Relationship between Wheel Tread Wear and Stability of Bogie

Fig. 4 shows the relationship diagram of the surface wear consumption against 209 bogies and the equivalent conicity. From this diagram, the larger the surface wear consumption is, the larger the equivalent conicity is. Fig. 5 shows the relationship diagram of the equivalent conicity of 209 bogies and the bogies' stability, among the equivalent conicity of the wheel set is more than 0.5, 67% bogies are unstable; among the unstable bogies, 67% equivalent conicity of the wheel set are more than 0.5.

Thus, the surface wear consumption against 209 bogies can cause the equivalent conicity increasing, but it is easy to lead the bogie unstable because of the large conicity.

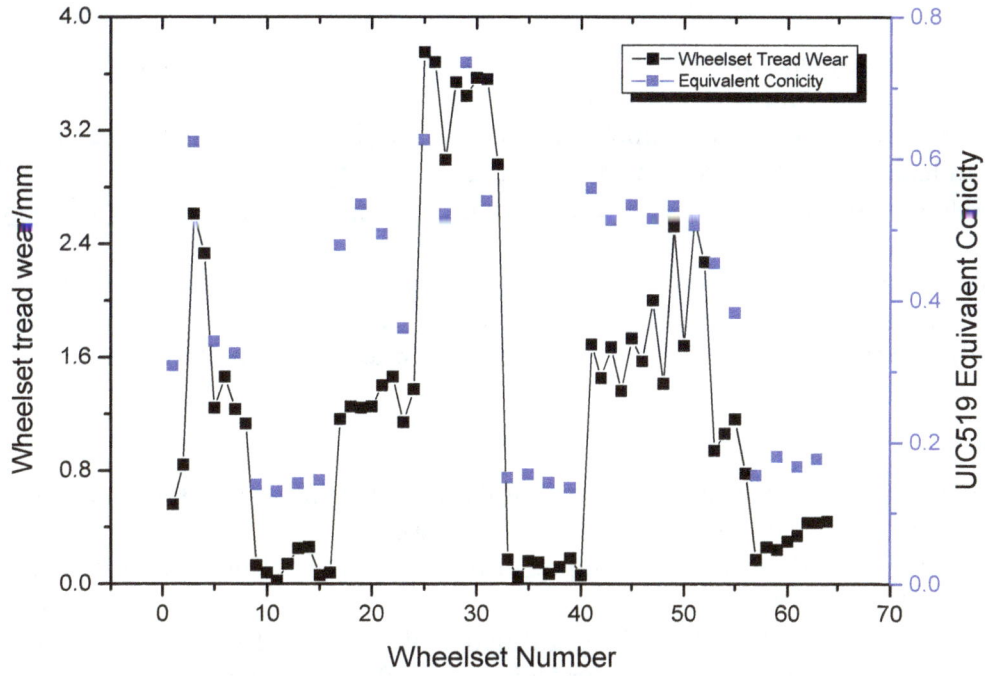

Fig. 4 The relationship between the wheel tread wear and equivalent conicity

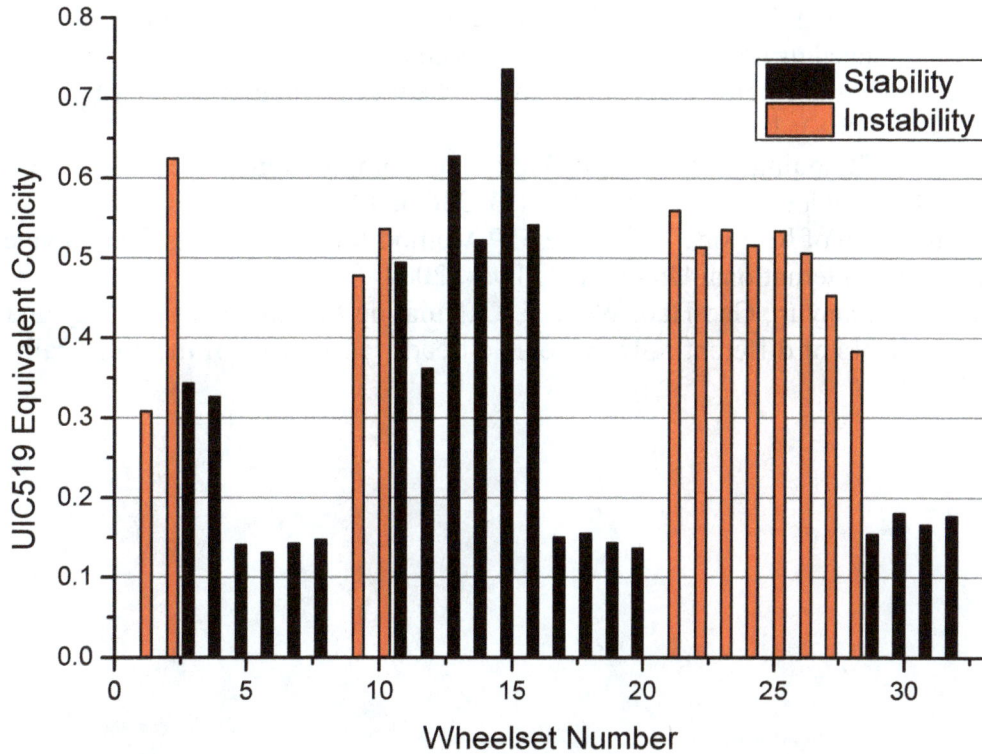

Fig. 5 The relationship between the equivalent conicity and stability of bogie

6. Conclusion

The surface wear consumption against 209 bogies with tread brake device is larger than that with disc brake and surface cleaning device.

The larger the surface wear consumption against 209 bogies is, the larger the equivalent conicity is. It is easy to lead the bogie to be unstable because of the large conicity.

It is suggested to choose UIC519 equivalent conicity 0.5 as the limited value of 209 bogies wheel set to the conicity management.

In the test, measure the vibration acceleration floor above bogie by the portable test device, the tested time is about 30s, the unstable signal may not be caught because of the wheel set's big conicity, so the more effective method can be adopted for testing in the future tests.

References

1. Huang Weizhao. Wheel treads wear and its influence on dynamic performance of vehicles [D]. Chengdu: Southwest Jiaotong University, 2012: 1-66.
2. KALAERJ J. Simulation of the development of railway wheel profile through wear[j]. Wear, 1991,0020150(1): 355-365.
3. Li Yan, Zhang Weihua, Zhou Wenxiang. Influence of wheel profile on dynamic performance of EMU [J]. Journal of Southwest Jiaotong University, 2010, 45(4): 549-554.
4. Wang Yijia Zeng Jing, Luo Ren, Wu Na. Wheel profile wear and wheel/rail contact geometric relation for a high-speed train [j]. Journal of vibration and shock, 2014, 33(7): 45-50.
5. Chen Rong, Wang Ping, Song Yang. Wheel/rail contact geometry of different wheel tread profile in railway high-speed turnout [j]. Advanced materials search, 2011, 255-260.
6. Polach O. Characteristic parameters of nonlinear wheel/rail contact geometry [j]. Vehicle system dynamics, 2010, 48(S): 19-36.
7. Wu Ning, Dong Xiaoqing, Lin Fengtao, Wen Bin. Computation and evaluation of equivalent conicity [j]. Railway locomotive & CAR, 2013, 33(1): 49-52.
8. International Union of Railway. UIC Code 519 Method for determining the equivalent conicity [S]. 1st ed. Paris: International Union of Railway, 2004.
9. Gan Feng, Dai Huanyun, Gao Hao, Wei Lai. Calculation of equivalent conicity and wheel-rail contact relationship of different railway vehicle treads [j]. Journal of the china railway society, 2013, 35(9): 19-24.

Structural Reliability Assessment Based on Adaptive Active Learning of Aviation Oxygen Flow Indicator

Yan-Xin Lin[1], Sa Wu[1,*], Yan-Sen Lin[2], Yi-Ping Ni[3]

[1]Sch Reliabil & Syst Eng, Beihang University, Peoples R China
[2]Sch Environm & Chem Eng, Shenyang Ligong University, Peoples R China
[3]SICHUAN YAMEI POWER TECHNOLOGY CO. LTD., Peoples R China
*Email: wusa@buaa.edu.cn,

A reliability assessment strategy named Structural System-oriented Active learning reliability method combining Kriging and Monte Carlo Simulation (SSo-AK-MCS) was proposed, the number of calls to the limit-state function of SSo-AK-MCS is 0.000121% of Monte Carlo Simulation (MCS), 4.3478% of Subset Simulation (SS), 11.2768% of Importance Simulation (IS). In the meantime, calculation precision of SSo-AK-MCS is 100.2930 times of SS, 11.5431 times of IS. The cost of achieving that outstanding performance is the uplift of the degree of intelligence w.r.t. reliability assessment strategy. At this moment, the reliability assessment strategy is no longer static, but dynamic, which has been equipped with the adaptive active learning strategy. Under the circumstances that uncertainty must be taken into account, the structural reliability assessment w.r.t. complex heterogeneous structure which is extremely sensitive to reliability, such as the duct of two augmented turbofan engines named RD-93 of fighter J-31.

Keywords: Reliability assessment; aviation oxygen flow indicator; stochastic modeling; adaptation; active learning; dynamic strategy.

1. Introduction

With the main force of the new generation fighter being equipped with and the promotion of the degree of automation, the pilots who are the supervisor of man-machine systems are undertaking more and more complex tasks w.r.t. information processing. Aviation oxygen flow indicator is the key device that pilots obtain the operating state of oxygen supply system, once the aviation oxygen flow indicator shows inaccurate reading, or even inverse one, the ability of risk assessment and decision disposal of pilots will be significantly reduced, even be seriously interfered with. Consequently, the non-tactical combat effectiveness of air force will be weakened inevitably. The home and abroad literature w.r.t. aviation oxygen flow indicator is pitiful. One w.r.t. the reliability assessment is even measlier.

In more and more extensive modern engineering background, uncertainty quantification is becoming a more and more crucial field. The model based on deterministic scheme is gradually replaced by the stochastic modeling because the inevitable physical phenomena and uncertainty in measurement is beyond the former. To cope with the rapid expansion of the amount of information, repeated execution of calculation to limit-state function, which is time consuming, and high computational burden is still the current solution to major reliability researchers[1,2].

A reliability assessment strategy named SSo-AK-MCS which is a combination of SSo and AK-MCS has been proposed, the trustworthy structural reliability index of aviation oxygen flow indicator has been shown with mild number of calls to the limit-state function, benefits shorter lead time and lower computational expense. The adaptive learning intelligent function has been introduced to our strategy, improvement of calculation precision by Kriging meta-model and rapid convergence has been accomplished synchronously. The cost of achieving that outstanding performance is the uplift of the degree of intelligence w.r.t. reliability assessment strategy. At this moment, the reliability assessment strategy is no longer static, but dynamic, which has been equipped with the adaptive active learning strategy.

*Corresponding author

641

The paper is organized as follows. 'Stochastic modeling and solving strategy' section describes three types of strategy for stochastic modeling. The proposed strategy SSo-AK-MCS which is used in adaptive active learning strategy for the structural reliability assessment of the air flow oxygen indicator will be established in 'Adaptive active learning strategies for structural system SSo-AK-MCS' section. 'Structural reliability assessment of aviation oxygen flow indicator' section describes the application of the proposed strategy SSo-AK-MCS in aviation oxygen flow indicator. The paper concludes with a summary of the main findings and respective outlooks.

2. Stochastic Modeling and Solving Strategy

The main task of stochastic modeling is to calculate the value of the limit-state function under stochastic input. In general, when entailing the evaluation of numerical model P_f based on vector X, the computational expense is very high. In order to avoid the nearly violent computational expense, we have excogitated the following three types of strategies:

1) Approximation method: SORM, FORM.
   The efficiency of these methods is very high, but when faced with complex and nonlinear limit-state function, they become very unreliable [3].
2) Simulation method: SS, IS.
   These methods are typically characterized by good convergence and can be used to calculate the consequent confidence interval of P_f, but they are in exchange for the cost of the extremely expensive computation expense [4].
3) Meta-model adaptive method: PCE, Kriging, AK-MCS.
   These methods are used to estimate the failure probability accurately with the surrogate models which are gradually improved with iterations [5, 6].

SSo is defined as a specific functional structure that is required to be specified in a set of well-defined security constraints. These constraints are considered in system design stage considering the expected environmental loading or operational loading that it will be subjected to. Uncertainty exists in the system of physical properties, such as manufacturing tolerances, environmental loading, such as the expected environmental conditions, service loading, such as transportation, the factors above can lead to structure deviation from the normal range of operation conditions. In these cases, our system encounters failure event.

The quantitative assessment of the probability w.r.t. failure event (failure probability) is analyzed by the structural reliability assessment, the uncertainty model w.r.t. structure, environment and loading parameters have been given [10, 11]. Gauss distribution $f(x)$ submitted to location parameter μ and scale parameter σ were selected to describe the parameters of structural system.

3. Adaptive Active Learning Strategies for Structural System SSo-AK-MCS

A reliability assessment strategy named SSo-AK-MCS, which is a combination of SSo and AK-MCS, has been proposed for the trustworthy structural reliability index of aviation oxygen flow indicator.

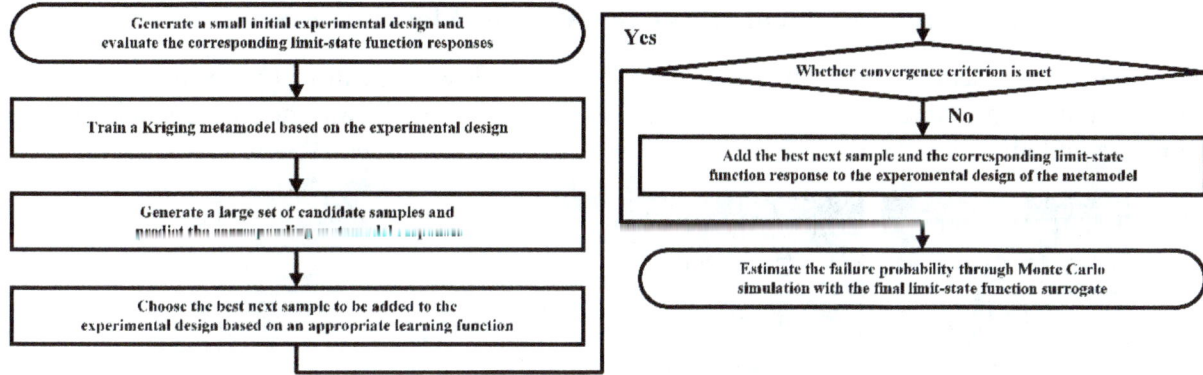

Fig. 1 DOE flowchart of AK-MCS

The basic framework of the AK-MCS reliability assessment strategy is the integration of MCS and Kriging (also known as Gauss process modeling) meta-model based on adaptive techniques. MCS and its variants are usually in distress, which is caused by heavy computational burden of limit-state function evaluation. The advantage of introducing Kriging meta-model instead of limit-state function in AK-MCS lies in the fact that without sacrificing the calculation precision, the problem is obviously out of the heavy burden of MCS computing expense. Fig. 1 shows the DOE flowchart of AK-MCS [7, 8].

Learning function measures attraction of candidate samples x, which have been added to DOE to improve the estimation precision of failure probability. The U function based on the concept of false classification is still the first choice for most researchers. Nevertheless, the expected feasibility function, which is shown in equation (1) is getting more and more attention and recognition.

$$
\begin{aligned}
\text{EFF}(\boldsymbol{x}) = {} & \mu_{\hat{g}}(\boldsymbol{x}) \left[2\Phi\left(\frac{-\mu_{\hat{g}}(\boldsymbol{x})}{\sigma_{\hat{g}}(\boldsymbol{x})}\right) - \Phi\left(\frac{-\epsilon - \mu_{\hat{g}}(\boldsymbol{x})}{\sigma_{\hat{g}}(\boldsymbol{x})}\right) - \Phi\left(\frac{\epsilon - \mu_{\hat{g}}(\boldsymbol{x})}{\sigma_{\hat{g}}(\boldsymbol{x})}\right) \right] \\
& - \sigma_{\hat{g}}(\boldsymbol{x}) \left[2\varphi\left(\frac{-\mu_{\hat{g}}(\boldsymbol{x})}{\sigma_{\hat{g}}(\boldsymbol{x})}\right) - \varphi\left(\frac{-\epsilon - \mu_{\hat{g}}(\boldsymbol{x})}{\sigma_{\hat{g}}(\boldsymbol{x})}\right) - \varphi\left(\frac{\epsilon - \mu_{\hat{g}}(\boldsymbol{x})}{\sigma_{\hat{g}}(\boldsymbol{x})}\right) \right] \\
& + \epsilon \left[\Phi\left(\frac{\epsilon - \mu_{\hat{g}}(\boldsymbol{x})}{\sigma_{\hat{g}}(\boldsymbol{x})}\right) - \Phi\left(\frac{-\epsilon - \mu_{\hat{g}}(\boldsymbol{x})}{\sigma_{\hat{g}}(\boldsymbol{x})}\right) \right]
\end{aligned}
\tag{1}
$$

Where Φ represents the standard normal cumulative distribution function and φ denotes the standard normal density function

The expected feasibility function is built with $\epsilon = 2\sigma_{\hat{g}}^2$.

Schobi [9] argues that the convergence criteria based on the U function is excessively conservative, and the alternative convergence criteria, which is shown in equation (2) w.r.t. the uncertainty of failure probability estimation of its own, is usually more efficient in the context of structural reliability.

$$
\frac{\hat{P}_f^+ - \hat{P}_f^-}{\hat{P}_f^0} \leq \epsilon_{\hat{P}_f}
\tag{2}
$$

Where $\epsilon_{\hat{P}_f} = 5\%$, $\hat{P}_f^0 = \mathbb{P}\big(\mu_{\hat{g}}(\boldsymbol{x}) \leq 0\big)$, $\hat{P}_f^{\pm} = \mathbb{P}\big(\mu_{\hat{g}}(\boldsymbol{x}) \mp k\sigma_{\hat{g}}(\boldsymbol{x}) \leq 0\big)$, $k = \Phi^{-1}\big(1 - \frac{\alpha}{2}\big)$.

4. Structural Reliability Assessment of Aviation Oxygen Flow Indicator

In this paper, the research object is the aviation oxygen flow indicator. The basic structure and assembly method of which is given in Fig. 2, while Fig. 3 shows the relationship between temperature and altitude.

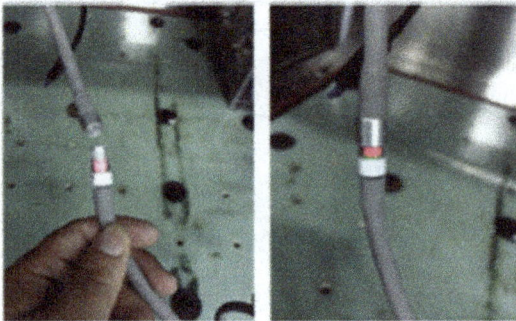

Fig. 2 Structure and assembly method of aviation oxygen flow indicator

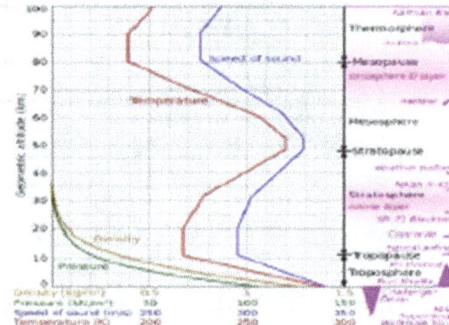

Fig. 3 Relationship between temperature and altitude

Table 1 shows photographs w.r.t. major high performance fighter in today's world, codename and ceiling. The ceiling range of high performance fighter is [15000m, 20000m], we can see from Fig. 2 that the corresponding temperature is 228K, namely, -45.15°C.

Table 1 Ceiling of high performance aircraft

photograph								
codename	F22	EF2000	F35	Su-35BM	Rafale	Mig-35	JAS-39	J-31
ceiling[m]	19812	18000	18288	18500	16800	16800	15240	18000

Table 2 presents the design parameters of the air oxygen indicator. Input random variable of limit-state function, say, parameter of dimension L_1, L_2, thermal expansion coefficient L_{1CTE}, L_{2CTE} are mutual independent, Fig. 4 shows the distribution of L_1, L_2 and $L_{1_{CTE}}$, $L_{2_{CTE}}$ respectively.

Table 2 Design parameters of air oxygen indicator (20°C)

PHA (Peg-in-Hole-Assembly)	Peg Part	Hole Part
Material	AS	2024T4
$L_n \sim N(\mu, \sigma)[mm]$	(9.410,0.010)	(9.390,0.010)
$L_{nCTE} \sim N(\mu, \sigma)[10^{-6}K^{-1}]$	(78,2)	(23,1)

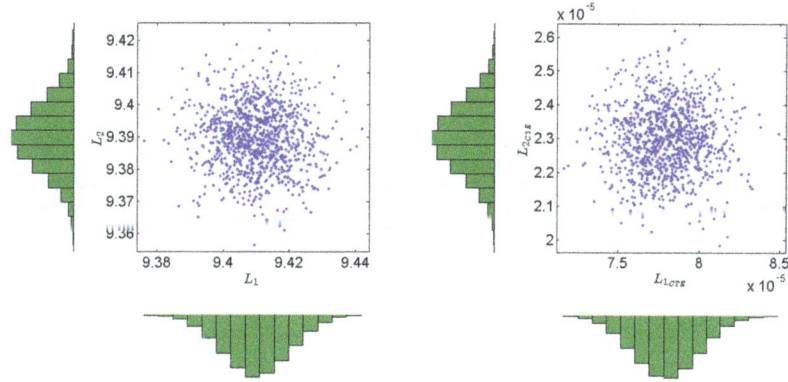

Fig. 4 Distribution of Input L_1, L_2 and $L_{1_{CTE}}, L_{2_{CTE}}$

Following assumptions have been made: 1) coefficient of thermal expansion $L_{n_{CTE}}$ does not change with temperature; 2) aviation oxygen flow indicator connecting structure is the ideal rigid thin-walled cylindrical shell with frictionless surface; 3) mutual extrusion effect of structure caused by deformation process under the change of temperature can be neglected.

The limit-state function for reliability assessment is:

$$g\left(L_1, L_2, L_{1_{CTE}}, L_{2_{CTE}}, t_0, t_1\right) = L_1 \cdot \left[1 - L_{1_{CTE}} \cdot (t_0 - t_1)\right] - L_2 \cdot \left[1 - L_{2_{CTE}} \cdot (t_0 - t_1)\right] \qquad (3)$$

Where $g\left(L_1, L_2, L_{1_{CTE}}, L_{2_{CTE}}, t_0, t_1\right)$ denotes the limit-state function, L_n is the parameter of dimension, $L_{n_{CTE}}$ represents the coefficient of thermal expansion respectively. t_0 is the initial temperature, while t_1 is the specified one.

At the time $g\left(L_1, L_2, L_{1_{CTE}}, L_{2_{CTE}}, t_0, t_1\right) \leq 0$, the aviation oxygen flow indicator encounters the connection release, which is predicated as structure failure.

For the sake of without loss of generation and convenience of application for aviation equipment researchers in aeronautical model research project as soon as possible, straightforward temperatures $t_0 = 20°C$, $t_1 = -10°C$ were chosen here, other temperatures can be set according to one's actual need.

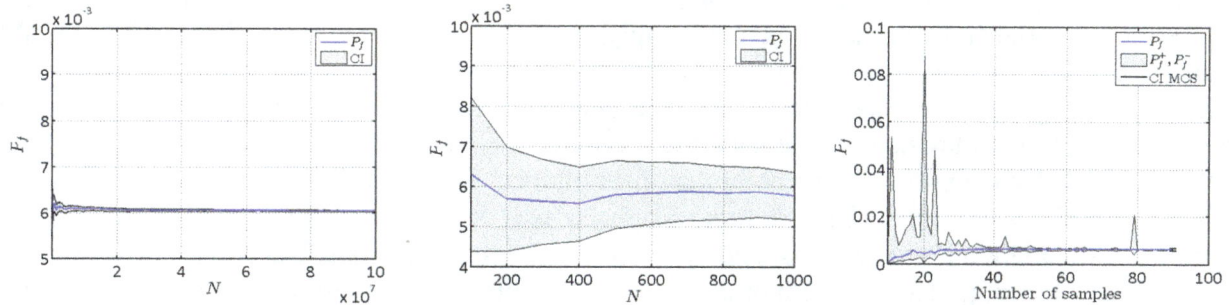

Fig. 5 Convergence trend of P_f by MCS, IS, SSo-AK-MCS

Fig. 5 shows the calculation precision and convergence trend of MCS, IS, SSo-AK-MCS and P_f.

Table 3 shows MCS, SS, IS, SSo-AK-MCS w.r.t. failure probability P_f, reliability measurement of index β convergence, namely, coefficient of variation CoV, confidence interval of failure probability $P_f CI$ with $\alpha = 0.05$ confidence interval of reliability index βCI with $\alpha = 0.05$ deviation ϵ_β compared with MCS, number of calls N_{call} to the limit-state function.

Table 3 Performance of MCS, SS, IS, SSo-AK-MCS

Method	$P_f[10^{-3}]$	β	CoV[10^{-3}]	$P_f CI[10^{-3}]$	βCI	ϵ_β	N_{call}
MCS	6.0368	2.5100	1.30	[6.02, 6.05]	[2.51, 2.51]	----------	1e8
SS	3.7100	2.6774	1.90	[2.33, 5.09]	[2.57, 2.83]	38.5435%	2783
IS	5.7690	2.5260	52.7	[5.17, 6.36]	[2.49, 2.56]	4.4361%	1073
SSo-AK-MCS	6.0600	2.5086	40.5	[5.58, 6.54]	[2.48, 2.54]	0.3843%	121

It can be perceived that the number of calls to the limit-state function of SSo-AK-MCS is 0.000121% of MCS, 4.3478% of SS, 11.2768% of IS. In the meantime, calculation precision of SSo-AK-MCS is 100.2930 times of SS, 11.5431 times of IS. The trustworthy structural reliability index of aviation oxygen flow indicator has been shown with mild number of calls to the limit-state function, benefits shorter lead-time and lower computational expense under the adaptive active learning strategy of SSo-AK-MCS.

5. Conclusion

The research results show that with the introduction of adaptive learning intelligent function to the proposed strategy SSo-AK-MCS, improvement of calculation precision by Kriging meta-model and rapid convergence has been accomplished synchronously.

1. The number of calls to the limit-state function of SSo-AK-MCS is 0.000121% of MCS, 4.3478% of SS, 11.2768% of IS.
2. The calculation precision of SSo-AK-MCS is 100.2930 times of SS, 11.5431 times of IS.

The cost of achieving that outstanding performance is the uplift of the degree of intelligence w.r.t. reliability assessment strategy. At this moment, the reliability assessment strategy is no longer static, but dynamic, which has been equipped with the adaptive active learning strategy. In the context uncertainty must be taken into consideration, we obtain the structural reliability assessment of aviation oxygen flow indicator based on proposed strategy SSo-AK-MCS, which reduces the potential risk of cockpit oxygen supply effectively, guarantees the flight safety of high performance fighter pilots, improves the combat effectiveness of pilots.

Under the circumstances that uncertainty must be taken into account, the structural reliability assessment w.r.t. complex heterogeneous structure which is extremely sensitive to reliability, such as the duct of two augmented turbofan engines named RD-93 of fighter J-31. Foreseeable application outlook is broad. The strategy proposed will be a strong impetus to the structural reliability assessment and the level of quality control of critical areas, such as manned space flight, deep space exploration, deep sea exploration, nuclear industry, automobile industry, shipbuilding industry, etc.

References

1. Bao, Jie, Zhijie Xu, and Yilin Fang. Uncertainty quantification for the reliability of the analytical analysis for the simplified model of CO2 geological sequestration. Greenhouse Gases: Science and Technology 5.2 (2015): 141-151.

2. Xuejun Zhang, Yanxin Lin, Sa Wu et al. Review on Probabilistic Design Method for Structural Reliability, Equipment Environmental Engineering, 2016, 03: 161-168. (in Chinese)

3. Low, B. K. FORM, SORM, and spatial modeling in geotechnical engineering. Structural Safety 49 (2014): 56-64.

4. Tong, Cao, et al. A hybrid algorithm for reliability analysis combining Kriging and subset simulation importance sampling. Journal of Mechanical Science and Technology 29.8 (2015): 3183-3193.

5. Schöbi, Roland, and Bruno Sudret. Application of Conditional Random Fields and Sparse Polynomial Chaos Expansions to Geotechnical Problems. Geotechnical Safety and Risk V (2015): 445.

6. Young, Michael T., et al. Satellite-Based NO2 and Model Validation in a National Prediction Model Based on Universal Kriging and Land-Use Regression. Environmental science & technology 50.7 (2016): 3686-3694.

7. S. Marelli, R. Schobi and B. Sudret Uqlab user manual–structural reliability. Chair of Risk, Safety & Uncertainty Quantification, ETH Zürich, 0.9-104 edition (2015).

8. Marelli, Stefano, and Bruno Sudret. UQLab: a framework for uncertainty quantification in MATLAB. ETH-Zürich, 2014.

9. Schöbi, R., B. Sudret, and S. Marelli. Rare Event Estimation Using Polynomial-Chaos Kriging. ASCE-ASME Journal of Risk and Uncertainty in Engineering Systems, Part A: Civil Engineering (2016): D4016002.

10. Yingjun Shen, Sa Wu. New Scheduling Heuristic for the Permutation Flowshop Problem. Journal of Beijing University of Aeronautics and Astronautics, 1998, 24(1): 83-87. (in Chinese)

11. Heidary-Torkamani, Hamid, et al. Fragility estimation and sensitivity analysis of an idealized pile-supported wharf with batter piles. Soil Dynamics and Earthquake Engineering 61 (2014): 92-106.

Chapter 12
Resource and Environment

Sustainable Energy Consumption in Beijing Rural Households

Lu Wang[a], Mian Liu[b], Shu-Yan Cao[c],*

Beijing Institute of Petro-chemical Technology, Beijing China, 102617

[a]wanglulu@bipt.edu.cn, [b]liumian@bipt.edu.cn, [c]caoshuyan@bipt.edu.cn

The energy consumption at rural household in china has accounted for significant portion of our power usage. In Beijing alone, the capital of China, although the rural household accounts for only about 14% of the inhabitants, they responsible for nearly 30% of the energy consumption, so much as the issue of sustainable energy has taken on extra urgency when Beijing were covered in smog. The article studies the energy usage pattern in rural Beijing households, and base on the finding to put forward plan for sustainable energy policy to manage the countryside to develop a sustainable energy model.

Keywords: Rural Beijing; residential energy use; energy policy.

1. Introduction

As the capital of the People's Republic of China, Beijing population grew by 70% in the last 15 years, and energy consumption rose by 77%, according to the report of A Review of Air Pollution Control in Beijing: 1998-2013 published by the UN Environment Program (UNEP) [1].

Beijing has a total population of above 20 million permanent residents as of 2012 with more than 86% of them living in urban areas. The residential energy consumption accounted for about 20% of total final energy consumption in Beijing, more than 1.5 times of its 1985 level and about double the national average level. On the other hand, however, Beijing countryside with only 14% of the inhabitants, responsible for more than 20% of the total energy consumption. In other words, carbon emission from rural households, which include burning of crop residues and firewood, obviously contribute to green house gases in Beijing.

2. Energy Consumption

The total amount of energy consumption in rural Beijing household (1990-2012) in last 20 year, is shown in Figure 1, which include solid fossil energy, coal, has dominated the energy consumption but showing sign of coming down, although it still account for about two-third of the total in 2012 energy consumption.

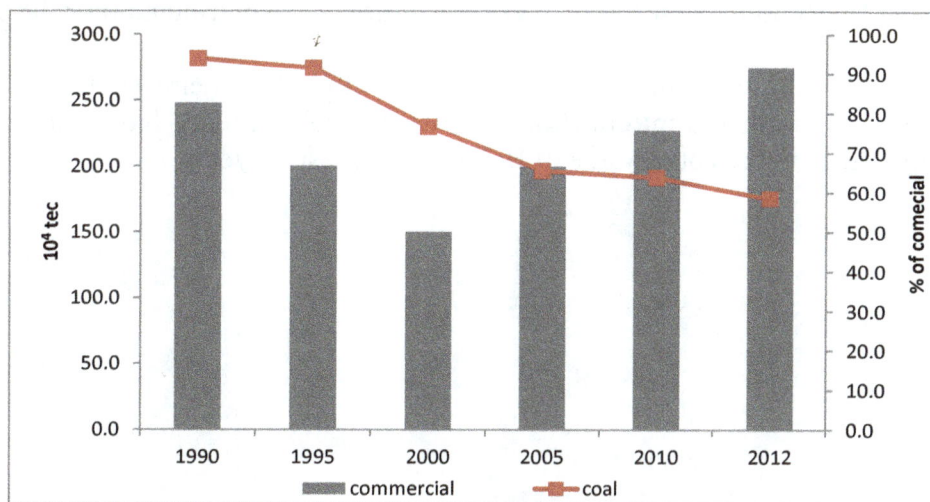

Fig. 1 Commercial energy consumption by household in rural Beijing

*Corresponding author

On the hands, non-grid based energy consumption, which include biomass energy (crops residue and forest residues), solar energy and biogas, has shown sign of moving from burning of biomass to solar energy, as indicated in Figure 2.

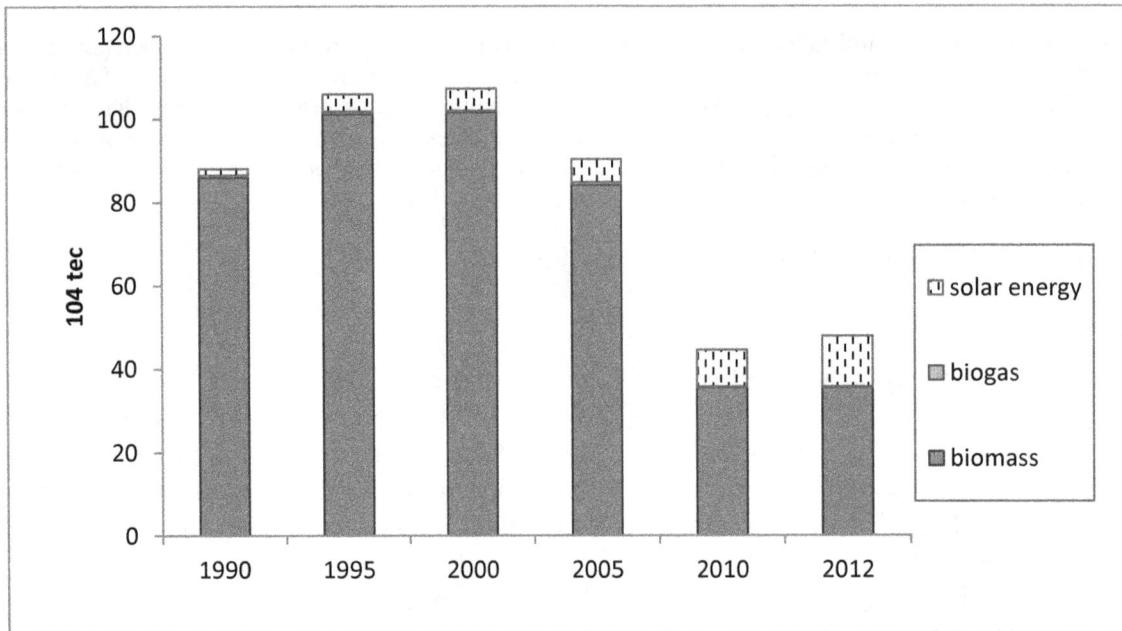

Fig. 2 Non-grid based energy consumption by household in rural Beijing

As a whole, the rural household energy usage has shown sign of moving away from high carbon emission to clean energy sources as shown in Figure 3, and the first U curve occurred during the period of 1990-2005, and the second during 2005-2012. The transformation of household energy usage in rural Beijing has been remarkable, moving away from burning of straw and firewood as the key sources of energy, for the use of clearer energy, covering electricity, gases and solar energy, has risen to 67% of total consumption, 21 and 36 percent point higher than that of 1990 and 2000 respectively. While the dirty energy, covering solid form of energy such as coals, crops residues and firewood in perspective of consumer use, still contributed near 30% of in house energy use.

Although there are more we could do to further reduce the carbon emission from rural area of Beijing, transformation with the introduction of low cost solar energy has great improved and altered the energy usage in the countryside of Beijing in the last 20 years.

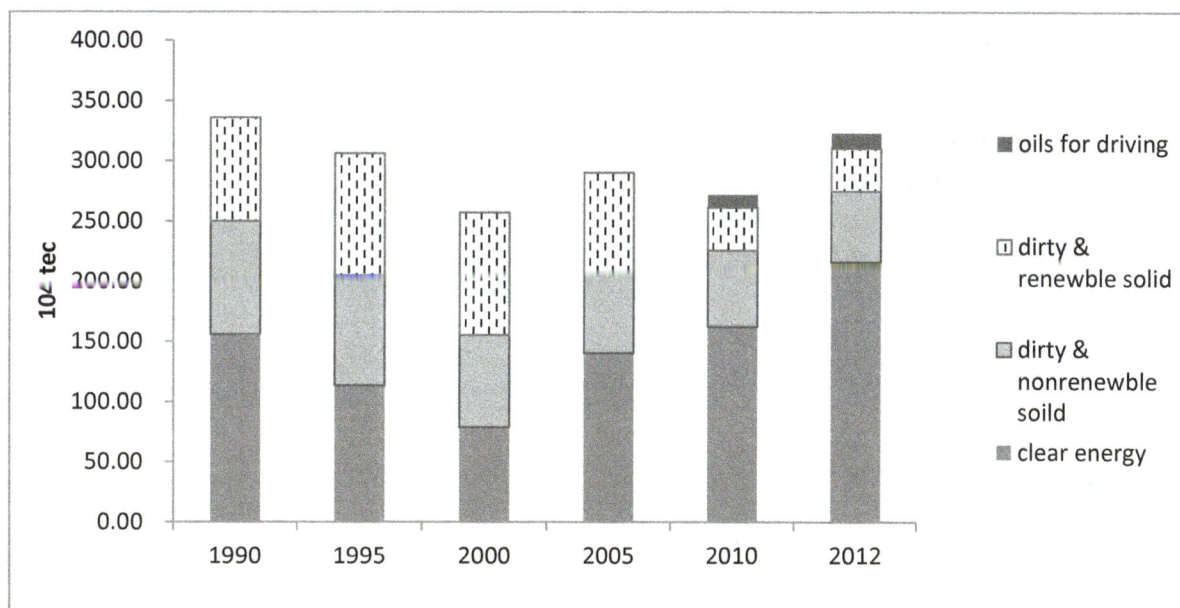

Fig. 3 Total energy consumption by household in rural Beijing

3. Challenges to Sustainable Energy Consumption

From **statistic** obtained, there are still 8000 rural household has biogas pool in 2012, reduced from more than 50,000 pools recorded in 1990. However, effort can not be spared to its future deduction. The records also shown that the heat energy utilization efficiency of burning solid biomass energy is very low, only 10% or so for traditional stoves and 20%-25% for improved stoves, with nutrition element and organic matter saying goodbye to nature and farming.

The heat energy utilization efficiency by directly burning coals is also very low. Rural area residents' awareness in the use of energy is still to improve; there are still 4.2% of rural household using crop/firewood residues and coals as the main energy for cooking. Greater effort would be needed to educate them to move away from these usages, price incentive and technology should be deployed to improve the situation.

During our investigation, a new energy problem occurs in rural Beijing. Only those middle-aged and elderly residents usually living at home, with younger members of the family only home once a while for a short stay at home, hence, energy usage per capital house hold is relatively large, the 2012 level for average rural resident is about 50 square meters, while 70 square meters for those 20% family with highest income. For those families with children working away from home, the heating or cooling house become uneconomical, as a result, the per capital coal consumption in the rural is three times that of the city in Beijing, the pattern of energy consumption in rural household should be transformed as a matter of urgency.

4. Efforts to Improve Rural Energy Consumption Pattern

In promoting sustainable energy, the Beijing municipal government has taken a series of effort to promote renewable energy sources, clear energy carriers and energy use efficiency in order for energy conservation and carbon emission reduction.

Transformation project of heating by coal to electricity, launched by Beijing governments originally for historic reserve areas in 2000, has extended to rural areas since 2013, as Beijing 2013-2017 Clean Air Action Plan Focused on the Task Decomposition in 2015, Beijing has declared that it will wipe out coal usage in its most rural areas by 2020. About 74500 rural households has replaced coal-fired heating stoves with those powered by electricity until to 2015 under the support of local governments. Local government has already cut electricity rates for heating by half during winter to encourage rural residents to discard their coal-heated stoves for electric heaters. Heating by electricity is a convenient and clean and improves indoor air quality and interior heat comfort and welcomed by rural population. Some rural family spontaneously use electric heater in the absence of subsidies when built new house. Coal-fired heating stoves would phase out of the rural household, according to local governmental plan.

The use of solar energy has been improved. Beijing locates in the second richest zone of China's solar energy resources, and has more mature solar technology [2]. Solar heaters and solar house have been valued in rural residents, as indicated in Fig.2. Local government also have incentives of supporting village using solar energy for heating rural schools and collective bathing houses, and lighting rural at night. A few of family have equipped solar power system to power their houses.

Switching from cooking using coal to gas has been facilitated. Local government has extended natural gas network into some rural areas.

Policies and efforts have facilitated more rural resident using gases. For those villages near the natural gas pipelines, coal to gas project has been implemented and local government subsided 30 percent of pipeline cost to plants. For those rural areas of rich subterranean heat, coal to heat project has been carried out by subsides. Coal to CNG (compressed natural gas) or LNG (liquefied natural gas) have widely adopted by subsides [2]. But gases have yet be limit in covered area and used quantities. And further support should be provided to accelerate the development base station, sub- station, network and user terminal building.

5. Conclusion

From our study, it is shown that the rural household energy usage has shown sign of moving away from high carbon emission to clean energy sources as shown in Figure 3. However, in order to further accelerate the process, nationwide policy is urgently needed, achievement in Beijing could provide evidence as how comprehensive policy could go a long way to address the issue of high carbon emission [3]

- **Education:** The issue of air pollution caused by coal-fired heating stoves, and burning biomass, and the health hazards it brings, have to be explained to the inhabitants living in the countryside.
- **The feasibility** of switching to cleaner energy is an infrastructure effect that have to be address at the government level to provide funding for bringing gas pipeline, and electricity into the home in rural country-side, to provide them with their energy need, failure to do that will result in encouraging them to look for alternative.
- **Incentive** to switching to cleaner alternative, subsidy should be made available for installation of solar panels, gas-fire cooker, and heating system in the home. The policy for subsidy for installation of solar panels were very successful introduced in England local councils to encourage their residents to switch over to cleaner alternative.
- **Regulation** should also introduce for homes in rural area to install fire safety equipment if they insist to hold on to the traditional way of life.

This stick and carrot approach could only be possible if national or local government take an active role in promoting the use of cleaner energy. As China has signed up to the **Paris Agreement** (French: *Accord de Paris*), which is an agreement within the United Nations Framework

Convention on Climate Change (UNFCCC) dealing with greenhouse gases emissions mitigation, adaptation and finance starting in the year 2020. The problem of sustainable energy consumption in Beijing rural households, is no more a local problem, it has become a national, and international problem, our study has provided a piece of evidence to demonstrate that sustainable energy is a collective efforts; both government and her inhabitants, together with all in the world to manage the greenhouse gases in the Paris agreement [4, 5, 6] , which went into effect on 4 November 2016.

Acknowledgments

The research funded by Beijing Municipal Program of Undergraduate Research & Training (Grant No. 2015X00010) and Key Projects in Social Sciences Pillar Program of Beijing Municipal Education Commission: (Grant NO. SZ201310017009).

References

1. UNEP. A Review of Air Pollution Control in Beijing: 1998-2013. United Nations Environment Programme (UNEP), Nairobi, Kenya. 2016.
2. S. Jiang, X. Liu. Subsequent Management Problems and countermeasures of new energy construction in Beijing Rural areas. Modern Agricultural Sciences and Technology. 12 (2010) 399-340.
3. Z. Zeng, W. Zhang. A Study of the Energy Problem in Rural Areas of China, Journal of Guizhou University (Social Science). 3 (2005) 105-108.
4. "Paris climate talks: France releases 'ambitious, balanced' draft agreement at COP21". ABC Australia. 12 December 2015.
5. "'Today is an historic day,' says Ban, as 175 countries sign Paris climate accord". United Nations. 22 April 2016.
6. "Paris Agreement to enter into force as EU agrees ratification". European Commission. 4 October 2016. *Retrieved 5 October 2016.*

Occurrence State of Scandium in Rare Earth at Yingjiang, Yunnan

Wei-Hua Yong[1, 4], Wen-Qi Gong[1], Jun-Hui Xiao[2,*], Yu-Shu Zhang[3]

[1]Wuhan University of Technology, Wuhan, Hubei, 430070, China

[2]School of Environment and Resource, Southwest University of Science and Technology, Mianyang, Sichuan, China

[3]Institute of Multipurpose Utilization of Mineral Resources, Chinese Academy of Geological Sciences, Chengdu, China

[4]Ministry of Land and Resources of Mineral Resources Reserves Evaluation Center, Beijing, China

*xiaojunhui33@163.com

In this article, a study on scandium bearing minerals species, content and dissemination characteristics of the rare earth mine at Yingjiang, Yunnan undertaken using chemical analysis, X-diffraction analysis, polarized light microscopy analysis, electron probe spectroscopy, and EDS analysis are presented. The results shows that scandium mainly distributes in the alteration of clay minerals (montmorillonite, talc) by ion adsorption, could be abstracted with direct leaching method. While another scandium mainly distributes in hornblende in the form of isomorphism, which could be abstracted by hornblende enrichment

Keywords: Scandium; occurrence state; isomorphism; ion absorption.

1. Introduction

Scandium is an important strategic resource, which is widely used in many important industrial fields [1-6]. With its increase popluarity and new applications [7-10]. Research and development in Scandium is gaining momentum. The average content of scandium in the Earth's crust is 36×10^{-4}%, which is close to the content of beryllium, boron, strontium, tin, germanium and tungsten. However, the distribution of which is extremely decentralized. Scandium is a kind of sparsity and spread around element. In nature, it mostly exists in other minerals in the form of intermingle [11-14]. Due to very decentralized, complex geological environment and mineral composition, diversity of composition and occurrence state of scandium bearing ore, extraction of scandium becomes very difficult. Using what kind of scandium extraction process is determined by occurrence of scandium in minerals. Therefore, ascertaining of scandium bearing minerals species, content and dissemination characteristics is the first step to recovery scandium resource.

For effective and reasonable developing scandium resources in Yingjiang, Yunnan, chemical analysis, phase analysis, X-diffraction analysis, electron probe spectroscopy, scanning electron microscopy analysis was used to study scandium bearing minerals species, content, occurrence and dissemination characteristics of scandium from the perspective of process mineralogy. The results obtained in this paper are of great significance for the development of reasonable scandium beneficiation process.

2. Experimental

The sample was taken from the scandium bearing ore of Yingjiang in Yunnan province of china. Ore sample was broken into powder, and was dry-ground in a porcelain ball mill. The products were then dry sieved to obtain various size fractions for various testing needs.

Chemical analysis was carried out using a QL-S3000C QL-S3000C joint measurement of infrared versatile multi-element analyzer. X-diffraction analysis using X'Pert PRO type X-diffractometer. The micro-area analysis of ore samples were carried out using Shimadzu EPMA-1720 electron probe. Spectrum analysis using Thermo Fisher's NORAN 7 type spectrum

---

*Corresponding author

analyzer. Polarized microscope analysis was carried out using BK-POL type polarizing microscope.

3. Results and Discussion

3.1. *Multielement and component analysis of raw ore*

3.1.1. *Multielement analysis of raw ore*

Multielement analysis results of scandium ore sample are presented in Table 1. As shown in Table 1, the grade of scandium in raw ore is 0.02436% with high value. The content of TiO2 in raw ore is 4.03%, which can be recovered comprehensively.

Table 1 Multielement analysis results of scandium ore (%)

CaO	MgO	TFe	SiO$_2$	Al$_2$O$_3$	S	P	Sc	TiO$_2$	K$_2$O	Na$_2$O
4.09	8.36	12.56	46.69	10.62	0.01	0.12	0.02436	4.03	1.23	1.14

3.1.2. *Component analysis of raw ore*

By means of microscope, X-diffractometer and chemical phase analyzer, there are 20 kinds of minerals in raw ore, which can be classified into four categories, such as sulfide, carbonate, oxide and silicate. proportion of silicate in raw ore is 89%, followed by oxide 7%, carbonate 3%, sulfide 0.5%. X- diffraction pattern of raw ore is shown in Fig. 1, and mineral species is shown in Table 2.

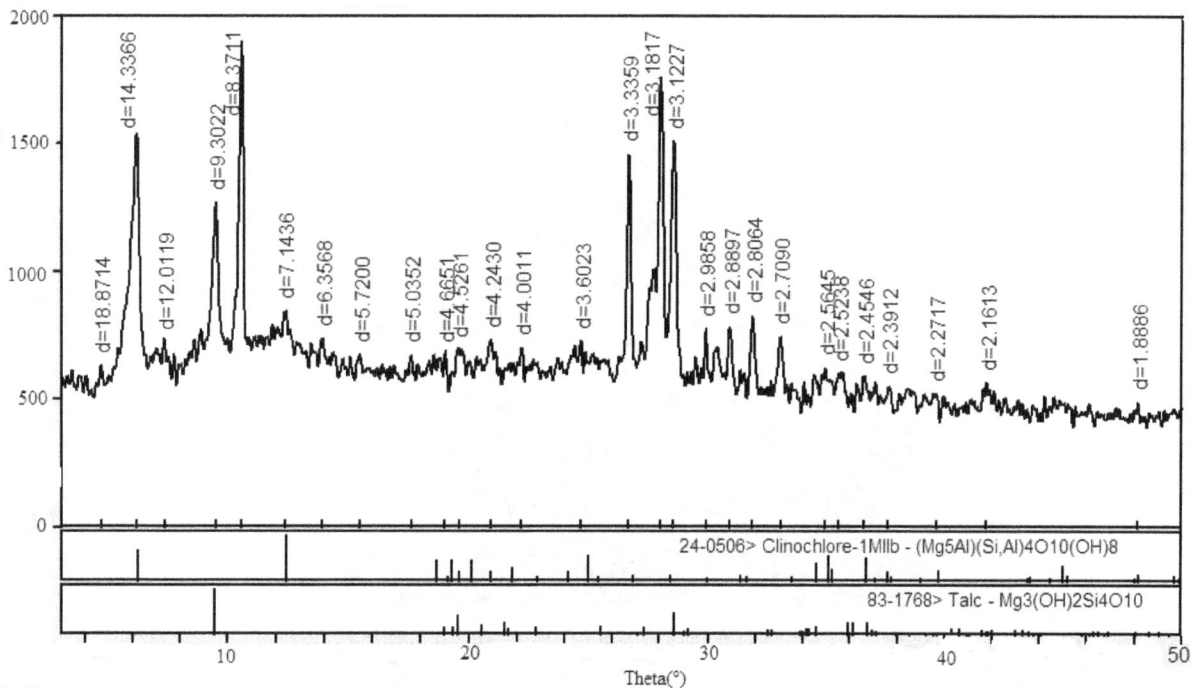

Fig. 1 X- diffraction pattern of raw ore

Table 2 Composition, dissemination size and content of raw ore (%)

pyrite	limonite	calcite	apatite	quartz	magnetite
0.2	0.6	3	0.5	4	2
hornblende	albite	orthoclase	diopside	tremolite	sericite
23	14	2	10	7	2.5
hematite	biotite	clay mineral	serpentine	fluorapatite	total
little	4	22	2	2	98.8

3.2. *Dissemination characteristics of scandium bearing minerals*

By means of polarized light microscopy, X-diffraction and electron microprobe analysis, there is no independent minerals of scandium, which exists mainly in the form of isomorphism in other minerals, such as magnetite, limonite, hornblende, plagioclase, diopside, clay minerals (montmorillonite, talc), fluorapatite. Therefore, dissemination characteristics of these scandium bearing minerals were explored in this paper.

3.2.1. *Magnetite*

The content of magnetite is about 2% in raw ore. Polarized optical images of magnetite is shown in Fig. 2. As shown in Fig. 2, most of which were xenomorphic granular textures, and a few were hypidiomorphic textures. Magnetite (black in image) coexists with hornblende and plagioclase closely. Particle size of magnetite is generally about 0.01-0.04mm. Magnetite is wrapped by albite to form inclusion texture. Part of magnetite is automorphic granular textures with large size. By means of electron microprobe analysis, scandium contained in magnetite is of only 0.00324%.

Fig. 2 Polarized optical image of magnetite(black in image)

3.2.2. *Limonite*

Limonite (Lm) with less content results from alteration of pyrite (Py). As shown in Fig. 3, Limonite is mainly composed of aphanitic or colloidal hematite, goethite, lepidocrocite and other aggregates, a few of which distributes as vein or disseminated.

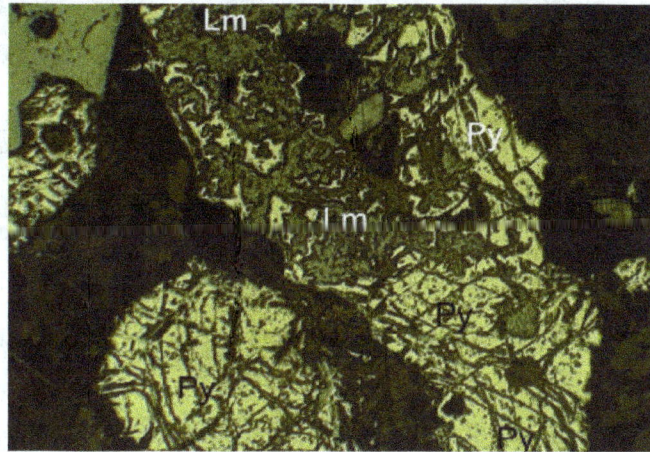

Fig. 3 Polarized optical images of limonite

Energy spectrum of limonite is shown in Fig. 4. Limonite contains small amounts of scandium locally, which is ion adsorption type.

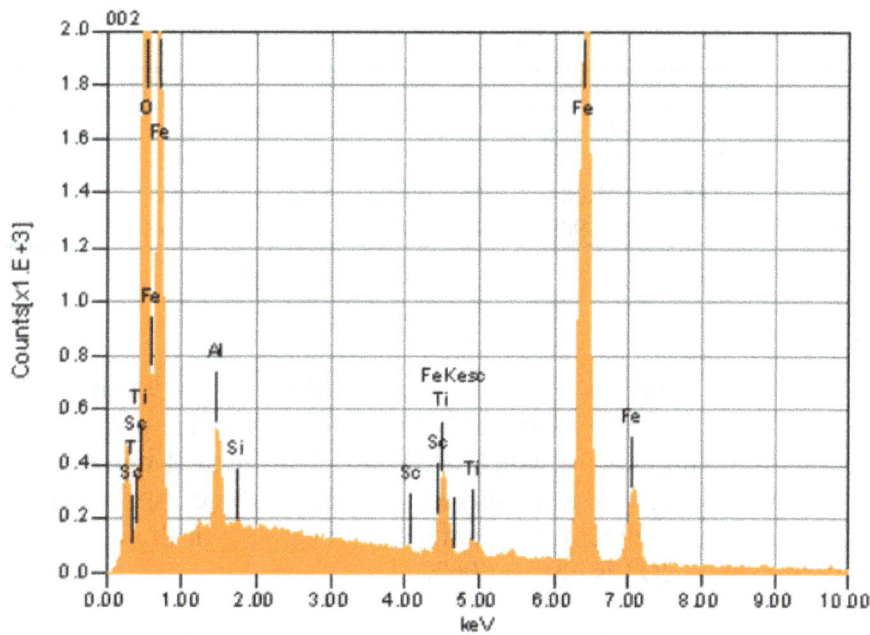

Fig. 4 Energy spectrum of limonite

3.2.3. *Hornblende*

The content of hornblende(Ho) in raw ore is of 23%, which mostly shows idiomorphic or hypidiomorphic granular structure, and often distributes between plagioclase(Pl) and chlorite(Ch) . Particle size of hornblende is 0.05-4mm, and most of which distributes between 0.1 and 1.5 as shown in Fig. 5. By means of electron microprobe analysis, the results are shown in Table 3, from which can be seen that the content of Sc2O3 reaches up to 0.025%.

Fig. 5 Polarized optical images of hornblende

Table 3 Electron microprobe analyses of hornblende /%

Position	Na$_2$O	Al$_2$O$_3$	SiO$_2$	MgO	CaO	TiO$_2$	FeO	K$_2$O
1	2.751	12.343	42.468	15.147	11.694	4.237	10.015	0.988
2	0.494	3.223	53.774	17.321	11.702	0.629	12.473	0.208
3	2.959	12.188	41.684	15.379	11.917	4.441	10.425	0.984
Average	2.068	9.251	45.909	15.949	11.711	3.102	10.971	0.727

3.2.4. *Plagioclase*

The content of plagioclase in raw ore is 14%, which mostly shows hypidiomorphic granular structure, and often closely coexists with hornblende, diopside and hematite. Particle size of plagioclase is 0.05-1mm, and most of which distributes between 0.1 and 0.5 as shown in Fig.5. Spectroscopy and electron microprobe analysis results of plagioclase are shown in Fig. 6 and table 4, respectively. From the results, plagioclase contains scandium, but the content is very low, and the average concentration of Sc2O3 is only 0.006%.

3.2.5. *Diopside*

The content of diopside in raw ore is 10%, which mostly shows hypidiomorphic granular structure, and often closely coexists with plagioclase and hornblende. Biotite (Bt) constitutes a larger size phenocrysts. Plagioclase (Pl) and diopside (Px) are fine granularity, and shows hypidiomorphic granular structure. Diopside is wapped by biotite, indicating that these two minerals formed during the same period to constitute porphyritic-like texture. Particle size of diopside is 0.05-0.8mm, and most of which distributes between 0.1 and 0.3 as shown in Fig.7. By means of spectroscopy and electron microprobe analysis, the results are shown in Fig.8 and table 5, respectively, from which can be seen that the average content of Sc2O3 is 0.021%.

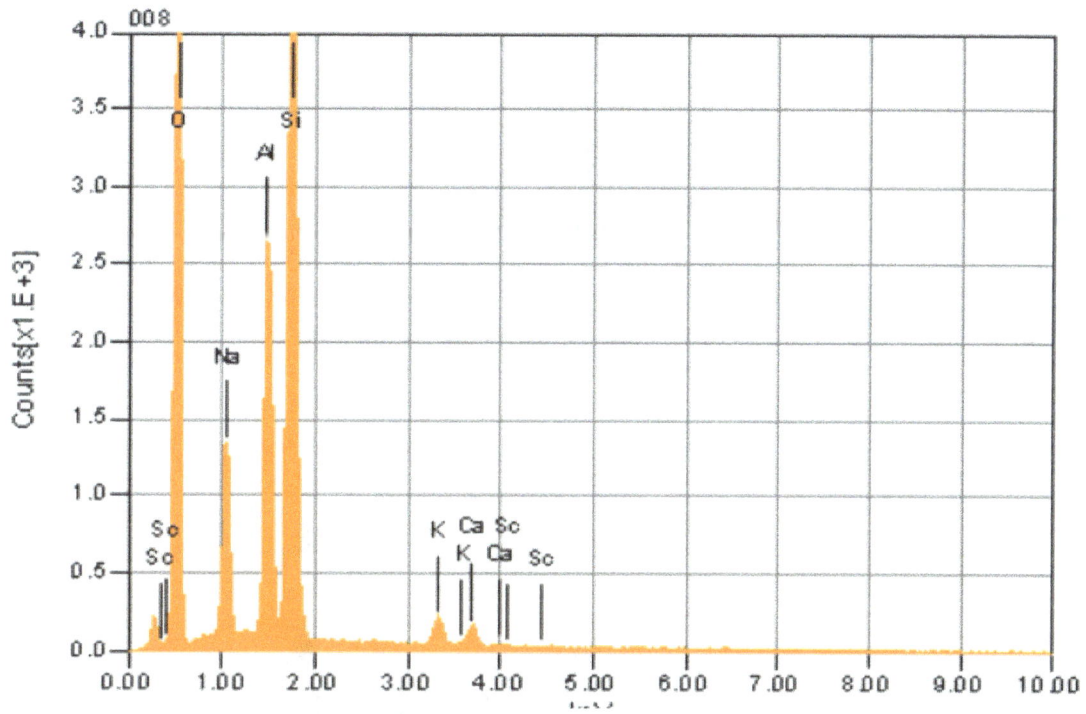

Fig. 6 Energy spectral analysis diagram of plagioclase

Table 4 Electron microprobe analyses of plagioclase (%)

Position	Na_2O	Al_2O_3	SiO_2	MgO	CaO	TiO_2	FeO	K_2O	Sc_2O_3
1	12.113	20.392	66.477	0.040	0.617	0.037	0.110	0.183	0.012
2	12.210	20.113	66.562	0.014	0.817	0.000	0.170	0.066	0.000
Average	12.162	20.253	66.520	0.027	0.717	0.019	0.140	0.412	0.006

Fig. 7 Polarized optical images of diopside

661

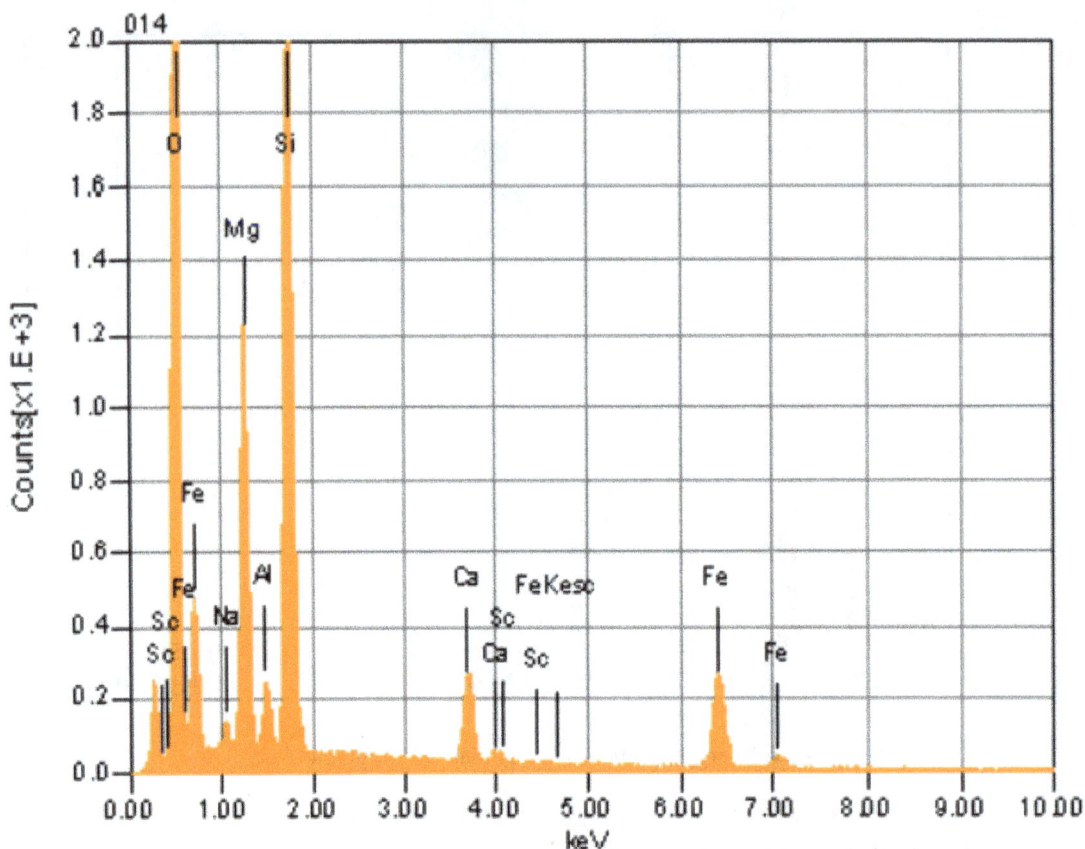

Fig. 8 Energy spectral analysis diagram of diopside

Table 5 Electron microprobe analyses of diopside(%)

Position	Na$_2$O	Al$_2$O$_3$	SiO$_2$	MgO	CaO	TiO$_2$	FeO	K$_2$O	Sc$_2$O$_3$
1	0.327	2.711	51.810	16.522	21.567	0.831	5.532	0.005	0.003
2	0.243	1.548	52.818	17.057	21.798	0.697	5.266	0.003	0.013
3	0.93	3.03	47.3	23.62	6.89	0.000	17.76	0.000	0.048
Average	0.500	2.430	50.643	19.066	16.752	0.509	9.519	0.003	0.021

3.2.6. *Clay minerals (montmorillonite, talc)*

Clay minerals is one of the main gangue minerals, and the content of which is about 22%. This kind of mineral results from alteration of other silicate minerals, such as diopside, hornblende, biotite, feldspar and olivine. Clay minerals is mainly composed of montmorillonite (Mnt) and talc (Tlc). Talc presents aggregates of microscopic scaly, and montmorillonite presents aggregates of cryptocrystalline. Particle size of clay mineral is 0.001-0.1mm, and most of which distributes between 0.001 and 0.05 as shown in Fig. 9. Chemical analysis shows that the content of Sc2O3 in clay minerals is 0.093%. Energy spectrum analysis result is shown in Fig.10. As shown in Fig.10, scandium distributes in clay minerals by ion adsorption.

662

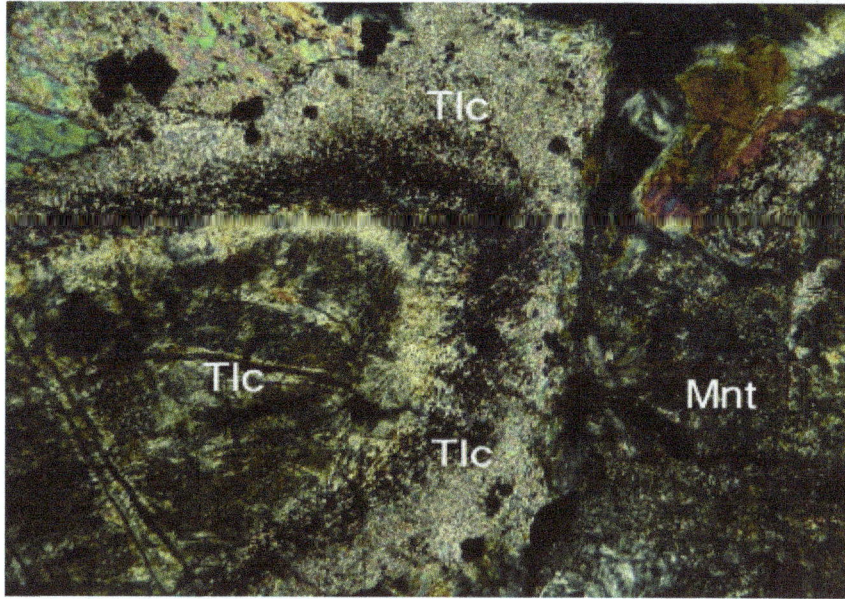

Fig. 9 Polarized optical images of talc and montmorillonite

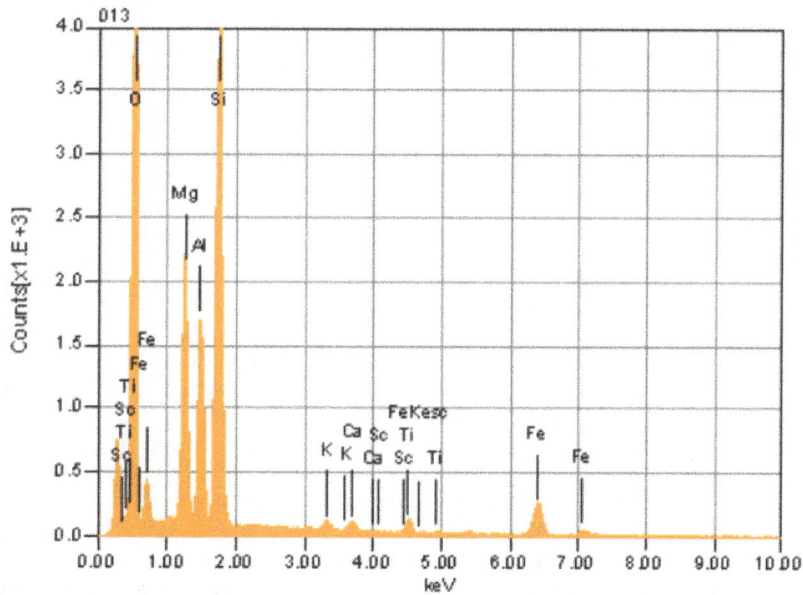

3.2.7. Fluorapatite

The content of fluorapatite in raw ore is 2%, which mostly shows idiomorphic or hypidiomorphic columnar structure, and often be wrapped by feldspar and biotite. Particle size of fluorapatite is 0.05-1.5mm, and most of which distributes between 0.1 and 0.3 as shown in Fig. 11. Electron microprobe and spectrum analysis shows that the content of Sc2O3 in fluorapatite is 0.1305% as shown in Table 6 and Fig. 12, respectively.

663

Fig. 11 Polarized optical images of fluorapatite

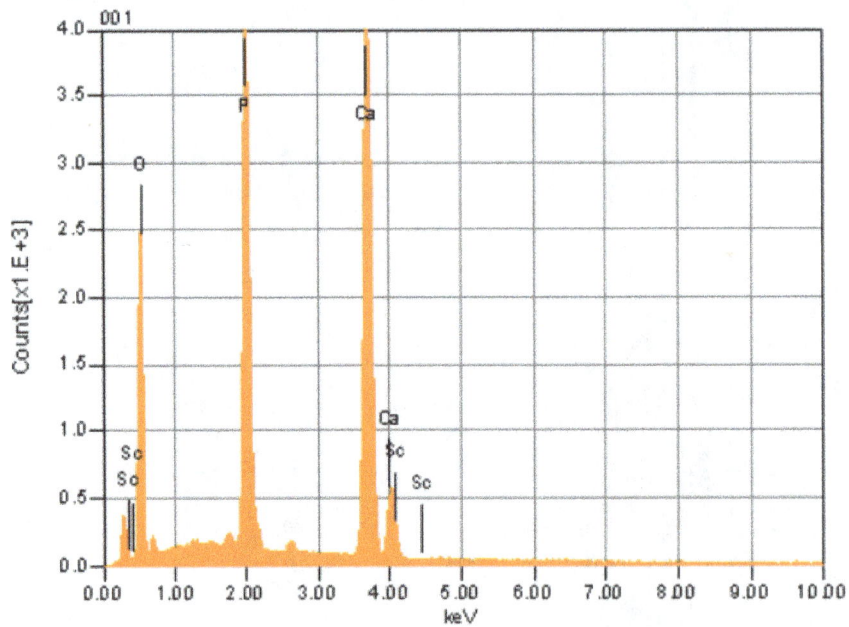

Fig. 12 Energy spectral analysis diagram of fluorapatite

Table 6 Electron microprobe analyses of fluorapatite (%)

Position	F	MgO	CaO	Cl	Ce_2O_3	P_2O_5	Sc_2O_3
1	2.451	0.015	54.862	0.577	0.401	41.569	0.126
2	3.161	0.022	55.045	0.624	0.098	40.858	0.130
3	2.785	0.039	54.507	0.596	0.313	41.605	0.114
4	0.000	0.000	57.36	0.000	0.000	39.12	0.152
Average	2.576	0.033	55.136	0.568	0.301	41.232	0.1305

3.3. *Occurrence state and beneficiability evaluation of scandium*

By means of chemical analysis, polarizing microscope, X- diffraction and electron microprobe analysis, no independent scandium minerals is found in raw ore, which exists mainly in the form of isomorphism in other minerals. Scandium distributes in each mineral, and content of which varies widely as shown in Table 7.

Table 7 Distribution rate of scandium in major minerals

Mineral	Mineral content (%)	Percentage of Sc_2O_3 in mineral (%)	Amount of Sc_2O_3 in mineral (10^{-6})	Distribution rate of scandium in minerals (%)
Diopside	10	0.021	21	6.6
Limonite	0.6	0.00258	0.15	0.05
Magnetite	2	0.00324	0.65	0.20
Hornblende	23	0.025	57.5	18.06
Plagioclase	14	0.006	8.4	2.64
Clay mineral	22	0.093	204.6	64.25
Fluorophosphate	2	0.1305	26.1	8.20
Total	73.6	/	318.4	100

64.25 percent of scandium mainly distributes in the altered clay minerals (montmorillonite, talc) by ion adsorption. This part of scandium can be leached, but the size of which is too fine to be pre enriched by physical methods.

Other part of scandium mainly distributes in hornblende in the form of isomorphism, and a small amount of scandium distributes in fluorapatite and diopside. Enriching hornblende can enrich scandium. However, it is not easy to separate hornblende from biotite and chlorite in the beneficiation process, which brings difficulty for scandium enrichment. There is a small amount of ion-adsorption type scandium distributes in the clay minerals (chlorite, sericite, etc.), which results in low scandium concentrate grade.

4. Conclusion

1) There is no independent minerals of scandium, which mainly exists in magnetite, limonite, hornblende, plagioclase, diopside, clay minerals and fluorapatite.
2) 64.25 percent of scandium distributes in clay minerals by ion adsorption, which can be recovered by direct leaching.
3) 18.06% percent of scandium mainly distributes in hornblende in the form of isomorphism, which can be enriched by hornblende enrichment.

Acknowledgment

This project was funded by Technical Standard System Construction Project of National Mineral Resources Reserves (CB2015) and China Postdoctoral Science Foundation (No2014M560734)

References

1. Gao Likun, Chen Yun. A study on the rare earth ore containing scandium by high gradient magnetic separation [J].   Journal of rare earths, 2010, 2010, (28): 4,622-626

2. Wu Horng yu, Gao Zhen wei, Lin Jia yu, etal. Effects of minor scandium addition on the properties of Mg-Li-Al-Zn alloy [J]. Journal of Alloys and Compounds.2009, 474(1-2):158–163

3. Reno R C, Rasera R L, Schmidt G. Temperature dependence of the electric quadrupole interaction at scandium impurities in titanium [J].Physics Letters A.1974, 50(4):243-244

4. Racka, K Avdonin,A Sochacki M, et al. Magnetic, optical and electrical characterization of SiC doped with scandium during the PVT growth[J].Journal of Crystal Growth.2015, 413:86-93

5. Kalashnikova A O, Yakovenchuka V N, Pakhomovskya Y A, etal. Scandium of the Kovdor baddeleyite–apatite–magnetite deposit (Murmansk Region, Russia): Mineralogy, spatial distribution, and potential resource [J].Ore Geology Reviews.2016, 72(1):532–537

6. Horovitz C T. Scandium, Its Occurrence, Chemistry, Physics, Metallurgy, Biology and Technology [M].London, Academic Press,1975, 70

7. Raade Gunnar. Scandium [J].Chemical and Engineering News, 2003, 81(36):68.

8. Zhang Zonghua, Zhang Guifang, Gao Likun, et al. Study on scandium separation from rare earth ore in Yunnan Province [J]. Journal of Rare Earths, 2005, 23(3):531-535.

9. Feijoo P C, Del Prado A, Toledano-Luque M, et al. Scandium oxide deposited by high-pressure sputtering for memory devices: Physical and interfacial properties[J]. Journal of Applied Physics, 2010, 107(8)

10. Shcheglov A D, Moskaleva V N, Markovskiy B A.Scandium hydrogeochemistry in waters of technically transformed media [J]. Earth science sections, 1994, (9):473-177.

11. Arbuzov S I, Volostnov A V, Mezhibor A M, et al. Scandium (Sc) geochemistry in coals (Siberia, Russian Far East, Mongolia, Kazakhstan, and Iran) [J].International Journal of Coal Geology, 2014, 125(22-35).

12. Arbuzov S I, Maslov S G, Volostnov A V, et al. Modes of occurrence of uranium and thorium in coals and peats of Northern Asia [J]. Solid Fuel Chemistry, 2012, 46(1):52-56.

13. Arbuzov S I, Rikhvanov L P, Volostnov A V, et al. Radioactive elements in Paleozoic coals of Siberia [J]. Geochemistry International, 2005, 43(5):478-492.

14. Arbuzov S I, Rikhvanov L P, Volostnov A V, et al. Radioactive elements in paleozoic coals of Siberia [J]. Geokhimiya, 2005, 43(5):527-541.

Effect of NO_3^-/NH_3 Ratio on Anaerobic Nitrogen Removal of Coal Gasification Wastewater

Shi-Dong Yang[1,*], Li-Qiang Yao[1], Lu-Hua Liao[2]

[1]*School of Civil Engineering and Architecture, Northeast Dianli University, Jilin132012, China*
[2]*Guangxi Polytechnic of Construction, Nanning530007, China*
Email: 15981105115@163.com

The effect of anammox coupling with heterotrophic denitrification process was investigated in an anaerobic reactor with an influent at the temperature of 35 °C, pH of 7.0, hydraulic detention time of 30h and the influent dilution ratio at 75%, the effects of different concentrations of nitrite on total nitrogen and organic matter removal were studied by changing NO_3^--N/NH_4^+-N ratio in the anaerobic reactor with the ammonia concentration of 140±5mg/L and COD of 900±5mg/L. The removal rate of ammonia nitrogen, nitrite, TN, COD were 54.71%, 81.49%, 73.58, 81.61%, respectively, with a NO_2^--N/NH_4^+-N ratio of 1.6. The optimum stoichiometric ratio can enhance anaerobic ammonium oxidation and denitrification, improving the removal efficiency of nitrogen carbon simultaneously.

Keywords: Coal gasification wastewater; anaerobic ammonium oxidation; denitrification; stoichiometric ratio; removal of carbon and nitrogen.

1. Introduction

Coal gasification wastewater is hard to treat with high concentration of ammonia and phenol [1-3]. Anaerobic technology has obvious advantages in the treatment of refractory wastewater due to its high BOD load, adaptability to different wastewater, less residual sludge, energy conservation, low operating cost and good treatment effect, etc [4-6]. The traditional anaerobic technology also has disadvantages in which there is high concentration of ammonia nitrogen in effluent. The discovery of a new biological nitrogen removal technology—anaerobic ammonia oxidation (ANAMMOX) is expected to solve this problem. Under anaerobic or anoxic conditions, autotrophic bacterium use nitrite as electron acceptor, directly convert ammonia nitrogen into nitrogen and produce a small amount of nitrate [7-9]. Under the condition of high concentration of organic carbon source, the activity of anammox bacteria will be inhibited due to the autotrophic bacteria competition, which has a serious impact on the removal effect of nitrogen[10]. So the coupling of denitrification and anammox can be taken into account in the process of anaerobic carbon removal.

A number of studies showed that the coupling of denitrification and anammox could be achieved [11-14]. In the coupling reactors, the removal of total nitrogen is the result of the interaction between anammox and heterotrophic denitrification, both of which have a synergistic effect. There is also a competition about the matrix—nitrite nitrogen between them. The effect of organic concentration on nitrogen removal performance by anammox has been investigated. Lv [15] founded that an anammox reactor could be operated steadily while the organic carbon source is low. The activity of anammox were inhibited while adding a large amount of organic matter while the denitrification activity was obvious. Increased concentration of nitrite can improve the removal performance of total nitrogen and COD. When the NO_2^--N was sufficient, the denitrification process didn't have a significant impact on the activity of anammox. Fu Jixiang,et al. [16] also confirmed that in the high concentration of COD, NO_2^--N can be added to mitigate the inhibitory effect on anammox with a ratio of m(NH_4^+-N):m(NO_2^--N)=1.92 in the influent. The usual used NH_4^+-N/NO_2^--N ratio is 1:(1~1.8), in which the researchers' interest is focused on 1:(1.4~1.5). Actual consumption of NO_2^--N in the former researches is higher than theoretical consumption

---

*Corresponding author

(1:1.32), some of which own a lower ratio of m(NH$_4^+$-N):m(NO$_2^-$-N)<1.32. No exac agreement has been achieved so far.

In this paper, with a optimum influent ration at 75% [17], the effects of different concentrations of nitrite on total nitrogen and organic matter removal were studied by changing the concentration of nitrite in anaerobic reactor through changing NO$_2^-$-N/NH$_4^+$-N ratio. The contribution of nitrogen removal by anammox and heterotrophic denitrification, the effect of different stoichiometric ratio on the organic compounds removal were studied at the same time. The optimum influent NO$_2^-$-N /NH$_4^+$-N ratio in an anaerobic ractor was finally got under high concentration of COD.

2. Materials and Methods

2.1. Experimental installation

The effective volume of anaerobic reactor was 2.5L with a height of 40cm, diameter of 10cm. The flow rate was controlled by the creeping pump. The reactor was stored in dark place and constant temperature water bath box was controlled at 35°C. The HRT of reactor was 30h. The influent pH was justed around 6.5-7.5 by NaOH and HCl.

2.2. Simulated wastewater composition

Simulated waste water was used in the experiment with an composition of : glucose 230-250mg/L, volatile phenol 150-250mg/L, ammonia 140-145mg/L, sulforho-anide 20-50mg/L, sulphide 20-50 mg/L, pyridine compounds 20-40mg/L, furans compounds 20-40mg/L, indoles compounds 20-30 mg/L, benzene compounds 100-150mg/L; KH$_2$PO$_4$ 27mg/L, CaCl$_2$·2H$_2$O 180mg/L, MgSO$_4$·7H$_2$O 300mg/L, NaHCO$_3$ 0.5g/L, trace element concentration I and II were 1mg/L, respectively.

2.3. Analysis items and methods

Water quality parameters were monitored accroding to the Standard Methods for Water and Waterwater Quality Monitoring. Sealed catalytic fast digestion method, Nessler's reagent spectrophotography, N-(1-naphthyl) ethylenediamine spectrophotometry, ultraviolet spectrophotometry, 4-aminoantipyrine direct spectrophotometric method, ultraviolet spectrophotometry were used to determine concentration of COD, NH$_4^+$-N, NO$_2^-$-N, NO$_3^-$-N, TN, volatile phenol and total phenol, respectively. All charts and data statistics analysis were processed by Origin 8.5.

2.4. Experimental methods

The anaerobic reactor has been operated for up to six months. The optimum operating conditions were determined through the orthogonal experiments: the HRT of 30h, the temperature of 35°C and the pH value of 6.5-7.5. The reactor was operated for 40d under different influent ratio R (0~100%). The removal rates of total nitrogen and COD were 64.56%, 80.1% respectively at the influent dilution ratio R of 75%. Under the above conditions, the content of nitrite in anaerobic influent was changed by changing the stoichiometric ratio (NO$_2^-$-N/NH$_4^+$-N ratio). The ammonia concentrations of simulated wastewater were kept as constants, and the contents of nitrite/nitrate were changed according to the stoichiometric ratio. Five different nitrite/ammonia ratios of 0.5, 0.8, 1.0, 1.32, 1.6, and 2.0 were investigated. The anaerobic reactor was operated for 10d under different stoichiometric ratio. The ammonia nitrogen, phenol, COD, nitrite nitrogen, nitrate nitrogen and total nitrogen content in influent and effluent were measured every day. For different stoichiometric ratio, the average value of the measured data in the corresponding running time were used to esitimate the content of pollutants in influent and effluent water.

The influent dilution ratio of R was the influent flow ratio of nitrite, nitrate nitrogen solution added to the influent in the anaerobic stage. Nitrite and nitrate in the simulatedl water were provided by sodium nitrite and sodium nitrate, the concentration of which was pumped into the bottom of he anaerobic reactor from the bottom through a peristalsis pump.

3. Results and Discussion

3.1. *Effect of stoichiometric ratio on nitrogen removal*

In Fig. 1, the stoichiometric ratio of 53~59d, 61~66d, 43~51d, 23~29d, 67~72d, 74~79d are 0.5, 0.8, 1.0, 1.32, 1.6, 2.0 respectively. The concentration of NH_4^+-N, NO_2^--N, NO_3^--N, TN in influent and effluent water were shown in Fig. 1.

From Fig. 1 we can see that, the effluent ammonia nitrogen was relatively stable and remained at about 32mg/L at the stoichiometric ratio of 0.8~1.6, while at both ends it was higher at about 40 mg/L. The trend of nitrite nitrogen removal was similar to that of total nitrogen when the influent nitrogen load was small. When the stoichiometric ratio was 0.5 and 0.8, the effluent nitrite nitrogen content could be maintained at below 10mg/L, the total nitrogen concentration was 48±5.0mg/L. The effluent concentration of nitrite and total nitrogen increased with the increase of influent nitrogen load. Removal rate of nitrite nitrogen and total nitrogen reached the maximum value when the stoichiometric ratio was 1.6. No obvious trend of nitrate concentration were shown in effluent.

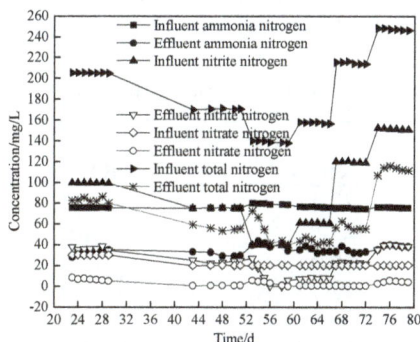

Fig. 1 NH_4^+-N, NO_2^--N, NO_3^--N, TN concentrations in influent and effluent

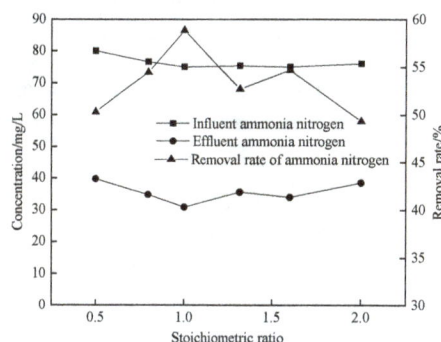

Fig. 2 Stoichiometric ratio on NH_4^+-N removal

In order to analyze the effect of nitrogen removal in anaerobic stage, the removal rate of NH_4^+-N, NO_2^--N, NO_3^--N, TN was investigated at the different influent ratio. As it can be seen from Fig. 2, when the influent ammonia was maintained at about 75mg/L, the effluent ammonia was about 35mg/L, showing a stable removal rate of about 55%. The ammonia nitrogen removal rate increased rapidly from 50.3% to 58.83% while the stoichiometric ratio increased from 0.5 to 1 below the theoratical ratio of 1.32 for NO_2-N/NH_3. The ammonia removal increased slightly while the stoichiometric ratio increased to 1.6 and showed a significantly decrease to 49.33% at 2.0.

669

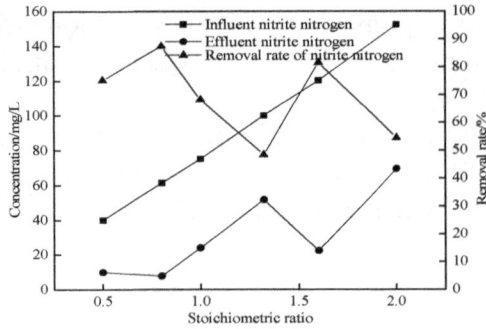

Fig. 3 Stoichiometric ratio on NO$_2^-$-N removal

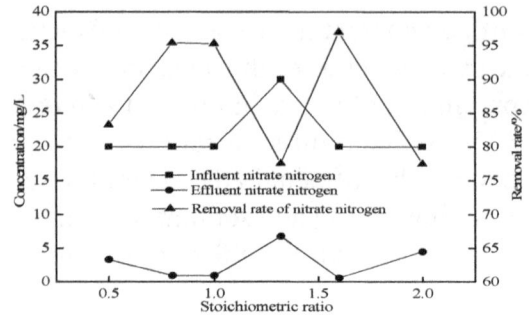

From Fig. 3 and Fig. 5, it could be seen that the effect of the stoichiometric ratio change on the removal rate of NO$_2^-$-N and TN was similar. The theoretical ratio of 1.32 was also seen as the dividing point, on both side of which the removal rate of NO$_2^-$-N, TN in both sides was symmetrical. The phenomenon showed that a optimum ration of NO$_2$-N/NH$_3$, as that found in former references. The higher ration caused inhibition of ANNOMAX, which might suggest the existance of anerobic denitrification.

As can be seen from Fig. 6, with the increased stoichiometric ratio, the total nitrogen removal increased as well. When NO$_2$-N/NH$_3$ ratio increased to 2.0, the total nitrogen removal rate decreased by 23mg/L than that at the ratio of 1.6.

The theoretical amount of nitrite consumed and nitrate produced by anaerobic ammonium oxidation could be calculated by the amount of ammonia nitrogen removed. With the NO$_2^-$-N, NO$_3^-$-N consumption listed in Fig. 6, the amount of NO$_2^-$-N consumption in anaerobic ammonium oxidation, heterotrophic denitrification and denitrification removal of NO$_x^-$-N can be caculated, contribution rate of anaerobic ammonium oxidation and heterotrophic denitrification to total nitrogen removal was also calculated with the data. The calculation result was shown in Table 1.

Table 1 Calculation results of anaerobic ammonium oxidation and denitrification on nitrogen removal (mg/L)

Stoichiometry	NH$_4^+$-N	Theoriatical NO$_2^-$-N	NO$_3^-$-N Produced	Totla NO$_3^-$-N removal	Denitrification of NO$_x^-$-N
0.5	40.24	53.12	10.46	27.11	4.12
0.8	41.71	55.06	10.84	29.92	28.46
1.0	44.12	58.24	11.47	30.52	23.38
1.32	42.02	55.47	10.93	27.78	42.29
1.6	41.10	54.25	10.69	30.09	73.49
2.0	37.56	49.58	9.77	25.12	58.39

Note: the denitrificaton NO$_x^-$-N in Table 1 indicates the part that generates the N$_2$

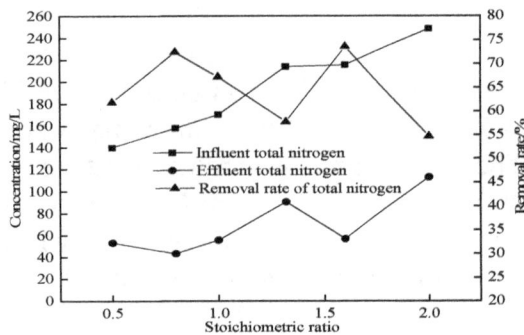

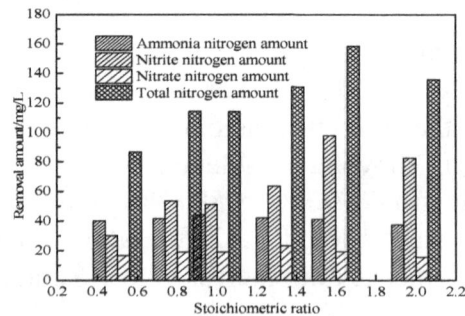

Fig. 5 Stoichiometric ratio on TN removal

Fig. 6 Stoichiometric ratio on removal amount of NH4+-N, NO2--N, NO3--N, TN

It was shown in Fig. 6 that when the stoichiometric ratio was below the critical ration of 1.32, the rem oval amount of nitrite nitrogen was lower than the elector acceptor that ammonia nitrogen required in the theoretical anaerobic ammonium oxidation. This showed that the insufficient part of nitrite might be compensated by incomplete conversion of nitrate denitrification. As can be seen in Table 1, when the stoichiometric ratio was 0.5, the removal amount of nitrite was 30.13mg/L, which was 22.99mg/L lower than the elector acceptor that ammonia nitrogen required in the theoretical anaerobic ammonium oxidation. This showed that the inadequate theoretical nitrite demand was compensated by incomplete conversion of nitrate denitrification, and the amount of nitrate nitrogen converted to N_2 was only 4.12mg/L. The contribution rate of anaerobic ammonium oxidation and heterotrophic denitrification total nitrogen removal at this ratio was 95.25%、4.75%, respectively.

By analyzing the theoretical and actual amount of nitrite removed, it could be seen that there was still a nitrate removal even the stoichiometric ratio was not at theoretical value of 1.32. When the ration was less than 1.32, the concentration of nitrite in the influent was insufficient, and the deficient part compensated through partial denitrification of nitrate. It was found that the introduction of nitrate can make up the negative effect of the influent nitrite nitrogen deficiency. The influent nitrite nitrogen can be satisfied with the demand of the substrate competition between heterotrophic denitrification and anaerobic ammonium oxidation while the stoichiometric ratio was greater than or equal to 1.32. And when the NO_2-N/NH_3 ratio ratio reached 1.6, the total nitrogen removal achieved the maximum value, when the contribution of anaerobic ammonium oxidation to total nitrogen removal was equal to denitrification.

3.2. Effect of stoichiometric ratio on organic matter removal

In Figs. 7 and 8, the stoichiometric ratio was 0.5,0.8,1.0,1.32,1.6,2.0 at 53~59d, 61~66d, 43~51d, 23~29d, 67~72d, 74~79d respectively, the influent and efflent concentrations were taken from the average values of the measured data. The volatile phenol(VP), total phenol(TP) and COD concentrations in influent and effluent were shown in Figs. 7 and 8.

From Figs. 7 and 8, it can be seen that the VP, TP COD removal were almost the same under the same stoichiometric ratio. The effluent concentrations owned a larger fluctuation range at 1.5. The effluent concentrations of VP, TP COD were minimum with value of 19.89mg/L, 36.71mg/L, 165.89 mg/L, respectively, at 1.6.

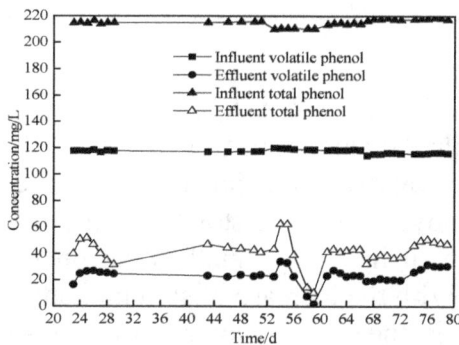

Fig. 7 Content change curve of influent and effluent phenol

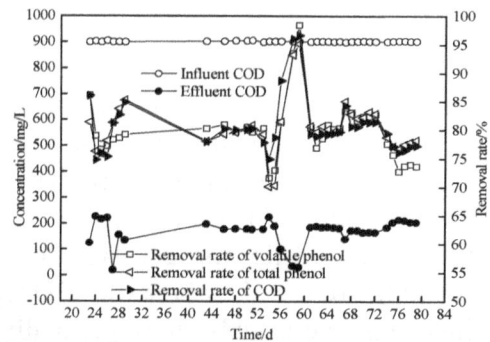

Fig. 8 Phenol & COD removal characteristics under different stoichiometric ratio

From Figs. 9, 10 and 11, the removal pattern of VP, TP and COD was alomost the same. The removal rates decreased with the stoichiometric ratio increased from 0.5 to 0.8 and from 1.6 to 2.0. The removal rate of organic compounds was stable at 0.8~1.32 and increased from 79.3% to 81.61%. At the ratio 1.6, the removal rate could achived an peak value, when the removal rate of total phenol could reach 83.4%.

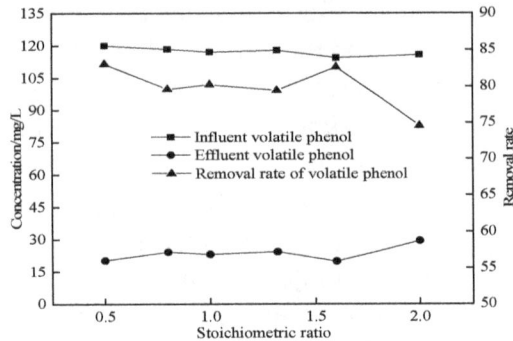

Fig. 9 Stoichiometric ratio on volatile phenol removal

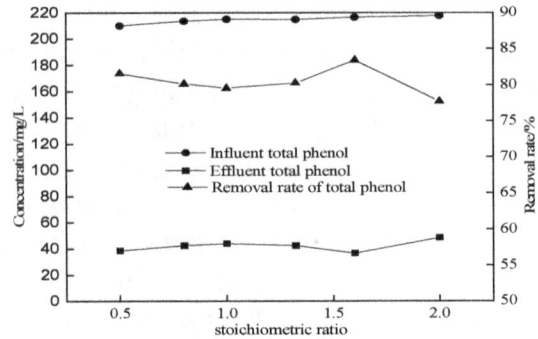

Fig. 10 Stoichiometric ratio on total phenol removal

The removal of organic matter was the synthetic result of heterotrophic anaerobic oxidation combined with heterotrophic denitrification, so the removal of COD by denitrification can be calculated through the NO_x-N that removed by heterotrophic denitrification process. From the theoretical equation of denitrification reaction, NO_2^--N 0.58mg and NO_3^--N 0.35mg are required to decompose COD 1mg COD. Through the analysis of the of nitrogen removal, the amount of COD by denitrification can be calculated, the results of which were shown in Table 2.

Table 2 COD removal (mg/L)

Stoichiometry	nitrite consumed	nitrate consumed	COD consumed by nitrite	COD consumed by nitrate	COD consumed by denitrification
0.5	0	27.11	0	36.85	36.85
0.8	0	29.92	0	82.96	82.96
1.0	0	30.52	0	74.89	74.89
1.32	8.10	34.19	13.97	97.69	111.66
1.6	43.7	30.09	75.34	85.97	161.31
2.0	33.27	25.12	57.36	71.77	129.13

As can be seen in Table 2, the removal of COD through heterotrophic denitrification increased with the increase of stoichiometric ratio from 0.8 to 1, while the amount of COD by denitrification was reduced by 7.97mg/L in contrast. The removal of COD by denitrification was maximun at the stoichiometric ratio of 1.6. From the analysis of COD removal under the stoichiometric ratio, the concentration of nitrite nitrogen was only 40mg/L at 0.5, it had very small effect on methane bacteria and the removal amount of COD was the biggest. With the increase of stoichiometric ratio, the influence on methane bacteria gradually increased, that is, the contribution rate of heterotrophic denitrification to COD removal rate increaesed. The results showed that when the stoichiometric ratio was 1.6, the contribution of heterotrophic denitrification to COD removal was the largest, but the total COD removal was only less than 35mg/L compared with the stoichiometric ratio was 0.5.

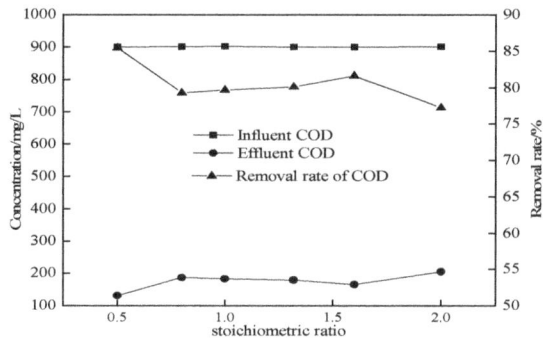

Fig. 11 Stoichiometric ratio on COD removal

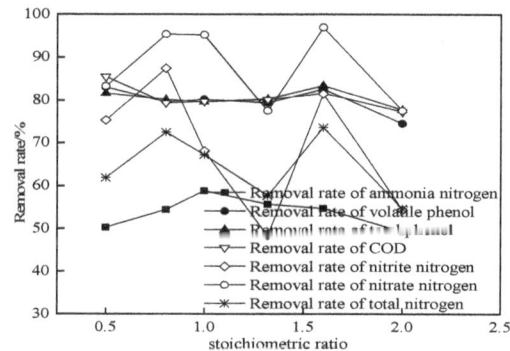

Fig. 12 Stoichiometric ratio on the removal of pollutants

4. Conclusion

The following conclusions can be drawn from the study:

1. While the stoichiometric ratio less than 1.32, the nitrite nitrogen in the influent was insufficient, and the deficient part compensated was supplied by partial denitrification of nitrate. the results showed that the introduction of nitrate nitrogen could strengthen the effect of anaerobic ammonium oxidation, which improve the ideal coupling anaerobic ammonium oxidation and aerobic denitrification. When the stoichiometric ratio was 1.32 or greater, the influent nitrite nitrogen can satisfy with the demand of the substrate competition between heterotrophic denitrification and anaerobic ammonium oxidation.

2. With the increase of stoichiometric ratio, the influence of heterotrophic anaerobic bacteria increased gradually. Denitrification bacteria and anaerobic heterotrophic bacteria can also achieve better coupling effect below the ration of 1.6. Methane bacteria was predominant in coupling, the contribution rate of anaerobic heterotrophic bacteria on COD removal was 78.04%.

3. The change of stoichiometric ratio can affect the interaction of anaerobic ammonium oxidation bacteria, heterotrophic denitrification bacteria and heterotrophic anaerobic bacteria. The result showed that the removal of nitrogen and carbon was best at the stoichiometric ratio of 1.6, the removal rates of TN and COD were 73.58% and 80.1%, respectively. The contribution rate of heterotrophic denitrification to COD, TN removal was 46.57%, 21.36%, respectively.

4. When the stoichiometric ratio was 1.6, the effluent NH_4^--N, TN was to 56.89mg/L, 34.03mg/L, respectively, which can be remove in the advanced treatment.

Acknowledgment

The project was funded by the Science, and Technology Department of Jilin Province (Provincial Science Foundation Project No.201510190JC).

References

1. Wang W, Han H J, Recovery strategies for tackling the impact of phenolic compounds in a UASB reactor treating coal gasification wastewater, J. Bioresource Technology. 2012, 103: 95–100.

2. Wang Z X, Xu X C, Gong Z, et al, Removal of COD, phenols and ammonium from Lurgi coal gasification wastewater using A2O-MBR system, J. Journal of Hazardous Materials. 2012, 235: 78-84.

3. Guo X L, Wang J, Wang Y, et al, Research of phenols adsorption from simulated coal gasification wastewater by resin, J. Procedia Environmental Sciences. 2012, 12: 152–158.

4. Xiao Benyi, Qu Jiuhui, Lin Xinkan, Anaerobic biological treatment at room temperature, J. Water Supply and Drainage. 2007, 33: 174-177.

5. Lv Jianguo. Development and latest status of wastewater anaerobic biological treatment technology, J. Environmental Science and Management. 2012, 37(12): 87-92.

6. Zhao Lijun, Teng Dengyong, Liu Jinling, et al, Review and research progress of wastewater anaerobic biological treatment technology, J. Technology and Equipment for Environmental Pollution Control. 2001, 2(5): 59-65.

7. Xu X C, Xue Y, Wang D, et al, The development of a reverse anammox sequencing partial nitrification process for simultaneous nitrogen and COD removal from wastewater, J. Bioresource Technology. 2014, 155: 427-431.

8. Wu X, Liu S T, Dong G L, et al, The starvation tolerance of anammox bacteria culture at 35°C, J. Journal of Bioscience and Bioengineering. 2015, 120(4): 450-455.

9. Catarina T, Catarina M, Samantha B J, et al, The contribution of anaerobic ammonium oxidation to nitrogen loss in two temperate eutrophic estuaries, J. Estuarine Coastal and Shelf Science. 2014, (143): 41-47.

10. Cao Shenbin, Wang Shuying, Wu Chengcheng, et al, Impact of organic compounds on anaerobic ammonium oxidation system, J. China Environmental Science. 2013, 33(12): 2164~2169.

11. Liu Changjing, Li Zebing, Zheng Zhaoming, et al, Effect of different organic compounds on anaerobic ammonium oxidation coupled denitrification, J. China Environmental Science. 2015, 35(1): 87-94.

12. Wang Chunxiang, Liu Changjing, Zheng Linxue, et al, Study on denitrification bacteria in anaerobic ammonium oxidation coupled denitrification system, J. China Environmental Science. 2014, 34(7): 1878~1883.

13. Zhou Shaoqi. Stoichiometric analysis of the synergistic effect of anaerobic ammonium oxidation and denitrification, J. Journal of South China University of Technology (natural science edition). 2006, 34(5): 1-4.

14. Huang Xiaoxiao, Chen Zhongjun, Zhang Rui, et al, Research progress in the coupling reaction of anaerobic ammonium oxidation and denitrification, J. Journal of Applied Ecology. 2012, 23(3): 849-856.

15. Lv Yongtao, Chen Zhen, Wu Yahong, et al, Experimental study on the effect of organic matter concentration on the performance of anaerobic ammonium oxidation to nitrogen removal, J. Journal of Environmental Engineering. 2009, 3(7): 1189-1192.

16. Fu Jinxiang, Tong Ying, Yu Pengfei, et al, Bidirectional effects of organic compounds on anaerobic ammonium oxidation and its inhibition, J. Industrial water treatmen. 2014, 34(7): 19-23.

17. Yang Shidong, Liao Luhua, Effect of influent ratio on anaerobic treatment efficiency of coal gasification wastewater, J. Silicate Bulletin. 2016 (Accepted).

The Adsorption Characteristic Research of Sulfamethoxazole in Albic Soil

Xin-Shuang Wang, Li Li[*], Yi-Bo Yang

Jilin Agricultural University, Jilin Changchun 130118
[*]*E-mail: jlnydxll@163.com*

This research aims to study the adsorption kinetics and isothermal adsorption features, as well as how different background pH, cation and various organic matter amount may affect on the adsorption effect in Albic soil of sulfamethoxazole (SMZ) by OECD guideline 106 batch equilibrium method. The results indicated that: when the initial concentration was 5 mg·L^{-1}, the absorption process of SMZ in the Albic soil followed the Particle Diffusion Equation. The adsorption process can be divided into fast and slow stages, and the equilibrium time is 24 hours. The isothermal adsorption behavior of the sediment SMZ could be better fitted with Freundlich and Langmuir adsorption isotherm, correlation coefficient (r) of which could be as high as 0.9890 and 0.9967. The adsorption isotherm of Freundlich equation is categorized as "L" type, and the adsorption capacity of K_f is 1.8616. With the solution pH increased, the adsorption ability of SMZ was weakened. The isothermal adsorption curves of SMZ in the Albic soil made based by various background solutions with different cations showed a better imitative effect on the Freundlich equation and Langmuir equation. The higher the ionic states are, the less K_f, Q_m are contained. The absorption ability of SMZ increases if more organic matters exist.

Keywords: Sulfamethoxazole; adsorption; soil; pH; cationic type; organic matter.

1. Introduction

Generally, sulfonamides refers to any drug contains amino benzene sulfonamide structure, which is widely used in prevention and treatment of bacterial infectious diseases. Due to the broad spectrum of antibiotics and low prices, it's one of the most commonly used additives in veterinary clinical and animal husbandry, which contributes to a large quantity of production and consumption [1, 2]. Because most antibiotics consumed by livestock are excreted through the feces and urine, which cause potential pollution on soils, surface and ground water [3, 4, 5, 6]. In addition, the livestock and poultry manure recycled in agricultural use can also lead to a high sulfonamides concentration in soil, which could be ranged from a few dozens to a few hundred grams [1, 7]. According to the wide use and long-term persistence of sulfonamide antibiotics, much attention has been attracted to its behavior in environment [8, 9, 10, 11].

In this research, sulfamethoxazole (SMZ) was studied as an example, in which the absorption, pH, cation species, organic matters and other various conditions were tested to study how the actual environmental condition may affect sulfamethoxazole adsorption. As a result, evaluation basis can be provided for environmental impacts studies.

2. Materials and Methods

2.1. *Soil samples*

The soil samples were Albic soil collected from Yongji County, Jilin Province, the town of Laxi; Soil samples were mixed soils of 0~20cm below the surface. Soil gravel, plant roots, straw and other debris were removed; the samples were grinded after natural air dry process, and then selected by 0.25 mm screen. The physical and chemical properties of the soil samples are shown in Table 1.

---

[*]Corresponding author

Table 1 Basic chemical properties of the Soil samples

Soil samples	pH	Organic matter /%	Cation exchange capacity /cmol·kg^{-1}	Particles（%）		
				Clay	Particles	Grit
Albic soil	4.2	2.54	22.42	35.8	39.7	24.5

2.2. Instruments and reagents

Primary instrument: Agilent 1260 high performance liquid chromatography, TDL-5-A centrifuge, Constant temperature oscillator, KQ-500E type ultrasonic cleaner, electronic balance, 0.22μm organic membrane.

Main reagents: Sulfamethoxazole($C_{10}H_{11}N_3O_3S$), the purity of more than 99.9, from Dimma technologies; Methanol as chromatographic grade reagent; Formic acid is pure class distinctions; Other reagents were analytical pure; Test water for pure water.

2.3. Chromatographic conditions

Agilent 1260 high performance liquid chromatography, UV detector, The column was 150 mm× 4.6 mm of ZORBAX SB-C$_{18}$, The injection volume 20μL, Column temperature 30°C, 0.1% formic acid: methanol = 7: 3 (V: V) as the mobile phase, The flow rate was 1.0 mL·min^{-1}, The detection wavelength was 270 nm.

2.4. Experiment design

2.4.1. Adsorption kinetics test

Adsorption tests were carried out by referring to the guideline OECD 106 batch equilibrium method [12]. 4.0000 g (±0.0005 g) of soil samples were dropped in polyethylene centrifuge tube (50 mL), 10 mL SMZ background electrolyte solution of 5 mg·L^{-1} were added. The background electrolyte solution was 0.01 mol·L^{-1} CaCl$_2$ (containing 0.01 mol·L^{-1} NaN$_3$). After being sealed, the samples were kept shaking in warm water (25°C±1°C) at the speed of 200 r·min^{-1}; the shaking duration time were 5 min, 10 min, 15 min, 30 min, 1 h, 2 h, 4 h, 8 h, 12 h, 24 h. Then hold the samples for 10 min (4000 rpm·min^{-1}) to get centrifugal. Supernatant were filtered by 0.22μm filter membrane. Equilibrium concentration of SMZ in the supernatant was tested by HPLC. The above test was repeated three times, the one without SMZ was blank group, and the one without soil sample was control group.

2.4.2. Adsorption isotherm test

4.0000 g (±0.0005 g) of soil samples were dropped in polyethylene centrifuge tube (50 mL), 10 mL SMZ background electrolyte solutions of different concentrations were added. The background electrolyte solution was 0.01 mol·L^{-1} CaCl$_2$. The initial SMZ concentration were set as 1, 2, 5, 7, 10 mg·L^{-1}. After being sealed, the samples were kept shaking in warm water (25°C±1°C). Then hold the samples to get equilibrium concentration. Supernatant were filtered by 0.22μm filter membrane. Equilibrium concentration of SMZ in the supernatant was tested by HPLC. The above test was repeated three times, the one without SMZ was blank group, and the one without soil sample was control group.

2.4.3. Effect of different factors on the adsorption of SMZ in Albic soil

(1) Effect of different pH on adsorption
Use HCl or NaOH solution to adjust the pH background, making the background solution pH as 5, 6.8, 9, referring to the test method in 1.4.2 to repeat the test, then determine the equilibrium concentration of SMZ.

(2) Effect of cation types on adsorption
Use $0.01 mol \cdot L^{-1}$ KCl and $AlCl_3$ as the background solution, referring to the test method in 1.4.2 to repeat the test, then determine the equilibrium concentration of SMZ.

(3) Effect of different organic matters on adsorption
In this study, peat containing 40% organic matter were used as an additive, which contributed to the organic matter of 0, 1%, 3%, 5%, The other conditions were referred to adsorption isothermal test.

3. Results and Discussion

3.1. *Adsorption kinetics of SMZ in Albic soil*

Fig. 1 describes how the adsorption capacity of SMZ changes with the time. It can be seen from Fig. 1 that when the initial concentration is 5 mg·L^{-1}, the adsorption capacity of SMZ is up to 82.5% of the initial addition. The adsorption processes can be divided into two stages, one is fast and the other is slow. In the whole adsorption process, at first 0~30 min, the aqueous solution proportion of SMZ decreased dramatically, which indicated the adsorption capacity of SMZ was the fastest; after that the slow stage started, then the concentration of SMZ reached to equilibrium. At this point, the adsorption was much slower. This was because the adsorption points on the soil surface are fixed. In the early stage, the adsorption points are sufficient, so the adsorption rate is high; with the increase of the adsorption time, the adsorption points were insufficient, and the adsorption rate was decreased. This process lasted until the adsorption equilibrium reached. After 24 hours, the adsorption amount was stable, indicating the equilibrium was satisfied.

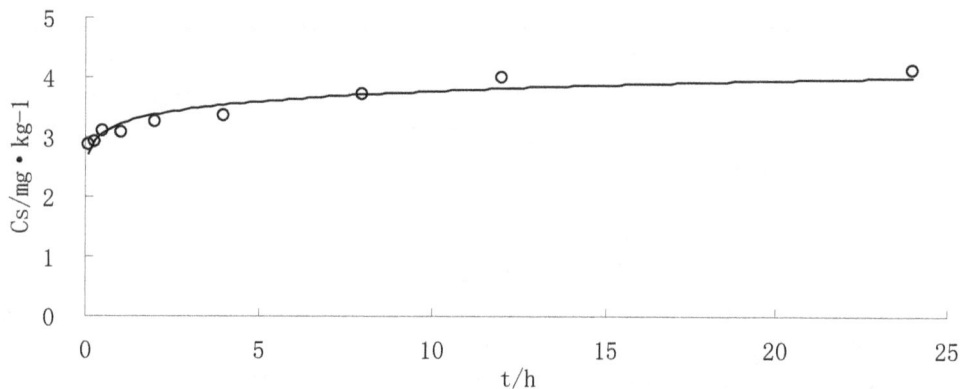

Fig. 1 Adsorption kinetics of SMZ in Albic soil

In general, the control of pollutant adsorption processes includes mass transfer, diffusion control, chemical reaction, particle diffusion, etc. Elovich equation, particle diffusion equation and double constant equation were used to test SMZ adsorption kinetics in white pulp soil.

Table 2 Fitting parameters of adsorption kinetic models

Soil samples	Elovich parameter			Particle diffusion parameter			Double constant parameter		
	A	K_t	r	B	K_{id}	r	C	K_s	r
Albic soil	3.2438	0.2339	0.9364**	2.8198	0.2950	0.9787**	1.1673	0.0724	0.9526**

Note: $n=10$, $r0.01=0.765$, ** indicates that the correlation coefficient is highly significant, ($p<0.01$).

As shown in Table 2, the fitting correlation coefficient (r) indicated that the particle diffusion equation was best on describing the kinetics of SMZ adsorption in white pulp soil. This supports that the adsorption behavior of SMZ includes external liquid membrane, the internal diffusion of the particles and the surface adsorption.

3.2. *Isothermal adsorption properties of SMZ in Albic soil*

The adsorption isotherm of SMZ in the Albic soil was fitted by Langmuir equation, Freundlich equation and linear equation. Adsorption isotherm is shown in Figure 2, Adsorption constants and correlation coefficients are shown in Table 3.

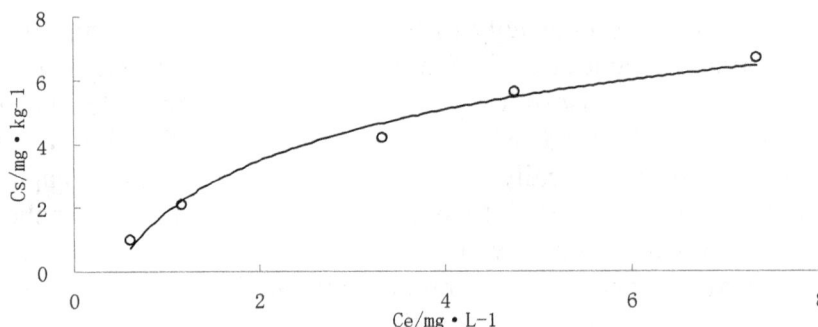

Fig. 2 Adsorption isotherms of SMZ in Albic soil

Table 3 Fitting parameters of adsorption models

Soil samples	Freundlich parameter			Langmuir parameter			Linear parameter	
	K_f	$1/n$	r	K_L	Q_m	r	K_d	r
Albic soil	1.8616	0.6621	0.9890**	0.1613	12.4729	0.9967**	1.0404	0.9372*

Note: $n=5$, $r_{0.01}=0.959$, $r0.05=0.878$,* indicates that the correlation coefficient is significant; ** indicates that the correlation coefficient is highly significant, ($p<0.05$).

According to Fig. 2 and Table 3, Freundlich equation, Langmuir equation and linear equation correlation coefficients (r) are 0.9372, 0.9890 and 0.9967. Therefore, the Freundlich equation and Langmuir equation are more suitable to fit the adsorption behavior of SMZ in the Albic soil. This also supports that the adsorption of SMZ on the Albic soil is a combined action of various adsorption points. In Freundlich equation, K_f is 1.8616, $1/n<1$, indicating that with the increase of adsorbent, the free energy of adsorption becomes stronger. According to the relationship between the $1/n$ value and the shape of the isothermal adsorption curve [13], the adsorption isotherm is "L" type. Based on this view, when the SMZ concentration is lower, the main limitation of adsorption may be the competition from water molecule. The adsorption ratio of SMZ is larger if the initial concentration is higher [14]. In this experiment, the effects from different solution pH, cation types and followed this rule.

The change of adsorption free energy is an important basis for judging the soil adsorption mechanism. According to the level of the changes, the adsorption mechanism can be determined. When the free energy difference is less than 40 $kJ \cdot mol^{-1}$, then the adsorption is physical; or the adsorption is chemical. The adsorption free energy value of SMZ in the Albic is -11.71 $kJ \cdot mol^{-1}$; ΔG is less than 0, indicating that the adsorption process is spontaneous. In addition, the absolute value is less than 40 $kJ \cdot mol^{-1}$, indicating the adsorption is physical.

3.3. *Effects of different factors on the adsorption of SMZ in Albic soil*

3.3.1. *Effects of pH on adsorption capacity*

The effects of pH on sorption amount of SMZ in the Albic soil were illustrated in Fig. 3.

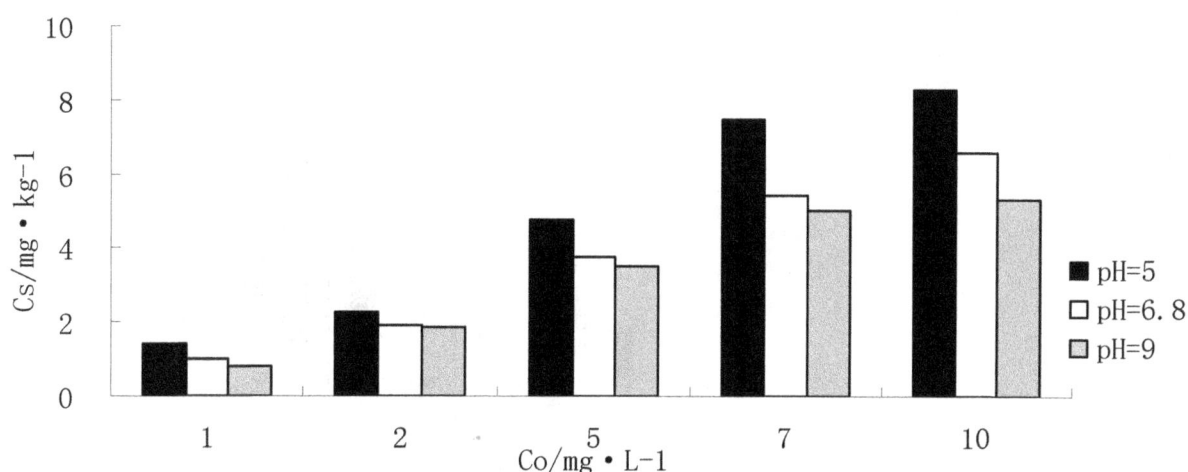

Fig. 3 Effect of pH on sorption amount of SMZ in the Albic soil

According to Fig. 3, in this experiment, the adsorption capacity of SMZ in the Albic soil decreased with the increasing of pH value. The SMZ contains NH_2 basic group and SO_2 acidic group (pKa values respectively were 1.8 and 5.6) which can be combined respectively with the solution of H^+ and OH^-, then existing as the form of cationic, anionic or zwitterionic. The Albic soil used in this experiment had a pH value of 4.2. When a background solution of pH 5 was added, SMZ existed mainly in the form of $SMZH^+$ in the Albic soil. When the background solution pH value was between 5 and 9, SMZ would exist as the form of $SMZH^{\pm}$. When the background solution had a pH of 9, SMZ existed mainly as the form of $SMZH^{\pm}$ or SMZ^-. Therefore, pH can affect the type of SMZ, as well as the adsorption of SMZ.

Table 4 Adsorption model parameters under different pH

Soil samples	pH	Freundlich parameter			Langmuir parameter			Linear parameter	
		K_f	$1/n$	r	K_L	Q_m	r	K_d	r
	5	2.4958	0.6594	0.9618**	0.1677	16.1211	0.9691**	1.4367	0.9173*
Albic soil	6.8	1.6238	0.7132	0.9915**	0.1121	14.6226	0.9928**	0.9883	0.9610**
	9	1.5829	0.6263	0.9604**	0.1861	9.3704	0.9794**	0.8117	0.8923*

According to Table 4, the mean correlation coefficient (r) values from Freundlich equation, Langmuir equation and linear fitting equation were 0.9712, 0.9804, and 0.9235, indicating Freundlich equation and Langmuir equation are better to fit the adsorption of SMZ under background solutions of different pH values. In those equations, K_f is adsorption constant, and K_f decreased with the increasing of pH values, indicating the binding ability weakened at higher pH values. The adsorption of SMZ on the Albic soil showed an obvious nonlinear tendency, and $1/n$ were less than 1, indicating the adsorption tendency decreased at higher initial concentration. This maybe results from the location of adsorption points on the soil surface. At the beginning, SMZ was first adsorbed by high energy adsorption points on the surface of the Albic soil, when these high energy adsorption points were fully occupied, then adsorption points with low energy started. Therefore, the adsorption is the result of multiple forces. In Langmuir equation, Q_m describes the theoretical saturation adsorption capacity of SMZ. From Table 4, the saturated adsorption capacity decreased with the increasing of pH values. In the linear equation, K_d is also decreased with the increasing of pH values. All the above parameters showed that the adsorption capacity of SMZ in the Albic soil decreased with the increasing of pH values.

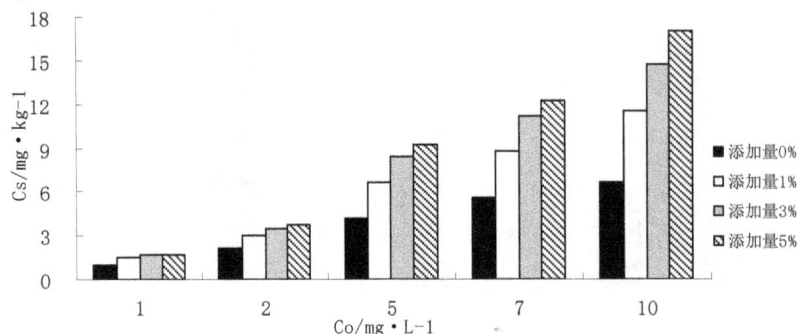

Fig. 4 Effect of organic matter on sorption amount of SMZ in the Albic soil

3.3.2. *Effects of cation types on adsorption capacity*

In the presence of different ions, the adsorption processes can be well fitted by Freundlich equation, Langmuir equation and linear equation, and the correlation coefficient (r) was significantly correlated. In Freundlich equation, the adsorption capacity of K_f was $K^+ > Ca^{2+} > Al^{3+}$, indicating adsorption capacity of the K^+ is the largest. In Langmuir equation, the adsorption capacity of single molecule layer (Q_m) was $K^+ > Ca^{2+} > Al^{3+}$. As a result, K^+ also has the most obvious effects on adsorption. In the linear equation, the adsorption parameter K_d is also enhanced with the decrease of the ionic valence state. In this study, SMZ exists at the form of zwitterion or anion.

Table 5 Adsorption model parameters under different cation types

Soil samples	Cation type	Freundlich parameter			Langmuir parameter			Linear parameter	
		K_f	$1/n$	r	K_L	Q_m	r	K_d	r
Albic soil	K^+	1.9142	0.6873	0.9978**	0.1393	14.4495	0.9991**	1.1263	0.9570**
	Ca^{2+}	1.8616	0.6621	0.9890**	0.1613	12.4729	0.9967**	1.0404	0.9372**
	Al^{3+}	1.7388	0.6643	0.9911**	0.1556	11.9358	0.9947**	0.9691	0.9435*

It's easier for anions to form stable complex goods with metal cations, which causes impacts on the adsorption function. Therefore, under different media, the adsorption of SMZ varies. This difference results from how hard it is to form compounds with different soil constituents.

3.3.3. *Effect of organic matter on the adsorption capacity*

From Table 6, we can see that the Freundlich equation, Langmuir equation and linear equation are better fit for SMZ, The average correlation coefficient r is 0.9888, 0.9931, 0.9609. Illustrated Freundlich equation, Langmuir equation and linear equation are suited to fit the adsorption behavior of different organic matter added amount of the SMZ in Albic soil. K_f With the increase in the amount of organic matter added increases; Q_m With the increase in the amount of organic matter added increases; K_d is also with the increase in the amount of organic matter added increases; All the above parameters showed that the adsorption capacity of SMZ increased with the increase of organic matter. With the increase of organic matter content, the increase of negative charge on the surface, the adsorption sites in the soil increased, make the adsorption capacity increased.

Table 6 Adsorption model parameters under different organic matter content

Soil samples	Additive amount/%	Freundlich parameter			Langmuir parameter			Linear parameter	
		K_f	$1/n$	r	K_L	Q_m	r	K_d	r
Albic soil	0	1.8616	0.6621	0.9890**	0.1613	12.4729	0.9967**	1.0404	0.9372*
	1	3.5451	0.7107	0.9973**	0.1630	24.5358	0.9992**	2.3507	0.9650**
	3	5.4886	0.7233	0.9810**	0.2138	31.8035	0.9844**	4.0013	0.9597**
	5	6.5372	0.8351	0.9882**	0.1369	55.7669	0.9921**	5.6318	0.9817**

4. Conclusion

(1) The adsorption process of SMZ in the Albic soil is divided into two stages: fast and slow adsorption. Reached adsorption equilibrium in 24 hours, the particle diffusion equation was best on describing the kinetics of SMZ adsorption in white pulp soil. This supports that the adsorption behavior of SMZ includes external liquid membrane, the internal diffusion of the particles and the surface adsorption.

(2) The adsorption ratio of SMZ is larger if the equilibrium concentration is higher. The isothermal adsorption behavior of the sediment SMZ could be better fitted with Freundlich and Langmuir adsorption isotherm, correlation coefficient (r) of which could be as high as 0.9890 and 0.9967, All reached extremely significant correlation level ($p<0.01$).

(3) In the presence of different ions, the adsorption processes can be well fitted by Freundlich equation, Langmuir equation and linear equation. The effect of different cations on the adsorption capacity of SMZ in the Albic soil was decreased with the increase of the ionic valence state.

(4) In the presence of different organic content matter, the adsorption processes can be well fitted by Freundlich equation, Langmuir equation. The average correlation coefficient r is 0.9888, 0.9931, and 0.9609. With the increase of organic matter content, the adsorption capacity of SMZ increased.

References

1. Zhang-liu Chen.Trends in the development of chemical drugs for animals [J]. Chinese Journal of Veterinary Drug, 2005, 39(7): 1-6. In Chinese

2. Zhen Li, Wen-hai Duan, Rong Shao. Analysis on the present situation of the use of antibiotics in China [N]. International Medicine and Health Guidance News, 2005, 21: 81-82. In Chinese

3. Chee-Sanford J.C., Mackie R.L, Koike S. Fate and Transport of Antibiotic Residues and Antibiotic Resistance Genes following Land Application of Manure Waste [J]. 2009, 38(3): 1086-1108.

4. Tian-tian Ren, Yin-bao Wu, et al. Research Progress on the environmental behavior of the veterinary drug [J]. Animal Husbandry and Veterinary Medicine, 2013, 45(5): 97-101. In Chinese

5. Ying-qin Wu, Huang-feng Jiang, Ming-guang Ma, et al. Study on adsorption behavior of insoluble humic acid nitroaniline to water [J]. Journal of Northwest Normal University. Natural Science Edition, 2006, 42 (3): 62-65. In Chinese

6. Xue-huan Li, soil chemistry [M]. Beijing Higher Education Press, 2001. In Chinese

7. Burkhardt M, Stamm C. Depth distribution of sulfonamideantibiotics in pore water of an undisturbed loamy grassland soil [J]. Journal of Environmental Quality, 2007, 36(2): 588-596.

8. Yu-xia Liu, Yan-yu Bao. Advances in research on the contamination of tetracycline antibiotics in soils [J]. Environmental Pollution and Control, 2011, 33(8): 81-86, 91. In Chinese

9. Hao Chen, Jin-qiang Zhang, Zhong Ming, et al. Adsorption characteristics of the typical paddy soil in Taihu area [J]. China Environmental Science, 2008, 28(4): 309-312. In Chinese

10. Figueroa RA, Leonard A, Mackay AA. Modeling tetracycline antibiotic sorption to clays [J]. Environ. Sci. Technol., 2004, 38(2): 476-483.

11. Sarmah AK, Meyer MT, Boxall AB. A global perspective on the use, sales, exposure pathways, occurrence, fate and effects of veterinary antibiotics (VAs) in the environment. Chemosphere, 2006, 65(5): 725-759.

12. OECD. OECD Guide lines for Testing of Chemicals, Test Guide line 106: Adsorption/ desorption Using a Batch Equilibrium Method. Paris: OECD, 2000. 45.

13. Calvet R. Adsorption of organic chemicals in soils [J]. Environmental Health Perspectives, 1989, 83: 145-177.

14. Sukul P, Lamshöft M, Zühlke S, et al. Sorption and desorption of sulfadiazine in soil and soil-manure systems [J]. Chemosphere, 2008, 73(8): 1344-1350.

15. Bin Li, Zhi-chun Wang. The soda alkaline soil exchangeable cation and correlation analysis [J]. Soil agricultural Bulletin, 2008, 24(6): 271-275. In Chinese

16. Juan Zhang, Wen-lu Guo, Cheng Sun, et al. Adsorption behavior of SD in black soil and different particle size fractions [J]. Journal of Agricultural Environmental Science, 2011, 30(2): 301-306. In Chinese

17. Ying Lu, Shuo-kui Han. Sorption of herbicide mefenacet in soils [J]. Environmental Chemistry, 2000, 19(6): 513-517. In Chinese

An Environmental-friendly and Efficient Configuration Plan of Marble Processing Equipment

Sen-Wei Zheng*, Wei Liu, De-Hong Sun, Hui-Fen Tong

Minnan University of Science and Technology, Quanzhou, China 362700
Email: 84246354@qq.com

With the rapid development of high-precision measuring equipment in manufacturing industry, the demands of reference platform are becoming higher and higher. With the features of long-term strength, high hardness and good stability, marble platform becomes one of the most ideal high-precision measuring equipment. However, there are many problems during the process of polishing with the marble platform, for example, big noise, heavy dust and complex processing procedures and so on. Therefore, this article mainly discusses the marble platform processing equipment, which can solve the above-mentioned problems about environmental protection and integrate milling, grinding and polishing at the same time.

Keywords: Environmental protection; marble platform; integrated processing.

1. Introduction

Marble platform is a precision reference measurement tools and it is made of natural stone materials. Its advantages are as follows: even texture, good stability, high strength, high hardness, no rust, acid and alkali-resistance, no magnetization, no deformation, good abrasion resistance. Besides, accuracy can be maintained under heavy load. Even after an impact or scratch, the working surface can be only with a pit, without any wale and burr, no influence on the measurement accuracy. Because of its unique characteristics in high-precision measurements, the traditional cast iron plate dwarfs. Marble platform as a high-precision instruments, precision tools, mechanical parts of the test over reference plane, low cost, high value-added products can be widely used in industrial production and laboratory measurements.

During the process of polishing, either manual or automatic, it will cause environmental pollution and generate a lot of dust, which can be very harmful for the health of polishing operators [1].

Therefore, this article mainly discusses a marble platform processing equipment, taking efficiency and quality into account, so that marble can be polished in a more efficient and unpolluted way.

2. Structure of Polishing Machine

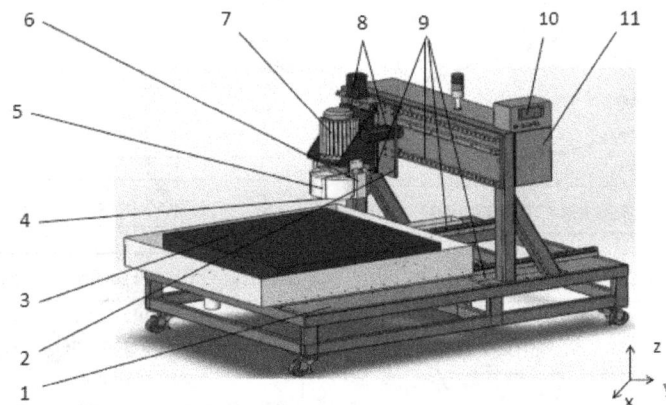

Fig. 1 Marble polishing machine

1 bed; 2, the mobile panel; 3, table; 4, polishing disc; 5, the protective cover; 6, water circulation system; 7, the electric motor; 8, motor; 9, guide way of three degrees of freedom & rectangular coordinates; 10, a control panel; 11, a control box

As shown in Fig. 1, which contains 1. bed; 2. moving panel; 3, work table; 4. polishing disc; 5. protective cover; 6. water circulation system; 7. electric motor; 8. motor; 9. guide way of three degrees of freedom & rectangular coordinates; 10. control panel; 11. control box, marble polishing grinding block should be fixed on the polishing disc, polishing disc and a motor should be connected to the motor, which fixed on the mobile panel. The moving panel is moving on the guide way and moving towards the Y and Z directions of 3 - DOF rectangular coordinates, the control panel is set above the control box, the table is moving towards X direction.

The guide way is set on the bed, the protective cover is fixed on the moving panel, and the water system is fixed on the protective cover. It can be used in polishing the components of flat stones. The polishing can be in different precision. The polishing disc can move towards Y and Z directions and process different platforms of different sizes. The work piece can be fixed on the working table and move together towards the X direction. According to different precision machining marble platform, grinding blocks on the polishing disc can be changed and replaced. In the process of processing, the noise and pollution of lime can be reduced through a protective cover and water circulation system.

3. The Control System Settings

The machine is equipped with PLC so the operators can use PLC to control the equipment. The operation becomes easy because of the touch screen on the operation panel. The speed of rotation and the press data of spindle motor, the moving distance of X, Y, Z axes are all displayed. The spindle is connected directly to electric motor.

Add an inverter to control the change of stepless speed. A spring can be added to Z-axis, in addition to its own gravity of Z-axis, achieved by changing the elastic force. This should be set according to the parameters of elastic effect. The control flow of X, Y, Z-axis stepper motor is shown in Fig. 2. For the grinding processing efficiency and surface finish, the same pressure is applied to each location to ensure flatness, the need for control of pressure.

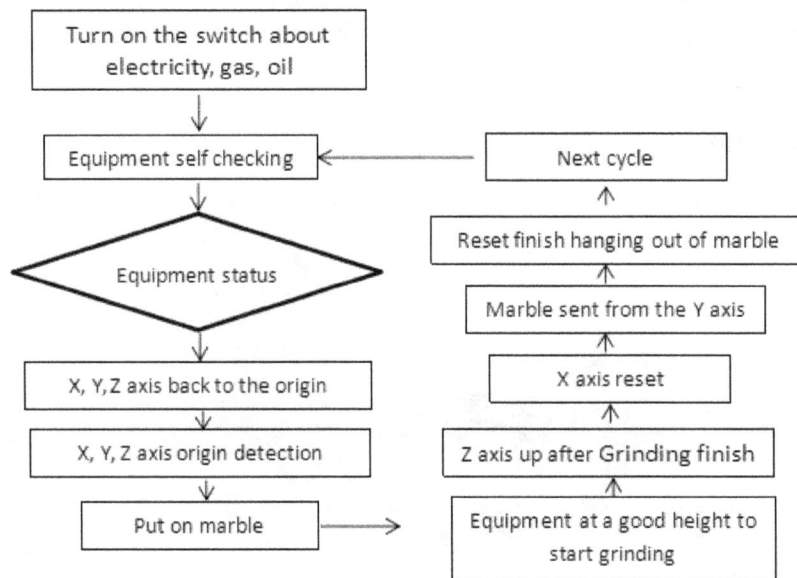

Fig. 2 Control flow chart

To increase the accuracy of the surface, in the process, a variety of trace trajectory is used to achieve random trajectory. Thus the path is set, the movement locus in the XY direction are as shown in Fig. 3. Two kinds of trajectory are used to produce [2].

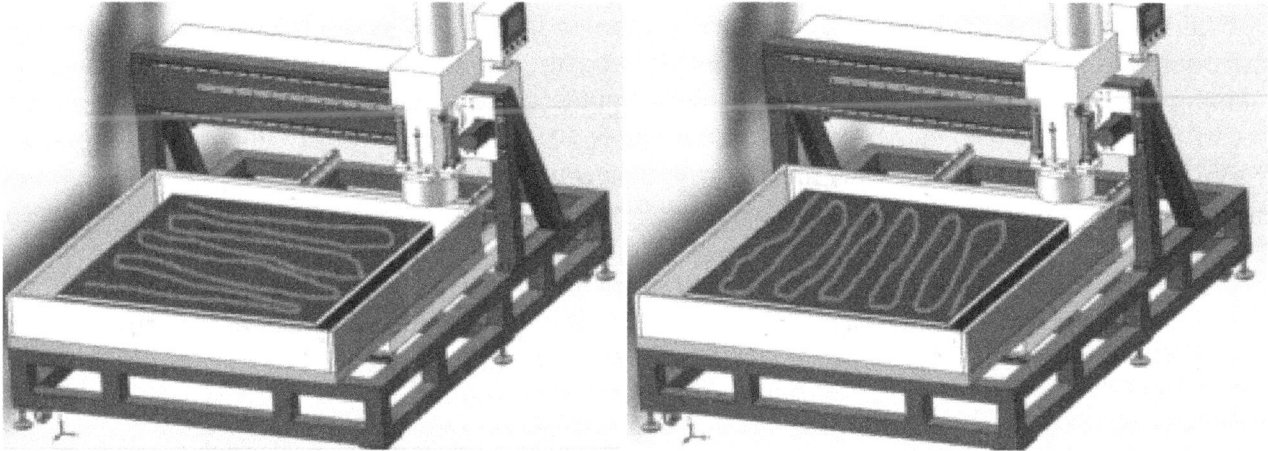

Fig. 3 Marble processing locus

4. Setting of Polishing Shield

In the process, the disc, with high speed rotation, could make the grinding block contact with the surface of marble, so that it can be grinded and polished. Due to friction and rotation of polishing disc, small blocks will come off and fly off in all directions, amount of dust and frication noise will also appear, the working environment will be polluted unavoidably to some extent. Therefore, it is quite necessary to add a protective cover around the polishing disc, as shown in Fig. 4, the shield covering the entire polishing dis. When the noise is spreading out during the processing, the shield could decrease the noise because of its acrylic material. Meanwhile, the shield could also prevent dust and tiny marble blocks from spreading out, so the pollution can be decreased to a certain degree.

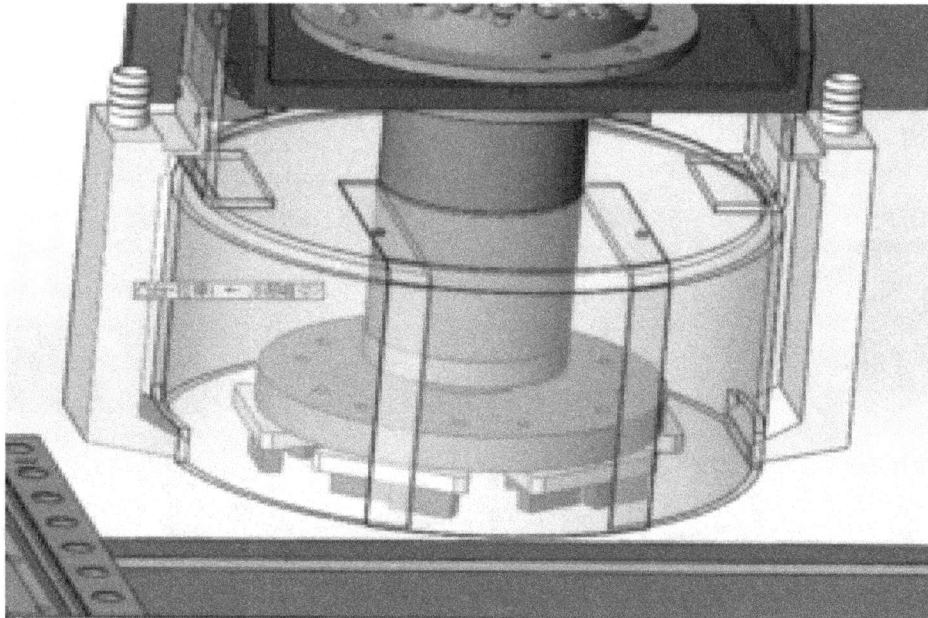

Fig. 4 Polishing guard

5. Setting of Water Circulation System in Polishing Equipment

Marble processing, due to friction with the surface of grinding marble block, heat and dust are generated [3]. Therefore, there must be a water circulation device in processing part, set up a water circulation system in the polishing apparatus, as shown in Fig. 5, it could have a cooling effect on the one hand, and prevent the dispersion of dust on the other hand. During the processing, the running water at room temperature washes the manufacturing surface. Water circulation system is set in the tank and the inlet of water circulation is fixed on both sides of the shield. After the washing, the water will return to the water tank through the sink, thus a recycled water supply system is formed. The system could offer several major benefits: less dust spreading, less frication because of the wet manufacturing surface, less loss of grinding block, less noise and higher processing efficiency. Besides, some tiny blocks will be produced during the manufacturing process and they will scatter on the surface of marble blocks, but the running water could wash them away, so the marble surface becomes clean and smooth, the producing quality can be improved in this way.

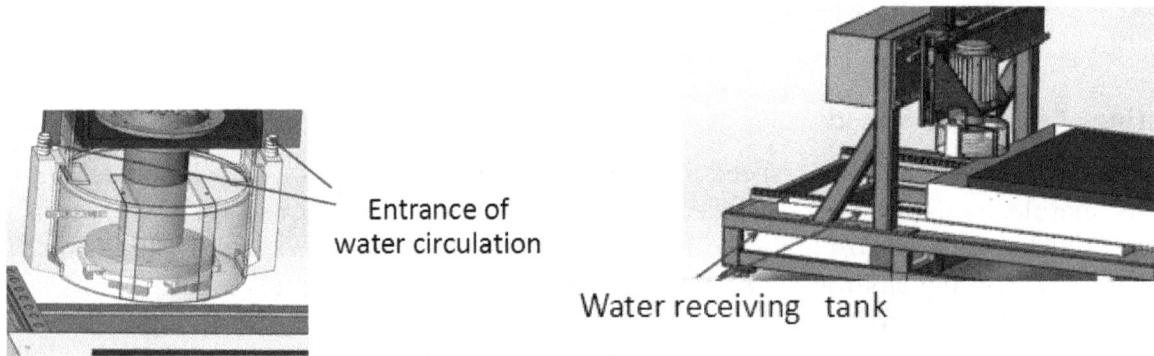

Fig. 5 Water circulation system setting

6. Setting of Processing and Polishing Parts

Fix the polishing grinding block, six groups of grinding blocks are fixed with screws, the blocks can be replaced and changed by other different blocks after two fastening screw becoming loose, as shown in Fig. 6.

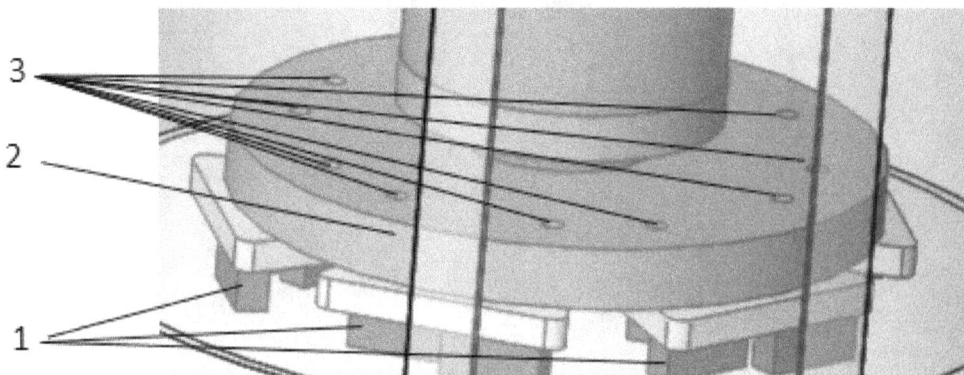

Fig. 6 Connection diagram of grinding, polishing block & grinding disc
1, the module; 2, polishing; 3, screw

The grinding blocks are easily damaged because of the friction during the process. After a period of use, they need to be replaced by other different grinding blocks. If the grinding disc assembles with the block, then replacing the blocks will be more troublesome, the cost of grinding blocks will also increase. In order to solve these problems, the grinding disc and the blocks can be connected by bolts. The material of the blocks is the super hard diamond and the diamond will help the processing of different precisions [4]. To meet the demand of different precisions, the turning speed of the grinding disc should be adjustable. In order to complete speed CVT, an inverter can be added to the spindle motor of the grinding disc.

7. Conclusion

The problems of big noise, more dust and many manufacturing procedures can be solved by improving the configuration scheme of processing equipment. Through offering a convenient and practical solution, we could get the following conclusions:

1. The traditional marble processing, through multi-channel processes multiple mounting clip marble to complete the grinding and polishing of marble, result as the process complex and the efficiency low, and our processing equipment polishing parts of the set with replaceable type grinding head of the grinding and polishing, all the marble grinding and polishing processes can be completed with only once clip of marble, so the processing steps can be less and the processing efficiency can be higher.
2. The traditional marble processing equipment, processing will generate a lot of dust, causing environmental pollution, serious damage to the health of the operating personnel, and our equipment through the protective cover can prevent the diffusion of dust and noise and realize the polishing process of no dust generation. The result is no environmental pollution, low noise and protecting the health of workers.
3. The traditional marble polishing, the operator from the polishing wheel is very close, more likely to cause personal safety hazards. Worker set the NC program settings by the control system. Because of the automatic operation, people do not have to close to the polishing wheel, so the worker can work efficient without dangerous.
4. The traditional marble polishing in the polishing process, the polishing liquid will be discharged directly or stored in a pool. This processing method cannot be recycling and environmental protection. Set the water circulation system in our equipment, can be achieve recycling and environmental protection.

Acknowledgments

This research was funded by Elitist Training foundation of Quanzhou (2012Z126) and Tow Items of expenditure of Shishi (2014SK14), (2012SD6), (2013SD18).

References

1. High Stone Group Shenzhen factory. Stone polishing technology introduction [J]. Stone, 2004 (6).
2. Conley Stone Group. Stone processing technology to the product quality [J]. Stone, 2009. Section 11.
3. I B.G. Koeke et al, 1977, Grinding Damage in ceramics, presented at SEM's International Tool & Manufacturing Conference.
4. Zhang Cao, Zhu Mei-chu. Stone grinding and polishing abrasive experimental study [J]. Stone, 1994 (6).

Thermal Performance Analysis for Heat Storage Container Used in Solar Thermal Power Generation

Hai-Ting Cui[*], Ning Li, Chang-Le Yi

School of Mechanical Engineering, Hebei University of Science and Technology,
Shijiazhuang Hebei 050018, China
[*]Email: cuiht@126.com

In this article, a numerical simulation is performed to investigate both the charging and discharging process of high-temperature phase change thermal energy storage container used in solar thermal power generation using the solidification/melting model. The effects of the several impact factors such as Fourier number, Stephen number and Reynolds number on the phase change process are discussed in detail. As the rule of the phase change process of high-temperature phase transition thermal energy storage container in solar thermal power generation was obtained, it provides an important reference value and theoretical basis for its optimization design.

Keywords: Solar thermal power generation; high temperature thermal energy storage container; phase change material; FLUENT; numerical simulation.

1. Introduction

The deployment of solar thermal for power generation has been gather pace [1]. It is recognized that the latent heat storage can provide a much higher heat storage density compared to the conventional sensible heat storage technologies. This can be achieved by using the latent heat during the solid-liquid phase transition which can store and release the heat at constant phase transition temperature. In addition, its reversible phase change processes would also allow for repetitive uses [2-3]. These advantages of the latent heat storage are of paramount importance to utilize the intermittent heat sources such as solar energy. Molten salts with melting points over 500°C and large latent heat was previously reported as a high-temperature phase change material (PCM) [4]. Molten salt has significant disadvantages in terms of its low thermal conductivity and high volume expansion. The low thermal conductivity of the PCM would lead to a low charging and discharging rate of the latent heat storage system. A high volume expansion ratio would make the design of the heat storage container, tank and heat exchanger difficult and decrease the heat storage density of the total system [5-6]. Therefore, molten salt is not the best PCM for high temperature applications, using metals or alloys as PCMs could be an innovative option for high-temperature latent heat storage. As a result, the use of metallic PCMs has now attracted great attention as an alternative to molten salts. Development of new metallic PCMs [7–8] and design of systems with these PCMs have been reported.

In this paper, an Al-12Si alloy was selected as the phase change material due to its advantages of thermal stability, slow decay, steady temperature change and little volume change during the heat storage and release processes. A solidification/melting model is developed to investigate the charging and discharging process of high-temperature phase change thermal energy storage container used in solar thermal power generation using commercial software FLUENT. The effects of the Fourier number, Stefan number and Reynolds number on the heat storage container and storage properties is examined in a systematic manner. In addition, the heat transfer criterion is fitted and the obtained results will be beneficial to improve the performance of heat storage device as well as its storage efficiency.

---

[*]Corresponding author

2. The Heat Transfer Model of Heat Storage Container

2.1. *Physical model for heat storage container*

Fig. 1 shows the schematic diagram of the solar thermal power generation used by the heat storage container, which is mainly composed of PCM storage unit. The length, width and height of the container are 1200mm, 970mm and 650mm, respectively. The inlet is on the lower left side while the outlet is on the upper right side. The inlet and outlet pipes are tubes with 100mm in length and 60mm in radius. The heat storage container is packaged by a 20 PCM storage unit. The storage unit adopts composite materials of ceramic steel, the radius and height are 90 mm and 600 mm respectively, and the heat transfer fluid is hot air. Thermophysical properties of Al-12Si are shown in Table 1[9].

Analysis was performed based on the following assumptions [7]:
1) Phase change material is isotropic;
2) Specific heat capacity, thermal conductivity, density of the phase change material is constant;
3) Heat loss due to the outer surface and the wall thickness effect is ignored;
4) Meet Boussinesq assumption, and consider the fluid density changes only in the buoyancy items;
5) The phase change material is melted incompressible Newtonian fluid;
6) Consider the impact of natural convection, and the natural convection is laminar flow

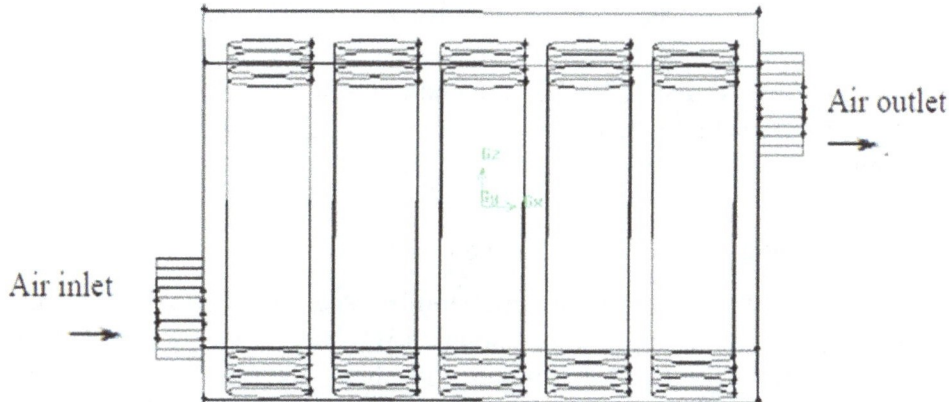

Fig. 1 The heat storage container used in the solar thermal power generation

Table 1 Thermophysical properties of Al-12Si

Type	component	phase change temperature (K)	Solution heat (kJ/kg)	density (kg/m³)	specific heat (kJ/(kg*k))	Thermal conductivity (W/(m*k))
PCM	Al-12Si	850	560	2700	1.148	160

2.2. *Mathematical model for heat storage container*

In the current study, the solidification/melting model with the enthalpy- approach [10] will be taken into account. The governing equations for the PCM–air system are:

$$\frac{\partial(\rho H)}{\partial t} + \nabla(\rho \vec{v} H) = \nabla(\lambda \nabla T) + S \tag{1}$$

$$H = \beta L + h_{ref} + \int_{T_{ref}}^{T} c_p dT \tag{2}$$

where ρ is the density, kg /m3; λ is the thermal conductivity, W / (m * k); H is the specific enthalpy, kJ / kg; href is the standard enthalpy, kJ / kg; L is the latent heat of phase change, kJ / kg; t is the phase change time, s; T is the temperature, K; Tref is the reference temperature, K; Cp is the constant pressure specific heat, kJ / (kg *K).

β is the liquid fraction, in the process of phase change, the value of β changes between 0 and 1. When the temperature of the PCM is less than the melting temperature, the value of β is 0, and PCM becomes solid phase. When the temperature of the PCM is equal to the melting temperature, the value of β is between 0 and 1, and PCM becomes solid-liquid two-phase. When the temperature of the PCM is higher than that of melting temperature, the value of β is 1, and PCM becomes liquid phase.

For the purpose of clearness, the following dimensionless parameters are introduced [10-11]:

$$Fo = \frac{\alpha t}{l^2} \quad Ste = \frac{C_p |T_{in} - T_m|}{L} \quad Re = \frac{ul}{v} \quad Pr = \frac{v}{\alpha} \quad Nu = \frac{hl}{\lambda}$$

where α is the thermal diffusivity, m2/s; l is the characteristic length, m; Tin is inlet temperature, K; Tm is the average phase change temperature, K; v is the kinematic viscosity, m2/s; h is the heat transfer coefficient of convection heat surface, W / (m * k).

3. The Numerical Simulation of Heat Storage Container

In the current study, the heat storage container model is established and the mesh is generated using GAMBIT software. A three-dimensional (3D) separated unsteady solver with the solidification/melting model is used to simulate the phase change process. For the heat storage container, the governing equation is 3D transient heat conduction equation. Boundary conditions are set as follows: the velocity inlet boundary and free outflow boundary that pressure gradient is zero are adopted, the heat storage unit is set as coupling interface, and the heat storage container as adiabatic, while the boundary is set as Heat flux to select a default. PCM melting phase is calculated by using Enthalpy-porosity, the coupled pressure and velocity fields are processed by applying Simple algorithm. Along the direction of time, difference scheme uses first order implicit scheme. In the iterative calculation process, appropriate adjustments to the relaxation factor and the time step can ensure stable convergence in the maximum number of iteration steps [11].

3.1. *Effect of Fourier number on heat storage performance*

Fig. 2a shows the effect of Fo on the liquid fraction in the heat melting process with Re is 33523, St is 0.1377 and Fourier is 4.471. Fig. 2b shows that the PCM liquid fraction varies with Fourier in solidification process with Re is 21384.4 and St is 0.3278, while the liquid fraction and Fourier number have the relationship of a decreasing function, i.e. the liquid fraction decreases when the Fourier number is 0.774. The sensible heat, while the solidification process requires a shorter process to reach the phase change temperature.

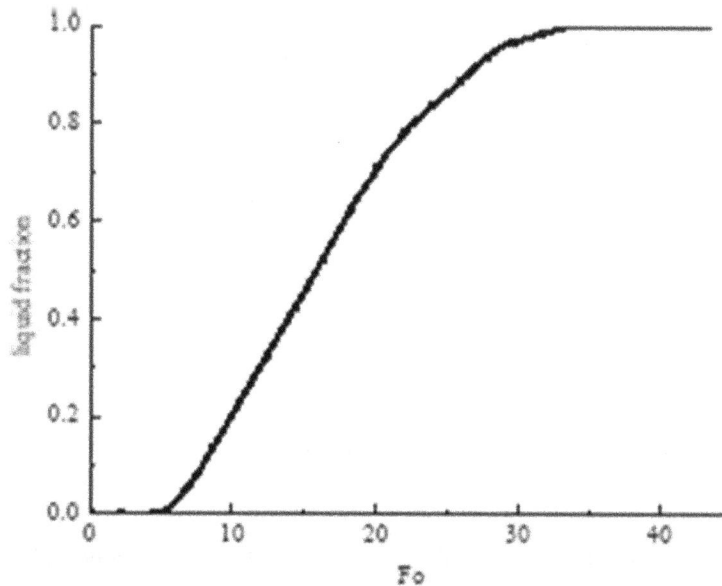

Fig. 2a Effect of Fo on the liquid fraction in the heat melting process

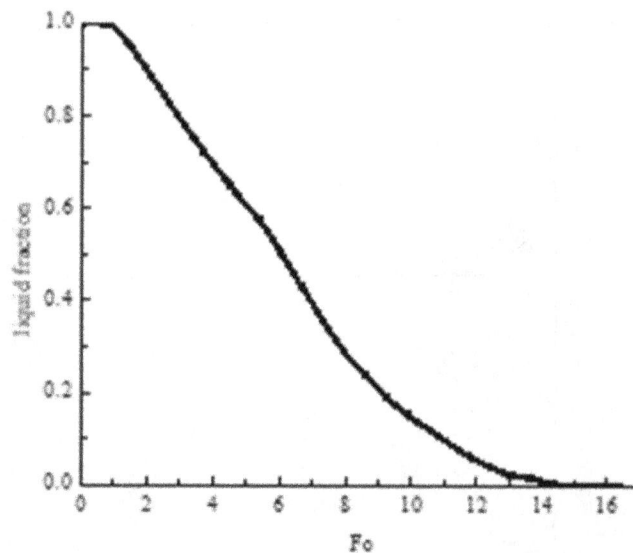

Fig. 2b Effect of Fo on the liquid fraction in the solidification process

And with the change of Fourier, the change rate of liquid fraction in the two processes have obvious varieties, while Fourier less 6, liquid fraction is 0, while Fourier in the range of 6-32, liquid fraction to speed up rapidly, while Fourier more than 32, liquid fraction nearly not change. This is because of heat conduction in the heat storage effect gradually weakened, natural convection heat transfer strengthened, and the transformation rate accelerated.

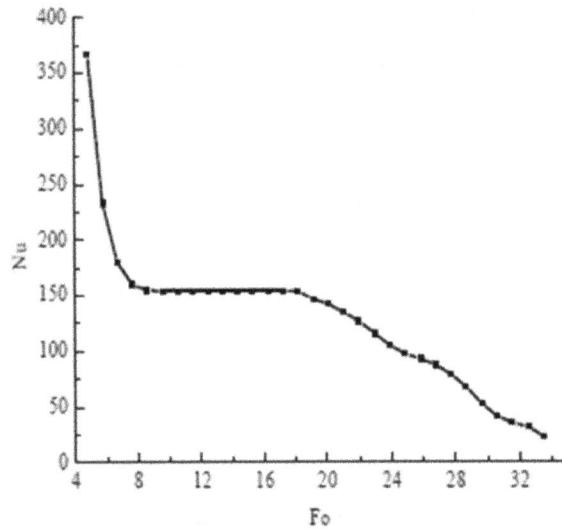

Fig. 3a Effect of Fo on the average Nu in the melting process

Fig. 3a and Fig. 3b demonstrate the change of the average Nusselt with Fourier during the process of melting and solidification process. It is noted that the Nusselt performs a sharp decreasing and constant tendency followed by a gentle decreasing tendency with the increase of the Fo. This could be the reason that the initial state is in the maximum unbalanced state, and the influence of the initial state distribution is gradually weakened, then the initial state is in a nearly stable phase change process.

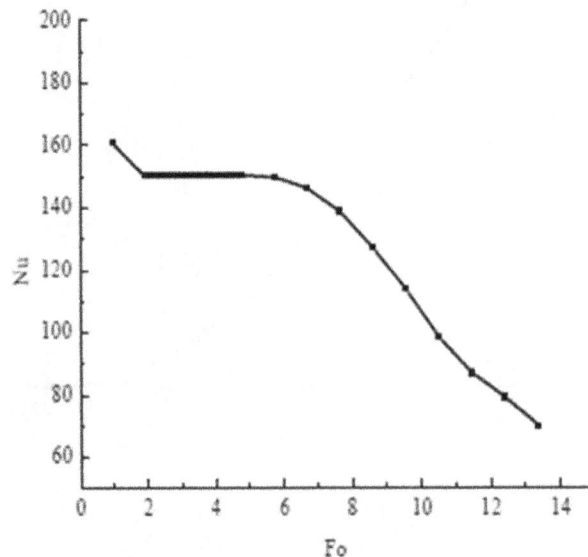

Fig. 3b Effect of Fo on the average Nu in the solidification process

3.2. *The effect of Reynolds number on the performance of heat storage*

Fig. 4a shows the variation of the change of the melting rate of PCM with Fourier under different Reynolds at Stefan is 0.1377.Furthermore, it also shows that the melting time of PCM is shortened with the increment of the Reynolds. When Reynolds is 50284.8, PCM first begins to melt, and the time is the shortest. When Reynolds is 16761.6 and Fourier is 50.122, PCM melts completely, while PCM has melted completely when Reynolds is 50284.8 and Fourier reduced by 52.4% is 26.167. Meanwhile, Fig. 4b shows the melting rate of the PCM with Fourier under different Reynolds when Stefan is 0.3278. When the Reynolds increases, the solidification time of PCM is shortened.

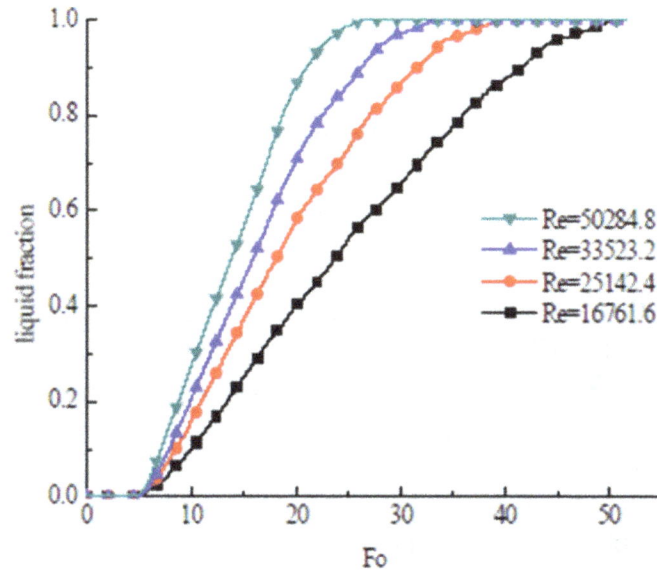

Fig. 4a Effect of Re number on the liquid fraction in the melting process

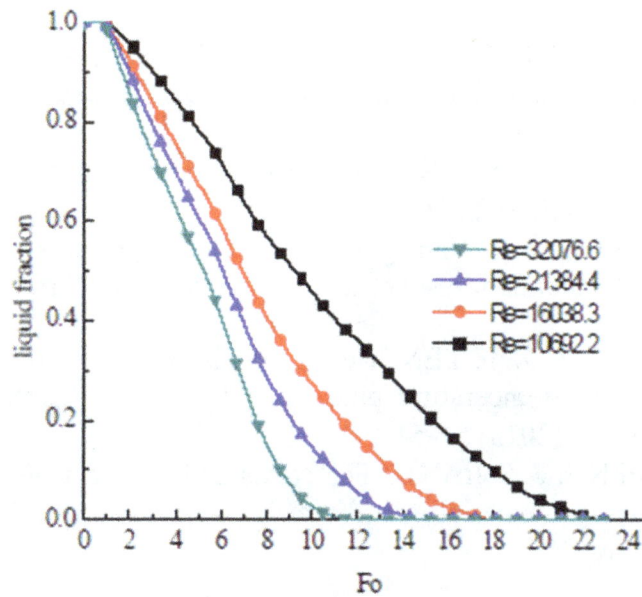

Fig. 4b Effect of Re number on the liquid fraction in the solidification

4. Conclusion

In this paper, a high temperature phase change thermal storage device of solar thermal power generation system is studied numerically.

1. The phase transformation processes of different Fo, Ste and Re were numerically analyzed by using FLUENT solidification/melting model.
2. The influence of the heat storage performance was obtained, and the change law was summarized.
3. It provides an important reference for the future research and application of thermal storage device into solar thermal power generation systems.

Acknowledgment

This work is funded by Natural Science Foundation of Hebei Province (E2014208005), Corresponding Author

References

1. Y.T. WU, L.N. ZHANG, C.F. MA. High temperature heat storage technology in solar thermal power, Solar Energy. 3 (2007) 23-25.
2. H.T. CUI, Z.H. WANG, Y.S. GUO, et al. Experimental study on heat performance of new phase change thermal energy storage unit, Acta Energiae Solaris Sinica. 30（2009）1188-1192. "In Chinese"
3. H.T. CUI, G. ZHANG, J.Z. JIANG. Numerical simulation on melting and solidification process of aluminum-silicon alloy,Journal of Hebei University of Science and Technology. 33(2012) 453-458. "In Chinese"
4. Z.Y. ZUO. Thermal process characteristics on high-temperature thermocline hybrid thermal energy storage with molten salt materials．South China University of Technology, 2010. "In Chinese"
5. G.S. CHEN, R.Y. ZHANG, F. LI, et al. The numerical simulation of the Cylindrical aluminum silicon alloy heat storage body at horizontal, ActaEnergiae Solaris Sinica. 33 (2012) 2093-2097.
6. R.Y. ZHANG, J.Q. SUN, X.F. KE, et al. Heat storage properties of Al-Si alloy, Journal of Materials Research. 20 (2006) 156-160.
7. N. Gokon, S. Nakammura, T. Yamaguchi, et al. Cyclic properties of thermal storage/discharge for Al-Si alloy in vacuum for solar thermochemical fuel production, Energy Procedia. 69 (2015) 1759-1769
8. J.F. CHEN, X.M. CUI, X.Y. PENG, et al. Research on the microstructure and property of aluminum-silicon high-temperature phase transformation thermal storage materials, Hunan Nonferrous Metals. 29 (2013) 51-54．
9. H.T. CUI, P.Y. PENG, J.Z. JIANG. The status and prospect on Al-Si alloy and heat storage unit as phase change material for thermal energy storage, Materials Review.28 (2014) 72-75．"In Chinese"

10. Z.Z. Han, J. Wang, X.P. LAN. The engineering examples and applications of the FLUENT fluid simulation calculation. Beijing: Beijing Institute of technology press, 2004. "In Chinese"

11. C.X. GUO, H.D. XIONG, X.J. WEI, et al. Numerical research on solidification of cooling storage ball by FLUENT, Energy Conservation Technology. 23 (2005) 488-491.

Study on the Change of the Energy Reflectivity with the Incident Angle for Multilayer Optical Anti-counterfeiting Film

An-Ling Wang, Fu-Ping Liu[*]

Beijing Institute of Graphic Communication, Beijing 102600 China

[*]Email: fupingliu60@sina.com

Based on the inconsistency of inhomogeneous wave between the propagation direction of phase shift and the amplitude attenuation direction in conductive medium, by the solution of the light wave equation, with the boundary conditions of electromagnetic field, the recurrence formula of optical wave propagation has been derived for non-uniform vertical polarized light in composite film of multilayer ultrathin metal and transparent medium. We have also given the calculation examples of the energy reflectivity for 12 layers with different thickness of metal film under different wavelength, analyzed for each calculation example, and obtained the different incident angles that can be seen in the same color, which provide a theoretical basis for the printing of the optical variable ink.

Keywords: Multilayer anti fake film; energy reflectivity; incident angle.

1. Introduction

The printing of color changing anti fake ink has gradually become an important part and means of modern society [1-4], because the color printing ink has a dynamic color effect, has a very good concealment, when the angle of view changes, it is obvious that the color of the original pattern has changed, which have an intuitive with an eye to be able to recognize [7-9], and cannot be copied by color copying machine and electronic scanner. Because of the ink manufacturing complex, high investment, difficult to counterfeit, with anti-counterfeiting effect is very strong [7-13], so the printing technology of color anti-counterfeit ink has important applications in banknotes, checks, such as bonds and stocks demanding high securities to prevent forgery. In order to improve the narrow band high reaction of optical thin film, the thin metal film and the transparent film are used to form a multilayer composite film system [1-4], which has a film of high conductivity in this film. Due to propagation of the light in conductive medium, the wave vector changed into complex wave vector [8-11], in most cases the amplitude attenuation direction and phase shift direction is inconsistent. In recent years, people began to pay attention to the propagation characteristics of inhomogeneous waves, not only gives the schema that the amplitude attenuation direction and phase shift propagation direction of the light is not consistent on the interface of the reflection and refraction, but also gives the quantitative relationship between the phase shift constant and the attenuation constant [7-11]. In this paper, considered the propagation properties of the general inhomogeneous wave, we have derived the equation of the wave propagation in the hybrid film of the metal film and the transparent film, given the variation curves of energy reflectivity with incident angle, analyzed the characteristics of the curve, and gotten some new rule understanding

2. Energy Reflectivity of Multilayer Anti-False Film

The optical film is composed of $J+1$ layer, by the wave field recurrence formula of multilayer film, we have [12, 13]

$$\begin{pmatrix} E_j^+ \\ E_j^- \end{pmatrix} = \begin{pmatrix} E_1^+ Q_{j,11}/Q_{11} \\ E_1^+ Q_{j,21}/Q_{11} \end{pmatrix} \tag{1}$$

---

[*]Corresponding author

Here, $Q_j = M_j^{-1} N_j \cdots M_J^{-1} M_J$, $M_j = \begin{bmatrix} e^{-ik_{jz}z_j} & e^{ik_{jz}z_j} \\ \xi_j e^{-ik_{jz}z_j} & -\xi_j e^{ik_{jz}z_j} \end{bmatrix}$,

$$N_j = \begin{bmatrix} e^{-ik_{j+1z}z_j} & e^{ik_{j+1z}z_j} \\ \xi_{j+1} e^{-ik_{j+1z}z_j} & -\xi_{j+1} e^{ik_{j+1z}z_j} \end{bmatrix}, \xi_j = \frac{1}{\sqrt{\alpha_j - i\beta_j}} \sqrt{\frac{\varepsilon_j}{\mu_j}} (\mathbf{k}_j^+ \cdot \mathbf{n}).$$

Where $\boldsymbol{\alpha}_j$ is the phase shift constant vector of uniform plane wave incident, $\boldsymbol{\beta}_j$ is the attenuation constant vector, ε_j and μ_j are the dielectric permittivity and permeability of j-th layer medium, respectively, $\mathbf{k}_j = \boldsymbol{\alpha}_j - i\boldsymbol{\beta}_j$ is the incident wave complex wave vector, z is the coordinates for the z direction. The field in each layer can be represented by the incident field E_1^+ in the first layer, here E_1^+ and E_1^- respectively represent the wave of the forward and backward propagation.

Equation (1) can be used to calculate the reflection transmission coefficient of each layer and the total reflection coefficient of multilayer films, so by equation (1) we can obtain the total amplitude reflection coefficient of a multilayer film

$$C_r = \frac{E_1^-}{E_1^+} = \frac{Q_{21}}{Q_{11}} \tag{2}$$

Amplitude transmission coefficient of multilayer film

$$t = \frac{E_{J+1}^+}{E_1^+} = \frac{1}{Q_{11}} \tag{3}$$

Light energy reflectivity of multilayer films

$$R = C_r \cdot C_r^* = \frac{T_{21}}{T_{11}} \cdot \left(\frac{T_{21}}{T_{11}}\right)^* \tag{4}$$

3. The Example Analysis

As a result of the general situation about the medium parameters are given in the form of index of refraction [12, 13]. In order to calculate conveniently we give the calculating method of refraction index of the metal film. For general non-magnetic optical waveguide $\mu \approx \mu_0$, the refractive index of the light is defined as

$$n = \sqrt{\varepsilon_r} = \sqrt{\frac{\varepsilon}{\varepsilon_0}}$$

Due to complex permittivity $\varepsilon^* = \varepsilon - \frac{i\sigma}{\omega}$

Then the complex refractive index is

$$n = \sqrt{\frac{\varepsilon^*}{\varepsilon_0}} = \sqrt{\frac{\varepsilon}{\varepsilon_0} - \frac{i\sigma}{\omega\varepsilon_0}} = \xi - i\tau$$

So the dielectric constant and refractive index of dielectric permittivity can be calculated by the complex refractive index, where σ is the medium conductivity, ω is the angular frequency of the light wave.

The following membrane system is A/yMxL(LH)^P(LH)^PzLdN/G, a 12-layer system, where A medium is one of the most top, G is the lowest medium, M represents a chromium film, N denotes the silver film, L and H are the low and high refractive index dielectric film. The number in front of the letter indicates the thickness of the dielectric film (a multiple of the center wavelength of the film).

In this film system, $n_H = 2.35$, $n_L = 1.32$, $P = 2$, $n_g = 1.52$, $n_A = 1.52$. Chromium film thickness $d_{cr} = [2nm, 4nm, 7nm]$, Ag film thickness $d = 100nm$, the central wavelength of 520nm, wavelength $\lambda = [516.5nm, 460nm, 480nm]$. The conductivity of chromium film $\sigma = 7.752 \times 10^6 s/m$, Electrical conductivity of silver film $\sigma = 6.06 \times 10^7 s/m$, Chromium film complex refractive index $n_{cr} = 2.91 - 3.33i$, complex refractive index of silver film $n_{ag} = 0.13 - 3.07i$. The energy reflectivity curve of the film system varies with the incident angle are given from Figure 1 to Figure 3, where $d_{cr} = [2nm, 4nm, 7nm]$, $\lambda = [516.5nm, 460nm, 480nm]$. The figures show that the thicker the chromium film, the smaller the energy reflectivity.

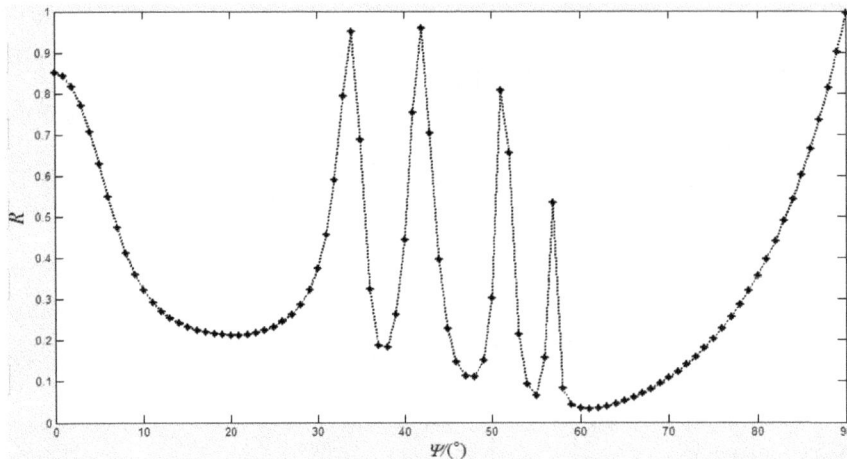

Figure 1 The variation of the energy reflectivity of multilayer optical film with the incident angle
$(d_{cr} = 2nm, \lambda = 516.5nm)$

When the main wavelength is $\lambda = 520nm$ the variation curve of the energy reflectivity with the incident angle is shown in Figure 1 for the different thickness of the 12 layer film, where the Chromium film thickness $d_{cr} = 2nm$. From the figure we can find that the curve has 6 peaks, and the peak value becomes small at $\psi = 53°, 58°$, where the peak value points are corresponding to the incident angles $0°$, $37.6°, 42.5°, 53°, 58°, 90°$. This shows that in those the reflection angles we can see the same color reflected light, there is a strong reflection on the light of main wavelength.

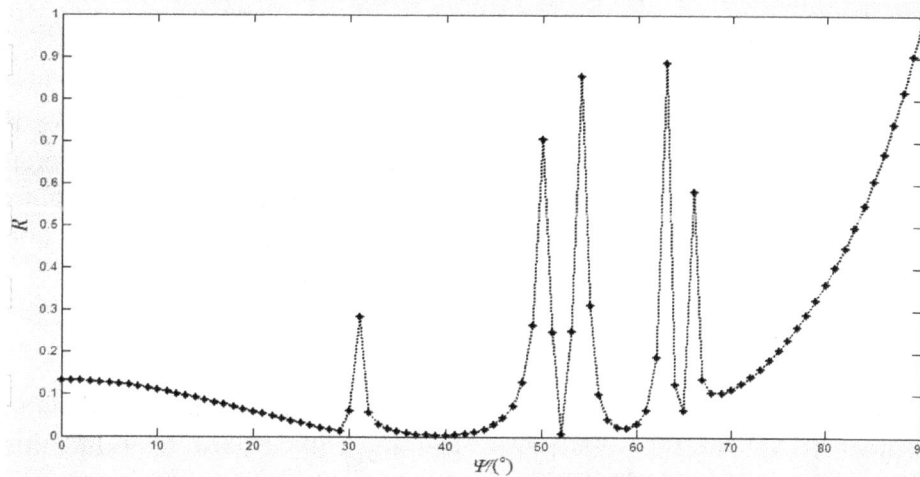

Figure 2 The variation of the energy reflectivity of multilayer optical film with the incident angle
$(d_{cr} = 4nm, \lambda = 460\text{nm})$

Figure 2 shows the variation curve of the energy reflectivity with the incident angle for the different thickness of the 12-layer film, where the Chromium film thickness $d_{cr} = 4nm$. The curve has 7 peaks, and the peak value becomes small at $\psi = 55°, 63°$, where the peak value points are corresponding to the incident angles $0°$, $32°$, $50°$, $54°$, $64°$, $67°$, $90°$.

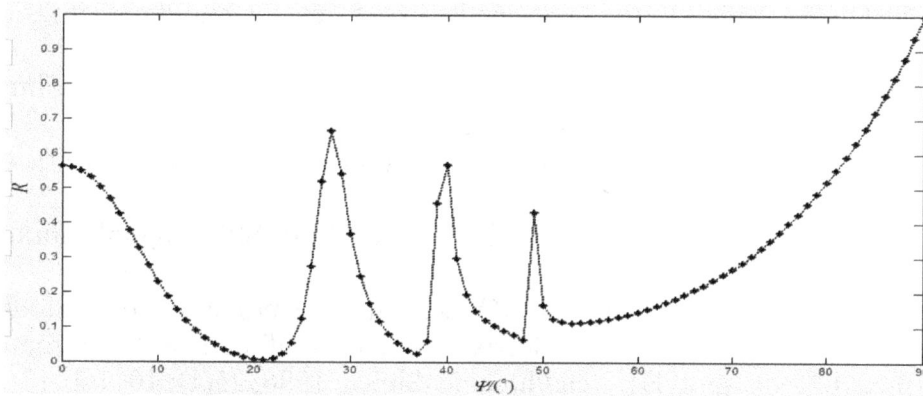

Figure 3 Variation of the energy reflectivity of multilayer optical film with the incident angle $(d_{cr} = 7nm, \lambda = 480\text{nm})$

Figure 3 gives the variation curve of the energy reflectivity with the incident angle for the different thickness of the 12 layer film, where the Chromium film thickness $d_{cr} = 7nm$. The curve has 5 peaks, where the peak value points are corresponding to the incident angles $0°, 30°, 40°, 50°$ and $90°$. The peak value becomes small at $\psi = 30°$, $40°$.

4. Conclusion

The considered inconsistency between the direction of the phase shift and the amplitude attenuation for the inhomogeneous optical wave, with the boundary conditions of electromagnetic field, we have derived the recurrence formula of optical wave propagation for non-uniform vertical polarized light in the composite film of multilayer ultrathin metal and transparent medium, and given three numerical calculating examples for the energy reflectivity of multilayer optical anti-counterfeiting film, the results show that the thicker the chromium film, the smaller the energy reflectivity; energy reflectivity curves have multi-peak points in different angles, This shows that in those the reflection angles we can see the same color reflected light, on there, there is a strong reflection for the light of main wavelength. This paper provides a theoretical basis for us to study the appeared color of the printing of the optical variable ink.

Acknowledgments

This work was funded by Funding Project for the Beijing City Board of Education Science and technology project (KM201510015009); The Beijing City Board of Education Science and technology key project (KZ201510015015).

References

1. ZHANG Jing-fang, LIU Li-min, YE Ye-dong, et al. Optical anti-counterfeiting technology and its application [M]. Beijing: National Defense Industry Press, 2011: 175-216.
2. ZHANG Yi-xin, CHEN Jie, SHI Gong-cheng. Color Prediction Model of Optically Variable Anti-forgery Ink [J]. Packaging Engineering, 2011, 32(5): 94-96.
3. WANG San. Theoretic designing and its optical characteristics of multilayer high reflectivity film [J]. Science & technology information, 2008, 32:148.
4. YU Hua, SHONG Li-min. A designing method of X-ray for multilayered reflection film [J]. Laboratory Science, 2010, 13(2): 42-44.
5. YIN Zhong-wen, XU Ai-hua. Calculation of the reflection index in optics film [J]. Journal of Nanyang NormalUniversity, 2007, 3(6): 24-27.
6. PHILIPS R W, BLEIKOLM Λ F. Optical Coatings for Document Security [J]. Applied Optics, 1996, 35(28): 5529-5534.
7. BLEKOLM A F. New Design Opportunities with OVI [J]. SPIE Digital Library, 1998, 3314: 223-230.
8. TAN Man-qing, LIN Yong-chang, ZHAO Da-zun. The properties of periodic symmetrical coatings and the Design of high reflectivity coatings of narrow band for containing the ultra-thin metallic film (n=k) [J]. ActaPhotonicaSinica, 1996, 25(1): 1011-1017.
9. TANG Jin-fa, GU Pei-fu, LIN Xu. Modern Optic technology [M]. Hangzhou: Zhejiang University press, 2006
10. PHILIPS R W. Paired Optically Variable Article with Paierd Optically Variable Structures and Ink, Paint and Foil Incorporating the Same and Method: US, 5766738 [P] 1998:06-16.
11. TAN Li, XIANG Qian-yong, LIU Yu-lin. Theoretical and experimental studies of color correction in laser digital photo finishing [J]. Acta Photonica Sinica, 2004, 33 (6) , 765-767.
12. XU Jian-hua. The Electromagnetic Field and Electromagnetic Wave in Layer Media [M]. Beijing: Petroleum Industry Press, 1997: 38-59.
13. WANG An-ling, LIU Fu-ping, ZHU Xiao-feng, YANG Chang-chun. The Reflection Properties of Multilayer Anti-counterfeiting Optical films for Vertical-polarized wave ACTA PHOTONICA SINICA. 2014, 43(8): 1002-1~8.

Research on Three Types of Spring Mattresses Burning Behavior

Xia Zhao[1, 3], Jia-Qi Huhang[2], Ai-Hua Yi[1, 3,*], Jian-Yong Liu[1, 3]

[1]Guangdong Province Enterprise Key Laboratory of Materials and Elements Fire Testing Technology, China
[2]Rensselaer Polytechnic Institute, Troy, NY, US
[3]Guangzhou Building Material Institute Limited Company, Guangdong, Guangzhou, China
*Email: yiaihua92751@163.com

Through the investigation on flame-retardant properties of three kinds of inner-spring mattresses, it is found that without flame retardant, the heat release rate (HRR) of spring mattress is up to 600 kW, the concentration of carbon monoxide (CO) and carbon dioxide (CO_2) is 0.08% and 1.5% respectively. Thus, mattress has a very high fire risk. It could cause injuries and deaths through spreading of fire and emitted gases. The mattress with only fabric treated with flame retardant could be ignited, but did not cause the flame spread across a large area, the flame retardant performance satisfied grade B_1 in GB 8624-2012. For the mattress which has both fabric and filler being flame retardant, the flame extinguished soon after the ignition sources were removed. Therefore, flame retardant performance for the last mattress was very well.

Keywords: Mattress; flame retardant; combustion.

1. Introduction

Most of mattresses are made in a large size by foam and sponge. Mattress without flame-retardant treatment is very flammable, it could release large amount of heat and lead to fire spreading. In recent years, fire caused by mattress burning happened more frequently, and it's highly dangerous to life and property [1, 2]. In general, spring mattress is composed of three major components: mattress frame (spring), filler, and fabric. Among all components, filler and fabric lack of flame-retardant ability in particular. In this research, three types of mattresses including mattress with no treatment of flame-retardant, mattress with flame-retardant fabric, mattress with flame-retardant fabric and filler were chosen to be tested by following the standard GB 8624-2012. The oxygen index method and vertical method, which is in 5.2.2 of GB 8624-2012, was followed for testing fabric; the oxygen index method was also followed for testing filler; The open flame burning test stated in appendix A of GB 8624-2012 was followed for testing the whole mattress [3].

2. Flame Resistance Analysis of Filler and Fabric

2.1. Instruments and methods

Three samples used in this experiment are mattress with no flame retardant treatment (mattress 1), mattress with flame retardant fabric (mattress 2), mattress with flame retardant fabric and filler (mattress 3). The filler of these three mattresses is made of polyurethane foam, and the fabric cover is made of cotton.

The oxygen method experiment was carried by following GB 5454-1997 assigned in GB 8624-2012, the oxygen index meter was made by Nanjing Jiangning Apparatus Factory; The experiment about vertical burning properties was carried by following GB 5454-1997 assigned in GB 8624-2012, the horizontal and vertical burning tester was also made by Nanjing Jiangning Apparatus Factory.

Table 1 below collects flame retardant properties of filler and fabric for all types of spring mattresses, it shows that the fabric of mattress 2 and 3, which is treated with flame retardant, both

---

*Corresponding author

has an limit oxygen index greater than 32.0%, damaged lengths of vertical burning are also less than 150mm, afterflame time and afterglow time are all less than 5s. Therefore, these two fabrics satisfies grade B₁ in GB 8624-2006 for their fair flame retardant properties. For mattress without treatment of flame retardant (Mattress 1), the limit oxygen index of its filler and fabric are18.8% and 21.3% respectively.

Table 1 Flame retardant properties of filler and fabric for three types of spring mattresses

	Mattress 1		Mattress 2		Mattress 3	
	Filler	Fabric	Filler	Fabric	Filler	Fabric
Limit Oxygen Index(%)	18.8	21.3	19.1	32.2	24.5	32.8
Damaged Length（mm）	—	256	—	35	—	31
Afterflame Time（s）	—	1.2	—	0	—	0
Afterglow Time（s）	—	5.6	—	0.5	—	0.7

This mattress can burn steadily in air, indicating a high fire hazard, i.e. high possibility to cause a fire. For mattress 2, filler's limit oxygen index is only 19.1%, meaning it can barely resist burning. But just as mentioned before, fabric of mattress 2 has good flame retardant properties. For mattress 3, the limit oxygen index of filler is 24.5%, indicating a partial flame retardant ability. Same as fabric of mattress 2, fabric of mattress 3 also has a good flame retardant properties. Further experiments of burning the entirety of mattress 2 and mattress 3 are required to see if they satisfy grade B₁ in GB 8624-2012.

3. Analysis of Open Flame properties of Entire Mattresses

3.1. *Apparatus and methods*

The open flame flammability test was carried by following methods provided in Appendix A, GB 8624-2012. To be more specific, two propane burners was placed on the top and by the side of mattress, they were ignited simultaneously. The upper burner continued burning for 70s, while the side burner also continued burning for 50s. During the experiment, data such as composition of smoke and gases, HRR, etc. were recorded. The furniture calorimeter used in the experiment was made by Fire Testing Technology Ltd.

3.2. *Analysis of testing results*

3.2.1. *Heat Release Rate (HRR)*

According to GB 8624-2012, if the peak HRR of mattress is less than 200kW within 1800s, the mattress will satisfy grade B₁ in the standard. After the combustion experiment was carried for about 550s, flame on mattress 1 burst into big fire because of absence of fire retardant. Considering the safety of experimenters and apparatus, the experiment was ended by extinguishing the fire manually. Fig. 1, Fig. 2, and Fig. 3 below are three spring mattresses after experiment, Fig. 4 is the curves of HRR during experiment.

Fig. 1 Mattress 1

It can observe fire spread continually after the burners were removed. At about 450s, the flame expanded rapidly and soon burst into big fire, at the same time, the peak value of HRR also increase to 600kW drastically, which is way larger than 200kW, which is required in GB 8624-2012. So without treating with flame retardant, mattress could be a high fire hazard for being burnt easily, and soon transform into big fire easily.

Fig. 2 Mattress 2

For mattress 2, the fire was still present after the burners were removed, the flame spread along the fabric and kept expanding. However, since the fabric was treated with flame retardant, and has a limit oxygen index of 32.2%, it protected the filler from being burnt. After about 800s, fire expanded gradually, its HRR reached to around 58kW at 1800s, which is well below 200kW specified in GB 8624-2012.

Fig. 3 Mattress 3

For mattress 3, since both fabric and filler were treated with flame retardant, the fire went off soon after the burners were removed. The HRR is 0, indicating that the mattress has a good resistibility to fire.

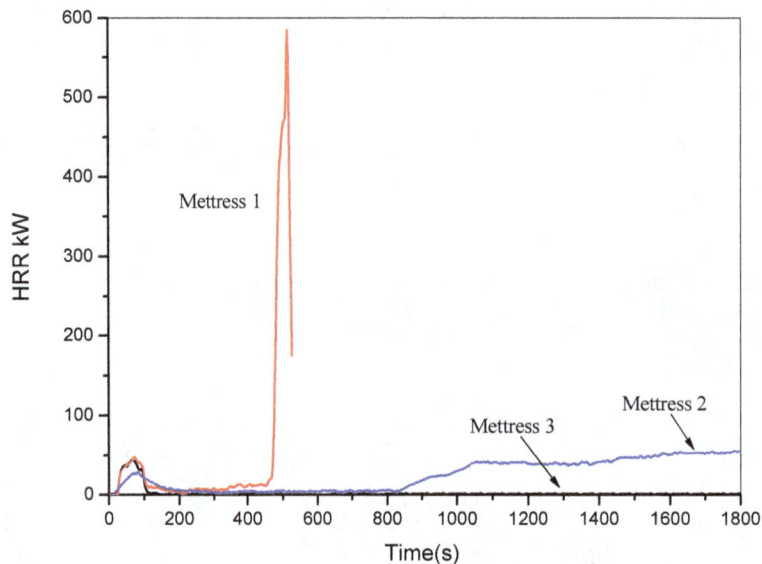

Fig. 4 HRR curves of three mattresses

3.2.2. CO and CO2

Previous research shows that most of the injuries and deaths in the fire results from the toxicity of smoke. According to relative animal experiment, $2300 \sim 5700 mg/m^3$ of CO, or converting to volume concentration, $0.184 \sim 0.456\%$ of CO could kill all the mice [4, 5]. Hence, CO could lead to deaths at a relatively low concentration. As required in GB 8624-2012, the sampling point was located in the exhaust pipe which was certain distance away from samples, so the concentration of the collected gases were diluted at some degree. Fig. 5 and Fig. 6 below are variation curves of CO and CO_2 in the combustion gases.

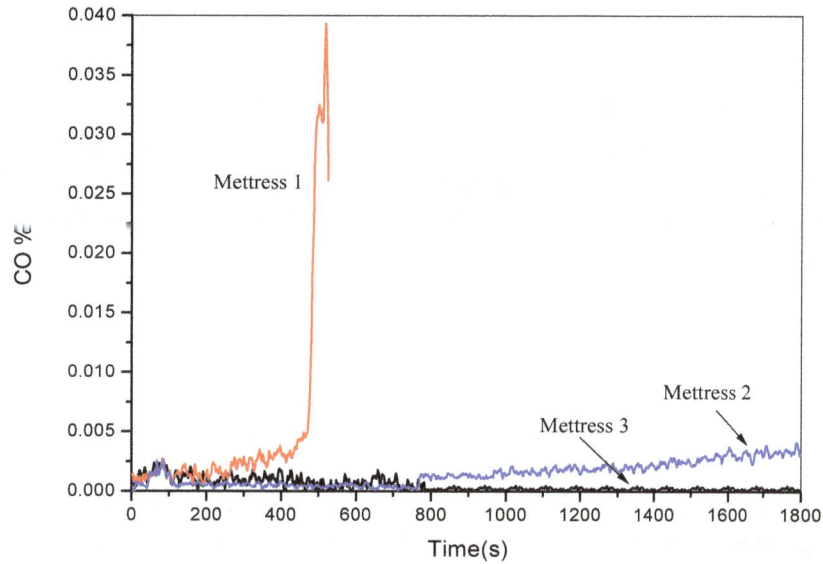

Fig. 5 Variation curve of CO in smoke

From Fig. 5, it can be observed that the highest value of CO is 0.08%, although this concentration doesn't reach to the lethal concentration of 0.184%, the actual concentration is greater than the recorded because of the setting of apparatus. Therefore, mattress 1 could be more dangerous in an actual fire. During combustion, CO concentration of mattress 2 is about 0.008%, which is only one tenth of CO concentration of mattress 1, indicating a much lower toxicity. Since mattress 3 was so retardant to fire, the CO concentration is almost 0. CO_2 in the combustion gases is not toxic, but it could cause anoxia at high concentration, people might be injured or dead for lacking the ability to escape from fire. From Fig. 6, peak value of CO_2 concentration reached to around 1.5% during combustion, again, the actual concentration is also expected to be higher, showing a high fire hazard. The peak values of CO_2 concentration for mattress 2 and mattress 3 are well less than that of mattress 1, so it can be concluded that CO_2 concentration of mattress 2 and 3 are too low to become a hazard.

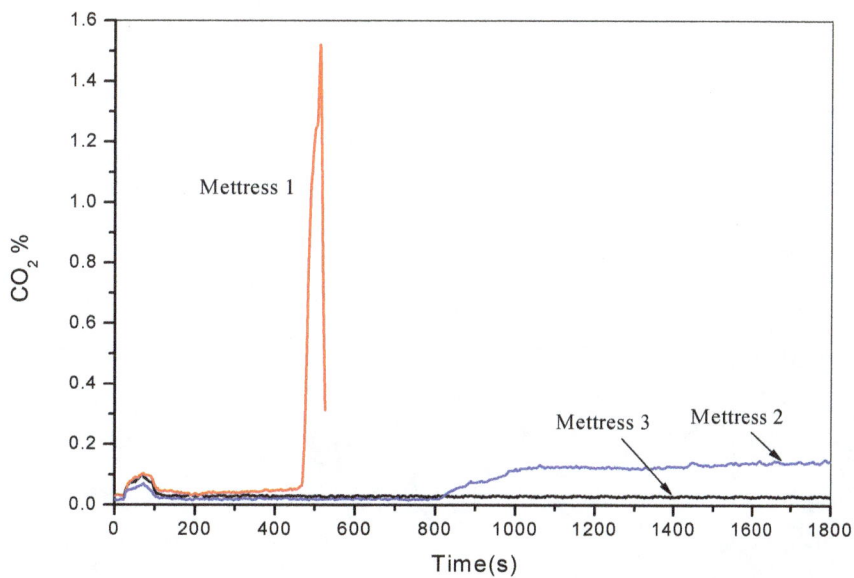

Fig. 6 Variation curve of CO_2 in smoke

4. Conclusion

This research shows that without treated with flame retardant, mattress has high fire hazard, it could be easily ignited and cause fire spreading. For mattress with flame retardant fabric, the fabric stopped fire from spreading to filler. Although the mattress was ignited, fire didn't spread across a big area, showing a good flame retardant properties that satisfy grade B_1 in GB 8624-2012. For mattress with both fabric and filler treated with flame retardant, fire extinguished by itself after the burners were removed, showing a very good flame retardant properties.

References

1. Jianyong Liu, Zhan Yang, Xia Zhao, et al. Summarization on the assessment method of upholstered furniture's combust ion performance. Fire Science and Technology. 28 (2009): 889-892.
2. Shilong Zou, Caisheng Yan, Suqin Tian, et al. Review on experimental regulations for combust ion performance of upholstered mattress. Fire Science and Technology. 27 (2008): 473-476.
3. Chenggang Zhao, The understanding of the standard 《GB 8324-2012 Fire classification of construction products and building elements. Thermal Insulations & Energy-Saving Technology. 2013 (5): 6-11.
4. Jie Zhao, Mingxue Zhu, Yiming Lu. The toxic gases and poisoning mechanism of the fire smoke [J]. Journal of Emergency Medicine. 13 (2004): 497-498.
5. Huang Rui, Yang Lizhong, Fang Weifeng, et al. Progress in study of Hazard Analysis of Fire Smoke [J]. Engineering Science. 4 (2002): 80-85.

Chapter 13
Simulation and Environment

Calculation of Allowable Stress for Ultrahigh Strength Sucker Rod

Yuan-Gang Xu, Fan-Fan Hao[*]

Xi'an Shiyou University, Xi'an, Shaanxi, 710065, China

[*]*Email: Fan_FanHao@126.com*

In recent years, the application of ultrahigh strength sucker rod has been increasing gradually in domestic oil fields, but its reasonable use is restricted by the lack of theoretical research on mechanical properties. By analyzing modified Goodman-stress diagram, the general formulas of allowable stress for ultrahigh strength sucker rod based on fatigue damage theory and API methods are given, which have considered the strength of sucker rod, loads bearing and working conditions. According to the results of tensile tests, a computing method of maximum allowable stress for grade HY and HL sucker rod is presented, which can provide theoretical references for optimal design of sucker rod string.

Keywords: Allowable stress; ultrahigh strength sucker rod; Goodman-stress diagram.

1. Introduction

The size of stress at any cross section of sucker rod changes over time when working. Its change is asymmetrically cyclic. Sucker rod will have fatigue failures when reciprocating motion, thus the designing always is on the basis of fatigue strength [1]. The API method, one of major ways used to design sucker rod string, requires that maximum stress less than the allowable stress when working. The maximum allowable stress is not an intrinsic property of sucker rod [2], but reflects its ultimate strength under different loading and working conditions, and its size relates to materials, strength, load bearing and working conditions, et al. As the important data for designing sucker rod string, the accuracy of maximum allowable stress affects the rationality of designing. In recent years, the application capability of ultrahigh strength sucker rod has been increasing gradually in domestic oil fields, but its rational use is restricted by the lack of theoretical research on mechanical properties. Hence computing methods of allowable stress for ultrahigh strength sucker rod will be studied here.

2. The Analysis of Modified Goodman-stress Diagram

Goodman had assumed that the safety fatigue limit was half (point D in Fig. 1) of tensile strength, indicated by T, when pulsating tensile cycle. And simplified Goodman linear, ADK, can be deemed as fatigue limit of sucker rod.

---

[*]Corresponding author

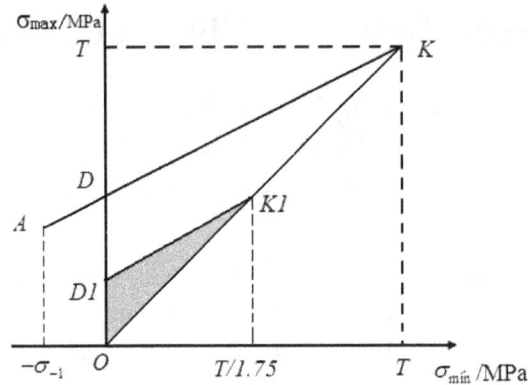

Fig. 1 Modified Goodman-stress diagram

According to the API method [3, 4], if safety factor is 2, maximum allowable stress is *T/4* (point *D1* in Fig. 1) when pulsating tensile cycle, and maximum allowable stress is *T/1.75* (point *K1* in Fig. 1) with static load. API specifies linear *D1K1* as maximum allowable stress linear with different stress radio.

Based on modified Goodman-stress diagram, considering the corrosive fluid effect, a computing formula of allowable stress for sucker rod is,

$$\sigma_{all} = \left(0.25T + 0.5625\sigma_{min}\right) \cdot \overline{SF} \tag{1}$$

Where σ_{all} is maximum allowable stress, T is ultimate minimum tensile strength, σ_{min} is minimum stress, $\overline{SF}$ is service factor shown in Table 1.

Table 1 The data of service factor

Rop type	Properties of working medium			Suitable condition
	Non-corrosive	Saline	H$_2$S-containing	
C	1.0	0.65	0.5	Light to medium load
K	1.0	< 1.0	< 1.0	Light to medium load
D	1.0	0.9	0.7	Medium to weight load
H	1.0	< 1.0	< 1.0	Weight to overweight load

API recommended the shadow area *OD1K1* in modified Goodman-stress diagram as safety operating area. If the stress point plots within area *OD1K1*, sucker rod will not have fatigue damage.

3. Calculation Methods of Allowable Stress for Ultrahigh Strength Sucker Rod

Mechanical properties of sucker rod, shown in Table 2, vary by its materials and processing technology [5], and the tensile strength and yield strength in the different grades are obviously different.

Table 2 Mechanical properties and materials of sucker rod

Rop type	Materials	Tensile strength (MPa)	Yield strength (MPa)
C	Quality carbon or alloy steel	621~793	≥ 414
K	Ni-Mo alloy steel	621~793	≥ 414
D	Quality carbon or alloy steel	793~965	≥ 586
KD	Ni-Mo alloy steel	793~965	≥ 586
HL	Alloy steel	965~1195	≥ 793
HY	Alloy steel	965~1195	-

Table 3 Foreign formulas of maximum allowable stress for different ultrahigh strength sucker rod

Rop type	Formulas of maximum allowable stress
EL	$\sigma_{all} = (379 + 0.2143\sigma_{min}) \cdot \overline{SF}$
S88	
T66/XD	$\sigma_{all} = (T/2.8 + 0.375\sigma_{min}) \cdot \overline{SF}$
HD	
UHS	

Note: EL, S88, T66/XD and HD are Weatherford's products, and UHS is Tenaris' products

4. General Formula of Allowable Stress for Ultrahigh Strength Sucker Rod

Let α be the radio of minimum tensile strength and yield strength. It is defined as Eq. 2:

$$\alpha = T / \sigma_s \tag{2}$$

and

$$\sigma_s = T / \alpha \tag{3}$$

Let K_0 be the safety factor when pulsating tensile cycle. According to API method, in stress diagram, point $D1$ will move to $D2$, coordinate for $(0, T/2K_0)$, and $OD2 = OD/K_0 = T/(2K_0)$. Similarly, the allowable stress is required less than yield strength in order to ensure the normal work of sucker rod, that is $\sigma_{all} \leq \sigma_s = T/\alpha$, so point $K1$ will move to point $K2$, coordinate for $(T/\alpha, T/\alpha)$. Finally this paper obtains linear $\overline{D2K2}$ called maximum allowable stress line of ultrahigh strength sucker rod is shown in Fig. 2.

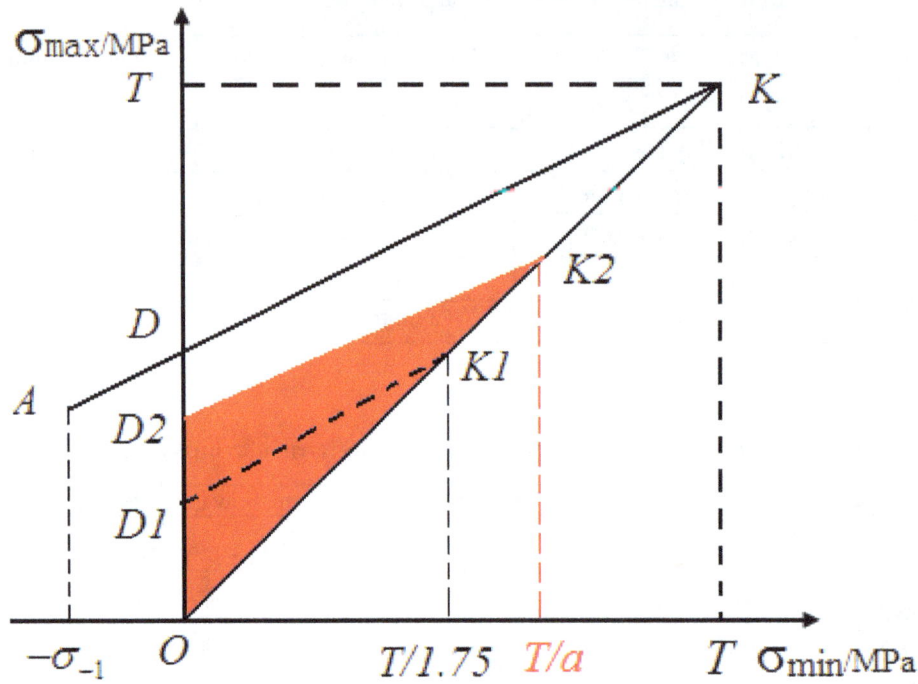

Fig. 2 Allowable stress diagram of ultrahigh strength sucker rod

The slope of linear $D2K2$ is,

$$\frac{T/\alpha - T/(2K_0)}{T/\alpha - 0} = 1 - \frac{\alpha}{2K_0} \tag{4}$$

And the equation of linear $D2K2$ is,

$$\sigma_{all} = \frac{T}{2K_0} + (1 - \frac{\alpha}{2K_0})\sigma_{min} \tag{5}$$

Considering fluid corrosion, the equation of linear $D2K2$ is,

$$\sigma_{all} = \overline{SF}\left[\frac{T}{2K_0} + \left(1 - \frac{\alpha}{2K_0}\right)\sigma_{min}\right] \tag{6}$$

Eq. 6 is the general formula of maximum allowable stress for ultrahigh strength sucker rod, and K_0, safety factor when pulsating tensile cycle, is 1.5~2.0 [8].

In order to get the value of α, we conducted tensile tests with ultrahigh strength sucker rod respectively produced by 5 manufacturers (A, B, C, D, E), and tests are done in the laboratory by electro-hydraulic servo universal testing machine. Results of tests are shown in Table 4.

Table 4 Tensile strength and yield strength of ultrahigh strength sucker rod from different manufacturers

Grade	Tensile strength (MPa)			Yield strength (MPa)		
	A	B	E	A	B	E
HL	1028	1099	1055	972	1007	995
	Average tensile strength is 1061 MPa			Average yield strength is 991 MPa		
	A	C	D	A	C	D
HY	1081	1085	1053	884	871	933
	Average tensile strength is 1073 MPa			Average yield strength is 896 MPa		

From Table 4, for grade HL, α=1.04; for grade HY, α=1.18. And considering general case, if K_0=2, $\overline{SF} = 0.8$,

For grade HL sucker rod,

$$\sigma_{all} = 206 + 0.592\sigma_{min} \tag{7}$$

For grade HY sucker rod,

$$\sigma_{all} = 211 + 0.564\sigma_{min} \tag{8}$$

5. Conclusion

Based on fatigue damage theory and results of tensile tests, this paper proposes reasonable general formula of maximum allowable stress for ultrahigh strength sucker rod, and provides theoretical references for its scientific use. However, it may result in a slight error because of experimental error.

Acknowledgments

This work was funded by Fund Project of Excellent Master Degree Thesis Cultivation of Xi'an Shi You University (2015yp140104), and Innovation Fund Project of Xi'an Shiyou University Graduate Students (2015cx140104).

References

1. Qi ZHANG. The Principle and Design of Production Engineering [M]. Dongying: Publishing House of China Petroleum University (east China), 2000 (In Chinese).
2. Jian-Ming GAO, Xiao-Min CHEN, et al. Material Mechanical Properties [M]. Wuhan: Publishing House of Wuhan University of Technology, 2003 (In Chinese).
3. Mao LIU. Analysis of Goodman-stress Diagram to Ensure Allowable Stress of Sucker Rod [J]. Oilfields Machinery, 1986, 15(2): 7-10 (In Chinese).

4. Rui-Dian LV, Ming XU. Modified Goodman-stress Diagram and Allowable Safety Factor [J]. Acta Petrolei Sinica, 2001, 22(6): 86-90 (In Chinese).

5. SY/T 5029-2013. Sucker rod [S]. Beijing: Publishing House of Oil Industry, 2014 (In Chinese).

6. http://www.weatherford.com/

7. http://www.tenaris.com/

8. Peng GAO. Mechanical Properties and Designing of Methods Grade H Sucker Rod String [D]. Xi'an: Xi'an Shi You University, 2014 (In Chinese).

A Hierarchical Genetic Algorithm for Sheet Metal Cutting Path Optimization

De-Zhong Qi[*], San-Qiang Zhang

Hubei Agricultural Machinery Engineering Research and Design Institute,
Hubei University of Technology, Wuhan, 430068, China
[*]Email: derek@mail.hbut.edu.cn

In order to ensure all contours are cut in turn and the cutting path is shortest in sheet metal cutting process, the cutting path optimization problem is proposed. It can be simplified to the cutting sequenced and the piercing point of the contour selected. An optimization model for the cutting path optimization problem is constructed, and a hierarchical genetic algorithm is developed for the better solution. A hierarchical structure with two chromosomes is designed in this algorithm. It can be used to simultaneously solve the cutting sequenced and the piercing point selected. The computational result and comparison prove that the presented approach is quite effective for the considered problem.

Keywords: Hierarchical genetic algorithm; path optimization; sheet metal cutting.

1. Introduction

Sheet metal cutting [1] is a getting part process which involves nesting a series of shapes of parts on plate in accordance with some methods, finding out the cutting sequence and selecting the piercing point of the contour, cutting these parts from the sheet according to a sequence. It's not hard to see the cutting path optimization is the key problem of sheet metal cutting optimization. Up to now, many scholars have made a lot of research for the cutting path optimization problem. Reginald [2] presents a review of the literature on generating cutting paths for laser cutting machines. Trends in research in cutting path generation are given in this paper.

The cutting path optimization problem can be attributed to the traveling salesman problem (TSP). Su et al. [3] converted the cutting path optimization problem into a classic traveling salesman problem by adding a node, and solved by the genetic algorithm. Lou et al. [4] transformed the cutting path optimization problem into traveling salesman problem by forming air travel optimization mathematical model, and solved by the adaptive genetic algorithm. Lee [5] proposed a two-step genetic algorithm combining global search for piercing point optimization and local search for part sequencing to solve the cutting path optimization. Jang [6] found optimal cutting paths based on an SA algorithm by allowing all the convex vertices of the parts to be piercing points compared with existing works that used only fixed piercing points. These researches step by step to achieve solving the cutting path optimization by adopting the idea of simplify the problem. Although the computational efficiency has improved in this way, it loss most optimal solution space because of the local search and neighboring local optimization algorithm have its own drawbacks. To better solve the cutting path optimization problem, a hierarchical genetic algorithm was designed in this paper. Hierarchical Genetic Algorithm (HGA) was proposed by Tang et al. according to hierarchical structure of chromosome genetic [7]. The hierarchical structure is consisted of parameter genes and control genes. The parameter genes are the lower level, and control genes are in the higher levels of the parameter genes [8]. The parameter genes are controlled by the control genes.

The rest of the paper is organized as follows. In section 2, a mathematic model for the cutting path optimization problem is constructed. In section 3, a process for solving the problem with a hierarchical genetic algorithm is demonstrated. Then the results of computational experiment are described in section 4. Finally, conclusions are given in section 5.

---

[*]Corresponding author

2. Mathematical Modeling

2.1. *Geometric representation*

Many cutting parts include outer contour and inner contour simultaneously. Each closed contour is composed by line, arc and round. A part only has an outer contour, but it may be have a plurality of inner contour. We can define the start and end points of the straight line as the polygon vertices, define the starting point (the piercing point) and end point of the arc as the polygon vertices, and define any point on the arc as the polygon vertices. The polygon vertices can be the start point or the end point of the cutting. The part geometric representation is shown in Fig. 1, the part n ($1 \leq n \leq N$) has K_n inner contour, and the outer contour of part n has V_n vertices. $M_{n,k}$ represents inner contour k ($1 \leq k \leq K_n$) has $M_{n,k}$ vertices. $V_{n,l}$ represents the l_{th} ($1 \leq l \leq V_n$) vertice of part n. $v_{n,k,m}$ represents the m_{th} ($1 \leq m \leq M_{n,k}$) vertice of inner contour k of part n.

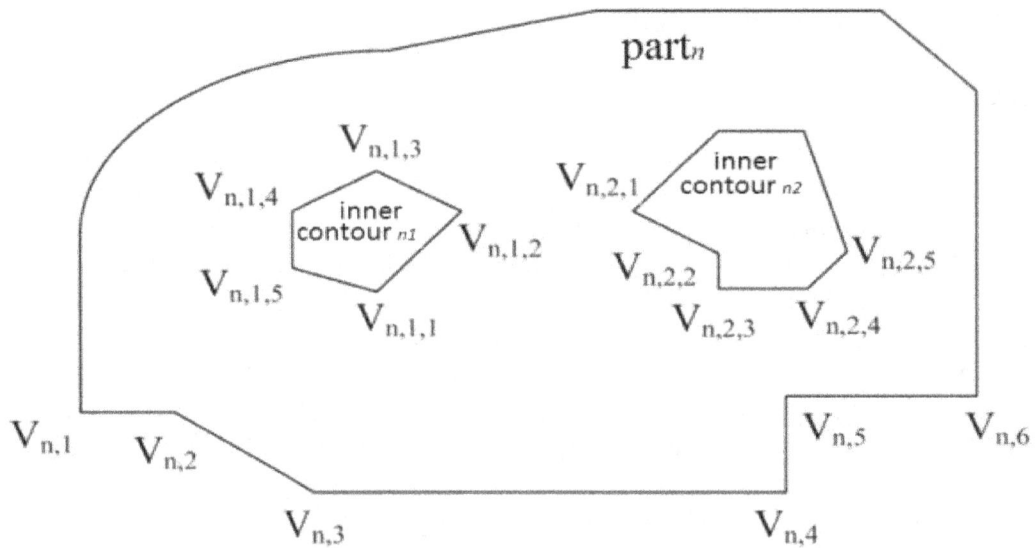

Fig. 1 A part geometric representation

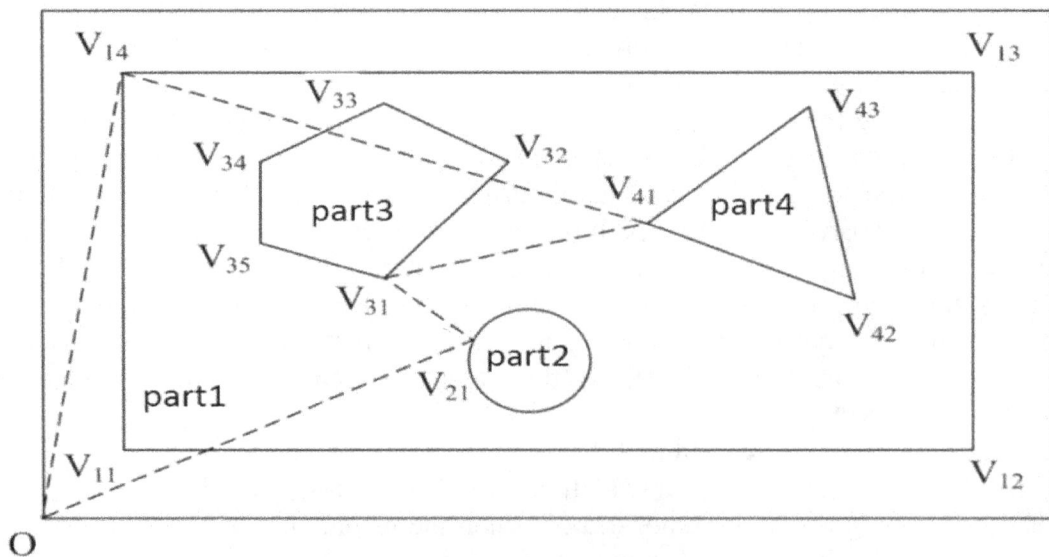

Fig. 2 The cutting sequence sketch

716

2.2. The mathematical model for selecting the piercing point

If the starting point of cutting is set on the part contours vertices, the number of the starting points is $\sum_{i=1}^{n} V_n + \sum_{i=1}^{n}\sum_{i=1}^{k} M_{n,k}$, $N + \sum_{i=1}^{n} K_n$ points take part in cutting path optimization calculation, and the shortest path optimization objective function can be defined as follow.

$$\min f(x) = I_{n+1} * L_{a,n+1} + (1 - I_{n+1}) * L_{b,n+1} \tag{1}$$

where, I_{n+1} represents inner contour of the part $(n+1)$, $L_{a,n+1}$ represents the total length of cutting path which includes the inner contour of the part $(n+1)$, $L_{b,n+1}$ represents the total cutting path length which doesn't include the inner contour of the part $(n+1)$.

$$L_{a,n+1} = \sum_{i=1}^{n} v_{n,1} v_{n+1,1,1} + \sum_{i=1}^{n}\sum_{i=1}^{k} v_{n+1,k,1} v_{n+1,k+1,1} + \sum_{i=1}^{n} v_{n+1,K_n,1} v_{n+1,1} \tag{2}$$

$$L_{b,n+1} = \sum_{i=1}^{n} v_{n,1} v_{n+1,1} \tag{3}$$

where, $\sum_{i=1}^{n} v_{n,1} v_{n+1,1,1}$ represents the length from the start cutting point of outer contour of part n to the start cutting point of inner contour of part $(n+1)$, $\sum_{i=1}^{n}\sum_{i=1}^{k} v_{n+1,k,1} v_{n+1,k+1,1}$ represents the total cutting path length which includes the inner contour of the part n and part $(n+1)$, $\sum_{i=1}^{n} v_{n+1,K_n,1} v_{n+1,1}$ represents the length from the last inner contour to the outer contour of part $(n+1)$.

2.3. The mathematical model for cutting sequencing

As shown in Fig. 2, the cutting sequence from part1 to part 4 is described as follow. Where, air move represents the move where the laser head is not cutting.

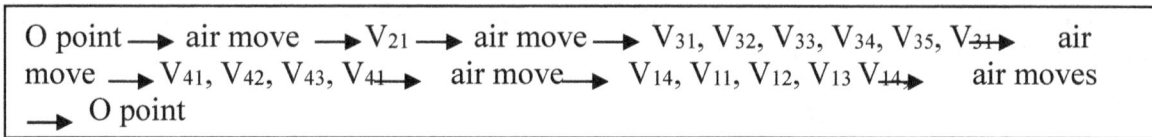

O point ⟶ air move ⟶ V21 ⟶ air move ⟶ V31, V32, V33, V34, V35, V31 ⟶ air move ⟶ V41, V42, V43, V41 ⟶ air move ⟶ V14, V11, V12, V13 V14 ⟶ air moves ⟶ O point

The mathematical model for cutting sequencing can be defined as follow.

$$D = \min \sum \left(\sqrt{(x_{ik} - x_{j1})^2 + (y_{ik} - y_{j1})^2} \right) \tag{4}$$

Where, x_{ik} represents the x coordinate of the k_{th} vertex points value of outer contour i, y_{ik} represents the y coordinate of the k_{th} vertex points value of outer contour i, x_{j1} represents the x coordinate of the first vertex points value of outer contour j, y_{j1} represents the y coordinate of the first vertex points value of outer contour j. The cutting sequence optimization problem can be converted into a problem to solve minimize air moves.

3. A Hierarchical Genetic Algorithm

To solving the cutting sequence optimization problem and the piercing point selecting optimization problem simultaneously, a hierarchical genetic algorithm is designed in this section.

The hierarchical structure consisted of parameter genes and control genes, where parameter genes can solve the piercing point selecting optimization problem, control genes can solve the cutting sequence optimization problem. The parameter genes are in the lowest level, and control genes are in the higher levels of the parameter genes. The parameter genes are controlled by the control genes.

3.1. Code

If the binary encoding is used for solving the cutting paths optimization problem, the chromosome will be very complex and difficult for the crossover operator and decoding. So, to overcome these difficulties, the natural number encoding couple with binary encoding is used.

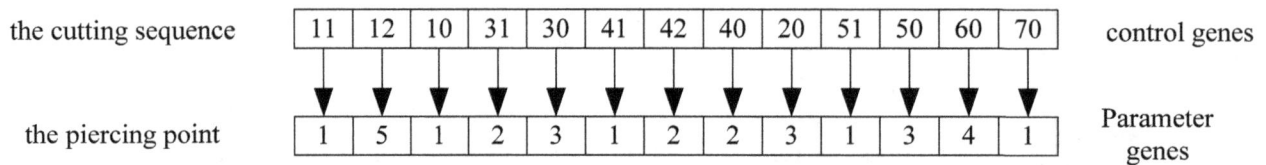

| the cutting sequence | 11 | 12 | 10 | 31 | 30 | 41 | 42 | 40 | 20 | 51 | 50 | 60 | 70 | control genes |

| the piercing point | 1 | 5 | 1 | 2 | 3 | 1 | 2 | 2 | 3 | 1 | 3 | 4 | 1 | Parameter genes |

Fig. 3 The process of natural number encoding

As shown in Fig. 3, thirteen parts were needed to be cut. The cutting sequence is showed by the control gens. The value of the first number except zero from the right represents inner contour number, the rest of number and represents outer contour number. For example, the number 11 represents the first inner contour of outer contour 1, the number 10 represents outer contour 1. The piercing point selected is showed by parameter genes. The value represents vertex point number.

3.2. Initial population

A set of feasible solutions for cutting sequenced question and piercing point selected question were chosen randomly. Before a outer contour is cut, the inner contour of this outer contour should be achieve its cutting.

3.3. Fit function

Minimizing air moves is objective function, and the fitness function can be obtained by the exponential transform of objective function.

$$f = a\exp(-b\sum(\sqrt{(x_{ik} - x_{j1})^2 + (y_{ik} - y_{j1})^2}) \tag{5}$$

Where, a is a positive real number, b is obtained by formula (6).

$$b = \begin{cases} N \times M \times \dfrac{f_{min}}{f_{min} + f_{max}} & f \geq f_{avg} \\[3mm] N \times M \times \dfrac{f_{max}}{f_{min} + f_{max}} & f \leq f_{avg} \end{cases} \tag{6}$$

Where, N is evolution generation, M is the number of individuals in the population, f_{ave} is the average fitness, f_{min} is the minimum fitness of the individual, f_{max} is the maximum fitness of the individual, and f is the individual fitness.

3.4. *Selection*

The hierarchical genetic algorithm allows the population progress from one generation to the next. The selection process is based on the fitness of the individuals, higher fitness results in more frequent selection. There are different selection rules such as the roulette wheel implementation, tournament selection, and elitism. Roulette wheel selection method is used.

Firstly, the fitness of individual i (f_i) is calculated by formula (5), then the selected probability of individual i can be calculated by the following formula.

$$p_i = f_i \Big/ \sum_{k=1}^{n} f_k \tag{7}$$

Secondly, the cumulative probability of each chromosome is calculated by formula (8).

$$q_i = \sum_{i=1}^{l} p_i \tag{8}$$

Where, l represents the iteration times.

Final, using roulette selection method selects the individual.

3.5. *Crossover*

Some of the genetic material of two individuals are swapped (i.e. crossover operator), creating new individuals (offspring), who are possibly better than their parents. For the control gene chromosome, partially mapping crossover method is used. The process of partially mapping crossover can be seen in Fig. 4. Firstly, two crossover points from parents' chromosome are selected. Secondly, the selected parents' chromosome fragments are exchanged. Thirdly, for the other genes, if the genes do not belong to the exchanged fragment of parents chromosome, the genes retain their value, otherwise, the value of those genes can be got using partially mapping method. In this process, if the position of gene for an outer contour is in the front of gene for the inner contour of this outer contour, the position of these two genes should be interchanged. Using this crossover method could avoid illegal individuals.

It is prone to generate illegal solution for crossover operation of the parameter genes. Therefore, the process of crossover operator applies only to control genes.

swap part 1

| 11 | 12 | 10 | 31 | 30 | 41 | 42 | 40 | 20 | 51 | 50 | 60 | 70 |

Chromosome of parent
individual A

| 20 | 11 | 12 | 31 | 30 | 41 | 42 | 40 | 10 | 51 | 50 | 60 | 70 |

Chromosome of parent
offspring B′

| 31 | 30 | 41 | 11 | 12 | 10 | 51 | 50 | 42 | 40 | 20 | 60 | 70 |

Chromosome of parent
offspring A′

| 20 | 31 | 30 | 11 | 12 | 10 | 51 | 50 | 42 | 41 | 40 | 60 | 70 |

Chromosome of parent
individual B

swap part 2

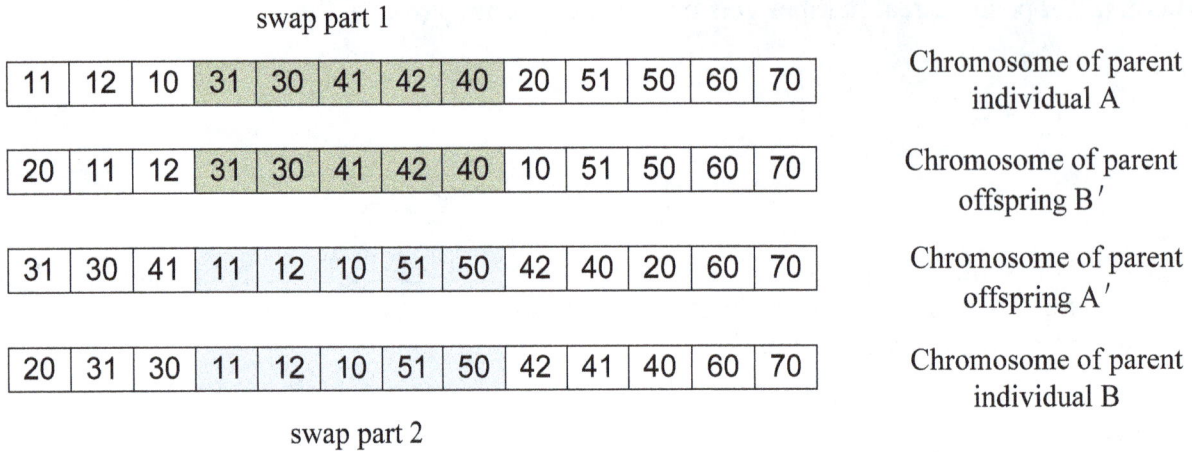

Fig. 4 The process of the first level control genes crossover operating

3.6. Mutation

In order to explore new areas of the search space, the mutation with introducing a variation in the population and avoid premature convergence is needed. For control genes of chromosome, changing sequence variation is used. Randomly two points of parent chromosome is selected, then the value of the selected point of parents chromosome is exchanged each other. If the position of gene for an outer contour is in the front of gene for the inner contour of this outer contour, the position of these two genes should be interchanged. For parameter genes of chromosome, the integer variation is used, that is, the parent parameter gene is replaced by an integer v ($v \in V$) with a certain probability.

3.7. Stop criteria

In this paper, setting a maximum iteration number has been used, and the algorithm will stop when the iteration reaches the setting maximum iteration.

4. Computational Experiments

In this section, the proposed optimization approach is proven to be available, and the performance of the solution strategy is evaluated by describing an experiment. There are eight part need to be cut, where three parts have inner contour, others have not inner contour. This instance was solved by the above proposed method.

Fig. 5 shows the cutting path 1 without using the optimization method and the cutting path 2 with using the above proposed method. As shown in Fig. 5, the air move can be shortened 17.8% by the above proposed method.

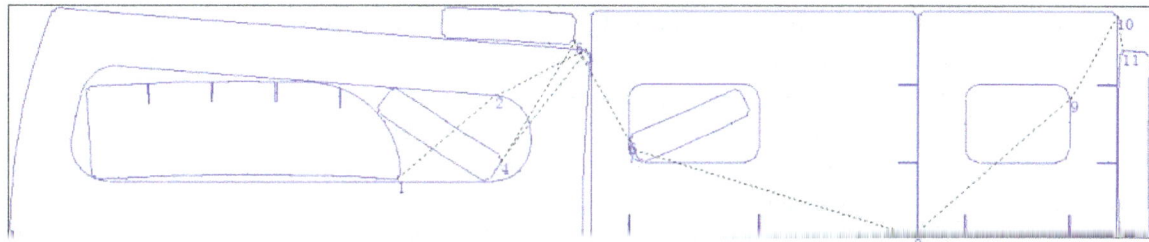

The cutting path 1(air move 1010 mm)

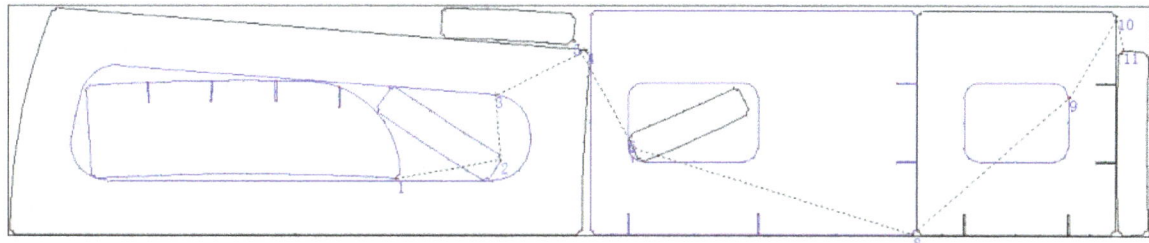

The cutting path 2 (air move 830 mm)

Fig. 5 The comparison of cutting path

5. Conclusion

A hierarchical genetic algorithm for the cutting paths optimization problem was discussed in this paper. An optimization model for the cutting paths optimization problem was constructed. In order to better solving this model, a hierarchical genetic algorithm was developed. This algorithm designed a hierarchical structure of two chromosomes which used to simultaneously solve the cutting sequence and the piercing point select. The sheet metal cutting path optimization can be effectively solved by this method. And a detailed case study was used to prove its effectiveness. Nevertheless, the computational efficiency of the paper proposed method is low. And it can be improved in the further research.

Acknowledgments

This research work was funded by the scientific research project of Hubei Province Education Department [grant number Q20151408] and the Hubei University of Technology High-level Talent Fund.

References

1. H. Zhu, F. Han and W. Lin, Generation of three-dimensional laser cutting path for sheet metal part. Zhongguo Jiguang/Chinese Journal of Lasers. 40(2013) DOI 10.3788/CJL201340. 0803007
2. R. Dewil, P. Vansteenwegen and D. Cattrysse, A review of cutting path algorithms for laser cutters. The International Journal of Advanced Manufacturing Technology. (2016) DOI 10.1007/s00170-016-8609-1
3. H.P. Sun, J. Li and W.G. Gou, Application of genetic algorithm on optimization of high power beam cutting paths. Transactions of the chinese society for agricultural machinery. 39 (2008) 158-160.

4. C.Y. Luo, B. LU and L. Han, Adaptive neighborhood method & GA for solving the vacancy route optimization of machining. Journal of chongqing university. 32 (2009) 1477-1481.

5. M.K. Lee, K.B. Kwon, Cutting path optimization in CNC cutting processes using a two-step genetic algorithm. International Journal of Production Research. 44 (2006) 5307-5326.

6. C.D. Jang, Y.K. Han, Approach to efficient nesting and cutting path optimization of irregular shapes. Journal of Ship Production. 15 (1999) 129-135.

7. K.S. Tang, K.F. Man, S. Kwong, et al. Design and optimization of IIR filter structure using hierarchical genetic algorithms. IEEE transaction on industrial electronics. 45 (1998) 481~487.

8. H. Zhou, W. Tang and B. Niu, Optimization of multiple traveling salesman problem based on hierarchical genetic algorithm [J]. Application Research of Computers. 26 (2009) 3754-3757.

9. M.J. Varnamkhasti L.S. Lee, A Fuzzy Genetic Algorithm Based on Binary Encoding for Solving Multidimensional Knapsack Problems. Journal of applied mathematics. (2012) DOI 10.1155/2012/703601.

10. P. Fattahi, V. Hajipour and A. Nobari, A bi-objective continuous review inventory control model: Pareto-based meta-heuristic algorithms. Applied soft computing. 32 (2015) 211-223.

Finite Element Analysis of the Crack Behavior of Rubber Material

Tian-Hua Zhang, Wei Wang*

Key Laboratory of Rubber-plastics, Ministry of Education/Shandong Provincial Key Laboratory of Rubber-plastics, Qingdao University of Science and Technology, Qingdao 266042, China.
Email: wdavid1@163.com

There are three finite element methods to solve the fracture problems, such as cohesive element, virtual crack closure technique (VCCT) and the extended finite element method (XFEM). In this paper, these methods were used to analyze the crack behavior of the same rubber specimen. Through the crack analyses, we found that the cohesive element is not suitable for rubber material. Results obtained from the other two methods agree very well by comparing the calculated results such as strain energy density and Mises stress. Hence, we consider the VCCT and XFEM can be applied to predict the crack behavior of rubber material and products.

Keywords: rubber; tensile crack; VCCT; XFEM.

1. Introduction

Rubber is used in various fields due to its excellent properties, such as high elasticity, low modulus, and fatigue resistance. Hence, the fatigue analysis and life prediction of rubber product have attracted the attention of many researchers. Mars and Fatemi [1] proposed that rubber components subjected to cyclic loading often fail due to the nucleation and growth of the crack. For rubber, the crack growth approach has been extensively studied and used. Generally, rubber products are made of many kinds of materials. Many factors such as materials diversity, structural complexity et al. make the study of rubber product's fatigue life fairly difficult. However, with the development of computer, numerical methods have been used extensively to study the rubber fatigue for its accuracy and great efficiency, especially the finite element method (FEM). For metal material, the FEM has been widely used to study the fatigue damage. But for rubber, the analysis of fatigue and fracture using FEM still an open research topic. In 1970s, Lindley [2] calculated tearing energy by using FEM for the first time and he analyzed the relationship between simple shear rubber's tearing energy and crack size.

The objective of this paper is to study the crack behavior of a rubber sample using the finite element method. The cohesive element, virtual crack closure technique (VCCT), and extended finite element method (XFEM) included in ABAQUS software are used to analyze the crack behavior of rubber in this study.

2. Cohesive Element

The cohesive element is often used to simulate the connection between two parts. It requires that the material size and strength of the bond are less than the parts. Generally, the local cohesive element acts as the crack in this method. In most situations, this method is applied to simulate the brittle material, for example the reinforced concrete. According to our simulation results，it is shown that the cohesive element is not applicable to rubber material. However, this method may be applied to calculate the rubber composite products, such as bonding of rubber and metal parts in shock absorber. So, we don't compare the results calculated by the cohesive element with the other two methods in the present paper.

3. Virtual Crack Closure Technique (VCCT)

VCCT was proposed according to Irwin's energy theory. Its principle was that the energy released during the crack propagation was equal to the energy absorbed during the crack closure. The VCCT

---

*Corresponding author

has been used for the rigid material. However, this method also gradually began to be applied in the rubber materials. Kim et al. [3] combined this method with cracking energy density to predict the fatigue life of the tires. They calculated the strain energy for every crack increment using VCCT. We also adopted the similar method to Kim et al. described in [3]. For simplification of this paper, the detailed description of VCCT method can be found in [3].

4. Extended Finite Element Method (XFEM)

In 1999, Belytschko et al. [4, 5] proposed a new calculation method, *i.e.*, the extended finite element method (XFEM) which was used to deal with the discontinuous problems. The XFEM may be utilized to simulate the crack propagation without defining the propagation path. Generally, this method is used to analyze the crack of rigid material. As a new method, the XFEM has begun to be applied to analyze the crack of elastic material. Duarte et al. [6] used XFEM to analyze the rubber concrete crack propagation behavior.

Based on the studies of many researchers, the three methods mentioned above can be used to explore the fatigue fracture problems. But for the rubber materials, the similarities and differences of VCCT and XFEM will be discussed in this paper.

5. Model and Material

A two-dimensional model of the rubber sample with an initial crack is constructed in this study, which is length 100 mm and width 30 mm, as shown in Fig. 1.

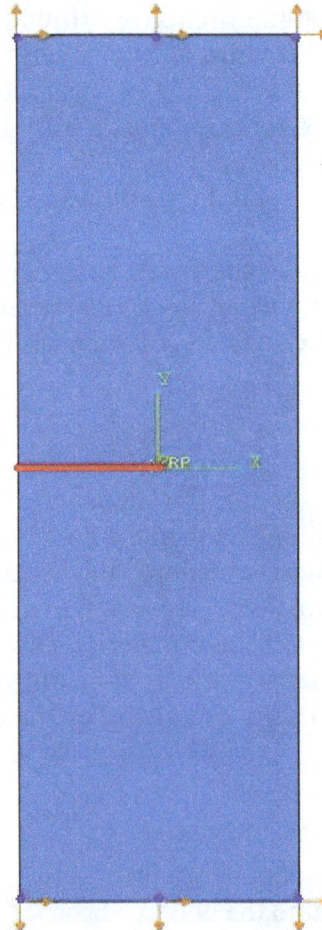

Fig. 1 The model for the rubber sample

The horizontal line in the middle of the sample denotes the location of precrack. The precrack length is half of the specimen width as the red line shown in Fig. 1. We selected the rubber of sidewall in a tire, because it easily appears fatigue crack failure. The material parameters of Yeoh model used to describe the constitutive behavior of rubber material are shown in Table 1.

Table 1 Material parameters of rubber sample

Rubber model	C_{10} (MPa)	C_{20} (MPa)	C_{30} (MPa)	E (MPa)
Yeoh	0.378013	-0.06452	0.012688	7.8

6. Results and Discussion

The calculated result contours of two methods are shown in Figs. 2 and 3.

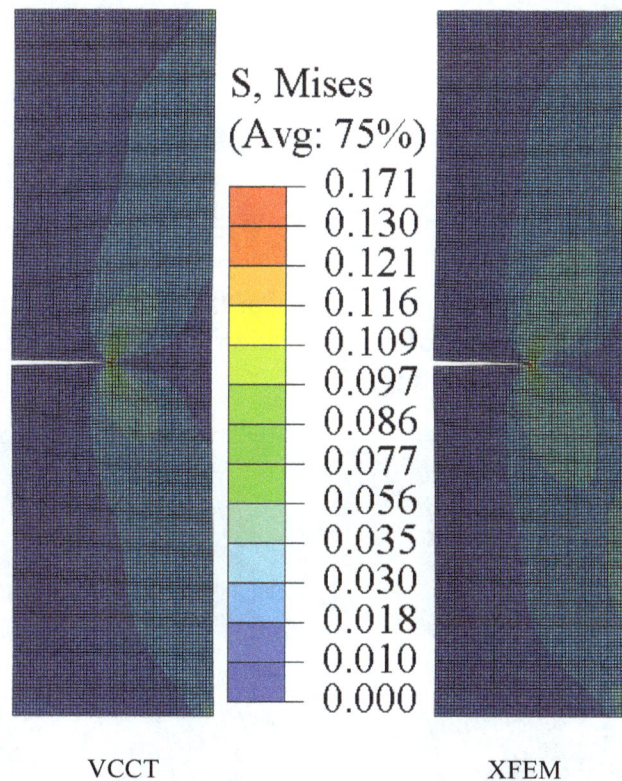

Fig. 2 The calculated Mises stress contours of VCCT and XFEM

From Fig. 2 and Fig. 3, we observe that the stress concentration occurs at the crack tip of the rubber specimen. This means that there is a higher tendency for the crack to grow toward the crack tip.

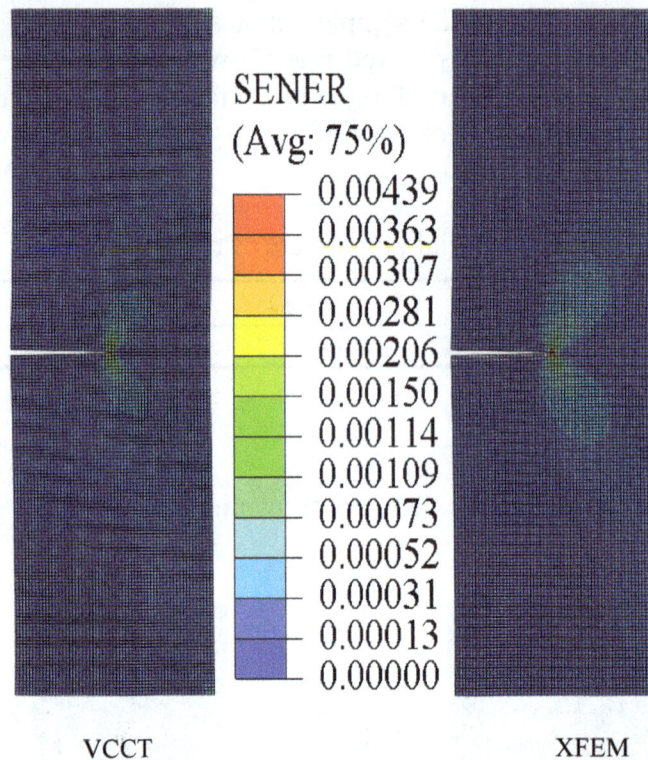

Fig. 3 The calculated strain energy density contours of VCCT and XFEM

Fig. 4 shows the crack path predicted by VCCT. For the VCCT and XFEM models, the sample size, the number of meshes, and the type of crack location is exactly the same. The selected path of Mises stress is also identical, as shown in Fig. 4.

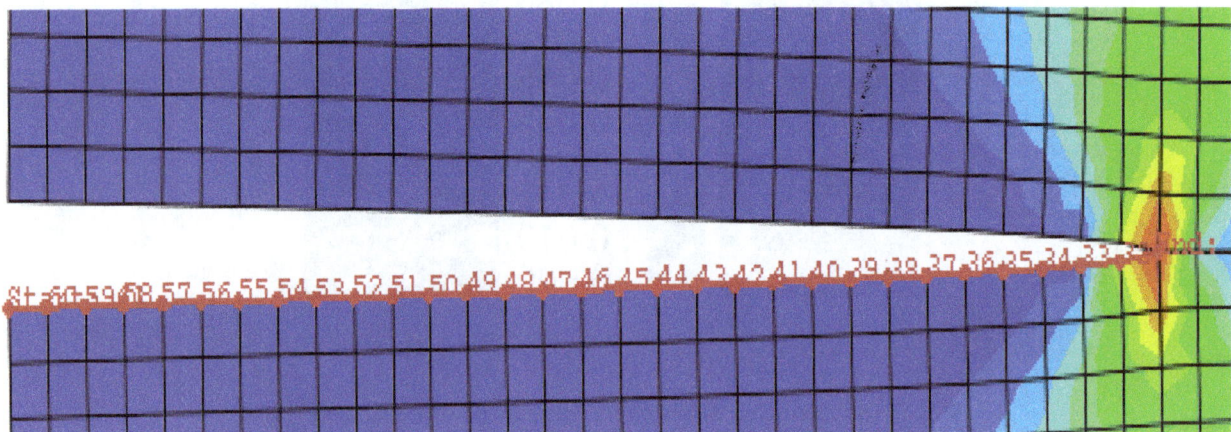

Fig. 4 Selected nodes along the crack path

We extract the strain energy density of node at the crack tip and Mises stress along the crack path respectively as shown in Figs. 5 and 6. It can be seen from the two figures that the results predicted by the mentioned two methods agree very well. It can be seen from Fig. 5 that the strain energy density (SED) shows good agreement between VCCT and XFEM methods. The SED is generally used to reveal the crack behavior of rubber material. The SED at crack tip of the rubber specimen increases as the time evolution goes on. Through the contour of calculated results we can see that the strain energy density is maximized at the crack tip. Similarly, under the dynamic situation, the strain energy density can be used to predict the fatigue life of rubber products.

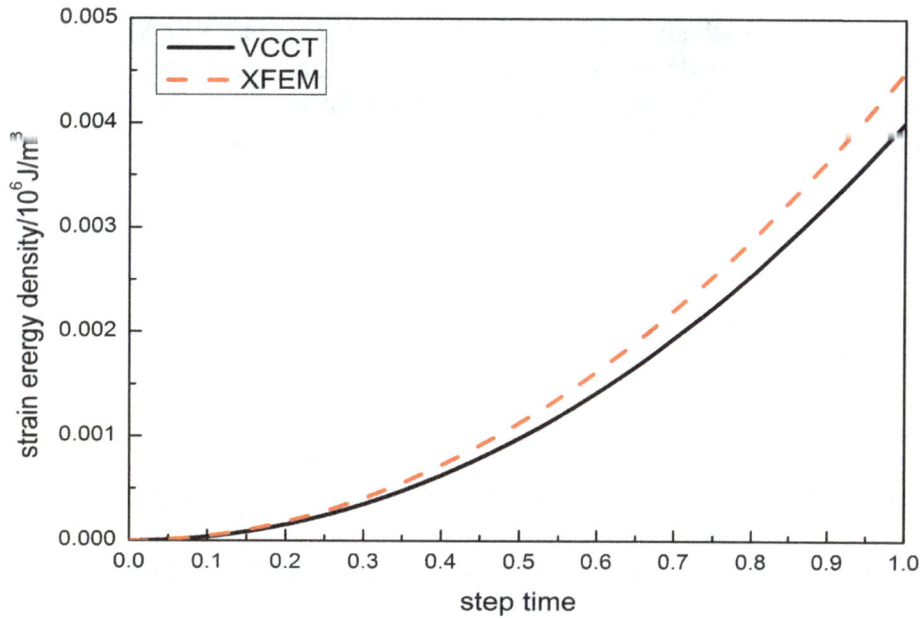

Fig. 5 Evolution of the crack tip strain energy density varies with the loading history

From Fig. 6, we can see that the variation tendency of Mises stress is basically the same. Away from the location of the crack tip, the variation tendency of Mises stress is slowly but Mises stress present the obviously up-trend near the crack tip, which indicates that the stress concentration occurred at the crack tip. Similarly Mises stress along the selected path between VCCT and XFEM is very similar. Hence, in the case of small strain the VCCT and XFEM methods could be applied to analyze simple rubber specimen's tensile crack behavior.

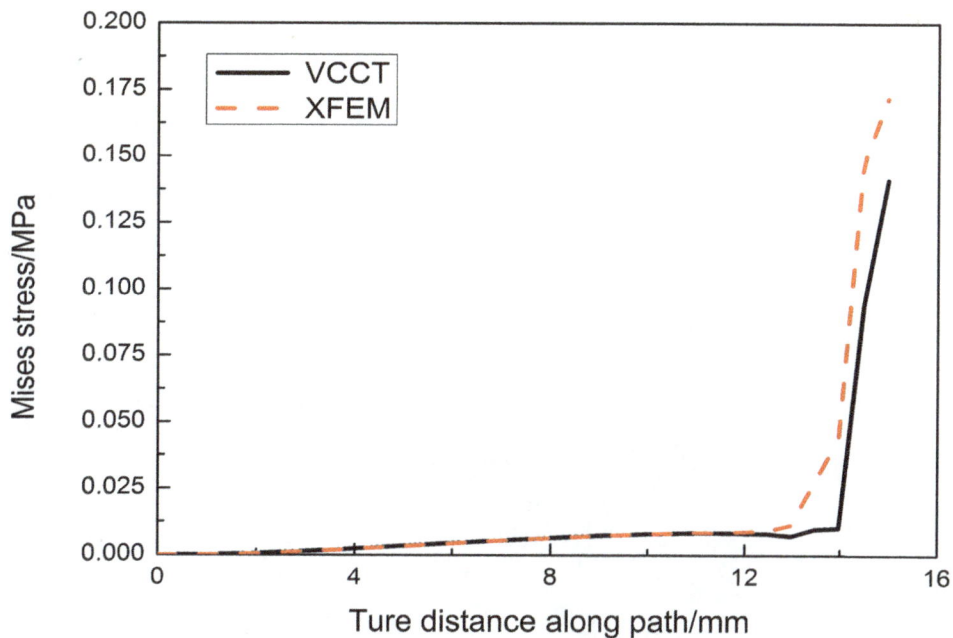

Fig. 6 Mises stress along the selected path

7. Conclusion

Through the comparative analysis, we consider that the VCCT and XFEM methods can be used to study the rubber sample's tensile crack behavior. In this paper, we chose the sidewall rubber as an object of study because the sidewall of tire often shows fatigue fracture. In our future studies, we will try to apply the VCCT and XFEM methods to reveal the fatigue fractures for the sidewall and shoulder of tire.

Acknowledgment

The authors gratefully acknowledge the funding from the National Natural Science Foundation of China through Contract/Grant numbers 51273099, 21274072.

References

1. M.V. Mars, A. Fatemi, A literature survey on fatigue analysis approaches for rubber. Int. J. Fatigue. 24 (2002) 949-961.
2. P. B. Lindley. Energy for crack growth in model rubber components. J. Strain Anal. 7 (1972) 132-140.
3. T.W. Kim, H.Y. Jeong, J.H. Choe, Y.H. Kim, Prediction of the fatigue life of tires using CED and VCCT, Key Eng. Mater. 297-300 (2005)102-107.
4. T. Belytsehko, T. Blaek, Elastic crack growth in finite elements with minimal remeshing. Int. J. Numer. Meth Eng. 45 (1999) 601-620.
5. N. Moes, J. Dolbow, T.A. Belytsehko, Finite element method for crack growth without remeshing. Int. J. Numer. Meth Eng. 46 (1999) 131-150.
6. A.P.C. Duarte, B.A. Silva, N. Silvestre, J. de Brito, E. Julio, Mechanical characterization of rubberized concrete using an image-processing/XFEM coupled procedure. Composites Part B. 78 (2015) 214-226.

Theoretical Study on Hydrogen Desorption Property of Cl-doped LiNH₂

Jing Zhao[1], Tian-Fu Gao[1], Chun-Hai Jiang[2], Ren-Zhong Huang[1,*]

[1]*College of Physical Science and Technology, Shenyang Normal University, Huanghe Street 253, Shenyang 110034, China*

[2]*School of Materials Science and Engineering, Xiamen University of Technology, Ligong Road 600, Xiamen 361024, China*

*Email: rzhuang09@163.com

The hydrogen storage property of Cl-doped LiNH₂ has been investigated by using first-principles method based on density functional theory. The calculated results show that Cl doping may result in the substitution of NH₂⁻ by Cl⁻ in the hydride lattice and accordingly, a favorable thermodynamics modification. The electron structure analysis shows that Cl doping induces the movement of Li-2s towards higher energy levels and weakens the interaction between Li and N. The increased interaction between Cl and Li benefit the early release of NH₃. The hydrogen desorption property is thus improved.

Keywords: Hydride; first-principles calculation; formation enthalpy; substitution enthalpy.

1. Introduction

Hydrogen energy has been considered as a promising secondary energy due to its high energy density. The key for applying hydrogen energy in the future is to develop hydrogen storage. Being regarded as the most ideal materials for automotive applications, solid state hydrogen storage materials should have large gravimetric, large volumetric storage capacity, suitable thermo-dynamic property and fast hydrogen sorption and desorption kinetics [1, 2]. Therefore, most experimental and theoretical works focus on the hydrogen sorption and desorption properties of metal hydrides and chemical hydrides such as LiNH₂ [3-5], LiBH₄ [6, 7] and Li₄BN₃H₁₀ [8, 9].

It has been reported by Chen *et al* [5] in experiments that the lithium nitride Li₃N can absorb/desorb a large number of hydrogen in the following two-step reaction:

$$Li_3N + 2H_2 \leftrightarrow Li_2NH + LiH + H_2 \leftrightarrow LiNH_2 + 2LiH \qquad (1)$$

Theoretically, 10.4wt% hydrogen can be stored in this reaction. However, nearly 6.5wt% hydrogen can be stored in the second step. Thus, many investigations have focused on the second reaction [10-12]:

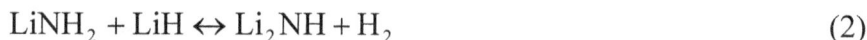

$$LiNH_2 + LiH \leftrightarrow Li_2NH + H_2 \qquad (2)$$

In fact, it has been shown that the reaction proceeds with two elementary steps, as defined by the equation (3) and (4):

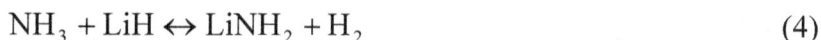

$$2LiNH_2 \leftrightarrow Li_2NH + NH_3 \qquad (3)$$

$$NH_3 + LiH \leftrightarrow LiNH_2 + H_2 \qquad (4)$$

Because of the slow reaction rate, a small amount of LiNH₂ starts to decompose near 300°C. Thus the release of a small portion of hydrogen occurs above 300°C accompanying with the emission of ammonia gas. To improve the reaction rate, many experiments have been done to search for the additives to enhance the dehydrogenation property, such as Ni, Fe, Co, VCl₃, Ti,

---

*Corresponding author

TiCl₃ and so on [13,14]. Among them, the most suitable catalyst is TiCl₃ which can reduce the desorption temperature to 220°C and avoid the emission of ammonia.

In this work, we aim to explore the mechanism of improved hydrogen desorption property of Cl-doped LiNH₂ by using plane-wave pseudopotential approach based on the density functional theory (DFT). It is shown that the Cl doping may result in the substitution of NH_2^- by Cl^- in the hydride lattice and facilitates the early release of NH_3. The hydrogen desorption property is thus improved for the hydride.

2. Calculation Method and Models

The first-principle method based on DFT is employed to investigate the effect of Cl doping on the dehydrogenation property of LiNH₂. All computations have been performed by using a plane wave pseudopotential method within the generalized gradient approximation (GGA) [15] to DFT. The Perdew-Burke-Ernzerhof (PBE) exchange-correlation potential [16] is adopted for the hydride. Cutoff energy of 330eV is used for the plane wave expansion in order to ensure the convergence of the total energy and forces acting on the atoms. The Brillouin zone integrations are performed by using a 3x3x1 k-point mesh according to the Monkhost-Pack scheme [17, 18]. The forces acting on the atoms are converged to 0.01eV/Å. The stress is converged to 0.05GPa and the system total energy is converged to 1x10⁻⁵eV. The atom positions and the lattice parameters are relaxed simultaneously to find the lowest energy state of the system.

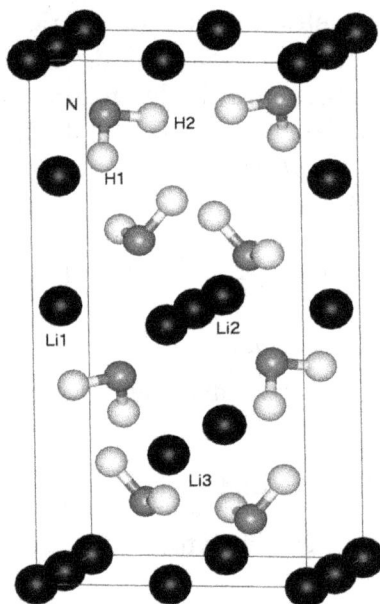

Fig. 1 Unit cell for LiNH₂: Li-black ball; N-grey ball; H-white ball.

LiNH₂ has a tetragonal crystal structure with space group
$$3Li_{30}B_8N_{24}H_{79}NiCl \rightarrow Li_9N_2Cl_3 + 24Li_3BN_2 + 3Li_2NH + 3LiNNi + 16NH_3 + 93H_2 \text{[19]}.$$

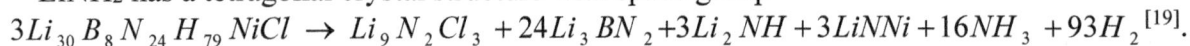

The primitive unit cell of LiNH₂ includes eight formula units (see Fig. 1(a)). Li ions are tetrahedrally coordinated by NH_2^-. There are two nonequivalent H ions and three nonequivalent Li ions in the LiNH₂ lattice. Our calculated lattice constants of a = 5.042Å and c = 10.435 Å are in good agreement with the experimental values of a = 5.034 Å and c =10.255 Å and other calculated results of a = 5.034 Å and c=10.348 Å [20]. As for doping, a 2x2x1 supercell of LiNH2 is chosen in order to avoid the interaction between doping ions in different supercells. The formal charge of Cl anion is -1 and thus substituting H^+ by Cl^- isn't allowed in the LiNH2 lattice. The substitution of Cl^- for NH_2^- in LiNH2 lattice is feasible and results in two additional H vacancies. The atom position and the lattice parameter of the proposed model are relaxed simultaneously to find their lowest energy state. The relaxed lattice parameters are listed in Table 1. It is seen from this Table that the substitution of Cl for NH_2^- results in the increase of lattice constants due to the difference of their ion radii.

Table 1 Formation enthalpy ΔH_{form}, substitution enthalpy ΔH_{sub} and lattice parameters of the Cl-doped hydride

Model	Dopant ion	Substituted ion	ΔH_{form} [KJ/mol atom]	ΔH_{sub} [KJ/mol atom]	a[Å]	b[Å]	c[Å]
Li32N32H64			-62.403		10.107	10.107	10.459
Li32N31H62Cl	Cl-	NH2-	-64.741	-1.042	10.160	10.143	10.482

3. Results and Discussion

Thermodynamic conventions are used to determine the stability of the dopant phase. Both negative formation enthalpy and negative substitution enthalpy indicate the stability of the dopant phase and the feasibility of the doping reaction. As a reasonable approximation, a substitution ratio of 31:1 has been selected, resulting in the doped hydride Li32N31H62Cl. Generally, the formation enthalpy ΔH_{form} and the substitution enthalpy ΔH_{sub} of the Cl-doped hydrides are given by the following formula:

$$\Delta H_{form}(Li_{32}N_{31}H_{62}Cl) = \frac{1}{N_{atom}}[E(Li_{32}N_{31}H_{62}Cl) - 32E(Li) - \frac{31}{2}E(N_2) - 31E(H_2) - \frac{1}{2}E(Cl_2) \qquad (5)$$

$$\Delta H_{sub}(Li_{32}N_{31}H_{62}Cl) = \frac{1}{N_{atom}}[E(Li_{32}N_{31}H_{62}Cl) + \frac{1}{2}E(N_2) + E(H_2) - E(Li_{32}N_{32}H_{64}) - \frac{1}{2}E(Cl_2)] \qquad (6)$$

Where $E(Li_{32}N_{31}H_{62}Cl)$ and $E(Li_{32}N_{32}H_{64})$ refer to the total energy of Li32N31H62Cl and Li32N32H64 crystal respectively. $E(N_2)$, $E(H_2)$ and $E(Cl_2)$ refer to the total energy of gaseous N2, H2 and Cl2 respectively. $E(Li)$ refers to the total energy of bcc Li.

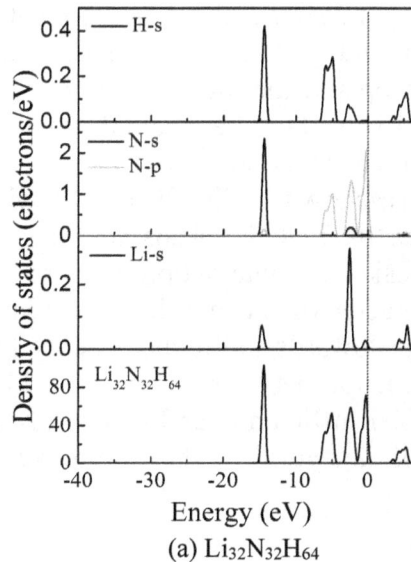

(a) Li32N32H64

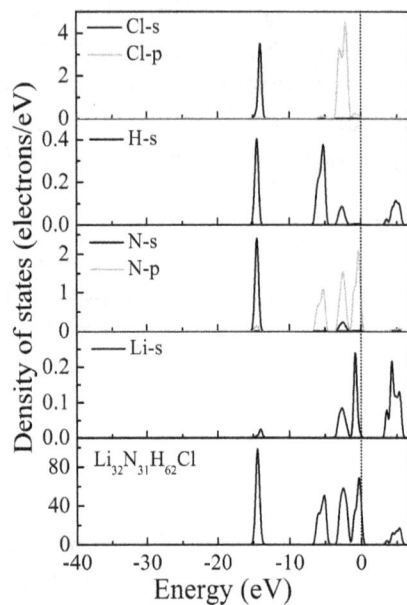

(b) $Li_{32}N_{31}H_{62}Cl$

Fig. 2 Total (TDOS) and partial density states (PDOS)

To evaluate the possibility of Cl doping, we have calculated the ΔH_{form} and the ΔH_{sub} of the proposed Cl-doped $LiNH_2$ according to the equations (5) and (6). It is seen from the results shown in Table 1, that the proposed model of substituting NH_2^- by Cl^- has a negative ΔH_{form} of -64.74 KJ/(mol atom) and a negative ΔH_{sub} of -1.04 KJ/(mol atom). In the viewpoint of thermodynamics, it is possible that Cl^- is incorporated into the main lattice of the hydride and substitutes for NH_2^-. It is also seen from this Table that the formation enthalpy of the Cl-doped hydride is lower than that of the pure one. It indicates that the Cl-doped hydride is more stable than the pure one. The negative ΔH_{sub} means that the substitution doesn't need any driving force and is a thermodynamic favorable reaction. Thus, it is inferred that the substitution of Cl^- for NH_2^- is a favorable substitution reaction and Cl doping will decrease the dehydrogenation temperature of the hydride.

To illustrate the bonding nature in undoped and doped hydride and reveal the micro-mechanism of improved hydrogen desorption property, the total density of states (TDOS) and partial density of states (PDOS) of $Li_{32}N_{32}H_{64}$ and $Li_{32}N_{31}H_{62}Cl$ are calculated and plotted in Fig. 2. Among these figures, the Fermi energy (E_F) is referenced as zero.

It is seen from the TDOS of $LiNH_2$ shown in Fig. 2(a) that a wide energy gap of about 3.1eV is indicative of an insulator character. Its PDOS reveals that a strong covalent interaction exists between N-2s and H-1s while a weak ionic interaction appears between Li-2s and N-2p. As for Cl-doped $LiNH_2$, the insulator character of the hydride still remains as shown in Fig. 2(b). Cl doping induces the movement of Li towards the higher energy level. The increasing interaction between Li and Cl reduces the interaction between Li and N and benefits the release of NH_3. The early released NH_3 will react with LiH sufficiently and no extra NH_3 releases at higher temperature. The hydrogen desorption property of the hydride is thus improved.

4. Conclusion

Our study shows that Cl doping on $LiNH_2$ may result in a substitution of NH_2^- by Cl^- in the hydride lattice and accordingly, a favorable thermodynamics modification. The electronic structure analysis indicates that the main peak of Li-2s moves close to the Fermi level. The increasing interaction between Li and Cl benefits the early release of NH_3 and no extra NH_3 releases at higher temperature. The hydrogen desorption property of the hydride is thus improved.

Acknowledgments

This work was funded by National Natural Science Fund of China: (Grant No. 11347003), Scientific Project of the Educational Department of Liaoning Province; (Grant No. L2012386), and Natural Science Foundation of Liaoning Province (Grant No. 20102208).

References

1. P. Chen, M. Zhu, Recent progress in hydrogen storage, Mater. Today 11 (2008) 36-43.
2. J. Yang, A. Sudik, C. Wolverton, D.J. Siegel, High capacity hydrogen storage materials: attributes for automotive applications and techniques for materials discovery, Chem. Soc. Rev. 39 (2010) 656-675.
3. T. Tsumuraya, T. Shishidou, T. Oguchi, Theoretical analysis of x-ray absorption spectra of Ti compounds used as catalysts in lithium amide/imide reactions, Phys. Rev. B 77 (2008) 235114.
4. Y.L. Teng, T. Ichikawa, Y. Kojima, Catalytic Effect of Ti−Li−N Compounds in the Li−N−H System on Hydrogen Desorption Properties, J. Phys. Chem. C 115 (2010) 589-593.
5. P. Chen, Z. Xiong, J. Luo, J. Lin, K.L. Tan, Interaction of hydrogen with metal nitrides and amides, Nature 420 (2002) 302–304.
6. T. He, H. Wu, G. Wu, J. Wang, W. Zhou, Z. Xiong, J. Chen, T. Zhang, P. Chen, Borohydride hydrazinates: high hydrogen content materials for hydrogen storage, Energy Environ. Sci. 5 (2012) 5686-5689.
7. X. Zheng, G. Wu, W. Li, Z. Xiong, T. He, J. Guo, H. Chen, P. Chen, Releasing 17.8 wt% H_2 from lithium borohydride ammoniate, Energy Environ. Sci. 4 (2011) 3593-3600.
8. X. Zheng, Z. Xiong, Y. Lim, G. Wu, P. Chen, H. Chen, Improving Effects of LiH and Co-Catalyst on the Dehydrogenation of $Li_4BN_3H_{10}$, J. Phys. Chem. C 115 (2011) 8840-8844.
9. Y.E. Filinchuk, K. Yvon, G.P. Meisner, F.E. Pinkerton, M.P. Balogh, On the Composition and Crystal Structure of the New Quaternary Hydride Phase $Li_4BN_3H_{10}$, Inorg. Chem. 45 (2006) 1433-1435.
10. T. Ichikawa, H.Y. Leng, S. Isobe, N. Hanada, H. Fujii, Recent development on hydrogen storage properties in metal–N–H systems, J. Power Sources 159 (2006) 126-131.
11. K. F. Aguey-Zinsou, J.H. Yao, Z.X. Guo, Reaction Paths between $LiNH_2$ and LiH with Effects of Nitrides, J. Phys. Chem. B, 111 (2007) 12531-12536.
12. L.P. Cheng, B.E. Xu, X.J Gong, X.Y. Li, Y.L. Zeng, L.P. Meng, First-principles study of hydrogen vacancies in lithium amide doped with titanium and niobium, Int. J. Hydrogen Energy, 38 (2013) 11303-11312.
13. S. Isobe, T. Ichikawa, N. Hanada, H.Y. Leng, M. Fichtner, O. Fuhr, H. Fujii, Effect of Ti catalyst with different chemical form on Li–N–H hydrogen storage properties, J. Alloys Comp. 404 (2005) 439-442.
14. T. Ichikawa, S. Isobe, N. Hanada, Lithium nitride for reversible hydrogen storage, J. Alloys Comp. 365 (2004) 271-276.

15. J.P. Perdew, J.A. Chevary, S.H. Vosko, K.A. Jackson, M.R. Pederson, D.J. Singh, C. Fiolhais, Atoms, molecules, solids, and surfaces: applications of the generalized gradient approximation for exchange and correlation, Phys. Rev. B 46 (1992) 6671–6687.

16. J.P. Perdew, K. Burke, M. Ernzerhof, Generalized gradient approximation made simple, Phys. Rev. Lett. 77 (1996) 3865-3868.

17. M.C. Payne, M.P. Teter, D.C. Allan, T.A. Arias, J.D. Joannopoulos, Iterative minimization techniques for *ab initio* total-energy calculations: molecular dynamics and conjugate gradients, Rev. Mod. Phys. 64 (1992) 1045-1098.

18. H.J. Monkhorst, J.D. Pack, Special points for Brillouin-zone integrations, Phys. Rev. B 13 (1976) 5188-5192.

19. J. B. Yang, X. D. Zhou, Q. Cai, W. J. James, W. B. Yelon, Crystal and electronic structures of $LiNH_2$, Appl. Phys. Lett. 88 (2006) 041914.

20. J. Wang, Y. Du, H. Xu, C. Jiang, Y. Kong, L. Sun, Z.K. Liu, Native defects in $LiNH_2$: a first-principles study, Phys. Rev. B 84 (2011) 024107.

Finite Element Analysis on Pull-out Test of Polypropylene Fiber

Wei-Hong Xuan[1], Yun-Hao Wu[2], Yu-Zhi Chen[1,*], Hui-Juan Jia[1], Xiao-Hong Chen[1]

[1]Architectural Engineering Institute, Jinling Institute of Technology, Nanjing 211169, China;
[2]Nanjing Qi Qiao Construction and Installation Engineering Co. Ltd, Nanjing, 210017, China
[*]Email: yuzhichen@jit.edu.cn

Numerical analysis is carried on for the pull-out test of single fiber by ANSYS. Using bond-slip relationship between fiber and cement matrix obtained by test, the bond stress and its distribution of the fiber and cement matrix is studied, and the characteristics of the bond and debond part is obtained by calculating. Finite element analysis shows that fiber stress is transferred to the fiber embedded side as the fiber carrying the pullout force. When the pullout force is smaller, bond stress of fiber-mortar takes the form of a triangle or trapezoid; when it is maximum, bond stress is at maximum. The maximum pullout force calculated by FEM differs by 9.02% from experimental data, and the calculated value of pullout displacement at the pullout side differs by 7.43%~12.37% from the experimental data. The calculated values are in good agreement with experimental.

Keywords: Finite element; pull-out test; fiber reinforced concrete.

1. Introduction

Normally, the performance of the composite material depends on the combination of each material, referred to as "Mixing Effects". But composite material can't always play a corresponding role, because the connection surface of two materials, called "interface", plays an important role. Composite material transmitted stress through the bonding of interface. If the bonding performance of interface is weaker, two materials whose strength is far from failure strength get relative movement as passing stress, thus the composite material is damaged. Therefore, the interfacial bond property is an important factor for influencing the performances of composite materials.

Despite the bonding performance between the low elastic modulus fibers and cement matrix is lower than steel fibers with high elastic modulus, it cannot be taken absolute. Full advantage of the tensile strength of the fiber is limited by the bonding performance between cement matrix and fiber [1]. Taking shear straight line shape steel fiber as an example [2-5], utilization rate of tensile strength of steel fiber is expressed by the ratio of maximum bond force to its the maximum tension, that is,

$$\rho = 2\tau l_f / (\sigma_b \cdot d_f) \tag{1}$$

In which is interfacial bond strength, value for 1.130; is the largest tensile stress of steel fiber, values for 380; is the length of steel fiber, averaging value for 30.3mm; equivalent diameter of steel fiber, value for 0.67mm.

Utilization rate of the tensile strength of steel fiber is only 26.9% by the calculation. This is mainly because interfacial bond strength reached critical value, strain of steel fiber is far from the yield point, and it is difficult to make tensile strength utilization of steel fiber improved significantly. Therefore, low elastic modulus fiber still has its application value of high cost performance by full advantage of material and economic concerns, worthy of further theoretical and applied research.

The pull-out tests on single fiber is numerically analyzed by ANSYS software. With the bond-slip relationships of fiber-cement matrix obtained by test, this paper studies the bond stress and the

*[*]Corresponding author

distribution, the characteristic of calculating the bond and the separation section, and compares the experimental with FEM results.

2. Finite Element Analysis

2.1. Element type

2.1.1. Cement matrix element

SOLID 65 element can simulate the strengthening rebar of concrete in ANSYS software, and the cracking and crushing phenomenon of material, as shown in figure 1. There are two cases for the application of the element. When finite element analysis is carried with the integral, SOLID 65 can be set up 1~3 kinds of reinforcing material as the material number, volume ratio and azimuth angle, etc. Reinforcing material (including creep and plastic) only has uniaxial stiffness and form into an integral with element. When it is carried with the separated, the several kinds of reinforced material are ignored. SOLID 65 element becomes plain concrete unit, and reinforced material was put in by other ways (such as fiber, etc.).

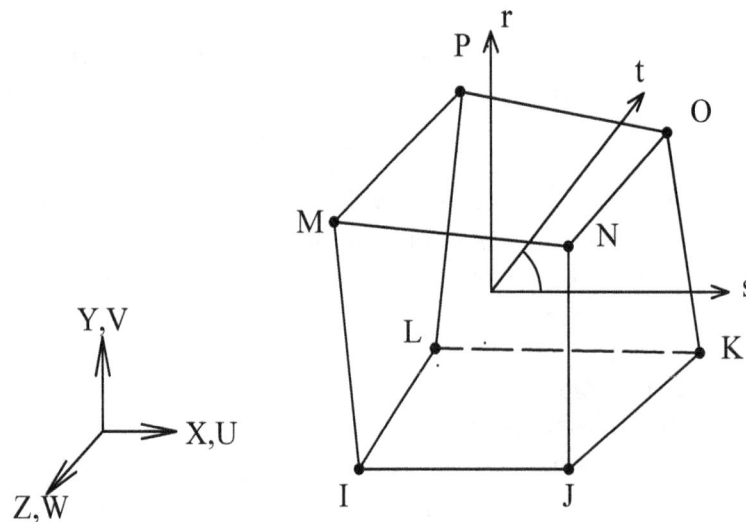

Fig. 1 Solid 65

2.1.2. Fiber element

Fiber is material with uniaxial stiffness, corresponding to LINK10 element in the ANSYS software, as shown in figure 2. LINK10 3d unit in each node has three degrees of freedom, and it can bear the uniaxial tension or compression. Initial strain of unit (ISTRN) is the difference between the unit length L and the initial zero strain. During the process of model establishment, the fiber is considered as line without section area, whose section size can be ignored. But the cross-sectional area of the fiber is limited and consideration in the specific application that material constant was defined. LINK10 element can limit the initial stress of the material, while the material constant is limited.

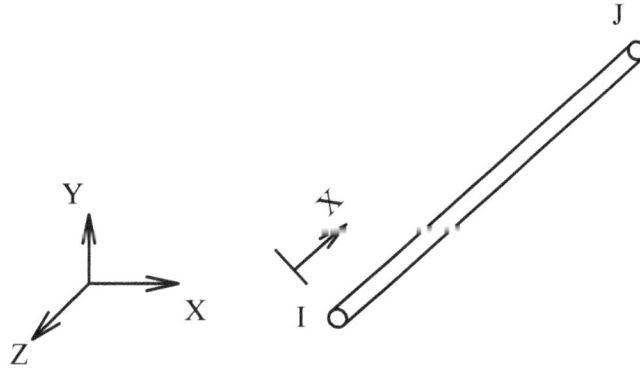

Fig. 2 Link10

2.1.3. Contact element

In order to simulate the bond-slip relationship between fibers and cement matrix, spring element is used. It is arranged in parallel on the interface between fiber and cement matrix to calculate the relative slip and bond force, without the actual geometry size, easy to set up, and unit division with no impact. COMBINE39 element selected within ANSYS is a nonlinear unit with two nodes. Each node could be defined the number and direction of degrees of freedom, and it can be a maximum of three degrees of freedom. Location and the shape of the unit are shown in figure 3. Bond force-deformation curve which is defined unit mechanics performance is shown in figure 4.

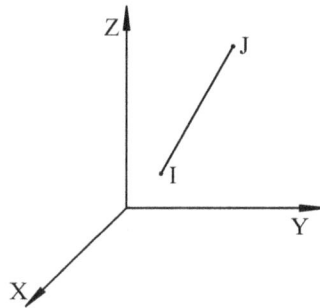

Fig. 3 Shape and positioning of element   Fig. 4 Force-deformation curve of element

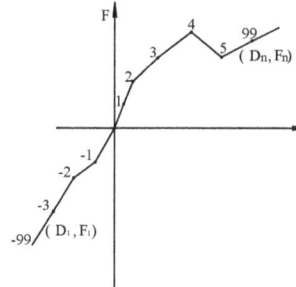

2.2. Calculation model

Based on the 30mm×30mm×50mm specimens, calculation model is shown in figure 5. Matrix use SOLID65 element, and the x, y direction to mesh with 3 mm, z direction to mesh with 1mm. Elastic modulus of cement matrix is 2.0×10^4. Thermal expansion coefficient is $1.1 \times 10^{-4}/°C$. Poisson's ratio is 0.167. Fiber use LINK10 element, mesh as 1mm, set in the centre of the mortar specimen along Z direction. Length of fiber respectively takes 5 mm, 10 mm and 15 mm, whose initial diameter is 0.63, that elastic modulus is 2647. Pull out force is applied at the end of the fiber, and fixed constraint is imposed on the other end of the matrix unit. The bond-slip relationships of fiber-cement matrix are obtained by test [6].

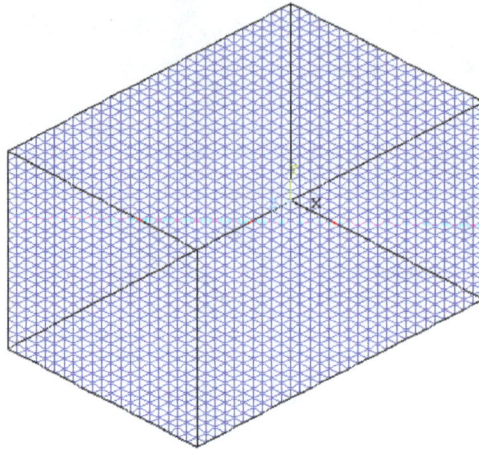

Fig. 5 Calculation Model of fiber pull-out test

2.3. Calculation results and analyses

2.3.1. Factor analysis on embedding length of fiber

Firstly, the influence of embedment length on the fiber stress is analyzed. Set extraction force F=1N. Length of fiber respectively takes 5mm, 10mm and 15mm. Distribution of fiber stress along the embedment length is calculated as different embedment length.

As shown in figure 6, the fiber stress, corresponding the three kinds of fiber length, decreases along embedment direction. Compared the date in Table 1, the fiber stress at the pullout end increases with increasing embedment length, while which at the embedment end decreases with increasing embedment length. Fiber embedment length is smaller, and its stress value is the more uniform as pullout force unchanged.

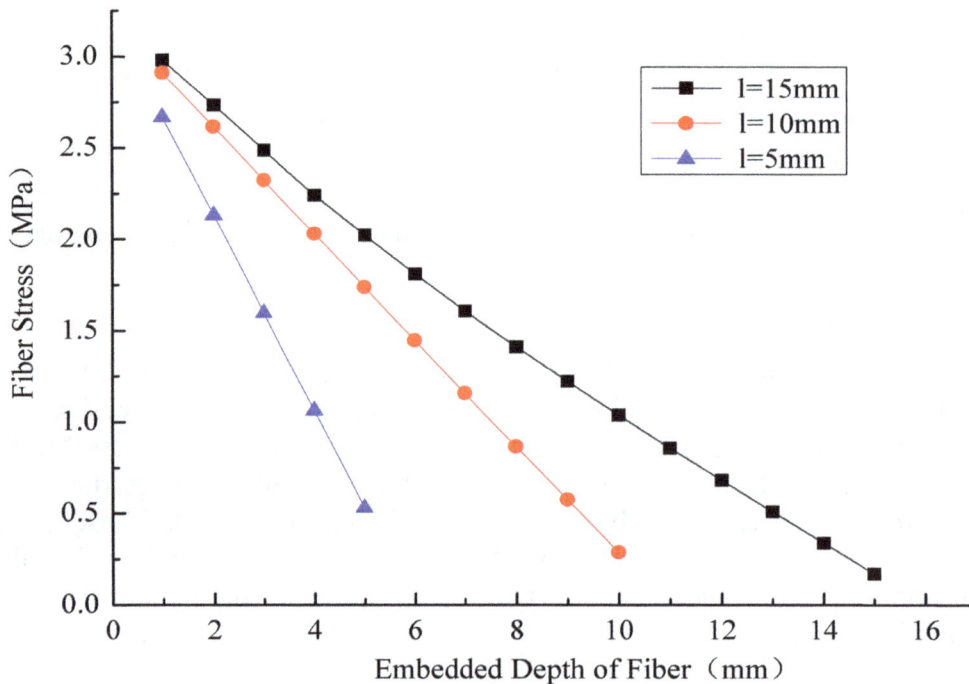

Fig. 6 Distribution of fiber stress with the different fiber embedded length

Table 1 Fiber stress with the different fiber embedded length (Mpa)

Distance/mm	l=5 mm	l=10 mm	l=15 mm	Distance/mm	l=10 mm	l=15 mm	Distance/mm	l=15 mm
1	2.666	2.900	2.982	6	1.420	1.919	11	0.939
2	2.128	2.596	2.760	7	1.133	1.718	12	0.749
3	1.593	2.297	2.544	8	0.848	1.520	13	0.561
4	1.061	2.001	2.332	9	0.666	1.321	14	0.373
5	0.530	1.709	2.123	10	0.282	1.130	15	0.186

According to fiber pull-out tests, Fiber passes pullout force to the cement matrix by the bond force of interface between the two, so the bond force need to research. Length of fiber respectively takes 5mm, 10mm and 15mm. The bond force is calculated as different embedment length, as shown in Table 2.

Table 2 Bond force with the different fiber embedded length (N)

Distance /mm	l=5 mm	l=10 mm	l=15 mm	Distance/mm	l=10 mm	l=15 mm	Distance /mm	l=15 mm
0	-0.169	-0.096	-0.070	6	-0.089	-0.063	11	-0.059
1	-0.168	-0.095	-0.069	7	-0.089	-0.062	12	-0.059
2	-0.167	-0.093	-0.068	8	-0.088	-0.061	13	-0.058
3	-0.166	-0.092	-0.066	9	-0.088	-0.060	14	-0.058
4	-0.165	-0.091	-0.065	10	-0.088	-0.060	15	-0.058
5	-0.165	-0.090	-0.064		Σ=-1N			Σ=-1N
	Σ=-1N							

Table 2 shows that fiber pullout force keeps balance with bond force between fiber and mortar. In the case of fiber and substrate material is constant, there is obvious relationship between the distribution of bond force and fiber embedment length. As shown in figure 7, distribution of bond force is uniform as the fiber embedment length of 5 mm, 10 mm; and when the fiber embedment length is 15 mm, the bond force decreases with distance from fiber pull-out side increasing. Both cases are consistent with the fiber pull out test result.

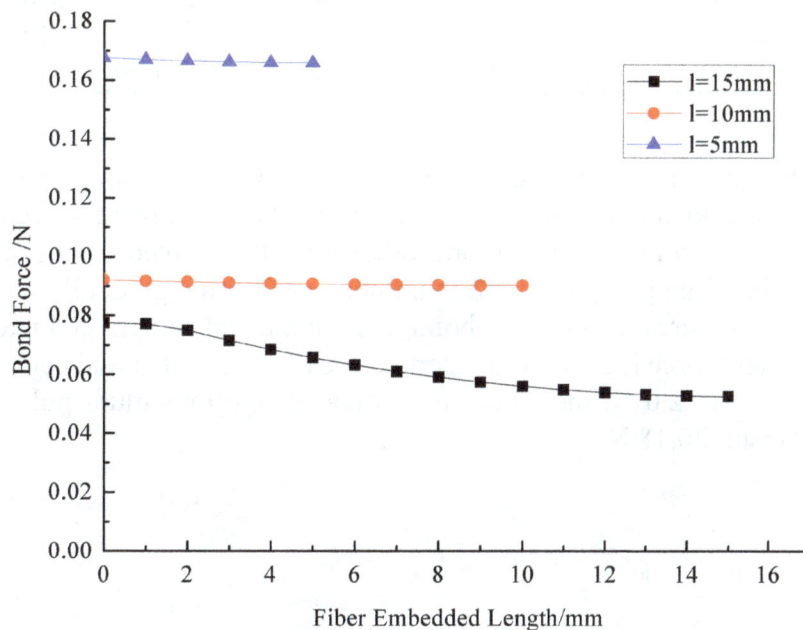

Fig. 7 Distribution of bond force with the different fiber embedded length

Based on the analysis to the impact of the embedding length of fiber on bond force between fiber and cement matrix, it can be considered that a pull-out force certainly corresponds to a minimum fiber embedment length, while the entire length of fiber participate in the work, bond stress between fiber and cement matrix reaches the maximum value. The minimum fiber embedment length is corresponding to fiber critical length of a pullout force.

2.3.2. *The comparative analysis of calculation results and the pull-out tests results*

According to the aforementioned finite element analysis method, comparing calculated results that the maximum pull-out displacement and force of fiber pull-out side corresponded to fiber pull-out force to the test, it is reasonable to use the calculation methods, and it can provide evidence for the fiber pull-out model.

Table 3 Bond force between fiber and cement matrix with the different pull-out force

Distance/mm	-3.344N	-7.82N	-10.971N	-16.78N	-20.18N	-21.18N	-22.0N
0	-0.236	-0.536	-0.739	-1.161	-1.317	-1.326	-1.326
1	-0.231	-0.527	-0.726	-1.142	-1.315	-1.326	-1.326
2	-0.226	-0.518	-0.714	-1.120	-1.312	-1.326	-1.326
3	-0.221	-0.510	-0.703	-1.101	-1.310	-1.326	-1.326
4	-0.217	-0.503	-0.692	-1.083	-1.305	-1.326	-1.326
5	-0.213	-0.497	-0.683	-1.067	-1.289	-1.326	-1.326
6	-0.210	-0.491	-0.678	-1.052	-1.274	-1.326	-1.326
7	-0.207	-0.485	-0.676	-1.039	-1.261	-1.326	-1.326
8	-0.204	-0.481	-0.674	-1.027	-1.250	-1.324	-1.326
9	-0.201	-0.477	-0.672	-1.017	-1.240	-1.323	-1.326
10	-0.199	-0.473	-0.671	-1.008	-1.231	-1.322	-1.326
11	-0.198	-0.470	-0.670	-1.001	-1.224	-1.321	-1.326
12	-0.196	-0.468	-0.669	-0.996	-1.218	-1.320	-1.326
13	-0.195	-0.466	-0.669	-0.991	-1.214	-1.320	-1.326
14	-0.195	-0.465	-0.668	-0.988	-1.211	-1.320	-1.326
15	-0.194	-0.464	-0.668	-0.987	-1.210	-1.319	-1.326
total/N	-3.344	-7.82	-10.971	-16.78	-20.18	-21.18	-22.0

Table 3 shows the bond force between fiber and cement matrix along the direction of fiber embedded corresponding to the different pull-out force. The bond force increases with pull-out force increasing, until reaching the maximum value and it remained almost unchanged as the pull-out force increasing. The position of the maximum bond force gradually move into the fiber, that is, the debonded part increases, until the bond force at the embedment end reach the maximum. When value of the pullout force is 21.18N, the bond force of partial fiber and cement matrix reached the maximum; When it is 22.0 N, the whole fiber debond, the maximum pullout force differ by 9.02% from the test result 20.18 N.

Table 4 Pull-out displacement with the different pull-out force (mm)

Distance /mm	3.344N	7.820N	10.971N	14.678N	16.780N
0	-0.166	-0.879	-1.223	-1.586	-1.756
1	-0.163	-0.870	-1.210	-1.569	-1.736
2	-0.159	-0.861	-1.198	-1.552	-1.717
3	-0.156	-0.854	-1.187	-1.537	-1.700
4	-0.153	-0.847	-1.177	-1.522	-1.681
5	-0.150	-0.841	-1.167	-1.510	-1.670
6	-0.148	-0.835	-1.159	-1.499	-1.657
7	-0.148	-0.830	-1.151	-1.488	-1.645
8	-0.146	-0.825	-1.144	-1.479	-1.635
9	-0.144	-0.821	-1.138	-1.471	-1.626
10	-0.142	-0.818	-1.133	-1.464	-1.618
11	-0.140	-0.815	-1.129	-1.458	-1.612
12	-0.139	-0.813	-1.126	-1.454	-1.607
13	-0.138	-0.811	-1.123	-1.450	-1.603
14	-0.138	-0.810	-1.121	-1.448	-1.600
15	-0.137	-0.810	-1.120	-1.447	-1.599

Table 4 shows the pullout displacement along the direction of fiber embedded corresponding to the different pull-out force. And the pullout displacement of fiber increases gradually with the pullout force increasing. Comparing the calculated results of the pull-out displacement as applied the different pull-out force to the test, the result is shown in Table 5, and the range of relative difference between calculated and experimental value is 7.40% ~ 12.37%.

Table 5 Test results of the pull-out displacement at the fiber pullout side as applied the different pull-out force

Pull-out force/N	3.344	7.82	10.971	14.678	16.78
Test/mm	0.194	0.969	1.356	1.744	1.938
Calculation/mm	0.166	0.879	1.223	1.586	1.756
Deformation of pullout fiber /mm	0.004	0.008	0.016	0.029	0.038
Total of calculated value /mm	0.170	0.887	1.239	1.615	1.794
Relative difference /%	12.37	8.46	8.63	7.40	7.43

3. Conclusion

Finite element analysis show that fiber stress is transferred to the fiber embedded side as the fiber carrying the pullout force. When the pullout force is smaller, bond stress of fiber-mortar takes the form of a triangle or trapezoid; when it is maximum, bond stress is at maximum. The maximum pullout force calculated by the bond-slip relationships of fiber and mortar, which obtained by the test, differs by 9.02% from the test, and the calculated value of pullout displacement in upward section pullout force-displacement curve differs by 7.43%~12.37% from the test. Therefore the fiber spanned across the crack of cement matrix is pulled out in the process of crack developing, which consumes energy and restrains crack developing.

Acknowledgments

This research was funded by the Prospective Study Projects of Jiangsu Industry-University-Research and Joint Innovation Funding: (BY2012039), Doctor Foundation of Jinling Institute of Technology (jit-b-201107, jit-b-201223).

References

1. Wang Zhi, Hu Xiao-bo, Bao Guang-yu. Review of the research of interfacial bond property between fiber and cement matrix [J]. Concrete, 2003, 168(11): 7-10. (In Chinese)
2. Yang Meng, Huang Cheng-kui. Bond Slip Characteristics between Steel Fiber and High Strength Mortar[J]. Journal of building Materials, 2004, 7(4): 384. (In Chinese)
3. M.J. Shannag, R. Brinker, W. Hansen, Pullout behavior of steel fibers from cement-based composites, Cem. Concr. Res. 27 (6) (1997) 925– 936.
4. Robins P, Austin S, Jones P. Pull-out behavior of hooked steel fibers. Mater Struct 2002; 35(7): 434-42.
5. Hamoush S, Abu-Lebdeh T, Cummins T, Zornig B. Pullout characterizations of various steel fibers embedded in very high-strength concrete. Am J Eng Appl Sci 2010; 3(2): 418-26.
6. Xuan Weihong, Liu Jianzhong, Li Xiaochun, Experimental Study on Bond Performance between Polypropylene Fiber and Cement Mortar [J]. Journal of Jiang Su University (Natural Science Edition), 2003, 31(6): 726-730. (In Chinese)

Study on Pressure-induced Structural and Magnetic Phase Transitions of Binary Half-metals: SrC and SrN

Hui Zhao, Yang Zhang[*], Yan Zhao, Yi-Ning Zhang

Department of Physics, Anshan Normal University, Anshan 114007, China
[*]Email: 460919991@qq.com

The phase transitions under hydrostatic pressure in SrC and SrN are investigated by first-principles calculation based on density functional theory. Four structures: rocksalt (B1), cesium chloride (B2), zinc-blende (B3), and nickel arsenide (B8$_1$) and two magnetic states: ferromagnetic (FM) and paramagnetic (PM) are considered for both SrC and SrN. The B1 to B2 structural transition, high-spin to low-spin magnetic transition, and half-metal to metal electronic topological transition are reported. We predict FM B1 $\xrightarrow{\text{4 GPa}}$ PM B2 for SrC and FM B1 $\xrightarrow{\text{14 GPa}}$ FM B2 for SrN. Phonon band dispersions, density of phonon states, and elastic stiffness constants are given to test the lattice dynamical and mechanical stability for pressure-induced phase.

Keywords: Binary ferromagnet; phase transition; pressure; density of phonon states; elastic stiffness constants.

1. Introduction

Since the first prediction in 1983 by de Groot et al. of half-metallic (HM) ferromagnetism in the half-Hesuler compounds NiMnSb and PtMnSb, many other systems have been found to be HM ferromagnets [1]. In 2004, Kusakabe and collaborators predicted that B3 Ca pnictides (i.e., CaP, CaAs, and CaSb) belong to the class of HM ferromagnets [2]. Then, Sieberer et al. and Volnianska et al. found HM ferromagnetism in many II^A-V^A B3 compounds [3-5]. Evidence of the growth of such sp compounds has been provided by Liu et al. who have reported successful self-assembly growth of high-quality ultrathin CaN in the B1 structure on top of Cu (001) [6]. In the past years, sp magnetic phenomena have attracted plenty of attentions in the academic circle. Nowadays, many works have been done on numerous sp magnets such as perfect and defect bulk solids, electrides, nanomaterials, surfaces and interfaces [7–14].

Especially, binary compounds with simple structures, which are compatible with important III-V and II-VI semiconductors are useful and meaningful for applied physical community. Many *sp*-band HM ferromagnets were intensively investigated in the compounds of I^A-IV^A [15], I^A-V^A [16], I^A-VI^A [17], II^A-III^A [18], II^A-IV^A [19], II^A-VI^A [20], and so on. Meanwhile much attention has been given to understand the physical mechanisms and possible applications. Geshi *et al.* proposed a synthesis process for these compounds using the high pressure and temperature environment [21]. Li *et al.* investigated effect of spin-orbital coupling on half-metallic magnetic properties of these compounds [22]. Laref *et al.* reported exchange interactions, spin waves, and Curie temperature in these compounds [23]. Özdoğan *et al.* studied robustness and stability of these compounds against doping and deformation [24, 25]. Tabatabaeifar *et al.* explored the surface properties of these compounds [26].

It is known that that density functional calculation can predict structural stability of these novel compounds. However, the structural stability of them has only rarely been explores [27-29]. By comparison thousands of literatures are only devoted to study the magnetic and bonding properties from the electronic structures. A lack of confirmation of the structural stability may discourage experimentalists to synthesize the proposed materials. To search for a new structure, we usually calculate Gibbs-free energy $G = E + PV - TS$ of possible structures and finally we recognize the structure with lowest Gibbs-free energy as a desired structure. Normally, density functional

---

[*]Corresponding author

calculations are performed at zero temperature, and therefore the Gibbs-free energy becomes equal to the enthalpy H = E + PV. A stable structure at a given pressure is one for which enthalpy has its lowest value, and thus transition from one phase to another is given by a pressure at which the enthalpies for the two phases are equal. So we obtained the phase transition pressure by matching enthalpies of all the phases. Therefore, at transition pressure the difference in enthalpies for the corresponding two phases becomes zero. Nevertheless, the thermodynamic stability may not necessarily guarantee the structural stability of the system. In order to confirm the structural stability of the proposed compounds, the elastic constants and phonon dispersion, at least, should be known. Additionally, magnetic materials may undergo a pressure-induced magnetic transition because of band widening or because of changes in the crystal field [30-33]. Since the cell volume and crystal structure controls the dispersion of electronic bands it has probably the largest influence on the formation of HM ferromagnetism. As the changes in the cell volume and crystal structure are controlled by pressure, the pressure-induced investigations on a HM ferromagnet should be a useful and meaningful way to understand its basic and important properties.

In this article, the compressibility behaviors of two sp compounds (SrC and SrN) are studied using first-principles calculations based on density functional theory. We choose B1, B2, B3 and B8$_1$ structures for the four considered modifications. The enthalpies, total energies, elastic stiffness constants, magnetic moments, volumes, bond lengths, electronic band structures, electronic total and partial density of states, phonon band dispersions, and density of phonon states at ambient and elevated pressures are calculated to help discuss the structural preferences, electronic properties, and magnetic behaviors.

2. Computational Methods

The computational studies presented here are carried out using the calculation module CAMBRIDGE SERIAL TOTAL ENERGY PACKAGE of Materials Studio 5.5 [34-36]. The generalized gradient approximation (GGA) in the scheme of Perdew, Burke and Ernzerhof (PBE) is used for the exchange-correlation functional [37, 38]. The cutoff kinetic energy for the plane waves is set to be 310 eV for SrC, and 280 eV for SrN. Good convergence was obtained with the energy Change to 5.0×10^{-6} eV/atom, maximum force to 0.01 eV/$\overset{\circ}{A}$, maximum stress to 0.02 GPa, maximum displacement tolerance to 5.0×10^{-4} $\overset{\circ}{A}$ and k-points separation to 0.02 $\overset{\circ}{A}^{-1}$[39]. The atomic configurations used to generate the ultrasoft pseudopotentials are $4s^2 4p^6 5s^2$ for Sr, $2s^2 2p^2$ for C, and $2s^2 2p^3$ for N [40].

Table 1 Calculated optimized equilibrium lattice parameters and magnetic moments per formula

Crystal structure	Lattice parameter ($\overset{\circ}{A}$)		Magnetic moment (μ_B)	
	SrC	SrN	SrC	SrN
FM B1	a = 5.675	a = 5.386	2	1
PM B1	a = 5.605	a = 5.363		
FM B2		a = 3.238		1
PM B2	a = 3.341	a = 3.231		
FM B3	a = 6.138	a = 5.813	2	1
PM B3	a = 6.035	a = 5.776		
FM B8$_1$	a = 3.885	a = 3.665	2	1
	c = 6.929	c = 6.678		
PM B8$_1$	a = 3.698	a = 3.665		
	c = 7.249	c = 6.575		

3. Results and Discussions

We performed calculations for two sp compounds: SrC and SrN. The calculated optimized equilibrium lattice parameters and magnetic moments per formula unit at zero pressure are calculated and presented in Table 1. FM B1→PM B2 phase transition for SrC and SrN are predicted by the enthalpy method. And then the common tangent method [41, 42], which is thought to be another approach to evaluate the transition pressure, is employed to test the results of the transition pressures. Fig. 1(a) shows that at ambient pressure SrC crystallizes in B1 structure and is ferromagnetically stable in FM state and FM B1 phase changes to PM B2 phase as pressure reaches 4 GPa. Similarly, for SrN, Fig. 1(b) indicates a phase transition from FM B1 to FM B2 at a pressure of 14 GPa.

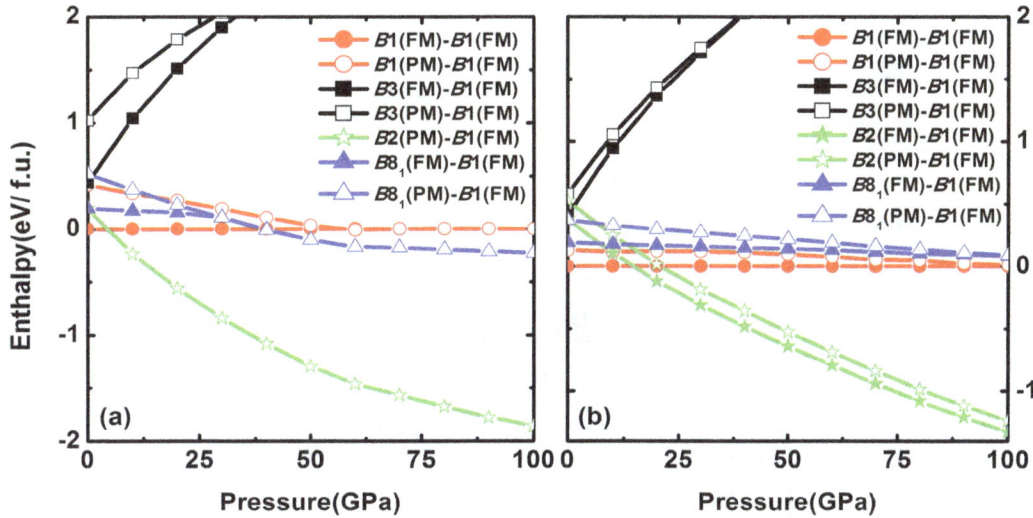

Fig. 1 Enthalpy-pressure curves for the FM and PM phases of (a) SrC and (b) SrN of B1, B2, B3 and B8$_1$ structures, referenced to FM B1 phase

The common tangent also give the transition pressure of 3.95 GPa and 13.96 GPa, for SrC and SrN, respectively, which agree well with the values obtained from the enthalpy method (see Fig. 2). The present structural transition may be understood based on Buerger mechanism or WTM mechanism [43].

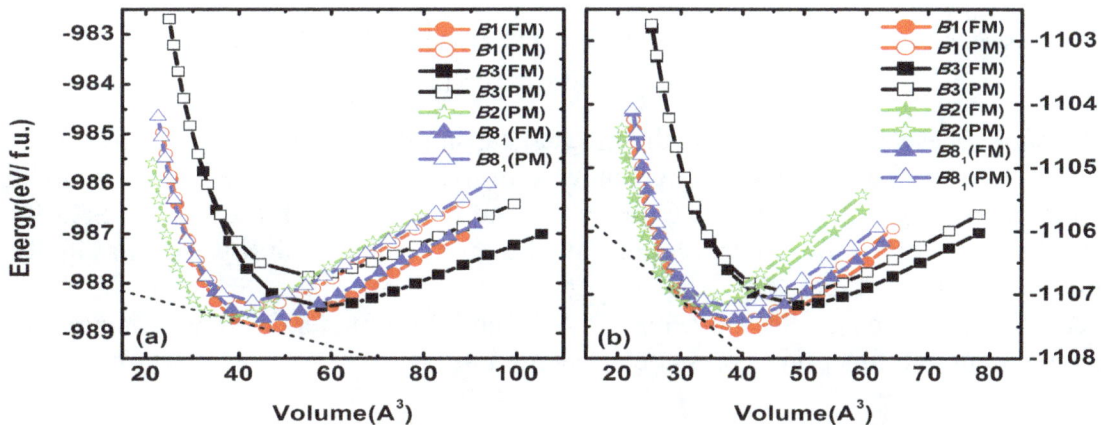

Fig. 2 Energy versus volume curves for the FM and PM phases of (a) SrC and (b) SrN of B1, B2, B3 and B8$_1$ structures (The transition pressure is derived from the common tangent method)

745

Phonon dispersion of SrC and SrN with B2 structure is calculated and the typical phonon band dispersion and density of phonon states are shown in Fig. 3 It is found that both SrC and SrN are stable lattice dynamically. The elastic stiffness constants for SrC and SrN with cubic B2 structure are tabulated in Table 2.

Table 2 Calculated Elastic stiffness constants C_{11}, C_{44}, and C_{12} for B2-SrC at 5 GPa (a=3.262 Å), and B2-SrN at 15 GPa (a=3.078 Å).

Compound	C_{11}	C_{44}	C_{12}
SrC	130.463	41.565	54.465
SrN	69.964	17.151	160.126

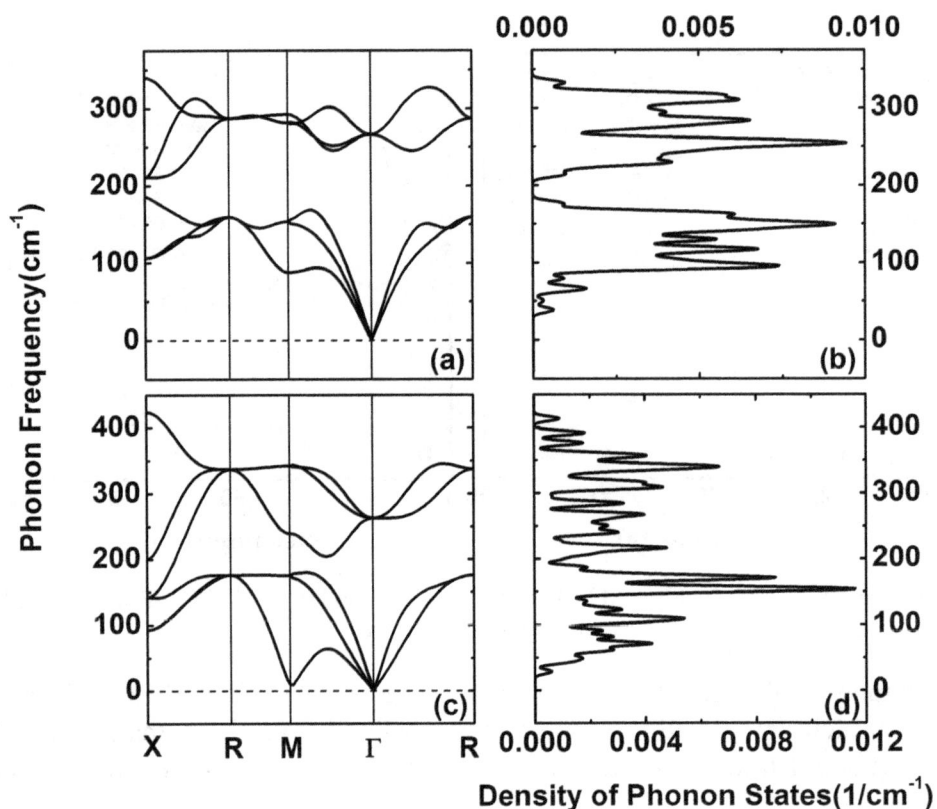

Fig. 3 Phonon band dispersion and density of phonon states of [(a) and (b)] B2-SrC under 5 GPa (a=3.262 Å) and [(a) and (b)] B2-SrN under 15 GPa (a=3.078 Å)

For the cubic phase, at zero pressure, the mechanical stability criteria are given by $C_{11}>0$, $C_{44}>0$, $C_{11}>|C_{12}|$, and $(C_{11}+2C_{12})>0$ [44]. In the special case of hydrostatic pressure, the new conditions for elastic stability are $C_{11} + 2C_{12} + P > 0$, $C_{44} - P > 0$, and $C_{11} - C_{12} - 2P > 0$ [45]. For SrC, it is stable mechanically. For SrN, the contravention of the mechanical stability criteria indicates the structural mechanical instability.

We calculated the magnetic moments of SrC and SrN under various compressions. The magnetic moment as a function of pressure are plotted in Fig. 4.

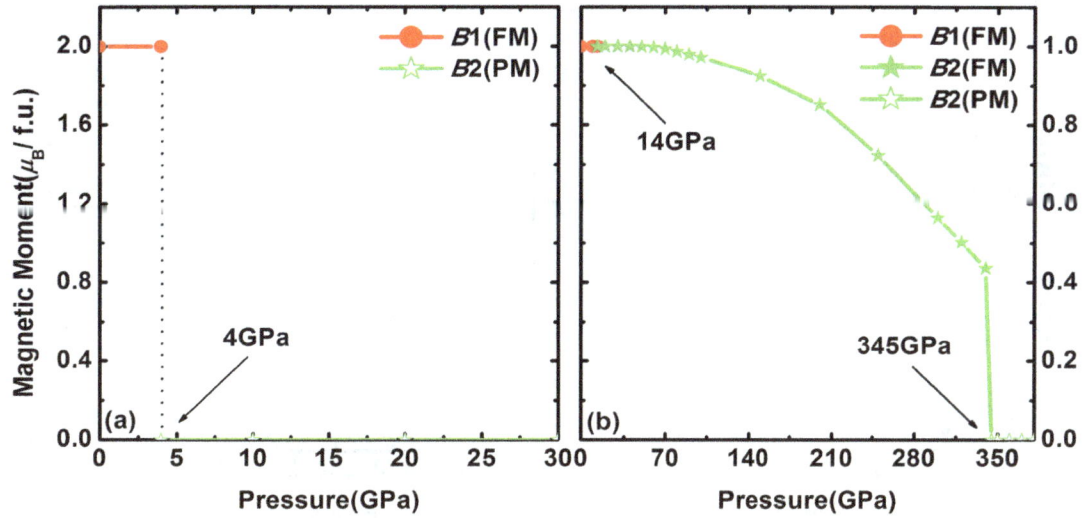

Fig. 4 The magnetic moment (in Bohr magneton μ_B) dependence on the pressure for (a) SrC and (b) SrN, respectively

At ambient pressure the magnetic moment of SrC is 2.00 μB, and it becomes zero rapidly after the phase transition pressure 4 GPa. For SrN, the magnetic moment decreases slowly, so it is much more stable than SrC. At 14 GPa the magnetic moment of SrN is continuous, which indicate that SrN transforms from B1 to B2 structure without change in magnetic moment. Volume collapses are predicted by plotting the pressure-volume relations in Fig. 5.

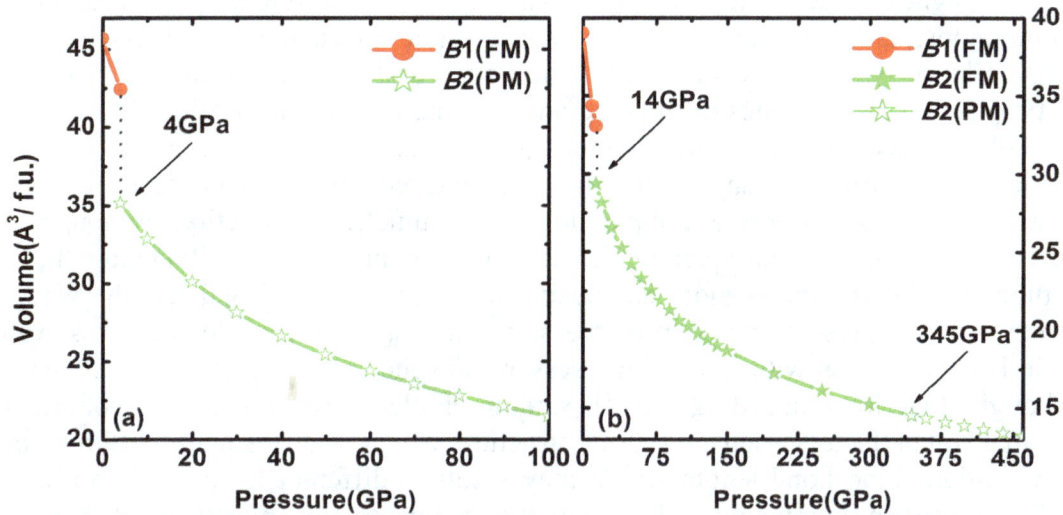

Fig. 5 Variation in volume with relative pressure for (a) SrC and (b) SrN, respectively

For SrC, FM B1 transforms into PM B2 phase at 4 GPa with a volume contraction of 17.1%. For SrN, it undergoes a phase transition from FM B1 to FM B2 at 14 GPa with 11.1% volume collapse. The bond length of SrC decreases from 2.84 to 2.77 Å when increasing the pressure from 0 to 4 GPa. At 4 GPa it rises back to 2.84 Å and then decreases monotonously. For SrN, the bond length decreasing from 2.70 to 2.55 Å when increasing the pressure from 0 to 14 GPa. At 14 GPa, it rises to 2.67 Å and then decreases monotonously (see Fig. 6).

747

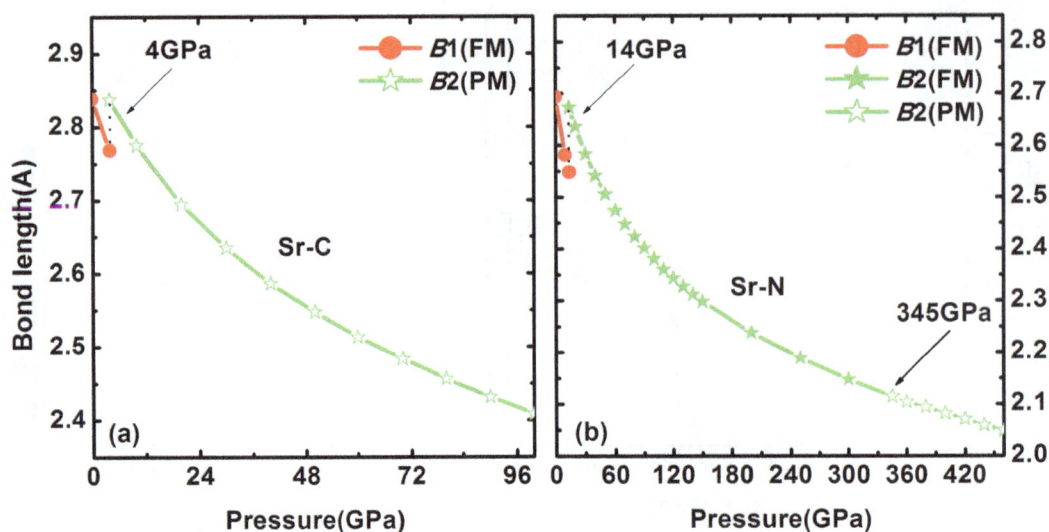

Fig. 6 Pressure dependence of the cation-anion bond length for (a) SrC and (b) SrN, respectively

4. Conclusion

We use both the enthalpy method and common tangent method to evaluate the transition pressures and the results are in good agreement. The phase transitions: FM B1 $\xrightarrow{4\text{ GPa}}$ PM B2 and FM B1 $\xrightarrow{14\text{ GPa}}$ M B2 for SrC and SrN, respectively, have been predicted by means of detailed structural and magnetic phase diagrams. We found that B2 structure is preferred for both the present compounds, although some of them are instable mechanically. It is found that the volumetric collapse always accompanies the structural phase transition. Nevertheless, the magnetic phase transition does not induce any volumetric collapse ordinarily. Moreover, the volumetric collapse accompanying the structural phase transition demonstrates that as pressure increases the crystal structure may prefer the close-packed arrangement. It is found that the magnetic sharply decrease occurs accompanying the volumetric contraction, although the bond length increases obviously. Consequently, it may be thought that for influencing the value of magnetic moment, the volume is more conclusive than cation-anion bond length. We generally find that, for both the present compounds, the spin exchange splitting decreases as well as the width of the band increases with increasing pressure, despite there is a structural transition. It is may be thought that for controlling the dispersion of electronic bands, the volume is more conclusive than other factors such as crystal structure which may result in different exchange splitting and cation-anion bond length which may result in different levels of hybridization and bonding. For these sp compounds, the magnetic moments are mostly contributed by the nondegenerate spin splitting of the degenerate 2p bands and the main reason for the pressure-induced magnetic phase transitions of them is the band widening of anion p states. We also find that SrN could maintain FM order after structural phase transition as well as undergoes FM-PM transition under relatively high pressure. Contrarily, SrC would undergo FM-PM transition under relatively low pressure when the structural phase transition occurs.

References

1. R. A. de Groot, F. M. Mueller, P. G. van Engen, and K. H. J. Buschow, Phys. Rev. Lett. 50, 2024 (1983).

2. K. Kusakabe, M. Geshi, H. Tsukamoto, and N. Suzuki, J. Phys.: Condens. Matter 16, S5639 (2004).

3. M. Sieberer, J. Redinger, S. Khmelevskyi, and P. Mohn, Phys. Rev. B 73, 024404 (2006).

4. O. Volnianska, P. Jakubas, and P. Bogusławski, J. Alloys Compd. 423, 191 (2006).

5. O. Volnianska and P. Bogusławski, Phys. Rev. B 75, 224418 (2007).

6. X. Liu, B. Lu, T. Iimori, K. Nakatsuji, and F. Komori, Surf. Sci. 602, 1844 (2008).

7. J. M. D. Coey, Solid State Sci. 7, 660 (2005).

8. J. J. Attema, G. A. de Wijs, and R. A. de Groot, J. Phys.: Condens. Matter 19, 165203 (2007).

9. O. Volnianska and P. Bogusławski, J. Phys.: Condens. Matter 22, 073202 (2010).

10. A. L. Ivanovskii, Phys.-Usp. 50, 1031 (2007).

11. C. J. Pickard and R. J. Needs, Phys. Rev. Lett. 107, 087201 (2011).

12. S. J. Dong and H. Zhao, Appl. Phys. Lett. 100, 142404 (2012).

13. G. Fischer, N. Sanchez, W. Adeagbo, M. Lüders, Z. Szotek, W. M. Temmerman, A. Ernst, W. Hergert, and M. C. Muñoz, Phys. Rev. B 84, 205306 (2011).

14. Y. Gohda and S. Tsuneyuki, Phys. Rev. Lett. 106, 047201 (2011).

15. W. X. Zhang, Z. D. Song, B. Peng, and W. L. Zhang, J. Appl. Phys. 112, 043905 (2012).

16. A. Lakdja, H. Rozale, and A. Chahed, Comp. Mater. Sci. 67, 287 (2013).

17. M. Moradi, M. Rostami, and M. Afshari, Can. J. Phys. 90, 531 (2012).

18. L. Adamowicz and M. Wierzbicki, Acta Phys. Pol. A 115, 217 (2009).

19. G. Y. Gao, K. L. Yao, E. Ş̧a̧sıoğ̌lu, L. M. Sandratskii, Z. L. Liu, and J. L. Jiang, Phys. Rev. B 75, 174442 (2007).

20. M. Geshi, K. Kusakabe, H. Nagara, and N. Suzuki, J. Phys. Soc. Jpn. 76, 074717 (2007).

21. M. Geshi, K. Kusakabe, H. Nagara, and N. Suzuki, Phys. Rev. B 76, 054433 (2007).

22. Y. Li and J. Yu, Phys. Rev. B 78, 165203 (2008).

23. A. Laref, E. Şaşioglu, and I. Galanakis, J. Phys. Condens. Matter 23, 296001 (2011).

24. K. Ōzdoğan, E. E. Şaşioglu, and I. Galanakis, J. Appl. Phys. 111, 113918 (2012).

25. K. Ōzdoğan, and I. Galanakis, J. Adv. Phys. 1, 69 (2012).

26. A. H. Tabatabaeifar, S. Davatolhagh, and M. Foroughpour, J. Appl. Phys. 114, 213705 (2013).

27. M. Geshi, Physica B 405, 517 (2010).

28. Z. Nourbakhsh, S. J. Hashemifar, and H. Akbarzadeh, J. Alloys Compd. 579, 360 (2013).

29. R. R. Palanichamy, G. S. Priyanga, A. J. Cinthia, A. Murugan, A. T. A. Meenaatci, and K. Iyakutti, J. Magn. Magn. Mater. 346, 26 (2013).

30. R. E. Cohen, I. I. Mazin, and D. G. Isaak, Science 275, 654 (1997).

31. J. F. Lin, V. V. Struzhkin, S. D. Jacobsen, M. Y. Hu, P. Chow, J. Kung, H. Liu, H. K. Mao, and R. J. Hemley, Nature (London) 436, 377 (2005).

32. J. Kuneš, A. V. Lukoyanov, V. I. Anisimov, R. T. Scalettar, and W. E. Pickett, Nat. Mater. 7, 198 (2008).

33. T. Kawakami, Y. Tsujimoto, H. Kageyama, X. Q. Chen, C. L. Fu, C. Tassel, A. Kitada, S. Suto, K. Hirama, Y. Sekiya, Y. Makino, T. Okada, T. Yagi, N. Hayashi, K. Yoshimura, S. Nasu, R. Podloucky, and M. Takano, Nat. Chem. 1, 371 (2009).

34. V. Milman, B. Winkler, J. A. White, C. J. Pickard, M. C. Payne, E. V. Akhmatskaya and R. H. Nobes, Int. J. Quantum Chem. 77, 895 (2000).

35. M. D. Segall, P. J. D. Lindan, M. J. Probert, C. J. Pickard, P. J. Hasnip, S. J. Clark, and M. C. Payne, J. Phys.: Condens. Matter 14, 2717 (2002).

36. S. J. Clark, M. D. Segall, C. J. Pickard, P. J. Hasnip, M. I. J. Probert, K. Refson, and M. C. Payne, Z. Kristallogr. 220, 567 (2005).

37. J. P. Perdew and Y. Wang, Phys. Rev. B 33, 8800 (1986).

38. J. P. Perdew, K. Burke, and M. Ernzerhof, Phys. Rev. Lett. 77, 3865 (1996).
39. J. Monkhorst and J. Pack, Phys. Rev. B 13, 5188 (1976).
40. D. Vanderbilt, Phys. Rev. B 41, 7892 (1990).
41. J. Cai and N. X. Chen, J. Phys.: Condens. Matter 19, 266207 (2007).
42. Y. Wang, T. J. Hou, S. Tian, S. T. Lee, and Y. Y. Li, J. Phys. Chem. C 115, 7706 (2011).
43. X. Zhou, J. L. Roehl, C. Lind, and S. V. Khare, J. Phys.: Condens. Matter 25, 075401 (2013).
44. Z. J. Wu, E. J. Zhao, H. P. Xiang, X. F. Hao, X. J. Liu, and J. Meng, Phys. Rev. B 76, 054115 (2007).
45. G. Grimvall, B. Magyari-Köpe, V. Ozoliņš, and K. A. Persson, Rev. Mod. Phys. 84, 945 (2012).

Equivalent Model and Simulation Analysis of Solar Wing Substrate

Wei-Wei Bian[1,*], Zong-Ze Xia[2], Fei Xia[3], Zhen-Fang Xin[1]

[1]*Beijing Institute of Mechanical Equipment, Beijing, 100854, China*
[2]*State Grid Liaoyang Electric Power Supply Company, Liaoning, 111000, China*
Email: karovie@163.com

As energy carrier for the satellite and spacecraft, solar wings are related to the works of them. Thus, in the early design stage, the modal parameters of the solar wings are usually classified as the focus of the study. Compared several common substrate modeling methods and ideas, an equivalent model and an approximation method are proposed to analysis the floor and honeycomb core in existing solar wing substrates. The results indicate that it will be a good approximation to the honeycomb core model based on geometry by using the method of finite element modeling for anisotropic material equivalent honeycomb core.

Keywords: Solar wing substrate; honeycomb core; modal analysis; equivalent model.

1. Introduction

A solar wing substrate is a sandwich structure that consists of the core layer with honeycomb core and the panel with carbon fiber. Generally, the honeycomb materials are Aluminum and Aramid papers. The key to modeling the substrate structure is to select an appropriate equivalent model of the carbon fiber layer model and the aluminum honeycomb. The existing simulation analysis methods have some differences in the complexity, precision and applicability [1]. How to select an appropriate equivalent model according to the analysis of the types and purposes needs comparative research based on the equivalent models. In order to ensure the reliability of the solar wings, the modal parameters are important items in the product acceptance, because the modal parameters are directly related to the dynamic response characteristics of the product. Therefore, in the early design stage, it is necessary to carry out the modal calculation and simulation analysis of the modal characteristics of the substrates, so as to ensure that the products are not destroyed in the mechanical environment [2].

In this paper, the mechanical properties of the surface layer and the honeycomb core are analyzed primarily, and two independent modeling ideas of the honeycomb core and the surface layer are presented. Then combined with the existing literatures, three kinds of integrated modeling schemes are presented, and the differences and similarities of them are put forward. Finally on the basis of theoretical analysis of existing equivalent models, investigate the differences of the substrates between the threes kinds schemes through the finite element simulation analysis. The results obtained can be used to guide the analysis and calculation of the honeycomb sandwich structure.

2. Equivalent Models of the Panel Layer and the Honeycomb Core

2.1. *Equivalent models of the panel layer*

The panel layers are located on both sides of the honeycomb core and in mirror symmetrical distribution. When designing the panel layer, the tension, compression and shear stress caused by the design load should be considered, and it needs to be able to prevent the panel from forming under the design load.

---

*Corresponding author

$$\begin{cases} \chi_2 = \dfrac{0.2}{1-v_m}\left(1.1-\sqrt{\dfrac{E_m}{E_{f2}}}+\dfrac{3.5E_m}{E_{f2}}\right)\left(1+0.22V_f\right) \\[4mm] \chi_{23} = 1.095+\left(0.8-V_f\right)\left[0.27+0.23\left(1-\dfrac{E_{f2}}{E_{f1}}\right)\right] \\[4mm] \chi_{12} = 0.28+\sqrt{\dfrac{E_m}{E_{f2}}} \\[4mm] E_1 = V_f E_{f1}+V_m E_m, \quad E_2 = \dfrac{E_{f2}E_m(V_f+\chi_2 V_m)}{V_f E_m+\chi_2 V_m E_{f2}}, \quad E_2 = E_3 \\[4mm] v_{12} = V_f v_{f12}+V_m v_m, \quad v_{12}=v_{13}, \quad v_{23}=(V_f v_{f12}+V_m v_m)\chi_{23} \\[4mm] G_{12} = \dfrac{G_f G_m(V_f+\chi_{12}V_m)}{V_f G_m+\chi_{12}V_m G_f}, \quad G_{12}=G_{13}, \quad G_{23}=\dfrac{E_2}{2(1+v_{23})} \end{cases} \tag{1}$$

Where, V_f is the volume fraction of fibers; V_m is the volume fraction of matrix; E_{f1} is the longitudinal elastic modulus of fiber; E_{f2} is the transverse elastic modulus of fibers; E_m is the matrix elastic modulus; v_{f12} and v_{f13} are the Poisson ratio of fibers; v_m is the Poisson ratio of matrix; G_f is the shear modulus of fibers; G_v is the shear modulus of matrix.

The panel layer is formed by stacking layers of composite materials in a certain direction and sequence. In order to facilitate the mechanical analysis, the following assumptions are made for the single layer composite material: the composite materials in a single layer are orthotropic materials and the mechanical properties of the lateral monolayer composites are the same. In other words, a single composite material is a transversely isotropic material, and the material parameters, such as E_1, E_2, E_3, G_{12}, G_{13}, G_{23}, v_{12}, v_{13}, and v_{23}, can be solved by following equations [3]:

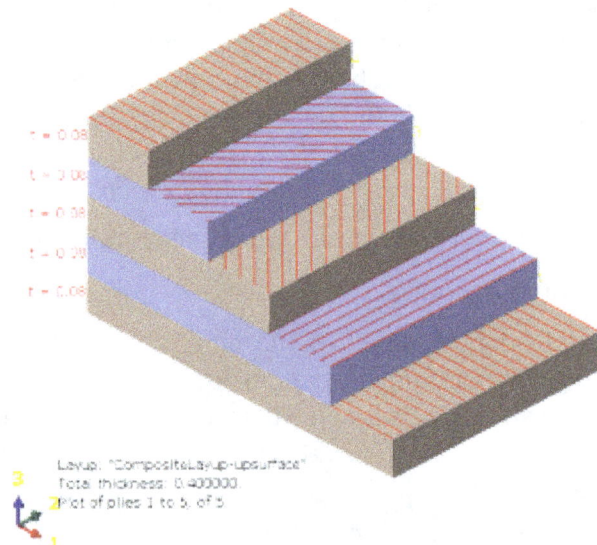

Fig. 1 Panel layer in ABAQUS

The existing finite element analysis software such as ABAQUS and ANSYS can complete multilayer panel layer modeling, can define direction, thickness, mechanical properties of each layer, and can realize the modeling work of multilayer materials composed of shell element and solid element. Figure 1 shows the result of the definition of the panel layer by using the finite element analysis software ABAQUS.

2.2. Equivalent model of the honeycomb core

According to the structure characteristics, the honeycomb core is equivalent to an orthotropic uniform continuous structure. In this way, the in-plane stiffness of the core layer can be considered. Assuming the shape of the aluminum honeycomb core is a regular hexagon, the elastic modulus, the shear modulus, the density, the Poisson ratio of material are E_s, G_s, ρ_s, v_s respectively, the wall thickness of honeycomb core is t_c, the side length is l, and the shear modulus, the density, the Poisson ratio of equivalent core layer are E_c, G_c, ρ_c, v_c respectively, the equivalent elastic parameters of the honeycomb core can be solved by following equations [4]:

$$\begin{cases} E_{cx} = E_{cy} = \dfrac{4}{\sqrt{3}}(\dfrac{t_c}{l})^3 E_s, \ E_{cz} = \dfrac{2}{\sqrt{3}}(\dfrac{t_c}{l})E_s, \ G_{cyz} = \dfrac{\sqrt{3}\gamma}{2}(\dfrac{t_c}{l})G_s, \\ G_{cyz} = \dfrac{\sqrt{3}\gamma}{2}(\dfrac{t_c}{l})G_s, \ G_{cyz} = \dfrac{\gamma}{\sqrt{3}}(\dfrac{t_c}{l})G_s, \ \rho_{ceq} = \dfrac{8}{3\sqrt{3}}(\dfrac{t_c}{l})\rho_s, v_c = v_s \end{cases} \quad (2)$$

Where, γ is the correction coefficient, and generally take 0.4-0.6.

In order to eliminate the longitudinal and transverse mechanical characteristics error caused by equivalent model, the actual geometric model can be used to simulate the honeycomb model, as shown in Figure 2.

Fig. 2 Honeycomb model based on actual geometric

3. Equivalent Model and Modeling Method of the Substrate

3.1. The mutual independence modeling method for the panel layer and the honeycomb core

Based on mechanical analysis, the finite element model of the solar wing substrate can be completed by using the mutual independence modeling method for the panel layer and the honeycomb core, combined with finite element modeling technology. The finite element model of the solar wing substrate can be set up in two ways according to Table 1.

Table 1 The realization approaches of the finite element model for the substrate

Number	Panel layer	Honeycomb core
1	Orthotropic (transverse isotropic)	Actual geometric
2	Orthotropic (transverse isotropic)	Orthotropic

3.2. *The integrated modeling method for the panel layer and the honeycomb core*

3.2.1. *Equivalent plate theory*

With equivalent plate theory, the whole honey comb sandwich plate is equivalent to isotropic plate of different thickness, and then the elastic modulus, thickness and density of the equivalent plate are derived according to the two conditions of equal stiffness and mass of the sandwich plate with the equivalent plate.

According to Hoff theory [5], the mechanical properties of the equivalent plate can be obtained as follows:

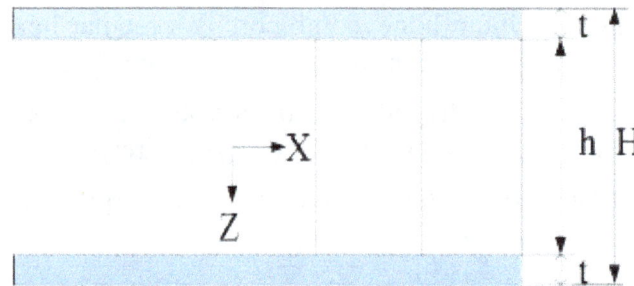

Fig. 3 The parameters of the substrate

$$
\begin{cases}
t_{eq} = \sqrt{t^2 + 3(h+t)^2}, & E_{eq} = \dfrac{2E_f t}{t_{eq}} \\[3mm]
v_{eq} = v_0, & \rho_{eq} = \dfrac{\rho_{ceq}h + 2\rho_0 t}{t_{eq}}k
\end{cases}
\tag{3}
$$

Where, t_{eq}, E_{eq}, v_{eq}, ρ_{eq} are the equivalent thickness, the elastic modulus, the Poisson ratio, and the density of the solar wing substrate respectively; ρ_{eq} is the equivalent density of the central honeycomb core; t is the panel thickness; h is the honeycomb core thickness; ρ_0, v_0 are the density and Poisson ratio of the panel respectively; E_f is the elastic modulus of the panel; k is the added-mass coefficient, and generally take 1.3-2.1.

The theory is easy to implement, but it is not able to reflect the influence of the shape of the honeycomb core and the characteristics of the panel on the whole performance of the sandwich plate. At the same time it is worth noticing that the calculated stress is not the actual stress, which should be converted according to the principle of equal torques or internal forces.

3.2.2. *Advanced allen theory*

Allen model [6] assumes that the honeycomb core extremely soft and could only resist transverse shear stress, ignoring the ability of the thin panel to resist transverse shear stress, and doesn't consider the flexural strength of the honeycomb core, that will lead to bigger error when the panel is thick relatively. Through introducing volume modulus of honeycomb sandwich, considering the bending stiffness, the advanced Allen theory designs the equivalent thickness of the honeycomb core, derives the bending stiffness of the sandwich plate, and then process the equivalent of the bending stiffness and the shear stiffness. The mechanical properties can be obtained as follows:

$$
\begin{cases}
E_{eq} = \dfrac{12(1-v_{eq}^2)}{(h+2t)}\left(D+\dfrac{1}{12}Kh^3\right) \\[4mm]
G_{eq} = G_{cxz}\dfrac{h+t}{h+2t}, \quad v_{eq} = \dfrac{Dv_f + \dfrac{1}{12}Kh^3}{D+\dfrac{1}{12}Kh^3}
\end{cases}
\tag{1}
$$

Where, $K = \delta E_c / 2\sqrt{3}l$ is the volume modulus of the honeycomb core; E_c is the elastic modulus of core material; l is the side length of the honeycomb core; δ is the thickness of the honeycomb core; $D = (E_f(h+t)^2 t)/2(1-v_f^2)$ is the bending strength of the sandwich plate; E_f is the elastic modulus of the panel layer material.

3.2.3. Honeycomb panel theory

With honeycomb panel theory [7], the whole honeycomb sandwich plate is equivalent to an orthotropic plate with equal stiffness and same size. The equivalent elastic parameters of the equivalent mechanical model are derived by considering the surface and outside mechanical properties of the sandwich plate, as shown in following equations:

$$
\begin{cases}
e_{11}^c = \dfrac{E_{cx}}{1-v_{cxy}^2}, \quad e_{22}^c = \dfrac{E_{cy}}{1-v_{cxy}^2}, \quad e_{12}^c = \dfrac{v_{cxy}E_{cx}}{1-v_{cxy}^2} \\[3mm]
e_{44}^c = G_{cxy}, \quad e_{55}^c = G_{cxz}, \quad e_{66}^c = G_{cyz}, \\[3mm]
e_{11}^f = \dfrac{E_f}{1-v_f^2}, \quad e_{12}^f = \dfrac{v_f E_f}{1-v_f}, \quad e_{22}^f = \dfrac{E_f}{1-v_f^2} \\[3mm]
e_{44}^f = G_f, \quad e_{55}^f = \alpha G_f, \quad e_{66}^f = \alpha G_f
\end{cases}
\tag{5}
$$

$$
\begin{bmatrix} e_{11} \\ e_{12} \\ e_{22} \\ e_{44} \\ e_{55} \\ e_{66} \end{bmatrix}
=
\begin{bmatrix}
\left[(H^3-h^3)e_{11}^f + h^3 e_{11}^c\right]/H^3 \\
\left[(H^3-h^3)e_{12}^f + h^3 e_{12}^c\right]/H^3 \\
\left[(H^3-h^3)e_{22}^f + h^3 e_{22}^c\right]/H^3 \\
\left[(H^3-h^3)e_{44}^f + h^3 e_{44}^c\right]/H^3 \\
(2te_{55}^f + he_{55}^c)/H \\
(2te_{66}^f + he_{66}^c)/H
\end{bmatrix}
;\quad
\begin{bmatrix} E_x \\ E_y \\ G_{xy} \\ G_{xz} \\ G_{yz} \\ v_{xy} \end{bmatrix}
=
\begin{bmatrix}
(e_{11}e_{22}-e_{12}^2)/e_{22} \\
(e_{11}e_{22}-e_{12}^2)/e_{11} \\
e_{44} \\
e_{55} \\
e_{66} \\
e_{12}/e_{22}
\end{bmatrix}
$$

Where, e_{ij}^f, e_{ij}^c are the stiffness coefficients of the panel layer and the honeycomb core respectively; ρ_f, ρ_c are the density of the panel layer and the honeycomb core respectively; $E_x, E_y, G_{xy}, G_{xz}, G_{yz}, v_{xy}$ are the equivalent parameters of the equivalent plate; E, G are the material parameters of the surface plate.

Stated thus, the equivalent methods can be used to complete the finite element modeling of the solar wing substrate by considering the modeling idea of the integrated modeling method for the panel layer and the honeycomb core. The realization approaches and mechanical performances are shown in Table 2.

Table 2 The realization approaches of the finite element model for the whole substrate

Number	Theory	Equivalent model
1	Equivalent Plate Theory	Isotropic
2	Advanced Allen Theory	Isotropic
3	Honeycomb Panel Theory	Orthotropic

4. Simulation and Analysis

According to the above modeling methods, the equivalent plate theory (marked as Model A), the equivalent honeycomb core model (marked as Model B), and the honeycomb model based on actual geometric (marked as Model C) are selected to process the modal analysis.

The size of the solar wing substrate is $1032\text{mm} \times 810\text{mm} \times 25\text{mm}$. Assume the materials of the panel layer and the honeycomb core are isotropic and homogeneous aluminum alloy material. The panel layers are distributed mirror symmetrical on both sides of the honeycomb core, each side has 5 layers, and each layer has a thickness of 5mm. Table 3 presents the calculation results of equivalent parameters for the three models.

Table 3 The parameters design of the equivalent models

Model	Material of honeycomb core	Material of panel layer
Model A	$E_{cx} = E_{cy} = 0.16\text{Mpa}, \quad E_{cz} = 808\text{Mpa},$ $G_{cxy} = 0.03\text{Mpa}, G_{cyz} = 303\text{Mpa}$ $G_{cxz} = 202\text{Mpa}, \rho_c = 4.16e-11\text{t/mm}^3,$ $v_c = 0.3$	$E = 70000\text{Mpa},$ $\rho_c = 2.7e-9\text{t/mm}^3,$ $v_c = 0.3$
Model B	$E = 70000\text{Mpa},$ $\rho_c = 2.7e-9\text{t/mm}^3,$ $v_c = 0.3$	$E = 70000\text{Mpa},$ $\rho_c = 2.7e-9\text{t/mm}^3,$ $v_c = 0.3$
Model C	$E = 1314\text{Mpa}, \rho_c = 8.05e-11\text{t/mm}^3, v_c = 0.3$	

In order to eliminate the error caused by the grid difference, the shape, number and location of the surface mesh grids should be consistent. The grids of the three models are shown in Figure 4.

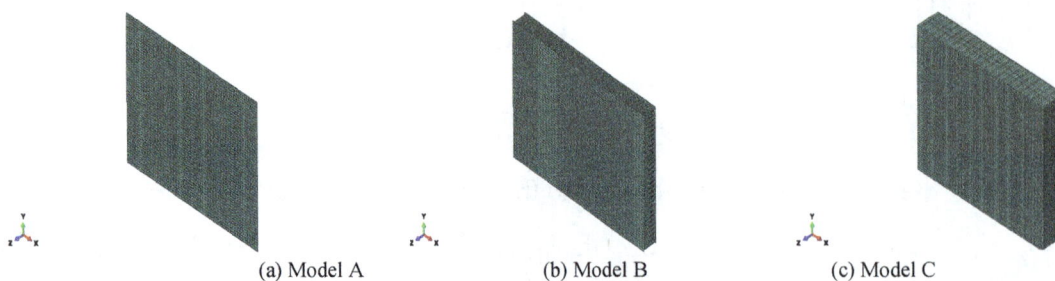

| (a) Model A | (b) Model B | (c) Model C |

Fig. 4 The finite element models of the three models

The statistics results of the first six order modal of the three models are shown in Table 4.

Table 4 The modal calculation results of the three models

Modal number	Model A	Model B	Model C
1	139.1	133.5	121.7
2	180.8	170.7	153.3
3	304.6	287.2	260.0
4	327.4	315.2	289.6
5	391.3	375.5	344.1
6	503.2	483.7	439.8

The calculation results show that the error of the first modal frequency value of the Model A and Model B is 4.3%, but the error of the Model C is 12.9%. In addition, the computation times of the three models are 110min, 26min, and 13min in turn. Therefore, the finite element modeling and analysis of the solar wing substrate can be carried out under the condition of full consideration of the calculation time and precision.

The modal shapes of the first order are shown in Figure 5 to Figure 10.

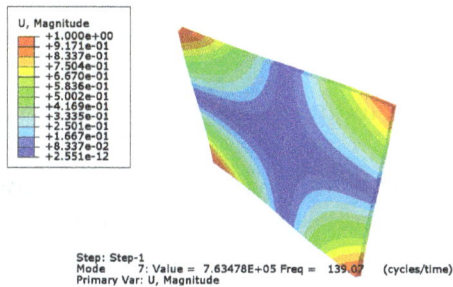

Fig. 5 The first order modal

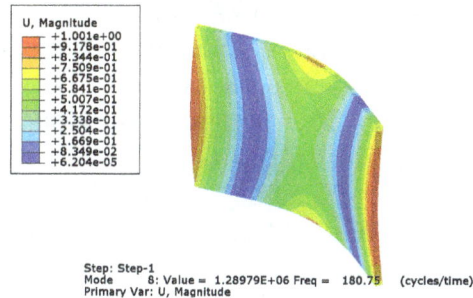

Fig. 6 The second order modal

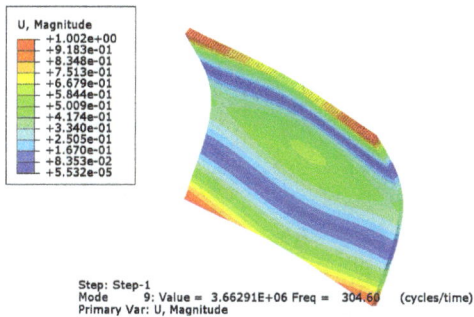

Fig. 7 The third order modal

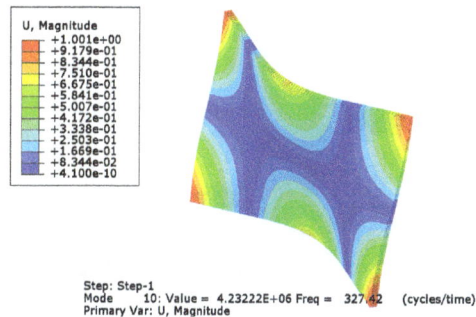

Fig. 8 The fourth order modal

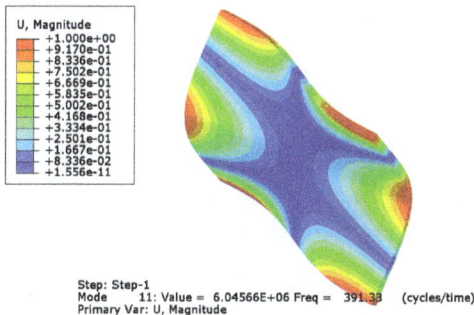

Fig. 9 The fifth order modal

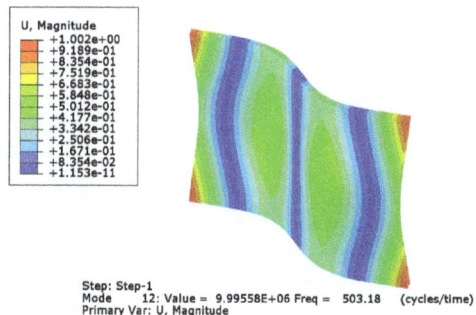

Fig. 10 The sixth order modal

5. Conclusion

The finite element calculation analysis illustrate that if take the honeycomb core equivalent as anisotropic materials to model when adopting the mutual independence modeling method for the panel layer and the honeycomb core, it not only can ensure the precision of the calculation, and also can greatly shorten the calculation time.

References

1. Fu Liying, Wang Weiyang. Equivalent Calculation and Experiment Research on the Honeycomb Sandwich Plates Used in Satellite[J]. Science Technology and Engineering, 2008, 8(23): 6429-6432.
2. Irenenuz Kreja. A Literature Review on Computational Models for Laminated Composite and Sandwich Panels [J]. Central European Journal of Engineering. 2011, 1(1): 59-80.
3. Reddy H N. Mechanics of Laminated Composite Plates: Theory and Analysis [M]. Boca Raton, FL: CRC Press, 1997: 752-754.
4. Zhu Tao, Wang Deyu. Nonlinear Equivalent Elastic Parameters of Honeycomb Core[J]. Aerospace Shanghai, 2008, (4): 113-118.
5. Du Zhengxing, Xue yingju, Liu Hongquan. General Stability research of honeycomb Sandwich Structure [J]. 2014, (4): 31-35.
6. Xia Lijuan, Jin Xianding, Wang Yangbao. Equivalent Analysis of Honeycomb Sandwich Plates for Satellite Structure [J]. Journal of Shanghai Jiaotong University, 2003, 37(7): 999-1001.
7. Deng Zongbai, Yan Jingyu. Dynamic Numerical Simulation of Aluminum Honeycomb Sandwich Plates Y Equivalent Model[J]. Machinery & Electronics, 2013, (4): 15-18.

Numerical Analysis of Cracks for Foam Metal Material

Zhen-Fang Xin[1,*], Fei Xia[2], Zong-Ze Xia[3], Wei-Wei Bian[1]

[1]Beijing Institute of Mechanical Equipment, Beijing, 100854, China
[2]State Grid Liaoyang Electric Power Supply Company, Liaoning, 111000, China
[*]Email: xzf340825@163.com

The numerical and analytical methods are hardly used to solve the cracks problems for foam metal materials. Only experiments currently are used for the crack tensile and compression, as well as the cracks study of them. The finite element analysis is used to solve the problems of the crack tips, based on the similarity of the TG model and DP model. The results of the crack tips under the two conditions of the plane stress and plane strain for single notched side tension specimen are presented.

Keywords: Cracks; foam metal; finite element.

1. Introduction

Compared with the dense material, due to the presence of sub structure-hole, foam metal material has a series of special performance, for example: light quality, good compressibility, good mechanical properties, and such advantages. It can be used for sound absorption, electromagnetic shielding and vibration damping, etc. also it has a wide range of application prospects in the field of comprehensive utilization of these properties, due to the integration of various characteristics.

In the aviation field, it is generally used as a lightweight, heat transfer support structure, such as the wing metal shell support body, the missile's shell temperature collapse support, radar mirror reflection material, etc. [1]. If the engine blades are made into porous structure by directional solidification method, the weight of the engine can be reduced greatly, the cooling capacity of the blades can be improved, and the performance of the engine can be advanced effectively. Figures 1 are the macro and STM micro structures for aluminium alloy Foam Metal Material.

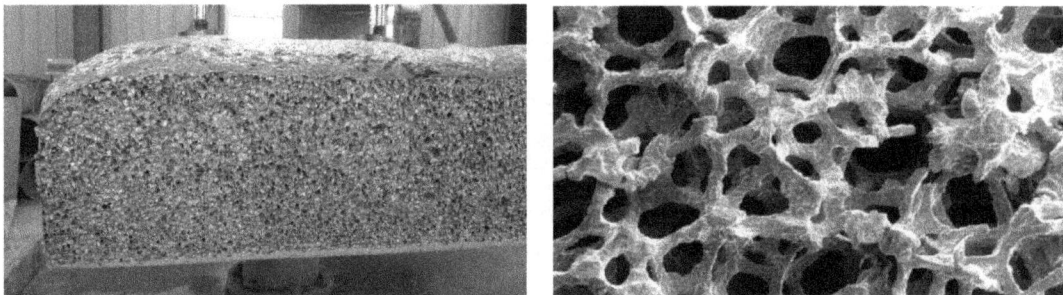

Fig. 1 Macro and STM micro structure for aluminium alloy Foam Metal Material

Because of the existence of large amount of cells and the proportion of material is small, the density distribution has become one of the most important structural characteristics. The relative density, the ratio of the density of the foam metal material to the density of the cell wall material, is about 0.02-0.2. In recent years, with the continuous improvement of manufacturing process, a lot of foam metal materials with better strength, elongation, processing performance and appropriate specific strength, have been developed. Take the dense aluminum alloy as an example, the elastic modulus $E = 71.4\text{GPa}$, the uniaxial yield strength $\sigma_Y = 258\text{MPa}$, but $E = 0.271\text{GPa}$ and $\sigma_Y = 0.811\text{MPa}$ of the INCO nickel foam material. Although the mechanical properties of cellular

---

[*]Corresponding author

material are much lower than those of the dense material, the density of the foam metal material is only a few percent of the dense material and the ratio σ_Y / E of the both are almost same.

2. Finite Element Equivalent Model

Because of the compressible plastic property of foam materials, there is no material model for direct study in commercial software. It has a form similar to the yield criterion of Drucker-Pragger (DP model), through the study on the plastic yield criterion of the foam metal materials. By analogy, the material constants of foam metal materials can be converted into the parameters needed for the DP model, which can help the elastic and plastic analysis for the foam metal materials by using DP model in commercial software, eliminating the need for independent development programming difficulties.

2.1. TG model

In Triantafillou and Gibson (TG) model [2], assuming that the effective stress has the following form:

$$\sigma = \sigma_e + 0.81 \frac{\rho^*}{\rho_s} \frac{\sigma_m^2}{\sigma_Y} \tag{1}$$

Where, σ_e is the von Mises effective stress; ρ^* / ρ_s is the relative density; σ_Y is the uniaxial yield strength; σ_m is the mean stress (equivalent to the static water pressure effect); σ_e can be defined as follows:

$$\sigma_e = \left(\frac{3}{2} s_{ij} s_{ij} \right)^{1/2} \tag{2}$$

Where, s_{ij} is the stress deviator tensor, which can be represented as the below:

$$s_{ij} = \sigma_{ij} - \sigma_{kk} \delta_{ij} / 3 = \sigma_{ij} - \sigma_m \delta_{ij} = \sigma_{ij} + p \delta_{ij} \tag{3}$$

Where, σ_{ij} is the stress tensor, and $\sigma_{kk} = \sigma_{11} + \sigma_{22} + \sigma_{33} = 3\sigma_m = -3p$.

According to the calculation of the given material constants, the Eq. (1) can be simplified as follows:

$$\hat{\sigma} = \sigma_e + 0.03 \frac{\rho^*}{\rho_s} \sigma_m \tag{4}$$

Combined with the yield criterion, there are some following expressions:

$$\Phi = \hat{\sigma} - Y = 0, \quad \sigma_e = \sqrt{3 J_2}, \quad \sigma_m = J_1 / 3$$

Therefore, the yield equation of foam metal materials can be derived after the simple mathematical deduction as follows:

$$0.03 \frac{\rho^*}{\rho} \frac{J_1}{3} + \sqrt{3 J_2} - \sigma_Y = 0 \tag{5}$$

2.2. DP model

The Kulun yield condition and flow rule of DP model in geotechnics are different approaches but equally satisfactory results with the TG model of foam materials. The DP model is suitable for the plastic deformation for materials under small scale yield condition. In general problems of rock and soil, with the increase of static water pressure, the yield stress and failure stress are greatly increased, which can use Kulun shear failure criterion [3] to indicate the beginning of the shear slip as follows:

$$\tau_n = C - \sigma_n \tan\phi \qquad (6)$$

Where, C is the cohesion; ϕ is the internal friction angle; σ_n is the normal stress on the fracture surface.

Taking the promotion of static water pressure into account, add a static water pressure factor on Mises case as follows:

$$f = \alpha J_1 + (J_2')^{\frac{1}{2}} - K = 0$$

$$(J_2')^{\frac{1}{2}} = \sigma_e$$

Considering

$$\alpha = \frac{\tan\phi}{(9+12\tan^2\phi)^{\frac{1}{2}}}, \qquad K = \frac{3C}{(9+12\tan^2\phi)^{\frac{1}{2}}}$$

The cohesion C and the internal friction angle ϕ can be solved.
The rock yield criterion can be represented as follows:

$$\alpha J_1 + \sqrt{J_2} - K = 0 \qquad (7)$$

α and K can be solved.

$$\begin{cases} \alpha = \dfrac{0.01}{\sqrt{3}} \dfrac{\rho^*}{\rho_s} = 9.24 \times 10^{-4} \\ K = \dfrac{\sigma_Y}{\sqrt{3}} = 0.468 \text{Mpa} \end{cases}$$

Therefore, ϕ and C can be solved.

$$\phi = 0.16, \quad C = 4.68 \times 10^5$$

The internal friction angle ϕ (unit: °), which used to control the number of volume expansion in physical meaning, is determined by the following three cases:

(1) $\phi_f = \phi$, the law of movement is defined as associated, and the volume expansion will be obviously;
(2) $\phi_f < \phi$, the law of movement is defined as no-associated, and the volume expansion will be little;

761

(3) $\phi_f = 0$, there will be no volume expansion because of the plastic flow is vertical to the yield surface.

The average stress of the foam materials causes compression but not expansion, therefore, under the drawing condition, assuming $\phi_f = 0$.

3. Calculation and Analysis

3.1. *Finite element pretreatment*

In ABAQUS, the material constants need to input for DP model are the internal friction angle, the flow stress ratio, the expansion angle and the yield limit. The material constants of INCO Nickel foam material can be calculated as follows [4]:

$$E = 0.271GPa, \quad v = 0.3, \quad \rho^* / \rho_s = 0.16, \quad \phi = 0.16, \quad \sigma_Y = 0.811MPa$$

The flow stress ratio is 1, and the expansion angle is 0.

The method of simulating the crack in ABAQUS is mainly preset a crack in the presence of the crack. Due to the function of displacement singularity simulation of the preset crack, it can simulate the displacement field of the crack tip, and then the stress intensity factor and stress displacement field distribution of the crack tip can be obtained by the calculation. Based on the theory of fracture mechanics, there will be singularity of $\dfrac{1}{\sqrt{r}} \dfrac{-b \pm \sqrt{b^2 - 4ac}}{2a}$ in the stress field

expression of the crack tip. The general unit cannot achieve this function, so the crack tip stress intensity factor cannot be calculated.

In the unit division of the model, the crack tip part needs special treatment, using distortion unit and element subdivision, and transforming the intermediate node of the equal parameter element to the 1/4 of the preset. The crack preset method can solve the problem of the stress singularity at the crack tip [5], and the calculated displacement field in ABAQUS can fit well with the displacement value of the theory. In the two dimensional finite element analysis of crack problems, singular distortion unit commonly uses six nodes triangular singular element and eight nodes quadrilateral singular element. According to the shape of the former network, the former is selected to define crack in special, simulate the singularity with collapse element and calculate the fracture mechanics parameters in the interaction module.

3.2. *Example and analysis*

Without needing 3D model, it will be effective to take 2D model for finite element calculation for the two kinds of stress state of the plane stress and plane strain, and the thickness can be set as 1mm.

3.3. *Unilateral stretching under plane stress condition*

The physical model is shown in Figure 2. $2H$ is the height of the specimen; a is the actual crack size; σ is the static loads of the upper and lower boundary of the specimen. The geometric dimensions used for the model are: $W = 100mm$, $H = 50mm$, $a = 10mm$, $20mm$, $30mm$, $40mm$; and the loads are: $\sigma = 0.15MPa$, $0.2MPa$, $0.25MPa$, $0.3MPa$, $0.35MPa$.

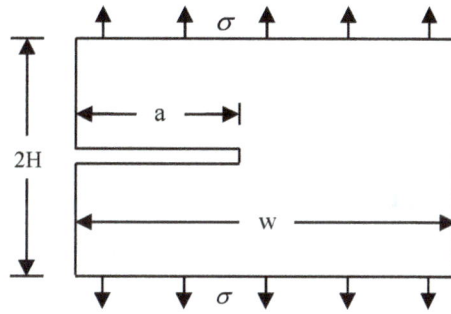

Fig. 2 Physical model of unilateral stretching specimen of finite size foam material

Taking $a = 30mm$ and $\sigma = 0.3MPa$ as an example, the results of the stress field and the displacement field are shown in Figure 2.

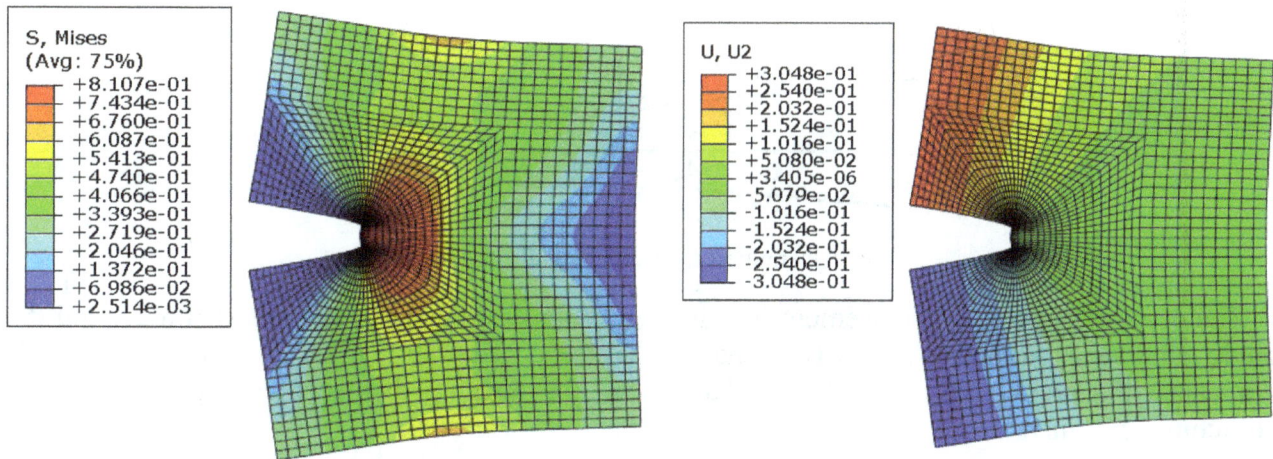

Fig. 3 Stress (left) and displacement (right) distribution near crack

From (a) and (b) of Figure 3, the unilateral stretching specimen of the finite size foam metal material under the plane stress state, the effective stress maximum of Von Mises appears at the crack tip, and the distribution of stress and displacement are axial symmetrically. The shape of plastic zone (red zone) near the crack tip is mainly concentrated in the vicinity of the crack tip. When the crack size and external load changes, the shape of the plastic zone is basically the same, but the size is different. Therefore, it would be better to take the length of the plastic zone as a measure on the crack surface to analyze the impact of external load on the plastic deformation of foam metal materials.

The size relationship between R and σ is shown in Figure 4. With a fixed crack length, the relationship between the size of the plastic zone and the external load is linear, which is consistent with the conclusion obtained from the classical fracture theory. When the size of the crack is bigger, the size of the plastic zone changes severely, the linear feature corresponded to small displacement is lost, and the relationship between R and σ is nonlinear.

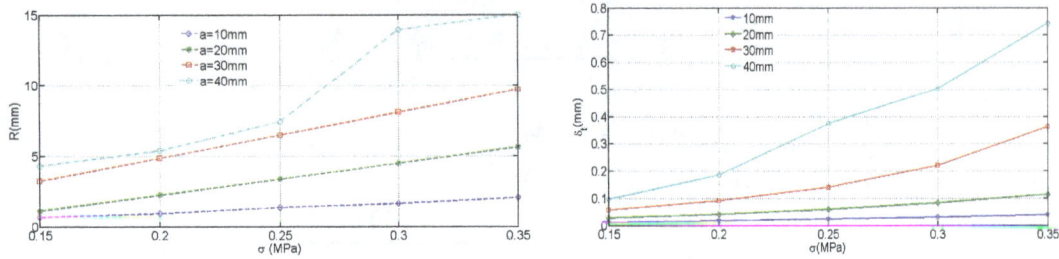

Fig. 4 The relationship between $R-\sigma$ (left) and $\delta_t - \sigma$ (right) under plane stress condition

For the definition of opening displacement of the crack tip, there are some different formulations in fracture mechanics. This study adopts the method proposed by Rice to draw two lines with 45° at current crack tip. The two lines intersect with crack surface at point A and point A', and the distance between the two points is used as the crack opening displacement, which is shown in Figure 5.

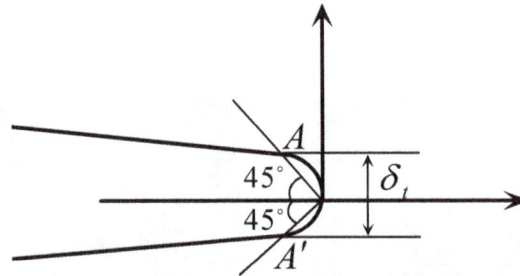

Fig. 5 Definition of the opening displacement

The crack tip opening displacement is consistent with the change of the external load, and the curve function is nonlinear similarly to that of the dense materials, which is shown in Figure 5. And also, the crack size has a great influence on the opening displacement of the unilateral stretching specimen.

3.4. *Unilateral stretching under plane strain condition*

Under plane strain condition, the geometric size and material constants of the specimen are the same of the plane stress state. Taking $a = 10mm$ and $\sigma = 0.3MPa$ as an example, the results of the stress field and the displacement field are shown in Figure 6.

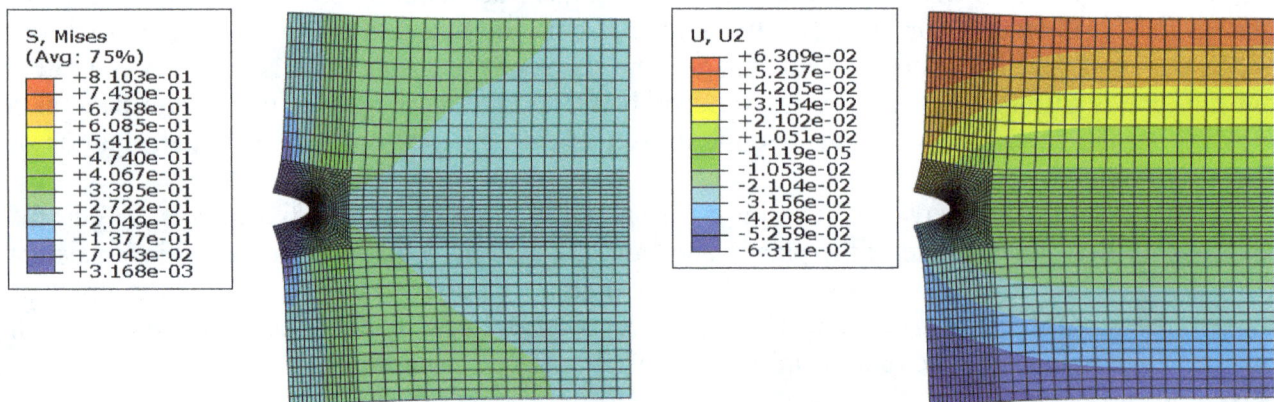

Fig. 6 Stress (left) and displacement (right) distribution near crack

The unilateral stretching specimen of the finite size foam metal material under the plane strain state, the effective stress maximum of Von Mises appears at the crack tip, and the distribution of stress is about x axial symmetrically. The displacement maximum appears at the two corners of the specimen, the displacement absolute is about y axial symmetrically, and the direction is opposite. The shape of plastic zone near the crack tip is shown in (a) of Figure 6.

The relationship between the size of the crack plastic zone and the applied boundary load is shown in Figure 7. R and σ are basically linear for different crack sizes.

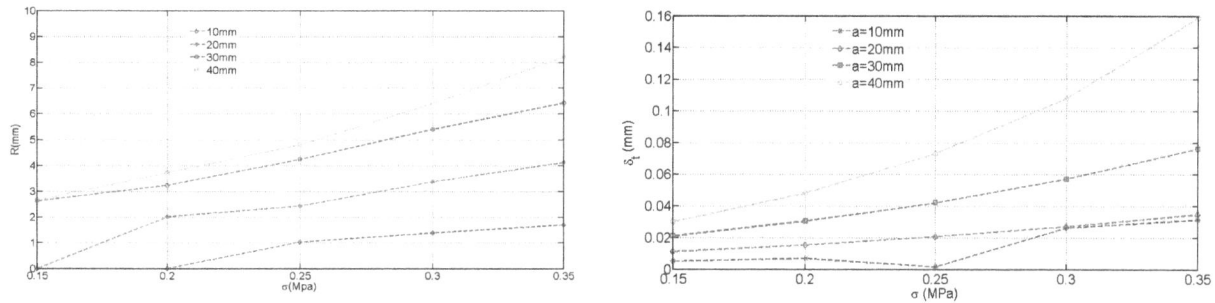

Fig. 7 The relationship between $R - \sigma$ (left) $\delta_t - \sigma$ (right) under plane strain condition

Compared with the plane stress state, the crack tip opening displacement is smaller. The relationship between $\delta_t - \sigma$ is also shown in Figure 7. The crack tip opening displacement is in line with the change of the external load, but with the increase of the load, the relationship between the two is nonlinear.

Compared the plastic zone size and the opening displacement between the plastic zone under the plane stress with the zone under the plane strain stress of the unilateral stretching specimen, it is not difficult to find that the plane stress is larger than that of the plane strain under the same load.

4. Conclusion

From the observation of numerical, the plastic zone length of the plane strain state is much smaller than that of plane stress state, and on the whole, the plastic deformation increases with the increase of the load. Under the condition of smaller load and same crack geometry sizes, the material of the plane stress state yields but the material of the plane strain state has not yet entered the plastic stage.

Compares with the stress states, the plastic zone of the plane strain state is much smaller than that of the plane stress state, which illustrates that the stress state has a great influence on the plastic deformation of the material. The plane stress state occurs in the interior of the thick object, and the thickness of the material specimen is larger, that is, the thickness of the specimen is larger in the z direction. For the common material, the front edge material of the crack under the plane strain condition is in the three direction tensile stress state. Tests show that the material is brittle failure but less plastic deformation. On the contrary, the plane stress state occurs in thin objects, and σ_{zz} is very small, even equal to 0. Therefore, in the crack front of thin materials, the material is in two directions stress state, and the deformation in the z direction is not restricted, so that the plastic deformation zone is larger.

References

1. Liu Huan. Energy Absorption and Explosion Proof Ability of Aluminum Foam [D]. Dongbei University. 2014.
2. Triantafillou T C, Gibson L J. Constitutive modeling of elastic-plastic open-cell foams [J]. Engng. Fract. Mech., 1990 (16): 2772-2778.
3. Yu Yuyin, Xi Feng, et al. Dynamic strength behavior of a Zr-based bulk metallic glass under shock loading [J]. Chinese Physics B. 2015 (06): 56-60.
4. Ruiping Guo. The analyzing solution for metel Foam Material crack expand and 3D dynamic fracture numerical analysis [D]. Beijing Inisitute of Technology. 2004.
5. Yingze Wang, Dong Liu, Qian Wang. Effect of fractional order parameter on thermoelastic behavious in infinite elastic medium with a cylindercal cavity [J]. Acta Mechanica Solida Sinica. 2015 (03): 89-95.

Microstructure Evolution and Its Finite Element Modelling of Al Matrix Composite Sheet during Wedge Pressing

Yi-Qiang He[1, 2,*], Hai-Sheng Zhou[2], Jun-Jie Li[2]

[1]*Jiangsu Marine Resources Development Research Institute, Lianyungang, Jiangsu 222005, China*
[2]*College of Mechanical Engineering, Huaihai Institute of Technology, Lianyungang, Jiangsu 222005, China*
*Email: ant210@126.com

Spray deposited SiC_P/Al-Fe-V-Si composite is characterized with pores and oxide films between deposited particles, therefore further densification is needed. Densification of the composite preforms in large size has become the research emphasis and challenge. Critical parameters of wedge pressing, density distribution, flow behavior, and densification mechanism and regularity were investigated by combining experiment and finite element modelling. Density of the external layer are higher than that of the inter layer, density distribution become uniformity with overall reduction. Materials in preformed area flows along thickness direction with decreasing thickness of the plate, while that in main deformation area flows along length direction with increasing length of the plate. Wedge pressing of the spray deposited Al matrix composite plate follows porous metal plastic deformation mechanism under the condition of plane strain. Practical relative density of the plate is higher than the calculated value in the initial stage of pressing, while the practical relative density become lower than the calculated value as true strain exceeds 0.55, which can be attributed to the pores that are difficult to be eliminated resulted from aggregation and breaking of SiC particles.

Keywords: Al matrix composite; spray deposition; wedge pressing; microstructure evolution; finite element simulation.

1. Introduction

In generally, relative density of spray deposited billet is 85% to 90%, pores in the blank recult in failling to obtain full metallurgical bonding between deposited particles, and the interface between particles and layers can not be eliminated completely, would result to a sharply decline in the mechanical properties and formability. Therefore study on subsequent densification of the spray deposited porous materials atrated many scholars' attention. Forging, extrusion, rolling, hot isostatic pressing and spinning, etc [1, 2, 3] are the dominant and eraditional densification technologies, but these technologies failed to effectively solve the problem of densification of the large size porous material, and systematic study of densification mechanisms are absence.

Densification and performance of spray deposited aluminum alloy have been investigated. Xiao Yude prepared heat-resistant aluminum alloy and composite materials with good mechanical properties by means of spray deposition [4]. Tan Dunqiang studied on spray deposited Al-Fe-V-Si, and improved the cooling rate of the alloy [5]. Zhang Hao used ceramic rolling to dense effectively sprays deposited SiC_P/Al composites and to improve the microstructures and mechanical properties [6, 7]. Chen Zhigang employed wedge pressing to dense spray deposited SiC_P/Al composites, and obtain large size dense ring blank [8, 9]. S. Devaraj studied on influence of hot isostatic pressing on the microstructure and mechanical properties of a spray-formed Al-4.5 wt.% Cu Alloy, after hot isostantatic, the porosity were reduced to less than 0.5% without any appreciable increase in grain size, while the porosity of the spray formed deposits varied between 5% to 12% [10]. Li Changhao, Li Wei worked on fatigue behavior of spray deposited SiC_P/Al-Si [11, 12]. The authors adopted rolling after extrusion and pressing to densify the composite, analysized SiC particle distribution, SiC-Al interface morphology and mechanical properties [13, 14, 15, 16]. However, restricted by the high deformation resistance at high temperature and limited by tonage of equipment and high processing costs, SiC_P/Al-matrix composite can not be densitied effectively. To solve this problem, wedge pressing is proposed by the authors to densify large

*Corresponding author

spray deposited slab, for industrial production of such a porous alloy sheet. Densification mechanisms of spray deposited SiCp/Al-matrix composite will be revealed and theoretical basis for industrial production will be provided.

2. Principle of Wedge Pressing

The wedge pressing device is mainly composed of three parts, namely a wedge shaped punch (die) in a hydraulic press with a capacity of 6300kN, a cast steel mould and a set of heating device, as it is shown in Fig. 1. The porous workpiece are encapsulated in steel mould. Several electrical heating bars are embedded in the holes of steel mould to heat the workpiece. The wedge shaped punch has an angle of 5~6° between the flat pressing surface and the pre-pressing one, and the latter provided a beneficial pre-deformed zone during deformation. Reduction of thickness during pressing process can be precisely controlled by an inhibiting device. The workpiece remained stationary during pressing. After each pressing, the punch was elevated and moved forward with a appropriate distance to the next position for another pressing. The pressing procedure is repeated until the necessary pressing reduction is obtained. The as-pressed workpiece is annealed for 20 minutes so as to continue the next cyclic of compaction. Thus, by means of accumulating multi-pass local small deformation, a nearly full density of spray deposited flat is expected to be obtained.

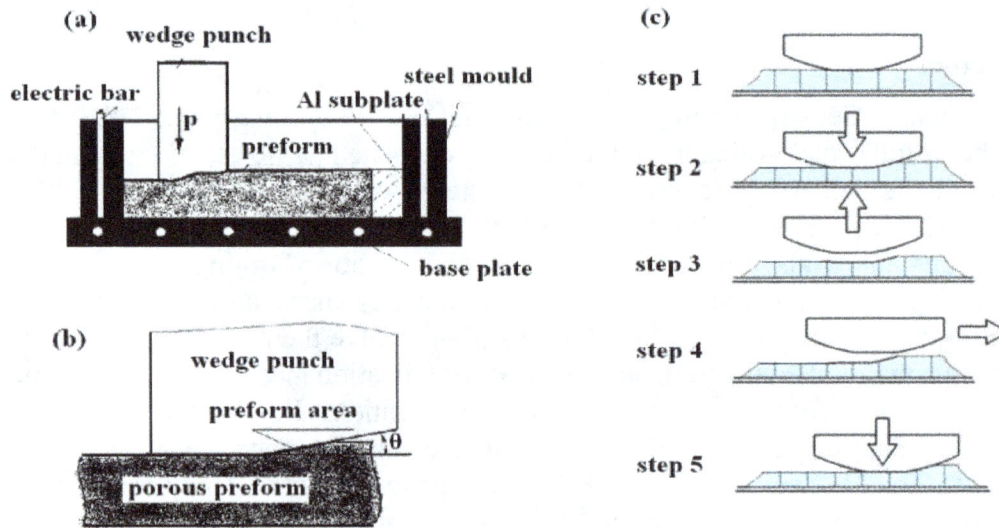

Fig. 1 Schematic diagram of wedge pressing: (a) device, (b) pressing theory, and (c) pressing process

3. Forming Principle and Pilot Programs

3.1. *Material preparation*

15vol.% SiCp/Al-8.5Fe-1.3V-1.7Si composite was prepared by means of spray deposition equipment, the average size of SiC particles is 2μm.

3.2. *Wedge pressing*

Temperature of the workpiece was monitored by a thermocouple during wedge pressing. The temperature was set at 480°C. Billet and the mould were preheated to 480°C for 60min before wedge pressing, each pressing reduction of 10~30%, and pressing rate of 1mm/s were adopted.

3.3. *Microstructure observation and density measurement*

Metallographic samples were observed by means of XJL-03 large Optical Microscope (OM), interface morphology was observed by means of JSM-3010 High Resolution Transmission Electron Microscopy (HRTEM), and fracture morphology of tensile specimen was observed through JSM-5600 Scanning Electron Microscope (SEM). The density and relative density were measured according to the principle of Archimedes.

3.4. *Finite element simulation*

Pressing process was simulated by Deform software. The initial size of the porous blank is 150mm×110mm×20mm, the initial relative density was about 87%, simulation temperature was setted to 480°C. Linear expansion coefficient of the blank was setted to $2.08×10^{-5}$/K, elastic modulus was setted to 107GPa, thermal conductivity was setted to 95W/m·K, and specific heat capacity was setted to 903J/Kg·K. The wedge shaped punch was regarded as rigid material, depression rate of 1mm/s, overall reduction of 1mm, and mould temperature of 480°C were setted. The heat exchange with environment was defined. All surfaces of workpiece were seemed as heat exchange surface with environment. Workpiece and punch were located, and then contact relationship was settled. Pastic shear friction model, friction coefficient was set to 0.3, and heat transfer efficient default was set to 11W/ $(m^2·K)$ was applied.

4. Results and Discussion

4.1. *Critical process parameters*

Appropriate pass reduction of wedge pressing was needed to elevate density and formability of the spray deposited SiCp/Al-Fe-V-Si billet. Ununiform deformation will be happened between the surface and the interior of the plate preform for too small pass reduction, while preform surface crack problem appeared because of too high pass reduction. Pass reduction of 10%~15% was utilized at the initial stage of wedge pressing because of poor bonding among deposited particles, and it can be elevated to 15% as a result of bonding improvement when overall reduction exceeds 25%. Space of 10mm between the end of the preform and the steel mould can be setted to prevent the preform from arch camber and cracking, due to extension of the preform along length direction were generated during wedge pressing.

4.2. *Density distribution*

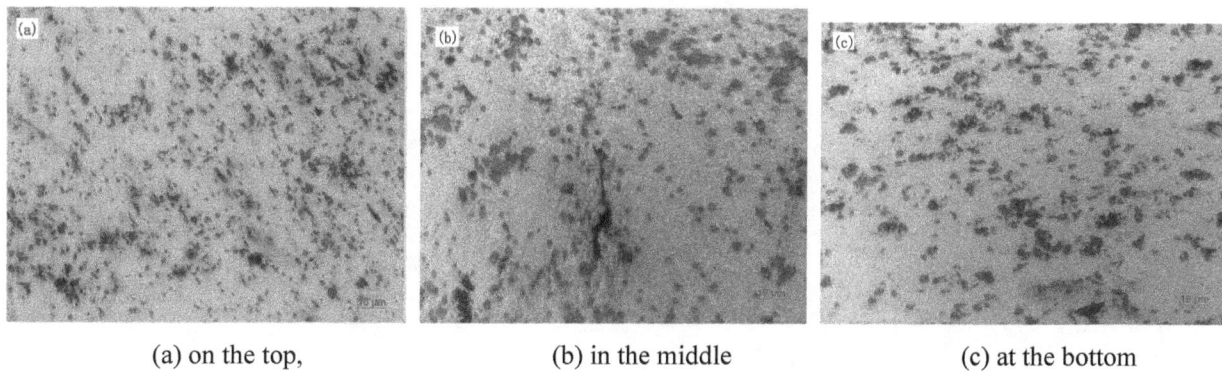

| (a) on the top, | (b) in the middle | (c) at the bottom |

Fig. 2 Microstructure of composite sheet with total reduction of 20%

Fig. 2 shows the density distribution in different positions with overall reduction of 20%. It can be seen from Fig. 2(a) and Fig. 2(c) that pores of large size collapse and fine pores disperse at the surface layer of the slab. Strip-type pores of 20μm in length distribute in the middle because of wedge pressing as shown in Fig. 2(b). It can be found that density decreases from the suface layer to the interior gradually.

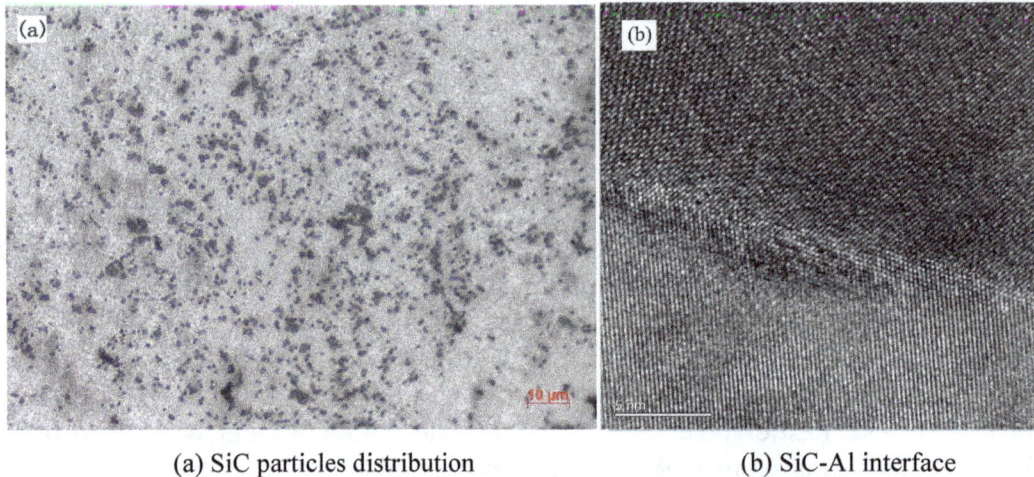

(a) SiC particles distribution   (b) SiC-Al interface

Fig. 3 Microstructure of the composite sheet with total reduction of 50%

This nonuniform density distribution is improved as overall reduction increases during wedge pressing. It can be seen from Fig. 3(a) that SiC particles distribute homogeneously and bond with the Al matrix well, and visible pores are eliminated in the interior region when the total reduction reaches 50%. Microstructure of the composite is improved further by rolling after wedge pressing as shown in Fig. 3(b), SiC particles distribute homogeneously, and microcracks between SiC particles and the Al matrix disappear.

The finite element simulation of density distribution of the preform with total reduction of 20% is shown in Fig. 4. It can be found that relative density in the surface layer of 0.891, which is higher than that in the centre of 0.874 during pressing process, and relative density decreases from the surface to the centre. This can be attibuted to that the pressure stress decreases from the surface to the interior. Higher-pressure stress brings more sufficient shearing deformation. Pores broke and collapse and then closed, while leads to a higher density.

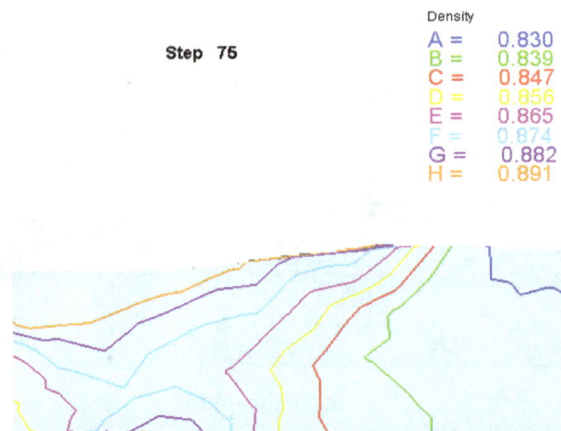

Density
A = 0.830
B = 0.839
C = 0.847
D = 0.856
E = 0.865
F = 0.874
G = 0.882
H = 0.891

Step 75

Fig. 4 Relative density distribution of the preform with total reduction of 20%

4.3. *Flow behavior*

It can be seen from Fig. 5 that flow rate decreases from the preformed area to the strained area.

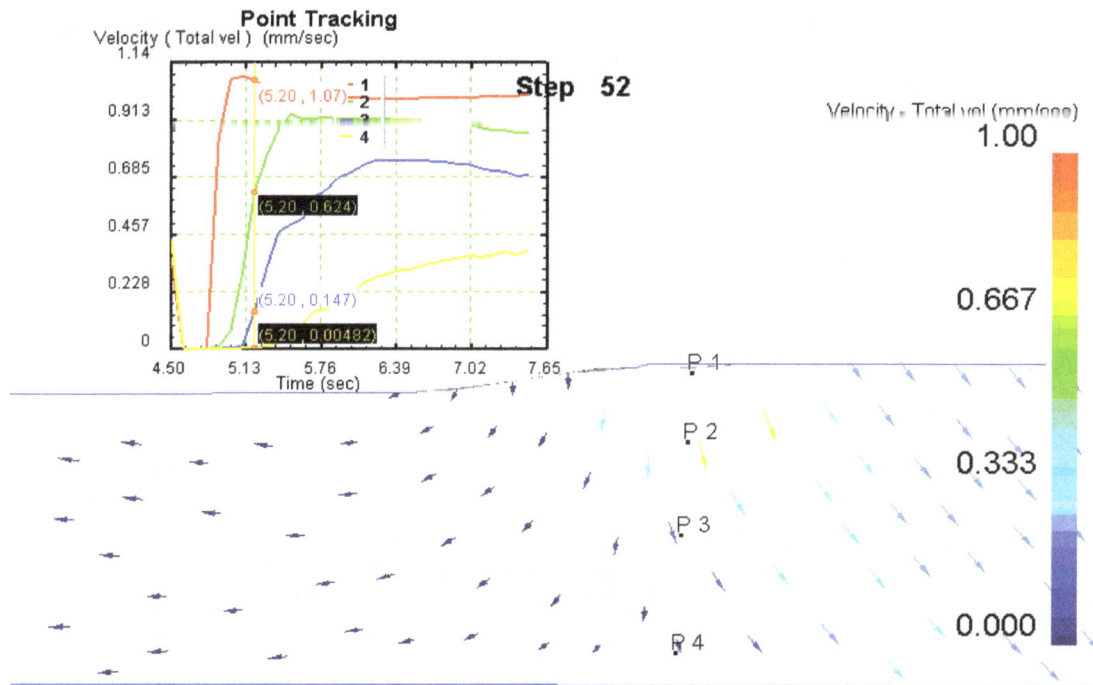

Fig. 5 Flow behavior of the sheet during wedge pressing

This can be attributed to that relative density and strength of the preformed area is lower than that of the strained area, densifying behavior of the preformed area under a strong pressing stress is stronger than that of strained area. And pores in this area collapse and then are closed. Materials in the preformed area flow is from up to down along thickness direction. The thickness of the slab is reduced. Materials in the strained zone flow from the preformed area to the stained area along the length direction and flow along thickness direction to a certain degree, and the length of the slab increases, while the thickness of this zone decreases to a certain extent.

Thickness of the slab decreases while length of the slab increases hardly because of the collapse of the pores at initial stage of pressing (the total reduction is less than 20%). Then length of the slab increases obviously when the overall reduction is more than 20% because of materials flow along length direction increasing.

4.4. *Densification mechanism and regularity*

Elongation in width direction of the preform is restricted by the steel mould, as a result, the preform almost remains the same in width, but length increases and thickness of the preform varies with overall reduction during wedge pressing. Therefore, materials deform under plate strain state during wedge pressing.

According to mass conservation of porous metal matrix material during plastic deformation, densification relation of porous metal material under plane strain condition can be expressed as Formula (1):

$$\varepsilon_1 = \frac{1}{4}\ln(1-\rho^2) - \ln\rho + C \tag{1}$$

Where ε_1 is principal strain, and ρ is density of the composite. Formula (1) can be used to describe densification behavior of porous metal material under plane strain condition, where integration constant C is decided by initial relative density. The calculated value of relative density of spray deposited SiC$_P$/Al-Fe-V-Si plate during pressing can be obtained by plugging the initial relative density into formula (1). Comparison between the calculated value and experimental value of relative density of the composite during pressing is shown in Fig. 6.

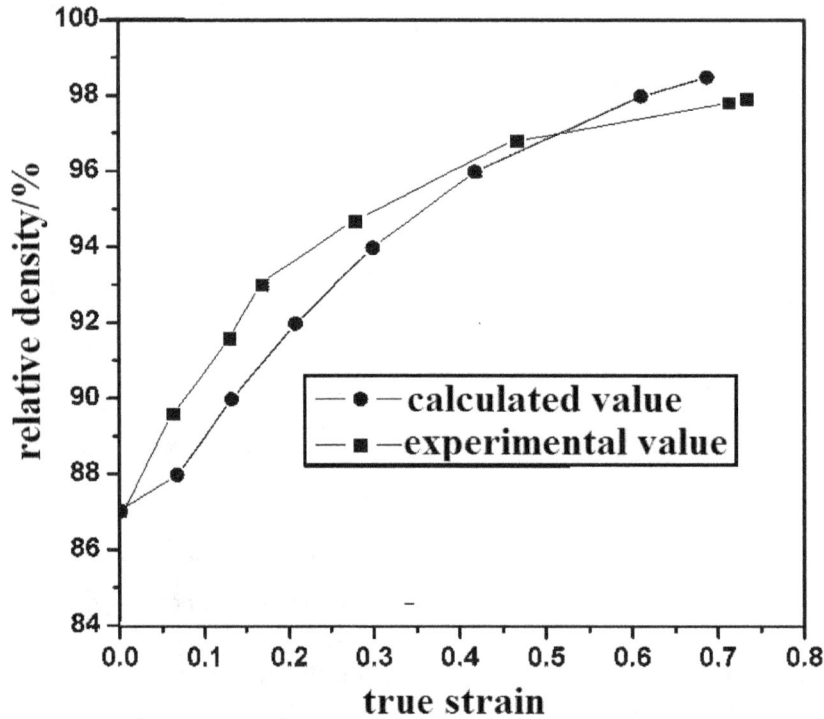

Fig. 6 Calculated value and experimental value of relative density of the composite during pressing

Relative density of the preform increases as the true strain during wedge pressing. The examined value of the relative density is higher than the caculated one, which means that the actual density and densification rate is higher than the calculated one at the initial stage of wedge pressing. The actual densification rate decreases as the ture stain ε exceeds 0.16, and the experimental relative density becomes lower than the calculated one when ε exceeds 0.55. It can be found that the actual densification rate decreases as the ture strain increases, while the experimental relative density increases. This can be explained as follow. On the one hand, the density increases sharply as result of collapse and close of pores in large size. On the other hand, the calculated value is based on the plate strain state, no strain exists in the width direction, actually, and strain in the width direction exists to a certain degree during wedge pressing.

SiC particles are ignored in the formula (1), but the densification rate at the final stage of wedge pressing will be reduced due to existence of SiC particles. This can be attributed to the three aspects as follow: ① Gathering of SiC particles were distributed slowly during pressing for low plastic flow of billet, the pores between the gathering particles is hard to be eliminated. Gathering pores can be seen in Fig. 7, densification rate decreases because of these pores; ② Relative sliding and rotating between the SiC particles and the matrix plastic during plastic flow, SiC particles and the matrix alloy are deboned and pores appear; ③ SiC particles with larger aspect ratio would be fracture to formate new pores. As a result, the experimental relative density became lower than the calculated one when ε being over 0.55.

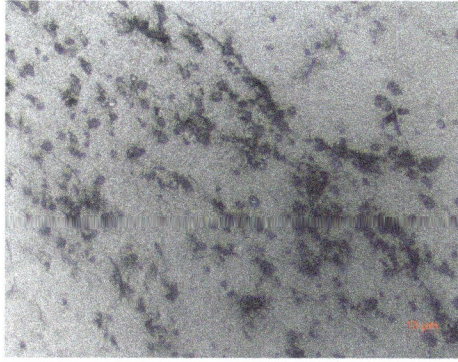

Fig. 7 Gathering and breakage of SiC particles (true strain ε=0.6)

5. Conclusion

A novel densification method, wedge pressing and then rolling was used to densify the spray deposited SiC$_P$/Al-Fe-V-Si sheet preforms with large size. The conclusions can be drawn as following:

1. The spray deposited SiC$_P$/Al-Fe-V-Si sheet preforms with large size can be effectively densified by wedge pressing. Wedge pressing can substantially reduce the porosity, and improve the metallurgical bonding between the as-deposited particles in the preforms effectively.
2. Pass reduction of 5% to 15% is appricate for spray deoposited SiC$_P$/Al-Fe-V-Si during wedge pressing. Materials in the deformed zone flow along the length direction, and length of this zone increases, while materials in preformed area flow along thickness direction, and thickness of this zone decreases to a certain extent.
3. Density distributes ununiformly from the suface layer to the centre, which is agree with its finite element simulation. This nonuniform density distribution is improved as overall reduction increases during wedge pressing.
4. Materials in the strained zone flow from the preformed area to the stained area along the length direction and flow along thickness direction to a certain degree.
5. Comply with densification behavior of porous metal material under plane strain condition, pores in spray deposited composite collapsed and closed increased rapidly. Actual relative density is higher than the calculated value when the ture stain ε is below 0.55; When the ture stain ε is over 0.55 during pressing, the experimental relative density becomes lower than the calculated one, which can be attributed that the actual densification rate decreases as the ture strain increases, while the experimental relative density increases.

Acknowledgments

The research described in this publication was funded by the Natural Science Foundation of China (No. 51004050), the Natural Science Foundation of Jiangsu Province (No. BK20141250), Natural Science Foundation of Jiangsu Colleges and Universities (No. 14KJB430005) and Natural Science Foundation of Lianyungang (No. CG1418, No. CXY1404), Key Research and Development Programs of Jiangsu Province (BE2015100), Postgraduate Research and Innovation Projects of Jiangsu Province (KYLX15_1485).

References

1. Feng-Xian LI, Yun-Zhong LIU, Y Jiang, Effect of processing parameters on the relative density of spray rolling 7050 aluminum alloy strip, The International Journal of Advanced Manufacturing Technology. 67 (2013) 2771-2778.

2. P.S. GRANT, Solidification in Spray Forming, Metallurgical and Materials Transactions A. 38 (2007) 1520-1529.

3. W.B. Shou, D.Q. Yi, H.Q. Liu, Effect of grain size on the fatigue crack growth behavior of 2524-T3 aluminum alloy, Archives of Civil and Mechanical Engineering. 16 (2016) 304-312.

4. Yu-De XIAO, W WANG, Wen-Xian LI, High temperature deformation behavior and mechanism of spray deposited Al-Fe-V-Si alloy, Transactions of Nonferrous Metals Society of China. 17(2007) 1175-1180.

5. Yi-Ping TANG, Dun-Qiang TAN, Wen-Xian LI, Preparation of Al-Fe-V-Si alloy by spray co-deposition with added its over-sprayed powders, Journal of Alloys and Compounds. 439 (2007) 103-108.

6. H. ZHANG, D. CHEN, Z.H. CHEN, Densification of Spray Deposited Aluminum Composite Sheets via Ceramic Rolling Technique, Materials and Manufacturing Process. 23 (2008) 479-483.

7. Ke Zhang, Guo-Yi Qin, Si-Yong Xu, Preparation of Ag-Ni-Cu Composite Material by Ultrasonic Arc Spray Forming and Accumulative Roll Bonding and the Evolution of Its Microstructure, Metallurgical and Materials Transactions A. 46 (2015) 880-886.

8. Zhi-Gang CHEN, Zhen-Hua CHEN, Ding CHEN, Microstructural evolution and its effects on mechanical properties of spray deposited SiCp/8009Al composites during secondary processing, Transactions of Nonferrous Metals Society of China. 19 (2009) 1116-112.

9. Zhi-Gang CHEN, Zhen-Hua CHEN, Guo-Ning Tang, Processing, Microstructure, and Mechanical Properties of Large Spray-Deposited Hypoeutectic Al-Si Alloy Tubular Preform, Journal of Materials Engineering and Performance. 20 (2011) 238-243.

10. S. Devaraj, S. Sankaran, R. Kumar, Influence of Hot Isostatic Pressing on the Microstructure and Mechanical Properties of a Spray-Formed Al-4.5 wt.% Cu Alloy, Journal of Materials Engineering and Performance. 23 (2014) 1440-1450.

11. C.H. LI, Z.H. CHEN, D. CHEN, Research on high cycle fatigue behavior of spray deposited SiC$_P$/Al-20Sicomposite, Journal of Mechanical Engineering. 48 (2012) 40-44.

12. W. LI, Z.H. CHEN, D. CHEN, Growth behavior of fatigue crack in spray formed SiC$_P$/Al-7Si composite, Acta Metallurgy Sinica. 47 (2011) 102-108.

13. Z.H. CHEN, Y.Q. HE, H.G. YAN, Ambient temperature mechanical properties of Al-8.5Fe-1.3V-1.7Si/SiC$_P$ composite, Materials Science and Engineering. 460-461 (2007) 180-185.

14. Yi-Qiang HE, Bin QIAO, Na WANG, A study on the interfacial structure of spray-deposited SiCP/Al-Fe-V-Si composite, Advanced Composite Letters. 18 (2009) 137-142.

15. Y.Q. HE, B. QIAO, N. WANG, Thermostability of Monolithic and Reinforced Al-Fe-V-Si Materials, Advanced Composite Materials. 18 (2009) 339-350.

16. Yi-Qiang HE, Hong TU, Bin QIAO, Tensile fracture behavior of spray deposited SiC$_P$/Al-Fe-V-Si composite sheet, Advanced Composite Materials. 22 (2013) 227-237.

17. H.A. KUHN, C.L. DOWNEY, Deformation characteristics and plasticity theory of sintered powder materials, International Journal of Powder Metallurgy. 7 (1971) 15-25.

The Cell Mechanical Response in Compression: A Finite Element Analysis

Na Li[1], Xu Cao[1], Si-Ying Chen[2], Kun Xiong[2,*]

[1]*Xiangya 3rd Hospital, Central South University, China*
[2]*Department of Anatomy and Neurobiology, School of Basic Medical Science, Central South University, China*
Email: xionghun2001@163.oom

This study was attempted to develop a new cell finite element model to present the cell mechanical response underlying the compression. The geometry and material properties of cytomembrane, cytoplasm, cytoskeleton, and the nucleus were obtained according the previous experimental data. 5Mpa compression loading was simulated in LS-DYNA solvers, the displacement, strain, stress contour of the cytomembrane, cytoplasm, cytoskeleton, and the nucleus were calculated in this study. According the prediction of this FE modeling, the mechanotransduction sequence of each part in cell was observed from cytomembrane to cytoskeleton and then the cytoplasm, finally to the nucleus. This phenomenon might indicate that the remodeling behavior of the cell cytoskeleton was influenced by the nucleus mechanical response.

Keywords: Finite element modeling; chondrocyte cell; compression; biomechanics.

1. Introduction

Studies have proved that cells show great sensitivity to their biomechanical environment. This is especially for chondrocyte because mechanics can directly influence its repair as a regulatory factor. A chronic lack of exercise leads to the atrophy of articular cartilage while degenerative joint disease, such as osteoarthritis, is associated with excessive loading and over use of joints. DENNIS DISCHE, a biomechanical professor of Pennsylvania University, considered that: 'Biomechanical analyses of different cell models have been developed in several organs (e.g., lung, heart, skeletal muscle, connective tissue) that quantitatively predict basic aspects of organ perfusion.' in his paper <Biomechanics: Cell Research and Applications for the Next Decade> [1]. Understanding the role biomechanical factor plays in both physical and pathological condition is of great significance for pathological study and clinical therapeutics. It helps to provide new ideas for clinical therapies. In addition, it is the key to explain the pathogenesis of many mechanical-based diseases such as osteoporosis, osteoarthritis and so on.

Based on cytoskeletal model, the present biomechanical models have made great development. However those previous models still have limitations in various degrees. The geometry and mechanical parameters of the Tensegrity Model [2] were not attained through biomechanical testing. Instead, they were derived from the hypothesis that microtubules are the main structure to balance the tension of microfilament. Subsequent studies proved this model cannot well reflect the characteristics of cell mechanics. The recent Coarse-grained Model [3] assumed the microfilament is linear elastic and uses Gauss-Seidel iterative method to adjust node position. This model simulated dynamic changes of cells under loading, but fails to take nuclear stress into consideration. Brownian dynamic model [4] gives full consideration to the concentration of actin monomer, the concentration ratio of cohesion to actin, the binding angel of cohesion, microfilament stiffness and the size of cohesion. This model well simulated the dynamic force acting on microfilament network structure. But it is a pity that the researchers still failed to consider the nucleus' biomechanical response. Therefore, the existing cell mechanical models are still insufficient to analyze and predict cell mechanical behavior. Nucleus' response to mechanical forces directly affects the metabolism, mechanotransduction and apoptosis of cells. After carefully considering the mechanical properties

---

*Corresponding author

of cells' internal structure and their concentration, our study attempt to develop a new biomechanical calculation method to predict the mechanical behavior of nucleus, aiming to provide a solid biomechanical basis for further study of cell behavior.

2. Method and Materials

The cytoskeleton plays a decisive role in determining mechanical properties of cells. The character of cell behavior under loading was determined by cytoskeleton, such as microtubules, microfilaments and intermediate filament. The microtubules are the main intracellular scaffold and help with the intracellular material transport; Intermediate filament is a component of cytoskeleton that plays the main supporting role. The tension and anti-shearing force of cell is determined by the mechanical property of the intermediate filament. Therefore, the cellular mechanical model could not be simplified as a homogeneous and isotropic mechanical material. Also, the extracellular matrix and cell fluid could not be simplified as Newton fluid.

As we all know, the accuracy of finite element model simulation is affected by many factors, such as acquisition of an accurate model of the external structure; measurement of material properties of cells; cell mechanical material constitutive model; selection of numerical method for fluid-structure interaction; the setting of boundary conditions of consistency with actual cell structure, the appropriate verification and validation of the experimental result. In order to correctly understand the mechanical behavior of cells and analysis and predict mechanical behavior of the cells, we propose to combine the cellular mechanics experiments with the cell-specific cellular mechanics finite model. The detailed research and modeling process are as follows:

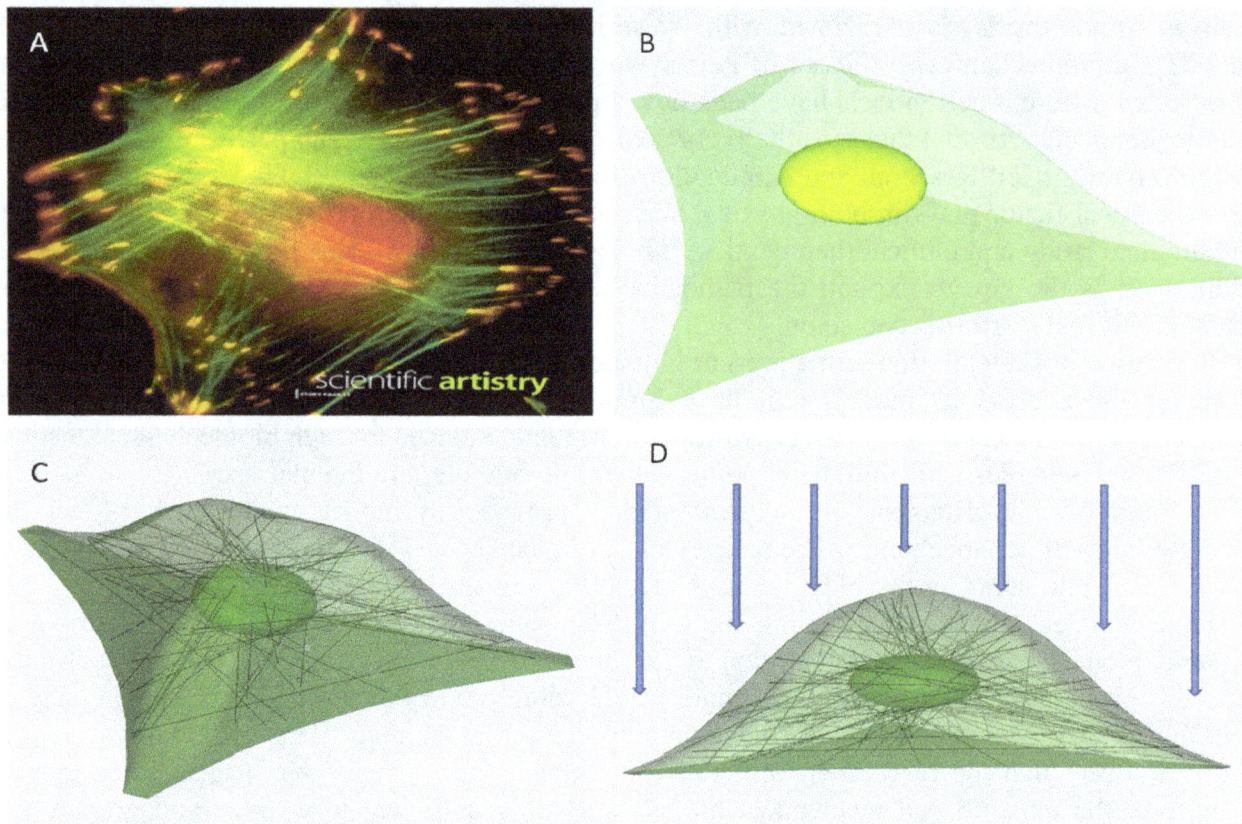

Figure 1A: Stress fibers (the filaments are red and microtubules are green) and nuclear structure in cultured chondrocytes; 1B: The geometry of the cell model; 1C: The FE model of cell cytoskeleton; 1D: the boundary condition of cell FE modeling.

3. Description of Cell Model

In this study, we begin with the acquisition of cell morphological structure. Accurate cell geometry can be obtained by laser confocal scanning microscopy's three-dimensional reconstruction, as shown in Figure 1A. Based on the structure characteristics of the actin filaments, a geometric model was established (Figure 1B). We used specific 3D software to establish geometric model. Based on the actual size of the cell, the gap between two angle is set as 55μm, the height of the highest point of the model is set as 17.15 microns [5]. The general image of the cell model was shown in Figure 1C.

The cytoskeleton is divided into two types: One type is connected to the cell membrane at both side, another type is connected to the cell membrane on one end while the other end is connected to nucleus.

1) The part of cytoskeletal that is connected to nucleus and cell membrane.[6]

   Divide the cell membrane into five parts: top, bottom, three lateral sides. Accordingly, the nucleus is also divided into five parts. Link the corresponding parts of cell membrane and nucleus. The connection points were chosen randomly using the Monte Carlo method as connection point of microtubules or actin filaments. The chosen probability is determined by the program. The probability for each point is fifty percent. Microtubule is black, filament is gray.

2) The type of cytoskeleton that is connected to the cell membrane at both sides.

   Randomly choose the connection point on the cell membrane using Monte Carlo method. Connect the chosen points. The probability of microfilament is 80%, the probability of microtubule is 20%. Microtubule is black, microfilament is gray.

3) Material properties.

   Cell membrane: thickness = 7.5 nm，linear elastic material, Young's modulus = 1000Pa, Poisson ratio = 0.45.[7, 8]; Cell fluid: Maxwell viscoelastic material, bulk modulus = $2.2*10^9$Pa, density = 1000g/m^3, τ = 1s, viscosity = 100Pa.s, elastic modulus = 100N/ square microns [9]; Cytoskeletal microtubule: diameter: 25 nm, linear elastic material, Young's modulus = 1.2Gpa, Poisson ratio = 0.3. Tension and compression are taken into consideration as well; Cytoskeletal microfilament: diameter = 7 nm, linear elastic material, Young's modulus = 2.6Gpa, Poisson ratio = 0.3; Only unidirectional tension was considered [10]. Nucleus karyolymph: Maxwell viscoelastic material, bulk modulus = $2.5*10^9$Pa, density =1000 g/m3, τ = 1s; Nucleus membrane: thickness = 40 nm, Maxwell viscoelastic material, Young's modulus = 1000Pa, Poisson ratio= 0.3, τ = 5s. Nuclear diameter = 12μm [11, 12].

4. Boundary Condition

In this finite element model, tetrahedral solid elements are used for karyolymph in nucleus and the cytoplasm. The solid elements were covered with shell element, representing the nuclear membrane and cell membrane. Connection points were set as co-nodes. Link element was used for cytoskeleton.

The loading mode was set according to the actual stress of cells. Perpendicular displacement was fixed at the bottom side of the model, as shown in Figure 1D: Exert a surface pressure of 5Mpa on the surface of the model. The compression started from zero, and linearly increased with time within one hour.

5. Result

Under the increasing compression, the cell height decreased, the displacement contour was shown in Figure 2A, the displacement trend to flat, the higher displacement was 0.43μm, and located on the edge of cell. As shown in Figure 2B, the displacement of three angles was not exactly symmetry, because of the random distribution of the cytoskeleton.

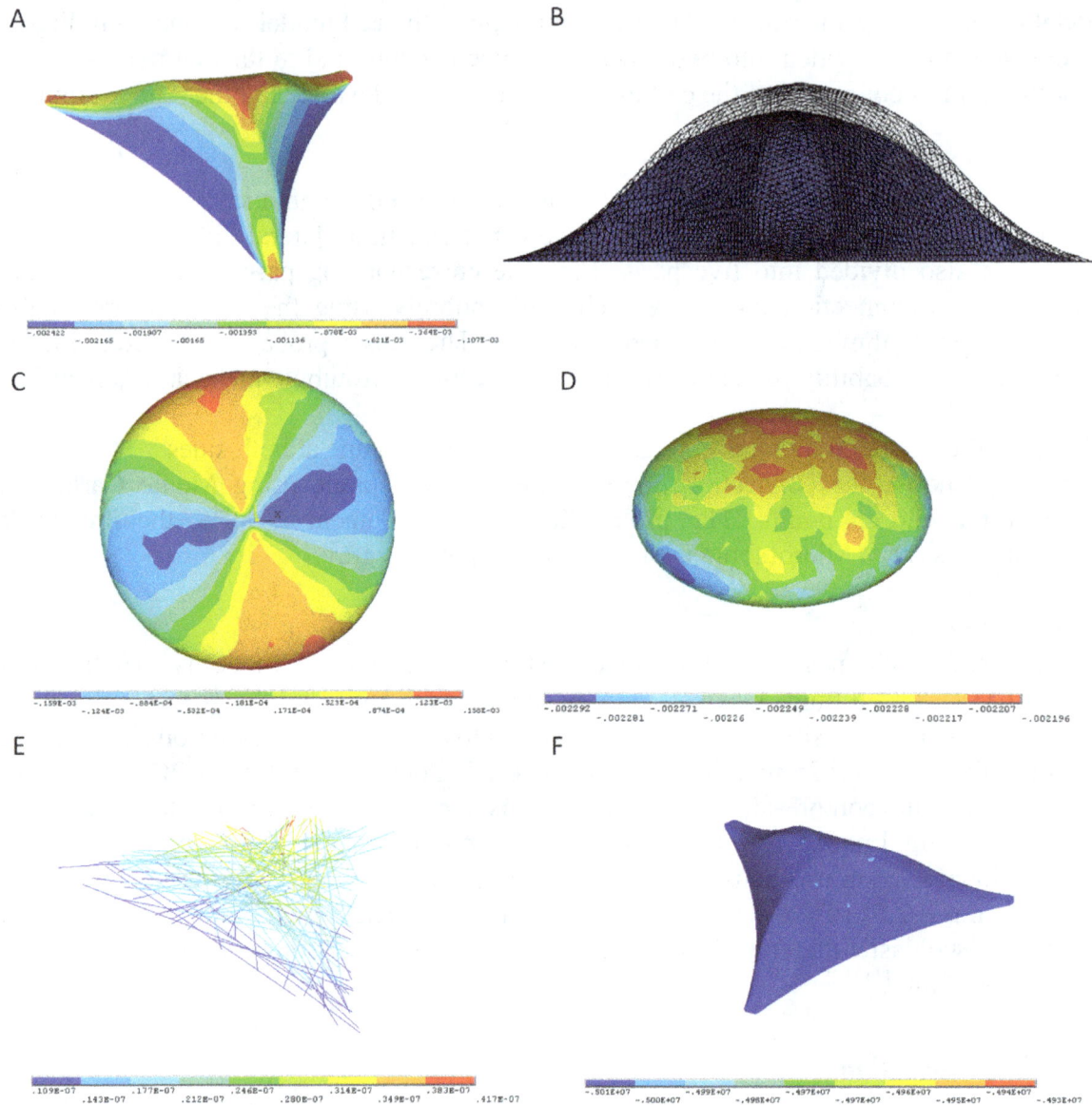

Figure 2A: the displacement contour of the cell, Figure 2B: the lateral view of the displacement of the cell; Figure 2C: the strain contour of the nucleus; Figure 2D: the stress contour of the nucleus; Figure 2E: the displacement contour of the cytoskeleton; Figure 2F: the stress contour of 1st primary stress of the cytoplasm.

In Figure 2C, the strain contour of the nucleus shown that the nucleus expanded in Z axis under the compression simulation. In the magnified 1000 times picture, we could observe the obvious the strain of nucleus, the peak value was 0.022.

In Figure 2D, the stress contour of nucleus shows that the stress distributed randomly due to the distribution of the cytoskeleton, and the peak stress was 0.278Mpa. In Figure 2E, the cytoskeleton displacement distributed on the upper part of the cell, and the peak value was similar to the cytomembrane displacement 0.43μm. Figure 2F was the 1st primary stress contour of the cytoplasm, the contour stress of the cytoplasm shows vanishingly small amounts.

6. Conclusion

According all these results, the cell height decreased under the compression, in this influence, the cytoskeleton will deform that follows the cell nucleus underlying the force. Under this mechanical stimulation, the cytoskeleton remodeling happened. Therefore, the mechanotransduction sequence of each part in cell was from cytomembrane to cytoskeleton and then the cytoplasm, finally to the nucleus. In this paper, we simulated this processing and quantitatively analyzed the mechanical response of cell. This study might explain the remodeling behavior of the cell cytoskeleton, at the same time, this model could validate the cell cytoskeleton remodeling, which was influenced by the nucleus mechanical response.

Acknowledgments

This research was funded by the National Science Foundation 51505137 and China Campus Creative Support Foundation 2012QNZT161.

References

1. Discher, D., Dong, C., Fredberg, J.J., Guilak, F., Ingber, D., Janmey, P., et al. Biomechanics: Cell Research and Applications for the Next Decade. Ann Biomed Eng. 2009, 37, 847-59.
2. Ingber, D.E. Cellular tensegrity: defining new rules of biological design that govern the cytoskeleton. Journal of cell science. 1993, 104, 613-.
3. Kang, J., Steward, R.L., Kim, Y., Schwartz, R.S., LeDuc, P.R., Puskar, K.M. Response of an actin filament network model under cyclic stretching through a coarse grained Monte Carlo approach. Journal of theoretical biology. 2011, 274, 109-19.
4. Kim, T., Hwang, W., Lee, H., Kamm, R.D. Computational analysis of viscoelastic properties of crosslinked actin networks. PLoS Comput Biol. 2009, 5, e1000439.
5. McGarry, J., Prendergast, P. A three-dimensional finite element model of an adherent eukaryotic cell. Eur Cell Mater. 2004, 7, 27-33.
6. Mijailovich, S.M., Kojic, M., Zivkovic, M., Fabry, B., Fredberg, J.J. A finite element model of cell deformation during magnetic bead twisting. Journal of Applied Physiology. 2002, 93, 1429-36.
7. Ronan, W., Deshpande, V.S., McMeeking, R.M., McGarry, P. Finite Element Analysis of Cytoskeletal Remodeling During Micropipette Aspiration. In: ASME 2010 Summer Bioengineering Conference, American Society of Mechanical Engineers, 2010, pp. 433-4.
8. Feng, F., Klug, W.S. Finite element modeling of lipid bilayer membranes. Journal of Computational Physics. 2006, 220, 394-408.
9. Or-Tzadikario, S., Gefen, A. Confocal-based cell-specific finite element modeling extended to study variable cell shapes and intracellular structures: the example of the adipocyte. J Biomech. 2011, 44, 567-73.
10. Barreto, S., Clausen, C.H., Perrault, C.M., Fletcher, D.A., Lacroix, D. A multi-structural single cell model of force-induced interactions of cytoskeletal components. Biomaterials. 2013, 34, 6119-26.

11. Jean, R.P., Chen, C.S., Spector, A.A. Finite-element analysis of the adhesion-cytoskeleton-nucleus mechanotransduction pathway during endothelial cell rounding: axisymmetric model. Transactions-American Society of Mechanical Engineers Journal of Biomechanical Engineering. 2005, 127, 594.
12. Vaziri, A., Mofrad, M.R.K. Mechanics and deformation of the nucleus in micropipette aspiration experiment. J Biomech. 2007, 40, 2053-62.

Atomic Decompositions of Two-parameter Weak Orlicz-Lorentz Martingale Spaces

Xue-Ying Zhang[1, 2,*], Pan Guo[1]

[1]*College of Science, Wuhan University of Science and Technology, Wuhan, 430065, China*
[2]*Hubei Province Key Laboratory of Systems Science in Metallurgical Process (Wuhan University of Science and Technology), Wuhan, 430081, China*
*Email: zhxying315@163.com

In this paper we present three atomic decomposition theorems of two-parameter weak Orlicz-Lorentz martingale spaces. These results generalize the known results of Orlicz martingale spaces and Lorentz martingale spaces. As an application, we get the boundedness of fractional integrals on $D_{F,\infty}$.

Keywords: Atomic decompositions; Orlicz-Lorentz spaces; two-parameter martingales.

1. Introduction

The idea of atomic decompositions in the martingale setting was derived from harmonic analysis [1]. Just as it does in harmonic analysis, the method of atomic decompositions is a key ingredient in martingale theory, such as in the study of martingale inequalities and of the duality theorems for martingale Hardy spaces, especially for small-index martingale and multi-parameter martingales. The technique of stopping times used in the case of one-parameter is usually unsuitable for the case of multi-parameter, but the method of atomic decompositions can deal with them in the same way. Weisz [2] gave some atomic decomposition on martingale Hardy spaces and proved many important theorems by atomic decompositions. Weisz [3] made a further study of the atomic decompositions of the weak Hardy space $w\Sigma_p$ consisting of Vilenkin martingales, and proved weak version of the Hardy-Littlewood inequality by atomic decomposition. Liu and Hou [4] studied the atomic decompositions of martingale for the vector-valued case and some atomic decompositions of Vector-valued martingale were obtained. Hou and Ren [5] considered the weak atomic decompositions and weak martingale inequalities of vector-valued martingales. Orlicz Hardy martingale spaces are also studied by some authors such as Miyamoto, Nakai, Sadasue, and Jiao [6-8]. At the same time, the Lorentz spaces are discussed [9-10]. For instance, the atomic decompositions of Lorentz martingales are first studied by Jiao in [9]. As the generalization of Orlicz and Lorentz spaces, the Orlicz-Lorentz spaces attract more attention. Montgomery-Smith [11] discussed the comparison of Orlicz-Lorentz spaces. Rajeev and Romesh [12] studied composition operators on Orlicz-Lorentz spaces. Echandia [13] discussed the interpolation of Orlicz-Lorentz spaces. Zhang [14] discussed the weak atomic decomposition of weak Orlicz-Lorentz martingales in one parameter case. Ho [16] discussed the atomic decompositions of martingale Hardy- Morrey spaces.

2. Preliminaries and Notations

Let us denote the set of integers and the set of non-negative integers by Z and N respectively. Let $N^2 = N \times N$ and define the partial ordering on N^2 as follows : $\forall m = (m_1, m_2), n = (n_1, n_2) \in N^2$, $m \leq n$ if $m_1 \leq n_1$ and $m_2 \leq n_2$; $m < n$ if $m \leq n$ and $m \neq n$; $m \prec n$ if $m_1 < n_1$ and $m_2 < n_2$.

---

*Corresponding author

Let (Ω, Σ, P) be a probability space and let Σ_n be a family of non-decreasing sub--algebras of Σ. We denote the conditional expectation operators relative to Σ_n by E_n.

A function v which maps Ω into the set of subspace of $N^2 \cup \{\infty\}$ is said to be a two-parameter stopping time relative to $(\Sigma_n, n \in N^2)$ if the elements of $v(\omega)$ are incomparable and for $\forall n \in N^2$, $\{\omega \in \Omega : n \in v(\omega)\} \in \Sigma_n$. The set of all two-parameter stopping times will be denoted by T_2.

2.1. Definition

An integrable sequence $f = (f_n, n \in N^2)$ is said to be a martingale if it is adapted to $(\Sigma_n, n \in N^2)$ and $E(f_m | \Sigma_n) = f_n$ for all $n \leq m$. The martingale $f = (f_n, n \in N^2)$ is regular if there exists a constant R such that $f_{n_1, n_2} \leq R f_{n_1-1, n_2}$ and $f_{n_1, n_2} \leq R f_{n_1, n_2-1}$ for all $n = (n_1, n_2) \in N^2$.

The stopped martingale $f^{(v)} = (f_n^{(v)}, n \in N^2)$ relative to a martingale $f = (f_n, n \in N^2)$ and stopping time $v \in T_2$ is defined by $f_n^{(v)} = \sum_{m \leq n} \chi(v \nmid m) d_m f$, , where $d_m f$ is martingale difference.

2.2. Definition

A Young function F is an even continuous and non-negative function in R^1, increasing on $(0, \infty)$, such that $\lim_{t \to \infty} F(t) = \infty, F(t) = 0$ if $t = 0$. A Young function F is said to satisfy the global Δ_2-condition if there is $c > 0$ such that $F(2t) \leq cF(t)$ for all $t \in R^1$. The function F is of lower type l for some $l \in [0,1]$ and upper type 1, denoted by $f \in \nabla$ i.e., there exists a constant $c \geq 1$ such that

$$F(tr) \leq c \max\{t^l, t\} F(r) \quad \text{for } t, r \in [0, \infty). \tag{2.1}$$

We know F is a concave function and $F \in \Delta_2$ if $F \in \nabla$. If F is a concave function, then F is subadditive, i.e. for

$$t \geq 0, r \geq 0, \quad F(t+r) \leq F(t) + F(r). \tag{2.2}$$

We define $\widetilde{F}(t) = 1/(F(1/t))$ if $t > 0$ and 0 if $t = 0$.

2.3. Definition

We define the weak Orlicz-Lorentz space $L_{F,\infty}$ as the set of all-measurable f's on Ω for which the Orlicz-Lorentz functional $\|f\|_{F,\infty} = \sup_{t \geq 0} \widetilde{F}^{-1}(t) f^*(t)$ is finite, where $f^*(t)$ is the non-increasing rearrangement of f defined by $f^*(t) = \inf\{s > 0, d_f(s) \leq t\}$.

2.4. Remark

If A is any measurable set, $\|\chi_A\|_{F,\infty} = \widetilde{F}^{-1}(P(A))$. Now we give some notations with respect to two-parameter martingales as follows.

$$M_n f = \sup_{i \le n} |f_i|, \qquad Mf = \sup_{n \in N^2} M_n f, \qquad S_n(f) = \left(\sum_{0 \le k \le n} |df_k|^2 \right)^{1/2},$$

$$S(f) = \sup_{n \in N^2} S_n(f), \qquad \sigma_n(f) = \left(\sum_{0 \le k \le n} E_{k-1} |df_k|^2 \right)^{1/2}, \quad \sigma(f) = \sup_{n \in N^2} \sigma_n(f)$$

Λ denotes the collection of all sequences (λ_n) of non-decreasing, non-negative and adapted functions. Set $\lambda_\infty = \sup_n \lambda_n$. Thus we can define some two-parameter weak Orlicz-Lorentz martingale spaces as follows:

$$H_{F,\infty}^\sigma = \{ f = (f_n) : \| \sigma(f) \|_{F,\infty} < \infty \},$$

$$Q_{F,\infty} = \{ f = (f_n) : \exists (\lambda_n) \in \Lambda, s.t. S_n(f) \le \lambda_{n-1}, \lambda_\infty \in L_{F,\infty} \}, \| f \|_{Q_{F,\infty}} = \inf_{(\lambda_n) \in \Lambda} \| \lambda_\infty \|_{L_{F,\infty}},$$

$$D_{F,\infty} = \{ f = (f_n) : \exists (\lambda_n) \in \Lambda, s.t. |f_n| \le \lambda_{n-1}, \lambda_\infty \in L_{F,\infty} \}, \| f \|_{D_{F,\infty}} = \inf_{(\lambda_n) \in \Lambda} \| \lambda_\infty \|_{L_{F,\infty}}.$$

2.5. Definition

A measurable function a is called a q-atom of the first category (or of the second category, of the third category, respectively) if there exists a stopping time τ such that

(i) $a_n = E_n a = 0$ if $\tau \prec$ n, (ii) $\| \sigma(a)_q \| \le P(\tau \ne \infty)^{1/q}$

(or $\| S(a)_q \| \le P(\tau \ne \infty)^{1/q}$, $\| M(a)_q \| \le P(\tau \ne \infty)^{1/q}$)

These three category atoms are briefly called q-1-atom, q-2-atom, and q-3-atom, respectively. Throughout this article, we use c to denote different constants at different occurrences.

3. Main Results

3.1. Theorem

Let function $\widetilde{F}^{-1} \in \Delta$. Then $f = (f_n) \in H_{F,\infty}^\sigma$ if and only if there exist a sequence $(a^k, k \in Z)$ of 2-1-atoms and the corresponding stopping times $(\tau_k, k \in Z)$ such that

(1) $\quad f_n = \sum_{k \in Z} u_k E_n a^k$, (2) $\| f \|_{H_{F,\infty}^\sigma} \sim \inf \sup_{k \in Z} 2^k \widetilde{F}^{-1}(P(\tau_k \ne \infty))$,

where the infimum is taken over all the preceding decompositions of f.

Proof: Assume $f = (f_n) \in H_{F,\infty}^\sigma$. Set $F_k = \{ \sigma(f) > 2^k \}$, and $\tau_k = \inf\{ n : E_n(\chi_{F_k}) > 1/2 \}$. Then the sequence of these stopping times is non-decreasing and $\tau_k \to \infty$. Let $(f_n^{(\tau_k)}, n \in N^2)$ be the stopped martingale, we have

$$f_n = \sum_{k \in Z} f_n^{(\tau_{k+1})} - f_n^{(\tau_k)} \tag{3.1}$$

Now let $u_k = 2^{k+3/2}$, $a_n^k = u_k^{-1}(f_n^{(\tau_{k+1})} - f_n^{(\tau_k)})$. Then for a fixed $k \in Z$, (a_n^k) is a martingale and

$$E(|f_n^{(\tau_{k+1})} - f_n^{(\tau_k)}|^2) = E(|\sum_{m \leq n} \chi_{(\tau_k \prec m \nsucc \tau_{k+1})} d_m f|^2) \leq E(\sum_{m \leq n} \chi_{(\tau_k \prec m \nsucc \tau_{k+1})} |d_m f|^2) \tag{3.2}$$

$$= E(\chi_{\sigma(f) \leq 2^{k+1}} E_{n-1}(\sum_{m \leq n} \chi_{(\tau_k \prec m \nsucc \tau_{k+1})} |d_m f|^2)) + cE(\chi_{\sigma(f) > 2^{k+1}} E_{n-1}(\sum_{m \leq n} \chi_{(\tau_k \prec m \nsucc \tau_{k+1})} |d_m f|^2)) := (A) + (B)$$

Obviously,

$$(A) \leq 2^{2(k+1)} P(\tau_k \neq \infty) \tag{3.3}$$

From the definition of τ_{k+1}, it follows that $E_{n-1}(\chi_{F_{k+1}}) \leq 1/2$ if $\tau_k \leq n$. so

$$(B) \leq \frac{1}{2} E(\sum_{m \leq n} \chi_{(\tau_k \prec m \nsucc \tau_{k+1})} |d_m f|^2). \tag{3.4}$$

Therefore, $E(\sum_{m \leq n} \chi_{(\tau_k \prec m \nsucc \tau_{k+1})} |d_m f|^2) \leq (A) + \frac{1}{2} E(\sum_{m \leq n} \chi_{(\tau_k \prec m \nsucc \tau_{k+1})} |d_m f|^2)$, which implies

$$E(|f_n^{(\tau_{k+1})} - f_n^{(\tau_k)}|^2) \leq 2(A) \leq 2^{(2k+3)} P(\tau \neq \infty). \tag{3.5}$$

Thus $\|M(a^k)\|_2 \leq c\|\sigma(a^k)\|_2 < \infty$ and $(a_n^k, n \in N^2)$ is L_2 bounded martingale. So there exists an integrable function a^k such that $a_n^k = E_n a^k$. If $n \leq \tau_k$, $E_n a^k = 0$ so we get a^k is really a 2-1-atom. Moreover, we have $f_n = \sum_{k \in Z} u_k E_n a^k$. Hence we get (1).

Now, we turn to the proof of (2). As

$$P(\tau_k \neq \infty) \leq 4E(\sup_n E_n(\chi_{F_k})^2) \leq 8P(\sigma(f) > 2^k) \tag{3.6}$$

We have $2^k \tilde{F}^{-1}(P(\tau_k \neq \infty)) \leq 2^k \tilde{F}^{-1}(8P(\sigma(f) > 2^k)) \leq c\sup_{t>0} t\tilde{F}^{-1}(P(\sigma(f) > t)) = c\|f\|_{H_{F,\infty}^\sigma}$, which implies

$$\sup_{k \in Z} 2^k \tilde{F}^{-1}(P(\tau_k \neq \infty)) \leq c\|f\|_{H_{F,\infty}^\sigma} \tag{3.7}$$

Conversely, assume that $f = (f_n)$ has a decomposition of the form (1).

Let $M = \sup_{k \in Z} 2^k \tilde{F}^{-1}(P(\tau \neq \infty))$. For any fixed $y > 0$ choose $j \in Z$ such that $2^j \leq y < 2^{j+1}$.

Let $f_n = \sum_{k \in Z} 2^{k+3/2} E_n a^k = \sum_{k \leq j-1} 2^{k+3/2} E_n a^k + \sum_{k \geq j} 2^{k+3/2} E_n a^k := g_n + h_n$. Thus by the fact $\sigma(f) \leq \sigma(g) + \sigma(h)$, we have $P(\sigma(f) > 2y) \leq P(\sigma(g) > 2y) + P(\sigma(h) > 2y)$. If $n \leq \tau_k$, $E_n a^k = 0$, thus $\sigma(a^k) = 0$ on the set $\{\tau_k = \infty\}$. Moreover $\sigma(h) \leq \sum_{k=j}^\infty 2^{k+3/2} \sigma(a^k)$ and $\{\sigma(h) > 0\} \subset \cup_{k=j}^\infty \{\tau_k \neq \infty\}$. Since $F^{-1} \in \nabla$, $\tilde{F}^{-1} \in \nabla$. Consequently,

$$\tilde{F}^{-1}(P(\sigma(h) > y)) \le \tilde{F}^{-1}(P(\sigma(h) > 0)) \le \sum_{k=j}^{\infty} \tilde{F}^{-1}(P(\tau_k \ne \infty)) \le My^{-1}. \tag{3.8}$$

$$\tilde{F}^{-1}(P(\sigma(g) > y)) \le \tilde{F}^{-1}(y^{-2} \|\sigma(g)\|_2^2) > y))$$

$$\le \tilde{F}^{-1}(\sum_{k=-\infty}^{j-1} y^{-2} 2^{2k+3} P(\tau_k \ne \infty)) \le \sum_{k=-\infty}^{j-1} y^{-2} 2^{2k+3} \tilde{F}^{-1}(P(\tau_k \ne \infty)) \le CMy^{-1}, \tag{3.9}$$

which implies $\|f\|_{H_{F,\infty}^\sigma} = \sup_{y>0} 2y \tilde{F}^{-1}(P(\sigma(f) > 2y)) \le c \sup_{k \in Z} 2^k \tilde{F}^{-1}(P(\tau_k \ne \infty))$. Thus we have proved Theorem 3.1.

Similarly, we get atomic decompositions theorems of the space $Q_{F,\infty}$ and $D_{F,\infty}$. We omit the proofs.

3.2. *Theorem*

Let function $F^{-1} \in \nabla$. Then $f = (f_n) \in Q_{F,\infty}$ if and only if there exist a sequence $(a^k, k \in Z)$ of 2-2-atoms and the corresponding stopping times $(\tau_k, k \in Z)$ such that

$$(1) \quad f_n = \sum_{k \in Z} u_k E_n a^k, \qquad (2) \quad \|f\|_{H_{F,\infty}^\sigma} \sim \inf \sup_{k \in Z} 2^k \tilde{F}^{-1}(P(\tau_k \ne \infty)),$$

where the infimum is taken over all the preceding decompositions of f.

3.3. *Theorem*

Let function $F^{-1} \in \nabla$. Then $f = (f_n) \in D_{F,\infty}$ if and only if there exist a sequence $(a^k, k \in Z)$ of 2-3-atoms and the corresponding stopping times $(\tau_k, k \in Z)$ such that

$$(1) \quad f_n = \sum_{k \in Z} u_k E_n a^k, \qquad (2) \quad \|f\|_{H_{F,\infty}^\sigma} \sim \inf \sup_{k \in Z} 2^k \tilde{F}^{-1}(P(\tau_k \ne \infty)),$$

where the infimum is taken over all the preceding decompositions of f.

4. Applications

It is well known that the fractional integrals have occupied a very important role in the classical harmonic analysis. We will discuss the fractional integral in martingale settings.

4.1. *Definition*

For $f = (f_n, n \in N^2)$, $\alpha > 0$, the fractional integral $I_\alpha f = ((I_\alpha f)_n, n \in N^2)$ of f is defined by $(I_\alpha f)_n = \sum_{k \le n} b_{k-1}^\alpha d_k f$, where b_k is a Σ_k-measurable function such that for all atom $B \in \Sigma_k, b_k = P(B)$ on B.

4.2. *Lemma*

Let $(\sum_n, n \in N^2)$ is regular and $\alpha > 0$. If there exists $A \in \sum$ such that $Mf \leq \chi_A$, then there exists a positive constant c independent of f and A such that $M(I_\alpha f) \leq cP(A)^\alpha \chi_A$ [15].

4.3. *Theorem*

Let $(\sum_n, n \in N^2)$ is regular, $\alpha > 0$ and $F^{-1} \in \nabla$. If $F^{-1}(x) \geq cx^{1-\alpha}$ for all $x \geq 1$, then there exists a constant c such that $\left\| I_\alpha a^k \right\|_{D_{F,\infty}} \leq c$ for each 2-3 atom a^k.

Proof: By the construction of atom in Theorem 3.1, we know $M(a^k) \leq \chi_{(\tau_k \neq \infty)}$. Then by lemma 4.2 we obtain that $M(I_\alpha(a^k)) \leq cP(\tau_k \neq \infty)^\alpha \chi_{(\tau_k \neq \infty)}$. Let $(\lambda'_n) = \left\| I_\alpha a^k \right\|_\infty \chi_{(\tau_k \leq n)}$. Then (λ'_n) is a nonnegative, non-decreasing and adapted sequence with $|(I_\alpha a^k)_{n+1}| \leq \lambda'_n$. Hence we have

$$\left\| I_\alpha a^k \right\|_{D_{F,\infty}} \leq cP(\tau_k \neq \infty)^\alpha \left\| \chi_{(\tau_k \neq \infty)} \right\|_{F,\infty} \leq cP(\tau_k \neq \infty)^\alpha P(\tau_k \neq \infty)^{1-\alpha} \leq c.$$

5. Conclusion

Three atomic decomposition theorems of two-parameter weak Orlicz-Lorentz martingale spaces have been discussed in the work. Proof is given and the potential application is proposed, which shows the boundedness of fractional integrals on $D_{F,\infty}$.

Acknowledgments

This research was funded by the National Natural Science Foundation of China under Grant 11201354, by Hubei Province Key Laboratory of Systems Science in Metallurgical Process (Wuhan University of Science and Technology) under Grant Y201511 and by Natural Science Foundation of Hubei Province under Grant 2015CFB602.

References

1. C. Herz, Hp-space of martingales, $0 < p \leq 1$, Z. Wahrscheinlichkeitstheor. Verw. Geb. 28 (1974) 189-205
2. F. Weisz, Martingale Hardy Spaces and Their Applications in Fourier Analysi, Springer, Berlin, 1994.
3. F. Weisz, Bounded operators on weak Hardy spaces, Acta Math. Hung. 80 (1998) 249-264.
4. P. D. Liu, Y. L. Hou L Yu, Atomic decompositions of Banach-space-valued martingales, Sci. China Ser. A 42 (1999) 38-47.
5. Y. L. Hou, Y. B. Ren, Weak martingale Hardy spaces and weak atomic decompositions, Sci. China Ser. A 49 (2006) 912-921.
6. T. Miyamoto, T, E. Nakai, G. Sadasue, Martingale Orlicz-Hardy spaces, Math. Nachr. 285 (5-6) (2012) 670-686.
7. Y. Jiao, L. Wu, Weak Orlicz-Hardy martingale spaces, arXiv:1304.3910.
8. Y. Jiao, Embeddings between weak Orlicz martingale spaces, J. Math. Anal. Appl., 378 (2011) 220-229.

9. Y. Jiao, L. P. Fan, P. D. Liu, Interpolation theorems on weighted Lorentz martingale spaces, Sci. China Math., 50(9) (2007) 1217-1226.

10. Y. Jiao, L. H. Peng, P. D. Liu, Atomic decompositions of Lorentz martingale spaces and applications, J. Funct. Spaces Appl., 7(2) (2009) 153-166.

11. S. J. Montgomery-Smith, Comparison of Orlicz-Lorentz spaces, Stud. Math., 103(2)(1992) 161-189.

12. K. Rajeev, K. Romesh, Composition operators on Orlicz-Lorentz spaces, Integral Equ. Oper. Theory, 60 (2008) 79-88.

13. V. Echandia, Interpolation between Hard-Lorentz-Orlicz spaces, Acta Math. Hung., 66(3) (1995) 217-221.

14. C. Z. Zhang, M. M. Zhong, X. Y. Zhang, Atomic decompositions of weak Orlicz-Lorentz martingale spaces, Journal of Inequalities and Applications, 1(66) (2014) 1-9.

15. G. Sadasue, Fractional integrals on martingale Hardy spaces for $0 < p \leq 1$, Mem. Osaka Kyoiku Univ, Ser. III, Nat. Sci. Appl. Sci., 60(2011) 1-7.

16. K. P. HO, Atomic decompositions of martingale Hardy-Morrey spaces, Acta Math. Hungar. 149 (1) (2016), 177-189.

Progressive Collapse Process Analysis of Steel Frame Under Fire Condition

Dong Chen[*], Xue Chen

Anhui Jianzhu University, Hefei 230601, China
Email: chenchenchu@163.com

Steel frame structures are prone to failure of local load-bearing components under fire, and the progressive collapse could occur. Exploring the mechanism and law of progressive collapse of steel frame under fire is significant in design and reinforcement of steel structure, and to avoid progressive collapse under fire. In this paper, the analysis of the specimens of six story three-span plane steel frame of the bottom column and side column under fire after the failure of the remaining elements of progressive collapse process by ANSYS finite element software. The analysis results well simulated the whole progressive collapse process of the steel frame under the fire, then analyzed the change of structure and shape of the progressive collapse of the steel frame in every moment, which lay the foundation for the subsequent collapse analysis.

Keywords: Fire; steel frame; progressive collapse; finite element.

1. Introduction

Steel structure is more and more used in high-rise buildings because of its unique properties and advantages, but the steel structure also has its fatal defects, and its physical and mechanical properties changed greatly under high temperature. Recently, there have been a number of high-rise steel structure building occur progressive collapse at high temperature under fire. The progressive collapse of the structure is that the structure is subjected to accidental loads to produce local initial damage, then caused damage to the frame members connected to the disabled component due to internal force redistribution, and formed a chain reaction, eventually formed a large range of damage and even collapse in a large scale compared with the initial local failure [1].The progressive collapse process of the structure is very complicated, and it is difficult to simulate the structure accurately. Sun Ruirui and others [2] used method of combining static analysis and dynamic analysis, and carried on analysis to the progressive collapse of steel frame under high temperature of fire, and structural failure mode and internal force variation of a single column and a plurality of columns were given at the same time under fire conditions. Xie Fuzhe and others [3] used pumping column method to carry out the progressive collapse simulation of the space frame and the frame support structure with different span and height, then the relationship between the dynamic effect of structures and the failure time of the columns and the displacement time history curves of each failure point were obtained. Li Yi and others [4-5] analyzed the progressive collapse resistance of reinforced concrete frame structure based on energy method and established progressive collapse resistance demand structure calculation theory method under bean and catenary mechanism. Lamont S [6] studied the effect of different fire duration and maximum fire temperature on the fire behavior of composite steel frame structure. Ju Zhu [7] used temperature curve of hydrocarbon combustion, and analyzed the failure process of the three-Story-bending steel frame through the explicit dynamic analysis. The internal force and deformation of internal forces and deformation at different locations in the structure were given. However, there are few literature about the description of the failure process of the steel frame structure at the temperature of the fire.

---

*Corresponding author

This paper used the finite element software ANSYS to establish and analyze a six-story three-span steel plane finite element model, and to simulate the failure of the bottom middle column and side column under fire, then removed them. To analyze the other parts of the progressive collapse process, then study the instantaneous structure and variation of inner force of the steel frame progressive collapse. The method of the analysis is that:

1) Static linear analysis of the model is analyzed before the column failure and the static internal force of the column is obtained;

2) Then the failure column is removed and static force is applied to the failure point, which make the residual structure and the original structure static equivalent.

3) In a certain period of time, the static internal force value of the reverse action is reduced to 0, and then the dynamic effect of the residual structure after the column is calculated.

2. Fire Temperature-Rise Curve

In this paper, the ISO834 standard fire heating curve is used to heat the structure and the expression of heating curve is:

$$T_g = 20 + 345\lg(8t+1) \tag{1}$$

Tg (°C) represents gas temperature and **t** (min) represents burning time. The standard heating curve is shown in Fig. 1:

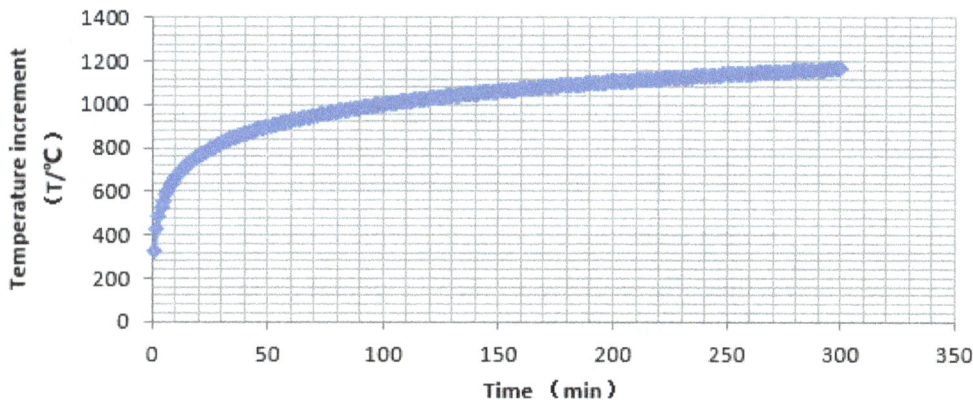

Fig. 1 ISO834 standard fire temperature curve

3. Steel Frame Structure Model

A six-story three-span plane steel frame model was set up as shown in Fig. 2, span is 6m, and floor height is 3.6m. All beams and columns were welded cross section, beams used 300mm × 300mm× 10mm× 16mm (height × width × web plate thickness × flange thickness) and side columns used 400mm× 300mm × 10mm × 16mm and middle column used 500mm × 300mm × 10mm × 16mm. Steel elastic modulus: E=2.06×105MPa, density: ρ=7.85×103kg/m3，yield strength is 345MPa. Assume that all of the same size of beam uniform load are 70kN/m. The columns on the A axis in the graph were defined as the column A1~A6 in the order of layers. The beams between the A axis and the B axis were defined as the beam A1~A6 and the remaining beams and columns are numbered in the same way. The fire locations of corner column and middle column were as shown in Fig. 2.

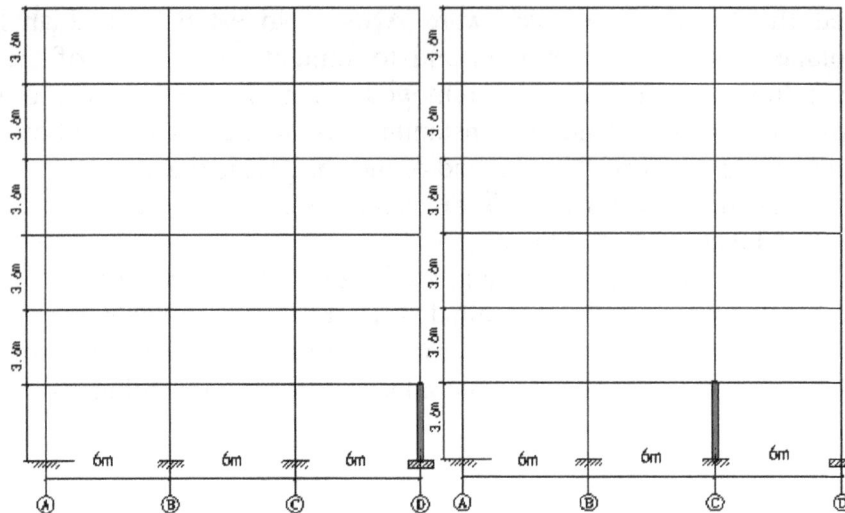

Fig. 2 Steel frame model

4. Analysis of Progressive Collapse Process

Using ANSYS finite element analysis software, to respectively analyze the basic form of progressive collapse of the disabled remaining components in different time, when the corner columns and middle columns which are in ground floor were under the effect of high temperature of fire. Because of the steel frame under the action of fire, the process of progressive collapse was very rapid, so took the 1s time after the beginning of the collapse process to analyze. In order to facilitate the analysis, the positive direction of the frame longitudinal (x) was defined along the axis of the frame, the direction of upward gravity was the vertical direction (z), and horizontal perpendicular to the axis of the frame was defined as the positive direction of the frame lateral (y).

4.1. *Progressive collapse process after failure of the middle column C1*

According to the analysis results, it was shown that the collapse process of the steel frame after C1 failure can be divided into the following stages:

(1)　Stage one : t=0.0s～0.21s

After the failure of C1 column which was under fire, the columns and beams associated with the C axis lost supporting function, and the larger deformation along the Z axis was gradually generated; the vertex of the D6 column and the D3 column had a larger deformation along the X axis. The deformation produced by this stage did not reach the ultimate strain of the steel, the deformation was mainly along the Z axis and X axis direction. As shown in Fig. 3.

(2)　Stage two : t=0.21s～0.35s

As time increased, the deformation continued to expand, the fracture occurred at the joints of B3 beam and B3 column in t=0.21s, and the deformation trend of frame was consistent with stage one; Fracture occurs at the joints of B2 beam and C2 column in t=0.22s, the deformation of the frame continued to increase; The joints which connected to C1 column fractured in succession in t=0.25s, The A1, B1 and D1 columns had a large deformation due to the redistribution of the internal force, and the deformation of D1

790

column was most serious. When t=0.30 ~ 0.35s, the deformation of B1 ~ B6 beam exceeded the limit strain, the joints of beam and column were broken, C1 ~ C6 beam occurred plastic hinge in succession, D1 column has completely collapsed, the deformation of the whole structure along the X axis and the Z axis was greatly increased. As shown in Fig. 4.

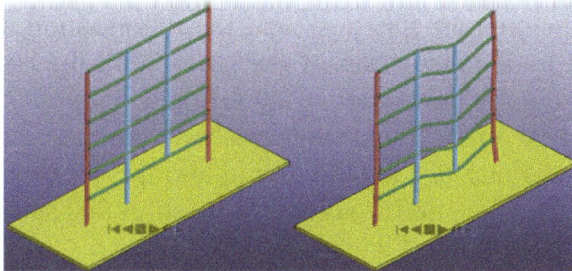

Fig. 3 Stage 1 deformation diagram

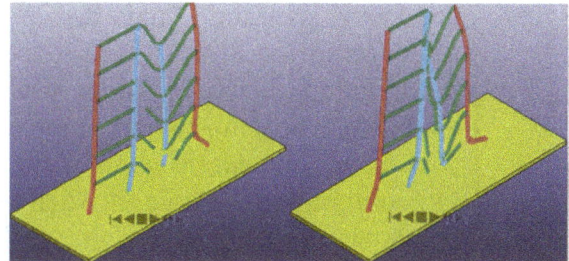

Fig. 4 Stage 2 deformation diagram

(3)　Stage three : t=0.35~0.55s

There was a very obvious plastic hinge in the A1 and B1 columns in this stage, the whole structure exhibited a very unstable state. Because of the failure of the D1 column, the columns on the C axis and the D axis and the beam which was connected to the column lost supporting function, and it collapsed along the direction of the X shaft. B1 column exited after failure when t=0.45s, A1 column was also completely out of work after failure when t=0.55s. Columns and beams on the A axis and the B axis collapsed along the X axis. As shown in Fig. 5.

(4)　Stage four : t=0.55~1.00s

In this stage, due to the columns of the bottom had been destroyed, all beams and columns were completely collapsed under the action of load and self-weight, columns and beams on the A axis and the B axis continued to collapse along the negative direction of the X axis, columns and beams on the C axis and the D axis continued to collapse along the positive direction of the X axis. As shown in Fig. 6.

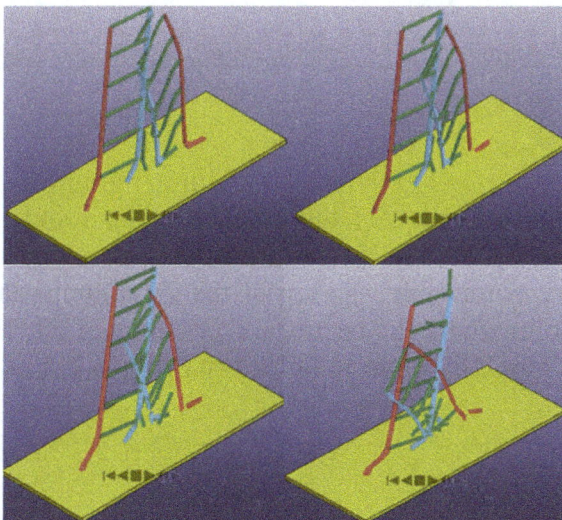

Fig. 5 Stage 3 deformation diagram

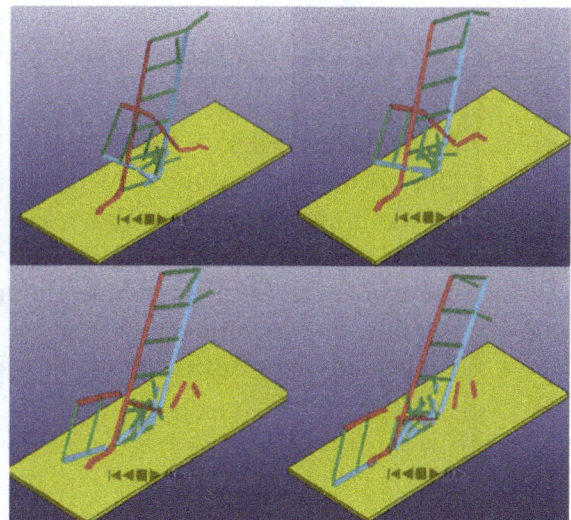

Fig. 6 Stage 4 deformation diagram

4.2. Progressive collapse process of side column D1 after failure

According to the analysis results, it was shown that the collapse of the steel frame can be divided into the following stages after the failure of the side column D1.

(1)   Stage one : t=0~0.26s

After the failure of the D1 column in the fire, the columns and the beams which were connected with the D axis had lost the support function, and the deformation along the direction of the Z axis was produced under the action of the load. With the increase of deformation, the plastic hinge was gradually generated at the left and right ends of C1 ~ C6 beam. The deformation of the column on the D axis was gradually increased along the X axis, and the deformation of the lower end of the D2 column and the upper end of the D5 column was the biggest when t=0.25s. The deformation produced in this stage did not reach the ultimate strain of the steel. As shown in Fig. 7.

(2)   Stage two : t=0.26~0.35s

As the deformation continued to increase, fracture occurred at the joint of D2 column and C1 beam in t=0.26s, deformation continued to increase, fracture occurs at the joint of C5 beam and C5 column in t=0.29s; fracture occurs at the joint of D3 column and D4 column in t=0.30s; beams and columns which were associated with D axis lost the constraint function, and the collapse occurred. As shown in Fig. 8.

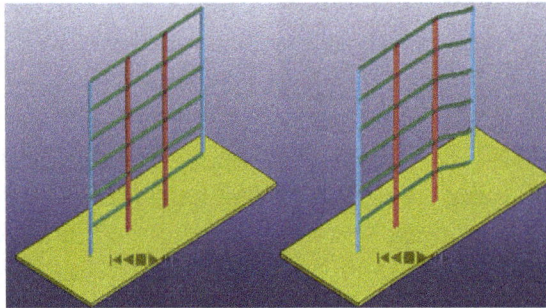

Fig. 7 Stage 1 deformation diagram   Fig. 8 Stage 2 deformation diagram

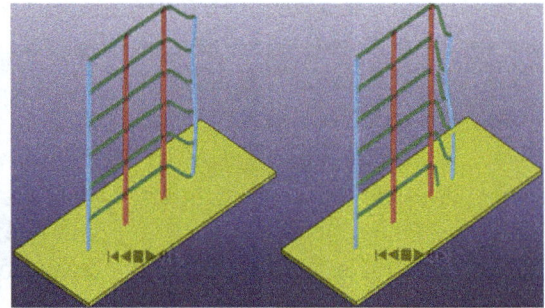

(3)   Stage three : t=0.35~0.60s

Due to the redistribution of internal force, A1, B1 and C1 columns gradually changed. A1 columns became invalid under the influence of the collapsed beam and column in t=0.53s; B1 column also failed in t=0.55s. Due to the failure of the A1 and B1 columns, the whole frame of the whole frame was unstable, the top of the C1 column damaged in t=0.60s, the fracture occurred at the joint where the beams and columns of B1, B2 and B6 connected, the whole framework was basically close to failure. As shown in Fig. 9.

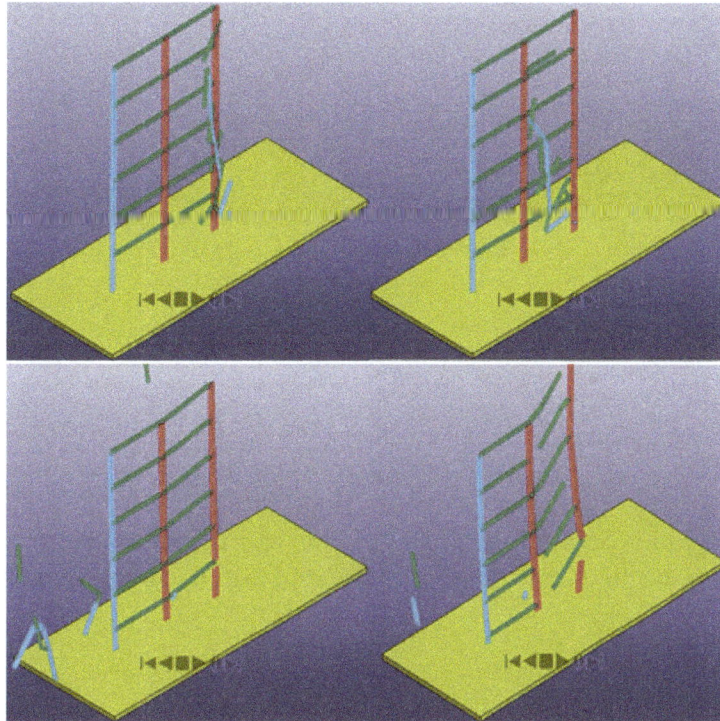

Fig. 9 Stage 3 deformation diagram

(4)   Stage four : t=0.60~1.00s
Because each column of the ground floor had been destroyed, the residual structure continued to deform along the respective collapse tendency under the load, the structure collapsed finally. As shown in Fig. 10.

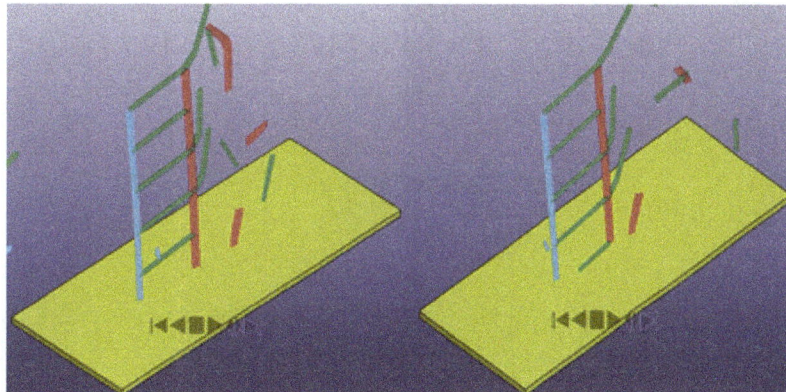

Fig. 10 Stage 4 deformation diagram

4.3. *Internal force and displacement curve*

According to the analysis results, when the middle column and corner column became invalid, the displacement curve of the top joint of the failure column was shown in Fig. 11.

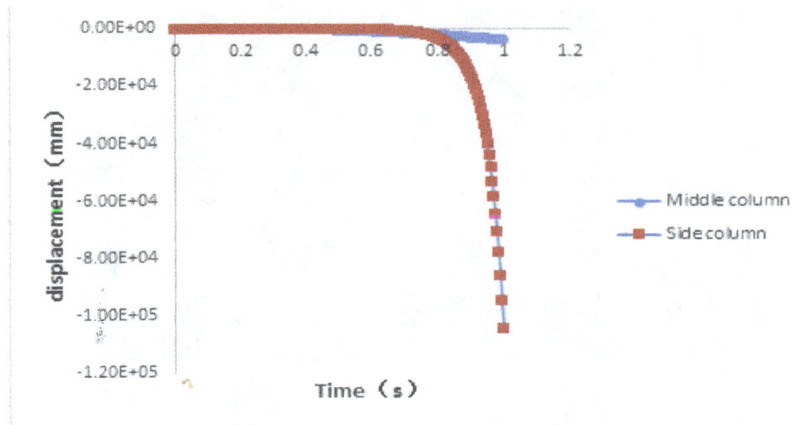

Fig. 11 Displacement curve

When the corner column became invalid, C1 beam right moment curve was shown in Fig. 12, when the middle column became invalid, C1 left beam bending moment curve was also shown in Fig. 12.

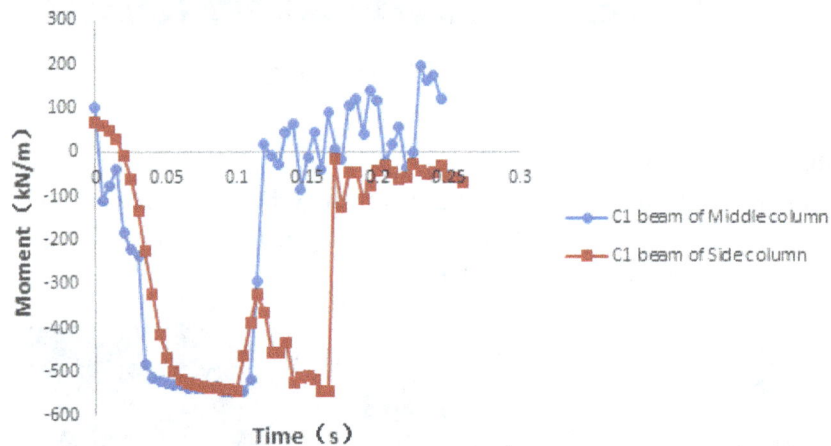

Fig. 12 Bending moment curve

5. Conclusion

The progressive collapse of steel frame under the high temperature of fire was studied in this paper:

(1) A finite element model of a six-story three-span plane steel frame had been built, respectively simulated and analyzed the whole process of the progressive collapse of the remaining members after the failure of the side columns and the middle columns under the fire.

(2) Describing the various stages of the progressive collapse process, a preliminary analysis of the structure and internal force of each moment of the progressive collapse of the steel frame.

(3) Compared with the collapse process of the middle columns and the corner columns, it can be seen that the collapse of the corner columns were more rapid than that of the middle column, and the form of the collapse was more complicated because of the complex characteristic of the force that corner columns suffered.

Acknowledgments

The authors sincerely thank the Nature Science Research Project of Anhui Province (Research Project No. 1408085QE96) and Anhui Universities Natural Science Research Project (Research Project No. KJ2015A046) for funding this work.

References

1. Ellingwood B R, Mitigating Risk from Abnormal Loads and Progressive Collapse, J. Journal of Performance of Constructed Facilities. 20(4) (2006) 315-323.
2. Sun Ruirui, Huang Zhaohui, Progressive collapse analysis of steel structures under fire conditions, J. Engineering Structures. 34 (2012) 400-413.
3. Xie Fuzhe, Shu Ganping, Feng Junmin, Progressive collapse analysis of steel frame structure using removing column method, J. Journal of Southeast University (Natural Science Edition). 40(1) (2010) 154-159. (In Chinese)
4. Li Yi, Ye Lieping, Lu Xinzheng, Progressive collapse resistance demand of RC frame structures based on energy method I: beam mechanism, J. Journal of Building Structures. 32(11) (2011) 1-8. (In Chinese)
5. Li Yi, Lu Xinzheng, Ye Lieping, Progressive collapse resistance demand of RC frame structures based on energy method II: catenary mechanism, J. Journal of Building Structures. 32(11) (2011) 9-16. (In Chinese)
6. Lamonta S, Usmanib A S, Gilliec M, Behaviour of a small composite steel frame structure in a "long-cool" and a "short-hot" fire, J. Fire Safety Journal. 39(5) (2004) 327-357.
7. Ju Zhu, Wang Zhenqing, Han Yulai, Liang Wenyan, Li Jialei, Structure collapse analysis of steel frame under fire condition, J. Engineering Mechanics, 31 (2014) 121-124. (In Chinese)

Mathematical Model Design of Materials Circulating System in Circulating Fluidized Bed Boiler

Yue Zhang, Yun-Fei Liu*, Yu-Han Men

Hebei Engineering Research Center of Simulation & Optimized Control for Power Generation (North China Electric Power University), Baoding, China, 071003
*Email: 1281385741@qq.com

The aim of this study is to investigate the concentration of solid particles inside the furnace. Starting from the material circulating system and distribution of concentration of solid particles in CFBB, the model of material circulating system was established. Finally, the system was simulated in Matlab. Through analysis of the simulation results: along the direction of the furnace, the concentration of solid particles inside the furnace shows in the distribution of 'the upper part is dilute and the under part is concentrated'. Moreover, in the dense phase zone, the concentration of solid particles decreases rapidly. But in the transition zone and dilute phase zone, it decreases slowly and tends to be stable at the exit of the furnace.

Keywords: Fluidized bed; materials circulation; particle concentration; mass balance.

1. Introduction

Circulating fluidized bed boiler has the advantages of high efficiency of combustion, low emissions, strong adaptability for fuels and large range of load adjustment, which makes it get rapid development and widespread application [1]. Currently, people for the study of circulating fluidized bed material circulation system focus on the flow characteristics of gas-solid flow and concentration distribution of solid particles, and has established a Euler-Euler equations based on two-fluid model, discrete particle model, turbulence model, drag models and cell models. Distribution of the concentration of material has a great influence on the economy and security of the fluidized bed boiler, and complex reactions within the bed makes the study of concentration distribution of the material very difficult [2].

Based on previous studies [3, 4], this paper establishes the models of particle combustion process and the quantity balance of the materials inside and outside the loop through mechanism analysis. This paper is no longer the modeling of the subsystem in the material circulation system. But as a whole, the flow and circulation of the process can be reflected. Finally through the MATLAB simulation, the simulation results are analyzed.

2. Modeling of Material Circulation System

This paper divided the combustion into the dense area, the transition zone and the dilute phase zone [5]. Each area has three solid materials: coal, ash and limestone. Sulfate in the formation of desulfurization reaction of limestone is classified into ash. Then the materials of three areas were modeled. The modeling process is shown in Figure 1.

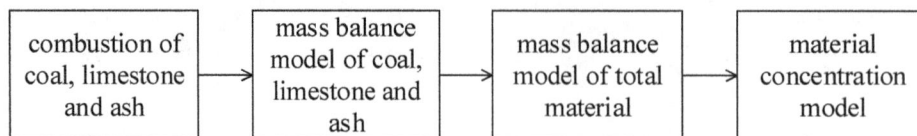

Fig. 1 Modeling flow diagram of materials circulating system

3. Modeling of Coal (Carbon Residue) Combustion

For the circulating fluidized bed boiler, due to the relatively large coal particles, the coal that has just been fed into the furnace is not completely burned, and part of which will be turned into coke. In this paper, the residual coke circulating in the furnace is called residual carbon [1]. The reaction of carbon combustion mechanism can be expressed by [6]:

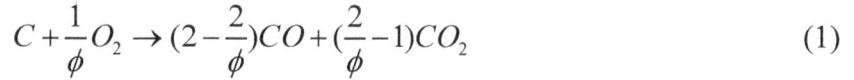

$$C + \frac{1}{\phi}O_2 \rightarrow (2 - \frac{2}{\phi})CO + (\frac{2}{\phi} - 1)CO_2 \tag{1}$$

Among them, ϕ is the mechanism factor, which determines the balance between CO and CO_2 [5, 7]. The expression of ϕ is as follows:

$$\phi = \begin{cases} \dfrac{2p+2}{p+2} & d_c < 0.05 \\[2ex] \dfrac{2p+2 - \dfrac{p}{0.095}(d_c - 0.0005)}{p+2} & 0.05 \leq d_c \leq 1 \\[2ex] 1 & d_c \geq 1 \end{cases} \tag{2}$$

Among them, d_c indicates that the average diameter of the carbon particles; p is the ratio of CO and CO_2, and the burning rate of a single char particle r_c is:

$$r_c = 12\pi d_c^2 k_c \frac{P}{RT_b} Y_{O_2} \tag{3}$$

Among them, k_c indicates that the burning rate constant of carbon particles, d_c indicates that the average diameter of the carbon particles, P indicates the bed pressure around the carbon particles, T_b indicates the bed temperature, and Y_{O_2} indicates the volume concentration of oxygen around the carbon particles.

Assuming that the accumulation of residual carbon in each area is composed of the particles with a diameter of d_c, the total combustion rate of carbon residue R_C is:

$$R_C = \frac{6M_c}{\pi d_c^3 \rho_c} r_c \tag{4}$$

4. Modeling of Limestone Combustion

The desulfurization process in circulating fluidized bed furnace is as follows: a certain amount of limestone is fed into the furnace, and the calcination reaction occurs at high temperature to produce porous CaO. Coal combustion generates the SO_2, which is absorbed by the CaO, then the reaction between SO_2 and CaO generates the $CaSO_4$.

The reaction rate of single limestone particle uses the following formula [5, 8]:

$$r_{ca} = \frac{\pi d_p^3}{6} k_v C_{SO_2} \tag{5}$$

$$k_v = 490 \exp(-\frac{17500}{RT}) S_R \lambda \tag{6}$$

Among them, λ is the activity coefficient of limestone. $\lambda = 0.035$; T indicates the temperature of the desulfurization agent particle. Assuming that the quantity of the desulfurization agent is M_{ca}, which is all composed of particles with a diameter of d_p, the consumption rate of the desulfurization agent is:

$$R_{ca} = \gamma \frac{56 M_{ca}}{\rho_{ca} d_p} \frac{P}{RT} k_v Y_{SO_2} \tag{7}$$

In the formula, γ is the correction factor; Y_{SO_2} is the volume concentration of SO_2 in the flue gas, and ρ_{ca} is the particle density of the desulfurization agent.

5. Distribution of Material Concentration in Bed

Solid particle entrainment rate model uses the relationship proposed by Wen and Chen. The formula is as follow:

$$E(h) = E_\infty + (E_0 - E_\infty) e^{-ah} \tag{8}$$

Among them, $E(h)$ is any high's solid entrainment rate above the dense phase zone surface; E_0 is the dense phase zone surface's uplift rate, E_∞ is the saturation entrainment rate; a is the attenuation index, which can be obtained through the experiment.

Saturated entrainment rate is calculated according to the following formula:

$$E_\infty = 23.7 \rho_g u_0 \exp\left(-5.4 \frac{u_t}{u_0}\right) \tag{9}$$

Among them, u_t is the ultimate velocity ρ_g is the gas density.

6. Mass Balance Equation of Coal, Limestone and Ash

Through literature [9, 10] review, the entrainment and mixing of regional solid materials in a fluidized bed is shown. E_i is the material entrainment between the regional interface, W_{ei} is the backmixing from the area above to the below, W_i is the net material amount from the area below to the above.

6.1. *Mass balance equation of coal (carbon residue)*

It is assumed that the reaction rate of carbon is the same in different regions.

The mass balance equation of dynamic residual carbon in dense phase zone is:

$$\frac{dM_{C1}}{dt} = \eta_1 W_C + W_{RC} + W_{H1} - R_C - W_{C1} \tag{10}$$

The mass balance equation of dynamic residual carbon in transition zone is:

$$\frac{dM_{C2}}{dt} = W_{C1} + W_{H2} - W_{H1} - R_C - W_{C2} \tag{11}$$

The mass balance equation of dynamic residual carbon in dilute phase zone is:

$$\frac{dM_{C3}}{dt} = W_{C2} - W_{H2} - R_C - W_{OC} \tag{12}$$

Among them, M_{C1}, M_{C2} and M_{C3} indicates residual carbon quantity of dense phase zone, transition zone and dilute phase zone; W_C indicates the coal feed rate; η_1 indicates the mass fraction of carbon in coal; W_{C1} and W_{C2} indicates the carbon quantity from dense phase zone to transition zone and from transition zone to dilute phase zone; W_{H1} and W_{H2} indicates the coke quantity from transition zone to dense phase zone and from dilute phase zone to transition zone; R_C indicates carbon reaction rate, which is equal to the quantity of the consumed carbon per unit time; W_{RC} indicates the quantity of the carbon returned from the return valve; W_{OC} indicates the quantity of carbon in the exit of dilute phase zone.

6.2. *Mass balance equation of limestone*

The mass balance equation of desulfurization agent in dense phase zone is:

$$\frac{dM_{ca1}}{dt} = 0.56W_{ca} + W_{sca} + W_{rca2-1} - W_{oca1-2} - W_{cca1} \tag{13}$$

The mass balance equation of desulfurization agent in transition zone is:

$$\frac{dM_{ca2}}{dt} = W_{oca1-2} + W_{rca3-2} - W_{cca2} - W_{oca2-3} - W_{rca2-1} \tag{14}$$

The mass balance equation of desulfurization agent in dilute phase zone is:

$$\frac{dM_{ca3}}{dt} = W_{oca2-3} - W_{rca3-2} - W_{cca3} - W_{oca} \tag{15}$$

Among them, M_{ca1}, M_{ca2} and M_{ca3} indicates desulfurization agent quantity of dense phase zone, transition zone and dilute phase zone; W_{oca}, W_{ca}, W_{sca}, W_{oca1-2}, W_{rca2-1}, W_{oca2-3} and W_{rca3-2} indicates desulfurization agent quantity from dilute phase zone to the separator inlet, added to the furnace , return from the material returning device , flowed into transition zone from dense phase zone, from transition zone to dense phase zone, from transition zone to dilute phase zone, from dilute phase zone to transition zone ; W_{cca1}, W_{cca2}, and W_{cca3} indicates consumed desulfurization agent quantity per unit time in dense phase zone, transition zone and dilute phase zone.

6.3. *Mass balance equation of ash*

The mass balance equation of ash in dense phase zone is:

$$\frac{dM_{ash1}}{dt} = W_{ash} + \eta_2 W_c + W_{sash} + W_{rash2-1} + W_{ash1} - W_{oash1-2} \tag{16}$$

The mass balance equation of ash in transition zone is:

$$\frac{dM_{ash2}}{dt} = W_{oash1-2} + W_{rash3-2} + W_{ash2} - W_{rash2-1} - W_{oash2-3} \tag{17}$$

The mass balance equation of desulfurization agent in dilute phase zone is:

$$\frac{dM_{ash3}}{dt} = W_{oash2-3} + W_{ash3} - W_{rash3-2} - W_{oash} \tag{18}$$

Among them, M_{ash1} , M_{ash2}, M_{ash3} and W_{ash} indicates the total ash quantity of dense phase zone, transition zone , dilute phase zone and added into the furnace when starting the boiler; η_2 indicates the quantity share of ash in coal feed; W_{sash} , $W_{oash1-2}$, $W_{rash2-1}$, $W_{oash2-3}$ and $W_{rash3-2}$ indicates the ash quantity from return valve to dense phase zone, from dense phase zone to transition zone, from transition zone to dense phase zone, from transition zone to dilute phase zone and from dilute phase zone to transition zone; W_{ash1} , W_{ash2} and W_{ash3} indicates the $CaSO_4$ quantity generated from desulfurization reaction in dense phase zone ($W_{ash1} = 2.43W_{cca1}$, W_{cca1} indicates the consumed desulfurization agent quantity in dense phase zone per unit time), transition zone and dilute phase zone; W_{oash} indicates the ash quantity in the exit of dilute phase zone.

6.4. *Mass balance equation of total material*

Material balance refers to that the net difference of material quantity importing and exporting a certain area and production and consumption per unit time, which is equal to changes of total material quantity in this area. The dynamic mass equation is the accumulation process of the materials in each area. The mass conservation equations of the different areas are as follows.

Don't consider the slag quantity.

The mass balance equation of materials in dense phase zone is:

$$\frac{dM_1}{dt} = W_f + W_{cc} + W_{ca1} + W_{r1} - W_{e1} - W_{rs1} \tag{19}$$

The mass balance equation of materials in transition zone is:

$$\frac{dM_2}{dt} = W_{ca2} + W_{e1} + W_{r2} - W_{r1} - W_{e2} - W_{rs2} \tag{20}$$

The mass balance equation of materials in dilute phase zone is:

$$\frac{dM_3}{dt} = W_{ca3} + W_{e2} - W_{r2} - W_{rs3} - W_o \tag{21}$$

Among them, M_1, M_2, M_3 and W_f indicates the total materials quantity of dense phase zone, transition zone, dilute phase zone and feeding into dense phase zone, including coal, ash and limestone; W_{cc} indicates cyclic materials quantity feeding into the furnace; W_{ca1}, W_{ca2} and W_{ca3} indicates the $CaSO_4$ quantity generated from desulfurization reaction in dense phase zone, which is classified into ash, transition zone and dilute phase zone; W_{e1}, W_{e2}, W_{r1} and W_{r2} indicates the materials quantity from dense phase zone to transition zone, from transition zone to dilute phase zone, from transition zone to dense phase zone and from dilute phase zone to transition zone; W_{rs1}, W_{rs2}, W_{rs3} and W_o indicates the consumed materials quantity in dense phase zone, including consumed coal and limestone, in transition zone, in dilute phase zone and of the exit of furnace.

7. Model of Concentration Distribution

LUO Zhong-yang et al [11] studied on the materials' concentration distribution by the experiments and established the expression of the materials concentration of circulating fluidized bed, which is as follow:

$$c = 14 \frac{Gs}{u_0} \exp\left(-2.61 \frac{h}{H}\right) \tag{22}$$

Among them, c indicates the solid particle concentration(kg/m^3); H indicates bed height(m); h indicates the height from under the bed(m); u_0 indicates the fluidizing velocity(m/s); Gs indicates the returned material quantity (kg/s).

In the formula, it can be seen that the concentration of solid particles is related to the returned material quantity Gs and the fluidizing velocity u_0. However, the experiment above did not consider the influence of the quantity of coal feed. In this paper, basing on this formula, the returned material quantity Gs is replaced by summation of actual returned material quantity, coal feed quantity and desulfurization agent quantity. Finally, the expression of the solid particle concentration can be obtained. It's as follow:

$$c = \frac{14Gt}{u_0}\exp(-2.61\frac{h}{H})\qquad(23)$$

Through literature [12] review, each parameter can be determined(the backmixings of each area are given to the constant value), which will be taken into the equilibrium equation to obtain the change of material quantity in dense phase, transition and dilute phase zone. At the same time, the returned material quantity can be obtained. And then the concentration distribution in the furnace can be obtained.

8. Simulation Results on MATLAB

8.1. *Simulation under rated condition*

This paper takes XinSu Power Plant's 220MW CFBB (DG725/13.75-II 1) as the research object. Boiler parameters are shown in Table 1, while MATALB initial conditions are shown in Table 2.

Table 1 XinSu Power Plant's boiler parameters

furnace height (m)	average particle size of coal (mm)	average particle size of limestone (mm)	fuel consumption (t·h⁻¹)	limestone consumption (t·h⁻¹)
38	1.2	0.45	116.93	43.74
static bed material height (mm)	furnace's sectional area of dense phase zone (m²)	furnace's sectional area of dilute phase zone (m²)	the number of grid plate's nozzle	
800	102	200	2000	

Table 2 Initial conditions of simulation

bed temperature /(°C)	mechanism factor φ	empirical coefficient y	ultimate velocity of carbon particles /(m·s⁻¹)	oxygen volume concentration in area 1/(%)	oxygen volume concentration in area 2/(%)	oxygen volume concentrati-on in area 3/(%)
908	1	0.8	8.4	7.2	11.3	5.5
void fraction	fluidizing velocity /(m·s⁻¹)	maximum height of dense phase zone /(m)	ultimate velocity of limestone particles /(m·s⁻¹)	SO₂ volume concentration in area 1 /(%)	SO₂ volume concentration in area 2/(%)	SO₂ volume concentrati-on in area 3/(%)
0.41485	5.4	38	3.0	15.3	10.5	6.8
attenuation index	furnace bottom's pressure /(pa)	carbon reaction rate constant	height percentage of dilute phase zone /(%)	height percentage of escape zone /(%)	Density of dense phase zone /(kg·m⁻²)	separation efficiency of cyclone separator /(%)
3.5	110	1	35	65	950	99

Programming in MATLAB, the obtained simulation results are as follows:

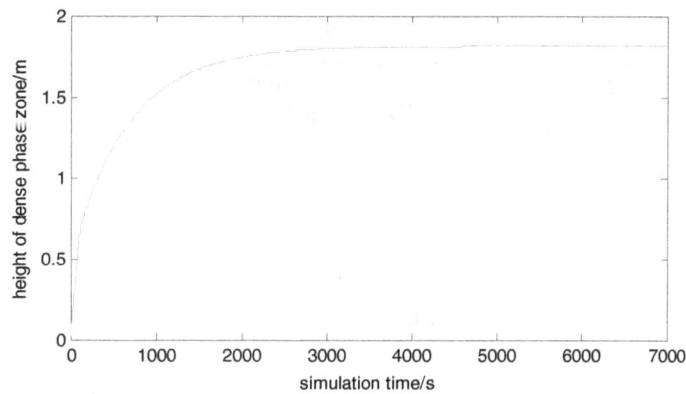

Fig. 2 Dense phase zone's height variation of 220MW CFB

Figure 2 shows the change of the height of the dense phase zone in CFB, With the material combustion, heat transfer and mass transfer in furnace,which stabilizes at 1.8m finally. It can be seen that the height of the dense phase zone increases rapidly in the initial stage. The trend tends to be stable with increasing reaction time.

Figure 3 shows the variation of particle concentration along the bed in CFB,which is obtained under the condition of different fluidizing velocity but same returned material quantity. What can be seen from the figure is that solid particle concentration showed a downward trend along the bed height direction. There is a dense phase zone at the bottom of the bed, whose particle concentration is relatively high. There is a transition zone at the middle of the bed, whose change of particle concentration is relatively slow. The upper area of the bed is dilute phase zone. In this area, the concentration change is little and basically constant because of the violent movement of particle swarm.

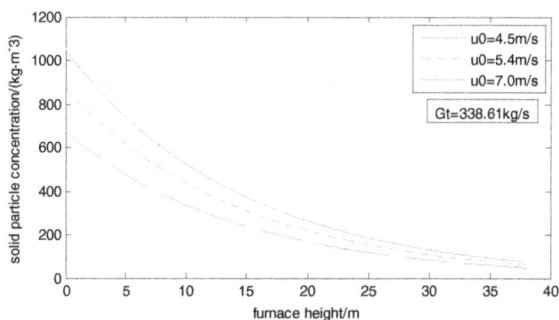

Fig. 3 Solids concentration distribution along the furnace under different streams of wind speed

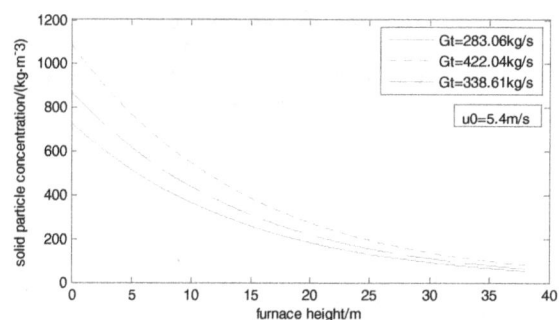

Fig. 4 Solids concentration distribution along the furnace under different returned material quantity

Figure 4 shows the variation of particle concentration along the bed in CFB,which is obtained under the condition of same fluidizing velocity but different returned material quantity. What can be seen from Figure 4 is that the particle concentration increases with the increase of the returned material quantity, which is due to the increase of the quantity of the particle in the bed.Also, because the wind speed is constant,which leads to the basically constant residence time of the particles, the concentration of solid particles in the bed will increase.

8.2. *Simulation comparison under three conditions*

Table 3 Simulation parameters under different conditions

parameter	coal feed quantity (t·h⁻¹)	limestone quantity (t·h⁻¹)	fluidizing velocity (m·s⁻¹)	oxygen volume concentration in area 1(%)
condition 1	58.5	21.7	4.3	4.3
condition 2	70.2	26.2	5.4	5.5
condition 3	102.9	38.5	7.0	7.4

By changing the coal feed quantity, fluidizing velocity and primary air flow and finding a suitable wind coal ratio to maintain the balance of the furnace materials, the simulation can be carried out. Three simulation conditions are shown in Table 3.

The simulation results are as follows.

Figure 5 is a contrast diagram of the dense phase zone height under three conditions. What can be seen from Figure 5 is that dense phase zone height increases with the increase of coal feed quantity and air flow ,which is due to increase solid particle concentration of dense phase zone resulting from increase coal feed quantity. More air flow is needed to maintain the balance of the furnace materials. However, to a certain extent,the increase of wind speed increases the height of dense phase zone.

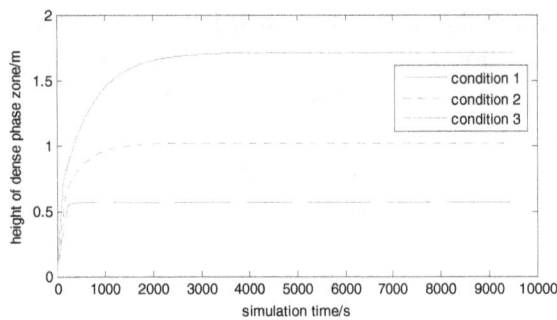

Fig. 5 Dense phase zone's height variation under different conditions

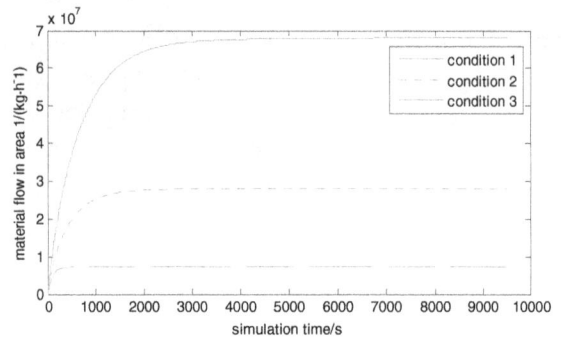

Fig. 6 Material flow variationof zone 1 under different condition

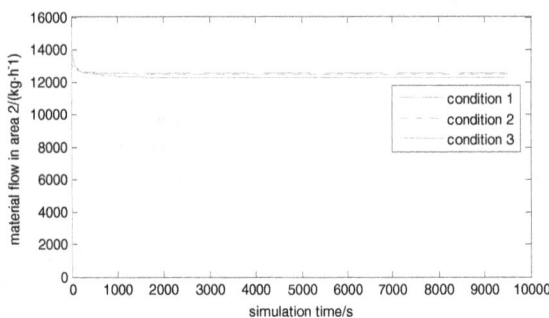

Fig. 7 Material flow variation of zone 2 under different conditions

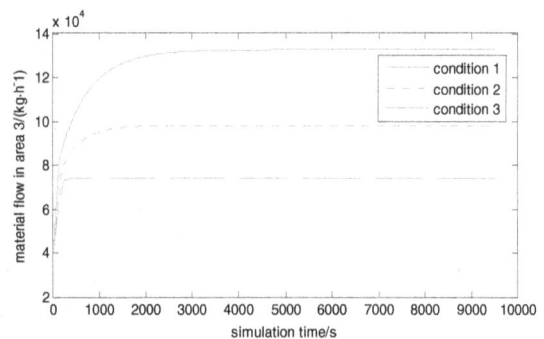

Fig. 8 Material flow variation of zone 3 under different conditions

Figures 6-8 show the comparison of the material flow in three areas under the three conditions. From the figures, we can see that the material flow of area 1 and 3 is increased with the increase of the coal feed quantity and air flow, but area 2 remains constant basically.The returned material quantity and solid particle concentration of all three areas will also increase accordingly.

In summary, the effect of coal feed quantity, primary air flow and fluidizing velocity on CFB is more obvious. At the same time, suitable wind coal ratio on the bed material balance and stability is essential.

9. Conclusion

This paper models the material circulation system of circulating fluidized bed basing on cell model, and then use the MATLAB to simulate the model. Finally the following conclusion can be obtained: Along the direction of the furnace, the concentration of solid particles inside the furnace shows in the distribution of 'the upper part is dilute and the under part is concentrated'. Moreover, in the bottom of the furnace that also called dense phase zone, the concentration of solid particles decreases rapidly. But in the upper furnace, also called transition zone and dilute phase zone, it decreases slowly and tends to be stable at the exit of the furnace. The height of the dense phase zone is flat with slightly lower secondary tuyere. At the same time, suitable wind coal ratio on the bed material balance and stability is essential. This shows that the model of the material circulation system established in this paper can correctly reflect the change of the material concentration in the furnace.

References

1. Ke-Fa CEN, Ming-Jiang NI, Zhong-Yang LUO *et al*. Theoretical Design and Operation of Circulating Fluidized Bed Boiler [M] .Beijing: China Electric Power Press. 1998. In Chinese.
2. Jia-Yi LU. Study on Particle Population Balance and Heat Balance in Large-Scale Circulating Fluidized Bed Boiler [D]. Chongqing: Chongqing University, 2012. In Chinese.
3. Xuan ZHANG, CHANG Taihua. Dynamic Bed Temperature Modeling of Large-scale CFBB[J]. Journal of Chinese Society of Power Engineering, 2013, 33(3):88-92. In Chinese.
4. Xing-Long ZHOU, Jian-Wen XIE, Sheng-Bin GAO, *et al*. Measurementof Solid Flux Distribution near Water Wall of a 330MW CFB Boiler [J]. Journal of Chinese Society of Power Engineering, 2014, 34(10): 753-758. In Chinese.
5. Chen TIAN. Research on the effect of furnace structural features on the hydrodynamics of CFB boiler [D]. Zhejiang: Zhejiang University. 2011. In Chinese.
6. W. Zhou, C S Zhao, L B Duan, *et al*. Two-dimensional computati-onal fluid dynamics simulation of coal combustion in a circulating fluidized bed combustor [J]. Chemical Engineering Journal, 2011, 166(1):306-314.
7. Cui-Jiu DUAN. Study on Oxy-fuel Combustion Characteristics and Pollutant Emission of Coal in Circulating Fluidized Bed [D]. Beijing: Graduate University of Chinese Academy of Sciences. 2012. In Chinese.
8. Liang-Min ZHONG. Research on Modeling and Optimization Control on the Bed Temperature of Large-Capacity Circulating Fluidized Bed Boiler [D]. Beijing: North China Electric Power University. 2014. In Chinese.

9. Su-Xia MA, Xian-Yong YANG. Experimental Research on Density of Matters in Suspension in Circulating Fluidized Bed Boilers Un-der Different Operational Conditions [J]. Journal of Power Engineering, 2005, 25(5):639-642. In Chinese.

10. Jian-Qiang GAO. Study of Real-time Simulation Model and Operation Property for Big Capacity Circulating Fluidized Bed Boiler Unit [D]. Baoding: North China Electric Power University. 2005. In Chinese.

11. Zhong-Yang LUO, Hong-Zhou HE, Qing-Hui WANG, et al. Status Quo-Technology of Circulating Fluidized Bed Boiler and Its Prospects of Development [J]. Journal of Power Engineering, 2005, 24(6):639-642. In Chinese.

12. Wei-Ming CHANG, Su-Xia MA, Ding-Ling LUO, et al. Experimental Study on Particle Concentration Distribution in a 1060t/h CFB Boiler [J]. Journal of Chinese Society of Power Engineering, 2014, 34(5):346-350. In Chinese.

Numerical Analysis on the Multi Layer Bonded Steel Butt Joint Under Charpy Impact Test

Min You[1, 3,*], Ya-Lan Zhao[2], Jian-Li Li[3], Ying-Ying Li[1]

[1]Hubei Key Laboratory of Hydroelectric machinery Design & Maintenance,
China Three Gorges University, Yichang 443002, China;
[2]Wuhan FiBK Technology Co., Ltd, Wuhan 430025, China
[3]Hubei Three Gorges Polytechnic, Yichang 443000, China
*Email: youmin@ctgu.edu.cn

The effect of the notch depth on the impact response of the steel butt joint bonded by multi layer under the Charpy impact test was studied using the elasto-plastic finite element method (FEM). The results obtained from numerical simulation show that both the elastic strain and plastic strain occurred at the point near the upper or lower surface decreased significantly when the notch depth increased from 0 mm (normal joint) to 6 mm. The value of the normal stress Sx and the von Mises equivalent stress $Seqv$ increased first when the notch depth is increased and then it decreased significantly when the notch depth is greater than 2 mm. There is a value fluctuation for stress $Seqv$ at the point 0.5 mm away from the upper surface when the notch depth reached 6 mm.

Keywords: Multilayer bondline; steel butt joint; notch depth; Charpy impact test; FEA.

1. Introduction

In recent years, some investigations related to the impact properties of the adhesively bonded metal joint have been down [1-5]. The effect of the adhesives thickness, elastic modulus of the adhesives on the impact response of butt joint [3] and the effect of the adhesive thickness on the impact toughness [4] was studied using finite element method. The influence of notch depth under Charpy [5] impact testing on the impact response of the adhesively bonded steel butt joint was investigated and it was found that the impact energy absorbed by unit area of joint is increased as the notch depth increased. The effect of multi layer on the stress distribution in the adhesively bonded single lap aluminum joint was significant [6]. This work is aimed to study the effect of notch depth on the steel butt joint bonded by the multi layer under the Charpy impact test.

2. Establishment of the Model

The diagram of Charpy impact test and a multi layer is shown in Fig. 1. The specimen for impact test (Fig. 2a) is in accordance with GB/T 19748 except the specimen was bonded by a sandwich (Phenolic resin-Epoxy-Phenolic resin, PEP for short) adhesive multi layer and with different notch depth as shown in Fig. 2b.

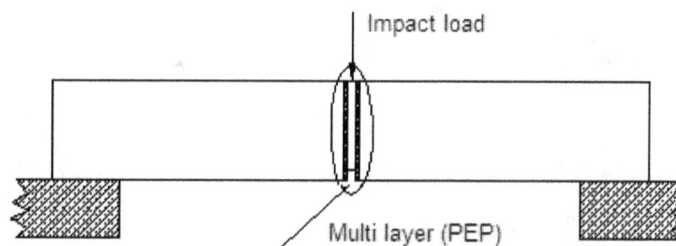

Fig. 1 Diagram of the impact testing

The adherends were made from Q235 structural steel and the two halves of specimen were bonded with an adhesive of a 0.4 mm thickness (0.1mm P - 0.2 mm E - 0.1 mm P). The area of adhesive layer was 10 mm by 10 mm for un-notched specimen. The properties of the materials are given in Table 1.

Table 1 Materials properties

	Materials	Elastic Modulus (MPa)	Poisson's Ratio	Yield Strength (MPa)	Tang Modulus (MPa)
	Steel Q235	203,000	0.27	235	6100
P	(Phenolic Resin)	2,875	0.42	90	500
E	(Epoxy Resin)	1,888	0.33	50	50

The impact time was set as 0.2 ms with an initial velocity of 3.2 m/s. The finite element model was built using the ANSYS/ LS-DYNA software as shown in Fig. 2. The eight-node hexahedral element Solid164 was used for both adhesive layer and adherend and it was restrained in Y direction by two supports. Two typical nodes are marked as P and Q in Fig. 2c (0.5 mm away from the upper surface or the notch tip respectively) for analyzing the response of the bondline under the impact test.

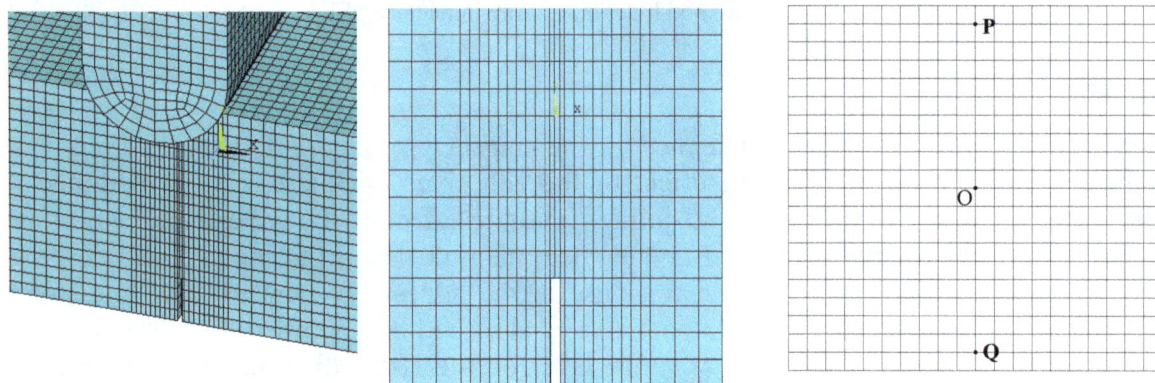

(a) Model of FEM   (b) Mesh for the multi-layer bonded butt joint   c) Node P and Q
Fig. 2 The impact test for specimen

3. Results and Discussion

Effect of notch depth on the impact response with time at the point P (0.5 mm away from the upper surface) and Q (0.5 mm away from the notch tip) under Charpy impact test is shown in Fig. 3.

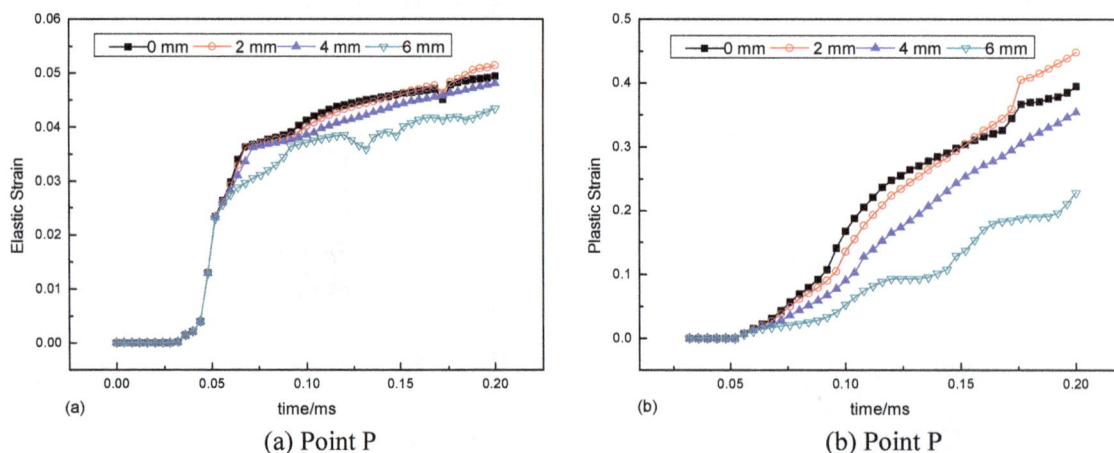

(a) Point P   (b) Point P

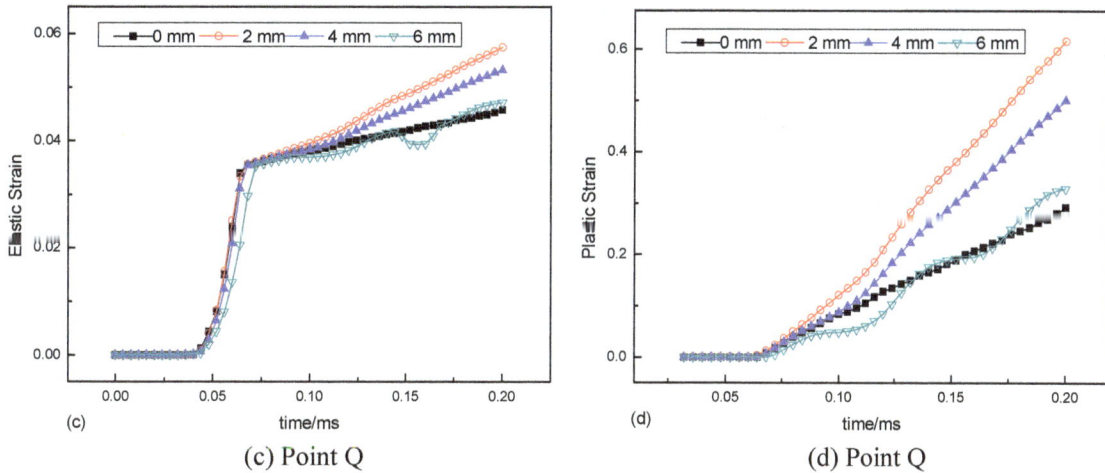

(c) Point Q   (d) Point Q

Fig. 3 The effect of notch depth and the node on the strain vs. time

It can be seen from the figure that the elastic strain occurred at node P or Q is increased rapidly after about 0.04 ms time lap to a certain value (about 0.07 ms the strain reached about 0.036) and then the increase becomes slow down (Fig. 3a and Fig. 3c) but the increase of the plastic strain is steadily with the time at node P or Q (Fig. 3b and Fig. 3d). In general, the strain (either elastic or plastic one) is decreased when the notch depth is increased except for the 2 mm depth notch, which is gradually greater than that of the normal joint at the node P and it is greater than the others after 0.07 ms at the node Q. In other words, the 2 mm depth notch is more sensitive to the impact test and in the standard GB/T 19748 it is the required size of the notch depth.

The effect of the notch depth on the normal stress Sx and von Mises equivalent stress $Seqv$ response with the time at the point at P and Q under Charpy impact test is shown in Fig. 4.

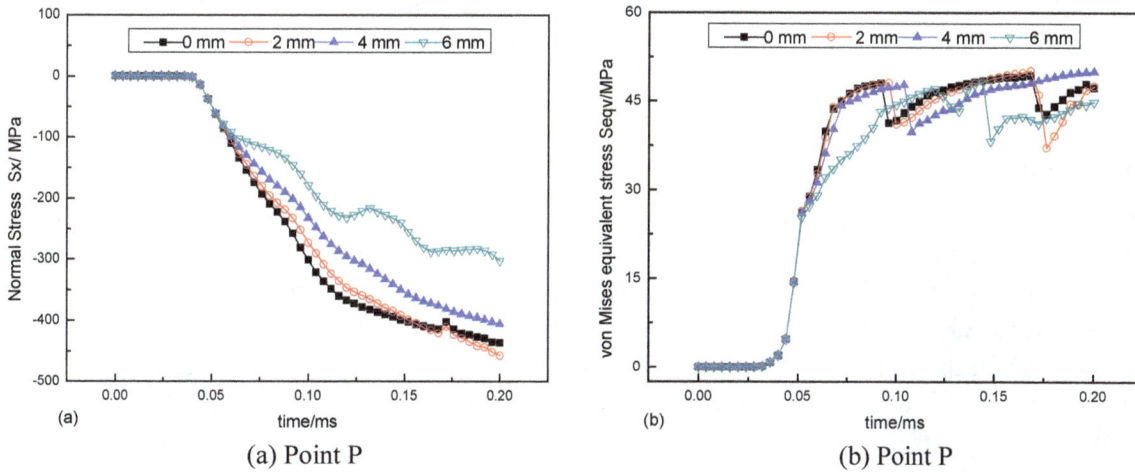

(a) Point P   (b) Point P

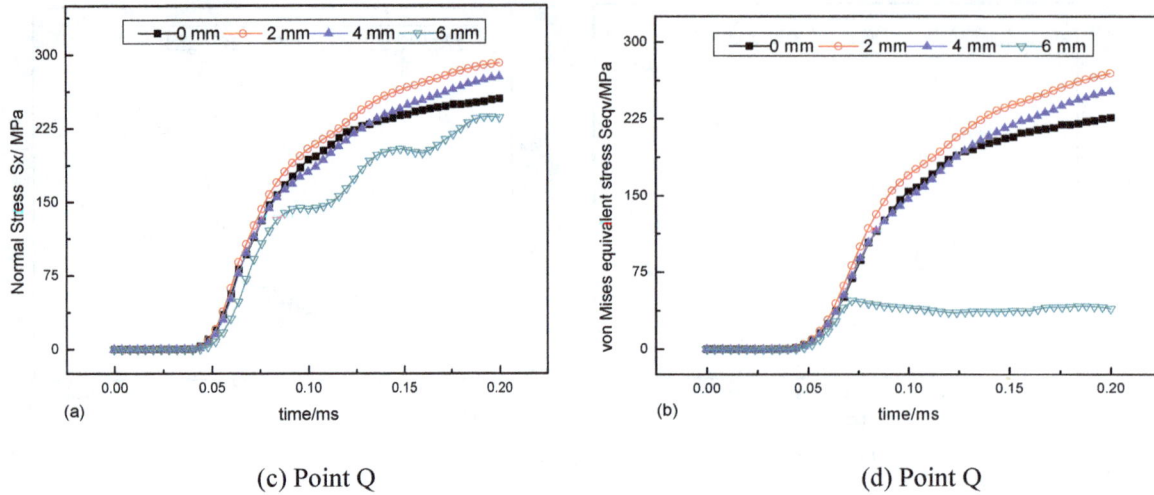

(c) Point Q   (d) Point Q

Fig. 4 The effect of notch depth and the node on the stress vs. time

It can be seen from the figure that the stress *Sx* and *Seqv* response at the point P and Q (0.5 mm away from the tip of notch) is significantly affected by the notch depth. When the notch depth increased from 0 to 6 mm, the value of the stress *Sx* response at node P increased and the response time became longer (Fig. 4a). The response tendency of the stress *Sx* at the node Q (Fig. 4c) is similar to the elastic strain (Fig. 3c) where the maximum value of the stress *Sx* response is in accordance with the 2 mm notch depth. Because the response tendency of the other two normal stress (*Sy* and *Sz*) is similar to that of the *Sx* either at node P (all negative) or at node Q (all positive), they are not presented. When the time is longer than 0.1 ms, the von Mises equivalent stress *Seqv* at node P presents a strong stress fluctuation (Fig. 4b). The von Mises equivalent stress *Seqv* occurred at node Q (Fig. 4d) is much different from the stress *Seqv* occurred at node P (Fig. 4b). The value of stress *Seqv* is increased first (till 2 mm notch depth) and then decreased as the depth increase. It is decreased evidently when the notch depth reached 6 mm to a value lower than 50 MPa (Fig. 4d).

4. Conclusion

Under the condition of this work, the results obtained from the numerical modeling show that the impact response of the butt joint bonded by multi layer bondline is evidently affected by the notch depth under the Charpy impact test. The value of the normal stress *Sx* and the von Mises equivalent stress *Seqv* increased first when the notch depth is increased and then it decreased significantly when the notch depth is greater than 2 mm. There is a value fluctuation for the stress *Seqv* at the point 0.5 mm away from the upper surface when the notch depth reached 6 mm meanwhile the value of the stress *Seqv* at the point 0.5 mm away from the notch tip is decreased significantly.

Acknowledgment

The authors would like to acknowledge the funding by the Hubei Province Natural Science Foundation of China under project no. 2014CFA123.

References

1. R. D. Adams, J. A. Harris, A critical assessment of the block impact test for measuring the impact strength of adhesive bonds, Int. J. Adhesion & Adhesives, 16 (1996) 61-71.
2. S. Xu, D. A. Dillard, Determining the impact resistance of electrically conductive adhesives using a falling wedge test, IEEE Trans. Components & Packaging Technol., 26 (2003) 551 562.
3. M. You, J. -L. Yan, X. -L. Zheng, et al, 3-D finite element analysis of bonded joints under impact loading, Advanced Materials Research, 97-101 (2010) 763-766.
4. X. -L Zheng, L. Wu, M. You, et al, Numerical analysis on the butt-joint under Izod impact test, Advanced Materials Research, 602-604 (2013) 2279-2282.
5. M. You, M. Li, J. -L Li, et al, Effect of notch depth on the adhesively bonded steel butt-joint under Charpy impact test, Applied Mechanics and Materials, 488-9 (2014) 538-541.
6. M. You, Y. -L Zhao, J. -L Li, et al, Numerical analysis of multi-layer on the stress distribution in adhesively bonded single lap aluminum joint, Advanced Materials Research, 1061-2 (2015) 471-474.

Effect of Inner Step Height on the Stress Distribution in Weld-bonded Single Lap Aluminum Joint

Jian-Li Li[1], Peng Wang[2], Min You[1, 3,*], Ying-Ying Li[3]

[1]Hubei Three Gorges Polytechnic, Yichang 443000, China
[2]Wuhan Tower Works, Power Construction Corporation of china, Wuhan 430011, China
[3]Hubei Key Laboratory of Hydroelectric Machinery Design & Maintenance,
China Three Gorges University, Yichang 443002, China
*Email: youmin@ctgu.edu.cn

The influence of the step height toward the bondline on the stress distribution in the aluminum single lap weld-bonded joint was investigated using elasto-plastic finite element method (FEM). The results from the numerical simulation show that the peak stress along the mid-bondline at the end of lap zone is decreased as the step height increased when a couple of inner steps were arranged in the adherend. The stress Sx in the region of nugget is decreased as the step height increased from 0.5 mm to 1.5 mm. The appropriate inner step height is 0.5mm to 1 mm for optimizing the stress in the single lap aluminum weld-bonded joint under the study condition.

Keywords: Weld-bonded joint; aluminum alloy; inner step height; stress distribution, FEA.

1. Introduction

The modification technologies have been developed in the recent years to uniform the stress distribution in adhesively bonded single lap joints [1-4]. Belingardi et al [1] investigated the effect of chamfer size on the stresses in metal/plastics adhesive joints. Sancaktare and Nirantar [2] founded that the strength of single lap joints of metal adherends could increased by taper minimization. Kaye and Heller [3] pointed out some shapes could reduce adhesive stress concentrations by evenly distributing the load transfer along the bond-line. The numerical and experimental study of the effect of the inner chamfer on the stress distribution in the adhesively bonded aluminum single lap joints was carried out by the authors [4]. And the effect of the outer chamfer angle in the adherends [5] as well as the inner step length [6] on the stress distribution in the weld-bonded aluminum single lap joints were investigated in recent years. The aim of this work is to study the effect of inner step height on the stress distribution in the weld-bonded single lap aluminum joint.

2. Finite Element Model and Mesh

The model and mesh were built using the ANSYS finite element software as shown in Fig. 1 and Fig. 2. The properties of the materials used in this study considered the non-linear behavior, bilinear isotropic hardening plasticity option (BISO) to describe the elastic-plastic behavior of material are listed in Table 1. The load applied was taken as 2 KN and the dimensions of the aluminum adherend were made in accordance with the Chinese standard GB 7124 (equivalent to ISO 4587). The triangular element was used for both bondline and the nugget and quadrilateral element for adherend (Fig. 2).The thickness of the bondline was 0.2 mm and divided into 10 layers. To investigate the effect of the inner step height thoroughly, four step heights were taken into account as 0.5 mm, 1.0 mm, 1.5 mm as well as 2 mm (standard joint) respectively and the length of step was kept as 3 mm according to Ref. [6]. The nugget was assumed in a shape of ellipsoid.

---

*Corresponding author

Table 1 Mechanical properties of the materials

Materials	Elastic Modulus (GPa)	Poisson's Ratio	Yield Strength (MPa)	Tang Modulus (MPa)
Aluminum alloy 2124	71	0.32	400	240
Nugget	102	0.29	800	210
Phenolic resin adhesive	2.875	0.42	90	500

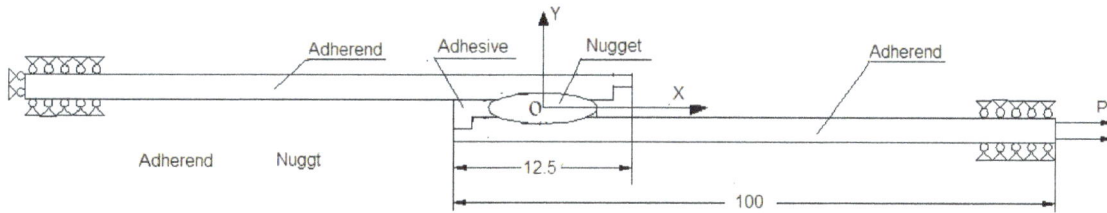

Fig. 1 Finite element model (unit: mm)

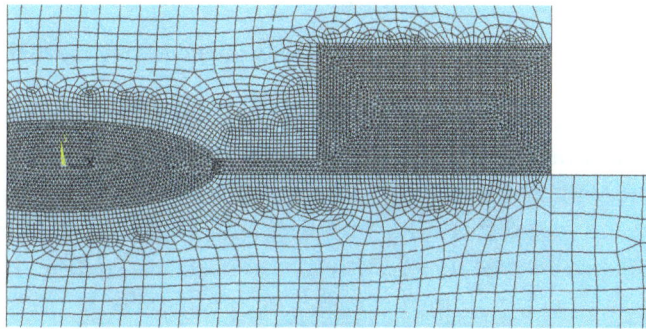

Fig. 2 Finite-element meshes for right half of over lap zone

3. Results and Discussion

The effect of the step height on the stress distribution along the mid-bondline (y = 0) is presented in Fig. 3. The results from the simulation showed that the peak values of the stress components Sx, Sy, Sxy and the von Mises equivalent stress $Seqv$ are all affected by the inner step height (Fig. 3).

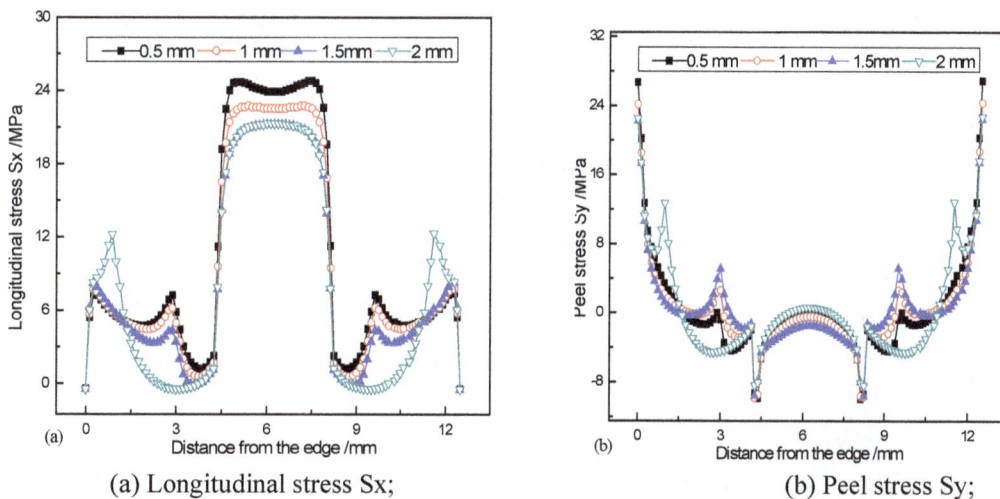

(a) Longitudinal stress Sx;

(b) Peel stress Sy;

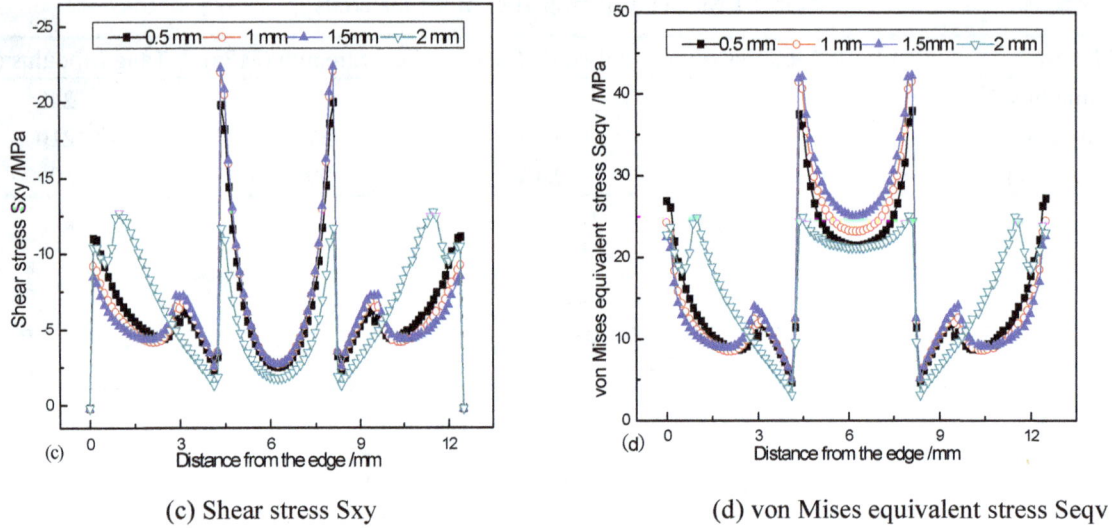

(c) Shear stress Sxy

(d) von Mises equivalent stress Seqv

Fig. 3 Effect of the step height on the stress distribution along the mid-bondline

For the peak value of the stress *Sy*, *Sxy* and *Seqv* occurred at the points near both ends of the over lap zone in the joint, it is decreased significant as the step height increased from 0.5 mm to 2 mm. For instance, the peak value of the peel stress *Sy* is decreased 15.8 % from 26.7 MPa to 22.5 MPa when the step height increased from 0.5 mm to 2 mm (Fig. 3b). For the von Mises equivalent stress *Seqv*, the decrease of peak stress is 15.2 % from about 26.9 MPa (H= 0.5 mm) to 22.8 MPa (H= 2 mm, Fig. 3d). But the peak value of stress longitudinal *Sx* is increased 12.4 % from 7.35 MPa to 8.26 MPa when the step height increased from 0.5 mm to 2 mm (Fig. 3a). The stress distribution tendency along the mid-bondline of the shear stress *Sxy* and the von Mises equivalent stress *Seqv* is similar to that of the longitudinal stress *Sx* and peel stress *Sy* in which the value of the peak stress is decreased and moved to the middle part of the over lap zone when the step height is increased from 0.5 mm to 1.5 mm. For the stress *Sx* occurred in the region of nugget, it is clearly that the peak stress decreased as the step height increased from 0.5 mm (about 24.7 MPa) to 1.5 mm (about 21.2 MPa) and then the value it is kept as the same (21.3 MPa for H= 2 mm, Fig. 3a). Higher peak stress in the nugget means that more load could be carried out by the nugget so that the load bearing capacity of the joint might be raised [6]. In Fig. 3b, the absolute value of stress *Sy* at the point corresponding to the right edge of the nugget is decreased as the step height increased such as -9.98 MPa (H= 0.5 mm), -9.75 MPa (H= 1 mm), -9.67 MPa (H= 1.5 mm) and -8.45 MPa (H= 2 mm) respectively. And the absolute value of the shear stress *Sxy* and the value of the von Mises equivalent stress *Seqv* in the region of the nugget is increased as the step height increased to 1.5 mm.

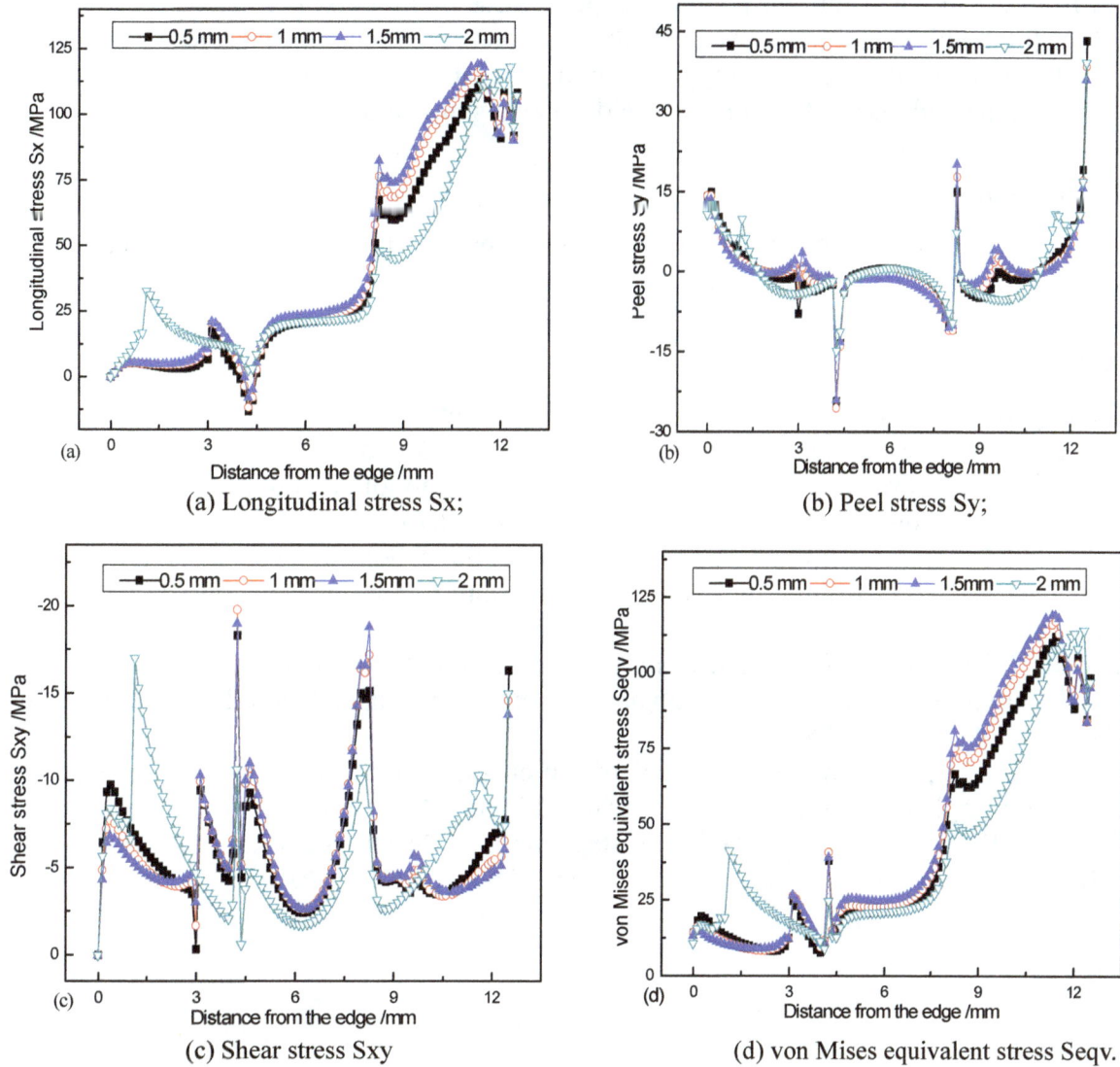

(a) Longitudinal stress Sx;

(b) Peel stress Sy;

(c) Shear stress Sxy

(d) von Mises equivalent stress Seqv.

Fig. 4 Effect of the step height on the stress distribution in adherend near the interface

The effect of the step height on the stress distributed in the adherend near the interface (y = -0.15 mm) is shown in Fig. 4. The results from the finite element analysis show that the stress distribution tendency of the *Sx* and that of the *Seqv* is similar to each other except there is a peak value at the point closed to the left edge of the nugget (Figs. 4a and 4d). And the highest value of the stress *Sx* and the *Seqv* is closed to 120 MPa occurred at the point near the right end of the lap zone. At the points corresponding to the edges of the nugget, there are a valley stress *Sy* occurred at left (-25.5 MPa, H = 1 mm) and a peak one at right (20.2 MPa, H = 1.5 mm) where the peak value of the stress *Sy* is increased as the step height increase until 1.5 mm (Fig. 4b).

4. Conclusion

The results obtained show that the peak values of the stress *Sx, Sy, Sxy* and *Seqv* along the mid-bondline are all affected by the inner step height. The peak value of the stress *Sy, Sxy* and *Seqv* occurred at the points near both ends of the over lap zone is decreased significant as the step height increased from 0.5 mm to 2 mm meanwhile the peak value of stress Sx is increased. Compared the results of the joint with the standard one, it is advantageous of reducing the peak stress near the both ends of the lap zone in weld-bonded single lap aluminum joints. Under the research conditions, the suitable inner step height is 0.5 to 1 mm for weld-bonded single lap aluminum joint.

Acknowledgment

The authors would like to acknowledge the funding by the Hubei Province Natural Science Foundation of China under project No. 2014CFA123.

References

1. G. Belingardi, L. Goglio, A. Tarditi, Investigating the effect of spew and chamfer size on the stresses in metal/plastics adhesive joints, Int. J. Adhesion & Adhesives, 22 (2002) 273-282.
2. E. Sancaktare, P. Nirantar, Increasing strength of single lap joints of metal adherends by taper minimization, J. Adhesion Sci. Technol., 17 (2003) 655-675.
3. R. Kaye, M. Heller. Through-thickness shape optimisation of typical double lap-joints including effects of differential thermal contraction during curing, Int. J. Adhesion & Adhesives, 25 (2005) 227–238.
4. M. You, Z.-M. Yan, X.-L. Zheng, et al, A numerical and experimental study of adhesively bonded aluminium single lap joints with an inner chamfer on the adherends, Int. J. Adhesion & Adhesives, 28 (2008) 71-76.
5. M. You, J.-L. Li, P. Wang, et al, Effect of the chamfer angle on the stress distribution in weld-boded single lap aluminum joint, Advanced Material Research, 1061 (2015) 450-453.
6. J.-L. Li, P. Wang, M. You, Effect of inner step length on the stress distribution in weld-bonded single lap aluminum joint, accepted by: Materials Science Forum, 2016.

Effect of Adhesive Thickness on the Stress Distribution in Weld-boded Single Lap DP 600 Steel Joint

Min You[1, 2,*], Jian-Jun Xu[1] Jian-Li Li[2], Ying-Ying Li[1]

Hubei Key Laboratory of Hydroelectric machinery Design & Maintenance,
China Three Gorges University, Yichang 443002, China
Hubei Three Gorges Polytechnic, Yichang 443000, China
*Email: youmin@ctgu.edu.cn

The effect of the adhesive thickness from 0.05 mm to 0.25 mm on the stress distribution in weld-bonded single lap DP 600 steel joint was investigated using elastic finite element method (FEM). The results obtained show that all the values of the peak stresses along the bondline at the points near the both ends of the lap zone decreased except Sx when the adhesive thickness increased. The peak stress of Sx and $Seqv$ in the region of the nugget is decreased a little as the adhesive thickness increased. It is suggested that an adhesive layer with the thickness of 0.15 mm to 0.2 mm be appropriate to optimize the stress distribution in the weld-bonded single lap DP 600 steel joint.

Keywords: Weld-bonded joint; DP 600 steel; adhesive thickness; stress distribution; FEA.

1. Introduction

The dual phase (DP) steel is one kind of higher yield stress metal used for manufacturing the light weight structure to establish a low-carbon economy and energy conservation system [1]. Davies et al [2] founded that numerical analysis of the stress state revealed larger stress concentration factors for tensile loading in thick bondline joints. Cognard et al [3] presented contributions of numerical modeling for the analysis of adhesively bonded assemblies for marine applications. Darwish and Al-Samhan [4] investigated the effect of the adherend thickness dissimilarity on the peel and shear strength of spot-welded dissimilar thickness joints. The experimental study on the effect of adhesive thickness on the impact toughness of adhesively bonded steel butt-joints [5] was carried out and it showed that the effect of the bondline thickness was evidently. And in authors' recent work the effect of nugget size on the stress distribution in weld-bonded single lap DP 600 steel joint has been down [6]. This work is aimed to study the effect of the adhesive thickness on the stress distribution in the weld-bonded single lap DP 600 steel joint.

2. Finite Element Model and Mesh

The model and mesh were built using the ANSYS finite element software as shown in Fig. 1 and Fig. 2. The properties of the materials used in this study are listed in Table 1.

Table 1 Materials properties of adhesive

Materials	Elastic Modulus (GPa)	Poisson's Ratio	Yield Strength (MPa)
DP 600 steel	207	0.3	370
Phenolic resin adhesive	2.875	0.42	90

*Corresponding author

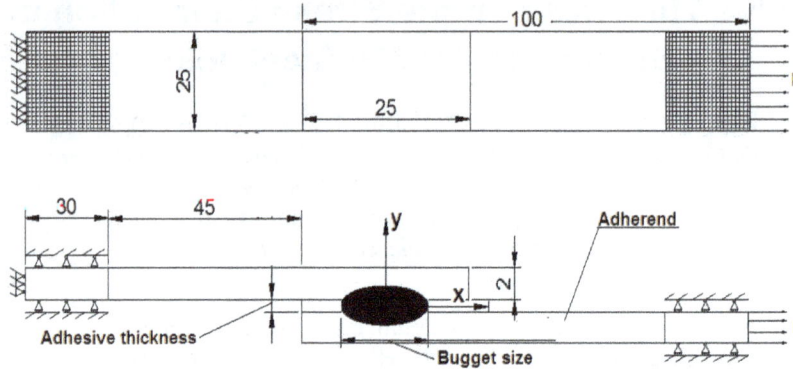

Fig. 1 Finite element model (unit: mm)

The load applied was taken as 2 kN and the dimensions of the DP 600 steel alloy adherend were made in accordance with the Chinese standard GB 7124 (equivalent to ISO 4587) but the length of the over lap zone is taken as 25 mm. The triangular element was used for both bondline and the nugget and the quadrilateral element for adherend (Fig. 2). To investigate the effect of the adhesive thickness thoroughly, five adhesive thicknesses were taken into account as 0.05 mm, 0.1 mm, 0.15 mm, 0.2 mm and 0.25 mm respectively. The nugget was assumed in a shape of ellipsoid.

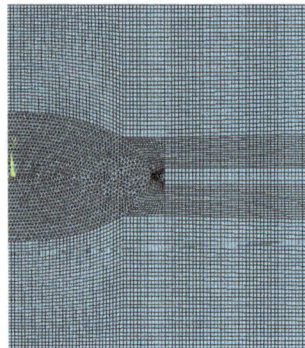

Fig. 2 Finite-element meshes for right half of over lap zone.

3. Results and Discussion

The effect of the adhesive thickness on the stress distribution along x axis (y = 0) is presented in Fig. 3. The results from the simulation show that the peak values of the stress components Sy, Sxy and the von Mises equivalent stress $Seqv$ are evidently affected by the adhesive thickness (Figs. 3b, 3c and 3d) but it is not significant for stress component Sx (Fig. 3a) as the thickness of the adhesive increased. And all the stresses distributed along the x-axis are symmetrical to the center of the overlap zone. For the peak value of the stress Sy, Sxy and $Seqv$ occurred at the points near both ends of the over lap zone in the joint, it is decreased significant as the adhesive thickness increased.

(a) Longitudinal stress Sx

(b) Peel stress Sy

(c) Shear stress Sxy

(d) von Mises equivalent stress Seqv

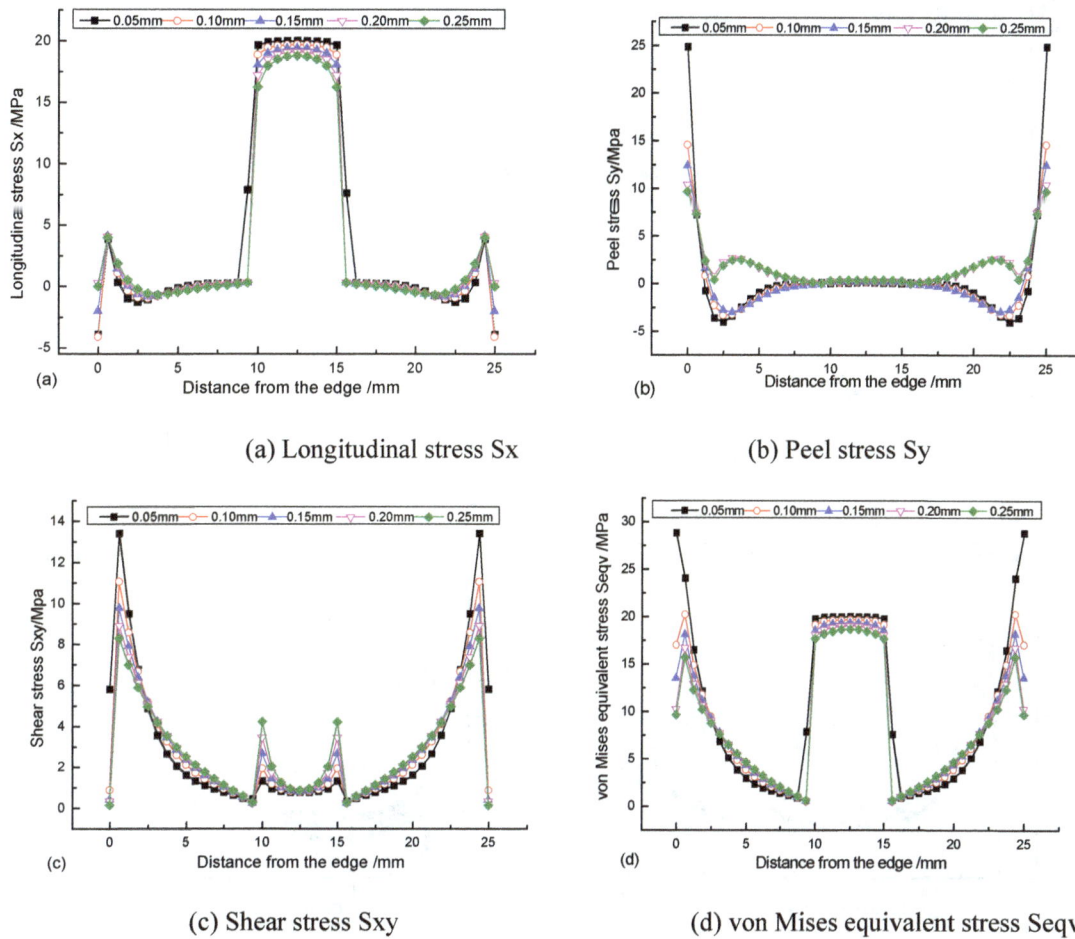

Fig. 3 Effect of the adhesive thickness on the stress distribution along the mid-bondline

For instance, the peak value of peel stress Sy is decreased about 60.7 % from about 24.7 MPa to about 9.7 MPa when the adhesive thickness increased from 0.05 mm to 0.25 mm (Fig. 3b). But it is nearly the same for peak value of the stress Sy, occurred in the region of the nugget (0.01 MPa to 0.39 MPa) as the thickness increased. It is clearly that the peak stress of the longitudinal stress Sx in the nugget decreased as the adhesive thickness increased (Fig. 3a) although for the peel stress Sy it is kept as the same value in the region of the nugget as the thickness of the adhesive increased from 0.05 mm to 0.25 mm (Fig. 3b). For the shear stress Sxy, the peak value at the point corresponding to the edge of the nugget is increased with the adhesive thickness increase. In other words, the effect of the thickness is opposite to that of the longitudinal stress Sx and the von Mises equivalent stress $Seqv$ in same region in which the peak value is decreased with the thickness increase. The stress distribution tendency of the longitudinal stress Sx (Fig. 3a) is similar to that of the von Mises equivalent stress $Seqv$ (Fig. 3d) except for the points near both ends of the lap zone.

819

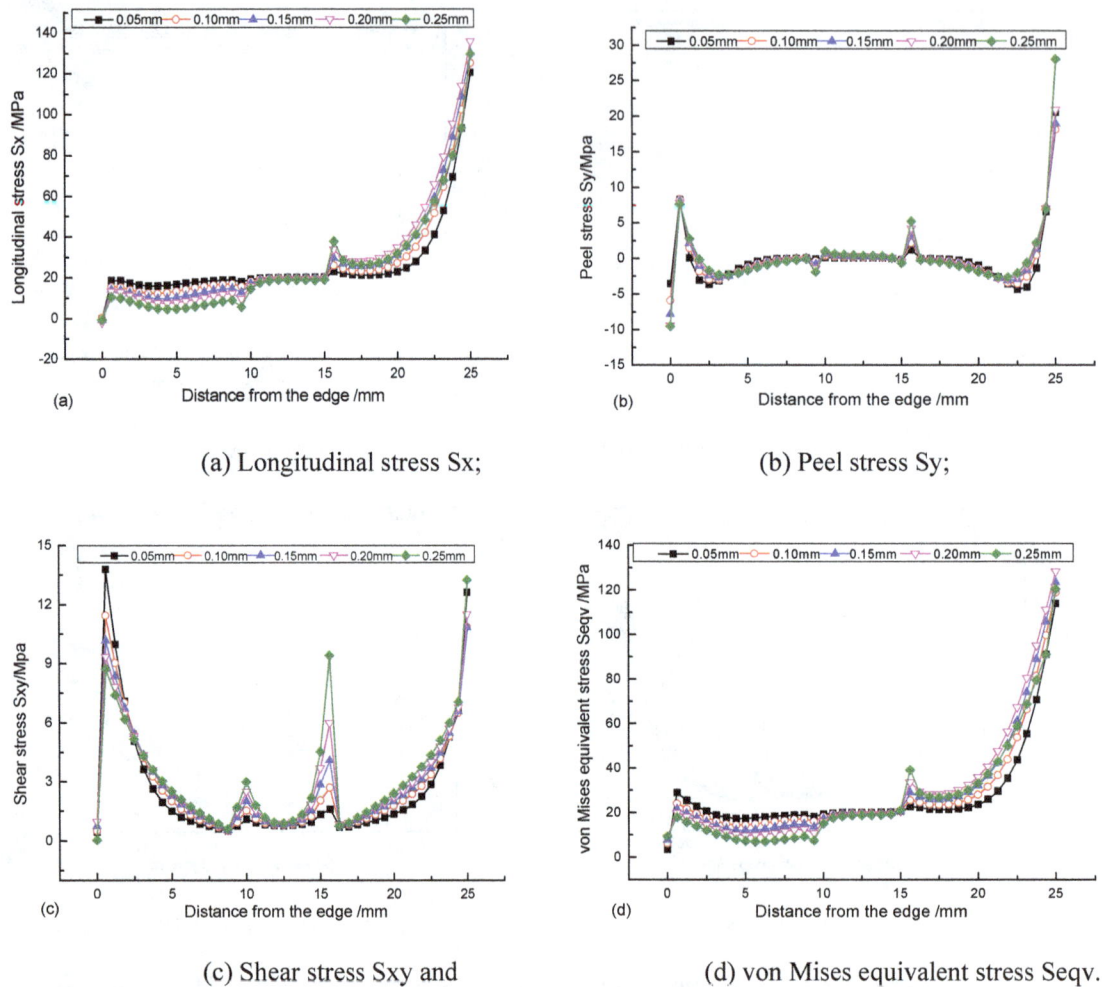

(a) Longitudinal stress Sx;

(b) Peel stress Sy;

(c) Shear stress Sxy and

(d) von Mises equivalent stress Seqv.

Fig. 4 Effect of the adhesive thickness on the stress distribution in adhered near the interface

When the conditions kept as the same, the effect of the adhesive thickness on the stress distributed in the adherend near the interface (0.05 mm below the interface, for a 0.2mm bondline thickness it is along the line $y = -0.15$ mm) is shown in Fig. 4. The results obtained are also showed that the values of the stress Sx and $Seqv$ are not varied distinguish as the adhesive thickness increased (Fig. 4a, 4d). For peel stress Sy and shear stress Sxy, the peak value at points corresponding to the right end of the overlap zone is increased when the adhesive thickness increased from 0.05 mm to 0.25 mm (Fig. 4b, 4c). And the same varying tendency is presented for the shear stress Sxy where a couple of the peak values are increased at the points corresponding to the edges of the nugget when the bondline thickness increased especially for the point at right edge of the nugget (Fig. 4c). So the stress Sy distributed in the adherend near the interface is approximately accorded with the one distributed along the x-axis (Fig. 3b).

4. Conclusion

The results show that all the peak values of the stress Sy, Sxy and $Seqv$ along the mid-bondline at the points near both ends of the over lap zone in the joint are decreased significant. But it is not significant for stress component Sx when the adhesive thickness is increased. The value of the peak stress of the longitudinal stress Sx in the nugget is decreased as the adhesive thickness increase. And the value of peel stress Sy is kept as the same in the region of the nugget as the thickness of the adhesive increased from 0.05 mm to 0.25 mm. For the shear stress Sxy, the peak value at the point corresponding to the edge of the nugget is increased with the adhesive thickness increase. The varying tendency of stress $Seqv$ in the nugget is similar to that of the stress Sx. It is recommended that the suitable adhesive thickness be 0.15 mm to 0.20 mm for weld-bonded single lap DP 600 steel joint according to its synthetic effects.

Acknowledgments

The authors would like to acknowledge the funding by the Hubei Province Natural Science Foundation of China under project No. 2014CFA123.

References

1. F. Hayat, Comparing properties of adhesive bonding, resistance spot welding, and adhesive weld bonding of coated and uncoated DP 600 steel, Int. J. Iron & Steel, 18 (n9, 2011) 70-78.
2. P. Davies, L. Sohier, J.-Y. Cognard, et al, Influence of adhesive bond line thickness on joint strength, Int. J. Adhesion and Adhesives, 29 (2009) 724-736.
3. J. Y. Cognard, R. Creachcadec, L. Sohier, et al, Influence of adhesive thickness on the behaviour of bonded assemblies under shear loadings using a modified TAST fixture, Int. J. Adhesion and Adhesives, 30 (2010) 257-266.
4. S. M. Darwish, A. Al-Samhan, Peel and shear strength of spot-welded and weld-bonded dissimilar thickness joints, Journal of Materials Processing Technology 147 (2004) 51–59.
5. M. You, J.-R. Hu, X.-L. Zheng, et al, Effect of adhesive thickness on the impact toughness of butt-joints, Advanced Materials Research, 230-232 (2011) 1350-1534.
6. M. You, J.-J Xu, J.-L. Li, et al, Effect of Nugget Size on Stress Distribution in Weld-boded Single Lap DP 600 Steel Joint: accepted by Materials Science Forum (2016).

Realization of the Homogenization Method in Meso Mechanics with ABAQUS Software

Chen-Xing Wei[1], Zhang-Hua Lian[2], Yan-Ru Guo[1,*], Tie-Jun Lin[2], Xiao-Jun Li[3], Pei-Lin Yu[4]

[1]*Research Institute of Bohai Drilling Engineering Technology, CNPC, Tianjin P. R. China 300280*

[2]*State Key Lab of Oil and Gas Reservoir Geology and Exploitation, Southwest Petroleum University, Chengdu P. R. China 610500*

[3]*Tubular Goods Research Institute of CNPC, Xi'an, P. R. China, 710077*

[4]*Shanghai Branch of CPPE, Shanghai, P. R. China, 200127*

[*]*Email: swpugyr@126.com*

Meso mechanics computation method is different from the macro mechanics calculation method and micro test methods in analyzing the drill string damage, which can denote the effect of micro crack propagation or hard inclusion on the macroscopic properties of the drill-stem material at the meso scale. The "homogenization" method is important to the meso mechanical computation. Based on description of the meso mechanics calculation process and data storage structure of ABAQUS software output database, the codes of achieving homogenization process of meso mechanics are finished by using ABAQUS scripting Python language. The calculation results of representative volume element model under the different load cases are treated by the homogenization code to acquire the parameters of elastic modulus, Poisson's ratio and stress-strain distribution curve, which are the same with the input of model. Verifying the integrity and correctness of the homogenization code provides the technical means for drilling tools' damage research on meso mechanics further.

Keywords: Drill string damage; meso mechanics; homogenization; ABAQUS software; python language.

1. Introduction

While drilling, the impact of the tension compression, torsion, bending, impact load, friction and erosion of carrying rock circulation medium, bottom hole assembly (BHA) is easy to fracture failure. As one of the important scientific research achievements in the field of mechanics in 20th century, meso mechanics computation method is different from the macro mechanics calculation method and micro test methods [1-4], which uses multiscale continuum mechanics theory and method, to study the quantitative relationship between microstructure and macroscopic properties of the material. At present this method is also introduced to analyze the failure mechanism and calculate working life for the drill string. The "homogenization" method of meso mechanics is a bridge between the meso scale mechanical parameters and the macro scale mechanical parameters. Based on description of the meso mechanics calculation principle and data storage structure of ABAQUS software, the codes of achieving homogenization process are finished by using ABAQUS scripting Python language, which is adopted to provide technical means to evaluate the influence of micro crack growth or hard inclusion on the macro performance of drilling stems in the latter work.

[*]Corresponding author

2. Meso Mechanics Calculation Method

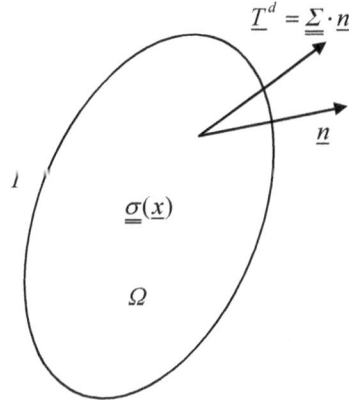

Fig. 1 Uniform stress boundary conditions for RVE

From the view of micro mechanics research, drill pipe material is composed of a large number of particles and disordered crystalline, so representative volume element (RVE) with the duality characterization can express the collection of the inhomogeneous and disordered materials [5], because it can meet the basic assumptions of continuous medium mechanics on the macro-scale, thereby, the equivalent homogeneous medium to replace the actual inhomogeneous materials, and ensure containing enough microstructure information on the meso-scale, thus accurately characterize the statistical average properties of the local continuous media [6, 7].

In Fig. 1, when the uniform stress $\underline{\underline{\Sigma}}$ exists on the boundary Γ of the REV, the surface load acting on the boundary Γ is $\underline{T}^d = \underline{\underline{\Sigma}} \cdot \underline{n}$, and the local stress $\underline{\underline{\sigma}}(x)$ is [8]:

$$\underline{\underline{\sigma}}(\underline{x}) = \underline{\underline{A}}_x (\underline{\underline{\Sigma}}, Y) \tag{1}$$

In Eq. (1), $\underline{\underline{A}}_x$ is the second order tensor function of stress localization, where the subscript x represents the local inhomogeneous characteristic of the material; Y is a geometric parameter associated with the micro structure geometry and mechanical properties of the RVE.

The constitutive relationship of local stress and strain is expressed as:

$$\underline{\underline{\varepsilon}}(\underline{x}) = \underline{\underline{F}}_x [\underline{\underline{\sigma}}(\underline{x})] \tag{2}$$

where $\underline{\underline{F}}_x$ is the second order tensor function of the local stress $\underline{\underline{\sigma}}$.

Eq. (3) represents the relationship between the average value of local strain $\underline{\underline{\varepsilon}}(x)$ in the region Ω and the macroscopic strain $\underline{\underline{E}}$, which is known as the "homogenization" method in the meso mechanics:

$$<\underline{\underline{\varepsilon}}> = \frac{1}{\Omega} \int_\Omega \underline{\underline{\varepsilon}}(\underline{x}) d\Omega = \underline{\underline{E}} \tag{3}$$

From Eqs. (1), (2) and (3), we find:

$$\underline{\underline{E}} = < \underline{\underline{\varepsilon}}(x) > = < \underline{\underline{F}}_x[\underline{\underline{\sigma}}(x)] > = < \underline{\underline{F}}_x[\underline{\underline{A}}_x(\underline{\underline{\Sigma}},Y)] > \tag{4}$$

The constitutive relationship of the macroscopic stress $\underline{\underline{\Sigma}}$ and the macroscopic strain $\underline{\underline{E}}$ can be obtained from Eq. (4):

$$\underline{\underline{E}} = \overline{\underline{\underline{F}}}_\sigma(\underline{\underline{\Sigma}}) \tag{5}$$

In Eq. (5), $\overline{\underline{\underline{F}}}_\sigma$ is the second order tensor function of the macroscopic stress $\underline{\underline{\Sigma}}$, which characterizes the influence of the micro-crack growth on the meso scale and the hard inclusion on the macroscopic properties of the drilling tool.

3. Storage Structure of ABAQUS Software Output Database

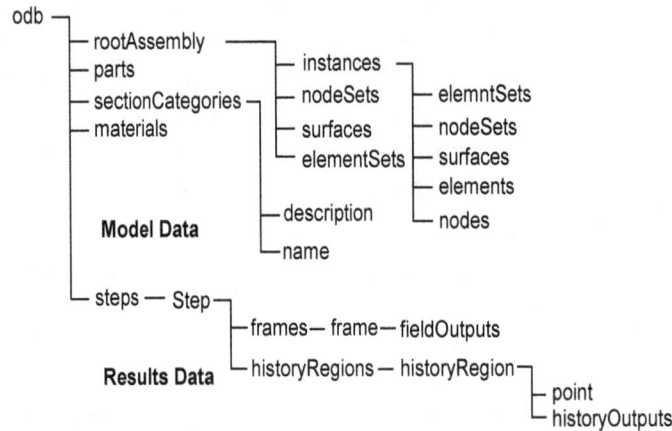

Fig. 2 The hierarchical relationship of ABAQUS output database

In Fig. 2, results composed of the model data and the results data can be calculated by using the software ABAQUS and stored in the output database (odb) [6], herein, which involves the results data of the output database on stress and strain by the "homogenization" method. While accessing to each field data object parameters in the output database using ABAQUS software scripting Python language, the "sequential method" should be followed, namely, accessing to all levels objects in turn according to the sequence created by them. Then obtain one of analysis steps of field distribution from certain parameter for one of frames.

4. Realization of the "Homogenization" Method

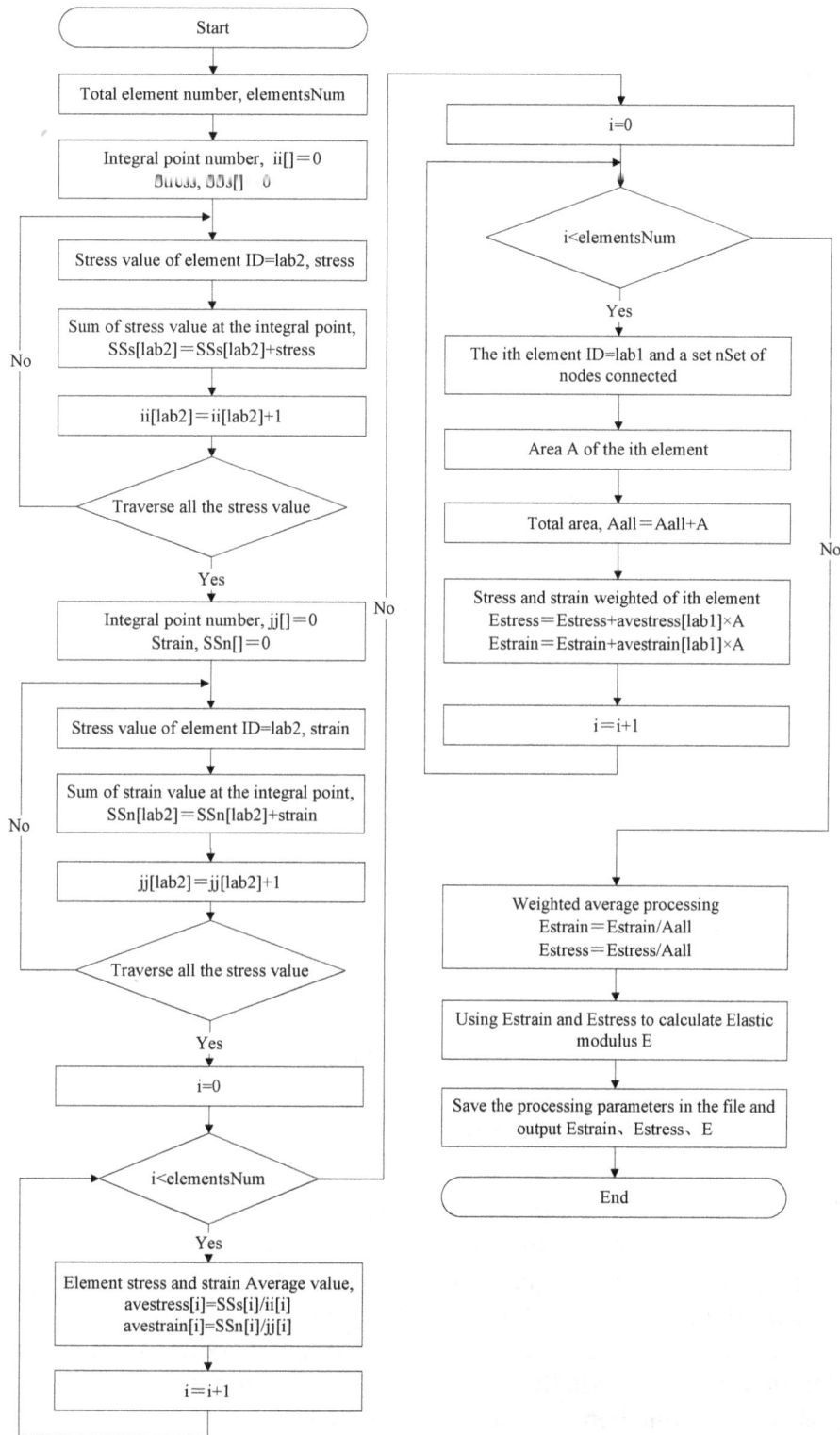

Fig. 3 Flow chart of "homogenization" program

In view of the meso mechanics homogenization method, the area-weighted average is used to process the calculation results from the ABAQUS software solver to achieve the "homogenization" process in this paper, in order to get the influence of microstructure information on macro mechanical properties. The flow chart shows the homogenization process of stress and strain parameters in Fig. 3, and the program is written in Python language.

Because there are several integral points in each element, firstly, the stress and strain values of the integral point are processed in each element one by one, and then the area-weighted average is used to process, and finally the macroscopic stress and macroscopic strain of the different directions are obtained and outputted by the homogenization process for the whole model, further the corresponding elastic modulus, Poisson's ratio parameters, are calculated and then stored in the output file.

4.1. Calculation of the element area

The four-node element and eight-node element, which are commonly used in the model analysis, combined with the order characteristics of the ABAQUS software storage nodes, are divided into a number of triangles to calculate the element area in Fig. 4.

For four-node element in Fig. 4a), the node order stored in ABAQUS software output database is (1, 2, 3, 4), and the element is divided into ①~② two triangle, and a triangle area can be obtained by Heron Formula [9].

For eight-node element in Fig. 4b), the node order stored in ABAQUS software output database is (1, 2, 3, 4, 5, 6, 7, 8). Because the eight-nodes element is not always regular quadrilateral, the element is divided into ①~⑥ six triangle.

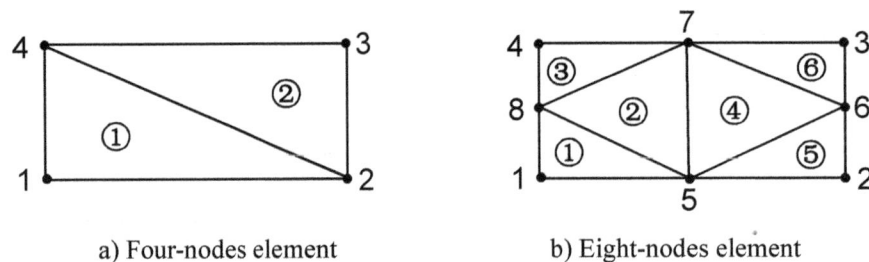

a) Four-nodes element   b) Eight-nodes element

Fig. 4 The area of different node elements

4.2. Calculation of the element stress and strain

In ABAQUS software, the nodes are employed to extract displacement and reaction force values, and integral points are employed to extract stress and strain values. Here, the integral points within each element and corresponding stress and strain values must be found.

All types of data objects are stored hierarchically in the ABAQUS software output database, so access to their data must also follow the hierarchy. For example, in order to extract the stress value in results data, the stress field of the different analysis steps and frames should be accessed at first, where arbitrary order stress value has the different element ID and integral point ID attributes, and the integral point ID are numbered from 1 in any element. According to the data storage characteristics of ABAQUS software output database, the average stress value of all integral points in each element can be obtained by traversing stress value of all integral points in the stress field, which is prepared to be called conveniently during weighted average processing of stress values in an element.

The "homogenization" process of the strain value is similar to the "homogenization" method of the corresponding stress value.

5. Validation Example

That is now using a finite element model of isotropic mechanics to verify the integrity and correctness of the program which has been compiled with the "homogenization" principle.

Meso mechanical model of 200 grains built by Voronoi diagram algorithm is shown in Fig. 5. Each Voronoi polygon represents a grain formatted during steel crystallization process [10], and material properties of each grain can be appointed individually or according to Weibull random algorithm. The external load σ_y applied on the model is the uniaxial tension, and the range of value is 1MPa~1125MPa. The data of stress-strain curve is derived from the uniaxial tensile test of drill pipe material, the elastic modulus is 2.1×10^5MPa, and the Poisson's ratio is 0.3.

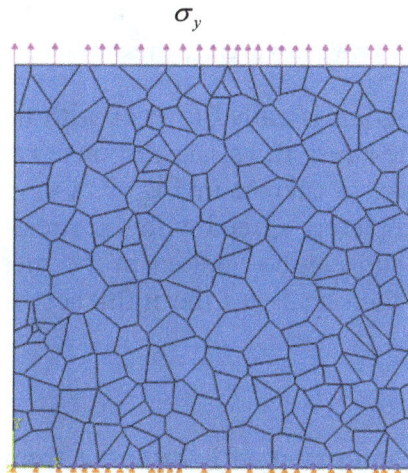

Fig. 5 Meso mechanical model of 200 grains

After solving the model in the ABAQUS software, the homogenization process program is called to obtain the area weighted average stress and strain values and the corresponding elastic modulus and Poisson's ratio. Through calculation, Y direction stress value obtained by using area weighted average is equivalent to the initial external load applied on the model boundary, which is consistent with characteristic of the meso mechanics calculation method, and the elastic modulus and Poisson's ratio obtained is also the same with the initial model parameters. As a result, the homogenization process program of the area weighted average is correct and feasible. The homogenization output results is shown in Fig. 6 when σ_y = 1 MPa.

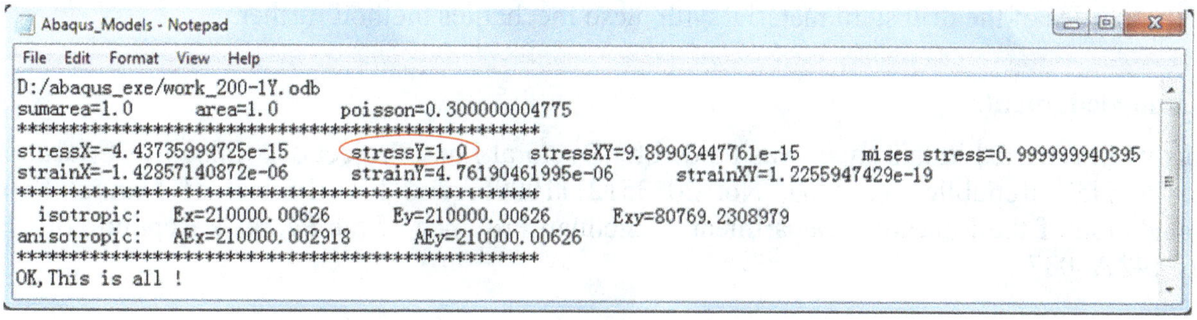

Fig. 6 "Homogenization" output file

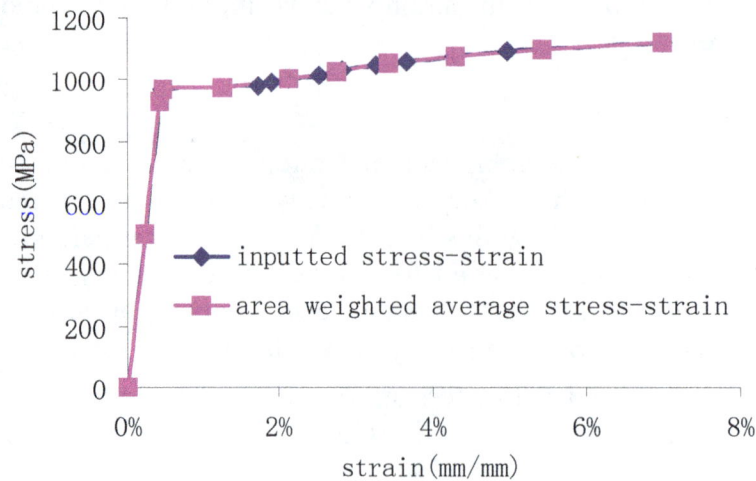

Fig. 7 The area weighted average stress-strain curves of the model

The homogenization process program is called to deal with the ABAQUS output database under the different external load, and the stress and strain response values are obtained, which are plotted in the model's stress-strain curve, as shown in Fig. 7. From the figure, the stress-strain curve obtained by homogenizing is completely agreed with the stress-strain curve inputted. It shows that this process method is suitable and can be applied in the meso mechanics model with the hard inclusions or micro cracks, and then analyzes the influence of microstructure on the mechanical properties of the material

6. Conclusion

(1) The meso mechanical calculation method is different from the macro mechanical calculation method and the microscopic test method, which has become the new method and direction of the research on the mechanism of drilling tool damage.

(2) Stress and strain values are stored in the results data of the ABAQUS software output database, which should be processed by traversing all integral points of different elements in the field before the area-weighted processing in order to speed up the calculation.

(3) The area of the four-node element and the eight-node element has been calculated separately, which can improve the calculation accuracy.

(4) According to the meso mechanics calculation method, the area-weighted process program has been compiled using the scripting Python language, which can achieve the homogenization process of the meso mechanics method, and provide an effective technical means to analyze the influence of the micro crack propagation or hard inclusions on the macroscopic mechanical properties of the drill stem material with meso mechanics method further.

Acknowledgments

This work reported in this paper was funded by Doctoral Fund Project of Ministry of Education of the People's Republic of China, No. 20135121110005, and Key Project of Natural Science Foundation of the Education Department of Sichuan Province of the People's Republic of China, No. 14ZA0037.

References

1. Wei YANG, Meso-mechanics and meso-damage mechanics, Advances in Mechanics. 22 (1992) 1-9, In Chinese.
2. Nemat-Nasser S and Hori M. Micromechanics: overall properties of heterogeneous materials. Elsevier, The Netherlands, 1993.
3. Yang W and Lee W. B. Mesoplasticity and its applications, Springer-Verlag, Berlin, 1993.
4. Shou-Wen YU, Xi-Qiao FENG. Damage mechanics, Tsinghua University Press, Beijing, 1997, In Chinese.
5. Hong-Jing FAN. Multiscale analysis of deformation and failure of materials, Science Press, Beijing, 2008, In Chinese.
6. Li NI. Finite element simulation for microstructure of heterogeneous materials, Master's degree thesis of Lanzhou University of Technology, Lan Zhou, 2005, In Chinese.
7. Feng-Lan HE, Xu-Dong LI, and Guo-Dong WANG, Finite element analysis of micro structure of three dimensional multi crystal materials based on DEFORM, Gansu Science and Technology. 25 (2009) 10-12, In Chinese.
8. Yan ZHANG, Zi-Ming ZHANG. The meso mechanics of materials, Science Press, Beijing, 2008, In Chinese.
9. Ji-Lan DONG. Handbook of practical mathematical formulas, Hebei Science & Technology Press, Shi Jia Zhuang, 1998, In Chinese.
10. Pei-De ZHOU. Computational geometry - algorithm design and analysis, Tsinghua University Press, Beijing, 2008, In Chinese.

Solution of Sparse Linear Systems with the Software Package LIS for Meso-scale Finite Element Simulation of Concrete Fractures

Jian-Ping Wu

Academy of Ocean Science and Engineering, National University of Defense Technology, Changsha, China
Email: wjp@nudt.edu.cn

Meso-scale simulation of concrete with finite elements is very time consuming, in which solution of the related linear systems occupies most of the time. In this paper, the software package LIS, which is designed to solve sparse linear systems and sparse linear eigenvalue problems in parallel, is adopted to accelerate the simulation. The flexural fracture of wet-sieved and three-graded concrete beams is simulated. For the linear systems considered are all symmetric positive definite, the preconditioned conjugate gradient is used all the time. And five symmetric preconditioners including ILU(0), ILU(1), SSOR, SAINV, and SAAMG are tested. The results show that, though the number of iterations with the smoothed aggregation based algebraic multigrid preconditioner SAAMG is less than that with the others, the average solution time is longer than that with the simple preconditioners ILU(0) and SSOR, and its efficiency is closely related to the selection of parameters.

Keywords: Concrete fracture; meso-scale simulation; sparse linear system; preconditioner; parallel computing.

1. Introduction

The meso-scale simulation with finite element approximation is an important way to investigate the performance of concrete, where the sample can be regarded as a three-phase composite material consisting of aggregate, mortar and interface between them. In this paper, the so-called Random Aggregate Random Parameter Model (RARPM) [1] is adopted to model the generation and extension of fractures.

In the RARPM, each node relates to three displacements, with each in one direction of the three-dimensional space. After the sample is approximated with finite elements, the damage model is applied to each element, and a simple elastic relation is applied to each node. Then we can derive the elemental mass, damping and stiff matrices, and the total stiffness matrix can be assembled. After the way and the quantity of loads are given, the displacement increments can be solved through non-linear equations, where many sparse linear systems should be solved.

Due to their projection properties, Krylov subspace methods are good choices to solve sparse linear systems. But the convergence rate is determined by the eigenvalue distribution of the coefficient matrix. If the eigenvalues are located in a relative small region, the iteration will converge very fast. And if they are distributed in a large area, the iteration will converge slowly, or even diverge. To improve the convergence rate, preconditioning schemes are widely used in recent years [2].

The most frequently used preconditioners include the classical iterations, incomplete LU factorizations [2, 3, 4, 5], sparse approximate inverses [2, 6] and multilevel preconditioners [7, 8, 9]. The classical iterations can be used as preconditioners, with a single iteration or several ones to approximate the inverse of the coefficient matrix, in which Jacobi and SSOR are the most widely used ones [2].

Incomplete factorizations factor the coefficient matrix into incomplete LU or QR factors. Based on the easier solution of triangular linear systems, the approximation of the inverse of the coefficient matrix is given implicitly and cheaper to implementation. In ILU(0), the incomplete LU factors have the same nonzero structures as the coefficient matrix, without any fill-ins [2]. ILU(1) is the extension of ILU(0), allowing the fill-ins within level 1 [2]. It is more effective for diagonally dominant systems, but is based on the matrix structures only, without any consideration to the magnitude of the elements. To solve this problem, Saad provide a method ILUT for general linear systems [4], and it is the most efficient preconditioner for general purposes. For the computation and the application of incomplete factors is serial in essence, their parallelization is not trivial, which attracts many attentions [5]. The block diagonal analog and additive Schwarz are two efficient parallelization techniques, where graph partitioning should be used in advance.

Sparse approximate inverses construct preconditioners explicitly, to approximate the inverse of the coefficient matrix directly, either column by column, or compute the inverse as a whole [2, 6]. The provided information in explicitly given preconditioner is less than that in implicit ones. Benzi presents a factorization type approximate inverse AINV, based on the LDU factorization, and a symmetrical version SAINV for symmetric definite matrices. These derived preconditioners are given as multiplication of some factors, thus, they are not completely explicit. The numerical results show that they can achieve similar performance as the ILU preconditioners [6].

In recent years, more and more researchers focus on the multilevel methods, either as single iterations solely, or as preconditioners of Krylov subspace methods, for their potential to achieve optimal convergence, and insensitivity to the grid scale [7, 8, 9]. Algebraic multigrid can be easily used to solve linear systems from un-structural problems or those with grid information unknown in advance, or even without any physical grid information. Among them, the aggregation based method is widely used for its cheap cost and easy implementation. But the convergence is limited for the simple grid transfer operators. SAAMG is an improvement to it, and smooth the interpolation operator with simple iterations such as Jacobi or damped Jacobi [9]. Many practices have proven the efficiency of this kind of multigrid.

In this paper, the software package LIS [10] is adopted to solve the sparse linear systems from meso-scale simulation of concrete samples. LIS is developed under the Scalable Software Infrastructure Project, and it is supported by the Development of Software Infrastructure for Large Scale Scientific Simulation Team of JST/CREST. To solve a linear system, one can use CG, BiCG, CGS, BiCGSTAB, GPBiCG, BiCGSafe, BiCGSTAB(l), Jacobi, Gauss-Seidel, SOR, IDR(s), CR, BiCR, CRS, BiCRSTAB, GPBiCR, BiCRSafe, TFQMR, Orthomin(m), GMRES(m), FGMRES(m), and MINRES. And there are ten preconditioners provided to users, including Jacobi, SSOR, ILU(k), ILUT, Crout ILU, I+S, SAAMG, Hybrid, SAINV, and additive Schwarz. Sparse matrices can be stored in CSR, CSC, MSR, DIA, ELL, JAD, BSR, BSC, VBR, COO, and DNS format.

In section 2, the outline of meso-scale simulation of concrete with static and dynamic loading will be given. Some numerical results will be given in section 3, and finally, in section 4 some conclusions are drawn.

2. Simulation of Concrete Fractures

In meso-scale simulation of concrete fractures, the sample can be regard as a three-phase composite material consisting of aggregate, mortar and interface between them and can be discretized with finite elements [1]. Usually, the meso-scale aggregate model in the middle regions and the end of the beam are discretized with hexahedron elements, and their transition region with tetrahedron ones. The discrete grids are shown in Figure 1.

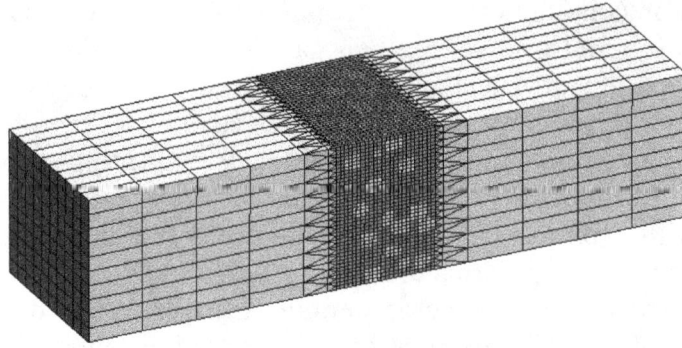

Figure 1 Finite mesh of a three graded concrete sample

In static loading, the loads are incrementally increased. Denote the elemental stiffness matrix, displacements, and static loads on the *i*-th loading step as $[K_i]$, $\{U_i\}$, and $\{P_i\}$ respectively, the balance equation on the *i*-th step is

$$[K_i]\{U_i\}=\{P_i\}, \tag{1}$$

and similarly on the (*i*+1)-th step, we have

$$[K_{i+1}]\{U_{i+1}\}=\{P_{i+1}\}. \tag{2}$$

Therefore

$$[K_i]\{\Delta U_i\}=\{\Delta P_i\}-[\Delta K_i]\{U_{i+1}\}, \tag{3}$$

where $\{\Delta P_i\}=\{P_{i+1}\}-\{P_i\}$ is the increment of loads vector, $\{\Delta U_i\}=\{U_{i+1}\}-\{U_i\}$ is the increment of displacement vector,

$$[\Delta K_i]= [K_{i+1}]-[K_i] =\sum_e[d^e(\varepsilon_i^e)-d^e(\varepsilon_{i+1}^e)][K_0^e]= -\sum_e[\Delta d^e(\varepsilon_i^e)][K_0^e] \tag{4}$$

is the stiffness increment, e and ε are element label and its strain, and d is the damage coefficient corresponding to the strain. Then, we can construct an iteration process as follows

$$[K_i^{(m)}]\{\Delta U_i^{(m)}\}=\{\Delta P_i\}+\sum_e[\Delta d^e(\varepsilon_i^{(e,m)})][K_0^e]\{U_{i+1}^{(m)}\}, \ m\geq0, \tag{5}$$

$$\Delta d^e(\varepsilon_i^{(e,m)})= d^e(\varepsilon_i^e)-d^e(\varepsilon_{i+1}^{m,e}), \ \varepsilon_{i+1}^{0,e}=\varepsilon_i^e, \ \{U_{i+1}^{(0)}\}=\{U_i\}. \tag{6}$$

In dynamic loading, the balance equations at time t is

$$[M]\{d^2U_d(t)/dt^2\} + [C]\{dU_d(t)/dt\} + [K_d(t)]\{U_d(t)\} + [K_s(t)]\{U_s(t)\} = \{P_d(t)\} + \{P_s\}. \tag{7}$$

The balance equation at time $t+\Delta t$ can be derived similarly, then, we have the equation of increment form

$$[M]\{\Delta(d^2U_d(t)/dt^2)\} + [C]\{\Delta(dU_d(t)/dt)\} + [K_d(t)]\{\Delta U_d(t)\} = R(t), \tag{8}$$

where

$$R(t) = \{\Delta P_d(t)\} - [\Delta K_d(t)]\{U_d(t+\Delta t)\} - [\Delta K_s(t)]\{U_s\},$$

$$\Delta Y = Y(t+\Delta t) - Y(t),$$

$[M]$, $[C]$, $[K_d]$, $[K_s]$ are the elemental mass, damping, dynamic stiffness, static stiffness matrices respectively, and $\{P_d\}$ and $\{P_s\}$ are the dynamic and static loads respectively. We solve equation (8) with Newmark formula [1], where we select $\delta \geq 0.5$, $\gamma = 0.25(0.5+\delta)^2$, and denote

$$a_0 = 1/(\gamma \Delta t^2),\ a_1 = \delta/(\gamma \Delta t),\ a_2 = 1/(\gamma \Delta t),\ a_3 = 1/(2\gamma),\ a_4 = \delta/\gamma,\ a_5 = (\Delta t/2)/(\delta/\gamma - 2). \tag{9}$$

The equation of increment form can be approximated by

$$[K_d(t) + a_0 M + a_1 C]\{\Delta U_d(t)\} = F(t), \tag{10}$$

where

$$F(t) = R(t) + [M]\{a_2 dU_d(t)/dt + a_3 d^2 U_d(t)/dt^2\} + [C]\{a_4 dU_d(t)/dt + a_5 d^2 U_d(t)/dt^2\}. \tag{11}$$

This equation is nonlinear, and can be approximated further as

$$[K_d(t) + a_0 M + a_1 C]\{\Delta U_d(t^{n+1})\} = F(t^n),\ n = 0,1,2,\ldots. \tag{12}$$

3. Numerical Experiments

In this section, some of the test results will be provided. All the given results are derived on a cluster of 32 Intel(R) Intel(R) Xeon(R) CPU E5-2692 v2@2.20GHz (cache 3.72MB), interconnected by infiniband. The operating system is Red Hat Linux 4.4.6-3. The message passing interface is MVAPICH2 1.8.1 and the compiler used is Intel 11.1.059. For the linear systems are all symmetric positive definite, the preconditioned CG iteration is exploited. In the iteration for each linear system, the initial guess is selected as the vector of all zeros, and the iteration is stopped when the Euclid norm of the residual vector is reduced by 1.0E-4.

The experiments are performed on two concrete samples, where one is of three-graded, and the other is wet-sieved. For each sample, there are two loading columns at the top, and two supporting columns at the bottom. The size of the three-graded sample is 1100mm x 300mm x 300mm. There are 44117 nodes and 53200 elements in all respectively. The two supporting columns are 0.1m away from left and right boundary respectively, while the two loading columns are 0.4m away from left and right boundary respectively. The size of the wet-sieved sample is 550mm x 150mm x 150mm. There are 71013 nodes and 78800 elements in all respectively. The two supporting columns are 0.05m away from left and right boundary respectively, while the two loading columns are 0.2m away from left and right boundary respectively.

Three experiments are done, with static loading test for the three-graded sample, and static and dynamic loading test for the wet-sieved sample. For the static loading test, the load added each step is 0.25kN. For dynamic loading test, the time step is 0.001 seconds, and the loads are increased linearly with increment of 0.8kN.

In the parallel simulation, the interface metis_PartGraphRecursive in the package METIS-4.0 [11] is used to partition the adjacent graph related to the coefficient matrix, and then, the computation related to each sub-graph is assigned to a processor. Each node of the graph is corresponding to a row of the coefficient matrix and a component of vectors.

For the derived linear systems are all symmetric positive, and the preconditioned conjugate gradient iterations are exploited, the preconditioners should be symmetric positive definite too. Thus, in the simulation with LIS, the relatively high efficient preconditioners of different type are tested, including SSOR, ILU(l), SAINV, and SAAMG. The less effective preconditioners such as Jacobi, and the preconditioner without symmetry property are not tested.

For the static loading of the three-graded sample, there occurs damaged elements at the 439-th step and the sample is completely damaged at the 567-th step. If there are damaged elements at a certain step, several linear systems need to be recursively solved to revise the displacements, which are induced by the degradation of the stiffness matrix. There are 697 linear systems to be solved in all. For the wet-sieved sample, in the static loading, there occur damaged elements at the 59-th step. The sample is completely damaged at the 94-th step and there are 178 linear systems in all. In the dynamic loading, besides the solution of linear systems corresponding to the incremental displacements, the computation of the matrices $[M]$ and $[C]$ require some of the eigenvalues of $[K]$, which needs to solve linear systems too. There are 221 linear systems in all and the number of loading steps is 38.

Table 1 Results of static loading for the three-graded sample,
where "-" means that the solver meets some problem and cannot be correctly used

		ILU(0)	ILU(1)	SAINV (0.05)	SSOR (1.0)	SAAMG			
						$\theta=0.18$	$\theta=0.19$	$\theta=0.2$	$\theta=0.3$
avgIts	NP=08	165.02	113.00	378.29	191.08	53.55	67.71	71.34	97.05
	NP=16	177.04	-	379.23	152.76	-	65.04	66.05	74.27
	NP=32	164.06	161.01	378.92	159.69	67.98	83.01	85.00	100.05
avgTme	NP=08	0.84	1.24	1.34	0.77	0.82	0.91	0.94	1.17
	NP=16	0.45	-	0.72	0.33	-	0.50	0.48	0.52
	NP=32	0.22	0.44	0.37	0.18	0.37	0.41	0.38	0.45

Some test results are given in Table 1 to Table 3, where avgIts and avgTme means average iterations and time used to solve each linear system. In preconditioner SAINV [6], the elements are less than 0.05 are dropped. If the linear system is $Ax=b$, and the strict lower triangular part, diagonal part, and the strict upper triangular part are denoted as L, D, and U respectively, the SSOR(ω) preconditioner [2] is given by $(D-\omega U)^{-1}D(D-\omega L)^{-1}$. In the tests, the parameter ω is selected as 1. In SAAMG, the nodes with strong connections are integrated into an aggregation, and if a_{ij}^2 is not less than $\theta a_{ii}a_{jj}$, we say that node i and node j is strongly connected [9]. In the tests, $\theta \leq 0.17$ always meets solution problem, and the systems cannot be correctly be solved, partly due to the large number of nonzeros in each row of the initial sparse linear systems, which is about 81 as the maximum. The selection $\theta=0.18$, 0.19, 0.2 and 0.3 are tested and the related SAAMG with θ as the classification parameter of strong connection is denoted as SAAMG(θ).

Table 2 Results of static loading for the wet-sieved sample

		ILU(0)	ILU(1)	SAINV (0.05)	SSOR (1.0)	SAAMG			
						θ=0.18	θ=0.19	θ=0.2	θ=0.3
avgIts	NP=08	229.49	184.09	602.84	284.24	112.02	123.06	128	150.02
	NP=16	240.02	203.49	603.13	292.33	113.02	126.1	131.07	154.02
	NP=32	244	207.09	603.11	300.58	115	132	136.02	154.04
avgTme	NP=08	2.55	3.53	6.04	1.89	2.9	2.7	2.60	2.92
	NP=16	1.3	1.95	3.09	1.01	1.59	1.5	1.46	1.55
	NP=32	0.7	1.03	1.46	0.55	0.98	0.95	0.87	0.84

Table 3 Results of dynamic loading for the wet-sieved sample,
where "-" means that the solver meets some problem and cannot be correctly used

		ILU(0)	ILU(1)	SAINV (0.05)	SSOR (1.0)	SAAMG			
						θ=0.18	θ=0.19	θ=0.2	θ=0.3
avgIts	NP=08	230.61	187.8	702.18	286.57	119.52	132	135.64	150.92
	NP=16	242.79	204.78	693.32	294.07	-	131.41	135.78	154.28
	NP=32	247.88	216.15	690.80	299.44	117.14	133.9	138.85	153.92
avgTme	NP=08	2.5	3.61	6.99	1.89	3.1	2.81	2.83	2.94
	NP=16	1.33	1.84	3.38	1.01	-	1.5	1.5	1.54
	NP=32	0.71	1.07	1.66	0.55	0.95	0.91	0.89	0.84

From the tables, we can see that the number of iterations with SAAMG is always less than the others. And for static loading of the three-graded sample, θ=0.18 is the best choice for SAAMG, and θ=0.2 is also a good choice, but in some cases, θ=0.18 meets solution problems. In addition, with the average solution time of each linear system in mind, SSOR is the best, and then ILU(0). SAAMG has no advantages even using the best choice of θ. For the simulation of the wet-sieved sample, θ=0.2 is the best for SAAMG in general. And in view of solution time, SSOR is the best, and ILU(0) is the second again. SAAMG has no advantages too. But the different between SAAMG and SSOR is not larger than that in the simulation of the three-graded sample, which may be caused by the scale of the problem. The three- graded sample has only 44117 nodes, thus the order of the linear systems is about 130000. While the wet-sieved sample has 71013 nodes and the order of the corresponding linear systems is about 210000, which is larger. It means that if the discrete scale of the sample increases further, the privileges of SAAMG preconditioner may occur, which is the focus of the future work. Another phenomenon is that when the number of processors increased step by step, the optimal θ is increasing correspondingly.

4. Conclusion

In this paper, the LIS software package is used to simulate the fracture of concrete samples, and five symmetric preconditioners including ILU(0), ILU(1), SSOR, SAINV, and SAAMG are tested. The results shows that though SAAMG has the potential to solve very large-scale problems, it has no advantages over ILU(0) or SSOR for the test samples, probably due to the relatively small discrete scale. It also illustrates that though the multigrid-like methods are efficient in many cases, the application to specific areas needs specific considerations to exert their privileges. Of course, the discrete scale of the current simulation is not very large. In future, large scale problems will be tested to verify the efficiency of SAAMG and other preconditioners further.

Acknowledgment

This research was funded by the National Science Foundation: (61379022).

References

1. Ma Huaifa, Chen Houqun, "Study on Dynamic Damage Mechanism of Full-graded Dam Concrete and its Meso-scale Mechanics Analysis", China WaterPower Press, Beijing, 2008
2. Y. Saad, "Iterative methods for sparse linear systems", PWS Publication Corporation, Boston, 1996
3. J. A. Meijerink and H. A. van der Vorst, "An iterative solution method for linear systems of which the coefficient matrix is a symmetric M-matrix", Mathematics of Computation, Vol. 31, No. 137, 1977, pp. 148-162
4. Y. Saad, "ILUT: A dual threshold incomplete ILU preconditioner", Numer. Lin. Alg. Appl., Vol. 1, No. 4, 1994, pp. 387-402
5. Wu Jian-ping, Zhao Jun, Song Jun-qiang, Zhang Wei-min, Li Xiao-mei, "A parallel preconditioner based on factors' combination for general sparse linear systems", SIAM J. Sci. Comput., Vol. 34, No. 4, 2012, pp. A2247-A2266
6. Benzi M., Cullum J. K., and Tüma M. Robust approximate inverse preconditioning for the conjugate gradient method. SIAM J. Sci. Comput., Vol. 22, No. 4, 2000, pp. 1318-1332
7. Henson, V. E., & Yang, U. M. (2002). Boomer AMG: A parallel algebraic multigrid solver and preconditioner. Applied Numerical Mathematics, 41(1), 155-177
8. Chartier, T., Falgout, R. D., Henson, V. E., Jones, J., Manteuffel, T., McCormick, S., Vassilevski, P. S. (2003). Spectral AMGe (rho AMGe). Siam Journal On Scientific Computing, 25(1), 1-26
9. R. Blaheta, Algebraic multilevel methods with aggregations: An overview, In I. Lirkov, S. Margenov & J. Wasniewski (Eds.), Large-Scale Scientific Computing, Vol. 3743, 2006, pp.3-14
10. The Scalable Software Infrastructure Project, Lis User Guide, Version 1.5.63, http://www.ssisc.org/lis/lis-ug-en.pdf, 2005
11. G. Karypis and V. Kumar, "MeTiS – A Software Package for Partitioning Unstructured Graphs, Partitioning Meshes, and Computing Fill-Reducing Orderings of Sparse Matrices – Version 4.0", Technical report, University of Minnesota, September 1998

Analysis and Cost Optimization of Desulfurizing Agent Consumption for KR

Yu Tang

CISDI Engineering Co., Ltd.

NO.1, Saidi Road, Jinyu Avenue, Yubei District 401122, Chongqing, China

Email: Yu.A.Tang@cisdi.com.cn

In order to control KR production cost, production data investigation was done on site at one steelmaking plant. The investigation is mainly about desulfurizing agent consumption. By comparison and analysis, it is found that the desulfurizing agent consumption in this plant can be improved further and major influencing factors are investigated such as initial [S] content, desulfurizing rate, hot metal temperature, impeller stirring status. By technical improvement, the desulfurizing agent consumption could be reduced with 0.5~1kg/t steel, saving 0.5~1RMB/t steel.

Keywords: Desulfurizing agent; KR; hot metal.

1. Introduction

Chinese iron and steel industry has entered the era of meagre benefit and adjustment. Under this circumstance, the steelmaking enterprise with lower cost and higher efficiency will survive or become the winner in this severe war.

In order to control the production cost, among many aspects, the raw material is a key factor. As the device of hot metal pretreatment, KR is now very popular in China. For KR, its raw material is mainly the desulfurizing agent. It would be a satisfactory result if KR can finish its task with low consumption and high desulfurizing rate. For this purpose, firstly, the influencing factor should be found out.

Previous papers [1-8] show the desulfurizing agent consumption would be influenced by some factors, such as hot metal temperature, stirring duration, stirring speed, initial [S] content, desulfurizing agent components and size, charging speed, desulfurizing depth according to different requirements of different steel grades.

2. Investigation and Results

2.1. *Investigation*

Site investigation was done at one steelmaking plant in Southern China in terms of production data and operation status. There are two sets of 210t KR, whose annual treatment capacity is 4.5 million ton per year.

All KR production data for 1.5 years from the data base of this steelmaking plant were fetched, investigated and analysed systematically. In this paper, we mainly analyse the desulfurizing agent consumption with methods of comparison and factor analysis.

2.2. *Results*

The average consumption of the desulfurizing agent is about 9.08 kg/t.

Fig. 1 shows the comparison of the desulfurizing agent consumption between several Chinese steelmaking plants. It is shown that the consumption for this plant is higher than other plants. It is 4.39kg/t higher than the lowest one (4.69kg/t). So, there should be more improving potential of the desulfurizing agent consumption for this plant.

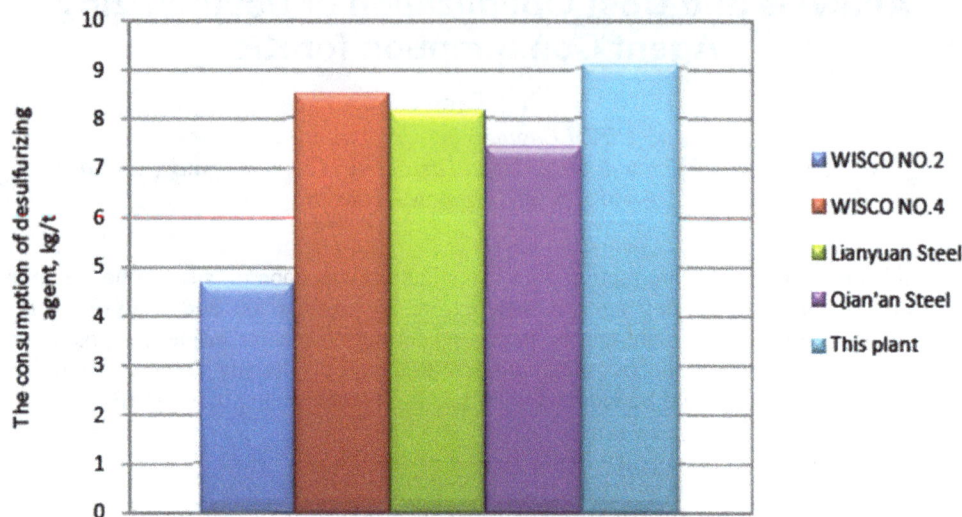

Fig. 1 Comparison of the desulfurizing agent between several Chinese steelmaking plants (kg/t)

Table 1 shows the desulfurizing agent consumption for each 0.001% [S] reduced during KR process, which can explain the desulphurization efficiency more reasonably than average data per ton steel.

Table 1 Desulfurizing agent consumption for each 0.001% [S] reduced during KR process

Initial [S],%	Final [S],%	Consumption for each 0.001%[S] reduced, kg/t
0.0100<x≤0.0150	0.0001<x≤0.0010	0.41
	0.0010<x≤0.0050	0.51
	0.0050<x≤0.0100	0.62
0.0150<x≤0.0200	0.0001<x≤0.0010	0.35
	0.0010<x≤0.0050	0.36
	0.0050<x≤0.0100	0.42
0.0200<x≤0.0230	0.0001<x≤0.0010	0.25
	0.0010<x≤0.0050	0.29
	0.0050<x≤0.0100	0.33

3. Discussion

Now, let's investigate the influencing factors of the desulfurizing agent consumption and the reason why it is higher than other plants.

3.1. *Initial [S] content in hot metal*

Fig. 2 shows the rate distribution of different initial [S] content in hot metal. It is shown that 0.0100 %< initial [S] ≤0.0230% takes 43.91%, and 0.0230 %< initial [S] ≤0.0500% takes 54.58%. So it is higher than normal level, leading to higher desulfurizing agent consumption.

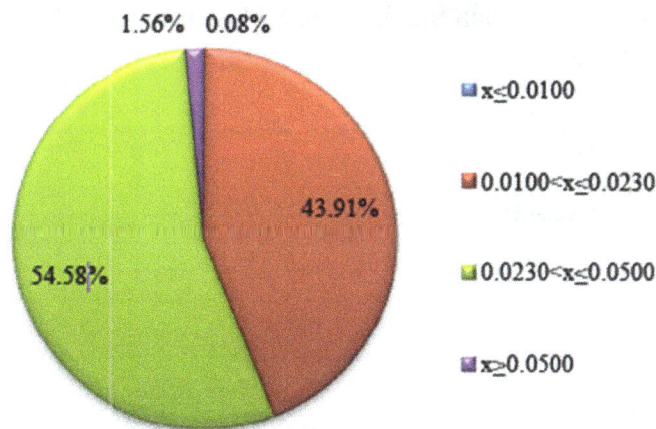

Fig. 2 Rate distribution of different initial [S] content in hot metal (%)

3.2. *Desulfurizing rate*

Fig. 3 shows the relationship between desulfurizing agent consumption (for every 0.001% [S] reduced) and desulfurizing rate.

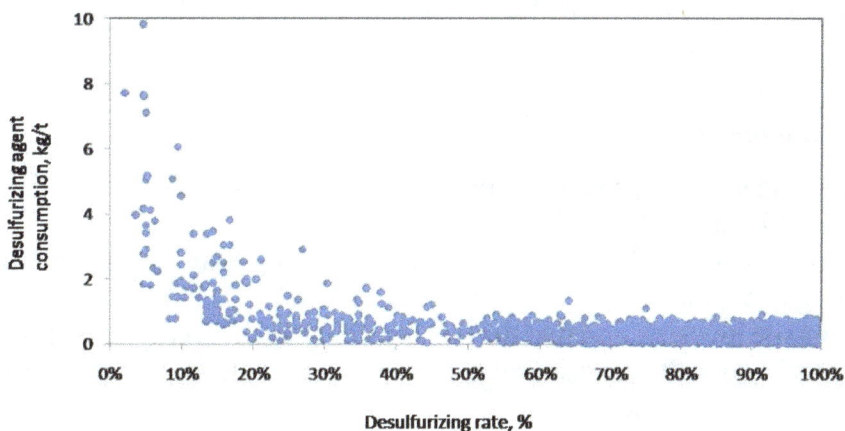

Fig. 3 Relationship between desulfurizing agent consumption
(for every 0.001% [S] reduced) and desulfurizing rate

It is shown that the desulfurizing agent consumption would decrease when the dusulfurizing rate increases. This rule is more obvious when the desulfurizing rate <40%.

3.3. *Hot metal temperature*

If hot metal temperature (when ladle car arrives at KR working position) is higher than proper level, the desulfurizing agent would agglomerate, reducing the reacting specific surface area, leading to lower desulfurizing rate and higher agent consumption.

Fig. 4 shows the comparison of hot metal temperature between Lianyuan Steel and this plant. For Lianyuan Steel, it also has two sets of 210t KR, so it is reasonable to compare the two plants. It is shown that the hot metal temperature of this plant is higher, there are 71.2% for the temperature >1390°C, much higher than Lianyuan Steel. The proper range for KR is 1300~1360°C.

There are 49% in this range for Lianyuan Steel, while 8.9% for this plant. So the consumption of this plant is higher than Lianyuan Steel.

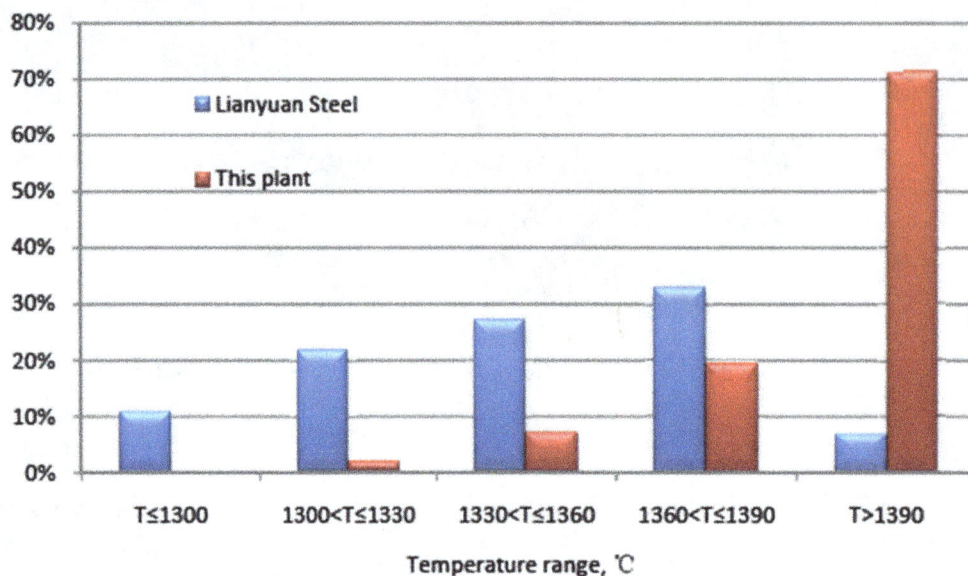

Fig. 4 Comparison of hot metal temperature between Lianyuan Steel and this plant

3.4. *Other influencing factors*

1) During desulfurizing process, production practice may be not completely on the basis of specific requirements of different steel grades;
2) There are 10% of ultra-low [S] steel grade, calling for more desulfurizing agent;
3) The desulfurizing agent may have comparatively lower taste, leading to more consumption;
4) It is shown by on-site investigation that the desulfurizing agent is easy to be lost during charging process because of too small size, leading to heavy dust and difficult reaction with the hot metal because of its floating on the surface.
5) During stirring process, the impeller's rotation is eccentric, influencing the flow field of the desulfurizing agent, lowering the desulfurizing efficiency.

4. Conclusion

Based on above, conclusions can be drawn that the desulfurizing agent consumption would be influenced by many factors, such as initial [S] content in hot metal, desulfurizing rate, hot metal temperature, desulfurization according to requirements, taste and particle size of desulfurizing agent and impeller stirring status.

In order to reduce desulfurizing agent consumption to control KR production cost, some measurement can be done, such as:

1) To control initial [S] content in hot metal by using fine charge such as iron ore, sinters or pellets for the blast furnace;
2) To improve production practice to enhance desulfurizing rate;
3) To control the hot metal temperature in proper range;
4) To desulfurize according to specific requirements of different steel grades;

5) To improve the taste and keep the proper particle size of the desulfurizing agent;
6) To optimize the impeller stirring status by optimized design of the impeller and strict control of its operating status.

By the measurements above, it is estimated that the desulfurizing agent consumption can be reduced 0.5~1kg/t at least. That is to say, the production cost can be reduced about 0.5~1 RMB/t steel. For this plant with 4.5 million ton treatment capacity per year, there will be 2.25~4.5 million RMB saved for KR every year.

References

1. Sun Liang, et al. The Application of 210t KR Desulfurization Technology. Technical Center Construction and New Products Seminar Proceedings, 2011: 88-92
2. Wang Wei, et al. Effect of KR Pretreatment Process Parameters on Hot Metal Desulfurization Results. Special Steel, 2006, 27 (4): 50-52
3. Yao Na, et al. Effect of Hot Metal Desulfurization Factors in KR Method. Journal of Materials and Metallurgy, 2010, 9 (3): 164-167
4. Yao Na, et al. Production Practice of Hot Metal Pretreatment Desulfurization by KR Mechanical Stirring Method. Special Steel, 2011, 32 (4): 34-35
5. Yuan Lifeng. Development and Application of Desulphurizer with KR Method in JIGANG. Journal of Anhui University of Technology, 2005, 22 (4): 576-579
6. Chen Junfu, et al. The Application of Desulfurization according to requirements for 200t KR process. Steelmaking and Continuous Casting Annual Conference Proceedings in Zhongnan & Zhusanjiao Zone, 2011: 122-124
7. Zhouyun, et al. Development and Application of KR Complex Desulfurizer in Jigang. Journal of Anhui University of Technology, 2007, 24 (4): 366-368
8. Zhou Jianfeng, et al. Research on Changing rules of Hot Metal Temperature during KR Process in Lianyuan Steel. Journal of Wuhan University of Science and Technology, 2010, 33 (6): 570-573

Based on SolidWorks Simulation Structure of Bridge Crane CAD/CAE Secondary Development Technology Research

Ming-Liang Yang[1,*], Rong-Zhen Lang[1], Lu-Yang Cheng[2], Ji-Sheng Wang[3], Wen-Jun Meng[1], Chao Wang[1]

*[1]School of Mechanical Engineering, Taiyuan University of Science and Technology
Taiyuan 030024, China
[2]Beijing Materials Handling Research Institute 100007, China
[3]Taiyuan Heavy Industry Co., Ltd
Email: 696718251@qq.com

The secondary development of SolidWorks simulation is very important in the research of overhead crane structure. The article is based on C# study on secondary development of SolidWorks. The emphasis is the secondary development of SolidWorks simulation method. The purpose is to realize integration of parametric modeling and finite element analysis in SolidWorks software about the overhead crane, which is helpful to improve the efficiency of product development and to reduce the repetitive work.

Keywords: The structure of overhead crane; Solidworks simulation; secondary development; CAD/CAE.

1. Introduction

With the rapid development of production, the quantity demand of market and customers about the overhead crane is growing. In order to improve the market competitiveness, enterprises need shorten the research and product development cycle. Interactive 3D software can't meet the needs of the professional machinery, so, it is necessary to develop appropriate professional CAD/CAE system.

In recent years, many scholars put the generalized modular and parametric design idea applied to the design of mechanical products. For example, Li Xin [1], who is from Dalian University of Technology chose VB.NET as tool and the platform of Solidworks2006, developed a parametric design system of overhead crane structure, and the 3D model can be automatically generated and bridge automatic assembly. For another example, Zongyan Wang, Shufang Wu, Huibin Qin [2] etc. come from North University of China, cooperated with enterprises such as Dalian Heavy Industry Group. SolidWorks was be used as the modeling tool, and have made remarkable achievements such as developed a parametric system of overhead crane, studied engineering drawing adjustment technology with the modular, parameters of product. But there is no one involved in parameterization of CAE aspect of the structure of overhead crane.

This article is in view of the crane structure parametric design. Using C# programming language in Visual Studio 2012 development platform, it connects the SolidWorks 3D modeling and performs finite element analysis, achieving a variational modeling and parameterized analysis integration. The application of parametric model reduces the repetitive work, improves work efficiency and is able to perform finite element analysis.

2. SolidWorks Secondary Development

2.1. *Principle*

The mainly method of SolidWorks secondary development is calling the built-in software API (Application Programming Interface) Interface functions; the SolidWorks API is a top-down multi-level tree network structure. The object relational hierarchy is shown in Fig. 1.

---

*Corresponding author

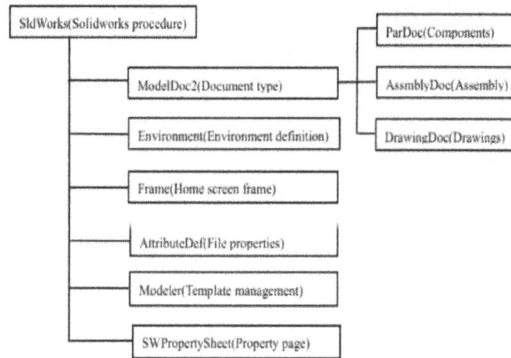

Fig. 1 Object relational hierarchy chart of SolidWorks API

2.2. *SolidWorks simulation API object model*

SolidWorks simulation objects include CWStudyManager, CWStudy, CWSolidManager, CWContactManager, CWShellManager, CWMesh, CWResults etc. as shown in Fig. 2. These objects can achieve the most basic operation of simulation program, such as invoking simulation, creating an example, defining the constraints, meshing, running the analysis etc. Also variables of SolidWorks simulation can be installed.

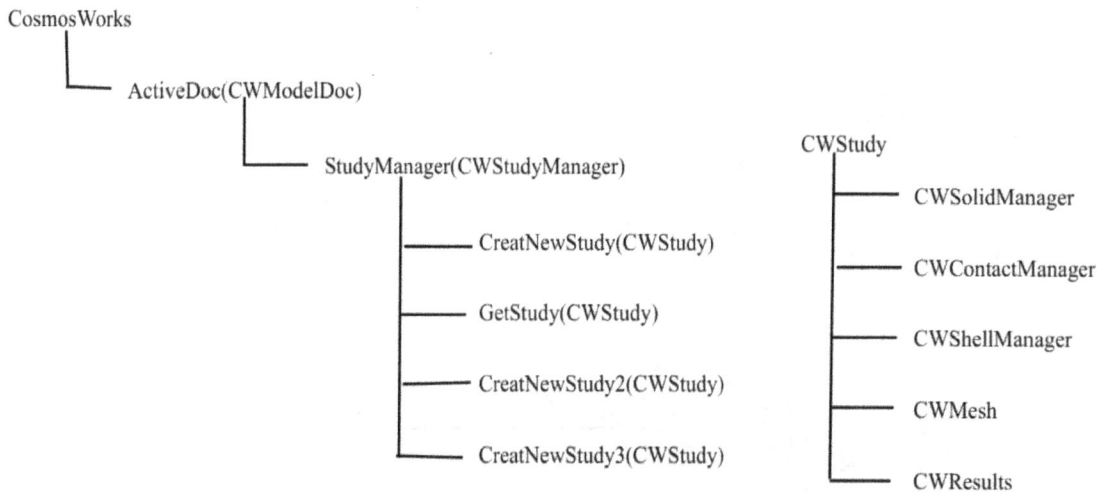

Fig. 2 Object relational hierarchy chart of SolidWorks simulation API

2.3. *Methods of SolidWorks secondary development*

There are two main types of secondary development methods for SolidWorks [3].

The first is the dimension-driven method. Components are built before hand and design variables which determine the actual size of the components are set. A series of components with the same structure but of various sizes can be obtained by assigning different values to design variables.

The second is model-driven method. With this method, models are built dynamically by using programs that describe the whole process of modeling, including creating a sketch, stretching resection, rotating characteristics, etc. The amount of programming required depends heavily on the complexity of the model. Therefore, this approach is often referred to as programming-driven method.

In this paper, the method that combines SolidWorks modeling with C# programming is adopted. Typical components, according to the relationship of design are selected and then built in

SolidWorks. First, according to the requirements of the design, analysis and determine the design variables of the model in the three dimensional modeling process. Second, calling the 3d model graphics file opened by API function, through the programming. Third, the design variables are modified to regenerate the model that is the change of model geometry has come true. Through the above three steps, the 3d variational modeling process of components have been done [4-7], then the CAE parametric control of the structure of overhead crane proceeds by calling the SolidWorks simulation function.

3. Secondary Development Process of SolidWorks Simulation

3.1. *Development process instance*

The Visual Studio 2012 is chosen as the development environment. C# is used as the programming language for secondary development of SolidWorks 2012, an executable application could be generated. The running workflow of the software is as shown in Fig. 3.

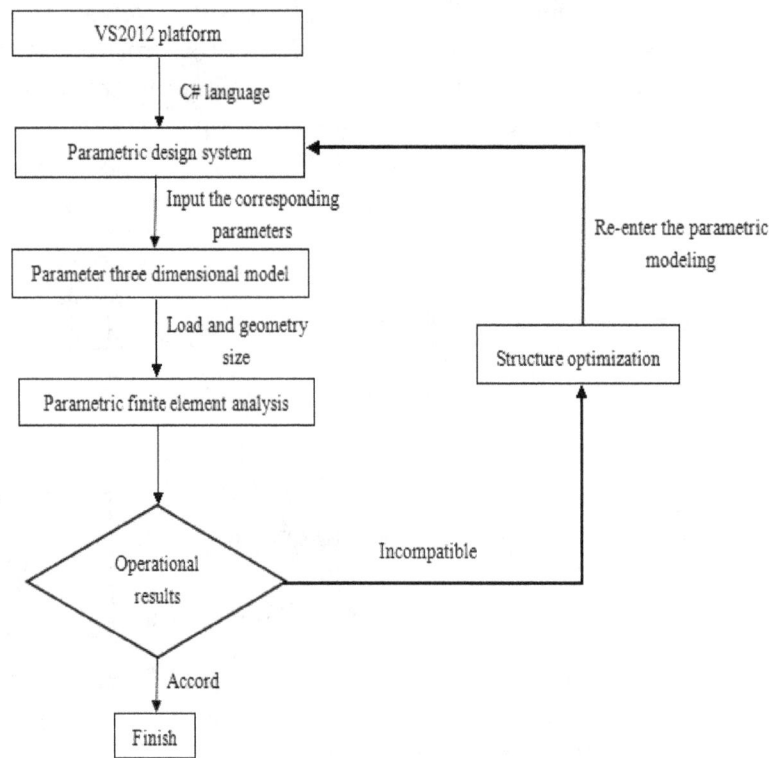

Fig. 3 The overhead crane development flowchart

3.2. *The establishment of the overhead crane model*

The three-dimensional model of double girder overhead crane is as shown in Fig. 4.

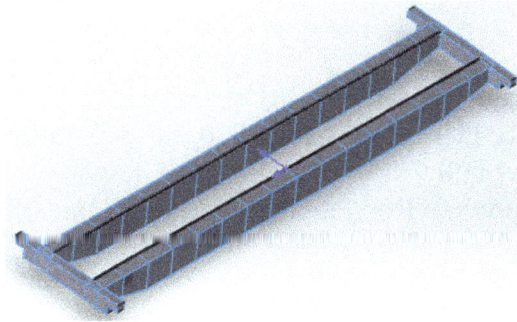

Fig. 4 Double girder overhead crane model

The program runs by determining the design variables of the model. Adding the necessary sizing constraints is shown in Fig. 5.

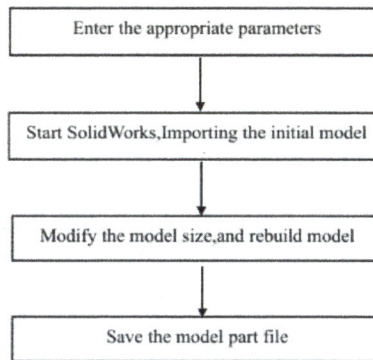

Fig. 5 Program flow

When using C# to call for secondary development of SolidWorks 2012, two reference namespaces are added to the new project, which are SolidWorks.Interop.sldworks and SolidWorks.interop [8]. Record the function by macro recording feature of SolidWorks. It is easy to know objects, properties and methods used by functions in SolidWorks. Saving the macro command after recording as SW VSTA C# Macro (*.csproj) format, forming the new function in CSldWorks.cs library through modifying the redundancy function code. A part code of CSldWorks.cs is shown as follows:

```
public void NewPart()
{
    swDoc = swApp.NewPart();//Create a new part
    swDoc.Visible = true;//Form visible
}
public void ModifyDimension1(string PartName, string SketchName, string DimenName, double
Value)//Modify the sketch dimension
{
    ModelDoc2 swDoc = null;
    bool boolstatus = false;
    swDoc = ((ModelDoc2)(swApp.ActiveDoc));
    boolstatus = swDoc.Extension.SelectByID2(SketchName, "SKETCH", 0, 0, 0, false, 0, null, 0);
```

```
        swDoc.EditSketch();
        swDoc.ClearSelection2(true);
        boolstatus = swDoc.Extension.SelectByID2 (DimenName + "@" + SketchName+ "@" +
        PartName + ".SLDPRT", "DIMENSION", 0,  0, 0, false, 0, null, 0);
        Dimension myDimension = null;
        myDimension = ((Dimension)(swDoc.Parameter(DimenName + "@" +  SketchName)));
        myDimension.SystemValue = Value / 1000;
        swDoc.ClearSelection2(true);
        swDoc.SketchManager.InsertSketch(true);
}
```

......

3.3. *SolidWorks simulation parameters of the call*

SolidWorks Simulation plug-in is added through the program calls. Adding quote named SolidWorks.Interop.cosworks into project built in Visual Studio 2012 platform [9]. The quote contains correlation function of SolidWorks simulation API. A part of code is shown as follows:

```
    swpath = swApp.GetExecutablePath();
    swApp.LoadAddIn(swpath+"\\simulation\\cosworks.dll");
```

SolidWorks Simulation analysis key steps in secondary development are shown in Fig.6:

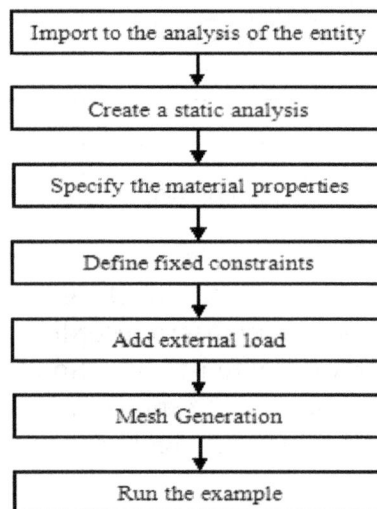

Fig. 6 SolidWorks simulation operating procedure

1. Part = ((ModelDoc2)(swApp.OpenDoc6(Path + @"\parts library\" + "overhead crane.SLDPRT", 1, 0, "", ref longstatus, ref longwarnings)));//Call the established double-girder overhead crane parts finite element analysis model.
2. CWObject = (CwAddincallback)swApp.GetAddInObject("SldWorks.Simulation");//Call the SolidWorks simulation objects,get SolidWorks simulation plug-in SolidWorks software.
3. ActDoc = (CWModelDoc)COSMOSWORKS.ActiveDoc;//Get documentation for the project.
4. StudyMngr = (CWStudyManager)ActDoc.StudyManager;//Create a new static calculation example.

5. SolidMgr = (CWSolidManager)Study.SolidManager;//Specify the material properties,set the material properties in the SolidWorks material library.
   Get entity face of information to determine the force constraining surface and the bearing surface required pidcollector.exe, can be found in the installation directory (SolidWorks\api).
6. LBCMgr=(CWLoadsAndRestraintsManager)Study.LoadsAndRestraintsManager;//Add the fixed constraint
7. CWForce = (CWForce)LBCMgr.AddForce(1, (varArray2), null, out errCode);//Put pressure on the select the face
8. CwMesh = (CWMesh)Study.Mesh;//Mesh Generation
9. errCode = Study.RunAnalysis();//Run the example

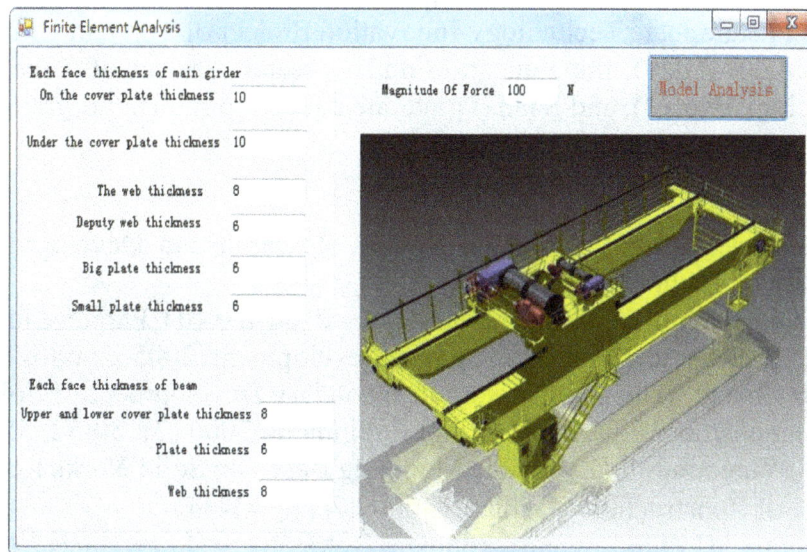

Fig. 7 Program interface

Generated window interface is shown in Fig. 7, with the results is shown in Fig. 8.

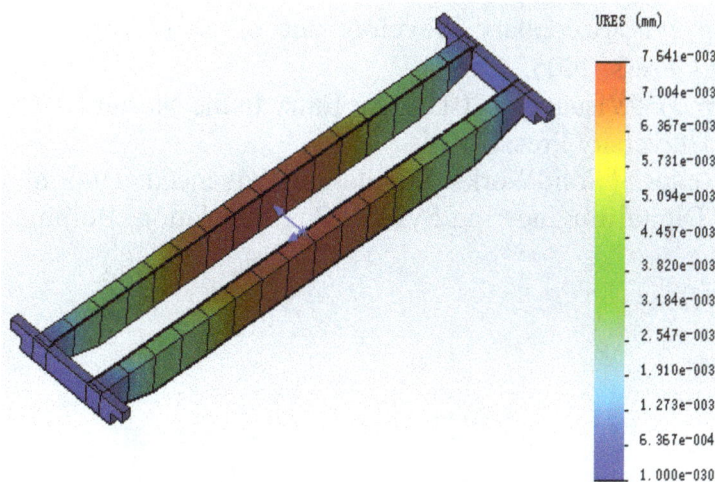

Fig. 8 SolidWorks simulation parameterization entity

847

4. Conclusion

In this paper, the CAD/CAE system of overhead crane structures is built based on the secondary development of SolidWorks and the platform of Visual Studio 2012. The design and engineering analysis of crane structures are integrated into a single software, which would greatly speed up the process of designing, improve the efficiency of product analysis shorten the design periods. In a word, this system can function as a helpful aid for engineering.

Acknowledgments

This paper is funded by the Postdoctoral Startup Foundation of Taiyuan University of Science and Technology (No.20142011), the Doctor Startup Foundation of Taiyuan University of Science (No.20122001), the Postgraduate Technology Innovation Project of Taiyuan University of Science and Technology (No.20145029), the Education and Research Project of Taiyuan University of Science and Technology (No.11), and Shanxi graduate education reform program (No.2015JG11).

References

1. Li Xin. "Steel Crane Parametric Design System Research and Development [D]." Dalian: Dalian University of Technology, 2008.
2. Wu ShuFang, Qin huiBin, Wang Zongyan. "Based on the 3D Parametric Crane Girder of SolidWorks [J]." Mechanical Management and Development, 2005, 2: 68-69.
3. Wang Xiaoli, Ji Zhong. "The Comparison of Secondary Development of SolidWorks Method [J]." Modern Manufacturing Technology and Equipment, 2006 (2): 50-52, 55.
4. Xu Gening, Zhu Yingdong. "3D Variable Modeling Based on SolidWorks Redevelopment [J] " Chinese Journal of Construction Machinery, 2009,7(1)：41-45
5. Zhao Liping, Qin Huibin, Wang Zongyan. "Mechanical Products Three-dimensional Parametric Design Variations Study and Application [J]." Mechanical Science and Technology (2008), 27 (10): 1154-1157.
6. Tao YuanFang. "Mechanical CAD Application Technology [M]." Beijing: Mechanical Industry Press, 2012, 5.
7. SolidWorks Company. "Secondary Development of SolidWorks API [M]" Beijing：Mechanical Industry Press, 2005.
8. (UK), Sharp (Sharp, J), "Visual C # 2012 from Entry to the Master [M]." Zhou Jing, translate. Beijing: Tsinghua University Press, 2014.
9. SolidWorks companies. "SolidWorks Simulation Advanced Tutorial: Edition 2011 [M]." Hangzhou Xin Di Digital Engineering System Co., translation. Beijing: Mechanical Industry Press, 2011, 4.

Crashworthiness Research of Sandwich Multi-cell Conical Tube Under Axial Impact

Xiao-Lin Deng

School of Mechanical & Material Engineer, Wuzhou University, Guangxi Wuzhou, China
Email: dengxiaolin3@163.com

Sandwich multi-cell conical tube was researched and a finite element model was build. Crashworthiness of sandwich multi-cell conical tube and ordinary conical tube, four-cell conical tube under axial impact were researched. The sandwich multi-cell conical tube deformation mode was analyzed. Energy absorption performance with different angles was studied. Specific energy absorption of sandwich multi-cell conical tube more than four cell conical tube increased by 35.79% and more than ordinary conical tube increased by 157%. Six different angles of sandwich cone tube with $\theta = 5°$, $\theta = 7.5°$, $\theta = 8.5°$, $\theta = 10°$, $\theta = 11°$ and $\theta = 12.5°$ were researched. The initial peak force increased with the increasing of cone angle became smaller. Energy absorption is the best when sandwich multi-cell conical tube with $\theta = 10°$. As angle increases or decreases, specific energy absorption declines.

Keywords: Sandwich multi-cell conical tube; axial impact; crashworthiness; energy absorbing.

1. Introduction

The energy absorption of material and structure under the impact is called crashworthiness. It is an important problem of safety studies in automobile. Experts and scholars conducted a lot of research about energy absorption properties of thin-walled structure, especially for round pipe, square tube, polygonal, and obtained many achievements. Tapered tubes relative to the round tubes, square tubes, and other thin-walled structure with better design, lower peak stress, and have better energy absorption ability under the oblique impact, so the research has the vital significance of tapered tubes. Theory and experiment of tapered tubes structure under the impact have carried on by Mamalis [1-3]. With the rapid development of finite element technology, the use of tapered tube technology of finite element simulation analysis is conducted, energy absorption characteristics of tapered thin-walled under the impact of got a lot of research, including axial impact [4-6], oblique impact [7-9]. In recent years, the related research of the tapered thin-walled structure mainly concentrated in the functionally graded tube and optimization design of thin-walled. Axial compression behavior has carried on for functional gradient tube and tapered tube by the finite element analysis and experimental verification by ZHANG [10]. The pipe is lower peak stress and higher energy absorption compared to common pipe. Li [11] contrast analysis of the function gradient tube and tapered tube in dynamic response under the oblique impact, through the analysis, found that the tapered tube has better oblique impact resistance compared to common pipe. Optimization is also the research focus in the thin-walled structure energy absorption in recent years. Song [12] with foam filled thin-walled structure as an example, build the surrogate model by response surface method, kriging model respectively and the radial basis function method, carry out optimized research with sequential quadratic programming method and the different methods such as particle swarm optimization algorithm. Hou [13] analyze a single tapered pipe, foam filling tapered and the crashworthiness of coaxial double tapered tube, and the response surface method was used to construct the surrogate model. Acar [14] analyze the crashworthiness of thin-walled structure, and also through the use of different surrogate model building method to build the corresponding surrogate model, with specific energy absorption and the average load and peak strength as the optimization goal, for thin wall pipe have optimized.

Although a lot of related researches on tapered tube have to carry out, but the research mainly focused on the different impact form, wall thickness, function gradient change and optimization design. Here, on the basis of previous studies, this paper research a new kind of sandwich multi-cell conical tube, has better crashworthiness compared with the traditional tapered tube by the comparative analysis.

2. Finite Element Model

2.1. *Model description*

Using finite element software ABAQUS to simulation analysis. Model structure diagram as showed in Fig. 1. $L = 200mm$, $d_1 = 150mm$, $d_2 = 40mm$, $t = 0.5mm$, θ is the angle of the tapered tube, by changing the cone model can get a different angle. The model is primarily composed of the rigid body of impact, model and the rigid body of fixed end, the model with four nodes reduced integral shell element to simulate. In order to effectively save computer time, after the convergence test, the unit average length is 4mm, with five integration points along the thickness direction. The fixed end of a rigid body with binding model, friction coefficient of 0.2.In order to avoid the deformation of thin-walled structure penetrate each other, the entire model applying single contact algorithm automatically. Impact of a rigid body is with a constant speed $v = 10m/s$. Impact is on 80% of the structure of the original length, which is 160mm.

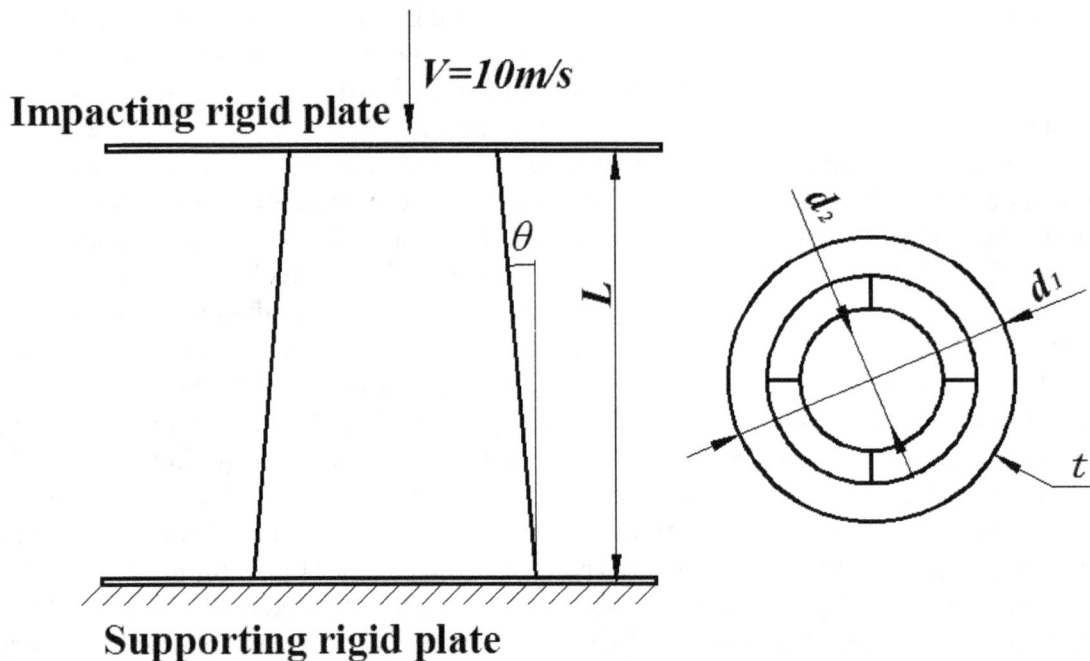

Fig. 1 Sandwich multi-cell conical tube model for the shock

AA6060T4 aluminum alloy material, material properties references [15-17], the density $\rho = 2700kg/m^3$, young's modulus $E = 68.2Gpa$ and Poisson's ratio $u = 0.3$, the initial yield stress $\sigma_y = 80Mpa$, ultimate stress $\sigma_u = 173Mpa$. Due to the aluminum alloy material is not sensitive to strain rate, here to ignore its strain rate effect.

2.2. Crashworthiness indicators

Crashworthiness indicators mainly include initial peak force, specific energy absorption, the average load and other indicators. Specific energy absorption is the most important indicator of crashworthiness, its expression as show in formula 1:

$$SEA = \frac{E_{total}}{m} = \frac{\int_0^{Se} Fds}{m} \qquad (1)$$

E_{total} refers to totally absorb energy of the structure;
m refers to the structure of the total mass;
Se refers to compression distance;
F refers to the structure load;
The average load formula F_m as shown in formula 2:

$$F_m = \frac{\int_0^{Se} Fds}{Se} \qquad (2)$$

3. Results and discussion

3.1. Comparative analysis

(a) Ordinary conical tube   (b) Four cell conical tube   (c) Sandwich multi-cell conical tube

Fig. 2 The deformation mode of ordinary conical tube, four cell conical tube and sandwich multi-cell conical tube

Fig. 2 is deformation of the ordinary taper tube, four-cell taper tube was proposed by literature [18] and sandwich multi-cell conical tube in shock to 80%. In Fig. 2, the ordinary conical tube has a very uniform progressive buckling mode, four cell taper pipe because of the existence of middle plate subjected to impact load ability is stronger, the diaphragm and conical wall contact part and the middle part of diaphragm, due to the mutual penetration, its deformation is more intense. Four cell conical tube and sandwich multi-cell conical tube have similar deformation mode. In the middle of the thin-walled rings also produced the axisymmetric deformation of the diamond model. Energy absorption capacity of four cell cone tubes is stronger in the middle of the cross diaphragm.

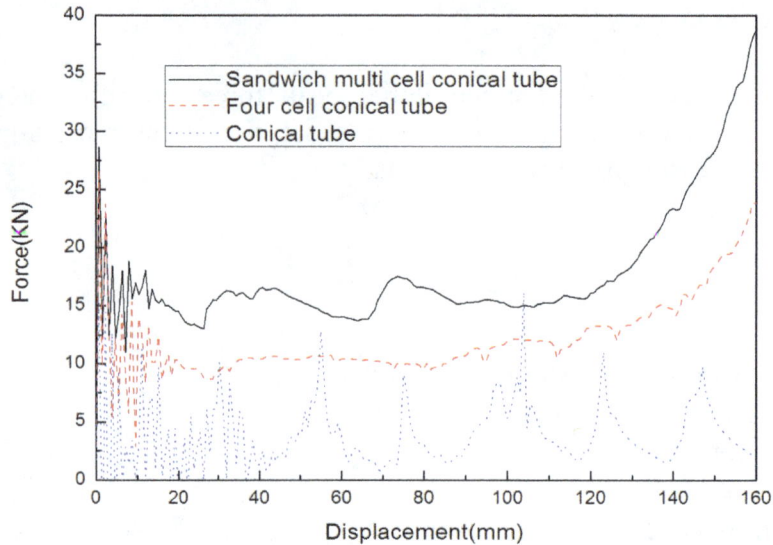

Fig. 3 Load displacement curve of different structure

Fig. 3 is displacement diagram of three different structures for the ordinary conical tube, four cell conical tube and sandwich multi-cell conical tube under impact load. Through analysis shows that different structures of three kinds of thin-walled structure have experienced three stages in the similar, namely the initial stage, the impact load of the average load fluctuations, and finally the rapidly rising densification stage. More than three different structures, sandwich multi-cell conical tube average impact load is largest, an ordinary conical tube is lowest.

Initial peak force is one of the important evaluation indicators of the crashworthiness, good energy absorption structure should be relatively constant reaction, avoid producing high initial peak stress. Three different structures of the initial peak stress are: $15.33KN$, $26.66KN$, $28.65KN$. Worth pointing out that sandwich multi-cell conical tube average impact load has improved significantly compared with the four cell cone tube, but the initial peak force relative to the four cell taper tube only increased by 7.5%, and illustrates the sandwich multi-cell conical tube in limiting initial peak force has a good advantage.

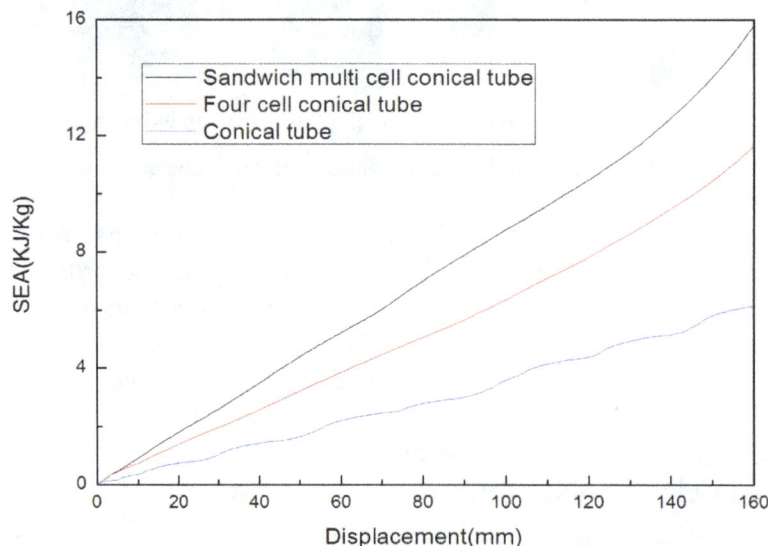

Fig. 4 SEA of different structure

Fig. 4 is specific energy absorption for the ordinary conical tube, four cell conical tube and sandwich multi-cell conical tube. Fig. 4 shows that the energy absorption of sandwich multi-cell conical tube has promoted compared with the ordinary conical tube and four-cell conical tube. Compressed to 80%, specific energy absorption of the ordinary conical tube, four cell conical tube and sandwich multi-cell conical tube respectively: $6.14 KJ/Kg$, $11.65 KJ/Kg$, $15.82 KJ/Kg$.

Specific energy absorption of sandwich multi-cell conical tube is relatively four cell conical tube increased by 35.79%. Relative the ordinary conical tube increased by 157%. Therefore, sandwich multi-cell conical tube relative the ordinary conical tube and four-cell conical tube has better energy absorption ability.

3.2. Deformation mode

Fig. 5 is deformation model of sandwich multi-cell conical tube. The figure shows that the collision at the first stage is mainly composed of impact end to the fixed end crushing in turn. With the increase of compression, the middle of the conical tube and the baffle plate by crushing, so that the deformation mode is more intense. The bionic tube presents a more stable fold contraction deformation mode.

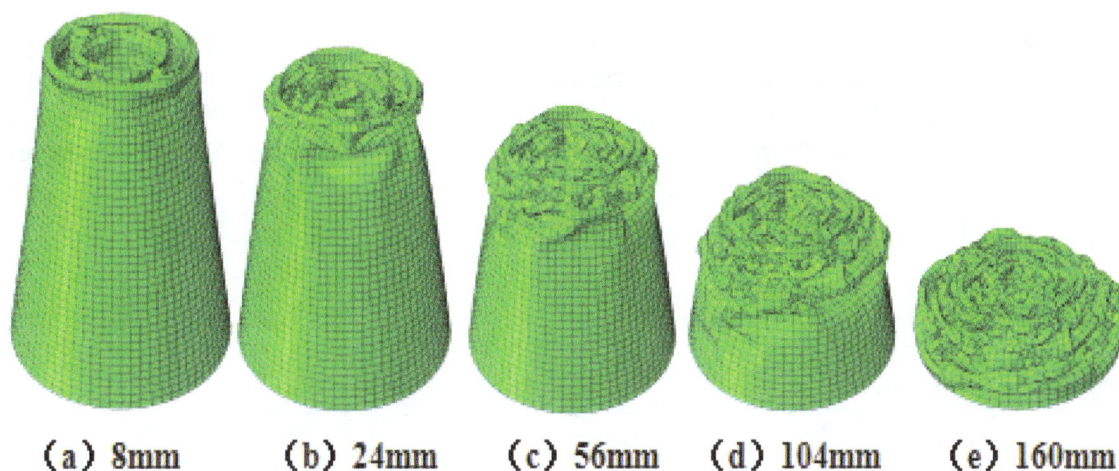

(a) 8mm   (b) 24mm   (c) 56mm   (d) 104mm   (e) 160mm

Fig. 5 Deformation mode of sandwich multi-cell conical tube

3.3. Effect of angle for crashworthiness

To analyze different cone angles of energy absorption performance of sandwich multi-cell conical tube, for $\theta = 5°$, $\theta = 7.5°$, $\theta = 8.5°$, $\theta = 10°$, $\theta = 11°$, $\theta = 12.5°$ a total of more than six different angles of sandwich multi-cell conical tube under axial impact crashworthiness has carried on. Fig. 6 shows that different angle of the load displacement curve. Different angles of the sandwich multi-cell conical tube showed the similar load displacement curve. It is mainly because the cone angle change is small, so the tube deformation mode is basically identical.

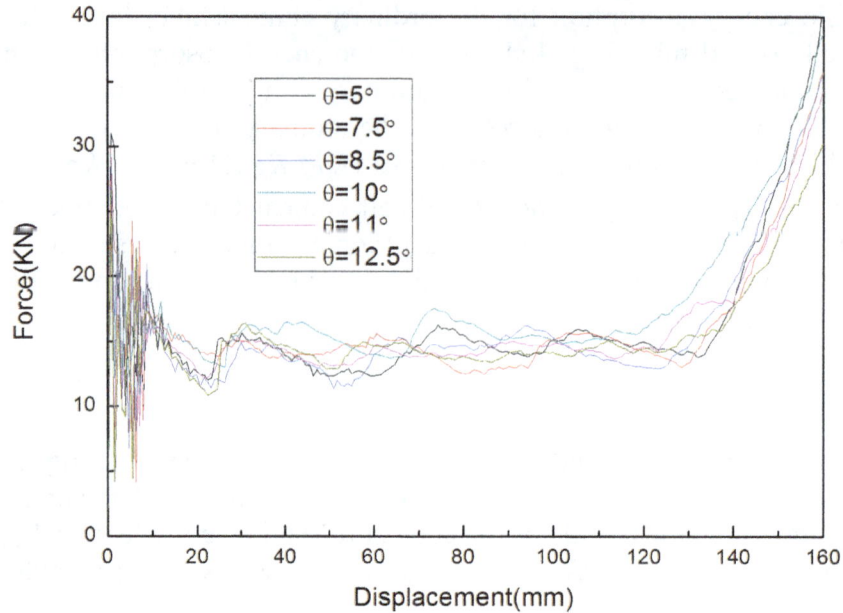

Fig. 6 Load displacement curve of different angles

Through the different angle of sandwich multi-cell conical tube is shown in Fig. 7 multi-cell conical tube average peak load and tried to clear the average load and the change of the peak force characteristics. Figure 7 shows the force displacement curve of sandwich multi-cell conical tube of different angle. Among them, the peak force along with the increase of the Angle became smaller, and the average load did not appear drab increases or decreases, the average load of sandwich multi-cell conical tube for $\theta = 10°$ is the largest.

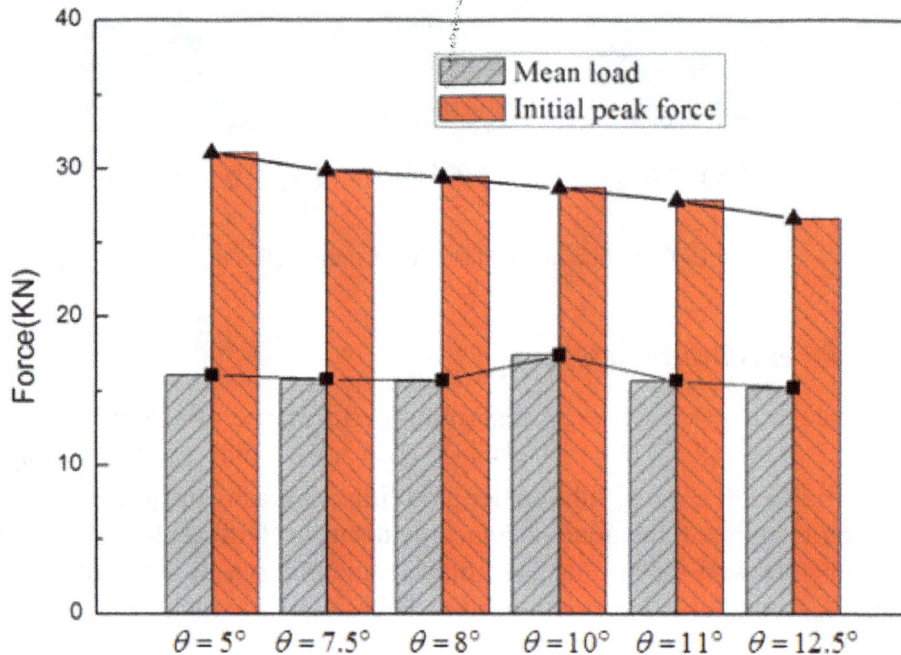

Fig. 7 F_m of different angles

The quality of sandwich multi-cell conical tube is different due to different angles, in order to more effectively compares and analyzes, the energy absorption diagram that different angles of sandwich multi-cell conical tube were shown Fig. 8.Through the analysis of the figure 8 shows that the energy absorption of sandwich multi-cell conical tube for $\theta = 10°$ is best, and with the angle of increasing or decreasing, energy capacity begin to decline, the energy-absorbing capacity of sandwich multi-cell conical tube for $\theta = 5°$ is the worst.

4. Conclusion

Energy absorption characteristics of sandwich multi-cell conical tube were studied using finite element numerical simulation method under axial impact. The research results show that the specific energy absorption of sandwich multi-cell conical tube relative to ordinary conical tube and four-cell conical tube has greatly improved. Six different angles of sandwich cone tube with $\theta = 5°$, $\theta = 7.5°$, $\theta = 8.5°$, $\theta = 10°$, $\theta = 11°$ and $\theta = 12.5°$ were researched. The initial peak force increased with the increasing of cone angle became smaller. Energy absorption is of the best when sandwich multi-cell conical tube with $\theta = 10°$. As angle increases or decreases, specific energy absorption decline. The results can provide reference for the optimization design of sandwich multi-cell conical tube.

Acknowledgments

This research was funded by the Guangxi Natural Science Foundation (No.2014jjBA60066; No.2016JJA110045) and The project of improving the basic ability of young teachers in universities of Guangxi (No. KY2016YB437).

References

1. Mamalis A G, Johnson W. The quasi-static crumpling of thin-walled circular cylinders and frusta under axial compression J]. International Journal of Mechanical Sciences, 1983, 25(9-10): 713-732.
2. Mamalis A G, Johnson W, Viegelahn G L. The crumpling of steel thin-walled tubes and frusta under axial compression at elevated strain-rates: Some experimental results [J]. International Journal of Mechanical Sciences, 1984, 26(84): 537–547.
3. Mamalis A G, Manolakos D E, Saigal S, et al. Extensible plastic collapse of thin-wall frusta as energy absorbers [J]. International Journal of Mechanical Sciences, 1986, 28(4): 219-229.
4. Nagel G M, Thambiratnam D P. A numerical study on the impact response and energy absorption of tapered thin-walled tubes[J]. International Journal of Mechanical Sciences, 2004, 46(2): 201–216.
5. Nagel G M, Thambiratnam D P. Computer simulation and energy absorption of tapered thin-walled rectangular tubes [J]. Thin-Walled Structures, 2005, 43(8): 1225–1242.
6. A G Mamalis, D E Manolakos, M B Ioannidis, et al. Numerical simulation of thin-walled metallic circular frusta subjected to axial loading [J]. International Journal of Crashworthiness, 2005, 10(5): 505-513.
7. Nagel G M, Thambiratnam D P. Dynamic simulation and energy absorption of tapered thin-walled tubes under oblique impact loading [J]. International Journal of Impact Engineering, 2006, 32(10): 1595-1620.

8. Nagel G. Impact and Energy Absorption of Straight and Tapered Rectangular Tubes [J]. Tapered Tubes, 2005.

9. Chang Q I, Dong F L, Yang S, et al. Energy-absorbing characteristics of a tapered multi-cell thin-walled tube under oblique impact [J]. Journal of Vibration & Shock, 2012.

10. Zhang X, Zhang H, Wen Z. Axial crushing of tapered circular tubes with graded thickness [J]. International Journal of Mechanical Sciences, 2015, 92:12-23.

11. Li G, Xu F, Sun G, et al. A comparative study on thin-walled structures with functionally graded thickness (FGT) and tapered tubes withstanding oblique impact loading [J]. International Journal of Impact Engineering, 2015, 77: 68–83.

12. Song X, Sun G, Li G, et al. Crashworthiness optimization of foam-filled tapered thin-walled structure using multiple surrogate models[J]. Structural & Multidisciplinary Optimization, 2013, 47(2): 221-231.

13. Hou S, Han X, Sun G, et al. Multiobjective optimization for tapered circular tubes [J]. Thin-Walled Structures, 2011, 49(7): 855–863.

14. Acar E, Guler M A, Gerçeker B, et al. Multi-objective crashworthiness optimization of tapered thin-walled tubes with axisymmetric indentations [J]. Thin-Walled Structures, 2011, 49(1): 94–105.

15. Santosa S P, Wierzbicki T. Experimental and numerical studies of foam-filled sections [J]. International Journal of Impact Engineering, 2000, 24(5): 509–534.

16. Acar E, Guler M A, Gerçeker B, et al. Multi-objective crashworthiness optimization of tapered thin-walled tubes with axisymmetric indentations [J]. Thin-Walled Structures, 2011, 49(1): 94–105.

17. Yin H, Wen G, Liu Z, et al. Crashworthiness optimization design for foam-filled multi-cell thin-walled structures [J]. Thin-Walled Structures, 2014, 75(2): 8–17.

18. Zou Meng, WEI Cangang, Xu Shucai, et al. Energy absorption characteristics simulation of cone thin-walled metal tube research in the automotive collision [J]. J Automotive Safety and Energy, 2012, (4): 326-331.

Optimization of Amplitude Hinge-point Based on Ideal Point Method

Jian-Feng Wang, Le-Feng Wang[*], Bing Zhang

[1]School of Mechanical Engineering, Xijing University, Xi'an, China
[*]Email: 769140379@qq.com

Taking the minimum stress of the dangerous section of the variable amplitude cylinder and the telescopic arm was used as the optimization goal of the high altitude working platform. On this basis, the mathematical model was established by the ideal point method. MATLAB optimization toolbox was used to seek the optimal solution of the problem.

Keywords: Amplitude hinge-point; the ideal point method; optimization algorithm.

1. Introduction

Structural parameters of the variable amplitude mechanism of aerial work platform, that is three hinge points made under the root hinge point of telescopic boom and the upper and lower hinge points of derricking cylinder, directing impact on the jib lubbing form, the structure of the compact and the stability of the system. In this paper, the MATLAB optimization toolbox is used to optimize the multi-objective optimization of the verifying nodes, and the optimal solution is processed by the ideal point method.

2. Ideal Point Method and MATLAB Toolbox

The problems encountered in practical engineering are very few of the single objective optimization problems with several fixed constraints [1]. Many times we need to pursue multiple goals $F(x) = \{F_1(x), F_2(x), \cdots, F_n(x)\}$ at the same time, is called multi-objective optimization.

The most basic method to solve the multi-objective optimization is to evaluate the function method. With the aid of geometric background, the multi-objective optimization problem is transformed into a single objective optimization problem, then the optimal value is obtained by the single objective optimization method, and it is considered as the optimal solution of the multi-objective optimization problem. The common methods of constructing evaluation function: Ideal point method, Linear weighted method, Minimax method. This article focuses on the introduction of the ideal point method.

2.1. *Ideal point method*

Firstly, the optimal solution of the single objective problem is obtained $\min f_i, (i = 1, 2, \cdots, n)$, suppose its value is $f_i^*(X)$, called $f^* = (f_1^*, f_2^*, \cdots f_n^*)$ as an ideal point in its range. Because it is difficult to achieve, it is expected that in a certain measure, to find f the nearest distance f^* as the approximate value. Construct the evaluation function $\phi(x) = \sqrt{\sum_{i=1}^{n}(z - f_i^*)^2}$ and minimum that is $\min \phi(f(X)) = \min \sqrt{\sum_{i=1}^{n}(f_i(X) - f_i^*)^2}$. The solution is taken as the optimal solution in this sense.

[*]Corresponding author

2.2. *MATLAB optimization toolbox*

Optimization toolbox in MATLAB contains a series of optimization algorithm function. These functions expand the processing capacity of MATLAB mathematical computing environment, solving many practical engineering problems. The steps are as follows:

(1) According to the optimization problem, the mathematical model of optimization is established;
(2) Specific analysis and research on the establishment of the model, selecting the appropriate optimization method;
(3) Find the answer.

3. Set up the Mathematical Model of the Jib Lubbing Mechanism

3.1. *Determine design variables*

Three hinge point positions of the jib lubbing mechanism, that is the geometric relationship between the A of the lower hinge point of the variable amplitude cylinder, the upper hinge point B and the point C of the movable arm is determined. Three variables are introduced in this paper: Position of the Lower hinge point A (X_1, X_2), Piston stroke of oil cylinder S (X_3). Express as $X = \{X_1, X_2, \cdots, X_3\}^T$, Δ is a non-working stroke hydraulic cylinder, h is known quantities. All parameters are shown in Fig. 1.

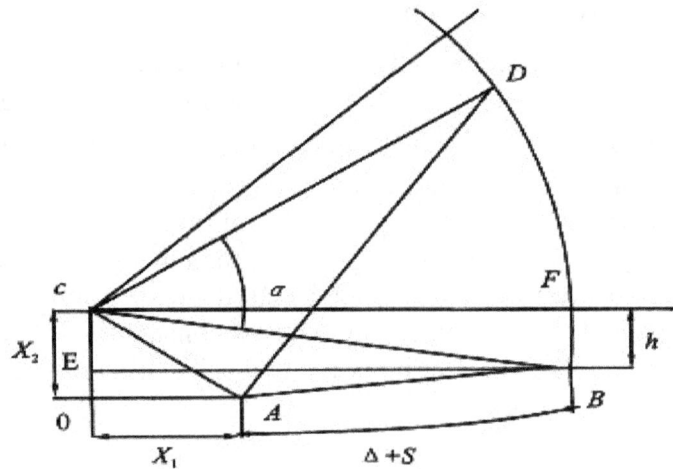

Fig. 1 Simplified graph of amplitude hinge-point

3.2. *The establishment of objective function*

At the total design stage, the working range $(0° \sim 86°)$ and working conditions of aerial working platform are established [2], as shown in Fig. 2. This article only considers: In the case of the lifting torque, the force of the variable amplitude cylinder is the least, The minimum stress of the dangerous section of the telescopic arm.

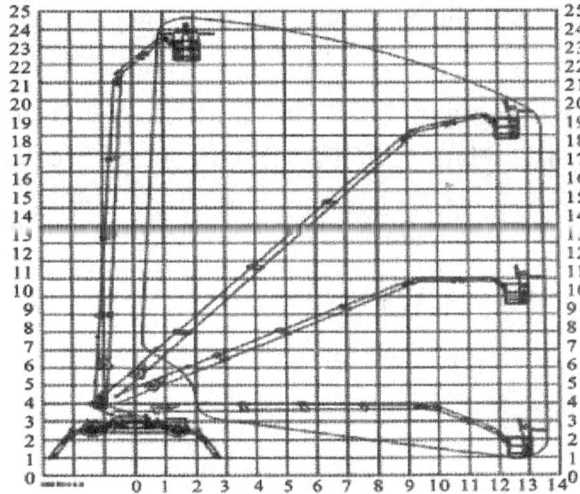

Fig. 2 Schematic graph of working condition of aerial work platform

3.2.1. *The force of the variable amplitude cylinder is the least*

The force analysis of variable amplitude cylinder is shown in Fig. 3.

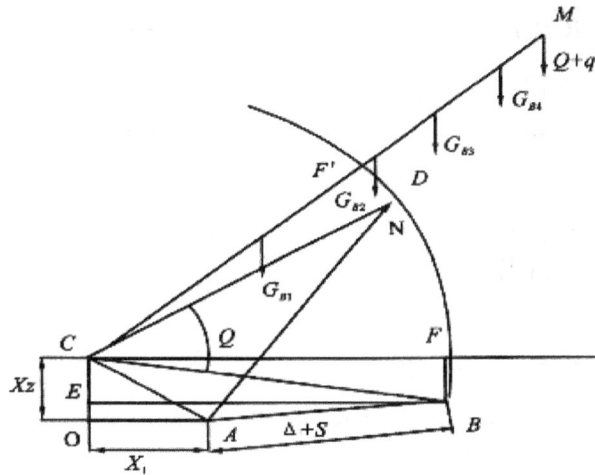

Fig. 3 Simplified stress graph

According to the principle of the balance torque, the torque of the C point is zero. That is $\sum M_c = 0$.

$$N = \left[\varphi_1\left(l_{B1} \times G_{B1} + l_{B2} \times G_{B2} + l_{B3} \times G_{B3} + l_{B4} \times G_{B4}\right) + \varphi_2\left(Q+q\right) \times l_B\right] \times \cos\alpha + M/l \qquad (1)$$

$$M = \frac{1}{2} \times Q \times l_{B5} + q \times l_{B5} \quad l = \frac{CA \times CD \times \sin\left(\alpha + \angle ACB\right)}{AD} \qquad (2)$$

In the formula: N ------the force of variable amplitude cylinder; M ------the torque arm and the handing basket; φ_1 ------lifting impact coefficient; φ_2 -----lifting force coefficient; l ------luffing cylinder arm; Q ------ the dead weight of the flying arm; α ------elevation angle of the power

859

arm; q ------the handing basket and the maximum load; $G_{B1}, G_{B2}, G_{B3}, G_{B4}$ ------estimation value of the weight of each node; $l_{B1}, l_{B2}, l_{B3}, l_{B4}$ ------the distance between the center of gravity of each arm and the hinge point of the moving arm. l_B ------ at work ,the length of the telescopic boom are becoming longer and longer. The relationship between the geometric parameters:

$$AC^2 = X_1^2 + X_2^2 \tag{3}$$

$$CB^2 = CD^2 = \left(X_1 + \sqrt{(\Delta + S)^2 - (X_2 - h)^2} \right)^2 + h^2 \tag{4}$$

$$AD^2 = AC^2 + CD^2 - 2 \times AC \times CD \times (\cos \angle ACB + \alpha) \tag{5}$$

$$\cos(\alpha + \angle ACB) = \frac{X_2 h + X_1 \left(X_1 + \sqrt{(\Delta + S)^2 - (X_2 - h)^2} \right)^2}{CA \times CB} \cos \alpha - $$
$$\frac{X_2 \left(X_1 + \sqrt{(\Delta + S)^2 - (X_2 - h)^2} \right) - X_1 h}{CA \times CB} \sin \alpha \tag{6}$$

$$\sin(\alpha + \angle ACB) = \frac{X_2 h + X_1 \left(X_1 + \sqrt{(\Delta + S)^2 - (X_2 - h)^2} \right)^2}{CA \times CB} \sin \alpha + $$
$$\frac{X_2 \left(X_1 + \sqrt{(\Delta + S)^2 - (X_2 - h)^2} \right) - X_1 h}{CA \times CB} \cos \alpha \tag{7}$$

As seen from the above formula, the force can be expressed by the design variable, that is $N = N(X)$. The objective function: $f_1(X) = N(X)$

3.2.2. Minimum bending moment of dangerous section of boom

The dangerous section of the telescopic arm is in the upper hinge point of the oil cylinder. According to the analysis：

$$\sum M_D = (Q + q) \times \left(L_B - X_1 - \sqrt{(\Delta + S)^2 - (X_2 - h)^2} \right) \times \cos \alpha + $$
$$G_{B4} \times \left(L_{B4} - X_1 - \sqrt{(\Delta + S)^2 - (X_2 - h)^2} \right) \times \cos \alpha + $$
$$G_{B3} \times \left(L_{B3} - X_1 - \sqrt{(\Delta + S)^2 - (X_2 - h)^2} \right) \times \cos \alpha + \tag{8}$$
$$G_{B2} \times \left(L_{B2} - X_1 - \sqrt{(\Delta + S)^2 - (X_2 - h)^2} \right) \times \cos \alpha + $$
$$G_{B1} \times \left(L_{B1} - X_1 - \sqrt{(\Delta + S)^2 - (X_2 - h)^2} \right) \times \cos \alpha + M $$

also have: $M(X) = \sum M_D$, objective function: $f_2(X) = M(X)$

3.3. *Establishment of constraint conditions*

Determine the scope of the variable according to the requirements of the actual problem:

$$X_{min} \le Xi \le X_{max} (i = 1,2) \tag{9}$$

$$S_{min} \le S \le S_{max} \tag{10}$$

Constraints between ΔCAB and ΔCAD:

$$CA + CB - AB > 0; CA + AB - CB > 0;$$
$$CB + AB - CA > 0;$$
$$CA + CD - AD > 0; CA + AD - CD > 0; \tag{11}$$
$$CD + AD - AC > 0_{\circ}$$

Meet the maximum lifting angle conditions, that is:

$$AD^2 = AC^2 + CD^2 - 2 \times AC \times CD \times \cos(\angle ACB + 86°) \tag{12}$$

3.4. *Optimization of variable amplitude hinge point*

In this paper, we used the ideal point method to transform the multi-objective processing into a single objective, using constrained nonlinear function in MATLAB Optimization toolbox Variable selection range $0 < X_1 < 600, 0 < X_1 < 600, 100 < S < 1100$. This paper only considers the dangerous conditions, that is, the $86°$ elevation of the working condition [3]. When applying the ideal point method, the objective function is simplified: $\min \varphi(f(X)) = \min \sum_{i=1}^{n} \left(f_i(X) - f_i^*\right)^2 (i = 1,2)$

f_i^* indicates that the optimal value of the objective function is within the same design variable range under the same constraints. The objective function after 102 iterations, the change has been a straight line, to achieve convergence [4]. The optimization results are verified by the geometric method, which meets the design requirements. At this time, the design variables and the objective function values are shown in Table 1.

Table 1 Design variables and objective function values

f	X_1	X_2	S
$4.5079e - 005$	519.136	434.356	922.346

According to the corresponding relationship between design variables and objective function in Table 1, the objective function is iterated. After numerous iterations, the graph of the objective function is almost close to a straight line, and the effect of convergence is reached.

4. Conclusion

Using the MATLAB optimization toolbox and the optimization algorithm based on the ideal point method, the optimization calculation of the three hinge points of a high altitude working platform is carried out. In practical engineering, there are more objectives and constraints to be considered, MATLAB optimization toolbox can make the problem more focused on how to establish a mathematical model to reflect the actual situation, not too much to consider the implementation of the specific algorithm. This paper presents a method for solving multi-objective optimization in order to solve similar problems as reference.

Acknowledgments

I would like to extend my sincere gratitude to the Scientific Research Fund of Shaanxi Provincial Education Department (16JK2244, 14JK2160) and the collage fund XJ150111 of XiJing University for funding this work.

References

1. Jianwen Wang, Jihua Xin. Mechanism design of crane bat [J], 2007 (2).
2. Haibo Kang, Zhongpeng Zhang. Design and Research of three hinge point optimization based on multi-objective hybrid genetic algorithm [J]. (2006) (4).
3. Fei Si technology product development center. MATLAB6.5 auxiliary optimization calculation and design [M]. Electronic Industry Publishing House 2003 (5).
4. Zhenghua Zhang. Optimization design of engineering machinery for three hinge point of telescopic boom type truck crane [J] .1989 (5).

Dynamic Simulation Analysis of Crank Connecting Rod Mechanism of Compressor

Yi-Jun Zhou[*], Xiao-Dong Zheng, Lei Zhang

School of Mechanical Engineering, Anhui University of Science and Technology, HuaiNan, 232001, PR China
[*]Email: zhy31130@163.com

ADAMS software was used to analyze the kinematics of the crank link mechanism of the compressor. The ADMAMS/Durability module of ADAMS was used to analyze the durability of the crank link mechanism of the compressor. Through the analysis of the durability of the crank connecting rod mechanism, the service life of the mechanism was predicted, which has a good reference value for the improvement of the mechanism.

Keywords: Crank connecting rod mechanism of compressor; kinematics simulation; ADMAMS/durability; durability.

1. Introduction

The use of the compressor is very wide in reality. Its working principle is mainly that the rapid rotation of the crankshaft to drive the piston reciprocating motion, and to complete the suction, compression, expansion, exhaust four processes with the intake valve and exhaust valve. At last, the low-pressure gas is changed into the high pressure gas to be transported out. The crankshaft is one of the most important parts of the piston compressor. However, it is also very easy to be damaged in the long run under the high speed. This determines the life of the compressor [1]. Therefore, the three-dimensional model of the compressor was established based on the theoretical analysis of the force of the crankshaft. The kinematics and durability of the crank link mechanism of the piston compressor were analysised using ADAMS software. The service life of the mechanism was predicted, and the theoretical basis was provided for the improvement design of the piston compressor.

2. Modeling and Simulation Analysis of Crank Connecting Rod Mechanism

ZW-0.8/10-16 liquefied petroleum gas piston compressor crankshaft was studied. We set up parametric solid modeling and analyzed dynamics through CAD/CAE software. Crank has been checked under normal operating conditions and the reliability of the safety factor calculation. Testing the reliability of the work of the crankshaft, the result of research carried out scientific analysis and got the expected results. Some structures, such as convex sets and oil holes, which had little effect on the results, were neglected in modeling. Otherwise, it would increase the workload and also lead to the increase of the cumulative error. The accuracy of the calculation would be reduced due to the more and more round and convex platform, as well as the existence of oil holes and other structures [2]. The assembly drawing of the crank connecting rod mechanism of the compressor is shown in Fig. 1. The compressor's working conditions are very complex due to the compressor in the work, friction, gas and other media of thermodynamic changes, the resistance of the lubricating oil, mechanical vibration, and the existence of factors. The influence of these factors on the static and dynamic characteristics of the crankshaft is not very large. So, in the study of dynamic simulation of crank link mechanism, the following assumptions are made [3]:

(1) The motion of the gas was regarded as an ideal cycle during the compression of the compressor;

(2) The influence of air leakage and deflection on the force of the piston was negated during the working process of the compressor;

(3) The lubrication effect of the crank link mechanism was good, the frictional resistance of the mechanism was very small, and so it could be ignored;

(4) The crank connecting rod mechanism was a rigid structure model, which could be ignored the influence of vibration on it.

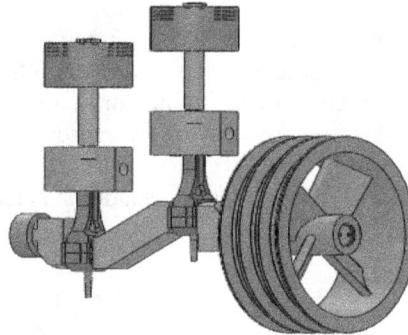

Fig. 1 Assembly drawing of crank linkage mechanism

The model which was built in Fig. 1, it was opened with ADAMS and displayed in the form of entity so that the ADAMS can automatically obtain some parameters and characteristics of the model [4]. The parameters of the model were set up, and then we could establish the structure model of the crank link mechanism in ADAMS, and added the constraint pair in the ADAMS according to the real situation of the crank link mechanism. After the constraint, the kinematics model of the crank link mechanism was obtained. Finally, in the ADAMS, an external drive to the crankshaft set the motor speed 550r/min, time was 1s, and the simulation step was 500.

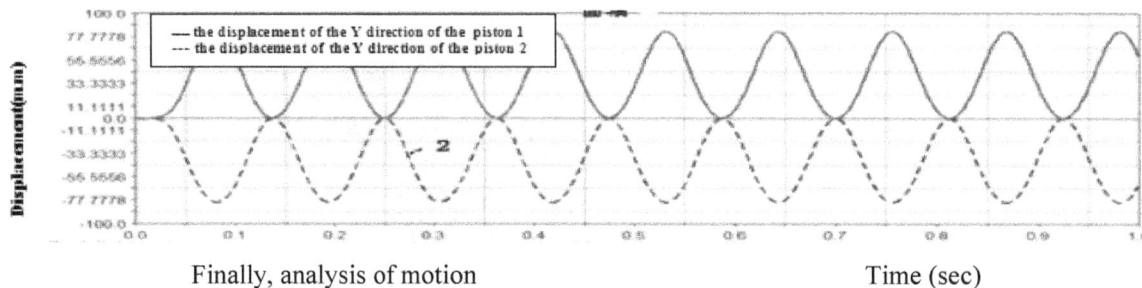

Finally, analysis of motion   Time (sec)

Fig. 2 Displacement time curve of piston Y

The piston was regarded as a first stage piston at one side of the flywheel, which was the zero point of the lower dead point, and the reference point for the Y direction. We could get the displacement time curve, velocity time curve and acceleration time curve of the first stage piston and the two-stage piston after the simulation was shown in Figs. 2, 3 and 4. It could be concluded when the first piston was in the upper and lower dead point position, the two pistons was in the lower dead point and the top dead point position, and the displacement of the two pistons was opposite. When the piston TDC down the stop motion, velocity increased gradually, arrived at the maximum value, and began to gradually decrease to zero; after speed began in the opposite direction from zero gradually was increased to a maximum and finally reduces to zero, to complete a work cycle.

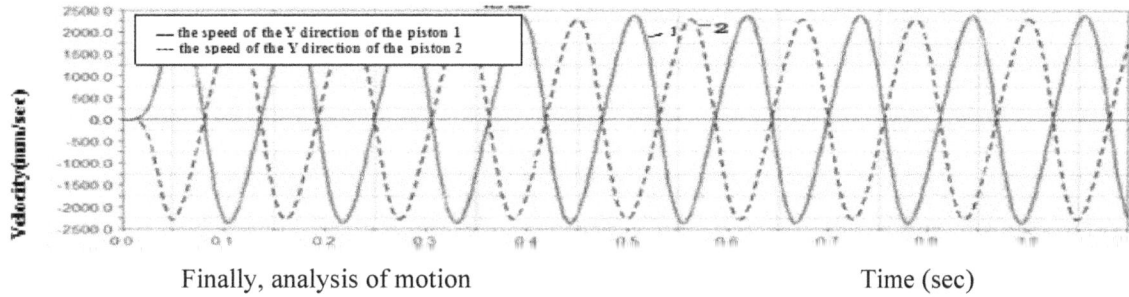

Finally, analysis of motion   Time (sec)

Fig. 3 Velocity time curve of piston Y

It could be seen from Fig. 4. There was a time difference between the maximum and minimum values of the acceleration of the first stage and the two stage piston. And the reason why was this phenomenon, mainly because of the effect of gas forces on the piston was different in the work process.

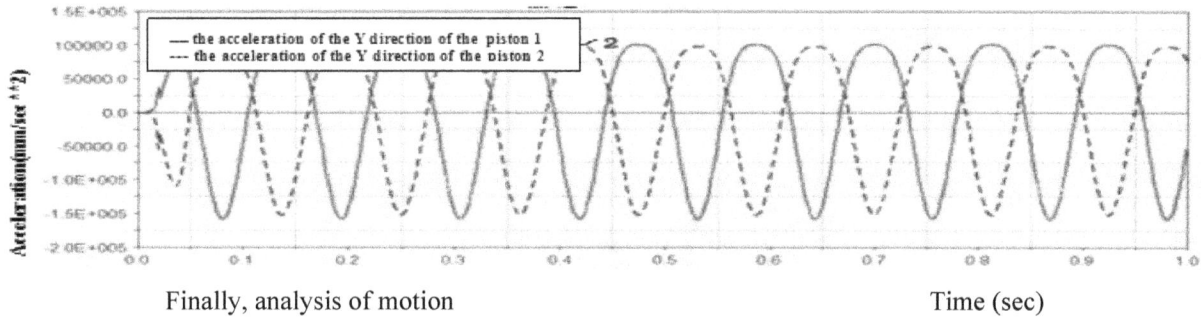

Finally, analysis of motion   Time (sec)

Fig. 4 Acceleration time curve of piston Y

3. Durability Analysis of Crank Link Mechanism

3.1. *Simulation and data processing*

Durability analysis purpose was to service life of the product to make accurate assessment and predict the early wear failure of parts, find out the reasons, so that we could find out the problems existing in the design, to provided a theoretical basis for the optimization design of the products, and make the products in the market competition more advantage [5]. The ADAMS/Durability module in ADAMS provided a way to analyze the durability of the mechanism. After the simulation has been finished, the data was entered into the post processing stage. The acceleration curve of the crankshaft could be obtained after the data was processed, and the Fourier transform and the FFT3D transform of the curve were carried out.

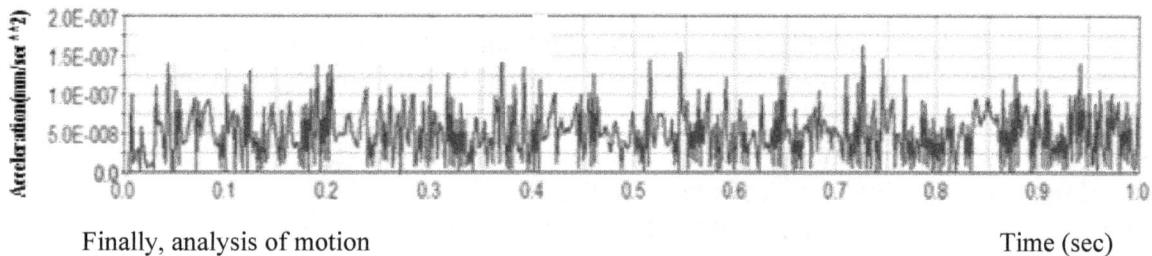

Finally, analysis of motion   Time (sec)

865

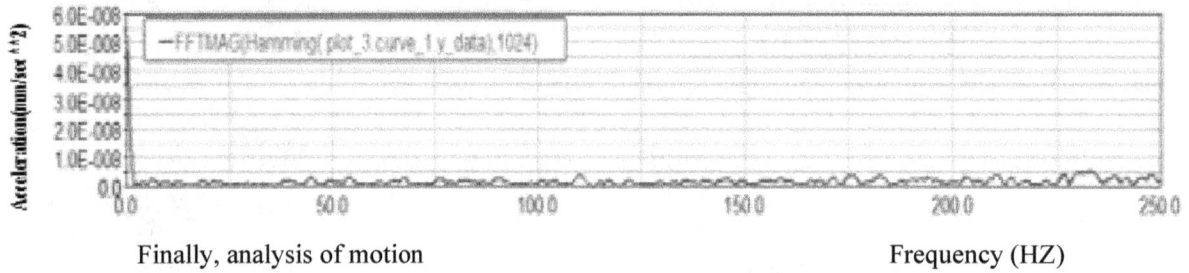

Finally, analysis of motion   Frequency (HZ)

Fig. 5 FFT curve

The corresponding FFT curve and FFT3D curve could be obtained as shown in Figures 5 and 6. This can be intuitive to show the trend of changes in the curve.

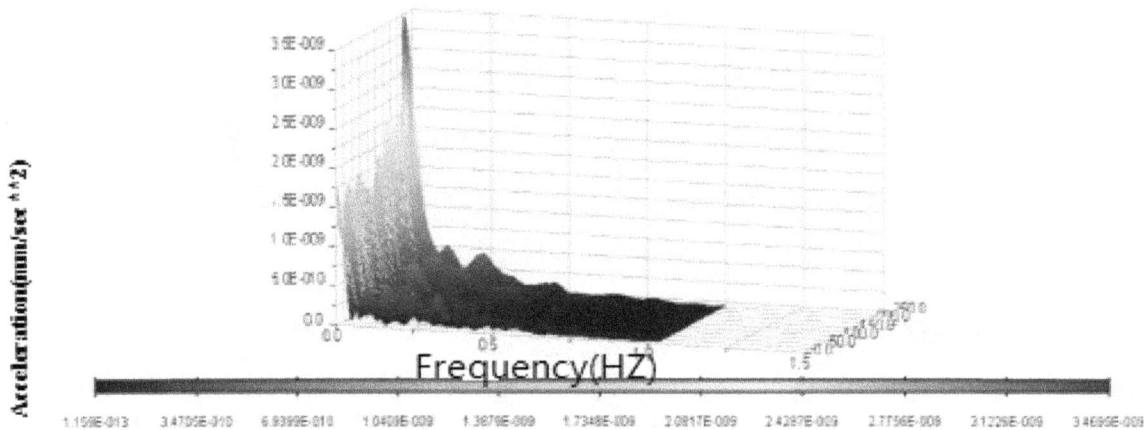

Fig. 6 FFT 3D curve

3.2. *Results analysis*

The relationship between the acceleration, the vibration frequency and the time of the mechanism has been obtained, through the analysis of the durability of the crank connecting rod mechanism. We could know that the vibration frequency of the crank connecting rod mechanism would increase with time, and the damage degree of the compressor would be increased through the analysis. Through the analysis of Fig. 5 we can predict when the compressor cumulative working hours were reached to 34600 days, the vibration will reach the maximum, at this time, the compressor can hardly work.

4. Simulation Results and Analysis

The fluid properties were definite water set the parameter in Fluent, and entrance velocity was 8m/s, the results are convergent. We got the static internal flow charts of different number of blades as shown in Figure 3. It can be seen from Figure 3; the static pressure value of impeller was gradually increasing from inlet to outlet and reached the maximum in the end of impeller blade working surface close to the rear. With the increasing of the number of impeller static appeared obvious changes, it is at least the internal flow static pressure of 12 blades impeller. The fluid is steady in middle layer and pressure change is uniform. The velocity distribution of flow field of centrifugal pump impeller was shown in Figure 4. It can be seen in figure, the velocity distribution was same as static pressure, and the velocity value of impeller was gradually increasing from inlet to outlet and reached the maximum in the end of impeller blade working surface close to the rear. With the increase of blade number the velocity distribution was more stable and uniform, high-speed area gradually reduced.

5. Conclusion

Firstly, it was introduced the specific function and the use method of the dynamic simulation software ADAMS, and the appropriate modules were selected for simulation and analysis. Secondly, the solid model was introduced into ADAMS, and the displacement, velocity, acceleration and time curve of the piston were obtained by kinematics simulation. Through the analysis, the motion law of the piston was obtained. Finally, the ADAMS/Durability module was used to analyze the durability of the crank link mechanism, and the simulation analysis was made to predict the life cycle of the mechanism, it would be 34600 days. And after 27756 days of work, the vibration would be gradually increased. Therefore, prior to this, in order to prevent damage to other parts of the compressor, the crankshaft must be maintained or replaced the crankshaft.

References

1. HE Bin-hui. System analysis and reference of compressor: Design [J]. Coal technology: 27 (04) 2012, 31 ~ 28.
2. WANG De-hai, Yin Jian-min, Yuan Yin-nan, et al. Three dimensional finite element method in the study of crankshaft strength of multi cylinder engine [J]. Journal of Jiangsu University of Science & Engineering, 1997, 18 (1): 7 ~ 11.
3. ZHANG HONG-jun. Dynamic characteristics of large scale industrial compressor crankshaft [D]. Nanjing: Nanjing University of Science and Technology, 2007.
4. ZHANG Guo-qing, HUANG Bo-chao, PU Geng-qiang, et al. Crankshaft fatigue life calculation based on dynamic simulation and finite element analysis [J]. Internal combustion engine engineering, 2006, 27 (1): 41 ~ 44.
5. WU Li-yan, Cao zhi, CHEN Yao-dong, et al. Analysis and case study on the durability of components [J]. Application engineering, 2010 (5): 73 ~ 74.

Fuzzy Self Learning Control of Glass Tempering and Annealing Temperature Based on Genetic Algorithm Optimal Approach

Xiao-Kan Wang[1,2,*], Qiong Wang[1]

[1]*Mechanical and Electrical Department, Henan Mechanical and Electrical Vocational College, Zhengzhou 450002, China*

[2]*School of Electronic and Information Engineering, Beijing Jiaotong University, Beijing 100044, China*
Email: wxkbbg@163.com

According to the problems of the time varying parameters and time lag characteristic in temperature control for glass tempering and annealing process, a kind of self-learning fuzzy controller based on improved genetic algorithm is put forward in this paper. Also, some of the strategies for improving genetic algorithm are stated. The improved algorithm can be used to fast search global optimal weighting factors. Thus the fuzzy control rules are perfected and corrected. The simulation results demonstrate that this kind of control method is suitable for systems with time varying parameters and time lag characteristics.

Keywords: Tempering and annealing; temperature; fuzzy control; self-learning; the optimized genetic algorithm.

1. Introduction

Many industrial objects have the characteristics of large time-delay and parameters time-varying, and it is difficult to establish mathematical models. The traditional Smith predictor and forecasting model can't obtain satisfactory control performance. Fuzzy control is a control technology that controlled object model does not rely on imitation of human thinking, which uses expert prior knowledge of certain field for approximate reasoning, and has achieved remarkable results in many fields [2]. In order to obtain good control results, the better perfect fuzzy control rules must be required for those time-varying parameters and large time-delay systems. However, controlled process's complexity will often cause fuzzy control rules rough or imperfect so that affect the control effect in varying degrees. To solve the problem, this paper presented a fuzzy self-learning controller based on improved genetic algorithm. Self-learning fuzzy controller design includes the following two steps: firstly, design two-dimensional fuzzy controller with many weighted factors; then use the genetic algorithm to the fuzzy controller's weighted factor to carry on the synthesis optimization based on the obtained online control information. Genetic algorithm can realize controller parameter's optimal process to achieve the control rules' learning and revision and improve controller performance [1, 3]. This makes the optimized parameters of the fuzzy controller can adapt time-varying and large time-delay control system.

2. Control Scheme Design

A Fuzzy control system based on genetic algorithm is showed in Figure 1. Adjustable fuzzy control rules are core components; the controller based on genetic algorithm has self-learning ability [4-6].
The main features of control system are as follows:

(1) Transformed accurate quantity to the corresponding fuzzy set universe by using the normalized fuzzy quantification method;
(2) The fuzzy control rule is easily described as many adjustment factor synthesis analytic expressions, so it is convenient for computer control and automatic adjustment the control rules;

(3) If the accurate mathematical model of controlled object is unknown, we could optimize many adjustment factors according to the obtained online input and output data by proposing the improved genetic.

By this way the self-learning and self-adaptive ability of the fuzzy controller was strengthened.

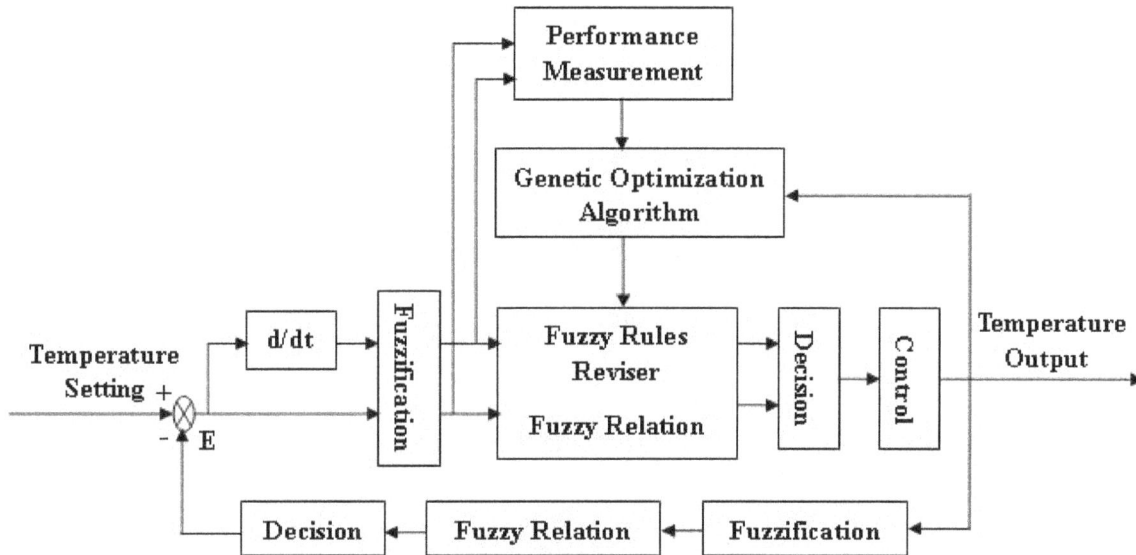

Fig. 1 Fuzzy control system based on genetic algorithm

The main features of control system are as follows:

(4) Transformed accurate quantity to the corresponding fuzzy set universe by using the normalized fuzzy quantification method;
(5) The fuzzy control rule is easily described as many adjustment factor synthesis analytic expressions, so it is convenient for computer control and automatic adjustment the control rules;
(6) If the accurate mathematical model of controlled object is unknown, we could optimize many adjustment factors according to the obtained online input and output data by proposing the improved genetic.

By this way the self-learning and self-adaptive ability of the fuzzy controller was strengthened.

3. Fuzzy Controller Design

3.1. *Fuzzy controller structure design*

This paper designed a two-dimensional fuzzy controller; its input error is e and error change rate is ec; the output is u; n is the sampling time.

$$e(n) = y(n) - r(n) \tag{3.1}$$

$$ec(n) = e(n) - e(n-1) \tag{3.2}$$

$$\Delta u = u(n) - u(n-1) \tag{3.3}$$

3.2. Inputs fuzzification

However, the quantification factor's choice has a great influence on control system's performance influence, between the interaction of *Ke* and *Kc*, so it is very difficult to choose a group that can suitable for control system's quantification factor.

Meanwhile it is very difficult to guarantee that the entire process of the controlled process is the optimizing control condition, if only has one kind of constant quantification factor which the corresponding long process with large inertial system; so that it reduces the fuzzy control system's robustness. Therefore some scholars realize the change quantification factors by using the array quantification factors, or carries on self-adjustment quantitative factor under system's different condition [3]. In this paper the normalized fuzzy method can avoid the choice of quantification factor; simultaneously it can also very conveniently transform the precise values of error and error change rate into the fuzzy universe [6-8].

The main step of normalized fuzzy method may divide into normalized processing and classify fuzzification. Supposed *e(n)* and *ec(n)* respectively is the *nT* time system's error and error change rate, and T is the sampling period. Through calculating *e/R* and *ec/R* may realize normalization for *e* and *ec*, and R is system's setting value. Dividing *e/R* and *ec/R* into certain ranks in [0, 1] the closed interval, that is classify fuzzification. We may obtain the fuzzy value of error *E* and the error change rate *EC* through classified fuzzification.

The universe of the error set *E* and the error change rate *EC* respectively is:

$$\{E\} = \{-3,-2,-1,0,1,2,3\} \tag{3.4}$$

$$\{EC\} = \{-2,-1,0,1,2\} \tag{3.5}$$

The corresponding language variables are respectively:

$$\{E\} = \{NB, NM, NS, ZE, PS, PM, PB\} \tag{3.6}$$

$$\{EC\} = \{NB, NS, ZE, PS, PB\} \tag{3.7}$$

The normalized fuzzy quantity is:

$$E = \begin{cases} 3\mathrm{sgn}(e), |e/R| \geq 0.5 \\ 2\mathrm{sgn}(e), |e/R| \geq 0.3 \\ 1\mathrm{sgn}(e), |e/R| \geq 0.001 \\ 0\mathrm{sgn}(e), \mathrm{others} \end{cases} \tag{3.8}$$

For the *E* & *EC* fuzzy set that needn't be evenly divided into sub-file, usually zero-profile is designed to the smallest interval in order that ensure the control precision, with the grade increases, the corresponding interval-valued in turn increasing.

$$EC = \begin{cases} 2\mathrm{sgn}(e), |e/R| \geq 0.3 \\ 1\mathrm{sgn}(e), |e/R| \geq 0.02 \\ 0\mathrm{sgn}(e), \quad \mathrm{others} \end{cases} \tag{3.9}$$

3.3. Fuzzy control rule design

The fuzzy control rule's adjustment is a key part of enhancing the control performance, and also the decision link for producing the control tables. Many fuzzy controllers have the different improvement in the language variable, the fuzzy quantification rank, membership function regulations of fuzzy set and fuzzy control output decision-making; fuzzy control quantity output decision-making have the different improvement, but all what are same in essentially, namely the fuzzy control rules expressed in tabular form. The main problem of this kind of controller is that the establishment of fuzzy control rules is difficult, many parameters can't be determined, and whether or not the right choice is essential. For the convenience of computing and the rule automatic control, many scholars have put forward the catalytical expression of control rules which lay a foundation for fuzzy controller design entering into automatic stage [7]. Presented below is a new catalytical expression on the basis of studying and describing a reasonable analytic expression for control rules.

$$\Delta U = s \bullet E + q \bullet EC \qquad (3.10)$$

Where, weighted factor s and the q value are separately determined by E and EC's value which corresponding to the normalized fuzzy quantification value for e and ec. Taking into account both the error message and the error changes message, so the controller can fully introduce the dynamic information of the control processing. Compared with the former scholar proposed expression, this method can reduce the number of adjustable factors and shorten the online self-learning time.

$$s = f(E) = s1, s2, ..., s7 \ (E \text{ separately is } -3, -2...3) \qquad (3.11)$$
$$q = f(EC) = t1, t2, ..., t5 \ (EC \text{ separately is } -2, -1... 2) \qquad (3.12)$$

In the above formula, the optimized parameters separately are *s1, s2, s3, s4, s5, s6, s7, t1, t2, t3, t4, t5*, total of 12 parameters.

4. The Genetic Algorithm Optimal Approach

Genetic algorithm is a searching algorithm based on natural selection and heredity genetic mechanism, which could be used to simulate natural selection and natural genetic process of reproduction, mating and mutation phenomena. It will encode each individual into a string form, appraisal each individual according to the predetermined objective function, and give a fitness value. Genetic algorithms always have some random individual when it starts iteration. We can make genetic operators operating for these individuals according to these individual fitness value, retains the superior individual and eliminates poor individuals, and get a new group of individuals. The new individuals inherit some fine characters of the previous generation, and surpass the previous generation obviously, so that the genetic algorithm could towards a more optimal direction to evaluating. In order to improve the operation accuracy, we encode 12 optimized parameters using the real number code. First, encoding optimized parameters in accordance with *s1, s2, s3, s4, s5, s6, s7, t1, t2, t3, t4, t5*, and then determined these parameters to be in the range [0, 1].Supposed he total of the initial population is 50 and the maximum genetic generation is 100. The fitness function is the error absolute integral performance IAE, namely:

$$J(IAE) = \int_0^\infty |e(t)| dt \qquad (4.1)$$

Use the improved genetic algorithm in order to speed up the global optimization of above parameter.

5. Fitness Function Calibration

Assign the individual fitness value in descending order location for the calibration method of the fitness value function. This method can solve the minimum optimization problem, and the fitness function is its objective function. So the calibration algorithm steps are as follows:

(1)Supposed population size of N, each individual fitness value of the population in descending order, that is, the highest individual fitness puts the first place; the smallest individual fitness is put in the lowliest place.

(2)Based on distribution location of the individual fitness value, calibration equation is as follows:

$$f(i) = 2 \times (i-1)/N \qquad (4.2)$$

Where, i represents each individual corresponding location, $1 \leq i \leq N$.

After calibrating the fitness function based on the above algorithm, the high fitness individuals' fitness value is large; the smaller ones of individual fitness are small.

6. Selection

Firstly, the sort of the current group according to the fitness value's information; secondly, retaining a larger individual so that make them directly into the next generation; finally, the selection of the rest of the individual could base on ranking selection mechanism [6]. Only by this way can expand the search space and also doesn't destroy the existing best solution, it always will make the evolutionary process moving in the optimal direction.

7. Reorganization

The genetic algorithm of this paper uses reorganization to replace crossover operation. The difference between the two is that: two parent individuals re-produced only one future generation, while the crossover operation will produce two future generations. The reorganization method is as follows:

Supposed two father generation and the descendant individual respectively are:

$$x(x_1, x_2, ..., x_n), \; y(y_1, y_2, ..., y_n), \; z(z_1, z_2, ..., z_n). \; z_i = \alpha_i(x_i - y_i) + x_i \qquad (4.3)$$

Where, α_i s the random belonging to the range [-0.25, 1.25], i = 1, 2, ... , n.

8. Variation

Supposed $x(x_1, x_2, ..., x_n)$ is on behalf of the father generation individual, the component x_i is the variation. If $x_i \in [a_i, b_i]$, the individual of adaptive descendant mutation yk is:

$$yk = xk + mut \times (bk - ak) \times \Delta \qquad (4.4)$$

The genetic algorithm of this paper uses reorganization to replace crossover operation. The difference between the two is that: two parent individuals re-produced only one future generation, while the crossover operation will produce two future generations. The reorganization method is as follows:

Where: mut is 0, 1, -1, if the assumed mutation rate is Pm, the mut 1 or -1 of the probability is Pm / 2, the remainder is 0; $\Delta = \sum_{i=0}^{m-1} \partial_i 2^{-i}$, where $\partial_i = 1$, the probability is $1/m$, m = 20, the others is 0.

If the variation rate is bigger and population diversity is better, the premature possibility is lower. But a big variation rate will result to genetic algorithms degenerate into pure random searching and the evolutionary rate will slow down. The smaller variation rate can maintain the quicker searching speed for the algorithm, but it is difficult to jump out when encountered local minimum point. So it is easy for premature phenomenon to occur. To better solve this problem, the mutation rate Pm must adaptive change according to the actual evolution [4-6].

When the evolutionary process is going smoothly, namely the variation rate maintenance will approach zero level when every generation appears a better solution; but if it does not appear the optimal solution without a longer evolutionary time, considered to rely on the existing groups is difficult to find the optimal solution, and increased mutation rate in order to broadening the searching space at this situation. When not-evolved generation achieves to a threshold which had set, it may believe that the group has been caught in premature state, and at this time through a very big variation rate separating from the premature state rapidly. All what above is the basis thinking of self-adaptive mutation operators in this paper.

Based on the above thought, mutation rate Pm is defined as:

$$Pmg = 0.001 + NG \bullet cof \tag{4.5}$$

Where: g is the current generation; NG is a continuous no-evolution generation from the last evolution to the current generation; *cof* is the enhancement coefficient of variation rate and usually its value is very small.

9. Simulation Research

Glass tempering and annealing temperature processing can be described as a second-order time-varying time delay system:

$$a(t)\dot{y}(t) + y(t) = b(t)u(t - L(t)) \tag{5.1}$$

In the Eq. 5.1, $a(t)$ and $b(t)$ are time-varying parameters, $L(t)$ is the variable time lag parameters. In the simulation experiment, the change rules of the three parameters are as follows:

$$a(t) = 55 + 0.25t, \ b(t) = 510/(53 + t),$$

$$L(t) = 120 + \exp(-250/t) \tag{5.2}$$

The input is a unit step signal, and the sampling period t_s =20s.

The system of the Eq.(5.1) as shown was controlled by using the designed control scheme in this paper. In the control processing, the improved genetic algorithm continuously optimized weighted factors of the fuzzy controller based on-line changing information of the error and error, so that it could realize the fuzzy control rules' revision and perfection. Weighted factors optimal processing is essentially a learning processing of genetic algorithm's control experience. The whole optimization process of the objective function value changes was shown in Figure 2. In order to see more clearly the changes trend of the objective function value, y-coordinate is expressed by the logarithm of the goal function value and x-coordinate is expressed by the evaluated generation when drawing Figure 2.

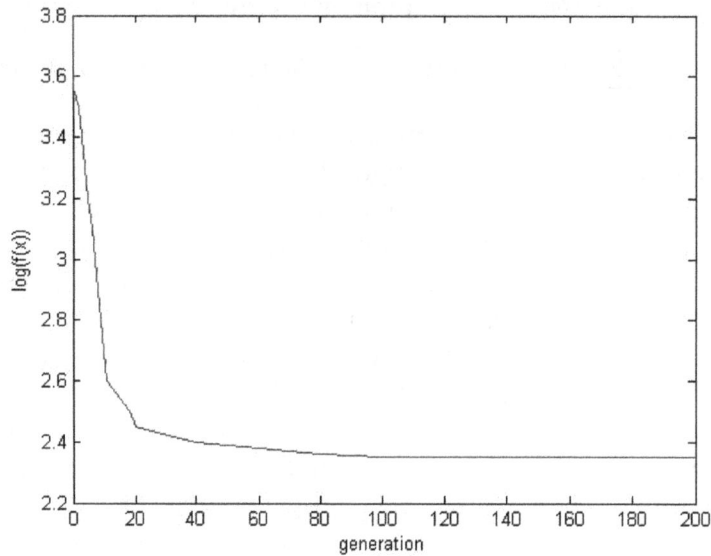

Fig. 2 Fitness value changes curve of optimized process

The system simulation curve of step response is shown in Figure 3 by using the optimized fuzzy controller to control the system in the equation (5.2).

Seen from Figure 3, the system unit step response curve has characteristics with no overshoot, no-oscillation, small steady-state error and the short rise time.

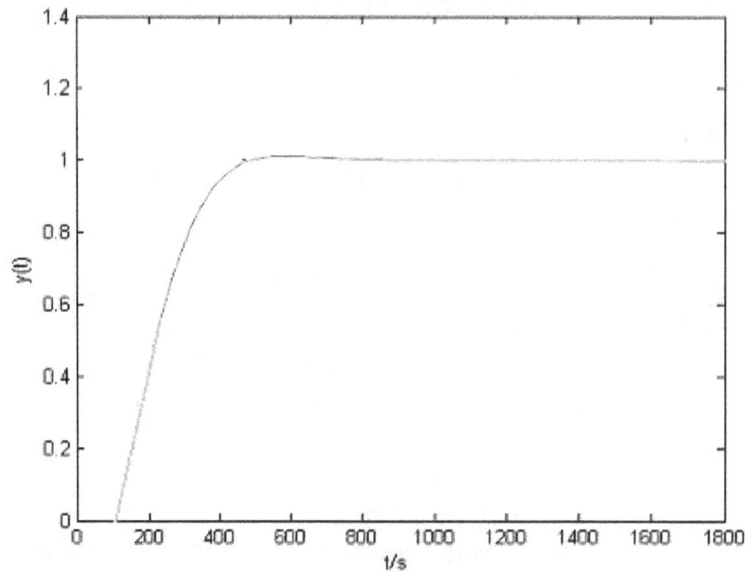

Fig. 3 System unit step response curves of time-varying and time lag parameters

With the ongoing of evaluated processing, the objective function value decreases continuously and the entire control system performance continues to increasing. The main reason lies in that amendments of the weighted factors make the fuzzy control rule to gradually in line with the actual control needs. Given the greatest evolutionary steps 100 in advance is genetic algorithm's terminal condition. Obtaining the optimized parameters after the optimal process has ended.

Evolutionary steps 100 in advance are genetic algorithm's terminal condition. Obtaining the optimized parameters after the optimal process has ended.

The system simulation curve of step response is shown in Figure 3 by using the optimized fuzzy controller to control the system in the equation (5.2).

Seen from Figure 3, the system unit step response curve has characteristics with no overshoot, no-oscillation, small steady-state error and the short rise time.

10. Conclusion

When using fuzzy control in the parameters time-varying and large time-delay systems, the main difficulty is how to obtain perfect control rule. Proposed fuzzy rule analytic expressions with has many adjustable factors in this paper, by introducing genetic algorithm to implement adjustable factor optimization and achieve the learning and revision of control rules that strengthened the Self learning function of fuzzy controller. All what indicated that the use of genetic algorithms supporting the controller design is a promising approach.

References

1. D.Q. Feng. "Integrated Intelligent Control Methods in Omethoate Synthetic Reaction Process and Application". Shanghai： Shanghai University, 2009

2. G.L. Sun, and Zh. Jin. "Simulation Research of Adaptive Fuzzy Temperature Controller Based on LMS". Journal of System Simulation, 2006 (26)11: 101-103

3. Luoyi Qian. "Study and Design of Float Glass Annealing Lehr". Chengdu: Chengdu University of Technology, 2009

4. Y.L. Cao. "Study on Intelligent Temperature Control System for Oil-Burning Annealing Furnace". Xi'an: Xi'an University of Technology, 2005

5. Zh. Y. Yin. "Temperature Control of Hot Water Boiler Based on New Fuzzy-PID Controller". Industrial Heating, 2008 (20)4: 66-68

6. X.K. Wang, and ZH.L. Sun, and L. Wang, and D.Q. Feng. "Design and Research Based on Fuzzy PID-parameters Self-tuning Controller with MATLAB". 2008 International Conference on Advanced Computer Theory and Engineering (ICACTE 2008), Phuket Thailand: IEEE CPS, 996-999.

7. X.K. Wang, and L. Wang, and Zh. L. Sun, and D.Q. Feng. "Short-Term Load Forecasting Based on RBF Adaptive Neural Fuzzy Inference". Proceedings of the 14th Youth Conference on Communication (2009), USA: Scientific Research Publishing, 220-224

8. X.K. Wang, and Zh. L. Sun, and L. Wang, and Sh. P. Huang. "Simulation and Optimization of Parameters on DC Motor Double Closed-Loop Control System Based on Simulink". 2009 International Conference on Intelligent Human-Machine Systems and Cybernetics (IHMSC09) Hangzhou: IEEE CPS, 253-256

First-principles Study of Structure, Elastic and Electronic Properties of Precipitates Al₃Fe, Al₆Mn and Mg₂Si in Al-Mg Alloys

Yuan-Chun Huang[1,2,a,*], Yin Li[2,b] and Zheng-Bing Xiao[1,2,c]

¹School of Mechanical and Electrical Engineering, Central South University, Changsha 410083, China

²Light Alloys Research Institute, Central South University, Changsha 410083, China

ᵃscience@csu.edu.cn, ᵇli.yin@csu.edu.cn, ᶜxiaozb@csu.edu.cn

The structural, elastic and electronic properties of Al₃Fe, Al₆Mn and Mg₂Si have been researched by first-principles calculations within the framework of generalized gradient approximation (GGA). The structural parameters and electronic structure such as lattice constant (a_0), shear modulus (G) and density of states (DOS) are in good agreement with the theoretical and experimental results available. It can be inferred from the negative cohesive energy and formation enthalpy that these compounds are structural stable, and the Al₃Fe phase is much steady from energetic point of view. The elastic constant of these phases obtained comply with the mechanical stability conditions. Then the shear modulus G, bulk modulus B, Young's modulus E and Poisson's ratio v of the studied compounds were concluded. Electronic structure of these compounds has been analyzed from electron density and density of states distribution.

Keywords: First-principles; Al-Mg alloys; DFT; elastic properties; electronic properties.

1. Introduction

Commercial Al alloys of 5xxx series are crucial important for transportation industries because they exhibit high strength, good weldability and excellent corrosion properties [1]. It is well known that these properties will definitely be affected by presence of precipitates [2]. Al₃Fe, Al₆Mn and Mg₂Si precipitates are the main intermetallics presented in the 5xxx series of aluminum alloy. Apart from the direct relation to the second phase strengthening in alloys, precipitates were connected to other physical and mechanical properties of this material. Although the importance, systematic information on the mechanical and physical properties of the precipitates in Al-Mg compounds is scarce.

As a powerful and helpful tool, first principles calculation is highly desirable due to it provide further stress on the crystalline structure and its physical property have relation to the electronic configuration of the metal [3]. Herein this works, the phase stability, elastic and electronic properties of binary Al₃Fe, Al₆Mn and Mg₂Si phases have been systematically studied through the first principles calculation. It is well known that the lattice parameters are difficult to be measured directly; however, the theoretical simulation grounded on accurate density functional theory can compensate the lack of experimental data. The elastic properties have closely relations with various physical fundamental properties, such as the information on the ability to keep up strength and stiffness of the material with temperature enhanced [4, 5]. Although the available theoretical and experimental values are limited, the outcomes of this work can provide the advantage of simple and easy relationships, and more importantly they could demonstrate trends which will be greatly helpful in developing new and tailor actual alloys and benefit to understand the microscopic strengthen mechanism.

2. Method of Computation

The calculations here were carried out grounded on the density functional theory (DFT) [6]. The electronic structures and elastic properties of Al₃Fe, Al₆Mn and Mg₂Si followed by geometry optimization were computed by the generalized gradient approximation, the so-called GGA-PBE. In

---

*Corresponding author

addition, ultrasoft pseudopotential [7] was used in our models. All calculations were totally relaxed in the case of entire degrees of freedom, cell vectors, volume and the internal atomic positions. The number of special k-points and the kinetic energy cut-off are the two parameters, which influence the exactitude of the valuation [8]. The plane-wave basis set cut-off energy was 330 eV for Al_3Fe, 310 eV for Al_6Mn, while 380 eV for Mg_2Si separately. The Monkhorst-Pack method by means of a mesh of 4×4×2, 5×5×3 and 7×7×7 employs the special k-points taking of samples synthesis onto the Brillouin zone [9]. Broyden-Fletcher-Goldfarb-Shanno (BFGS) minimization technique [10] is applied to geometry optimization within CASTEP (Cambridge Sequential Total Energy Package) code. The convergence criteria used in geometry optimization have been set as follows: the tolerance in total energy being within $5×10^{-6}$ eV/atom, the maximum stress within 0.02GPa, the maximum ionic Hellmann-Feynman force within 0.01 eV/Å, and the maximum ionic displacement within $5×10^{-4}$ Å. The calculations of Al_3Fe and Al_6Mn enforced in this work were in spin-polarized calculation (use formal spin as initial).

3. Results and Discussion

3.1. *Structure properties*

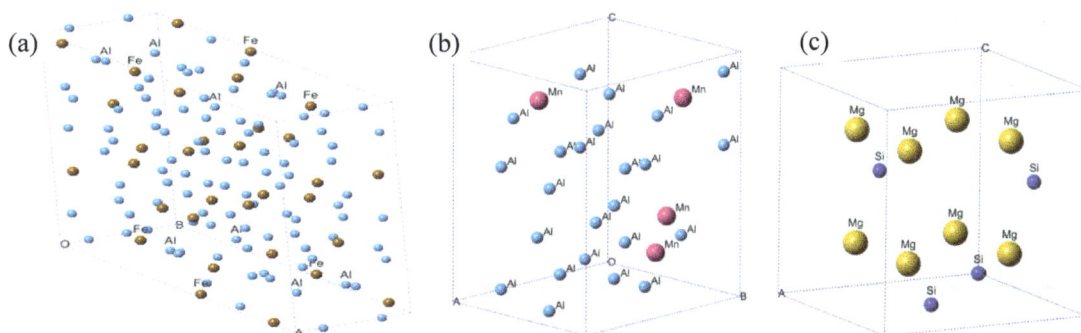

Fig. 1 The crystal structure of Al_3Fe (a), Al_6Mn (b), and (c) Mg_2Si

The relaxed crystal structure of Al_3Fe, Al_6Mn, Mg_2Si are shown in Fig. 1. A summary of equilibrium lattice constants and other matching theoretical results are assembled in Table 1. All calculated results are highly accurate and reliable as the corresponding measured values in good agreement between them. It indicates that the parameters and used calculation methods make the theoretical foundation for dependability of the followed study of the stability, anisotropy and charge density distributions of the important material.

The total energies E_{tot} of Al_3Fe, Al_6Mn and Mg_2Si intermetallic compounds have been obtained in the spin polarized calculations. The total energy is very imperative in calculating the heat of formation enthalpy ΔH and cohesive energy E_{coh} of compounds. The formation enthalpy and cohesive energy have been calculated of the crystal to check the thermodynamic stability of Al_3Fe, Al_6Mn, Mg_2Si structure [3, 11]. The formation enthalpy is well defined as the energy absorbed or released when a crystal forms from atoms. A negative value means the formation of the crystal is an exothermic process and a spontaneous reaction. As a basic property of the material, the cohesive energy measure the ability of the power that isolated atoms combined into solid [12]. Therefore, the lower formation enthalpy and cohesive energy is, the more stable the structure is. The cohesive energy E_{coh} and heat of formation enthalpy ΔH for a binary phase A_xB_y were calculated using the following expressions [12, 13]:

$$\Delta H = \frac{1}{x+y}\left(E_{tot} - xE_{solid}^A - yE_{solid}^B\right) \tag{1}$$

$$E_{coh} = \frac{1}{x+y}\left(E_{tot} - xE_{atom}^A - yE_{atom}^B\right) \tag{2}$$

In Eq. (1) and (2) E_{tot} is the total energy of the unit cell, E_{solid}^A and E_{solid}^B are the total energies of the solid states of the pure elements A and B, each in the equilibrium (zero-pressure) geometry. E_{atom}^A and E_{atom}^B are the total energies of the isolated A and B atoms, respectively, and x, y relate to the numbers of A and B atoms in the unit cell. The acquired formation enthalpy and cohesive energy are listed in Table 2. Grounded on these calculated results, it can be seen that formation enthalpy of Al_3Fe, Al_6Mn and Mg_2Si are all negative, which confirms that the structure of the compounds are able to exist and stable [14]. Due to the negative formation enthalpy of Al_6Mn, Mg_2Si and Al_3Fe are gradually decreased, it is deduced that Al_3Fe compound own the highest forming competence, then Mg_2Si, lastly Al_6Mn.

Table 1 Lattice constants (nm) and structural parameters of Al_3Fe, Al_6Mn, Mg_2Si

Phase	Group (No.)	Structure type	Lattice constants a, b, c		V_0/nm^3
Al_3Fe	$C12/m1$ (12)	monoclinic	Present	1.539, 0.802, 1.242	1.488
			Ref. [15]	1.548, 0.808, 1.248	1.462
Al_6Mn	Cmcm (63)	orthorhombic	Present	0.755, 0.647, 0.883	0.431
			Ref. [16]	0.755, 0.650, 0.887	0.436
Mg_2Si	$Fm\bar{3}m$ (225)	cubic	Present	0.638, 0.638, 0.638	260.304
			Ref. [17]	0.635, 0.635, 0.635	256.048

From Table 2, it can be observed that all the cohesive energy of the investigated phases is also negative, which means that this phase is energetically stable. By further analysis the investigated compound, it can be concluded that Al_3Fe compound is the most stable as the result of the highest E_{coh} at this condition, next is Al_6Mn, while the stability of Mg_2Si is weakest because of the lowest E_{coh}.

Table 2 Formation enthalpy (kJ mol^{-1}) and Cohesive energy (kJ mol^{-1}) of Al_3Fe, Al_6Mn, Mg_2Si phases

Phase	Formation enthalpy			Cohesive energy	
	Present	Ref	Exp	This work/eV·atom^{-1}	Ref
Al_3Fe	-35.58	-33.37[18]	–	-415.385	–
Al_6Mn	-21.06	-23.08[19]	–	-360.962	–
Mg_2Si	-25.64	-15.164[14]	-21.20 [20]	-281.6	-281.9 [14]

3.2. Elastic properties

The zero-pressure elastic properties of Al_3Fe, Al_6Mn and Mg_2Si were investigated in the present work. From elastic behavior, we can obtain valuable information to reflect the characteristics of materials [21]. The researched compounds in this part pertain to monoclinic, orthorhombic and cubic crystal class. Further information about the independent elastic constants and the relative restrictions for monoclinic, orthorhombic and cubic are listed in Table 3.

Table 3 Independent elastic constants, mechanical stability conditions, equations for monoclinic, orthorhombic & cubic

Phase	Structure type	Independent Elastic Constants	Stability Conditions
Al$_3$Fe	monoclinic	C_{11}, C_{22}, C_{33}, C_{44}, C_{55}, C_{66}, C_{12}, C_{13} C_{23}, C_{15}, C_{25}, C_{35}, C_{46} [26]	$C_{11}>0$, $C_{22}>0$, $C_{33}>0$, $C_{44}>0$, $C_{55}>0$, $C_{66}>0$, $[C_{11}+C_{22}+C_{33}+2(C_{12}+C_{13}+C_{22}>0]$, $(C_{33}C_{55}-C_{35}^2)>0$, $(C_{44}C_{46}-C_{46}^2)>0$, $(C_{11}C_{33}-2C_{13})>0$
Al$_6$Mn	orthorhombic	C_{11}, C_{22}, C_{33}, C_{44}, C_{55}, C_{66}, C_{12}, C_{13}, C_{23} [27]	$C_{11}>0$, $C_{22}>0$, $C_{33}>0$, $C_{44}>0$, $C_{55}>0$, $C_{66}>0$, $[C_{11}+C_{22}+C_{33}+2(C_{12}+C_{13}+C_{22}>0]$, $(C_{11}+C_{22}-2C_{12})>0$, $(C_{11}+C_{33}-2C_{13})>0$, $(C_{22}+C_{33}-2C_{23})>0$
Mg$_2$Si	cubic	C_{11}, C_{12}, C_{44} [28]	$C_{44}>0$, $C_{11}>|C_{12}|$, $C_{11}+2C_{12}>0$

Table 4 shows the derived quantities from elastic constants. As the bulk modulus B (average) characterizes the resistance to volume change under applied pressure, this implies that the typical value for bond strength in Al$_6$Mn is slightly stronger than in Al$_3$Fe, quite stronger than in Mg$_2$Si. The bulk moduli of Al$_3$Fe and Mg$_2$Si are comparatively low (smaller than 100 GPa) and, thus ought to be categorized as a relatively soft materials with high compressibility (higher than 0.01). In addition, shear modulus G, a measure of resistance to reversible deformation upon shear stress [22], so the larger shear modulus of Al$_3$Fe and Al$_6$Mn show that they have stronger resistance to reversible deformation upon shear stress than Mg$_2$Si. One of the widely used critical values that separate brittleness and ductility is around 1.75. If B/G < 1.75, the material behaves in a brittle manner; otherwise, a ductile is predicted [12]. Al$_3$Fe, Al$_6$Mn and Mg$_2$Si show brittleness, as the B/G ratios of the studied compounds are smaller than 1.75, thus their mechanic properties decrease quickly with increasing temperature. Another used to distinguish the brittleness from ductility is Poisson's ratio [23]. The critical value is about 0.57. In the present work, the values of Al$_3$Fe, Al$_6$Mn and Mg$_2$Si are 0.203, 0.209 and 0.1696, respectively, implying that the studied compounds are essentially brittle, and in Good agreement with the results estimated by the B/G ratio.

Table 4 Anisotropy factor A, bulk modulus B/GPa (average), shear modulus G/GPa, B/G, Possion's ratio v, and Young's modulus E/GPa at the GGA levels for Al$_3$Fe, Al$_6$Mn, Mg$_2$Si

Phase	Elastic constants					
	A	B	G	v	B/G	E
Al$_3$Fe	0.906	90.921	67.425	0.203	1.35	162.184
Al$_6$Mn	0.835	102.662	74.265	0.209	1.38	179.510
Mg$_2$Si	0.973	52.207	44.238	0.1696	1.18	103.485

From Table 4, the Young's modulus of Al$_3$Fe, Al$_6$Mn and Mg$_2$Si are found to be 162.184 GPa, 179.510 GPa and 103.485 GPa, respectively, it indicates that, Al$_3$Fe and Al$_6$Mn will show a rather strong stiffness than Mg$_2$Si. The elastic anisotropy of compounds has an important implication in engineering science, and it is highly correlated with the possibility of inducing micro-cracks in materials [24, 25]. Clearly, the computed anisotropy factor (A) of the herein studied compounds exhibits a weak anisotropy.

3.3. Electronic structure

The important band structure features of the herein studied materials can be obtained from the density of states (DOS). Fig. 2 illustrates the total density of states (TDOS) and the orbital resolved partial density of states (PDOS) projected onto Al, Fe, Mn, Si and Mg contributions. The calculated total DOS of Al3Fe is mainly occupied by Al-p orbitals and Fe-d orbitals. The hybridization between Al and Mn is clear in entire region and the valence bands (VBs) close proximity of the Fermi level, is mainly made of the occupied Mn-d states with small contributions from the Al-p states. From the total DOS of Al3Fe and Al6Mn, the precipitates exhibit metallic character since the density of states at Fermi energy is non-zero. There are quasigap which indicate the presence of the directional covalent bonding [29]. Compare with the pure metallic bonding, the covalent bonding would enhance the strength of material, therefore the Al3Fe and Al6Mn have pronounced stability. Different with quasigap, the gap is observed for Mg2Si, and it amounts to about 0.2eV. Compared to Al3Fe and Al6Mn, a covalent character in Mg2Si is predominantly decreased.

In order to further reveal the feature of bonding, the charge density distributions are also investigated. From Fig. 3(a), it can be seen clearly that there exists strong directional bonding between Al and its nearest Al atoms, while the bonding between Al and Fe atom is mainly ionic. Fig. 3(b) also shows directional bonding between Al atoms, and ionic Al-Mn bonding for Al6Mn. The computed net charges on Mg and Si are 0.68/0.68 and -1.36, respectively. The build-up of charge around silicon atom as shown in its charge density plot in Fig. 3(c), and the charge depletion around Mg atom essentially mark the ionic character in the bonds between Mg and Si. However, from the apparent hybridization between Mg(s) and Si(p) below the Fermi level as shown in Fig. 2(c), a covalent character in Mg2Si cannot be totally ignored. Therefore, the more stable phase are Al3Fe and Al6Mn followed by Mg2Si, which is in good accordance with the E_{tot}.

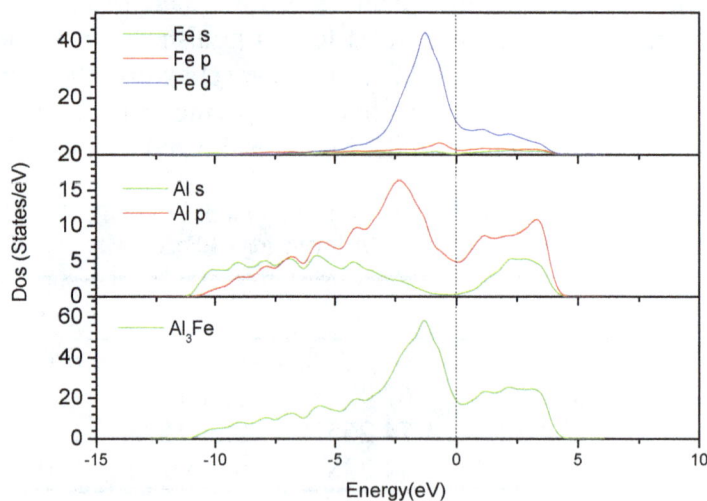

(a) Al₃Fe,

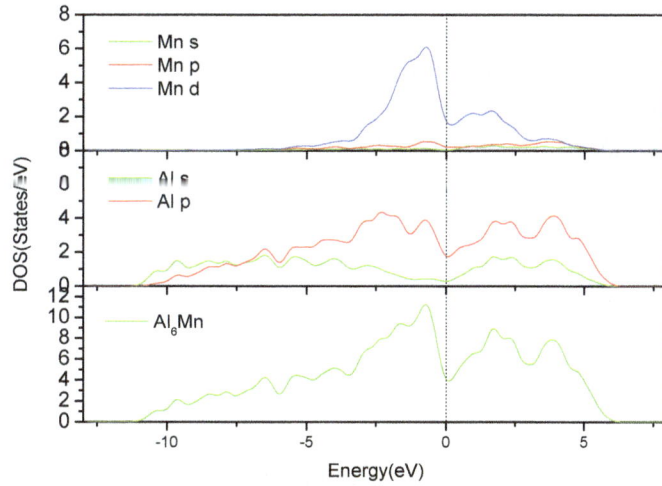

(b) Al$_6$Mn,

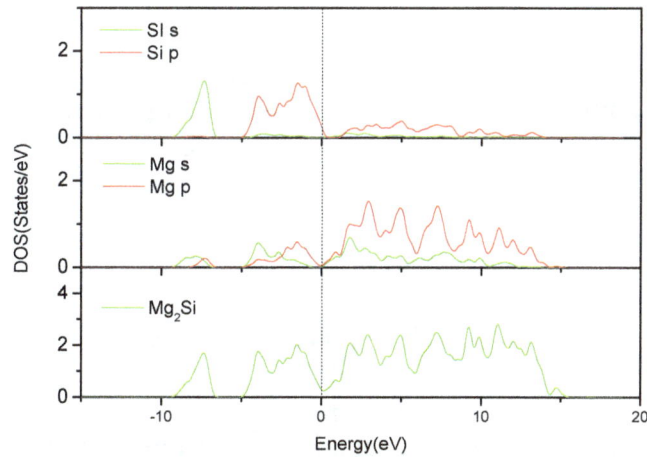

(c) Mg$_2$Si

Fig. 2. The partial and total densities of states (PDOS and TDOS, respectively)
The Fermi level is set at zero energy and marked by the vertical lines.

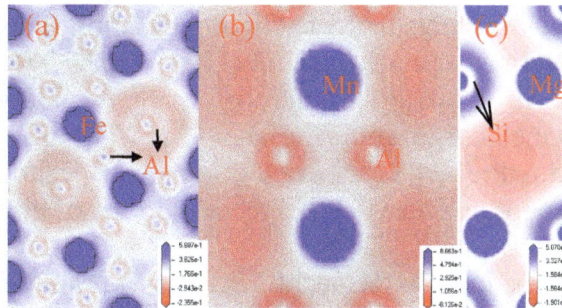

(a) (1-10) plane for Al$_3$Fe, (b) (110) plane for Al$_6$Mn, (c) (1-10) plane for Mg$_2$Si (c).

Fig. 3. Electron density distribution

4. Conclusion

In conclusion, by means of pseudopotential plane-wave method in the framework, we studied in details the structural, elastic, and electronic properties of Al_3Fe, Al_6Mn and Mg_2Si. The calculated equilibrium structural parameters are in good agreement with the available experimental measurements. From the negative formation enthalpy and cohesive energy, it is inferred that these compounds are mechanically stable. Compared with orthorhombic Al_6Mn and cubic Mg_2Si, monoclinic Al_3Fe exhibits considerable large cohesive energy and heat of formation enthalpy. Our analysis of the predicted elastic moduli shows that Al_3Fe and Mg_2Si are relatively high bond strength and soft materials with high compressibility. The studied intermetallics are essentially brittle, and characterized by a weak elastic anisotropy. From the analysis of the DOS and electron density characteristics, it is found that there are directional covalent bonding and ionic bonding in Al_3Fe and Al_6Mn, and a predominant ionic bonding characteristic in Mg_2Si. According to the calculated electronic structure of the studied phases, the higher stability of Al_3Fe and Al_6Mn may be attributed to the directional covalent bonding. These results would be significant for the further optimization and design of Al-Mg alloys with excellent mechanical properties.

Acknowledgment

This research was financially supported by the National Science Foundation. The project is supported by the Fundamental Research Funds for the Central Universities of Central South University (Contract No.: 2016zzts319) and the National Basic Research Program of China (Grant No.2012CB619504).

References

1. R. Goswami, G. Spanos, P.S. Pao, R.L. Holtz, Precipitation behavior of the ß phase in Al-5083, Materials Science and Engineering: A, 527 (2010) 1089-1095.
2. N. Acharya, B. Fatima, S.S. Chouhan, S.P. Sanyal, Ab-initio study of structural, electronic and elastic properties of cobalt intermetallic compounds, Computational Materials Science, 98 (2015) 226-233.
3. X. Hao, P. Jia, Y. Cui, J. Wang, F. Gao, Y. Qiao, Structural, electronic and elastic properties of Y_3AlC_3, YAl_3C_3 and Y_3AlC via first principles study, Journal of Alloys and Compounds, 658 (2016) 1025-1030.
4. E. Busso, F. McClintock, Mechanisms of cyclic deformation of NiAl single crystals at high temperatures, Acta metallurgica et materialia, 42 (1994) 3263-3275.
5. R. Jayaram, M. Miller, An atom probe study of grain boundary and matrix chemistry in microalloyed NiAl, Acta metallurgica et materialia, 42 (1994) 1561-1572.
6. M. Segall, P.J. Lindan, M.a. Probert, C. Pickard, P. Hasnip, S. Clark, M. Payne, First-principles simulation: ideas, illustrations and the CASTEP code, Journal of Physics: Condensed Matter, 14 (2002) 2717.
7. O. Boudrifa, A. Bouhemadou, N. Guechi, S. Bin-Omran, Y. Al-Douri, R. Khenata, First-principles prediction of the structural, elastic, thermodynamic, electronic and optical properties of Li4Sr3Ge2N6 quaternary nitride, Journal of Alloys and Compounds, 618 (2015) 84-94.
8. D. Zhou, J. Liu, S. Xu, P. Peng, Thermal stability and elastic properties of Mg_2X (X= Si, Ge, Sn, Pb) phases from first-principle calculations, Computational Materials Science, 51 (2012) 409-414.

9. P. Black, The structure of FeAl$_{3.1}$, Acta Crystallographica, 8 (1955) 43-48.

10. A. Kontio, P. Coppens, New study of the structure of MnAl$_6$, Acta Crystallographica Section B: Structural Crystallography and Crystal Chemistry, 37 (1981) 433-435.

11. J.G.Barlock, L.F. Mondolfo, Structure of Some Aluminum-Iron-Magnesium-Manganese-Silicon Alloys, Zeitschrift fur Metallkunde, 66 (1975) 605-611.

12. M. Mihalkovič, M. Widom, Structure and stability of Al$_2$Fe and Al$_5$Fe$_2$: First-principles total energy and phonon calculations, Physical Review B, 85 (2012) 014113.

13. G.T. de Laissardiere, D.N. Manh, L. Magaud, J. Julien, F. Cyrot-Lackmann, D. Mayou, Electronic structure and hybridization effects in Hume-Rothery alloys containing transition elements, Physical Review B, 52 (1995) 7920.

14. A. Nayeb-Hashemi, Phase diagrams of binary magnesium alloys, ASM International, Metals Park, Ohio 44073, USA, 1988. 370, (1988).

15. S. Daho, M. Ameri, Y. Al Douri, D. Bensaid, D. Varshney, I. Ameri, First-principles calculations of structural, elastic, thermodynamic, electronic and magnetic investigations of the filled skutterudite alloy UFe$_4$Sb$_{12}$, Materials Science in Semiconductor Processing, 41 (2016) 102-108.

16. W. Lin, J.-h. Xu, A. Freeman, Electronic structure, cohesive properties, and phase stability of Ni$_3$V, Co$_3$V, and Fe$_3$V, Physical Review B, 45 (1992) 10863.

Study on the Parameterized Model for Tunnel Lining Structure Safety Analysis with APDL and UIDL

Jian-Cong Xu[1,*], Hui-Hao Xue[2], Quan-Ji Zhou[3]

[1]*Key Laboratory of Geotechnical and Underground Engineering of Ministry of Education, Tongji University, Shanghai 200092, P.R. China*
[2]*Department of Geotechnical Engineering, Tongji University, Shanghai 200092, P.R. China*
[3]*Shanghai Construction Group Co., Ltd., Shanghai 200080, P.R. China*
Email: xjc1008@tongji.edu.cn

In tunnel engineering, the forms of lining structure usually are similar. In order to conveniently and quickly evaluate and analyze the tunnel lining structure safety, based on the general finite element software ANSYS, a parameterized model for tunnel lining structure safety analysis with APDL and UIDL was researched in this paper. The results show as follows: The developed parameterized model using the numerical method of limit analysis of strata structure may not only conveniently and quickly evaluate and analyze the tunnel lining structure safety, but also analyze and calculate the safety factor of tunnel lining structure taking into account the effect of tunnel excavation process. The applicability and simplicity of the developed parameterized model in the paper also support its usefulness.

Keywords: Parameterized model; ANSYS; secondary development; safety factor.

1. Introduction

In finite element analysis, if we were to model and analyze each tunnel structure, the efficiency of analysis will be much lower. It's a preferable way to enhance the efficiency of analysis by the secondary development based on the general finite element software. Parametric model analysis of tunnel lining structures will be more convenient.

Zhang (2006) developed a parameterized tunnel lining structure design system using the load structural method [1].

ANSYS is one of the most popular general finite element software in the world. It is widely used in civil engineering. ANSYS is open and customizable, and it also provides several ways of secondary development. Users can conveniently extend its function or integrate system on their request based on the standard ANSYS version. In this paper, a parameterized model for the tunnel lining structure safety analysis, which considers the features of tunnel engineering, is developed by the secondary development based on ANSYS.

2. Module Function and Structure

ANSYS Parametric Design Language (APDL) and User Interface Design Language (UIDL) are most commonly used secondary development functions in ANSYS.

At present, the load-structure method and the stratum-structure method are two main ways to calculate internal force and deformation of tunnel lining in tunnel engineering design. Using the APDL, UIDL technology of ANSYS, based on the Chinese Tunnel Engineering Design Codes and experiences, a TUNNEL_ANALYSIS module is developed. This module consists of SS_MAIN using the stratum-structure method in the paper, which may analyze the safety of tunnel lining.

Considering the features of tunnel engineering, the TUNNLE_ANALYSIS module has several characteristics as follows:

(1) Fine number control to tunnel finite element model, which brings very good versatility and expansibility and is convenient for the follow-up study.

---

*Corresponding author

(2) Highly parametric input of tunnel information. Finite element model is auto-generated by inputting geometry parameters of tunnel cross-section, stratum parameters and lining material parameters in the custom interface.

(3) Simulate lining structures by the asymmetric variable cross-section beam elements. It may simulate tunnel lining structures better compared to common beam elements.

(4) A safety factor calculator macro, which generates lining safety factor conveniently and intuitively.

(5) Analyze and evaluate the tunnel lining safety using the TUNNLE_ANALYSIS module with the onsite measurement data of lining quality.

The typical processes of the TUNNLE_ANALYSIS module are as follows: (1) define parameters; (2) create geometry model; (3) mesh generation; (4) load; (5) solve; (6) post-process. The structure of SS_MAIN module using the stratum-structure method is as Fig. 1. It defines a material library and a safety factor calculator macro in the SS_MAIN module.

After the SS_MAIN module is completed, it was integrated with the customized toolbar function of ANSYS. The TUNNLE_ANALYSIS toolbar structure of tunnel lining safety analysis is as Fig. 2.

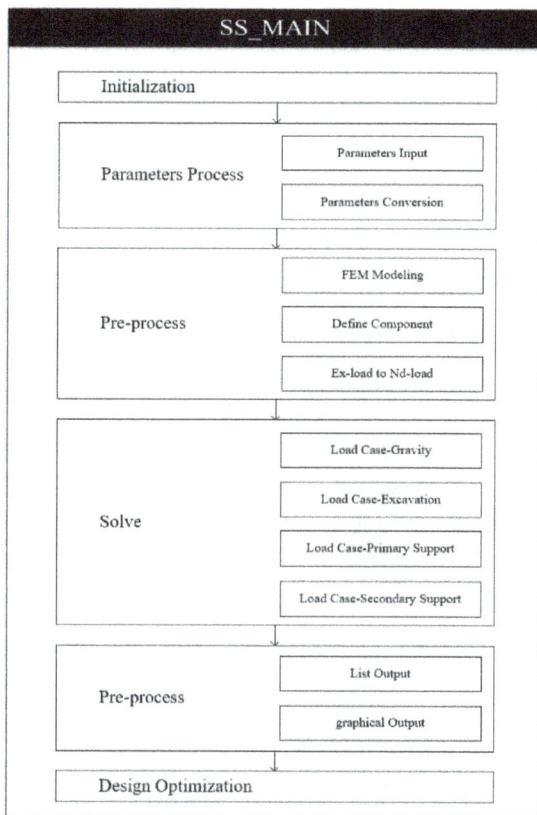

Fig. 1 Structure diagram of SS_MAIN module

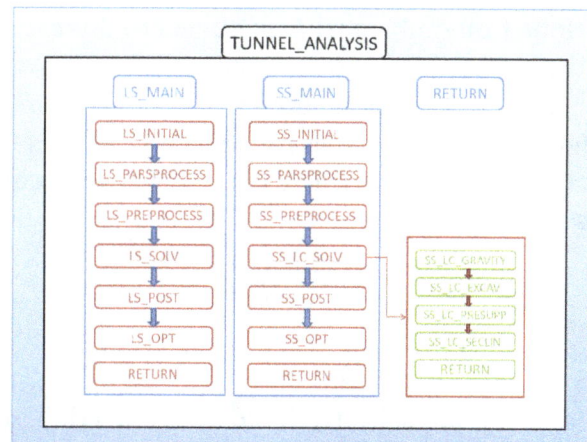

Fig. 2 Structure of TUNNLE_ANALYSIS toolbar

3. Calculation Principles of SS_MAIN Module

In the numerical simulation and analysis of tunnel engineering, not only the stability of surrounding rock but also the space-time effects of construction may be considered. The stress state of tunnel structure changes in the construction process, described in Fig. 3.

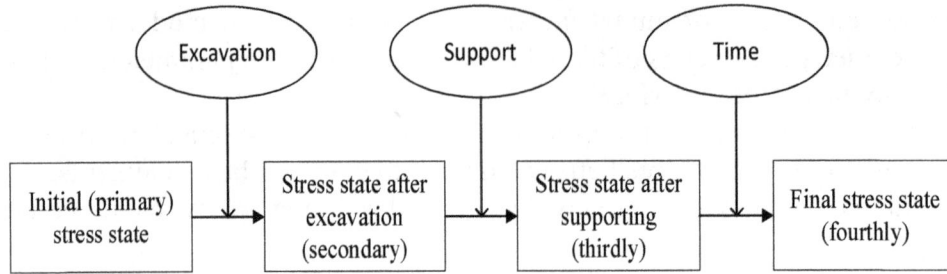

Fig. 3 Stress state changes in the construction process

To consider the space-time effects of construction, the virtual-supporting-force method is used in the finite element simulation. Ma (2007) applied virtual force on the border of tunnel excavation to simulate the gradual unloading of surrounding rock [2, 3], described in Fig. 4.

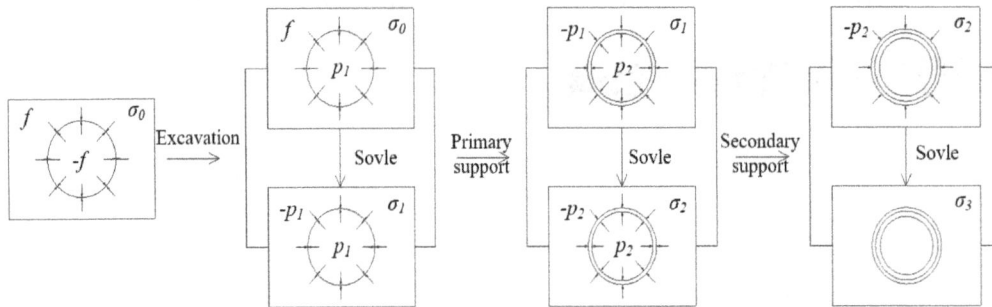

Fig. 4 The simulation of tunnel construction progress

Mohr-Coulomb equal-area circle D-P3 yield criterion is adopted in this paper.

The strength reduction method (SRM) is to acquire the safety factor, which cohesive force c and the tangent of internal friction angle $\tan \varphi$ are reduced continually by a coefficient ω until the failure of rock or soil mass occurs, described as Eq.1. The sliding face is calculated automatically by elastic-plastic numerical calculation. Meanwhile the failure information of rock or soil mass was got.

$$\begin{cases} c' = c / \omega \\ \tan \varphi' = (\tan \varphi) / \omega \end{cases} \tag{1}$$

Where, c' = the cohesive force after reduction, φ' = the internal friction angle after the reduction.

4. Case Study and Validate

The clear width of cross section of a railway tunnel is 13.3m and its clear height is 10.85m, shown in Fig. 5.

886

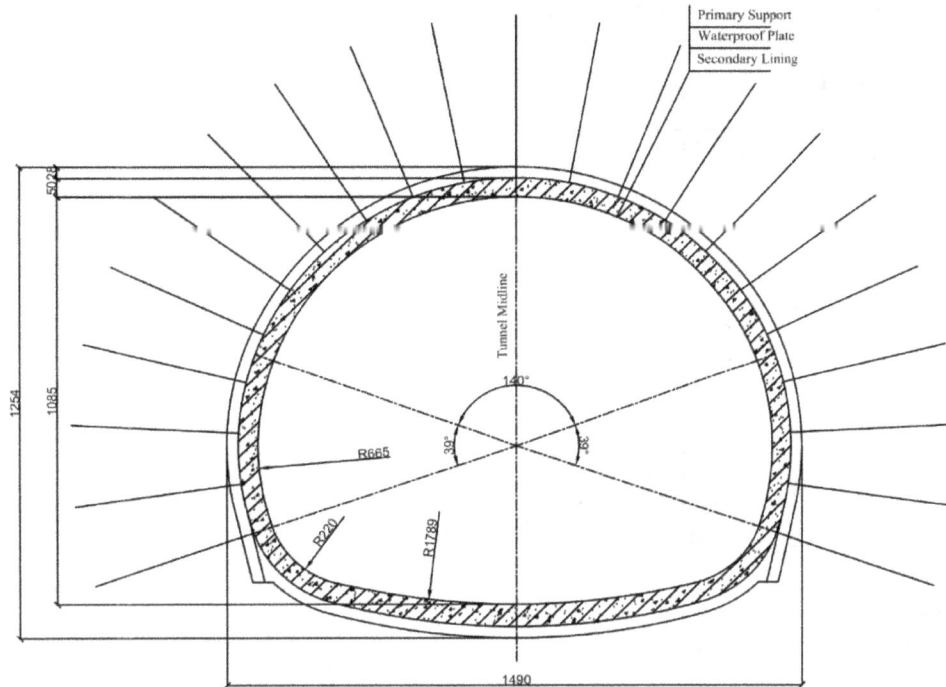

Fig. 5 Cross section of tunnel

The buried depth of tunnel is about 50m and the surrounding rock classification of tunnel is IV according to the Chinese Tunnel Design Codes [4, 5]. In order to simplify the calculation, the effect of unsymmetrical pressure and underground water are ignored. The calculation is done using the SS_MAIN module developed in the paper.

The necessary input parameters are introduced in the form of table, shown in Table 1-Table 6.

Table 1 Tunnel inner contour line geometry parameters

Parameters / Parts	X coordinate of center of circle (m)	Y coordinate of center of circle(m)	Radius of circle arc(m)	Central angle of circle arc(°)
Crown arch	0	0	6.65	140
Left side wall	0	0	6.65	38.6716667
Left arch springing	-4.2158	-1.4246	2.2	55.7438889
Invert arch	0	13.6905	17.8920	31.1688889

Table 2 Stratum parameters

Surrounding rock classification	Cohesive force of rock mass c (MPa)	Internal friction angle φ (°)	Strength reduction coefficient ω
IV	0.1	27	1.0

Notes: The surrounding rock classification was ascertained according to the Chinese Tunnel Design Codes.

Table 3 Release coefficient of surrounding rock stress to each load-step

Excavation	Primary support	Secondary lining
20%	40%	100%

887

Table 4 Primary support parameters

Concrete strength grade	Steel type	Interval of steel arch(m)	Cross-section height of steel arch(m)	Thickness of primary support(m)
C30	HRB335	0.5	0.22	0.28

Table 5 Anchor bolt parameters

Steel type	Number of anchor bolts	Length of anchor bolt(m)	Diameter of anchor bolt(m)	Strength increase coefficient of Reinforced area
HRB335	21	4	0.022	1.2

Table 6 Secondary lining parameters

Parameters / Position	Concrete strength grade	Steel type	Interval of main rebar(m)	Diameter of rebar(m)	Thickness of secondary lining(m)
Arch wall	C35	HRB335	0.2	0.02	0.5
Invert wall	C35	HRB335	0.2	0.02	0.6

Notes: Thickness of protective layer was 0.05m, symmetrical reinforcement.

A typical parameters input interface is as Fig. 6 showed.

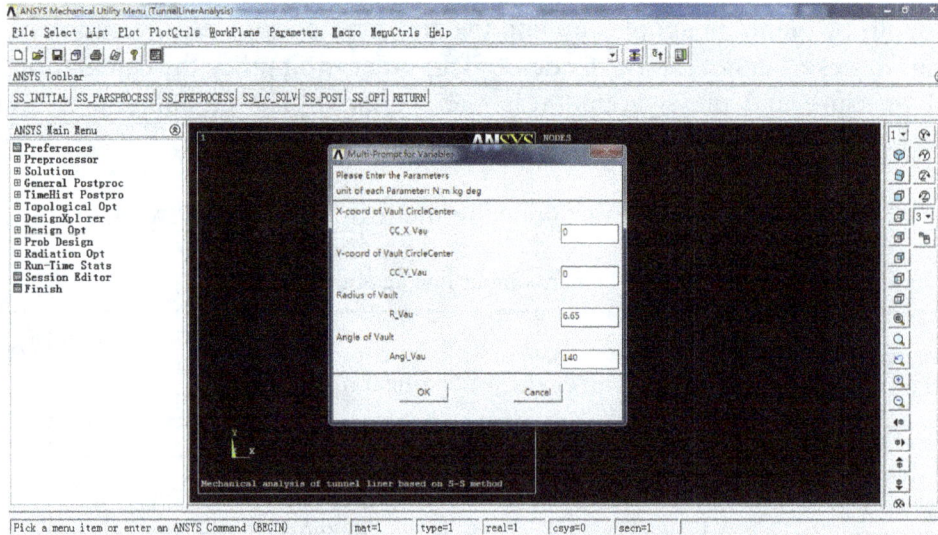

Fig. 6 Typical parameters input interface

The FEM model contains 12208 nodes, 11880 elements for stratum simulation, 144 elements among which turn into reinforced area. 144 beam elements are used to simulate lining structure and 184 link elements are used to simulate anchor bolts. Left boundary, right boundary and bottom are constrained completely in the FEM model.

The final internal forces of lining structure are shown as Fig. 7 and Fig. 8. The result is also output as a list file showed in Fig. 9.

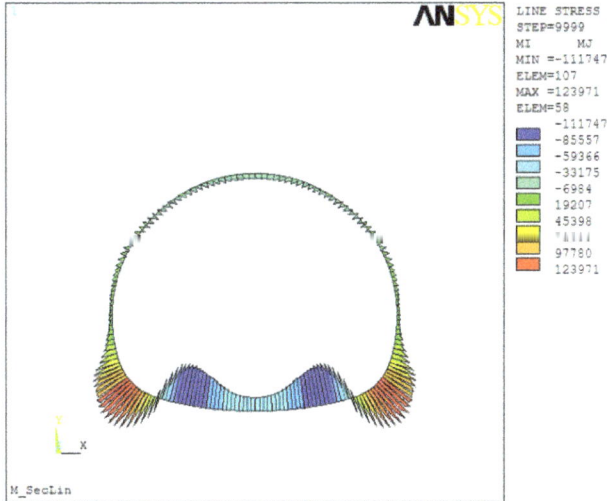

Fig. 7 Bending moment diagram of lining structure

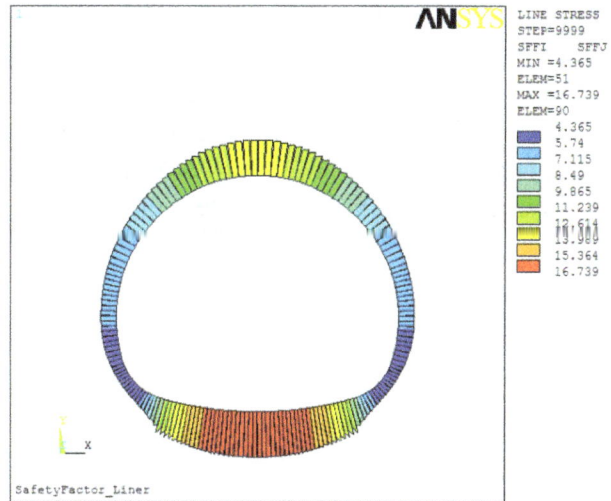

Fig. 8 Safety factor diagram of lining structure

Fig. 9 Lining Safety factors list output file

According to Fig. 7, the tunnel arch springing of both sides are the most dangerous parts of lining structure. The minimum safety factor of tunnel is 4.365 but still far above the standard value 2.4 specified by the Chinese Tunnel Design Codes.

5. Conclusion

(1) The TUNNEL_ANALYSIS module developed in the paper can conveniently calculate internal forces and safety factors of tunnel lining structure, which may quickly evaluate and analyze the tunnel lining structure safety, wherein the SS_MAIN module is based on the stratum-structure method.
(2) Results of analysis may be given in the form of graphics and list files automatically.

(3) Compared to the commercial software, this module is more expandable. Serial numbers of the finite element model are well controlled so that the model and the calculate result can be used in the follow-up study using ANSYS.

(4) In tunnel project, the thickness of lining is usually unequal because of the construction quality defect. The TUNNEL_ANALYSIS module may simulate tunnel lining structures using the asymmetric variable cross-section beam element.

References

1. J. Zhang, F. Yang, J. Qian, Redevelopment and Application of ANSYS Based on APDL and UIDL, J. Manufacture Information Engineering of China, 23 (2000) 79-81.
2. K. Ma, J. Xu, L. Wang, Application of virtual-supporting-force method numerical simulation technology, Western China Communications Science & Technology, 02 (2007) 43-47.
3. Y. Xu, L. Tao, W. Li, J. Fan, and W. Wang, A numerical simulation study on the settlement laws of the high-speed railway subgrade induced by the construction of twin shield tunnel, Chinese Journal of Beijing University of Technology, 36 (2010) 1618-1623.
4. Z. Wang, S. Du, W. Zhang, Y. Li, and X. Wu, Analysis of construction settlement of shallow railway tunnel under crossing the highway, Chinese Journal of Underground Space and Engineering, 5(2009) 531-535.
5. P. Zhang, Study on the surface settlement control standard of metro tunnel under the high-speed railway, Chinese Journal of Underground Space and Engineering, 10 (2014) 1699-1703.

Study on the Design Optimization Module for Tunnel Lining Structure with APDL and UIDL

Jian-Cong Xu[1,*], Cai-Hong Jin[2], Quan-Ji Zhou[3]

*1Key Laboratory of Geotechnical and Underground Engineering of Ministry of Education,
Tongji University, Shanghai 200092, P.R. China*
*2Department of Geotechnical Engineering, Tongji University,
Shanghai 200092, P.R. China*
3Shanghai Construction Group Co., Ltd., Shanghai 200080, P.R. China
*Email: xjc1008@tongji.edu.cn

In order to economically and quickly design the tunnel lining structure based on the general finite element software ANSYS, a design optimization module of the tunnel lining structure with APDL and UIDL was studied in this paper. The results show as follows: the TUNNEL_ANALYSIS module developed in the paper can conveniently calculate internal forces and safety factors of tunnel lining structure, which may quickly evaluate and analyze the tunnel lining structure safety; the design optimization module developed in the paper may obtained the results of optimization conveniently for the tunnel lining structure. The applicability and simplicity of the design optimization module developed in the paper also support its usefulness. The case validation shows that the module may satisfy the requirements of engineering and study.

Keywords: Tunnel engineering; ANSYS; secondary development; design optimization.

1. Introduction

In tunnel engineering, a series of lining structures are designed, which are similar in structure characteristics. Besides, most of lining designs need to be modified because of the variation of engineering geology, the construction method adjustment, the security management, the cost control and so on [1~3].

ANSYS Parametric Design Language (APDL) and User Interface Design Language (UIDL) are most commonly used secondary development functions in ANSYS [4, 5].

APDL consists of the program design language and ANSYS command stream. With the help of UIDL, users can extend the function of ANSYS meanwhile create corresponding graphical interfaces. For example, add an option menu to the main menu, and design a corresponding dialog box, and realize the input of parameters.

---

*Corresponding author

2. Module Function and Structure

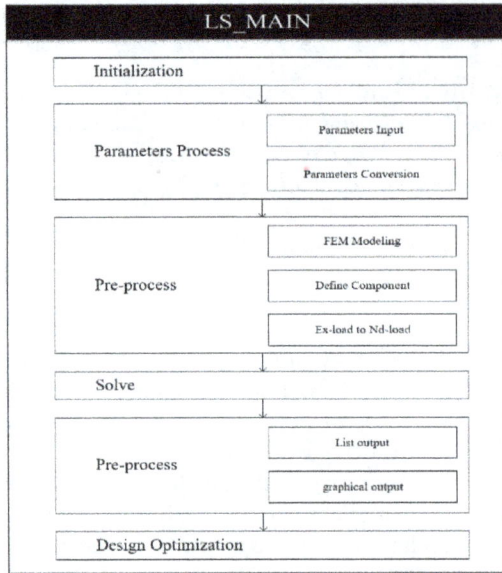

Fig. 1 Structure diagram of LS_MAIN module

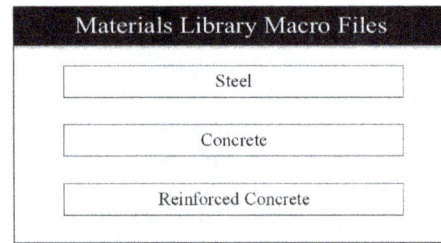

Fig. 2 Materials Library Macro Files

In this paper, a design optimization module for the tunnel lining structure developed by the secondary development based on ANSYS.

At present, the load-structure method and the stratum-structure method are two main ways to calculate internal force and deformation of tunnel lining in tunnel engineering design. Using the APDL, UIDL technology of ANSYS, based on Chinese Tunnel Engineering Design Codes and experiences, a TUNNEL_ANALYSIS module is developed. This module consists of LS_MAIN using the load-structure method, which may analyze the safety of tunnel lining and optimize the tunnel lining design by invoking the ANSYS DesignOPT module.

The typical processes of the TUNNLE_ANALYSIS module are as follows: (1) define parameters; (2) create geometry model; (3) mesh generation; (4) load; (5) solve; (6) post-process. The structure of LS_MAIN module using the load-structure method is as Fig. 1.

It defines a material library as Fig. 2 and a safety factor calculator macro as Fig. 3 in the LS_MAIN module. After the LS_MAIN module is completed, it is integrated with the customized toolbar function of ANSYS. The TUNNLE_ANALYSIS toolbar structure of tunnel lining safety analysis is as Fig. 4.

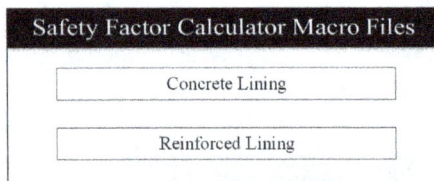

Fig. 3 Safety factor calculator macro files

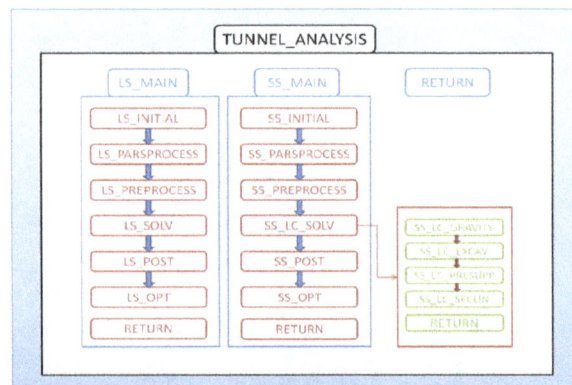

Fig. 4 Structure of TUNNLE_ANALYSIS toolbar

3. Calculation Principles of LS_MAIN Module

The calculation model is shown as Fig. 5.

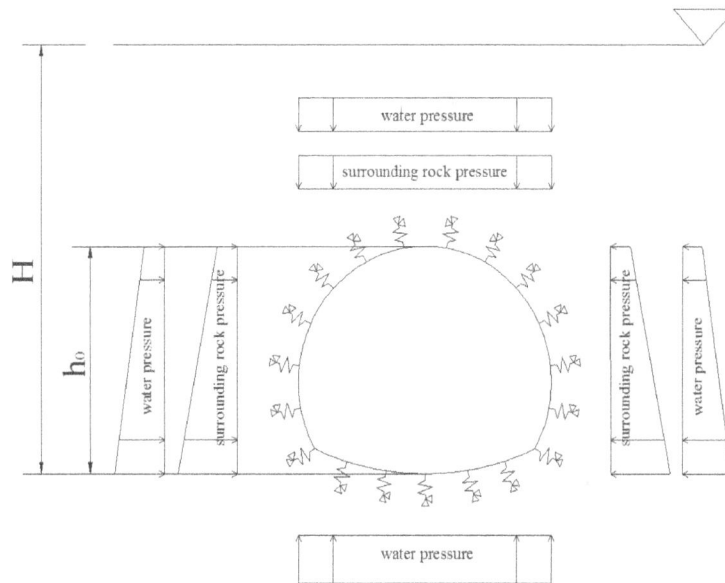

Fig. 5 Calculation model of load-structure method

LS_MAIN is based on the load-structure method. According to the Chinese Tunnel Design Codes and engineering experience, the basic calculation hypotheses are as follows:

(1) Tunnel is a slender structure, which may be analyzed by the plane strain model.

(2) It is presumed that lining structure is a small deformation elastic beam which is divided into a lot of dispersed variable thickness beam elements.

(3) The interaction between surrounding rock and structure is simulated by spring elements distributed on nodes of the model. The spring elements can only endure pressure but not endure tension. The elasticity coefficient of spring element is ascertained by the local deformation theory based on the Winkler hypothesis, which adopts the value of stratum elastic resistance coefficient.

4. Principle of Design Optimization Function

The built-in DesignOPT module of ANSYS may achieve the function of design optimization. There are two design optimization methods as follow: zero-order method and first-order method. They may achieve most optimization tasks. Usually there are three important factors in structure design optimization: design variable (variable to be optimized), objective function (assessment criteria of optimization plan) and state variables (the constraint conditions need to be satisfied). In ANSYS, the process of structure design optimization is shown as Fig. 6.

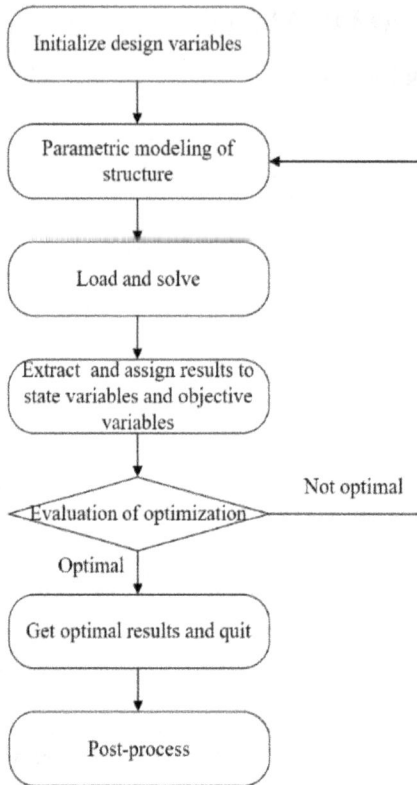

Fig. 6 Process of structure design optimization in ANSYS

Using the LS_MAIN module developed in the paper, it is convenient to get the necessary OPTfile for ANSYS DesignOPT module. In tunnel projects, the lining structure designed empirically usually have extra safety factors. The cost will be reduced after the optimization.

There are two optimization modes in TUNNEL_ANALYSIS module as follow.

(1) Two parameters optimization mode. Design variables: Arch wall thickness T_SecLin; Inver wall thickness T_InvLin. State variable: the minimum deviation value of safety factor sff_offs_min. Objective function: the volume of lining concrete volume_C.

(2) Four parameters optimization mode. Design variables: arch wall thickness T_SecLin; invert wall thickness T_InvLin; the rebar diameter of arch wall D_Rebar_Seclin; the rebar diameter of Invert wall D_Rebar_Inv. State variable: the minimum deviation value of safety factor sff_offs_min. Objective function: the volume of lining concrete cost_RC.

5. Case Study and Validate

The clear width of cross section of a railway tunnel is 13.3m and its clear height is 10.85m, shown in Fig. 7.

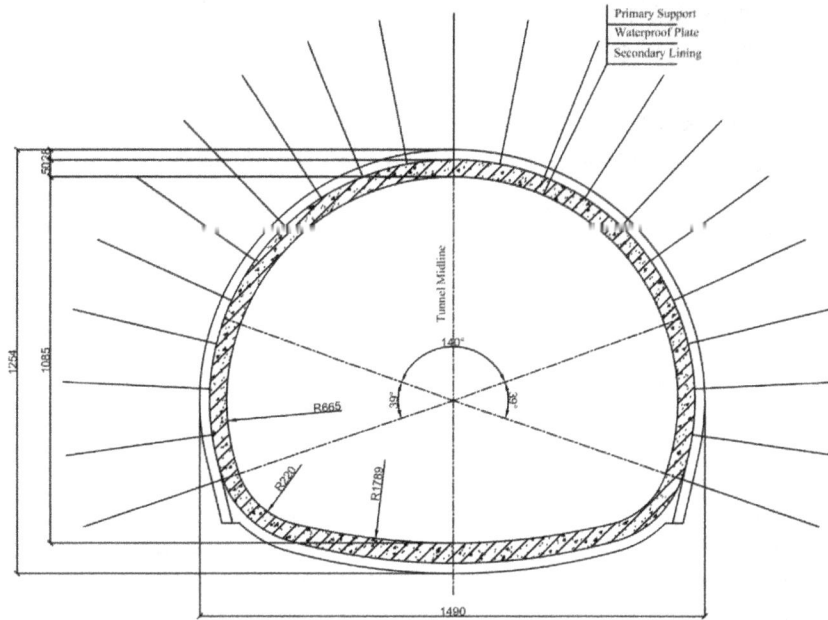

Fig. 7 Cross section of tunnel

The buried depth of tunnel is about 10m and the surrounding rock classification of tunnel is V according to the Chinese Tunnel Design Codes. In order to simplify the calculation, the effect of unsymmetrical pressure and underground water are ignored.

The necessary input parameters are introduced in the form of table, shown in Table 1~Table 3.

Table 1 Tunnel inner contour line geometry parameters

Parameters / Parts	X coordinate of center of circle (m)	Y coordinate of center of circle(m)	Radius of circle arc(m)	Central angle of circle arc(°)
Crown arch	0	0	6.65	140
Left side wall	0	0	6.65	38.6716667
Left arch springing	-4.2158	-1.4246	2.2	55.7438889
Invert arch	0	13.6905	17.8920	31.1688889

Table 2 Physical and mechanical indexes of surrounding rock

Surrounding rock classification	γ (kN/m³)	k (MPa/m)	E (MPa)	μ	φ(°)	c (MPa)	φ_c (°)
V	18.5	150	1.5	0.4	25	0.1	46

Notes: The surrounding rock classification was ascertained according to the Chinese Tunnel Design Codes.

Table 3 Secondary lining parameters

Parameters / Position	Concrete strength grade	Steel type	Interval of main rebar(m)	Diameter of rebar(m)	Thickness of secondary lining(m)
Arch wall	C35	HRB335	0.2	0.02	0.5
Invert wall	C35	HRB335	0.2	0.02	0.6

Notes: Thickness of protective layer was 0.05m, symmetrical reinforcement.

In Table 2, γ = unit weight, k= elastic resistance coefficient, E= deformation modulus, μ = Poisson's ratio; φ= internal friction angle; c = cohesion; φ_c = calculation friction angle.

The FEM model contains 320 nodes, 160 beam elements, 160 link elements. All nodes outsides the link elements are constrained completely in the FEM model.

The optimization function (four parameters mode) of LS_MAIN module is used to optimize the tunnel lining design. The result of optimization is shown as Fig. 8, which *SET 14* is the best solution got easily in the case, and the safety factor of the most dangerous part (vault) is 2.688 that is 0.119 only larger than the lower limit of tunnel lining structure safety factor (2.4) required in the Chinese Tunnel Design Codes. Parameters before/after the optimization are listed in Table 4.

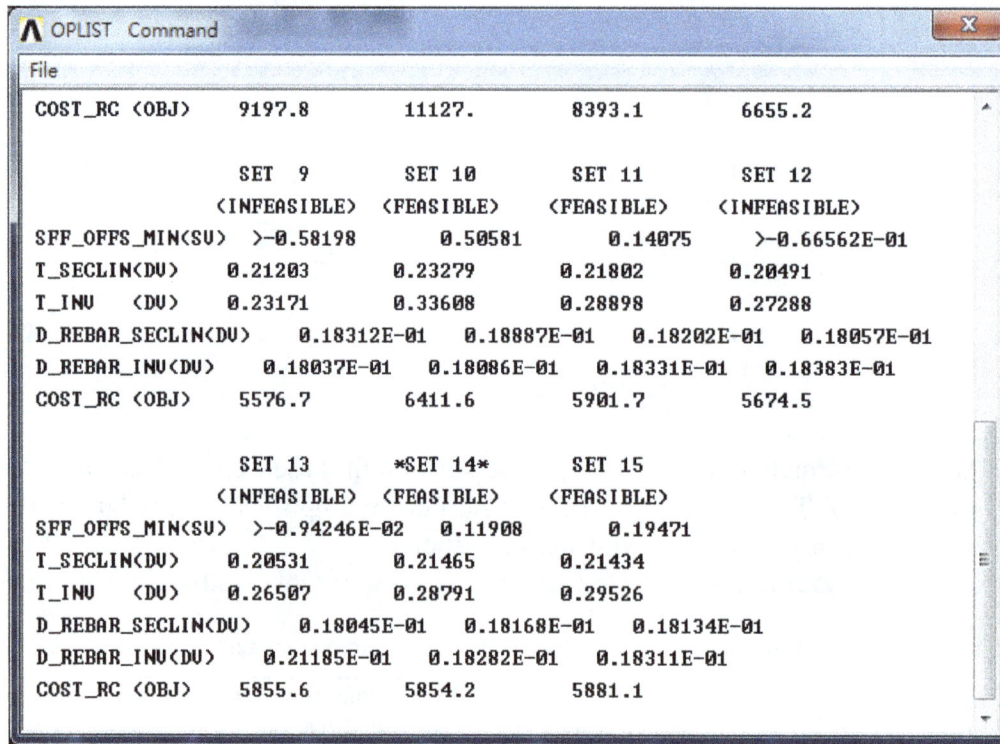

```
∧ OPLIST  Command                                                      X
File
COST_RC <OBJ>     9197.8          11127.          8393.1          6655.2

                  SET  9          SET 10          SET 11          SET 12
                <INFEASIBLE>    <FEASIBLE>      <FEASIBLE>      <INFEASIBLE>
SFF_OFFS_MIN<SV>  >-0.58198        0.50581         0.14075        >-0.66562E-01
T_SECLIN<DV>      0.21203         0.23279         0.21802         0.20491
T_INV   <DV>      0.23171         0.33608         0.28898         0.27288
D_REBAR_SECLIN<DV>    0.18312E-01     0.18887E-01     0.18202E-01     0.18057E-01
D_REBAR_INV<DV>       0.18037E-01     0.18086E-01     0.18331E-01     0.18383E-01
COST_RC <OBJ>     5576.7          6411.6          5901.7          5674.5

                  SET 13          *SET 14*        SET 15
                <INFEASIBLE>    <FEASIBLE>      <FEASIBLE>
SFF_OFFS_MIN<SV>  >-0.94246E-02    0.11908         0.19471
T_SECLIN<DV>      0.20531         0.21465         0.21434
T_INV   <DV>      0.26507         0.28791         0.29526
D_REBAR_SECLIN<DV>    0.18045E-01     0.18168E-01     0.18134E-01
D_REBAR_INV<DV>       0.21185E-01     0.18282E-01     0.18311E-01
COST_RC <OBJ>     5855.6          5854.2          5881.1
```

Fig. 8 The result of optimization

Table 4 Comparison of parameters before/after optimization

Parameters	Thickness of arch wall T_{SecLin} (m)	Thickness of invert wall T_{Inv} (m)	Diameter of rebar in arch wall $D_{Rebar\ Seclin}$ (m)	Diameter of rebar in invert wall $D_{Rebar\ Inv}$ (m)
Before optimization	0.5	0.6	0.02	0.02
After optimization	0.22	0.29	0.018	0.018

To validate whether the result of optimization meet the requirement for the Chinese Tunnel Design Codes, the parameters after the optimization are input into the LS_MAIN module developed in the paper, which the safety factor of lining structure after optimization is obtained, shown as Fig. 9.

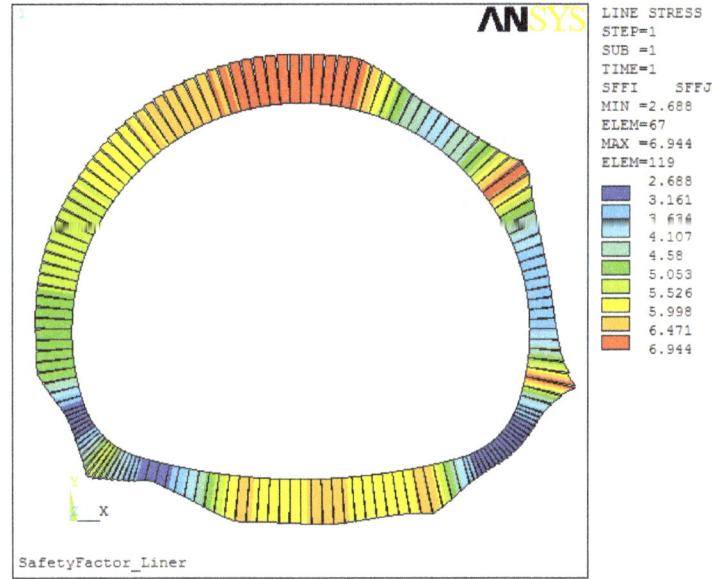

Fig. 9 Safety factor of lining structure after the optimization

6. Conclusion

(1) The TUNNEL_ANALYSIS module developed in the paper can conveniently calculate internal forces and safety factors of tunnel lining structure, which may quickly evaluate and analyze the tunnel lining structure safety, wherein the LS_MAIN module is based on the load-structure method.

(2) Results of analysis may be given in the form of graphics and list files automatically.

(3) The design optimization module developed in the paper may obtain results of optimization conveniently for the tunnel lining structure.

(4) The applicability and simplicity of the design optimization module developed in the paper also support its usefulness. The case validation shows that the module may satisfy the requirements of engineering and study.

References

1. N. Dolzhenko, P. Mathieu, Experimental study of the phases of settlement of a tunnnel with the aid of a scaled-down two-dimensional model, Can. Geotech. J., 42 (2005) 352–364.

2. S.S. Vardakos, M.S. Gutierrez and N.R. Barton, Back-analysis of Shimizu Tunnel No. 3 by distinct element modeling, Tunnelling and Underground Space Technology, 22 (2007) 401-413.

3. A. Fakhimi, D. Salehi and N. Mojtabai, Numerical back analysis for estimation of soil parameters in the Resalat Tunnel project, Tunnelling and Underground Space Technology, 19 (2004) 57-67.

4. J. Zhang, F. Yang, J. Qian, Redevelopment and Application of ANSYS Based on APDL and UIDL, J. Manufacture Information Engineering of China, 23 (2000) 79-81.

5. J. Zhou, Research and development of highway tunnel lining design system [Master dissertation], Tongji University, Shanghai, China, 2006.

Forecast of β-phosphogypsum Strength based on General Regression Neural Network

Wen-Jia Zhang, Zhi-Man Zhao*, Yi-Hui Yao

Faculty of Civil Engineering and Architecture, Kunming University of science and Technology, Kunming 650100, P.R China

Email: lzd2005@126.com

β-phosphogypsum is prepared from phosphogypsum in industrial waste. With characteristics and influence factors determined, the General Regression Neural Network (GRNN) model was established for the forecast of β-phosphogypsum strength. 15 groups of laboratorial data, which were trained by the network as learning samples. With the high precision of predicted results, the ability of GRNN which includes nonlinear mapping function, fault tolerance and self-study, was demonstrated efficaciously in predicting β-phosphogypsum's strength. With laboratorial level and efficiency improved, large quantities of aimless proportioning test and material waste were avoided.

Keywords: β-phosphogypsum; general regression neural network; strength forecast.

1. Introduction

Phosphogypsum, a kind of industrial waste and including large amounts of harmful substances, has harmful effects on human health and ecological environment, whose vast emission and stack not only immensely increased the cost and the land of piling up ,but also destroyed the balance of soil and water[1-5]. According to the data conversion of IFA (International Fertilizer Industry Association), by 2009, global phosphogypsum production was about 175 million tons, while the utilization rate of only 4.5% [6]. As for China, the average annual emissions of phosphogypsum had gone up to 50 million ton, the highest of the whole world; the utilization rate was no more than 10% yet in 2015[7]. Researches of β-phosphogypsum can effectively solve the difficulty of dealing with industrial residue and recycle the waste, which is beneficial to social economy and environment and of great significance to the construction of environment-friendly society.

Artificial neural network was a system model concerning with managing complex information, modelled after the pattern of information transmission between cerebral neurons, which made people constantly learn the generation process of cognitive and construction of principle from the outside world.[14] Compared with the traditional model, it was able to deal with a complex logic operations and nonlinear mapping. The BP neural network and RBF neural network were used most widely in plentiful artificial neural models. But when analyzing forecast information, there were some drawbacks, low convergence speed of error learning and tiny local error, which needed a large number of sample data as the foundation to improve precision. Therefore, it was not advisable when facing with problems of insufficient sample data and redundant interferes [8]. By contrast, when processing data, general regression neural network (GRNN) converged to a optimized regression surface with samples gathering more ,and precedes much in the ability of learning speed, error approximation ability and data processing or classification. It was suitable to apply the case of short of data samples and the unsteady data, and that usually got good predictions [9,10]. β-phosphogypsum's preparation was influenced by many factors and needed a lot of repeated ratio test, consuming time and big intensity labor. General Regression Neural Network (GRNN) can effectively reduced test times and cost, avoided the waste of resources, improved the level and efficiency of experiment, found and trained the relationship among various factors and predict preparation strength.

*Corresponding author

2. Basic Theory and Network Construction

2.1. *Analysis of basic theory*

General regression neural network (GRNN) was first proposed in 1991 by Sprecht in *Neural Network*. As shown in Fig.1, GRNN comprises two layers of artificial neurons, radial basis layer and special linear layer, where R is the number of elements in input vector, Q is the number of neurons of output samples.

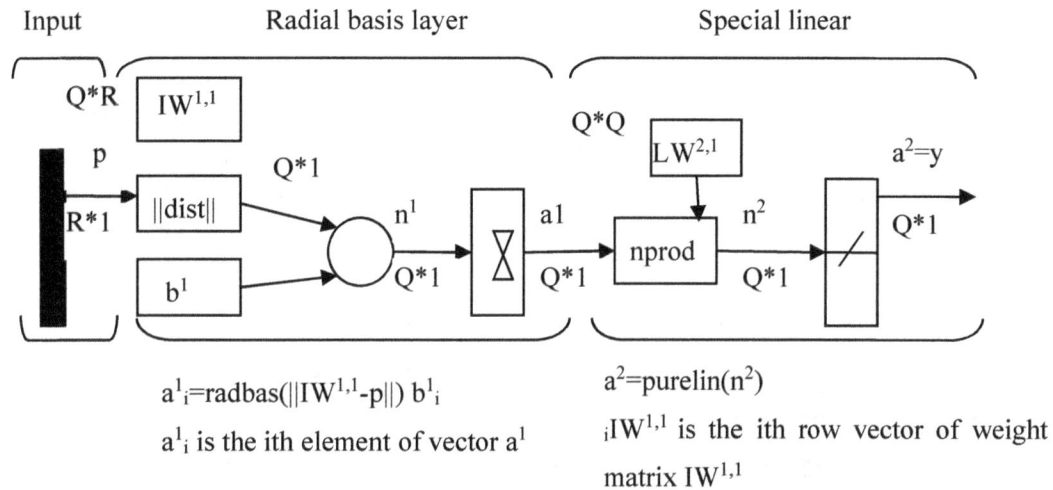

$$a^1_i = radbas(\|IW^{1,1}-p\|)\, b^1_i$$

a^1_i is the ith element of vector a^1

$$a^2 = purelin(n^2)$$

$_iIW^{1,1}$ is the ith row vector of weight matrix $IW^{1,1}$

Fig.1 Structure of generalized regression neural network (GRNN)

Input signals are introduced by the first layer containing neurons of Group Q and this layer uses a Gaussian as the node transfer function of the first layer, as Fig.1 shown. The variable of Gaussian function—smoothing factor σ_i —determines the shape of basic functions of the *i*th hidden layer, the higher σ_i is, the smoother the basic function shows[11,12]. Weight is the distance between network input vector P and the row vector of input weight $IW^{1,1}$, which is calculated by Euclidean distance function module (expressed in $\|dist\|$), and then the threshold value b_1 is used to control output, the output result is passed to a_1. Usually, network output layer processes the data by normalized dot product (expressed in nprod) firstly, then uses $a^2 = purelin\ (n^2)$, a linear transfer function, to calculate the network output, determines the output finally.

$$R_i(x) = \exp\left(-\frac{\| x - c \|^2}{2\sigma_i^2}\right) \tag{1}$$

The feed-forward neural network of BP neural network is still used to learn and correct the connect weight of generalized regression neural network, which makes the process of weight learning tend to be more precise. Gaussian function responses locally to the input signal. That is, when the input signals are close to the central range of basis function, the nodes of hidden layer will produce larger output [13]. So using Gaussian function as the radial basis node function can make the network learn faster than BP neural network and RBF network. In addition, the artificial parameters of GRNN are less, which can minimize effects of subjective factors on the prediction.

2.2. *Network construction*

The wasted phosphogypsum materials of this experiment were all from the industrial residue of Yunnan SanHuan chemical CO. LTD. After the measure of authority, phosphogypsum's color was grayish yellow, whose moisture content was 18.1%, PH value was about 2.3 after adding 50% water, main composition was $CaSO_4 \cdot 2H_2O$(content of 86.49%) and mineral composition was shown in Table 1. All phosphogypsum of experiment were pretreated, including sieving twice (100mu), watering, adding $Ca(OH)_2$ and crystal, drying. Finally, phosphogypsum conformed to national standard was prepared. Using the above preparation of phosphogypsum to implement experiment with the optimal mix proportion, in which water reducer adopts polycarboxylate superplasticizer, liquid, concentration of 60%. Retarder were citric acid, solid, content of 98%. Hydrated lime mixed with calcium oxide and water was field configuration, concentration of 20%. The experiments were implemented with different mixing ratio of experiment respectively and spot, whose phosphogypsum was moulded by casting with a mould of 160 mm x 40 mm x 40 mm. Demoulding after with 3h natural curing, and then curing again on the room temperature after 28d, phosphogypsum were able to used in compression and flexural experiments.

Table 1 Phophogypsum component

Raw materials	State	SiO_2	Al_2O_3	TFe_2O_3	MnO	MgO	CaO	K_2O	P_2O_5	SO_3	Cl^-	F^-	Organic
Phosphog-ypsum	Powdery	14.52	1.66	0.15	0.005	0.17	31.94	0.22	0.94	45.38	0.027	0.12	0.25

Based on the experimental results, with the analysis for factors of strength of phosphogypsum, the water-gypsum ratio(WG), PH value, hydrated lime content (HL), water reducing agent content(WR) and retarder(R) were determined as the main factors of phosphogypsum strength. Whereafter the compression and flexural experiments after 28d were measured with national standard and used as output factor and output of network. According to determined output factor and input of network, GRNN was built with the training samples that were made up of the matching data and test results' statistic data of 15 sets of laboratory preparation. Network model uses experimental data of Group 1-12 as training samples of network and experimental data of Group 13-15 as extrapolation forecast. Specific statistic of output and input samples were shown in Table 2.

Table 2 Experimental test sample data of phosphorus gypsum

number	Input sample					Target sample	
	WG	PH	HL(%)	WR(%)	R(%)	F(MPa)	C(MPa)
1	0.362	10.39	5.0	0.55	0.35	6.18	13.07
2	0.362	11.01	4.8	0.70	0.33	6.17	15.17
3	0.355	9.78	5.0	0.90	0.30	5.66	13.64
4	0.412	10.05	4.5	0.55	0.35	5.67	9.27
5	0.523	10.03	5.0	0.40	0.32	5.23	8.52
6	0.351	8.89	6.0	0.65	0.30	5.41	7.64
7	0.278	9.44	5.0	0.71	0.35	6.25	16.91

(*Continued*)

Table 2 (*Continued*)

number	Input sample					Target sample	
	WG	PH	HL(%)	WR(%)	R(%)	F(MPa)	C(MPa)
8	0.484	10.01	4.5	0.65	0.33	6.15	9.23
9	0.362	9.77	4.0	0.90	0.32	4.73	6.79
10	0.413	8.55	5.5	0.35	0.35	5.04	6.58
11	0.401	10.12	6.0	0.35	0.31	5.57	7.92
12	0.297	11.11	5.0	0.76	0.35	6.96	16.33
13	0.374	9.67	5.5	0.85	0.33	6.02	14.61
14	0.375	10.21	5.9	0.74	0.36	5.87	10.54
15	0.401	9.49	5.5	0.50	0.32	5.62	7.95

PS: WG is the water-gypsum ratio, PH is PH value, HL is hydrated lime content, WR is water reducing agent content, R is retarder, C is compression strength and F is flexural strength.

A lot of experimental data show that phosphogypsum could achieve better compressive strength and flexural strength when water paste ratio ranges from 0.2 to 0.6; PH shown alkalinity; hydrated lime content was in the area of 4% to 6%; water reducing agent content was between 0.3% and 1%; dosage of citric acid retarder was between 0.3% and 0.4%. As shown in Table 2, between 15 groups of experimental data selected randomly, water paste ratio's maximum was 0.523, minimum was 0.278, average was 0.383; PH's maximum was 11.11, minimum was 8.55 and average was 9.901; hydrated lime content's minimum was 4.0%, maximum was 6.0% and average was 5.15%; water reducer minimum of dosage of water reducer was 0.35%, maximum was 0.90% and average was 0.637%; citric acid retarder content's minimum was 0.3%, maximum was 0.36% and average was 0.331%. Data in Table 2 are normalized to introduce to the area [0, 1] through the normalization formula(as shown in Formula 2). As for the data having been in the area [0, 1], such as paste ratio, water reducer content and retarder content, there was no need to normalize[11].

$$x^* = \frac{x - x_{min}}{x_{max} - x_{min}}$$

(2)

Where X is the actual value, X* is the normalized value; X_{max} and X_{min} are the maximum and minimum of corresponding actual value.

3. Testing Results of Data and Analysis

After normalization, Matlab was used to build network to train and test. Because the smoothing factor of network was an artificial parameter which could affect performance of network prediction and output error, the determination of best value of that needed constant trial according to the error distribution curve. In this work, the smoothing parameter σ=0.1, 0.2, 0.3, 0.4, 0.5 was chosen to train network, and the errors between actual values and forecasting values were shown in Fig. 2.

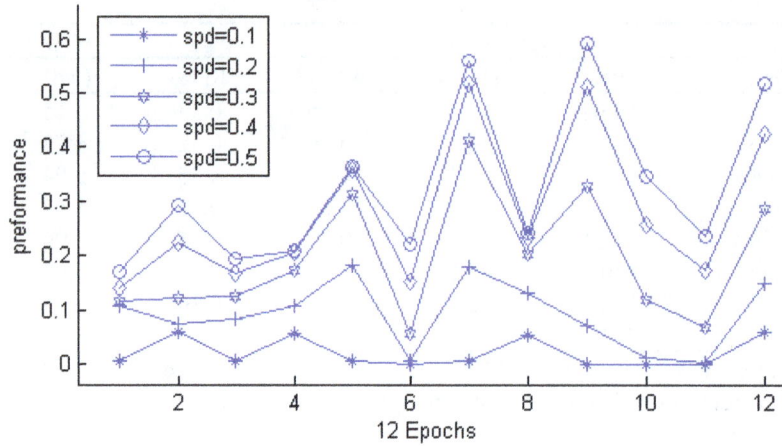

Fig. 2 The approximation error of network

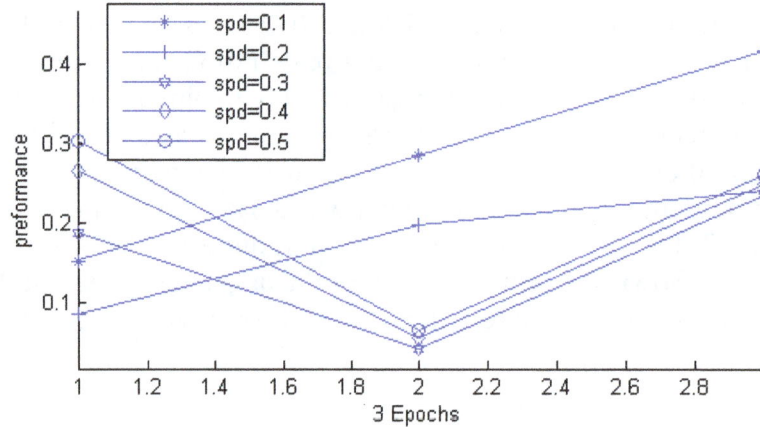

Fig. 3 The prediction error of network

It was concluded that when the smoothing parameter reduces from 0.5 to 0.1, errors of samples in network fluctuated weakly, that was to say errors were less. In the curve of prediction error of network shown in Fig. 3, when $\sigma=0.1$, prediction error was biggish and rising; when $\sigma=0.2$, prediction error tend gently, about 0.2; when $\sigma=0.3$, 0.4, 0.5, prediction errors fluctuate unsteadily in the area [0.1,0.6]. In conclusion, taking approximate error and prediction error into comprehensive consideration, 0.2 was chosen to the best smoothing parameter to train the network.

As shown in Table 3 was the calculated result and relative errors of fitting, when $\sigma=0.2$, the absolute value of greatest relative fitting error of flexural strength was 14.54%, the minimum absolute value is 0.03% and the absolute average of relative fitting error was 7.28%; the relative error of fitting compressive strength is 18.10%, the minimum absolute value was 0.32% and the absolute average of relative error was 9.33%; the fluctuation range of error was narrow and acceptable.

Table 3 Relative error of fitting sample (%)

items	Number											
	1	2	3	4	5	6	7	8	9	10	11	12
Relative error of flexure	6.87	7.63	7.90	8.90	14.54	0.55	8.56	11.88	6.11	0.96	0.03	13.38
Relative error of compression	10.90	7.65	8.56	10.96	18.23	0.55	18.10	13.21	7.23	1.23	0.32	14.98

After normalization to output results of network, as shown in Table.4 were calculation results and relative errors of extrapolation sample, the maximum relative error of flexural strength was 10.59%, the minimum was 0.26% and the absolute average of relative error was 5.96%; the maximum relative error of compressive strength was 24.13%, the minimum was 8.66% and the absolute average of relative error was 17.51%; the fluctuation range of prediction error of flexural strength was narrow and acceptable, while the compressive strength's was existent and also allowable. Through the analysis of compressive and flexural strength's fitting training and prediction, it was found that either the error between fitting value and measured value, or between prediction value and measured value, exist certain fluctuation, but the prediction values were close to the measured values. It was proved that the prediction of network to phosphogypsum's strength was applicative, the error conforms to the precision of fitting and prediction and GRNN was suitable for prediction to β- phosphogypsum strength.

Table 4 Calculation results and relative errors of extrapolation sample

Number	Measured flexural strength	Predicting flexural strength	Error value (%)	Measured compressive strength	Predicting compressive strength	Error value (%)
13	6.02	6.09	7.72	14.61	14.69	8.66
14	5.87	5.98	10.59	10.54	10.73	19.74
15	5.62	5.63	0.26	7.95	8.19	24.13

4. Conclusion

According to the experimental study, the phosphogypsum strength was deeply affected by water paste ratio and admixture. Under the circumstance that citric acid content was about 0.35%, polycarboxylic acid was about 0.7% and water paste ratio was about 0.3, phosphogypsum could achieve high strength and preparation with more stable strength. The process of network forecasting exists certain errors at the reason of limit of training samples, however, the errors of network forecasting are acceptable and the forecast is applicative. The research of β-phosphogypsum's mix proportion and forecasting of compressive and flexural strength is diverse, nonlinear, widely involved and high comprehensive. With the traits—high nonlinearity, fault tolerance, self-learning and real-time processing of GRNN, the network solves the problem of strength forecasting of phosphogypsum well, outputs more precisely, saves materials, reduces the times of experiment, provides scientific guidance for researches of phosphogypsum, meanwhile promotes the utilization of computer science and technology in experiments of building materials.

References

1. Wang Qiqing. Gypsum building materials application [M]. Chemical industry press, 2008.12: 16-17.

2. The ministry of industry and information technology guidance about comprehensive utilization of industrial by-product gypsum [J]. Journal of ministry of no. [201173]

3. B. N. Estevinho, E. Ribeiro, A. A. Santos, A preliminary feasibility study for pentachlorophenol column sorption by almond shell residues, Chem. Eng. [J]. 136 (2008) 188-194.

4. V.C. Srivastava, I.D. Mall, I.M. Mishra, Adsorption thermodynamics and isosteric heat of adsorption of toxic metal ions onto bagasse fly ash (BFA) and rice husk ash (RHA), Chem. Eng. [J].132(2007) 267-278.

5. Ceng Ming Ruan Yan, Chen Jing Chong-Ying Wang, Zhang Bing, Zhou Zichen. Phosphogypsum is different from the effect of the treatment comparison [J]. Journal of Building Materials in the World. 2011.6: 18-21.

6. Wang Xinlong, Cheung Chi. The phosphorus gypsum utilization way. Phosphate fertilizer and compound fertilizers [J], 2010, 25(3):61—66.

7. Ye Xuedong. "twelfth five-year" has a long way for comprehensive utilization of phosphogypsum [J]. Journal of phosphate fertilizer and compound fertilizers. 2012, 1(27): 7-9.

8. Ge Zhexue, Sun Zhiqiang. Neural network and MATLAB R2007 Application [M]. Publishing House of Electronics Industry of China 2007.9

9. Yan Pingfan, Zhang Changshui. Artificial neural network and simulated evolutionary computation [M] Beijing: Tsinghua University Press, 2000.

10. Shi Zhongzhi. Knowledge discovery [M], Beijing: Tsinghua University Press, 2002.

11. Sprecht D F. A General Regression Neural Network [J]. IEEE Tran Neural Network, 1991, 2:568-576.

12. Sprecht D F. A General Regression Neural Network Rediscovered [J]. Neural Network, 1993, 6:1033-1034.

13. Zhao Chuang, Liu Kai, Freight Volume Forecast [J]. Journal of the China Railway Society, 2004, 2, 26(1): 12-15.

14. Jiang Shaofei. Artificial neural network is used to data processing method in the field of construction engineering [J]. Journal of Harbin University of Civil Engineering and Architecture, 1999, 32(5): 24-28.

Analysis of Problem for Secondary Lining Contact of Highway Tunnel in Cold Regions Based on ANSYS

Xiao-Wan Zhang[1,*], Yi-Feng Zheng[1], Fan Liu[2]

[1]*College of Construction Engineering, Jilin University, Changchun 130026, China*
[2]*Jilin Provincial Communication Planning and Design Institute, Changchun 130021, China*
*Email: 944748003@qq.com

In this paper, we exploit the contact unit (TARGE170 and CONTA174) in ANSYS analysis software to simulate the secondary lining structure and waterproof material of tunnels. Then the longitudinal deformations of the secondary lining structure in air temperature field for cold-region are computed based on instances under different temperatures respectively. Finally, we compare the computation with the data field observed and analyze the results.

Keywords: Finite element; ANSYS; contact analysis; secondary lining of the tunnel; cold regions.

1. Introduction

Highway tunnel construction is developing rapidly in China. However, due to the crack disease, tunnel lining is affected in different degrees and there is no exception in cold regions. In cold areas, cracking not only leads to degradation of performance but brings hidden dangers to pedestrians and vehicles, which has caused great losses to the country's economy. Therefore, it is of great significance to study the longitudinal deformations of the highway tunnels in cold regions, which provides the design of highway tunnels with reference frame.

There is an enormous amount of study conducted by researchers worldwide. Zhang *et al.* make a comprehensive analysis of reasons for crack of tunnel lining. They exploit finite element model to simulate the tunnel linings' stresses and it demonstrated that crack make more significant variation of stresses [1]. Huang *et al.* analyze the monitoring data and field geological condition of a tunnel in Zhejiang province and build a finite element model of composite lining based on spring-contact. It shows that asymmetry pressure and variable temperatures have great influence on the lining structure [2]. Yang *et al.* simulate crack of the secondary lining of tunnel under different external force and get the regularities of development and distribution of crack [3].

In this paper, a tunnel in Tonghua county, Jilin province is studied. We analyze the longitudinal deformations rule of the secondary lining of tunnel in cold regions and get some useful conclusions.

2. Project Profile of a Certain Tunnel and Experimental Research

The tunnel we studied, locates in Tonghua county, Jilin province, is a separate type single lane double-track tunnel. The left tunnel is 988 meters in length and its stake number is ZK44+366∼ZK45+354. And the right tunnel is 980 meters in length and its stake number is YK44+342.116∼YK45+322.116. By field measurement and experimental research, we got the experimental parameter of a certain tunnel as following:

Difference in temperature: The temperature difference of concrete material is computed by temperature variation (T_0), hydration heat temperature difference (T_r) and shrinkage temperature difference of concrete material (T_s).

In some period from January 2010 to July 2010, our group member Li observed the air temperature changes in the tunnel thus measured data of temperature in winter and summer are

---

**Corresponding author

obtained. The highest mean temperature in summer of the tunnel is 20°C~26°C while the lowest mean temperature in winter is -13°C~-17°C. Then the temperature changes in the tunnel range from 33°C~43°C.

In addition, Professor Wang Tiemeng works out the relationship between hydration heat temperature difference T_r and wall thickness h of the structure on the basis of a large number of measured data in China [8]. Based on this relationship, we obtain the revised hydration heat temperature difference of the tunnel: $T_r' = k_1 k_2 k_3 k_4 T_r = 1.2 \times 1.0 \times 1.56 \times 1.0 \times (4.8 \sim 6.0) = (9.0 \sim 11.2)\,°C$.

In this paper, we set $T_r'=10°C$. Finally, considering the actual situation of the tunnel, we estimate the concrete shrinkage referring to the EU standards. Then we compute the concrete shrinkage temperature difference of the secondary lining structure of the tunnel: $T_s = -\varepsilon_s / \alpha_t \approx -9°C$

Where ε_s denotes the shrinkage strain limit of the secondary lining structure of the tunnel. Generally, we set $\varepsilon_s = 3.03 \times 10^{-4}\,mm$. In this paper, we consider the humidity condition during construction of the secondary lining structure and take a correction factor 0.3 like [11]. Then we have $\varepsilon_s = 0.3 \times 3.03 \times 10^{-4} = 9.09 \times 10^{-5}$.

Therefore, we can compute the temperature difference of the secondary lining structure of the tunnel by temperature variation, hydration heat temperature difference and shrinkage temperature difference of concrete material:

$$\Delta T = -(T_r + T_0 - T_s) \ldots\ldots\ldots\ldots\ldots\ldots\ldots\ldots (1-1)$$

Take the relevant data into (1-1), we get the temperature difference of the tunnel: $\Delta T = -(52 \sim 62)\,°C$.

Surrounding rock pressure: According to the calculation methods for surrounding rocks of deep and shallow buried tunnels [10], we compute the vertical uniform pressure of IV level surrounding rocks of the tunnel. The result is 126.0846 (kn/m^2) and this will not be introduced in detail here.

Resistance coefficient of waterproof material: In [4], our research group member Men Yang obtained the relationships between shear stress and deformation on the contact surface of the waterproof material and lining structure under different normal stress by shear test. According to the relationship, we derive the friction coefficient between waterproof material (EVA and PVC) and lining structure. Thereby, we get the restraint effect of waterproof material on the free deformed secondary lining of the tunnel.

Test analysis of EVA type specimen:

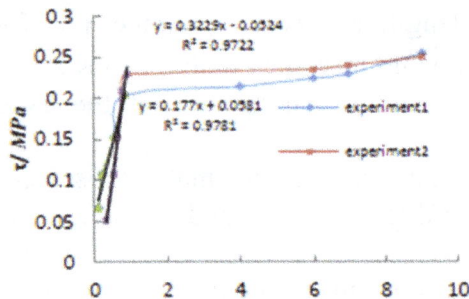

Fig. 2.1 τ-u curve for EVA+geotextile ($\sigma_n=1MPa$)   Fig. 2.2 τ-u curve for EVA+geotextile ($\sigma_n=2MPa$)

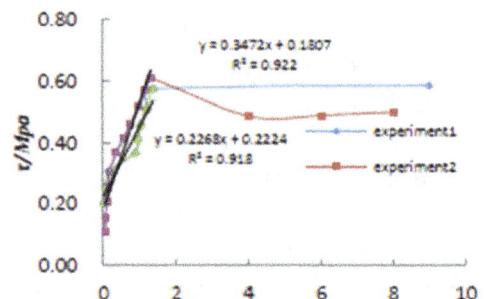

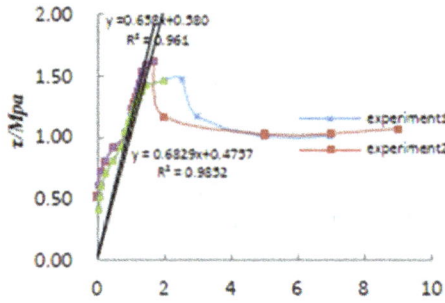

Fig. 2.3 τ-u curve for EVA+geotextile (σ_n=3MPa)   Fig. 2.4 τ-u curve for EVA+geotextile (σ_n=4MPa)

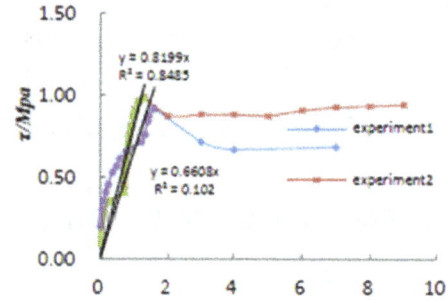

In Figure 2.1~ Figure 2.8, we show the graph of relation between the shear stress of test specimen (EVA and PVC) and displacement (τ-u) under four different normal stresses respectively. We set the normal stress $\sigma_n = 1MPa$、$2MPa$、$3MPa$、$4MPa$. The curves in these figures conform to relationship in the elastic coulomb model figure, which proves that our test data is reasonable.

Test analysis of PVC type specimen:

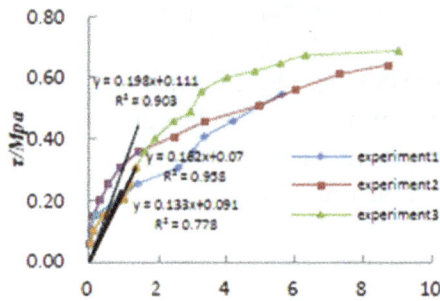

Fig. 2.5 τ-u curve for PVC+geotextile (σ_n=1MPa)   Fig. 2.6 τ-u curve for PVC+geotextile (σ_n=2MPa)

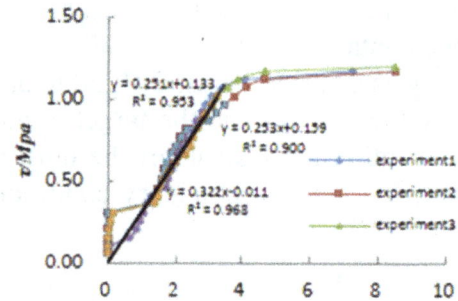

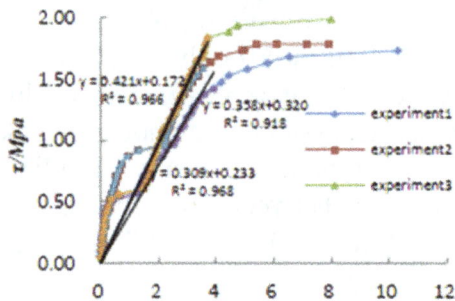

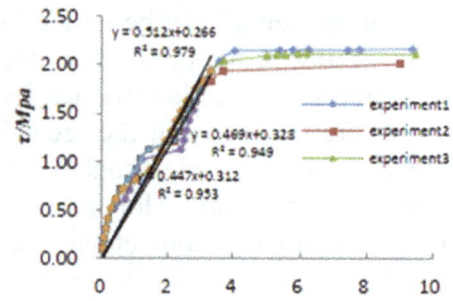

Fig. 2.7 τ-u curve for PVC+geotextile (σ_n=3MPa)   Fig. 2.8 τ-u curve for PVC+geotextile (σ_n=4MPa)

In accordance with the τ-u curve and the computational formula of elastic coulomb model, we derive the resistance coefficient values between the secondary lining and waterproof material of each test specimen group. According to the above test results, the friction coefficients between waterproof material and the secondary lining under different load in the supporting structure of the tunnel are as follows: PVC+geotextile: $\mu = 0.55$ EVA+geotextile: $\mu = 0.26$

When we study the contact of waterproof material and the secondary lining, the longitudinal deformation of the secondary lining will experience two stages, which are elastic and plastic deformation respectively. Our research group member Men Yang worked out the relationship between shear stress and deformation of waterproof material and the secondary lining in the elastic stage by experiment. Based on this relationship, we compute the shear modulus between waterproof material and the secondary lining structure: $PVC + geotextile: C_x = K_s = 0.102\sigma_n + 0.065 = 0.0777$ $EVA + geotextile: C_x = K_s = 0.150\sigma_n + 0.058 = 0.0766$

3. Introduction to Critical Theories of ANSYS Modeling of the Tunnel

Contact problem means the problem that stress and strain are produced between local areas of two objects under the action of outer load. In the supporting structure of tunneling, the contact problem of waterproof material and the secondary lining is a highly nonlinear problem.

There are two basic types of contact problem in ANSYS: the contact between rigid body and soft body, the contact between soft body and soft body [9]. According to [6], the contact has three forms: point to point, point to surface and surface to surface. The model in this paper is coulomb friction model, which is controlled by parameters such as TAUMAX, FACT, DC, COHE. Contact unit includes CONTA171 ~ CONTA178, while target unit includes TARGE169 and TARGE170 like [2]. When studying the contact problem of waterproof material and the secondary lining in cold-region highway tunnels, we define the secondary lining and waterproof material as "target" surface and "contact" surface respectively and simulate the actual contact problem with surface to surface contact form [5].

Analysis steps contains: 1 Build geometric model and divide mesh; 2 Identify the contact pair; 3 Specify the contact and the target surface; 4 Define the contact and the target surface; 5 Set key options and real constants of the unit; 6 Define the movement of target surface for rigid-flexible contact; 7   Impose boundary conditions; 8 Define solution options and load steps; 9 Solve and get results [7].

ANSYS Modeling Analysis of the Tunnel and Comparison of Results

At first, we build a 12 meters long solid model. Along the tunnel longitudinal, we take 1 meter as a unit so that the tunnel can be divided into 12 parts. Along the transverse, for the reason that the secondary lining of the tunnel is connected by multiple arcs, it is difficult to divide the mapping mesh. Therefore, we divide meshes for the 7 cm thick area of waterproof material + buffer layer and the 40 cm thick area of the secondary lining structure separately. We choose Drucker-Prager elastic-plastic model, which contains 9383 joints and 8908 units. Solid65 unit is applied to divide the supporting structure. Moreover, we simulate the contact between the secondary lining and waterproof material by contact pair built by contact unit TARGE170 and CONTA174, as shown in Figure 3.1.

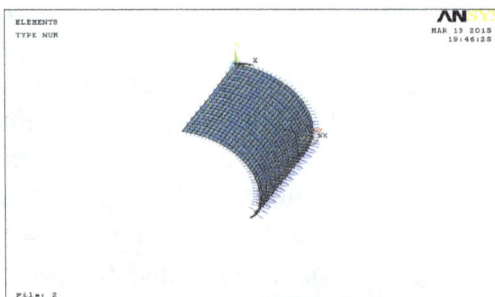

Figure 3.1 Sketch map of contact surface

While modeling, half part of the tunnel structure is built. Then we build the whole tunnel model by symmetry command, as shown in Figure 3.2 and Figure 3.3.

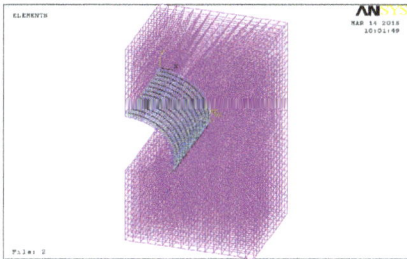

Figure 3.2 Sketch map of 1/2 model   Figure 3.3 Sketch map of the whole model

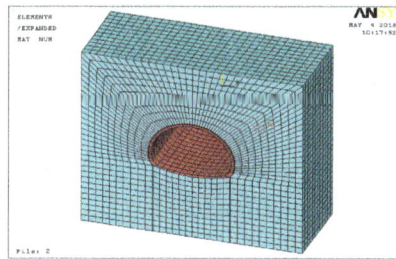

We obtain the distribution of deformation and stress of the secondary lining structure under a -30°C temperature difference load case through nonlinear iteration calculation conducted by ANSYS modeling. As shown in Figure 3.4, the deformation and stress values of point A (a series points along the tunnel longitudinal) are extracted for analysis.

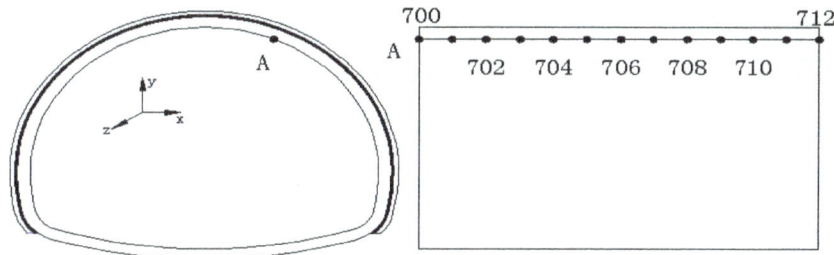

Figure 3.4 Sketch map of point A's location

Based on the extracted computational data, we can plot the deformation curve and stress curve under a -30°C temperature difference load case, as shown in Figure 3.5 and Figure 3.6.

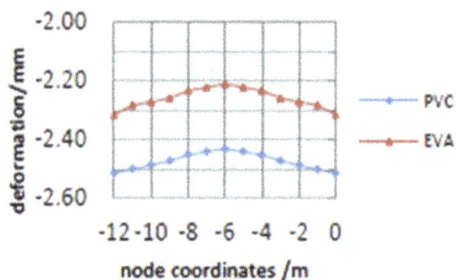

Fig. 3.5 Deformation curve of nodes along   Fig. 3.6 Stress curve of nodes along
the tunnel longitudinal   the tunnel longitudinal

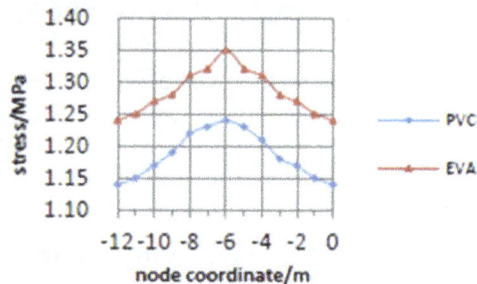

From Figure 3.5 and Figure 3.6, the maximum values of deformation and stress are all obtained at node 706 (Z=-6m), which locates at the middle part of the built model. This result is consistent with the computation of Men Yang [4]. Then we select node 706 as the research object, and study the longitudinal deformation law and the corresponding stress change at this node under various temperature conditions in the temperature field.

We simulate the contact processes between the secondary lining and two kinds of waterproof material (EVA and PVC) respectively and output the results. Then we plot the deformation curve in Figure 3.7 and stress curve in Figure 3.8 under various temperature, which is from 25°C to -21°C (the temperature difference is from -14°C to -60°C).

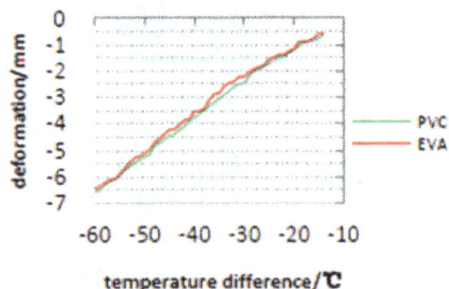

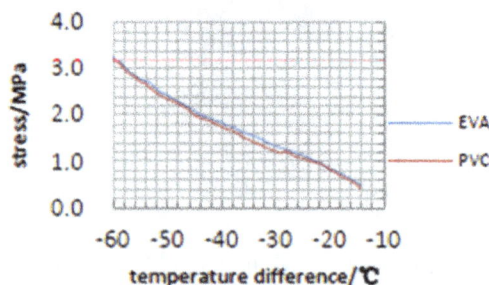

Figure 3.7 Temperature-deformation curve   Figure 3.8 Temperature-stress curve

Comparing the output of our model with the field observations, we can get the crack width between them, as shown in Table 3.1.

Table 3.1 Comparison of field observations and ANSYS simulated result

	Winter crack width(mm)	Summer crack width(mm)
Field observations	2.5~5.3	0~1.2
ANSYS simulated result	PVC: 4.65 EVA: 4.49	0

4. Conclusion

(1) Analyzing the comparison of the model output and the results field observed, we can see that the longitudinal deformation of the secondary lining under different temperature is in accordance with the theoretical change rule. During the process of cooling in winter, the crack size of the secondary lining is basically consistent with the results field observed.

(2) By comparing the effects of EVA and PVC to the secondary lining, we concluded that the longitudinal deformation of the secondary lining under the contact of PVC waterproof material is larger than the deformation under the contact of EVA waterproof material, which demonstrates the restriction effect to the second lining of PVC is smaller than the effect of EVA.

References

1. Zhang Wei, Li Xi bing, Gong Feng qiang, Li Di yuan. Cause analysis and numerical simulation study of lining cracking of highway tunnel [J]. Engineering construction, 2007, 39(1): 26-29.
2. HUANG ZhiYi YU WeiDa, WANG JinChang. Numerical Simulation of Highway Tunnel Composite Lining Temperature Stress [J]. Highway Engineering, 2010, 35(4); 54-57.
3. Yang Hua, Zhang Wenzheng. Numerical Simulation of a Longitudinal Lining crack of a Highway Tunnel [J]. MODERN TUNNELLING TECHNOLOGY, 2011, 48(5); 46-51.
4. Men Yang. Mechanism study on longitudinal deformation of secondary lining of highway tunnel in cold regions[D]. JiLin University, 2013.

5. Guo Xiaoming, Zhao Huilin. Study on contact problems in engineering structures [J]. JOURNAL OF SOUTH EAST UNIVERSITY (Natural Science Edition), 2003, 33 (5); 577-582.

6. Li Yan. The contact element analysis and application in engineering based on the ANSYS software [D]. JiLin University, 2004.

7. Wang Xu cheng, Finite element method [M]. Beijing: Tsing Hua University Publishing House, 2003,7.

8. Wang Tie meng. Engineering structure crack control [M]. Beijing: China Building Industry Press, 1997.

9. He Ben Guo. Application Cases of Civil Engineering with ANSYS Software Beijing: China Water Power Press, 2011.10

10. Mashimo, H. & Isago, N et al. 2001. Experimental study on static behavior of road tunnel lining[C]. Proc. of Modern Tunneling Science and Technology 1: 451-456.

11. Xiao TongGang. The Safety Analysis of Tunnel Surrounding Weak Rock-Supporting System and Its Engineering Application [D]. Shanghai: Tongji University, 2007.

Numerical Simulation for Lost Foam Casting of Diamond Grinding Wheels

Qiu-Lian Dai[a,*], Can-Bin Luo[b], Fang-Yi You[c]

MOE Engineering Research Center for Brittle Materials Machining, Huaqiao University
Xiamen, Fujian, 361021, P.R China
[a]lisadai@hqu.edu.cn, [b]lchrobin@hqu.edu.cn, [c]fangyiy@hqu.edu.cn
*lisadai@hqu.edu.cn

The Al-based matrix diamond grinding wheels manufactured by the Lost Foam Casting (LFC) process was explored in this research. Software of Pro-CAST for numerical simulation casting process was used to simulate the flow fields and temperature fields of aluminum-based diamond grinding wheels during the LFC process. The influences of different gating system designs and process parameters on the filling and solidification behaviors of the wheels were simulated. Casting defects were also predicted. By analyzing the simulation results of the casting process, the gating system and process parameters were optimized as follows: The runner and gating system was located at the center of the cast; The pouring temperature was 740°C and the vacuum degree was 30kPa; The foam densities of wheel substrate and wheel grinding layer were 15kg/m^3 and 30kg/m^3 respectively. By using the optimized results, diamond wheels were fabricated successfully.

Keywords: Al-based matrix; diamond grinding wheels; lost foam casting; numerical simulation; gating system; casting parameters.

1. Introduction

Lost foam casting (LFC) or Vacuum evaporative pattern casting (V-EPC) utilizes a polystyrene pattern embedded in dry sand which is not removed before pouring the melt. The polystyrene pattern is decomposed after introduction of the melt and is substituted by the molten metal [1,2]. LFC has been focused in the past years because of numerous advantages such as the ability to produce complex shapes, no mold parting line, no cores, and more accurate dimensions, high infiltration capability due tosucking force of the vacuum degree, no environmental pollutants in the foundry practice [3].

One of the other benefits of this process is the viability of inserting different inserts into the foampattern before pouring the melt. It is, therefore, feasible to fabricate metal matrix composites in near-net shape by this process. At present, lost foam casting (LFC) infiltration processing is an attractive process to produce metal matrix surface composites reinforced with hard ceramic particles, which is applied to wear-resistant parts [4]. In nature, metal-bonded super-hard abrasive grinding wheels can be regarded as a kind of surface-layer composites reinforced with diamond or CBN particles in the metal matrix. Therefore, it is possible to use LFC infiltration processing to fabricate the metal-bonded super-hard abrasive grinding wheels.

The traditional metal-bonded multilayer super-hard abrasive grinding wheels are manufactured by powder metallurgy. Sufficient alloying cannot develop under the hot pressing conditions, where the hot pressing temperature is low and the time is short. Accordingly, interfacial reaction between the alloy elements and the diamond grits is limited and bonding strength between metal matrix and diamond grits is weak. Compared to the traditional manufacturing method of powder metallurgy, the following advantages can be expected by using LFC infiltration processing to fabricate the metal-bonded super-hard abrasive grinding wheels: (1) The retention of diamond grits can be enhanced by the increasing of the mechanical locking and the chemical bonding. The increase of the mechanical locking is due to the contraction of the matrix around the diamond grits when it solidifies from liquid to solid during the LFC infiltration processing. The chemical bonding is expected because the metal matrix is in liquid state during the process and chemical bonding can be promoted due to higher reaction rate existing at the interface. (2)Grinding substrate and its grinding layer is a once-molded product formed in one piece. (3)The cost of the products is lower in mass production due to the lower cost of the raw materials and higher production rates of the casting process.

The quality of the LFC casting is influenced by many parameters [5]. This study aims to investigate the effects of the gating system design, the pouring temperature, the density of polystyrene foam and the vacuum degree on the filling and solidification characteristics, as well as porosity on the Al-based matrix diamond grinding wheels during the LFC casting process. Software of Pro-CAST for numerical simulation casting process was used to simulate the flow field and temperature field as well as the temperature of the diamond grits during the LFC process.

2. Methodology

Casting process analysis

The shape and size of the diamond wheel was shown in Figure 1. It can be seen that the wheel has a disc shape consisting of grinding substrate and grinding layer. In order to ensure the casting quality, especially the quality of the grinding layer, the optimization criteria for the casting wheel were defined as:

(1) It has a short filling time;

(2) Liquid metal fills smoothly; and

(3) No obvious shrinkage cavities are formed at the grinding layer. Based on this consideration, the gate should be designed at the center of the wheel.

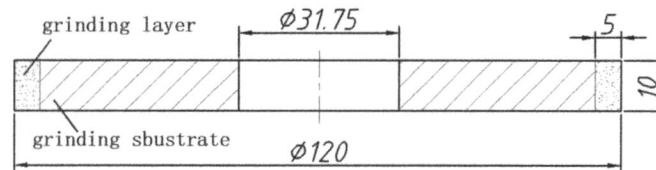

Fig. 1 Schematic illustration of the diamond wheel

Settings of boundary conditions and simulation parameters

The metal matrix for the diamond wheel in this study was Aluminum alloy(ADC12). The compositions of the alloy were shown in Table 1. Its liquids and solids are 625°C and 476°C respectively. The thermo-physical properties of polystyrene, diamond and sand were presented in Table 2. The heat transfer coefficient between mould and casting, mold and sand were set to be 500 W/(m²·K) . The thickness of coating was set to be 0.5mm and the air permeability was set to be 1×10^{-9} cm²/(Pa·s). The initial temperature of sand and foam pattern was set to be 20°C.

Table 1 Chemical composition (wt%) of Aluminum alloy ADC12

Constituent element	Si	Cu	Mg	Zn	Fe	Mn	Ni	Sn	Al
Percentage	11.0	2.5	0.3	1.0	0.9	0.5	0.5	0.3	Bal.

Table 2 Thermo-physical properties of polystyrene, diamond and sand

Materials	Thermal-conductivity (W/m·K)	Specific-heat (kJ/kg·K)	Latent-heat (kJ/kg)	Density (kg/m³)	Air-permeability cm²/(Pa·s)	Liquids (°C)	Solids (°C)
Polystyrene	0.15	3.7	100	350		330	
Diamond	2000	0.5	—	3515			
Sand	0.53	1.22	—	1520	1×10^{-7}		

The establishment of the geometric model

The three-dimensional entity model for the casting was established by using Solid-works. The number of diamond grits in the grinding layer was set to be 990.The diamond grits were supposed to be spherical and had a size of 0.6mm. The model was inputted into the Pro-CAST system and then translated into the FE(finite-element) model.

3. Results and Discussion

Numerical simulation to optimize runner and gating system design

According to the casting process analysis of the grinding wheel above, four models were proposed to investigate the effects of different gating system designs on their filling behaviors. The pouring temperature was set to be 740°C. The vacuum degree of 30kPa was used. The foam density of the wheel substrate and grinding layer were set to be 15kg/m³ and 30kg/m³ respectively. Figure 2 was the simulation results of the filling time distributions of the four models. It can be seen that the melt filled the spur and runner first and then filled the whole casting from the central portion of the part to the outside layer. In Model IV, three equal risers were added to the top of the casting part.

However, the filling behavior of Model IV was more turbulent than in the other models. Also, the outside grinding layer could not be filled simultaneously, which may enhance the gathering of diamond grits. The filling behaviors of Model I (three runners design) and Model II (a boss at the center) as well as Model III were similar. The metal filled generally and layer by layer from the central portion of the part to the outside grinding layer, which was essential for ensuring the quality of the grinding layer of the diamond wheel. By comparing the three models, the filling time of model III was the shortest.

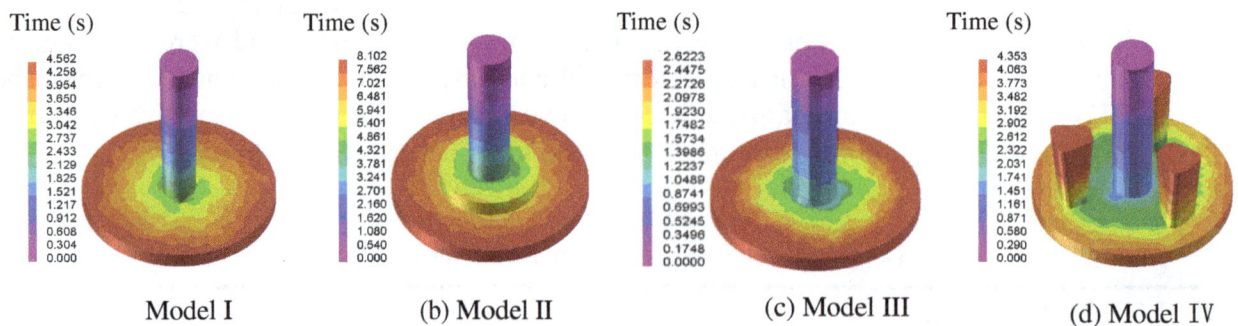

| Model I | (b) Model II | (c) Model III | (d) Model IV |

Fig. 2 Filling time distributions of different gating system design

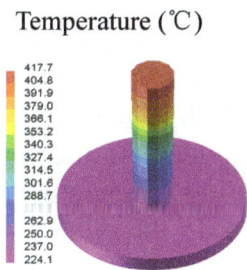

Fig. 3 Solidification time distributions

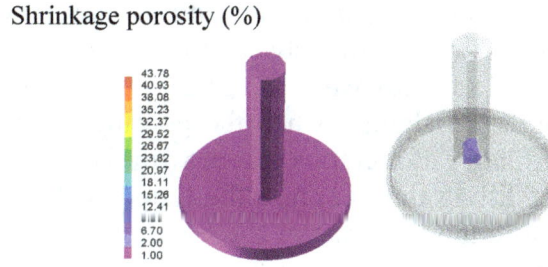

Fig. 4 Shrinkage cavities distributions

The simulation results revealed that the solidification behaviors of the metal and the tendency of formation of shrinkage cavities in the casting were similar in Model I, Model II and Model III. The simulation results of Model III were shown in figure 3 and figure 4. It can be seen from figure 3 that the solidification of the whole casting was from outside grinding layer to the spure. The whole casting had small shrinkage cavities with the size of 1-2%. The size of the shrinkage cavity was defined as the volume ratio of the shrinkage cavity and the casting. Shrinkage cavities with the size more than 2% only appeared in the intersection of the central portion of the part and the spure. The size of the grinding wheel center hole was bigger than the spure diameter. Hence the shrinkage cavities in the intersection of the central portion of the part and the spure will not affect the casting quality. The small shrinkage cavities in the cast cannot be avoided because the metal used has a wide solidification range. Based on above simulation results, the Model III was therefore considered as the optimized design of the gating system, in which riser was not necessary.

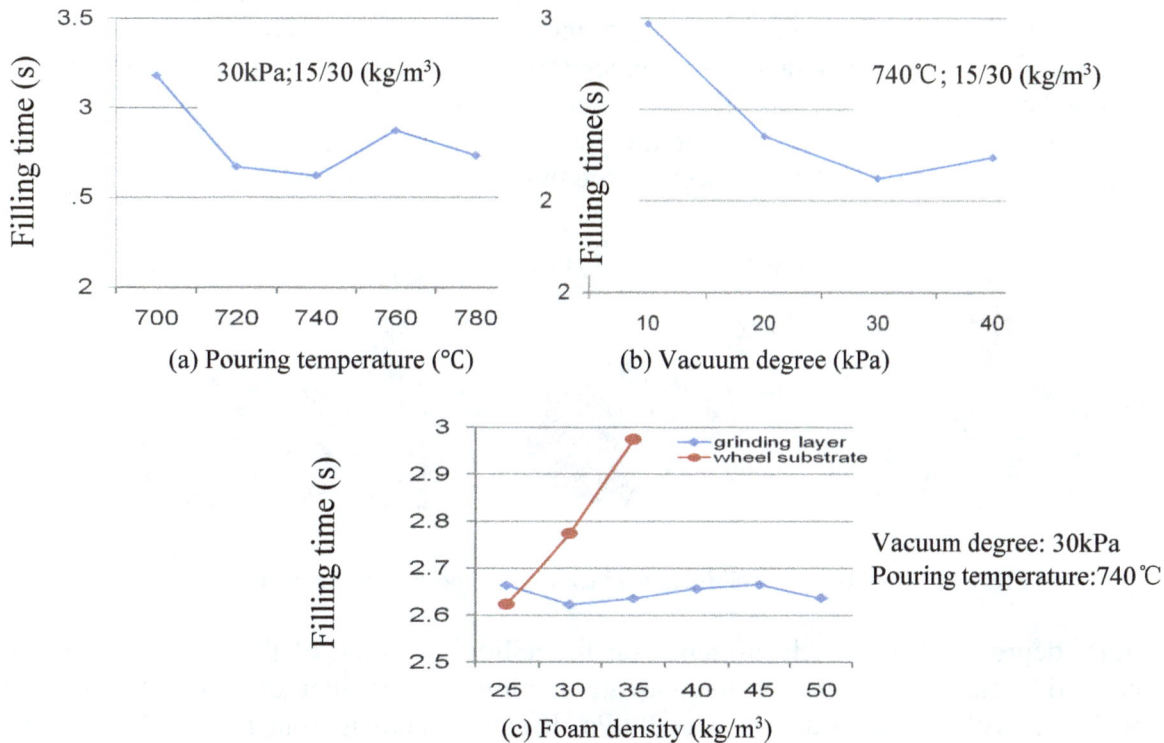

(a) Pouring temperature (°C)

(b) Vacuum degree (kPa)

(c) Foam density (kg/m^3)

Fig. 5 Relations between casting parameters and the filling time of the meta

Numerical simulation to optimize casting parameters

Figure 5(a) shown the relation between the filling time and pouring temperature. The filling time decreased from 3.1s to 2.6s when the pouring temperature increased from 700°Cto 740°C. But the filling time increased slightly when the pouring temperature was further increased. The results indicated that fluidity of metal increased with the increase of pouring temperature which leads to the decrease of filling time because the liquid with the higher pouring temperature has a lower viscosity therefore the flow rate of the liquid metal can be increased. In the other hand, the flow rate of liquid metal at high temperature will decrease because the volume of the gas produced by polystyrene foam increases significantly [6].

The influence of vacuum degree on the filling time was shown in the figure 5(b). It can be known that the filling time decreased from 3.5s to 2.6s when the vacuum degree increased from 10kPa to 30kPa. When the vacuum degree was further increased to 40kPa, the filling time increased slightly instead. The reason is that higher vacuum degree will be in favor of the eliminating of the emission gas, which is produced by the polystyrene foam combustion. However, too high vacuum degree will lead to turbulence of the liquid metal. The turbulence of the liquid metal may cause air trapped in the liquid. Consequently the flow rate of the liquid metal will be decreased.

The simulation results revealed that the foam density of wheel substrate affected the filling time significantly while the variation of filling time with the increase of the foam density of grinding layer was negligible. This is due to the reason that higher foam density will result in more gas, which will slow down the flow rate of the liquid metal. Because the volume percentage of grinding layer in the whole cast is small, hence the effect of the foam density of grinding layer is negligible.

The influence of pouring temperature on the solidification behavior of the metal was shown in figure 6. The time for solidifying the whole cast increased with the increase of the pouring temperature because casting that poured on a higher temperature means it has higher superheat. The higher superheat will make the casting take longer time to solidify from liquid to solid. The diamond wheel solidifies simultaneously when the pouring temperature is 740°C. This will reduce stress and distortion of the diamond wheel because the thickness of the wheel is small.

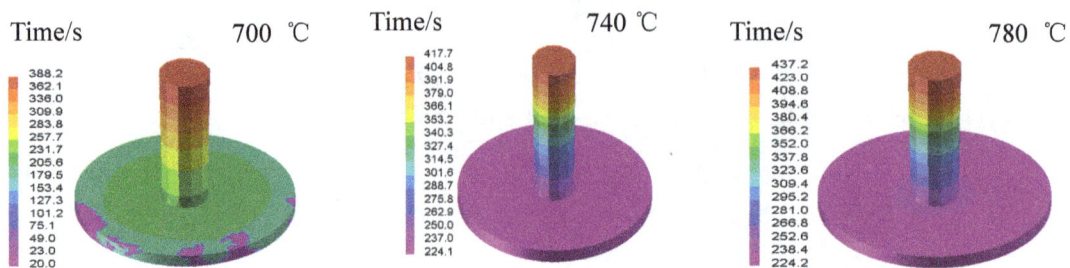

Fig. 6 Relations between solidification behavior and pouring temperature

The vacuum degree had not much influence on the solidifying time of the whole cast but it affected the solidification behavior of the casting obviously. The diamond wheel solidified simultaneously when the vacuum degree was 30kPa. It can be known from figure 5(b) that the filling time was longer when the vacuum degree was lower or higher than 30kPa. Longer filling time will cause the liquid metal to consume more heat which leads to a greater difference of metal temperature in different part of the wheel.

The foam density of wheel substrate had not much influence on the solidifying time of the whole cast but it affected the solidification characteristics of the casting. When the foam density of wheel substrate was set to be 15kg/m³, the diamond wheel will solidifies simultaneously. However, the influence of foam density of grinding layer is negligible.

The peak temperature of diamond grits in the grinding layer
Thermal damage to diamond grits must be considered due to the tendency of graphitization of diamond grits at high temperature. In order to study the temperature variation of the diamond grits during the LFC process, twenty diamond grits in the grinding layer were selected randomly for Numerical simulation. The results indicated that the temperature variation of the diamond grits under different casting parameters had the same regulation as shown in Figure 7.

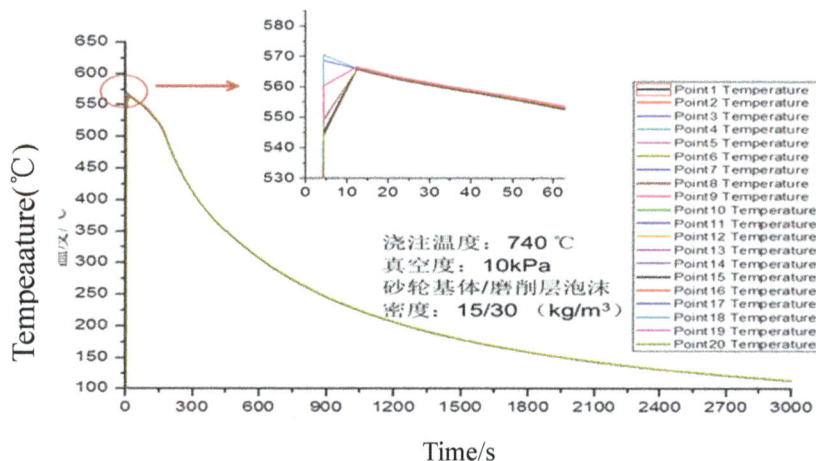

Fig. 7 Temperature variation of the diamond grits during the LFC process

Table 3 shows the peak temperatures of diamond grits under different pouring temperatures. It could be found that the peak temperatures of diamond grits were fluctuant and they raise with the increase of the pouring temperature. However, the peak temperatures of diamond grits were always much lower than the pouring temperature. When the pouring temperature was set to be 740°C, the maximum temperature of the diamond grits was around 707°C, which was much lower than that of the graphitization temperature of diamond. It is known that graphitization of diamond begins at about 800°C in the air [7].

Table 3 The peak temperatures of diamond grits under different pouring temperatures

Pouring temperature (°C)	Vacuum degree (kPa)	Foam density (kg/m³) wheel substrate/ grinding layer	Peak temperature of twenty diamond grits (°C)
700			563.6~598.5
720			568.4~617.5
740	30	15/30	572.3~707.4
760			639.4~719.8
780			667.5~738.8

Based on the above analysis, optimized gating system design and casting parameters were set. By using the optimized results, a diamond wheel as shown in Figure 8 was fabricated using the LFC process. The topography of the diamond wheel under a 3D video system was presented in Figure 9. It exhibited that the diamond grit could be tightly attached to the metal matrix and there was not any thermal damage on the diamond grit.

Fig. 8 The cast diamond wheel   Fig. 9 Topography of the diamond wheel

4. Conclusion

Model III is considered as the optimized design, in which the runner and gating system is located at the center of the cast. No additional riser is necessary. The optimized pouring temperature is 740°C and the vacuum degree is 30kPa. The optimized foam densities of wheel substrate and wheel grinding layer are 15kg/m^3 and 30kg/m^3 respectively. The simulation results match the experimental results well.

Acknowledgment

This study was supported by Grant No. 51175192 from the Natural Science Foundation of China.

References

1. D.A. Caulk: 'A foam melting model for lost foam casting of aluminum', *Int. J. of Heat and Mass Transfer*, 2006, 49, 2124–2136.
2. A. Charchi, M. Rezaei, S. Hossainpour, J. Shayegh, S. Falak: 'Numerical simulation of heat transfer and fluid flow of molten metal in MMA–Stcopolymer lost foam casting process', *J. of Mater. Pro. Tech.*, 2010, 210, 2071–2080.
3. J. Shayegh, S. Hossainpour, M. Rezaei, A. Charchi: 'Developing a new 2D model for heat transfer and foam degradation in EPS lost foam', *Int.COMM. in Heat and Mass Transfer*, 2010, 37, 1396–1402.
4. L. Zulai, J. Yehua, Z. Rong, L. Dehong, Z. Rongfeng: 'Dry three-body abrasive wear behavior of WC reinforced iron matrix surface composites produced by V-EPC infiltration casting process', Wear, 2007, 262, 649–654.

5. Suyitno, Sutiyoko: 'Effect of Pouring Temperature and Casting Thickness on Fluidity, Porosity and Surface Roughness in Lost Foam Casting of Gray Cast Iron', Procedia Engineering, 2012, 50, 88 – 94.

6. M. Khodai, N. Parvin: 'Pressure Measurement and Some Lost Foam Casting in Observation', *J. of Mater. Pro. Tech.*, 2008, 206, 1-8.

7. S. Wang, T.H. Zhang: 'Analysis on the Researches on Diamond Thermal Stability and Some Propositions', Diamond & Abrasives Engineering, 2001, 5, 36~39.

Simulation Research of Temperature Field for Heat Transfer in Cu/Sn-58Bi-0.05Sm/Cu Micro Solder Joints

Ning Zhang[1,*], Hao-Bin Shao[1], Chun-Hong Zhang[2], Ming-Fan Zhu[1], Li-Bin Yin[1]

[1]Jiangsu Key Laboratory of Large Engineering Equipment Detection and Control, School of Mechanical & Electrical Engineering, Xuzhou Institute of Technology, Xuzhou, 221018, China

[2]Xuzhou Engineering Technology Research of green and clean composite solder, Department of Mechanical & Electrical Engineering, Xuzhou Bioengineering Technical College, Xuzhou, China

*Email: sprningzn@126.com

The method of composite brazing induced by ultrasonic vibration is used in preparing Sn-58Bi-0.05Sm micro braze welding joint. Heat transfer experiments are carried out using a self-made multi-field coupling aging device. Temperature changes at the ends of the specimen and the ends of solder are detected by a thermocouple thermometer. ABAQUS, an finite element analysis software, is used to draw the three-dimensional model of the micro tensile specimen, bring into the parameters and the measured temperature value, simulate the temperature at both ends of the solder, and calculate the temperature gradient. The results of ABAQUS simulation show that: the cold end and hot end temperature of Cu/ Sn-58Bi-0.05Sm /Cu micro solder joint are 66.95°C and 73.05°C, the temperature gradient being 305°C /cm. When the temperature of the hot end and cold end reaches 40°C and 180°C, the temperature gradient at the ends of the solder joint is as high as 1068°C / cm. Such a high temperature gradient can be sufficient to induce heat transfer of metal atoms and cause serious problems of reliability.

Keywords: Temperature field; heat transfer; Sn-58Bi solder; rare earth Sm.

1. Introduction

The heat transfer of metal atoms is controlled by diffusion under a certain driving force [1]. Heat transfer studies began in 1879, the study found, the temperature difference across the tube will make the salt solution concentration nonuniform and the hot end of the salt solution concentration is lower than the cold end. Therefore, it can be concluded that the salt transport flux is caused by the temperature gradient. Similarly, the composition of the alloy in a certain temperature gradient will become uneven [2]. While the phenomenon of interaction of the heat exchange and atomic diffusion to de alloying (known as SORET, also known as heat transfer or atomic diffusion) driven by temperature gradient [3]. Micro electronic packaging solder interconnection is mainly two alloys or multicomponent alloy based on Sn, which will also show heat transfer phenomenon under a certain temperature gradient [4].

Electronic products continue to pursue high density, high performance and miniaturization, the size of solder joints continues to decrease, and the Joule heat generated by electronic devices in service has become one of the main problems faced by microelectronic technology [5]. As the microelectronics industry gradually entered the era of post Moore's law of integrated circuit (IC), IC 3D packaging has become one of the ways to solve the physical limit of super large scale IC [6]. The stacking of the chip in the IC 3D package makes the problem of Joule heat more serious and requires the introduction of a heat sink for heat dissipation. In the paper, based on the self-made multi-field coupling aging device ABAQUS, an finite element analysis software is used to simulate the temperature of Cu/Sn-58 Bi-0.05 Sm/Cu solder at the cold and heat end and calculate the temperature gradient of heat transfer, which provides data support for the study on reliability of lead-free solder joint.

2. Experimental Materials and Methods

2.1. *Experimental materials*

The matrix solder used in the test is Sn-58Bi particles, the size of particle is 40~55μm; the doped rare earth element is Sm, the purity is 99.9%, the particle size is 70~75μm; the thickness of the pure copper plate is 1mm.

2.2. *Experimental methods*

Before brazing, 50×15×1mm pure copper in acetone and alcohol in the ultrasonic should be cleaned for 20 minutes to remove surface impurities. Then take the required Sn58Bi paste into the mortar and again that take a certain amount of Sm, a few into the mortar by many times, with a glass rod mechanical stirring mixed edge addition, until the rare earth element Sm was uniformly dispersed in the paste with composite solder. The whole process lasted 30 minutes. Then, use the self-made brazing fixture and ultrasonic vibration welding equipment for brazing. Brazing temperature is 185°C, ultrasonic power is 800W and ultrasonic time is 15s. After the welding, it is naturally cooled in the air.

The welding parts are processed into micro tension samples by using wire cutting machine tools, as shown in Fig. 1. Then the specimen is loaded and placed in a multi-field coupling aging device. Tensile stress of the 300g, current 1A enters the copper wire. A group of 6 high temperature magnets is placed in symmetrical groove. Open the heating table and heating sink, the heating temperature is set to be 130°C, so that the device turns into the work state. After the device is stably heated, the temperature changes of both ends of the sample and the two ends of the solder are measured by the thermocouple temperature measuring instrument. Then, by using ABAQUS to draw the three-dimensional model of mini-tensile-specimen, bringing into the parameters and the measured temperature value, simulating the solder ends of the temperature, finally the temperature gradient can be calculated.

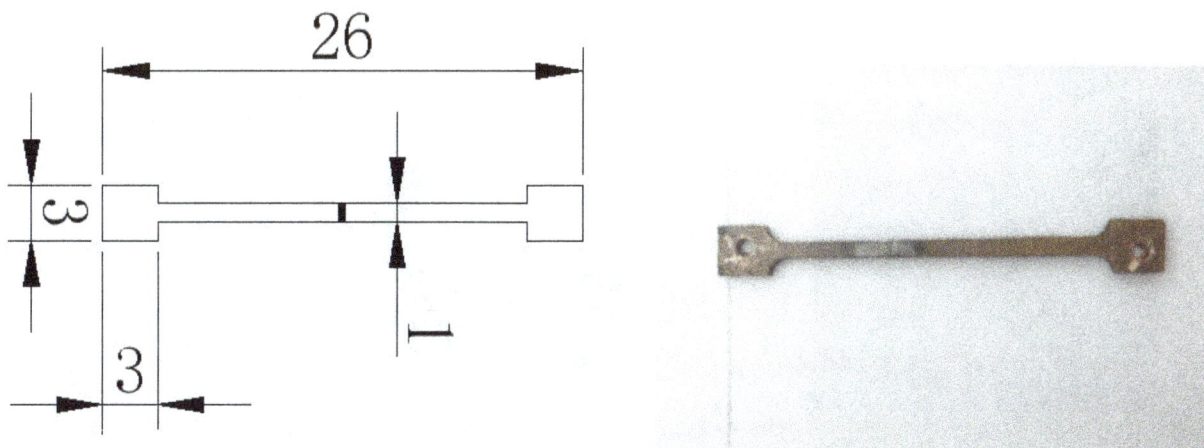

Fig. 1 Micro tensile specimen

3. Experimental Results and Discussion

3.1. *The main parameters of Cu/Sn-58Bi-0.05Sm/Cu*

Thermal physical properties of the material, but also referred to as thermal properties, it is a kind of material properties [7]. According to the second law of thermodynamics, heat spontaneous transfers from the high temperature object to the body at a lower temperature, the heat transfer process has a great relationship with properties of the material, usually considered the heat transfer process and object containing heat change is directly related to the nature of known as the thermal properties of the material. Specifically, the thermal properties of materials include density, thermal conductivity, specific heat, thermal expansion coefficient, melting point, boiling point. ABAQUS heat conduction simulation can be used to calculate the temperature distribution and thermal physical parameters of a system or component, such as the heat, thermal gradient, heat flux and so on. Heat transfer includes heat conduction, radiation and convection [8]. This model mainly involves transmission modes, heat conduction and heat radiation, without considering the influence of convection. The ABAQUS heat conduction is mainly used for the specimen in the study on the process of the cold end and hot end heat conduction. The distance measured by actual measurement process for solder at both ends is only 0.2mm. Because of the influence of measuring accuracy, the thermometer used to measure the solder at both ends of the temperature will have some errors. So compared with the actual measured value, the use of heat conduction of ABAQUS to simulate the heat conduction process of the hot end to the cold end has more reliability.

Using ABAQUS finite element analysis software to draw the 3D model of the micro tensile specimen, as shown in Fig. 2, then confer a specimen of the material properties and different parts of the section properties, Cu: Conductivity:401, Density:8960, Specific Heat:380.6; Sn58Bi: Conductivity :21, Density :8750, Specific Heat:170.2, without considering the material deformation. For each attribute definition of contact surface, the solder part, the heat radiation rate is set to 0.21, the hot end and the cold end of the sample copper substrate, the heat radiation rate is set to 0.05, the ambient temperature is set to 20°C, as shown in Fig. 3.

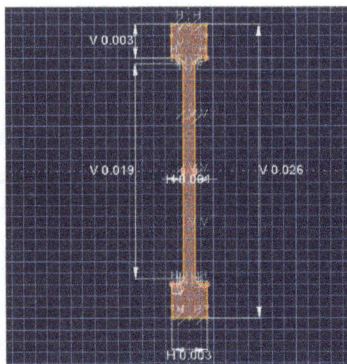

Fig. 2 3D Model of sample

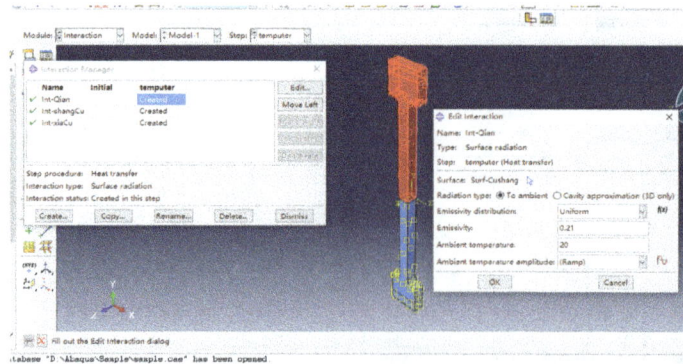

Fig. 3 Definition of contact property

As shown in Fig. 4, when boundary conditions are formulated, both ends of up and down part of the sample model need to have restriction on the parameters, such as temperature. The temperature of a part of upper end of the specimen (as the hot end) is set to 90°C, while the lower end (as the cold end) of the field temperature is set to 50°C. The ambient temperature is set to 20°C.

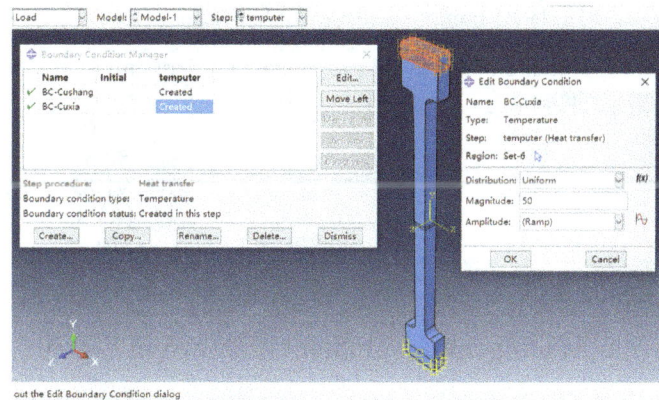

Fig. 4 Definition of boundary conditions

3.2. *Simulation results of the temperature field and heat transfer*

In this model, the 8 node block element is used to divide the grid, and then the operation is generated. The simulation results of the temperature field are shown in Fig. 5.

In the actual experiment, the surface temperatures of the sample solder at hot end and cold end measured by thermocouple are 67.7°C and 69.2°C respectively. The temperatures of solder joint at the cold and hot end simulated by ABAQUS are 66.95°C and 73.05°C, the temperature gradient (TG) is 305°C / cm.

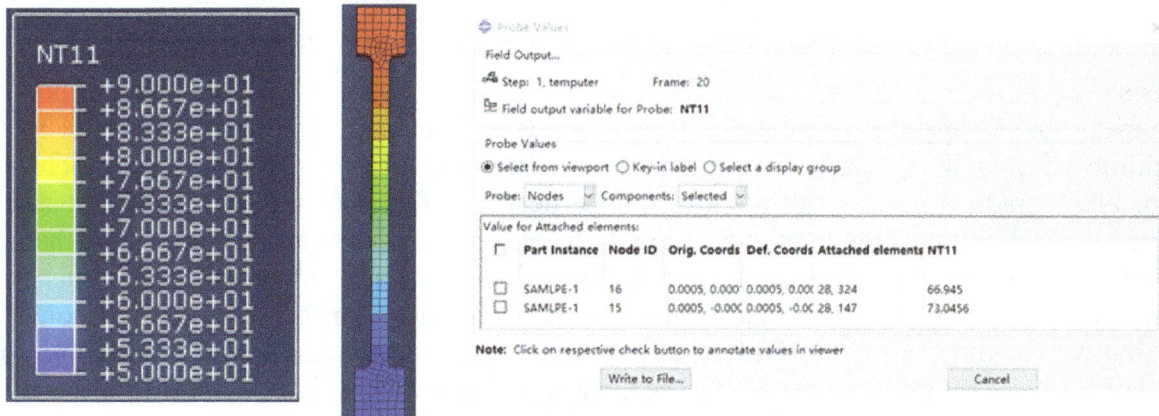

Fig. 5 Simulation results of temperature field

3.3. *Discussion of heat conduction simulation*

Because Sn-Bi solder used in this experiment is a solder with low temperature, by the DSC experiment testing, adding rare earth Sm of the composite solder and the original Sn-58Bi solder melting point has little difference. The heat transfer temperature at the hot end is only set to 90°C and less than the melting point of the Sn-58Bi (138°C). If a mesothermal solder with higher melting point temperature (such as Sn-Zn) or a high-temperature solder (such as Sn-Cu and Sn-Ag) is adopted to test, temperature gradient will be greater, as Table 1 shows.

Table 1 Simulation results of heat conduction

Sample	Cold end of sample(°C)	Hot end of sample(°C)	Cold end of solder joint(°C)	Hot end of solder joint(°C)	Temperature gradient(°C /cm)
1	50	90	66.95	73.05	305
2	40	110	69.65	80.32	533.5
3	40	130	78.13	91.85	686
4	40	150	86.60	103.38	836
5	40	180	99.30	120.66	1068

The comparison of the sample temperature gradient is shown in Fig. 6.

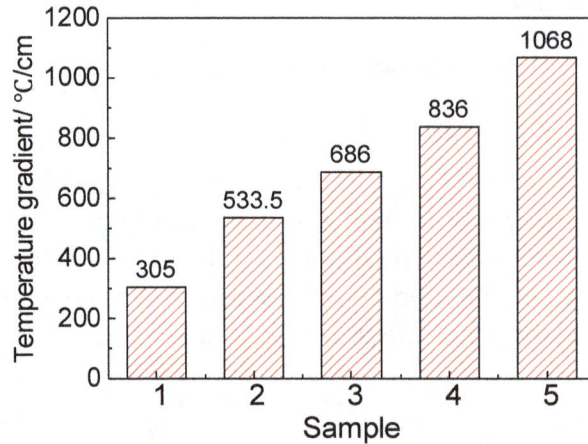

Fig. 6 Comparison of temperature gradient of each sample

As shown in the figure, with the increase of the difference between the cold end and hot end of specimens, the temperature gradient of solder joints at both ends is also increasing. Under the condition of sample 5, temperature gradient has been as high as 1068°C / cm, such a high temperature gradient can be sufficient to induce heat transfer of metal atoms and cause serious problems of reliability. This is because the interfacial reaction between the solder and the copper substrate is the key to influence the reliability of the package interconnection. Intermetallic compound (IMC) formatted by interfacial reaction is necessary for the realization of solder joint metallurgy, but the brittle nature of IMC interface makes the thickness and morphology of the interface must be controlled effectively. The growth and evolution of interfacial IMC are controlled by the diffusion of filler metal and based metal atoms. The temperature gradient in the micro solder joints increases the directional diffusion capacity of the metal atoms in the solder, which causes the heat transfer, and makes the element redistribution, which will significantly affect the growth of the interface IMC, and then affect the reliability of the micro interconnections.

4. Conclusion

(1) The method of composite brazing induced by ultrasonic vibration is used in preparing Sn-58Bi-0.05Sm micro braze welding joints. Heat transfer experiments are carried out using a self-made multi field coupling aging device. The measured surface temperatures of the solder cold end and the hot end of the test sample are 67.7°C and 69.2°C respectively.

(2) With the aid of ABAQUS finite element analysis software, the temperature of Cu/Sn-58Bi-0.05Sm /Cu micro solder joints at cold end and hot end are simulated and the results are 66.95°C and 73.05°C, and the temperature gradient is 305°C /cm.

(3) With the increase of difference between specimen's cold end and the hot end, solder joints at both ends of the temperature gradient is also increasing. Under the condition of sample 5, temperature gradient has been as high as 1068°C /cm, such a high temperature gradient can be sufficient to heat transfer induced by metal atoms and cause serious problems of reliability.

Acknowledgments

This work was funded by the Open Project of Jiangsu Key Laboratory of Large Engineering Equipment Detection and Control; (Grant No. JSKLEDC201510), the Science and Technology Project of Xuzhou (Grant No. KC15SM041), the College Students' practical and innovative Project of Jiangsu Province (Grant No. 201611998034Y), the Science and Technology Planning Project of Xuzhou Bioengineering Technical College (Grant No. 2015B01), and sponsored by Qing Lan Project of University of Jiangsu Province in 2014.

References

1. Desmarest S.G. Reliability of Pb-free solders for harsh environment electronic assemblies. Mater Sci Technol, 2012; 28: 257-73.
2. Li Yang, Chengchao Du, Jun Dai, et al. Effect of nanosized graphite on properties of Sn–Bi solder. J Mater Sci: Mater Electron, 2013, 24: 4180-4185.
3. Ning Zhang, Chunhong Zhang, Juli Li. Research of in-situ synthesis Ti(C, N)-WC particle reinforced Ni60A composite coating by argon arc cladding. Journal of Xuzhou Institute of Technology (Natural Sciences Edition), 2015, 30(1): 47-51.
4. Pang X Y, Shang P J, Wang S Q, et al. Weakening of the Cu/Cu3Sn (100) Interface by Bi Impurities. Journal of electronic materials, 2010, 39(8): 1277-1282.
5. Zhang Ning, Qiang Yinghuai, Zhang Chunhong, et al. Microstructure and property of WC/steel matrix composites[J]. Emerging Materials Research, 2015, 4(2):149-156.
6. Li Yang, Jun Dai, Yaocheng Zhang. Influence of BaTiO3 Nanoparticle Addition on Microstructure and Mechanical Properties of Sn-58Bi Solder. Journal of ELECTRONIC MATERIALS, 2015, 44(7): 2473-2478.
7. Dekun Zhang, Junjie Duan. On the Siding-rolling Friction and Wear Properties of Point Contact Friction Couple between GCrl5 Steel Ball and GCrl5 Steel Disc. Journal of Xuzhou Institute of Technology (Natural Sciences Edition), 2014, 29 (4): 7-12.
8. Chen C, Hsiao H Y, Chang Y W, et al. Thermomigration in solder joints [J]. Materials Science and Engineering: R: Reports, 2012, 73(9): 85-100.

Finite Element Analysis for Thermal Stress of WC Particulates Reinforced Steel Matrix Composites

Ning Zhang[1,*], Chun-Hong Zhang[2], Jian Fu[1]

[1]*Jiangsu Key Laboratory of Large Engineering Equipment Detection and Control, School of Mechanical & Electrical Engineering, Xuzhou Institute of Technology, Xuzhou, 221018, China*
[2]*Department of Mechanical & Electrical Engineering, Xuzhou Bioengineering Technical College, Xuzhou, China*
Email: sprningzn@126.com

Based on the software ANSYS of the finite element analysis, a material model with bilinear elastic-plastic matrix and elastic reinforcement was adopted to establish the finite element model of stress simulation for the matrix and reinforcement which included radiation heat transfer and heat transfer by oil convection. The simulation results show that the difference of thermal expansion coefficient exists between 5CrNiMo steel matrix and WC particulate reinforcement. In the process of oil quenching treatment, there is a large stress gradient near the interface. The stress of the composites increases with the decreasing of the particle pointedness. When the pointedness is less than 60 degrees, the stress of the particle and matrix is very large, and the probability of microscopic damage is greater. The stress rises with the increase of particle diameter, while the change is relatively small when the particle size of the reinforcement is in the range of 0.2~0.8mm, but the stress value is larger when it is greater than 1.0mm.

Keywords: WC particulates reinforced steel matrix composites; thermal stress; finite element analysis.

1. Introduction

In the process of studying the properties of the composite materials, the factors that affect the mechanical properties can be classified into two types: one kind is the elastic constants of each phase in the composite, and the other is the microstructure characteristics of the composite, including the geometry, size, distribution and interaction of the inclusions, particles, holes, and cracks [1]. This problem is due to the establishment of an appropriate micro mechanical model, and to simplify the factors that affect the microstructure and select the appropriate material model and so on. Scholars in this area have also made some valuable progress [2].

2. Theoretical Model and Numerical Algorithm

The particle reinforced steel matrix composite material produced by electroslag melting and casting and the heat treatment process are very complex. It is very difficult to establish a strict scientific analysis system for the thermal stress analysis of materials science, thermodynamics, elastic-plastic mechanics, physical chemistry and material processing engineering and other disciplines [3]. On the whole, the thermal stress is mainly the numerical simulation of the temperature field and the stress field for the temperature field (no phase transition, so there is no phase change stress). The numerical simulation of temperature field is the basis of that of the stress field, and the accuracy of the temperature field is directly affected by the stress field [4].

In view of the above situation, on the basis of proper simplification, a finite element method based on the model of particle elasticity and matrix elastic and plastic was adopted to simulate and calculate the different models by software ANSYS of commercial finite element analysis. The influence of particle parameters on the internal stress distribution was obtained when the composite materials were produced and heat treatmented. It is hoped that the above work can provide some references for the preparation of the related particle reinforced steel matrix composites.

In this study, based on the analysis characteristics of the thermal stress coupled field of software ANSYS, the work flow chart is designed as illustrated in Fig. 1.

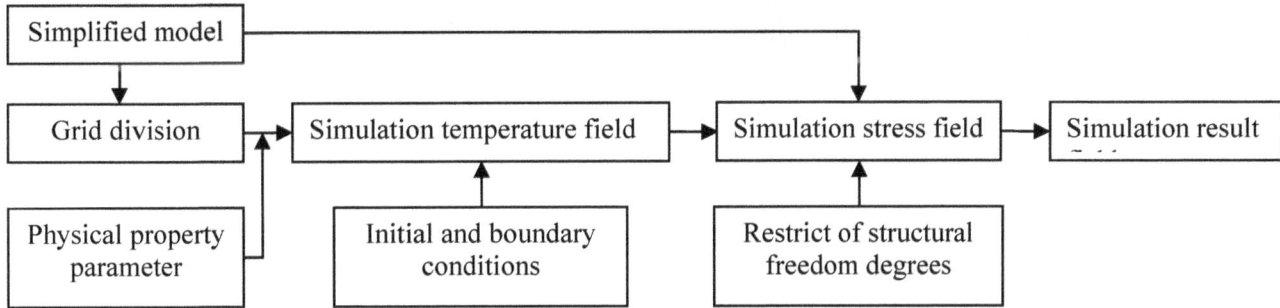

Fig. 1 Working flow of thermal stress simulation

3. Analysis of Numerical Simulation Results

In order to compare the particle model of the different shape, the equivalent diameter is defined for the premise to ensure that the same volume of different pointedness. Fig. 2 is a section of rotary particles. The contour line is represented as function y=f (x), R is the equivalent diameter. The definition is as follows [5]:

$$r = 3\sqrt{\frac{\int_a^b \pi f^2(x)dx}{\frac{4}{3}\pi}} = 3\sqrt{\frac{3}{4}\int_a^b f^2(x)dx} \tag{1}$$

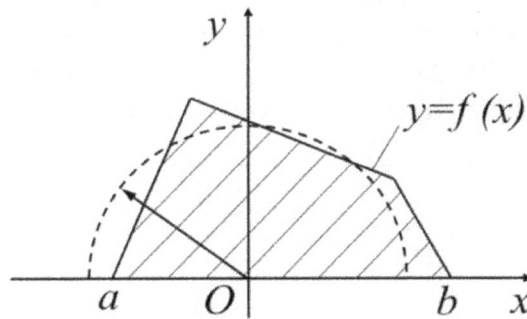

Fig. 2 Definition of equivalent diameter

3.1. Relationship between stress and time of near interface matrix

The difference of thermal expansion coefficient between the reinforcement and the matrix is more influence on the stress at the interface in the oil quenching process. The region first appears the plastic zone, and it is also the most likely region of crack initiation and propagation. But the equivalent stress of different points in this region is different from the time of occurrence. Fig. 3 is the schematic diagram of the three typical locations for this region, of which point 1 is near the particle in the vicinity of the heat transfer interface, point 3 is near the material close to the interior of the material, point 2 is located in the side of the particle. As can be seen from the figure, because the point 1 is closer to the heat transfer boundary, the equivalent stress of point 1 is the maximum at the beginning of a period time, while point 3 is the minimum. At the later stage of cooling, the equivalent stress of point 2 is the maximum, which is mainly caused by the stress of the surface layer due to the cooling and contraction of the inner. The difference of equivalent stresses between point 1 and point 3 are not very large.

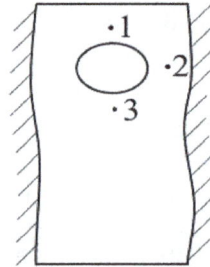

Fig. 3 Schematic diagram of three typical locations near the particle

3.2. *Effect of reinforced body pointedness on the strength*

From knowledge of elastic plastic mechanics, the place with the presence of sharp angles will appear the stress concentration, which is the most likely parts of the crack initiation, expansion and cracking [6]. In order to investigate the effect of the reinforced body pointedness on the strength, for the premise of keeping all other conditions under the same, the reinforced body pointedness are set as 30 degrees, 60 degrees, 90 degrees, 120 degrees and 180 degrees (corresponding to the spherical reinforcement) to investigate this effect. The equivalent diameter of particles for different pointedness is 0.6mm.

The effect of size of the reinforcement pointedness on the tip of the body is shown in Fig. 4. As can be seen from the graph, the stress in the two directions of the reinforcement tip is all compressive stress, and gradually becomes smaller with the angle of pointedness increases. The spherical reinforcement is the minimum. When the pointedness is less than 60 degrees, the change of stress is greater. This shows that the sharper the pointedness is, in the oil quenching process, the greater the possibility of strengthening the reinforced body damage.

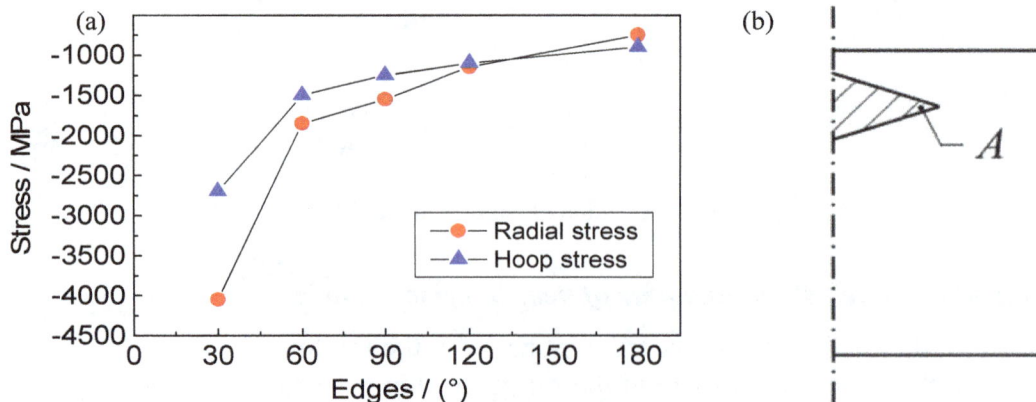

(a) Relationship between the radial of point A and the surrounding stress with the pointedness
(b) Location of point A

Fig. 4 Effect of pointedness on the stress at the tip of the reinforced body

The simulation results are also shown that the reinforced body pointedness has great influence on the stress concentration of the matrix near the substrate. After the matrix stress value reaches the yield limit of the material, the micro plastic zone begins to form, and continues to expand with the continuous growth of the stress. As can be seen from the Fig. 5, when the reinforced body pointedness is the same, the equivalent stress increases with the enlargement of cooling time, and the residual stress reaches the maximum at the end of oil quenching treatment. At different times, the equivalent stress value is smaller with the enhancement of reinforced body pointedness. The simulation results of composite cooling time of 150s, 20s, 40s and 5S are given in Fig. 5.

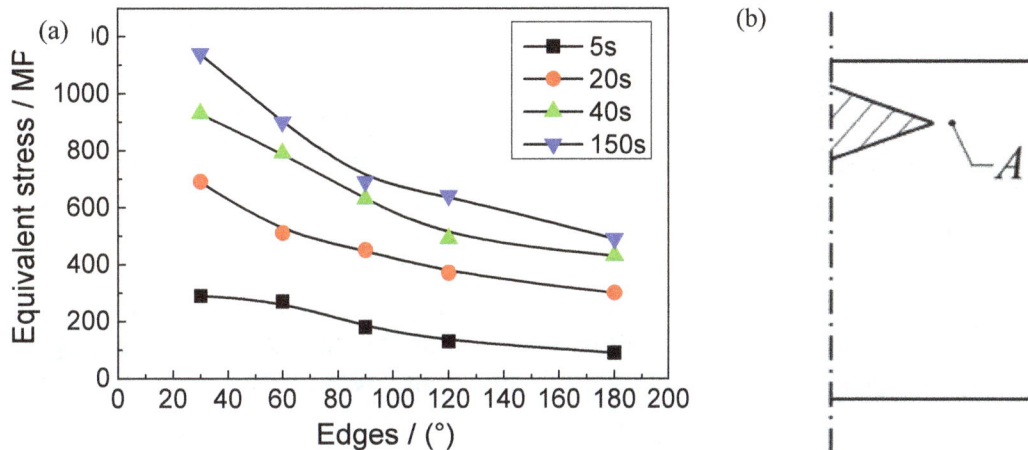

(a) Relationship between the equivalent stress of point A with the pointedness at different time
(b) Location of point A

Fig. 5 Effect of pointedness on the equivalent stress of the matrix

3.3. *Effect of reinforced body equivalent diameter on the strength*

According to the definition of equivalent diameter, when the reinforced body pointedness is selected to 90 degrees, and the variation of equivalent diameter is in the range of 0.2~1.6mm, to study the influence law of the equivalent diameter to the equivalent stress of the reinforcement and matrix at the corners of both sides of the interface for the particles.

The simulation results of the effect of the equivalent diameter on the stress at the tip of the reinforced body are shown in Fig. 6. As can be seen from the figure, in the process of oil quenching, the influence of particle size effect on the equivalent stress at the tip of the reinforcement is very large. With the increase of the equivalent diameter of the reinforced body, the radial and circumferential stresses of the tip gradually become larger. That is, the smaller the equivalent diameter of the reinforced particles is, the smaller the possibility of particle breakage caused by thermal stress during the process of oil quenching is.

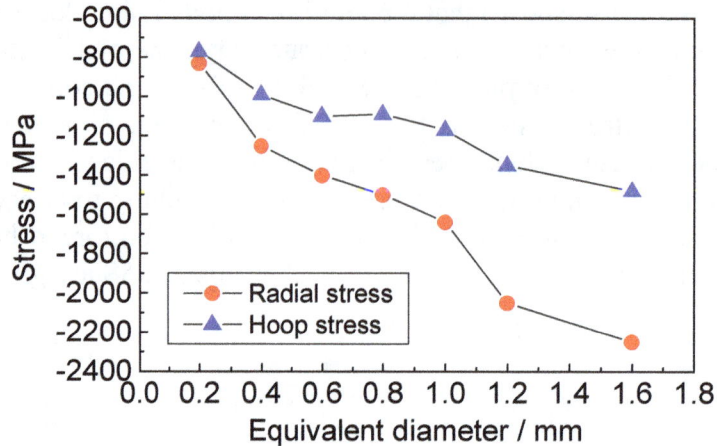

Fig. 6 Effect of equivalent diameter on the stress of the tip of the reinforced body

The simulation results of the effect of reinforcement size on the equivalent stress of in the matrix near the particle tip are shown in Fig. 7. As can be seen from the figure, the variation range of the equivalent stress in the matrix at different cooling time is not very large when the equivalent diameter is 0.2~0.8mm. While when the equivalent diameter continues to become larger in case of greater than 1.0mm, the increasing amplitude of equivalent stress in the matrix is relatively larger.

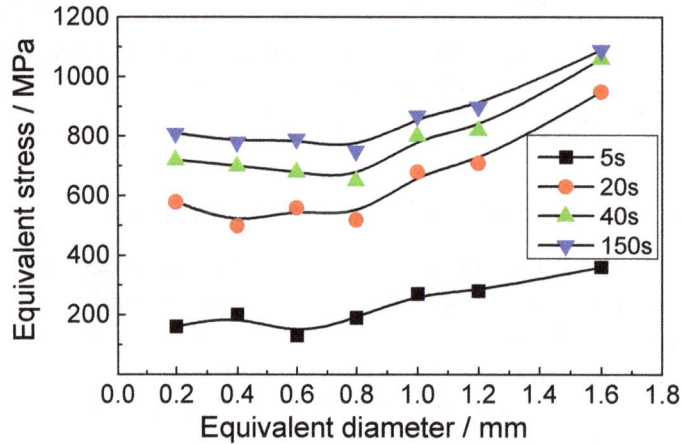

Fig. 7 Effect of equivalent diameter on the equivalent stress of the matrix near tip

4. Conclusion

1. The difference of thermal expansion coefficient exists between 5CrNiMo steel matrix and WC particulate reinforcement. In the process of oil quenching treatment, there is a large stress gradient near the interface.
2. The stress of the composites increases with the decreasing of the particle pointedness. When the pointedness is less than 60 degrees, the stress of particle and matrix is very large, and the probability of microscopic damage is greater.
3. The stress rises with the increase of particle diameter, while the change is relatively small when the particle size of the reinforcement is in the range of 0.2~0.8mm, but the stress value is larger when it is greater than 1.0mm.

Acknowledgments

This work was funded by Natural Science Foundation of colleges and universities of Jiangsu Province in 2015 (Grant No. 15KJB430030), General project of Xuzhou Institute of Technology (Grant No. XKY2014318), and the Science and Technology Project of Xuzhou (Grant No. KC14SM100).

References

1. Zhong L, Yan Y, Ovcharenko V E, et al. Microstructural and Mechanical Properties of In Situ WC-Fe/Fe Composites[J]. Journal of Materials Engineering & Performance, 2015, 24(11):4561-4568.
2. Nutthita Chuankrerkkul; Parinya Chakartnarodom. Fabrication of injection moulded 304L stainless steels reinforced with tungsten carbide particles, Materials Science Forum, 2012, 706-709, 638–642.
3. Zhang Ning, Qiang Yinghuai, Zhang Chunhong, et al. Microstructure and property of WC/steel matrix composites[J]. Emerging Materials Research, 2015, 4(2):149-156.
4. Nutthita Chuankrerkkul, Parinya Chakartnarodom. Fabrication of Injection Moulded 304L Stainless Steels Reinforced with Tungsten Carbide Particles [J]. Materials Science Forum, 2012.
5. Dekun Zhang, Junjie Duan. On the Siding-rolling Friction and Wear Properties of Point Contact Friction Couple between GCrl5 Steel Ball and GCrl5 Steel Disc. Journal of Xuzhou Institute of Technology (Natural Sciences Edition), 2014, 29 (4): 7-12.
6. Ning Zhang, Chunhong Zhang, Juli Li. Research of in-situ synthesis Ti(C, N)-WC particle reinforced Ni60A composite coating by argon arc cladding. Journal of Xuzhou Institute of Technology (Natural Sciences Edition), 2015, 30(1): 47-51.

Study on the Noise Reduction of Sound Absorption Noise Barrier

Wen-Jun Luo, Gong-Yu Liu

East China Jiao Tong University Engineering research center of Railway Environment Vibration and Noise, Ministry of Education, 330013

Absorptive sound barrier is an effective way to reduce the damage of vehicle driving noise to the surrounding environment. The characteristics of sound absorbing materials are important factors to determine the noise reduction effect. This paper presents two different sections of the noise spectrum, combined with a variety of sound absorption materials, by calculation the results, the contribution of the sound absorption coefficient of the sound absorption type barrier to the overall sound absorption and noise reduction is discussed, and obtained the influence degree of the change of the sound absorption coefficient on the whole noise reduction, dependent on sections of the noise spectrum distribution.

Keywords: Absorptive sound barrier; sound absorbing material; noise reduction effect; sound absorption coefficient.

1. Introduction

In engineering, we usually use absorptive sound barrier instead of reflecting sound barrier for the purpose of reducing the effect of noise barrier on the reflection of sound waves, adding noise barrier sound absorption material is to further reduce noise radiated by the roadside. There is an important factor to influence the noise reduction effect is Sound absorption characteristics of absorptive sound barrier. Domestic scholars have made some research and Analysis on the noise reduction measures of absorptive sound barrier. Yang Lu [1] had used the boundary element method to calculate the total reflection and the loss of the sound barrier, in order to explore the influence of the sound absorptive performance on the insertion loss of the road noise barrier. Xinhua Zhang [2] through the RAYNOISE software simulation analysis in order to explore the relationship between with sound barrier in different regions of the noise reduction effect and the installation of sound absorptive material location. According to relevant research, people not only lack of research on the sound absorptive characteristics of the sound barrier but also the simulation analysis method is simpler. The boundary element method is not economical to solve the large size model at high frequency. Statistical energy principle was extracted from random description of the total to research object and ignored the details parameters, concerned about time domain, frequency domain and space statistical average value, adopt "energy" view uniformed deal with vibration and acoustic field distribution, the advantages of which was calculated very fast and can obtain more accurate results in high frequency band [3]. At present, the statistical energy method is widely used in the field of aerospace, ship and automobile, but the application of the noise is less in the railway system. In this paper, the statistical energy method is used to combine the theoretical calculation formula, applied research on this aspect.

1.1. Sound absorptive characteristics of absorptive sound barrier

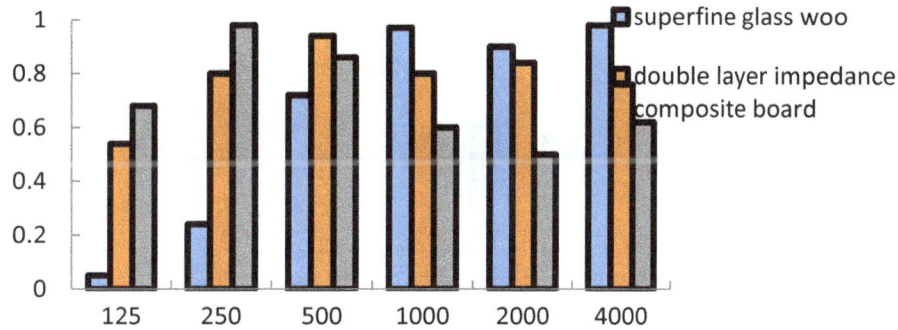

Fig. 1 Sound absorption coefficient of sound absorption material

Sound absorbing materials of sound barrier are often use porous sound absorbing materials. The parameters that represent the sound absorptive ability of the sound absorptive material are the sound absorptive coefficient and the acoustic impedance. The main factors affected the sound absorptive coefficient of the porous material is flow resistance, porosity and structure factor, thickness, bulk density and so on. This paper calculates adopt sound absorption coefficient of superfine glass wool, impedance composite plate and micro perforated plate, as shown in Figure 1.

2. Method for Calculating the Noise Reduction Effect of Absorptive Sound Barrier

2.1. Statistical energy model of absorptive sound barrier

Each subsystem energy is obtained by calculating, and the vibration velocity and sound pressure can be calculated. According to the basic principle of the statistical energy method, this paper establishes a statistical energy model of the sound space acoustic barrier[4]. As shown in Figure 2, the total length of the sound barrier is 200m.

$$\begin{cases} P_{Lin} = \omega\eta_1 E_1 + \omega\eta_{12} E_1 - \omega\eta_{21} E_2 \\ 0 = \omega\eta_2 E_2 + \omega\eta_{21} E_2 - \omega\eta_{12} E_1 + \left(\dfrac{\rho_a c_a}{\rho}\right) E_2 \end{cases} \tag{1}$$

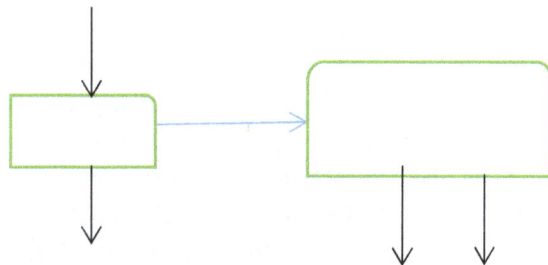

Fig. 2 Prediction SEA model of sound absorption noise inrailway sound barrier

933

Plug in each parameter, have

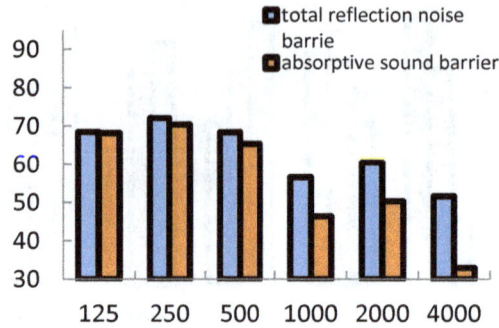

(a) Theoretical calculation formula gets the sound pressure level

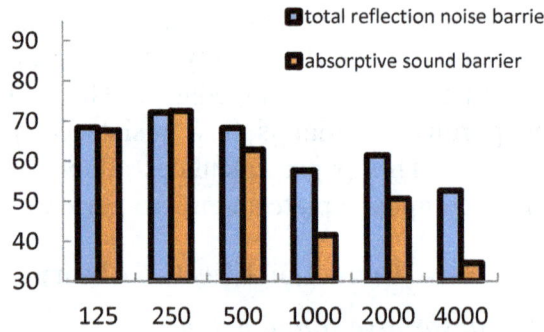

(b) Statistical energy method calculation gets the sound pressure level

Fig. 3 Compared with statistical energy method and the theoretical calculation results

Among them: η_1--internal loss factor of sound space, η_2 --Internal loss factor in sound barrier of sound absorption type, η_{21} --Coupling loss factor of sound absorption type sound barrier to sound space , ρ_a --Air density, c_a --Sound speed in the air, ρ --Density of sound barrier .

In the model, the acoustic space was approximated to the closed space,energy involved in the coupling of sound barrier,When the SEA parameters were estimated,the acoustic space handled into a one-dimensional acoustic field,and the sound barrier radiation was treated as an energy outflow pathway, in the calculation of the sound field,   it wasn't considerate a variety of terrain, climate, environmental and so on. In the calculation, by applying different sound absorbing materials, it is compared the influence of sound absorbing material to sound absorptive characteristics of sound barrier. The wheel noise spectrum was selected as a sound source when the speed at 250km/h and was input. At the wheel rail rolling noise, the superfine glass cotton screen suction noise reduction effect of sound barrier was obtain when the location is away from the track center line 4.2m and the train at the running speed of 250km / h. Compared with the noise reduction effect of the type (1) as shown in Figure 3.

We can draw from Figure 3 that the statistical energy method calculates result is roughly consistence with the theoretical results. The error is not more than 2dB. Therefore, the model can be used as a model of railway noise prediction model of absorptive sound barrier.

3. Calculation Result Analysis

3.1. *Analysis calculate result by use statistical method*

The sound pressure level difference was calculated between the absorption sound barrier and the total reflection noise barrier from the rail centerline 4.2m, as shown in Figure 4.

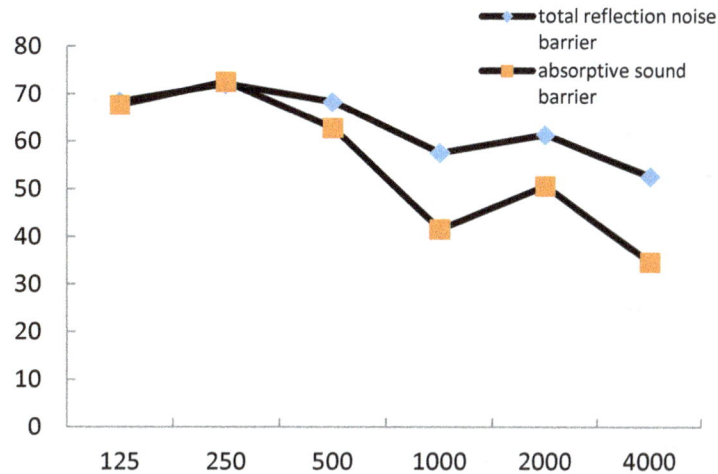

Fig. 4 Noise reduction effect of sound barrier with different sound absorption type

It can be seen from the calculation results that the different sound absorptive materials of sound barrier have obvious effect on the noise reduction. Impedance composite plate and micro perforated plate in the middle and low frequency noise reduction effect is more obvious; At high frequencies band，superfine glass wool noise reduction effect is obvious .In addition, Because the sound source only considers the wheel / rail noise and the sound source is located at the bottom of the vehicle, the noise reduction effect of the sound barrier may be have large noise reduction effect. In this paper, the contribution distribution map was made by using the wheel / rail noise source spectrum value as parameter and the formula, as shown in Figure 5:

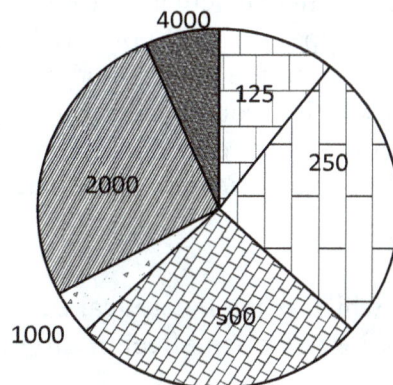

Fig. 5 Contribution distribution map (the number in the area of pie chart represent corresponding frequency)

935

The calculation results of Figure 7 shows that the sound absorptive coefficient at low frequencies increase can significantly improve the noise reduction, the impedance composite panels and micro perforation plate compare with superfine glass wool for low frequency sound absorptive performance is good. Therefore the noise reduction result is better than superfine glass wool.

3.2. *Analysis calculate result by use statistical method*

According to the calculation of statistical energy method in need of modal density, coupled loss factor, damping loss factor and put into parameter, In the range of calculated frequency bands, at the situation of the sound barrier of different sound absorbing materials and the whole reflecting sound barrier, the sound pressure level difference is obtained away from the center of rail about 4.2m, as shown in Figure 6.

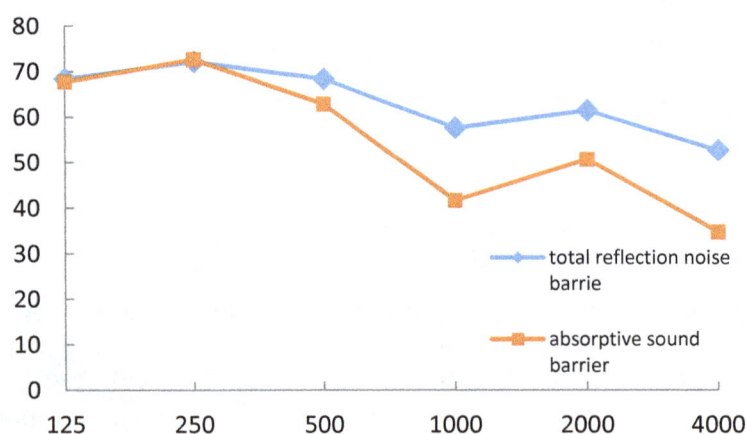

Fig. 6 Different absorption material noise reduction effect of sound barrier

As can be seen from the calculation result of statistical energy method, the noise reduction effect of various type of absorptive sound barrier is more obvious than that of the total reflect sound barrier. In addition, because statistical energy method obtained the average value in the scope of space, compared with the theoretical calculation formula results is come from one point, slightly different, but the basic close type acoustic noise reduction regularity of noise barrier. The sound source is located at the bottom of the vehicle and only considers the wheel / rail noise, so the noise reduction effect of absorptive sound barrier is slightly larger. In general, sound absorptive material has certain effect on noise reduction effect of sound barrier, the difference is between 8 and 10dB compared with the total reflection noise barrier.

4. Conclusion

In this paper, two methods are used which is the theoretical calculation formula and the statistical energy method, and evaluation the noise reduction effect of different absorptive sound barrier. The conclusions are as follows:

(1) In sound absorptive noise reduction measures, there has a certain effect on the noise reduction effect of the sound barrier accompany with the sound absorbing material join, and the difference between them is about 5 ~ 6dB.

(2) At the running speed of 250km / hit can significantly increase the amount of noise reduction through improve the low frequency sound absorptive coefficient, impedance composite panels and micro perforation plate compare with superfine glass wool in low frequency sound absorptive performance is better, therefore, the noise reduction effect is better than superfine glass wool.

(3) Statistical energy method has fast computing speed in the calculation of high frequency. The statistical energy method is adopted to predict the amount of noise reduction, and the error is not more than 2dB. This model can be as railway noise prediction model of absorption sound barrier.

Acknowledgments

This research partly supported by National Natural Science Foundation (51468021), Jiangxi Province Natural Science Foundation (20161BAB206160).

Reference

1. Yang Lu, Zhongrui Jiang, Effect of sound absorption performance on the insertion loss of road noise barrier [J]. Environmental Impact Assessment, 2015. (1): 92-96.
2. Xinhua Zhang, Dianyong Wang,leiming Song.Numerical simulation analysis between the relationship of sound-absorbing material position and noise reduction effect of sound barrier.Vibration and Noise Academic Exchange Meeting of the Environmental Protection Committee of China Railway Society, 2008.
3. Deyuan Yao, Qizheng Wang, Principle and application of statistical energy analysis [M]. Beijing: Beijing Institute of Technology Press, 1995.
4. Shaojia Wang, ShuyingGao, Prediction of noise reduction effect of railway sound barrier by statistical energy method [J]. Noise and Vibration Control, 2001. (6): 645-647.

Chapter 14
Measurements and Characterization

Research on Dynamic Detection Technology of Mineral Aggregate Gradation in HMA Based on Digital Image

Chao Geng, Hong-Fei Yao*

*Key Laboratory for Special Area Highway Engineering of Ministry of Education,
Chang'an University, Xi'an, Shannxi, P.R.China 710064*
*Email: 869590050@qq.com

In order to solve the lagging problem of mixture material gradation test in HMA, the image detection and analysis technology is studied, which achieves the dynamic grading test in construction. In this paper, firstly, the platform of image detection and analysis equipment is set up in a road construction project; secondly, the conversion of image information and gradation data is realized by using computer technology; finally, the test gradation curve and design gradation curve are drawn and compared. The measurement and analysis show that the image acquisition technology can quickly get a large amount of aggregate gradation information, and the grading curve can meet the requirements of construction accuracy.

Keywords: Asphalt mixture; gradation; image detection.

1. Introduction

Reasonable aggregate mixture ratio is an important factor to ensure the performance of asphalt pavement [1], which impacts on asphalt concrete performance in the strength, durability, structural adaptability and surface features, etc. However, the mineral mixture segregation is common in construction. Researchers [2] at Tongji University, conducted the drill core sampling and extraction screening test for nine high grade road pavement in a certain area, discoverded that almost all of the core sample size was outside the scope of the design and the actual construction generally didn't meet the design requirements.

One of the main reasons for asphalt pavement aggregate segregation is the problem of detection means. Existing gradation detection methods comprise the nucleon method, extraction method (centrifugal or return type), combustion method etc. which centrifugal extraction has been used as the industry standard test method [3]. These methods mostly are carried out after concrete mixing, so that these tests are time consuming and can not guide the construction in advance.

Aiming at these problems, a kind of real-time image acquisition system has been paid attention to. Wang Linbing [4] of the Louisiana State University in the United States in 2004 provided a program to calculate the volume and inertia moment of inertia between particles. Sha Aimin and others [5] found that the use of image processing technology to detect aggregate gradation of asphalt mixture is feasible, and it can be used as an effective alternative to the screening method.

This study focuses on the construction of image detection grading equipment, the introduction of detection methods, and the drawing analysis of test data.

---

*Corresponding author

2. Test Equipment

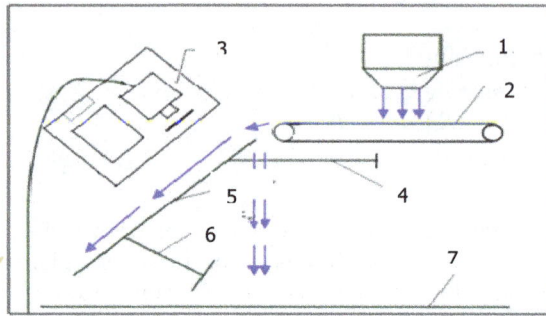

1. Cold silo
2. The first conveyor belt
3. Image acquisition device
4. Flow control lever
5. Plate for separating aggregate
6. Rod for adjusting angle
7. Second conveyor belts

Fig. 1 Schematic diagram of image detection equipment installation

Fig. 1 shows the schematic diagram of the testing platform for the construction site. The system is mainly composed of image acquisition equipment, aggregate control equipment and computer processing equipment. The anti vibration and dust removal measures should be paid attention to in the detection process.

Fig. 2 shows the equipment installed in the physical map. By installing and debugging, the system runs smoothly.

Fig. 2 Field installation diagram

3. Detection Method

The mineral aggregate drops from the cold silo (1), by the first conveyor belt (2) being transmitted to the plate for separating aggregate (5). The flow and angle-adjusting rod (4, 6) should be adjusted until the aggregate on the plate meets the requirements of the shoot number and divergence. At this time, the image acquisition system (3), including digital cameras and supplementary light source) acquires the aggregate image information on the plate. Finally, the image information is transmitted through the data transmission line to the computer terminal. Then the computer extracts the aggregate gradation information from the image through the specific program. A set of gradation data is obtained every 3 minutes (adjustable) in this study.

4. Results and Discussion

In this project, the gradation of the upper layer (SMA-13) and the middle surface layer (AC-20) of a project is measured, and a large number of field gradation data are obtained.

Gradation curve of SMA-13 asphalt mixture

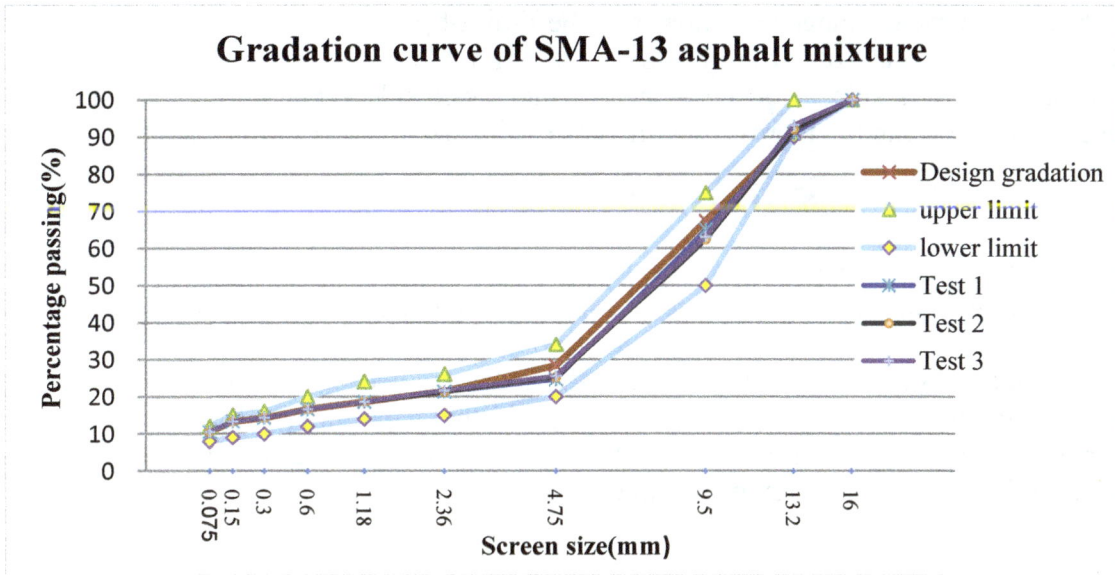

Fig. 3 SMA-13 asphalt mixture image detection grading curve

Fig. 3 is the gradation curve of SMA-13 asphalt mixture using the image detection method. 30 groups of original test data (each with 3min) are selected, which are divided into three groups and averaged respectively.

The average test values of the three groups are located between the grades of the assembly line, and located near the design level. Therefore, the image method is used to detect the gradation of mineral mixture. The three test groups were lower than the design-grading curve at 4.75mm, therefore, the 2.36mm file is less.

Fig. 4 is the AC-20 asphalt mixture gradation curve with image detection technology. The test values of the three groups were the mean of the ten groups of the original test data.

Gratation curve of AC-20 asphalt mixture

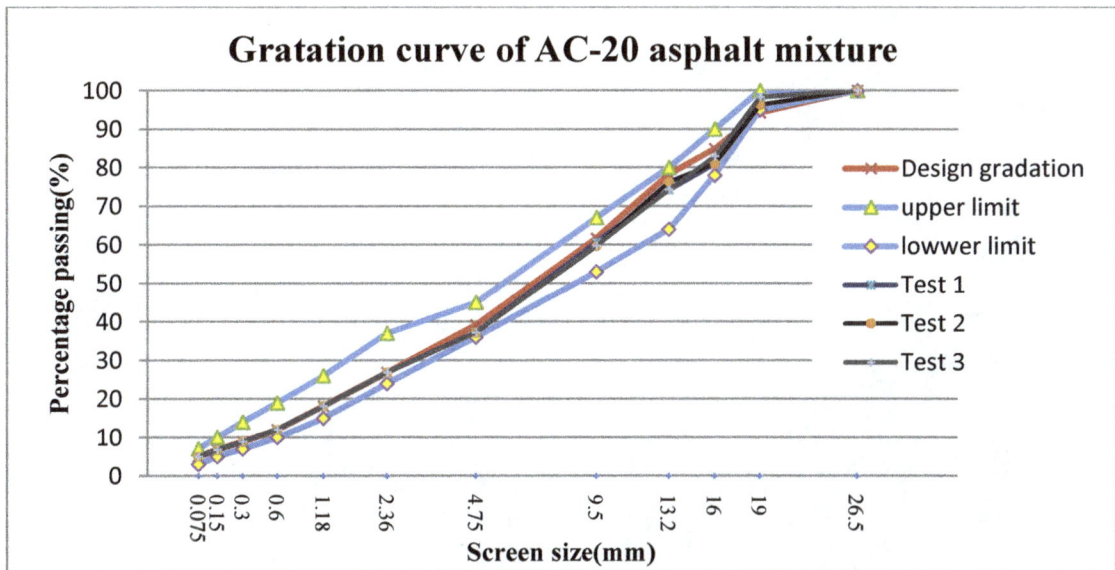

Fig. 4 AC-20 asphalt mixture image detection grading curve

The gradation curve of image detection is in the limit of gradation, which are similar rules with SMA-13 grading. In addition, from 19mm to 26.5mm file, three groups of test grading curve change from the lower side of design curve to the upper side. This shows that the 19mm ore material is less, and it can be improved by increasing the size of the 19mm gear.

5. Conclusion

The mineral aggregate gradation detection technology based on image acquisition technology is implemented in the real project, which has has the characteristic of high efficiency and accuracy.

The following conclusions are obtained through the research. Firstly, image detection techniques used for gradation detection were shown to be feasible in the construction process. Secondly, the method achieved rapid detection, and in this paper, a set of test data was obtainted every three minites. Thirdly, based on the new equipment, the grading curves were drawn quickly and used to guide the construction in real time.

References

1. Jiaji Yan. Road construction materials [M]. The third edition. Beijing: China Communications Press, 2004: 45-78
2. Famao Tan. Ways and methods to improve the performance of asphalt pavement [D]. Engineering master's degree thesis, Tianjin University, 2002
3. Chaofan Wang. Research on gradation detection technology of asphalt mixture based on digital image [D]. Xi'an: Chang'an University, 2007
4. Wang L B, Frost J D, Lai J S. Three-dimensional digital representation of granular material micro-structure from X-ray tomography imaging [J]. Journal of Computing in Civil Engineering, 2004,18 (1): 28-35.
5. Aimin Sha, Chaofan Wang, Zhaoyun Sun. A method of asphalt mixture gradation detection based on image [J]. Journal of Chang'an University, Natural Science Edition. 2010, 09:30-5.

Characterization of Hot Deformation Behavior of TC4-DT Titanium Alloy

Qun-Lan Wu[1, *], Shao-Yang Wang[1], Fu-Hua Peng[1], Wei Gu[1], Wen-Jun Yu[1], Li-Wei Zhu[2]

[1]*AVIC Chengdu Aircraft Industrial (Group) Co. Ltd., Chengdu 610092, China*
[2]*Beijing Institute of Aeronautical Materials, Beijing 100095, China*
Email: wuqunlan132@163.com

The hot deformation behavior of TC4-DT alloy was investigated using a hot Gleeble-1500D thermal simulator in the temperature range of 908 °C ~1038 °C and at constant strain rate from $0.01s^{-1}$ to $10s^{-1}$. Flow behavior of TC4-DT alloy was discussed. The deformation activation energy was calculated. Processing map of TC4-DT alloy was established. The results indicate that the hot deformation behavior of TC4-DT alloy is sensitive to the deformation temperature and strain rate. The peak flow stress decreases with the increase of the test temperature and decrease of the strain rate. A constitutive equation was constructed as a function of temperature and strain rate, and the activation energy of deformation was estimated to be 685KJ/mol in the α+β phase region and 242KJ/mol in β phase region, respectively. The processing map is generated at the true strain of 0.7, which shows that the peak efficiency domain appears at the temperature of 965°C and the strain rate of 0.01^{-1} with a peak efficiency of 54%.

Keywords: Titanium alloy; TC4-DT; flow behavior; constitutive equation; processing map.

1. Introduction

In recent years, titanium and titanium alloys have been widely used in the aerospace industry as structural materials due to low density, high strength, good corrosion resistance and fatigue properties, such as jet engines and airframe components [1-4]. TC4-DT alloy is an α+β titanium alloy based on the chemical composition of TC4 (Ti-6Al-4V) with the extra low interstitial (ELI) grade with the contains of oxygen in the range of 0.09%~0.13%, which is to attain high fracture toughness and a good combination of strength, plastic and toughness [5, 6]. It is normally used at a medium tensile strength of 900MPa with good ductility, high fatigue and damage tolerance properties. According to its specific, TC4-DT alloy is particularly suitable to manufacture the integrated critical force-bearing component parts such as large frame, beam and joints in order to meet the damage tolerance design for long life, security and dependability.

It is well known that it is difficult for titanium alloys to deform because of their poor deformability, high resilience and low plasticity. Many factors could affect the compressive behavior of titanium alloys during hot working, such as deformation temperature, strain rate and deformation degree. And the constitutive equation of materials could express the relationship among the deformation temperature, flow stress and strain rate. Furthermore, the evolution of microstructures is very sensitive to process parameters such as temperature, strain rate and train. The application of processing map to optimize the workability of a wide variety of materials is compiled in detail, which are useful for numerical analysis and simulation of deformation processes. Prasad *et al*. characterized the hot working behavior of titanium and its alloys using the approach of processing maps [7]. Kim *et al*. studied the flow behavior, microstructure evolution and dynamic globularization during the hot working of Ti-6Al-4V alloy with widmanstäten microstructure [8]. Zhu *et al*. analyzed the hot deformation behavior of Ti-4.5Al-3V-2Fe-2Mo alloy, and the instability region was discussed according to the processing map [9]. Therefore, careful process control and profound knowledge of the influence of processing parameters on hot working behavior are significant important for the manufacturing of titanium alloys.

---

*Corresponding author

In this work, the hot deformation behavior of TC4-DT alloy was investigatied based on the results of hot compression tests. Hence, the influence of deformation temperature and strain rate on flow behavior is considered, and then the constitutive equation and the processing map are constructed. The approach of hot deformation behavior of TC4-DT alloy is focused on to analyze the instability region and optimize the manufacturing parameters for processing.

2. Experimental

2.1. *Material*

The TC4-DT alloy was triple melted by consumable vacuum arc remelting (VAR) process. The finished ingots were processed into bars of 210mm in diameter with two steps. The ingots was initially break down around 1150°C followed by forging into plate in α+β field. The chemical composition in weight percent is 6.26Al, 4.18V, 0.03C, 0.03Fe, 0.012N, 0.004H, 0.11O, and the balance Ti. The β transus temperature is approximately 985°C. The initial microstructure of the plate of TC4-DT alloy is shown in Figure 1.

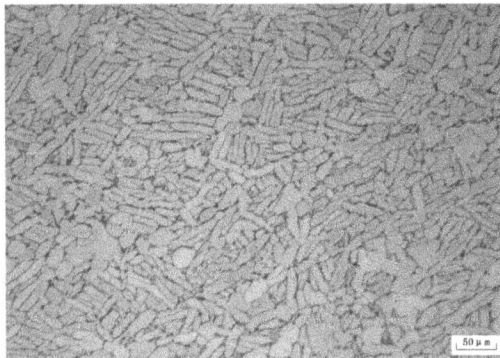

Fig. 1 Initial microstructure of TC4-DT alloy   Fig. 2 Schematic diagram of hot compression test

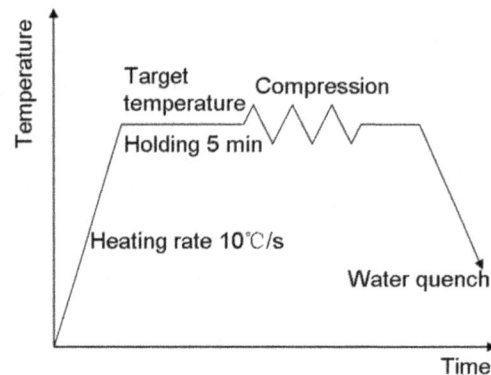

2.2. *Compression tests*

The isothermal hot compression tests were carried out on the Gleeble-3800 simulator at temperature from 908°C to 1038°C, and at constant strain rate of 0.01, 0.1, 1 and 10s^{-1}, respectively. Schematic diagram is shown in Fig. 2. The cylindrical specimens of 10 mm in diameter and 12 mm in height were machined from the plate along the axial direction. The specimens were reheated to the target temperature, holding for 5 min and then deformed at the selected temperature with a constant strain rate to a height reduction. The as-deformed specimens were immediately water quenched to room temperature. The load-stroke curves obtained from the compression tests are converted into true stress-strain curves by standard equations.

2.3. *Microstructure observation*

The deformation microstructure was observed the after tests. The hot compression specimens were cut into two parts as the metallographic specimens along the compression axial by wire cutting. The metallographic specimens were corroded using the solution containing 10% HNO$_3$, 15% HF and 75% H$_2$O. And the microstructure was observed using conventional optical microscopy.

3. Results and Discussion

3.1. *True stress-true strain curves*

The typical true stress-strain curves obtained of 908°C ~1038°C and strain rates of 0.01s⁻¹~10s⁻¹ are presented in Fig. 3, which are representative of deformation behavior in three regions of α+β region, near β transus, and the β region. The curves show that the flow stress behavior of TC4-DT alloy is sensitive to strain rates and compression temperature. All the true stress-strain curves display a peak stress at the beginning of deformation, followed by a continuous flow softening till the end of hot compression to a true strain of 1.0. After the peak stress, the true stress decreased gradually with increasing the true strain, and then tended to a steady state. In general, flow stress was observed to decrease with increasing the deformation temperature and increasing the strain rate. Such continuous decrease in the flow stress with increasing strain has been previously reported for Ti-6Al-4V alloy, IMI834 alloy [10, 11].

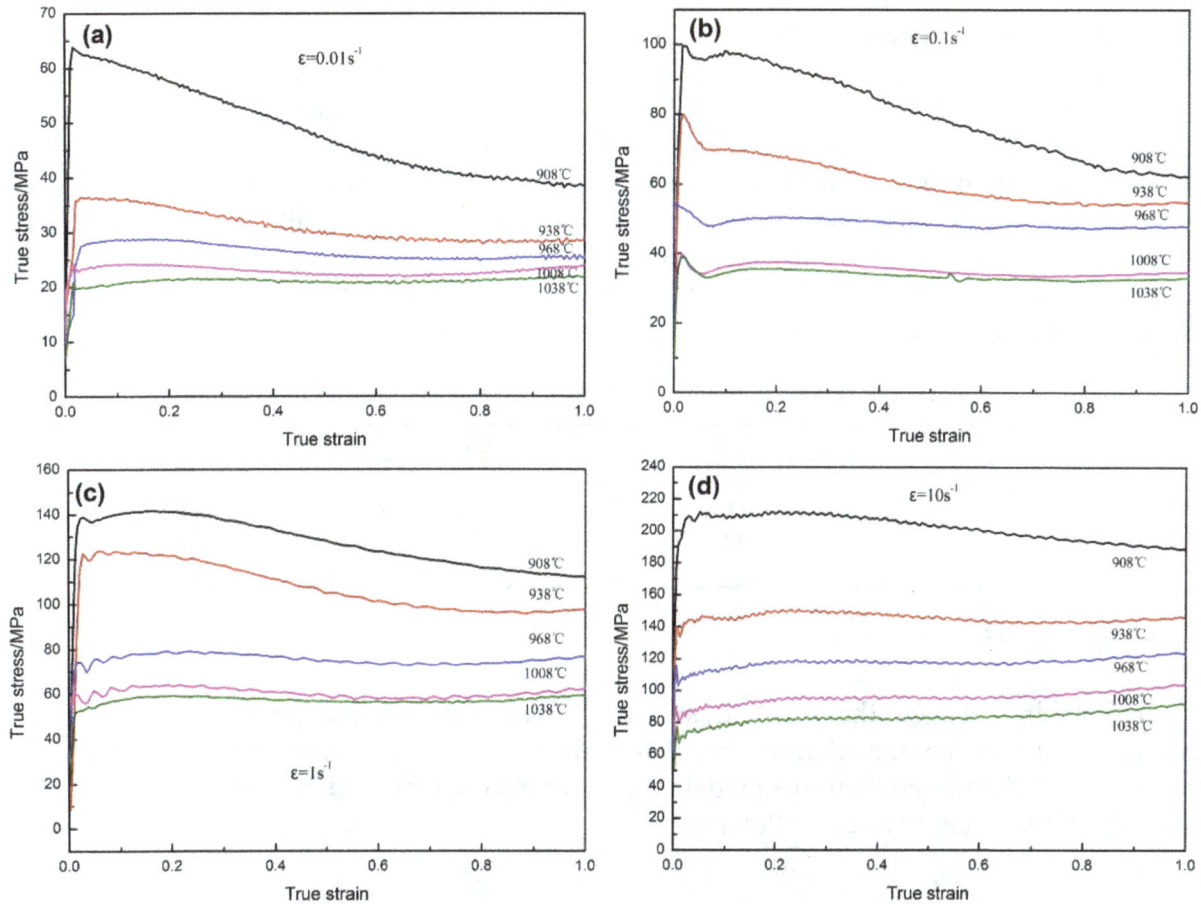

Fig. 3 True stress-true strain curves of TC4-DT alloy obtained under different strain rate at (a) 650°C;
(b) 850°C; (c) 900°C; (d) 950°C, with the temperature from 908°C and 1038°C

As illustrated in Fig. 3, there is a significant difference in the true stress-true strain curves observed for deformation in the α+β region as compared to the β region. At the temperature below β transus, especially at 908°C, the true stress-true strain curves show an initial sharp peak, and then reach a steady-state stress condition at strain rate of 0.01s⁻¹ and 0.1s⁻¹ or decrease gradually at strain rate of 1s⁻¹ and 10s⁻¹. Consequently, when TC4-DT alloy is deformed in single β region, the true

947

stress reaches a steady state or increase a little as deformation proceeds without showing a yield drop.

It suggests that the strain hardening was greater than the dynamic recovery or dynamic recrystallization softening process, therefore the stress increased rapidly with the strain ascending in the early period of deformation. The flow stress would reach an initial sharp peak value, when hardening process was approximately equal to softening process. The phenomenon of initial sharp peak can be concluded that α phase that acts as hard particals plays an important role in dislocation pinning. Therefore, the true stress increase abruptly duo to a fast generation of dialocations in early state of deformation. However, when the becomes large enough, dislocation starts to slip from the pinning sites. Consequently, the true stress decrease sharply. Then, with the deformation continuing, the true stress almost remained constant [12]. when the deformation temperature above β transus, the strain hardening was a little higher than the dynamic recovery or dynamic recrystallization softening process at high strain rate of 10s⁻¹.

In addition, the results of peak flow stress in different conditions are listed in Table 1. when the strain rate are 0.01s⁻¹ and 10s⁻¹ at the temperature of 908°C, the values of peak flow stress are 63MPa and 212MPa, respectively, with the difference of 149MPa; at the temperature of 1038°C, the values of peak flow stress are 20MPa and 83MPa, respectively, with the difference of 63MPa. It can be seen that the larger the strain rate is, the larger the peak stress is, especially at low deformation temperature of 908°C. The peak stress value decreased with the increase of deformation temperature. When the strain rate of 0.01s⁻¹, the stress-strain curves have little change with the strain increasing, almost showing a horizontal line, only existing a tiny change between 908 °C and 938°C . There may be several suggests to consider this reasons [13, 14]. This is mainly due to the increase of strain rates, the dynamic recovery or dynamic recrystallization could not supply in time and in sufficient, consequently the stress increases with increasing the strain rates.

Table 1 The peak flow stress with different testing condtions (MPa)

strain rate [s⁻¹]	908 °C	938°C	968°C	1008°C	1038°C
0.01	63	36	28	24	20
0.1	99	79	54	39	38
1	142	123	79	64	59
10	212	150	119	96	83

3.2. Kinetic analysis

The relationship between the flow stress, strain rate and temperature can be described by constitutive equation. For the relationship between the flow stress and the strain rate, There are three expression of Arrhenius constitutive models that have been widely reported in literatures to model the material flow behavior as the following [15]:

$$\dot{\varepsilon} = A_1 \sigma^{n_1} \exp(-Q/RT) \tag{1}$$

$$\dot{\varepsilon} = A_2 \exp(\beta\sigma) \exp(-Q/RT) \tag{2}$$

$$\dot{\varepsilon} = A[\sinh(\alpha\sigma)^n] \exp(-Q/RT) \tag{3}$$

Where $\dot{\varepsilon}$ is the strain rate; σ is the flow stress; Q is the average apparent activation energy of deformation; R is the ideal gas constant (8.314 J·mol^{-1}·K^{-1}); T is the temperature in Kelvin; n_1, n are the stress exponent; $A_1, A_2, A, \beta,$ and α are material constants. The relationship among α, β and n_1 is

$$\alpha = \beta/n_1 \tag{4}$$

The power law, Eqn. (1) and the exponential law, Eqn. (2), are suitable for a low stress and a high stress, respectively. The law of Eqn. (3) is generally used to describe the flow stress and deformation behavior over a wide range of temperature and strain rate.

By taking natural logarithm, Eqn. (1) and Eqn. (2) can be written as

$$\ln \sigma = \frac{\ln \dot{\varepsilon}}{n_1} - \frac{\ln A_1}{n_1} \tag{5}$$

$$\sigma = \frac{\ln \dot{\varepsilon}}{\beta} - \frac{\ln A_2}{\beta} \tag{6}$$

Because of that the material constants of $A_1, A_2, Q, R,$ and T is invariable, Eqn. (5) and Eqn. (6) by taking derivative can be written as

$$n_1 = \frac{\partial \ln \dot{\varepsilon}}{\partial \ln \sigma} \tag{7}$$

$$\beta = \frac{\partial \ln \dot{\varepsilon}}{\partial \sigma} \tag{8}$$

The relation of $\ln \sigma - \ln \dot{\varepsilon}$ when the stress is low is shown in Fig. 4(a), and the relation of $\sigma - \ln \dot{\varepsilon}$ when the stress is high is shown in Fig. 4(b). Since the transus temperature of TC4-DT alloy is about 986°C, the kinetic parameters may be evaluated separately into the two-phase region (908-968°C) and the signal-phase region (998-1038°C). By linear regression of the relations of $\ln \sigma - \ln \dot{\varepsilon}$ and $\sigma - \ln \dot{\varepsilon}$ at different temperature, the values of n_1 and β are obtained for the $\alpha+\beta$ region and the β region. Then taking the values into Eqn. (4), the value of α in the $\alpha+\beta$ region and the β region can be inferred to be 0.0117 and 0.0205, respectively.

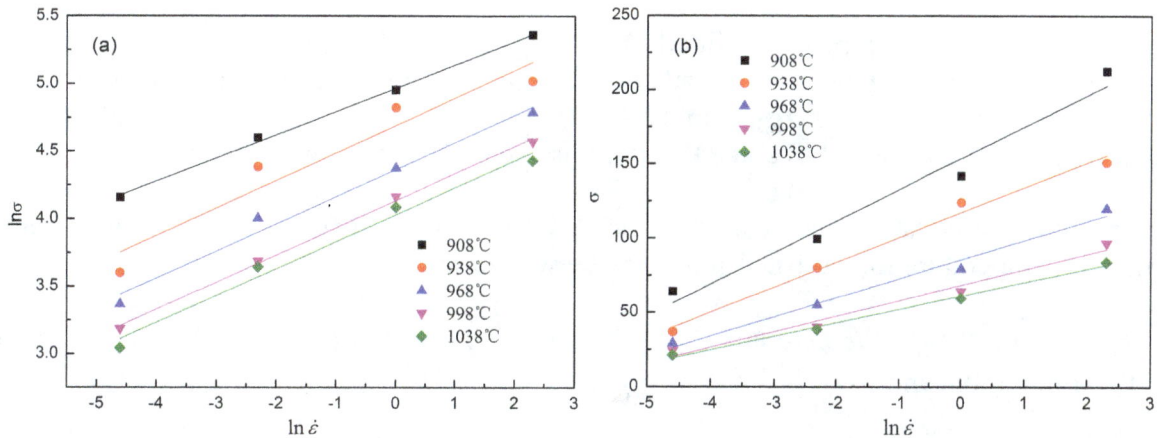

Fig. 4 Plots of (a) $\ln \sigma - \ln \dot{\varepsilon}$ and (b) $\sigma - \ln \dot{\varepsilon}$

949

If the activation energy (Q) does not change with deformation temperature (T), the Eqn. (3) takes into logarithm forms as

$$\ln \dot{\varepsilon} = \ln A + n \ln[\sinh(\alpha\sigma)] - Q/RT \tag{9}$$

Taking partial derivative for Eqn. (9), n and Q can be obtained as

$$n = \left. \frac{\partial \ln \dot{\varepsilon}}{\partial \ln\left[\sinh(\alpha\sigma)\right]} \right|_{T} \tag{10}$$

$$Q = Rn \left. \frac{\partial \ln\left[\sinh(\alpha\sigma)\right]}{\partial(1000/T)} \right|_{\varepsilon} = R \left. \frac{\partial \ln \dot{\varepsilon}}{\partial \ln\left[\sinh(\alpha\sigma)\right]} \right|_{T} \left. \frac{\partial \ln\left[\sinh(\alpha\sigma)\right]}{\partial(1000/T)} \right|_{\varepsilon} \tag{11}$$

According to the value of α and the true stress-true strain curves, the stress exponent n and activation energy Q for deformation is calculated in $\alpha+\beta$ region and the β region. The plot of $\ln\dot{\varepsilon}$ versus $\ln[\sinh(\alpha\sigma)]$ and the plot of $\ln[\sinh(\alpha\sigma)]$ versus T^{-1} in both phase fields are shown in Fig. 5.

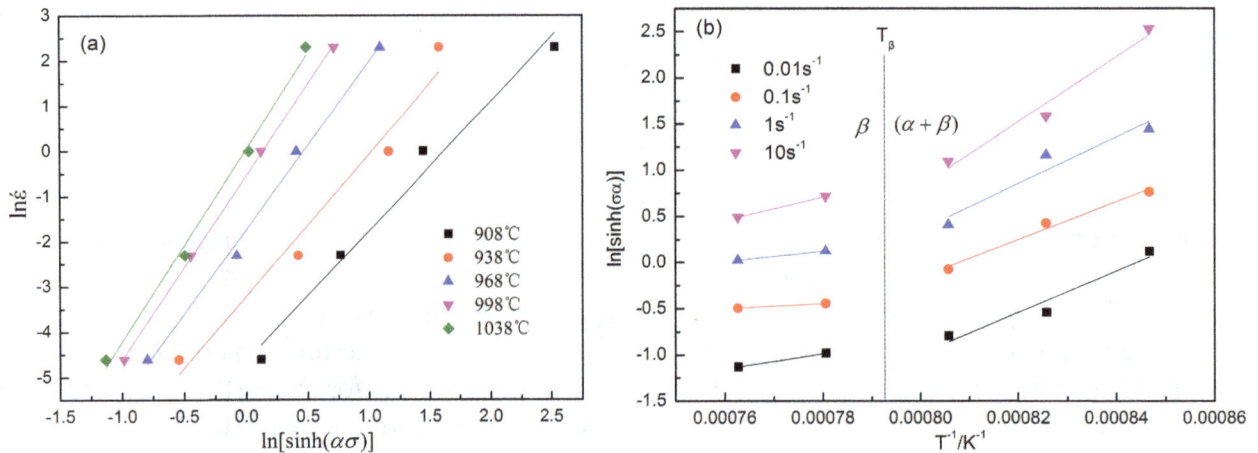

Fig. 5 Plots of (a) $\ln\dot{\varepsilon} - \ln[\sinh(\alpha\sigma)]$ and (b) $\ln[\sinh(\alpha\sigma)] - T^{-1}$

It can be seen that the slope of the fitted line in Fig. 5(b) is changing which indicates different deformation mechanism. By regression analyzing the curves shown in Fig. 5, the values of the stress exponent n and the activation energy Q of TC4-DT alloy is determined to be 3.185 and 685 kJ/mol in the $\alpha+\beta$ phase region and 4.148 and 242 kJ/mol in the β phase region, respectively.

The constitutive equation that desicribes the flow behavior as a function of the strain rate and deformation temperature for TC4-DT alloy may be written as

$$\dot{\varepsilon} = e^{65.688}[\sinh(0.00117\sigma)]^{3.185} \exp(-6.85 \times 10^{5}/RT) \tag{12}$$

for the $\alpha+\beta$ region, and

$$\dot{\varepsilon} = e^{23.165}[\sinh(0.0205\sigma)]^{4.148} \exp(-2.42 \times 10^{5}/RT) \tag{13}$$

for the β region, respectively.

3.3. Processing map

Processing maps are useful in explaining the hot deformation behavior, and describing the manner in which power is dissipation through microstructural evolution during hot deformation for a large number of metals alloys. Based on the principles of dynamic materials model [16, 17], the processing map of TC4-DT alloy in the deformation temperatures from 908°C to 1038°C and strain rates from 0.01 to 10s^{-1} for a strain of 0.7. has been established shown in Fig. 7, which indicated the safe domain and the unsafe domain during plastic processing. As seen in Fig. 7, the material is under safe region through all the tested temperatures in this test at strain rate lower than 1.5s^{-1}, whereas there are two instable fields in the processing maps: one is in the temperatures from 908°C to 945°C and strain rates from 1.5s^{-1} to 10s^{-1}, and another is in the temperatures from 1005°C to 1038°C and strain rates from 2.5s^{-1} to 10s^{-1} showing the instable processing conditions of TC4-DT alloy. In the safe area, it can be seen that in Fig. 7, there are three regions with high values of power dissipation and ther are occurring at the temperature and strain rate of (908~995°C, 0.01~1.5 s^{-1}), (960~1000°C, 3.16~10 s^{-1}) and (1000~1080°C, 0.1~2.5 s^{-1}). The most peak efficiency domain with a peak efficienty of 54% appears at the temperature of 965°C and strain rate of 0.01s^{-1}. The large safe region and the small unsafe region exhibit the good processing ability for TC4-DT alloy.

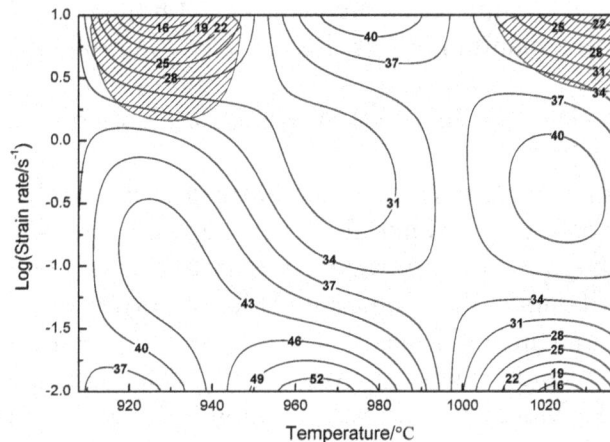

Fig. 7 Processing map of TC4-DT alloy (ε=0.7)

4. Conclusion

Hot compression tests of TC4-DT alloy in the temperature range of 908°C ~1038°C and strain rate range of 0.01s^{-1}~10s^{-1} were carried out. Based on the flow curves and the constructed processing map, the hot deformation behaviors of TC4-DT alloy were investigated. The following principal conclusions can be drawn from the present study:

1. The flow behavior of TC4-DT alloy is sensitive to the strain rate and the compression temperature. The peak flow stress decrease with increasing of compression temperature and decreasing of strain rate.
2. The deformation activation energy Q of TC4-DT alloy is determined to be 685 kJ/mol and 242 kJ/mol for the two-phase α+β field and single-phase β field, respectively. The constitutive equation for TC4-DT alloy are formulated to describe the dependence of the flow stress on deformation temperature and strain rate.

3. The stability domain of processing map appears at the temperatures ranging from 908°C to 1038°C and strain rates ranging from $0.01s^{-1}$ to $1.5s^{-1}$. The most peak efficiency domain with a peak efficienty of 54% appears at the temperature of 965°C and strain rate of $0.01s^{-1}$.

4. Two instability deformation fields are obtained, in which the first field is at 908°C to 945°C and $1.5s^{-1}$ to $10s^{-1}$, and another is at 1005°C to 1038°C and $2.5s^{-1}$ to $10s^{-1}$. These two fields must be avoided because of the flow localization in the deformation of TC4-DT alloy.

References

1. D. C. James, P. C. Larry, R. P. Herry.Titanium alloys on the F-22 fighter airframe. Advanced Materials and Processes. 160 (2002) 25-28.

2. Lane Lineberger. Titanium aerospace alloy. Advanced Materials & Process. 1998 (5):45 -46.

3. D. W.D. Brewer, R.K. Bird, T.A. Wallace. Titanium alloys and processing for high speed aircraft. Mater. Sci. Eng. A 243 (1998) 299-304.

4. C. Leyens, M. Peters. Titanium and titanium alloys fundamental and applications. Wiley-VCH Gmbh & Co., 2003, pp. 5–10.

5. Z. S. Zhu. Recent research and development of titanium alloys for aviation application in China. Journal of Aeronautical materials. 34 (2014) 44-50.

6. L.W. Zhu, W. X. Wang, Z. S. Zhu, et al. Near-threshold fatigue crack propagation behavior of TC4-DT damage tolerance titanium alloys. Rare Metal Materials and Engineering. 43 (2014) 1342-1346.

7. Y. V. R. K. Prasad, T. Seshacharyulu. Processing maps for hot working of titanium alloys. Mater. Sci. Eng. A243 (1998): 82-88.

8. J. H. Kim, S. L. Semiatin, C. S. Lee. High temperature deformation behavior of Ti-6Al-4V alloy with widmanstäten microstructure. Materials Science Forum. 426-432(2003) 689-694.

9. L. W. Zhu, W. X. Wang, F. Yue, et al. Characterization of hot deformation behavior of Ti-4.5Al-3V-2Mo-2Fe titanium alloy. Materials Science Forum. 849 (2016) 309-316.

10. S.L. Semiatin, V. Seetharaman, and I. weiss. Flow stress and globularization kinetics during hot work of Ti-6Al-4V with a colony alpha microstructure. Mater. Sci. Eng. A 263 (1999) 257-259.

11. P. Wanjara, M. Jahazi, H. Monajati et al. Hot working behavior of near-α alloy IMI834. Mater. Sci. Eng. A 396 (2005) 50-60.

12. W.J. Jia, W.D. Zeng, Y.G. Zhou, et al. High-temperature deformation behaviour of Ti60 titanium alloy. Mater. Sci. Eng. A 528 (2011) 4068-4074.

13. G.F. Zhang, S.Z. Chen. Hot deformation behavior of Ti-6.5Al-3.5Mo-1.5Zr-0.3Si alloy with acicular microstructure. J Cent South Univ Technol. 18 (2011) 296-302.

14. SVS. Narayana Murty, R.B. Nageswara. Instability criteria for hot deformation of materials. J Mater Process Technol. 104 (2000) 103–109.

15. S. Guo, D. Li, X. Wu, et al. Characterization of hot deformation behavior of a Zn–10.2Al–2.1Cu alloy using processing maps. Mater. Des. 41 (2012) 158–166.

16. V.V.Balasubrahmanyam, YVRK. Prasad. Deformation behavior of beta titanium alloy Ti-10V-4.5Fe-1.5Al in hot upset forging. Mater. Sci. Eng. A 336 (2002) 150-158.

17. YVRK. Prasad, H.L. Gegel, S.M. Doraivelu, et al. Modeling of dynamic material behavior in hot deformation: forging of Ti-6242. Metall. Mater. Trans. A 15 (1984) 1883–1892.

Facile Preparation and Characterization of VO$_2$(M) on Thermal Decomposition of Ammonium Perchlorate (AP)

Yi-Fu Zhang*, Qiu-Shi Wang, Ji-Qi Zheng

School of Chemistry, Dalian University of Technology, Dalian 116024, PR China
Email: yfzhang@dlut.edu.cn

VO$_2$(M) nanobelts were successfully synthesized using commercial V$_2$O$_5$, ethanol and water as the starting materials by a template-free hydrothermal method and subsequent calcination. The as-obtained products were characterized by X-ray photoelectron spectroscopy (XPS), X-ray powder diffraction (XRD) and transmission electron microscopy (TEM). Furthermore, the study on the influence of the as-prepared VO$_2$(M) nanobelts on the thermal decomposition of ammonium perchlorate (AP) were evaluated by Thermo-Gravimetric Analysis and Differential Thermal Analysis (TGA/DTA), which showed the thermal decomposition temperatures of AP in the presence of the as-prepared VO$_2$(M) nanobelts were reduced to 421 °C (decreased by 35 °C). The results indicated that VO$_2$(M) nanobelts had great influence on the thermal decomposition temperature of AP.

Keywords: VO$_2$(A) nanobelts; hydrothermal process; ammonium perchlorate; catalytic activity.

1. Introduction

Ammonium perchlorate (AP) is the most common oxidant in composite solid propellants. AP's thermal decomposition characteristics directly influence the combustion behavior of the solid propellant [1-6]. Reducing the particle size of AP can improve the combustion behavior of the solid propellant, however, this method is restricted and superfine AP is dangerous. In the past decades, many materials scientists investigated many transition metal oxides and their related composites on thermal decomposition of AP to improve the combustion behavior of solid propellant [2, 3, 7-9]. For example, Liu and his coworkers [10] reported that Cu, Al and Ni particles can respectively lower the thermal decomposition temperature (T$_d$) of AP to 347, 425 and 364 °C. Although lots of materials have been investigated to decrease the T$_d$ of AP, to find new catalysts applied in the thermal decomposition of AP is still a great challenge for material scientists who are engaging in the field of catalyst [2]. Recently, vanadium oxides as the additives on the thermal decomposition of AP have attracted some attention. Vanadium pentoxide (V$_2$O$_5$) nanowires [7], vanadium oxide hydrate (V$_3$O$_7$·H$_2$O) nanobelts and V$_3$O$_7$·H$_2$O@C [8], vanadium sesquioxide (V$_2$O$_3$) particles and V$_2$O$_3$@C [11], V$_6$O$_{13}$ nanobelts [2] and vanadium dioxide with a B phase (VO$_2$(B)) nanobelts [12] were respectively reported to reduce the T$_d$ of AP. To our best knowledge, the influence of vanadium dioxide with a M phase (VO$_2$(M)) on the thermal decomposition of AP has not been reported.

As a class of vanadium oxides, vanadium dioxide (VO$_2$) is a representative binary compound with different polymorphs, including VO$_2$(M), VO$_2$(R), VO$_2$(B), VO$_2$(A), VO$_2$(C), VO$_2$(D), etc. Among them, the monoclinic vanadium dioxide VO$_2$(M) is the most important because it shows a fully reversible first-order metal-to-insulator transition (MIT) with the phase transition temperature (T$_c$) at about 68 °C, accompanied by a crystallographic transition between a low temperature monoclinic phase (M) and a high temperature tetragonal rutile phase (R) [13-16]. On warming through the transition, drastic changes occur in both optical and electrical properties [17, 18]. For example, the change in electrical resistivity is in the order of 10^5 and its infrared transmission characteristics dramatically change over the phase transition. These features make VO$_2$(M) and its related compounds to be suitable for various applications [19-24], such as smart window coatings, optical switches, storage medium, temperature-sensing devices, laser protection and so on. However, to the best of our knowledge, there is no report on the investigations of the influence of VO$_2$(M) on the

---

*Corresponding author

953

thermal decomposition of AP. In this contribution, we first synthesize VO₂(M) nanobelts by a facile one-pot hydrothermal approach and combination of calcination. Then the as-obtained VO₂(M) nanobelts were explored as the additive to the thermal decomposition of AP, the key component of composite solid propellants.

2. Experimental Section

2.1. *Synthesis*

In a typical procedure, 0.455 g of commercial V₂O₅ was dispersed into 31 mL of redistilled water with magnetic stirring, then 2.0 mL of H₂O₂ (30 wt%) and 2.0 mL of absolute ethanol were continuously added into the above solution. The mixture was stirred for about 1 h to obtain the brown liquid at room temperature. The above mixture was transferred into a 50 mL Teflon Lined stainless steel autoclave, then sealed and maintained at 180 °C for 48 h. When the reaction was finished, the precipitates VO₂(B) were filtered off, washed with distilled water and anhydrous alcohol several times, respectively, and dried in vacuum at 75 °C. VO₂(M) nanobelts was synthesized by the transformation from VO₂(B) under the inert atmosphere [19, 24]. The hydrothermal products VO₂(B) nanobelts were heated in a tube furnace with 5 °C/min heating rate under a high purity Ar (99.999%) atmosphere at 700 °C for 2 h, and cooled to room temperature in the Ar flow to prevent the oxidation of VO₂(M) nanobelts.

2.2. *Characterization*

X-ray photoelectron spectroscopy (XPS) was used to investigate the composition of the products and confirm the oxidation state of vanadium preformed on ESCALAB250Xi, Thermo Fisher Scientific. X-ray powder diffraction (XRD) was carried out on D8 X-ray diffractometer equipment with Cu *Kα* radiation, $\lambda = 1.54060$ Å. The morphology and dimension of the as-obtained products were observed by scanning electron microscopy (SEM, Quanta 200) and transmission electron microscopy (TEM, JEM-2100). Thermo-Gravimetric Analysis and Differential Thermal Analysis (TG/DTA) was performed on SETSYS-1750 (AETARAM Instruments). About 5 mg of the sample was heated in an Al₂O₃ crucible in nitrogen atmosphere from ambient temperature to 500 °C at a constant rise of temperature (10 °C/min). AP and the products were mixed and ground to make them dispersed homogeneously to prepare the target samples for thermal decomposition analyses.

3. Results and Discussion

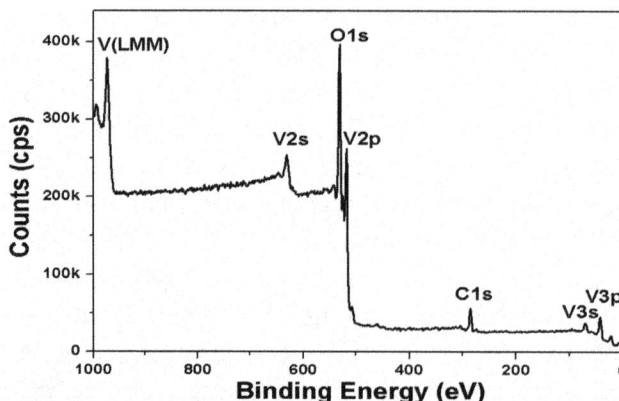

Fig. 1 XPS curve of the product

The composition of the sample was confirmed by XPS and XRD. Fig. 1 represents the typical XPS spectra of the as-obtained sample, which reveals that there are two elements: V and O. It is noted that the C peak is used as the standard value. XRD pattern was used to identify the phase of the sample. Fig. 2 shows the XRD patterns of the as-obtained $VO_2(M)$ nanobelts. All the diffraction peaks can be readily indexed to the monoclinic crystalline phase (space group: $P_{21/c}$) of $VO_2(M)$, in agreement with the literature value (JCPDS No. 72-514) [25], as shown in Fig. 2c. No impurity phases, such as V_2O_5, V_3O_7, $VO_2(B)$, $VO_2(A)$ and V_2O_3, are detected, indicating the as-obtained $VO_2(M)$ with high purity and well crystallization.

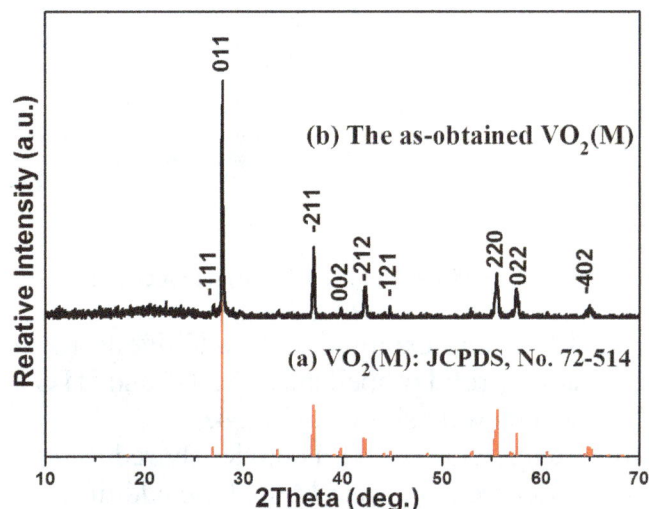

Fig. 2 XRD pattern of the product

Fig. 3 shows the morphology of the as-obtained $VO_2(M)$ nanobelts. The TEM image shows that the as-obtained $VO_2(M)$ contains lots of 1D nanobelts. The result confirms that $VO_2(M)$ nanobelts grow well separated and with high crystallinity, and also show that they are with length in the range of several to tens of micrometers, typically 100-260 nm wide and 10-25 nm thick.

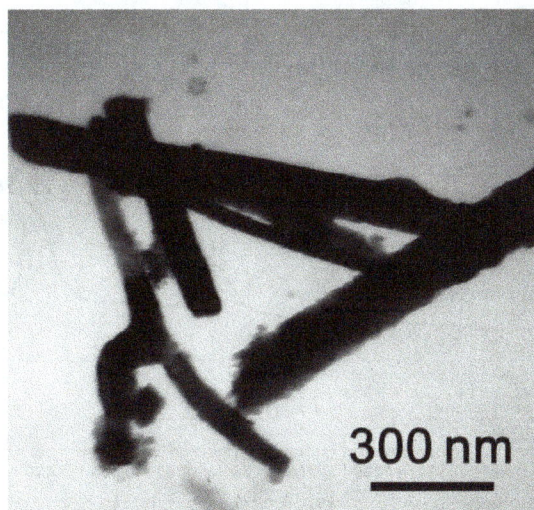

Fig. 3 The morphology of the as-obtained $VO_2(M)$ nanobelts

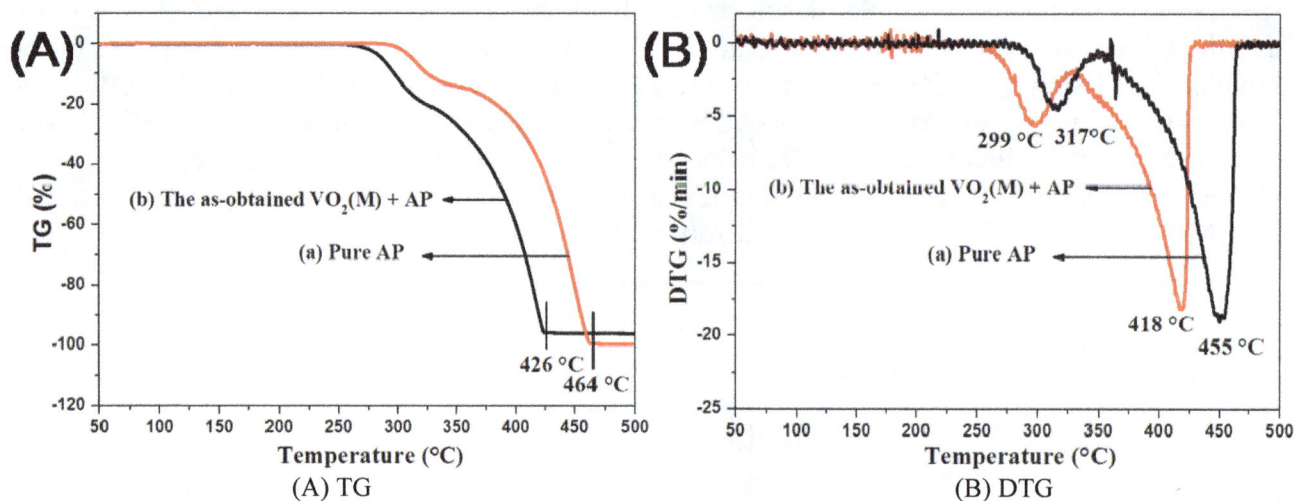

Fig. 4 Curves for the influence of as-obtained $VO_2(M)$ nanobelts on decomposition of AP

Furthermore, $VO_2(M)$ nanobelts were explored as an additive to the thermal decomposition of AP, the key component of composite solid propellants. The TG and DTG curves of pure AP and AP in the presence of $VO_2(M)$ nanobelts were shown in Fig. 4.

The starting decomposition temperature of AP is obviously reduced by the addition of $VO_2(M)$ nanobelts. It could be clearly observed from Fig. 4A that the addition of $VO_2(M)$ nanobelts in AP led to a significant reduction of the ending decomposition temperature of AP (pure AP: 464 °C and $VO_2(M)$ nanobelts + AP: 426 °C). Fig. 4B shows the corresponding DTG curves and two peaks can be seen in each curve, which reveals that the thermal decomposition of AP contains two steps. Further information of the influence of $VO_2(M)$ nanobelts on decomposition of AP was provided by heat flow curves, as shown in Fig. 5. The exothermic peaks at 318 °C and 456 °C in pure AP (Fig. 5a) are assigned to the partial decomposition of AP to form some intermediate products such as NH_3 and $HClO_4$ and then complete decomposition to volatile products [26], respectively, in agreement with the TG and DTG curves in Fig. 4.

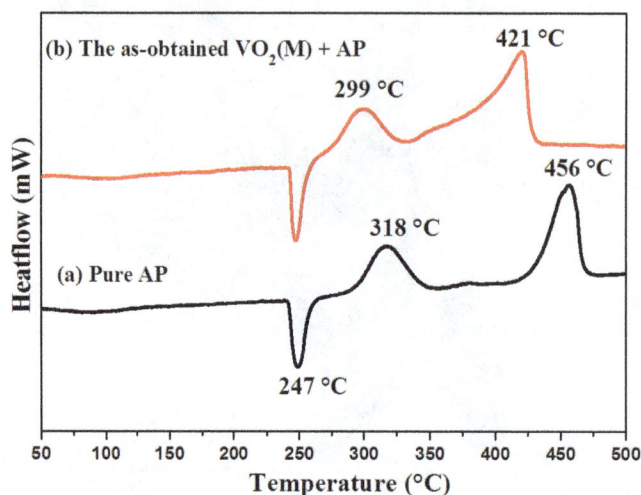

Fig. 5 Heat flow curves for the influence of the as-obtained $VO_2(M)$ nanobelts on decomposition of AP

956

However, obvious changes can be observed (Fig. 5) with the addition of $VO_2(M)$ nanobelts in AP. The thermal decomposition temperature of AP was lowered to 421 °C (decreased by 35 °C). Thus, we can conclude that $VO_2(M)$ nanobelts has great influence on the thermal decomposition of AP, which may be used as the promising additives in the future. Compared Fig. 4 with Fig. 5, there is no weight loss at the endothermic peak located at 247 °C, which is due to the transition from orthorhombic to cubic AP [1] and is not influenced by the additives. On the basis of the above results, it reveal that $VO_2(M)$ nanobelts have high catalytic activity towards the thermal decomposition of AP.

According to previous literatures [1, 3, 27, 28], the decomposition of AP consists of essentially of three steps:

(1) 240-250°C: the crystal transformation from orthorhombic to cubic phase;

(2) 300-330°C: the first decomposition step — a solid-gas multiphase reaction including decomposition and sublimation shown as follows: $NH_4ClO_4 \rightarrow NH_4^+ + ClO_4^- \rightarrow NH_3(g) + HClO_4(g)$;

(3) 450-480°C: the second decomposition step — NH_3 and $HClO_4$ react after entering the gas phase and the products are N_2O, O_2, Cl_2, H_2O and few NO.

From our experimental results, we can deduce that the additives can accelerate the thermal decomposition of AP via the third step, because the ending decomposition of temperature was greatly reduced during this process. Up to now, the thermal decomposition mechanism of AP is not yet fully understood because the decomposition process is a complex hetero-phase process involving coupled reactions in the solid, adsorbed and gaseous phases. There still remain some unsolved issues. According to the traditional electron-transfer theory [1], the presence of partially filled 3d orbit in vanadium atom provides help in an electro-transfer process. Positive hole in vanadium atom can accept electrons from AP ion and its intermediate products, by which the thermal decomposition of AP is accelerated.

4. Conclusion

In summary, a facile route was developed to successfully synthesize $VO_2(M)$ nanobelts using bulk V_2O_5, ethanol and water as the starting materials. The as-obtained products were characterized by XPS, XRD and TEM. Furthermore, the catalytic properties of the as-prepared $VO_2(M)$ nanobelts on the thermal decomposition of AP were evaluated by TGA/DTA. The results showed the thermal decomposition temperatures of AP in the presence of the as-prepared $VO_2(M)$ nanobelts were reduced to 421 °C (decreased by 35 °C). The results indicated that $VO_2(M)$ nanobelts had great influence on the thermal decomposition temperature of AP. Finally, the thermal decomposition mechanism of AP was proposed based on the traditional electron-transfer theory, and positive hole in vanadium atom can accelerate the thermal decomposition of AP.

Acknowledgments

This work was partially funded by the Fundamental Research Funds for the Central Universities (DUT16LK37), Science research project of Liaoning Province Education Department (L2015123) and the National Natural Science Foundation of China (Grant No. 21271037).

References

1. V.V. Boldyrev, Thermal decomposition of ammonium perchlorate, Thermochim. Acta, 443 (2006) 1-36.
2. Y. Zhang, C. Huang, C. Meng, Controlled synthesis of V_6O_{13} nanobelts by a facile one-pot hydrothermal process and their effect on thermal decomposition of ammonium perchlorate, Mater. Express, 5 (2015) 105-112.
3. Y. Zhang, X. Liu, J. Nie, L. Yu, Y. Zhong, C. Huang, Improve the catalytic activity of α-Fe_2O_3 particles in decomposition of ammonium perchlorate by coating amorphous carbon on their surface, J. Solid State Chem., 184 (2011) 387-390.
4. Z. Jia, D. Ren, Q. Wang, R. Zhu, A new precursor strategy to prepare ZnCo2O4 nanorods and their excellent catalytic activity for thermal decomposition of ammonium perchlorate, Appl. Surf. Sci., 270 (2013) 312-318.
5. H. Zhao, L. Guo, S. Chen, Z. Bian, Synthesis, complexation of 1,2,3-(NH)-triazolylferrocene derivatives and their catalytic effect on thermal decomposition of ammonium perchlorate, RSC Advances, 3 (2013) 19929-19932.
6. J. Lu, J. Zhu, Z. Wang, J. Cao, X. Zhou, Rapid synthesis and thermal catalytic performance of N-doped ZnO/Ag nanocomposites, Ceram. Int., 40 (2014) 1489-1494.
7. Y. Zhang, N. Wang, Y. Huang, W. Wu, C. Huang, C. Meng, Fabrication and catalytic activity of ultra-long V_2O_5 nanowires on the thermal decomposition of ammonium perchlorate, Ceram. Int., 40 (2014) 11393-11398.
8. Y. Zhang, X. Liu, D. Chen, L. Yu, J. Nie, S. Yi, H. Li, C. Huang, Fabrication of $V_3O_7 \cdot H_2O$@C core-shell nanostructured composites and the effect of $V_3O_7 \cdot H_2O$ and $V_3O_7 \cdot H_2O$@C on decomposition of ammonium perchlorate, J. Alloys Compd., 509 (2011) L69-L73.
9. Y. Zhang, C. Meng, Facile fabrication of Fe_3O_4 and Co_3O_4 microspheres and their influence on the thermal decomposition of ammonium perchlorate, J. Alloys Compd., 674 (2016) 259-265.
10. L.L. Liu, F.S. Li, L.H. Tan, L. Ming, Y. Yi, Effects of nanometer Ni, Cu, Al and NiCu powders on the thermal decomposition of ammonium perchlorate, Propell. Explos. Pyrot., 29 (2004) 34-38.
11. Y. Zhang, J. Zhang, J. Nie, Y. Zhong, X. Liu, C. Huang, Facile synthesis of V_2O_3/C composite and the effect of V_2O_3 and V_2O_3/C on decomposition of ammonium perchlorate, Micro Nano Lett., 7 (2012) 782-785.
12. Y. Zhang, X. Tan, C. Meng, The influence of $VO_2(B)$ nanobelts on thermal decomposition of ammonium perchlorate, Mater Sci-Pol, 33 (2015) 560-565.
13. J.B. Goodenough, The two components of the crystallographic transition in VO_2, J. Solid State Chem., 3 (1971) 490-500.
14. Y. Zhang, J. Zhang, X. Zhang, C. Huang, Y. Zhong, Y. Deng, The additives W, Mo, Sn and Fe for promoting the formation of $VO_2(M)$ and its optical switching properties, Mater. Lett., 92 (2013) 61-64.
15. Y. Zhang, J. Zhang, X. Zhang, Y. Deng, Y. Zhong, C. Huang, X. Liu, X. Liu, S. Mo, Influence of different additives on the synthesis of VO_2 polymorphs, Ceram. Int., 39 (2013) 8363-8376.
16. Y. Zhang, M. Fan, F. Niu, W. Wu, C. Huang, X. Liu, H. Li, X. Liu, Belt-like $VO_2(M)$ with a rectangular cross section: A new route to prepare, the phase transition and the optical switching properties, Curr. Appl. Phys., 12 (2012) 875-879.
17. F.J. Morin, Oxides Which Show a Metal-to-insulator Transition at the Neel Temperature, Phys. Rev. Lett., 3 (1959) 34-36.
18. A. Zylbersztejn, N.F. Mott, Metal-insulator transition in vanadium dioxide Phys. Rev. B, 11 (1975) 4383-4395.

19. Y. Zhang, W. Li, M. Fan, F. Zhang, J. Zhang, X. Liu, H. Zhang, C. Huang, H. Li, Preparation of W– and Mo–doped $VO_2(M)$ by ethanol reduction of peroxovanadium complexes and their phase transition and optical switching properties, J. Alloys Compd., 544 (2012) 30-36.

20. Y. Zhang, J. Zhang, X. Zhang, S. Mo, W. Wu, F. Niu, Y. Zhong, X. Liu, C. Huang, X. Liu, Direct preparation and formation mechanism of belt-like doped $VO_2(M)$ with rectangular cross sections by one-step hydrothermal route and their phase transition and optical switching properties, J. Alloys Compd., 570 (2013) 104-113.

21. Y. Zhang, Y. Huang, J. Zhang, W. Wu, F. Niu, Y. Zhong, X. Liu, X. Liu, C. Huang, Facile synthesis, phase transition, optical switching and oxidation resistance properties of belt-like $VO_2(A)$ and $VO_2(M)$ with a rectangular cross section, Mater. Res. Bull., 47 (2012) 1978-1986.

22. I.P. Parkin, T.D. Manning, Intelligent Thermochromic Windows, J. Chem. Educ., 83 (2006) 393-400.

23. Y. Zhang, C. Chen, J. Zhang, L. Hu, W. Wu, Y. Zhong, Y. Cao, X. Liu, C. Huang, Fabrication of belt-like $VO_2(M)$@C core-shell structured composite to improve the electrochemical properties of $VO_2(M)$, Curr. Appl. Phys., 13 (2013) 47-52.

24. Y. Zhang, C. Chen, W. Wu, F. Niu, X. Liu, Y. Zhong, Y. Cao, X. Liu, C. Huang, Facile hydrothermal synthesis of vanadium oxides nanobelts by ethanol reduction of peroxovanadium complexes, Ceram. Int., 39 (2013) 129-141.

25. G. Andersson, Studies on vanadium oxides II. The crystal structure of vanadium dioxide, Acta Chem. Scand., 10 (1956) 623-628.

26. X.Q. Shi, X.H. Jiang, L.D. Lu, X.J. Yang, X. Wang, Structure and catalytic activity of nanodiamond/Cu nanocomposites, Mater. Lett., 62 (2008) 1238-1241.

27. L.M. Song, S.J. Zhang, B. Chen, J.J. Ge, X.C. Jia, A hydrothermal method for preparation of alpha-Fe_2O_3 nanotubes and their catalytic performance for thermal decomposition of ammonium perchlorate, Colloids Surf. A, 360 (2010) 1-5.

28. L.J. Chen, L.P. Li, G.S. Li, Synthesis of CuO nanorods and their catalytic activity in the thermal decomposition of ammonium Perchlorate, J. Alloys Compd., 464 (2008) 532-536.

Characteristics of Road Dust Emission in Autumn of Shijiazhuang City

Shuo Guo, Jie-Ying Xiao*, Sai An, Juan Liu, Wen-Xia Zhao

Pollution Prevention Biotechnology Laboratory of Hebei Province, School of Environmental Science and Engineering, Hebei University of Science and Technology, Shijiazhuang 050018, Hebei, China
Email: jyxiao2014@126.com

Road dust samples were collected from 8 typical paved roads of Shijiazhuang city in autumn, 2014, for analyzing the characteristic of silt loading and finding the difference according to road types. Correlation analyses of silt loading of vehicle lane and bicycle lane from four types of road were conducted. Results show that, silt loading of expressway, main trunk road, secondary trunk road and branch road of Shijiazhuang city are $2.14g/m^2$, $2.28g/m^2$, $3.70 g/m^2$ and $0.65 g/m^2$ respectively. And its sequence is secondary trunk road>main trunk road>expressway>branch road. The differences of silt loading between vehicle lane and bicycle lane are large. Linear correlations are significant between the silt loading of vehicle lane and bicycle lane. Results can provide suggestion for the Shijiazhuang city road dust management.

Keywords: Urban paved road; sampling method; silt loading.

1. Introduction

In recent years, with the rapid economic development of Shijiazhuang City, the amount of motor vehicles increases daily, road area expands constantly, which lead to a gradually increasing contribution to urban air pollution from road dust [1]. According to the source apportionment reported on 2014: dust is the main pollution source to PM_{10} in Shijiazhuang, its contribution rate was slightly higher than that to $PM_{2.5}$. At the same time, road dust emissions are at low height, and locate at both sides of transportation line or intersection, closer to the dense population and human activity places, has the direct impact on air quality and human health [2]. Therefore, the road traffic dust pollution problems and related research has been widely concerned.

AP-42 emissions factors model, which proposed by the U. S. EPA, is accepted widely currently in global scale. In this paper, according to Huang's [3] road dust loading measurement method, road dust samples from vehicle lane and bicycle lane, different types of road were collected in Shijiazhuang City. The dust loading relationship among different seasons and relationship among different types of road were studied for providing reference to find more convenient and efficient dust load detection methods in the future.

2. Materials and Methods

2.1. *Materials and laboratory equipment*

Vacuum cleaner (horizontal cyclone G1007 multi-stage vacuum dust collecting device) and quadrat, sampling car, generator (3000 Watt), standard sieve (20 mesh and a 200 mesh), supporting paper bags, electric vibration sieve machine, electronic balance (AB204-S), plastic brush and so on.

2.2. *Sampling method*

The safety coefficient is very low and the feasibility is very poor when sampled in the reality urban road pavement in accordance with the AP-42 road dust sampling method. In order to ensure

---

*Corresponding author

the feasibility and sample collector's safety, sampling method of this paper, according to Chinese dust load detection method standard, was applied for sample collection in vehicles lane and bicycle lanes.

The specific sampling method is introduced in Fig. 1. At first, sampling path, the length of the sampled road and the location of the sampling points were determined. Sample area is a 1m² rectangle made by delineation of 1m² quadrats sampling range. Collecting area of vehicle lane is 4m², while collecting area of bicycle lane is 2m². Secondly, uniform coverage sweep in horizontal and vertical direction in the sampling area with vacuum cleaner were conducted carefully and make it sure that no road dust left in the edge of quadrats. Thirdly, the dust collecting boxes were cleared, and road dust was collected into the sample bag with a fine brush. Then the sample bags were sealed, labeled, brought back to laboratory for analysis. At the same time the portable GPS was used for recording latitude and longitude position information of each sampling point. Finally, the surrounding environment situation of each point was noted.

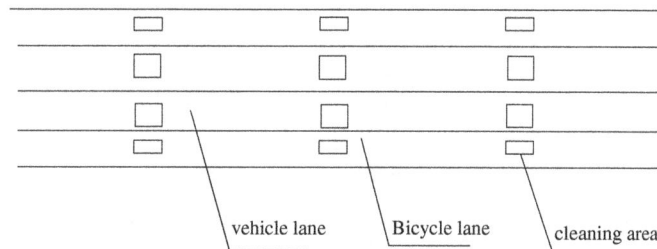

Fig. 1 Vacuum cleaner method for paved road dust

2.3. Sampling road

Urban roads are divided into expressway, main trunk road, secondary trunk road and branch according to Chinese urban road design standard. In this paper, 8 city paved roads were selected according to the standard and the characteristics of roads in Shijiazhuang City. One representative section was selected in each road. 128 dust load samples were collected at both vehicle lane and bicycle lane in each road on October 2014.

2.4. Silt loading calculation of urban paved road

Silt loading of paved road is defined by the particulate mass per unit area on the road under the 200 mesh sieve (sieving grain diameter less than 74μm). Larger weed, cigarette butts, garbage and other impurities were removed from the collected samples in laboratory. The samples were dried under the condition of 105°C, and were balanced for 3 days in the dryer, then were sealed and stored. We need to record the particulate mass after 20 mesh sieves, and 200 mesh sieves, then we can obtain the silt loading by the formula (1).

$$sL = \frac{W - W_{20} - W_{200}}{S} \tag{1}$$

Where, sL was silt loading in g m⁻², and W was total sample weight in g, W_{20} was oversize fraction weight of 20 meshes sieve in g, W_{200} was oversize fraction weight of 200 meshes sieve in g, and S was sampling area value in m².

3. Results and Analysis

3.1. *Silt loading comparison of different types of road*

Silt loading of each lane was calculated according to formula (1), and the average values of silt loading from different types of road were then calculated (Fig. 2).

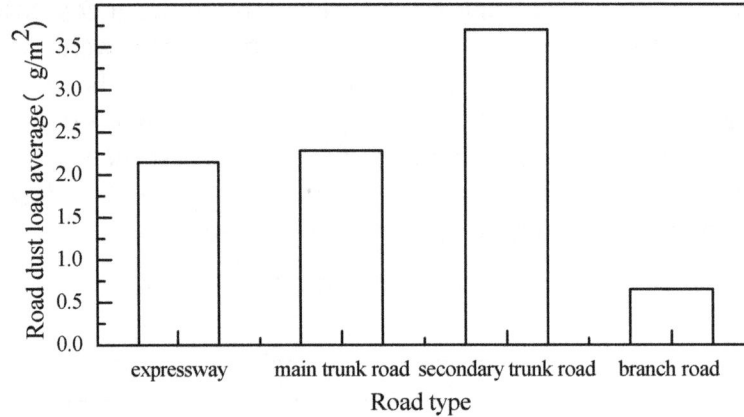

Fig. 2 The average value of road dust load from different types

The silt loading expressway, main trunk road, secondary trunk road and branch road of autumn of Shijiazhuang city are 2.14g/m², 2.28g/m², 3.70g/m², 0.65g/m². And its sequence is secondary trunk road>main trunk road>expressway>branch road. The silt loading of branch road is the least. The research results of remaining three roads are consistent with Xu Yan's and Fan Shoubin's result.

3.2. *Silt loading comparison between vehicle lane and bicycle lane*

Silt loading was calculated according to formula (1) in each lane. The silt loading of each road type of vehicle lane and bicycle lane were summarized, for comparing the silt loading of vehicles to bicycle lane, results shown in Fig. 3.

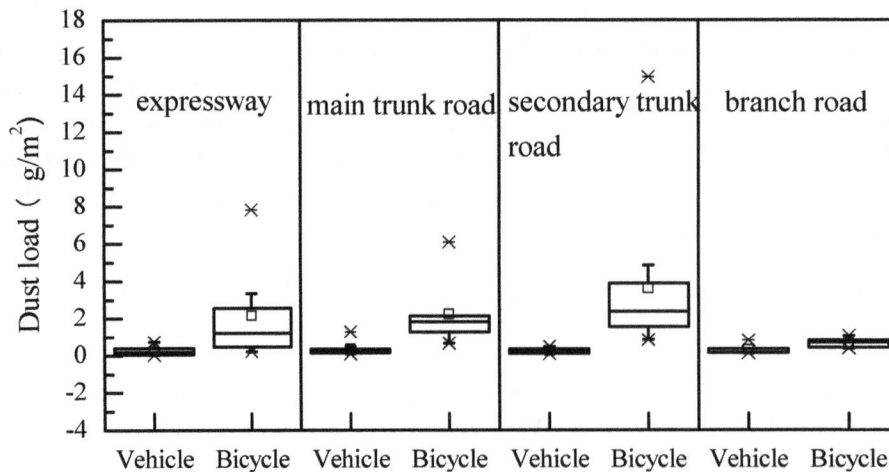

Fig. 3 Silt loading of vehicles and bicycle lane

Fig. 3 shows that the silt loading of bicycle lane is significantly higher than vehicle lane. The main reason of this phenomenon is caused by traffic flow that the bicycle's is higher than motorized. There are bicycles, tricycles and electric vehicles driven on the bicycle lane, and they are traveling on a slower speed than vehicles. Most of dust deposited on pavement road is because of little soil rolled up by bicycle on bicycle lane. In addition, there is higher frequency of sweeping about vehicle lane than bicycle lane. At the same time, the surface of bicycle lane is slightly lower than vehicle lane, which makes dust rushed to bicycle lane from vehicle lane by precipitation or sprinkler and deposited on bicycle lane. That explains why the silt loading of bicycle lane was significantly higher than that of vehicle lane. It can be seen that vehicle types, vehicle flow and vehicle speed are the important factors which influence the pavement of silt loading.

3.3. *Correlation analysis of silt loading from vehicle lane and bicycle lane*

In order to explore the correlation of silt loading from different types of road , silt loading of different types of road were divided into vehicle lane and bicycle lane, and correlation analysis (Pearson test) were carried on. The test results are shown in Table 1, all the correlation coefficient of Pearson about expressway, main trunk of road,secondary trunk roads and branch road of the vehicle lane and bicycle lane are more than 0.5.And two-tailed test P values is less than 0.05, it indicates that their is a linear correlation.

Table 1 The correlation of silt loading from various types of vehicle lane and bicycle lane

Road type	Expressway	Main trunk road	Secondary trunk road	Branch road
Pearson Correlation	0.795**	0.848**	0.868**	0.894**
Sig. (2-tailed)	0.001	0.000	0.000	0.000

Note: Sig. (2-tailed) said the establishment of probability 2-tailed correlation coefficient in the assumptions and values of P, when $P<0.05$, is associated with * representation; $P<0.01$, significantly correlated with * *; $P<0.001$, significant correlation

3.4. *Linear fitting of silt loading from vehicle lane and bicycle lane*

The silt loading data from vehicle lane and bicycle lane of Shijiazhuang city in winter were calculated and then linear fitting relationship analyses were conducted by using origin software. Detail results were shown in Fig. 4, which SL means silt loading.

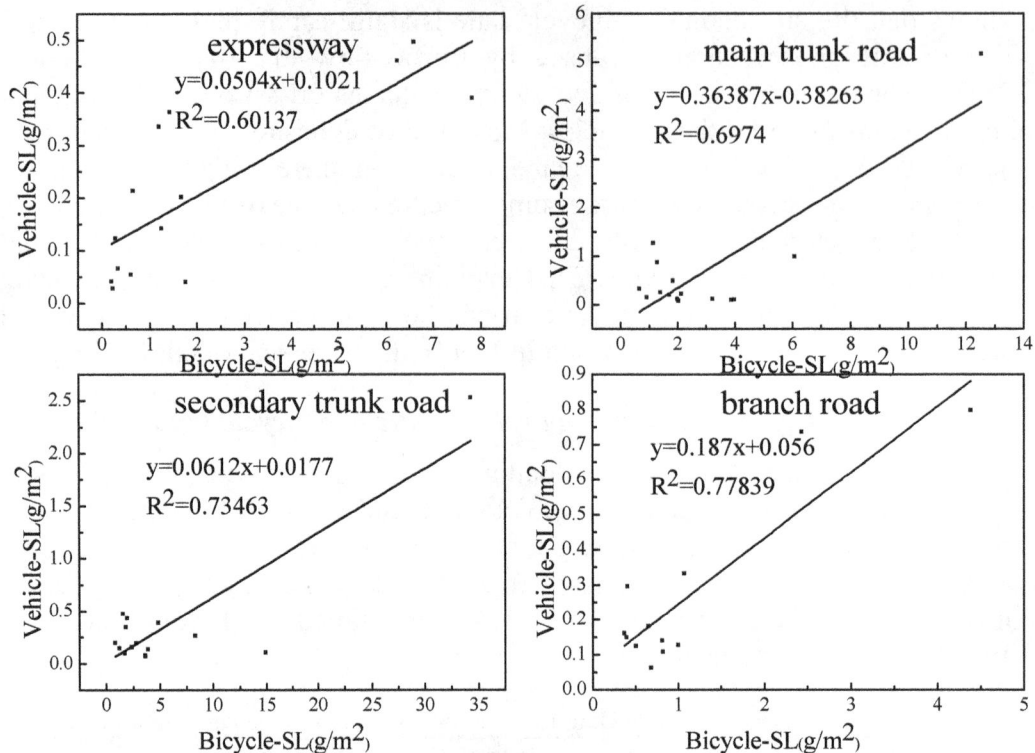

Fig. 4 Correlation analysis of bicycle silt loading and vehicles silt loading in autumn

From Figure 4, we can find that the fitting function R^2 of all types of road, are beyond 0.6. Also we can find that there is significant linear relationship between vehicle lane and bicycle lane from the same type of road. These results can provide us a easy and safety way for obtaining not only silt loading of vehicle lane from only owing silt loading data of bicycle lane by linear relationship formulas, but also silt loading from whole road. It can not only reduce greatly the workload of sample collecting, but also can avoid the potential safety problem during sample collecting process.

4. Conclusion

Road dust is one of the important sources of atmospheric particulates, 128 road duat samples were collected according to the traditional road dust sampling methods and were treated and analyzed in laboratory, the main conclusions are as follows

(1) The pollution intensity sequence of paving road silt loading in autumn in Shijiazhuang City is: secondary trunk road>main trunk road>expressway>branch road; silt loading of bicycle lane is significantly greater than that of vehicles lane in the same type of road;

(2) Silt loading is also affected by the impact of external environment factors, such as secondary trunk road, has a high silt loading value among all types of road because of the demolition nearby, and has maximum silt loading in bicycle lane.

(3) By comparing the silt loading of vehicles lane and bicycle lane, significant linear relationship of silt loading between vehicles lane and bicycle lane from four types of road were found; and we can predict the silt loading of one of them by the linear relationship expression.

References

1. Shou-bin FAN, Gang TIAN, Jian-ping QIN, et al. Emission characteristics of road dust-fall in Beijing [J]. Chinese Journal of Environmental Engineering, 2010,4(3):629-632. (in Chinese)
2. Yu SHI, Jian-hui ZHANG, Hai-jiang LUO. Analysis of characteristics of atmospheric particulate matter pollution in Beijing during the fall and winter of 2012 to 2013 [J]. Ecology and Environmental Sciences, 2013, 22(9): 1571-1577. (in Chinese)
3. Yu-hu HUANG, Gang LI, Tao YANG, et al. Establishment and Comparison of Evaluation Methods for Fugitive Road Dust [J]. Research of Environmental Sciences, 2011, 24(1):27-32. (in Chinese)

Oxygen Transmission Rate Measurement for Plastic Film with Equal-pressure Method

Ji-Yan Zhang[1,*], Jin-Xin Dong[2], Jun-Cheng Dai[1], Ming-Liang Gao[1]

[1]Division of Medical and Biological Measurements, National Institute of Metrology, Beijing, China
[2]College of mechanical and electrical engineering, Harbin Institute of Technology, Harbin, China
*Email: zhangjy@nim.ac.cn

In order to develop national reference material and solve valuation method for oxygen transmission rate, apparatus performance and measurement factors involved in valuation should be assessed first. According to the definition of the oxygen transmission rate, based on the equal-pressure method with coulometric sensors, experimental study on oxygen transmission rate was carried out. Repeatability and stability were measured under normal conditions with a typical commercial equal-pressure apparatus, and effects of different measurement temperatures were analyzed. The results show that the relative measurement repeatability can reach about 1% and the relative stability is about 5% for the equal-pressure method apparatus at 23°C and 0% relative humidity. Moreover, the oxygen transmission rate generally increases as the temperature increases. The temperature is proved to be an important factor to the oxygen transmission rate measurement. If the transmission cell temperature at 23°C is controlled within ±0.5°C, about 2% relative error of the oxygen transmission rate may be caused.

Keywords: Oxygen transmission rate; equal-pressure method; coulometric sensor; barrier property; packaging material; plastic film.

1. Introduction

People cannot live without food and medicine, and whether it is safe or not is closely related to the health of people. Food and drug safety refers to the quality of not only food and drug itself but also packaging materials used. Because packaging materials contact with food and drug directly, and are very important for quality assurance during shelf-life.

There are many parameters to evaluate the property of food and drug packaging materials, such as barrier property, mechanical property, solvent residue, sealing performance and so on. Among them, barrier property refers to the barrier effect of packaging materials on gas and liquid permeation, while the oxygen transmission rate is one of the most important parameters. Through the oxygen transmission rate measurement of packaging materials, oxidative deterioration and other issues are analyzed due to oxygen-sensitive, which is very useful to develop and evaluate the product shelf-life. Currently two main test methods are usually used for measurement of oxygen transmission rate of packaging materials. One is the equal-pressure method and the other is the differential-pressure method. Correspondingly national and international standards are drafted based on these two methods, which are GB/T 19789 [1], GB/T 1038 [2], ISO 15105-1 [3] and ISO 15015-2 [4]. Equal-pressure method usually uses coulometric sensors [5]. Oxygen permeated through materials will be taken to the coulometric sensor and electrons will be released due to the electrolysis reaction. The number of oxygen molecules can be detected according to the size of the current, and then oxygen transmission rate can be calculated. While for differential-pressure method, it is commonly based on pressure sensors. By measuring the amount of pressure change, the number of oxygen molecules permeated is estimated and then oxygen transmission rate is obtained. Comparison with the differential-pressure method, equal-pressure method is used in a wide range of applications due to the less dependence of the ambient temperature and the flow rate of the carrier

*Corresponding author

gas, especially suitable for low oxygen transmission rate measurement of higher barrier materials. Therefore equal-pressure method is widely used in the industry.

In this paper, on the basis of the definition of oxygen transmission rate and the principle of equal-pressure method, experimental study on oxygen transmission rate measurement is conducted. With the experimental results obtained, measurement repeatability, stability, temperature influence and other factors are analyzed and discussed for equal-pressure method. The results will be useful for reference material development and valuation method research.

2. The Oxygen Transmission Rate

Permeation property of a material is related to the specific penetration object. Oxygen transmission rate means the property of a material permeable to the oxygen. The entire oxygen permeation process for the test specimen (e.g., film) can be divided into four parts: adsorption, dissolution, diffusion and desorption. The oxygen from a high concentration side enters the surface area [6], and then diffuses within the material. Finally desorbs from the other side with a low concentration of oxygen [7], shown in Fig. 1.

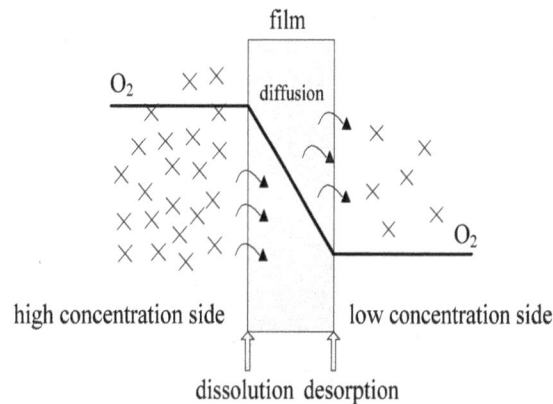

Fig. 1 Diagram of permeation process

In GB and ISO standards [1-4], Oxygen transmission rate is defined as the volume of oxygen passing through a plastic material per unit area and unit time. Under the standard atmospheric pressure, the unit is $cm^3/(m^2 \cdot 24h)$. Oxygen permeance is defined as the volume of oxygen passing through a plastic material, per unit area and unit time, under unit partial-pressure difference between the two sides of the material. Its unit is $cm^3/(m^2 \cdot 24h \cdot 0.1MPa)$. For oxygen permeability coefficient, it is defined as the volume of oxygen passing through a plastic material of unit thickness, per unit area and unit time, under unit partial-pressure difference between the two sides of the material. The unit is $cm^3/(m \cdot 24h \cdot 0.1MPa)$. It can be seen that oxygen permeability coefficient is a physical property of a material and it does not vary with the particular shape of the material.

3. Measuring Principle of Equal-pressure Method

The measuring principle of equal-pressure method is shown in Fig. 2.

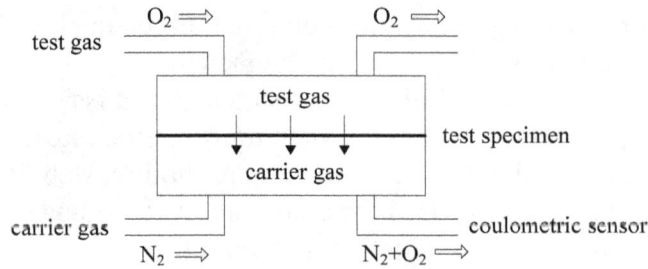

Fig. 2 Principle of equal-pressure method

A test specimen is mounted in a gas-transmission cell so as to form a sealed barrier between two chambers. One chamber is slowly swept with a carrier gas. The second chamber is fed with the test gas. The total pressure is identical in each chamber but, since the partial pressure of the test gas is higher in the second chamber, the test gas permeates through the barrier into the carrier gas in the first chamber. The test gas which permeates through the specimen is carried by the carrier gas to a sensor the nature of which will depend on the test specimen and the test gas used. In Fig. 2, the test gas is oxygen and the carrier gas is nitrogen [1, 4, 5].

An example of oxygen transmission rate apparatus based on equal-pressure method is shown in Fig. 3. When the oxygen permeability reaches a steady state, the output voltage will reach a constant value. Then oxygen transmission rate can be calculated.

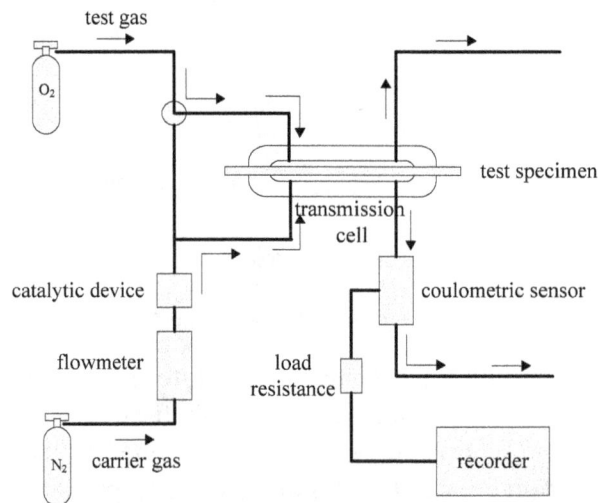

Fig. 3 An example of oxygen transmission rate measurement based on equal-pressure method

The coulometric sensor is a very important and essential part in equal-pressure method for oxygen transmission rate measurement, which consists of a cadmium anode and a graphite cathode. It is a consumption sensor which is accordant with the Faraday's law. Each molecule of oxygen permeated into the coulometric sensor will release four free electrons and form a current. The current size is proportional to the number of oxygen molecules into the sensor. With Eq. 1 the oxygen transmission rate can be calculated [1].

$$GTR = (E_e - E_0) \times Q / (A \times R).$$ (1)

Wherein, GTR is the oxygen transmission rate of the test specimen, unit is $cm^3/(m^2 \cdot 24h)$; E_e is the voltage in the steady state, unit is mV; E_0 is the zero-value voltge, unit is mV; A is the effective permeation area, unit is m^2; Q is the calibration constant for the apparatus, unit is $cm^3 \cdot \Omega/(mV \cdot 24h)$; R is the load resistance, unit is Ω.

It can be seen that, for oxygen transmission rate apparatus based on equal-pressure method, when A, R and Q are known, the oxygen transmission rate GTR can be calculated with Eq. 1 only by accurately measuring E_e and E_0.

4. Measurement and Analysis of Experiments

4.1. *Preparations*

Prior to the use of equal-pressure method to take oxygen transmission rate measurement, a most important work is adequate pipeline purge. Use mixed gas of (97%-98%) nitrogen and (2%-3%) hydrogen as a carrier gas to purge pipes inside the instrument and oxygen remaining on the coulemetric sensor with the flow rate of (5-10) ml/min. Purge time is dependent on the actual case. Normally the longer the instrument downtime is the longer the purge time.

Inorder to evaluate the repeatability and stability of the equal-pressure method apparatus, test specimen used must be reliable. A kind of Polyethylene Terephthalate (PET) film is chosen to take the measurement because it has the advantage of low linear expansion coefficient, low mold shrinkage and good chemical stability. Its density is $(1.29-1.40)g/cm^3$, and the internal molecular structure is constant from $-70°C$ to $+70°C$. Moreover, the gas permeability of the PET film is hardly affected by the ambient humidity [8]. In the measurement three PET films A, B and C are used, and nominal oxygen transmission rate is 48.9, 10.16 and $1.98cm^3/(m^2 \cdot 24h)$ respectively at $23°C$ and 0% relative humidity, which is claimed to be tracebale to the National Institute of Standard and Technology (NIST) by the manufacturer. Therefore, it can be used to evalute the repeatability and stability of the apparatus itself.

4.2. *Measurement repeatability*

Repeatability is used to evaluate the capability of a measuring instrument to provide similar indicated values for repeated measurements of the same measurand under the same conditions.

At $23°C$ and 0% relative humidity, repeatability measurements are taken on a commercial equal-pressure apparatus with plastic film A, B and C of different oxygen transmission rate. Each film is measured at least six times under the same test conditions, results are shown in Table 1 below. It is shown that the relative repeatability for this equal-pressure apparatus is within 1% and relative error is within 5% under the normal condition of $23°C$ and 0% relative humidity.

Table 1 Repeatability results

Film	Unit：[cm³/(m²·24h)]						Average value	Standard deviation	Relative repeatability
	Oxygen transmission rate								
	1	2	3	4	5	6			
A	50.40	50.42	50.45	50.43	50.47	50.52	50.448	0.043	0.1%
B	10.55	10.59	10.65	10.70	10.72	10.73	10.657	0.075	0.7%
C	1.93	1.92	1.92	1.92	1.92	1.93	1.923	0.005	0.3%

4.3. *Measurement stability*

Stability is used to evaluate the capacity of a measuring instrument to maintain its metrological characteristics be constant with time.

In order to evaluate the stability of the apparatus with equal-pressure method, measurements are taken at 23°C and 0% relative humidity in different time periods. With the same film A, B and C described above, stability measurements are taken every 20 days. Experimental results are listed in Table 2. It is shown that the relative stability for this equal-pressure method apparatus is within 5% under the normal conditions of 23°C and 0% relative humidity.

Table 2 Stability results

Film	Unit: [cm³/(m²·24h)]							Relative stability
	Oxygen transmission rate				(Maximum -minimum)	Average value	Standard deviation	
	1	2	3	4				
A	48.43	48.06	50.45	50.53	2.47	49.368	1.199	2.4%
B	9.76	10.45	10.65	10.77	1.01	10.408	0.491	4.7%
C	1.96	1.93	1.93	1.89	0.07	1.928	0.034	1.8%

4.4. *Analysis on effect of different temperatures*

The transmission cell temperature is a very important factor for oxygen transmission rate measurement in equal-pressure method. Usually, the temperature accuracy is required within ±0.5°C. In order to evaluate the influence of temperature, experiments are taken at different temperatures in the range of (13-33)°C for the same plastic film. Detailed results are listed in Table 3 and corresponding curve is shown in Fig. 4.

Table 3 Experimental results at different temperature

Temperature [°C]	Oxygen transmission rate [cm³/(m²·24h)]		
	Film A	Film B	Film C
13	31.15	6.91	1.23
18	38.63	8.65	1.54
23	46.47	10.65	1.92
28	57.60	13.05	2.32
33	70.76	15.93	2.82

From the experimental results it can be concluded that the oxygen transmission rate generally increases as the temperature of transmission cell rises. For the PET film with higher oxygen transmission rate measured in the experiment, the oxygen transmission rate increases about (1.5-2.5)cm³/(m²·24h) when temperature changes 1°C. While for medium and low oxygen transmission rate film, it increases about (0.3-0.6)cm³/(m²·24h) and (0.06-0.1)cm³/(m²·24h) when temperature changes 1°C. Generally the temperature accuracy is required within ±0.5°C, which may introduce about 2% relative measurement error of oxygen transmission rate at the set temperature of 23°C. Therefore, the temperature is proved to be a very important factor for oxygen transmission rate measurement in equal-pressure method and it should be controlled accurately.

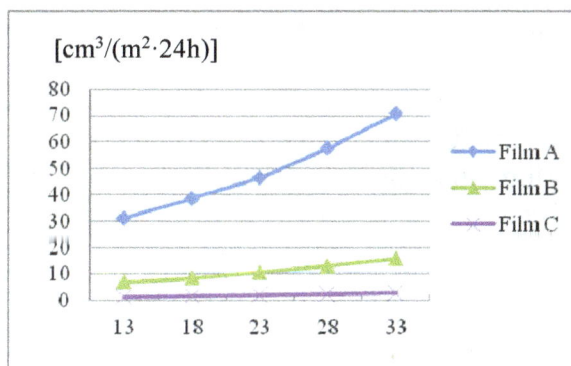

Fig. 4 Temperature-oxygen transmission rate curve

5. Conclusion

Whether the oxygen transmission rate is accurate or not is directly related to the assessment of the effective shelf-life, therefore it is very important. However, the apparatus quality for oxygen transmission rate measurement cannot be evaluated with unified method due to the lack of national reference materials.

In this paper, first the definition of oxygen transmission rate is introduced, and measurement principle of equal-pressure method based on coulometric sensor is described. Then experiments are taken out on the measurement of oxygen transmission rate for PET films. Analysis and calculations are made for the repeatability and stability under normal conditions for equal-pressure method. Moreover, affects of different temperatures are analyzed. The results show that the relative measurement repeatability of this typical apparatus is about 1% which is good. The relative stability is about 5% at 23°C and 0% relative humidity, which mainly due to the different sealing conditions, film and environmental conditions in different measurement periods. By measuring the oxygen transmission rate for one same film in the temperature range from 13°C to 33°C, it is found that the oxygen transmission rate generally increases as the temperature of the transmission cell rises. If the temperature accuracy was set about ±0.5°C at 23°C, it may cause about 2% relative error of the oxygen transmission rate. Therefore, the transmission cell temperature should be controlled accurately in equal-pressure method.

By the above measurement and analysis, apparatus performance and measurement factors with equal-pressure method are assessed, which provides the statistical data for valuation method research of oxygen transmission rate. Next performance and factors with differential-pressure method will be measured and analyzed, and finally valuation method will be developed.

Acknowledgments

This research was funded by the State Administration of Quality Supervision, Inspection and Quarantine.

We thank the IWMSE 2016 organizing committee for providing this template and all colleagues who previously provided technical supports.

References

1. GB/T 19789, Packaging material-Test method for oxygen gas permeability characteristics of plastic film and sheeting-Coulometric sensor, 2005. "In Chinese"

2. GB/T 1038, Plastics-Film and sheeting-Determination of gas transmission-Differential-pressure method, 2000. "In Chinese"

3. ISO 15105-1, Plastics-Film and sheeting-Determination of gas-transmission rate-Part 1: Differential-pressure methods, 2007.

4. ISO 15105-2, Plastics-Film and sheeting-Determination of gas-transmission rate-Part 2: Equal-pressure method, 2003.

5. ASTM D 3985, Standard test method for oxygen gas transmission rate through plastic film and sheeting using a coulometric sensor, 2002.

6. Zhi-Gang TU, Zeng-Qing WU, Kan-Cheng MAI, Development of barrier plastic film for packaging, Packaging engineering. 24(6):22-24, 2003. "In Chinese"

7. Group with practical techniques textbook, Modern packaging materials and Technology, Guangdong science and technology press, Guangzhou, 2004. "In Chinese"

8. Qin-Jian YIN, Preparation and properties of functional polyester (PET) composite, Sichuan University, 2007. "In Chinese"

The CBR Test of Compacted Red Clay

Kai-Sheng Chen[*], Shuang Deng

School of Civil Engineering, Guizhou University, Guizhou Guiyang, China
[*]Email: chen_kaisheng@163.com

Red clay has high liquid limit, high plasticity, high void ratio and other special engineering properties. Whether the red clay can be used as road material? How does the CBR properties present? These problems are worthy of being studied. No soaking water, soaking water 1, 2, 4, day-night and under the dry-wet circulation of CBR tests are made by subgrade pavement material strength test system. Study results show that the CBR value of soil samples without soaking water decreases, when moisture content increases, and more than the highway specification requirement of each parts for CBR value. The CBR value which soaking water after 4 day-night cannot satisfy the CBR value requirement of highway specification, each parts of first grade roads. CBR value decreases with the increase of soaking water time, soaking water after 1 day-night and CBR value reduction is large. Soil samples' CBR value decreased with increasing cycles, the first cycle reduced amplitude is larger than other cycles, in which reduced amplitude becomes gentle. Soil samples of different compaction degree and moisture content experience more dry-wet circulations, their CBR value becomes closer.

Keywords: Red clay compaction; CBR test; dry-wet circulation.

1. Introduction

CBR (California Bearing Ratio) proposed by the highways agency, California, is used to assess subgrade soil and indexes of pavement material strength. Foreign standards prefer to use CBR as the design parameters of pavement materials and subgrade soil [1]. Although domestic design parameters of the pavement and subgrade are the resilience modulus [2] according to asphalt concrete pavement design specifications, there are specific provisions [3] about the highway subgrade compaction degree and the CBR value of filling in the" Specification for design of highway subgrade" (JTJ D30-2015), which becomes the basis of subgrade filling selection.

2. Basic Physical Properties of Red Clay

This test soil sample is taken from a foundation pit in western campus, Guizhou University. Soil characteristics are as follows: palm red, wet, homogeneous, compact and there is a small amount of gravel and plant roots, grain composition is given in advance with silty clay. The basic physical indexes of soil sample are shown in Table 1 [4].

Table 1 Basic physical indexes of red clay

According to highway test procedures (JTG E40-2007)	>0.075mm/%	0.075 ~ 0.002mm/%		<0.002mm /%	
	19.36	64.30		16.34	
Optimum water content /%	Maximum dry unit weight g/cm^3	Plastic Limit /%	Liquid limit /%	Nonuniform coefficient	Coefficient of curvature
30.8	1.47	31.2	68.4	16.7	1.5

[*]Corresponding author

3. Testing Program

3.1. *Sample preparation*

Test sample preparation using heavy compaction, are shown in Table 2

Table 2 Sample preparation

Water content /%	Degree of Compaction /%		
20	93	94	96
30	93	94	96
35	93	94	96
40	93	94	96

3.2. *Test method*

3.2.1. *Soaking water test*

To simulate the material in the process of the most unfavorable condition, send samples respectively soaking water 4 day- night, 2 day-night, 1 day-night and 0 day-night before loading, at the top of the sample surface load board, to simulate pavement structure for the additional stress of subgrade. Then study on soaking water time impact on CBR value through penetration test.

3.2.2. *Dry-wet circulation test*

To keep material consistency in site condition and use process, samples respectively under dry-wet cycles test before loading. Drying process: the initial sample in the oven to 105°C, measure the quality of the sample every 1 hours, until sample quality change no longer, dry finish. Humidification process: add water until the initial moisture content using a syringe, braised materials in wet cylinder one day [5] [6]. That completes a dry-wet circulation, which in turn to finish 2 times, 3 times of the dry-wet cycle test, and then studies the influence of dry-wet cycles for CBR value through penetration test.

3.2.3. *Penetration test*

In penetration test, the higher bearing capacity of the material, the larger load material bearing when penetrating certain depth. CBR value is the ratio which the unit pressure when sample penetration quantity is 2.5 mm or 5 mm and standard load intensity (7 MPa and 10.5 MPa) when standard rubble into the same penetration quantity, is the evaluation index of subgrade engineering characteristics. If the CBR value of penetration quantity to 5 mm greater than the CBR value of penetration quantity to 2.5 mm, the test should redo, if still get this result, the CBR value of penetration quantity to 5 mm is accepted.

4. Test Result

4.1. *Relationship between the CBR Value and water content*

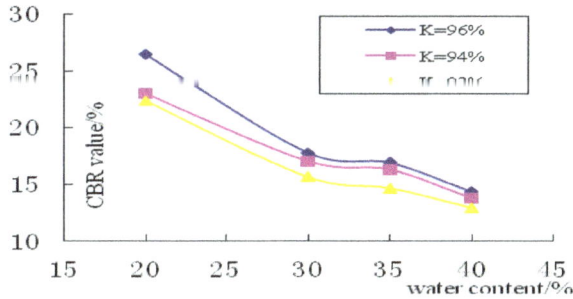

Fig. 1 Relationship between the CBR value of no 1 day-night soaking water soil sample with water content

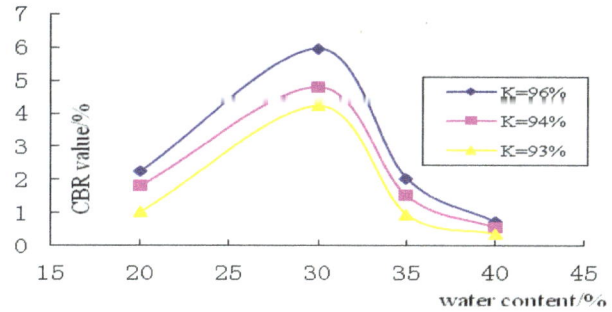

Fig. 2 Relationship between the CBR value of soaking water soil sample with initial water content

Fig. 1 shows that the CBR value decreases with the increase of initial water content. Secondly, when soil samples no soaking water, the CBR value is greater than the requirement of the CBR values in highway specification of each part.

Fig. 2 shows that, with the increase of initial water content, CBR value firstly increases, then decreases, and reaches the maximum in the optimum water content. Secondly, under the optimum water content condition, soil samples after 1 day-night soaking water and its CBR value can not meet the specification requirements of highway and highway subgrade, but meet the specification requirements of under subgrade and embankment.

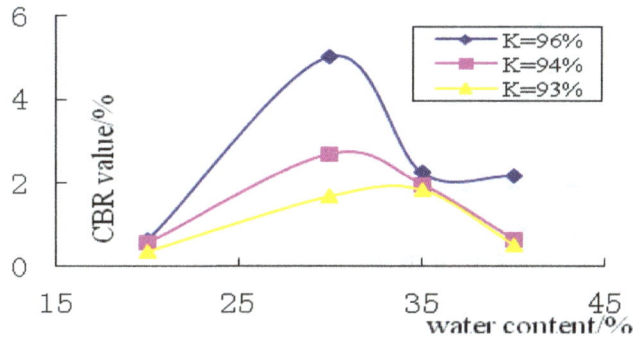

Fig. 3 Relationship between the CBR value of 2 day-night soaking water soil sample with water content

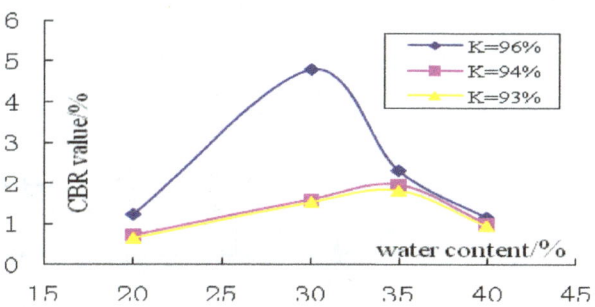

Fig. 4 Relationship between the CBR value of 4 day-night soaking water soil sample with water content

Fig. 3 shows that, with the increase of initial water content, CBR value firstly increases, then decreases, and reaches the maximum in the optimum water content. Secondly, under the optimum water content condition, soil samples after 2 day-night soaking water and its CBR value can not meet the specification requirements of highway and highway subgrade, also can not meet the specification requirements of under subgrade and embankment.

Fig. 4 shows that, with the increase of initial water content, CBR value firstly increases, then decreases, and reaches the maximum in the optimum water content. Secondly, under the optimum water content condition, soil samples after 4 day-night soaking water and its CBR value can not meet the specification requirements of each part of highway.

4.2. Relationship between the CBR Value and degree of compaction

Fig. 5 shows that CBR value increased with the increase of degree of compaction. CBR value of soil sample not soaking water is much larger than of CBR value after soaking water.

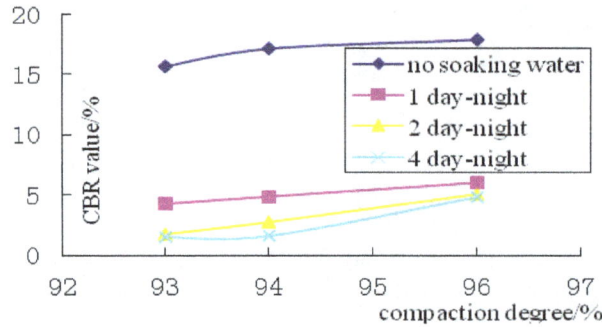

Fig. 5 Relationship between the CBR value and degree of compaction under the optimum water content condition

4.3. Relationship between the CBR Value and soaking time,

The relationship between the CBR Value and Soaking Time are Shown in Fig. 6.

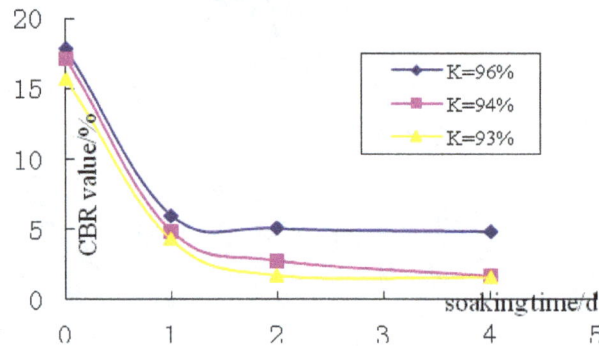

Fig. 6 Relationship between the CBR value and soaking time under the optimum water content condition

4.4. Relationship between the CBR Value and the amount of cycles

Figs. 7-10 show that CBR value decreases with the increase of cycles times, the first cycle reduced amplitude is large, then flatten out. The same water content soil samples with different degree of compaction experience more dry-wet circulation, their CBR values would be closer.

4.4.1. Relationship between the CBR value and the amount of cycles with different degree of compaction

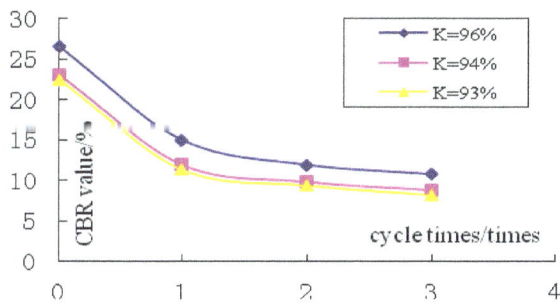

Fig. 7 w=20% Relationship between the CBR value and the amount of cycles

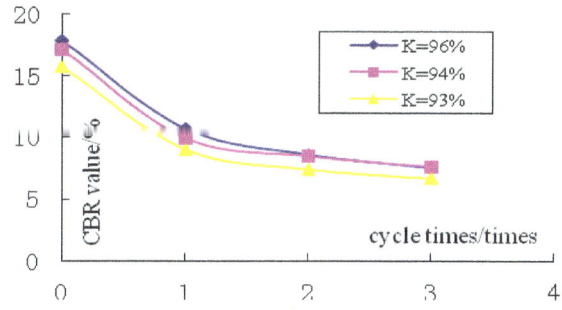

Fig. 8 w=30% Relationship between the CBR value and the amount of cycles

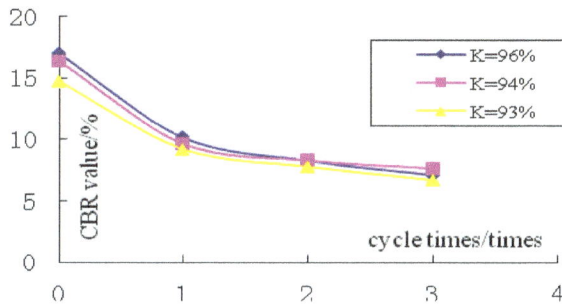

Fig. 9 w=35% Relationship between the CBR value and the amount of cycles

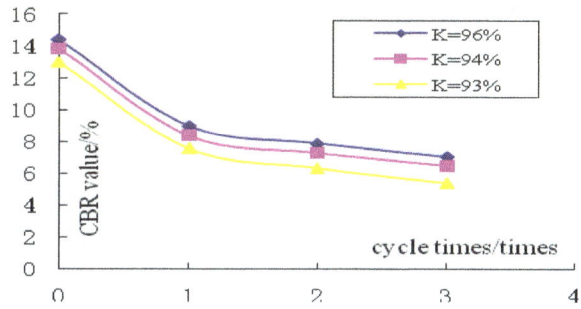

Fig. 10 w=40% Relationship between the CBR value and the amount of cycles

4.4.2. Relationship between the CBR Value and the amount of cycles with different water content

The Figs. 11-13 show that CBR value decreases with the increase of cycles times, the first cycle reduced amplitude is large, then flatten out. The same degree of compaction soil sample with different water content experience more dry-wet circulation, thus their CBR values will be closer.

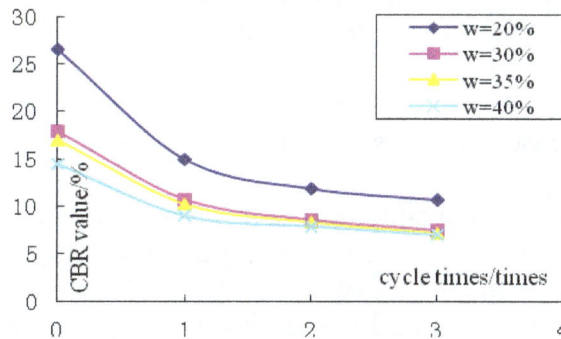

Fig. 11 K=96% Relationship the CBR value and the amount of cycles

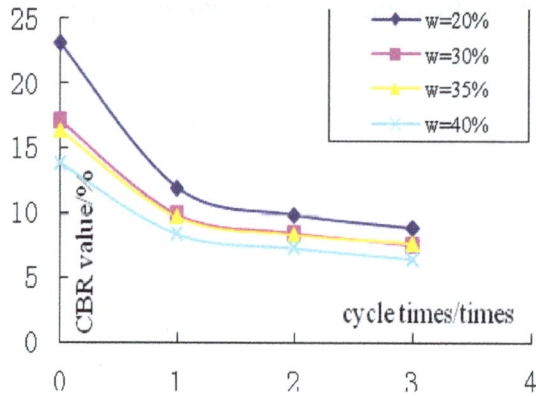

Fig. 12 K=94% Relationship between the CBR value and the amount of cycles

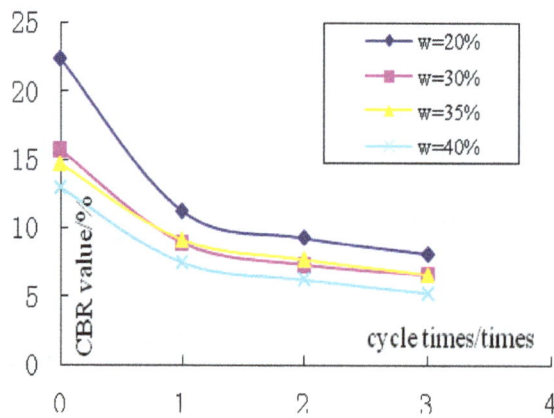

Fig. 13 K=93% Relationship between the CBR value and the amount of cycles

5. Conclusion

1. When soil samples no soaking water, the CBR value decreases with the increase of initial water content. The CBR value is greater than the requirement of the CBR values in highway specification of each part.
2. When soil samples soak water 1, 2, 4 day-night, with the increase of moisture content, the CBR value firstly increases, reaches the maximum in the optimum water content then decreases.
3. Under the optimum water content condition, soil samples after 1 day-night soaking water and its CBR value can not meet the specification requirements of highway and highway subgrade, but meet the specification requirements of under subgrade and embankment. Soil samples after 2 day-night soaking water and its CBR value can not meet the specification requirements of highway and highway subgrade, also can not meet the specification requirements of under subgrade and embankment.
4. Under the optimum water content condition, soil samples after 4 day-night soaking water and its CBR value can not meet the specification requirements of each part of highway.
5. The CBR value increases with the increase of degree of compaction. The CBR value of sample no soaking water is much larger than the CBR value of samples after soaking water. With the

optimum water content, the CBR value decreases with the increase of soaking water time. Among samples soaking water 1 day-night, the CBR value reduction of is large, then curve flattens, the reduction amplitude is small.

6. The CBR value of soil sample decreases with the increase of cycle time, the first cycle reduced amplitude is large, then flatten out. The same degree of compaction soil sample with different water content experience more dry-wet circulation, their CBR value more and closer.

Acknowledgments

This work was funded by National Natural Science Foundation of China (Approval number: 51368010), Guizhou science and Technology Department—Guizhou University Foundation (Qiankehe Lh zi[2014]7663), Natural science research project of Guizhou Provincial Department of Education (Qianjiaohe KY zi[2015]488).

References

1. Zhiduo Zhu, Jianxin Hao, Liping Huang. CBR test influencing factors and several problems that should be paid attention to in engineering [J]. Rock and soil mechanics, 2006, 27 (9) : 1593-1595160

2. Kaisheng Chen, Yuan Yin. Guiyang - Qingzhen Expressway red clay strength indexes performance test [J]. Journal of Highway and Transportation Research and Development, 2011, 28 (3) : 61-66

3. Design code for highway subgrade (JTJ D30-2015)

4. Highway geotechnical test code (JTG E40-2007)

5. Xianjie Mu, Xiaoping Zhang. Study of expansive soil mechanics performance test under the condition of dry-wet circulation [J]. Rock and soil mechanics, 2008, 28 (suppl) : 580-582

6. Wen-hua Liu, Qing Yang, Xiao-wei Tang et al. Study of silty clay with cycle loading under dry-wet circulation condition dynamic characteristics experimental [J], Journal of Hydraulic Engineering, 2015 46-48 (4) : 425-432

Carbon Fibre Microstructure Characterization

Si-Yu Jin[1], Shi Wen[1], Wei-Wei Du[2], Lyes Douadji[1,*]

[1]*Chongqing Academy of Science & Technology, Research Center for Advanced Materials, Chongqing 401123, China*
[2]*Southwest University, College of Material Science and Engineering, Chongqing 400715, China*
Email: ldouadji@yahoo.fr

Carbon fibre has been widely used as reinforced fibres in composites for auto and civil engineering industry. The characterizations of this fibre inform us more about the interface behaviour and help us to improve the properties of the composites using these fibres. In this paper, methods including Scanning Electron Microscopy (SEM), Optical Microscopy have been used to determine the surface morphology of carbon fibres. The results suggested that: Firstly, the number of filaments of a carbon fibre bundle can be well obtained by using optical microscope. Second, the carbon fibre with PAN precursor has ribbon like shallow furrows along the longitudinal direction. Finally, the carbon fibre with higher modulus shows more clearly skin-core structure.

Keywords: Morphology; carbon fiber; SEM.

1. Introduction

There are three types of structures that a carbon fibre could formed; namely crystalline, amorphous and partly crystalline [1]. A typical microstructure for a carbon fibre is shown has presented in introduction of our results [2]; graphite is one structural type which forms crystalline domains. This graphite structure is made of *sp2* hybridized carbon atoms, which are organized in a two-dimensional honeycomb structure in the *x-y* plane. Carbon atoms in one layer are bonded by covalent bonding and together with metallic bonding. The distance between the hexagonally bound carbon atoms, which are part of the layers, is 1.42Å. Nevertheless the distance between the layers is 3.35Å and this gives rise to the anisotropy of properties [3]. The covalent bonds are formed through overlap of the *sp2* hybridized orbitals, and metallic bonds are created by the delocalization of the *Pz orbitals*. This is the reason why graphite is a good electric conductor and a good thermal conductor in the *x-y* plane. The layers are connected by van der Waals' bonds; therefore the carbon layers can easily slide past each other. In the direction perpendicular to the layers, graphite is an electrical insulator and a thermal insulator. As a result, graphite is highly anisotropic which is attributed to the difference between the in-plane and out of plane bonding. Moreover, a higher modulus is obtained in the direction parallel to the plane; on the contrary a lower modulus is obtained perpendicular to the plane. The bond strength within the layer plane is about 600 kJ/gm atoms, whereas the interlayer binding energy is estimated to be only 5.4 kJ/gm atom. Young's modulus parallel to the layers can be 30 times of that perpendicular to them [4].

Similarly, for the structure of a high modulus carbon fibre, the graphene layers should be parallel to the fibre axis in order to obtain a higher modulus. This crystallographic favoured orientation is called fibre texture. In other words, for the same reason, in the direction of loading along the fibre axis it has a higher elastic modulus, higher electrical and thermal conductivities, and lower coefficient of thermal expansion also exist. When the degree of alignment of carbon layers parallel to the fibre axis is high, the fibre texture turns out to be stronger, and furthermore, the fibre density, the carbon content, the *c*-axis crystallite, fibre modulus, fibre internal strength, electrical conductivity and thermal expansion are all increased.

*Corresponding author

2. Materials and Methods

The major materials used in this project are a high modulus carbon fibre from Toray with the product code of M46J, and a low modulus carbon fibre from Toho Tenex with the code of IMS60. Methods including Scanning Electron Microscopy (SEM) and optical microscopy have been used to determine the microstructure of these materials.

3. Results and Discussion

3.1. *Carbon fibre morphology characterization by optical microscopy*

To examine the surface morphology by optical microscopy a single carbon fibre was firstly fixed on a paper card to secure it in place. The paper card has a window in the middle, and is the same as the sample used for the Raman spectroscopy and tensile testing studies. The fibre card sample was then placed under the optical microscope to analyse the surface morphology and fibre diameter.

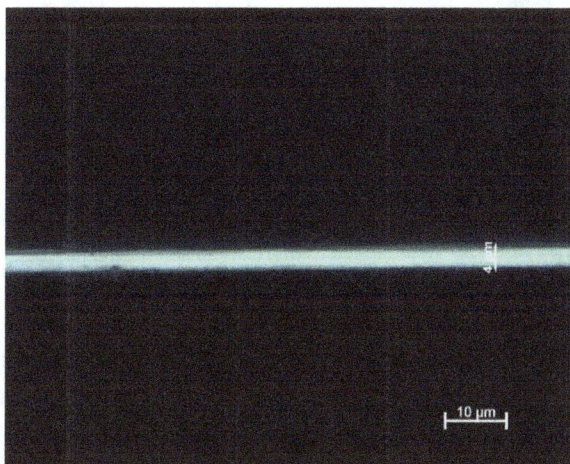

Fig. 1 Optical micrograph of a high modulus carbon fibre

Figure 1 shows a typical optical micrograph of a model carbon fibre in the longitudinal direction. An objective lens of ×100 is used to record the image. It can be seen that the contrast of the micrograph is poor, with the edge of the carbon fibre hard to discern. Due to the wavelength of an electron is much smaller than that of visible light, the measurement of the fibre diameter by optical microscopy is not as accuracy as that measured by SEM.

3.2. *Filaments count for commercial carbon fibres by optical microscopy*

Both types of carbon fibres were provided by the manufacturers in a bundle form. The ends of these bundles have been imaged under the optical microscope. An example of this is shown in Fig. 2.

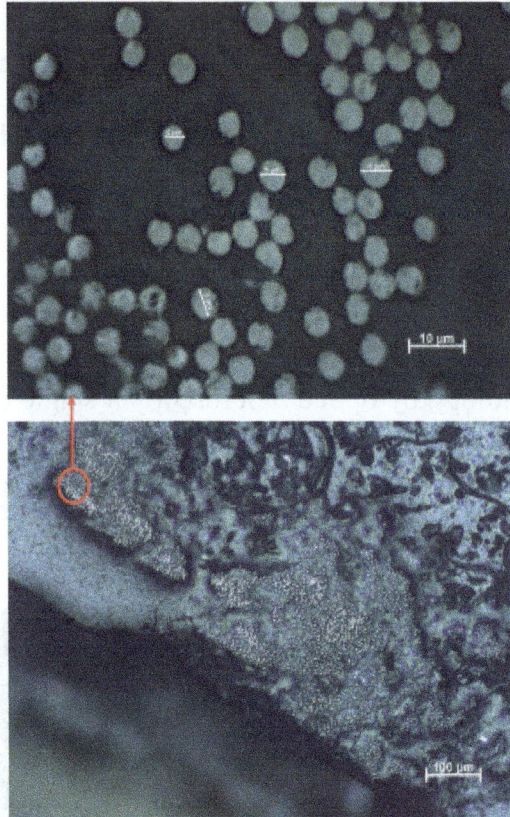

Fig. 2 Optical micrographs of bundles of commercial carbon fibres

Before the test bundles of high modulus carbon fibres and low modulus carbon fibres were embedded in polyester stubs separately. These stubs were then grinded until the end of each bundle of fibres was exposed at the top of the sample. The cross sectional area at the surface of each polyester stub was then polished in order to be observed clearly in an optical microscope. The average numbers of carbon fibre filaments in each bundle were 5900±200 and 11300 ±1000 for the high modulus carbon fibre and low modulus carbon fibres respectively. More importantly, as shown in Figure 2 it has confirmed the circular cross-section which is important for the next step of SEM analysis. The fibre diameters measured across the cross sections show their variability, which are generally between 5 μm and 6 μm.

3.3. *Carbon fibre morphology characterization by SEM*

In this study both types of carbon fibres have been examined in the longitudinal direction, along the fibre surface. The cross section area was determined using Scanning Electron Microscopy (SEM), assuming a circular cross-section. Typical SEM micrographs are shown in Figure 3 and Figure 4.

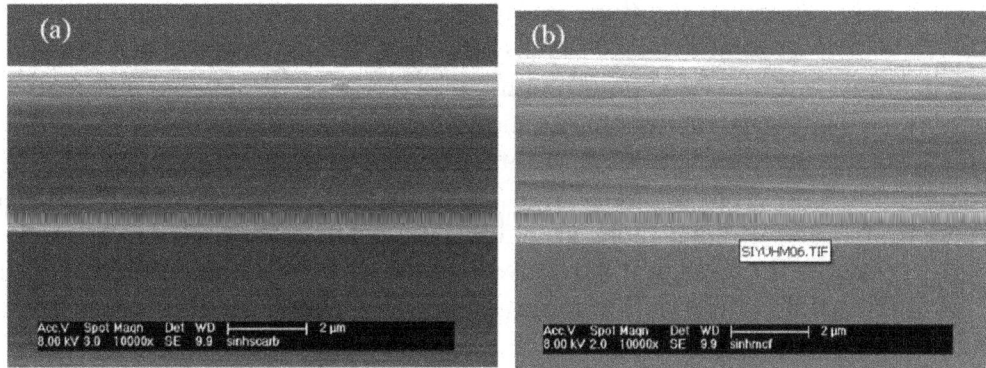

(a) Low modulus carbon fibre (b) High modulus carbon fibre

Fig. 3 SEM micrographs in the longitudinal direction

According to Figure 3 there are no significant differences between the surface morphologies of the two types of fibres; both surfaces have some shallow „furrows" present, parallel to the fibre axis. This could be due to the fact that both types of fibres are produced from a PAN precursor. These „furrows" are thought to be created during the fibre spinning process [4] (Smiley, 1993). In the literature these furrows are also described as wrinkled and distorted ribbons [5]. Because they are considered to be defects on the carbon fibre surface and detrimental to the fibre strength, it is easy to understand that high strength carbon fibres have comparatively fewer furrows [5]. The degree of surface roughness normally decreases with an increase in fibre modulus [1]. Both types of carbon fibres have been surface treated with a sizing agent by the manufacturer. None of the sizing can however be observed in the SEM images.

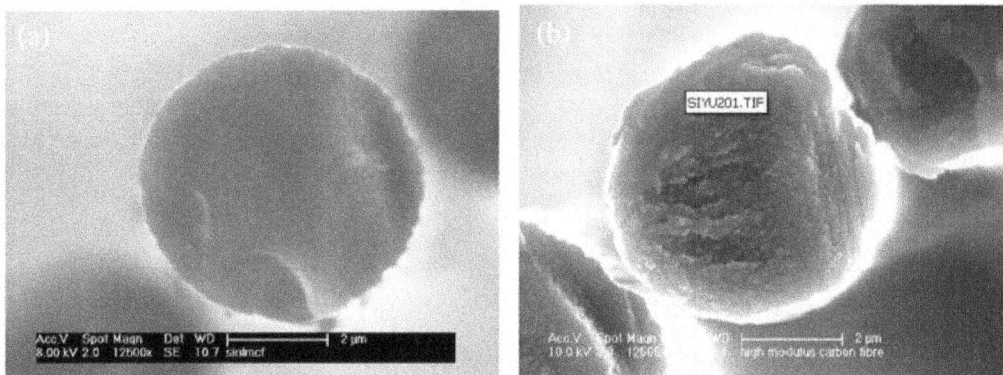

(a) Low modulus carbon fibre (b) High modulus carbon fibre

Fig. 4 SEM micrographs of the cross-sectional areas

The cross section areas for fibres fractured during the tensile test have also been measured under SEM. According to Figure 4 both types of carbon fibre have mostly a circular cross section, but the low modulus carbon fibre exhibits a rather smooth fracture surface compared to the high modulus one which shows clearly a skin-core shape as the graphite basal planes are created turbostratically into a layered structure during the graphitization process [6]. Furthermore it can be observed that the outer layers (skin) are more organized than the interior (core).

Paris indicated that PAN-based carbon fibres usually exhibit random cross sectional areas [7]. The fundamental building block of carbon fibres are graphitic crystallites, which are generally made up from the layers of graphite basal planes. The size of the graphitic crystallites and their plane layers are highly reliant on the time and temperature conditions during the heat treatment procedure; larger planer graphitic basal planes and layered crystallites are created when the treatment time is longer and the temperature is higher [8]. These crystallites are generally parallel to the fibre axis in a filamentary or ribbon like morphology type (the „furrows" discussed earlier). During manufacturing of the fibre, these crystallites become more parallel to the fibre axis if strain and higher temperatures are both applied. As a result of these mechanisms, the surface of high modulus carbon fibre tends to have more graphitic basal planes than the lower modulus carbon fibres which are produced in comparatively lower graphitization temperatures. This is why the cross section of high modulus carbon fibre indicates the presence of graphitic basal planes, and the low modulus fibres have a comparatively smoother cross section [9].

3.4. *Carbon fibre characterization by atomic force microscopy*

In this work AFM was used to characterise the surface morphology of carbon fibres.

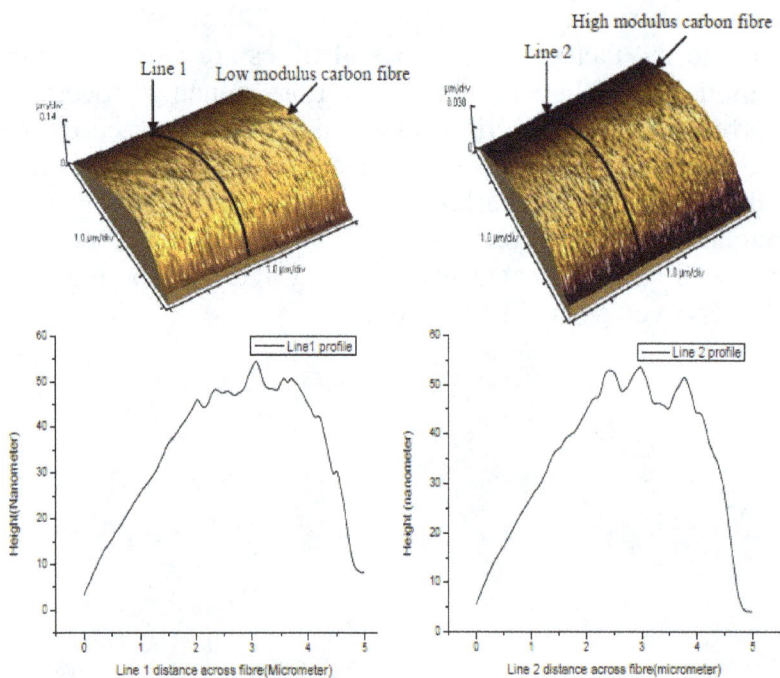

Fig. 5 The surface morphology micrograph of carbon fibre characterised by AFM

Figure 5 shows the surface morphology of the two types of carbon fibres; their surface roughness has been quantified as well. It can be seen that the shallow furrows which used to be detected from SEM images cannot be observed entirely clearly in the AFM image, this could be due to resolution limitation that a tip with higher aspect will give a better resolution and the radius of curvature of the probe results in different tip convolution. However as the cantilever tip is able to touch the fibre surface, the depth of the shallow furrows can be determined through the line profile function as is shown in Figure 5. According to the crossed line profile the determined average roughness are 6.17nm and 9.32nm for the low modulus carbon fibre and high modulus carbon fibre respectively.

4. Conclusion

Carbon Fibre Microstructure Characterization has been investigated with Scanning Electron Microscopy (SEM), Optical Microscopy. The results presents in this paper suggested that:

1) The number of filaments of a carbon fibre bundle can be well obtained by using optical microscope

2) The Carbon fibre with PAN precursor has ribbon like shallow furrows along the longitudinal direction. Finally, the Carbon fibre with higher modulus shows more clearly skin-core structure.

Acknowledgment

The authors thank Ministry of Science and Technology of the People's Republic of China (contract grant number 2015DFA51330) for funding this work.

References

1. Fitzer, E. (1990) *Carbon Fibres Filaments and Composites*, edited by Figueredo, J.L., Bernardo, C.A., Baker, R.T.K. and Huttinger, K.J., Dordrecht: Kluwer Academic, pp. 405-439
2. Siyu jin, PhD theses, 2013
3. Bacon, R. and Tang, M.M. (1964) Carbonization of cellulose fibres.2. physical properties study. *Carbon*, 2(3), PP. 221.
4. Peters, D.M. (1970) Carbon Fibre, Breakthrough and Early Development, pp. 3-11.
5. Bascom, W.D., Drzal, L.T. (1987) *The surface properties of carbon fibre and their adhesion to Organic Polymers*. NASA Contractor Report 4084, pp. 35.
6. Diefendorf, R.J. and Tokarsky, E. (1975) High Performance Carbon Fibres, *Polymer Engineering Science*, 15(3), pp. 150-159.
7. Paris, O., Loidl, D., Muller, M., Lichtenegger, H., and Peterlik, H. (2001) *Cross sectional texture of carbon fibres analysed by scanning micro beam X-ray diffraction*, Journal of Applied Crystallography,34, pp. 473-479.
8. Badami, D. U., Joiner, J. C. and Jones, G. A. (1967) Microstructure of High Strength High Modulus Carbon Fibres, *Nature*, 215, pp. 386-387.
9. Chuang, D.D.L. (1994) *Carbon fibre composites*. Newton USA: Butterworth-Heinemann.

Study of Consolidation Deformation Characteristic for Compacted Red Clay

Kai-Sheng Chen[*], Shuang Deng

School of Civil Engineering, Guizhou University, Guizhou Guiyang, China
[*]*Email: chen_kaisheng@163.com*

Based on the laboratory one-dimensional consolidation test, the paper analyses the initial state and dry-wet circulation state consolidation deformation characteristics of red clay, and improves the moisturizing method in the process of consolidation test. The results show that the moisturizing method of compression test supports previous hydrating & wet towel cover measures, and indicate how much water needs to be replenished according to the loss of water. When the water content is between optimum water content minus 2% and the optimum water content, the variation, regarding void ratio, compression deformation coefficient and compression coefficient, is not obvious. When the water content is greater than the optimum value, void ratio, compression deformation coefficient and the amplitude of compression coefficient are getting larger with the increase of water content. From the view point of controlling the sedimentation of subgrade, it suggests that the water content of compacted subgrade is between optimum water content minus 2% and the optimum water content. The stress-strain relationship of compacted red clay is available to be expressed by $\dfrac{\varepsilon}{p} = Kp^{\,n}$. Compression coefficient of red clay is obviously larger than the initial state after dry-wet circulation. With the increase of cycles, compression coefficient increases. The three previous cycles' compression coefficient increases are larger than other cycles, while the fourth and the fifth cycle compression coefficients tend to be stable. Suggested by subgrade settlement calculation using long-term compression coefficient index (the fourth or fifth cycle compression coefficient), the result of calculation tallies better with the actual situation. The study results provide good technical supports on subgrade settlement calculation.

Keywords: Compacted red clay; consolidation deformation; compression coefficient; dry-wet circulation.

1. Introduction

Red clay is mainly distributed in the southern part of China. Due to the high liquid limit, high plasticity and high water content, some problems [1] are possibly associated with red clay subgrade, such as subsidence, cracking, sliding and relevant pavement distress. Red clay subgrade damage problem is still not well solved, seriously affecting the highway economic and social benefits. The main reason of such problems is that red clay shrinks during the dry season due to desiccation, while expands during the rainy season because of water saturation. This reciprocation leads to the change of soil structure, soil strength gradual reduction and increasing deformation, resulting in the destruction of the subgrade [2-4]. At present, the study achievements of red clay are abundant, mainly concentrating in the highway performance, improved technology, construction technology and its treatment of engineering properties [5-8]. However, fundamental study, such as the deformation characteristics of red clay, is lacking. The study about the deformation characteristics of red clay under dry-wet circulation is especially lacking. Those characteristics are important causes of subgrade damages. Through a series of laboratory consolidation tests, this paper studies the consolidation deformation characteristics of red clay under initial normal state and dry-wet circulation state, thus improving the method of moisture in the consolidation test process.

---

[*]Corresponding author

2. Basic Physical Properties of Red Clay

Table 1 Basic physical indexes of red clay

According to highway test procedures (JTG E40-2007)	>0.075mm/%	0.075-0.002mm/%		<0.002mm /%	
	18.2	61.3		16.34	
Optimum water content/%	Maximum dry unit weight g/cm³	Plastic limit /%	Liquid limit /%	Nonuniform coefficient	Coefficient of curvature
33.8	1.81	38.1	62.7	6.39	1.50

This test soil sample is taken from a foundation pit in western campus, Guizhou University. Soil characteristics are as follows: the palm red, wet, homogeneous, compact and there is a small amount of gravel and plant roots, grain composition is given priority to with silty clay. The basic physical indexes of soil sample are shown in Table 1 [4].

3. Testing Program

3.1. *Sample preparation*

Test sample preparation using heavy compaction, are shown in Table 2.

Table 2 Sample preparation

Water content /%	Degree of Compaction /%				
85	28	31	33.8	36	40
90	28	31	33.8	36	40
93	28	31	33.8	36	40
96	28	31	33.8	36	40

3.2. *Dry-wet circulation test*

Drying: under the condition of 105°C drying, drying until the water content is below 1%. Humidification: calculation water content for the initial water content need to add water, and then slowly with a syringe injection samples. Sealed materials: in the 24 hours moisturizing cylinder, so that uniform water permeability, complete a dry-wet circulation test, in turn, 2, 3, 4 and 5 times the dry-wet circulation test, as shown in Fig. 1.

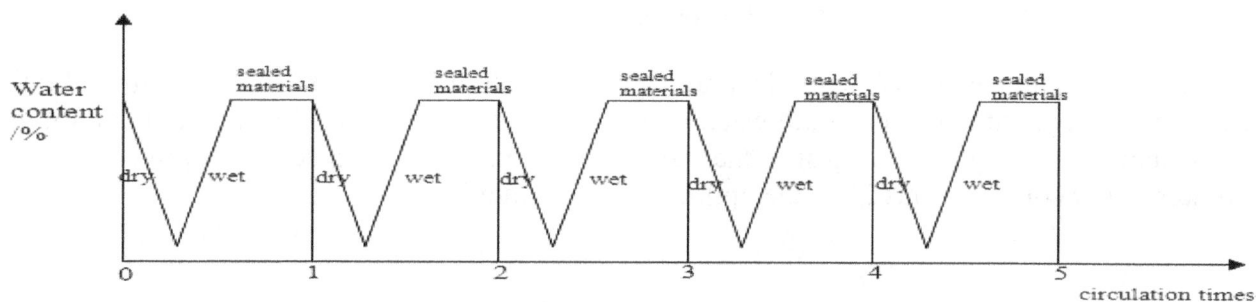

Fig. 1 Test methods of drying-wetting cycles drying-wetting cycles

3.3. *Consolidation test*

Uniaxial compression test is performed on initial samples and samples after dry-wet circulation, with pressure of 25kpa, 50kpa, 100kpa, 200kpa, 400kpa and 800kpa. Stable standard under per level pressure is that per hour deformation is not more than 0.01 mm.

4. Test Result

4.1. *Improvement of moisture methods*

Highway test procedures (JTG E40-2007) [9] refer: for unsaturated samples, press wet cotton yarn around the plate to avoid moisture evaporation. Actual situation shows this method can not meet the requirements of moisture, especially in the summer, moisture evaporation be fast. Because floating ring compression apparatus at the bottom shows dentate, water not only can evaporate from the sides of the pressure plate, there are many pores also can make water abstraction, means that, the whole tank is water abstraction channel. Therefore, in the process of the test, previous hydrating + wet towel cover in order to reduce water loss. Water supplement based on the amount of water loss. Practice proved (Table 3), previous hydrating + wet towel cover measures make error greatly reduce.

Table 3 Moisturizing method

Wet towel cover				Previous hydrating + wet towel cover				
Water content /%	Water content after compression /%	Water loss/ %	Error /%	Water content /%	Filling water in advance /%	Water content after compression /%	Water loss /%	Error /%
28	26.41	1.59	5.68	28	1.59	27.93	0.07	0.25
31	29.57	1.43	4.61	31	1.43	30.77	0.23	0.74
33.8	31.73	2.07	6.12	33.8	2.07	33.56	0.24	0.71
36	33.19	2.81	7.81	36	2.81	35.82	0.18	0.50
40	37.68	2.32	5.80	40	2.32	39.84	0.16	0.40

4.2. *Initial state compacting consolidation deformation characteristics of red clay e-p line*

Fig. 2 shows that:
   (1) With the pressure increasing, void ratio decreasing, the greater the water content, the smaller the void ratio.
   (2) When the water content between optimum water content minus 2% ~ the optimum water content, the variation of void ratio is not obvious.

When the water content is greater than the optimum water content, the amplitude of void ratio increasing is large with the increase of water content. Therefore, from the view point of control the sedimentation of subgrade, suggested that the water content of compacted subgrade between optimum water content minus 2% ~ the optimum water content.

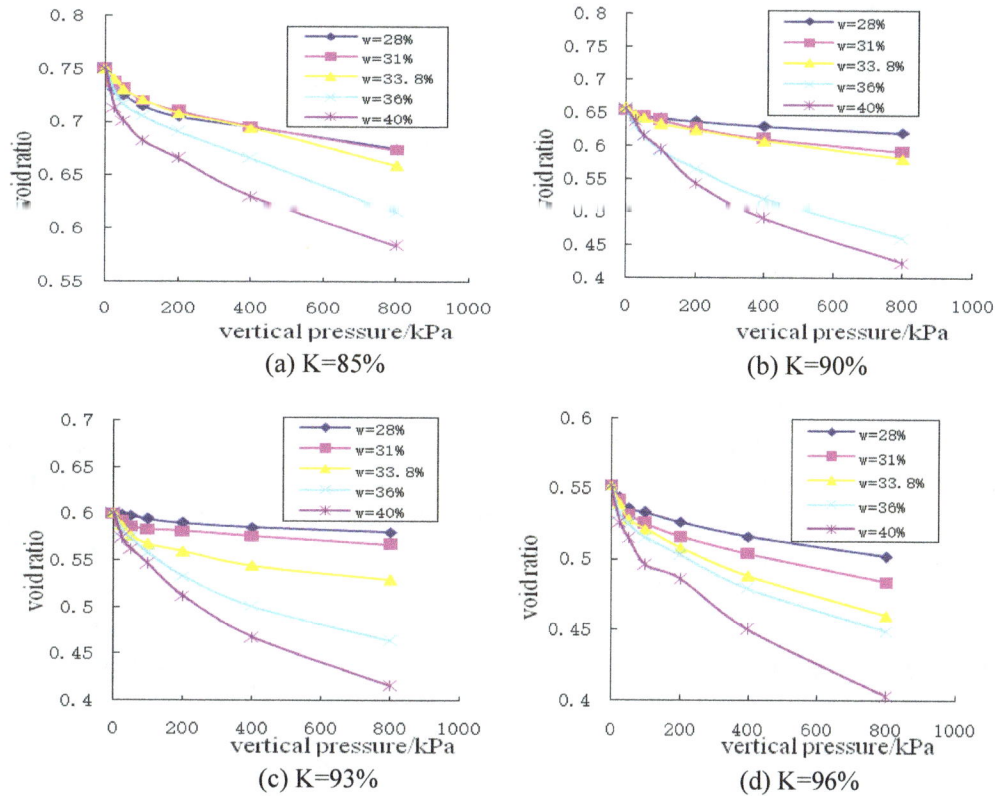

Fig. 2 The e-p curve under different water content

4.2.1. *Relationship between compression deformation coefficient and water content*

Definition: $\delta_p = \dfrac{h_0 - h_i}{h_0}$, δ_p for compression deformation coefficient, h_0 for initial sample height, h_i for the sample stable height after the level i pressure. The definition of compressive deformation coefficient refers, the greater the compression deformation coefficient δ_p, the greater compressibility of soil samples.

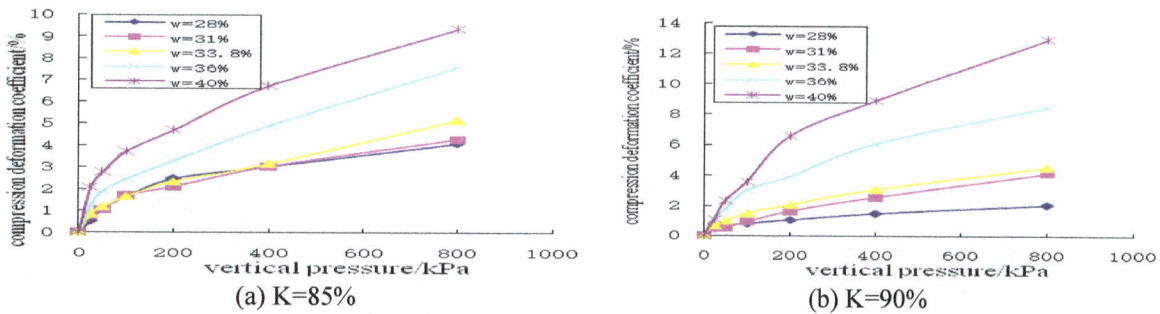

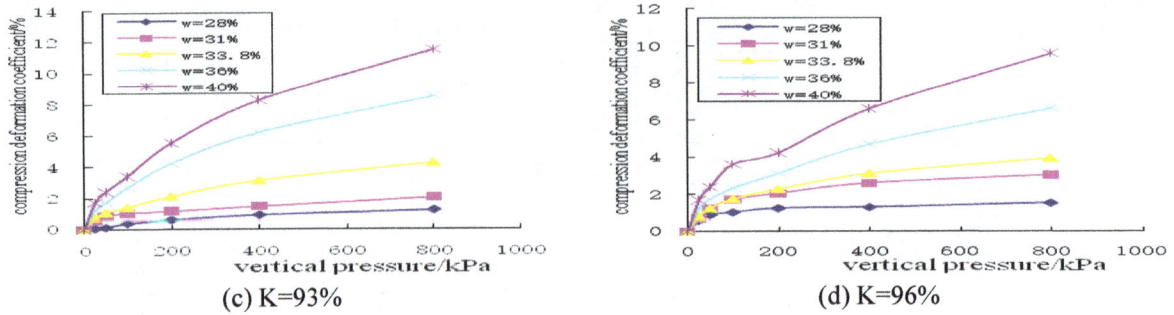

(c) K=93%   (d) K=96%

Fig. 3 Relationship between compression deformation coefficient and vertical pressure under different water content

Fig. 3 shows that (1) compression deformation coefficient increases as the pressure; increases with the increase of water content; (2) When the water content is greater than the optimum water content, the increase of compression coefficient is obvious. When the water content between optimum water content minus 2% ~ the optimum water content, the amplitude of compression deformation coefficient is small. Suggest that the water content of compacted subgrade between optimum water content minus 2% ~ the optimum water content.

4.2.2. Relationship among compression coefficient a_{1-2}, water content and degree of compaction

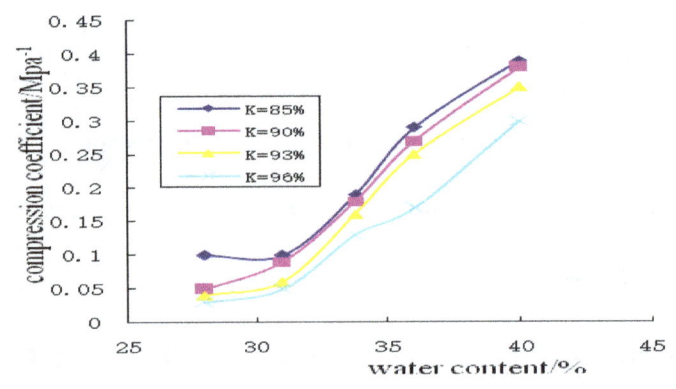

Fig. 4 Relationship between compression coefficient and water content

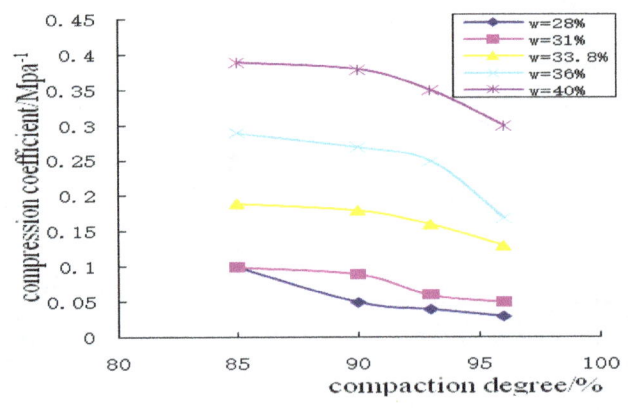

Fig. 5 Relationship between compression coefficient and compaction degree

Fig. 4 and Fig. 5 show that the compression coefficient increases with the increase of water content. When the water content between optimum water content minus 2% ~ the optimum water content, with the increase of compaction degree, the amplitude of compression deformation coefficient is small. When the water content is greater than the optimum water content, with the increase of compaction degree, the increase of compression coefficient is obvious. The compression coefficient decreases with the increase of the degree of compaction. With low water content, the relationship line of compression coefficient and compaction degree curve smooth. With high water content, the compression coefficient decreased significantly with the increase of the degree of compaction.

4.2.3. *Express compression curve of strain*

Organize compacted red clay $\lg\dfrac{\varepsilon_{si}}{P_i} \sim \lg P_i$ test data [10]. The results are shown below in Fig. 6:

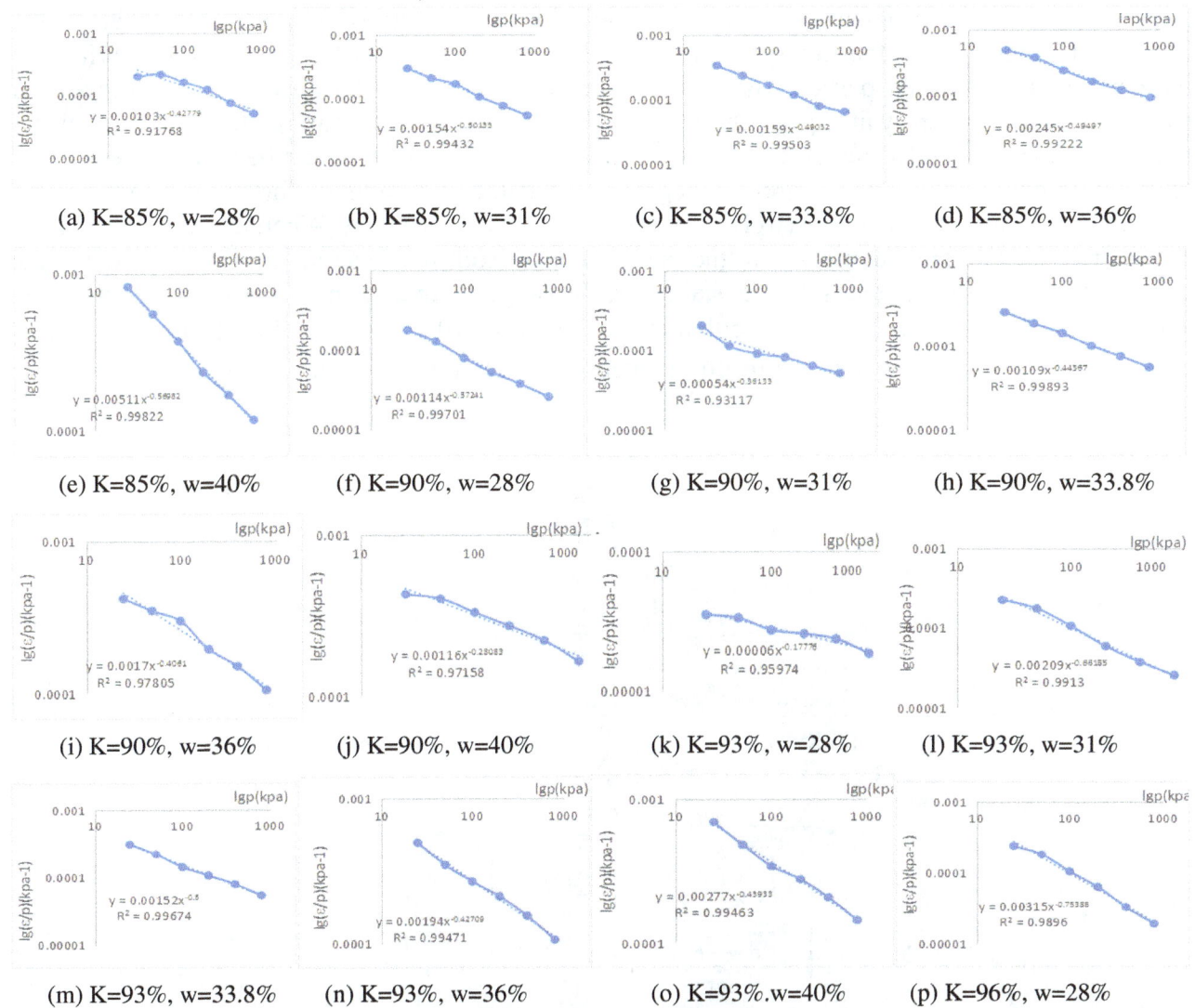

(a) K=85%, w=28%   (b) K=85%, w=31%   (c) K=85%, w=33.8%   (d) K=85%, w=36%

(e) K=85%, w=40%   (f) K=90%, w=28%   (g) K=90%, w=31%   (h) K=90%, w=33.8%

(i) K=90%, w=36%   (j) K=90%, w=40%   (k) K=93%, w=28%   (l) K=93%, w=31%

(m) K=93%, w=33.8%   (n) K=93%, w=36%   (o) K=93%.w=40%   (p) K=96%, w=28%

Fig. 6 The curve of $\lg\dfrac{\varepsilon_{si}}{P_i} \sim \lg P_i$

Fig. 6 shows that linear fitting on the relationship of compacted red clay $\lg\frac{\varepsilon_{si}}{P_i} \sim \lg P_i$ is ideal,

the correlation coefficient is high. Its expression is: $\frac{\varepsilon}{p} = Kp^n$

In this expression, K and n are test constants.

4.3. Relationship between compression coefficient of compacted red clay a_{1-2} and cycles of dry-wet circulation

Fig. 7 shows that red clay compression coefficient is obviously larger than that of the initial state after dry-wet circulation, and with the increase of cycles, compression coefficient increases. The three previous cyclic compression coefficient increases are large, while the fourth and the fifth cycle compression coefficients tend to be stable with gentle curves. The reason is that, in the process of water loss for the first dry-wet circulation, the water loss of soil causes dry shrinkage, the cementation between particles is reduced, the soil surface cracks appear. In the process of soaking humidification, water goes through cracks into the interior soil so that the pore water pressure increases. Under the larger pore water pressure, the bonding effect between cohesive soil particles is destroyed. Large aggregate begins to disperse, resulting in further increase of soil porosity. Soil carrying capacity sharply attenuates after water softening. When dry-wet cycles keep increasing, cracks are stabilized in the soil, interior soil reaches an equilibrium state, the compression coefficient tends to be stable [3]. It suggests that using long-term compression coefficient index a_{1-2} in subgrade settlement calculation (that is, the fourth or fifth cycle compression coefficient) makes the result of calculation better tallies with the actual situation.

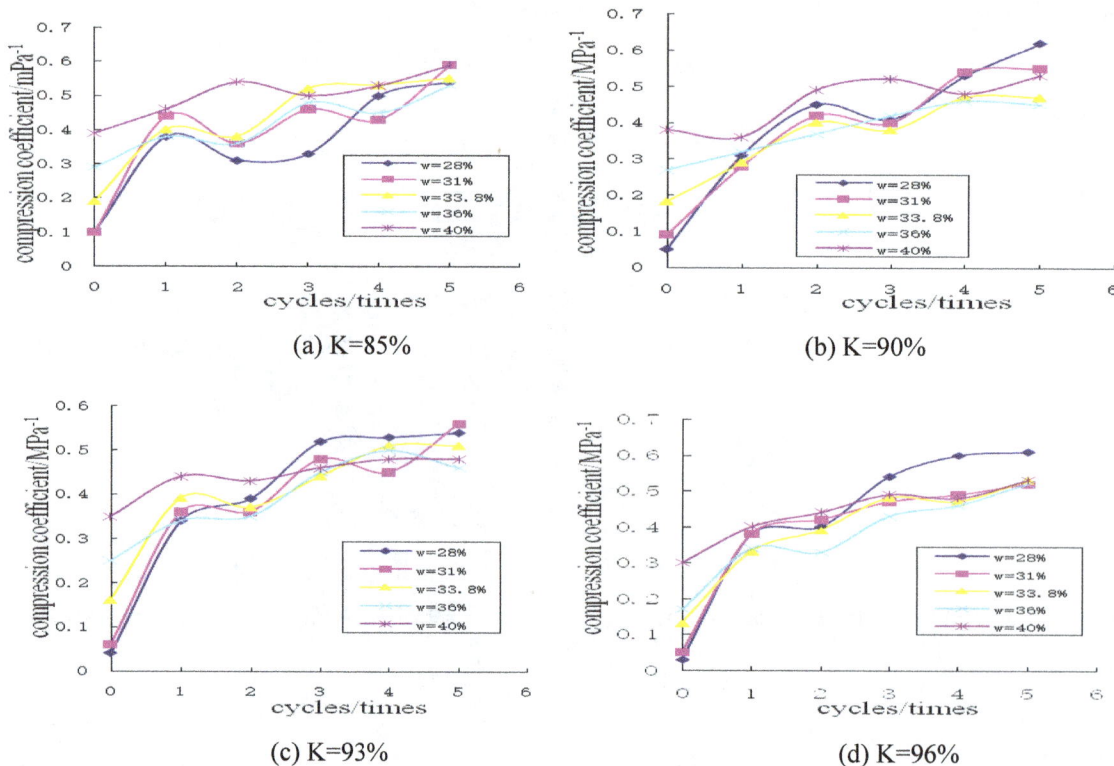

(a) K=85%

(b) K=90%

(c) K=93%

(d) K=96%

Fig. 7 Relationship between compression coefficient and cycles numbers

5. Conclusion

1. Moisturizing method of compression test recommended previous hydrating + wet towel cover. Water supplement based on the amount of water loss.

2. With pressure, void ratio is reduced, the greater water content, the smaller void ratio Compression deformation coefficient increases as the pressure, also increases as water content. Compression deformation coefficient and void ratio basically presents linear relationship. With the increase of void ratio, compressive deformation coefficient decreases. The compression coefficient increases as water content.

3. When the water content between optimum water content minus 2% ~ the optimum water content, the variation of void ratio, compression deformation coefficient, compression coefficient is not obvious. When the water content is greater than the optimum water content, the amplitude of void ratio, compression deformation coefficient and compression coefficient is larger with the increase of water content. Therefore, from the view point of control the sedimentation of subgrade, suggested that the water content of compacted subgrade between optimum water content minus 2% ~ the optimum water content.

4. Stress-strain relationship of compacted red clay can express as $\frac{\varepsilon}{p} = Kp^n$. In this expression, K and n are test constants.

5. Red clay compression coefficient obviously is larger than that of the initial state after dry-wet circulation, and with the increase of cycles, compression coefficient increases. The three previous cyclic compression coefficients have large increase, the fourth and the fifth cycle compression coefficients tend to be stable, gentle curve. It suggests that using long-term compression coefficient index a_{1-2} in subgrade settlement calculation (i.e. the fourth or fifth cycle compression coefficient) makes the result of calculation more tally with the actual situation.

Acknowledgments

This research was funded by the National Natural Science Foundation of China (51368010), Science and Technology Department of Guizhou Province- Alliance Fund Projects of Guizhou University (Science and technology alliance of Qian [2014] 7663) , Natural science research project of Guizhou Provincial Department of Education (Qianjiaohe KY zi[2015]488).

References

1. ZHANG Yong-ting, WANG Bao-tian, SHANG Shu. Experimental Study on Deformation Characteristics of Red Clay with High Liquid Limit under Different Initial Conditions [J]. Journal of Water Resources and Architectural Engineering, 2013, 11 (3): 111-115

2. LI Cong, DENG Wei-dong, CUI Xiang-kui. The study on resilient modulus of remolded loess in moisture circulation process [J]. Traffic science and Engineering, 2009, 25 (2) : 8-12

3. LIU Hong-tai, ZHANG Ai-jun, DUAN Tao et al. The influence of alternate dry-wet on the strength and permeability of remolded loess [J]. HYDRO-SCIENCE AND ENGINEERING, 2010 (4): 38-42

4. MU Xian. jie, ZHANG Xiao-ping. Research on mechanical properties of expansive soil under wetting-drying cycle [J]. Rock and Soil Mechanics, 2008, 28(Suppl): 580-582. (in Chinese)
5. ZHANG Qizhe. Discussion on Construction Techniques of Red Clay [J]. Journal of Water Resources and Architectural Engineering, 2007, 5(1): 83-84.
6. LIU Longwu, YANG Heping, KANG Shilei, et al. Research on Nature of Red-Clay-Filling Material Used in Road[J]. Highway, 2002, (6): 125-128.
7. ZHU Tianzhang, HUANG Xiangjin. The Project Characteristic Properties and Problems of Laterite in Hunan Highway are discussed with the Countermeasure [J]. Highway Engineering, 2008, 33(5): 24-28.
8. ZENG Jing, DENG Zhibin, LAN Xia, et al. Experimental Study on Properties of High Liquid
9. Limit Soil and Red Clay of Zhucheng Highway [J]. Rock and Soil Mechanics, 2006, 27 (1): 89-92, 98.
10. P. R. China. Ministry of Communications. JTG E40-2007 Specification of Highway Soil Test [S]. Beijing: China Communications Press, 2007.
11. Liu Baojian, Zhang Junli. Application and Analysing Method for Soil Compression Test [J]. China Journal of Highway and Transport, 1999, 12 (1): 37-41, 100.

Crystal Structure and Microwave Dielectric Properties of $Ca[(Li_{1/3}Nb_{2/3})_{0.95}Zr_{0.05}]O_{3-\delta}$ Ceramics with Pr^{3+} Substitution at A-Site

Gang Xiong[*], Han-Fen Zhang

Department of Electronic and Information Engineering, Xianning College, Xianning 437100, China.
[*]Email: xgang68@aliyun.com

Crystal structure and microwave dielectric properties of $(Ca_{1-x}Pr_x)[(Li_{1/3}Nb_{2/3})_{0.95}Zr_{0.05}]_{3-\delta}$ ($0.0 \leq x \leq 0.2$, CPLNZ) ceramics were investigated. A single phase with orthorhombic perovskite structure was obtained at $x=0.0 \sim 0.09$. With an increase of Pr^{3+} content, the quality factor value firstly increased and then began to decrease at x=0.06 due to a decrease of the B-site 1:2 ordering degree. The variation of τ_f with tolerance factor was discussed. When x=0.07, the optimum microwave dielectric properties: the permittivity is 30.1, the quality factor is 25010GHz, and the temperature coefficient of resonator frequency is $-10.6 \times 10^{-6}/°C$.

Keywords: Microwave dielectric ceramics; microwave dielectric properties.

1. Introduction

Among those low-temperature-cofired ceramics (LTCC), $Ca(Li_{1/3}Nb_{2/3})O_{3-\delta}$-based ceramics were newly developed and widely investigated because of its excellent microwave dielectric properties and low sintering temperature of about 1150°C. J.W.Choi et al. [1,2] have reported Ti^{4+}, Sn^{4+} and Zr^{4+} B-site substitution in $Ca(Li_{1/3}Nb_{2/3})O_{3-\delta}$ ceramics. By their means, microwave dielectric properties of $Ca(Li_{1/3}Nb_{2/3})O_{3-\delta}$ ceramics were improved. In this study, we investigated the substitution effect of Pr^{3+} to Ca^{2+} at A-site on the microwave properties, particularly properties-structure relationships in $Ca(Li_{1/3}Nb_{2/3})O_{3-\delta}$-based. Based on our preliminary work, $Ca[(Li_{1/3}Nb_{2/3})_{0.95}Zr_{0.05}]O_{3-\delta}$ ceramics was selected as the matrix since it possesses good dielectric properties: $\varepsilon_r = 28.5$, $Q_f = 23920$ GHz and $\tau_f = -14.7$ ppm/°C after sintering at 1150°C for 4h. Crystal structure and its relationships with microwave dielectric propertyes in $(Ca_{1-x}Pr_x)[(Li_{1/3}Nb_{2/3})_{0.95}Zr_{0.05}]O_{3-\delta}$ ($0.0 \leq x \leq 0.2$) were studied as well.

2. Experimental

The $(Ca_{1-x}Pr_x)[(Li_{1/3}Nb_{2/3})_{0.95}Zr_{0.05}]O_{3-\delta}$ ($0.0 \leq x \leq 0.2$) powder compositions were synthesized by the conventional solid-state reaction method. Specimens were sintered in a closed Al_2O_3 crucible at 1170°C for 4h. The bulk densities of sintered specimens were measured by Archimede method. Phase formation and microstructure were examined by X-ray diffractometer (X'Pert PRO) using CuKa radiaiton. The measurement of microwave dielectric properties were performed on TE_{011} mode at the resonant frequency from 5 to 8 GHz by the Hakki-Coleman's dielectric resonator method using a network analyer (ADVANTEST R3767C). The temperature coefficient of resonator frequency (τ_f) was calculated at the range between 20 and 80 °C.

3. Results and Discussion

Fig. 1 shows the XRD patterns of CPLNZ specimens sintered at 1170°C for 4h with the various values of x. A single perovskite phase with the orthorhombic structure was obtained from x =0.00 to 0.07, however, secondary phase was detected when X≥0.09.With increasing Pr^{3+} content, the peak positions firstly shifted to higher 2θ angle indicating a increasing in the lattice parameter of the solid solution because of the relatively larger ionic radius of Pr^{3+} (0.117nm, coordination number CN=12) than the ionic radius of Ca^{2+} (0.134nm, CN=12) entered the A-site of the ABO_3 perovskite structure[3]. Then as x further increased from 0.09 to 0.2, the 2θ angle began to shift to higher angle, which was due to the appearance of second phase. The relative density of the sintered specimens was above 96% of x-ray density.

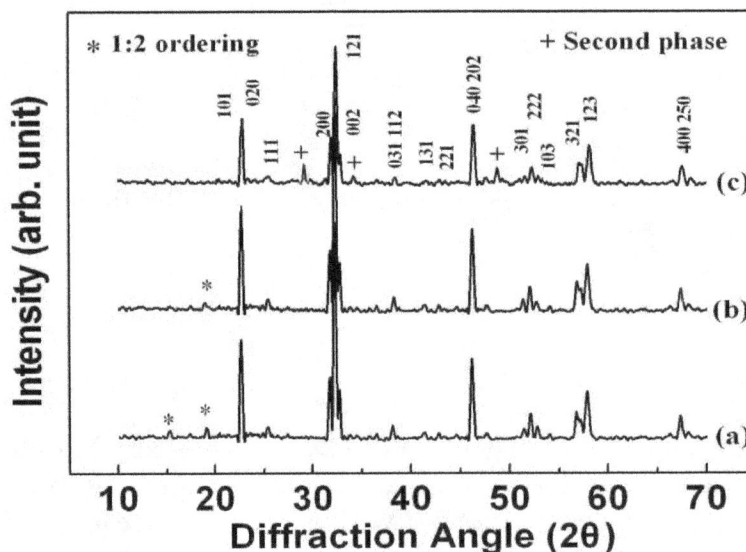

Fig. 1 XRD patterns of CPLNZ specimens sintered at 1170°C for 4h with the x value of (a) 0.00, (b) 0.07 and (c) 0.09

Fig. 2 shows dielectric constant ε_r of CPLNZ specimens sintered at 1170°C for 4h with the various values of x. The ε_r gradually increased from 28.5 to31.2 as Pr^{3+} content increased from 0.0 to 15mol%. This phenomenon is considered that the ionic polarizability of Pr^{3+} (5.32$Å^3$)is higher than Ca^{2+}(3.16 $Å^3$) and a increase of unit-cell volume in CPLNZ ceramics caused by A site Pr^{3+} substitution[4]. When Pr^{3+} content become greater than 15mol%, the ε_r steadily decreased from 31.2 to 30.5, which was due to the formation of secondary phase.

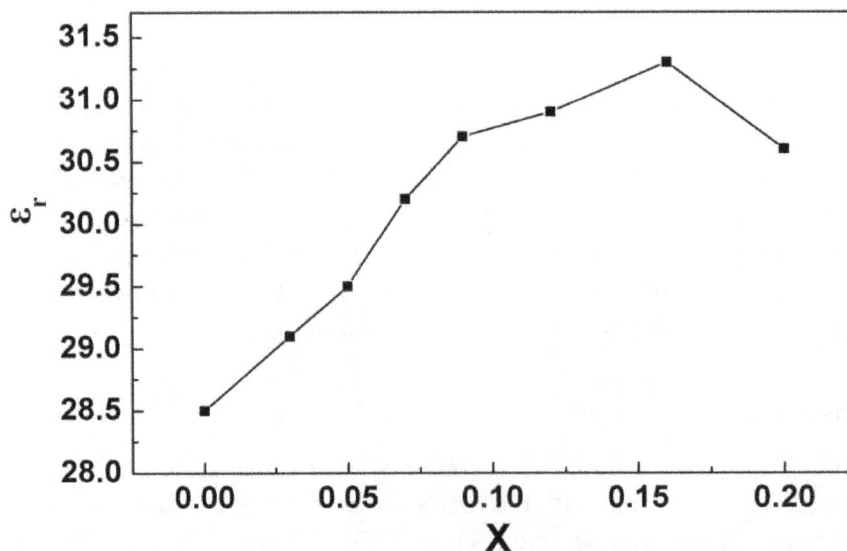

Fig. 2 Dielectric constant (εr) patterns of CPLNZ specimens sintered at 1170°C for 4h with the various values of x

It is observed in Fig. 1 and in Fig. 3 that when x≤5mol%, with x increases, the peaks of superlattice diffractions of specimen 1:2 increase, the degree of B-site 1:2 ordering increase, Qf value increase greatly; when x>5mol%, the peaks of superlattice diffractions of specimen 1:2 decreases until disappear, the degree of B-site 1:2 ordering will decrease, second phase starts to appear. Therefore, the Qf value of the specimen is dramatically reduced to 25010GHz.With the further increase of Pr supersedes, the contents of second phase increase, the extent of deviation from original ceramic composition increase, Qf value decrease from 23260GHz when x=9mol% to 17560GHz when x=20mol%.

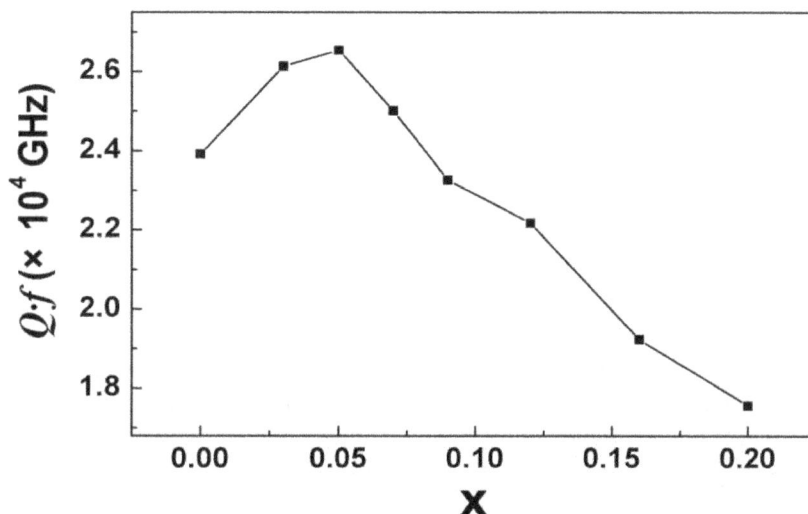

Fig. 3 Quality factor (Qf) patterns of CPLNZ specimens sintered at 1170°C for 4h with the various values of x.

Fig. 4 shows the temperature coefficient of resonant frequency (τ_f) and the Tolerance factor (t) of CPLNZ specimens sintered at 1170°C for 4h with the various values of x. The τ_f value firstly increase from −14.7 to -9.6 ppm /°C with increasing Pr^{3+} content from 0.0 to 12mol%, then decrease to -10.5×10^{-6}/°C. The τ_f value of CPLNZ specimens increased with the decrease of the Tolerance factor (t).The τ_f can be related to the Tolerance factor (t) of the ABO_3 perovskite structure according to the work of Reaney al [5].

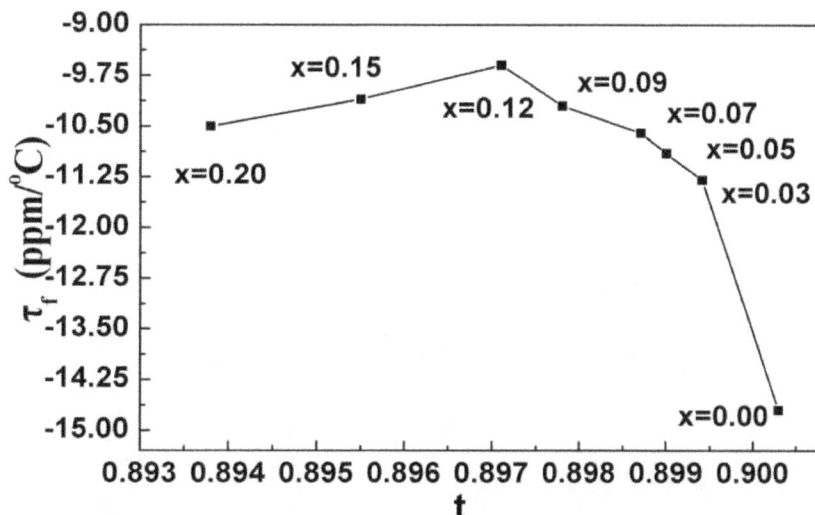

Fig. 4 Relationship between τ_f and the Tolerance factor (t) of CPLNZ.

4. Conclusion

A single phase with orthorhombic perovskite structure was obtained at x=0.0~0.07 in $(Ca_{1-x}Pr_x)[(Li_{1/3}Nb_{2/3})_{0.95}Zr_{0.05}]O_{3-\delta}$ ceramics. Other specimens with Pr^{3+} content x>7mol% possessed second phase. A site Pr^{3+} substitution made the tolerance factor value decrease. Therefore, the τ_f value increased from−14.7 to -9.6 ppm /°C with increasing Pr^{3+} content from 0.0 to 12.0mol%. At x=0.12, the τ_f value began to moved to the negative direction. When *x*=0.07, the optimum microwave dielectric properties: ε_r=30.2, Qf=25010GHz, τ_f = −10.6 ppm/°C.

Acknowledgments

This work was funded by China Undergraduate Scientific and Technological Innovation Project (No.201410927006), the Natural Science Foundation of Hubei Province (No. 2014CFC1084), Foundation of Bureau of Human Resources and Social Service of Xianning (No.xnrsg-3) and the Natural Science Foundation of Scientific office of Xianning (No.XNKJ-1302).

References

1. J.W. Choi, C.Y. Kang, S.J. Yoon, H.J. Kim and H.J. Jung, Microwave dielectric properties of $Ca[(Li_{1/3}Nb_{2/3})_{1-x}M_x]O_{3-\delta}$(M=Sn, Ti)ceramics. J. Mater. Res. 14 (1999): 3567-3570.
2. J.W. Choi, J.Y. Ha, S.J. Yoon, H.J. Kim and K.H. Yoon, Microwave dielectric properties of $Ca[(Li_{1/3}Nb_{2/3})_{1-x}Zr_x]O_{3-\delta}$ ceramics. Jpn. J. APPL. Phys. 43 (2004): 223-225.
3. R.D. Shannon, Revised effective ionic radii snd systematic study of interatomic distances in halides and chalcogenides.Acta Crystallogr. (secA.), 1976, 32 (1976): 751-767.
4. R.D. Shannon, Dielectric poparizabilities of ions in oxides and flucorides. J. Appl. Phys., 73 (1993): 348-366
5. Reaney I M., Colla E L, Setter N., Dielectric and structural characteristics of Ba- and Sr-based complex perovskites as a function of tolerance factor. Jpn. J. Appl. Phy., 33(1994): 3984-3992

Crystal Structure and Microwave Dielectric Properties of Ca[(Li$_{1/3}$Nb$_{2/3}$) $_{0.93}$Zr$_{0.07}$]O$_{3-\delta}$–xTiO$_2$ Ceramics

Gang Xiong*, Han-Fen Zhang

Department of Electronic and Information Engineering, Xianning College, Xianning 437100, China.
Email: xgang68@aliyun.com

Crystal structure and microwave dielectric properties of Ca[(Li$_{1/3}$Nb$_{2/3}$) $_{0.93}$Zr$_{0.07}$]O$_{3-\delta}$–xTiO$_2$(0≤x≤0.2) ceramics were investigated. A single phase with orthorhombic perovskite structure was obtained at x=0.0~0.1. With an increasing of Ti^{4+} content, the Qf value decreased due to a decrease of the degree of B-site 1:2 ordering. However, the τ_f value increased from-14.2ppm /°C to -6.3ppm/°C. When x=0.04, the optimum microwave dielectric properties: ε_r=31.6, Qf =17160GHz and τ_f=−8.5 ppm/°C.

Keywords: Microwave dielectric ceramics; microwave dielectric properties.

1. Introduction

Recently, Ca(Li$_{1/3}$Nb$_{2/3}$)O$_{3-\delta}$ ceramics have been newly developed and widely investigated because of its excellent microwave dielectric properties and low sintering temperature of about 1150°C [1, 2]. Ca[(Li$_{1/3}$Nb$_{2/3}$)$_{0.93}$Zr$_{0.07}$]O$_{3-\delta}$ shows ε_r of 28.9 and Qf of 22020 GHz, while τ_f is a large negative value (−14.4ppm/°C). It is important to adjust τ_f to near zero for application of the microwave device. In order to adjust τ_f to near zero, two or more compounds with positive and negative τ_f value are employed to form a solid solution or mixed phases [3]. TiO$_2$ was reported to possesses good microwave dielectric properties: ε_r =100, τ_f = 400ppm/°C and Qf $\approx$ 40000 GHz. Therefore, it is expected that the microwave materials with Near-Zero temperature coefficient, good sinterability and dielectric properties were obtained by combining Ca[(Li$_{1/3}$Nb$_{2/3}$)$_{0.93}$Zr$_{0.07}$]O$_{3-\delta}$ with TiO$_2$. In the present study, the influence of TiO$_2$ addition on microwave dielectric properties and crystal structures of Ca[(Li$_{1/3}$Nb$_{2/3}$)$_{0.93}$Zr$_{0.07}$]O$_{3-\delta}$ ceramics have been investigated.

2. Experimental

The Ca[(Li$_{1/3}$Nb$_{2/3}$)$_{0.93}$Zr$_{0.07}$]O$_{3-\delta}$−xTiO$_2$ (0.0≤x≤0.1) powder compositions were synthesized by the conventional solid-state reaction method. Specimens were sintered in a closed Al$_2$O$_3$ crucible at 1170 °C for 4h. The bulk densities of sintered specimens were measured by Archimede method. Phase formation and microstructure were examined by X-ray diffractometer (X'Pert PRO) using CuKa radiaiton. The measurement of microwave dielectric properties were performed on TE$_{011}$ mode at the resonant frequency from 5 to 8 GHz by the Hakki-Coleman's dielectric resonator method using a network analyer (ADVANTEST R3767C). The temperature coefficient of resonator frequency (τ_f) was calculated at the range between 20 and 80 °C.

3. Results and Discussion

Fig. 1 shows part of powder XRD patterns of the Ca[(Li$_{1/3}$Nb$_{2/3}$)$_{0.93}$Zr$_{0.07}$]O$_{3-\delta}$−xTiO$_2$ (0.0≤x≤0.1) ceramics varied with TiO$_2$ content sintered at 1170°C for 4h. With x increases, the peaks of superlattice diffractions of specimen 1:2 decreases until disappear, the degree of B-site 1:2 ordering will decrease, second phase disappearence, which means Ti^{2+} entered the B-site of the ABO$_3$ perovskite structure, and then formed complete solid solution with single perovskite phase. TiO$_2$

addition accelerated the composition of main phase in ceramic systems, but inhibited the degree of 1:2 ordering in $Ca[(Li_{1/3}Nb_{2/3})_{0.93}Zr_{0.07}]O_{3-\delta}$-$xTiO_2$ ceramics.

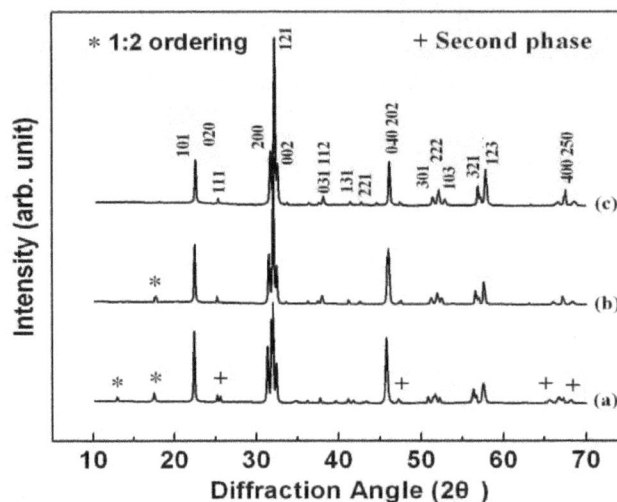

Fig. 1 XRD patterns of CPLNZ specimens sintered at 1170°C for 4h with the x value of (a) 0.00, (b) 0.04 and (c) 0.08

Fig. 2 shows the relationship between the dielectric constant ε_r and the composition (x). With the increasing content of TiO_2 ranged from 0.06 to 1.0, the dielectric constant ε_r increased almost linearly from 28.9 to 34.7 by the compensation effect, which was due to the formation of complete solid solution with single perovskite phase, $CaTiO_3$ has a high ε_r of 170 [4]. Although the samples are not the mixtures of pure $Ca[(Li_{1/3}Nb_{2/3})_{0.93}Zr_{0.07}]O_{3-\delta}$ and TiO_2, the measured values could still be compared with the theoretical values calculated using the logarithmic mixing rule $\ln\varepsilon_r = V_1\ln\varepsilon_{r1} + V_2\ln\varepsilon_{r2}$ where V_i and ε_{ri} are the volume fraction and dielectric constant of each component [5].

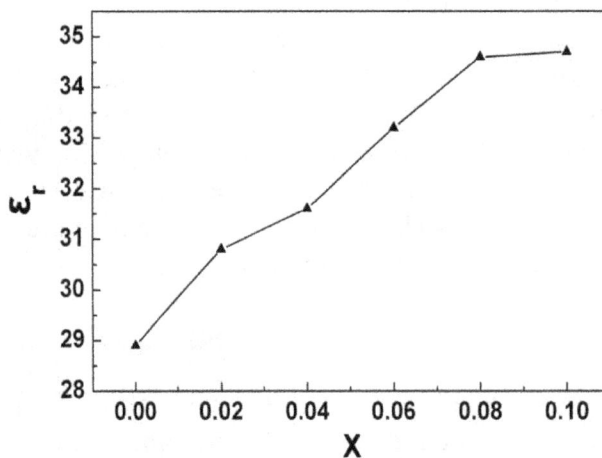

Fig. 2 ε_r values of $Ca[(Li_{1/3}Nb_{2/3})_{0.93}Zr_{0.07}]O_{3-\delta}$-$xTiO_2$ specimens with the various values of x

Fig. 3 shows the effect of TiO$_2$ addition on the quality factor.

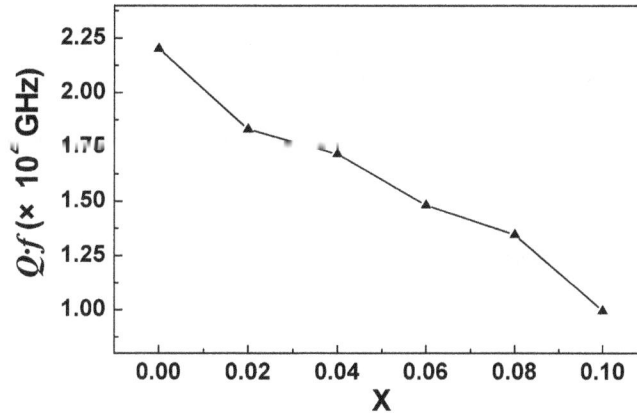

Fig. 3 Qf values of Ca[(Li$_{1/3}$Nb$_{2/3}$)$_{0.93}$Zr$_{0.07}$]O$_{3-\delta}$−xTiO$_2$ specimens with the various values of x

The quality factor decreases with TiO$_2$ addition. This is expected since TiO$_2$ addition inhibited the degree of 1:2 ordering in Ca[(Li$_{1/3}$Nb$_{2/3}$)$_{0.93}$Zr$_{0.07}$]O$_{3-\delta}$-xTiO$_2$ ceramics and thus cause the decreases of the quality factor [6, 7].

The relationship between the temperature coefficient of the resonant frequencies and the TiO$_2$ content in Ca[(Li$_{1/3}$Nb$_{2/3}$)$_{0.93}$Zr$_{0.07}$]O$_{3-\delta}$-xTiO$_2$ ceramics was plotted in Fig.4. As indicated τ$_f$ values varied almost linearly with TiO$_2$ content which was quite similar to that of dielectric constant. It ranged from negative value of -14.4ppm/°C to negative value of −6.3ppm/°C when TiO$_2$ content increased from 0 to 10 mol%.The results indicated that by proper choose of two compounds with opposite temperature coefficients, it is possible to achieve temperature stable material.

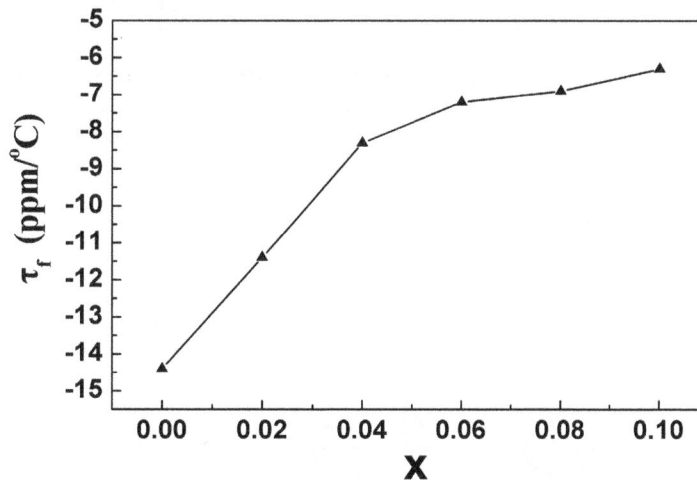

Fig. 4 τ$_f$ values of Ca[(Li$_{1/3}$Nb$_{2/3}$)$_{0.93}$Zr$_{0.07}$]O$_{3-\delta}$−xTiO$_2$ specimens with the various values of x

4. Conclusion

TiO$_2$ addition accelerated the composition of main phase in ceramic systems, but inhibited the degree of 1:2 ordering in Ca[(Li$_{1/3}$Nb$_{2/3}$)$_{0.93}$Zr$_{0.07}$]O$_{3-\delta}$-xTiO$_2$ ceramics. When 6mol%≤x≤10mol%, the systems changed into single perovskite phase. The ε$_r$ increased from 28.9 to34.7 with an increase of Ti content. Which was due to the ε_r value of $CaTiO_3$ is 170, higher than that of $Ca[(Li_{1/3}Nb_{2/3})_{0.93}Zr_{0.07}]O_{3-\delta}$ ceramics. The Qf value decreased due to a decrease of the degree of B-site 1:2 ordering and an change of the matrix. With an increasing of Ti^{4+} content, the τ_f value increased from -14.4 ppm/ °C to -6.3 ppm/°C, When $x=0.04$, the optimum microwave dielectric properties: $\varepsilon_r=31.6$, Qf $=17160$GHz and $\tau_f=-8.5$ ppm/°C.

Acknowledgments

This work was funded by China Undergraduate Scientific and Technological Innovation Project (No.201410927006), the Natural Science Foundation of Hubei Province (No. 2014CFC1084), Foundation of Bureau of Human Resources and Social Service of Xianning (No.xnrsg-3) and the Natural Science Foundation of Scientific Office of Xianning (No.XNKJ-1302).

References

1. J.W. Choi, C.Y. Kang, S.J. Yoon, H.J. Kim and H.J.Jung, Microwave dielectric properties of $Ca[(Li_{1/3}Nb_{2/3})_{1-x}M_x]O_{3-\delta}$(M=Sn, Ti) ceramics. J. Mater. Res.14 (1999): 3567-3570.

2. J.W. Choi, J.Y. Ha , S.J. Yoon, H.J. Kim and K.H. Yoon, Microwave dielectric properties of $Ca[(Li_{1/3}Nb_{2/3})_{1-x}Zr_x]O_{3-\delta}$ ceramics. Jpn. J. APPL. Phys. 43 (2004): 223-225.

3. C.L. Huang, H.L. Chen, C.C. Wu, Improved high Q value of $CaTiO_3$-Ca $(Mg_{1/3}Nb_{2/3})O_3$ solid solution with near zero temperature coefficient of resonant frequency. Mater. Res. Bull., 36 (2001): 1645-1652.

4. O. Hirotaka, P. Liu, E.S. Kim, et al., Low-temperature sintering and microwave dielectric properties of $Ca(Li_{1/3}Ta_{2/3})O_{3-\delta}-CaTiO_3$ ceramics. J. Eur. Cer. Soc., 23 (2003): 2417-2421.

5. K. Lichtenecker, Dielectric constants of cubic ionic compounds. Physik Zeits., 27 (1926): 833-835.

6. I.T. Kim, Y.H. Kim, S.J. Chung, Order-disorder transition and microwave dielectric properties of $Ba(Zn_{1/3}Ta_{2/3})O_3$ ceramics. Jpn. J. Appl. Phys., 34 (1995), : 4096-4101.

7. P.K. Davies, J. Tong, T. Negas, Effect of ordering-induced domain boundaries on low-loss $Ba(Zn_{1/3}Ta_{2/3})O_3$-$BaZrO_3$ perovskite microwave dielectrics. J. Am. Ceram. Soc., 80 (1997): 1727-1740.

Development and Application of Multi Dimensional Deformation Monitor

Shi-Yu Wei[1,2], Chuan Li [1,2,*], Chao Li[1,2]

[1]*Chongqing Engineering Research Center of Automatic Monitoring for Geological Hazards, Chongqing 400042, China*

[2]*Chongqing Institute of Geology and Mineral Resources, Chongqing 400042, China*

Email: 641419444@qq.com

In order to further enhance the capability of engineering monitoring, considering the deformable body, whose deformation mechanism is ambiguity and the destruction form is complex. Through the study of sensing principle and adopting the technology integration method, this paper developed a set of multidimensional deformation monitors with the function of three dimensional displacement monitoring, Dutch roll and inclination. The application results show that the displacement observation accuracy is better than 1 mm, and dip angle observation accuracy is better than 0.01°. This device can meet the demand of engineering monitoring.

Keywords: Multidimensional; PSD displacement sensing; inclination sensing; integration.

1. Introduction

At present, as the deformable body's deformation mechanism is unknown and the failure mode is complex, monitoring methods mostly rely on GNSS, total station, crack meter, guyed displacement meter and other equipment to achieve absolute or relative displacement, or are based on pressure sensing devices for pressure monitoring. In general, existing monitoring methods pay more attention to displacement monitoring. Different monitoring data are relatively isolated. There is a lack of rotation quantity monitoring and multidimensional monitoring data analysis. It is difficult to build a set of combination space domain with time domain of complete monitoring system. Therefore, it is necessary to import new technologies, new methods, through the sensor technology integration means, the research and development of new intelligent monitoring equipment, breaking through the existing technical bottlenecks and effectively strengthening the monitoring security efforts.

2. The Research on Sensor Technology

To achieve the purpose of the 3d displacement and space rotation integration observation, research should start from the principle of sensor, laser ranging, communication technology, the technical implementation and integration of the three aspects. The research technical route is shown in Fig. 1.

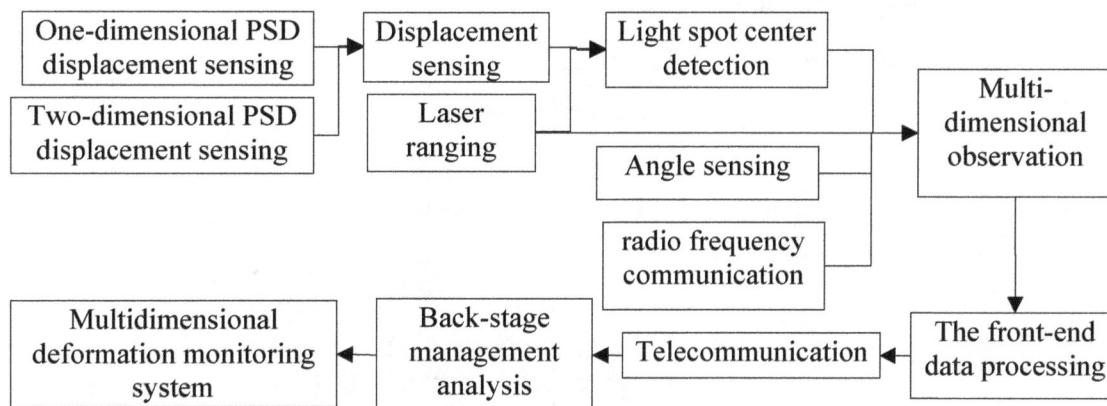

Fig. 1 The research on technical route

*Corresponding author

3. Research on PSD Displacement Sensor Technology

3.1. *Sensing principle*

Photoelectric PSD with high sensitivity is a new type of photoelectric device, or called coordinate cell. It is a kind of non-split type device. light spot on the photosensitive surface can be transferred to electrical signals. Sensor is composed of three layers of P, I, N. And P is the photosensitive surface, and both ends have I1, I2, as the output. There is a public side on the back of the N layer on the photosensitive surface that is connected to bias voltage [1]. As shown in Fig. 2.

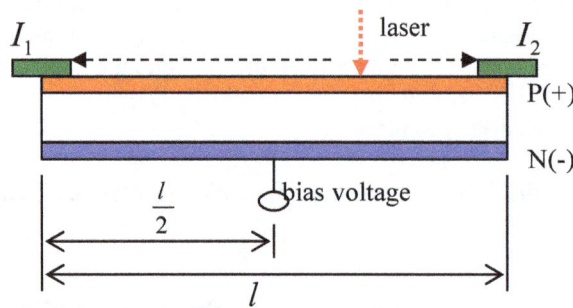

Fig. 2 Sensing principle

When lights, PIN photodiode produce photocurrent, and the relationship between facula center position and the output photocurrent is shown in Eq. 1.

$$X_a = \frac{L}{2} \times \frac{I_2 - I_1}{I_2 + I_1}.$$

(1)

In Eq. 1 : X_a—Distance of facula center to PSD center distance, mm;

   L—Photosensitive surface length, mm;

   I1, I2—Photocurrent on two output sides.

Two-dimensional square PSD device has independent photosensitive layer in X, Y, directions, which respectively percept the change of light spot position in X and Y direction. And its principle diagram is shown in Fig. 3. Based on the same working principle of one-dimensional PSD, relational expression can be exported between the location parameters of the two-dimensional square PSD device and the electrode current (the origin of coordinates as device center). The formula is shown in Eq. 2 and Eq. 3.

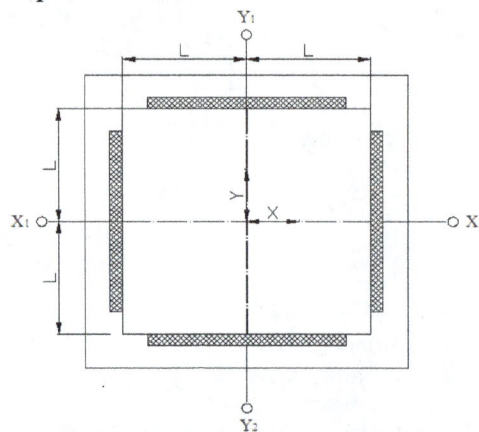

Fig. 3 Working principle diagram of two-dimensional PSD

$$\frac{X_2 - X_1}{X_2 + X_1} = \frac{X}{L}. \tag{2}$$

$$\frac{Y_2 - Y_1}{Y_2 + Y_1} = \frac{Y}{L}. \tag{3}$$

When sighting position is within the measuring range of mobile, the distance between facula and PSD two electrodes change, which makes the output current of two electrodes change as the position of light spot change, so by measuring the size of the output current of sensor two electrodes, displacement changes of the object tested can be perceived [2].

3.2. *The core circuit design*

PSD output signal I1 and I2 in the form of photocurrent. Because the sighting object plane has a certain distance with the PSD, and photosensitive surface of PSD receives low light energy, so that the PSD output photocurrent is weak; thus adapter circuit design is crucial for the extraction of photocurrent. According to statement above, the origin of coordinates is located in the center of photosensitive surface of PSD. When devices work in reverse bias and small signal condition, in general, when the intensity of light spot is stable, Fig. 4 the circuit can be used (Figure shows the circuit schematic diagram in the X direction, Y direction and X direction). The output is connected to A/D and makes conversion; the circuit mainly includes the current voltage converter (U1A, U2A) and voltage amplifier (U1B, 2 b).

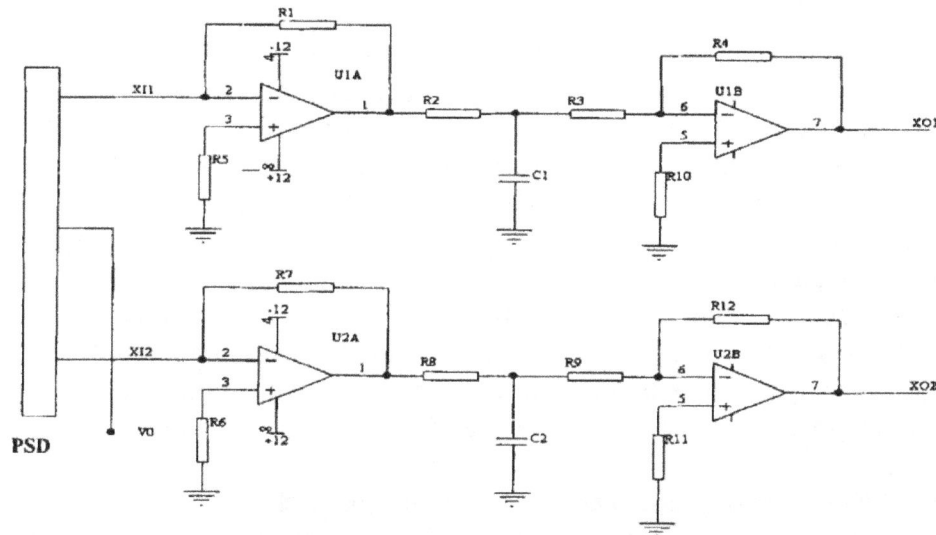

Fig. 4 Amplifying circuit design in X direction

3.3. *Research on angle sensing technology*

Based on the Newton's second law, speed cannot be measured within a system, but its acceleration can be measured. If initial velocity of the overall system has been known, the linear velocity can be calculated through the integral method so that calculate linear displacement. When the angle sensor is static, namely there is no lateral and vertical direction acceleration, so only the acceleration speed of gravity works on it. Included angle between the vertical axis of gravity and acceleration sensor sensitive axis is the system angle. The above basic principles can be used to measure dip angle changes relative to the horizontal plane or vertical plane [6].

With the development of MEMS technology, inertial sensor, one the most widely used MEMS, has become the most successful in recent years, and micro accelerometer (micro - accelerometer) is an excellent representative of inertial sensor. As the most sophisticated inertial sensor, the current MEMS accelerometer has a very high level of integration, namely sensing system and the interface circuits are integrated on a chip [7].

Inclinometer highly integrates MCU, MEMS accelerometer, modulus conversion circuit, communication unit, to realize two-axis oblique observation. The sensor's features include: silicon micro mechanical sensor (MEMS) has the biaxial angle change in horizontal plane. The output angle takes the level surface as reference, and datum can be calibrated again. It is output in the form of data, and the interface forms include RS232 and RS485. In addition, it has strong resistance to outside electromagnetic interference.

4. Research on Laser Ranging and Facula Detection Technology

4.1. *Laser ranging technique*

To solve the objective demands of three dimensional displacement observation, projects use laser ranging principle to achieve space vector observations. Working methods of laser ranging can be divided into: pulse laser ranging and continuous wave laser ranging. As the laser ranging technology has matured, here there is no need to do explanation.

4.2. *Detection technology in facula center*

In the process of displacement observation, due to the existence of laser speckle, surface reflection characteristics changes of the object tested, the surface tilt and external disturbance, the light intensity of image facula signal obtained by CCD is not distributed uniformly, and relative centroid of the geometric center of facula offset, thus seriously affect the accuracy of the measuring system. Therefore, accurate positioning of facula center position in laser speckle image is one of the key technologies needed to be solved for displacement observation. When hardware measures reduce the distortion of the facula, it can also improve the localization algorithm method of facula center from software aspect, so as to reduce the measurement error, and improve the accuracy of measurement. This study first uses iterative threshold method to calculate the optimal threshold value of the facula image segmentation, intercepts facula feature point imaging area; And use bilinear interpolation method to accurately calculate facula center coordinates [3-5].

1) **The sub pixel localization principle:** Related studies have shown that the premise of sub pixel localization is: the target is not isolated individual pixels, and it must be compose of a series of pixel in gray level distribution and shape distribution, for example the targeted features are dot, angular point, "cross intersection, a straight line, curve, etc., and they have obvious gray level variance and certain area. Laser facula in this study is nearly circular geometric shape, so corresponding algorithm can be adopted to realize sub-pixel positioning.

2) **Interception of facula image feature point area with iterative threshold method:** When determine the facula area at image feature points (coarse location), first filter the speckle,

eliminate the stray light and noise; then iterative method is used to calculate the optimal threshold, to split facula. The realization steps of the iterative threshold method are as followed:

Step 1: Calculate the facula gray image's minimum grey value G_{min} and maximum grey value G_{max} after filtering the wave and noise, and then the initial threshold value is:

$$T^k = \frac{G_{min} + G_{max}}{2}. \tag{4}$$

Step 2: Based on threshold value TK, segment image into target image and background image, according to the Eq. 5 and Eq. 6, calculate the average gray value, G_O and G_B

$$G_O = \frac{\sum_{g(i,j)<T^k} g(i,j) \times k(i,j)}{\sum_{g(i,j)<T^k} g(i,j)}. \tag{5}$$

$$G_B = \frac{\sum_{g(i,j)>T^k} g(i,j) \times k(i,j)}{\sum_{g(i,j)>T^k} g(i,j)}. \tag{6}$$

In Eq. 5 and Eq. 6: g(i, j) is the grey value of (i, j) in image. k (i, j) is the weighting coefficient of point (i, j), and it is generally taken 1.

Step 3: Calculate the new threshold

$$T^{k+1} = \frac{G_O + G_B}{2}. \tag{7}$$

Step 4: If $T^k=T^{k+1}$, iteration ends; Otherwise let T^k equal to the T^{k+1}, then go to step 2 to continue the iteration, until meet the conditions of $T^k=T^{k+1}$.

Step 5: Image segmentation. After dealing with the iterative threshold method, separate the grey value greater than the threshold value of image region, namely the facula image feature point area.

In the process of actual application processing, step 4 is difficult to meet. To avoid endless iteration computation, a limit value of iterations can be set. There are 50 times of limit value of iteration in the process of processing facula image. Fig. 5 (a) is part of color facula images collected by CCD, (b) is the gray image after filtering and (c) is intercepted facula area when T = G_{max} / 3, and (d) is the facula area intercepted by iterative threshold method.

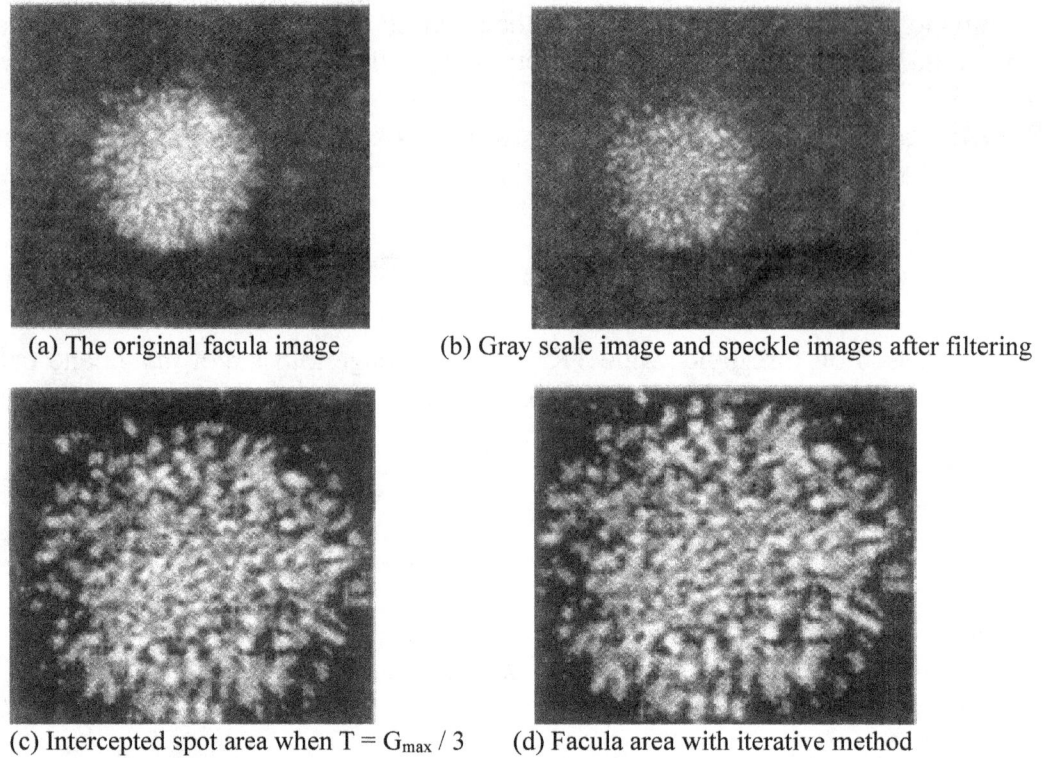

(a) The original facula image   (b) Gray scale image and speckle images after filtering

(c) Intercepted spot area when $T = G_{max} / 3$ (d) Facula area with iterative method

Fig. 5 The determination of laser facula feature point area

As can be seen from Figs. 5 (c) and (d), the proposed algorithm can well realize preliminary positioning of facula image feature points area, which is very important for the accurate positioning of facula center coordinates and will greatly reduce the number of pixels to improve the running speed of the program.

3) **The sub pixel localization of facula center location**: In laser triangulation measurement, image obtained by CCD camera is a typical symmetric highlight facula, and its light intensity is in line with the gaussian distribution. In this case, calculation accuracy based on gray centroid algorithm method is better than that of the polynomial fitting method. Gray centroid algorithm calculation formula is

$$X_C = x_0 + \sum_{i=1}^{n} x_i \cdot g(x_i, y_i) / \sum_{i=1}^{n} g(x_i, y_i). \tag{8}$$

$$Y_C = y_0 + \sum_{i=1}^{n} y_i \cdot g(x_i, y_i) / \sum_{i=1}^{n} g(x_i, y_i). \tag{9}$$

In Eq. 8 and Eq. 9: X_C and Y_C are facula center coordinates; x_i and y_i are coordinates of point (x_i, y_i); x_0 and y_0 are the starting point coordinates in facula area; $g(x_i, y_i)$ is the gray value of point (x_i, y_i) .

To improve the facula center coordinates' calculation precision and stability, before calculating facula center coordinates, we first use the bilinear interpolation method to make interpolation processing on intercepted facula image.

Set $0 < x < 1$, $0 < y < 1$, the grey value of the interpolation point $(i + x, j + y)$ can be obtained through the following formula

$$g(i+x,j)= g(i,j)+x[g(i+x,j)-g(i,j)] \tag{10}$$

$$g(i,j+y)= g(i,j)+y[g(i,j+y)-g(i,j)] \tag{11}$$

$$g(i+x,j+y)=x[g(i+1,j)-g(i,j)]+y[g(i,j+1)-g(i,j)]+xy[g(i+1,j+1)+g(i,j)-g(i+1,j)-g(i,j+1)]+g(i,j) \tag{12}$$

The value of x, y both are 0.5, namely interpolate value in the middle of every two pixel.

4.3. *Communication technology*

According to the research and development needs, research needs to solve the problems of equipment on-site communication and telecommunications. The on-site communicate includes information interaction among on-site communication equipments, and telecommunication is responsible for monitoring data teletransmission. The integrated study respectively chooses nRF24L01 chips and M6310 IOT private network communication module to carry out integration study [8, 9].

4.4. *Equipment integration*

Based on the research on two-dimensional PSD sensor, obliquity sensor, laser ranging, RF communications and wireless communications technology , supplemented by the corresponding peripheral circuit, has realized the 3D displacement, revolution of plane (roll) and vertical surface plane rotation (pitching) posture observe five parameters observation, the conceptual model shown in Fig. 6, technical parameters are shown in Table 1.

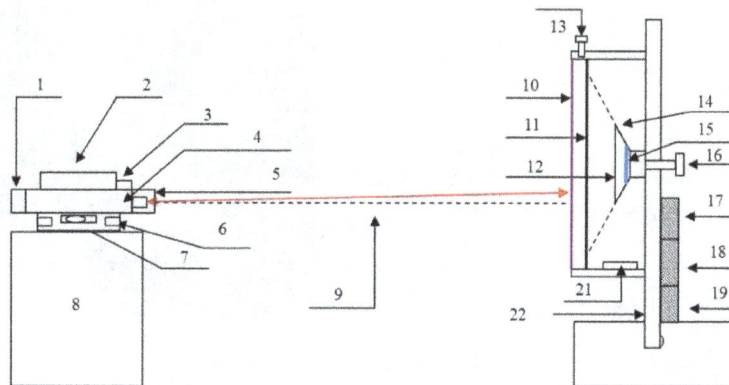

Fig. 6 Structure design and principle of multidimensional deformation monitor

Digital light detector; 6. Substrate; 7. Level tube; 8. Control point; 9. Laser beam; 10. Filter; 11. Imaging plate; 12. Wide-angle lens; 13. Adjustable screw; 14. Two-dimension CCD; 15. Lens corrective lenses; 16. Range adjusting screw; 17. Wireless communication module; 18. CCD driving circuit; 19. Pilot circuit; 20. monitoring point; 21. Dual-axis tilt sensor; 22. Bracket.

Table 1 The technical parameters of multidimensional deformation monitor

Accuracy	0.2mm 0.001°	Power input	DC12V（Battery）+12VSolar panels
Measuring range	70m, 200m, 500m	Average power consumption	100mW, 500mW, 2W
Range	200×200mm/20×20mm	Operating temperature	Industrial grade（-40~+85℃）
Transmission mode	GSM, GPRS, RF	Working humidity	＞90%

5. Application Demonstration

To verify the research and development of multidimensional space deformation monitoring in the engineering practice of environmental adaptability, observation precision and sensitivity, the reliability of the remote communication and the correctness of data analysis algorithm, the research projects selected a landslide in Chongqing Yunyang for field observation and application demonstration.

According to the *building deformation measurement specification* (JGJ8-2007), the corresponding requirements, considering the canteen deformation status and unfavorable deformation of uneven settlement and building incline, so in the leading edge and east and west sides of canteen and the east of trailing edge layout three sets of multidimensional deformation monitor which is shown in Fig. 7.

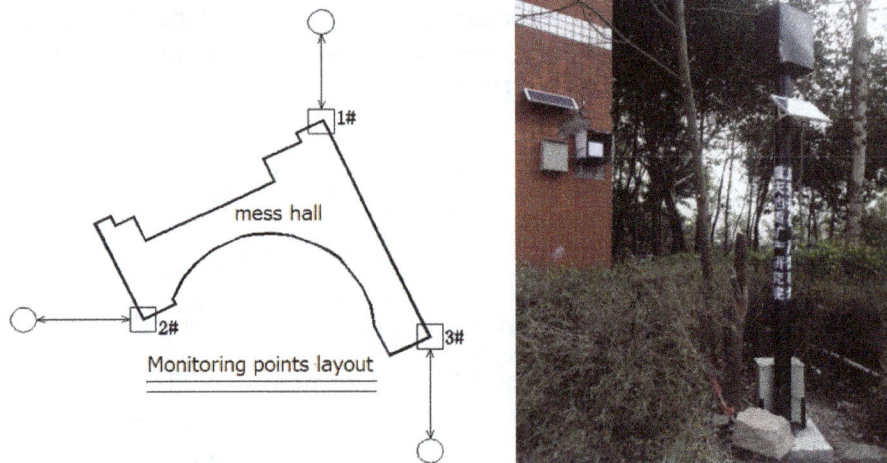

Fig. 7 Monitoring points layout

According to the equipment observed function and monitoring requirements, can focus on observation settlement displacement, point to the slope displacement, vertical surface rotation (that is, housing tilt) three parameters.

The application demonstration starts since April 30, 2015, end on October 31, 2015, lasted six months. The monitoring data shows that during the period of emergency drainage construction, 2 # point produced significant deformation (excerpts from monitoring data are shown in Table 2), the monitoring data in the table shows that for the settlement deformation and two-dimensional space reverse during the period, integrated representation shown in Fig. 8, namely the deformation is more complex multidimensional deformation, further analysis show deformation and building infrastructure form, geological conditions and drainage construction is closely related, the multidimensional deformation situation feedback to the construction side in a timely manner during monitoring, the construction side immediately slow drainage construction speed, to avoid the further deformation.

Table 2 The displacement statistical table of 2 # monitoring point (excerpts)

Time	X	Y	Z	ΔX	ΔY	ΔZ	ΔΔX	ΔΔY	ΔΔZ
2015/4/30 18:01	100.05	84.53	3433.21	0.00	0.00	0.00	0.00	0.00	0.00
2015/5/1 19:17	100.13	84.64	3433.23	0.08	0.11	0.02	0.08	0.11	0.02
2015/5/2 19:48	100.22	85.48	3433.23	0.09	0.84	0.00	0.17	0.95	0.02
2015/5/3 8:47	101.94	86.46	3433.25	1.72	0.98	0.02	1.89	1.93	0.04
2015/5/8 16:36	102.31	86.50	3433.24	0.37	0.04	-0.01	2.26	1.97	0.03
2015/5/10 15:11	102.33	86.50	3433.23	0.02	0.00	-0.01	2.28	1.97	0.02
2015/5/12 17:01	102.45	86.61	3433.25	0.12	0.11	0.02	2.40	2.08	0.04
2015/5/13 23:21	102.56	86.77	3433.25	0.11	0.16	0.00	2.51	2.24	0.04
2015/5/22 19:21	102.78	86.78	3433.23	0.22	0.01	-0.02	2.73	2.25	0.02
2015/5/2 16:55	103.01	87.00	3433.23	0.23	0.22	0.00	2.96	2.47	0.02
2015/5/10 17:44	104.26	93.78	3433.23	1.25	6.78	0.00	4.21	9.25	0.02

Table 2 (continued)   The rotation statistical table of 2 # monitoring point(excerpts)

Time	α	β	Δα	Δβ	ΔΔα	ΔΔβ
2015/4/30 18:01	89.08	91.15	0.00	0.00	0.00	0.00
2015/5/1 19:17	89.09	91.15	0.02	0.00	0.02	0.00
2015/5/2 19:48	89.08	91.15	-0.02	0.00	0.00	0.00
2015/5/3 8:47	89.07	91.15	-0.01	0.00	-0.01	0.00
2015/5/8 16:36	89.12	91.15	0.05	0.00	0.05	0.00
2015/5/10 15:11	89.11	91.15	-0.01	0.00	0.03	0.00
2015/5/12 17:01	89.15	91.15	0.04	0.00	0.08	0.00
2015/5/13 23:21	89.18	91.13	0.03	-0.02	0.11	-0.02
2015/5/22 19:21	89.23	91.13	0.05	0.00	0.16	-0.02
2015/5/2 16:55	89.25	90.98	0.02	-0.15	0.17	-0.17
2015/5/10 17:44	89.36	89.03	0.11	-1.95	0.28	-2.12

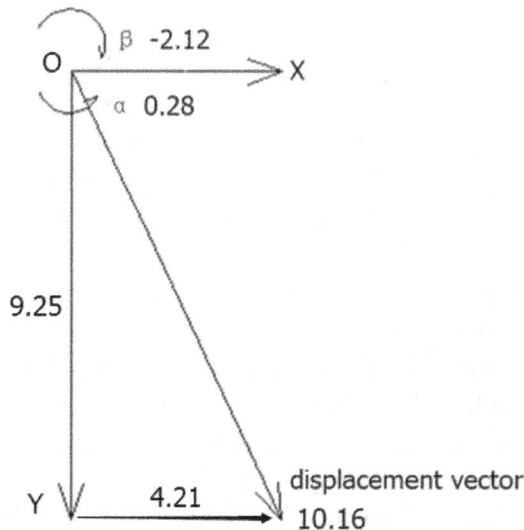

Fig. 8 Multidimensional deformation phenogram

Due to large students traffic in monitoring area, teaching and accommodation areas are relatively dense, to timely grasp the deformation situation of the region, using multidimensional deformation monitor to conduct canteen real-time monitoring, to capture significant deformation timely during the period, to achieve the original intention of monitoring design, at the same time, indicates that the equipment is different from the conventional displacement or oblique observation equipment, engineering practice can observe the complex synthesis deformation, monitoring achievement is reliable.

6. Conclusion

Use sensing technology research and integrated means, supplemented by the corresponding of outside power supply, communications, lightning protection circuit, which can realize multidimensional space deformation observation function, through application demonstration can get the following conclusion:

1. The multidimensional deformation monitor has established function of research, through the observation of component can analysis and get the complex multidimensional deformation information of object;
2. The data continuity is good during the application demonstration, which shows that the equipment has strong adaptability in the field environment, and the communication function is more reliable;
3. The application results show that displacement observation accuracy is better than 1 mm, space rotation Angle precision is better than 0.01 °, meet the needs of engineering practice;
4. Devices able to capture the actual micro deformation during the monitoring, which shows that the equipment observation sensitivity is good, can provide safeguard for engineering practice.

Acknowledgment

This research was funded by the Science and Technology Planning Project of Chongqing Land Bureau: Research and demonstration of multidimensional deformation monitor for geological disaster (CQGT-KJ-2014044).

References

1. ZENG Chao, LI Feng, XU Xiangdong. Characteristics of optoelectronic position sensitive detector and its application[J]. OPTICAL INSTRUMENTS, 2002, 24(Z1): 30-33.
2. LIU Jiaoyue, YANG Juqing, DONG Dengfeng. Application and research of laser tracker's optoelectronic aiming technology based on PSD [J]. Instrument Technique and Sensor, 2015 (7): 98-100.
3. CHEN Tingzheng, LV Haibao. CCD subdivision method and application [J]. ACTA OPTICA SINICA, 2002, 22(11): 1396-1399.
4. SU Li. The numerical method of quantitative applies in CCD sub-pixel subdivision [J]. SCIENCE & TECHNOLOGY INFORMATION, 2008 (28): 98-101.
5. ZHOU Hongfeng, GONG Ailing. The research of CCD facula center sub-pixel positioning in image [J]. Measurement Technique, 2007 (11): 21-23.
6. WANG Chao, XU Xiaohui, GUO Tao, et al. Design of wireless dual-axis tilt sensor [J]. Research and Exploration in Laboratory, 2015, 34(08): 56-60.

7. CHEN Fengjun, WANG Jiawei, XU Tao, et al. Design of high precision inclinometer used for adjusting verticality based on MEMS sensor [J]. Building Construction, 2016, 38(4): 504-505.
8. SHANG Xiaofeng, XU Pengfei. Wireless communication technology based on radio frequency [J]. Modern Electronics Technique, 2010, 33(19): 29-31.
9. TANG Wei. Research of IOT system based on communications network [D]. Beijing: Beijing University of Posts and Telecommunications, 2012.

Chapter 15
Other Important Subjects for Material Science and Engineering

Surface Modification of Aramid Fiber-III by Ultrasonic Vibration

Feng-De Wang[1,2], Yong-Lei Lv[1], Xu Wei[2], Guo Ling[2], Zhe-Wen Han[1,*]

1. Key Laboratory of Specially Functional Polymeric Materials and Related Technology, Ministry of Education, School of Materials Science and Engineering, East China University of Science and Technology, Shanghai, China.
2. China Bluestar Chenguang Chemical Research Institute Co., Ltd., High-tech Organic fibers Key Laboratory of Sichuan Province, Chengdu, China
*Email: zhewhan@ecust.edu.cn

Aramid fiber III has been treated by ultrasound to enhance the adhesion between aramid fiber III and epoxy. The effects of different ultrasonic systems and conditions were studied. The XPS and SEM indicated the obvious physical and chemical corrosion in the surface of fiber after ultrasonic treatment. The yarn pull-out method was used to evaluate the effect of surface modification. The evaluation results show the tensile strength and NOL ILSS of treated aramid fiber III/epoxy composite increased by 12.5% and 15.2%, respectively. In summary, the effect of ultrasonic treatment is obvious and has potential in industrial application.

Keywords: Aramid fiber; ultrasound; surface modification.

1. Introduction

Aramid fiber III is one kind of novel organic fiber, which is prepared from terephthaloyl chloride, p-phenylenediamine and the third monomer via low-temperature solution poly condensation, wet spinning and special post treatment. As a kind of high-performance fiber, aramid fiber III which has high strength and high modulus is widely used as the reinforcement of composites. Aramid fiber III possesses excellent mechanical properties, composite performance and resistance to combustion performance due to its special chemical structure and process. Hence aramid fiber III can be used in wide fields and can meet high requirements. However, its application is limited due to its smooth surface, chemical inertness and low interface strength, which result in poor interfacial adhesion between fibers and epoxy resin matrix [1-4]. It is well known that the mechanical properties of fiber reinforced polymer composites depend not only on the mechanical properties of fibers and matrix, but also on the interfacial adhesion property of fibers and matrix which transfer the load from matrix to reinforced fiber. To improve the performance of fiber/resin composite and related products, it is essential to modify the surface of aramid fiber III can improve surface property, enhance interfacial bond strength. Many surface modification methods have been studied in previous literatures, including plasma treatment, chemical treatment, coupling agent treatment, ray irradiation treatment, and so on.

Although these methods can enhance the interfacial adhesion of composites, they still have negative effects on the tensile strength of aramid fibers. For example, it is difficult to control the chemical treatment on the surface of fiber. With increase the power is used for the chemical treatment, the aramid fiber will suffer serious damage. Therefore, new strategies are widely explored to avoid this problem.

In the past decades the ultrasound is widely applied in the industrial process. As a mean to improve the interfacial adhesion between aramid fiber and epoxy matrix, the ultrasonic treatment act on each component in composites. [3-5]

---

*Corresponding author

In this paper, the effects of ultrasonic treatment on the resin system are studied in details. The physicochemical changes of aramid fibers under the ultrasonic treatment also been discussed in this paper. The surface characteristics of aramid fibers before and after the ultrasonic treatment are analyzed. The aim of this study is to establish an ultrasonic treatment system and determine the optimum ultrasonic parameters to improve the interfacial performance and the mechanical properties of aramid fiber/epoxy composites.

2. Experiment

2.1. *Materials*

The uncoated aramid fiber III, used in this experiment is produced according to previous work. [2] The Bisphenol A epoxy (E51), acid anhydride curing agent and accelerating agent purchased in the commercial market and used as received.

2.2. *Ultrasonic treatment*

The transducer of ultrasound with frequency of max 22 kHz (the max power is 1200W) was applied to the fiber. The frequency and power are adjusted to study the effect of ultrasonic parameter on the fiber.

2.3. *Analysis and test*

The XPS analysis of aramid fiber III samples was carried out by XSAM800 multifunctional XPS equipment (Kratos Corporation, the United Kingdom, X-ray source: Al Ka (hv =1486.6eV; scan range: 0-1100eV, 150mS/0.65eV) to measure element components and the differences of electron binding energy of aramid fiber III. The scanning electric microscope (SEM) morphologies of the aramid fiber III were observed by JSM-5900 (JEOL Corporation, Japan, scan voltage: 0-20 Kv). The NOL ring interlaminar shear strength (NOL ILSS) of Aramid fiber III was determined using the Yarn pull-out method. The tensile strength of the NOL ring was tested on an Instron testing machine at a rate of 5 mm/min. More than 30 specimens were tested to ensure precise testing.

3. Results and Discussion

3.1. *The effect of ultrasonic treatment on the Aramid fiber III*

3.1.1. *The effect of ultrasonic treatment system*

The effect of ultrasonic treatment system on the Aramid fiber III was studied. As shown in Table 1, when the ultrasonic treatment system is pure water, no improvement can be found in pull out strength. When the ultrasonic treatment system is Dimethylacetamide (DMAc) or DMAc + Chloroacetyl chloride (CAC), the improvement is obviously.

Table 1 The effect of ultrasonic treatment system

Ultrasonic treatment system	Fracture strength (cN/dtex)	Fracture modulus (cN/dtex)	Elongation at break (%)	Pull out strength (N/mm·tex)
No	28.98	930.11	3.22	0.469
Water	28.95	927.31	3.20	0.459
DMAC	28.82	900.72	3.25	0.570
Phosphoric acid (20%)	28.90	912.77	3.21	0.488
CAC+DMAc (20%)	28.21	899.72	3.18	0.545

3.1.2. *The effect of ultrasonic treatment time*

The effect of ultrasonic treatment time on the Aramid fiber III was studied. The ultrasonic treatment system is DMAc. As shown in Table 2, the best ultrasonic treatment time is 5 min.

Table 2 The effect of ultrasonic treatment time

Ultrasonic treatment time (min)	Fracture strength (cN/dtex)	Fracture modulus (cN/dtex)	Elongation at break (%)	Pull out strength (N/mm·tex)
0	30.81	934.6	3.47	0.604
5	30.36	876.3	3.76	0.671
10	30.12	880.5	3.78	0.659
20	29.45	865.3	3.74	0.585

3.1.3. *The effect of ultrasonic treatment power*

The effect of ultrasonic treatment power on the Aramid fiber III was studied. The ultrasonic treatment system is DMAc. As shown in Table 3, the best ultrasonic treatment power is 600 W.

Table 3 The effect of ultrasonic treatment power

Ultrasonic treatment power (W)	Fracture strength (cN/dtex)	Fracture modulus (cN/dtex)	Elongation at break (%)	Pull out strength (N/mm·tex)
0	30.81	934.6	3.47	0.604
300 W	30.48	921.6	3.41	0.625
600 W	30.36	876.3	3.76	0.671
900 W	28.62	860.1	3.54	0.580
1200 W	26.25	870.3	3.30	0.512

3.2. *XPS analysis*

Fig. 1 shows the entire XPS scanning spectra of native HMPBO and treated HMPBO. According to Fig. 1, various elements contents can be calculated by sensitivity factor, and corresponding results are listed in Table 4.

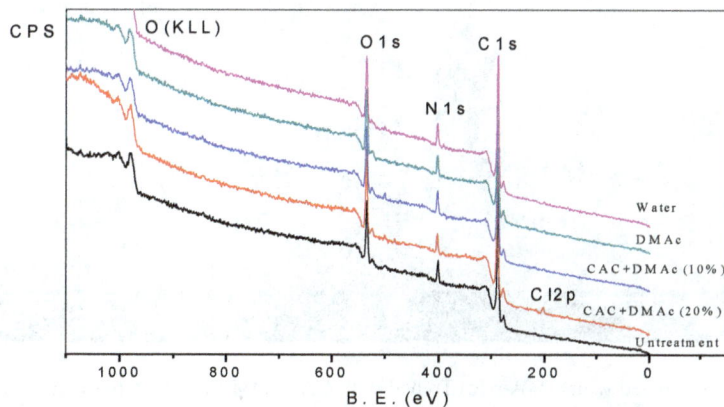

Fig. 1 XPS spectra of native aramid fiber-III and treated aramid fiber-III

Table 4 The XPS data of fiber surface before and after ultrasonic treatment

	Element	CPS	S.F.	CO.IN (%)
Original fiber	O	2500	0.66	16.12
	N	1000	0.42	10.13
	C	4300	0.25	73.19
Treaed with CAC+DMAc (20%)	O	2850	0.66	19.42
	N	800	0.42	8.57
	C	3900	0.25	70.16
	Cl	300	0.73	1.85
Treaed with CAC+DMAc (10%)	O	2500	0.66	17.81
	N	900	0.42	10.08
	C	3800	0.25	71.47
	Cl	100	0.73	0.64
Treaed with DMAc	O	2500	0.66	17.67
	N	1000	0.42	11.11
	C	3750	0.25	69.97
Treaed with water	O	2300	0.66	18.11
	N	850	0.42	10.52
	C	3400	0.25	70.67

3.3. *SEM analysis*

The SEM morphologies of the surface of the original fiber and the treated fiber are shown in Fig. 2.

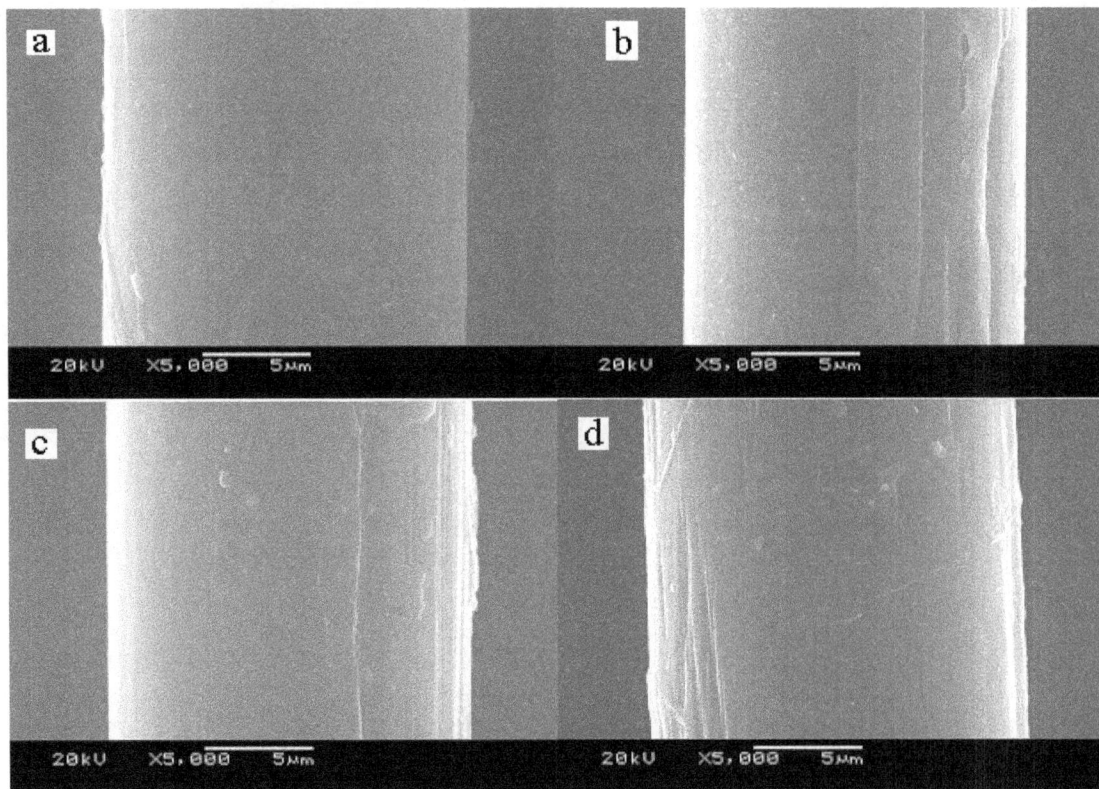

(a): original fiber; (b) Treaed with DMAc; (c) Treaed with CAC+DMAc (10%); (d) Treaed with CAC+DMAc (20%))

Fig. 2 The SEM of fiber surface before and after ultrasonic treatment

The surface of native fibers is smooth Fig. 2(a). After ultrasonic treatment, the surface of aramid fiber becomes rough obviously Fig. 2(b-d), which would be good for enhancing the mechanical interlocking of fiber surface and epoxy resin, and then to improve the interface bonding strength. The surface morphology also shows that under the ultrasonic treatment in DMAc+CAC, the aramid fiber surface appeared obvious physics and chemical corrosion, which is more obvious with the increasing of CAC concentration in DMAc.

3.4. *Tensile properties*

Table 5 The Mechanical properties of fiber during yarn pull-out test

	Yarn tensile strength(cN/dtex)	Wetted Yarn tensile strength (MPa)	NOL ILSS (MPa)
Original fiber	28.65	4800	46
Treaed with CAC+DMAc (10%)	27.96	5100	50
Treaed with CAC+DMAc (20%)	27.51	5400	53

The tensile strength and single fiber pull-out strength of aramid fiber III before and after ultrasonic treatment are shown in Table 5. The results show that the aramid fiber III has a tensile strength of 28.65cN/dtex. However, the treated aramid fiber III has an average value of 27.73cN/dtex. The tensile strength of fiber is maintained well. Meanwhile, the NOL ILSS of treated aramid fiber III/epoxy composite increased by 12.5% and 15.2%, respectively.

4. Conclusion

Aramid fiber III has been treated by ultrasound during the winding process to enhance the adhesion between aramid fiber III and epoxy. The effect of different ultrasonic systems and conditions was studied. The XPS and SEM indicate the obvious physics and chemical corrosion in the surface of fiber after ultrasonic treatment, which is helpful for its use in the aramid fiber/epoxy composites. The SEM results also illustrate that the binding force between fiber and resin is enhanced. The yarn pull-out method was used to evaluate the effect of surface modification. The evaluation results show the tensile strength and NOL ILSS of treated aramid fiber III/epoxy composite increased by 12.5% and 15.2%, respectively. In summary, the effect of ultrasonic treatment is obvious and has potential industrial application prospects.

Acknowledgments

The authors express their gratitude to the Shanghai Natural Science Foundation for funding this work (12ZR1407900). This project is also funded by the Basic Innovation Research Program of Science.

References

1. L. Liu, Y. D. Huang, Z. Q. Zhang, X. B. Yang, Effect of Ultrasound on Wettability Between Aramid Fibers and Epoxy Resin, J. Appl. Poly. Sci. 99 (2006) 3172–3177.

2. L. Liu, Y. D. Huang, Z. Q. Zhang, B. Jiang, J. Nie, Ultrasonic Modification of Aramid Fiber–Epoxy Interface, J. Appl. Poly. Sci. 81 (2011) 2764–2768.

3. H. J. Dong, J. Wu, G. Y. Wang, Z. G. Chen, G. Y. Zhang, The Ultrasound-Based Interfacial Treatment of Aramid Fiber/Epoxy Composites, J. Appl. Poly. Sci. 113 (2009) 1816–1821.

4. L. Liu, Y.D. Huang, Z.Q. Zhang, Z.X. Jiang, L.N. Wu, Ultrasonic treatment of aramid fiber surface and its effect on the interface of aramid/epoxy composites, Appl. Surf. Sci. 254 (2008) 2594–2599.

5. J. W. Gu, J. Dang, W. C. Geng, Q. Y. Zhang, Surface Modification of HMPBO Fibers by Silane Coupling Agent of KH-560 Treatment Assisted by Ultrasonic Vibration, Fib. Poly. 13 (2012) 979-984.

Thermal Degradation Kinetics of Purified Natural Rubber Treated with Adsorption Precipitation Method

Ming-Zhe Lv[1], Si-Dong Li[2], Lei Fang[1], Zi-Ming Yang[1], Fan Zhang[1], Pu-Wang Li[1,*]

[1]Agricultural Product Processing Research Institute, Chinese Academy of Tropical Agricultural Sciences, China
[2]College of Science, Guangdong Ocean University, Zhanjiang 524088, China
*Email: puwangli@163.com

Relative to the conventional enzymatic treatment of deproteinization, purified natural rubber treated with adsorption precipitation method has better thermal stability because of the removal the water-soluble proteins instead of binding proteins in NR. Thermal degradation mechanisms of purified natural rubber (PNR) were studied by thermogravimetric analysis (TGA) and differential thermal analysis (DSC). The results show that, the removal of water-soluble proteins and the other non-rubber components from the NR leads to a remarkable change in the degradation stabilization. The total activation energy of PNR calculated by Coats-Redfern method was 167.99 kJ/mol, which is 35.05 KJ/mol higher than that of NR. For the whole thermos-oxidative degradation stage, the thermal decomposition mechanism of PNR is similar to that of NR. The thermal decomposition mechanism of PNR corresponds to a three-dimensional diffusion (Jander equation) at the first stage, and a phase boundary controlled reaction (one-dimensional movement) at the second stage. Kinetic analysis showed that activation energy (Ea), activation entropy (ΔH) and activation Gibbs energy (ΔG) values are all positive, indicating that the first thermos-oxidative degradation process of PNR is non-spontaneous.

Keywords: Natural rubber; purified natural rubber (PNR); water-soluble proteins; aluminium hydroxide; thermo-oxidative degradation.

1. Introduction

Natural rubber (NR) containing more than 92% cis-1, 4-polyisoprene is an important biopolymer and strategic material for national economy and national defense construction. Compared to synthetic rubber based on petroleum, NR has excellent comprehensive properties. It has widely used in industry, agriculture, national defense, transportation, machinery manufacturing, medical and health fields and daily life [1-3].

The fresh latex collected from Hevea brasiliensis, the commercial source of natural rubber, is composed of about 30-35 % of rubber fraction, 5% of non-rubber components such as protein, lipid, sugar, heavy metal ion, which were dispersed in water [4]. Although significant quantities of non-rubber substances are removed during concentrating the latex to about 60% rubber content by creaming or centrifugation, a good deal of that are kept in the NRL [5]. Non-rubber components present in NRL are presumed to be distributed in the serum fraction as well as surrounding the rubber particle surface. Non-rubber substances are the most important factor for NRL to maintain stable. The NRL starts to deteriorate or coagulase as a result of non-rubber substances changing. Due to these non-rubber components, the internal molecular structure of NR is more complex than synthetic rubber and make the NR has superior comprehensive performance. NR could not be replaced owing to the importance in rubber industries especially in high-end tires and aircraft tire manufacturing.

In fact, non-rubber components (mainly proteins and lipids) in NR play important roles for controlling the properties of NR. A hygroscopic nature because of the non-rubber components impair the electrical properties of the vulcanizate has long been known [6]. The ordinary natural rubber presents shorter curing time than PNR because there are some phospholipids and proteins which are natural accelerators for curing reaction. The presence of non-rubber components seems to play a major role on cross-linking density [7].

*Corresponding author

1023

Some people are sensitive to latex products and cause allergic symptoms such as difficult breathing. It can reduce the protein allergic reaction on humans by removal water-soluble proteins of natural rubber [8]. NR was purified to remove protein by many approaches including the usage of several times centrifugation with high speed, the dry gel leaching method [9, 10], adsorption and replacement method [11], proteolytic enzyme treatment [12, 13] and radiation process [14, 15]. And the technique of enzymatic enzyme treatment is the most widely used in producing the commercial DPNR. However, enzymatic treatment is indiscriminate hydrolysis one kind of protein and cannot differentiate between water-soluble proteins and rubber binding proteins. Furthermore, the appropriate amount of binding protein can promote on accelerating vulcanization, enhance stress at definite elongation, and improve the aging resistance performance of rubber.

To overcome the shortage of NR purification process above-mentioned, aluminium hydroxide was used successfully to remove water soluble proteins for reducing allergic and retain binding proteins in our work.

However, there are few reports on the thermal degradation of purified natural rubber. Thermal analyses play a vital role in studying the structure and properties of any material [16]. Thermogravimetric analysis has been widely used to investigate the decomposition characteristics of polymeric materials [17, 18]. Thermal decomposition properties of polymeric materials are important for their practical applications. The most specific interest in thermal decomposition of polymeric materials is the decomposition kinetics as well as the activation energy required during the thermal decomposition.

In the present work, thermos-oxidative degradation of PNR treated by adsorption precipitation method was examined with TGA and differential scanning calorimeter (DSC) analysis to understand the thermal degradation kinetics and mechanism of PNR.

2. Experimental

2.1. Materials

Natural Rubber Latex (NRL) with a dry rubber content of 60% was obtained from Guangken Rubber Company Maoming Branch (Guangdong, China). Sodium Dodecylsulphate (SDS) and Aluminium Hydroxide (AlOH3) were of analytical grade and supplied by Tianjin Kermel Chemical Reagent Co., Ltd (Tianjin, China).

Aluminium Hydroxide has been used as deproteinization agent to treat NRL [19]. Purified natural rubber (PNR) was obtained by treatment of high ammonia NR latex with Aluminum hydroxide for 2 h, followed by double centrifugation. The NRL was diluted with water to 30% at first and stabilized with 0.5% (w/v) of SDS for 0.5 h at room temperature. And then 2% (w/w) Aluminum hydroxide was put into NRL and treated for 2 h under stirring. After stewing for 24h, the NRL was centrifuged at 12000 rpm and the top creamy fraction of rubber particles was collected to obtain PNR latex. The PNRL and NRL were cast on glass plates and dried at room temperature.

2.2. Measurements

A STA449C thermogravimetric analyzer (NETZSCH, Germany) was used for thermal and thermos-oxidative decomposition measurement. To record the TGA, DTG and DSC curves, as much as 10 mg of samples were heated from 30 to 700°C with different heating rates (B) of 10, 15, 20, 25 and 30°C /min, using an empty aluminum oxide pan as reference. Air was used as carrier gas and the flow rate was 50 ml/min.

The molecular weight of raw natural rubber dissolved in tetrahydrofuran was measured by PL-GPC 220 High Temperature Gel Permeation Chromatography (Agilent Technologies, USA) at

50°C. FT-IR spectra were recorded by using a Perkin–Elmer GX-1 infrared spectrometer (Perkin Elmer, USA). The chemical structure of PNR were confirmed by using Attenuated total reflection (ATR) equipped with KBR ATR accessory in the wave number range of 4 000-650 cm $^{-1}$ with a resolution of 4 cm^{-1}.

2.3. *Kinetic methods for thermogravimetric analysis*

The activation energies of degradation of the PNR and NR were determined by applying Coats-Redfern method as follows [19]. Coats-Redfern's method is a model-fitting method based on a single heating rate [24, 25] and is represented by Eq. (1):

$$\ln \frac{g(X)}{T^2} = \ln[\frac{AR}{BE_a}(1-\frac{2RT}{E_a})] - \frac{E_a}{RT}$$ (1)

The g(X) has different expressions for different types of mechanisms and kinetic models, and these are tabulated in Table 1 [19]. E_a is calculated from the slope (-E_a/R) of the straight line of plot $\ln[g(X)/T^2]$ against 1/T.

Table 1 Algebraic Expressions for g(X) for the most frequently used mechanisms of solid-state processes [23]

Code	g(X)	Mechanisms
		Sigmoidal curves
A2	$[-\ln(1-X)]^2$	Nucleation and growth (Avrami equation (1))
A3	$[-\ln(1-X)]^3$	Nucleation and growth (Avrami equation (2))
A4	$[-\ln(1-X)]^4$	Nucleation and growth (Avrami equation (3))
		Deceleration curves
R1	X	Phase boundary controlled reaction(one-dimensional movement)
R2	$2[1-\ln(1-X)^{1/2}]$	Phase boundary controlled reaction(contracting area)
R3	$3[1-\ln(1-X)^{1/3}]$	Phase boundary controlled reaction(contracting area)
D1	X^2	One-dimensional diffusion
D2	$(1-X)\ln(1-X)+ X$	Two-dimensional diffusion (Valensi equation)
D3	$[1-\ln(1-X)^{1/3}]^2$	Three-dimensional diffusion (Jander equation)
D4	$[1-(2/3)X]-(1-X)^{2/3}$	Three-dimensional diffusion(GinstlingeeBrounshtein equation)
F1	$-\ln(1-X)$	Random nucleation with one nucleus on the individual particle
F2	$1/(1-X)$	Random nucleation with two nuclei on the individual particle
F3	$1/(1- X)^2$	Random nucleation with three nuclei on the individual particle

Another thermal decomposition parameter, activation entropy, was determined from the equation [20]:

$$A = \frac{kT_p}{h}\exp\left(\frac{\Delta S}{R}\right)$$ (2)

Where A is pre-exponential factor and was obtained from the intercept (ln[(1-2RT/E)]AR/βE]) of the straight line of plot ln[g(α)/T2] against 1/T, k is the Boltzmann constant, h is the Planck constant and Tp is the temperature of the DTG peak maximum value determined from Fig. 3. A third thermal degradation parameter, activation enthalpy (ΔH), was obtained from the equation Eα = Δ H - RTp. Finally, the activation Gibbs energy was calculated from the equation:ΔG = ΔH - TmaxΔS [21-23] .

3. Results and Discussion

3.1. *Characterization of PNR*

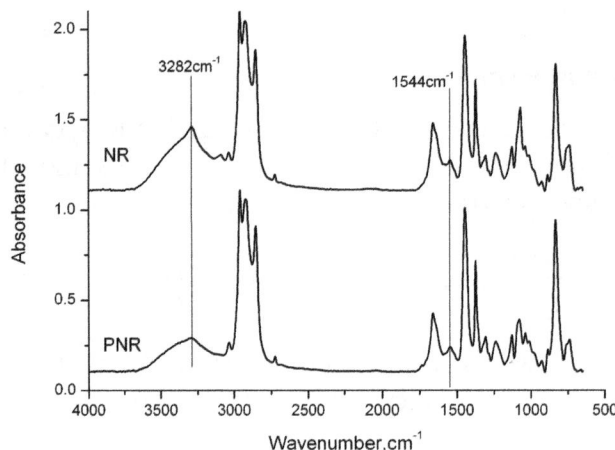

Fig. 1 FT-IR of natural rubber (NR) and purified natural rubber (PNR)

Fig. 1 shows that the appearance of PNR of characteristic amide bands at 3282 (-NH2 stretching) and 1544 cm^{-1} (amide II vibration), which can be attributed to the signals of proteins, smaller than that of NR. It confirms that a part of proteins in NR were eliminated and the total protein content decreased by introducing aluminum hydroxide.

It was been shown that from Table 2 the number-average molecular weight (Mn) after the adsorption of aluminum hydroxide increases and the polydispersity index (PDI) decreases. It indicates that removal of small molecules such as water-soluble proteins lead to the molecular weight distribution of PNR narrow down.

Table 2 The molecular weight of natural rubber was measured by GPC

Sample	M_n	M_w	PDI
NR	273300	1717000	6.283
PNR	282200	1686000	5.976

3.2. *Degradation process*

The thermos-oxidative ageing resistance of PNR and NR can be assessed from the investigation of thermos-oxidative decomposition. Fig. 2 shows the TG and DTG curves of the PNR and NR, respectively. The thermal degradation temperatures of PNR and NR are listed in Table 3, the initial temperature (T_0) and the final temperature (T_f) are obtained with a bitangent method. T_p is the temperature at maximum weight loss rate, which can be obtained from the peak of the DTG curves. The thermal degradation processes of PNR and NR are very similar, and there are two distinct and well-separated turns (220-450°C and 450-600°C) in the TG curves and two corresponding weight-loss peaks in the DTG curves. Therefore, the thermal degradation of PNR and NR can be roughly regarded as a two-step-degradation. The initial temperature (T_0) of PNR is 326.6°C, which is 9.8°C higher than that of NR. Compared with NR, the thermal degradation curves of PNR show a shift toward high temperatures, meaning that the thermal stability of PNR is better than that of NR.

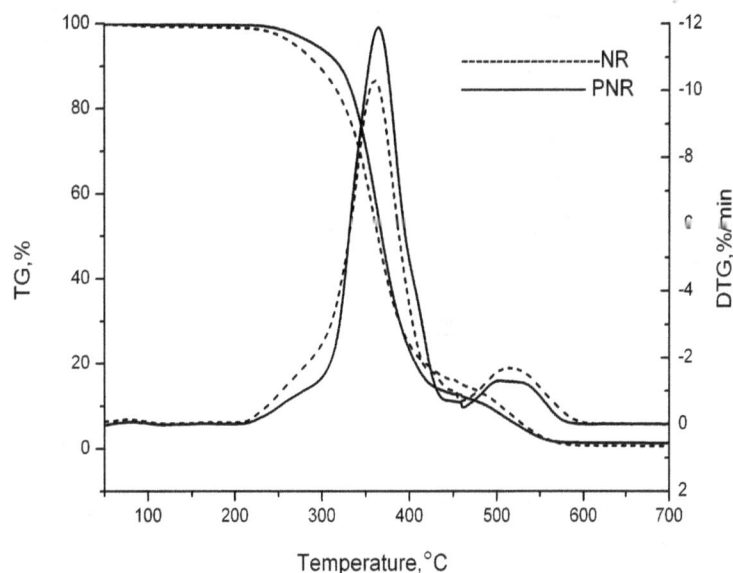

Fig. 2 Thermogravimetric analysis (TG) curves of natural rubber (NR) and purified natural rubber (PNR)

Table 3 Thermal degradation temperatures of natural rubber (NR) and purified NR (PNR)

Samples	The first stage (220-450°C)			The second stage (450-600°C)		
	T_0, (°C)	T_p (°C)	T_f(°C)	T_0 (°C)	T_p(°C)	T_f(°C)
NR	316.8	360.0	391.9	488.1	515.5	560.3
PNR	326.6	364.5	396.7	483.9	501.6	557.0

3.3. Degradation kinetic and mechanism analysis

The assignment of the mechanism of thermal decomposition is based on the assumption that the form of g (X) depends on the reaction mechanism. In this investigation, 13 forms of g(X) for PNR and NR were used (Table 1), in order to enunciate the mechanism of thermal decomposition in each stage. According to Ea using Coats-Redfem method, correlation coefficients for all of these forms were calculated. The results are summarized in Table 4.

The thermal decomposition mechanism of PNR was similar to that of NR at the whole thermos-oxidative degradation stage. It can be seen that when it has the best value of correlation coefficient Ea of PNR is 348.96 kJ/mol, which is much more than that obtained by Friedman and Flynn–Wall–Ozawa methods (It was about160 kJ/mol). When the Ea was close to 160 kJ/mol, the best value of correlation coefficient for the first stage for both materials decomposition was obtained using the function: g(X)= [1-ln(1-X)1/3]^2, which corresponds to a mechanism involving three-dimensional diffusion (Jander equation). However, for the second stage, the best value of correlation coefficient of NR decomposition was obtained using the function: g (X) = X, which corresponds to a mechanism involving phase boundary controlled reaction (one-dimensional movement).

1027

Table 4 Correlation coefficients calculated using nine forms of g (X) for natural rubber (NR) and purified natural rubber (PNR)

| Functions g(X) | The first stage (220-450°C) | | | | The second stage (450-600°C) | | | |
| | NR | | PNR | | NR | | PNR | |
	E_a (kJ/mol)	R	E_a (kJ/mol)	R	E_a (kJ/mol)	R	E_a (kJ/mol)	R
$[-\ln(1-X)]^2$	135.12	0.9949	169.52	0.9946	80.302	0.9631	84.01	0.9589
$[-\ln(1-X)]^3$	207.63	0.9951	259.24	0.9948	127.05	0.9663	132.63	0.9623
$[-\ln(1-X)]^4$	280.15	0.9953	348.96	0.9949	126.20	0.9420	181.24	0.9640
X	50.957	0.9837	66.963	0.9865	6.35	0.9839	7.99	0.9904
$2[1-\ln(1-X)^{1/2}]$	0.066	0.0173	0.8459	0.1649	15.62	0.8827	18.37	0.8916
$3[1-\ln(1-X)^{1/3}]$	2.758	0.7155	2.05	0.4853	11.08	0.8347	13.76	0.8574
X^2	111.83	0.9870	143.84	0.9886	0.50	0.2326	2.76	0.8096
$(1-X)\ln(1-X)+ X$	118.53	0.9903	151.09	0.9912	10.48	0.9647	6.10	0.8999
$[1-\ln(1-X)^{1/3}]^2$	126.59	0.9931	160.00	0.9934	37.93	0.9782	37.33	0.9793
$[1-(2/3) X]-(1-X)^{2/3}$	121.19	0.9914	154.03	0.9920	18.29	0.9806	13.89	0.9642
$-\ln(1-X)$	62.599	0.9939	79.80	0.9938	33.55	0.9496	35.40	0.9447
$1/(1-X)$	16.839	0.8059	20.125	0.7694	145.82	0.9185	178.27	0.9220
$1/(1-X)^2$	43.595	0.8657	50.168	0.8287	304.85	0.9243	369.75	0.9266

As previously mentioned, the removal of water-soluble proteins led to a remarkable change in the degradation stabilization. PNR degrades with much more energy than that of NR at the two degradation steps. The total activation energy (Ea) of PNR is 167.99 kJ/mol, which is 35.05 kJ/mol higher than that of NR. It indicates that the removal of small molecule substances, such as water-soluble proteins, increases the thermal stability of the NR. Kinetic analysis from Table 5 showed that activation energy (Ea), activation entropy (ΔH) and activation Gibbs energy (ΔG) values are all positive at the first stage, indicating that the thermos-oxidative degradation process of PNR is non-spontaneous. At the second stage the result of formula (1-2RT/Ea) is negative, activation entropy (ΔS) cannot Calculated by this way.

Table 5 Kinetic parameters for the first thermos-oxidative degradation stages of natural rubber (NR) and purified natural rubber (PNR)

Samples	E_a (kJ/mol)	A (s^{-1})	Δ H (kJ/mol)	Δ S(J/mol·K)	Δ G (kJ/mol)
The first thermos-oxidative degradation stage					
NR	126.59	2.60×10^8	131.85	36.35	108.84
PNR	160.00	1.23×10^{11}	165.30	87.63	109.43

3.4. DSC analysis

The DSC curves of two samples recorded in air from 50°C to 700°C are shown in Figure 3. It can be seen that the two thermal oxidative degradation stages of NR are exothermic, peak temperatures are 340.1°C and 491.1°C, respectively. It is clear for PNR that peak temperatures of the two thermal oxidative degradation stages are 344.7°C and 512.5°C, which are 4.6°C and 21.4°C higher than that of NR, respectively. It indicates that the removal of small molecule substances, such as water-soluble proteins, not only leads to increase of oxidative instability, but increase the thermal stability of the NR. These results are in good agreement with the TGA results.

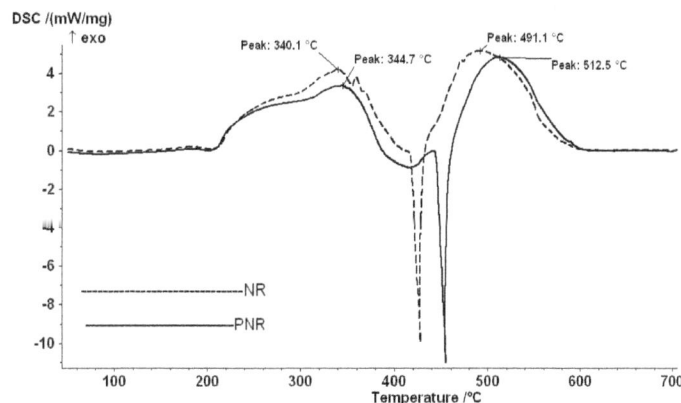

Fig. 3 The differential scanning calorimeter (DSC) curves of thermo-oxidative degradation of natural rubber (NR) and purified natural rubber (PNR)

4. Conclusion

The thermal degradation of PNR and NR is a two-step-degradation. Compared to NR, the TG curve of PNR shifted towards higher temperatures, indicate that the thermal stability of the PNR is better than that of NR and improved by the purification with adsorption and precipitation method. The thermal decomposition mechanism of PNR corresponds to a one-dimensional diffusion at the first stage, and a phase boundary controlled reaction (one-dimensional movement) at the second stage. Kinetic analysis showed that activation energy (Ea), activation entropy (ΔH) and activation Gibbs energy (ΔG) values are all positive, indicating that the first thermos-oxidative degradation process of PNR is non-spontaneous.

Acknowledgments

The work was funded by the Natural Science Foundation of Hainan Province (NO. 20165203) and Fundamental Research Funds for Rubber Research Institute, CATAS (No. 1630022015022).

References

1. Angellier H, Molina-Boisseau S, Dufresne A. Macromolecules, 2005, 38, 9161-9170.
2. Sato S, Honda Y, Kuwahara M, et al. Biomacromolecules, 2004, 5, 511-515.
3. Sanguansap K, Suteewong T, Saendee P, et al. Polymer, 2005, 46, 1373-1378.
4. Wiwat Pichayakorn, Jirapornchai Suksaeree, Prapaporn Boonme, et al. Industrial & Engineering Chemistry Research, 2012, 51, 13393−13404.
5. Frank W. Perrellaa and Anthony A. Methods, 2002, 27, 77–86.
6. J. R Scott.. Rubber Chemistry and Technology, 1948, 21(3), 711-726.
7. Wirasak Smitthipong, Rattana Tantatherdtam, Kanokwan Rungsanthien, et al. Advanced Materials Research, 2013, 844, 345-348.
8. Ng K P, YiP E, Mok K L. Journal of Natural Rubber Research, 1994, 9(2), 87-95.
9. A.H. Eng. International Rubber Conference, Seoul, Korea, 1999.
10. K.L. Mok. 4th Annual International latex Conference, Akron, Ohio, USA, 2001.
11. Seiichi Kawahara, Warunee Klinklai, Hirofumi Kuroda, et al. Advances in Polymer Technology, 2004,15,181-184.

12. Frank W. Perrella, and Anthony A. Methods, 2002, 27, 77–86.
13. Lucksanaporn Tarachiwin, Jitladda Sakdapipanich, Koichi Ute, et al. Biomacromolecules, 2005, 6, 1851-1857.
14. Siby V, Yosuke K, Kioz M,et al. Rubber Chemistry and Technology, 2000, 73(1), 80-88.
15. Sizue O.Rogero, Ademar B.Lug, Fumio Yoshii, et al. Radiation Physics and Chemistry, 2003, 67, 501-503.
16. A.M. Donia. Thermochimica Acta, 1998, 320, 187-199.
17. Wilkie C.A., Thomson J.R., Mittleman M.L. Journal of Applied Polymer Science, 1991, 42, 901-909.
18. Al Shawabkeh A.F., Al Wahab H.A., Shahab Y.A. Journal of Optoelectronics and Advanced Materials, 2007, 9, 2075-2077.
19. He Canzhong, Wang, Yueqiang, Luo Yongyue, et al. Journal of Polymer Engineering, 2013, 33(4), 331-335.
20. Gabal MA. Thermochimica Acta, 2003, 402, 199-208.
21. Straszko J, Olszak-Humienik M, Mozejko J. Thermochimica Acta, 2000, 59, 935-942.
22. Prado AGS, Torres JD, Martins PC, et al. Journal of Hazardous Materials, 2006, 136, 585-588.
23. Militky J, Sesták J. Thermochimica Acta, 1992, 203, 31-42.

Cutting Temperature and Cutting Forces in High-speed Cutting of Titanium Matrix Composites with PCD Tool

Ying-Fei Ge[1,*], Hai-Qin Fan[1], Hai-Xiang Huan[2], Jiu-Hua Xu[2]

[1]School of Mechanical Engineering, Nanjing Institute of Technology, Nanjing 211167, China
[2]College of Mechanical & Electrical Engineering, Nanjing University of Aeronautics & Astronautics, Nanjing 210016, China

*Email: yingfeige@163.com

High-speed machining tests were performed on vol.(0-10%) TiC_p/Ti6Al4V and vol.10% (TiC_p+TiB_w)/Ti6Al4V composite in the speed range of 15-150 m/min using Polycrystalline Diamond (PCD) tools to investigate the cutting temperatures and cutting forces. The results showed that the cutting temperature range was 260-590 °C under the cutting parameters used. Cutting speed and tool wear had a significant effect on cutting temperature. For all the machining tests, the cutting temperature for the new tool was 60-100 °C higher than that for the worn tool (VB=0.1 mm) at all cutting speed levels used. The cutting temperature for the titanium matrix composites was 40-90 °C higher than that for the Ti6Al4V matrix. However, it slightly decreased when relatively higher volume fraction composites were used. The cutting forces increase by 15% with the increasing cutting speed at the range of 15-100 m/min but they increase by up to 35% when the cutting speed is between 100 m/min and 150 m/min. The cutting forces increased slightly or even decreased with the increasing cutting speed when using a worn tool, also they moderately increased when machining the workpiece with higher reinforcement volume fraction. The cutting forces significantly increased with the increasing tool flank wear especially for the peripheral force and feed force which increased 300-500 percentages.

Keywords: Titanium matrix composite; high speed cutting; PCD tool; cutting temperature; cutting force.

1. Introduction

Titanium matrix composites (TMCs) have the higher specific stiffness/strength, wear/corrosion resistance, anti-fatigue performance and better high temperature properties and hence are recognized as one of the new materials to replace titanium alloys. However, due to the low plasticity and fracture toughness, non-uniformity and abrasive nature of the reinforcement, machinability of this kind of material is poor for the excessive tool wear, poor surface finish, low productivity and high machining cost even cut with diamond tools, which severely restricts their wide application [1-3]. Numerous studies have been done on the machining of Aluminum matrix composites [4-6]. However, investigation on the machining of TMCs is still on its initiate stage. Sharon M. Hayes et al. [7] studied the machinability (chip formation, tool wear, machined surface) of continuous fiber reinforced titanium metal matrix composites using high speed steel and carbide tools under dry cutting condition. They reported that severe tool flank wear by abrasion was noted for all the cutting conditions, which led to the poor surface quality immediately. They concluded that the carbide tool only can be used for rough machining. The optimization of cutting parameters was carried out by Aramesh et al. to find the optimum cutting conditions for surface roughness and tool wear rate [8]. Tool life of PVD coated grades was also estimated by Aramesh et al. when turning vol.(10-12)%TiC_p/ Ti6Al4V composites at the cutting speed of 60 m/min [9]. R. Bejjani et al. [10] investigated the laser assisted machining (LAM) of vol.(10-12)% particle reinforced Ti6Al4V matrix composites using Polycrystalline Diamond (PCD) tool. The results showed that tool life increased by up to 180% using LAM. However, it was found that surface roughness increased by up to 15%. What's more, under high cutting speed and high laser power, tool diffusion wear was promoted and accordingly tool life was decreased. Ge et al. [11] studied the cutting force, cutting

*Corresponding author

temperature, and tool life when turning vol.10%(TiC$_p$+TiB$_w$)/ Ti6Al4V composites with PCD and carbide tools.

The literatures review indicates that merely several investigations have been done on the machining of TMCs. Conventional cutting research on this type of material with PCD tools has not been done yet. Moreover, our current understanding on the machinability in cutting of the TMCs has not well been understood. The present study intends to enhance the understanding of the high-speed turning and milling machinability of (TiC$_p$+TiB$_w$) and TiC$_p$ reinforced titanium matrix composites by investigating the cutting forces and cutting temperature using PCD tool.

2. Experimental Work

2.1. *Materials and machine tools*

The machined materials in the experiments were vol.10% (TiC$_p$+TiB$_w$)/Ti6Al4V ((TiC$_p$+TiB$_w$)/ TC4), vol.5% TiC$_p$/Ti6Al4V and vol.8% TiC$_p$/Ti6Al4V (TiC$_p$/TC4) composites which were produced through vacuum self-consuming electro-arc melting and hot forging technology. The mean size of the reinforcements of vol.10% (TiC$_p$+TiB$_w$)/TC4 is 1.5-20 μm in diameter for the particles and 35-80 μm in length for the fibers. Supower^{TM} PCD cutters (Fig. 1(a)) with 2-30 μm tool grain size were used in the machining tests.

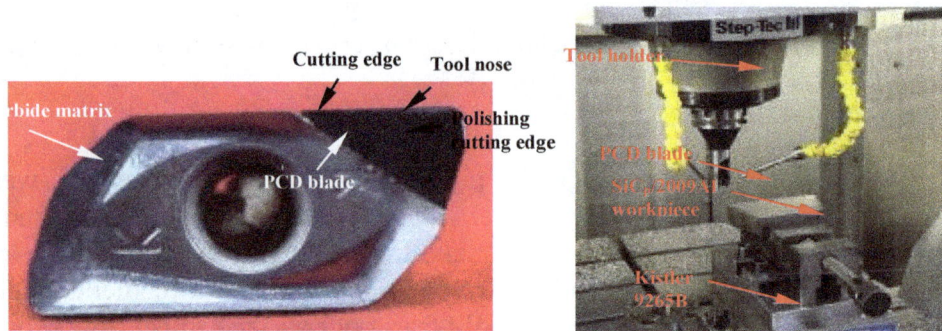

(a) Image of PCD tool   (b) High speed milling with PCD tools

Fig. 1 PCD tool image and the high speed milling photograph

The geometries of the PCD tips were listed in Table 1. The cutting parameters were shown in Table 2. Turning and milling tests were performed on SK50P CNC lathe and MICRON UCP710 high-speed five-coordinate vertical machine centre (Fig. 1(b)), respectively. The milling tests were conducted in down milling mode and under water-based emulsion cooling condition.

Table 1 The actual working angles of PCD tips

Machining method	Rake angle [°]	Clearance [°]	Inclination angle [°]	Tool nose radius[mm]	Cutting edge angle [°]	Side angle [°]	Negative chamfer[mm]
Turning	5	8	4	0.8	45	45	-
Milling	0	8	4	0.8	-	-	0.15

Table 2 The cutting parameters for turning and milling

Machining method	Cutting speed v [m/min]	Feed rate f [mm/rev]	Radial depth of cut a_w [mm]	Axial depth of cut a_p [mm]	Depth of cut a_e [mm]
Turning	15,30,45,60	0.08	-	-	0.5
Milling	60,80,100,150	0.08	1	3	-

2.2. Experimental scheme

The workpiece (TMC)-tool (PCD) natural thermocouple was used to measure the turning temperature while the workpiece (TMC)-constantan thermocouple was used to measure the milling temperature. The electromotive force signals of the thermocouple were recorded using a HP3562A dynamic signal analyzer. The cutting temperature can be calculated through the electromotive force of the workpiece-tool and workpiece-constantan thermocouple after calibration of the measured electromotive force using a special calibration system. Calibration of the workpiece-tool and workpiece-constantan thermocouples was done by heating two sets of thermocouples, the standard NiCr-constantan thermocouple and the calibration one, simultaneously. The two sets of thermocouples have a common hot junction which ensures that they share the same temperature at any moment during the calibration. The cold junctions of both thermocouples were kept at room temperature. The temperature that corresponding to the measured electromotive force of the hot and cold junctions can be obtained. An electromotive force-temperature curve can be gained after several times of readings of the temperature and the measured electromotive force. According to this curve, the corresponding temperature can be confirmed when a certain electromotive force was given. The fitting formulas of the electromotive force-temperature curves are as follows:

For the vol.5% TMC −PCD thermocouple:

$$Y=-7.374+45.436X+0.989X^2-0.052X^3 \tag{1}$$

For the vol.10% TMC−PCD thermocouple:

$$Y=-6.943+43.012X+1.078X2-0.064X3 \tag{2}$$

For the constantan −TC4 thermocouple:

$$Y=-12.238+24.2151X-0.0025X2+0.0009X3 \tag{3}$$

For the constantan − vol.5% TMC thermocouple:

$$Y=-26.27+27.741X-1.343X2+0.0637X3 \tag{4}$$

For the constantan − vol.8% TMC thermocouple:

$$Y=-23.543+26.469X-0.999X2+0.0513X3 \tag{5}$$

Cutting force was measured with a Kistler 9265B three component piezoelectric dynamometer and associated 5019A charge amplifiers connected to the PC employing Kistler Dynoware force measurement software. F_p, F_c and F_f are the peripheral force, main cuttting force and feed force for turning, respectively. F_t, F_r and F_a are the tangential force, raidial force and axial force for milling, respectively.

3. Results and Discussion

3.1. Cutting temperature

As shown in Fig. 2, the cutting temperature increases significantly with the increasing cutting speed because more cutting power will be needed and more friction heat will generate at higher cutting speed.

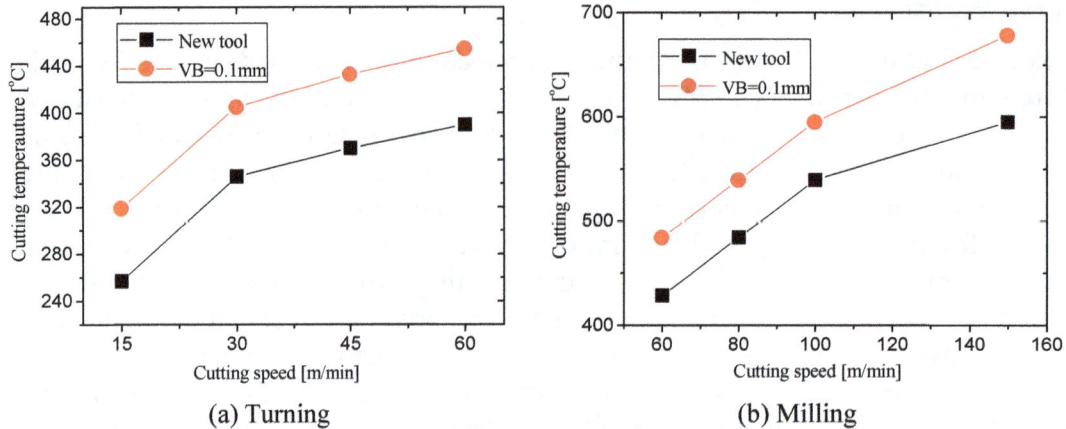

(a) Turning

(b) Milling

Fig. 2 Cutting temperature under different cutting speed and tool wear (vol.5% TiC$_p$/TC4)

The cutting temperature increased from 260 °C to 370 °C (Fig. 2 (a)) for the turning tests when cutting speed increased from 15m/min to 60 m/min while it increased from 430 °C to 590 °C (Fig. 2 (b)) for the milling tests when cutting speed increased from 60 m/min to 150m/min. From Fig. 2, it also can be observed that the cutting temperature moderately increased with the tool flank wear VB. For all the machining tests, the cutting temperature for the new tool was 60-100 °C higher than that for the worn tool (VB=0.1 mm) at all cutting speed levels used.

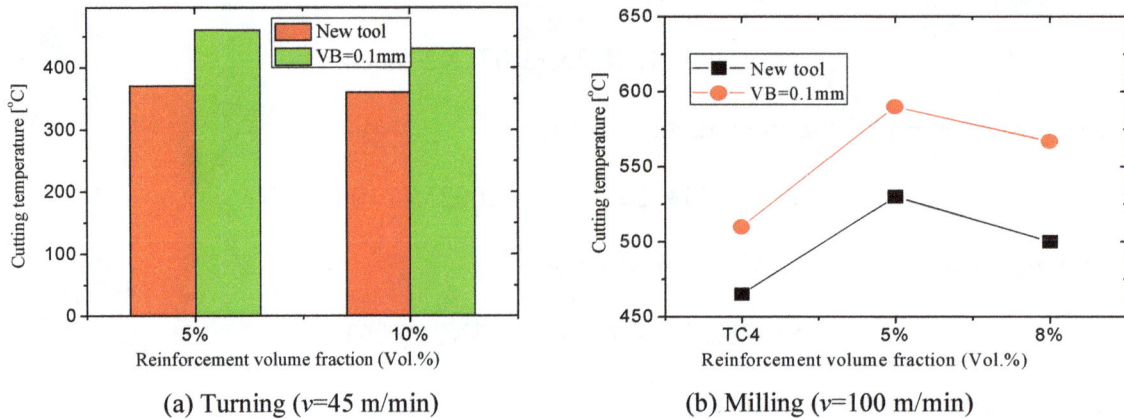

(a) Turning (*v*=45 m/min)

(b) Milling (*v*=100 m/min)

Fig. 3 Cutting temperature for different workpiece materials and tool wear

Fig. 3 (b) shows that the cutting temperature for the titanium matrix composites is 40-90 °C higher than that for the TC4 matrix. This attributes to the fact that a higher strength will be obtained when the reinforcement is added into the TC4 matrix. Furthermore, the strength under the higher temperature for the titanium matrix composites is also substantially increased compared to its matrix material, as shown in Table 3.

Table 3 The physical and mechanical properties of the titanium matrix composites

Materials	Tensile strength [MPa]	Yield strength [MPa]	Elongation [%]	Modulus of elasticity [GPa]	Hardness [HRC]
Vol.10%(TiC$_p$+TiB$_w$)/TC4	1100	1000	2.2	135	35-38
Vol.8%TiC$_p$/TC4	1050	970	8.6	140	36-37
Vol.5%TiC$_p$/TC4	1020	930	9.3	120	34-36
TC4	950	830	10	110	30

However, for the vol.5% and vol.8% composites, the cutting temperature slightly decreased when relatively higher volume fraction composites were used, as shown in Fig. 3. This is because the thermal conductivity coefficient substantially increased with the increasing volume fraction especially when the temperature is high, as shown in Table 4.

Table 4 The thermal conductivity coefficient for titanium matrix composites with different volume fraction

Temperature [°C]	25	100	300	500	700
Vol.5% TiC_p/TC4 [W/m·K]	5.708	5.924	7.482	9.982	12.597
Vol.8% TiC_p/TC4 [W/m·K]	5.858	6.686	9.546	11.740	14.221
Vol.10% (TiC_p+TiB_w)/TC4 [W/m·K]	5.968	7.005	10.363	13.408	15.675

3.2. Cutting forces

As can be seen from Fig. 4 (a), when turning titanium matrix composites, the main cutting force F_c is much bigger than the peripheral force F_p and feed force F_f. For milling, the tangential force F_t is the biggest one, as shown in Fig. 4 (b). It also can be observed from Fig. 4 that the cutting forces increase moderately (by 15%) with the increasing cutting speed at the range of 15-100 m/min. However, they increase significantly (by up to 35%) when the cutting speed used is between 100 m/min and 150 m/min, which induced by the sever deformation and workhardening of the material in the shear zone under the higher cutting speed. It should be noted that the milling forces are high due to the interrupt cutting effect, which would result in more and sever chipping on the cutting edge of the PCD tool.

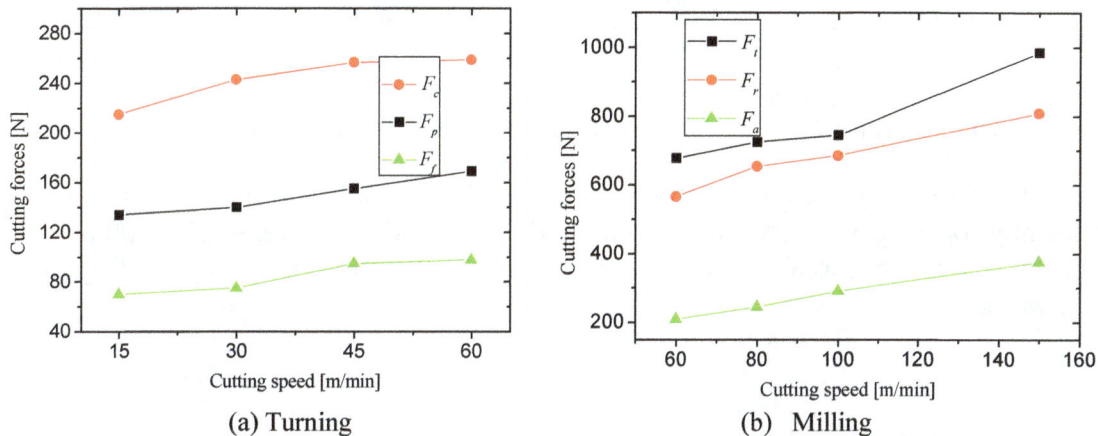

(a) Turning   (b) Milling

Fig. 4 Cutting forces under different cutting speed (new tool, vol.5% TiC_p/TC4)

Fig. 5 shows that the cutting forces slightly increase or even decrease with the increasing cutting speed when using a worn tool. This is due to the matrix softening effect by the high cutting temperature generated when using a worn tool combined with a high cutting speed.

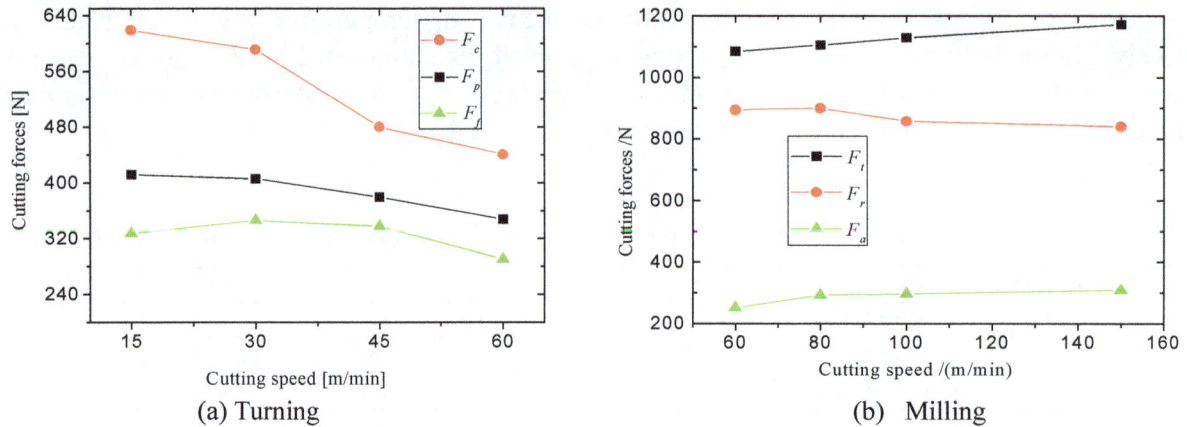

(a) Turning   (b) Milling

Fig. 5 Cutting forces under different cutting speed using worn tool (VB=0.1 mm, vol.5% TiC_p/TC4)

Fig. 6 (a) and Fig. 7 show that cutting forces moderately increase when machining the workpiece with higher reinforcement volume fraction using a new tool but the difference of the cutting forces is trivial when machining with a worn tool (Fig. 6 (b)).

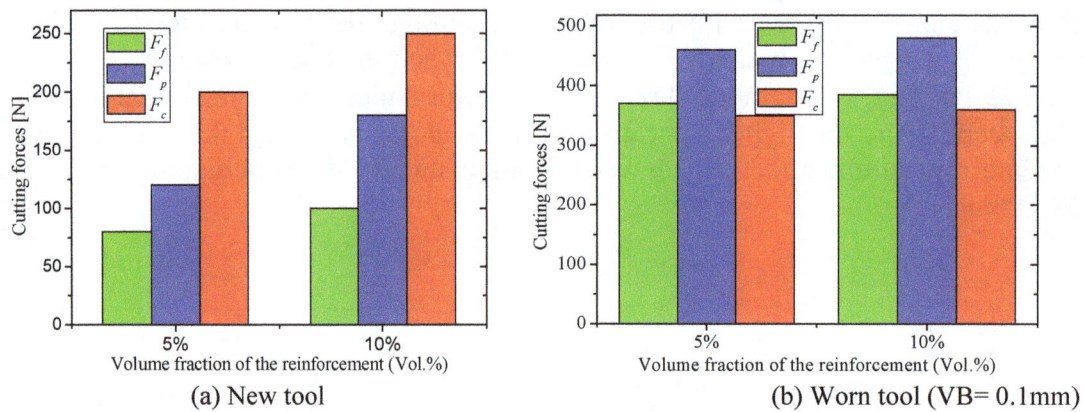

(a) New tool   (b) Worn tool (VB= 0.1mm)

Fig. 6 Cutting forces for different workpiece materials when turning (v=45 m/min)

Fig. 6 also shows that the cutting forces significantly increase with the increasing tool flank wear especially for the peripheral force F_p and feed force F_f which increase 300-500 percentages. This indicates that the tool wear rate will increases rapidly due to the very high cutting forces when the tool is worn.

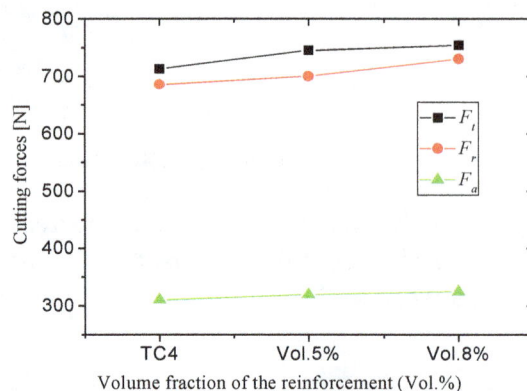

Fig. 7 Cutting forces for different workpiece materials when milling (new tool, v=100 m/min)

Therefore, the relatively small tool failure criterion should be selected when machining titanium matrix composites. Due to the high temperature and cutting forces, PCD tool generally presents obvious crater wear and chipping in the rake face and abrasive wear in the flank (Fig. 8).

(a) Turning   (b) Milling

Fig. 8 Tool wear appearance when cutting titanium matrix composites (vol.5% TiC_p/TC4, cutting 4 min)

4. Conclusion

1. The cutting temperature range is 260-590 °C under the cutting parameters used. The cutting temperature increased from 260 °C to 370 °C for the turning tests when cutting speed increased from 15 m/min to 60 m/min while it increased from 430 °C to 590 °C for the milling tests when cutting speed increased from 60 m/min to 150 m/min. The cutting temperature for the new tool was 60-100 °C higher than that for the worn tool (VB=0.1 mm) at all cutting speed levels used. The cutting temperature for the titanium matrix composites was 40-90 °C higher than that for the TC4 matrix. However, it slightly decreased when relatively higher volume fraction composites were used.

2. The cutting forces increased slightly with the increasing cutting speed at the range of 15-100 m/min. However, they began to increase significantly when the cutting speed was bigger than 100 m/min. The cutting forces slightly increased or even decreased with the increasing cutting speed when using a worn tool. Cutting forces moderately increase when machining the workpiece with higher reinforcement volume fraction using a new tool, however, the difference of the cutting forces is trivial when machining with a worn tool. The cutting forces significantly increase with the increasing tool flank wear especially for the peripheral and feed forces which increase 300-500 percentages.

Acknowledgments

This project was funded by the National Natural Science Foundation of China (51275227), Nanjing Science and Technology Development Plan (201306024), Qinglan Project of Jiangsu Province (2014) and Students Innovation Fund of Nanjing Institute of Technology (TB20161705).

References

1. S. Ranganath, Review on particulate-reinforced titanium matrix composites, J. Mater. Process. Technol. 32(1) (1997) 1-16.

2. M.S. Thompson, V.C. Nardone, In-situ-reinforced titanium matrix composites, Mater. Sci. Eng. A. 144(1-2) (1991) 121-126.

3. S. Abkowitz, S.M. Abkowitz, H. Fisher, P.J. Schwartz Cerm, Ti discontinuously reinforced Ti-matrix composites: manufacturing, Properties and Applications, JOM 56(5) (2004) 37-41.

4. K. Venkatesan, R. Ramanujam, J. Joel, P. Jeyapandiarajan, M. Vignesh, D.J. Tolia, R.V. Krishna, Study of Cutting force and Surface Roughness in machining of Al alloy Hybrid Composite and Optimized using Response Surface Methodology, Procedia Eng. 97 (2014) 677-686.

5. G. Chaudhary, M. Kumar, S. Verma, A. Srivastav, Optimization of drilling parameters of hybrid metal matrix composites using response surface methodology, Procedia Mater. Sci. 6 (2014) 229-237.

6. T. Wang, L.J. Xie, X.B. Wang, L. Jiao, J.W. Shen, H. Xu, F.M. Nie, Surface integrity of high speed milling of Al/SiC/65p aluminum matrix composites, Procedia CIRP 8 (2013) 475-480.

7. S. M. Hayes, M. Ramulu, W.E. Pedersen, Machining characteristics of a titanium metal matrix composite, NAMRC (2001) 189-196.

8. M. Aramesh, B. Shi, A.O. Nassef, H. Attia, M. Balazinski, H.A. Kishawy, Meta-modeling optimization of the cutting process during turning titanium metal matrix composites (Ti-MMCs), Procedia CIRP 8 (2013) 576-581.

9. M. Aramesh, Y. Shaban, M. Balazinski, H. Attiac, H.A. Kishawy, S.Yacout, Survival life analysis of the cutting tools during turning titanium metal matrix composites (Ti-MMCs), Procedia CIRP 14 (2014) 605-609.

10. R. Bejjani, B. Shi, Laser assisted turning of Titanium Metal Matrix Composite, CIRP Ann. Manuf. Techol. 60 (2011) 61-64.

11. GE Yingfei, XU Jiuhua, FU Yucan. High-speed turning of titanium matrix composites with PCD and carbide tools, Mater. Sci. Forum, Vol. 770 (2014) 39-44.

Experimental Research of High Speed Cutting Difficult-to-cut Material Based on Fractal Theory

Qiang Wu[1], Yong Mao[2], Lan-Ying Xu[1,*]

[1]*School of Automotive Engineering, Guangdong Polytechnic Normal University, Guangzhou 510635, China*
[2]*Chinese Communist Party School of Guangzhou Railway (Group) Corporation, Guangzhou 510600, Guangdong, China*
[*]Email: xulanying2012@126.com

High speed adaptive cutting experiment of high strength alloy steel based on fractal theory has been carried out. The impacts of cutting parameters on the cutting force, fractal dimension and surface roughness were analyzed. Based on the W-M function, the relationship between the fractal parameters and the traditional surface precision index is established. Meanwhile, the relationship between the cutting parameters and the fractal parameters of the cutting surface is also determined. Experimental results indicate that the main cutting force decreased by about 6% when the cutting speed increased from 100m/min to 240m/min under the condition of the cutting depth being 3.5mm and the feeding being 0.25 mm/r. The surface roughness decreases with the increase of the cutting speed, while it increases with the increase of the cutting feed and cutting depth. The biggest influence on the surface roughness is the cutting speed. The biggest influential factor is the cutting speed on the surface roughness. The impacts of cutting feeding and cutting depth are smaller.

Keywords: Fractal theory; difficult-to-cut material; high-speed machining; surface integrity.

1. Introduction

High-speed machining is the development direction of modern manufacturing technology, with the characteristic of high efficiency, high machining accuracy and high surface quality, so it is widely used in the manufacture of difficult-to-cut material [1, 2, 3]. However, high speed machining is a complicated and nonlinear process. The friction and wear resistance of cutting tool can be improved by high-speed adaptive cutting to improve the work-piece machining surface quality. But the traditional parameters can only reflect the local characteristics in the evaluation parameters for the processing quality.

B. Mandelbrot puts forward the concept of fractal in 1975. Each part is similar to the overall in some way, namely the parameters of the scale invariance and scale invariance of characterization are the fractal dimensions, so fractal dimension can completely be described. In traditional turning and milling, it is found that the surface topography has fractal characteristics. But the surface analysis of the advanced compound machining shows that the surface texture can be expressed by fractal dimension [4, 5]. Recently, scholars from domestic and overseas research and characterize surface microstructure, surface roughness, contact mechanics and etc. In these areas, they are increasingly using the powerful mathematical tools of fractal geometry theory [6, 7, 8].

In the process with high speed and high efficiency, we also found that the carbide coated cutting tool appears "self-organization" phenomenon in high speed dry cutting process [8, 9], it forms complex "amorphous" structure of the oxide film or secondary structure in friction interface, which can have the effect of protection, lubrication and heat insulation, so it improves the friction and wear resistance of a tool to reduce the adhesion on the surface of cutting tool, which enlarges the tool life and increases the surface quality of the machined surface [10, 11, 12]. This paper will use the fractal theory to analyze the fractal characteristics of machined surface integrity of high strength alloy steel in the process of high speed cutting. The prediction model of machined surface integrity is established to reveal the formation mechanism of processing surface.

---

[*]Corresponding author

2. The Application of Fractal Theory

The surface roughness is the main parameter to evaluate the machining surface quality. The past researches reveal that machining surface is not usually accurate self-similarity but the statistical sense of self similarity. As a result, the surface roughness can only use statistical fractal theory to study. The mathematical characteristic of rough surface is continuous, non-differentiable and self affine. Non-differentiable concept is that the more and more elaborate structures appear constantly of the roughness of certain point when the rough surface is amplified repeatedly, and the tangent line or tangent plane cannot be described at any point of the surface. The self-affinity is that the magnification of horizontal and vertical coordinates is not different in the course of the actual measurement. The W-M function can meet these characteristics in the fractal geometry. The contour height of rough surface [12]:

$$Z(x) = G^{D-1} \sum_{n=n_1}^{\infty} \frac{\cos(2\pi r^n x)}{r^{(2-D)n}}; 1 < D < 2, r > 1 \tag{1}$$

Where: x is measurement coordinates of rough surface; G is the characteristic length scale coefficient of the rough surface; D is the fractal dimension; r is the space frequency mode.

There are many kinds of the definition of fractal dimension. In general, the box dimension is used to express surface roughness. Now the double logarithmic diagram of power spectral density function will be used to calculate the fractal dimension. The power spectrum is got by the fast Fourier transform from Eq.1:

$$P(f) = \frac{G^{2(D-1)}}{2 \ln r} \frac{1}{f^{5-2D}} \tag{2}$$

Where: f is spatial frequency; G is the intercept of least squares line on the power shaft.

From Eq.2 it is found that the power spectrum obeys the power law, so the slope k of the least squares line can be calculated in the double logarithmic power spectrum diagram. So the fractal dimension of the work-piece surface:

$$D = \frac{5-k}{2} \tag{3}$$

3. Test Conditions

The experiment was carried out on the type of CAK6140VA numerical control lathe, the cutting force is measured by the turning dynamometer and charge amplifier, the sampling frequency is 5.5 kHz, sampling points is 16384. The Work-piece material is high strength steel, the diameter is 100mm, its length is 250 mm and its hardness is HRC60. The composition of the material is shown in Table 1, its mechanical properties are shown in Table 2.

Table 1 Chemical composition of test materials /%

C	Si	Mn	S	P	Cr	Ni	Cu	V	Mo
0.39-0.42	0.19-0.39	0.65-0.75	≤0.015	≤0.015	0.75-0.85	1.67-1.85	≤0.025	0.001	0.21-0.25

Table 2 Mechanical properties of materials (quenching temperature at 850°C)

Tensile strength σ_b	Yield strength σ_s	Elongation $\delta_5[\%]$	Section shrinkage $\varphi[\%]$	Impact work $Akv[J]$	Impact toughness value $\alpha kv[J/cm^2]$
≥980	≥835	≥12	≥55	≥78	≥98

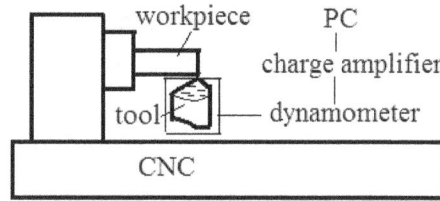

Fig. 1 The schematic diagram of the experiment

The tools are the coated cemented carbide tools TiCN, the schematic diagram of the test is shown in Fig. 1, the surface roughness of the workpiece is measured by the roughness tester TR200 after cutting every time.

4. The Test and Analysis

4.1. *The relationship of cutting speed, cutting force and fractal dimension*

The influence of cutting speed on the cutting force and the fractal dimension is studied by changing the cutting speed under the condition of the cutting depth is 3.5mm and the feeding is 0.25 mm/r. Fig. 2 and Fig. 3 are the relationship of the cutting speed, cutting force and the fractal dimension.

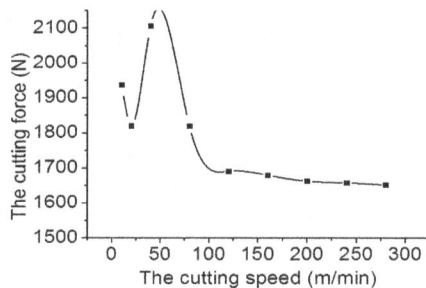

Fig. 2 Relationship between cutting force and cutting speed

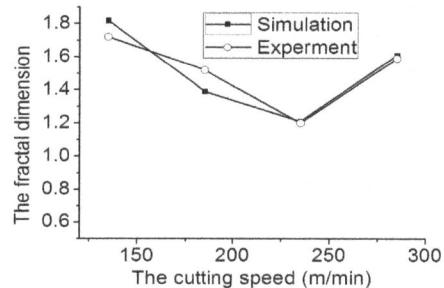

Fig. 3 Relationship between fractal dimension of cutting force and cutting speed

From the Fig. 1 we found that the main cutting force began to decline if the cutting speed is improved when cutting high strength alloy steel, then the main cutting force rose, however, it began to drop in a linear relationship after reaching a maximum. It can be seen that the main cutting force decreased by about 6% when the cutting speed increased from 100m/min to 240m/min.

As can be seen from Fig. 3, the change trend of the fractal dimension of the cutting force is basically the same as that of the experimental value. When the cutting speed is low, the fractal dimension of the cutting force is larger, which indicates that the cutting force is large and the cutting process is not stable. This is because when the cutting speed is low, the efficiency of cutting tool is low, the chip form is broken, which leads to high frequency change of cutting force. When the chip is broken, the surface of the work-piece will form a bump, which will affect the surface quality. The cutting force change exerts an alternative excitation force on the tool and machine, causing the vibration of the machine system. If the cutting force frequency is close to the natural frequency of the machine system, the vibration will increase, which will affect the machining precision, surface roughness and tool life. With the increase of the cutting speed, the cutting force decreases, the number of chip breaking is decreased, the cutting process is stable, and the fractal dimension of cutting force decreases quickly. But when the cutting speed continues to increase, the fractal dimension of the cutting force appears to increase rapidly.

4.2. *The relationship of feeding, cutting force and its fractal dimension*

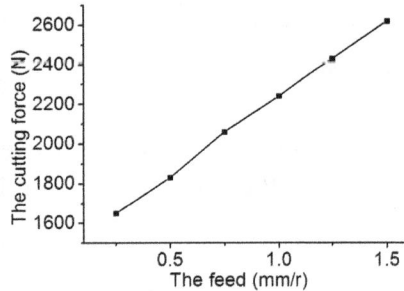

Fig. 4 Relationship between cutting force and the feed

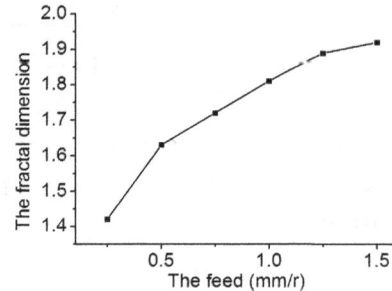

Fig. 5 Relationship between fractal dimension of cutting force and the feed

The impact that the feeding influence on cutting force and its fractal dimension is studied by changing the feeding under the condition of the cutting speed is 150m/min and the cutting depth is 0.3mm. Fig. 4 and Fig. 5 are the relationship curve of the feeding, cutting force and its fractal dimension.

The Figures show that the cutting force increases linearly with the increase of the feed, and the fractal dimension D changes significantly, and then with the increase of the feed its growth rate slowed down when the feed increases from 0.4 mm/r to 1.2 mm/r.

When the feed is small, the shear slip plastic deformation is less, the cutting force is small, the cutting process is relatively smooth, the roughness value of the surface roughness is small, and the fractal dimension of the cutting force is small. However, due to the influence of various random factors, the waveform of the surface profile is short, which may lead to large fractal dimension. When the feed increases, the plastic deformation of the processed material increases, the stability of the cutting becomes worse, and the roughness value of the machined surface increases. At this point, the tool path has little effect, the contour of the machined surface is longer and the frequency component is reduced. The growth rate of the fractal dimension of the cutting force decreases and tends to be stable.

4.3. *The relationship among cutting depth and cutting force and its fractal dimension*

When the cutting speed is 150m/min, the feed is 2.5mm/r, the test is carried out. Fig. 6 expresses the relationship between the cutting depth and the cutting force meanwhile Fig. 7 expresses the relationship between the cutting depth and the fractal dimension respectively. The Figure shows that cutting force increases linearly with the increase of cutting depth. The fractal dimension of cutting force increases with the increase of cutting depth at first, then it reaches maximum when the cutting depth is 1.9mm, afterwards it decreases with the increase of cutting depth in the future.

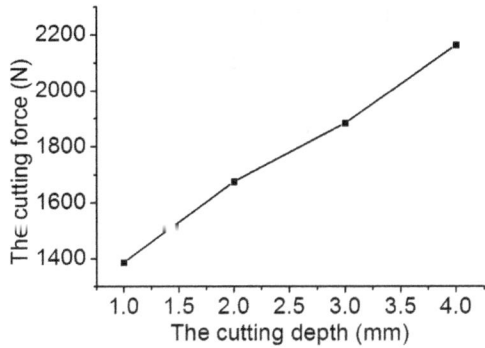

Fig. 6 Relationship between cutting force
and cutting depth

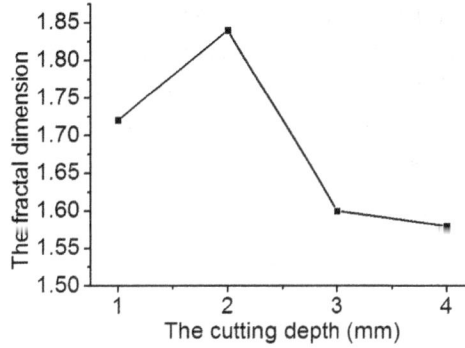

Fig. 7 Relationship between fractal dimension
of cutting force and cutting depth

The cutting depth is small, the material removal of the unit time is less, so the cutting force is small, the cutting process is relatively stable, and the fractal dimension of cutting force is small. As the cutting depth increases, the cutting force increases with the increasing of material removal, the stability of the cutting becomes worse, and the fractal dimension of the cutting force increases to the maximum. After the stability of the cutting process, the fractal dimension of the cutting force decreases quickly and tends to be stable. This is due to the distortion and fine structure appearing on the machined surface and gradually accumulates and forms. When the equilibrium state is reached, the fractal dimension is in stable condition.

4.4. The relationship between surface roughness and cutting parameter

Fig. 8 shows the relationship between surface roughness and cutting speed. With the increase of cutting speed, the surface roughness value shows a decreasing trend. This is due to the cutting heat generated too fast to be passed to the work-piece then it has been taken away by chip during the high speed cutting, which leads to reduce the surface temperature of the work-piece, so the work-piece has to be squeezed to produce a smaller uplift, and the surface roughness decreased.

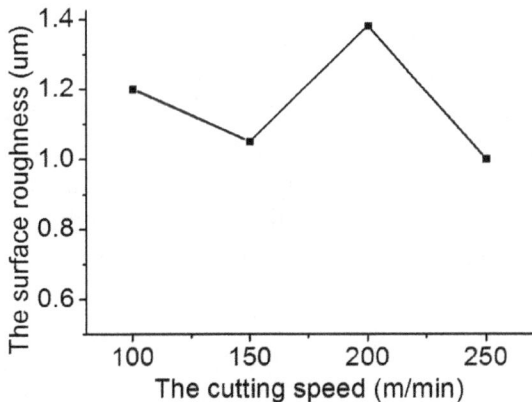

Fig. 8 Variation of surface roughness with the
cutting speed

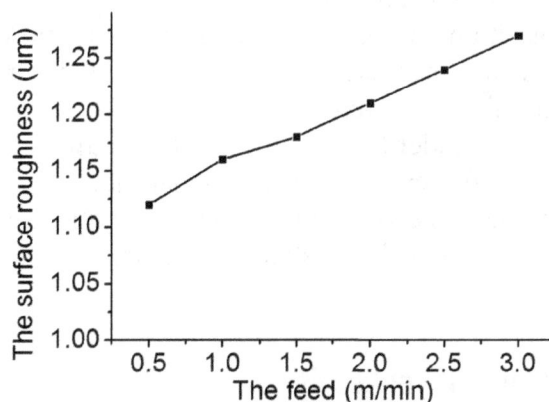

Fig. 9 Variation of surface roughness with the feed

Fig. 9 reveals the relationship between the surface roughness and the feed. The surface roughness increases linearly with the increase of the feed. There are two main reasons: firstly, from the residual area height theories tell us that the height of the residual area is also increased with the increase of the feed. Second, the vibration generated during cutting process enhances with the increase of the feed, and the radial force is also increased. This reflects in the front cutting force analysis, so the cutting vibration intensifies. Both of these effects lead to the increase of surface roughness.

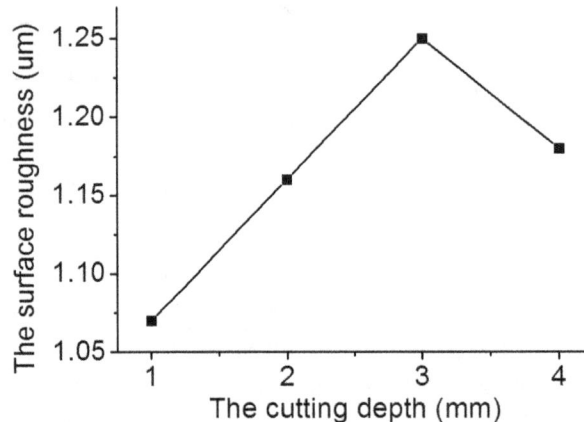

Fig. 10 Variation of surface roughness with depth of cut

The Fig. 10 shows that the surface roughness increase linearly with the increase of cutting depth, just after cutting depth is 2.9 mm, surface roughness fell slightly. This is because the cutting tool has negative chamfer, when cutting depth is low; mainly the negative edge bears the cutting task, with the increase of cutting depth, contact area between tool and chip raises, removal of metal materials increases, which leads to high cutting temperature. The bulge of surface is bigger because of the extrusion, so the surface roughness value increases. After cutting depth increases to 2.9mm, negative chamfer is no longer responsible for cutting task, and at this time mainly the cutting edge bears cutting task, which leads to a slight decrease in the surface roughness.

5. Conclusion

The optimized process parameters are high cutting speed, low the feed and small cutting depth. Choose high cutting speed, feed and small back turning for processing.

With the increase of cutting speed, the surface roughness value is on the decline. The main cutting force decreased by about 6% when the cutting speed increased from 100m/min to 240m/min under the condition of the cutting depth is 3.5mm and the feeding is 0.25 mm/r.

With the increase of feed, the surface roughness value rises linearly, and the fractal dimension of cutting force is getting bigger. With the increase of cutting depth, the surface roughness raises linearly, finally drops slightly. Fractal dimension increases at first and then decrease the cutting force.

Acknowledgments

This project is funded by National Natural Science Foundation of China (No. 51375101), Natural Science Foundation of Guangdong Province of China (No. 2014A030313638 and 2015A030313673) and Guangdong Science and Technology Program (No. 2014A010104014).

References

1. G.M. Zheng, J. Zhao, X.Y. Song, Q.Y. Cao and Y.E. Li: Ultra High Speed Turning of Inconel 718 with Sialon Ceramic Tools, Advanced Materials Research, Vol. 126-128 (2010), pp. 653-657.
2. H. Schulz, E. Abele and N. He: *Theory and Application of High Speed Machining* (Science Press, China 2010).
3. X. Ai: *High Speed Machining Technology* (National Defence Industry Press, China 2003).
4. Z.Y. Duan: *Research on Surface Texture of Turning Milling Based on Fractal Theory.* (MS.,Shenyang Ligong University, China 2011), pp.28-32.
5. M. Jumdar and T. Ien: Fractal Characterizat Ion and Simulation of Rough Surfaces, Wear, Vol. 136 (1990), pp.313-327.
6. C.G. Li and G.X. Zhang: The Relationship between Fractal Dimension and Surface Roughness Parameters, Tool technology, Vol. 12 (1997), pp. 36-37.
7. Y.L. Pan, L.Q. Wu and Y.D. Zhang: Research on the Relationship between Fractal Model Parameters and Roughness Parameter Ra, Engineering Science in China, Vol. 6 (2004), pp. 49-51.
8. G.S. Rabinovich, K. Yamomoto and S.C. Veldhuis: Tribological adaptability of TiAlCrN PVD coatings under high performance dry machining conditions, Surface and Coatings Technology, Vol. 200 (2005), pp. 1804-1813.
9. S. Li, Y. Liu, R. Zhu, H. Li, W. Ding: Heating Current Control in Electric Hot Minipore drilling hard-to-cut material. Applied Mechanics and Materials Vols. 34-35 (2010) pp. 1605-1608.
10. Lanying Xu, Qiang Wu, Yong Tang et al, Experimental Study on Force of Electric Heating Drilling to Hard-to-Cut Materials, Materials and Manufacturing Processes, 2015, 30: 263–271.
11. Zeng Quanren,Liu Geng,Liu Lan. Quantitative Description Model of Surface Integrity for Machined Parts[J]. China Mechanical Engineering. 2010, 21(24): 2995-2999.
12. G.S. Rabinovich, K. Yamomoto, K. Yamamoto and A.I. Kovalev: Wear Behavior of Adaptive Nano-multilayered TiAlCrN/NbN Coatings under Dry High Performance Machining Conditions, Surface and Coatings Technology, Vol. 202 (2008), pp. 2015-2022.

Experimental Research on the Micro Droplet Spray for 3D Printing Material

Qiang Wu[1], Fang-Zheng Wu[2], Lan-Ying Xu[1,*]

School of Automotive Engineering, Guangdong Polytechnic Normal University, Guangzhou 510635, China
Department of Computer Science and Technology, Central South University, Changsha, 410083, China
Email: xulanying2012@126.com

In order to analyze the forming law of micro droplet spray material, a series of experiments was carried out on the micro droplet spray forming technology by the self-designed test platform. Firstly, using water as a jet of liquid, the information of the injection process was collected by the high speed camera. This paper analyzed the mechanism of droplet injection and the change law of the liquid zone length with the fracture time. Meanwhile the influences of the trapezoidal wave amplitude, the high level holding time and the driving frequency on the micro liquid droplet were investigated. Finally, the various problems encountered during the injection process were discussed, and the micro droplet spray head was evaluated comprehensively. Experimental results indicate that the conical spray chamber needs less voltage amplitude in the case of same liquid length. When driving frequency is 25Hz and high level keep time is 1100μs, the liquid length of cylindrical cavity is biggest that the voltage is 130V or so.

Keywords: 3D printing; micro droplet injection; piezoelectric nozzle; incremental manufacturing.

1. Introduction

Incremental manufacturing technology is a typical advanced manufacturing technology, and it is also an important development direction of digital manufacturing [1, 2, 3]. Micro droplet free jetting forming is the key technology of incremental manufacturing. The various micro droplets by nozzle jet, with volume from microlitre to femtoliter [4, 5], are gradually accumulated into various solid structures on the substrate. These products are widely used in various fields of industry [6, 7]. Therefore, it is of great significance to develop the micro droplet spray forming technology.

Based on independent design of the micro droplet spray forming 3D motion experimental platform and the micro droplet observation system, we developed a piezoelectric nozzle which is suitable for micro droplet spray forming, and carried out a series of experiments on the jet dynamics. In addition, several nozzles with different cavity configurations were designed. The influence of the cavity shape on the droplet ejection was studied by experiments, which lays the foundation for the further research of the incremental manufacturing technology.

2. Test Conditions and Methods

The micro droplet ejection test system is shown in Fig. 1, which mainly includes the micro liquid drop nozzle, liquid supply module, piezoelectric drive power supply, liquid drop observation module, working platform, driving system and control system, and so on.

Test platform adopts MJ-AB-01 type nozzle and Jet-Drive III driver produced by Microfab Company. The spray nozzle is suitable for spraying water soluble substances or tiny droplets produced by related solutions; it can spray the liquid with a viscosity less than 20cp and the surface tension at 20-70dynes/cm. The spray chamber and the nozzle are made of special glass material, which can observe the formation of liquid and liquid droplets in the nozzle. The inner diameter of the nozzle is 10-100μm, and the temperature of the spraying liquid is 55°C. The maximum output voltage of the driver is 150V, the maximum frequency is 32 KHz, and the static voltage ripple is less than 20mV. Drive can provide triangular wave, square wave, sine wave and

*Corresponding author

other waves. It connects the host computer control program through the serial port protocol [7, 8]. The driver sets a stroboscopic light source to drive the output channel and sends out a high-level trigger signal of 5V to the light source controller to control the high-speed camera when the liquid drop was observed.

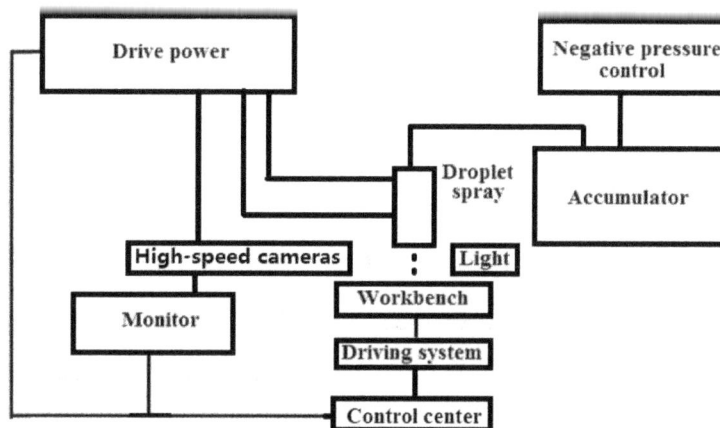

Fig. 1 Test system

The tube wall of the nozzle can produce impurities and attachments after using for a period of time, and even block the flow channel, it needs regular cleaning. The cleaning liquid can be stored in the liquid storage tank, through adjusting the negative pressure controller, the cleaning liquid can be repeatedly vibrating in the nozzle, and the purpose of cleaning is realized.

In this experiment it uses ultra high speed camera, its frame rate reaches one hundred thousand frames per second. Although a micro droplet jetting process is usually only a few milliseconds, even a few microseconds, and the size of micro droplet is only a few microns, but we can still use the ultra high speed camera to direct the process of shooting, and get the information of liquid drop.

To ensure relative motion with at least three freedoms between the micro liquid drop spray head and the forming working table, we designed the droplet spray head to be stationary and the working table can move freely in three directions of XYZ. Adopting motion control card and stepping driver to control the stepper motor, then to drive the movement of the three axes in the way of driving the synchronous belt, which achieves the motion control of X, Y and Z axes in three directions.

3. Experiment and Analysis

Independent design of piezoelectric nozzle consists of a piezoelectric diaphragm and a piezoelectric resonator and glass nozzle. Piezoelectric cavity has cylindrical shape and conical type; glass nozzle can be changed freely. By the different pressure injection chambers and nozzles can be combined into different types of nozzle, and it can get different nozzle jet test data, thus we obtain the relevant injection law by comparative analysis. When the selected driving waveform is trapezoidal wave, its amplitude, high level holding time and frequency have a great influence in test, as shown in Fig. 2.

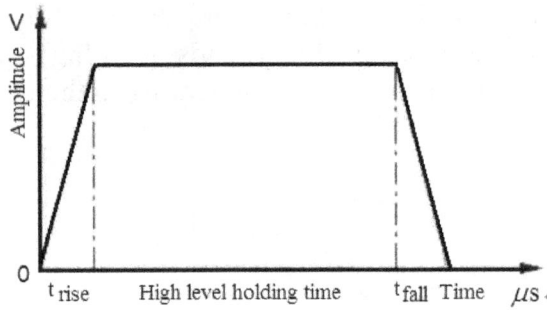

Fig. 2 Driving waveform

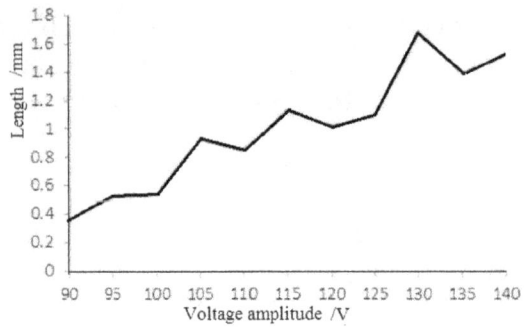

Fig. 3 Variation of maximum length of liquid zone

3.1. *Amplitude effect*

A cylindrical spray chamber is selected to test, at this time, the outlet diameter of the glass nozzle is 415 μm, the inside diameter is 90μm, and mark it as No.1 nozzle. When driving frequency is 25Hz and high level keep time is 1100μs, we change the voltage amplitude and get the curve of the maximum length of the liquid zone with the voltage amplitude as shown in Fig. 3, meanwhile, we also obtains the curve of the fracture time with the change of voltage amplitude, as shown in Fig. 4.

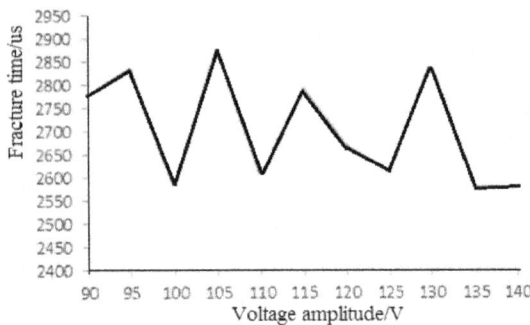

Fig. 4 Variation of the fracture time with voltage amplitude

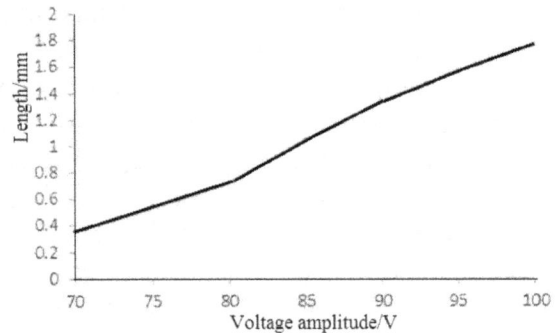

Fig. 5 Variation of maximum length of liquid zone

From Fig. 3 and Fig. 4, we can see that the maximum length of the liquid zone increases with the increase of the voltage amplitude, and the fracture time and the voltage amplitude of the liquid zone show the nonlinear relation of oscillation type. In case of the same glass nozzles, a fixed driving frequency of 25 Hz, high level keep time for 1400μs, we use conical spray chambers to carry on the same test, the maximum length of liquid zone is found to show a linear relationship with the change of the voltage amplitude, as shown in Fig. 5.

It is found that the high level holding time is different under different test schemes. This is due to the different shape of the nozzle and the spray chamber, as well as the different interval of the high level holding time. Comparing Fig. 3 with Fig. 5 it can be seen that the conical spray chamber needs less voltage amplitude in the case of the same liquid length. Under the same experimental conditions, the variation of fracture time with the voltage amplitude is shown in Fig. 6. The relationship between them is nonlinear. At the beginning, the fracture time decreases with the increase of voltage, then gradually increases, and last decreases gradually after reaching the maximum value.

Selecting the conical spray chamber to test, we change the size of the glass nozzle, at this time, the outlet diameter of the glass nozzle is $700\,\mu\text{m}$, the inside diameter is $210\,\mu\text{m}$, and mark it as No.2 nozzle. Under the same experimental conditions, the maximum length of the liquid band with the change of the voltage amplitude is shown in Fig. 7.

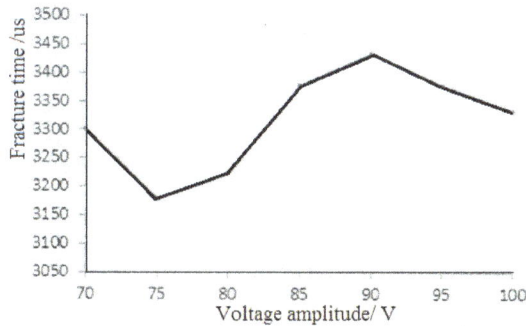

Fig. 6 Variation of the fracture time of liquid with

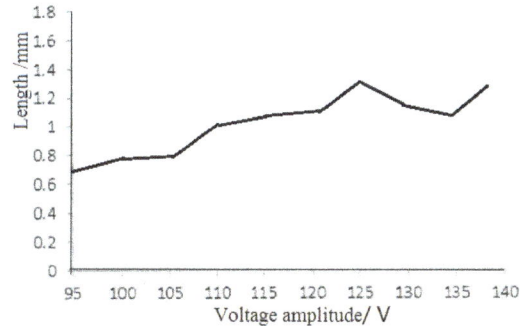

Fig. 7 Variation of maximum length of liquid zone zone voltage amplitude

Fig. 7 shows that the maximum length of the liquid zone increases with the increase of the voltage amplitude, comparing Fig. 7 with Fig. 5 it can be found that the voltage amplitude of Fig. 7 increases. This shows that the larger the nozzle diameter, the larger the voltage amplitude of the nozzle is needed. From Fig. 3, Fig. 5 and Fig. 7, we can find that the larger the amplitude of the driving voltage, the longer the liquid zone, but the maximum length of the liquid zone is not related to the selection of the spray chamber and glass nozzle. Under the same test conditions, the fracture time of the liquid zone with the change of voltage amplitude is obtained by using No. 2 nozzle, as shown in Fig. 8. In view of this kind of glass nozzle, the fracture time of the liquid zone gradually decreases with the increase of the voltage amplitude. From Fig. 4, Fig. 6 and Fig. 8, it is found that the forms of spray chamber and the glass nozzle have a great influence on the fracture time, and they have strong correlation.

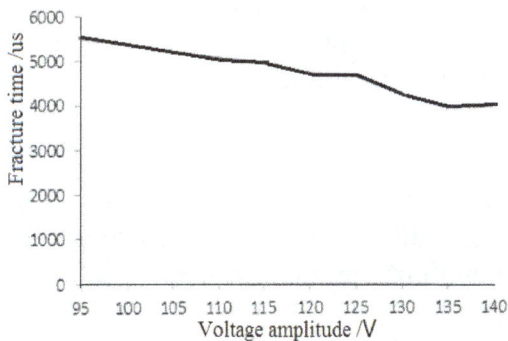

Fig. 8 Variation of the fracture time of liquid zone with voltage amplitude

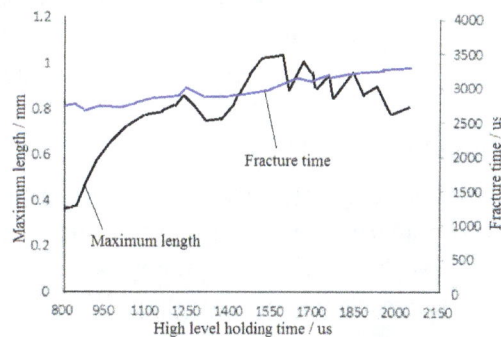

Fig. 9 Variation of the maximum length and the fracture time of the liquid zone

3.2. High level holding time

In the experiment voltage amplitude and drive frequency are fixed the same, we change the high level holding time and record the change of liquid zone. Under the test conditions in Table 1, the change of the maximum length and the fracture time of the liquid zone with high level were analyzed.

Table 1 Technological conditions of the test

Test scheme	Spray head form	Nozzle number	Voltage [V]	Frequency [Hz]
1	Cylinder type	1	90	25
2	Cone type	1	90	25
3	Cone type	2	150	25

The results obtained in accordance with the test scheme No.1 is shown in Fig. 9, the results obtained in accordance with the test scheme No.2 is shown in Fig. 10, the results obtained in accordance with the test scheme No.3 is shown in Fig. 11.

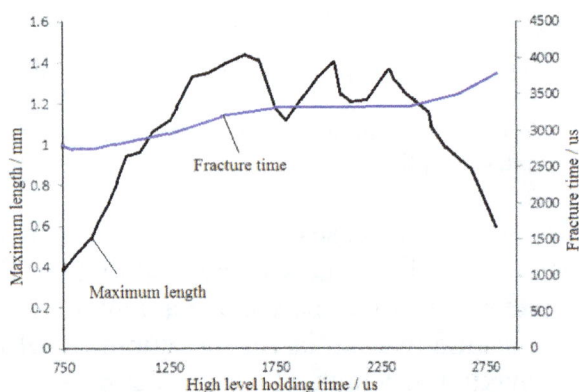

Fig. 10 Variation of the maximum length and the fracture time of the liquid zone

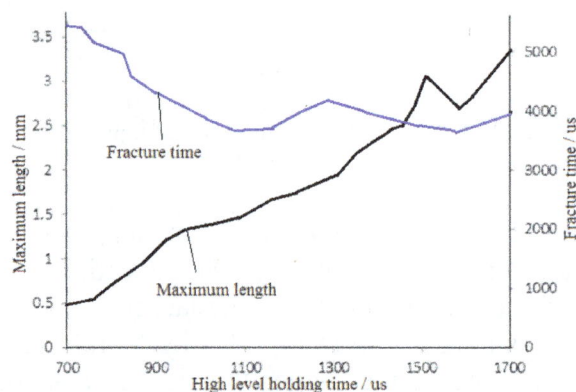

Fig. 11 Variation of the maximum length and the fracture time of the liquid zone

Comparing Fig. 9 with Fig. 10 it shows that the maximum length of the liquid zone begins to increase and then decreases when the high level holding time increases, for No. 1 glass nozzle, the fracture time of the liquid zone increases slowly with the increase of high level hold time. Comparing Fig. 10 with Fig. 11 it can be seen that glass nozzle diameter increases with the increase of liquid zone maximum length when they are driven by the same trapezoidal wave, but the shape of the cavity is not sensitive to the fracture time of the liquid zone.

3.3. Frequency effect

When the voltage amplitude and the high level holding time are constant, the injection test is carried out according to the technological parameters of Table 2. To change the driving frequency, the effect of the driving frequency on the maximum length of the liquid zone and the fracture time of the liquid zone was investigated.

Table 2 Technological conditions of the test

Test scheme	Spray head form	Nozzle number	Voltage [V]	High level holding time [μs]
1	Cylinder type	1	150	900
2	Cone type	1	100	1300
3	Cone type	2	150	1700

The results obtained in accordance with the test scheme No.1 is shown in Fig. 12, the results obtained in accordance with the test scheme No.2 is shown in Fig. 13, the results obtained in accordance with the test scheme No.3 is shown in Fig. 14.

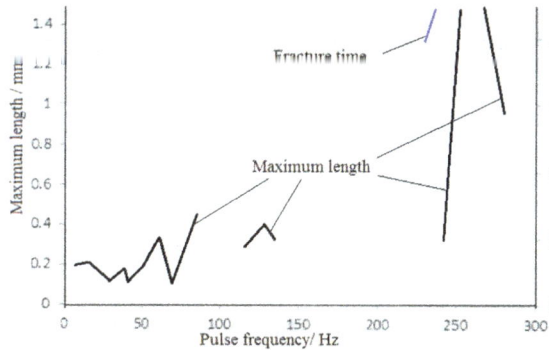

Fig. 12 Variation of the maximum length and the fracture time of the liquid zone

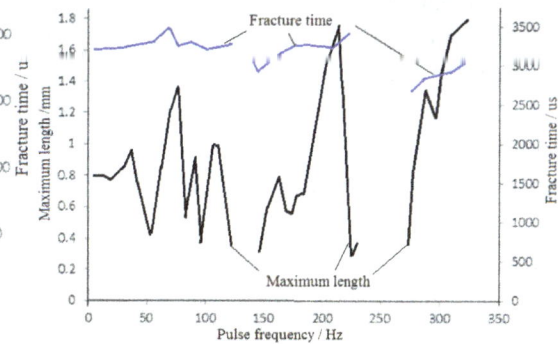

Fig. 13 Variation of the maximum length and the fracture time of the liquid zone

Because piezoelectric diaphragm type spray chamber can't work normally under all driving frequency, that is to say, under the condition of certain frequency it can't spit out micro droplets, so the test data is not continuous, as shown in Fig. 12 and Fig. 13. It can be found that the piezoelectric micro droplet injection system appears resonance phenomenon under certain frequency condition, at the same time, the length of the liquid zone has been received the maximum value. When the cone type spray chamber is used, there are several resonance frequencies in the low frequency band. Meanwhile, it also has several great values.

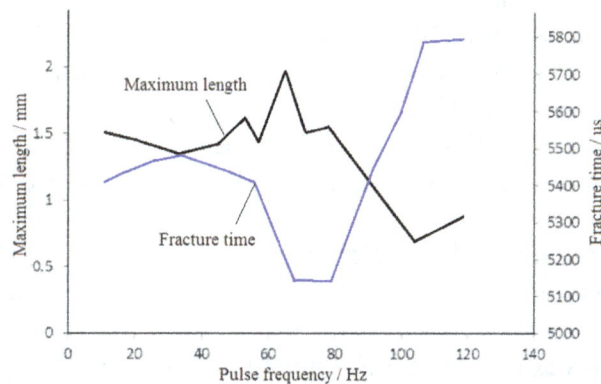

Fig. 14 Variation of the maximum length and the fracture time of the liquid zone

As shown in Fig. 14, the injection frequency has a maximum value, once the frequency exceeds this maximum frequency, the test platform cannot be sprayed out of micro droplets normally and then the test data can no longer be obtained. From Fig. 13, we can find that the larger the diameter of the glass nozzle, the greater the impact of the pulse frequency on the injection system, the smaller the frequency range of the normal injection. Comparison of Fig. 12, Fig. 13 and Fig. 14 can also be found that the stability of the No. 1 nozzle is higher and the liquid zone of the fracture time of the No. 1 nozzle is smaller when the frequency is changed.

4. Conclusion

Adopting self-developed piezoelectric diaphragm micro droplet spray head, with water for injection liquid and a high-speed camera for recording droplet ejection process, we collected a large number relevant information of droplets. Through the different pressure injection chambers and nozzles can be combined into different types of nozzle, the influence of injection parameters on the droplet ejection was investigated by comparing the experimental data with various injections. The experiment results show that the nozzle is not only convenient to change but also has good injection effect. When driving frequency is 25Hz and high level keep time is 1100μs, the maximum liquid length of the cylindrical cavity is biggest that the voltage is 130V or so.

Acknowledgments

This project is funded by National Natural Science Foundation of China (No. 51375101), Natural Science Foundation of Guangdong Province of China; (No. 2014A030313638 and 2015A030313673) and Guangdong Science and Technology Program (No. 2014A010104014).

References

1. Grau J, Cima NJ and Sachs E: Alumina Molds Fabricated by 3-Dimensional Printing for Slip Casting and Pressure Slip Casting. Ceramic Industry. Vol. 23 (1998) No. 3, pp. 22-24.
2. Michaels S, Sachs EM and Cima MJ: Metal Parts Generation by Three-Dimensional Printing. Proc. of SFF Symposium, University of Texas at Austin, Texas.(2002) No. 7, pp. 244-246.
3. Singh. M., H.M. Haverinen, P.Dhagat and G.E. Jabbour: Inkjet printing-process and its applications. Adv Mater. Vol. 22 (2010) No. 4, pp. 673-674.
4. Srivastava VC, Upadhyaya A and Ojha SN: Microstructural Features Induced by Spray Forming of a Ternary Pb-Sn-Sn Alloy. Bull. Mater. Sci. Vol. 23 (2015) No. 8, pp. 73-75.
5. Malone E: Freeform Fabrication and Characterization of ZincAir Batteries. Rapid Prototyping Journal. Vol. 14 (2013) No. 5, pp.128.
6. J.A. Penoyer, G. Burnett, D.J. Fawcett and S.-Y. Liou: Knowledge based product life cycle systems: principles of integration of KBE and C3P. Computer-Aided Design. Vol. 32 (2014) No. 9, pp. 311-312.
7. P. Rossbacher, M. Hirzand and A. Harrich: The Potential of 3D-CAD Based Process Optimization in the Automotive Concept Phase. SAE International. Vol.2 (2009) No. 6, pp. 250-254.
8. K.P. Karunakaran, P. Vivekananda and S. Ganathan: Rapid Prototyping of Metallic Parts and Moulds. Journal of Materials Processing Technology. (2000) No. 7, pp. 371-375.

Study on the Gray Level Distribution in the Vicinity of Weld Pool Under Different Current

Xiao-Gang Liu[1, 2], Tian-Yuan Liu[1,*], Shi Huang[1]

[1]*Department of Mechanical Engineering, Guangxi University of Science and Technology, Guangxi Liuzhou 545000, China*

[2]*Department of Mechanical Engineering, Guilin University of Aerospace Technology, Guangxi Guilin 541000, China*
Email: 707837094@qq.com

The arc in the vicinity of the molten pool in the process of CO_2 welding has a complex distribution. Changes of the arc limit the development of seam tracking technology based on vision seriously. Therefore, exploring and studying the image's gray distribution, which is captured by the CCD camera in the vicinity of the molten pool, can provide the basic conditions for the selection of suitable welding current and the irradiated area, thus preparing for getting a clear picture. Firstly, image acquisition system which can be flexibly adjusted is established. Secondly, the numerical model of gray value in the area near the welding pool under different welding currents is built with the toolbox called cftool in Matlab. Experimental results show that: with the increase of welding current, the gray value of the image, which is captured by the CCD camera near the weld pool, is increased slightly; with the increase of distance, the gray value of the image, which is captured by the CCD camera near the weld pool, is reduced. When the distance between torch and the area is less than 10mm, the gray value decreased sharply. When the distance between torch and the area is more than 10mm, the gray value decreased slowly. The fitting results conform to the objective laws. It lays the foundation for the welding seam automatic tracking technology.

Keywords: Welding current; the vicinity of welding pool; gray value; numerical model.

1. Introduction

With the development of machinery, shipbuilding, aerospace, power, materials and other industries, welding has become the third largest industry after assembly and machining in the metal manufacturing industry. In view of the complexity of the weld and deformation of weld deviation, in order to ensure the welding stability of product and to improve labor productivity and working conditions of welding workers, the weld seam tracking technology must be studied.

At present, the sensors used in the automatic tracking technology of welding seam in China and abroad include mechanical sensors, electromagnetic induction sensors, arc sensors, vision sensors and so on. Among them, the visual sensor is favored by the users because of its high sensitivity and precision, strong anti electromagnetic interference ability, and no contact with the workpiece [1-4]. By using the visual sensor and microcomputer, it realizes the automatic seam tracking and calculation of the gap size, recognition the joint and a variety of functions. What's more, it realizes the intelligent welding process in the process of automatic seam tracking and on the welding quality of real-time control [5]. The CO_2 arc welding droplet short circuit transition process of molten pool near the regional splash metal, light illumination intensity distribution and complex variation have serious impact on the timeliness and reliability of weld feature extraction. The application of seam tracking technology based on visual CO_2 arc welding is severely restricted. Therefore, the study of different welding current CO_2 welding droplet short circuiting transfer processes, the molten pool area in the vicinity of the arc light distribution and the change law is very meaningful. In this paper, the image of the area near the weld pool of CO_2 welding under different welding current is obtained. Using the size of gray value reflects the strength of the arc. Numerical modeling and in-depth analysis of gray distribution and variation of the area near the weld pool in CO_2 welding are good for laying a foundation of finding the appropriate welding current and arc specific basis.

---

*Corresponding author

2. Establishment of Model

2.1. *Create an image acquisition system*

An excellent image acquisition system plays an important role in obtaining the better quality of the image in the vicinity of the weld pool. CO_2 welding vision image acquisition system has been developed for several decades, including the system of the collection principle has not changed, the overall structure is not changed, but the ability to upgrade the system hardware. The image acquisition system has a complete optical path is composed by welding arc column (source), of base metal (metal reflector), a filter, a CCD camera. Requirements for the components in the optical path have a high positioning accuracy. In order to ensure that the vibration error of each component in the system is in the allowable range, it is required to have the characteristics of convenient adjustment and reliable performance. In this system, welding torch and a CCD camera with L plate connected and fixed by nuts. In the experiment, the CCD camera is fixed in the direction perpendicular to the welding base material, and the center axis of the CCD camera is parallel to the direction of welding wire, and the corresponding length of the unit pixel in the image is consistent. During the welding process, the area around the weld pool which is captured by the CCD camera is collected by the image capture card and then transferred to the industrial computer to convert to digital image. The visual image acquisition system is shown in Fig. 1:

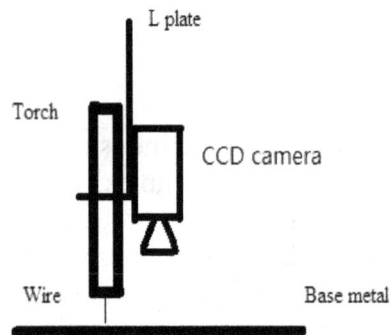

Fig. 1 Image acquisition system

2.2. *Test parameters*

The NBC-350 inverter CO_2 gas protection welder was chosen for this experiment. Welding torch protective gas is the flow of pure CO_2 of 10 L/min. Stainless steel wire has a diameter of 1.2mm. The length of welding wire is 10 mm. Welding base material is low carbon steel. Welding speed is 550mm/min. The industrial CCD camera model is MV1-D1312I. Sampling frequency is 485 frames per second. AcutEye high speed image long-time storage system is used to collect the image of the area near the weld pool and the real-time current and voltage data. It is welded by plate welding. According to a large number of test cases, the center wavelength of narrow band filter used in this experiment is 550nm. Welding voltage is 21V. Camera exposure time is 1.8ms. The form of droplet transfer is short circuit transition. The range of welding current test is 100-140A.

3. The Establishment of Numerical Model

3.1. *Calibration in this research system*

When dealing with the image, it is carried out in the unit of the pixel. Therefore, when we study the relationship between the gray value and the distance after the treatment, we must solve the problem of calibration, namely, what is the actual distance of each pixel of the image. In a system of this study, the CCD camera obtained for each image pixel value is 544'544.In order to reflect the relationship between the pixel grid and the actual distance, we use coordinate paper as a transfer medium, each large grid graph paper is one square centimeter, each small grid graph paper is one square millimeter. The resulting graph paper image is shown in Fig. 2. Set the origin of the coordinates at the 544th rows and the 272 columns of the image's pixel grid. The welding torch itself determines the camera optical axis and the attitude angle of the welding plane. Camera under different gesture images obtained from different location to the corresponding pixels per unit length is different. To this end, welding wire is perpendicular to the welding plane in the image acquisition system, and the camera lens optical axis is parallel to the wire axis. At this point, the unit pixel corresponds to the same length. The distance m corresponding to a single pixel lattice along the X axis can be obtained by the formula: $m = \dfrac{m_1}{544}$ .m_1 is the small square lattice's number contained along the axis of symmetry of the X axis. The distance n corresponding to a single pixel lattice along the Y axis can be obtained by the formula: $n = \dfrac{n_1}{544}$. n_1 is the small square lattice's number contained along the axis of symmetry of the Y axis.

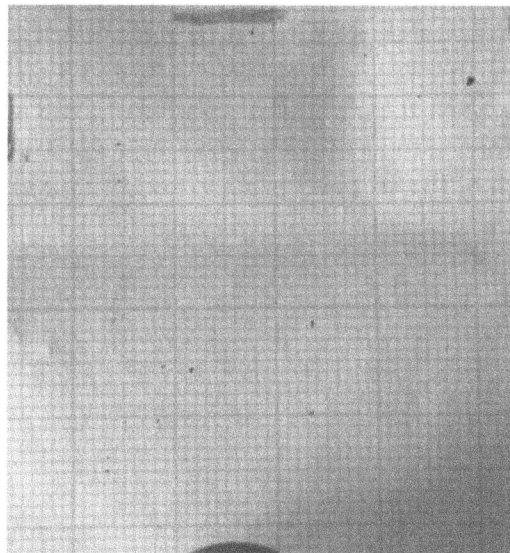

Fig. 2 Mesh of processing area

From Fig. 2 can be estimated m_1=50.5mm, n_1=50.5mm, the m_1 and n_1 were substituted into the formula obtained m=0.093mm, n=0.093mm, namely the actual distance of each pixel grid for 0.093mm.

3.2. Profile of cftool Matlab

Matlab has a powerful curve fitting toolbox cftool, which has the characteristics of easy to use and can achieve a variety of types of linear, non-linear curve fitting. This curve fitting toolbox is used to build the Matlab graphical user interface (GUI) and the integration of M file functions. Using the toolbox can fit parameters (when trying to find the regression coefficients and the physical meanings behind them), or using smoothing splines or various other interpolation methods for non parametric fitting (when the regression coefficient does not have a physical meaning and don't care about them). Using this toolbox interface, you can quickly and effectively achieve many of the basic curve fitting in an environment which is simple and easy to use [6]. The fitting types provided by the toolbox are: Custom Equations; Exponential: there are 2 types: $a \times e^{bx}$, $a \times e^{bx} + c \times e^{dx}$; Fourier: there are 7 types and the foundation type is $a_0 + a_1 \times \cos(\omega x) + b_1 \sin(\omega x)$; Interpolant: there are 4 types: linear, nearest neighbor, cubic spline, shape-preserving; Polynomial: there are 9 types:linear,quadratic,cubic,4-9th degree; Power: there are 2 types: $a \times x^b$, $a \times x^b + c$; Rational: the numerator and denominator of the common type is linear,quadratic,cubic,4-5th degree. In addition, the molecule also includes the constant type; Smooth Spline; Sum of Sin Functions: there are 8 types and the foundation type is $a_1 \times \sin(b_1 \times x + c_1)$; Weibull: there is only one type: $a \times b \times x^{b-1} \times e^{-ax^b}$

3.3. Experimental results and analysis

In the experiment, it is assumed that the arc column is a point light source, which is consistent with the illumination distribution of each direction around the weld pool at this time. Therefore, this experiment selected along the vertical direction of the torch light distribution situation as the research object. Through a lot of experiments can be found that when the distance from the welding torch about 2.5cm, toward the direction away from the welding torch, the gray value almost not any change. Therefore, a total of 27 gray value data were selected from the beginning of the 543rd pixel grid and each of the 10 pixel grids to select a pixel grid of gray information. The scattered point distribution of the gray level information is shown in Fig. 3.

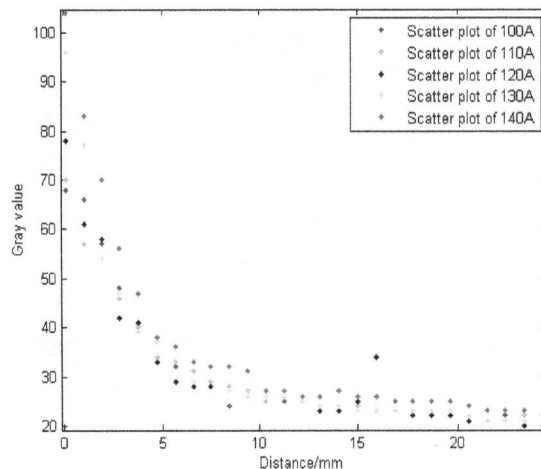

Fig. 3 Scatter plots of gray value distribution in the vicinity of weld pool under different welding current

According to the scatter plots of the gray value distribution, the numerical model of gray distribution in the vicinity of the weld pool is established by the exponential approximation of the MATLAB curve fitting toolbox. Test data at different currents are shown in Table 1.

Table 1 Gray level information in the vicinity of molten pool under different current

100A	110A	120A	130A	140A	Actual distance （mm）
68	70	78	96	104	0.093
66	57	61	77	83	1.023
57	54	58	54	70	1.953
48	46	42	47	56	2.883
41	40	41	39	47	3.813
33	34	33	37	38	4.743
32	33	29	36	36	5.673
29	31	28	29	33	6.603
28	29	28	32	32	7.533
24	28	27	27	32	8.463
26	27	26	26	31	9.393
25	25	26	26	27	10.323
25	26	26	26	27	11.253
25	25	25	25	26	12.183
24	24	23	24	26	13.113
24	23	23	24	27	14.043
24	24	25	23	26	14.973
23	23	34	23	26	15.903
23	23	25	23	25	16.833
23	22	22	23	25	17.763
23	22	22	23	25	18.693
22	22	22	23	25	19.623
22	22	21	22	24	20.553
21	21	21	21	23	21.483
22	21	21	21	23	22.413
22	22	20	21	23	23.343
21	21	20	21	22	24.273

When we choose the welding current at 100A, we can get the fitting results of distance and gray value:

$$f(x) = 46.69 \times e^{-0.3218x} + 12.91 \times e^{-0.01419x} \tag{1}$$

When we choose the welding current at 110A, we can get the fitting results of distance and gray value:

$$f(x) = 45.88 \times e^{-0.2828x} + 25.1 \times e^{-0.007161x} \tag{2}$$

When we choose the welding current at 120A, we can get the fitting results of distance and gray value:

$$f(x) = 52.34 \times e^{-0.3942x} + 28.26 \times e^{-0.01187x} \tag{3}$$

When we choose the welding current at 130A, we can get the fitting results of distance and gray value:

$$f(x) = 69.48 \times e^{-0.4629x} + 30.43 \times e^{-0.01624x}$$

(4)

When we choose the welding current at 140A,we can get the fitting results of distance and gray value:

$$f(x) = 78.76 \times e^{-0.3691x} + 29.05 \times e^{-0.009524x}$$

(5)

The fitting curves of the distance and the gray value under different welding current are shown in Fig. 4.

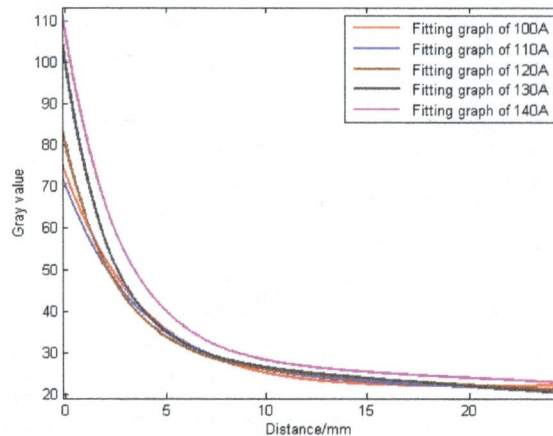

Fig. 4 Fitting curves of distance and gray value under different welding current

We can see from Fig. 4, with the increase of welding current, the image gray value of the area near the weld pool obtained by CCD camera is slightly increased, but the enlargement is not very large. This is because with the increase of welding current, arc voltage increases, arc radiation intensity increases. But the welding current is not the main factor that causes the CCD camera to obtain the image of the area near the weld pool. Therefore, the increase is not large. With the increase of the distance, gray value gradually decreased. This is because without taking into account the absorption, the intensity of light emitted by a point source is inversely proportional to the square of the distance. Therefore, with the increase of distance, the base material is reduced by the illumination, the illumination intensity of the CCD camera is also reduced, and the performance of the image gray value is reduced. When the distance between torch and area is less than 10mm, the gray value decreased sharply. The distance between torch and area is more than 10mm, the gray value decreased slowly. This is because when the distance from the light source 1cm, attenuation of 75% per cm, when the distance from the light source 10cm, attenuation per cm 17%, when the distance from the light source 100cm, attenuation per cm 2%, away from the light source 1000cm, attenuation per cm 0.2%.

4. Conclusion

1. The image acquisition system in the vicinity of the weld pool which can be flexibly adjusted is established, and the numerical model of the gray value of the image in the vicinity of the weld pool is established with the cftool Matlab toolbox.

2. The results show that, with the increase of welding current, the image gray value of CCD

camera is increased, but not very obvious. With the increase of distance, the image gray value of the area near the weld pool is decreased.

3. When the distance between torch and area is less than 10mm, the gray value decreased sharply. The distance between torch and area is more than 10mm, the gray value decreased slowly.

Acknowledgment

This research was funded by the Guangxi Natural Science Foundation (2014GXNSFAA118310).

References

1. Guo-jun Song, Liu-mei Zhu, Wei Wang, et al. Research on real time image processing of vision-based seam tracking system [J]. Welding Technology, 2003, 32(1): 10-12. (In Chinese)
2. Fu-Qiang Liu. Feature Recognition and Accurate Measurement of Welding Seam [D]. Harbin Engineering University, 2013. (In Chinese)
3. S. Lin, Y. Li, S. B. Kang, X. Tong, and H.Y. Shum. Diffuse-specular separation and depth recovery from image sequences. In proceeding of Europe Conference of Computer Vision, pages 210–224, 2002.
4. Ping Huo, Ning-Ning Li, Hai-Wang Zhang, et al. Analysis and research on the model of structured light visual sensor. [J]. Electric Welding Machine. 2014, 44(4): 108-111. (In Chinese)
5. Peng-Jun Mao, CO2 welding visual image morphological processing and welding seam tracking [D]. South China University of Technology, 2003. (In Chinese)
6. Qing-Wan Hu, Curve fitting using MATLAB curve fitting toolbox [J]. Computer Knowledge and Technology. 2010, 06(21): 5822-5823. (In Chinese)

Thermomechanical Modeling Research of Workpiece in End Milling Based on Moving Heat Source Methods

Guo-Dong Yan[*], Yong Feng, Bing-Hui Jia, Xiao-Lin Jia

School of Mechanical Engineering, Nanjing Institute of Technology, Nanjing 211167, China
[*]*Email: ygd0537@126.com*

A method of dynamically solving cutting force and cutting temperature in end milling is established based on moving heat source methods. Also, the distribution of cutting temperature and cutting forces of AISI1045 in a certain operating condition is studied through combining with the experimental data. Analysis results indicate that the maximum of F_x, F_y and F_z are 198.2 N, 50.8 N and 90.4 N respectively, the deviation between calculated and measured value are 2.1%, 20.3% and 10.2% respectively. The maximum temperature of tool-chip area showing three nonlinear phases is about 150°C. The deviation between the calculated and the measured value is about 10%, which shows good consistency. It indicates that the established method can be used to accurately investigate the thermo-mechanics of workpiece in milling process.

Keywords: Thermomechanical modeling; cutting temperature; cutting force; moving heat source methods.

1. Introduction

Modelling and study of the thermomechanical model during the end milling process has become a research hot point. It is also the main basis of high speed cutting process analysis and optimization, which is very important for tool wear, productivity and machining accuracy. MOUFKI et.al [1] took fully account of thermal-dynamic couple effect which would affect cutting formation, and put forward the calculation method of flow cutting angle. Oxley [2] proposed the thermomechanical model of orthogonal cutting process, and forecast analysis on oblique cutting force was carried out under the assumption that the cutting force has no relation to the cutting bevel and the chip flow angle. Song et.al [3] presented a general model for predicting the cutting forces in oblique cutting operation with respect to vector transformation derived from orthogonal cutting process. WU [4] gave theoretical formation of oblique cutting force by coordination transfer method. However, most former researchers mainly focused on cutting force and temperature of oblique cutting, the research on milling thermomechanical characteristics has been fairly rare. Furthermore, the existing research cannot accurately reflect the dynamic changing process of milling force and temperature with the blade rotation angle because the organic connection of cutting force and temperature has not been established. Draw lessons from the original modeling method based on oblique cutting, by introducing moving heat source method, the organic link between temperature prediction model and cutting force prediction model has been built in this paper, which provides accurate basis for high speed milling work analysis and optimization.

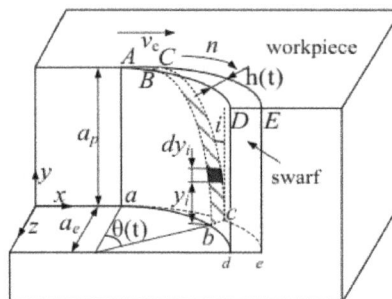

Fig. 1 Schematic drawing of helical tooth end milling

[*]Corresponding author

Due to the milling process containing complex heat, force, machinery and its coupling effect, it is difficult to establish the thermomechanical model accurately. Therefore, the milling process must be properly simplified for attaining the thermomechanical model, and then more accurate analysis of the real time physical quantities such as temperature, stress and strain can be done. As for the helical tooth end milling, the cutting process could be simplified as the oblique cutting thermomechanical process in the cutting plane by a single helical tooth. The simplified milling model is shown in Fig. 1, where v_c is the cutting speed, n is the milling rotation speed, a_p is the axial cutting depth, a_e is the radial cutting depth, Aa stands for the initial cut-in line of single tooth, DdEe denotes the cut-out plane of single tooth, BbCc denotes the shear plane of single tooth, h(t) and θ(t) indicate the cutting thickness and the rotation angle of single tooth under single cutting respectively, and i is the nominal helical angle of the milling cutter.

2. Thermomechanical Modeling Description

2.1. *Cutting force model of oblique cutting*

According to the cutting force model during oblique cutting as shown in Fig. 2, in the equivalent plane, friction force F_f and shear force F_s can be expressed as Eqs 1 and 2, respectively [5].

$$F_f = F \sin \beta_e \tag{1}$$

Here, F is cutting resultant force in the cutting process (N), βe is equivalent friction angle in the equivalent plane (rad), and it can be acquired by tgβe=tgβn/cosηc.

$$F_s = \tau A_{sh} \tag{2}$$

Where, τs is flow stress of the workpiece material (Pa).

The mechanical model of oblique cutting is more complex than that of orthogonal cutting, but in the normal plane, the two models are the same at all.

$$F_{tn} = \tau_s wt_c \frac{\cos \eta_s \cos(\beta_n - \gamma_n)}{\sin \phi_n \cos \lambda_s \cos(\phi_n + \beta_n - \gamma_n)} \tag{3}$$

$$F_{an} = \tau_s wt_c \frac{\cos \eta_s \sin(\beta_n - \gamma_n)}{\sin \phi_n \cos \lambda_s \cos(\phi_n + \beta_n - \gamma_n)} \tag{4}$$

$$F_{rn} = F_s \sin \eta_s = \tau_s wt_c \frac{\sin \eta_s}{\sin \phi_n \cos \lambda_s} \tag{5}$$

Considering the influence of tool cutting edge inclination λs, after coordinate transformation, the cutting forces in the main plane can be deduced as

$$\begin{bmatrix} F_x \\ F_y \\ F_z \end{bmatrix} = \begin{bmatrix} \cos \lambda_s & 0 & \sin \lambda_s \\ 0 & 1 & 0 \\ -\sin \lambda_s & 0 & \cos \lambda_s \end{bmatrix} \begin{bmatrix} F_{tn} \\ F_{an} \\ F_{rn} \end{bmatrix} \tag{6}$$

During oblique cutting, the cutting forces in the normal plane are shown in Fig. 2, and the force components are as follows

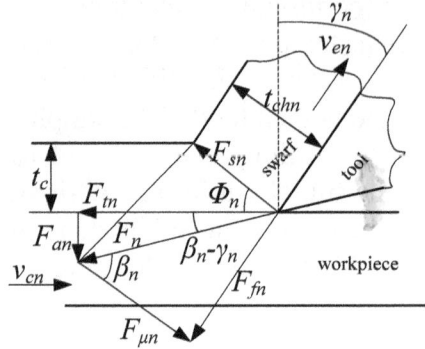

Fig. 2 Cutting forces in the normal plane of oblique cutting

2.2. *Temperature field model of steady oblique cutting based on moving heat source method*

In grinding and orthogonal cutting processes, moving heat source method has been proved to be an effective method for predicting cutting temperature, and in equivalent plane Pe the oblique cutting mechanism can be regarded as the accumulation of two-dimensional cutting state. Therefore, just putting the classical formula for moving heat source method into the equivalent plane, an approximate temperature calculation model describing oblique cutting can be obtained. With shear heat source and moving heat source working together, the temperature calculation model of the equivalent plane is shown in Fig. 3. Here, the origin of coordinates is at the end of tool/chip contact line, z axis is perpendicular to the rake face, and x axis coincides with the tool/chip contact line. O'A' and A'B' stand for the mirroring heat sources of friction heat source and shear heat source respectively, and after setting the mirroring heat sources, the temperature at any point of the object is the superposition of the two temperatures caused by real and mirror strong heat source respectively [6]. And thus during oblique cutting, the temperature field calculation formulas for workpiece and tool in the equivalent plane are respectively expressed as Eqs.7 and 8.

$$
\begin{aligned}
T_{M,chip} &= \frac{q_{pls}}{2\pi\lambda} \int_{w_i=0}^{t_c/\cos(\phi-\alpha)} e^{-(x-x_i)v_e/2a} \\
&\times \left[K_0\left(\frac{v_e}{2a}R_{is}\right) + K_0\left(\frac{v_e}{2a}R'_{is}\right) \right] dw_i + \frac{q_{plf}B_{chip}}{\pi\lambda} \\
&\times \int_{l_i=0}^{l_c} e^{-(x-l_i)v_c/2a} \left[K_0\left(R_{if}v_c/2a\right) + K_0\left(R'_{if}v_c/2a\right) \right] dl
\end{aligned}
\tag{7}
$$

Where, a is thermal diffusion coefficient of material (m2·s-1); q_{plf} is density of friction heat flow (W·m-2), and $q_{plf} = F_{fve}/(l_c \times w_c)$; q_{plf} is density of shear heat flux (w. m-2), and $q_{plf} = F_{svs}/A_{sh}$; K_0 is zero order Bessel function of second kind, and the others are as follows:

$$
R_{if} = \sqrt{(x-x_i)^2 + z^2}, \quad x_i = l_c - \omega_i \sin(\phi-\gamma_n), \quad R'_{if} = \sqrt{(x-x_i)^2 + (z-2t_{ch})^2}, \quad z_i = \omega_i \cos(\phi-\alpha)
$$

$$
\omega_i = 0 \sim t_{ch}/\cos(\phi-\alpha), \quad R_{ix} = \sqrt{(x-x_i)^2 + (z-z_i)^2}, \quad R'_{ix} = \sqrt{(x-x_i)^2 + (2t_{ch}-z-z_i)^2}
$$

$$
T_{M,tool} = \frac{q_{plf}}{2\pi\lambda} B_{tool} \int_{y_i=-w_c/2}^{w_c/2} dy_i \int_{x_i=0}^{l_c} \left(R_i^{-1} + R_i'^{-1}\right) dx_i
\tag{8}
$$

1062

Where, B_{tool} and B_{chip} are distribution ratios of friction heat source between tool and chip (B_{tool} + B_{chip}=1), and while calculating, the two parameters can be dynamically determined according to the condition that at both sides of the tool/chip contact line the average temperatures caused by any heat source are exactly equal [7]; and the others are as follows

$$R_i = \sqrt{(x-x_i)^2 + (y-y_i)^2 + z^2} \, , \quad R_i' = \sqrt{(x-2l+x_i)^2 + (y-y_i)^2 + z^2}$$

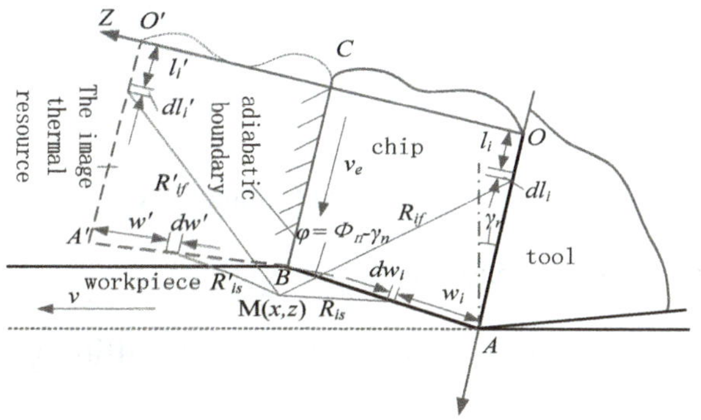

Fig. 3 Temperature calculation model of equivalent plane

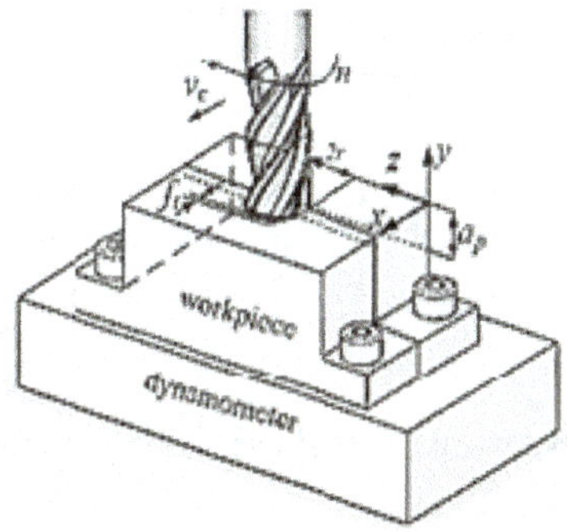

Fig. 4 Cutting test principle

3. Experiments and Thermomechanical Analysis

Experimental Principle. Measuring principle of milling force and temperature is shown in Fig. 4 [8], the coordinate origin in the system is located in the position of lines in the top right the artifacts. x, z axis is shown in the figure, y axis is determined by the right-hand rule. Kistler 9257B dynamometer is used to measure the three-dimensional cutting force in milling process. One constantan thermocouple wire is put in about two-thirds location of the axial cutting depth between the two workpiece pieces, and another constantan thermocouple wire is fixed as thermocouple cold end, away from the cutting point position. In milling process, once cutting temperature detected by the clamped constantan wire the millivoltmeter will output voltage values. The voltage-temperature relationship measured by constantan thermocouple can be acquired, and after calibration it can be expressed as Eqs 9. In calibration test, K type standard thermocouple made of nickel-chrome and nickel-silicon has been used, which has wide measuring range (0-1300°C), high sensitivity, high precision,

$$T = 6.41 + 33.87U \tag{9}$$

Where, T is pending temperature (°C), U is test voltage (mV).

Experimental Results. Electric potential signal curve collected by test is shown in Fig. 5(a), and filtering processing is made in order to get clearer changes of electric potential signal during the process of milling cutter cut-in to cut-out, as is shown in Fig. 5(b). It can be seen from the graph that the maximum of F_x is around 160 N; Cutting force fluctuation change, after removal not ideal signal curve, the minimum basic cutting force appears near the 0 N, accords with the basic change law of milling forces.

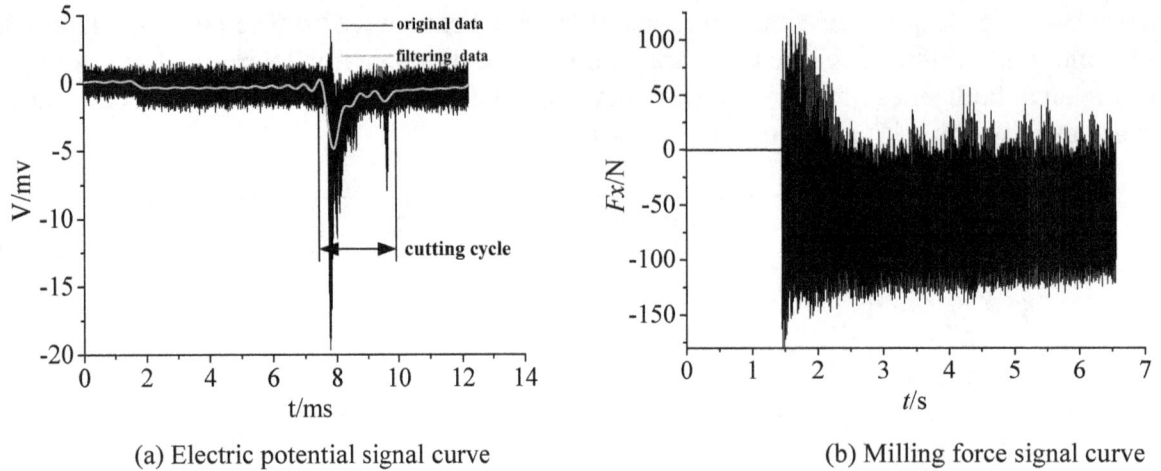

(a) Electric potential signal curve   (b) Milling force signal curve

Fig. 5 Experimental curves measured

Thermomechanical Analysis. Thermomechanical calculation steps of vertical milling processing are shown as follows:

(1) Determine the point of integral step length $\Delta\theta$, integration step Δap of the blade in axial, and the cutting out angle Φ_{ex} (Φ_{st}=acos(ae/r-1), Φ_{ex}=π).
(2) When the milling cutter rotation to arbitrary Angle $\theta(t)$, calculate the cutting force and temperature distribution of blade micro element by Eqs.3-.6 and Eqs. 7-8.
(3) Composite $\theta(t)$, Φ_{st} and Φ_{ex} geometric relations, it can ensure the angle $\theta(t)$in the k_{th} blade, then calculate the cutting force and temperature for integral processing in their respective axial cutting depth. After this step, it can obtain the total cutting force and temperature of the k_{th} blade.
(4) Change $\theta(t)$, repeat step 2 and 3, it can get the cutting force and temperature changes when the blades rotating.

According to the above steps, the cutting force with blade rotation within one circle is obtained as Table 1 shown. According to the comparison of cutting force curves produced by theoretical and experimental results, the maximum values of theoretical analysis on F_x, F_y, F_z (about 198.2N, 50.8N and 90.4N respectively) all occur on the time of milling cutter rotating to $\pi/4$, and have errors of 2.1%, 20.3% and 10.2% with the experimental results respectively, but both change trends are basically consistent. Errors are calculated as Eqs.10:

$$\text{Error= (test value} - \text{theoretical value) / test value} * 100\% \tag{10}$$

Table 1 Calculation results compared with test results

Cutting force	F_{xmax}	F_{ymax}	F_{zmax}
Theoretical value	202.3N	40.5N	99.6N
Test value	198.2N	50.8N	90.4N
Errors	2.1%	20.3%	10.2%

The maximum temperature T_{max} in tool-chip area as shown in Fig. 6 is also obtained from above steps, which shows that in the early processing stage (about 0-$\pi/4$) T_{max} has a sharp increase with a peak value of about 150°C, and the error of the highest temperature between theoretical and experimental results is around 15°C, and both the overall trends are nearly consistent. In addition, through the theoretical results compared with the experimental results. It can be concluded that due to the combined effects of working environment, measuring method and facilities, and other factors, in some locations there exist bigger differences between theoretical analysis and experimental results.

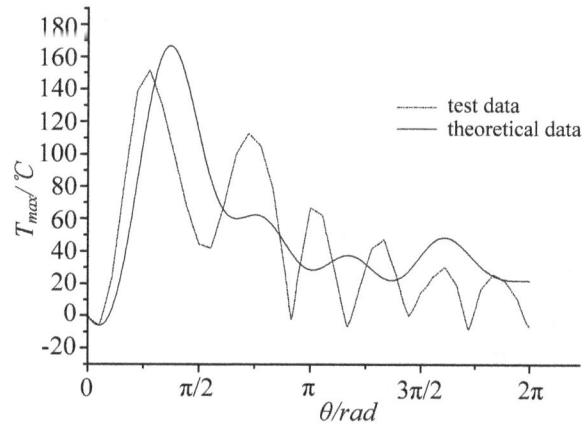

Fig. 6 T_{max} in tool/chip area

4. Conclusion

In this work, an intensive study on thermomechanical modelling for end milling workpiece has been carried out based on moving heat source method. The distributions of cutting force and temperature of AISI1045 steel under certain milling conditions has been presented according to theoretical analysis and experimental results. Based on the results obtained, the main conclusions can be extracted as follows:

1. The thermomechanical model of steady oblique cutting has been established. Based on the above model established, the dynamical solving method for cutting force and temperature of helical tooth end milling has been proposed.
2. For whole cutting process, the maximum cutting force Fx, Fy and Fz are 198 N, 40 N and 90 N respectively, the error between the test data are 2.1%, 20.3% and 10.2% respectively, basically meet the requirements of thermal error of problem solving.
3. As to the method proposed in this work, not only the material flow stress under thermomechanical effect is taken into account, but also the influence of chip thickness on cutting force and temperature is considered, and the changes of intermediate parameters can be calculated based on a small number of cutting conditions, which reduces the dependence on test data.

Acknowledgments

This work was funded by the Natural Science Foundation of Jiangsu Province of China (NO: BK20131341 and BK20150728) and the National Natural Science Foundation of China (No. 51405234)

References

1. Moufki A, Dudzinski D, Molinari A, Rausch M (2000) Thermoviscoplastic modeling of oblique cutting: forces and chip flow predictions. International Journal of Mechanics Science 42:1205-1232
2. Oxley PLB (1989) Mechanics of Machining. Ellis Horwood press, Chichester, UK
3. Song G, Sui SC, Tang LM (2015) Precision prediction of cutting force in oblique cutting operation. International Journal of Advanced Manufacturing Technology DOI 10.1007/s00170-015-7206-z
4. WU W.G. Study on high efficiency precision cutting and its vibrating characteristic. Jiang Su University. (2007). (in Chinese)
5. Moufki A, Devillez A, Dudzinski D, Molinari A (2004) Thermomechanical modelling of oblique cutting and experimental validation. International Journal of Machine Tools & Manufacture 44:971-989
6. Komanduri R, Hou ZB (2001) Thermal modeling of the metal cutting temperature process part III: temperature rise distribution due to shear plane heat source at the too-chip interface. International Journal of Mechanical Sciences 43:89-107
7. Dudzinski D, Molinari A (1997) A modeling of cutting for viscoplastic materials. International Journal of Mechanics Science 39:369-389
8. Maurel-Pantel A, Fontaine M, Michel G, Thibaud S, Gelin JC (2013) Experimental investigations from conventional to high speed milling on a 304-L stainless steel. International Journal of Advanced Manufacturing Technology 69:2191-2213

Temperature Field Control of 35CrMo Steel During the Process of Temperature-controlled Die Forging

Yuan-Chun Huang[1, 2,*], Jing-Jing Wang[2], Yu-Tian Huang[2]

1School of Mechanical and Electrical Engineering, Central South University, Changsha 410083, China
2Light Alloys Research Institute, Central South University, Changsha 410083, China
Email: science@csu.edu.cn

By the means of OM, Hot-tensile test and SEM, we study the austenitic phase transformation kinetics and thermoplastic behaviors of 35CrMo steel during the process of temperature-controlled die forging. At the same time, the coarsening temperature of austenite grain growth and the better thermoplastic temperature range of steel are proven. The experimental results show that, when the holding time is certain, the austenite grain size of 35CrMo steel is growing exponentially with the increase of heating temperature. The austenite grain coarsening temperature is 950 °C. When the heating temperature is certain, the austenite grain size is approximately of parabola growth with the prolonging of holding time. In order to obtain the uniform size of austenite grain, the holding time should be controlled in 1h or so. Meanwhile, the true stress-true strain curves of the hot tensile, the changes of the area reduction at different temperature and the fracture morphology of SEM are comprehensively considered to show that the heating temperature should be controlled in 950 °C ~1000 °C to ensure good thermoplastic.

Keywords: 35CrMo steel; heating temperature; holding time; austenite grain; thermoplastic.

1. Introduction

For steel, the grain size of austenite at high temperature directly affects the characteristics of austenite transformation and the microstructure and mechanical properties of the transformation products during the subsequent cooling [1, 2]. Grain refinement is currently the only enhanced way to improve the strength and toughness of steel at the same time. The smaller the austenite grains at high temperature, the smaller the microstructure of austenite transformation at room temperature, so the higher the strength and toughness of the steel [3].

Thermoplastic behavior of steel reflects resistance to various stresses in steel under high temperature plastic deformation. In general, high-temperature tensile section shrinkage and tensile strength are the index of high temperature mechanical properties of steels. It is the basis of order and perfect manufacturing technology to study the thermoplastic behavior and its change rules of steel [4]. It is necessary to study the high temperature thermoplastic behavior of 35CrMo steel, to ensure that the stress of 35CrMo steel under high temperature will not exceed its allowable stress, in order to avoid the internal crack [5, 6, 7]. The heating temperature and the holding time of austenitizing have a great influence on the grain size of the forging, which affects the final microstructure and properties [8]. Therefore, for 35CrMo steel at high temperature, the austenitic phase transformation kinetics and thermoplastic behaviors were studied, proven the coarsening temperature of austenite grain growth and the better thermoplastic temperature range of steel, so as to provide a theoretical basis for the subsequent hot working process and heat treatment process of 35CrMo steel, as well as, is of important significance to improve the strength and toughness of 35CrMo steel.

2. Experimental Materials and Procedures

The chemical compositions (wt. %) of 35CrMo steel in electromagnetic casting are shown in Table 1.

---
*Corresponding author

Table 1 Chemical compositions of 35CrMo steel (wt %)

	C	Si	Mn	Cr	Mo	S	P	Al	Fe
Standard	0.32~0.4	0.17~0.37	0.4~0.7	0.8~1.1	0.15~0.25	≤0.03	≤0.03	≤0.03	Bal.
Measure	0.34	0.21	0.56	0.95	0.19	0.0051	0.019	0.0032	Bal.

Austenite grain growth: The samples were divided into 6×6 groups. The box-type resistance furnace KBF1400 was employed to heat the samples. The samples were heated to different temperatures (850 °C, 900 °C, 950 °C, 1000 °C, 1050 °C, 1100 °C and 1150 °C) with the heading rate of 10 °C /min. After the heating treatment, the samples were quenched into water immediately. Then, the samples were grinding and polishing. Next , the samples were corroded with the corrosive liquid ratio of picric acid 2g + 1-2g detergent + 50ml of distilled water, and kept for 5min-10min. The austenite grains were observed by optical microscope Olympus DSX500 for the grain morphology. According to GB/T 6394-2002 "Metal average grain size determination methods", the average grain size of austenite grain can be assessed by concentric round measurement [9].

Thermoplastic: The hot-tensile tests of the samples with a size of φ6mm×120mm were carried on Gleeble-1500 thermal simulation machine at a strain rate of $0.1s^{-1}$. The samples were heated to different temperatures (850 °C, 900 °C, 950 °C, 1000 °C, 1050 °C, 1100 °C and 1150 °C) with the heating rate of 10 °C /s, and held at the temperature for 120s. When the samples are fractured, the samples were quenched into water to room temperature immediately. The high-temperature tensile fracture morphology and longitudinal section of tissue were observed by using the SEM. We analyzed the impact of high temperature mechanical deformation temperature steel properties and fracture mechanisms.

3. Results and Discussion

3.1. *The austenitic phase transformation kinetics of 35CrMo steel*

Quenching the samples after heated to over Ac_1 in different heating temperatures and holding time, the austenite grain size is increasing with the increase of heating temperatures and holding time [10]. Table 2 summarizes the average austenite grain size of 35CrMo steel at different heating temperatures and holding time.

Table 2 Average austenite grain size at different temperatures and holding time (μm)

Temperature/°C	Holding time/min					
	0	30	60	90	120	150
850°C	16.92	17.86	19.45	20.14	20.56	20.72
900°C	17.63	18.81	22.33	23.18	24.05	24.84
950°C	24.60	29.83	31.86	32.58	33.13	33.48
1000°C	54.68	61.28	69.77	72.89	74.76	76.57
1050°C	78.27	86.19	103.72	112.53	117.31	120.56
1100°C	137.39	150.42	176.59	187.93	194.03	199.27

3.1.1. *The effect of heating temperature on austenite grain growth behavior*

Figure 1 shows the austenite grain morphology of 35CrMo steel at different temperatures holding for 30 min. And the variation of austenite grain size with temperature holding for 30 min is shown in Figure 3(a). It can be seen from Figure 3(a), when the heating temperature is 850~950 °C, the austenite grains size increases with the heating temperature slowly increases, within the temperature range to maintain a relatively uniform grain size; when the temperature rises to 950~1000 °C, a portion of the austenite grains grow rapidly with increasing temperature, another portion of the grains remain small; when the temperature continues to rise to over 1050 °C, small grains are basically disappeared. From the experimental results, the coarsening temperature of austenite grains is 950 °C. When the holding time is constant, usually obey Arrhenius relationship between the large angle grain boundary mobility and temperature [11]:

$$M = M_0 \exp(-Q / RT). \tag{1}$$

Where, M_0 constant related to the material, Q for the grain boundary migration of surface activation energy, J / mol; R is the ordinary gas constant, 8.31J / (mol • K); T is the heating temperature, K. From the formula (1), austenite grain growth rate increases with heating temperature increase exponentially relationship. Figure 3(a) curve has better response to this rule.

With the austenitizing temperature and the superheat increases, the critical nucleus radius decreases, and to accelerate the formation of austenite speed; in addition, austenite grain growth process is the annexation process of large grain to small grain. The total number of grains is dropping, leading to reduction in grain boundary area, the total surface energy decreases accordingly, and the grain growth is accelerating.

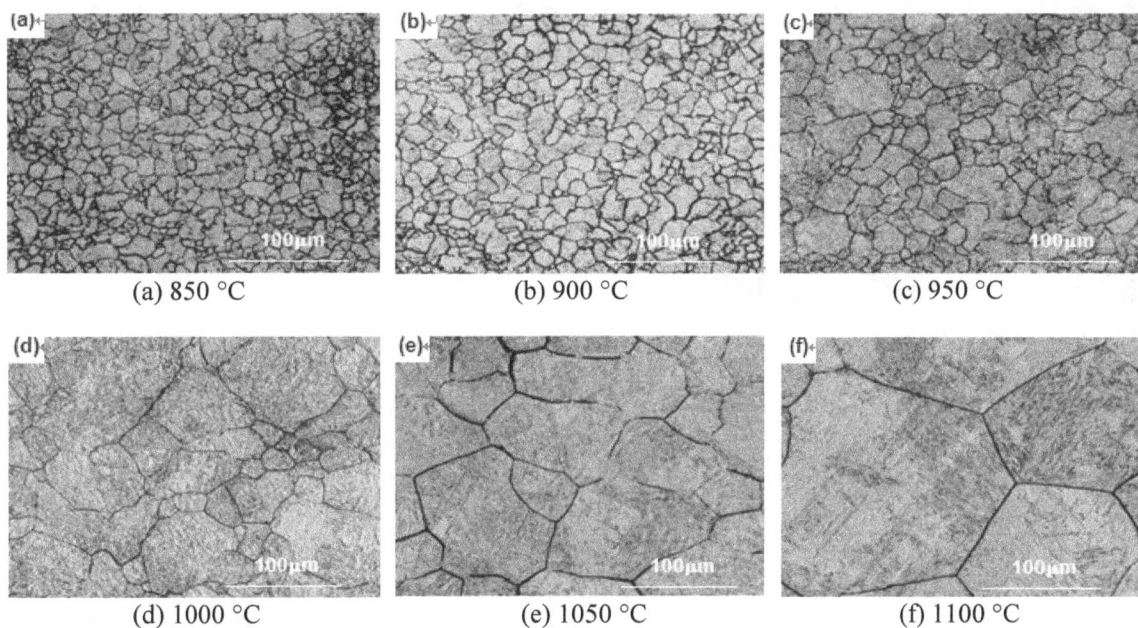

| (a) 850 °C | (b) 900 °C | (c) 950 °C |
| (d) 1000 °C | (e) 1050 °C | (f) 1100 °C |

Fig. 1 Austenite grain morphology of 35CrMo steel at different temperatures holding for 30 min

3.1.2. *The effect of holding time on austenite grain growth behavior*

Austenite grain morphology is shown at different holding time at 950 °C in Figure 2. The change curve of average austenitic grain size with the holding time at 950 °C is shown in Figure 3(b). According to Figure 2 and Figure 3(b), the original austenite grain size increases with the prolonging of holding time, when the austenite grain grows up to a certain size, the growth process slows down until it stops growing. The effect of holding time on the growth of austenite grain is much less than that of the heating temperature, as compared with Figure 3(a) and Figure 3(b).

When normal grain growth in isothermal conditions, the average grain size of austenite and holding time is approximately a parabola, the process can be described with Beck equation [3]:

$$D = Kt^n \qquad (2)$$

Where, D is the average grain size of austenite grain growth at time t, μm; K, n for the constants related to material and temperature; t is the holding time, S.

It can be seen from Figure 2, in the early stage of thermal insulation, the initial austenite grain size is not uniform, which provides the driving force of the austenite grain growth, lead to grains growing up quickly. However, with the increase of holding time, the austenite grain size tends to be uniform, the grain boundaries become more and more straight, the driving energy is obviously decreasing, and the grain growth rate also tends to be smooth. Then, in order to obtain the uniform size of austenite grains, at the same time, to ensure the uniform diffusion of elements in the heating process, the holding time should be controlled at about 1h.

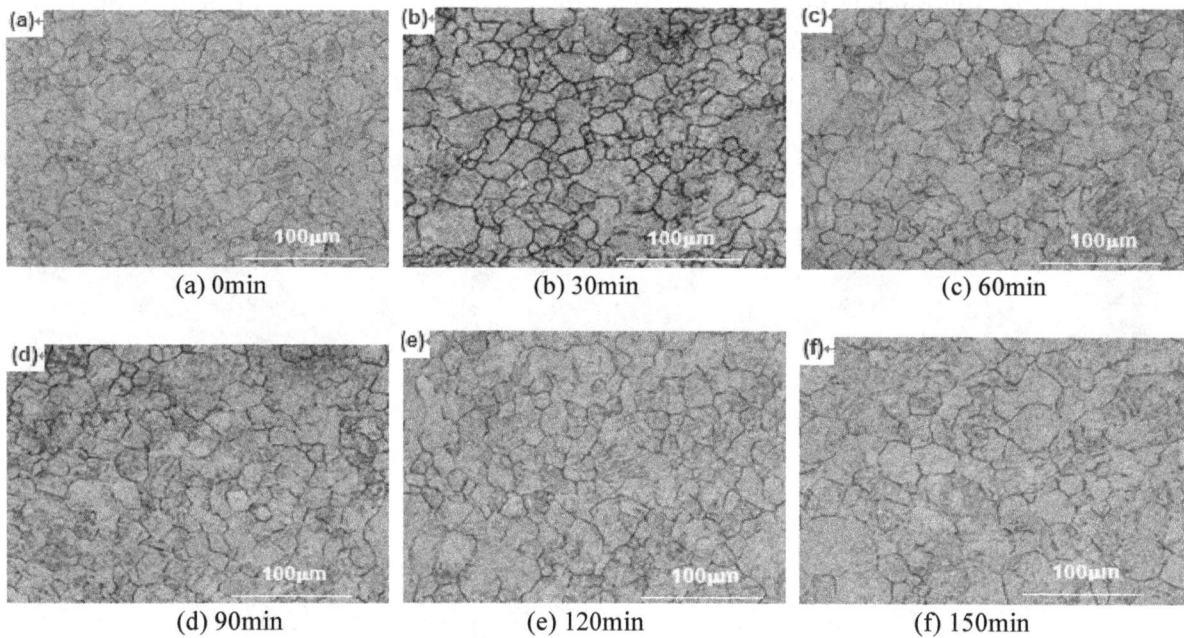

(a) 0min   (b) 30min   (c) 60min

(d) 90min   (e) 120min   (f) 150min

Fig. 2 Morphology of austenite grains heated at 950 °C with different holding time

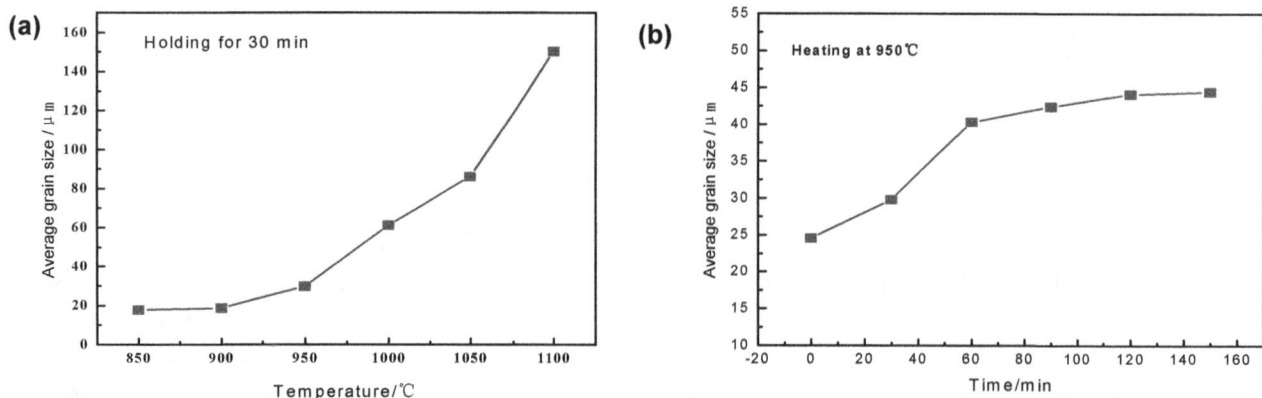

Fig. 3 (a) Variation of austenite grain size with temperature holding for 30 min; (b) Curve of average austenitic grain size with the holding time at 950°C

In the austenitizing process of 35CrMo steel, considering the effects of heating temperature, holding time and initial austenite grain size, we deduces the austenite grain growth model of 35CrMo steel. The calculated results of model are in agreement with the experimental data.

3.2. 35CrMo steel thermoplastic behavior at high temperature

3.2.1. The effect of deformation temperature on thermoplastic

Figure 4(a) shows that the high-temperature tensile curves of 35CrMo steel under different deformation temperatures at a strain rate of $0.1s^{-1}$. By the figure can be seen, the true stress-true strain curves show obvious peak characteristics at different deformation temperatures, the peak stress gradually decreases with the increase of deformation temperature. The thermoplastic curves of 35CrMo steel are shown in Figure 4(b), 35CrMo steel has good plasticity in the temperature range of 850~1150 °C. With the increase of temperature, the area increases is decreasing gradually, and are more than 90%. The tensile strength decreases approximately linearly with the increase of temperature. Usually, the section shrinkage rate of 60% is defined as the dividing line between the low and high plasticity. Therefore, the 35CrMo steel showed good thermoplastic in the experimental temperature range. Especially when the deformation temperature up to 1000 °C above, the reduction of area reached more than 99%, as the temperature continues to rise, the reduction of area is close to 100%, showing a fairly good plasticity. When the temperature rises to 1150 °C, the plastic deformation is reduced. mainly because, with the increase of temperature, the grain coarsening, and grain boundary strength gradually decreasing, and the initiation crack at the grain boundary in tensile process, which makes tensile samples break down as the lower deformation [12]. The tensile strength and deformation temperature are linearly fitted by Origin, and the relationship between them is as follows:

$$\sigma_b = -0.487T + 612.57 \qquad (3)$$

Where, σb is the tensile strength, MPa; T is the deformation temperature, °C.

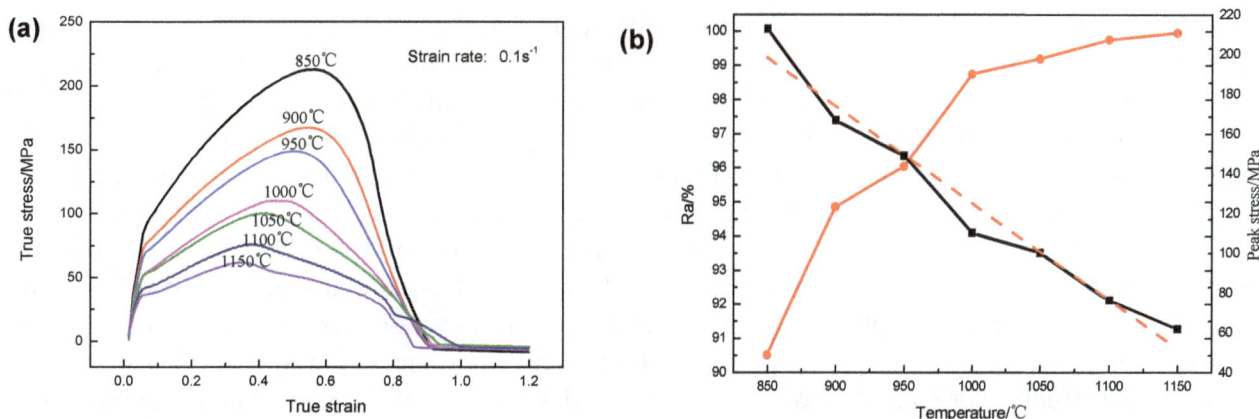

Fig. 4 (a) high temperature tensile fracture morphology; (b) the thermoplastic curves of 35CrMo steel

3.2.2. *Fracture morphology of SEM at different deformation temperatures*

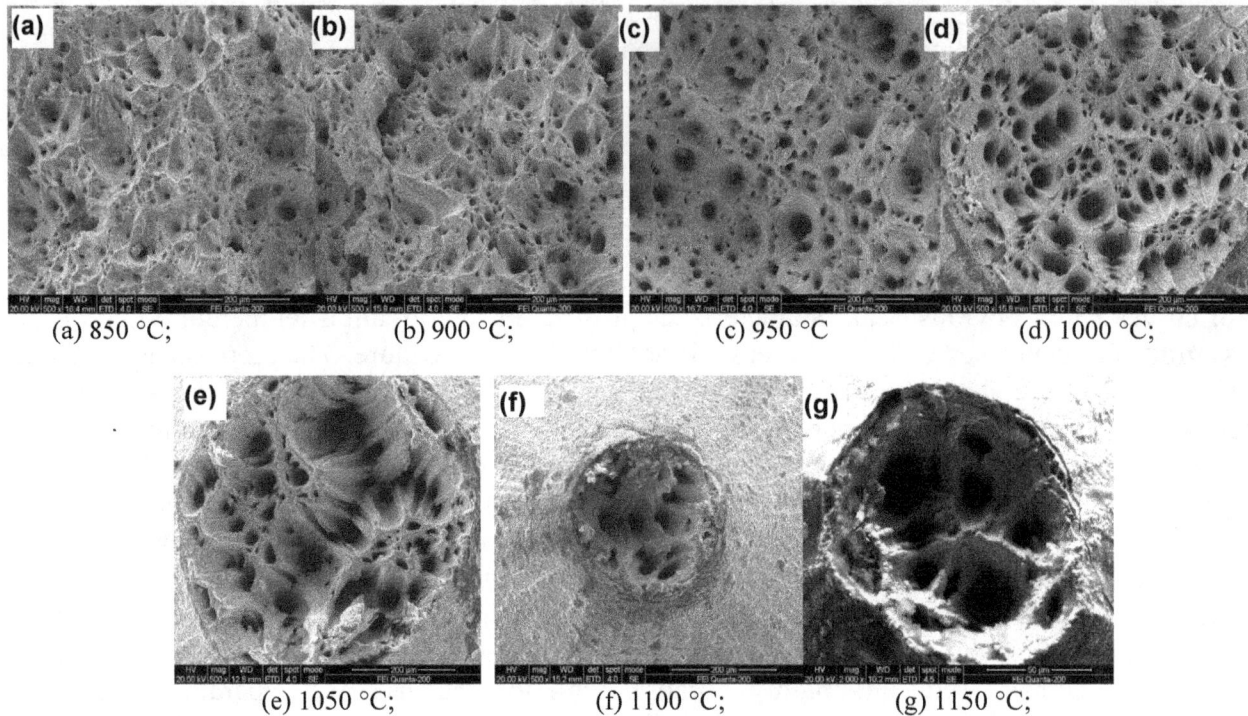

(a) 850 °C; (b) 900 °C; (c) 950 °C (d) 1000 °C;

(e) 1050 °C; (f) 1100 °C; (g) 1150 °C;

Fig. 5 The fracture morphology of SEM at the different deformation temperatures

Figure 5 shows that the Fracture morphology of SEM of high-temperature tensile samples under different temperatures at a strain rate of $0.1s^{-1}$. The figure shows, the fracture of 35CrMo steel samples in the temperature range of 850 °C ~1150 °C is basically all of the dimple fracture, which showed ductile fracture. With the increase of the deformation temperature, the depth of the dimple increases, the diameter increases, the fracture area is reduced gradually. Through the analysis of the morphology and size of the dimple, the plasticity of the material can be better assessed. The deeper the fracture dimple, the larger the diameter, shows that the better the plasticity of the material.

As shown in Figures 5 (a)~(c), when the deformation temperature is in the range of 850 °C ~950 °C, you can see the different diameter size of equiaxed dimples, which is the typical ductile fracture, and larger plastic deformation is found around the dimples, but the depth of dimples is shallow. As shown in Figures 5 (d)~(f), when the deformation temperature rises to the range of 1000 °C ~1150 °C, the obvious plastic deformation is found around the dimples, the fracture is fibrous, and fiber area of the center don't see the second crack and ductile tearing, shows the better morphology of equiaxed dimples. With the increase of deformation temperature, recovery and recrystallization occurred in the internal metal, the critical shear stress is reduced, slip systems is increased, the structure organization of metal is changed, the effect of thermoplastic is strengthened, and the effect of grain boundary sliding to strengthen. It can be known that the plasticity of the 35CrMo steel sample increases with the increase of temperature in the temperature range of 850 °C ~1150 °C.

Comprehensive the true stress-true strain curves of the hot tensile, the changes of the reduction of area at different temperature, and the analysis of the fracture morphology of SEM, when the deformation temperature is 850 °C ~1150 °C and the strain rate is $1\ s^{-1}$, 35CrMo steel has good plasticity, and there is no brittle temperature range. When the temperature is above 1000 °C, the section shrinkage rate is above 99%, and the 35CrMo steel has excellent thermoplastic in the temperature range of 1000 °C ~1150 °C. In the process of the temperature-controlled die forging, it is not only to prevent the excessive growth of original austenite grain caused by excessive temperature in the early heating process, but also to prevent the crack generation because of too large forging stress in the process of the temperature-controlled die forging. Considering the kinetics of austenite transformation and the changing rule of the thermoplastic behavior of 35CrMo steel, when the temperature rises above 1000 °C, the increase of plasticity is not obvious, but, when the temperature is higher than 950 °C, the austenite grain coarsening significantly. Therefore, in order to obtain the microstructure of homogeneous fine-grain after forging, the temperature should be controlled in 950 °C ~1000 °C in the process of the temperature-controlled die forging.

4. Conclusion

1. When the holding time is certain, the austenite grain size of 35CrMo steel is growing exponentially with the increase of heating temperature, and the austenite grain coarsening temperature is 950 °C. When the heating temperature is certain, the austenite grain size of 35CrMo steel is approximately parabola grew up with the prolonging of holding time. But the effect of holding time on austenite grain growth is much less than that of temperature. In order to obtain the uniform size of austenite grain, the holding time should be controlled in 60min or so.
2. In the austenitizing process of 35CrMo steel, considering the effects of heating temperature, holding time and initial austenite grain size, we deduces the austenite grain growth model of 35CrMo steel. The calculated results of model are in agreement with the experimental data.
3. The temperature range of good plasticity is 1000~1150 °C. When the temperature is above 1000 °C, the section shrinkage rate is above 99%.
4. Considering the effects of temperature on phase transformation kinetics and thermal plasticity of 35CrMo steel, in order to obtain the homogeneous fine-grain of austenite grain and good thermoplastic, the heating temperature should be controlled at 950 °C ~1000 °C in the process of the temperature-controlled die forging.

Acknowledgment

The authors are grateful for the funding by the National Program on Key Basic Research Project of China (No.2014CB046702).

References

1. Jun-Ru LI, Le-Yu ZHOU, Chen GONG. Growth behavior of austenite grain in heat resistant steel 10Cr12Ni3Mo2VN, Transactions of Materials and Heat Treatment, in Chinese, S1, (2014) 106-111.
2. Bin-Feng LU. The surface composite layer and wear resistance of chromium carbide by high energy beam in situ synthesis, Shanghai: Shanghai Jiao Tong University, 2010.
3. Li-Guang HU, Xi-Wen XIE. Heat treatment of steel. Fourth Edition, Xi'an: Northwestern Polytechnical University Press, (2014) 14-37.
4. Yong-Ming YAN, Ya-Zheng LIU, Sheng XU. Hot ductility behavior and fracture mechanism of 23CrNi3Mo steel, Transactions of Materials and Heat Treatment, in Chinese, 06 (2014) 80-84.
5. Qing-Yun Sha, Zu-Qing Sun. Grain growth behavior of coarse-grained austenite in a Nb–V–Ti microalloyed steel, Materials Science and Engineering: A, 523 (2009) 77-84.

6. Zeng-Min Shi, Kai Liu, Mao-Qiu Wang. Thermo-mechanical properties of ultra high strength steel 22SiMn2TiB at elevated temperature, Materials Science and Engineering: A, 528 (2011) 3681-3688.

7. GW Fan, J Liu, PD Han. Hot ductility and microstructure in casted 2205 duplex stainless steels, Materials Science and Engineering: A, 515 (2009) 108-112.

8. Le-Fei SUN, Fu-Ming WANG, Su-Fen TAO. Austenite grain growth behavior of locomotive wheel steel, Journal of University of Science and Technology Beijing, 36 (2014) 301-307.

9. Wen-Tao ZENG, Yan LUAN, Qiang Gu, Zhong-Ping LIU. Metal-methods for estimating the average grain size, Fushun: China Metallurgical Information and Standardization Institute, (2002) 2-22.

10. Zhe JIN. Study on microstructure properties and corrosion resistance mechanism of weathering bridge steel, Shenyang: Northeastern University (2013).

11. Yong-Ning YU. Principle of Physical Metallurgy, Beijing: Metallurgical Industry Press, 366 (2000).

12. HJ McQueen, S Yue, ND Ryan, Hot working characteristics of steels in austenitic state, Journal of Materials Processing Technology, 53 (1995) 293-310.

Research on the Solution of Insufficient Stretching in Shallow-drawing Process of Car Roof Panel

Cheng-Yong Wang*, Yuan-Yuan Yao, Zi-Ren An, Ming Cheng, Cheng Dai, Fang-Fang Guo

School of Materials Science and Engineering, Hefei University of Technology,
Hefei, P.R. China
Email: wangchengyong@hfut.edu.cn

Based on the shallow-drawing analysis of car roof panel, the problem of insufficient stretching during the drawing process was studied. Reasons and solutions were discussed. The finite element software was used for the numerical modeling of the drawing process. Three parameters, namely distribution of the lock force percentage, distance of the drawbeads and blank holder force, were testified with the orthogonal experimental design method. The influence of each parameter on insufficient stretching of the blank was analyzed with other two parameters unchanged. According to the results of orthogonal experiment, the process parameters were optimized to control the defects of panel during drawing process. The research can be a valuable reference for solving the deficient stretch in the process of drawing.

Keywords: Orthogonal test; finite element method; sheet forming.

1. Introduction

Shallow-drawing small-curvature car cover panel accounts for a considerable proportion of automobile panel parts, which has structural features such as large dimensions, small depth and small curvature. Thus, this kind of parts is easy to generate the issue of insufficient stretching, which leads to lacking of stiffness of parts and to make great noise [1-2].

Car roof panel is a typical shallow-drawing small-curvature cover part. Due to a larger number of other covers linked to it in the process of assembling, smooth appearance is required and defects such as cracking, wrinkling, drawing marks, depressions and ripples are definitely not allowed. The drawing process of car roof is prone to having forming quality problems in the car roof stamping process. Whether it is available or not relates to the appearance quality and application performance of the car roof [3].

This paper analyzed the problem of insufficient stretching of car roof panel in the process of drawing and several available solutions were examined via using the finite element software DYNAFORM, such as adopting the addendum to increase the drawing depth in design of drawing parts, reasonable distribution of lock force percentage in the die design, proper blank holder force.

2. Structure Analysis and CAE Modeling of Car Roof Panel

The 3D schematic of the car roof is shown in Figure 1.

Fig. 1 Schematic of car roof panel

The material of the car roof panel is SPCD36, and the thickness is 1mm. The outline dimension is about 2445mm×1287mm.

*Corresponding author

As a large shallow-drawing part, there is a window in the front of it, and technology holes lie in the center and surrounding of it. There are also flanges around the window and the two sides, and the tail shape models complexly. The undercut exits both in the front and the tail of the car roof. On the surface of car roof also evenly distribute decorated ribs. Due to its shallow drawing depth and complex space curved surface, car roof is easy to form some defects, such as material insufficient stretching, lacking of stiffness, spring back and poor dimension accuracy [4-5]. In order to meet the requirements of the drawing process, it is necessary to modify the undercut, unfold flange, create binder and add inner and outer addendum in DYNAFORM. The schematic of modification of drawing part through adding technology is shown in Figure 2.

Fig. 2 Drawing part of car roof

As shown in Figure 2, in the most regions of the car roof, the contact pressure is small compared with the flow stress of the sheet and it is usually acceptable to assume plane stress deformation. In the zone A, because the curved surface is gently curved, and the friction force imposed on the sheet is great, the tension increases slowly along the cross section line. Due to this, the region of zone A is inadequately plastically deformed. At the corner radius, namely the zone B, both tension and compression stress act on it, and the smaller the corner radius is, the smaller the plastic deformation zone is. According to the level that tension and compression account for, rupture or wrinkle will occur. The zone C is a straight side, the material is drawn inwards, which generates radial deformation and small tangential deformation. Thus, it can be ruptured easily. The zone D is the flange of the drawing part, whose material partly transforms into straight side. The material is exerted on both tensile stress in radial direction and compressive stress in tangential direction when imposed on binder. Thus it is elongated in radial direction and is compressed in tangential direction, leading to the increase of the thickness and the appearance of wrinkles.

3. Numerical Model Setup

3.1. *Material properties*

In the numerical simulation, considering the anisotropy of sheet metal, SPCD36 steel is selected. Detailed mechanical properties of the material are shown in Table 1.

Table 1 Main mechanical properties of SPCD36 steel

Elastic modulus E/GPa	Yield ratio σ_s/σ_b	Poisson's ratio μ	Hardening rule n	Thick anisotropy index r
207	0.70	0.28	0.233	1.65

3.2. *Drawbead design*

In the drawing process of the car roof, the feed resistance of different parts is different. Due to the large area of car roof and the shallow drawing depth, it is prone to causing insufficient drawing. Thus, drawbeads were settled to adjust the feeding resistance by changing the parameters of the different parts and control the feeding quantity and the speed of material flowing into the die. The blank of the drawbeads outside undergoes a series of bending or unbending deformation when it glides on the drawbeads, so the blank generates severe plastic deformation and strain hardening, enhancing the stiffness and the anti-instability capability. In the finite element modeling, drawbeads are divided into two categories, real drawbeads and equivalent drawbeads. If adopt the former, there were a lot of work to do, such as surface meshing. On the contrary, the computing scale can be reduced remarkably [6].

The size of equivalent resistance and the distribution of drawbeads are identified by some factors such as the shape of drawing part, drawing depth and material flowing. It is evident that the flow resistance in corner is greater than in straight side and there is local shape forming in the end. Various factors were taken into consideration, it is assumed that open drawbeads are adopted, drawbeads of different parts are distributed with different percentage of resistance and different drawbead distance (the distance between drawbead centerline and trimming line) [7]. The drawbeads design is shown in Figure 3.

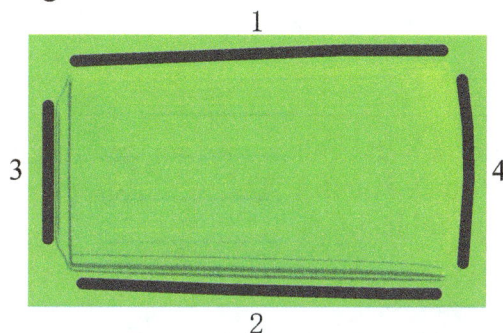

Fig. 3 Drawbead distribution

3.3. *Blank holder force loading*

In the process of drawing, the blank holder force is calculated by formula (1).

$$F = AF_q \qquad (1)$$

Where F_q is the unit blank holder force, A is the effective area of the contact between the plate and the blank holder. It can be obtained that A is 1.17m2. The unit blank holder force is 2 MPa. Therefore, 2000kN, 2250kN and 2500kN are selected as the three levels of the blank holder force, respectively.

4. Process Parameter Optimization

4.1. *Orthogonal experimental design*

In order to investigate the influence of the distribution of drawbeads resistance and the drawbead distance impose on the blank forming, three different levels of them are designed, respectively. (d1, d2, d3, d4 represent the distance between the drawbead 1, 2, 3, 4 and trimming line, respectively). As shown in Table 2 and Table 3.

Table 2 Distribution of drawbead resistance level

Process conditions	Lock force percentage of drawbead 1, 2(%)	Lock force percentage of drawbead 3(%)	Lock force percentage of drawbead 4(%)
A1	25	22	28
A2	30	27	33
A3	35	32	38

Table 3 Drawbead distance level

Process conditions	Drawbead distance of 1, 2(mm)	Drawbead distance of 3, 4(mm)
B1	60	40
B2	70	50
B3	80	60

With the other parameters constant, the distribution of drawbead resistance, the drawbead distance and the blank holder force are regarded as three factors of related three levels. As shown in Table 4.

Table 4 Orthogonal test factors and levels

Items	A	B	C
1	A1	B1	2000(C1)
2	A2	B2	2250(C2)
3	A3	B3	2500(C3)

The minimum thickness after the sheet deformed, inadequate drawing area and the ruptures, are defined as the evaluation indexes of quality. The results of orthogonal experiment arrangement and evaluation index are shown in Table 5.

Table 5 Orthogonal test and the results evaluation

Number	Factors			Results		
	A	B	C	Minimum thickness(mm)	Area of deficient stretching	Fracture
1	A1	B1	C1	0.785	46%	no
2	A1	B2	C2	0.823	48%	no
3	A1	B3	C3	0.824	35%	no
4	A2	B1	C2	0.824	35%	no
5	A2	B2	C3	0.642	8%	yes
6	A2	B3	C1	0.775	12%	no
7	A3	B1	C3	0.742	4%	yes
8	A3	B2	C1	0.691	0	yes
9	A3	B3	C2	0.749	6%	no

4.2. *Simulation result analysis*

According to the results of orthogonal experiment, the influence of the factors on the minimum thickness of the parts is analyzed by using the minimum thickness data. The summation of the minimum thickness of the blank are defined as I_j, II_j, III_j (j = A, B, C), which are the corresponding factors to the level of the factors (1, 2, 3). Similarly, $\underline{I_j}$, $\underline{II_j}$, $\underline{III_j}$ is the average value of the minimum thickness of the blank. K indicates the range of the j column, k=max ($\underline{I_j}$, $\underline{II_j}$, $\underline{III_j}$) -min (I_j, II_j, III_j). Range calculation is shown in Table 6.

Table 6 Range computing

Items	A	B	C
I j	2.432	2.351	2.251
II j	2.241	2.156	2.396
III j	2.182	2.348	2.208
I j	0.811	0.784	0.750
II j	0.747	0.719	0.799
III j	0.727	0.783	0.736
k	0.084	0.065	0.063

From the Table 6, it can be obtained that the effect of each factor imposing on the minimum thickness of the plate is ordered by their extent, as follows: A(the distribution of drawbead resistance)>B(the drawbead distance)>C(the blank holder force). According to the result of orthogonal experiment, the combination of the smallest thickness in the safe deformation area is A3B2C3. By regulating the parameters of the combination of the smallest thickness slightly, namely A3B2C3, according to the result of orthogonal test, we acquired a set of optimization parameters as follows, the drawbead restraining force distribution is 34% (1, 2drawbead) ,31% (3 drawbead), and 37% (4 drawbead), drawbead distance is 70mm, 50mm, respectively, blank holding force is 2490kN. With input of the optimized parameters into FE model, forming limit diagram for the nodes in the part and thinning distribution are shown in Figure 4 and Figure 5, respectively.

Fig. 4 Forming limit diagram with optimized parameters

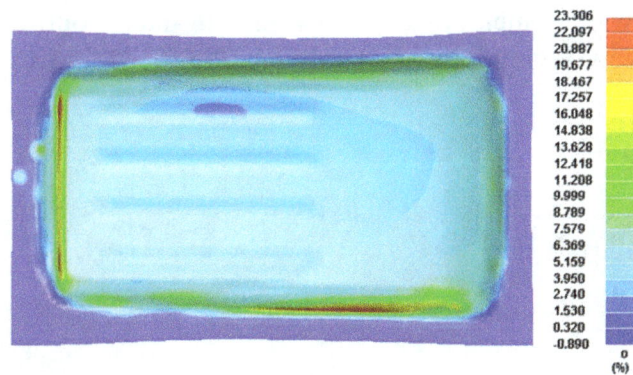

Fig. 5 Thinning distribution under optimized parameters

As Figure 4 and Figure 5 shown, blank deformation is sufficient without fractures after the parameters are optimized and the maximum thinning rate reaches 23.3% which meet with the quality requirements.

5. Conclusion

This paper analyzed the deformation characteristics of the car roof panel, which is the typical shallow-drawing small-curvature car cover. It has been shown that insufficient stretching is one of the main defects in the process of drawing. Based on the mechanism of sheet forming and die design, some plans such as adopting the addendum to increase the drawing depth of draw parts, setting up reasonable distribution of lock force percentage in the die design and selecting proper blank holder force in stamping, were introduced to solve the problem of insufficient stretching.

The finite element software DYNAFORM was used to establish drawing model and to identify the reliability of the solutions above. Additionally, the orthogonal experiment was used to obtain the influence degree of three factors, such as the distribution of drawbead resistance, the drawbead distance and the blank holder force. According to the result of experiment, a set of optimization was obtained. The result from this work can be a guidance for the process of drawing of car roof panel for auto industrial manufacturing.

References

1. Lamei Shi, Leigong Wang, Yue Zhang, Formability stiffness improvement of shallow-drawing small-curvature car cover panel based on numerical simulation, Forging & Stamping Technol. 6, 37 (2012). In Chinese.
2. Xueyan Ren, Deforming characteristics and quality measures of shallow-drawing small-curvature car cover, Die & Mold Industry 11, 237(2000). In Chinese.
3. Lingjiang Cui, Stiffness of auto-cover and its control. Metal Forming Machinery. 6, 36 (2001). In Chinese.
4. Qinsheng Li, Wei Chen, Zhenrong Huang, Research on forming process of automobile cover and technological parameter, Forging& Stamping Technol. 5, 36(2011). In Chinese.
5. Ishigaki H, Okamoto I, Nakagawa N, Analysis of growth and disappearance of surface deflection in press forming of large-size autobody panels, International Journal of Vehicle Design. 56, 240 (1985).
6. Wang N M, A mathematical model of drawbead forces in sheet metal forming, Journal Of Applied Metalworking. 9, 193 (1982).
7. Xiao Sun, Front wall inner plate drawing die-face optimum design based on ETA/DYNAFORM, New Technology and Craft. 35, 79 (2013). In Chinese.

Study on Preparation of Porous Beta-tricalcium Phosphate Scaffold by In-situ Decomposition Method

Lin Yao[1], Gui-Lin Yu[1,*], Su Cheng[1], Shu-Cheng Mu[2], Nan Li[2]

[1]College of Medicine, Wuhan University of Science and Technology, Wuhan, China

[2]The Key State Laboratory of Refractories and Metallurgy, Wuhan University of Science and Technology, Wuhan, China

*Email: yuguililn@wust.edu.cn

This paper is to prepare porous beta tricalcium phosphate scaffolds by in-situ decomposition method. The particles of $CaCO_3$ and $NH_4H_2PO_4$ with different sizes were used to react at the different temperatures. Their phase and microstructure were detected. The results of XRD and SEM energy spectrum analysis showed that $CaCO_3$ and $NH_4H_2PO_4$ reacted and generated β-$Ca_3(PO_4)_2$ at high temperature. SEM analysis showed that a lot of pores are formed in the scaffold materials, and the pores are connected and irregular. The average size of pores of $CaCO_3$ powder groups is from 20μm to 90μm, while that of Nano $CaCO_3$ groups is from 30μm to 110μm.

Keywords: Porous scaffold materials; in-situ decomposition method; beta-tricalcium phosphate; solid state reaction.

1. Introduction

The biological scaffold as extracellular matrix of seed cells, which formed a new organizational framework, is one of the three important factors in the tissue engineering. Its properties directly affect the seed cells' adsorption, proliferation, differentiation and other biological characteristics, and at last affect the tissue construction [1]. The ideal scaffold should simulate the morphology, structure, and function of tissues with good biocompatibility and biodegradability, and also have a certain strength and porous structure because the appropriate pore size, porosity and large surface area are contributed to cell's adhesion, proliferation, differentiation, and provide sufficient space for cell's growth.

The traditional production methods of the biological scaffold (such as fiber bonding, solvent casting-particulate leaching, gas foaming method, and phase separation method) are to join the admixture and produce the porous structure, then remove additive by the different ways. Disadvantages of the traditional production methods are getting the same pore size and stomatal pores, and too complicated process. In recent years, with the development of three-dimensional printing technology, there are also some defects in forming the scaffold, for example, the pore size is too small, and the mechanical properties and forming accuracy need to be improved. If the scaffold can be produced the interconnected pores with the different size and porosity by decomposition of the raw materials, and formed in one time, then there will be an ideal method for the preparation of the porous scaffold materials [2-3].

Beta tri-calcium phosphate has good biocompatibility and good biomechanical properties, is a biodegradable material, and is suitable for the biological scaffold materials of tissue engineering. But the beta calcium phosphate powder prepared by previous dry method is coarse, its composition is not uniform, and its purity is not high with miscellaneous phase. The experimental conditions of the wet method and the hydrothermal method are complex, and it is difficult to operate the experiment. The product is just powder, which needs to be re-shaped, so it is difficult to control the three-dimensional porous structure and the strength of scaffold materials. Due to the biological scaffold materials needing high strength to bear certain pressure, the porous structure and high strength itself are contradicting. How to unify the two is a difficult problem in current research [4-7].

*Corresponding author

The in-situ decomposition method is based on the characteristics of decomposition reaction of raw materials with the releases of gas under a certain temperature. The process is as follows: after a variety of raw materials mixed evenly and heated to a certain temperature, the decomposition reaction is taken with gas releasing and left pores in situ, and at the same time the reaction products formed the porous ceramic materials to the required in situ. Its advantages are that pore size and distribution are easy to control, has a simple production process, has a one-time molding, and has no other impurities. In this study, phosphate, calcium carbonate and calcium hydroxide are used as raw materials with the different granularities. The scaffolds are prepared under the different process conditions by in situ decomposition method, in which there are different size pores connected and different porosity. In the previous study, there are few people do research of influence of granularity of raw materials and process conditions on the pore structure and the mechanism of aperture. So it is necessary to study the influence of the granularity of raw materials and process conditions on the pore size and porosity of the scaffold, and to elucidate the mechanism of pore formation [8-9].

We hope that we can change the granularity of raw materials and process conditions to control the pore size and distribution, according to the requirements of different pore structure, in order to make an ideal scaffold material.

2. Materials and Methods

2.1. *Experimental materials*

The granularity of fine $CaCO_3$ powder is 28 μm (purity is 98%, made in Chuanying Powder Plant), the granularity of $CaCO_3$ powder is 40nm (purity is 99%, made in Xi'an Fuda Industrial Co Ltd), and the granularities of $NH_4H_2PO_4$ powder is 0.2mm (purity is 98%, made in Wuhan Chemical Reagent Factory).

2.2. *Experimental procedure*

There are two groups, which are A and B. A is fine $CaCO_3$ powder while B is nano $CaCO_3$ powder. Each group is 390g of $CaCO_3$ and 300g of $NH_4H_2PO_4$ by molar ratio of 3:2, and is placed into an alumina ball grinding tank and mixed grinding for half an hour. Then 30g of mixed material was put into the circle model with diameter of 36mm and height of 30mm, using a hydraulic machine with 100MPa pressure to make 20 for each group. All specimens were placed into a drying oven at 110°C for 24 hours. All specimens of each group were heated at 900°C, 1100°C, 1300°C, and 1500°C for 3h, respectively, which is marked as number 1, number 2, number 3, and number 4 respectively.

2.3. *Detection methods*

2.3.1. *Phase analysis*

Some samples of group A and group B were grinded into powder with a mortar were selected to analysis phase by X'Pert Pro X-ray diffractometer machine made in Philip Company.

2.3.2. *Microstructure analysis*

The microstructure, composition, and distribution of the elements of the prepared samples were examined by a Scanning Electron Microscopy (SEM; Quanta 400; FEI, Hillsboro, OR) and an Electron Microprobe Analysis (EPMA; JX A8800; JEOL, Tokyo, Japan).

3. Results and Discussion

3.1. *Phase analysis*

3.1.1. *XRD analysis*

Results showed that β-$Ca_3(PO_4)_2$ was generated, especially in group A under four different temperatures formed with high purity. In this experiment, $CaCO_3$ and $NH_4H_2PO_4$ powder reacted under the high temperature, and reaction was as follows [10]:

$$3CaCO_3 + 2NH_4H_2PO_4 \rightarrow Ca_3(PO_4)_2 + 2NH_3\uparrow + 3CO_2\uparrow + 2H_2O\uparrow \tag{1}$$

β-$Ca_3(PO_4)_2$ had the characteristic diffraction peak when 2θ is $17°,25°,28°,31°,34°$,and $53°$.There was forming of β-$Ca_3(PO_4)_2$in two groups at 900°C. There was little $Ca_2P_2O_7$ generated in group B (using nano $CaCO_3$ powder), which may be intermediates in the reaction process. There was a lot of hydroxyapatite generated at 1300°C andα-$Ca_3(PO_4)_2$ formed at 1500°Cin Group B, those two phases were calcium phosphate and were commonly used as biomaterials. Why those two phases could be formed in group B, while they could not be formed in group A, the reason is maybe that the chemical reaction is more unstable because of the high activity of nano $CaCO_3$ powder. In the X - ray diffraction pattern, the higher intensity and the more compact, the crystallization degree is better. The crystallization of group A was gradually enhanced from 900°Cto 1300°C, and the optimum condition was reached at 1300°C. While in group B, the optimum condition was reached at 1100°C, which shows that temperature of the reaction of the nano powder can be as low as 1100°C.

The diffraction patterns were shown in Figure 1 and Figure 2.

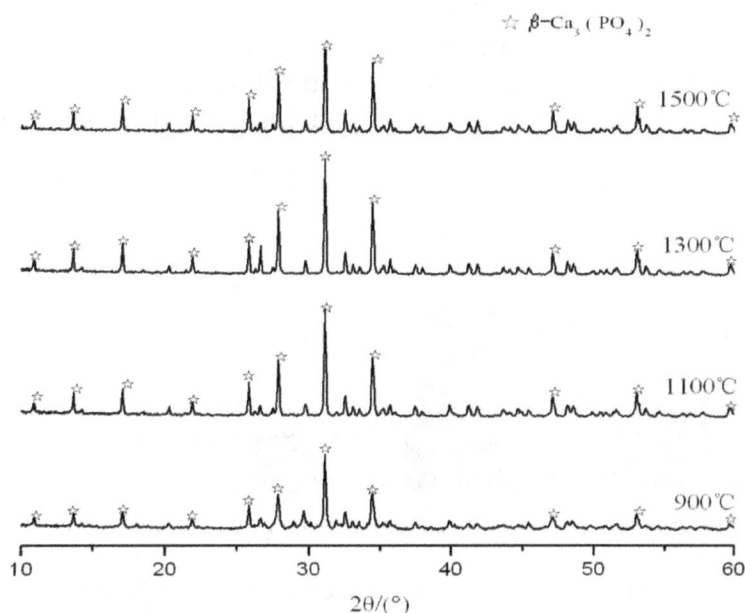

Fig. 1 X- ray diffraction of group A at different temperatures

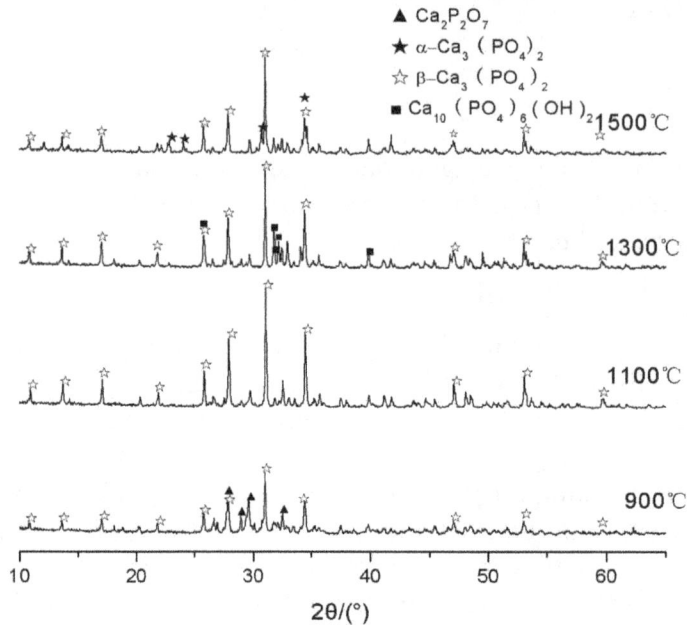

Fig. 2 X- ray diffraction of group B at different temperatures

3.1.2. *SEM energy spectrum analysis*

Results of analysis from scanning electron microscopy with energy dispersive X-ray spectrometer (EDS) also prove that products of the chemical reaction were mostly $Ca_3(PO_4)$.

One point was selected in Figure 3 and got the energy spectrum diagram as shown in the right illustration of Fig. 3. Table 1 was the element composition of the mark points, whose results showed that there were elements of Ca, P, O, and the element composition was similar to $Ca_3(PO_4)_2$.

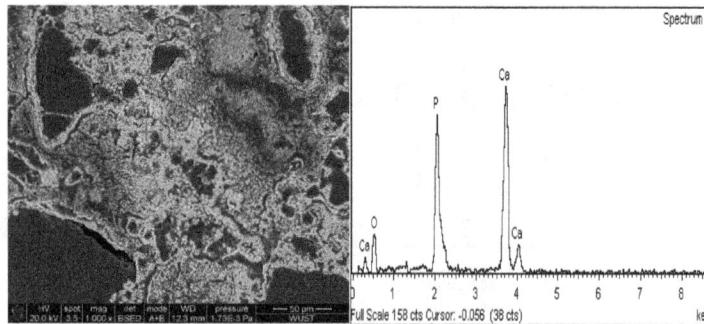

Fig. 3 The SEM diagram of group B at 1300°C and the energy spectrum diagram of the marked point

Table 1 Element composition of the mark points in Figure 3

Element	Weight (%)	Atomic (%)
O K	46.86	66.76
P K	18.04	13.28
Ca K	35.10	19.96
Totals	100.00	

The results of XRD and SEM energy spectrum analysis showed that $CaCO_3$ and $NH_4H_2PO_4$ reacted and generated β-$Ca_3(PO_4)_2$ at high temperature

3.2. *Microstructure of porous materials*

From Figure 4 and Figure 5, it can be found that a lot of pores are formed in the scaffold materials, and are mutually connected and irregular. Pores in group B were significantly larger than that of group A. A large amount of gas is generated in the solid state reaction. A part of the gas is not discharged and formed the air holes so that the scaffold material can have many pores. Because nano $CaCO_3$ has a higher activity, the reaction with $NH_4H_2PO_4$ is very intense, and instantly produce a lot of gas, which is not discharged in time, so there are the larger pores in group B than that of group B.

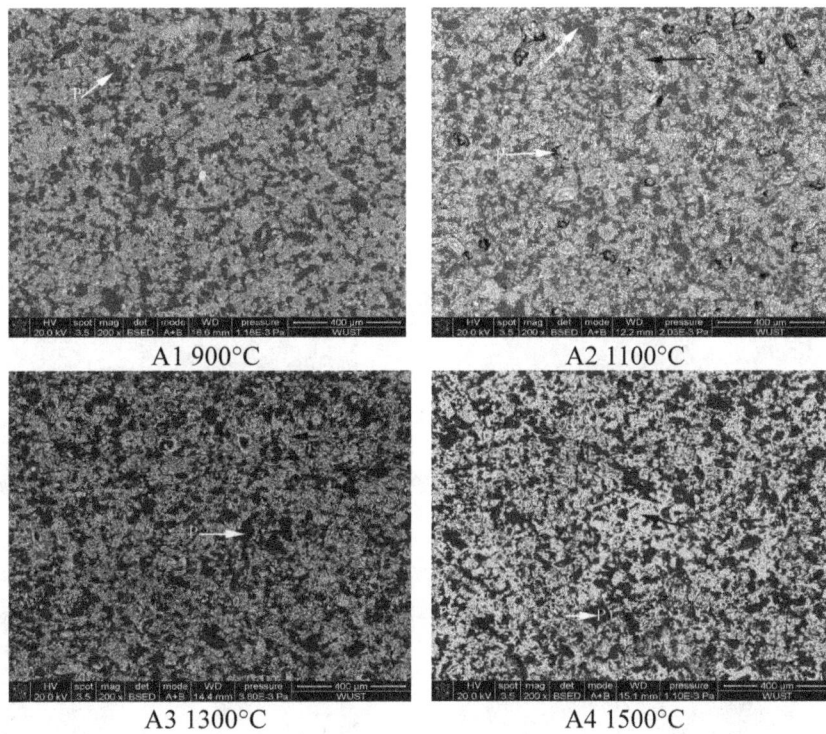

A1 900°C

A2 1100°C

A3 1300°C

A4 1500°C

Fig. 4 SEM of group A under 200X at the different temperatures (the white arrow P represents pore, the black arrow S represents $Ca_3(PO_4)_2$)

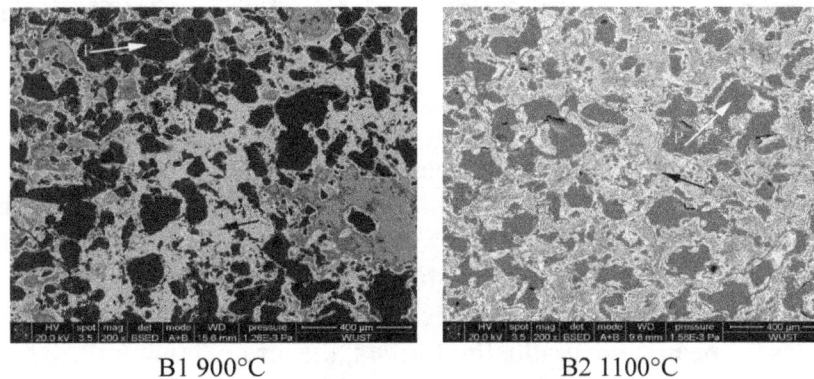

B1 900°C

B2 1100°C

B3 1300°C

B4 1500°C

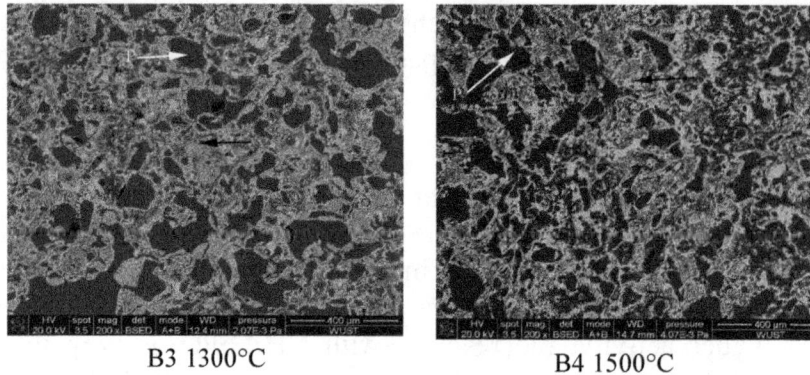

Fig. 5 SEM of group B under 200X at the different temperatures (the white arrow P represents pore, the black arrow S represents $Ca_3(PO_4)_2$)

From SEM under 1000X in Figure 6 and Figure 7, it can be found that the size of the pores is different, and the average size of large pores is from 30μm to 110μm, there are some holes greater than 130μm.

A1 900°C

A2 1100°C

A3 1300°C

A4 1500°C

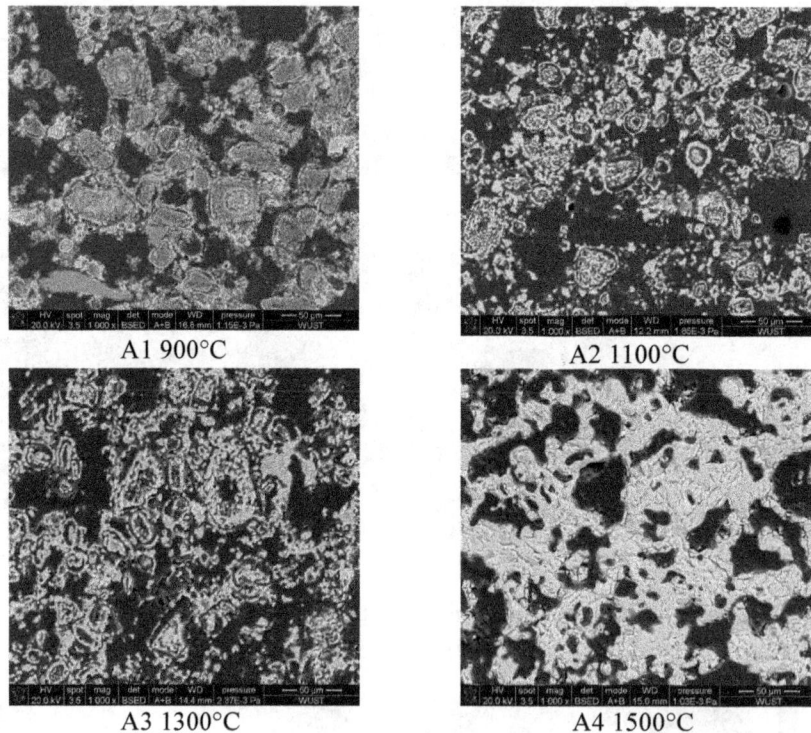

Fig. 6 SEM of group A under 1000X at the different temperatures

Pores in group B was significantly larger than that of group A, which are mutually connected. With the increase of sintering temperature, the crystallization of $Ca_3(PO_4)_2$ is more and more perfect, so a large number of closed pores are formed in group B at 1500°C. The reason is maybe that with the increase of sintering temperature: the contact area of particles is expanded; connection of particles is changed from the neck to the interface; particles are aggregated; center distance of particles is near, and gradually formed the grain boundary. The shape of pore is changed with the volume reduced from the connected holes to the independent closed holes. The size of pore is from 30μm to 110μm, which can provide enough space for the survival of the cells, and provide a channel for the flow of blood and body fluid [11].

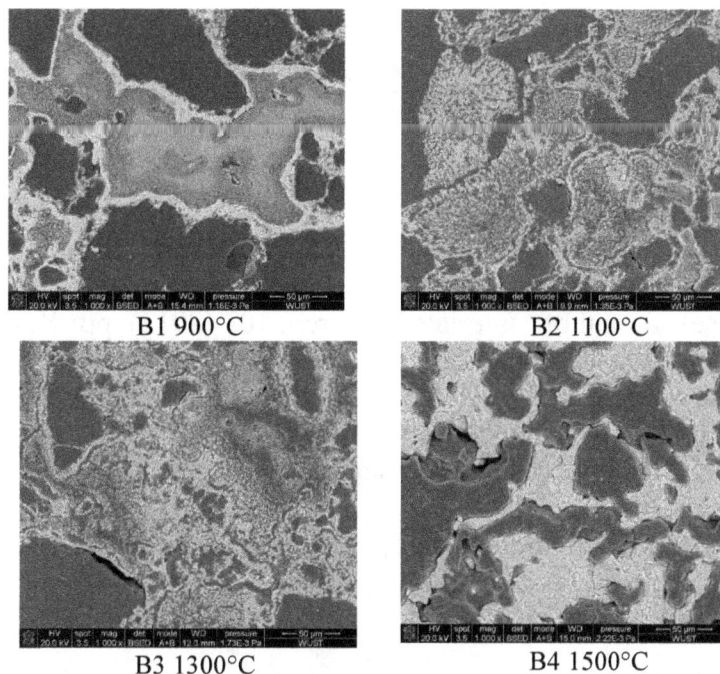

Fig. 7 SEM of group B under 1000X at the different temperatures

4. Conclusion

Porous beta-tricalcium phosphate was prepared by in-situ decomposition method at high temperature using CaCO$_3$ and NH$_4$H$_2$PO$_4$ as raw materials. The average pore size of porous scaffold materials generated by nano CaCO$_3$ reaction was significantly higher than that of micro powder CaCO$_3$.

Acknowledgment

Thank you very much for a lot of guidance and useful discussion of Guilin Yu professor and the senior sister apprentice Shuangyue Zhou of Wuhan University of Science and Technology. Thanks for the leadership's care and support of the School of Medicine, Wuhan University of Science and Technology. Thank you for lots of persistence of the training program (project number: 201310488039) all members!

References

1. Xu Xiao Yu. Biological materials [M]. Beijing: Science Press, 2005. 21-22 (in Chinese).
2. K.S. Jaw. Preparation of a biphasic calcium phosphate from Ca(H$_2$PO$_4$)$_2$·H$_2$O and CaCO$_3$[J]. Journal of Thermal Analysis and Calorimetry, 83 (2006), pp: 145-149.
3. Cijun Shuai, Jingyu Zhuang, Huanlong Hu. In vitro bioactivity and degradability of β-tricalcium phosphate porous scaffold fabricated via selective laser sintering, Biotechnology and applied biochemistry, 60 (2013), pp: 266-273.
4. Zairani, Nur Amyra Shazni. Fabrication and characterization of porous beta-tricalcium phosphate scaffolds coated with alginate, Ceramics international, 42 (2016), pp: 5141-5147.

5. Elena I. Oprita, Lucia Moldovan, Oana Craciunescu. In vitro behavior of osteoblast cells seeded into a COL/β-TCP composite scaffold. Central European Journal of Biology, 3 (2008), pp: 31-37.

6. Masatomo Yashima, Atsushi Sakai, Takashi Kamiyama, Akinori Hoshikawa. Crystal structure analysis of β-tricalcium phosphate $Ca_3(PO_4)_2$ by neutron powder diffraction [J]. Journal of Solid State Chemistry, 175 (2003), pp: 272~277.

7. Y Dang, Li Wang, Pengcheng Wu, Hao Zhao, Jin Long. Preparation and Performance Research of PLA/HA Porous Scaffold Materials Date, Plastics Science and Technology, 42(2014), pp: 87-91 (in Chinese).

8. Wen Yan, Hao Luo, Jun Tong, Nan Li. Effects of sintering temperature on pore characterization and strength of porous cordierite–mullite ceramics by a pore-forming in-situ technique, International Journal of Materials Research, 103 (2012), pp: 1239-1243.

9. Wen Yan, Nan Li, Bingqiang Han, Jun Liu, Guangping Liu. Preparation and Characterization of Porous Cordierite Ceramics with Well-distributed Interconnected Pores, Transactions of the Indian Ceramic Society, 70 (2011), pp: 65-69.

10. Lu Peiwen. Physical chemistry of silicate. Wuhan: Wuhan University of Technology press [M], 2005. 250~251(in Chinese).

11. H. Janik, M. Marzec. A review: Fabrication of porous polyurethane scaffolds, Materials Science and Engineering: C, 48 (2015), pp: 586–591.

Analysis of RC Rectangle Beams Reinforced with High Performance HFRP Sheets

Jian Zhang[1,*], Tai-Yun Fang[2], Xin Ye[3]

1 Nanjing University of Aeronautics and Astronautics, Jiangsu Nanjing, China
2 Quanlity Supervision Bureau of Transportation Department of Jiangsu Province, Jiangsu Nanjing, China
3 Southeast University ChengXian College, Jiangsu Nanjing, China
*Email: jianzhang78@126.com

In the past achievements of a single FRP or HFRP reinforced concrete beam, the most are focused on a certain amount of the experimental study of the specimens, because of the high nonlinear performance of hybrid fiber reinforced concrete beam element and the complexity of three dimensional nonlinear programming. The theoretical calculation and analysis are conducted by using the commercial finite element software. According to the theory of the solid degradation element, the main steels are managed with three degrees of freedom of spatial bar element. The contribution matrix of the main steels to the stiffness matrix of the hybrid composite element is completed on basis of the node displacement coordination of the main steel element. It is shown from the examples that the derived hybrid composite element can be accurately used for the whole process of analysis of the RC rectangle beams reinforced with the high performance HFRP sheets.

Keywords: HFRP sheets; RC rectangle beam; element.

1. Introduction

The reinforced technology by using FRP sheet has a certain application in practical engineering. It basically adopts the passive paste of non-prestressed FRP sheet to raise the bearing capacity of components [1-3]. The weight of bridge is often a significant proportion of the total load, and passive paste FRP is only used for resistance, the influence on the application effect on flexural member is not satisfactory. Therefore, the beam strengthened with prestressed HFRP sheet paste is better than that of non-prestressed HFRP sheet paste method. In order to improve the effectiveness of reinforcement, the HFRP sheet has a certain amount of stress by using the appropriate tension machine and then paste on the structure [4-5]. This kind of reinforcement technology can make use of FRP materials more effectively. It is able to solve the above problems in all aspects. Compared with the study of non-prestressed HFRP sheet, the study of prestressed HFRP sheet only appears in recent years. Like traditional prestressed structure, the initial prestress can be used to balance structure of self-respect and part of the load, delay the development of the cracks, reduce the width of the crack and deflection of structure, alleviate stress of the steel, and improve the ultimate bearing capacity of concrete beams. In this paper, the main steels are managed with three degrees of freedom of spatial bar element [6-8]. The contribution matrix of the main steels to the stiffness matrix of the hybrid composite element is completed on basis of the coordination of the displacement of the nodes of the main steel element. And then the analysis is completed.

2. Mechanical Model of RC Rectangle Beam Reinforced with High Performance HFRP Sheets

RC rectangle beam is simulated by entity degradation element. The midsection has nine controlling nodes in Fig. 1, which is used to describe the degradation element of information about the corresponding element decreased significantly. Concrete, stirrups and structural reinforcement in the concrete beams can take concrete layer element to simulate. For vertical reinforced (main), using

---

*Corresponding author

layer element simulating is too approximate, this paper uses composition element to simulate concrete beams. The displacement of 9 nodes degradation element is as follows:

$$\boldsymbol{\delta}_i = [u_i \quad v_i \quad w_i \quad \beta_{1i} \quad \beta_{2i}]^T \tag{1}$$

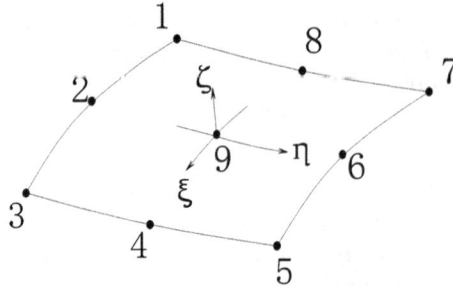

Fig. 1 Local coordinate of the mid surface of the element

In this formula: $[u_i \; v_i \; w_i]$ is the line displacement of the node i in the overall system, $[\beta_{1i} \; \beta_{2i}]$ is the angle displacement of the i node in the node coordinates. Displacement field by interpolation can be expressed as the shape function:

$$u = \sum_{i=1}^{n} N_i u_i + \sum_{i=1}^{n} N_i \frac{h_i}{2} \zeta (v_{1i}^x \beta_{1i} - v_{2i}^x \beta_{2i}) \tag{2}$$

$$v = \sum_{i=1}^{n} N_i v_i + \sum_{i=1}^{n} N_i \frac{h_i}{2} \zeta (v_{1i}^y \beta_{1i} - v_{2i}^y \beta_{2i}) \tag{3}$$

$$w = \sum_{i=1}^{n} N_i w_i + \sum_{i=1}^{n} N_i \frac{h_i}{2} \zeta (v_{1i}^z \beta_{1i} - v_{2i}^z \beta_{2i}) \tag{4}$$

Where N_i is the shape function of node i, h_i is the element thickness of node i, v_{1i}^x is a cosine between the node coordinate $\mathbf{v}_1$ and the overall system x of node i (the same to others). The main reinforcement in the combination element in concrete beams starts with point A and ends with point B. By using the displacement interpolation Eqs. (2) to (4), the node displacement in main reinforcement elements can be expressed as:

$$
\begin{bmatrix} u_A \\ v_A \\ w_A \\ u_B \\ v_B \\ w_B \end{bmatrix} = \sum_{i=1}^{n}
\begin{bmatrix}
N_i^A & & & N_i^A \frac{h_i}{2}\zeta_A v_{1i}^x & -N_i^A \frac{h_i}{2}\zeta_A v_{2i}^x \\
& N_i^A & & N_i^A \frac{h_i}{2}\zeta_A v_{1i}^y & -N_i^A \frac{h_i}{2}\zeta_A v_{2i}^y \\
& & N_i^A & N_i^A \frac{h_i}{2}\zeta_A v_{1i}^z & -N_i^A \frac{h_i}{2}\zeta_A v_{2i}^z \\
N_i^B & & & N_i^B \frac{h_i}{2}\zeta_B v_{1i}^x & -N_i^B \frac{h_i}{2}\zeta_B v_{2i}^x \\
& N_i^B & & N_i^B \frac{h_i}{2}\zeta_B v_{1i}^y & -N_i^B \frac{h_i}{2}\zeta_B v_{2i}^y \\
& & N_i^B & N_i^B \frac{h_i}{2}\zeta_B v_{1i}^z & -N_i^B \frac{h_i}{2}\zeta_B v_{2i}^z
\end{bmatrix}
\begin{bmatrix} u_i \\ v_i \\ w_i \\ \beta_{1i} \\ \beta_{2i} \end{bmatrix} = \sum_{i=1}^{n} \mathbf{R}_i \boldsymbol{\delta}_i \tag{5}
$$

Where N_i^A is the value of node A about the main reinforcement element in the shape function, N_i^A is the value of node B; ζ_A is the value of element's starting point A in the local ζ coordinates, ζ_B is the value of element's end point B in the local ζ coordinates, δ_i is the array of shell element node displacement, $\mathbf{R}_i$ is the transformation matrix. Take $\delta_s = [u_A \ v_A \ w_A \ u_B \ v_B \ w_B]^T$, $\mathbf{R} = [\mathbf{R}_1 \ \mathbf{R}_2 \ \cdots \ \mathbf{R}_n]$, $\delta_c = [\delta_1^T \ \delta_2^T \ \cdots \ \delta_n^T]^T$. Then Eq.(5) can be written as :

$$\delta_s = \mathbf{R}\delta_c \tag{6}$$

Application of virtual work principle, the winner muscle cell's contribution can be pushed to the composition element stiffness matrix:

$$\mathbf{K}_S = \mathbf{R}^T \overline{\mathbf{K}}_S \mathbf{R} \tag{7}$$

Where $\mathbf{K}_S$ is the element's contribution to the composition element stiffness matrix; $\overline{\mathbf{K}}_S$ is the steel stiffness matrix of the overall coordinate system. The contribution of the concrete layer, the stirrup layer and the reinforcement layer to the stiffness matrix can use Gaussian formula to calculate. Hybrid reinforcement of structure with HFRP fiber sheet thickness is small, single layer thickness is about 0.25 mm. The HFRP fiber sheet can use the shell element shown in Fig. 2. Initial stress equivalent load method is adopted to realize hybrid the space prestressing effect of HFRP sheet. Because of adopting advanced steel plate bolt anchorage technique and paste for prestressed HFRP sheet, the failure mode of test reinforcement beam is broken in the form of fiber blah or rib fracture, the failure mode of fiber sheet strip damage does not happen in this paper. Therefore, this article does not consider the interface bond-slip effect, the hybrid fiber sheet element with concrete beam element use the node processing.

3. Criterion for the Concrete Beam Reinforced with Prestressed C /G Fiber Sheet

The Owen triaxle yielding criterion is utilized as:

$$f(I_1, J_2) = (\alpha I_1 + 3\beta J_2)^{1/2} = \sigma_0 \tag{8}$$

Where I_1 is the first invariant stress tensor, J_2 is the second invariant stress tensor, σ_0 is equivalent stress, f_c is the concrete compressive strength, α and β are the material parameters which are respectively determined according to the uniaxial compression and biaxial compressive test calibration.

$$\alpha = \frac{1-t^2}{2t-t^2}\sigma_0, \quad \beta = \frac{2t-1}{2t-t^2} \tag{9}$$

Where $t = f_{cc}/f_c$, f_{cc} is bidirectional compressive strength.

Hardening model determine the subsequent yield surface movement in the process of plastic deformation. It determines the loading surface and the relationship between the cumulative plastic strains. By using the effective stress and effective plastic strain, it makes the description of the mechanical behavior of uniaxial test about concrete to uniaxial test. The parameters H' of the elastic-plastic matrix can be represented by a slope of effective stress σ on the effective plastic strain ε_p, the expression shows as follows:

$$H' = \frac{d\sigma}{d\varepsilon_p} \tag{10}$$

The expression of the relation between uniaxial effective stress σ and effective plastic strain ε_p based on the Madrid parabolic shows as follows:

$$\sigma = E_0(\varepsilon_e + \varepsilon_p) - \frac{1}{2}\frac{E_0}{\varepsilon_0}(\varepsilon_e + \varepsilon_p)^2 \tag{11}$$

Where E_0 is the initial elastic modulus, ε_0 is the total strain at the uniaxial compressive strength f_c and its value is $2f_c/E_0$. ε_e is the elastic strain, and its value is σ/E_0. The expression can be determined by uniaxial effective stress and effective plastic strain equation:

$$\sigma = -E_0\varepsilon_p + \sqrt{2E_0^2\varepsilon_0\varepsilon_p} \qquad (0.3f_c < \sigma \le f_c) \tag{12}$$

Thus the parameter H' can be solved by Eqs.(9-11).

4. Analysis of the Example

The size of prestressed HFRP fiber reinforced concrete beam is 150mm×250mm×250mm, longitudinal reinforcement is 2φ14, stirrup isφ8@150, symmetrical reinforcement, the thickness of protective layer is 25 mm. The thickness of fiber sheet is 0.23mm. According to the different ratio of HFRP hybrid fiber performance by composite material production elements, and after comprehensive comparison, seven kinds of hybrid fiber are chosen as a theoretical analysis, which is based on both strength and ductility index calculation analysis of prestressed HFRP fiber sheet. For convenient, carbon fiber can be abbreviated to C, carbon-glass 1:1 can be abbreviated to CG11, carbon-glass 1:2 can be abbreviated to CG12, carbon-glass 1:3 can be abbreviated to CG13, carbon-glass 2:1 can be abbreviated to CG21, carbon-glass 3:1 can be abbreviated to CG31, glass fiber can be abbreviated to G. The finite element model of the RC rectangle beam is shown in Fig. 2. Each load point is loaded as 3kN in the whole process of analysis.

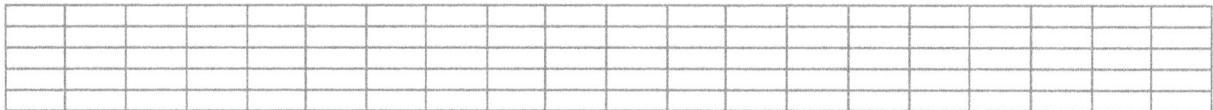

Fig. 2 The finite element model

The optimal mixture ratio is studied in the whole process of calculation about prestressed HFRP fiber reinforced RC rectangle beams in Fig. 3. The concrete strength level is C25, HFRP prestressed degree is 0.1-0.4. The crack load, yield load, ultimate load and the load level at the time of the prestressed HFRP sheet, mid-span deflection of the beams are shown in Fig. 3.

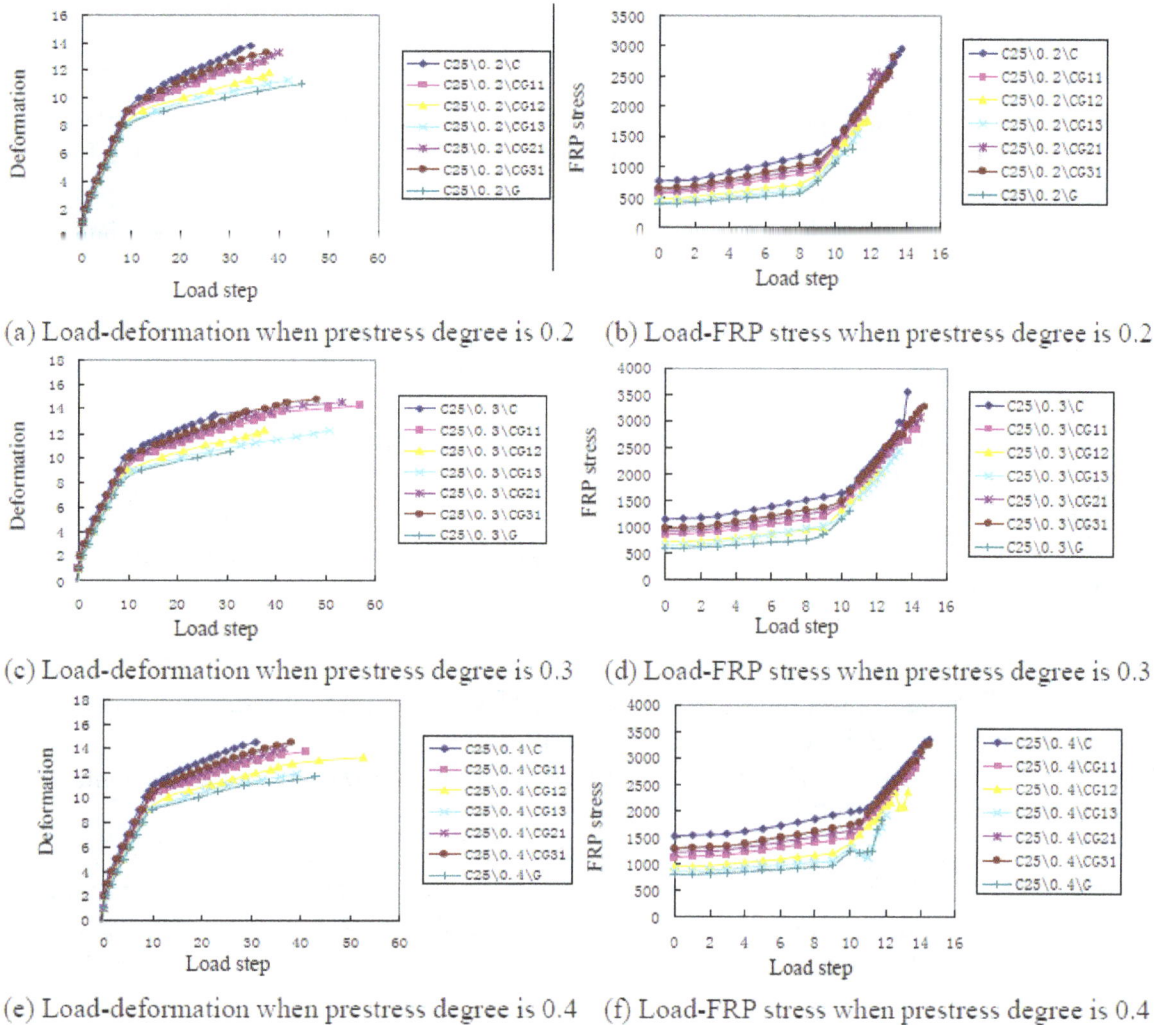

(a) Load-deformation when prestress degree is 0.2   (b) Load-FRP stress when prestress degree is 0.2

(c) Load-deformation when prestress degree is 0.3   (d) Load-FRP stress when prestress degree is 0.3

(e) Load-deformation when prestress degree is 0.4   (f) Load-FRP stress when prestress degree is 0.4

Fig. 3 Computational results of beams reinforced with prestressed HFRP sheet

It is shown in Fig. 3 that the results of the analysis, with the load rating, the load deflection curves of different fiber sheet reinforced concrete beam and the development curve of stress of fiber sheet change in rule; At the beginning of loading, deflection and fiber stress develop more gently, but to yield load, the mid span deflection and fiber stress develop rapidly until the structure damage.

5. Conclusion

RC rectangle beams are studied in the whole process, which are reinforced by carbon fiber, carbon-glass 1:1, carbon-glass 1:2, carbon-glass 1:3, carbon-glass 2:1, carbon-glass 3:1 and glass fiber. Characteristics of prestressed GFRP reinforcement values are relatively low, thus prestressed GFRP reinforcement is not appropriate. Characteristic value of load strengthened with prestressed CFRP sheets have improved, compared with the hybrid fiber cracking load which increase a little while the residual strength caused by the excessive waste and the ductility index both decline. Therefore, the hybrid with CFRP and GFRP sheet can give full play to their respective advantages.

Acknowledgments

Project is funded by the National Natural Science Foundation of China (Nos. 11232007 and 11272147), the Fundamental Research Funds for the Central Universities (NS2014003), Science Project of Jiangsu Yangtze River Bridge Co., Ltd. (GCJS2014-37), and Natural Science Foundation of Jiangsu Province (BK20130787).

References

1. J. Zhang, J.S. Ye, C.W. Zhou, Powell's optimal identification of material constants of thin-walled box girders based on Fibonacci series search method, Applied Math. Mech., 32 (2011), 97-106.
2. X.J. Wei, J.J. Zhang, S.M. Zhang, Group for shield tunnel construction induced ground maximal settlement, Rock Soil Mech., 29 (2008) 445-448.
3. J. Zhang, C.W. Zhou, W.G. Lan, Nonlinear dynamical identification of displacement parameters of multi-cell curve box based on Markov error theory, Chinese J. Applied Mech., 27 (2010), 746-750.
4. M. S. Khaled, B. K. John, Literature review in analysis of box-girder bridges, J. Bridge Eng., 7 (2002), 134-143.
5. K. Babu, M. Devdas, Correction of errors in simplified transverse bending analysis of concrete box-girder bridges, J. Bridge Eng., 10 (2005), 650-657.
6. R. A. Ghani, U. Hangang, Thin-walled multicell straight box-girder finite element, J. Struct. Eng., 117 (1991), 2953-2971.
7. Q.Z. Luo, Q.S. Li, J. Tang, Shear lag in box girder bridges, J. Bridge Eng., 7 (2002), 308-313.
8. J. Zhang, J.S. Ye, C.Q. Wang, Dynamic Bayesian estimation of displacement parameters of continuous thin-walled straight box with segregating slab based on CG method, Chinese J. Comp. Mech., 25 (2008), 574-580.

Application of the Image Processing Based on Metal Microelectrode in the Experimental Teaching

Yu-Hong Liu[1,2], Jian-Jun Chen[1,2,*]

[1]*College of Science, Huazhong Agricultural University, Wuhan 430070, PR China*
[2]*Institute of Applied Physics, Huazhong Agricultural University, Wuhan 430070, PR China*
Email: chenjianjun@mail.hzau.edu.cn

Research on the image processing based on the metal microelectrodes is widely popular in the experimental teaching of the physical electronics subject. In this research, the microelectrodes were designed using the gold electrodes. The number of the microelectrodes is 8, with the width of 1000μm. A method of image processing based on the system was demonstrated. Filtering result is optimal when the neighborhood window size is 3×3 by using median filtering algorithm. Applying the edge detection of Roberts's operator, the result is qualified when the threshold value is 0.05.

Keywords: Microelectrode, microscopic image, physical electronics.

1. Introduction

Microelectrodes have been in common usage in various fields, such as environmental analysis and monitoring, biology [1], medical science [2, 3]. Accurate resistivity mouse brain mapping is used by microelectrode arrays [4]. The extracellular glucose in brain is tested by high resolution microelectrode array biosensor [5]. Uric acid was detected by reversible nanostructured thin-film microelectrode [6].

Development of microelectrode technique benefits the progress of all other crossing disciplines, like material science [7, 8], electronics [9], microsystem, micromachining technology, and microfluidic technique and computer science. Hence, the potential of the microelectrode development is significant [10]. In the experiments of physical electronics subject, the study about microelectrodes is often involved. As a result, it is necessary to introduce the image processing of metal microelectrode in the experimental teaching of the physical electronics subject.

2. Experimental

2.1. *Preparation of metal microelectrode*

Fig. 1 Metal microelectrode physical map

*Corresponding author

The first step is to design the mask plate in the procedure of manufacturing the metal microelectrodes. The scheme diagram is printed onto the chip by photolithography using the mask plate in fabrication process, thus the accuracy of the mask plate determines the total accuracy of the system.

The glass substrate is dealt with ultrasonic wave, acetone and deionized water in sequence. It is then dried by N_2 to assure the material clean and also improve its adhesion on the metallic coating. The photo resist (~3μm) is spun and then baked on the surface of glass. It is followed by UV exposure and development. The glass substrate is coated with the Au in the magnetron sputtering system. Afterwards the photo resist layer is dissolved by using the acetone. The entire process of the chip fabrication is shown above.

The final step is to check the circuit in microscope and implement the electrical tests. The circuit is intact and has no crossing. The electrode is smooth, as shown in Fig. 1. In Fig. 1, the number of the microelectrodes is 8, with the width of 1000μm. Size of the electric pad on the right edge is 5000μm × 2500μm.

3. Results and Discussion

3.1. *Mage standard deviation*

For vector x_i, where $i=1, 2\cdots, n$, the standard deviation is Eq. 1[11].

$$s = \sqrt{\frac{1}{n-1}\sum_{i=1}^{n}(x_i - x)^2} \tag{1}$$

Among them $x = \frac{1}{n}\sum_{i=1}^{n}x_i$, The length of the vector is n.

3.2. *Image correlation coefficient*

The pixel of the gray image is a two-dimensional matrix. Two dimensional matrix of equal size, and its correlation coefficient is calculated. Its formula is shown in Eq. 2.

$$r = \frac{\sum_{m}\sum_{n}(A_{mn} - \overline{A})(B_{mn} - \overline{B})}{\sqrt{\left(\sum_{m}\sum_{n}(A_{mn} - \overline{A})^2\right)\left(\sum_{m}\sum_{n}(B_{mn} - \overline{B})^2\right)}} \tag{2}$$

Where A_{mn} and B_{mn} for the size of m row n column of the gray image, $\overline{A}$ is mean2 (A), $\overline{B}$ is mean2 (B).

3.3. *Sequential statistical filtering*

In the coordinates (x, y), the size of the window for the $m\times n$ said S_{xy}, median filter is selected window S_{xy} is interference image $g(x, y)$ of the median, as the output of the coordinates (x, y), the formula is Eq. 3.

$$f(x, y) = \underset{(s,t)\in S_{xy}}{Median}[g(s,t)] \tag{3}$$

Max value filter can remove salt and pepper noise, but it will remove some of the melanin from the edge of the black object, the formula of maximum filter is shown in Eq. 4.

$$f(x, y) = \underset{(s,t) \in S_{xy}}{Max}[g(s,t)] \tag{4}$$

The minimum filter is similar to the maximum filter, but it can remove some white pixels from the edge of the white object. The formula of the minimum filter is shown in Eq. 5.

$$f(x, y) = \underset{(s,t) \in S_{xy}}{Min}[g(s,t)] \tag{5}$$

3.4. *Roberts operator*

For discrete image f (x, y), the edge detection operator is to use the vertical and horizontal gradient of the image to approximate the gradient operator, that is shown in Eq. 6 [12].

$$\nabla f = \left(f(x,y) - f(x-1,y), f(x,y) - f(x,y-1) \right) \tag{6}$$

In edge detection, for each pixel in the image ∇f calculation, then seeks the absolute value and the threshold operation can be achieved. The calculation formula of Roberts's operator is shown in Eq. 7.

$$R(i, j) = \sqrt{\left[f(i, j) - f(i+1, j+1) \right]^2 + \left[f(i, j+1) - f(i+1, j) \right]^2} \tag{7}$$

The Roberts operator is composed of two templates.

$$\begin{bmatrix} 1 & 0 \\ 0 & -1 \end{bmatrix} \begin{bmatrix} 0 & 1 \\ -1 & 0 \end{bmatrix}$$

Figure 2 shows that salt and pepper noise is added to metal microelectrode grayscale image. Noise density is 0.05.

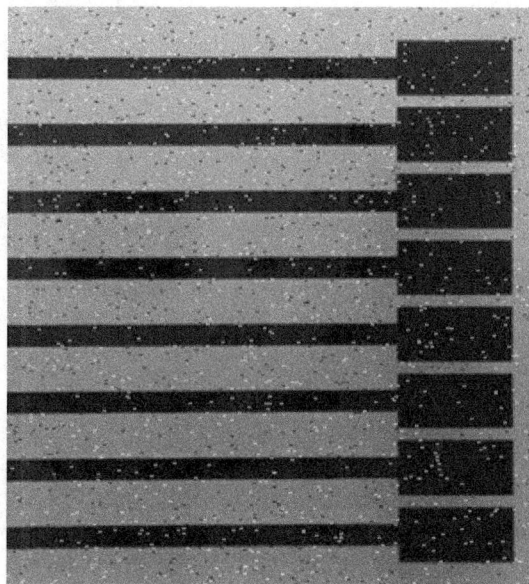

Fig. 2 Contains salt and pepper noise of metal microelectrode image

Fig. 3 shows that 3×3 Neighborhood window filter in metal microelectrode image. The number of image correlation coefficient is 0.9987 between Fig. 3 and metal microelectrode grayscale image. The number of mage standard deviation is 0.2410 in Fig. 3.

Fig. 4 shows that 7×7 Neighborhood window filter in metal microelectrode image. The number of image correlation coefficient is 0.9945 between Fig. 4 and metal microelectrode grayscale image. The number of mage standard deviation is 0.2397 in Fig. 4.

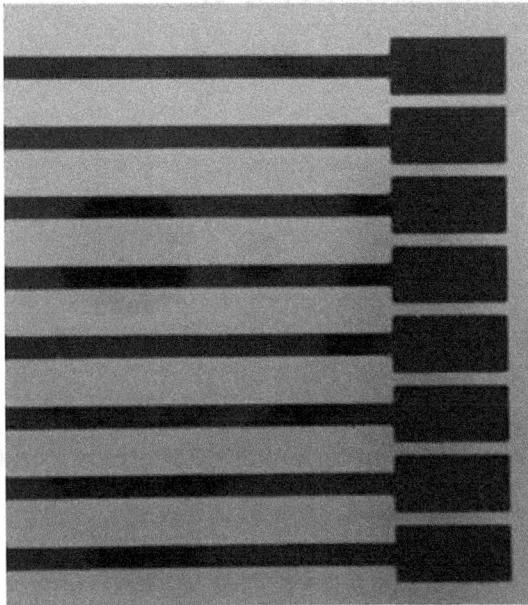

Fig. 3 3×3 Neighborhood window filter in metal microelectrode image

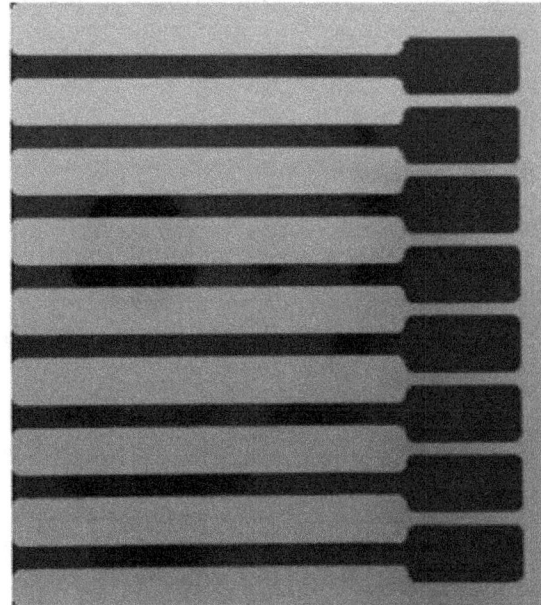

Fig. 4 7×7 Neighborhood window filter in metal microelectrode image

Fig. 5 shows that the automatic threshold Roberts operators for edge detection of the metal microelectrode image. Fig. 6 shows that threshold of 0.05 Roberts operators for edge detection of the metal microelectrode image.

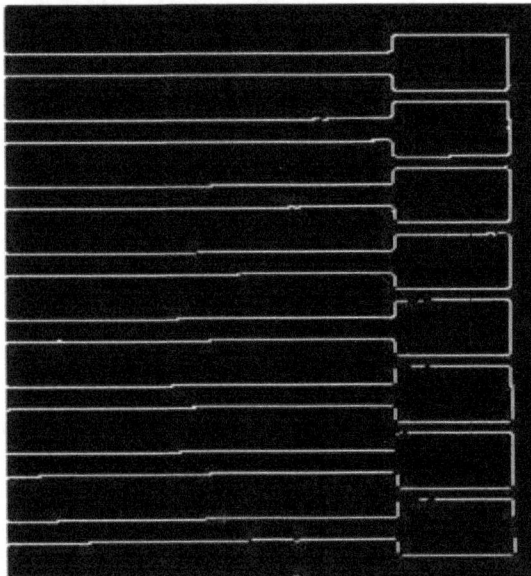

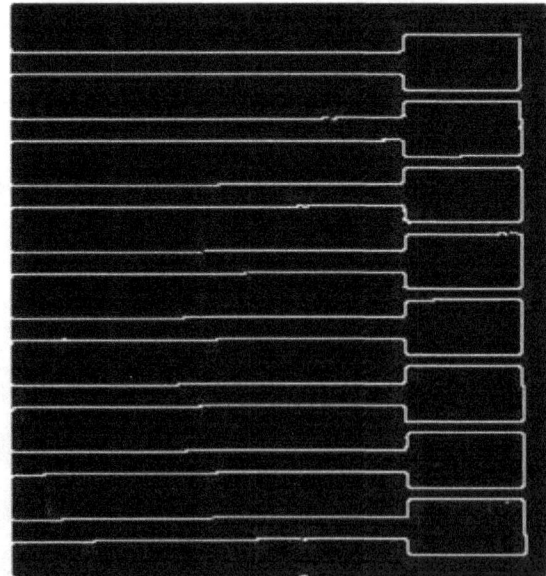

Fig. 5 Automatic threshold Roberts operators for edge detection of metal microelectrode image

Fig. 6 Threshold of 0.05 Roberts operators for edge detection of metal microelectrode image

4. Conclusion

In this paper, a method of image processing to the microelectrodes is demonstrated. The filtering result is optimal when the window size is 3×3 by using median filtering algorithm. Applying the edge detection of Roberts's operator, the result is qualified when the threshold value is 0.05. The research on the image processing of the metal microelectrodes is widely useful in the experimental teaching of the physical electronics subject.

Acknowledgment

This research was funded by the project grant from China Scholarship Council for studying in the United States of America as a visiting scholar ([2014]3012, File No. 201406765042).

References

1. B.X. Gu, Z. Liu, X.Y. Wang, X.X. Dong, RF magnetron sputtering synthesis of carbon fibers/ZnO coaxial nanocable microelectrode for electrochemical sensing of ascorbic acid, Materials Letters, 181 (2016) 265-267.
2. H.Y. Bai, F. J.D. Campo, Y.C. Tsai, Scanning electrochemical microscopy for study of aptamer–thrombin interfacial interactions on gold disk microelectrodes, Journal of Colloid and Interface Science 417 (2014) 333–335.
3. T. Anh-Nguyen, D. T. Tran, U. Pliquett and G. A. Urbana, Behavior and the response of cancer cells on anticancer drug treatment monitored with microelectrode array, Procedia Engineering 120 (2015) 928 – 931.
4. A. Beduer, P. Joris, S. Mosser, V. Delattre, P.C. Fraering, P. Renaud, Accurate resistivity mouse brain mapping using microelectrode arrays, Biosensors and Bioelectronics 60 (2014) 143–153.
5. C. F. Lourenco, A. Ledo, J. Laranjinha, G. A. Gerhardt, R. M. Barbosa, Microelectrode array biosensor for high-resolution measurements of extracellular glucose in the brain, Sensors and Actuators B: Chemical, 237 (2016) 298-307
6. Z. Herrasti, F. Martinez, E. Baldrich, Detection of uric acid at reversibly nanostructured thin-film microelectrodes, Sensors and Actuators B: Chemical, 234 (2016) 667-673
7. A. Khalifa, Z.L. Gao, A. Bermak, Y.Wang, Leanne Lai Hang Chan, A novel method for the fabrication of a high-density carbon nanotube microelectrode array, Sensing and Bio-Sensing Research 5 (2015) 1–7.
8. A. M.H. Ng, Kenry, C. T. Lim , H. Y. Low , K. P. Loh, Highly sensitive reduced graphene oxide microelectrode array sensor, Biosensors and Bioelectronics 65 (2015) 265–273.
9. C.A. Smith, Thermochemical and physical properties of printed circuit board laminates and other polymers used in the electronics industry, Polymer Testing, 52 (2016) 234-245
10. M. Bauch, T. Dimopoulos, Design of ultrathin metal-based transparent electrodes including the impact of interface roughness, Materials & Design, 104 (2016) 37-42
11. D. Yang, H.B. Zhao, Z. Long. Matlab image processing examples. Beijing, 2013. (In Chinese)
12. J. W. Wang, Y. J. Li. Matble7.0 graphic image processing. Beijing, 2006. (In Chinese)

Analytic Model of Stress Distribution on the Interference Fit Surface Between Roll Sleeve and Roll Mandrel in Four-High Strip Cold Rolling

Yong-Gang Dong*, Guo-Ling Luo, Jian-Feng Song

College of Mechanical Engineering, Yanshan University, Qinhuangdao 066004, China
Email: d_peter@163.com

For building the mathematical models of stress distribution on the interference fit surface between the roll sleeve and roll mandrel by the shrinkage fit method, the roll sleeve and roll mandrel were discretized segment by segment. Based on the green function of for biharmonic equation of circular domain, the mathematical models for the stress acted on the interference fit surface during strip cold rolling were derived. These models were transformed from polar coordinates to orthogonal coordinates. Moreover, the summation of stress during strip cold rolling and hot charging stress on the interference fit surface was calculated to obtain the radial stress, tangential stress and shear stress on the interference fit surface and flattening zone respectively. Therefore, it built up an important theoretical foundation to improve the manufacturing technology and practical application of the composite back-up roll.

Keywords: Composite back-up roll; roll sleeve; roll mandrel; fit surface; stress distribution; strip cold rolling.

1. Introduction

At present the domestic back-up roll for the plate or strip cold rolling mill is manufactured by the traditional method generally, such as unit cast method, solid forging method or compound forging method [1, 2]. Since the whole back-up roll have to be changed for the roll radius is less than a minimum working radius after being ground some times, the consumption of back-up roll is tremendous and the cost of products is increased. A novel composite back-up roll, which is assembled by the roll sleeve and roll mandrel by the shrinkage fit method, have been studied to decrease the consumption of back-up roll and the roll changing periodic. In the process of plate or strip cold rolling, the roll sleeve of this composite back-up roll is changed individually when it is ground to reach a minimum working radius, and the roll mangle can be used a longer time and it is not necessary to be changed frequently [3-5].

It is optimistic and worth to be desired for the composite back-up roll by the shrinkage fit to be applied in the practice. However, the processing parameter of shrinkage fit for the composite back-up roll, such as the physical dimension and the shrink range, is obtained by the empirical formula or practice experience. Moreover, the influence of the structure size, material performance parameter and the crown of roll sleeve and roll mangle on the elastic deformation of roll system in the rolling has not been studied.

In plate or strip cold rolling process by four-roll mill with a solid back-up roll, the elastic deformation mathematical model of rolls system had been studied profoundly by some researchers [6, 7]. But this model can't be applied to solve the elastic deformation of rolls system with a composite back-up roll directly for the difference of structure and manufacturing technology. So it is very necessary to build a proper mathematical model to obtain the the elastic deformation of rolls system with a composite back-up roll.

---

*Corresponding author

2. The Loads on the Composite Back-Up Roll

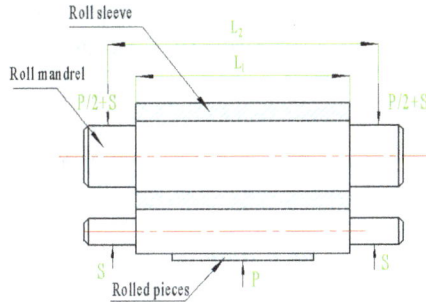

Fig. 1 The size and force of rolls system

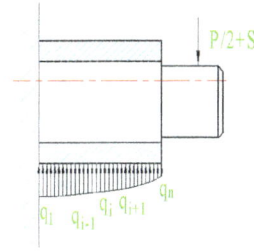

Fig. 2 The load simplification of working roll

As can be seen in Fig. 1, the size of roll radius and roll body length about one half of lower rolls system in strip rolling process was shown. According to the characteristic of composite back-up roll, the composite back-up roll can be separated two parts to study its deflection respectively.

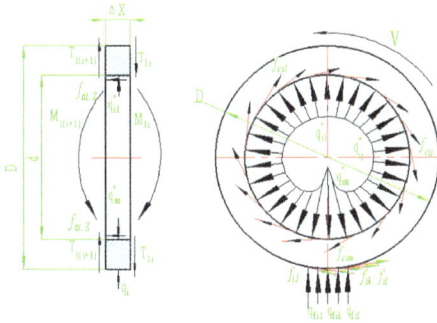

Fig. 3 Loads on the discretized element of roll sleeve

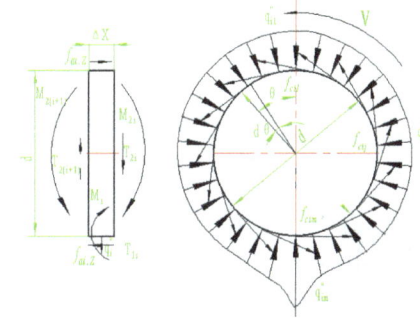

Fig. 4 Loads on the discretized element of roll mandrel

As shown in Fig. 3 and Fig. 4, $T_1(i)$, $T_1(i+1)$ are the shear stress on the right side and left side of discretized element of roll sleeve respectively. $T_2(i)$, $T_2(i+1)$ are the shear stress on the right side and left side of discretized element of roll mandrel respectively. $M_1(i)$, $M_1(i+1)$ are the torques on the on the right side and left side of discretized element of roll sleeve respectively. q_i is the contact force between the working roll and back-up roll of ith element of roll sleeve. q_{ij}'' is the radial contact force on the interference fit surface between the roll sleeve and roll mandrel. f_{aij}, f_{cij} are the axial friction force and circumferential friction force on the interference fit surface respectively. f_{i1} is the friction force on the contact zone between the working roll element and back-up roll element.

For discretized elements of roll sleeve, the boundary conditions can be shown as

$$T_1(1) = T_2(1) = 0, M_1(1) = 0 \qquad (1)$$

For discretized elements of roll mandrel, the boundary conditions can be shown as

$$T_2(1) = P/2 + S \qquad (2)$$
$$M_2(1) = (P/2 + S) \cdot (L_2 - L_1)/2 \qquad (3)$$

3. Solving the Stress Distribution on the Interference Fit Surface

According to the transformation relations between the orthogonal coordinates system and the polar coordinates, the stress distribution equation of polar coordinates system under the plane force can be shown as

$$
\begin{cases}
\sigma'_r = \dfrac{1}{2}(\sigma_x + \sigma_y) + \dfrac{1}{2}(\sigma_x - \sigma_y)\cos 2\theta + \tau_{xy}\sin 2\theta \\[2mm]
\sigma'_\theta = \dfrac{1}{2}(\sigma_x + \sigma_y) - \dfrac{1}{2}(\sigma_x - \sigma_y)\cos 2\theta - \tau_{xy}\sin 2\theta \\[2mm]
\tau'_{r\theta} = \tau_{xy}\cos 2\theta - \dfrac{1}{2}(\sigma_x - \sigma_y)\sin 2\theta
\end{cases}
\tag{4}
$$

The stress distribution equation of polar coordinates system under the body force can be shown

$$
\begin{cases}
\sigma_{r1} = \dfrac{1}{2}\left(\sigma_x + \sigma_y - \dfrac{L_1}{2n}\rho g y\right) + \dfrac{1}{2}\left(\sigma_x - \sigma_y + \dfrac{L_1}{2n}\rho g y\right)\cos 2\theta + \tau_{xy}\sin 2\theta \\[2mm]
\sigma_{\theta 1} = \dfrac{1}{2}\left(\sigma_x + \sigma_y - \dfrac{L_1}{2n}\rho g y\right) - \dfrac{1}{2}\left(\sigma_x - \sigma_y + \dfrac{L_1}{2n}\rho g y\right)\cos 2\theta - \tau_{xy}\sin 2\theta \\[2mm]
\tau_{r\theta 1} = \tau_{xy}\cos 2\theta - \dfrac{1}{2}\left(\sigma_x - \sigma_y + \dfrac{L_1}{2n}\rho g y\right)\sin 2\theta
\end{cases}
\tag{5}
$$

Therefore, the transformation relation between the stress equation of plane force and body force can be shown as

$$
\begin{cases}
\sigma_{r1} = \sigma'_r - \dfrac{L_1}{4n}\rho g y + \dfrac{L_1}{4n}\rho g y \cos 2\theta \\[2mm]
\sigma_{\theta 1} = \sigma'_\theta - \dfrac{L_1}{4n}\rho g y - \dfrac{L_1}{4n}\rho g y \cos 2\theta \\[2mm]
\tau_{r\theta 1} = \tau'_{r\theta} - \dfrac{L_1}{4n}\rho g y \sin 2\theta
\end{cases}
\tag{6}
$$

Where g is the acceleration of gravity, y and ρ can be obtained by

$$
y = \frac{D}{2}\cos\theta + \frac{D}{2}
\tag{7}
$$

$$
\rho = \frac{\left(T_{1i} + T_{2i}\right) - \left(T_{1(i+1)} + T_{2(i+1)}\right)}{\dfrac{L_1}{2n}\left(\dfrac{D}{2}\right)^2 \pi}
\tag{8}
$$

According to the equations of boundary integral and natural integral, the equations for the plane problem can be expressed as the equation of plane problem.

$$
\begin{cases}
\Delta\Delta\varphi(r,\theta) = 0 \\[2mm]
\varphi\big|_\Gamma = \varphi\big|_{r=d} = \varphi_0(\theta) \ , \ \dfrac{\partial\varphi}{\partial n}\big|_\Gamma = \varphi\big|_{r=d} = \varphi_n(\theta)
\end{cases}
\tag{9}
$$

Substituting the Green function $G(r,\theta;r',\theta')$ for biharmonic equation of circular domain into Eq.9 yields the Poisson's equation for stress function

$$\varphi(r,\theta) = \int_0^{2\pi} \left\{ \frac{(D^2 - 4r^2)[D - 2r\cos(\theta - \theta')]}{\pi[D^2 + 4r^2 - 4Dr\cos(\theta - \theta')]^2} \varphi_o(\theta') - \frac{(D^2 - 4r^2)^2}{16\pi[D^2 + 4r^2 - 2Dr\cos(\theta - \theta')]^2} \varphi_n(\theta') \right\} d\theta' \quad \left(0 \le r < \frac{D}{2} \right) \tag{10}$$

According to the references [8], the $\varphi_o(\theta')$ and $\varphi_n(\theta')$ can be solved and the stress equation can be shown as

$$\varphi(r,\theta) = -\frac{q_i(D^2 - 4r^2)}{4\pi} \int_0^{\frac{3\pi}{2}} \frac{\sin\theta'}{[D^2 + 4r^2 - 4rD\cos(\theta - \theta')]} d\theta' + \frac{\rho\frac{L_1}{n}D^2(D^2 - 4r^2)^3}{128} \int_{\frac{\pi}{2}}^{\frac{3\pi}{2}} \frac{\left[1 - \cos\left(\theta' - \frac{\pi}{2}\right)\right]\sin\theta'}{[D^2 + 4r^2 - 4Dr\cos(\theta - \theta')]^2} d\theta' \tag{11}$$

Integrating Eq.11 and simplifying it yields $\varphi(r,\theta) = Aq_i + B$

$$A = -\frac{(D^2 - 4r^2)}{\pi} \frac{Dr(\sin\theta + \cos\theta)}{(D^2 + 4r^2 - 4Dr\cos\theta)(D^2 + 4r^2 + 4Dr\sin\theta)}$$

$$B = \frac{\rho\frac{L_1}{n}D^2(D^2 - 4r^2)^3}{128} \int_{\frac{\pi}{2}}^{\frac{3\pi}{2}} \frac{\left[1 - \cos\left(\theta' - \frac{\pi}{2}\right)\right]\sin\theta'}{[D^2 + 4r^2 - 4Dr\cos(\theta - \theta')]^2} d\theta' \tag{12}$$

Therefore the general solution of stress can be expressed as

$$\sigma_r'(r,\theta,q_i) = \frac{\partial^2 \varphi}{\partial\theta^2} = \frac{\partial^2(Aq_i)}{\partial\theta^2} + \frac{\partial^2(B)}{\partial\theta^2} \tag{13}$$

$$\sigma_\theta'(r,\theta,q_i) = \frac{\partial^2 \varphi}{\partial r^2} = \frac{\partial^2(Aq_i)}{\partial r^2} + \frac{\partial^2(B)}{\partial r^2} \tag{14}$$

$$\tau_{r\theta}'(r,\theta,q_i) = \frac{\partial^2 \varphi}{\partial r\theta} = \frac{\partial^2(Aq_i)}{\partial r\theta} + \frac{\partial^2(B)}{\partial r\theta} \tag{15}$$

According to reference [8], the stress on the fitting surface can be expressed as

$$\sigma_{r1} = \int_{\theta - \arcsin\frac{b}{D}}^{\theta + \arcsin\frac{b}{D}} \sigma_r(r,\theta',q_i(\theta' - \theta)) d\theta' \tag{16}$$

$$\sigma_{\theta1} = \int_{\theta - \arcsin\frac{b}{D}}^{\theta + \arcsin\frac{b}{D}} \sigma_\theta(r,\theta',q_i(\theta' - \theta)) d\theta' \tag{17}$$

$$\tau_{r\theta1} = \int_{\theta - \arcsin\frac{b}{D}}^{\theta + \arcsin\frac{b}{D}} \tau_{r\theta}(r,\theta',q_i(\theta' - \theta)) d\theta' \tag{18}$$

4. Summation of Hot Charging Stress and Stress on the Fitting Surface

Stress distribution of roll sleeve generated by interference fit with a roll mandrel can be obtained by [9]

$$\sigma_{r2} = -\frac{Pa^2}{D^2-d^2}\left(\frac{D^2}{r^2}-1\right) = \frac{Pd^2}{D^2-d^2}\left(1-\frac{D^2}{r^2}\right) \tag{19}$$

$$\sigma_{\theta2} = \frac{Pd^2}{D^2-d^2}\left(1+\frac{D^2}{r^2}\right) \tag{20}$$

The stress on the interference fit surface can be expressed as

$$P = \frac{\delta}{k \cdot d} \tag{21}$$

$$k = [1-v_1]\frac{1}{E_1} + \left[\frac{D^2+d^2}{D^2-d^2}+v_2\right]\frac{1}{E_2} \tag{22}$$

Where δ is the magnitude of interference fit, mm.

According to the superposition principle, plane stress of roll sleeve can be obtained by calculating the summations of contact stress and hot charging stress

$$\sigma_r = \sigma_{r1} + \sigma_{r2} \tag{23}$$

$$\sigma_\theta = \sigma_{\theta1} + \sigma_{\theta2} \tag{24}$$

$$\tau_{r\theta} = \tau_{r\theta1} \tag{25}$$

Transforming polar coordinates system into rectangle coordinates system yields

$$\begin{cases} \sigma_x = \frac{1}{2}\left[\sigma_r + \sigma_\theta - 2\tau_{r\theta}\sin2\theta + (\sigma_r - \sigma_\theta)\cos2\theta\right] \\ \sigma_y = \frac{1}{2}\left[\sigma_r + \sigma_\theta + 2\tau_{r\theta}\sin2\theta - (\sigma_r - \sigma_\theta)\cos2\theta\right] \\ \tau_{xy} = \frac{1}{\cos2\theta}\left[\tau_{r\theta} - \tau_{r\theta}\sin2\theta + \frac{1}{2}(\sigma_r - \sigma_\theta)\cos2\theta\right] \end{cases} \tag{26}$$

5. Results and Discussions

Thickness of Incoming work-piece and outgoing work-piece are 3.0mm and 2.0mm respectively, width of incoming work-piece is 900mm, total rolling force 800ton, diameter of working roll is 520mm, diameter of back-up roll 1200mm,length of work roll and back-up roll are 1200mm, distance between two housing screw is 2456mm.

Material of roll sleeve is 70Cr3NiMo, and its elastic modulus is 2.1×1011Pa, its Poisson ratio is 0.26. Material of roll mandrel is 45, and its elastic modulus is 1.9×1011Pa, its Poisson ratio is 0.3.

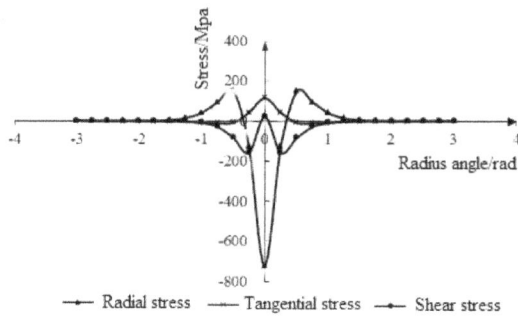

Fig. 5 Stress of the edge element on the fitting surface

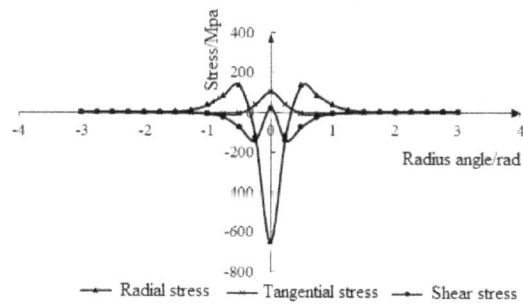

Fig. 6 Stress of the middle element

Circumferential distribution of radial stress on the interference fit surface were shown in Fig. 5 and Fig. 6, and it reaches its maximum value at the center of flattening zone and its direction changes at the outside of flattening zone since the force state of roll sleeve changes from compression to tension. Moreover, the positive radial stress reaches its maximum value where radial angle equals $\pi/6$. The radial stress at non-flattening zone is less than that of flattening zone greatly, and its value is dominated by the hot charging stress between the interference fit surface of roll sleeve and roll mandrel. Compared Fig. 5 with Fig. 6, the maximum radial stress of edge element is greater than that of middle element. The tangential stress on the interference fit surface decreases with the rise of distance from the symmetry plane, and the maximum value of tangential stress exists at the center of flattening zone. The plane shear stress on the interference fit surface reaches its maximum value at the edge of flattening zone and it approaches 0 at the center of flattening zone.

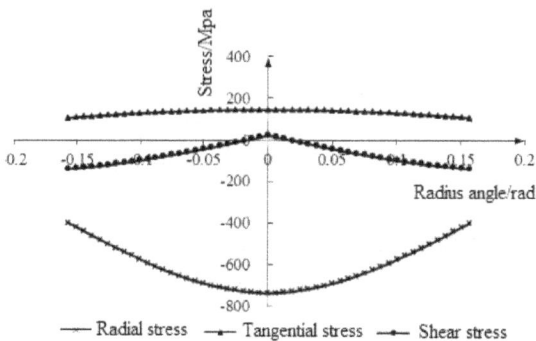

Fig. 7 Stress of edge element on flattening zone

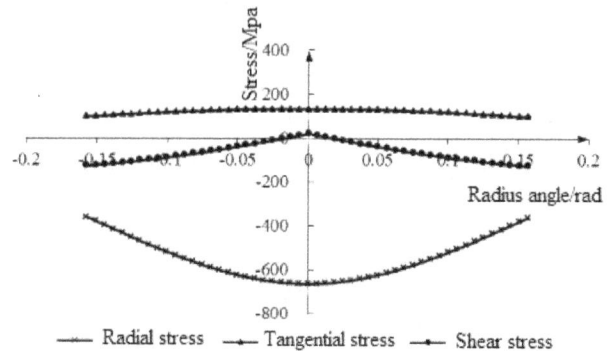

Fig. 8 Plane stress of middle element on flattening zone

As shown in Fig. 7 and Fig. 8, distribution curve of radial stress on the flattening zone is parabolic shape and it accords with the distribution law of flattening zone between two rolls. The tangential stress just changes a little from the edge to the center of flattening zone, and the direction of plane shear stress changes. Compared Fig. 7 and Fig. 8, the maximum value of radial stress on the edge element is greater than that on the middle element, and the tangential stress and shear stress doesn't change greatly.

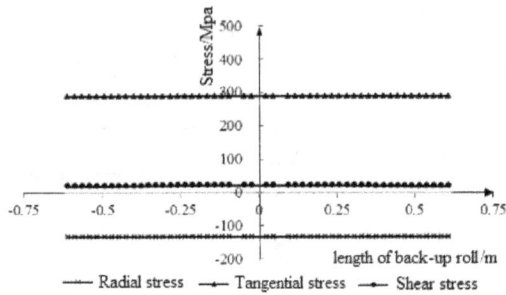

Fig. 9 Stress on the topmost part of fitting surface

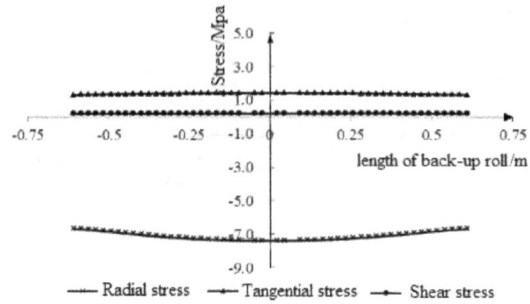

Fig. 10 Stress on the lowest part of fitting surface

As shown in Fig. 9 and Fig. 10, the radial stress on the topmost part of interference fit surface is less than that on the lowest of interference obviously, and the axial distribution of radial stress, tangential stress and shear stress is essentially uniform and it does not change steeply.

6. Conclusion

1. Radial stress on the interference fit surface reaches its maximum value at the center of flattening zone, and its direction changes at the outside of flattening zone since the force state of roll sleeve changes from compression to tension. Moreover, the positive radial stress reaches its maximum value where radial angle equals $\pi/6$;
2. The radial stress at non-flattening zone is less than that of flattening zone greatly, and its value is dominated by the hot charging stress between the interference fit surface of roll sleeve and roll mandrel;
3. Distribution curve of radial stress on the flattening zone is parabolic shape and it accords with the distribution law of flattening zone between two rolls. The tangential stress just changes a little from the edge to the center of flattening zone, and the direction of plane shear stress changes.

Acknowledgment

Foundation Item: Item funded by Natural Science Foundation of China (50775196); Natural Science Foundation of Hebei Province (E2015203431); First-class Foundation of Post Doctoral Research of Hebei Province

References

1. H.Y. Kima, C. Kimb, W.B. Bae, Development of optimization technique of warm shrink fitting process for automotive transmission parts (3D FE analysis), J. Journal of Materials Processing Technology. 187 (2007) 458-462.
2. Shimizu M, Shitamura O, Matsuo S, Development of high performance new composite roll, J. ISIJ International. 32 (1992) 1244−1249.
3. Fu Hanguang, Xing Jiandong, Zhao Aimin, Centrifugal casting of high speed steel/ nodular cast iron compound roll collar, J. Journal of Iron and Steel Research. 9 (2002) 32-35.
4. X Chen, R. Balendra, Y. Qin, A new approach for the optimization of the shrink-fitting of cold-forging dies, J. Journal of Materials Processing Technology. 145 (2004) 215-223.

5. Abdelkhalek S, Montmitonnet P, Potier-Ferry M, Strip flatness model including buckling phenomena during thin strip cold rolling, J. Iron-making and Steelmaking. 37 (2010) 290-297.
6. Abdelkhalek, S, Montmitonnet, P, Potier-Ferry, M,. Coupled approach for flatness prediction in cold rolling of thin strip, J. International Journal of Mechanical Sciences. 53 (2011) 661-675.
7. Yu Meng, Zhang Qingdong, Wang Bo, Research on rolling model based on non-circular contact arc for cold strip temper rolling, J. Advanced Materials Research. 145 (2011) 223-229,
8. The Iron and Steel Institute of Japan, Theory and Practice in strip or plate rolling (translated by Wang Guodong), M. Chinese Railway Press, Beijing, 1990, pp. 115-119.
9. Zhu Aihua, Zhu Chengjiu, Zhang Weihua, Calculation and analysis of friction torque of rolling bearing, J. Bearing, 6 (2008) 15-18.

Blasted Material Compaction Results Analysis in Dam

Feng-Yi Tan[1,2,*], Zheng-Yong Zhang[3], Yong-Xia Bao[3], Fang-Jun Zhang[3], Ying-Biao Liu[3]

[1]Yangtze River scientific Research Institute, Wuhan 430015, China

[2]Wuhan Changke Engineering Construction Supervision Co., Wuhan 430015, China

[3]SINOHYDRO BUREAU 5 CO., LTD, Chengdu 610065, China

*Email: tfy2003@126.com

By in-situ compaction test in the construction site for the blasted material with the 80.0cm thickness of placing and spreading and different compacted times, the compaction construction technology of blasted material in dam was studied and related construction parameters was suggested. The study showed that: when the watering content increased from 0.0% to 10.0%, the indexes, including average compacted settlement, porosity and dry density, of the blasted material was influenced obviously with the increasing of compaction times. But when the watering content further increased from 10.0% to 15.0%, there was no obvious change of the above indexes with the increasing of compaction times, particularly, when the watering content increased from 0.0% to 15.0%, it seemed that there was a little influence of the water content and indexed of granular with the change of compaction times. It satisfied with the design requirement with the following suggested construction technique and related parameters: when the blasted material is as the compacted materials, it is placed and spread by the step-forward technique , the thickness of placing and spreading is 80.0cm, the watering content is 5.0% or 10.0% (watering content is controlled by volumetric method), then it is compacted 10 times or 8 times according to different watering contents by the smooth drum vibratory roller with the 32 tons at the velocity of 2.8 km/h, during the compaction, the forward and backward method with lapped joint width is about 1/10 vibration wheel's width is adopted.

Keywords: Blasted material; compaction; construction technology; construction parameter.

1. Introduction

There was the concrete face sandy gravel-rockfill dam for the multipurpose hydraulic project of Aertashi. Along the downstream part of dam, it was the blasted material compacted zone with the following mechanical requirements after compaction: the content of particle with diameter less than 0.075mm was less than 5.0%, the content of particle with diameter less than 0.1mm was less than 5.0%, the content of particle with diameter less than 5.0mm was less than 15.0%, porosity was less than 19.0% and D_{max} was no more than 600mm [1].

The blasted material was artificial one, which had the engineering properties, such as high compaction property, strong permeability, high density, strong shear strength, small settlement, high bearing capacity, etc., meanwhile it could be obtained raw material in construction site, which indicated that the transport distance was shortened and the cost was saved. But the engineering property of blasted material was depended on the change of its grade, particularly the content of fine stuff. The study showed that to obtain the construction technology and parameters

*Corresponding author

of compacted materials, their compaction property should be studied [2~4], therefore there were a series of in-situ full scale compaction test of blasted material.

2. Test Plan

For the compaction test of blasted material, to obtain the optimal compaction times and the optimal watering content, the smooth drum vibratory roller with the 32 tons was used, and the thickness of placing and spreading was 80.0cm [5], shown in Table 1.

Table 1 Test parameters [6~8]

Material	Thickness of placing and spreading	Watering content (%)	Compacted times	Testing item
Blasted material	80.0cm	0.0,5.0,10.0,15.0	6,8,10	Porosity, settlement, dry density, water content, gradation

Note: the watering content was controlled by volumetric method.

The mechanical of blasted material was shown in Table 2, the plan of placing and spreading, compaction was shown in Table 3.

Table 2 Blasted material's mechanical parameters

Saturated compressive strength (Mpa)	Softening coefficient	Freeze thawing loss rate	Dry density (g/cm^3)
23.3~172.0	0.51~0.71	<1.0%	2.400~2.790

The compaction machine was the YZ32Y2 smooth drum vibratory roller with the 32 tons, which was the stepless control with frequency between 0.0~28 Hz. The nominal amplitude was 1.83mm, the vibration force was 590KN.

Table 3 Placing spreading and compaction

Placed and spread	Watering	Step forward and backward method
	Non-watering	Step forward method
Compaction	Preparation	After placing and spreading, bulldozer grading, compaction one without vibration
	Velocity	2.8km/h
	Lapped joint width	10.0~20.0cm
	Running mode	Step forward and backward method

3. Test Results Analysis

The thickness of placing and spreading of blasted material was 80.0cm, when the compacted times was 6, 8, 10, the following indexes were tested, porosity, settlement, dry density, water content and grade, and analyzed.

3.1. *Porosity*

When the placing and spreading thickness was 80.0cm, the relationship between the porosity of blasted material with different watering content and the compacted times was shown in Table 4.

Table 4 Relationship between porosity &compacted times

Watering content	Compacted times(times)		
	6	8	10
0.0	23.3	21.4	19.7
5.0	20.6	19.3	18.3
10.0	19.4	18.4	18.2
15.0	19.1	18.4	18.1

Under the condition of same watering content, when the compacted times was from 6 to 10, the porosity of blasted material decreased gradually. When it increased from 6 to 8, there was the evident decreasing of porosity that indicated it play an important role for the compaction effect. But with compacted times increasing to 10, the porosity decline of blasted material was a little that indicated there is a little influence for its compaction effect.

Under the condition of same compacted times, when the watering content was from 0.0 to 15.0%, the porosity of blasted material decreased gradually. When it increased from 0.0 to 10.0%, there was the evident decreasing of porosity that indicated it play an important role for the compaction effect. But with watering content increasing to 15.0%, the porosity decline of blasted material was a little that indicated there is a little influence for its compaction effect.

3.2. *Settlement*

When the placing and spreading thickness was 80.0cm, the relationship between the settlement of blasted material with different watering content and the compacted times was shown in Table 4.

Table 5 Relationship between settlement & compacted times

Watering content(%)	Compacted times(times)				
	2	4	6	8	10
0.0	22.0	42.0	54.0	65.0	72.0
5.0	31.0	54.0	67.0	76.0	79.0
10.0	42.0	67.0	78.0	84.0	86.0
15.0	50.0	74.0	83.0	88.0	89.0

Under the conditions of 80.0cm thickness of placing and spreading and the same watering content, the settlement of blasted material increased with the increasing of compacted times.

Under the condition of the watering content with zero:

(1) The velocity of settlement was 8.1 mm/times when the compacted times was between 1 ~8, the settlement was 65.0mm at the end of compacted times with 8,

(2) The velocity of settlement was 3.5 mm/times when the compacted times was between 8 ~ 10, the settlement was 72.0mm at the end of compacted times with 10.

Under the condition of the watering content with 5.0%:

(1) The velocity of settlement was 9.5 mm/times when the compacted times was between 1 ~ 8, the settlement was 676.0mm at the end of compacted times with 8,

(2) The velocity of settlement was 1.5 mm/times when the compacted times was between 8 ~ 10, the settlement was 79.0mm at the end of compacted times with 10.

Under the condition of the watering content with 10.0%:

(1) The velocity of settlement was 13.0 mm/times when the compacted times was between 1 ~ 6, the settlement was 78.0mm at the end of compacted times with 6,

(2) The velocity of settlement was 3.0 mm/times when the compacted times was between 6 ~ 8, the settlement was 84.0mm at the end of compacted times with 8,

(3) The velocity of settlement was 1.0 mm/times when the compacted times was between 8 ~ 10, the settlement was 86.0mm at the end of compacted times with 10.

Under the condition of the watering content with 15.0%:

(1) The velocity of settlement was 13.8 mm/times when the compacted times was between 1 ~ 6, the settlement was 83.0mm at the end of compacted times with 6,

(2) The velocity of settlement was 2.5 mm/times when the compacted times was between 6 ~ 8, the settlement was 88.0mm at the end of compacted times with 8,

(3) The velocity of settlement was 0.5 mm/times when the compacted times was between 8 ~ 10, the settlement was 89.0mm at the end of compacted times with 10.

When the watering content increased from 0 to 5.0%, the settlement of blasted material increased evidently that indicated that it play an important role for compaction effect. Under the condition of watering content with 5.0%, the settlement of blasted material was stable at the end to compacted times with 10. When the watering content further increased to 15.0%, though the settlement of blasted material increased, its application decreased, the settlement of blasted material was stable at the end to compacted times with 8.

3.3. Dry density

When the placing and spreading thickness was 80.0cm, the relationship between the dry density of blasted material with different watering content and the compacted times was shown in Table 6.

Table 6 Relationship between dry density & compacted times

Watering content (%)	Compacted times (times)		
	6	8	10
0.0	2.09	2.14	2.18
5.0	2.16	2.20	2.22
10.0	2.19	2.22	2.23
15.0	2.20	2.22	2.23

Table 6 showed that:

Under the conditions of 80.0cm thickness of placing and spreading and the same watering content, the dry density of blasted material increased with the increasing of compacted times.

(1) When the watering content was zero, the dry density of blasted material had the further increasing tendency with the compacted times from 6 to 8.
(2) When the watering content was 5.0%, its tended to be stable.
(3) When the watering content were 10.0% and 15.0%, its tended to be stable, and there was

A little amplification of dry density with compacted times further increasing to 10 times.

Under the conditions of 80.0cm thickness of placing and spreading and the same watering content, the dry density of blasted material increased with the increasing of watering content.

(1) When the compacted times was 6, the dry density of blasted material was increased substantially with the watering content increasing from zero to 15.0%.
(2) When the compacted times was 8, the dry density of blasted material was increased substantially with the watering content increasing from zero to 10.0% and there was a little change for its density with the watering content further increasing to 15.0%.
(3) When the compacted times was 10, the dry density of blasted material was increased substantially with the watering content increasing from zero to 5.0% and there was a little change for its density with the watering content further increasing from 5.0% to 15.0%.

3.4. *Water content*

When the placing and spreading thickness was 80.0cm, the relationship between the water content of blasted material with different watering content and the compacted times was shown in Table 7.

Table 7 Relationship between water content & compacted times

Watering content (%)	Compacted times (times)		
	6	8	10
0.0	0.2	0.3	0.2
5.0	1.5	1.5	1.4
10.0	1.6	1.4	1.5
15.0	1.8	1.7	1.6

Table 7 showed that:

Under the conditions of smooth drum vibratory roller with the 32 tons and 80.0cm thickness of placing and spreading, there was a little relationship between the water content of blasted material with different watering content and the compacted times. It was perhaps that the value of water content of blasted material was small and the content of fine stuff of blasted material was small.

3.5. Particle gradation

Before the compaction tests, there were the gradation tests to obtain the contents of different small grain sizes: the content of particle with diameter less than 5.0mm was 8.3%, the content of particle with diameter less than 0.075mm was 0.79%, the uniformity coefficient (C_u) was 28, the curvature coefficient (C_c) was 1.5, the above indexes was satisfied with the design requirement, but the content of particle with diameter between 200.0mm and 400.0mm was higher than the design. Under the conditions of 80.0cm thickness of placing and spreading and different watering contents, the gradation of blasted material before and after compaction tests was shown in Table 8.

Table 8 Content of particle

Watering content(%)	Diameter (mm)	Compacted times (times)		
		6	8	10
0	5.0	9.5	10.5	9.3
	0.075	0.4	0.5	0.5
5.0	5.0	11.4	11.7	10.9
	0.075	1.8	0.7	0.7
10.0	5.0	12.7	10.8	11.6
	0.075	1.5	1.2	0.9
15.0	5.0	10.0	11.2	11.2
	0.075	1.2	1.6	1.9
Before test	5.0	8.3		
	0.075	0.7		

Table 8 showed that after the compaction test, though the content of particle with diameter less than 5.0mm increased, the value was less than 15.0%. Meanwhile, there was a little change of content of particle with diameter less than 0.1mm, which is satisfied with the design requirement.

4. Conclusion

By in-situ compaction test in the construction site for the blasted material with the 80.0cm thickness of placing and spreading and different compacted times, the compaction construction technology of blasted material was studied and related construction parameters was suggested. The study showed that: when the watering content increased from 0.0% to 10.0%, the indexes, including average compacted settlement, porosity and dry density, of the blasted material was influenced obviously with the increasing of compaction times. But when the watering content further increased from 10.0% to 15.0%, there was no obvious change of the above indexes with the increasing of compaction times, particularly, when the watering content increased from 0.0% to 15.0%, it seemed that there was a little influence of the water content and indexed of granular with the change of compaction times. It satisfied with the design requirement with the following suggested construction technique and related parameters: when the blasted material is as the compacted materials, it is placed and spread by the step-forward technique, the thickness of placing and spreading is 80.0cm, the watering content is 5.0% or 10.0% (watering content is controlled by volumetric method), then it is compacted 10 times or 8 times according to different watering contents by the smooth drum vibratory roller with the 32 tons at the velocity of 2.8 km/h, during the compaction, the forward and backward method with lapped joint width is about 1/10 vibration wheel's width is adopted.

Acknowledgments

This research was cooperated by the Aertashi site supervision team of Wuhan Changke Engineering Construction Supervision Co. and Aertashi dam Project management team of SINOHYDRO BUREAU 5 CO.LTD

References

1. Construction bid documents of dam engineering of multipurpose hydraulic project of Aertashi [R], 2015.
2. YUAN Yao-yu. THE DESIGN OF THE FILTER IN SIPING SOIL AND ROCK DAM OF HUBEI [J], Resources Environment & Engineering, 2006, 20(5): 531-537.
3. TAN Feng-yi, ZOU Rong-hua, GONG Bi-wei, et al. Study of Different Treatment Measures' Compaction in Swelling Rock Slope [J]. South-to-North Water Transfers and Water Science& Technology, 2009, 7(6): 178-181.
4. TAN Feng-yi, ZOU Rong-hua, GONG Bi-wei, et al. Construction Technology of Treatment M easures in Swelling Rock Slope [J]. Journal of Yangtze River Scientific Research Institute, 26(11): 75-80.
5. Fengyi Tan, Ronghua Zou, Hanbing Hu, et al. Construction Technology of Treatment Measure of Swelling Rock Slope-Replaced Backfilling Clay [J]. Advanced Materials Research, 2011, 168-170: 2334-2339.
6. DL/T5129-2013 Specification for rolled earth-rockfill dam construction [S], 2013.
7. NB/T 35016-2013, Testing specification on material compaction for earth and rock-fill dams [S], 2013.
8. SL237-1999, Specification for soil test [S], 1999.

Study on Fracture Process of Concrete Under Uniaxial Compression

Xin-Yu Liang

School of Civil Engineering and Architecture, Xian Technology University, Xi'an 710032, Shaanxi Province, P.R China

Email: key_xinyu@163.com

A series of CT images were obtained by using CT to scan the meso fracture process of concrete under uniaxial compression. Through the gray image and numerical simulation for comparative analysis, it is clarified that, due to uniaxial compression, the concrete experiences dense pressure, expansion, CT scale crack initiation, expansion, coalescence and concrete macroscopic failure stage. Studies have shown that CT analysis is fine view damage detection and an effective analysis method. It provides an important basis for further quantitative description of the concrete fine view of failure process, and confirms the feasibility of calculation using numerical simulation.

Keywords: Meso concrete; numerical calculation; CT.

1. Introduction

Concrete is a kind of complex synthetic material, and it is considered that the concrete is a three-phase material consisting of aggregate, cement mortar and the bond between the two. The failure of concrete is actually the process of crack initiation, propagation and coalescence. The nonlinear behavior and softening process of concrete materials are largely related to the behavior of crack evolution.

CT technology can detect the internal structure characteristics of the materials dynamically and without damage [1-3]. In recent years, CT technology has been widely used in the exploration of rock meso fracture process [4-6], and has achieved many significant results. In this paper, the CT dynamic observation of the damage process of concrete under uniaxial compression is carried out by using the newly developed portable concrete CT test loading equipment. Of concrete CT image, CT number in the difference image changes were analyzed by analysis before and after cracking of concrete CT image change characteristic, mechanism of the fracture process of concrete research, in order to explore the law of concrete meso fracture.

2. CT Test Results

2.1. Test conditions

Specimens are used for concrete specimens, cylindrical concrete test specifications: specimens with a diameter of 60mm * 120mm, strength C15, aggregate for Banduo Hydropower Station aggregate field of natural gravel aggregate and natural sand, particle size 5-20 mm continuous gradation.

2.2. Test equipment

Using M8000 Marconi spiral CT scanner, the resolution of 24 lines on the /cm, the image size of 1024 x 1024, the fastest imaging speed for the 0.5s within the 4 layer CT images.

2.3. Test results

Figure 1 is the different stress stages of each scanning layer image, white as high density area, brown as low density area. Each slice of CT number average value increased, the occurrence of this phenomenon is the concrete test piece of each layer of the initial void and micro crack (damage) appear closed, the concrete is compacted and the density increases. Strength has improved, this stage is the compaction stage, in fact, a damage occurs in the weakening stage.

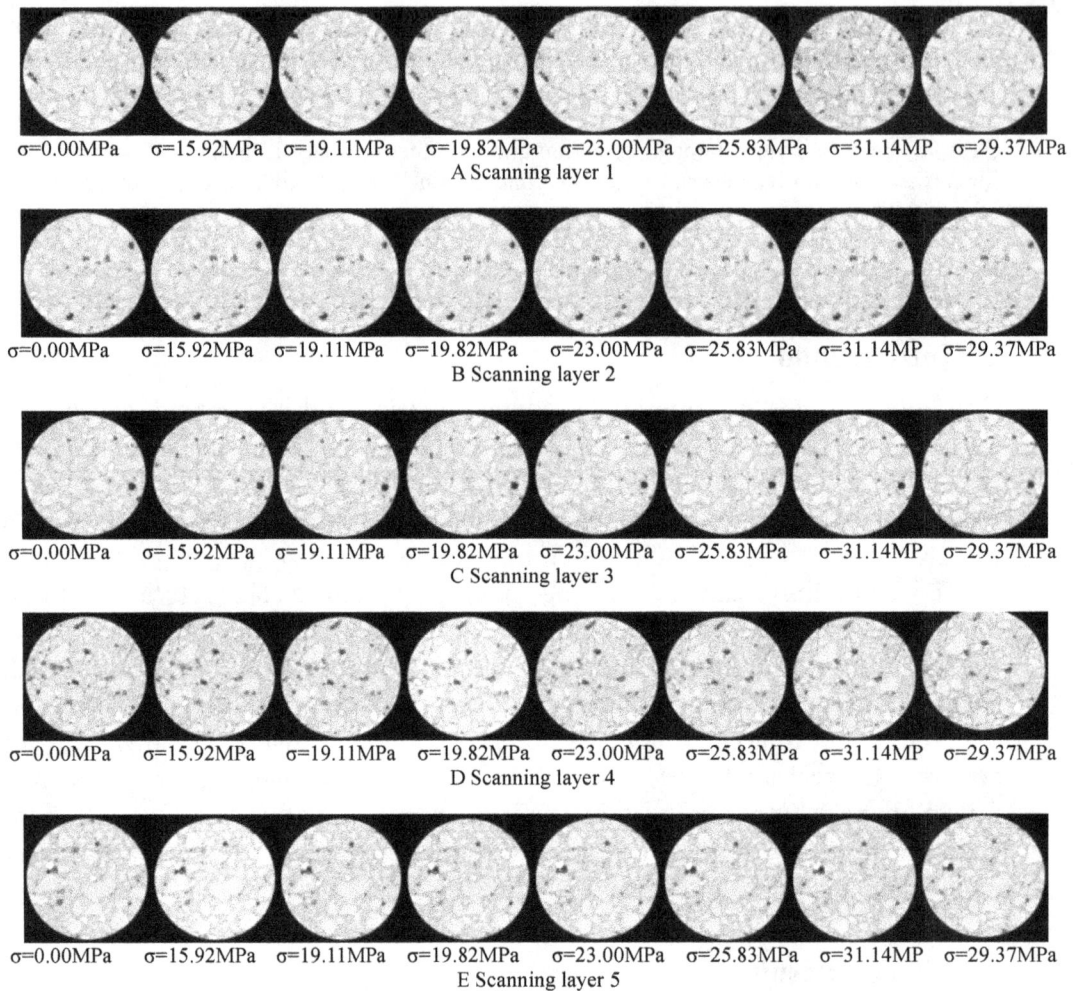

σ=0.00MPa   σ=15.92MPa   σ=19.11MPa   σ=19.82MPa   σ=23.00MPa   σ=25.83MPa   σ=31.14MP   σ=29.37MPa
A Scanning layer 1

σ=0.00MPa   σ=15.92MPa   σ=19.11MPa   σ=19.82MPa   σ=23.00MPa   σ=25.83MPa   σ=31.14MP   σ=29.37MPa
B Scanning layer 2

σ=0.00MPa   σ=15.92MPa   σ=19.11MPa   σ=19.82MPa   σ=23.00MPa   σ=25.83MPa   σ=31.14MP   σ=29.37MPa
C Scanning layer 3

σ=0.00MPa   σ=15.92MPa   σ=19.11MPa   σ=19.82MPa   σ=23.00MPa   σ=25.83MPa   σ=31.14MP   σ=29.37MPa
D Scanning layer 4

σ=0.00MPa   σ=15.92MPa   σ=19.11MPa   σ=19.82MPa   σ=23.00MPa   σ=25.83MPa   σ=31.14MP   σ=29.37MPa
E Scanning layer 5

Fig. 1 Cross-section under different stress of CT image

Load from the 15.92MPa 19.11MPa, CT number of each layer of the average value changed little, CT observation image has a little change, this stage test of meso damage was not significant, can think, this stage concrete fine damage, no big changes in the concept, concrete in the elastic deformation stage.

When load is the 19.82MPa, 1, 5 layer scanning of CT numbers continue to increase (compaction), the rest of the slice CT number average value began to decline, indicating that the meso damage evolution, inside of the micro crack initiation and expansion. Under the action of load, the initial damage micro pores, micro holes and micro crack initiation and propagation, which is based on the initial damage, has a new damage, the specimen density decreases, resulting in the average CT number decreases. This stage is an expansion of the stage, the specimen density decreased, the strength decreased.

When stress continues to increase to 25.83MPa CT number of each layer began to decline, the micro crack in the aggregate of weak AC interface initiation and micro crack appears elongation, bifurcation, phenomenon of linking up, but due to the small crack locations and CT number variations, also human on the gray resolution rate is low, from 19.82 MPa to 25.83MPa this stage is the initiation of the damage evolution and the stage of stable development.

When stress continues to increase to 31.14MPa, it reaches peak load. At each scanning section, CT number average increase rate decreases. This stage is the failure precursor stage, that is, damage evolution of the fastest stage at this time will have a number of new crack, making this drastically reduced the density test, direct manifestation of its microscopic is the number of the scanning section CT average decreases sharply, namely test significant expansion in the scanning section.

Finally, the stress starts to unload. When the process is unloading to the 29.37MPa, the specimen turns into the residual deformation stage. The scanning section CT average value decreased sharply and bearing capacity decreases gradually, a rapid expansion, section further crack coalescence, through the main crack spread quickly to become wider, through the macroscopic crack formation, specimens collapse completely. From after the destruction of the photos can see that extended the crack direction and the principal compressive force direction is parallel or has a small angle, and accompanied by shear diagonal cracks and crack number has multiple small spacing parallel cracks and eventually formed a run through of the main crack and the emergence of phenomenon of cubical expansion.

3. Numerical Test Results

In the numerical test, the random aggregate model is established, and the model size is referenced to the CT test specimen. The double broken line damage evolution model is adopted in the calculation. The model uses the bottom surface Z direction constraint. The midpoint of the full constraint, the crack initiation, development and through the simulation, each model is used to step by step loading, displacement control for the loading method, the calculation.

3.1. Numerical results

When the stress is 28.326MPa, stress showed a linear variation with strain and shall thereafter stress - strain curves of the slope decreases, interface unit and the mortar element damage, damage elastic modulus attenuation, stress to 29.96MPa strain greatly increased. This unit is damaged, the strength of the specimen to failure, the strain to increase until the test failure.

From the above analysis, we can see that the damage evolution of concrete under uniaxial.

3.2. Compression in several stages

(1) At the stage of damage weakening, the density of concrete increases and the strength increases.
(2) The stage of damage initiation and stable development shows that the meso damage evolution, this stage is a stage of expansion, the density of the specimen decreases and the strength decreases.

(3) Crack initiation expansion stage, the stress peak, this stage is failure precursor stages: injury the fastest stage evolution. This greatly reduced the density and the scanning section appears significant expansion.

4. Conclusion

In this paper, the damage degree of concrete under different loads is obtained by using CT test and numerical test, and the following conclusions can be obtained by the gray images and numerical results of the concrete CT:

Concrete specimens under uniaxial compression are in the process of compaction, expansion, crack initiation, propagation, coalescence and macroscopic failure stage, which reflect the failure process of concrete meso damage.

Acknowledgments

This research was funded by the headmaster fund of Xian Technology University (No. XAGDXJJ1317) & the fund of Shanxi Education Department (No. 15JK1349).

References

1. CHEN Hou-qun, DING Wei-hua, PU Yi-bin, DANG Fa-ning. Real time observation on meso fracture process of concrete using X-ray CT under uniaxial compressive condition [J]. SHUILIXUEBAO, 2006, 37(9): 1044-1050. (in chinese)
2. GE Xiu-run, REN Jian-xi, PU Yi-bin, et al. Real-in time CT test of the rock meso-damage propagation law [J]. SCIENCE IN CHINA (Series E) 2000, 30(2): 104-111. (in chinese)
3. GE Xiu-run, YANG geng-she. CT real-time testing on damage propagation microscopic mechanism of rock under uniaxial compression [J]. Rock and Soil Mechanics, 2001, 22(2): 130-133. (in chinese)
4. TIAN Wei, DANG Faning, LIANG Xinyu, et al. CT image analysis of meso-fracture process of concrete [J]. Engineering Journal of Wuhan University, 2008, 41(2): 69-72. (in Chinese)
5. DING Weihua, WU Yanqing, PU Yibin, et al. Measurement of crack width in rock interior based on X-ray CT[J]. Chinese Journal of Rock Mechanics and Engineering, 2003, 22(9):1 421–1 425. (in Chinese)
6. DING Wei-hua, WU Yan-qing, PU Yi-bin, et al. The Density Damage Increment and Its Digital Image of Rock in Compression [J]. Journal of Xi'an University of Technology, 2000, 16(1): 45-48. (in Chinese)

AFFF Spread Over Liquid Surface with Different Foam Expansion

Bao-Wei Wang[1, 2], Shen-Shi Huang[1], Qi-Ze He[1], Shou-Xiang Lu[1,*]

[1]*University of Science and Technology of China, Hefei, Anhui 230029, China*
[2]*Fire Department of Ministry of Public Security, Beijing 100054, China*
Email: sxlu@ustc.edu.cn

Aqueous film forming foam (AFFF) is an effective agent for combating two-dimensional liquid fuel fires. The fire suppression effect of AFFF extinguishant behaves differently on different burning liquids. Previous work indicates that, for most liquid fuel, the fire suppression effect depends on the foam spreading process and its coverage area. In this study, the foam spreading process was investigated, with different foam expansion of agent. Firstly, based on theoretical fluid dynamic analysis, an axisymmetric foam spread model was established. Then foam spreading experiments were carried out to obtain coefficients in the spread model. It was found that, in the stable spreading stage, the spreading speed of AFFF is well consistent with the axisymmetric model. Finally, by fitting the coefficients, the relation of foam coverage and time is obtained, which may provide theoretical guidance for fire suppression.

Keywords: AFFF; foam expansion; spread; foam area.

1. Introduction

Foam is one of the most effective extinguishant to put out liquid fire [1]. According to foam expansion, it can be divided into three types [2]: low-expansion, medium-expansion, and high-expansion. Low-expansion foam contains adequate water, and it can from a high altitude to a lower altitude due to gravity. This foam applies to two-dimension oil fire cases, such as flowing kerosene fire over warship deck and spilling oil fire in petrochemical factories. In contrast, high-expansion foam contains more air, thus it has the ability to accumulate in a room space.

AFFF is the agent of choice for many military and civilian applications requiring rapid fire extinguishment, because it can be applied using traditional water-discharge devices. Williams et al. [3] found that the AFFF agent has different effect on fire suppression over different burning liquid. For the liquid fuel whose flash point is higher than environment temperature (such as aviation kerosene), the fire suppression area is equals to the foam coverage. And for whose flash point is higher than the boiling point of water (such as bio-diesel), the fire suppression area is slightly larger than the foam coverage due to the evaporation of water. Persson et al. [4, 5] indicated that the foam spread process is independent with the foam mass loss of heat radiation from fire. Therefore, the foam spread, which directly affects the efficiency of fire suppression, can be considered independently.

In this study, to investigate to fire suppression effect of AFFF, the foam spread process of a typical low-expansion AFFF extinguishant is discussed. Based on the axisymmetric foam spread theory, the foam spread speed is analyzed. To obtain the coefficients in the spread model, a series of small scale experiments are conducted. This study will provide further understanding of the AFFF foam spread process on liquid surface and provide theoretical guidance for fire suppression using AFFF extinguishant.

2. Axisymmetric Foam Spread Theory

When fire fighters putting out oil fire, stream of AFFF spray from the hose and spread from a point source on fuel surface. Thus compared to traditional two-dimension spread theory [6], the axisymmetric foam spread theory is more accordant with the actual situation. Based on oil slicks spread model, axisymmetric foam spread theory was developed by Persson et al. in 1994 [4, 5], as shown in Fig. 1. The foam extinguishant of interest has a thickness of $h(x,t)$ and is flowing over a liquid bed. It is assumed that the foam is supplied at $r = R_0$ with a constant volume flow rate V. The thickness of foam at $r = R_0$ is h_0 and the front of foam at the time t is $r = R(t)$.

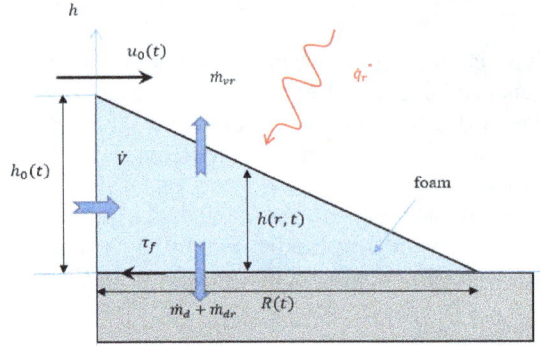

Fig. 1 Illustration of foam extinguishant spread process

The foam spread model without radiation exposure can be summarized as

$$\begin{cases} \dfrac{\partial}{\partial t}(rh) + \dfrac{\partial}{\partial r}(ruh) = 0 \\[2mm] u = -\beta h \dfrac{\partial h}{\partial r} \\[2mm] 2\pi R_0 u_0 h_0 = \dot{V} \end{cases} \qquad (1)$$

where r denotes the radius, h is the thickness of the foam, and u is the velocity of the foam. β is the friction parameter, which is depend on the physical property of the foam. The radius $r = R_0$ is the boundary condition where the foam flow is supplied.

It is found that foam layer approximately have a linear shape during the spread process, so it can be assumed as a cone shape in axisymmetric spread.

$$h(r,t) = h_0(t)\left(1 - \frac{r}{R(t)}\right) \qquad (2)$$

In this case, the thickness at the front of foam spread is zero, and it is possible to integrate equations along $r = R_0$ to $r = L(t)$. Therefore, the thickness of foam at $r = R_0$ and the front of foam can be obtained

$$\begin{cases} h_0 = \left(\dfrac{27\dot{V}^3}{4\pi^3 \beta^2 R_0^2}\right)^{1/7} t^{1/7} \\[4mm] R = \left(\dfrac{18}{\pi^2} \beta R_0 \dot{V}^2\right)^{1/7} t^{3/7} \end{cases} \qquad (3)$$

Hence, the thickness h_0 and front radius R of foam is determined by the volume flow rate V and the friction parameter β.

3. Foam Spread Experiment

Foam spread experiments on small scale were conducted to obtain the coefficients in the axisymmetric model. AFFF solution with concentrations of 3% was used in the experiment, as it is widely applied to military and domestic fire scenarios.

Compressed air foam generating system [7] is adopted in this experiment. AFFF generating system consists of high pressure nitrogen cylinders, foam solution tank, gas flowmeter, fluid flowmeter, mixing chamber, valves, transporting tubes and foam nozzle, shown as Fig. 2. Nitrogen cylinders is the driving source of system. One provided the inlet gas for mixing chamber, which is the important component of foam. The other is the power source of the foam solution, pushing the fluid flow from the tank to mixing chamber. There are flow control valves and flowmeters on both gas transporting tube and fluid transporting tube, so as to control the foam generation rate. When the foam generated by system reaches a stable state, then it would be collected in a vessel with volume of 1635mL. A stopwatches, an electronic balances are used to measure the density and volume flow rate of foam.

Stopwatch records the time cost for the foam fill up the constant volume vessel t_b , thus the volume flow rate of foam V can be calculated as

$$V = \frac{1635ml}{t_b} \tag{4}$$

Electronic balances weights the mass of the foam in vessel m_b , so that expansion of the foam S can be obtained as

$$S = \frac{1635ml}{m_b} \tag{5}$$

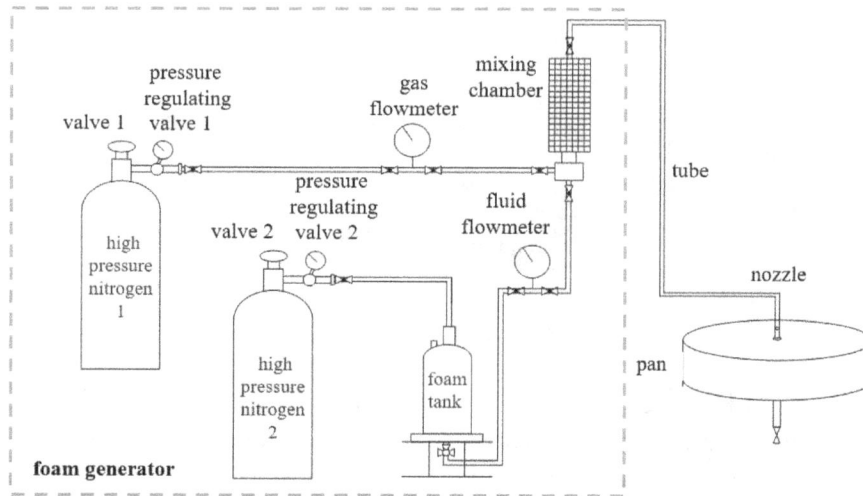

Fig. 2 Foam generator in experiment

Generated foam flow through the tube and ejected from the nozzle into the round pan, on which we can observe the spread performance of the foam. In this experiment, foam spread above the bottom water cushion is investigated. Based on foam spread theory, the spread process on water cushion is expected to be roughly the same to that on fuel cushion.

4. Result and Discussion

The experiment cases and results are shown in Table 1. The first four cases focus on the effect of foam flow rate on spread performance, and keep the foam expansion unchanged. The other focus more on the effect of foam expansion, in which both foam flow rate and expansion are changed at the same time.

Table 1 Experiment cases and results of AFFF spread

No.	Foam solution flow rate $\dot{V}_s$ (L/h)	Gas flow rate $\dot{V}_g$ (L/h)	Result				
			Fulfill time t_b (s)	Foam flow rate $\dot{V}$ (mL/s)	Foam mass m_b (g)	Expansion S	Spread time t_f (s)
1	15	60	38	43.03	156.88	10.42	27
2	25	100	23	71.09	146.43	11.17	26
3	35	140	15	109.00	130.98	12.48	23
4	45	180	11	148.64	126.36	12.94	22
5	15	75	32	51.09	130.95	12.49	30
6	15	90	25	65.40	100.49	16.27	50
7	15	100	22	74.32	89.24	18.32	43
8	15	120	18	90.83	65.72	24.88	58

The basic process of AFFF spread on liquid cushion is shown in Fig. 3. A compact water film is formed quickly between AFFF and liquid surface due to the effect of the fluorocarbon surfactant. As the surface tension of the fluorocarbon surfactant is much lower than other liquids, the thin foam film extended very fast at the first stage of the spread process. Then ejected foam piled up on the film, foam layer start to spread around due to the effect of gravity. Finally, the foam layer covered the whole pan surface which means the goal of fire extinguishing was attained. The shape of foam layer was approximately cone as mentioned in the axisymmetric foam spread theory, so that the model was considered to be reasonable.

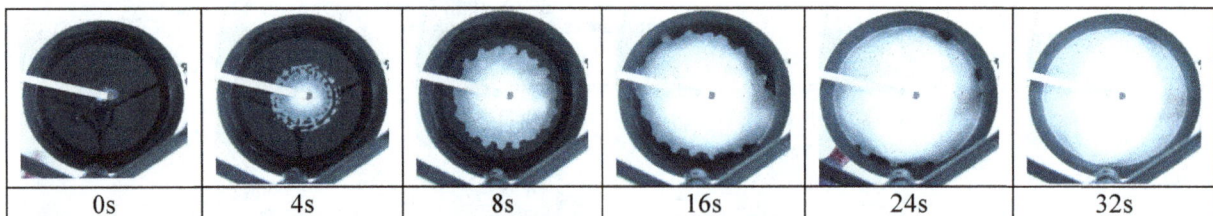

| 0s | 4s | 8s | 16s | 24s | 32s |

Fig. 3 Illustration of AFFF spread process (S =12.49, $\dot{V}$ =51.09)

The cover area by AFFF layer was the most important parameter that we care about, as the foam spreading area equals to fire suppression area when calculating the heat release rate during the fire extinguishing process. According to the axisymmetric foam spread model, the cover area of foam can be expressed as

$$A = \pi r^2 = \left(18\beta R_0\right)^{2/7} \pi^{3/7} \dot{V}^{4/7} t^{6/7} \tag{6}$$

Image recognition technique was used to capture the plane shape of foam layer. By dealing with foam spread videos, the change curves of foam spread area against time can be obtained. Fig. 3. illustrates an example of foam spread process, in which case foam expansion S=12.49 and volume flow rate $\dot{V}$ =51.09mL/s. It is found that a foam spread process can be mainly divided into three stage: (1) Fast extension. Due to the effect of surface tension, AFFF film extend very fast on the bottom water cushion in this stage. (2) Stable spread. Foam layer began to pile up and continue to spread forward. (3) Limited spread. When the foam gradually covered the water cushion surface and the front of foam layer reached the wall, the area of foam was limited to the pan size.

From observation of AFFF spread, foam fast extension stage lasts only a short time after foam ejection. Foam stable spread stage is the major process if the pan is large enough. Foam spread behavior in this stage consisted well with the axisymmetric model in theoretical analysis. Foam spread area A is proportional to six-seventh power of time $t^{6/7}$. Linear function is used to fit the foam area data in the second stage. The fitting result is shown as the red line in Fig. 4, which describes the foam stable spread stage properly.

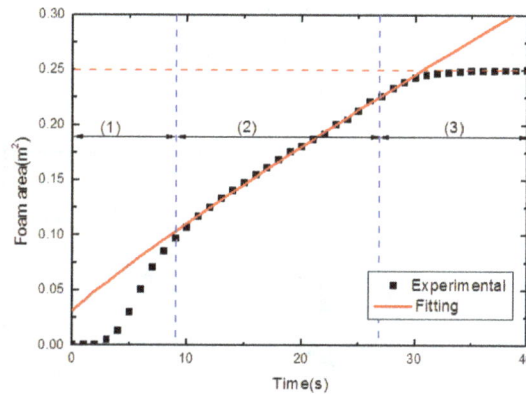

Fig. 4 Change curve of foam spread area against time (S =12.49, $\dot{V}$ =51.09)

To investigate into the relationship between foam spread speed and volume flow rate, cases with similar foam expansion are selected for comparison. Fig. 5 shows the change of spread area against time different foam flow rate. Foam flow rate is a decisive parameter to foam spread speed. For example, spread speed increases significantly with larger foam flow rate.

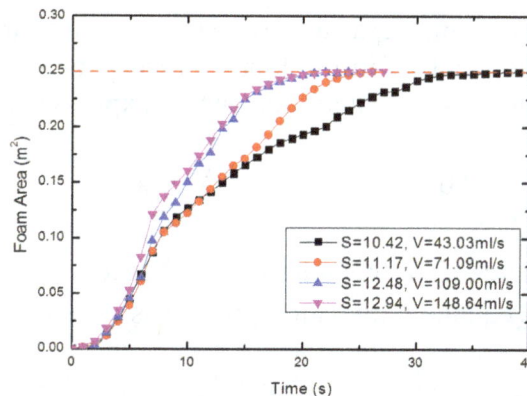

Fig. 5 Change of spread area against time with different foam flow rate

Based on the model analysis, the effect of both time and foam flow rate were considered in normalization. Linear function was used to fit the stable spread stage of these case, and the proportional coefficients a are very close, shown as in Table 2.

Table 2 Linear fitting result of fixed foam expansion cases

No.	Foam flow rate $\dot{V}$ (mL/s)	Foam expansion S	Fitting result ($Area = aV^{4/7}t^{6/7} + b$)	
			a ($\times 10^3$)	b ($\times 10^3$)
1	43.03	10.42	2.79	0.8
2	71.09	11.17	3.56	-60.1
3	109.00	12.48	3.68	-107.6
4	148.64	12.94	3.09	-79.7

In the last four cases, the effect of the foam expansion on spread performance was investigated. Fig. 6 shows the Change of spread area with changed foam expansion. It shows that the speed of foam spread was almost equal when the expansion S is less than 16.3, whereas the speed started to decline with the increase of foam expansion. The effect of density on parameters like β became lager when the foam continued to expansion grow.

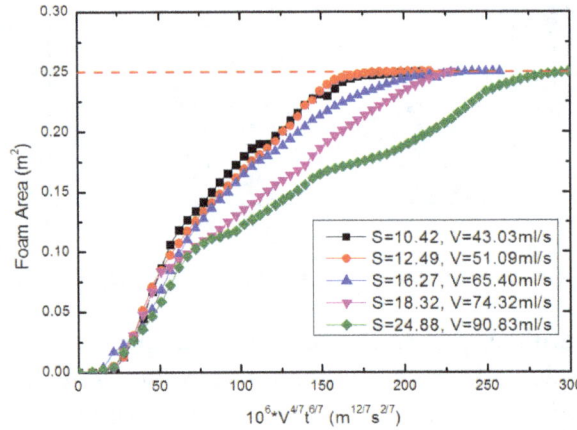

Fig. 6 Change of foam area against normalization of foam flow rate and time

In practical applications, the foam expansion of AFFF is controlled at about 10.0. Thus, we focused more on expansion around 10.0 in the foam spread experiment. By averaging the fitting coefficients, the relation of foam coverage and time can be expressed as

$$A = 0.0033 \left[10^{-24/7} \times (m/s)^{2/7} \right] \dot{V}^{4/7} t^{6/7} \tag{7}$$

where the unit of area is square meter, the unit of time of second, and the unit of foam flow rate is milliliter per second. From the equation, the value βR_0 was calculated as $2.06 \times 10^{-5} \, m/s$.

5. Conclusion

An axisymmetric foam spread model was established based on theoretical fluid dynamic analysis. This model was further validated through foam spreading experiments. It is remarkable that the foam spread speed is the function of both foam ejecting rate and time. The cofficient, βR_0, was determined as $2.06 \times 10^{-5} \, m/s$.

References

1. Kuchta, Joseph M., and Robert G. Clodfelter. Aircraft mishap fire pattern investigations. GREEN.

2. National Fire Protection Association. (2005). NFPA 11 Standard for Low. Medium-, and High-expansion Foam.

3. Williams, B. A., Sheinson, R. S., & Taylor, J. C. (n.d.) Regimes of Fire Spread Across an AFFF-Covered Liquid Pool, 1–10.

4. Persson, B., & Dahlberg, M. (1994). A Simple Model For Predicting Foam Spread Over Liquids. Fire Safety Science, 4, 265–276.

5. Persson H., "Fire Extinguishing Foams - Resistance Against Heat Radiation", SP Report 1993-54, 1993.

6. Dahlberg M., "Foam Spread Experiments on a Liquid Surface", SP Report 1994: 16, 1994.

7. Dlugogorski B Z, Kennedy E M. What properties matter in fire-fighting foams [C]. Proceedings of the National Research Institute of Fire and Disaster. 2002.

Author Index

Lu, Shou-Xiang, 1119
Luo, Can-Bin, 912
Luo, Guo-Ling, 1100
Luo, Jing-Wan, 287
Luo, Wen-Jun, 932
Luo, Xidan, 528
Luo, Xue, 299, 389
Luo, Zu-Wei, 287
Lv, Bing-Yang, 628
Lv, Ming-Zhe, 1023
Lv, Yong-Lei, 1017

Ma, Hong, 54
Ma, Wei, 339
Mao, Xiao-Nan, 122, 139
Mao, Yong, 1039
Men, Yu-Han, 796
Meng, Wen-Jun, 842
Mi, Wen-Zhong, 228
Mu, Run-Hong, 149
Mu, Shu-Cheng, 1081

Nguyen, Huu-That, 379
Ni, Li-Jie, 235
Ni, Yi-Ping, 641
Nie, Ming, 299, 389

Ouyang, Jian-Ming, 161, 167

Pan, Ming-Yu, 235
Pei, Hong-Jie, 537
Peng, Bo, 516
Peng, Fu-Hua, 945
Peng, Guo-Yan, 578
Pu, Shou-Zhi, 462

Qi, De-Zhong, 715
Qi, Wen-Jun, 330
Qi, Yun-Lian, 122, 139
Qin, Jiang-Lei, 195
Qin, Meng-Yang, 558

Qiu, Guo-Xing, 431
Quan, Heng, 235

Rao, Mei-Juan, 305
Ren, Shang-Kun, 585

Shang, Hui-Chao, 241
Shang, Shuo, 415
Shao, Hao-Bin, 920
Shen, Can-Duo, 456
Shen, Xiao-Qin, 423
Shi, Jin-Ming, 174
Shi, Ying, 54
Song, Dong-Fu, 330
Song, Jian-Feng, 1100
Song, Jie-Guang, 494
Song, Yong-Lun, 603
Su, Hang-Biao, 139
Su, Yao-Dong, 471
Sun, De-Hong, 683
Sun, Hua-Mei, 122
Sun, Yue, 403

Tan, Feng-Yi, 1108
Tang, Wen-Xian, 578
Tang, Xiu-Jian, 318
Tang, Xu, 207
Tang, Yu, 837
Tao, Ran, 312
Teng, Hua-Xiang, 398
Tian, Xin-Li, 318
Tong, Hui-Fen, 683

Wan, Mu-Hua, 161
Wang, An-Ling, 696
Wang, Bao-Wei, 1119
Wang, Bi-Li, 279
Wang, Bo, 423
Wang, Chao, 61, 842
Wang, Cheng-Yong, 1075
Wang, Dong, 449

www.ingramcontent.com/pod-product-compliance
Lightning Source LLC
Chambersburg PA
CBHW081346190326
41458CB00018B/6095